INSTRUCTOR'S SOLUTIONS
MANUAL

CALCULUS

INSTRUCTOR'S SOLUTIONS MANUAL

Henri Feiner

CALCULUS

GERALD L. BRADLEY & KARL J. SMITH

Prentice Hall, Upper Saddle River, NJ 07458

© 1995 by **PRENTICE-HALL, INC.**
Simon & Schuster/A Viacom Company
Upper Saddle River, NJ 07458

10 9 8 7 6 5 4

ISBN 0-13-178732-2

Printed in the United States of America

CONTENTS

PREFACE

This manual provides the solution for all problems in the textbook *Calculus* by Gerald L. Bradley and Karl J. Smith. Even though we did not show every step to every problem, we did show all of the important or tricky steps so that you assess the difficulty of each problem before assigning the problem. The *Student Solution Manual* offers some review, survival hints, and some added explanations for many of the odd problems.

There are several places in this manual where we used computer software. Even though such software is, of course, optional, it can help us through much of the tedium. The output shown is from *Converge* 4.0 available from JEMware, The Kawaiahao Plaza Executive Center, 567 South King Street, Suite 178, Honolulu, Hawaii 96813. Phone: 808-523-9911.

CHAPTER 1

Preview of Calculus: Functions and Limits

1.1 What Is Calculus?, Page 8

1. A mathematical model is a mathematical framework whose results parallel the real world situation. It involves abstraction, predictions, and then interpretations and comparisons with real world events. An excellent example of real world modeling from *Scientific American* (March 1991) is mentioned in the margin on page 7.

2. A real world problem can use up resources, may be nonrenewable, expensive and time consuming. A mathematical model just wastes the electrical energy required by executing a computer program (beyond the energy required to make up the model in the first place and that of the button-pusher to operate the equipment).

3. Sorry, ladies, but you cannot reach the wall, unless your hand extends beyond your foot. Your foot will get closer and closer to the wall, without reaching it.

4. If close enough means one inch (or whatever constant you choose), that number will be reached sooner or later. Problems such as this can sometimes lead to an interesting classroom discussion about the history of mathematical thought leading to calculus.

5. $0.3, 0.33, 0.333, 0.3333, \cdots$ Assumption: this is a repeating decimal. The limit appears to be $N = \frac{1}{3} = 0.\overline{3}$.

6. $0.2, 0.27, 0.272, 0.2727, \cdots$ The limit appears to be $\frac{3}{11}$.

7. $3, 3.1, 3.14, 3.141, 3.1415, 3.14159, 3141592, \cdots$. The limit appears to be π.

8. $1, 1.4, 1.41, 1.412, 1.4121, 1.414213, 1.4142135, \cdots$. $N = 1.414213\cdots$ could be an approximation to $\sqrt{2}$.

In Problems 9-12, the students may have very little background for working these problems. Any reasonable guess should be acceptable. (Some possibilities are shown in the solution to Problem 9.) A standard procedure, however, is shown for the other answers.

9. $\lim\limits_{n\to+\infty} \dfrac{2n}{n+4}$; Suppose n is very large, then adding the finite number 4 is negligible. For the sake of approximating, the denominator

can be treated as n since $n + 4 \approx n$ for large n, so we might guess the limit to be 2. Another possible solution is to notice for

$$n = 1, 10, 100, 100; \quad L = \frac{2}{5}, \frac{20}{14}, \frac{200}{104}, \frac{2000}{1004}.$$

It appears that $L = 2$. Finally, notice that as n gets very large the 4 becomes negligible, and the numerator is always twice the denominator. Mathematically, we might present the following argument:

$$\lim\limits_{n\to+\infty} \frac{2n}{n+4} = \lim\limits_{n\to+\infty} \frac{2n}{n+4} \cdot \frac{1/n}{1/n}$$
$$= \lim\limits_{n\to+\infty} \frac{2}{1+4/n} = 2$$

10. $\lim\limits_{n\to+\infty} \dfrac{2n}{3n+1} = \lim\limits_{n\to+\infty} \dfrac{2n}{3n+1} \cdot \dfrac{1/n}{1/n}$
$$= \lim\limits_{n\to+\infty} \frac{2}{3+1/n} = \frac{2}{3}$$

11. $\lim\limits_{n\to+\infty} \dfrac{3n}{n^2+2} = \lim\limits_{n\to+\infty} \dfrac{3n}{n^2+2} \cdot \dfrac{1/n^2}{1/n^2}$
$$= \lim\limits_{n\to+\infty} \frac{\frac{3}{n}}{1+\frac{2}{n^2}} = 0$$

12. $\lim\limits_{n\to+\infty} \dfrac{3n^2+1}{2n^2-1} = \lim\limits_{n\to+\infty} \dfrac{3n^2+1}{2n^2-1} \cdot \dfrac{1/n^2}{1/n^2}$
$$= \lim\limits_{n\to+\infty} \frac{3+\frac{1}{n^2}}{2-\frac{1}{n^2}} = \frac{3}{2}$$

13. a.   b.

    c.   d. There is no unique tangent line.

14. A_3 is an inscribed triangle. Connect the center with each vertex on the circle; since it is a unit circle, each of these lengths is 1. Three triangles are formed, each with central angle $2\pi/3$. Draw a perpendicular from the center of the circle to a side of the triangle, thus forming 6 triangles. If θ is the central angle for this small right triangle, $\theta = \pi/3$. The total area of A_3 is the area of the 6 triangles:

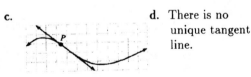

$$A_3 = 6[\tfrac{1}{2}(\cos \tfrac{\pi}{3})(\sin \tfrac{\pi}{3})] \approx 1.2990$$

A_4 is an inscribed square. Connect the center with each vertex on the circle; since it is a unit circle, each of these lengths is 1. Four triangles are formed, each with central angle $2\pi/4 = \pi/2$. Draw a perpendicular from the center of the circle to a side of the square, thus forming 8 triangles. If θ is is the central angle for this small triangle, $\theta = \pi/4$. The total area of A_4 is the area of the 8 triangles:

$$A_4 = 8[\tfrac{1}{2}(\cos \tfrac{\pi}{4})(\sin \tfrac{\pi}{4})] = 2$$

Continue the same process for A_5, A_6, $\cdots$ to obtain:

$$A_5 = 10[\tfrac{1}{2}(\cos \tfrac{\pi}{5})(\sin \tfrac{\pi}{5})] \approx 2.3776$$
$$A_6 = 12[\tfrac{1}{2}(\cos \tfrac{\pi}{6})(\sin \tfrac{\pi}{6})] \approx 2.5981$$
$$A_7 = 14[\tfrac{1}{2}(\cos \tfrac{\pi}{7})(\sin \tfrac{\pi}{7})] \approx 2.7364$$
$$\vdots$$
$$A_{100} = 200[\tfrac{1}{2}(\cos \tfrac{\pi}{100})(\sin \tfrac{\pi}{100})] \approx 3.1395$$

15. By drawing radii to each of the vertices of a regular inscribed n-gon we get n congruent triangles. Each of these triangles has a central angle of $2\pi/n$. A perpendicular to a side gives a right triangle with central angle π/n . So in this triangle

$$\tfrac{s}{2} = r \sin \tfrac{\pi}{n}, \text{ and } h = r \cos \tfrac{\pi}{n} .$$

The area of each triangle is

$$A = r^2 \sin \tfrac{\pi}{n} \cos \tfrac{\pi}{n}$$

or more simply (using the double angle identity)

$$A = \tfrac{r^2}{2} \sin \tfrac{2\pi}{n}$$

The area of a regular inscribed n-gon is:

$$A_n = \tfrac{nr^2}{2} \sin \tfrac{2\pi}{n}$$

Now $A_3 \approx 5.196152$, $A_4 = 8.000000$, $A_5 \approx 9.510565$, $A_6 \approx 10.392305$, $A_{100} \approx 12.558104$, $A_{1000} \approx 12.566288$. Since we are dealing with a circle, let's put these results in terms of π.
$$A_6 \approx 3.307973\pi,$$
$A_{100} \approx 3.997369\pi$, $A_{1000} \approx 3.999974\pi$. We will, therefore, assume that $L = 4\pi$.

16. There are 8 rectangles, each with a width of $\tfrac{1}{8}$

and a height of the square of the right

endpoint of the rectangle.

$$A = \tfrac{1}{8}[(\tfrac{1}{8})^2 + (\tfrac{2}{8})^2 + (\tfrac{3}{8})^2 + \cdots + (\tfrac{8}{8})^2]$$

$$= 0.3984375$$

17. There are 16 rectangles, each with a width of $\tfrac{1}{16}$ and a height of the square of the right endpoint of the rectangle.

$$A = \tfrac{1}{16}[(\tfrac{1}{16})^2 + (\tfrac{2}{16})^2 + (\tfrac{3}{16})^2 + \cdots + (\tfrac{16}{16})^2]$$

$$= \tfrac{1}{16^3}(1^2 + 2^2 + 3^2 + \cdots + 16^2)$$

Using the summation formula for squares we obtain:

$$A = \tfrac{1}{16^3}\left(\tfrac{16(17)(33)}{6}\right) = \tfrac{187}{512} \approx 0.3652$$

1.2 Preliminaries, Page 19

1. **a.** $(-3, 4)$ **b.** $3 \leq x \leq 5$ **c.** $-2 \leq x < 1$
   d. $(2, 7]$

2. **a.** $x < -2$ **b.** $\tfrac{\pi}{4} \leq x \leq \sqrt{2}$ **c.** $(-3, \infty)$
   d. $[-1, 5]$

3. **a.** This interval is closed on the left and open on the right.

   b. This interval is closed on both the right and left.

   c. This interval is open on both the right and left.

   d. This interval is closed on the left and open on the right.

   4. **a.**

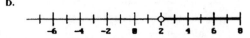

   b.

   c.

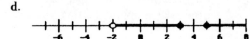

   d.

5. **a.**

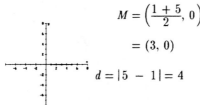

$$M = \left(\frac{1+5}{2}, 0\right)$$

$$= (3, 0)$$

$$d = |5 - 1| = 4$$

b.

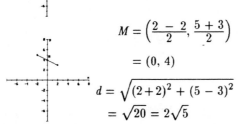

$$M = \left(\frac{2-2}{2}, \frac{5+3}{2}\right)$$

$$= (0, 4)$$

$$d = \sqrt{(2+2)^2 + (5-3)^2}$$

$$= \sqrt{20} = 2\sqrt{5}$$

6. **a.**

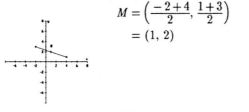

$$M = \left(\frac{-2+4}{2}, \frac{1+3}{2}\right)$$

$$= (1, 2)$$

$$d = \sqrt{(4+2)^2 + (3-1)^2} = \sqrt{40} = 2\sqrt{10}$$

b.

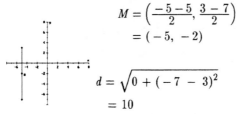

$$M = \left(\frac{-5-5}{2}, \frac{3-7}{2}\right)$$

$$= (-5, -2)$$

$$d = \sqrt{0 + (-7-3)^2}$$

$$= 10$$

7. **a.**

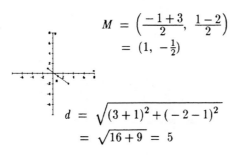

$$M = \left(\frac{-1+3}{2}, \frac{1-2}{2}\right)$$

$$= (1, -\tfrac{1}{2})$$

$$d = \sqrt{(3+1)^2 + (-2-1)^2}$$

$$= \sqrt{16+9} = 5$$

b.

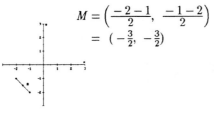

$$M = \left(\frac{-2-1}{2}, \frac{-1-2}{2}\right)$$

$$= (-\tfrac{3}{2}, -\tfrac{3}{2})$$

$$d = \sqrt{(-1+2)^2 + (-2+1)^2} = \sqrt{2}$$

8. **a.**

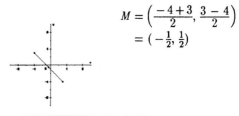

$$M = \left(\frac{-4+3}{2}, \frac{3-4}{2}\right)$$

$$= (-\tfrac{1}{2}, \tfrac{1}{2})$$

$$d = \sqrt{(3+4)^2 + (-4-3)^2} = 7\sqrt{2}$$

b.

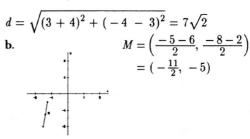

$$M = \left(\frac{-5-6}{2}, \frac{-8-2}{2}\right)$$

$$= (-\tfrac{11}{2}, -5)$$

$$d = \sqrt{(6-5)^2 + (8-2)^2} = \sqrt{1+36} = \sqrt{37}$$

9.
$$x^2 - x = 0$$

$$x(x - 1) = 0$$

$$x = 0, 1$$

10. $2y^2 + y - 3 = 0$

$$y = \frac{-1 \pm \sqrt{1 + 24}}{2(2)}$$

$$= 1, -\tfrac{3}{2}$$

11.
$$y^2 - 5y - 14 = 0$$

$$(y - 7)(y + 2) = 0$$

$$y = 7, -2$$

12. $x^2 + 5x + a = 0$

$$x = \frac{-5 \pm \sqrt{25 - 4a}}{2}$$

13.
$$3x^2 - bx = c$$

$$3x^2 - bx - c = 0$$

$$x = \frac{-(-b) \pm \sqrt{b^2 - 4(3)(-c)}}{2(3)}$$

$$= \frac{b \pm \sqrt{b^2 + 12c}}{6}$$

14.
$$4x^2 + 20x + 25 = 0$$

$$(2x + 5)^2 = 0$$

$$x = -\tfrac{5}{2}$$

15. $2x + 4 = 16$ or $-(2x + 4) = 16$

$$2x = 12 \quad \text{or} \quad -2x = 20$$

$$x = 6, -10$$

16. $5y + 2 = 12$ or $-(5y + 2) = 12$

$$5y = 10 \quad \text{or} \quad 5y = -14$$

$$y = 2, -\tfrac{14}{5}$$

17. $3 - 2w = 7$ or $-(3 - 2w) = 7$

$$2w = -4 \quad \text{or} \quad 2w = 10$$

$$w = -2, 5$$

18. $5 - 3t = 14$ or $-(5 - 3t) = 14$

$$3t = -9 \quad \text{or} \quad 3t = 19$$

$$t = -3, \tfrac{19}{3}$$

19. $\emptyset$; (The empty set; an absolute value can never be equal to a negative number.)

20. $\emptyset$

21. $\sin x = -\tfrac{1}{2}$ on $[0, 2\pi)$. The reference angle (in Quad I) is $\pi/6$. The sine is negative in Quad III and Quad IV, so $x = 7\pi/6, 11\pi/6$.

22. $(\sin x)(\cos x) = 0$ on $[0, 2\pi)$.

$$\sin x = 0 \qquad\qquad \cos x = 0$$
$$x = 0, \pi \qquad\qquad x = \tfrac{\pi}{2}, \tfrac{3\pi}{2}$$

$$x = 0, \pi, \tfrac{\pi}{2}, \tfrac{3\pi}{2}$$

23. $2\cos x + \sqrt{2} = 0 \qquad\qquad 2\cos x - 1 = 0$

$$\cos x = -\frac{\sqrt{2}}{2} \qquad\qquad \cos x = \tfrac{1}{2}$$
$$x = \tfrac{3\pi}{4}, \tfrac{5\pi}{4} \qquad\qquad x = \tfrac{\pi}{3}, \tfrac{5\pi}{3}$$
$$x = \tfrac{3\pi}{4}, \tfrac{5\pi}{4}, \tfrac{\pi}{3}, \tfrac{5\pi}{3}$$

24. $3\tan x + \sqrt{3} = 0 \qquad 3\tan x - \sqrt{3} = 0$

$$\tan x = -\frac{\sqrt{3}}{3} \qquad\qquad \tan x = \frac{\sqrt{3}}{3}$$
$$x = \tfrac{5\pi}{6}, \tfrac{11\pi}{6} \qquad\qquad x = \tfrac{\pi}{6}, \tfrac{7\pi}{6}$$
$$x = \tfrac{\pi}{6}, \tfrac{5\pi}{6}, \tfrac{7\pi}{6}, \tfrac{11\pi}{6}$$

25.

$$\cot x + \sqrt{3} = \csc x$$
$$\cot^2 x + 2\sqrt{3}\cot x + 3 = \csc^2 x$$
$$\cot^2 x - \csc^2 x + 2\sqrt{3}\cot x + 3 = 0$$
$$-1 + 2\sqrt{3}\cot x + 3 = 0$$
$$\cot x = -\frac{\sqrt{3}}{3}$$
$$x = \tfrac{2\pi}{3}, \tfrac{5\pi}{3}$$

A check is necessary because we squared both sides: $x = 5\pi/3$ is extraneous, so the solution is: $x = \tfrac{2\pi}{3}$

26.

$$\sec^2 x - 1 = \sqrt{3}\tan x$$
$$(\tan^2 x + 1) - 1 = \sqrt{3}\tan x$$

$$\tan^2 x - \sqrt{3}\tan x = 0$$
$$\tan x(\tan x - \sqrt{3}) = 0$$
$$\tan x = 0 \qquad \tan x - \sqrt{3} = 0$$
$$x = 0, \pi \qquad\qquad \tan x = \sqrt{3}$$
$$x = \tfrac{\pi}{3}, \tfrac{4\pi}{3}$$
$$x = 0, \tfrac{\pi}{3}, \pi, \tfrac{4\pi}{3}$$

27. $3x < -5$

$$x < -\tfrac{5}{3}$$

Answer: $(-\infty, -\tfrac{5}{3})$

28.
$$5(3 - x) > 3x - 1$$
$$15 - 5x > 3x - 1$$
$$-8x > -16$$
$$x < 2$$

Answer: $(-\infty, 2)$

29. $-5 < 3x < 0$ or $-\tfrac{5}{3} < x < 0$

Answer: $(-\tfrac{5}{3}, 0)$

30. $3 \le y < 8$ or $-8 < y \le 3$

Answer: $(-8, -3]$

31. $-3 < y - 5 \le 2$ or $2 < y \le 7$

Answer: $(2, 7]$

32. $-5 \le 3 - 2x < 18$

$$-8 \le \quad -2x \quad < 15$$
$$-\tfrac{15}{2} < \quad x \quad \le 4$$

Answer: $(-\tfrac{15}{2}, 4]$

33.
$$t^2 - 2t \le 3$$
$$t^2 - 2t - 3 \le 0$$
$$(t + 1)(t - 3) \le 0$$

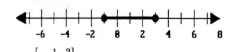

$[-1, 3]$

34. $s^2 + 3s - 4 > 0$

$$(s + 4)(s - 1) > 0$$

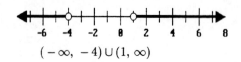

$(-\infty, -4) \cup (1, \infty)$

35. Put each of the three factors on a sign line:

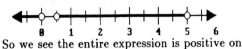

So we see the entire expression is positive on $(-\infty, 0) \cup (\frac{1}{2}, 5)$

36. Put each of the five factors on a sign line:

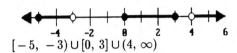

$[-5, -3) \cup [0, 3] \cup (4, \infty)$

37. Read this problem as a distance function: The distance between x and 8 is less than or equal to 0.001. The interval is $[7.999, 8.001]$.

38. Read this as a distance function: $(4.99, 5.01)$.

39. a. $\frac{\pi}{6}$ **b.** $\frac{\pi}{4}$ **40. a.** 1.17 **b.** 4.80

41. a. $60°$ **b.** $30°$ **42. a.** $229°$ **b.** $-57°$

43. The length of arc is a fraction of the circumference of the circle. In radian measure

$$s = \frac{\theta}{2\pi}(2\pi r) = r\theta$$

In this case, $s = 2(2) = 4$ cm

44. Write $30° = \frac{\pi}{6}$ so that $s = 10(\frac{\pi}{6}) = \frac{5\pi}{3} \approx 5$ in.

45. a. $\frac{\sqrt{2}}{2}$ **b.** $-\sqrt{2}$ **c.** $-\sqrt{3}$

46. a. -0.9093 **b.** 0.0707 **c.** 0.5463

47. a. 1.5574 **b.** -0.0584 **c.** 0.9656

48. a. $\frac{1}{2}$ **b.** $\sqrt{3}$ **c.** $\frac{\sqrt{3}}{3}$

49. $(x+1)^2 + (y-2)^2 = 9$

50. $(x-3)^2 + y^2 = 4$

51. $x^2 + (y-1.5)^2 = 0.0625$

52. $(x+1)^2 + (y+5)^2 = 16.81$

53. $x^2 - 2x + y^2 + 2y + 1 = 0$

$(x^2 - 2x + 1) + (y^2 + 2y + 1) = -1 + 1 + 1$

$(x-1)^2 + (y+1)^2 = 1$

Circle with center at $(1, -1)$ and $r = 1$

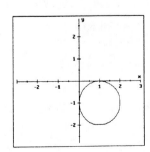

54.
$$4x^2 + 4y^2 + 4y - 15 = 0$$
$$x^2 + y^2 + y = \frac{15}{4}$$
$$x^2 + (y^2 + y + \frac{1}{4}) = \frac{15}{4} + \frac{1}{4}$$
$$x^2 + (y + \frac{1}{2})^2 = 4$$

Circle with center $(0, -\frac{1}{2})$ and $r = 2$

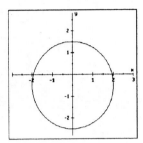

55.
$$x^2 + y^2 + 2x - 10y + 25 = 0$$
$$(x^2 + 2x + 1) + (y^2 - 10y + 25) = 1$$
$$(x+1)^2 + (y-5)^2 = 1$$

Circle with center $(-1, 5)$ and $r = 1$

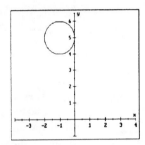

56.
$$2x^2 + 2y^2 + 2x - 6y - 9 = 0$$
$$x^2 + x + y^2 - 3y = \frac{9}{2}$$
$$(x^2 + x + \frac{1}{4}) + (y^2 - 3y + \frac{9}{4}) = \frac{9}{2} + \frac{1}{4} + \frac{9}{4}$$
$$(x + \frac{1}{2})^2 + (y - \frac{3}{2})^2 = 7$$

Circle with center $(-\frac{1}{2}, \frac{3}{2})$ and $r = \sqrt{7}$

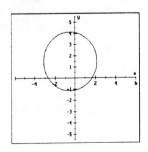

57.
$$\sin\left(-\tfrac{\pi}{12}\right) = \sin\left(\tfrac{\pi}{4} - \tfrac{\pi}{3}\right)$$
$$= \sin\tfrac{\pi}{4}\cos\tfrac{\pi}{3} - \cos\tfrac{\pi}{3}\sin\tfrac{\pi}{3}$$
$$= \frac{\sqrt{2} - \sqrt{6}}{4}$$

58.
$$\cos\left(\tfrac{7\pi}{12}\right) = \cos\left(\tfrac{\pi}{4} + \tfrac{\pi}{3}\right)$$
$$= \cos\tfrac{\pi}{4}\cos\tfrac{\pi}{3} - \sin\tfrac{\pi}{4}\sin\tfrac{\pi}{3}$$
$$= \frac{\sqrt{2} - \sqrt{6}}{4}$$

59.
$$\tan\tfrac{\pi}{12} = \tan\left(\tfrac{\pi}{4} - \tfrac{\pi}{6}\right)$$
$$= \frac{\tan\tfrac{\pi}{4} - \tan\tfrac{\pi}{6}}{1 + \tan\tfrac{\pi}{4}\tan\tfrac{\pi}{6}}$$
$$= \frac{1 - \tfrac{\sqrt{3}}{3}}{1 + \tfrac{\sqrt{3}}{3}}$$
$$= 2 - \sqrt{3}$$

60.
$$\sin 165° = \sin(135° + 30°)$$
$$= \sin 135°\cos 30° + \cos 135°\sin 30°$$
$$= \frac{\sqrt{2}}{4}(\sqrt{3} - 1)$$

61. If $ax^2 + bx + c = 0, a \neq 0$, then
$$x = \frac{-b \pm \sqrt{b^2 - 4ac}}{2a}$$

62. If $|x| = a$, then $x = a$ or $x = -a$.

63. If $|x| \leq a$, then $-a \leq x \leq a$.

64. Use identities to get all of the functions to be functions of the same angle (say $x, 2x, 3x, \cdots$). Solve for the function, then solve for the angle, and finally find all solutions in the given domain.

65. If $a > 0$, the solution is positive if $b < 0$, 0 if $b = 0$, and negative if $b < 0$. If $a < 0$, the solution is positive if $b > 0$, 0 if $b = 0$, and negative if $b < 0$.

66. Change c affects the discriminant, $d = b^2 - 4ac$. If $d < 0$, then there is no solution. If $d = 0$, there is a double root, and if $d > 0$, then the solutions

$$x_1 = \frac{-b + \sqrt{b^2 - 4ac}}{2a} \text{ and } x_2 = \frac{-b - \sqrt{b^2 - 4ac}}{2a}$$

approach $-\tfrac{b}{2a}$ as the $d \to 0$, or further always as d increases.

67. If $a = 1$, the period is 2π; if a increases then the period decreases and vice-versa. The period is $2\pi/a$.

68. Suppose $c > 0$ and $|x| < c$.

If $x \geq 0$, then $|x| = x \leq c$.
If $x < 0$, then $|x| = -x \leq c$ or $x \geq -c$ so that
$$-c \leq x \leq c$$

Suppose $c > 0$ and $-c \leq x \leq c$.
If $c \leq x < 0$, then $-x = |x| \geq c$.
If $0 \leq x \leq c$, then $x = |x| \leq c$ so that
$$-c \leq |x| \leq c \text{ or } |x| \leq c$$

69. If $x \geq 0$, then $|x| = x$. Thus,
$$-|x| \leq 0 \leq x \leq |x|$$
If $x < 0$, then $|x| = -x$. Thus,
$$-|x| \leq x \leq 0 \leq |x|$$
Thus, $-|x| \leq x \leq |x|$.

70. Suppose $a = b$ or $a = -b$.

If $a = b > 0$, then $|a| = |b|$.
If $a = b < 0$, then $-a = -b > 0$, so $|a| = |b|$.
If $a = -b > 0$, then $|a| = |b|$.

Suppose $|a| = |b|$.

If $a > 0$, $b > 0$, then $a = b$.
If $a < 0$, $b < 0$, then $-a = -b$.
If $a > 0$, $b < 0$, then $-b > 0$ $a = -b$

If $a = b = 0$, the proof is trivial.

71. $|a| < b$ and $b > 0$ implies $-b < a < b$.
If $a \geq 0$, then $|a| = a$ and $|a| < b$ means $a < b$.
If $a < 0$, then $|a| = -a$ and $|a| < b$ means
$$-a < b \text{ or } -b < a.$$
Combining these results leads to the conclusion $-b < a < b$.

72. By Problem 69, $-|x| \leq x \leq |x|$ and $-|y| \leq y \leq |y|$. Adding these inequalities leads to
$$-(|x| + |y|) \leq x + y \leq (|x| + |y|)$$
By Problem 68, this is equivalent to
$$|x + y| \leq |x| + |y|$$

73. By Problem 69, $y \leq |y|$ and $x \leq |x|$. Subtraction leads to
$$x - y \leq |x| - |y| \text{ or } y - x \geq |y| - |x|$$
By Problem 69, $-|y| \leq y$ and $-|x| \leq x$. Subtraction leads to
$$|x| - |y| \leq y - x \text{ or } |y| - |x| \geq -(y - x)$$
Combining these results leads to
$$-(y - x) \leq |y| - |x| \leq y - x \text{ or}$$
$$\big||y| - |x|\big| \leq |y - x| = |x - y|$$

This can be written as

$$||x| - |y|| \leq |x - y|$$

74. The horizontal distance between P and Q is $|x_2 - x_1|$. The x-value of the midpoint is

$$x = x_1 + \frac{x_2 - x_1}{2} = \frac{x_1 + x_2}{2}.$$

Similarly, $y = \dfrac{y_1 + y_2}{2}$.

1.3 Lines in the Plane, Page 24

1. If $y = 0$, solve for x and draw the vertical line $x = c$. If $y \neq 0$, solve for $y = mx + b$. Plot the y-intercept $(0, b)$ and then count out the slope, m. Draw the line passing through the y-intercept and the slope point.

2. $m = \dfrac{6 - 4}{3 - 1} = 1$; Use either point in the point-slope form:

$$y - 4 = (1)(x - 1)$$
$$x - y + 3 = 0$$

3. Using the point-slope form :

$$y - 7 = \frac{2}{-1}(x + 1)$$
$$2x + y - 5 = 0$$

4. $y + 5 = 0$

5. horizontal line; $y = \frac{1}{2}$ or $2y - 1 = 0$

6. Using the slope-intercept form: $y = 2x + 5$ or $2x - y + 5 = 0$.

7. vertical line; $x = -2$ or $x + 2 = 0$

8. Using the point-slope form:

$$y - 0 = -3(x - 5)$$
$$3x + y - 15 = 0$$

9. Use the intercept form:

$$\frac{x}{7} + \frac{y}{-8} = 1$$
$$8x - 7y - 56 = 0$$

10. Use the intercept form:

$$\frac{x}{4.5} + \frac{y}{-5.4} = 1$$
$$5.4x - 4.5y - 24.3 = 0$$

You could also write this as $6x - 5y - 27 = 0$.

11. Any line parallel to $3x + y = 7$ has the same slope, namely -3. Now use the point-slope form:

$$y - 8 = -3(x + 1)$$
$$3x + y - 5 = 0$$

12. $4x - 3y + 2 = 0$ has slope $m = \frac{4}{3}$, so a perpendicular line has slope $-\frac{3}{4}$. Now use the point-slope form:

$$y + 2 = -\frac{3}{4}(x - 3)$$
$$4y + 8 = -3x + 9$$
$$3x + 4y - 1 = 0$$

13. The line passing through the given points has slope $m = \dfrac{9 - 1}{5 - 2} = \dfrac{8}{3}$. Now use the point-slope form:

$$y - 5 = \frac{8}{3}(x - 4)$$
$$3y - 15 = 8x - 32$$
$$8x - 3y - 17 = 0$$

14. The desired line has slope $m = -2$. Using point-slope form:

$$y - 6 = -2(x + 1)$$
$$2x + y - 4 = 0$$

15. $x - 4y + 5 = 0$ has $m = \frac{1}{4}$ so our line must have $m = -4$. Solve the two given equations simultaneously to find their intersection at $(-1, 1)$. Now use the point and slope to find the line:

$$y - 1 = -4(x + 1)$$
$$4x + y + 3 = 0$$

16.
$$5x + 3y - 15 = 0$$
$$3y = -5x + 15$$
$$y = -\frac{5}{3}x + 5$$

The slope is $m = -\frac{5}{3}$ and the y-intercept is $(0, 5)$. The x-intercept is $(3, 0)$.

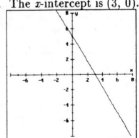

17. In slope-intercept form: $y = -\frac{3}{5}x - 3$. The slope is $m = -\frac{3}{5}$ and the y-intercept is $(0, -3)$. Let $y = 0$ to find x-intercept 5: $(-5, 0)$.

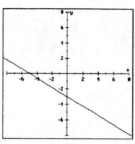

18. $y = -\frac{3}{2}x + 6$; $m = -\frac{3}{2}$, with intercepts $(0, 6)$ and $(4, 0)$.

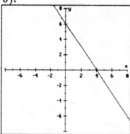

19. $y = \frac{3}{2}x - 3$; $m = \frac{3}{2}$, with intercepts $(0, -3)$ and $(2, 0)$.

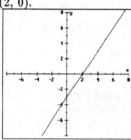

20. $m = 2$ with intercept $(0, 0)$.

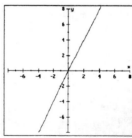

21. $m = \frac{1}{5}$ with intercept $(0, 0)$.

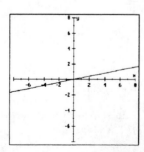

22. This is a horizontal line with $m = 0$ and

y-intercept $(0, 5)$. There is no x-intercept.

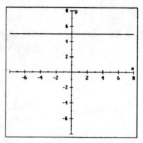

23. This is a vertical line with slope not defined and no y-intercept. The x-intercept is is $(-3, 0)$.

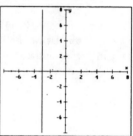

24. Let $x = k$, then $2y = 3k + 6$ and the height of the triangle at $x = k$ is

$$y = \frac{3}{2}(k + 2)$$

The area of the triangle is

$$A = \frac{1}{2}(2 + k)\frac{3}{2}(k + 2)$$

Thus,

$$3 = \frac{3}{4}(k + 2)^2$$
$$4 = (k + 2)^2$$
$$k + 2 = \pm 2$$
$$k = 0, -4$$

There are two such vertical lines:

$$x = 0, \text{ and } x = -4.$$

25. The opposite sides of a parallelogram are parallel and equal. $(3, -2)$ is 5 units down and 2 units right of $(1, 3)$, so the fourth vertex is 5 units down and 2 units right of $(4, 11)$, which is $(6, 6)$, or 5 units above and 2 units to the left of $(4, 11)$, which is $(2, 16)$. Using the first choice:
Let A:$(1, 3)$, B:$(3, -2)$, C:$(4, 11)$, D:$(6, 6)$
$\overline{AB}$: $y - 3 = -\frac{5}{2}(x - 1)$ or
$\quad\quad\quad 5x + 2y - 11 = 0$
$\overline{AC}$: $y - 3 = \frac{8}{3}(x - 1)$ or
$\quad\quad\quad 8x - 3y + 1 = 0$
$\overline{CD}$: $y - 11 = -\frac{5}{2}(x - 4)$ or
$\quad\quad\quad 5x + 2y - 42 = 0$

$\overline{BD}$: $y + 2 = \frac{8}{3}(x - 3)$ or
$$8x - 3y - 30 = 0$$
Using the second choice:
Let A:(1, 3), B:(3, −2), C:(4, 11), D:(2, 16)
$\overline{AB}$: $y - 3 = -\frac{5}{2}(x - 1)$ or
$$5x + 2y - 11 = 0$$
$\overline{AD}$: $y - 3 = 13(x - 1)$ or
$$13x - y - 10 = 0$$
$\overline{CD}$: $y - 11 = -\frac{5}{2}(x - 4)$ or
$$5x + 2y - 42 = 0$$
$\overline{BC}$: $y + 2 = 13(x - 3)$ or
$$13x - 3y - 41 = 0$$
This solution is not unique.

26. Let $y = E$ and $x = A$. Then
$$m = \frac{50 - 20}{24 - 60} = -\frac{5}{6}$$
and $y = 20 - \frac{5}{6}(x - 60) = 70 - \frac{5}{6}x$.
Thus, $E = -\frac{5}{6}A + 70$.

   a. If $E = 30$, $A = 48$

   b. If $A = 0$, then $E = 70$

   c. If $E = 0$, then $A = 84$

27. $m = \frac{212 - 32}{100 - 0} = \frac{9}{5}$ so that $F = \frac{9}{5}C + 32$

   a. If $C = -39$, then $F = -38.2$

   b. If $F = 0$, then $C = -\frac{160}{9} \approx -17.8$

   c. If $F = C$, then $F = C = -40$

28. a. Take the year 1987 as $t = 0$ and 1992 as $t = 5$. Then the slope is
$$m = \frac{575 - 545}{5} = -6$$
   The scores are related to t as
$$S = -6t + 575$$

   b. In 2000, $t = 13$. Then
$$S = -6(13) + 575 = 497$$

   c. $527 = -6t + 575$ or $t = 8$; the SAT scores should be 527 in 1995.

29. $C(0) = 5000$ and $m = 60$ so
   $C = 60x + 5000$

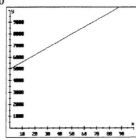

30. Let x denote the number of miles driven and

$C(x)$ the corresponding cost (in dollars). If $x \le 100$, then $C(x) = 40$. If $x > 100$, then $C(x) = 0.34(x - 100) + 40$.

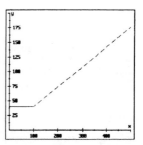

   a. A 50 mile trip costs \$40.

   b.
$$92.36 = 0.34(x - 100) + 40$$
$$0.34(x - 100) = 52.36$$
$$x = 254$$

31. Let t denote the age in years of the machinery and V be a linear function of t. At the time of purchase, $t = 0$ and $V(0) = 200,000$. Ten years later, $t = 10$ and $V(10) = 10,000$. The slope of the line through $(0, 200\,000)$ and $(10, 1000)$ is
$$m = \frac{10,000 - 200,000}{10 - 0} = -19,000$$
Thus, $V(t) = -19,000t + 200,000$.

In particular,

$V(4) = -19,000(4) + 200,000 = 124,000$

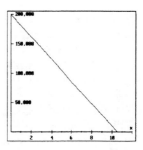

32. $A(1, 3)$, $B(-1, 2)$. Let d_1 and d_2 be the distances from $P(x, y)$ to A and B, respectively. Then
$$d_1{}^2 = d_2{}^2$$
$$(x - 1)^2 + (y - 3)^2 = (x + 1)^2 + (y - 2)^2$$
$$-2x + 1 - 6y + 9 = 2x + 1 - 4y + 4$$
$$4x + 2y - 5 = 0$$

33. a. The three possible parallelograms are:

   $P_1P_2P_3A$, $P_1P_3P_2B$, $P_1P_3CP_2$. Using

the fact that the slopes and lengths of opposite sides are equal (see Problem 25)

A:(3, 1),   B:(1, 11),   C:(-3, -5)

b. For $P_1P_2P_3A$, P_1P_3 joins (2, 6) to (0, -2) and has midpoint (1 , 2). $P_2P_4 = P_2A$ joins (-1, 3) to (3, 1) and has midpoint (1, 2). Having the same midpoint they bisect each other.

34. Let x be the number of days. The slope is
$$m = \frac{200 - 164}{12 - 21} = -4$$
Thus, $y = -4x + 248$. In particular, for the 8th month, $y = -4(8) + 248 = 216$.

35. Let x denote the number of days since the reduced rate went into effect and $N(x)$ the corresponding number of vehicles qualifying for the reduced rate. Since the number of qualifying vehicles is increasing at a constant rate, N is a linear function of x. Since $N(0) = 157$ and $N(44) = 289$ fourteen days from now,
$$m = \frac{247 - 157}{30 - 0} = 3$$
$N(x) = 3x + 157$ for $x \geq 0$. In 14 days, $x = 44$ so $N(44) = 289$.

36. a.

Age:	0	10	20	30	40
Value:	3	6	12	24	48

At the end of 30 years the book is worth $24 and at the end of 40 years it will be worth $48.

b. Since the slopes between pairs of points are not equal, the relationship is not linear.

37. Let $P(h, k)$ and $Q(x, y)$ be two points on a line. Consider a right triangle with vertices P, Q, and (x, k). If ϕ is the angle between the line and the horizontal line segment, then
$$m = \tan \phi = \frac{y - k}{x - h}$$
Thus, $y - k = m(x - h)$.

38. $A(x_1, y_1)$, $B(x_2, y_2)$ are two points on a line whose equation is $y = mx + b$. Then, $y_1 = mx_1 + b$ and $y_2 = mx_2 + b$.
$$mx_1 - y_1 = mx_2 - y_2$$
$$m(x_1 - x_2) = y_1 - y_2$$
$$m = \frac{y_2 - y_1}{x_2 - x_1}$$

The point (0, b) is on the line since
$$b = m(0) + b$$

39. $Ax + By + C = 0$ or $y = -\frac{A}{B}x - \frac{C}{B}$, $B \neq 0$

(If $B = 0$, the line is horizontal and the distance is trivial.) The slope of this line is $-A/B$. The normal line L_n has slope $m_n = B/A$. The equation of the normal is
$$y = \frac{B}{A}x + b$$
The point $P(x_0, y_0)$ lies on L_n. Then
$$y_0 = \frac{B}{A}x_0 + b \text{ or } b = y_0 - \frac{B}{A}x_0 = \frac{Ay_0 - Bx_0}{A}$$
The line L_n has equation
$$y = \frac{B}{A}x + \frac{Ay_0 - Bx_0}{A}$$
The point of intersection, I, of the lines can be found by simultaneously solving the system of equations:
$$-\frac{A}{B}x - \frac{C}{B} = \frac{B}{A}x + \frac{Ay_0 - Bx_0}{A}$$
$$A(-Ax - C) = B(Bx + Ay_0 - Bx_0)$$
$$-A^2x - AC = B^2x + ABy_0 - B^2x_0$$
$$x(-A^2 - B^2) = ABy_0 + AC - B^2x_0$$
$$x = \frac{B^2x_0 - ABy_0 - AC}{A^2 + B^2}$$

This is the first component of the point I. Substitute to find the second component:
$$y = -\frac{A}{B}x - \frac{C}{B}$$
$$= -\frac{1}{B}\left(\frac{AB^2x_0 - A^2By_0 - A^2C}{A^2 + B^2} + \frac{A^2C + B^2C}{A^2 + B^2}\right)$$
$$= -\frac{ABx_0 - A^2y_0 + BC}{A^2 + B^2}$$

Now use the distance formula to find s:
$$s^2 = \left(x_0 - \frac{B^2x_0 - ABy_0 - AC}{A^2 + B^2}\right)^2$$
$$+ \left(y_0 + \frac{ABx_0 - A^2y_0 + BC}{A^2 + B^2}\right)^2$$
$$= \frac{A^2(Ax_0 + By_0 + C)^2 + B^2(Ax_0 + By_0 + C)^2}{(A^2 + B^2)^2}$$

Thus, $s = \dfrac{|Ax_0 + By_0 + C|}{\sqrt{A^2 + B^2}}$

40. $\phi = \alpha_2 - \alpha_1$, so

$$\tan(\alpha_2 - \alpha_1) = \frac{\tan \alpha_2 - \tan \alpha_1}{1 + \tan \alpha_2 \tan \alpha_1}$$
$$= \frac{m_2 - m_1}{1 + m_2 m_1}$$

41. Rewrite $y - 8 = m(x + 1)$ as
$mx - y + m + 8 = 0$. Using the distance formula to find

$$5 = \frac{|2m - 4 + m + 8|}{\sqrt{m^2 + 1}}$$
$$5\sqrt{m^2 + 1} = 3m + 4$$
$$25(m^2 + 1) = 9m^2 + 24m + 16$$
$$16m^2 - 24m + 9 = 0$$
$$(4m - 3)^2 = 0$$
$$4m - 3 = 0$$
$$m = \frac{3}{4}$$

1.4 Functions and Their Graphs, Page 39

1. $D = $ all reals or $D = \mathbb{R}$ or $D = (-\infty, \infty)$
$$f(-2) = 2(-2) + 3 = -1$$
$$f(1) = 2(1) + 3 = 5$$
$$f(0) = 2(0) + 3 = 3$$

2. $D = (-\infty, \infty)$
$$f(0) = -(0)^2 + 2(0) + 3 = 3$$
$$f(1) = -(1)^2 + 2(1) + 3 = 4$$
$$f(-2) = -(-2)^2 + 2(-2) + 3 = -5$$

3. $D = (-\infty, \infty)$
$$f(1) = 3(1)^2 + 5(1) - 2 = 6$$
$$f(0) = 3(0)^2 + 5(0) - 2 = -2$$
$$f(-2) = 3(-2)^2 + 5(-2) - 2 = 0$$

4. $D = (-\infty, 0) \cup (0, \infty)$
$$f(-1) = -1 + \frac{1}{-1} = -2$$
$$f(1) = 1 + \frac{1}{1} = 2$$
$$f(2) = 2 + \frac{1}{2} = \frac{5}{2} \text{ or } 2.5$$

5. $D = (-\infty, -3) \cup (-3, \infty)$.
$$f(2) = 2 - 2 = 0$$
$$f(0) = 0 - 2 = -2$$
$$f(-3) \text{ is undefined}$$

6. $D = (\frac{1}{2}, \infty)$

$$f(1) = [2(1) - 1]^{-3/2} = 1$$
$$f(\tfrac{1}{2}) = [2(\tfrac{1}{2}) - 1]^{-3/2} \text{ is undefined}$$
$$f(13) = [2(13) - 1]^{-3/2} = 25^{-3/2} = \frac{1}{125}$$

7. $D = (-\infty, 2)] \cup [0, \infty)$
$$f(-1) = \sqrt{(-1)^2 + 2(-1)} \text{ is undefined}$$
$$f(\tfrac{1}{2}) = \sqrt{(\tfrac{1}{2})^2 + 2(\tfrac{1}{2})} = \frac{\sqrt{5}}{2}$$
$$f(1) = \sqrt{(1)^2 + 2(1)} = \sqrt{3}$$

8. $D = (-\infty, -3] \cup [-2, \infty)$
$$f(0) = \sqrt{0^2 + 5(0) + 6} = \sqrt{6}$$
$$f(1) = \sqrt{1^2 + 5(1) + 6} = \sqrt{12} = 2\sqrt{3}$$
$$f(-2) = \sqrt{(-2)^2 + 5(-2) + 6} = 0$$

9. $D = (-\infty, \infty)$
$$f(-1) = \sin[1 - 2(-1)] = \sin 3 \approx 0.1411$$
$$f(\tfrac{1}{2}) = \sin[1 - 2(\tfrac{1}{2})] = \sin 0 = 0$$
$$f(1) = \sin[1 - 2(1)] = \sin(-1) \approx -0.8415$$

10. $D = (-\infty, \infty)$
$$f(0) = \sin 0 - \cos 0 = -1$$
$$f(-\tfrac{\pi}{2}) = \sin(-\tfrac{\pi}{2}) - \cos(-\tfrac{\pi}{2}) = -1$$
$$f(\pi) = \sin \pi - \cos \pi = 1$$

11. $D = (-\infty, \infty)$
$$f(3) = 3 + 1 = 4$$
$$f(1) = -2(1) + 4 = 2$$
$$f(0) = -2(0) + 4 = 4$$

12. $D = (-\infty, \infty)$
$$f(-6) = 3 \text{ since } -6 < -5$$
$$f(-5) = -5 + 1 = -4 \text{ since } -5 \le -5 \le 5$$
$$f(16) = \sqrt{16} = 4 \text{ since } 16 > 5$$

13. $$\frac{f(x+h) - f(x)}{h} = \frac{[9(x+h) + 3] - (9x + 3)}{h}$$
$$= \frac{9h}{h} = 9$$

14. $$\frac{f(x+h) - f(x)}{h} = \frac{5 - 2(x+h) - (5 - 2x)}{h}$$
$$= \frac{5 - 2x - 2h - 5 + 2x}{h}$$
$$= \frac{-2h}{h}$$
$$= -2$$

15. $$\frac{f(x+h) - f(x)}{h} = \frac{5(x+h)^2 - 5x^2}{h}$$

$$= \frac{5x^2 + 10xh + 5h^2 - 5x^2}{2}$$

$$= \frac{10xh + 5h^2}{h}$$

$$= 10x + 5h$$

16. $\dfrac{f(x+h) - f(x)}{h} = \dfrac{6xh + 3h^2 + 2h}{h}$

$$= 6x + 3h + 2$$

17. $\dfrac{f(x+h) - f(x)}{h} = \dfrac{-x - h - (-x)}{h}$

$$= \frac{-h}{h} = -1$$

18. $\dfrac{f(x+h) - f(x)}{h} = \dfrac{x + h - x}{h} = 1$

19. $\dfrac{f(x+h) - f(x)}{h} = \dfrac{\frac{1}{x+h} - \frac{1}{x}}{h}$

$$= \frac{-\dfrac{h}{x(x+h)}}{h}$$

$$= \frac{-1}{x(x+h)}$$

20. $\dfrac{f(x+h) - f(x)}{h} = \dfrac{\frac{x+h+1}{x+h-1} - \frac{x+1}{x-1}}{h}$

$$= \frac{1}{h}\left[\frac{x^2 + hx + x - x - h - 1 - x^2 - x - hx - h + x + 1}{(x-1)(x+h-1)}\right]$$

$$= \frac{-2h}{h(x-1)(x+h-1)} = \frac{-2}{(x-1)(x+h-1)}$$

21. **a.** Not equal. $f(x) = g(x);\ x \neq 0$

   b. Equal, since $g(x)$ restricts the domain.

22. **a.** $f(x) = \dfrac{2x^2 - x - 6}{x - 2} = \dfrac{(2x+3)(x-2)}{x-2}$

   $$= 2x + 3,\ x \neq 2$$

   $f = g$ since they are equal and have the same domains.

   b. $f(x) = \dfrac{3x^2 - 7x - 6}{x - 3} = \dfrac{(3x+2)(x-3)}{x-3}$

   $$= 3x + 2,\ x \neq 3$$
   $f = g$ since they are equal and have the same domains.

23. **a.** $f(x) = \dfrac{3x^2 - 5x - 2}{x - 2} = \dfrac{(3x+1)(x-2)}{x-2}$

   $$= 3x + 1,\ x \neq 2$$
   $f \neq g$ since the domains are not the same.

   b. $f(x) = \dfrac{(3x+1)(x-2)}{x-2},\ x \neq 6$

$$= 3x + 1,\ x \neq 6,\ x \neq 2$$

$$g(x) = \frac{(3x+1)(x-6)}{x-6},\ x \neq 2$$

$$= 3x + 1,\ x \neq 2,\ x \neq 6$$

   $f = g$ since they are equal and have the same domains.

24. **a.** $f_1(-x) = (-x)^2 + 1 = x^2 + 1 = f_1(x)$

   f_1 is even

   b. $f_2(-x) = \sqrt{(-x)^2} = \sqrt{x^2} = f_2(x)$

   f_2 is even

25. **a.** $f_3(-x) = \dfrac{1}{3(-x)^3 - 4} = \dfrac{1}{-3x^3 - 3}$

   f_3 is neither

   b. $f_4(-x) = (-x)^3 + (-x) = -x^3 - x$

   $$= -f_4(x);\ f_4 \text{ is odd}$$

26. **a.** $f_5(-x) = \dfrac{1}{[(-x)^3 + 3]^2} = \dfrac{1}{(-x^3 + 3)^2}$

   f_5 is neither

   b. $f_6(-x) = \dfrac{1}{[(-x)^3 + (-x)]^2} = \dfrac{1}{(-x^3 - x)^2}$

   $$= \frac{1}{(x^3 + x)^2} = f_6(x);\ f_6 \text{ is even}$$

27. **a.** $f_7(-x) = |-x| = |x| = f_7(x)$

   f_7 is even

   b. $f_8(x) = |-x| + 3 = |x| + 3 = f_8(x)$

   f_8 is even

28. $(f \circ g)(x) = f(2x) = (2x)^2 + 1 = 4x^2 + 1$

   $(g \circ f)(x) = g(x^2 + 1) = 2x^2 + 2$

29. $(f \circ g)(x) = f(1 - x^2) = \sin(1 - x^2)$

   $(g \circ f)(x) = g(\sin x) = 1 - \sin^2 x = \cos^2 x$

30. $(f \circ g)(t) = f(t^2) = \sqrt{t^2} = |t|$

   $(g \circ f)(t) = g(\sqrt{t}) = t$

31. $(f \circ g)(u) = f\left(\dfrac{u+1}{1-u}\right) = \dfrac{\frac{u+1}{1-u} - 1}{\frac{u+1}{1-u} + 1}$

   $$= \frac{u + 1 - 1 + u}{u + 1 + 1 - u} = \frac{2u}{2} = u$$

   $(g \circ f)(u) = g\left(\dfrac{u-1}{u+1}\right) = \dfrac{\frac{u-1}{u+1} + 1}{1 - \frac{u-1}{u+1}}$

   $$= \frac{u - 1 + u + 1}{u + 1 - u + 1} = \frac{2u}{2} = u$$

32. $(f \circ g)(x) = f(2x + 3) = \sin(2x + 3)$

$(g \circ f)(x) = g(\sin x) = 2 \sin x + 3$

33. $(f \circ g)(x) = f(\tan x) = \dfrac{1}{\tan x} = \cot x$

$(g \circ f)(x) = g(\frac{1}{x}) = \tan \frac{1}{x}$

34. **a.** $u(x) = 2x^2 - 1; \; g(u) = u^4$

b. $u(x) = 5x - 1; \; g(u) = \sqrt{u}$

35. **a.** $u(x) = \tan x; \; g(u) = u^2$

b. $u(x) = x^2; \; g(u) = \tan u$

36. **a.** $u(x) = \sqrt{x}; \; g(u) = \sin u$

b. $u(x) = \sin x; \; g(u) = \sqrt{u}$

37. **a.** $u(x) = \dfrac{x+1}{2-x}; \; g(u) = \sin u$

b. $u(x) = \dfrac{2x}{1-x}; \; g(u) = \tan u$

38.
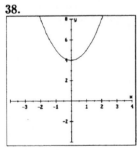

39.
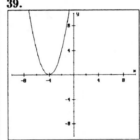

40.
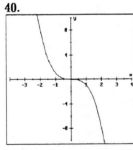

41.
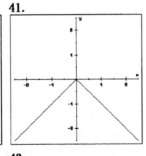

42.
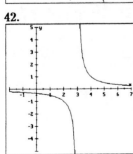

43.
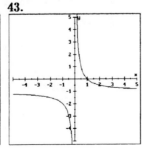

44.
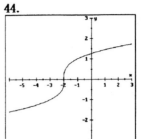

45.
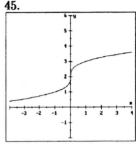

46.
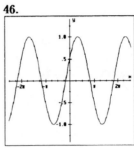

47.
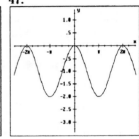

48.
49.
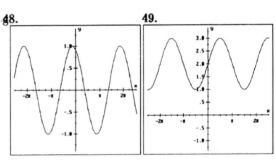

50.
51.
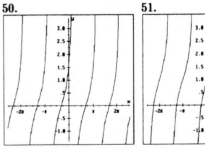
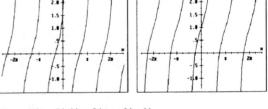

52. $P(5, f(5)); \; Q(x_0, f(x_0))$

53. $R(a, g(a)); \; S(x_0, g(x_0))$

54. **a.** $s(0) = -16(0)^2 + 96(0) + 144 = 144$

The height of the cliff is 144 ft.

b. $s(t) = 0$ if

$$-16t^2 + 96t + 144 = 0$$

$$t^2 - 6t - 9 = 0$$

$$t = \frac{6 \pm \sqrt{36 - 4(-9)}}{2} = 3 \pm 3\sqrt{2}$$

The ball hits the ground when

$t = 3 + 3\sqrt{2} \approx 7.2$ (reject the negative t)

  c. The maximum height occurs at the vertex of the parabola $s = -16t^2 + 96t + 144$.

$$s - 144 = -16(t^2 - 6t)$$

$$s - 144 + (-16)(9) = -16(t^2 - 6t + 9)$$

$$s - 288 = -16(t - 3)^2$$

It takes 3 seconds for the ball to reach its highest point.

55. a. $C(20) = 20^3 - 30(20)^2 + 400(20) + 500$
$= 4,500$; the cost is \$4,500

  b. $C(20) - C(19) = 4,500 - 4,129 = 371$
The cost of the 20th unit is \$371.

56. a. $f(2) = -(2)^3 + 6(2) + 15(2)^2 = 64$

  b. $f(2) - f(1) = 64 - 20 = 44$

57. a. $I = f(t) = \dfrac{K}{s^2} = \dfrac{K}{(6t - t^2)^2}$

    $= \dfrac{30}{t^2(6 - t)^2}$

  b. $I(1) = \dfrac{30}{1^2(6 - 1)^2} = \dfrac{6}{5}$ candles

    $I(4) = \dfrac{30}{4^2(6 - 4)^2} = \dfrac{15}{32}$ candles

58. a. $P(9) = 20 - \dfrac{6}{9 + 1} = 19.4$
This is 19,400 people.

  b. $P(9) - P(8) = 20 - \dfrac{3}{5} - (20 - \dfrac{2}{3}) = \dfrac{1}{15}$
This is about 67 people.

  c. The population will tend to 20,000 people in the long run. The denominator will have increased to the point of making $6/(t + 1)$ negligible.

59. a. $D = (-\infty, 0) \cup (0, \infty)$

  b. Since n represents the number of trials, n is a positive integer $[1, p]$, for some finite number p.

  c. For the third trial, $n = 3$ so

$$f(3) = 3 + \dfrac{12}{3} = 7 \text{ minutes}$$

  d. $f(n) \le 4$, so

$$3 + \dfrac{12}{n} \le 4$$

$$\dfrac{12}{n} \le 1$$

$$12 \le n \text{ since } n > 0$$

The rat first traverses the maze on the 12th trial.

  e. $12/n$ gets smaller and smaller as n increases. Thus $12/n \to 0$ as $n \to +\infty$ and $f(n)$ gets closer and closer to 3. Note: this mathematical result is not practical since n has an upper bound.

60. $S(r) = C(R^2 - r^2)$

$$= 1.76 \times 10^5(1.2^2 \times 10^{-4} - r^2)$$

  a. $S(0) \approx 25.344$ cm^3/s^2

  b. $S(0.6 \times 10^{-2}) \approx 19.008$ cm^3/s^2

61. a. $C(25t) = (25t)^2 + (25t) + 900$

$$= 625t^2 + 25t + 900$$

  b. $C(75) = 75^2 + 75 + 900 = \$6,600$.

  c. $11,000 = 625t^2 + 25t + 900$

$$0 = 625t^2 + 25t - 10,100$$

$$0 = 25t^2 + t - 404$$

$$0 = (25t + 101)(t - 4)$$

$$t = -\dfrac{101}{25}, 4$$

Negative values of time are not in the domain so $t = 4$ hours.

62. Rough crossings at $x \approx -2.6, 0.8, 3.8$

63. By long division obtain $x - 1$.

64. a. Yes, this function is "almost linear" locally about any x.

  b. $G(x) = (x + 2)(x^2 - 3)(x^2 + 3)$

  c. No, G/F behaves like a quadratic function for large x because G is quintic and F is cubic.

65. a. Hint: note the key angle θ satisfies

$$\dfrac{2\pi}{\theta} = \dfrac{2\pi r}{25}$$

  b. $r \approx 313.331$

66. Let $x = 11$, then

$$f(11 + 3) = \dfrac{f(11) - 1}{f(11) + 1} = \dfrac{10}{12} = \dfrac{5}{6}$$

Let $x = 14$, then

$$f(14 + 3) = \dfrac{f(14) - 1}{f(14) + 1} = \dfrac{\frac{5}{6} - 1}{\frac{5}{6} + 1} = -\dfrac{1}{11}$$

Let $x = 17$, then

$$f(17 + 3) = \dfrac{f(17) - 1}{f(17) + 1} = \dfrac{-\frac{1}{11} - 1}{-\frac{1}{11} + 1} = -\dfrac{6}{5}$$

Let $x = 20$, then

$$f(20 + 3) = \dfrac{f(20) - 1}{f(20) + 1} = \dfrac{-\frac{6}{5} - 1}{-\frac{6}{5} + 1} = 11$$

The pattern is for $f(3k + 2)$ for $k = 3, 4, 5, \cdots$

so that $3k + 2 = 2{,}000$ when $k = 666$. The pattern repeats every 4 elements, so $f(2000) = -\frac{6}{5}$.

1.4 The Limit of a Function, Page 47

1. 0   **2.** 2   **3.** 6   **4.** 7   **5.** 7   **6.** 10   **7.** 6

8. 4   **9.** 2   **10.** not defined   **11.** 2   **12.** 7

13.

x	2	3	4	4.5	4.9	4.99
$f(x)$	3	7	11	13	14.6	14.96

$\lim\limits_{x \to 5} f(x)$ appears to be 15.

14.

x	1	1.5	1.9	1.99	1.999	1.9999
$f(x)$	-1	-0.5	0.1	0.01	0.001	0.0001

$\lim\limits_{x \to 2} g(x)$ appears to be 0.

15.

x	1	1.9	1.99	1.999
$f(x)$	7	9.7	9.97	9.997

x	3	2.5	2.1	2.001
$f(x)$	13	11.5	10.3	10.003

$\lim\limits_{x \to 2} h(x)$ appears to be 10.

16.

x	1	0.5	0.1	0.01	0.001
$f(x)$	15.33	0.11	0.66	0.6666	0.666666

$\lim\limits_{x \to 2} g(x)$ appears to be $\frac{2}{3}$.

17. 8   **18.** 4   **19.** 2   **20.** 0   **21.** -1 **22.** 1

23. Finding a limit means "getting closer and closer to" without being at a particular point.

24. 0.00   **25.** 1.00   **26.** 0.00   **27.** 5.00   **28.** -0.50

29. -0.17 **30.** does not exist   **31.** does not exist

32. does not exist   **33.** -0.32 **34.** 0.24

35. 0.00   **36.** 0.00   **37.** 0.64   **38.** 8.00

39. does not exist   **40.** 5.00   **41.** 1.00

42. -2.00 **43.** 3.14   **44.** 0.17   **45.** 0.00

46. 0.25   **47.** 0.67   **48.** does not exist

49. -0.06 **50.** 2.00   **51.** 3.00   **52.** 1.00

53. 1.00   **54.** 2.72   **55.** 2.00   **56.** 0.00

57. 0.00   **58.** does not exist

59. a. $v(t) = \lim\limits_{x \to t} \dfrac{s(x) - s(t)}{x - t}$

$= \lim\limits_{x \to t} \dfrac{-16(x^2 - t^2) + 40(x - t)}{x - t}$

$= \lim\limits_{x \to t} \dfrac{(x - t)[-16(x + t) + 40]}{x - t}$

$= \lim\limits_{x \to t} (-16x - 16t + 40)$

$= -32t + 40$

b. $v(0) = \lim\limits_{x \to 0} (-16x - 0 + 40)$

$= 40$ ft/s

c. $s(t) = 0$ if $-16t^2 + 40t + 24 = 0$ or if

$2t^2 - 5t - 3 = 0$

$(t - 3)(2t + 1) = 0$

$t = 3, \ -\frac{1}{2}$

Reject the negative solution:

$v(3) = \lim\limits_{x \to 3} (-16x - 48 + 40)$

$= -56$ ft/s

d. At the highest point on the **trajectory**, the ball has stopped moving upward and has not yet started on its downward fall. This occurs at $-16t^2 - 16t + 40 = 0$ or $t = 1.25$ seconds.

60. Assume a linear relationship. The slopes between successive points should be approximately the same. Consider the slope, for example, between $(-1, 41.5)$ and $(1, 43.2)$:

$m = \dfrac{43.2 - 41.5}{2} \approx 0.85$

This is the miles per minute, so the speed is approximately

$\dfrac{0.85 \text{ miles}}{\text{minute}} \cdot \dfrac{60 \text{ minutes}}{1 \text{ hour}} = \ \approx 51 \text{ m/hr}$

61. 228   **62.** does not exist   **63.** 0   **64.** 1

65. a. values are all -1; might guess that limit is -1

b. values are all 0; might guess that limit is 0

c. Follow the instructions in **a.** and **b.** and make your conclusion about the limits. How can they be different? A look at the graph might help:

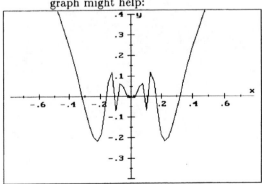

Even though we cannot determine the limit, we might guess that it is 0.

1.6 Properties of Limits, Page 54

1. $\lim\limits_{x \to -2} (x^2 + 3x - 7) = 4 - 6 - 7 = -9$

2. $\lim\limits_{t \to 0} (t^3 - 5t^2 + 4) = 0^3 - 5(0)^2 + 4 = 4$

3. $\lim\limits_{x \to 3} (x + 5)(2x - 7) = (3 + 5)(2 \cdot 3 - 7) = -8$

4. $\lim\limits_{x \to 4} \left(\dfrac{1}{x} + \dfrac{3}{x - 5} \right) = \dfrac{1}{4} + \dfrac{3}{-1} = -\dfrac{11}{4}$

5. $\lim\limits_{z \to 1} \dfrac{z^2 + z - 3}{z + 1} = \dfrac{1 + 1 - 3}{1 + 1} = -\dfrac{1}{2}$

6. $\lim\limits_{x \to 3} \dfrac{x^2 + 3x - 10}{3x^2 + 5x - 7} = \dfrac{3^2 + 3(3) - 10}{3(3^2) + 5(3) - 7} = \dfrac{8}{35}$

7. $\lim\limits_{x \to \pi/3} \sec x = \sec \dfrac{\pi}{3} = 2$

8. $\lim\limits_{x \to \pi/4} \dfrac{1 + \tan x}{\csc x + 2} = \dfrac{1 + \tan \frac{\pi}{4}}{\csc \frac{\pi}{4} + 2} = \dfrac{2}{2 + \sqrt{2}}$

$\qquad = 2 - \sqrt{2}$

9. $\lim\limits_{x \to \pi/3} \dfrac{x \sin \pi x}{1 + \cos \pi x} = \dfrac{\frac{1}{3} \sin \frac{\pi}{3}}{1 + \cos \frac{\pi}{3}} = \dfrac{\sqrt{3}/6}{3/2} = \dfrac{\sqrt{3}}{9}$

10. $\lim\limits_{x \to 6} \dfrac{\tan(\pi/x)}{x - 1} = \dfrac{\tan(\pi/6)}{6 - 1} = \dfrac{1/\sqrt{3}}{5}$

$\qquad = \dfrac{1}{5\sqrt{3}} = \dfrac{\sqrt{3}}{15}$

11. $\lim\limits_{x \to -2} \dfrac{4 - u^2}{2 + u} = \lim\limits_{x \to -2} \dfrac{(2 + u)(2 - u)}{(2 + u)}$

$\qquad = \lim\limits_{x \to -2} (2 - u) = 4$

12. $\lim\limits_{x \to 2} \dfrac{x^2 - 4x + 4}{x^2 - x - 2} = \lim\limits_{x \to 2} \dfrac{(x - 2)^2}{(x - 2)(x + 1)}$

$\qquad = \lim\limits_{x \to 2} \dfrac{x - 2}{x + 1} = \dfrac{2 - 2}{2 + 1} = 0$

13. $\lim\limits_{x \to 1} \dfrac{\frac{1}{x} - 1}{x - 1} = \lim\limits_{x \to 1} \dfrac{\frac{1 - x}{x}}{x - 1} = \lim\limits_{x \to 1} \dfrac{-1}{x} = -1$

14. $\lim\limits_{x \to 0} \dfrac{(x + 1)^2 - 1}{x} = \lim\limits_{x \to 0} \dfrac{(x + 1 + 1)(x + 1 - 1)}{2}$

$\qquad = \lim\limits_{x \to 0} \dfrac{(x + 2)(x)}{x} = \lim\limits_{x \to 0} (x + 2) = 2$

15. $\lim\limits_{x \to 1} \left(\dfrac{(x - 2)(x - 1)}{(x + 2)(x - 1)} \right)^2 = \lim\limits_{x \to 1} \dfrac{(x - 2)^2}{(x + 2)^2} = \dfrac{1}{9}$

16. $\lim\limits_{x \to 3} \sqrt{\dfrac{x^2 - 2x - 3}{x - 3}} = \lim\limits_{x \to 3} \sqrt{\dfrac{(x - 3)(x + 1)}{x - 3}}$

$\qquad = \lim\limits_{x \to 3} \sqrt{x + 1} = \lim\limits_{x \to 3} \sqrt{(x + 1)} = 2$

17. $\lim\limits_{x \to 1} \dfrac{\sqrt{x} - 1}{x - 1} = \lim\limits_{x \to 1} \dfrac{(\sqrt{x} - 1)(\sqrt{x} + 1)}{(x - 1)(\sqrt{x} + 1)}$

$\qquad = \lim\limits_{x \to 1} \dfrac{x - 1}{(x - 1)(\sqrt{x} + 1)}$

$\qquad = \lim\limits_{x \to 1} \dfrac{1}{\sqrt{x} + 1} = \dfrac{1}{2}$

18. $\lim\limits_{y \to 2} \dfrac{\sqrt{y + 2} - 2}{y - 2}$

$\qquad = \lim\limits_{y \to 2} \dfrac{(\sqrt{y + 2} - 2)(\sqrt{y + 2} + 2)}{(y - 2)(\sqrt{y + 2} + 2)}$

$\qquad = \lim\limits_{y \to 2} \dfrac{y + 2 - 4}{(y - 2)(\sqrt{y + 2} + 2)}$

$\qquad = \lim\limits_{y \to 2} \dfrac{1}{\sqrt{y + 2} + 2} = \dfrac{1}{\sqrt{2 + 2} + 2} = \dfrac{1}{4}$

19. $\lim\limits_{x \to 0} \dfrac{1 - \sin x}{\cos^2 x} = \dfrac{1 - 0}{\cos^2 0} = 1$

20. $\lim\limits_{x \to 0} \dfrac{1 - 2\cos x}{\sqrt{3} - 2\sin x} = \dfrac{1 - 2\cos 0}{\sqrt{3} - 2\sin 0} = \dfrac{-1}{\sqrt{3}}$

21. $\lim\limits_{x \to 0} \dfrac{\sin 2x}{x} = \lim\limits_{x \to 0} 2\left(\dfrac{\sin 2x}{2x} \right) = 2(1) = 2$

22. $\lim\limits_{x \to 0} \dfrac{\sin 3x}{2x} = \lim\limits_{x \to 0} \dfrac{3}{2}\left(\dfrac{\sin 3x}{3x} \right) = \dfrac{3}{2}(1) = \dfrac{3}{2}$

23. $\lim\limits_{x \to 0} \dfrac{\tan x}{x} = \lim\limits_{x \to 0} \dfrac{\frac{\sin x}{\cos x}}{x} = \lim\limits_{x \to 0} \dfrac{\frac{\sin x}{x}}{\cos x}$

$\qquad = \dfrac{\lim\limits_{x \to 0} \frac{\sin x}{x}}{\lim\limits_{x \to 0} \cos x} = \dfrac{1}{1} = 1$

24. $\lim\limits_{x \to 0} \dfrac{\sin^2 x}{x^2} = \left(\lim\limits_{x \to 0} \dfrac{\sin x}{x} \right)^2 = (1)^2 = 1$

25. $\lim\limits_{x \to 0} \dfrac{\frac{1}{x + 3} - \frac{1}{3}}{x} = \lim\limits_{x \to 0} \dfrac{3 - x - 3}{3x(x + 3)}$

$\qquad = \lim\limits_{x \to 0} \dfrac{-1}{3(x + 3)} = -\dfrac{1}{9}$

26. $\lim\limits_{x \to 3} \dfrac{\frac{1}{x} - \frac{1}{3}}{x - 3} = \lim\limits_{x \to 3} \dfrac{3 - x}{3x(x - 3)}$

$\qquad = \lim\limits_{x \to 3} \dfrac{-1}{3x} = -\dfrac{1}{9}$

27. $\lim\limits_{x \to 0} \dfrac{\sec x - 1}{x \sec x} = \lim\limits_{x \to 0} \dfrac{1 - \cos x}{x}$

$\qquad = \lim\limits_{x \to 0} \dfrac{(1 - \cos x)(1 + \cos x)}{x(1 + \cos x)}$

$\qquad = \lim\limits_{x \to 0} \dfrac{\sin^2 x}{x(1 + \cos x)}$

$\qquad = \lim\limits_{x \to 0} \dfrac{\sin x}{1 + \cos x} \left(\lim\limits_{x \to 0} \dfrac{\sin x}{x} \right) = 0$

28. $\lim\limits_{x \to \pi/4} \dfrac{1 - \tan x}{\sin x - \cos x}$

$\qquad = \lim\limits_{x \to \pi/4} \dfrac{\cos x - \sin x}{\cos x(\sin x - \cos x)}$

$\qquad = \lim\limits_{x \to \pi/4} \dfrac{-1}{\cos x} = -\sqrt{2}$

29. $\lim\limits_{x\to 0} \dfrac{\tan^2 x}{x} = \lim\limits_{x\to 0}\left(\dfrac{\sin x}{x}\right)\!\left(\dfrac{\sin x}{\cos^2 x}\right) = 1(0) = 0$

30. $\lim\limits_{x\to\pi} \dfrac{\tan x}{1 + \sec x} = \lim\limits_{x\to\pi} \dfrac{(\tan x)(1 - \sec x)}{(1 + \sec x)(1 - \sec x)}$

$\qquad = \lim\limits_{x\to\pi} \dfrac{(\tan x)(1 - \sec x)}{1 - \sec^2 x}$

$\qquad = \lim\limits_{x\to\pi} \dfrac{(\tan x)(1 - \sec x)}{-\tan^2 x}$

$\qquad = -\lim\limits_{x\to\pi} \dfrac{1 - \sec x}{\tan x}$

which is not defined since the numerator approaches 2 and the denominator approaches 0

31. The limit of a polynomial can be found by substitution; that is

$$\lim\limits_{x\to a} f(x) = f(a)$$

32. If replacing the variable by the limiting value does not give an undefined form, then evaluate the limit by substitution; that is,

$$\lim\limits_{x\to a} \dfrac{f(x)}{g(x)} = \dfrac{\lim\limits_{x\to a} f(x)}{\lim\limits_{x\to a} g(x)}$$

provided $\lim\limits_{x\to a} g(x) \neq 0$. If replacing the variable by the limiting value leads to $0/0$ or ∞/∞, then try to rewrite the fraction in such a way that 0 or ∞ reduces out of the fraction.

33. $\lim\limits_{x\to 0} \dfrac{\sin ax}{x} = \lim\limits_{ax\to 0} a\!\left(\dfrac{\sin ax}{ax}\right) = a(1) = a$

34. $\lim\limits_{x\to 2^-} (x^2 - 2x) = 2^2 - 2(2) = 0$

35. $\lim\limits_{x\to 1^+} \dfrac{\sqrt{x - 1} + x}{1 - 2x} = \dfrac{0 + 1}{1 - 2} = -1$

36. $\lim\limits_{x\to 0^-} \dfrac{|x|}{x} = \lim\limits_{x\to 0^-} \dfrac{-x}{x} = -1$

37. $\lim\limits_{x\to 0^+} \dfrac{|x|}{x} = \lim\limits_{x\to 0^+} \dfrac{x}{x} = 1$

38. $\lim\limits_{x\to 2^-} (3 - 2x) = 3 - 4 = -1$

39. $\lim\limits_{x\to 2^+} (x^2 - 5) = 4 - 5 = -1$

40. $\lim\limits_{s\to 1^-} \dfrac{s^2 - s}{s - 1} = \lim\limits_{x\to 1^-} s = 1$

41. $\lim\limits_{s\to 1^+} \sqrt{1 - s} = \sqrt{1 - 1} = 0$

42. Not defined because of 0 in the denominator.

43. Not defined because of 0 in the denominator.

44. $\lim\limits_{t\to 0} \dfrac{t^2 - 4}{t^2 - 4t + 4} = \lim\limits_{t\to 2} \dfrac{(t + 2)(t - 2)}{(t - 2)^2}$

$\qquad = \lim\limits_{t\to 2} \dfrac{t + 2}{t - 2}$

Not defined because of the 0 in the denominator.

45. Does not exist because the limit from the right is 1 and the limit from the left is -1.

46. Does not exist because the limit from the right is 2 and the limit from the left is -5.

47. Does not exist because the limit from the left is 5 and the limit from the right is -1.

48. Does not exist because $\tan x$ increases without bound as $x \to \pi/2$.

49. Does not exist because $\csc \pi x$ increases without bound as $x \to 1$

50. $\lim\limits_{x\to 3} \dfrac{x^2 - 9}{x - 3} = \lim\limits_{x\to 3} \dfrac{(x + 3)(x - 3)}{x - 3}$

$\qquad = \lim\limits_{x\to 3} (x + 3) = 6$

51. $\lim\limits_{x\to 0} \dfrac{x - \sin x}{x^3} \approx 0.166667$ by table

52. $\lim\limits_{x\to 1} \dfrac{x^5 - 1}{x - 1} = \lim\limits_{x\to 1} (x^4 + x^3 + x^2 + x + 1)$

$\qquad\qquad$ by synthetic division

$\qquad = 5$

53. $\lim\limits_{x\to 1} \dfrac{\sqrt{x} - 1}{x - 1} = \lim\limits_{x\to 1} \dfrac{\sqrt{x} - 1}{(\sqrt{x} - 1)(\sqrt{x} + 1)}$

$\qquad = \lim\limits_{x\to 1} \dfrac{1}{\sqrt{x} + 1} = \dfrac{1}{2}$

54. $\lim\limits_{x\to 1} \dfrac{\frac{1}{x} - 1}{\sqrt{x} - 1} = \lim\limits_{x\to 1} \dfrac{1 - x}{x(\sqrt{x} - 1)}$

$\qquad = \lim\limits_{x\to 1} \dfrac{(1 - x)(\sqrt{x} + 1)}{x(\sqrt{x} - 1)(\sqrt{x} + 1)}$

$\qquad = \lim\limits_{x\to 1} \dfrac{(1 - x)(\sqrt{x} + 1)}{x(x - 1)}$

$\qquad = (-1) \lim\limits_{x\to 1} \dfrac{\sqrt{x} + 1}{x} = -2$

55. $\lim\limits_{x\to 0}\left(\dfrac{x - 1}{x^2}\right) = \dfrac{-1}{0}$

This fraction approaches $-\infty$ as $x \to 0$, so the limit does not exist.

56. $\lim\limits_{x\to 5} f(x) = 8$

57. $\lim\limits_{t\to 2} g(t) = 4$ since the left and right limits are identical.

58. $\lim\limits_{x\to 2} f(x) = \lim\limits_{x\to 2} 2(2 + 1) = 6$

59. $\lim\limits_{x\to 3} f(x) = 8$ since the left and right limits

are identical.

60. $\lim\limits_{x\to 0}\left[x^2 - \dfrac{\cos x}{1,000,000,000}\right] = 10^{-9}$

On many calculators, the result appears to be 0.

61. $Q = \dfrac{f(x + \Delta x) - f(x)}{\Delta x}$

$= \dfrac{3(x + \Delta x) - 5 - (3x - 5)}{\Delta x}$

$= \dfrac{3\Delta x}{\Delta x} = 3$

$\lim\limits_{\Delta x\to 0} Q = 3$

62. $Q = \dfrac{f(x + \Delta x) - f(x)}{\Delta x}$

$= \dfrac{x^2 + 2x\Delta x + (\Delta x)^2}{\Delta x}$

$= 2x + \Delta x$

$\lim\limits_{\Delta x\to 0} Q = 2x$

63. $Q = \dfrac{f(x + \Delta x) - f(x)}{\Delta x}$

$= \dfrac{\dfrac{3}{x + \Delta x} - \dfrac{3}{x}}{\Delta x}$

$= 3\left[\dfrac{-1}{x(x + \Delta x)}\right]$

$\lim\limits_{\Delta x\to 0} Q = \dfrac{-3}{x^2}$

64. $Q = \dfrac{f(x + \Delta x) - f(x)}{\Delta x}$

$= \dfrac{\sqrt{x + \Delta x} - \sqrt{x}}{\Delta x}$

$= \dfrac{(\sqrt{x + \Delta x} - \sqrt{x})(\sqrt{x + \Delta x} + \sqrt{x})}{\Delta x(\sqrt{x + \Delta x} + \sqrt{x})}$

$= \dfrac{x + \Delta x - x}{\Delta x(\sqrt{x + \Delta x} + \sqrt{x})}$

$= \dfrac{1}{\sqrt{x + \Delta x} + \sqrt{x}}$

$\lim\limits_{\Delta x\to 0} Q = \dfrac{1}{2\sqrt{x}}$

65. $\lim\limits_{x\to 0} \dfrac{\sin ax}{\sin bx} = \lim\limits_{x\to 0}\left(\dfrac{\dfrac{a \sin ax}{a}}{\dfrac{b \sin bx}{b}}\right)$

$\lim\limits_{x\to 0}\left(\dfrac{\dfrac{a \sin ax}{ax}}{\dfrac{b \sin bx}{bx}}\right) = \dfrac{a \lim\limits_{x\to 0}\dfrac{\sin ax}{ax}}{b \lim\limits_{x\to 0}\dfrac{\sin bx}{bx}} = \dfrac{a}{b}$

66. $\lim\limits_{x\to 0}\dfrac{\tan ax}{\tan bx} = \dfrac{a}{b}\,\lim\limits_{x\to 0}\dfrac{\sin ax}{ax}\,\dfrac{bx}{\sin bx} = \dfrac{a}{b}$

67. $f(x) = \dfrac{1}{x^2};$ if $|x| < \dfrac{1}{10\sqrt{L}}$, then $x^2 > 100 L$.
As L grows, $|x|$ shrinks, and $f(x)$ grows beyond all bounds. This means $\lim\limits_{x\to c} f(x)$ does not exist.

68. $\lim\limits_{x\to c} f(x) = \lim\limits_{x\to c}\dfrac{1}{x - c}$ is not defined.

$\lim\limits_{x\to c} g(x) = \lim\limits_{x\to c}\dfrac{-1}{x - c}$ is not defined.

However, $f(x) + g(x) = 0$ regardless of what value x approaches.

69. a. $\lim\limits_{x\to c} f(x) = \lim\limits_{x\to c}\dfrac{1}{x^2 - c^2}$ does not exist.

$\lim\limits_{x\to c} g(x) = \lim\limits_{x\to c}(x - c) = 0$

$\lim\limits_{x\to c}[f(x)g(x)] = \lim\limits_{x\to c}\dfrac{x - c}{x^2 - c^2}$

$= \lim\limits_{x\to c}\dfrac{1}{x + c} = \dfrac{1}{2c}$ which exist for $c \neq 0$.

b. $\lim\limits_{x\to 0} f(x) = \lim\limits_{x\to 0}\dfrac{1}{x}$ does not exist.

$\lim\limits_{x\to 0} g(x) = \lim\limits_{x\to 0}\dfrac{1}{\sin x}$ does not exist.

$\lim\limits_{x\to 0}\dfrac{f(x)}{g(x)} = \lim\limits_{x\to 0}\dfrac{\sin x}{x} = 1$

70. $\lim\limits_{x\to 0}[f(x) + g(x)] = L_1$ and $\lim\limits_{x\to 0} f(x) = L_2$.

Then $\lim\limits_{x\to 0} g(x) = \lim\limits_{x\to 0}\{[f(x) + g(x)] - f(x)\}$

$= L_1 - L_2 = L$

71. To find $\lim\limits_{x\to x_0} \cos x$, let

$h = x - x_0$ or $x = h + x_0$

$\lim\limits_{x\to x_0} \cos x = \lim\limits_{x\to x_0}\cos(h + x_0)$

$= \lim\limits_{x\to x_0}[\cos h \cos x_0 - \sin h \sin x_0]$

$= 1(\cos x_0) - 0 = \cos x_0$

72. To find $\lim\limits_{x\to x_0} \tan x$, let

$h = x - x_0$ or $x = h + x_0$

$\lim\limits_{x\to x_0} \tan x = \lim\limits_{x\to x_0}\tan(h + x_0)$

$= \lim\limits_{x\to x_0}\dfrac{\tan h + \tan x_0}{1 - \tan h \tan x_0}$

$= \dfrac{\tan x_0}{1} = \tan x_0$

1.7 Continuity, Page 66

1. Temperature is continuous, so TEMPERATURE = f(time) would be a continuous function. The domain would be midnight to midnight; $0 \leq t < 24$.

2. The humidity on a specific day at a given location is continuous.

3. The selling price of ATT stock is not continuous. The domain is the set of positive rational numbers that can be divided evenly by 8.

4. The number of unemployed people in the US is not continuous. The domain is a subset of positive integers.

5. The charges (range of the function) consist of rational numbers only (dollars and cents to the nearest cent), so the function CHARGE = f(MILEAGE) would be a step function (that is, not continuous). The domain would consist of the mileage from the beginning of the trip to its end.

6. The charges to mail a package as a function of weight are not continuous. The price displayed on the postage meter (converted to cents) is a positive integer.

7. No suspicious points and no points of discontinuity with a polynomial.

8. $x = \frac{1}{2}$ is suspicious (causes division by 0). This is also a point of discontinuity.

9. The denominator factors to $x(x - 1)$, so suspicious points would be $x = 0, 1$. There will be a hole discontinuity at $x = 0$ and a pole discontinuity at $x = 1$.

10. The function f is a polynomial function so there are no suspicious points and no points of discontinuity.

11. $x = 0$ is suspicious and is a point of discontinuity, since the denominator vanishes. $x < 0$ are not in the domain since square roots of negative numbers are not defined in the set of real numbers.

12. There are no suspicious points. We are not dealing with an even root and there is no denominator. There are no points of discontinuity.

13. We have suspicious points where the denominators are 0 at $t = 0, -1$. There are pole discontinuities at each of these points.

14. Suspicious points $x = -2$ and $x = 1$ are also points of discontinuity (since the denominator is 0).

15. $x = 1$ is a suspicious point; there are no points of discontinuity.

16. $t = 1$ and $t = 3$ are suspicious points; $t = 3$ is a point of discontinuity.

17. The sine and cosine are continuous on the reals, but the tangent is discontinuous at $x = \pi/2 + n\pi$. Each of these values will have a pole type discontinuity.

18. $g(x) = \dfrac{\cot x}{\sin x - \cos x} = \dfrac{\cos x}{\sin x(\sin x - \cos x)}$
The denominator vanishes for a multiple for π because of $\sin x$. It also vanishes at $x = (2n + 1)\pi/4$, an odd multiple of $\pi/4$, because that is where $\sin x = \cos x$.

19. $h(x) = \csc x \cot x = \dfrac{\cos x}{\sin^2 x}$

Suspicious points are $x = n\pi$ (n an integer), and these are also the points of discontinuity.

20. $f(x) = \dfrac{\sec x}{\sin x \tan x} = \dfrac{\frac{1}{\cos x}}{\frac{\sin^2 x}{\cos x}} = \dfrac{1}{\sin^2 x}$
Suspicious points are $x = n\pi$ (n an integer), and these are also the points of discontinuity.

21. For continuity, $f(2)$ must equal
$$\lim_{x \to 2} f(x) = \lim_{x \to 2} \frac{(x - 2)(x + 1)}{x - 2}$$
$$= \lim_{x \to 2} (x + 1) = 3$$

22. For continuity, $f(2)$ must equal
$$\lim_{x \to 2} f(x) = \lim_{x \to 2} \sqrt{\frac{x^2 - 4}{x - 2}}$$
$$= \lim_{x \to 2} \sqrt{\frac{(x - 2)(x + 2)}{x - 2}}$$
$$= \lim_{x \to 2} \sqrt{x + 2} = 2$$

23. For continuity, $f(2)$ must equal
$$\lim_{x \to 2} f(x) = \lim_{x \to 2} \frac{\sin(\pi x)}{x - 2}$$
$$= \pi \quad \text{(by table or graphing)}$$

24. For continuity, $f(2)$ must equal
$$\lim_{x \to 2} f(x) = \lim_{x \to 2} \frac{\cos \frac{\pi}{2}}{x - 2}$$
$$= \frac{\pi}{4} \quad \text{(by table or graphing)}$$

25. The function is not defined at 2, and since $\lim_{x \to 2^-} f(x) = 11$ and $\lim_{x \to 2^+} f(x) = 9$

No value can be assigned to $f(2)$ to "tie together" the two pieces. This is sometimes called an "essential" discontinuity. Only hole

type discontinuities are "removable".

26. This function cannot be continuous at $x = 2$ because of the pole at this point. No finite value for $f(2)$ will circumvent the effect of the pole.

27. No suspicious points on $[1, 2]$; continuous

28. Suspicious point $x = 0$. Discontinuous on $[0, 1]$ since the pole $x = 0$ is in the domain. If the interval had been $(0, 1]$, the function would be continuous on the interval.

29. The function has a pole type discontinuity at $t = 0$, but 0 is not in the interval. The function is continuous on $[-3, 0)$.

30. Suspicious point $x = 1$; discontinuous on $[0, 7]$, since the pole $x = 1$ is in the domain.

31. Suspicious point $x = 2$; the limit from the left is 4 while that from the right is 7. The function is discontinuous at $x = 2$.

32. Suspicious point $t = 0$; the limit from the left is 15 while that from the right is 0. The function is discontinuous at $t = 0$.

33. No suspicious points; $y = x$ and $y = \sin x$ are continuous on the reals, so $f(x) = x \sin x$ will be continuous on $(0, \pi)$.

34. Suspicious point $x = 0$; discontinuous on $[0, \pi]$ because the pole $x = 0$ is in the interval.

35. Both f and g have jumps at $x = 0$, but
$$f(x) + g(x) = \begin{cases} 4x \text{ if } x \neq 0 \\ 0 \text{ if } x = 0 \end{cases}$$
is a function which removes the jump, and is therefore continuous at $x = 0$.

36. f/g and g are continuous at $x = c$. This means
$$\lim_{x \to c} \frac{f(x)}{g(x)} = L_1 = \frac{f(c)}{g(c)} \quad \text{and}$$
$$\lim_{x \to c} g(x) = L_2 = g(c)$$
Now,
$$\lim_{x \to c} \frac{f(x)}{g(x)} = \frac{\lim_{x \to c} f(x)}{\lim_{x \to c} g(x)} = \frac{\lim_{x \to c} f(x)}{L_2} = L_1$$
or $\lim_{x \to c} f(x) = L_1 L_2$.

Also, $\dfrac{f(c)}{g(c)} = \dfrac{f(c)}{L_2}$ or $f(c) = L_1 L_2$.

Thus, $\lim_{x \to c} f(x) = f(c)$.

37. $f(x) = \begin{cases} \dfrac{1}{x-1} & \text{if } x \neq 1 \\ -2 & \text{if } x = 1 \end{cases}$

and

$g(x) = \begin{cases} x^2 - 1 \text{ if } x \neq 1 \\ -1 \text{ if } x = 1 \end{cases}$

f is discontinuous at $x = 1$, but

$$f(x)g(x) = \begin{cases} x + 1 \text{ if } x \neq 1 \\ 2 \text{ if } x = 1 \end{cases}$$

is continuous everywhere.

38. The function f is continuous at one $x = 0$ if
$$f(x) = \begin{cases} 1 + x \text{ if } x \text{ is rational} \\ 1 \text{ if } x = 0 \\ 1 - x \text{ if } x \text{ is irrational} \end{cases}$$

39. $f(x) = \sqrt[3]{x} - x^2 - 2x + 1$ is continuous on $[0, 1]$ and $f(0) = 1$, $f(1) = -1$ so the hypotheses of the intermediate value theorem are met, and we are guaranteed that there is at least one number c on $[0, 1]$ such that $f(c) = 0$.

40. $f(x) = \dfrac{1}{x+1} - x^2 + x + 1$ is continuous on $[1, 2]$ and $f(1) = \frac{3}{2}$, $f(2) = -\frac{2}{3}$ so the hypotheses of the intermediate value theorem are met, and we are guaranteed that there is at least one number c on $[1, 2]$ such that $f(c) = 0$.

41. $f(x) = \sqrt[3]{x - 8} + 9x^{2/3} - 29$ is continuous on $[0, 8]$ and $f(0) = -31$, $f(8) = 7$ so the hypotheses of the intermediate value theorem are met, and we are guaranteed that there is at least one number c on $[0, 8]$ such that $f(c) = 0$.

42. $f(x) = \tan x - 2x^2 + 1$ is continuous on $[-\frac{\pi}{4}, 0]$ and $f(-\frac{\pi}{4}) = -\frac{\pi^2}{8}$, $f(0) = 1$ so the hypotheses of the intermediate value theorem are met, and we are guaranteed that there is at least one number c on $[-\pi/4, 0]$ such that $f(c) = 0$.

43. $f(x) = \cos x - \sin x - x$ is continuous on the reals and $f(0) = 1$, $f(\pi/2) = -1 - \pi/2$. Since the hypotheses of the intermediate value theorem have been met, we are guaranteed that there exists at least one number c on $[0, \pi/2]$ such that $f(c) = 0$.

44. $f(x) = \cos x - x^2 + 1$ is continuous on

$[0, \pi]$ and $f(0) = 2$, $f(\pi) = -\pi^2$ so the hypotheses of the intermediate value theorem are met, and we are guaranteed that there is at least one number c on $[0, 1]$ such that $f(c) = 0$.

45. $\lim\limits_{x \to 2^-} f(x) = 2 + 1 = 3 = f(2)$

$\lim\limits_{x \to 2^+} f(x) = 2^2 = 4 \neq f(2)$

46. $f(0) = f(2)$, so $b = 2a + b$

Also, $\lim\limits_{x \to 1^-} f(x) = 1 + b = f(1) = 4$

Thus, $a = 0$ and $b = 3$.

47. At 12:00 the hands coincide. For any other hour h the minute hand at $f(h)$ is at a position $5h$ minutes *less than* the position of the hour hand, and at $f(h+1)$ the minute hand is at a position $5(12 - h)$ minutes *greater than* the position of the hour hand. Since the position of the minute hand is a continuous function, we have met the hypotheses of the intermediate value theorem and are guaranteed of the existence of some time t on the interval $[h, h+1]$ where the distance between the hands is 0.

48. **a.** The colony dies out when $-8t + 66 = 0$ or when $t = 33/4$. This is about 8 min 15 sec.

b. $P(2) = 5$ thousand; the function is continuous at $t = 5$, and $P(7) = -56 + 66 = 10$ thousand. By the intermediate value theorem, there is at least one solution on $[2, 7]$.

49. $f(5) = 8$; from the right, $5a + 3 = 8$, so $a = 1$. From the left, $25 + 5b + 1 = 8$, so $b = -18/5$.

50. $f(2) = b$; if $t \neq 2$ and $a = 2$, then

$\dfrac{2t - 4}{t - 2} = 2$, so $b = 2$

51. This function is continuous on $[0, \infty)$ except possibly at $x = 1$. For continuity it is required that $\lim\limits_{x \to 1} f(x) = f(1)$. So we need:

$\lim\limits_{x \to 1} \dfrac{\sqrt{x} - a}{x - 1} = b$ Since this expression is undefined, we rationalize the numerator:

$\lim\limits_{x \to 1} \dfrac{(\sqrt{x} - a)(\sqrt{x} + a)}{(x - 1)(\sqrt{x} + a)}$

$= \lim\limits_{x \to 1} \dfrac{x - a^2}{(x - 1)(\sqrt{x} + a)}$.

Now if we let $a = 1$, this will reduce to

$\lim\limits_{x \to 1} \dfrac{1}{\sqrt{x} + 1} = \dfrac{1}{1 + 1} = \dfrac{1}{2} = b$.

Therefore $a = 1$, $b = \frac{1}{2}$ will make this function continuous at $x = 1$.

52. $g(-5) = -5b + 2$; if $x < -5$ and $a = \pm\frac{3}{25}$ we have

$$\dfrac{ax^2 - 3}{x + 5} = \dfrac{\pm\frac{3}{25}x^2 - 3}{x + 5}$$

$$= \pm\dfrac{3}{25}\left(\dfrac{x^2 - 25}{x + 5}\right)$$

$$= \pm\dfrac{3}{25}\left[\dfrac{(x + 5)(x - 5)}{x + 5}\right]$$

$$= \pm\dfrac{3}{25}(x - 5)$$

Now, $\lim\limits_{x \to -5} g(x) = \pm\dfrac{3}{25}(-10)$

$$= \pm\dfrac{6}{5} = -5b + 2$$

$b = \dfrac{16}{25}$ if $a > 0$, $b = \dfrac{4}{25}$ if $a < 0$.

53. $g(0) = 5$; if $x < 0$,

$\lim\limits_{x \to 0^-} \dfrac{\sin ax}{x} = a \lim\limits_{x \to 0^-} \dfrac{\sin ax}{ax} = a = 5$

Similarly, $\lim\limits_{x \to 0^+} g(x) = 0 + b = 5$, $b = 5$

54. $f(0) = 4$; if $x < 0$,

$\lim\limits_{x \to 0^-} \dfrac{\tan ax}{\tan bx}$

$= \dfrac{a}{b} \lim\limits_{x \to 0^-} \left[\left(\dfrac{\sin ax}{ax}\right)\left(\dfrac{bx}{\tan bx}\right)\left(\dfrac{\cos bx}{\cos ax}\right)\right]$

$= \dfrac{a}{b} = 4$

Similarly,

$\lim\limits_{x \to 0^+} f(x) = 0 + b = 4$, so $a = 16$

and $b = 4$.

55. $f(x) = x^2 - x - 1 = 0$ on $[1, 2]$

Left: $x_{l_0} = 1$ and right: $x_{r_0} = 2$ so that $f(x_{l_0}) = -1$, $f(x_{r_0}) = 1$

$x_1 = \dfrac{1 + 2}{2} = \dfrac{3}{2}$; $f(x_1) = -0.25$, so

$x_{l_1} = \dfrac{3}{2}$ and $x_{r_1} = 2$

$x_2 = \dfrac{\frac{3}{2} + 2}{2} = \dfrac{7}{4}$; $f(x_2) = 0.32$, so

$x_{l_2} = \dfrac{3}{2}$ and $x_{r_2} = \dfrac{7}{4}$

$x_3 = \dfrac{\frac{3}{2} + \frac{7}{4}}{2} = \dfrac{13}{8}$; $f(x_3) = 0.016$ so

$x_{l_3} = \dfrac{3}{2}$ and $x_{r_3} = \dfrac{13}{8}$

$$x_4 = \frac{\frac{3}{2} + \frac{13}{8}}{2} = \frac{25}{16} = 1.5625$$

Continue with a computer to find
$1.618 < x < 1.619$.

56. Use a computer to show $0.381 < x < 0.382$.

57. Use a computer to show $-0.415 < x < -0.414$.

58. Use a computer to show $0.754 < x < 0.755$.

59. Use a computer to show $1.116 < x < 1.117$.

60. Use a computer to show $-2.549 < x < -2.547$.

61. With $f(x) = \frac{\sin x - x}{x^3}$, $\lim_{x \to 0} \frac{\sin x - x}{x^3} = -\frac{1}{6}$
Define

$$f(x) = \begin{cases} \dfrac{\sin x - x}{x^3} & \text{if } x \neq 0 \\[2mm] -\dfrac{1}{6} & \text{if } x = 0 \end{cases}$$

62. With $f(x) = \dfrac{x^4 - 2x^3 + 3x^2 - 5x - 2}{|x - 2|}$

$$\lim_{x \to 2} \frac{x^4 - 2x^3 + 3x^2 - 5x - 2}{|x - 2|}$$

$$= \lim_{x \to 2} (x^3 + 3x + 1) = 15$$

Define

$$f(x) = \begin{cases} x^3 + 3x + 1 & \text{if } x \neq 2 \\ 15 & \text{if } x = 2 \end{cases}$$

63. f is continuous on $[a, b]$ and $\lim_{x \to c} f(x) = f(c)$
on $[a, b]$. By the intermediate value theorem,
there is a number c such that $f(c) = L$ for all
$f(a) < L < f(b)$. Assume $f(a) < 0$, $f(b) > 0$.
Pick $L = 0$ and c will be the root.

64. $f(x)$ is continuous at $x = c$, so it must be
continuous from the left and the right (where
c is not an endpoint of an interval). $\lim_{x \to c} f(x)$

exists if and only if the right and left hand
limits exist. Conversely, assume that
continuity from the left and from the right
exist. Then

$$\lim_{x \to c^+} f(x) = f(c) = \lim_{x \to c^-} f(x)$$

This implies continuity at $x = c$.

65. $\lim_{x \to 0} f(u(x)) = \lim_{x \to 0} f(x) = 0$, since $f(x) = 0$
for $x \neq 0$.

$$f\left[\lim_{x \to 0} u(x)\right] = f\left[\lim_{x \to 0} x\right] = f(0) = 1$$

1.8 Introduction to the Theory of Limits, Page 74

1. For any chosen value of ϵ we want the
distance between $f(x)$ and L to be less than ϵ.
In absolute value notation this is

$$|f(x) - L| < \epsilon$$

For our function and limit we need

$$|(2x - 5) - (-3)| < \epsilon$$
$$|2x - 2| < \epsilon$$
$$2|x - 1| < \epsilon$$

So the distance between x and 1 needs to be
less than $\frac{\epsilon}{2}$. Therefore, let $\delta = \frac{\epsilon}{2}$.

2. $|3x + 7 - 1| = 3|x + 2| < \epsilon$ so that

$$|x + 2| < \frac{\epsilon}{3}$$

Pick $\delta = \frac{\epsilon}{3}$.

3. $\lim_{x \to 1} (3x + 1) = 4 \neq 5$; the interval on the y-
axis in the neighborhood corresponding to
$x = 1$ can be arbitrarily restrictive.

$$0 < |x - 1| < \delta$$

must apply and the given statement is false.
The actual limit is 4.

4. $\lim_{x \to 2} (x^2 - 2) = 5$ implies $|x^2 - 2 - 5| < \epsilon$.
For $|x - 2| < \delta$, $2 - \delta < x < 2 + \delta$. The
doubter picks $\epsilon = 1$. The ϵ inequality
becomes

$$-1 < x^2 - 7 < 1, \ 6 < x^2 < 8 \text{ or}$$

$2.45 < \sqrt{5} < 2.83$. The believer has the
impossible task of finding a number around
$x = 2$. The given statement is false. The
actual limit is 2.

5. For any chosen value of ϵ we want the
distance between $f(x)$ and L to be less than ϵ.
In absolute value notation this is

$$|f(x) - L| < \epsilon$$

For our function and limit we need

$$|(x^2 + 2) - (6)| < \epsilon$$
$$|x^2 - 4| < \epsilon$$
$$|x - 2||x + 2| < \epsilon$$

As we approach 2, $|x + 2|$ is less than 5.
If the distance between x and 2, written as
$|x - 2|$, is to be less than δ, we need $|x - 2|$
$(5) < \epsilon = \delta$. Therefore, let $\delta = $ minimum

of $\{1, \frac{\epsilon}{5}\}$.

6. $|x^2 - 3x + 2| = |(x - 2)(x - 1)|$

Let $|x - 1|$ be the maximum of $\frac{1}{2}$ or δ. Then $\frac{3}{2} < x < \frac{5}{2}$ and $x - 2 < \frac{1}{2}$. Thus,

$$|x - 2||x - 1| < \frac{1}{2}\delta = \epsilon \text{ or } \delta = 2\epsilon$$

The statement is true; no matter what number the doubter selects, the believer will always be able to find $|x^2 - 3x + 2| < \epsilon$.

7. $|f(x) - L| = |(x + 3) - 5| = |x - 2| < \delta$

Choose $\delta = \epsilon$.

8. $|f(x) - L| = |(3t - 1) - 0| < |3t - 1|$

The statement if false. Choose $\epsilon = 0.3$ and it is not possible to find a delta.

9. $|f(x) - L| = |(3x + 7) - 1| = |3x + 6|$

$= 3|x + 2| < 3\delta$; choose $\delta = \frac{\epsilon}{3}$.

10. $|f(x) - L| = |(2x - 5) + 3| = 2|x - 1|$

$< 2\delta$; choose $\delta = \frac{\epsilon}{2}$.

11. $|f(x) - L| = |(x^2 + 2) - 6| = |x^2 - 4|$

$= |x - 2||x + 2|$; choose $\delta = \min(1, \frac{\epsilon}{5})$.

12. $|f(x) - L| = \left|\frac{1}{x} - \frac{1}{2}\right| = \frac{|x - 2|}{2|x|}$;

choose $\delta = \min(1, \frac{\epsilon}{5})$.

13. In order for $f(x)$ to be continuous at $x = 0$,

$\lim_{x \to 0} f(x)$ must equal $f(0)$. To show that $\lim_{x \to 0} \sin \frac{1}{x} = 0$ we need to show that for any $\epsilon > 0$ there exists a $\delta > 0$ such that

$|f(x) - L| < \epsilon$ when $|x - 0| < \delta$.

Arbitrarily letting $\epsilon = .5$ we need to find a δ-interval about 0 such that $\left|\sin \frac{1}{x}\right| < .5$.

However any interval about 0 contains a point $x = \frac{2}{\pi n}$ (with n odd) where $\sin \frac{1}{x}$

$= \sin \frac{\pi n}{2} = \pm 1$. Therefore, there does not exist a δ-interval about 0 such that $\left|\sin \frac{1}{x}\right| < 0.5$, and $f(x)$ must be discontinuous at $x = 0$.

14. By hypotheses, $\lim_{x \to c} f(x) = L$ which means there exists $\epsilon_1 > 0$ such that $|f(x) - L| < \epsilon$ with δ such that $0 < |x - c| < \delta$. Let $g(x) = kf(x)$ so that

$|g(x) - kL| = |kf(x) - kL|$

$= |k||f(x) - L| < |k|\epsilon_1$

This says that for $\epsilon = |k|\epsilon_1 > 0$, where ϵ is arbitrary because ϵ_1 was arbitrary, a δ exists with

$$0 < |x - c| < \delta$$

This inequality is restated as $\lim_{x \to c} g(x) = kL$ or $\lim_{x \to c} kf(x) = kL$.

15. $\lim_{x \to c} f(x) = L_1$ which means there exists an $\epsilon_1 > 0$ such that $|f(x) - L| < \epsilon$ with δ such that $0 < |x - c| < \delta$. Similarly $\lim_{x \to c} g(x) = L_2$ which means there exists an $\epsilon_2 > 0$ such that $|g(x) - L_2| < \epsilon_2$ with δ such that $0 < |x - c| < \delta$. Let $\frac{\epsilon}{2} = \max(\epsilon_1, \epsilon_2)$. Then $|f(x) - g(x) - (L_1 - L_2)|$

$= |[f(x) - L_1] - [g(x) - L_2]|$

$\leq |f(x) - L_1| + |g(x) - L_2|$

$< \epsilon_1 + \epsilon_2$

$< \frac{\epsilon}{2} + \frac{\epsilon}{2} = \epsilon$

This says that for any $\epsilon > 0$ there exists a $\delta > 0$ such that $\lim_{x \to c} |f(x) - g(x)| = L_1 - L_2$ whenever $0 < |x - c| < \delta$.

16. By hypotheses, $\lim_{x \to c} f(x) = L_1$ which means there exists an $\epsilon_1 > 0$ such that $|f(x) - L_1| < \epsilon_1$ with $0 < |x - c| < \delta$. Similarly $\lim_{x \to c} g(x) = L_2$ which means there exists an $\epsilon_2 > 0$ such that $|g(x) - L_2| < \epsilon_2$ with $0 < |x - c| < \delta$. Let $\frac{\epsilon}{2} = \max(a\epsilon_1, b\epsilon_2)$. Then,

$|a[f(x) - L_1] + b[g(x) - L_2]| < \frac{\epsilon}{2} + \frac{\epsilon}{2} = \epsilon$

This says that for any $\epsilon > 0$ there exists a $\delta > 0$ such that $\lim_{x \to c} [af(x) + bg(x)]$

$= a \lim_{x \to c} f(x) + b \lim_{x \to c} g(x)$ whenever $0 < |x - c| < \delta$.

17. By hypotheses, $\lim_{x \to c} f(x) = 0$ which means there exists an $\epsilon_1 > 0$ such that $|f(x) - 0| < \epsilon_1$ with $0 < |x - c| < \delta$.

Similarly $\lim_{x \to c} g(x) = 0$ which means there exists an $\epsilon_2 > 0$ such that $\left| g(x) - 0 \right| < \epsilon_2$ with $0 < |x - c| < \delta$.

Let $\frac{\epsilon}{2} = \max\left(\sqrt{\frac{\epsilon_1}{2}}, \sqrt{\frac{\epsilon_2}{2}} \right)$. Then,

$$\left| f(x) - 0 \right| \left| g(x) - 0 \right| < \sqrt{\frac{\epsilon_1}{2}} \sqrt{\frac{\epsilon_2}{2}} < \epsilon$$

This says that for any $\epsilon > 0$ there exists a $\delta > 0$ such that $\lim_{x \to c} f(x)g(x) = 0$ whenever $0 < |x - c| < \delta$.

18. $0 \le \left| f(x) - 0 \right| \le \left| g(x) - 0 \right| < \epsilon$ whenever $0 < |x - c| < \delta$. Now, $\left| f(x) - 0 \right| < \epsilon$ and $0 < |x - c| < \delta$ imply $\lim_{x \to 0} f(x) = 0$.

19. Let $h(x) = f(x) - g(x)$. Since $f(x) \ge g(x)$ throughout an open interval containing c, the limit limitation theorem guarantees that

$$\lim_{x \to c} h(x) = \lim_{x \to c} [f(x) - g(x)]$$
$$= \lim_{x \to c} f(x) - \lim_{x \to c} g(x) \ge 0 \text{ or}$$

$\lim_{x \to c} f(x) \ge \lim_{x \to c} g(x)$.

20. $0 \le \left| f(x) - |L| \right| \le \left| f(x) - L \right| < \epsilon$ whenever $0 < |x - c| < \delta$. Thus, $\left| |f(x)| - |L| \right| < \epsilon$ implies $\lim_{x \to c} |f(x)| = |L|$.

21. By hypotheses, $\lim_{x \to c} f(x) = L$, and by Problem 20, $\lim_{x \to c} |f(x)| = |L|$, or there exists an $\epsilon = \frac{|L|}{2} > 0$ so that $\left| f(x) - L \right| < \frac{|L|}{2}$ with a δ such that $0 < |x - c| < \delta$. Thus,

$$-\frac{|L|}{2} < \left| f(x) \right| - |L| < \frac{|L|}{2} \text{ and}$$
$$\frac{|L|}{2} < \left| f(x) \right| < \frac{3|L|}{2}.$$

22. **a.** $\left| f(x) \right| < \frac{3|L|}{2}$,

$$\left| f(x) - L \right| \le \left| f(x) \right| + |L| < |L| + \frac{3|L|}{2} = \frac{5|L|}{2}$$

Now, if L is replaced by $-L$,

$$\left| f(x) - (-L) \right| = \left| f(x) + L \right| < \frac{5|-L|}{2} = \frac{5|L|}{2}$$

b. $\left| [f(x)]^2 - L^2 \right| = \left| f(x) + L \right| \left| f(x) - L \right|$

$$\le \left[|f(x)| + |L| \right] \left| f(x) - L \right| < \frac{5|L|}{2} \epsilon_2;$$
$$\left| f(x) - L \right| < \epsilon_2 \text{ for } 0 < |x - c| < \delta_2$$

because $\lim_{x \to c} f(x) = L$, by hypotheses.

c. Let $\epsilon = \frac{5|L|}{2} \epsilon_2$, then $\left| [f(x)]^2 - L^2 \right| < \epsilon$ for $0 < |x - c| < \delta$ where $\delta = \min(\delta_1, \delta_2)$

23. $\frac{1}{4} [(f + g)^2 - (f - g)^2]$
$$= \frac{1}{4} [f^2 + 2fg + g^2 - f^2 + 2fg - g^2]$$
$$= \frac{1}{4} [2fg + 2fg]$$
$$= fg$$

Thus,

$$\lim_{x \to c} fg = \lim_{x \to c} \frac{1}{4} [(f + g)^2 - (f - g)^2]$$
$$= \frac{1}{4} \lim_{x \to c} [(f + g)^2 - (f - g)^2]$$

by Problem 14. By Problem 15, it also equals

$$\frac{1}{4} \left[\lim_{x \to c} (f + g)^2 - \lim_{x \to c} (f - g)^2 \right]$$

and by Problem 22 it becomes

$$\frac{1}{4} \left[(L + M)^2 - (L - M)^2 \right] = LM$$

24. Let $\epsilon = \frac{|f(c)|}{2}$. $\lim_{x \to c} f(x) = f(c)$ implies

$$\left| f(x) - f(c) \right| < \frac{|f(c)|}{2}$$
$$-\frac{f(c)}{2} < f(x) - f(c) < \frac{f(c)}{2}$$
$$0 < \frac{f(c)}{2} < f(x) < \frac{3f(c)}{2}, \ 0 < f(x)$$

25. **a.** Since f is continuous at L, we have

$\lim_{w \to L} f(w) = L$ which means that there exists a δ_1 such that $\left| f(w) - L \right| < \epsilon_1$ whenever $|w - L| < \delta_1$.

b. Since $\lim_{x \to c} g(x) = L$, by hypotheses, $\left| g(x) - L \right| < \epsilon_2$ whenever $|x - c| < \delta_2$. Let $w = g(x)$ in part **a.** Then

$$\left| f[g(x)] - L \right| = \left| f[g(x)] - g(x) - L \right|$$
$$\le \left| f[g(x)] - g(x) \right| + \left| g(x) - L \right| < \epsilon$$

whenever $|x - c| < \delta$ where $\epsilon = \frac{1}{2}(\epsilon_1 + \epsilon_2)$ $\delta = \max(\delta_1, \delta_2)$.

26. Since (as it turns out) the slope at $x = 3$ is 26; picking $K = 26$ or larger works.

CHAPTER 1 REVIEW

Proficiency Examination, Page 75

1. $\mathbb{N} = \{1, 2, 3, \cdots\}$; $\mathbb{W} = \{0, 1, 2, 3, \cdots\}$; $\mathbb{J} = \{\cdots, -3, -2, -1, 0, 1, 2, 3, \cdots\}$; $\mathbb{Q} = \{p/q$ so that p is an integer and q is a nonzero integer$\}$;

$\overline{\mathbb{Q}} = \{$nonrepeating or nonterminating decimals$\}$;

$\mathbb{R} = \mathbb{Q} \cup \overline{\mathbb{Q}}$

2. $|a| = a$ if $a \geq 0$; $|a| = -a$ if $a < 0$

3. $|x + y| \leq |x| + |y|$

4. $d = \sqrt{(x_2 - x_1)^2 + (y_2 - y_1)^2}$.

5. $m = \tan \theta$ where θ is the angle of inclination

6.   **a.** $Ax + By + C = 0$ **b.** $y = mx + b$

    c. $y - k = m(x - h)$ **d.** $y = k$ **e.** $x = h$

7.   Lines are parallel if they have the same slope, and perpendicular if their slope are negative reciprocals.

8.   A function is a rule that assigns to each element x of the domain D a unique element of the range R.

9. $(f \circ g) = f[g(x)]$

10.   The graph of a function f consists of all points whose coordinates (x, y) satisfy $y = f(x)$ for all x in the domain of f.

11.   **a.**

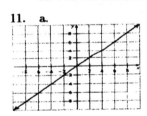

  b.

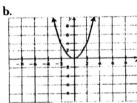

  c.

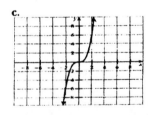

  d.

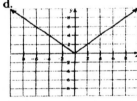

  e.

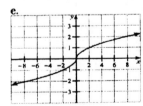

  f.

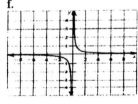

  g.

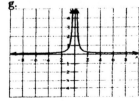

  h.

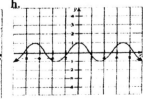

i.   **j.**

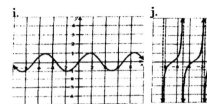

k.   **l.**

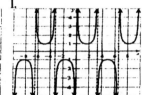

m.

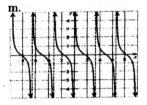

12. $f(x) = a_n x^n + a_{n-1} x^{n-1} + \cdots + a_2 x^2 + a_1 x + a_0$

13. $f(x) = \dfrac{P(x)}{D(x)}$, $D(x) \neq 0$

14. $\lim\limits_{x \to 0} f(x) = L$ means that the function values $f(x)$ can be made arbitrarily close to L by choosing x sufficiently close to c.

15.   **a.** $\lim\limits_{x \to c} k = k$ for any constant k

    b. $\lim\limits_{x \to c}[sf(x)] = s \lim\limits_{x \to c} f(x)$

    c. $\lim\limits_{x \to c}[f(x) + g(x)] = \lim\limits_{x \to c} f(x) + \lim\limits_{x \to c} g(x)$

    d. $\lim\limits_{x \to c}[f(x) - g(x)] = \lim\limits_{x \to c} f(x) - \lim\limits_{x \to c} g(x)$

    e. $\lim\limits_{x \to c}[f(x)g(x)] = [\lim\limits_{x \to c} f(x)][\lim\limits_{x \to c} g(x)]$

    f. $\lim\limits_{x \to c} \dfrac{f(x)}{g(x)} = \dfrac{\lim\limits_{x \to c} f(x)}{\lim\limits_{x \to c} g(x)}$ if $\lim\limits_{x \to c} g(x) \neq 0$

    g. $\lim\limits_{x \to c}[f(x)]^n = \left[\lim\limits_{x \to c} f(x)\right]^n$ n is a rational number and the limit on the right exists.

16.   A function f is continuous at a point $x = c$ if

    (1) $f(c)$ is defined

    (2) $\lim\limits_{x \to c} f(x)$ exists

    (3) $\lim\limits_{x \to c} f(x) = f(c)$

17.   If f is a polynomial, rational, power, or trigonometric function, then f is continuous at any number $x = c$ for which $f(c)$ is defined.

18.   If f is a continuous function on $[a, b]$ then there exists at least one number c on $[a, b]$

such that $f(c) = L$.

19. Determine a closed interval $[a, b]$ for which $f(a)$ and $f(b)$ have opposite signs. Let $c = (a + b)/2$ and then pick $[a, c]$ or $[c, b]$ and repeat the procedure.

20. $\lim_{x \to c} f(x) = L$ means that for each $\epsilon > 0$ there exists a number $\delta > 0$ such that

$|f(x) - L| < \epsilon$ whenever $0 < |x - c| < \delta$

21. If $g(x) \le f(x) \le h(x)$ for all x on an open interval containing c and if

$$\lim_{x \to c} g(x) = \lim_{x \to c} h(x) = L$$

then $\lim_{x \to c} f(x) = L$.

22. **a.**
$$y - 5 = -\tfrac{3}{4}[x - (-\tfrac{1}{2})]$$
$$y - 5 = -\tfrac{3}{4}(x + \tfrac{1}{2})$$
$$4y - 20 = -3x - \tfrac{3}{2}$$
$$6x + 8y - 37 = 0$$

b. $m = \dfrac{2 - 5}{7 + 3} = \dfrac{-3}{10}$
$$y - 5 = -\tfrac{3}{10}[x - (-3)]$$
$$10y - 50 = -3x - 9$$
$$3x + 10y - 41 = 0$$

c.
$$\frac{x}{4} + \frac{y}{-\frac{3}{7}} = 1$$
$$\frac{x}{4} - \frac{7y}{3} = 1$$
$$3x - 28y = 12$$
$$3x - 28y - 12 = 0$$

d. If we write the given equation in slope-intercept form, $y = -\tfrac{2}{5}x + \tfrac{11}{5}$,

we see that the slope is $-2/5$. A parallel line must have the same slope. Now use the point-slope form.
$$y - 5 = -\tfrac{2}{5}(x + \tfrac{1}{2})$$
$$5y - 25 = -2x - 1$$
$$2x + 5y - 24 = 0$$

e. Find the slope of $\overline{PQ}$. A perpendicular line will have a slope which is the negative reciprocal. Find the midpoint of $\overline{PQ}$. Then use the point-slope form for the equation of the line. The slope of $\overline{PQ}$ is $-3/4$. The midpoint of $\overline{PQ}$ is $(1, 4)$.
$$y - 4 = \tfrac{4}{3}(x - 1)$$

$$3y - 12 = 4x - 4$$
$$4x - 3y + 8 = 0$$

23. **a.** Probably the easiest way to graph this equation is by finding the intercepts. If $x = 0$, then $y = 6$, and if $y = 0$, then $x = 4$. Using $(0, 6)$ and $(4, 0)$:

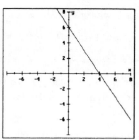

b. Complete the square in order to find the vertex of the parabola: $y = (x - 2)^2 - 14$, so the vertex is at $(2, -14)$.

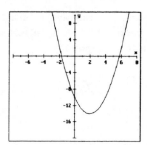

c. This is an absolute value function that has been translated one unit to the left and 3 units up. It has a vertex at $(-1, 3)$.

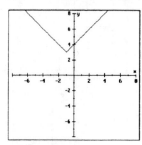

d. This is a cosine curve that has an amplitude of 2 and has been translated 1 unit to the right.

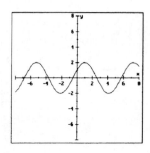

e. Remove the x coefficient from the parentheses in order to determine the phase shift and the change in frequency: $y + 1 = \tan 2(x + \frac{3}{2})$. This is a tangent curve that has been translated $\frac{3}{2}$ units to the left, 1 unit down, and has a change in frequency from π to $\frac{\pi}{2}$.

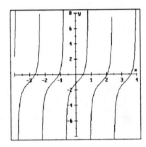

24. If $f(x) = \dfrac{1}{x + 1}$, then

$$f\left(\frac{1}{x + 1}\right) = \frac{1}{\frac{1}{x + 1} + 1}, \text{ and}$$

$$f\left(\frac{2x + 1}{2x + 4}\right) = \frac{1}{\frac{2x + 1}{2x + 4} + 1}$$

So we need to find the values of x for which:

$$\frac{1}{\frac{1}{x + 1} + 1} = \frac{1}{\frac{2x + 1}{2x + 4} + 1}.$$

Multiply numerator and denominator on the left by $(x + 1)$, and on the right by $(2x + 4)$:

$$\frac{x + 1}{1 + x + 1} = \frac{2x + 4}{2x + 1 + 2x + 4}$$

$$\frac{x + 1}{x + 2} = \frac{2(x + 2)}{4x + 5}$$

$$4x^2 + 9x + 5 = 2x^2 + 8x + 8$$

$$2x^2 + x - 3 = 0$$

$$(2x + 3)(x - 1) = 0$$

$$x = -\frac{3}{2}, 1$$

25. The root of a quotient equals the quotient of the roots when they have the same domain. The domain of $f(x)$ is $(-\infty, 0] \cup (1, \infty)$. The domain of $g(x)$ is $(1, \infty)$. The two functions are not the same.

26. $f \circ g = \sin(\sqrt{1 - x^2})$

$g \circ f = \sqrt{1 - \sin^2 x} = |\cos x|$

27. Since the cost is the number of worker-hours times the rate per hour,

$$C = 25T = \frac{25N(3N + 4)}{2N - 5}.$$

28. a. Since this function is defined at the limiting value, $\lim\limits_{x \to 3} f(x) = f(3) = \frac{3}{2}$

b. Factor the denominator and reduce:

$$\lim_{x \to 4} \frac{\sqrt{x} - 2}{(\sqrt{x} - 2)(\sqrt{x} + 2)}$$

$$= \lim_{x \to 4} \frac{1}{\sqrt{x} + 2} = \frac{1}{4}$$

c. Factor and reduce:

$$\lim_{x \to 2} \frac{(x - 2)(x - 3)}{(x + 2)(x - 2)} = \lim_{x \to 2} \frac{x - 3}{x + 2} = -\frac{1}{4}$$

d. $\lim\limits_{x \to 0} \dfrac{1 - \cos x}{2 \tan x} = \lim\limits_{x \to 0} \dfrac{(1 - \cos x)(1 + \cos x)}{2 \frac{\sin x}{\cos x}(1 + \cos x)}$

$$= \lim_{x \to 0} \frac{\sin^2 x}{2 \frac{\sin x}{\cos x}(1 + \cos x)}$$

$$= \lim_{x \to 0} \frac{\sin x \cos x}{2(1 + \cos x)} = \frac{0}{4} = 0$$

29. Polynomials are everywhere continuous, so the only problem is at $x = 1$. We need

$$\lim_{x \to 1^-} (Ax + 3) = 2, \text{ and } \lim_{x \to 1^+} (x^2 + B) = 2$$

$$A + 3 = 2, \quad A = -1 \text{ and } 1 + B = 2, \quad B = 1$$

30. Let $f(x) = x + \sin x - \dfrac{1}{\sqrt{x} + 3}$. This function is continuous on $[0, \infty)$, and

$$f(0) = -\frac{1}{3}, \quad f(\pi) = \pi - \frac{1}{\sqrt{\pi} + 3} \approx 2.93$$

So by the root location theorem there must be some value c on $[0, \pi]$ where $f(c) = 0$.

Supplementary Problems, Page 76

1. $d_1 = \sqrt{(-1 + 1)^2 + (8 - 3)^2} = \sqrt{25} = 5$

$d_2 = \sqrt{(11 + 1)^2 + (8 - 8)^2} = \sqrt{144} = 12$

$d_3 = \sqrt{(11 + 1)^2 + (8 - 3)^2} = \sqrt{169} = 13$

$P = 5 + 12 + 13 = 30; A = \frac{1}{2}(5)(12) = 30$

2. $L_1: y = 5; L_2: y = \frac{4}{3}x + \frac{11}{3};$

$L_3: -\frac{12}{5}x + 5$

Intersection of L_1 and L_2 is $(1, 5)$

Intersection of L_1 and L_3 is $(0, 5)$

Intersection of L_2 and L_3 is $(\frac{5}{14}, \frac{29}{7})$

$d_1 = \sqrt{(1 - 0)^2 + (5 - 5)^2} = 1$

$d_2 = \sqrt{(\frac{5}{14} - 0)^2 + (\frac{29}{7} - 5)^2} = \frac{13}{14}$

$$d_3 = \sqrt{(\tfrac{5}{14} - 1)^2 + (\tfrac{29}{7} - 5)^2} = \tfrac{15}{14}$$

$$P = 1 + \tfrac{13}{14} + \tfrac{15}{14} = 3$$

$$A = \tfrac{1}{2} \begin{vmatrix} 1 & 5 & 1 \\ 0 & 5 & 1 \\ \tfrac{5}{14} & \tfrac{20}{7} & 1 \end{vmatrix} \approx 1.07$$

3. $P = 8 + 2\sqrt{9 + 64} + 2 = 2(5 + \sqrt{73}) \approx 27$

$A = \tfrac{1}{2}(8)(2 + 8) = 40$

4. The curves intersect at

$$2x - 7 = x^2 + 6x - 3$$
$$x^2 + 4x + 4 = 0$$
$$(x + 2)^2 = 0$$
$$x = -2$$

If $x = -2$, then $y = 2(-2) - 7 = -11$

The point of intersection is $(-2, -11)$. The equation of the vertical line is is $x + 2 = 0$.

5. $r = 4; (x - 5)^2 + (y - 4)^2 = 16$

6. The circle contains $(0, 0)$, is tangent to $3x + 4y - 40 = 0$, and has center $(0, K)$. Thus, the desired equation has the form

$$x^2 + (y - K)^2 = K^2$$

The radius is the distance from $(0, 0)$ to the point of intersection of the tangent line and the line through the center normal to the tangent. For this normal, the slope is $m = 4/3$, and the equation is

$$y - K = \tfrac{4}{3}x$$
$$4x - 3y + 3K = 0$$

The intersection is $\left(\dfrac{120 - 12K}{25}, \dfrac{160 + 9K}{25}\right)$. The radius is

$$K^2 = \left(\dfrac{120 - 12K}{25}\right)^2 + \left(\dfrac{160 + 9K}{25} - K\right)^2$$
$$25^2 K^2 = 12^2(10 - K)^2 + 16^2(10 - K)^2$$
$$25^2 K^2 = 20^2(10 - K)^2$$
$$\pm 25K = 20(10 - K)$$
$$\pm 5K = 4(10 - K)$$
$$K = -40, \tfrac{40}{9}$$

There are two circles:

$$x^2 + (y + 40)^2 = 1{,}600$$
$$x^2 + (y - \tfrac{40}{9})^2 = \tfrac{1{,}600}{81}$$

7. $(x^2 + y^2 - 5x + 7y) - (x^2 + y^2 + 4y) = 3 - 0$

$$-5x + 3y = 3$$

The line is $5x - 3y + 3 = 0$.

8. **a.** $\tan(x + \tfrac{\pi}{3}) = \dfrac{\tan x + \tan \tfrac{\pi}{3}}{1 - \tan x \tan \tfrac{\pi}{3}} = \dfrac{A + \tan x}{1 + B \tan x}$

Thus, $A = \sqrt{3}$ since $\tan \tfrac{\pi}{3} = \sqrt{3}$ and $B = -\sqrt{3}$

b. Notice

$\sin 3x = \sin(2x + x)$

$= \sin 2x \cos x + \cos 2x \sin x$

$= 2 \sin x \cos^2 x + (1 - 2\sin^2 x)\sin x$

$= 2 \sin x(1 - 2\sin^2 x) + \sin x - 2\sin^3 x$

$= 2 \sin x - 4\sin^3 x + \sin x - 2\sin^3 x$

$= 3 \sin x - 6 \sin^3 x$

Thus, $-6 \sin^3 x = \sin 3x - 3 \sin x$

$$\sin^3 x = -\tfrac{1}{6}\sin 3x + \tfrac{1}{2} \sin x$$

Thus, $A = -\tfrac{1}{6}$ and $B = \tfrac{1}{2}$.

9. $A(-2, 1)$, $B(5, 6)$, and $C(3, -2)$.

a. $m_{AB} = \tfrac{5}{7}$, so the perpendicular line has slope $-\tfrac{7}{5}$. L_C:

$$y + 2 = -\tfrac{7}{5}(x - 3)$$
$$5y + 10 = -7x + 21$$
$$7x + 5y - 11 = 0$$

$m_{AC} = -\tfrac{3}{5}$; perpendicular has $m = \tfrac{5}{3}$. L_B:

$$y - 6 = \tfrac{5}{3}(x - 5)$$
$$3y - 18 = 5x - 25$$
$$5x - 3y - 7 = 0$$

$m_{BC} = 4$; perpendicular has $m = -\tfrac{1}{4}$. L_A:

$$y - 1 = -\tfrac{1}{4}(x + 2)$$
$$4y - 4 = -x - 2$$
$$x + 4y - 2 = 0$$

b. The last two equations in part **a** lead to

$$\tfrac{5}{3}(x - 5) + 6 = -\tfrac{1}{4}(x + 2) + 1$$
$$20x - 100 + 72 = -3x - 6 + 12$$
$$20x - 28 = -3x + 6$$
$$23x = 34$$
$$x = \tfrac{34}{23}$$

Substitute this value into the first equation to obtain $y = \frac{3}{23}$. The point is $\left(\frac{34}{23}, \frac{3}{23}\right)$.

10.
$$f(2x) = f(3x)$$
$$(2x)^2 + 5(2x) - 9 = (3x)^2 + 5(3x) - 9$$
$$4x^2 + 10x = 9x^2 + 15x$$
$$5x^2 + 5x = 0$$
$$x(x + 1) = 0$$
$$x = 0, -1$$

11. $s(t) = 256t - 16t^2 = -16t(t - 16)$

The projectile is at ground level when $t = 0$ and $t = 16$. It seems reasonable that the height point is the vertex of the parabola, or midway at $x = 8$. In fact, $s(8) = 1{,}024$.

12. The current population is when $t = 0$:
$$P(0) = \frac{11(0) + 12}{2(0) + 3} = 4 \quad \text{(thousand people)}$$

In 6 years,
$$P(6) = \frac{11(6) + 12}{2(6) + 3} = \frac{78}{15} = 5.2 \quad \text{(thousand)}$$

For $P = 5$ we have
$$5 = \frac{11t + 12}{2t + 3}$$
$$10t + 15 = 11t + 12$$
$$t = 3$$

The population will reach 5,000 in 3 years.

13. $\lim\limits_{x \to 1^+} \sqrt{\dfrac{x^2 - x}{x - 1}} = \sqrt{\lim\limits_{x \to 1^+} \dfrac{x(x - 1)}{x - 1}} = \lim\limits_{x \to 1^+} x = 1$

14. $\lim\limits_{x \to 1} \dfrac{x^3 - 1}{x^2 - 1} = \lim\limits_{x \to 1} \dfrac{(x - 1)(x^2 + x + 1)}{(x + 1)(x - 1)}$

$= \lim\limits_{h \to 0} \dfrac{x^2 + x + 1}{x + 1} = \dfrac{3}{2}$

15. $\lim\limits_{x \to 1} \dfrac{x^2 - 3x + 2}{x^2 - 1} = \lim\limits_{x \to 1} \dfrac{(x - 1)(x - 2)}{(x + 1)(x - 1)}$

$= \lim\limits_{x \to 1} \dfrac{x - 2}{x + 1} = -\dfrac{1}{2}$

16. $\lim\limits_{x \to 1/\pi} \dfrac{1 + \cos \frac{1}{x}}{\pi x - 1}$

$= \lim\limits_{x \to 1/\pi} \dfrac{(1 + \cos \frac{1}{x})(1 - \cos \frac{1}{x})(x - \frac{1}{\pi})}{\pi(x - \frac{1}{\pi})(1 - \cos \frac{1}{x})(x - \frac{1}{\pi})}$

$= \lim\limits_{x \to 1/\pi} \dfrac{\sin^2 \frac{1}{x}(x - \frac{1}{\pi})}{\pi(x - \frac{1}{\pi})^2(1 - \cos \frac{1}{x})}$

$= \lim\limits_{x \to 1/\pi} \dfrac{\sin^2[(\frac{1}{x} - \pi) + \pi](x - \frac{1}{\pi})}{\pi(x - \frac{1}{\pi})^2(1 - \cos \frac{1}{x})}$

$= \lim\limits_{x \to 1/\pi} \dfrac{[-\sin(\frac{1}{x} - \pi)]^2(x - \frac{1}{\pi})}{\pi(x - \frac{1}{\pi})^2(1 - \cos \frac{1}{x})}$

$= \lim\limits_{x \to 1/\pi} \left[\dfrac{\sin(\frac{1}{x} - \pi)}{x - \frac{1}{\pi}}\right]^2 \left[\dfrac{x - \frac{1}{\pi}}{\pi}\right]\left[\dfrac{1}{1 - \cos \frac{1}{x}}\right]$

Now we know
$$\lim\limits_{x \to 1/\pi} \left(1 - \cos \frac{1}{x}\right) = 1 - \cos \pi = 2$$
and
$$\lim\limits_{x \to 1/\pi} \dfrac{\sin(\frac{1}{x} - \pi)}{x - \frac{1}{\pi}} = 1$$

Thus,
$$\lim\limits_{x \to 1/\pi} \left[\dfrac{\sin(\frac{1}{x} - \pi)}{x - \frac{1}{\pi}}\right]^2 \left[\dfrac{x - \frac{1}{\pi}}{\pi}\right]\left[\dfrac{1}{1 - \cos \frac{1}{x}}\right]$$
$$= 1 \cdot 0 \cdot \dfrac{1}{2} = 0$$

17. $Q = \dfrac{f(x + \Delta x) - f(x)}{\Delta x}$

$= \dfrac{3(x + \Delta x) + 5 - (3x + 5)}{\Delta x}$

$= \dfrac{3\Delta x}{\Delta x} = 3$

$\lim\limits_{\Delta x \to 0} Q = 3$

18. $Q = \dfrac{f(x + \Delta x) - f(x)}{\Delta x}$

$= \dfrac{\sqrt{2x + 2\Delta x} - \sqrt{2x}}{\Delta x}$

$= \dfrac{(\sqrt{2x + 2\Delta x} - \sqrt{2x})(\sqrt{2x + 2\Delta x} + \sqrt{2x})}{\Delta x(\sqrt{2x + 2\Delta x} + \sqrt{2x})}$

$= \dfrac{2x + 2\Delta x - 2x}{\Delta x(\sqrt{2x + 2\Delta x} + \sqrt{2x})}$

$= \dfrac{2}{\sqrt{2x + 2\Delta x} + \sqrt{2x}}$

$\lim\limits_{\Delta x \to 0} Q = \dfrac{1}{\sqrt{2x}}$

19. $Q = \dfrac{f(x + \Delta x) - f(x)}{\Delta x}$

$= \dfrac{(x + \Delta x)(x + 1 + \Delta x) - x(x + 1)}{\Delta x}$

$= \dfrac{x^2 + 2x\Delta x + (\Delta x)^2 + x + \Delta x - x^2 - x}{2}$

$= 2x + \Delta x + 1$

$\lim\limits_{\Delta x \to 0} Q = 2x + 1$

20.

$$Q = \frac{f(x + \Delta x) - f(x)}{\Delta x}$$

$$= \frac{\frac{4}{x + \Delta x} - \frac{4}{x}}{\Delta x}$$

$$= \frac{1}{\Delta x}\left[\frac{4x - 4x - 4\Delta x}{x(x + \Delta x)}\right]$$

$$= \frac{-4}{x(x + \Delta x)}$$

$$\lim_{\Delta x \to 0} Q = \frac{-4}{x^2}$$

21. If $x > 2$:

$$\frac{x^2 - Ax - 6}{x - 2} = \frac{(x + 3)(x - 2)}{x - 2}$$

$$= \frac{x^2 + x - 6}{x - 2}; \text{ thus, } A = -1$$

$$\lim_{x \to 2^+} \frac{x^2 + x - 6}{x - 2} = \lim_{x \to 2^+} (x + 3) = 5$$

Thus, $\lim_{x \to 2^-} (x^2 + B) = 5$ so that $4 + B = 5$
or $B = 1$.

22. Let x denote the number of people on the tour and $R(x)$ the corresponding revenue.

  a. $R(x) = (\text{NUMBER OF PEOPLE})(\text{COST PER PERSON}) = (x + 100)(500 - 4x)$
$$= 50,000 + 100x - 4x^2$$

  b.

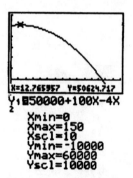

X=12.765957 Y=50624.717
Y₁=50000+100X-4X²
Xmin=0
Xmax=150
Xscl=10
Ymin=-10000
Ymax=60000
Yscl=10000

The maximum revenue is $50,625 at 12.8. Since the number of people must be an integer, we find that at $x = 12$ or $x = 13$ the maximum revenue is $50,624.

23. $x = 50 + \frac{y}{2} = 25y$ or $600 + 6y$ inches

$$x^2 + 50^2 = (50 + \tfrac{y}{2})^2 \text{ so}$$

$$x = \sqrt{(50 + \tfrac{y}{2})^2 - 2,500}$$

  a. $x \approx 15$ inches   **b.** $x \approx 21$ inches

  c. $x \approx 29$ inches   **d.** $x \approx 19$ inches

24. **a.** $h(t) = -16t^2 + 256$

$$h(2) = -16(2)^2 + 256 = 192 \text{ ft}$$

  b. $|h(3) - h(2)| = |-16(3)^2 + 256 - 192|$
$$= 80 \text{ ft}$$

  c. $h(0) = 256 \text{ ft}$

  d. $h(t) = 0$ when $-16t^2 + 256 = 0$ or when $t = 4$ (disregard $t = -4$). It will hit the ground in 4 seconds.

25. **a.** The domain consists of all $x \neq 300$.

  b. x represents a percentage, so $0 \leq x \leq 100$ in order for $f(x) \geq 0$.

  c. If $x = 50$, $f(50) = \dfrac{600(50)}{300 - 50} = 120$

  d. With $x = 100$,

$$f(100) = \frac{600(100)}{300 - 100} = 300$$

  e. With $f(x) = 150$, $\dfrac{600x}{300 - x} = 150$ or $x = 60$. The percentage of households should be 60%.

26. **a.**   **b.**

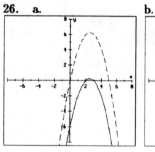 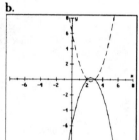

  c.

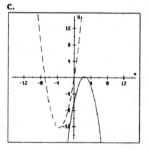

27. **a.** $r = 3$, so $A = 9\pi$

  b. $\ell = 3$ and $w = 2$, so $A = 2(3) = 6$

28. **a.** $r = 4$, $V = \frac{4}{3}\pi r^3 = \frac{256}{3}\pi$; $S = 4\pi r^2 = 64\pi$

  b. $\ell = 2$, $w = 3$, $h = 5$, so $V = 2(3)(5) = 30$;
$$S = 2(2)(3) + (2)(2)(5) + 2(3)(5) = 62$$

29. **a.** $r = 2$, $h = 4$; $V = 2(\pi r^2 h) = 16\pi$;
$$S = \pi r^2 + 2\pi rh = 8\pi + 16\pi = 24\pi$$

  b. $r = 3$, $h = 5$; $V = \frac{1}{3}r^2 h = \dfrac{5(3^2)}{3}\pi = 15\pi$

For the surface area, find the lateral area by cutting the cone open along an edge. The flattened surface is a sector of a circle of radius $r = \sqrt{9 + 25} = \sqrt{34}$. The circumference of the base of the cone, $2\pi(3) = 6\pi$ is the circular boundary of the sector. The full circle of such a radius has area 34π and circumference $2\pi\sqrt{34}$. The lateral surface area is

$$A = \frac{6\pi\sqrt{34}}{2} = 3\pi\sqrt{34}$$

The total surface area is $6\pi + 3\pi\sqrt{34}$.

30. The radius to the point of contact $P_0(x_0, y_0)$ is perpendicular to the tangent line. The slope of this radius is $m = y_0/x_0$. The slope of the tangent line is $-x_0/y_0$ and the equation of the tangent line is

$$y - y_0 = -\frac{x_0}{y_0}(x - x_0)$$

The desired equation is

$$y_0 y + x x_0 = x_0^2 + y_0^2 = r^2$$

31. a. The midpoints are $M_1(1, 3)$ and $M_2(1, -1)$. The line segment $M_1 M_2$ is horizontal and thus parallel to $\overline{BC}$. The length is 4 units, one-half the length of $\overline{BC}$, which is 8 units.

b. $A(0, a)$, $B(b, 0)$, and $C(c, 0)$ are any points in a plane, with the coordinate axes chosen very carefully. $M_1(\frac{b}{2}, \frac{a}{2})$, $M_2(\frac{c}{2}, \frac{a}{2})$. The slope is 0, so the line segment $\overline{M_1 M_2}$ is parallel to $\overline{BC}$. The length of $\overline{M_1 M_2}$ is $\frac{c - b}{2}$, one-half of the distance from B to C.

32. $A(0, a)$, $B(0, b)$, $C(c_1, c_2)$, $D(d_1, d_2)$ be arbitrary points of a quadrilateral. The corresponding midpoints are $P(\frac{a}{2}, \frac{b}{2})$, $Q(\frac{c_1}{2}, \frac{c_2 + b}{2})$, $R(\frac{d_1 + c_1}{2}, \frac{d_2 + c_2}{2})$, $S(\frac{d_1 + a}{2}, \frac{d_2}{2})$. The slopes $m_{PQ} = \frac{c_2}{c_1 - a}$ and $m_{RS} = \frac{c_2}{c_1 - a}$ are equal. Thus, $\overline{PQ}$ is parallel to $\overline{RS}$. Similarly, $m_{QR} = m_{PS} = \frac{d_2 - b}{d_1}$. This shows that $PQRS$ is a parallelogram.

33. $I(0, -5)$ generates $c = -\frac{4}{5}$. Thus,

$$x^2 + xy - \tfrac{4}{5}y = 4$$

For $(x_0, 0)$, $x^2 = 4$, so the intercepts are $(-2, 0)$ and $(2, 0)$.

34.

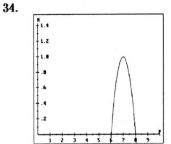

a. $N = -p^2 + 14p - 48 = 0$ if $p = 6$ or $p = 8$. The operation is profitable for $6 < p < 8$.

b. The maximum is likely to occur at the vertex or at the midpoint of the interval from 6 to 8. $N(7) = 1$ is the maximum profit.

35. a. The temperature is between 5.8 °C and 7.5 °C, the temperature change at 2.2 sec is approximately 6.2 degrees.

b. The temperature change at 8 °C is between 1 and 2 sec, leading to roughly 1.8 sec.

36. θ is the angle opposite the 7 ft mural and α is the angle opposite the 5 ft side of the right triangle.

$$\tan(\alpha + \theta) = \tfrac{12}{12} = 1$$

Thus, $\alpha + \theta = \frac{\pi}{4}$; $\alpha = \tan^{-1}\frac{5}{12}$ so that $\theta = \frac{\pi}{4} - \tan^{-1}\frac{5}{12}$.

37. Ship A cover a distance of $s_a = 9t$ hr after noon. Ship B covers a distance of $s_b = 7(t - 1)$ in $t - 1$ hr after noon. Using the law of cosines, the square of the distance s between A and B is

$$s^2 = s_a^2 + s_b^2 - 2s_a s_b \cos\tfrac{\pi}{3}$$
$$= 81t^2 + 49(t - 1)^2 - 2(63)t(t - 1)(\tfrac{1}{2})$$
$$= 81t^2 + 49(t^2 - 2t + 1) - 63t^2 + 63t$$
$$= 67t^2 - 35t + 49$$

After 4 hours, $s = 3\sqrt{109} \approx 31$ km.

38. Let x be the assessed value in dollars. Then taxes computed under the first bill will amount to $y = 0.08x + 100$, while the taxes under the second bill will be $y = 0.02x + 1{,}900$. The break-even point occurs at $0.06x = 1{,}800$ or $x = \$30{,}000$. Since it is safe to assume that the assessed value is greater than $\$30{,}000$, the homeowner should support the second bill.

39. Let x denote the time in hours the spy has been traveling. Then $x - 2/3$ is the time the smugglers have been traveling. The distance the spy travels is $72x$ km and the corresponding distance the smugglers travel is $168(x - 2/3)$ km. They reach the same point when:

$$72x = 168(x - \tfrac{2}{3})$$
$$72x = 168x - 112$$
$$x = \frac{112}{96}$$

This is 1 hr and 10 min. At the end of that time the spy, and thus the smugglers, have traveled

$$72(\tfrac{7}{6}) = 84 \text{ km}$$

The spy escapes, since freedom is reached at the border after only 83.8 km.

40. Let x be the number of hours since the second plane took off, and y the distance traveled in flight. The first plane is ahead by 0.5 hr or $0.5(550) = 275$ mi. The equations for the distances are $y = 550x + 275$ and $y = 650x$ for the first and second planes, respectively. They will have covered the same distance when

$$650x = 550x + 275$$
$$100x = 275$$
$$x = 2.75$$

This is 2 hr and 45 min after the second plane took off.

41. Let x be the number of additional days after today (the eight day) before the club takes all its glass to the recycling center. Assume that the same quantity of glass is collected daily, namely $2{,}400/8 = 300$ lbs. The daily price per pound is $15 - x$ cents. The club's revenue on day x is

$$(2{,}400 + 300x)(0.15 - 0.01x) = 3(8 + x)(15 - x)$$
$$= -3x^2 + 21x + 360$$

The maximum revenue occurs at the vertex; consider the graph:

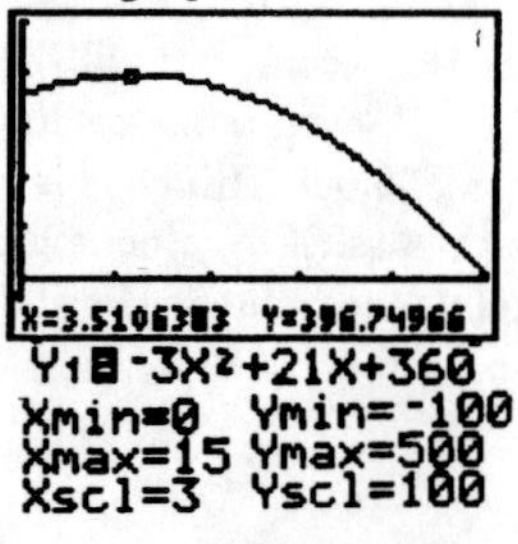
X=3.5106383 Y=396.74966
Y1■-3X²+21X+360
Xmin=0 Ymin=-100
Xmax=15 Ymax=500
Xscl=3 Yscl=100

It looks like the maximum revenue is \$397 when $x = 3.5$. However, since x must be an integer, we check $x = 3$ with revenue \$396 and $x = 4$ with revenue \$396. The glass should be taken in on the 11th or 12th day.

42. Let x x be the length of the side of the square base and y the height of the open box. The area of the base is x^2 m^2 and that of each side is xy. The total cost is $4x^2 + 3(4xy) = 48$. Solving for y in terms of x we find

$$12xy = 48 - 4x^2$$
$$3xy = 12 - x^2$$
$$y = \frac{12 - x^2}{3x}$$

The volume (in m^3) of the box is

$$V = x^2y = \frac{x^2(12 - x^2)}{3} = \tfrac{1}{3}(12x - x^3)$$

43. Form a triangle by connecting the endpoints of one side to the center of the circle. Let h be the altitude, $s = 2\ell$ the side, and $\theta = 2\pi/n$ the angle opposite the side. R is the radius of a circle.

a. $h = R\cos\frac{\pi}{n}$, $s = 2\ell = 2R\sin\frac{\pi}{n}$,

$$P(\theta) = 2ns = 2nR\sin\frac{\pi}{n} = \frac{4\pi R}{\theta}\sin\frac{\theta}{2}$$

b. $$\lim_{\theta\to 0} \frac{4\pi R}{2}\,\frac{\sin(\theta/2)}{\theta/2} = 2\pi R$$

c. $$A_n = \tfrac{1}{2}h(2\ell) = R^2\cos\frac{\pi}{n}\sin\frac{\pi}{n} =$$
$$R^2\cos\frac{\theta}{2}\sin\frac{\theta}{2} = \frac{R^2\sin\theta}{2}$$

Since $2\sin t\cos t = \sin 2t$, we see

$$A = \frac{R^2n}{2}\sin\theta = \pi R^2\,\frac{\sin\theta}{\theta}$$

As $\theta \to 0$, $A \to \pi R^2$ since $\dfrac{\sin\theta}{\theta} \to 1$.

44. The height in the cylinder decreases continuously through all nonnegative real numbers. Similarly, the height in the rectangular trough increases continuously through all real numbers from 0 to its maximum. Somewhere the two values are equal (by the intermediate value theorem).

45. The royalties for Publisher A are given by

$$P_A = \begin{cases} 0.12(5)x \text{ if } 0 < x \le 30{,}000 \\ 0.12(30{,}000)(5) + 17(5)(x - 30{,}000) \text{ if } x \ge 30{,}000 \end{cases}$$

Similarly, for Publisher B:

$$P_B = \begin{cases} 0 \text{ if } 0 < x \le 4{,}000 \\ 0.15(6)(x - 4{,}000) \text{ if } x \ge 4{,}000 \end{cases}$$

a. Publisher A certainly offers the better deal if $x \le 4{,}000$. If $x = 30{,}000$, then
$$P_A = 0.6(30{,}000) = 18{,}000$$

b. If $x = 100{,}000$, $P_A = 0.6(100{,}000) = 60{,}000$ and $P_B = 0.9(96{,}000) = 86{,}400$

c. In this case, Publisher B offers the better deal. Sign with Publisher A if
$$P(N) = P_A - P_B > 0$$

d. For x between 4,000 and 30,000 copies, the break-even point (if it exists) is
$$0.9(x - 4{,}000) = 0.6x$$
$$0.3x = 3{,}600$$
$$x = 12{,}000$$

For $x \ge 30{,}000$, Publisher B pays
$$23{,}400 + 0.9(x - 30{,}000)$$
while Publisher A pays
$$18{,}000 + 0.85(x - 30{,}000)$$

The graphs of these linear functions do not intersect for $x \ge 30{,}000$. The break-even point is 12,000.

46. Assume $x < c < y$. Show that
$$\lim_{x \to c} f(x) = f(c)$$
Let $\epsilon = 2M|x - y|$. Then,
$$\left| f(x) - f(c) - [f(y) - f(c)] \right|$$
$$= \left| f(x) - f(y) \right|$$
$$\le \left| f(x) - f(c) \right| + \left| f(y) - f(c) \right|$$
$$< \frac{\epsilon}{2} + \frac{\epsilon}{2} = \epsilon$$
Thus, $\quad \lim_{x \to c^-} f(x) = f(c)$ and
$$\lim_{y \to c^+} f(y) = f(c)$$
The limits from the left and the right are equal to $f(c)$ which implies continuity.

47.

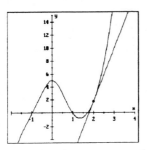

The following sequence of secant lines (graphs) were drawn using a computer.

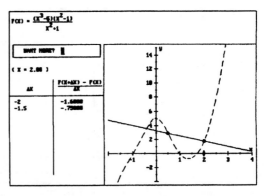

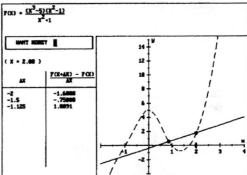

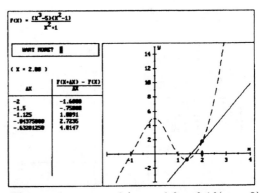

The tangent at $x = 2$ is $y = 1.8 + 8.16(x - 2)$.

48.

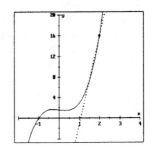

The following sequence of secant lines (graphs) were drawn using a computer.

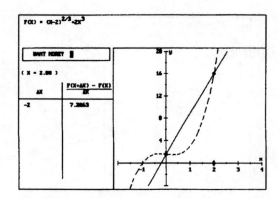

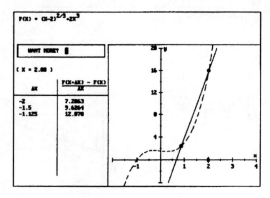

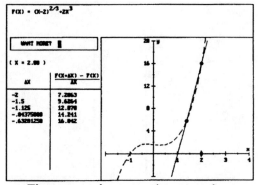

The tangent does not exist at $x = 2$.

49. Putnam competition problem solutions are copied with permission from *The William Lowell Putnam Mathematical Competition, Problems and Solutions: 1938-1964* by A. M. Gleason, R. E. Greenwood, and L. M. Kelly, published by The Mathematical Association of America. This is Problem 4 in the morning session for 1959.

$$1 = 1 - f(1) - g(1) + [f(1) + g(0)]$$
$$+ [f(0) + g(1)] - [f(0) + g(0)]$$

Now one of the numbers

$$\left|1 - f(1) - g(1)\right|, \left|f(1) + g(0)\right|,$$
$$\left|f(0) + g(1)\right|, \left|f(0) + g(0)\right| \text{ is at least } \frac{1}{4}.$$

Thus, the relation holds for at least one of the points $(1, 1)$, $(1, 0)$, $(0, 1)$, or $(0, 0)$.

50. This is Putnam Problem 5 in the morning session for 1960. Consider a constant function g, say $g(x) = a$. Then for $f[g(x)] = g[f(x)]$ becomes $f(a) = a$. Since this is true for all real a, f is the identity function; *i.e.* $f(x) = x$. Remark: the proof does not require the hypothesis that f be a polynomial function.

CHAPTER 2

Techniques of Differentiation with Selected Applications

2.1 **An Introduction to the Derivative:**
    Tangents, Page 93

1. Find the rate of change of $y = f(x)$ with
    respect to a change in x. Some describe it as
    a five-step process: (1) Find $f(x)$ (2) Find
    $f(x + \Delta x)$ (3) Find $f(x + \Delta x) - f(x)$

(4) Find $\dfrac{f(x + \Delta x) - f(x)}{\Delta x}$ (5) Finally

find $\lim\limits_{\Delta x \to 0} \dfrac{f(x + \Delta x) - f(x)}{\Delta x}$

2. The derivative of f at x is given by

$$f'(x) = \lim_{\Delta x \to 0} \frac{f(x + \Delta x) - f(x)}{\Delta x}$$

provided this limit exist.

3. $f'(x) = \lim\limits_{\Delta x \to 0} \dfrac{f(x + \Delta x) - f(x)}{\Delta x}$

$= \lim\limits_{\Delta x \to 0} \dfrac{f(-5 + \Delta x) - f(-5)}{\Delta x}$

$= \lim\limits_{\Delta x \to 0} \dfrac{3 - 3}{\Delta x} = 0$

4. $f'(x) = \lim\limits_{\Delta x \to 0} \dfrac{f(x + \Delta x) - f(x)}{\Delta x}$

$= \lim\limits_{\Delta x \to 0} \dfrac{f(2 + \Delta x) - f(2)}{\Delta x}$

$= \lim\limits_{\Delta x \to 0} \dfrac{2 + \Delta x - 2}{\Delta x} = 1$

5. $f'(x) = \lim\limits_{\Delta x \to 0} \dfrac{f(x + \Delta x) - f(x)}{\Delta x}$

$= \lim\limits_{\Delta x \to 0} \dfrac{f(1 + \Delta x) - f(1)}{\Delta x}$

$= \lim\limits_{\Delta x \to 0} \dfrac{2 + 2\Delta x - 2}{\Delta x} = 2$

6. $f'(x) = \lim\limits_{\Delta x \to 0} \dfrac{f(x + \Delta x) - f(x)}{\Delta x}$

$= \lim\limits_{\Delta x \to 0} \dfrac{f(1 + \Delta x) - f(1)}{\Delta x}$

$= \lim\limits_{\Delta x \to 0} \dfrac{2 + 4\Delta x + 2(\Delta x)^2 - 2}{\Delta x}$

$= \lim\limits_{\Delta x \to 0} \dfrac{\Delta x(4 + 2\Delta x)}{\Delta x}$

$= \lim\limits_{\Delta x \to 0} (4 + 2\Delta x) = 4$

7. $f'(x) = \lim\limits_{\Delta x \to 0} \dfrac{f(x + \Delta x) - f(x)}{\Delta x}$

$= \lim\limits_{\Delta x \to 0} \dfrac{f(0 + \Delta x) - f(0)}{\Delta x}$

$= \lim\limits_{\Delta x \to 0} \dfrac{2 - (\Delta x)^2 - 2}{\Delta x}$

$= \lim\limits_{\Delta x \to 0} -\Delta x = 0$

8. $f'(x) = \lim\limits_{\Delta x \to 0} \dfrac{f(x + \Delta x) - f(x)}{\Delta x}$

$= \lim\limits_{\Delta x \to 0} \dfrac{f(2 + \Delta x) - f(2)}{\Delta x}$

$= \lim\limits_{\Delta x \to 0} \dfrac{-4\Delta x - (\Delta x)^2}{\Delta x}$

$= \lim\limits_{\Delta x \to 0} \dfrac{\Delta x(-4 - \Delta x)}{\Delta x}$

$= \lim\limits_{\Delta x \to 0} (-4 - \Delta x) = -4$

9. $f'(x) = \lim\limits_{\Delta x \to 0} \dfrac{f(x + \Delta x) - f(x)}{\Delta x}$

$= \lim\limits_{\Delta x \to 0} \dfrac{5 - 5}{\Delta x} = 0$

Differentiable for all real x.

10. $g'(x) = \lim\limits_{\Delta x \to 0} \dfrac{g(x + \Delta x) - g(x)}{\Delta x}$

$= \lim\limits_{\Delta x \to 0} \dfrac{3x + 3\Delta x - 3x}{\Delta x}$

$= \lim\limits_{\Delta x \to 0} 3 = 3$

Differentiable for all real x.

11. $f'(t) = \lim\limits_{\Delta t \to 0} \dfrac{f(t + \Delta t) - f(t)}{\Delta t}$

$= \lim\limits_{\Delta t \to 0} \dfrac{3t + 3\Delta t - 7 - 3t + 7}{\Delta t}$

$= \lim\limits_{\Delta t \to 0} 3 = 3$

Differentiable for all real t.

12. $f'(u) = \lim\limits_{\Delta u \to 0} \dfrac{f(u + \Delta u) - f(u)}{\Delta u}$

$= \lim\limits_{\Delta u \to 0} \dfrac{4 - 5u - 5\Delta u - 4 + 5u}{\Delta u}$

$= \lim\limits_{\Delta u \to 0} (-5) = -5$

Differentiable for all real u.

13. $g'(r) = \lim\limits_{\Delta r \to 0} \dfrac{g(r + \Delta r) - f(r)}{\Delta r}$

$= \lim\limits_{\Delta r \to 0} \dfrac{3r^2 + 6r\Delta r + 3(\Delta r)^2 - 3r^2}{\Delta r}$

$= \lim\limits_{\Delta r \to 0} (6r + 3\Delta r) = 6r$

Differentiable for all real r.

14. $h'(x) = \lim_{\Delta x \to 0} \dfrac{h(x + \Delta x) - h(x)}{\Delta x}$

$\qquad = \lim_{\Delta x \to 0} \dfrac{2x^2 + 4x\Delta x + 2(\Delta x)^2 - 2x^2}{\Delta x}$

$\qquad = \lim_{\Delta x \to 0} \dfrac{\Delta x(4x + 2\Delta x)}{\Delta x}$

$\qquad = \lim_{\Delta x \to 0} (4x + 2\Delta x) = 4x$

Differentiable for all real x.

15. $f'(x) = \lim_{\Delta x \to 0} \dfrac{f(x + \Delta x) - f(x)}{\Delta x}$

$\qquad = \lim_{\Delta x \to 0} \dfrac{x^2 + 2x\Delta x + (\Delta x)^2 - x - \Delta x - (x^2 - x)}{2}$

$\qquad = \lim_{\Delta x \to 0} \dfrac{\Delta x(2x + \Delta x - 1)}{\Delta x}$

$\qquad = \lim_{\Delta x \to 0} (2x + \Delta x - 1) = 2x - 1$

Differentiable for all real x.

16. $g'(x) = \lim_{\Delta t \to 0} \dfrac{g(t + \Delta t) - g(t)}{\Delta t}$

$\qquad = \lim_{\Delta t \to 0} \dfrac{4 - t^2 - 2t\Delta t - (\Delta t)^2 - (4 - t^2)}{\Delta t}$

$\qquad = \lim_{\Delta t \to 0} \dfrac{\Delta t(-2t - \Delta t)}{\Delta t}$

$\qquad = \lim_{\Delta t \to 0} (-2t - \Delta t) = -2t$

Differentiable for all real t.

17. $f'(s) = \lim_{\Delta s \to 0} \dfrac{f(s + \Delta s) - f(s)}{\Delta s}$

$= \lim_{\Delta s \to 0} \dfrac{s^2 + (\Delta s)^2 + 1 + 2s\Delta s - 2s - 2\Delta s - s^2 + 2s - 1}{\Delta s}$

$= \lim_{\Delta s \to 0} \dfrac{\Delta s(\Delta s + 2s - 2)}{\Delta s}$

$= \lim_{\Delta s \to 0} (\Delta s + 2s - 2) = 2s - 2$

Differentiable for all real s.

18. $f'(x) = \lim_{\Delta x \to 0} \dfrac{f(x + \Delta x) - f(x)}{\Delta x}$

$\qquad = \lim_{\Delta x \to 0} \dfrac{1}{\Delta x}\left(\dfrac{1}{2x + 2\Delta x} - \dfrac{1}{2x}\right)$

$\qquad = \lim_{\Delta x \to 0} \dfrac{x - x - \Delta x}{2x\Delta x(x + \Delta x)}$

$\qquad = \lim_{\Delta x \to 0} \dfrac{-1}{2x(x + \Delta x)} = \dfrac{-1}{2x^2}$

Differentiable for $x \neq 0$.

19. $f'(x) = \lim_{\Delta x \to 0} \dfrac{f(x + \Delta x) - f(x)}{\Delta x}$

$= \lim_{\Delta x \to 0} \dfrac{\sqrt{5x + 5\Delta x} + \sqrt{5x}}{\Delta x}$

$= \lim_{\Delta x \to 0} \dfrac{\sqrt{5x + 5\Delta x} + \sqrt{5x}}{\Delta x}\left[\dfrac{\sqrt{5x + 5\Delta x} + \sqrt{5x}}{\sqrt{5x + 5\Delta x} + \sqrt{5x}}\right]$

$= \lim_{\Delta x \to 0} \dfrac{5x + 5\Delta x - 5x}{\Delta x(\sqrt{5x + 5\Delta x} + \sqrt{5x})}$

$= \lim_{\Delta x \to 0} \dfrac{5}{\sqrt{5x + 5\Delta x} + \sqrt{5x}} = \dfrac{5}{2\sqrt{5x}} = \dfrac{\sqrt{5x}}{2x}$

Differentiable for $x > 0$.

20. $f'(x) = \lim_{\Delta x \to 0} \dfrac{f(x + \Delta x) - f(x)}{\Delta x}$

$= \lim_{\Delta x \to 0} \dfrac{\sqrt{x + \Delta x + 1} - \sqrt{x + 1}}{\Delta x}$

$= \lim_{\Delta x \to 0} \dfrac{\sqrt{x + \Delta x + 1} - \sqrt{x + 1}}{\Delta x}\left[\dfrac{\sqrt{x + \Delta x + 1} + \sqrt{x + 1}}{\sqrt{x + \Delta x + 1} + \sqrt{x + 1}}\right]$

$= \lim_{\Delta x \to 0} \dfrac{x + \Delta x + 1 - x - 1}{\Delta x(\sqrt{x + \Delta x + 1} + \sqrt{x + 1})}$

$= \lim_{\Delta x \to 0} \dfrac{1}{\sqrt{x + \Delta x + 1} + \sqrt{x + 1}} = \dfrac{1}{2\sqrt{x + 1}}$

Differentiable for $x > -1$.

21. $f'(x) = 3 = m$ (see Problem 11)

$\qquad y - 2 = 3(x - 3)$

$\qquad 3x - y - 7 = 0$

22. $g'(t) = 6t; \; g'(-2) = -12 = m$ (see Problem 13)

$\qquad y - 12 = -12(t + 2)$

$\qquad 12t + y + 12 = 0$

23. $f'(s) = 3s^2$ (see Example 7)

$\qquad f'(-\tfrac{1}{2}) = 3(-\tfrac{1}{2})^2 = \tfrac{3}{4} = m$

$\qquad\qquad f(-\tfrac{1}{2}) = (-\tfrac{1}{2})^3 = -\tfrac{1}{8}$

$\qquad y - (-\tfrac{1}{8}) = \tfrac{3}{4}[s - (-\tfrac{1}{2})]$

$\qquad 3s - 4y + 1 = 0$

24. $h'(x) = -2x$ (see Problem 16)

$\qquad h'(0) = 0 = m$

$\qquad\qquad f(0) = 9 - x^2 = 9$

$\qquad y - 9 = 0$

25. $f'(x) = \lim_{\Delta x \to 0} \dfrac{f(x + \Delta x) - f(x)}{\Delta x}$

$\qquad = \lim_{\Delta x \to 0} \dfrac{1}{\Delta x}\left(\dfrac{1}{x + \Delta x + 3} - \dfrac{1}{x + 3}\right)$

$$= \lim_{\Delta x \to 0} \frac{1}{\Delta x}\left[\frac{x + 3 - x - \Delta x - 3}{(x + 3)(x + \Delta x + 3)}\right]$$

$$= \lim_{\Delta x \to 0} \frac{-1}{(x + 3)(x + \Delta x + 3)} = \frac{-1}{(x + 3)^2}$$

$$f'(2) = \frac{-1}{25} = m; \quad f(2) = \frac{1}{2 + 3} = \frac{1}{5}$$

$$y - \frac{1}{5} = \frac{-1}{25}(x - 2)$$

$$25y - 5 = -x + 2$$

$$x + 25y - 7 = 0$$

26. $g'(t) = \lim_{\Delta t \to 0} \dfrac{g(t + \Delta t) - g(t)}{\Delta t}$

$$= \lim_{\Delta t \to 0} \frac{\sqrt{t + \Delta t - 5} - \sqrt{t - 5}}{\Delta t}$$

$$= \lim_{\Delta t \to 0} \frac{\sqrt{t + \Delta t - 5} - \sqrt{t - 5}}{\Delta t}\left[\frac{\sqrt{t + \Delta t - 5} + \sqrt{t - 5}}{\sqrt{t + \Delta t - 5} + \sqrt{t - 5}}\right]$$

$$= \lim_{\Delta t \to 0} \frac{t + \Delta t - 5 - t + 5}{\Delta t(\sqrt{t + \Delta t - 5} + \sqrt{t - 5})}$$

$$= \lim_{\Delta t \to 0} \frac{1}{\sqrt{t + \Delta t - 5} + \sqrt{t - 5}} = \frac{1}{2\sqrt{t - 5}}$$

$$g'(9) = \frac{1}{4} = m; \quad g(9) = 2$$

$$y - 2 = \frac{1}{4}(t - 9)$$

$$4y - 8 = t - 9$$

$$t - 4y - 1 = 0$$

27. Since $f(x)$ has a slope of 5, the required equation will have a slope of $-1/5$ and must pass through $(3, 13)$. Using the point-slope equation:

$$y - 13 = -\frac{1}{5}(x - 3)$$

$$5y - 65 = -x + 3$$

$$x + 5y - 68 = 0$$

28. Since $g(t)$ has a slope of 2, the required equation will have a slope of $-1/2$ and must pass through $(0, -3)$. Using the point-slope equation:

$$y + 3 = -\frac{1}{2}(t - 0)$$

$$2y + 6 = -t$$

$$t + 2y + 6 = 0$$

29. $g'(1) = -\frac{1}{36}$ (see Problem 25); the required equation will have a slope of 36 and must pass through the point $(1, \frac{1}{6})$ since $g(1) = \frac{1}{6}$.

$$y - \frac{1}{6} = 36(t - 1)$$

$$6y - 1 = 216t - 216$$

$$216t - 6y - 215 = 0$$

30. $f'(1) = 1$ (see Problem 19); the required equation will have slope -1 and must pass through the point $(1, 2)$ since $f(1) = \sqrt{4(1)} = 2$.

$$y - 2 = -(x - 1)$$

$$x + y - 3 = 0$$

31. $\dfrac{dy}{dx} = \lim_{\Delta x \to 0} \dfrac{f(x + \Delta x) - f(x)}{\Delta x}$

$$= \lim_{\Delta x \to 0} \frac{2(x + \Delta x) - 2x}{\Delta x}$$

$$= \lim_{\Delta x \to 0} 2 = 2$$

$$\left.\frac{dy}{dx}\right|_{x = -1} = 2$$

32. $\dfrac{dy}{dx} = \lim_{\Delta x \to 0} \dfrac{f(x + \Delta x) - f(x)}{\Delta x}$

$$= \lim_{\Delta x \to 0} \frac{(4 - x - \Delta x) - (4 - x)}{\Delta x}$$

$$= \lim_{\Delta x \to 0} -1 = -1$$

$$\left.\frac{dy}{dx}\right|_{x = 2} = -1$$

33. $\dfrac{dy}{dx} = \lim_{\Delta x \to 0} \dfrac{f(x + \Delta x) - f(x)}{\Delta x}$

$$= \lim_{\Delta x \to 0} \frac{[1 - (x + \Delta x)^2] - (1 - x^2)}{\Delta x}$$

$$= \lim_{\Delta x \to 0} \frac{-2x\Delta x - (\Delta x)^2}{\Delta x}$$

$$= \lim_{\Delta x \to 0} (-2x - \Delta x) = -2x$$

$$\left.\frac{dy}{dx}\right|_{x = 0} = 0$$

34. $\dfrac{dy}{dx} = \lim_{\Delta x \to 0} \dfrac{f(x + \Delta x) - f(x)}{\Delta x}$

$$= \lim_{\Delta x \to 0} \frac{1}{\Delta x}\left(\frac{4}{x + \Delta x} - \frac{4}{x}\right)$$

$$= \lim_{\Delta x \to 0} \frac{4(x - x - \Delta x)}{x\Delta x(x + \Delta x)}$$

$$= \lim_{\Delta x \to 0} \frac{-4}{x(x + \Delta x)} = \frac{-4}{x^2}$$

$$\left.\frac{dy}{dx}\right|_{x = 1} = -4$$

35. **a.** $f(-2) = 4$ and $f(-1.9) = 3.61$

$$m_{sec} = \frac{4 - 3.61}{-2 + 1.9} = -3.9$$

b. $f'(x) = \lim\limits_{\Delta x \to 0} \dfrac{f(x + \Delta x) - f(x)}{\Delta x}$

$$= \lim\limits_{\Delta x \to 0} \frac{x^2 + 2x\Delta x + (\Delta x)^2 - x^2}{\Delta x}$$

$$= \lim\limits_{\Delta x \to 0} (2x + \Delta x) = 2x$$

$$m_{tan} = f'(-2) = -4$$

36. **a.** $f(1) = 1$ and $f(1.1) = 1.331$

$$m_{sec} = \frac{1.331 - 1}{1.1 - 1} = 3.31$$

b. $f'(x) = \lim\limits_{\Delta x \to 0} \dfrac{f(x + \Delta x) - f(x)}{\Delta x}$

$$= \lim\limits_{\Delta x \to 0} \frac{x^3 + 3x^2(\Delta x) + 3x(\Delta x)^2 + (\Delta x)^3 - x^3}{\Delta x}$$

$$= \lim\limits_{\Delta x \to 0} [3x^2 + 3x\Delta x + (\Delta x)^2] = 3x^2$$

$$m_{tan} = f'(1) = 3$$

37. $f(x) = x^2 - x$

$$f'(x) = \lim\limits_{\Delta x \to 0} \frac{f(x + \Delta x) - f(x)}{\Delta x}$$

$$= \lim\limits_{\Delta x \to 0} \frac{[(x + \Delta x)^2 - (x + \Delta x)] - [x^2 - x]}{\Delta x}$$

$$= \lim\limits_{\Delta x \to 0} \frac{x^2 + 2x\Delta x + (\Delta x)^2 - x - \Delta x - x^2 + x}{\Delta x}$$

$$= \lim\limits_{\Delta x \to 0} (2x + \Delta x - 1) = 2x - 1$$

The derivative is 0 when $x = \frac{1}{2}$; $f(\frac{1}{2}) = -\frac{1}{4}$, so the graph has a horizontal tangent at $(\frac{1}{2}, -\frac{1}{4})$.

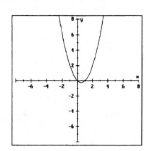

38. **a.** $f(x) = x^2 - 3x$

$$f'(x) = \lim\limits_{\Delta x \to 0} \frac{f(x + \Delta x) - f(x)}{\Delta x}$$

$$= \lim\limits_{\Delta x \to 0} \frac{[(x + \Delta x)^2 - 3(x + \Delta x)] - [x^2 - 3x]}{\Delta x}$$

$$= \lim\limits_{\Delta x \to 0} \frac{x^2 + 2x\Delta x + (\Delta x)^2 - 3x - 3\Delta x - x^2 + 3x}{\Delta x}$$

$$= \lim\limits_{\Delta x \to 0} (2x + \Delta x - 3) = 2x - 3$$

b. The derivative is 0 when $x = \frac{3}{2}$; $f(\frac{3}{2}) = -\frac{9}{4}$, so there is a tangent line through $(\frac{3}{2}, -\frac{9}{4})$. This tangent line is $y + \frac{9}{4} = 0$; this can be written as $4y + 9 = 0$.

c. Two lines are tangent if their slopes are the same. $m = 2x - 3 = -3$ if $x = 0$; $f(0) = 0$, so this tangent line has equation $y = -3x$.

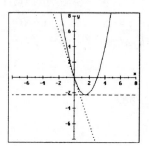

39. **a.** $f(x) = 4 - 2x^2$

$$f'(x) = \lim\limits_{\Delta x \to 0} \frac{f(x + \Delta x) - f(x)}{\Delta x}$$

$$= \lim\limits_{\Delta x \to 0} \frac{[4 - 2(x + \Delta x)^2] - (4 - 2x^2)}{\Delta x}$$

$$= \lim\limits_{\Delta x \to 0} \frac{-4x\Delta x - 2(\Delta x)^2}{\Delta x} = -4x$$

b. A horizontal line has $m = 0$, so $f'(x) = 0$, $-4x = 0$, $x = 0$. At $x = 0$, $f(0) = 4$, so

$$y - 4 = 0(x - 0)$$

$$y - 4 = 0 \text{ is the equation.}$$

c. $8x + 3y = 4$ has a slope of $-\frac{8}{3}$ so our line must have the same slope in order to be parallel. $f'(x) = -4x = -\frac{8}{3}$, $x = \frac{2}{3}$. $f(\frac{2}{3}) = \frac{28}{9}$. So the point at which the tangent is parallel to the given line is $(\frac{2}{3}, \frac{28}{9})$. If we wanted the equation of this tangent line, we would use the

point-slope equation:

$$y - \tfrac{28}{9} = -\tfrac{8}{3}(x - \tfrac{2}{3})$$

$$24x + 9y - 44 = 0$$

40. If $x < 0$, $f'(0) = \lim\limits_{\Delta x \to 0^-} [-2(0) - 2\Delta x] = 0$

If $x \geq 0$, $f'(0) = \lim\limits_{\Delta x \to 0^+} [2(0) + 2\Delta x] = 0$

Since the function is continuous at $x = 0$, and the limits are the same, the derivative exists at $x = 0$.

41. **a.** If $x < 1$, the limit from the right for the derivative is $f(1^+) = \frac{1}{2}$ (see Problem 19). If $x \geq 1$, the limit from the left for the derivative is $f(1^-) = -2$. This means that at the point $(1, -2)$ there are two tangent lines (depending on the direction of approach for the secant lines, as shown in following graph).

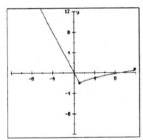

b. $f(-1) = -2$ and $\lim\limits_{x \to 1^-}(-2x) = -2$

and $\lim\limits_{x \to 1^+}(\sqrt{x} - 3) = -2$, we see that the function is continuous at $x = 1$.

42. $f(x) = |x - 5|$ is continuous on $(-\infty, \infty)$, but not differentiable at $x = 5$ (because of the corner point; see Example 6).

43. Let $\Delta x = 0.1$, $c = 1$, $f(x) = (2x - 1)^2$

$$f'(x) \approx \frac{f(x + \Delta x) - f(x)}{\Delta x}$$

$$= \frac{f(1.1) - f(1)}{0.1}$$

$$= \frac{1.44 - 1}{0.1} = 4.4$$

Let $\Delta x = 0.01$

$$f'(x) \approx \frac{f(x + \Delta x) - f(x)}{\Delta x}$$

$$= \frac{f(1.01) - f(1)}{0.01}$$

$$= \frac{1.0404 - 1}{0.01}$$

$$= 4.04$$

It appears that $f'(x) = 4$.

44. We could proceed as shown in the solution for Problem 43, but here we present data which could be generated using a graphing calculator, spreadsheet, or a computer program.

$f(x) = \dfrac{1}{x + 1}$, $c = 2$

Δx	$c + \Delta x$	$\dfrac{f(x + \Delta x) - f(x)}{\Delta x}$
0.5	2.5	-0.0952
0.125	2.125	-0.0671
0.03125	2.03125	-0.1111
0.00781	2.007181	-0.11104
0.00195	2.00195	-0.11108
0.00049	2.00049	-0.11109

We might guess that the derivative is

$$f'(2) = -\tfrac{1}{9} \approx 0.1111$$

45. We could proceed as shown in the solution for Problem 43, but here we present data which could be generated using a graphing calculator, spreadsheet, or a computer program.

$f(x) = \sqrt{x}$, $c = 4$

Δx	$c + \Delta x$	$\dfrac{f(x + \Delta x) - f(x)}{\Delta x}$
0.5	4.5	0.24264
0.125	4.125	0.24808
0.03125	4.03125	0.24851
0.00781	4.007181	0.24989
0.00195	4.00195	0.24997
0.00049	4.00049	0.24997

We might guess that the derivative is

$$f'(2) = \tfrac{1}{4} = 0.25$$

46. We could proceed as shown in the solution for Problem 43, but here we present data which could be generated using a graphing calculator, spreadsheet, or a computer program.

$f(x) = \sqrt[3]{x}$, $c = 8$

Δx	$c + \Delta x$	$\dfrac{f(x + \Delta x) - f(x)}{\Delta x}$
0.5	8.5	0.081652
0.125	8.125	0.082903
0.03125	8.03125	0.083225
0.00781	8.007181	0.083306
0.00195	8.00195	0.083323
0.00049	8.00049	0.083332

We might guess that the derivative is

$$f'(2) = \tfrac{1}{12} \approx 0.08333$$

47. $\dfrac{dy}{dx} = 2Ax$, $\dfrac{dy}{dx}\Big|_{x\,=\,c} = 2Ac$.

The point is $(c, f(c)) = (c, Ac^2)$

So the equation of the tangent line is:

$$y - Ac^2 = 2Ac(x - c).$$

The x-intercept occurs when $y = 0$:

$$-Ac^2 = 2Ac(x - c), \quad -c = 2(x - c), \quad x = \tfrac{c}{2}$$

The y-intercept occurs when $x = 0$:

$$y - Ac^2 = 2Ac(0 - c), \quad y = -Ac^2$$

48. **a.** From Example 1, $f'(x) = 2x$

For $g(x) = x^2 - 3$, we have

$$
\begin{aligned}
g'(x) &= \lim_{\Delta x \to 0} \frac{g(x + \Delta x) - g(x)}{\Delta x}\\
&= \lim_{\Delta x \to 0} \frac{[(x + \Delta x)^2 - 3] - [x^2 - 3]}{\Delta x}\\
&= \lim_{\Delta x \to 0} \frac{x^2 + 2x\Delta x + (\Delta x)^2 - 3 - (x^2 - 3)}{\Delta x}\\
&= \lim_{\Delta x \to 0} (2x + \Delta x) = 2x
\end{aligned}
$$

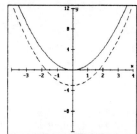

The graph of $y = x^2 - 3$ is the graph of $y = x^2$ lowered by 3 units. Thus, the graphs are parallel and their tangents have the same slopes for any value of x. This accounts geometrically for the fact their derivatives are identical.

b. The graph of $y = x^2 + 5$ is also parallel to the graph of $y = x^2$. Thus,

$$\frac{d}{dx}(x^2 + 5) = \frac{d}{dx}(x^2) = 2x$$

49. **a.**

$$
\begin{aligned}
f'(x) &= \lim_{\Delta x \to 0} \frac{f(x + \Delta x) - f(x)}{\Delta x}\\
&= \lim_{\Delta x \to 0} \frac{[(x + \Delta x)^2 + 3(x + \Delta x)] - (x^2 + 3x)}{\Delta x}\\
&= \lim_{\Delta x \to 0} \frac{x^2 + 2x\Delta x + (\Delta x)^2 + 3x + 3\Delta x - x^2 - 3x}{\Delta x}\\
&= \lim_{\Delta x \to 0} (2x + \Delta x + 3) = 2x + 3
\end{aligned}
$$

b.

$$
\begin{aligned}
g'(x) &= \lim_{\Delta x \to 0} \frac{g(x + \Delta x) - g(x)}{\Delta x}\\
&= \lim_{\Delta x \to 0} \frac{x^2 + 2x\Delta x + (\Delta x)^2 - x^2}{\Delta x}\\
&= \lim_{\Delta x \to 0} (2x + \Delta x) = 2x\\
h'(x) &= \lim_{\Delta x \to 0} \frac{h(x + \Delta x) - h(x)}{\Delta x}\\
&= \lim_{\Delta x \to 0} \frac{3(x + \Delta x) - 3x}{\Delta x}\\
&= \lim_{\Delta x \to 0} 3 = 3
\end{aligned}
$$

The derivative of $f(x) = x^2 + 3x$ is equal to the sum of the derivatives of $g(x) = x^2$ and $h(x) = 3x$.

c. $f'(x) = [g(x) + h(x)]' = g'(x) + h'(x)$

50. $f(x) = -x^2$, and $f'(x) = -2x$. Let $(c, f(c))$ be the point of contact for the tangent line. Then the equation of the tangent line is

$$y + c^2 = -2c(x - c)$$

For the line to pass through $P(0, 9)$,

$$
\begin{aligned}
9 + c^2 &= -2c(0 - c)\\
c^2 + 9 &= 2c^2\\
c &= \pm 3
\end{aligned}
$$

Then, $f(\pm 3) = -9$. Thus, the points are $(3, -9)$ and $(-3, -9)$.

51.

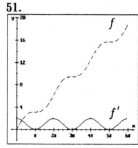

52.

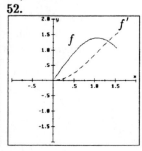

53.

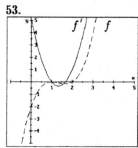

54.

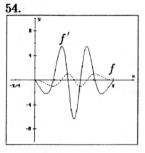

55. See the graphs of Problems 51-54. Note that the graph of $f'(x)$ is increasing where the derivative is positive, and $f'(x)$ is decreasing where the derivative is negative.

56. For example, using $b = 2.01$, we obtain

$$y = 1.8 + 8.22918(x - 2)$$

as opposed to the "real tangent"

$$y = 1.8 + 8.16(x - 2)$$

57. When you zoom *very* close to $x = 2$, you should see a sharp peak or "cusp" which is missed by most graphs. There is no tangent at $x = 2$.

58. For $f'(c) \neq 0$,

$$y - f(c) = -\frac{1}{f'(c)}(x - c)$$
$$f'(c)y - f(c)f'(c) = -x + c$$
$$f'(c)y - f(c)f'(c) + x - c = 0$$

If $f'(c) = 0$, $x = c$ is the equation of the vertical normal line.

59. $y = Ax^2$. Plot several points $(c, f(c))$ and $(\frac{c}{2}, 0)$. Draw the (very short) line segment of this tangent line to the parabola near the point of constant. Then draw a smooth curve tangent to all the plotted line segments.

2.2 Techniques of Differentiation, Page 102

1. (3) $f'(x) = 0$; $f'(-5) = 0$
      (4) $f'(x) = 1$; $f'(2) = 1$
      (5) $f'(x) = 2$; $f'(1) = 2$
      (6) $f'(x) = 4x$; $f'(1) = 4$
      (7) $f'(x) = -2x$; $f'(0) = 0$
      (8) $f'(x) = -2x$; $f'(2) = -4$

2. (9) $f'(x) = 0$ (10) $g'(x) = 3$
      (11) $f'(t) = 3$ (12) $f'(u) = -5$
      (13) $g'(r) = 6r$ (14) $h'(x) = 4x$

3. (15) $f'(x) = 2x - 1$
      (16) $g'(t) = -2t$
      (17) $f(s) = (s - 1)^2 = s^2 - 2s + 1$;
            $f'(s) = 2s - 2$

      (18) $f(x) = \frac{1}{2}x^{-1}$; $f'(x) = -\frac{1}{2}x^{-2}$

      (19) $f(x) = \sqrt{5}\,x^{1/2}$; $f'(x) = \frac{\sqrt{5}}{2x^{1/2}} = \frac{\sqrt{5x}}{2x}$

4. (21) $f'(x) = 3$; $f'(3) = 3$

      (22) $g'(t) = 6t$; $g'(-2) = -12$

(23) $f'(s) = 3s^2$; $f'(-\frac{1}{2}) = \frac{3}{4}$

(24) $h'(x) = -2x$; $h'(9) = -18$

5. **a.** $f'(x) = 3(4)x^3 - 0 = 12x^3$

      b. $g'(x) = 0 - 1 = -1$

6. **a.** $f'(x) = 5(2)x + 1 = 10x + 1$

      b. $g'(x) = 0$

7. **a.** $f'(x) = 3x^2 + 0 = 3x^2$

      b. $g'(x) = 1$

8. **a.** $-10t^{-2}$

      b. $g(t) = 7t^{-1}$; $g'(t) = -7t^{-2}$

9. $r(t) = t^2 - t^{-2} + 5t^{-4}$;

      $r'(t) = 2t + 2t^{-3} - 20t^{-5}$

10. $f'(x) = 0$

11. $f(x) = 7x^{-2} + x^{2/3} + C$;

      $f'(x) = -14x^{-3} + \frac{2}{3}x^{-1/3}$

12. $g(x) = \frac{1}{2}x^{-1/2} + \frac{1}{4}x^2 + C$

      $g'(x) = -\frac{1}{4}x^{-3/2} + \frac{1}{2}x$

13. $f(x) = x + 1 + x^{-1} - 7x^{-2}$

      $f'(x) = 1 - x^{-2} + 14x^{-3}$

14. $f(x) = 2x^2 - 3x^{-1} + 11x^{-3}$

      $f'(x) = 4x + 3x^{-2} - 33x^{-4}$

15. $f(x) = (2x + 1)(1 - 4x^3)$

      $\qquad = 2x - 8x^4 + 1 - 4x^3$

      $f'(x) = -32x^3 - 12x^2 + 2$

16. $g(x) = (x + 2)(2x^{1/2} + x^2)$

      $\qquad = 2x^{3/2} + x^3 + 4x^{1/2} + 2x^2$

      $g'(x) = 3x^{1/2} + 3x^2 + 2x^{-1/2} + 4x$

17. $f'(x) = \dfrac{(x + 9)(3) - (3x + 5)(1)}{(x + 9)^2}$

      $\qquad = \dfrac{22}{(x + 9)^2}$

18. $f'(x) = \dfrac{(x^2 + 5)(2x) - (x^2 + 3)(2x)}{(x^2 + 5)^2}$

      $\qquad = \dfrac{2x(x^2 + 5 - x^2 - 3)}{(x^2 + 5)^2}$

      $\qquad = \dfrac{4x}{(x^2 + 5)^2}$

19. $g(x) = x^2(x + 2)^2 = x^4 + 4x^3 + 4x^2$

$g'(x) = 4x^3 + 12x^2 + 8x$

20. $f(x) = x^2(2x + 1)^2 = 4x^4 + 4x^3 + x^2$

$f'(x) = 16x^3 + 12x^2 + 2x$

21. $f(x) = x^5 - 5x^3 + x + 12$

$f'(x) = 5x^4 - 15x^2 + 1$

$f''(x) = 20x^3 - 30x$

$f'''(x) = 60x^2 - 30$

$f^{(4)}(x) = 120x$

22. $f(x) = \frac{1}{4}x^8 - \frac{1}{2}x^6 - x^2 + 2$

$f'(x) = 2x^7 - 3x^5 - 2x$

$f''(x) = 14x^6 - 15x^4 - 2$

$f'''(x) = 84x^5 - 60x^3$

$f^{(4)}(x) = 420x^4 - 180x^2$

23. $f(x) = -2x^{-2}$

$f'(x) = 4x^{-3}$

$f''(x) = -12x^{-4}$

$f'''(x) = 48x^{-5}$

$f^{(4)}(x) = -240x^{-6}$

24. $f(x) = 4x^{-1/2}$

$f'(x) = -2x^{-3/2}$

$f''(x) = 3x^{-5/2}$

$f'''(x) = -\frac{15}{2}x^{-7/2}$

$f^{(4)}(x) = \frac{105}{4}x^{-9/2}$

25. $y = 3x^3 - 7x^2 + 2x - 3$

$\frac{dy}{dx} = 9x^2 - 14x + 2$

$\frac{d^2y}{dx^2} = 18x - 14$

26. $y = (x^2 + 4)(1 - 3x^3) = -3x^5 - 12x^3 + x^2 + 4$

$\frac{dy}{dx} = -15x^4 - 36x^2 + 2x$

$\frac{d^2y}{dx^2} = -60x^3 - 72x + 2$

27. $f(x) = x^2 - 3x - 5;\ f(-2) = 5$, so the point of tangency is $(-2, 5)$.

$f'(x) = 2x - 3$ and $f'(-2) = -7 = m_{\tan}$

Using the point-slope formula:

$$y - 5 = -7[x - (-2)]$$

$7x + y + 9 = 0$

28. $f(x) = x^5 - 3x^3 - 5x + 2;\ f(1) = -5$, so the point of tangency is $(1, -5)$.

$f'(x) = 5x^4 - 9x^2 - 5$ and

$f'(1) = -9 = m_{\tan}$

Using the point-slope formula:

$$y - (-5) = -9(x - 1)$$

$9x + y - 4 = 0$

29. $f(x) = (x^2 + 1)(1 - x^3) = -x^5 - x^3 + x^2 + 1$

$f(1) = (1^2 + 1)(1 - 1^3) = 0$, so the point of tangency is $(1, 0)$.

$f'(x) = -5x^4 - 3x^2 + 2x$ and

$f'(1) = -6 = m_{\tan}$

Using the point-slope formula:

$$y - 0 = -6(x - 1)$$

$6x + y - 6 = 0$

30. $f(x) = \frac{x + 1}{x - 1};\ f(0) = -1$, so the point of

tangency is $(0, -1)$.

$$f'(x) = \frac{(x - 1)(1) - (x + 1)(1)}{(x - 1)^2} = \frac{-2}{(x - 1)^2}$$

and $f'(0) = -2 = m_{\tan}$

Using the point-slope formula:

$$y + 1 = -2(x - 0)$$

$2x + y + 1 = 0$

31. $f(x) = \frac{x^2 + 5}{x + 5};\ f(1) = 1$, so the point of

tangency is $(1, 1)$.

$$f'(x) = \frac{(x + 5)2x - (x^2 + 5)(1)}{(x + 5)^2}$$

$$= \frac{x^2 + 10x - 5}{(x + 5)^2},$$

and $f'(1) = \frac{1}{6} = m_{\tan}$

Using the point-slope formula:

$$y - 1 = \frac{1}{6}(x - 1)$$

$x - 6y + 5 = 0$

32. $f(x) = 1 - x^{-1} + 2x^{-1/2};\ f(4) = \frac{7}{4}$, so

the point of tangency is $(4, \frac{7}{4})$.

$f'(x) = x^{-2} - x^{-3/2}$ and

$f'(4) = 4^{-2} - 4^{-3/2} = -\frac{1}{16} = m_{\tan}$

Using the point-slope formula:

$$y - \tfrac{7}{4} = -\tfrac{1}{16}(x - 4)$$

$$x + 16y - 32 = 0$$

33. $f'(x) = 6x^2 - 14x + 8$; solve

$$6x^2 - 14x + 8 = 0$$

$$2(3x - 4)(x - 1) = 0$$

$$x = \tfrac{4}{3}, 1$$

$f(\tfrac{4}{3}) = -\tfrac{1}{27}$ and $f(1) = 0$; the points are $(1, 0)$ and $(\tfrac{4}{3}, -\tfrac{1}{27})$.

34. $g(x) = (3x - 5)(x - 8) = 3x^2 - 29x + 40$

$g'(x) = 6x - 29$; $g'(x) = 0$ if $x = \tfrac{29}{6}$

$g(\tfrac{29}{6}) = -\tfrac{361}{12}$; the point is $(\tfrac{29}{6}, -\tfrac{361}{12})$.

35. $f(t) = t^{-2} - t^{-3}$

$f'(t) = -2t^{-3} + 3t^{-4}$; solve

$$-\frac{2}{t^3} + \frac{3}{t^4} = 0$$

$$-2t + 3 = 0$$

$$t = \tfrac{3}{2}$$

$f(\tfrac{3}{2}) = (\tfrac{3}{2})^{-2} - (\tfrac{3}{2})^{-3} = \tfrac{4}{27}$; the point is $(\tfrac{3}{2}, \tfrac{4}{27})$.

36. $f'(t) = 4t^3 + 12t^2 - 16t$; solve

$$4t^3 + 12t^2 - 16t = 0$$

$$4t(t^2 + 3t - 4) = 0$$

$$4t(t - 1)(t + 4) = 0$$

$$t = 0, 1, -4$$

$f(0) = 3$, $f(1) = 0$, and $f(-4) = -125$; the points are $(0, 3)$, $(1, 0)$, and $(-4, -125)$.

37. $f(x) = \sqrt{x}(x - 3) = x^{3/2} - 3x^{1/2}$

$f'(x) = \tfrac{3}{2}x^{1/2} - \tfrac{3}{2}x^{-1/2}$; solve

$$\tfrac{3}{2}x^{1/2} - \tfrac{3}{2}x^{-1/2} = 0$$

$$x - 1 = 0$$

$$x = 1$$

$f(1) = \sqrt{1}(1 - 3) = -2$; the point is $(1, -2)$.

38. $h(x) = \dfrac{x^2 - 2x + 1}{x - 1} = \dfrac{(x - 1)^2}{x - 1} = x - 1$

$h'(x) = 1 \neq 0$, so there are no horizontal tangents.

39. $h(x) = \dfrac{4x^2 + 12x + 9}{2x + 3} = \dfrac{(2x + 3)^2}{2x + 3} = 2x + 3$

$h'(x) = 2 \neq 0$, so there are no horizontal tangents.

40. a. $f'(x) = 4x - 5$

    b. $f(x) = (2x + 1)(x - 3)$

$$f'(x) = (2x + 1)(1) + 2(x - 3)$$

$$= 4x - 5$$

41. a. $f'(x) = \dfrac{x^3(2) - (2x - 3)(3x^2)}{x^6}$

$$= \dfrac{2x^3 - 6x^3 + 9x^2}{x^6}$$

$$= \dfrac{-4x + 9}{x^4}$$

    b. $f'(x) = x^{-3}(2) + (-3x^{-4})(2x - 3)$

$$= 2x^{-3} - 6x^{-3} + 9x^{-4}$$

$$= -4x^{-3} + 9x^{-4}$$

    c. $f'(x) = 2(-2)x^{-3} - 3(-3)x^{-4}$

$$= -4x^{-3} + 9x^{-4}$$

42. $f(x) = ax^2 + bx + c$; the curve passes through the origin, so $c = 0$ and $f(x) = ax^2 + bx$. The curve passes through the point $P(5, 0)$, so

$$25a + 5b = 0$$

$$b = -5a$$

$f'(x) = 2ax + b$ and $4a + b = 1$ since the slope is 1 when $x = 2$. Combining results for b leads to

$$b = -4a + 1 \text{ and } b = -5a$$

Thus, $-5a = -4a + 1$, or $a = -1$ and $b = 5$. Therefore, $f(x) = -x^2 + 5x$.

43. The given line $2x - y - 3 = 0$ has a slope of 2, so we want the points on $y = x^4 - 2x + 1$ that have a slope of 2. That is, $y' = 2$ so $4x^3 - 2 = 2$. $x^3 = 1$ implies that there is only one real root: $x = 1$. $f(1) = 0$, so we want the equation of a line through the point $(1, 0)$ with a slope of 2:

$$y - 0 = 2(x - 1)$$

$$2x - y - 2 = 0$$

44. $f'(x) = \dfrac{(1 + x)(3) - (3x + 5)(1)}{(1 + x)^2}$

$$= \dfrac{-2}{(1 + x)^2}$$

The given line $2x - y - 1 = 0$ has slope

$m = 2$, so the perpendicular line has slope $-1/2$. Let $(c, f(c))$ be the point of tangency. Then,

$$\frac{-2}{(1 + c)^2} = -\frac{1}{2}$$

$$-4 = -(1 + c)^2$$

$$1 + c = \pm 2$$

$$c = 1, -3$$

$$f(1) = \frac{3 + 5}{1 + 1} = 4; \quad f(-3) = \frac{-9 + 5}{-3 + 1} = 2$$

The desired equations for the tangent lines are

$$y - 4 = -\tfrac{1}{2}(x - 1) \qquad y - 2 = -\tfrac{1}{2}(x + 3)$$

$$x + 2y - 9 = 0 \qquad\qquad x + 2y - 1 = 0$$

45. a. $f(x) = (x^3 - 2x^2)(x + 2) = x^4 - 4x^2$

$$f(1) = 1^4 - 4(1)^2 = -3$$

$$f'(x) = 4x^3 - 8x; \ f'(1) = -4 = m$$

The equation of the tangent line is

$$y + 3 = -4(x - 1)$$

$$4x + y - 1 = 0$$

b. $f'(0) = 0$ so the slope is not defined and the normal line is horizontal with equation $x = 0$ (that is, the y-axis).

46. From Problem 45, $f'(x) = 4x^3 - 8x$. The slope of the given line is $1/16$, so the slope of the desired normal line is -16. Solve

$$f'(x) = -16$$

$$4x^3 - 8x = -16$$

$$x^3 - 2x + 4 = 0$$

$$(x + 2)(x^2 - 2x + 2) = 0$$

Use factor theorem and synthetic division or use graphing.

$$x = -2$$

Thus, $f(-2) = 0$, so the desired line has equation

$$y - 0 = 16(x + 2)$$

$$16x - y - 32 = 0$$

47. We are looking for particular points (x_0, y_0) on the graph of $y = 4x^2$ which have a tangent at that point which will pass through the point $(2, 0)$.

$$y' = 8x, \quad f(x_0) = 4x_0^2, \quad f'(x_0) = 8x_0.$$

So we have a point $(x_0, 4x_0^2)$ and the slope of the line at that point: $8x_0$. We can now

write the equation of the line:

$$y - 4x_0^2 = 8x_0(x - x_0)$$

This line must pass through the point $(2, 0)$, so

$$0 - 4x_0^2 = 8x_0(2 - x_0)$$

$$4x_0^2 - 16x_0 = 0$$

$$4x_0(x_0 - 4) = 0$$

$$x_0 = 0, 4$$

Therefore there are two points on the curve at which the tangent line will pass through $(2, 0)$; they are $(0, 0)$ and $(4, 64)$.

48. $f'(x) = 2x - 4; \ f'(c) = 2c - 4;$

$f(c) = c^2 - 4c + 25;$ the equations of the tangent lines are

$$y - (c^2 - 4c + 25) = (2c - 4)(x - c)$$

The origin is on these tangent lines so

$$-(c^2 - 4c + 25) = (2c - 4)(-c)$$

$$-c^2 + 4c + 2c^2 - 4c = 25$$

$$c^2 = 25$$

$$c = \pm 5$$

Thus,

$$y - (25 - 20 + 25) = (10 - 4)(x - 5)$$

$$y - 30 = 6x - 30$$

$$6x - y = 0$$

and

$$y - (25 + 20 + 25) = (-10 - 4)(x + 5)$$

$$y - 70 = -14x - 70$$

$$14x + y = 0$$

49. $f'(x) = 2x + 2; \ f''(x) = 2; \ f'''(x) = 0$

$$y''' + y'' + y' = 2x + 2 + 2 + 0 = 2x + 4$$

The equation is not satisfied.

50. $f'(x) = 3x^2 + 2x + 1; \ f''(x) = 6x + 2;$

$f'''(x) = 6$

$$y''' + y'' + y' = 3x^2 + 2x + 1 + 6x + 2 + 6$$

$$= 3x^2 + 8x + 9; \ \text{The equation is not satisfied.}$$

51. $f'(x) = x; \ f''(x) = 1; \ f'''(x) = 0$

$$y''' + y'' + y' = 0 + 1 + x = x + 1$$

This function satisfies the given equation.

52. $f'(x) = 4x + 1;\ f''(x) = 4;\ f'''(x) = 0$

$y''' + y'' + y' = 4x + 1 + 4 + 0 = 4x + 5$

The equation is not satisfied.

53. $P_2(x) = a_2 x^2 + a_1 x + a_0;\ P_2'(x) = 2a_2 x + a_1$

$\quad P_2''(x) = 2a_2$

$P_3(x) = a_3 x^3 + \cdots;\ P_3'(x) = 3a_3 x^2 + \cdots$

$\quad P_3''(x) = 3 \cdot 2a_3 x + \cdots$

$\quad P_3'''(x) = 3!a_3$

$\quad\quad \vdots$

$P_k(x) = a_k x^k + \cdots;\ P_k'(x) = ka_k x^{k-1} + \cdots;$

$\quad P_k''(x) = k(k-1)a_k x^{k-2} + \cdots;$

$\quad P_k'''(x) = k(k-1)(k-2)a_k x^{k-3} + \cdots;$

$\quad\quad \vdots$

$\quad P_k^{(k)}(x) = k!a_k$

The $(k+1)$st derivative is 0.

54. $F(x) = cf(x)$

$$F'(x) = \lim_{\Delta x \to 0} \frac{F(x + \Delta x) - F(x)}{\Delta x}$$

$$= \lim_{\Delta x \to 0} \frac{c[f(x + \Delta x) - f(x)]}{\Delta x}$$

$$= cf'(x)$$

55. $F(x) = f(x) + g(x)$

$$F'(x) = \lim_{\Delta x \to 0} \frac{F(x + \Delta x) - F(x)}{\Delta x}$$

$$= \lim_{\Delta x \to 0} \frac{f(x + \Delta x) + g(x + \Delta x) - [f(x) + g(x)]}{\Delta x}$$

$$= \lim_{\Delta x \to 0} \frac{f(x + \Delta x) - f(x)}{\Delta x}$$

$$+ \lim_{\Delta x \to 0} \frac{g(x + \Delta x) - g(x)}{\Delta x}$$

$$= f'(x) + g'(x)$$

56. $F(x) = f(x) - g(x)$

$$F'(x) = \lim_{\Delta x \to 0} \frac{F(x + \Delta x) - F(x)}{\Delta x}$$

$$= \lim_{\Delta x \to 0} \frac{f(x + \Delta x) - g(x + \Delta x) - [f(x) - g(x)]}{\Delta x}$$

$$= \lim_{\Delta x \to 0} \frac{f(x + \Delta x) - f(x)}{\Delta x}$$

$$- \lim_{\Delta x \to 0} \frac{g(x + \Delta x) - g(x)}{\Delta x}$$

$$= f'(x) - g'(x)$$

57. $F(x) = [f(x)]^2$

$$F'(x) = \lim_{\Delta x \to 0} \frac{F(x + \Delta x) - F(x)}{\Delta x}$$

$$= \lim_{\Delta x \to 0} \frac{[f(x + \Delta x)]^2 - [f(x)]^2}{\Delta x}$$

$$= \lim_{\Delta x \to 0} \frac{[f(x + \Delta x) - f(x)][f(x + \Delta x) + f(x)]}{\Delta x}$$

$$= \lim_{\Delta x \to 0} [f(x + \Delta x) + f(x)] \lim_{\Delta x \to 0} \frac{f(x + \Delta x) - f(x)}{\Delta x}$$

$$= 2f(x)f'(x)$$

58. $(fg)' = \frac{1}{2}[(f + g)^2 - f^2 - g^2]'$

$$= \frac{1}{2}[2(f + g)(f + g)' - 2ff' - 2gg']$$

$$= (f + g)(f' + g') - ff' - gg'$$

$$= ff' + fg' + fg' + gg' - ff' - gg'$$

$$= fg' + f'g$$

59. $q'(x) = \lim_{\Delta x \to 0} \dfrac{q(x + \Delta x) - q(x)}{\Delta x}$

$$= \lim_{\Delta x \to 0} \frac{1}{\Delta x}\left[\frac{f(x + \Delta x)}{g(x + \Delta x)} - \frac{f(x)}{g(x)}\right]$$

$$= \lim_{\Delta x \to 0} \frac{f(x + \Delta x)g(x) - f(x)g(x) - f(x)g(x + \Delta x) + f(x)g(x)}{\Delta x g(x)g(x + \Delta x)}$$

$$= \lim_{\Delta x \to 0} \frac{[f(x + \Delta x) - f(x)][g(x) - f(x)][g(x + \Delta x) - g(x)]}{\Delta x g(x)g(x + \Delta x)}$$

$$= \lim_{\Delta x \to 0} \frac{f(x + \Delta x) - f(x)}{\Delta x g(x + \Delta x)}$$

$$- \lim_{\Delta x \to 0} \frac{f(x)[g(x + \Delta x) - g(x)]}{\Delta x g(x)g(x + \Delta x)}$$

$$= \frac{f'(x)}{g(x)} - \frac{f(x)g'(x)}{[g(x)]^2}$$

Since $f(x) \neq 0$, $g(x) \neq 0$, $f'(x)$ and $g'(x)$ exist, so does $q'(x)$.

60. $\left(\dfrac{f}{g}\right)' = \lim\limits_{\Delta x \to 0} \dfrac{1}{\Delta x}\left[\dfrac{f(x + \Delta x)}{g(x + \Delta x)} - \dfrac{f(x)}{g(x)}\right]$

$\phantom{\left(\dfrac{f}{g}\right)'} = \lim\limits_{\Delta x \to 0} \dfrac{f(x + \Delta x)g(x) - f(x)g(x) - f(x)g(x + \Delta x) + f(x)g(x)}{\Delta x g(x)g(x + \Delta x)}$

$\phantom{\left(\dfrac{f}{g}\right)'} = \lim\limits_{\Delta x \to 0} \dfrac{[f(x+\Delta x) - f(x)][g(x) - f(x)][g(x+\Delta x) - g(x)]}{\Delta x g(x)g(x + \Delta x)}$

$\phantom{\left(\dfrac{f}{g}\right)'} = \lim\limits_{\Delta x \to 0} \dfrac{f(x + \Delta x) - f(x)}{\Delta x g(x + \Delta x)} - \lim\limits_{\Delta x \to 0} \dfrac{f(x)[g(x + \Delta x) - g(x)]}{\Delta x g(x)g(x + \Delta x)}$

$\phantom{\left(\dfrac{f}{g}\right)'} = \dfrac{f'(x)}{g(x)} - \dfrac{f(x)g'(x)}{[g(x)]^2}$

$\phantom{\left(\dfrac{f}{g}\right)'} = \dfrac{g(x)f'(x) - f(x)g'(x)}{[g(x)]^2}$

61. $r(x) = \dfrac{1}{f(x)}$; by the quotient rule, $r'(x) = \dfrac{f(x)1' - (1)f'(x)}{[f(x)]^2} = \dfrac{f'(x)}{[f(x)]^2}$

62. $(fgh)' = (fg)'h + (fg)h'$

$ = (fg' + f'g)h + (fg)h'$

$ = fg'h + f'gh + fgh'$

63. a. $g(x) = [f(x)]^3$

$ = f'(x)[f(x)]^2 + f(x)f'(x)f(x) + [f(x)]^2 f'(x)$ *See Problem 62 where $g = h = f$.*

b. $p(x) = [f(x)]^4$

$ = \{f(x)[f(x)]^3\}'$

$ = f'(x)[f(x)]^3 + \{[f(x)]^3\}'f(x)$

$ = f'(x)[f(x)]^3 + \{3[f(x)]^2[f'(x)]\}f(x)$

$ = 4[f(x)]^3 f'(x)$

64. $y' = 3Ax^2 + B;\ y'' = 6Ax;\ y''' = 6A;$

$y''' + 2y'' - 3y' + y = 6A + 2(6Ax) - 3(3Ax^2 + B) + Ax^3 + Bx + C$

If this is equal to x, we can equate the coefficients of x^3 to conclude $A = 0$. From the coefficients of x we see that $12A + B = 1$ or $B = 1$. The constants reveal that $C = 3B = 3$. Then $y = x + 3$.

65. $f(x) = a_n x^n + a_{n-1}x^{n-1} + \cdots + a_2 x^2 + a_1 x + a_0 \geq 0$, by hypothesis. Then

$$a_n x^n + a_{n-1}x^{n-1} + \cdots + a_2 x^2 + a_1 x + a_0 \geq 0$$

$$na_n x^{n-1} + (n-1)a_{n-1}x^{n-2} + \cdots + 2a_2 x + a_1 \geq 0$$

$$n(n-1)a_n x^{n-2} + (n-1)(n-2)a_{n-1}x^{n-3} + \cdots + 2 \cdot 3a_2 \geq 0$$

$$\vdots$$

$$n!a_n \geq 0$$

$$a_n \geq 0$$

The above rewritten $f(x) \geq 0$, $f'(x) \geq 0$, $f''(x) \geq 0$, $\cdots f^{(n)}(x) \geq 0$. Adding all the left members leads to

$$f(x) + f'(x) + f''(x) + \cdots + f^{(n)}(x) \geq 0$$

2.3 Derivatives of the Trigonometric Functions, Page 110

1. $\displaystyle\lim_{x \to 0} \frac{\sin 2x}{x} = \lim_{x \to 0} 2\left(\frac{\sin 2x}{2x}\right) = 2$

2. $\displaystyle\lim_{x \to 0} \frac{\sin 4x}{9x} = \lim_{x \to 0} \left(\frac{4}{9}\right)\left(\frac{\sin 4x}{4x}\right) = \frac{4}{9}$

3. $\displaystyle\lim_{x \to 0} \frac{3}{2} \frac{\frac{\sin 3x}{3x}}{\frac{\sin 2x}{2x}} = \frac{3}{2} \frac{\displaystyle\lim_{x \to 0} \frac{\sin 3x}{3x}}{\displaystyle\lim_{x \to 0} \frac{\sin 2x}{2x}} = \frac{3}{2}$

4. $\displaystyle\lim_{t \to 0} \frac{\sin 3t}{\cos t} = \lim_{t \to 0} (\sin 3t) \lim_{t \to 0}\left(\frac{1}{\cos t}\right) = 0\left(\frac{1}{1}\right) = 0$

5. $\displaystyle\lim_{t \to 0} \frac{\tan 5t}{\tan 2t} = \lim_{t \to 0} \frac{\sin 5t}{\cos 5t} \lim_{t \to 0} \frac{\cos 2t}{\sin 2t}$

$\displaystyle = \lim_{t \to 0} \frac{\sin 5t}{5t} \lim_{t \to 0} \frac{5}{\cos 5t} \lim_{t \to 0} \frac{\sin 2t}{2t} \lim_{t \to 0} \frac{\cos 2t}{2}$

$\displaystyle = (1\left(\frac{5}{1}\right)(1)\left(\frac{1}{2}\right) = \frac{5}{2}$

6. $\displaystyle\lim_{x \to 0} \frac{\cot 3x}{\cot x} = \lim_{x \to 0} \frac{\cos 3x}{\sin 3x} \lim_{x \to 0} \frac{\sin x}{\cos x}$

$\displaystyle = \lim_{x \to 0} \frac{\cos 3x}{\cos x} \lim_{x \to 0}\left(\frac{1}{3}\right) \frac{3x}{\sin 3x} \lim_{x \to 0} \frac{\sin x}{x}$

$\displaystyle = \left(\frac{1}{1}\right)\left(\frac{1}{3}\right)(1)(1) = \frac{1}{3}$

7. $\displaystyle\lim_{x \to 0} \frac{\frac{1 - \cos x}{x}}{\frac{\sin x}{x}} = \frac{\displaystyle\lim_{x \to 0} \frac{1 - \cos x}{x}}{\displaystyle\lim_{x \to 0} \frac{\sin x}{x}} = \frac{0}{1} = 0$

8. $\displaystyle\lim_{x \to 0} \frac{\sin^2 x}{2x} = \lim_{x \to 0} \frac{1 - \cos^2 x}{2x}$

$\displaystyle = \frac{1}{2} \lim_{x \to 0} \frac{(1 - \cos x)(1 + \cos x)}{x}$

$\displaystyle = \frac{1}{2}\lim_{x \to 0} \frac{1 - \cos x}{x} \lim_{x \to 0}(1 + \cos x)$

$\displaystyle = \frac{1}{2}(0)(1 + 1) = 0$

9. $\displaystyle\lim_{x \to 0} \frac{\sin(\cos x)}{\sec x} = \lim_{x \to 0} \cos x \sin(\cos x)$

Let $t = \cos x$; $t \to 1$ as $x \to 0$

$\displaystyle = \lim_{t \to 1} t \sin t = (1)(\sin 1) = \sin 1 \approx 0.8415$

10. $\displaystyle\lim_{x \to 0} \tan 2x \cot x$

$\displaystyle = \lim_{x \to 0}\left[2 \cdot \frac{\sin 2x}{2x} \cdot \frac{1}{\cos 2x} \cdot \cos x \cdot \frac{x}{\sin x}\right]$

$= 2(1)(\frac{1}{1})(1)(1) = 2$

11. $\displaystyle\lim_{x \to 0} \frac{\sin^2 x}{x^2} = \lim_{x \to 0} \frac{\sin x}{x} \frac{\sin x}{x} = (1)(1) = 1$

12. $\displaystyle\lim_{x \to 0} \frac{x^2\cos 2x}{1 - \cos x} = \lim_{x \to 0} \frac{x^2\cos 2x(1 + \cos x)}{(1 - \cos x)(1 + \cos x)}$

$\displaystyle = \lim_{x \to 0} \frac{x^2\cos 2x(1 + \cos x)}{\sin^2 x}$

$\displaystyle = \lim_{x \to 0}\left[\left(\frac{x}{\sin x}\right)^2(\cos 2x)(1 + \cos x)\right]$

$= (1)^2(1)(1 + 1) = 2$

13. $f'(x) = \cos x - \sin x$

14. $f'(x) = 2\cos x + \sec^2 x$

15. $g'(t) = 2t - \sin t$

16. $g(t) = 2\sec t \tan t + 3\sec^2 t$

17. $p'(x) = -x^2 \sin x + 2x \cos x$

18. $p'(t) = 2t \sin t + t^2 \cos t + 2\cos t$

19. $f'(t) = (\sin t)(\cos t) + (\sin t)(\cos t)$

$\quad = 2\sin t \cos t = \sin 2t$

20. $g'(x) = -2\sin x \cos x = -\sin 2x$

21. $f'(x) = \sqrt{x}\,(-\sin x) + (\cos x)\frac{1}{2}x^{-1/2}$

$\qquad + x(-\csc^2 x) + (\cot x)(1)$

$\quad = -\sqrt{x}\sin x + \frac{1}{2}x^{-1/2}\cos x$

$\qquad - x\csc^2 x + \cot x$

22. $f'(x) = 2x^3 \cos x + 2(3x^2)(\sin x)$

$\qquad - 3(1)\cos x - 3(x)(-\sin x)$

$= 2x^3 \cos x + 6x^2 \sin x + 3x \sin x - 3\cos x$

23. $q'(x) = \dfrac{x\cos x - \sin x}{x^2}$

24. $r'(x) = \dfrac{(\sin x)(1) - (x)(\cos x)}{\sin^2 x}$

$\quad = \dfrac{\sin x - x\cos x}{\sin^2 x}$

25. $h'(t) = \dfrac{(t)(\sec^2 t) - (\tan t)(1)}{t^2}$

$\quad = \dfrac{t\sec^2 t - \tan t}{t^2}$

26. $f'(\theta) = \dfrac{(\theta)(\sec \theta \tan \theta) - (\sec \theta)(1)}{\theta^2}$

$\quad = \dfrac{\theta \sec \theta \tan \theta - \sec \theta}{\theta^2}$

27. $f'(x) = \dfrac{(1 - 2x)(\sec^2 x) - \tan x(-2)}{(1 - 2x)^2} = \dfrac{\sec^2 x - 2x \sec^2 x + 2 \tan x}{(1 - 2x)^2}$

28. $g'(t) = \dfrac{\sqrt{t} \cos t - (1 + \sin t)(\frac{1}{2}\sqrt{t})}{t} = \dfrac{2t \cos t - \sin t - 1}{2t^{3/2}}$

29. $f'(t) = \dfrac{(t + 2)(\cos t) - (2 + \sin t)(1)}{(t + 2)^2} = \dfrac{t \cos t + 2 \cos t - \sin t - 2}{(t + 2)^2}$

30. $f'(\theta) = \dfrac{(2 + \cos \theta)(1) - (\theta - 1)(-\sin \theta)}{(2 + \cos \theta)^2} = \dfrac{2 + \cos \theta + \theta \sin \theta + \sin \theta}{(2 + \cos \theta)^2}$

31. $f'(x) = \dfrac{(1 - \cos x)(\cos x) - \sin x (\sin x)}{(1 - \cos x)^2} = \dfrac{\cos x - \cos^2 x - \sin^2 x}{(1 - \cos x)^2}$

$\qquad = \dfrac{\cos x - (\cos^2 x + \sin^2 x)}{(1 - \cos x)^2} = \dfrac{\cos x - 1}{(1 - \cos x)^2} = \dfrac{-1}{1 - \cos x} = \dfrac{1}{\cos x - 1}$

32. $f'(x) = \dfrac{(1 - \sin x)(1) - (x)(-\cos x)}{(1 - \sin x)^2} = \dfrac{1 - \sin x + x \cos x}{(1 - \sin x)^2}$

33. $f'(x) = \dfrac{(2 - \cos x)(\cos x) - (1 + \sin x)(\sin x)}{(2 - \cos x)^2} = \dfrac{2 \cos x - \cos^2 x - \sin x - \sin^2 x}{(2 - \cos x)^2}$

$\qquad = \dfrac{2 \cos x - \sin x - 1}{(2 - \cos x)^2}$

34. $g'(x) = \dfrac{(1 + \cos x)(-\sin x) - (\cos x)(-\sin x)}{(1 + \cos x)^2} = \dfrac{-\sin x - \cos x \sin x + \cos x \sin x}{(1 + \cos x)^2}$

$\qquad = \dfrac{-\sin x}{(1 + \cos x)^2}$

35. $f'(x) = \dfrac{(\sin x - \cos x)(\cos x - \sin x) - (\sin x + \cos x)(\cos x + \sin x)}{(\sin x - \cos x)^2}$

$\qquad = \dfrac{-(\sin x - \cos x)^2 - (\sin x + \cos x)^2}{(\sin x - \cos x)^2}$

$\qquad = \dfrac{-\sin^2 x + 2 \sin x \cos x - \cos^2 x - \sin^2 x - 2 \sin x \cos x - \cos^2 x}{(\sin x - \cos x)^2}$

$\qquad = \dfrac{-2}{(\sin x - \cos x)^2}$

36. $f'(x) = \dfrac{(3x + 2 \tan x)(2x + \sec^2 x) - (x^2 + \tan x)(3 + 2 \sec^2 x)}{(3x + 2 \tan x)^2}$

$\qquad = \dfrac{6x^2 + 3x \sec^2 x + 4x \tan x + 2 \tan x \sec^2 x - (3x^2 + 2x^2 \sec^2 x + 3 \tan x + 2 \tan x \sec^2 x)}{(3x + 2 \tan x)^2}$

$\qquad = \dfrac{6x^2 + 3x \sec^2 x + 4x \tan x + 2 \tan x \sec^2 x - 3x^2 - 2x^2 \sec^2 x - 3 \tan x - 2 \tan x \sec^2 x}{(3x + 2 \tan x)^2}$

$\qquad = \dfrac{3x^2 + 3x \sec^2 c + 4x \tan x - 2x^2 \sec^2 x - 3 \tan x}{(3x + 2 \tan x)^2}$

37. $g(x) = \sec^2 x - \tan^2 x + \cos x = 1 + \cos x$ so $g'(x) = -\sin x$

38. $g(x) = \cos^2 x + \sin^2 x + \sin x = 1 + \sin x = \cos x$

39. $f(\theta) = \sin \theta;\ f'(\theta) = \cos \theta;\ f''(\theta) = -\sin \theta$

40. $f(\theta) = \cos \theta;\ f'(\theta) = -\sin \theta,\ f''(\theta) = -\cos \theta$

41. $f(\theta) = \tan \theta;\ f'(\theta) = \sec^2\theta;\ f''(\theta) = 2 \sec \theta \sec \theta \tan \theta = 2 \sec^2\theta \tan \theta$

42. $f(\theta) = \cot\theta$; $f'(\theta) = -\csc^2\theta$;

$f''(\theta) = -2\csc\theta(-\csc\theta\cot\theta) = 2\csc^2\theta\cot\theta$

43. $f(\theta) = \sec\theta$; $f'(\theta) = \sec\theta\tan\theta$;

$f''(\theta) = \sec\theta(\sec^2\theta) + \tan\theta(\sec\theta\tan\theta)$

$f''(\theta) = \sec^3\theta + \sec\theta\tan^2\theta$

44. $f(\theta) = \csc\theta$; $f'(\theta) = -\csc\theta\cot\theta$

$f''(\theta) = -\csc\theta(-\csc^2\theta) - (-\csc\theta\cot\theta)(\cot\theta)$

$= \csc^3\theta + \csc\theta\cot^2\theta$

45. $f'(x) = \cos x - \sin x$; $f''(x) = -\sin x - \cos x$

46. $f'(x) = x\cos x + \sin x$;

$f''(x) = -x\sin x + \cos x + \cos x$

$= -x\sin x + 2\cos x$

47. $g'(y) = -\csc y\cot y - (-\csc^2 y)$

$= \csc y(\csc y - \cot y)$

$g''(x) = \csc y[-\csc y\cot y - (-\csc^2 y)]$

$\qquad + (\csc y - \cot y)(-\csc y\cot y)$

$= -\csc^2 y\cot y + \csc^3 y - \csc^2 y\cot y$

$\qquad + \csc y\cot^2 y$

$= \csc^3 y - 2\csc^2 y\cot y + \csc y\cot^2 y$

48. $g'(t) = \sec t\tan t - \sec^2 t$

$g''(t) = \sec t(\sec^2 t) + (\sec t\tan t)(\tan t)$

$\qquad - 2\sec t(\sec t\tan t)$

$= \sec^3 t + \sec t\tan^2 t - 2\sec^2 t\tan t$

49.

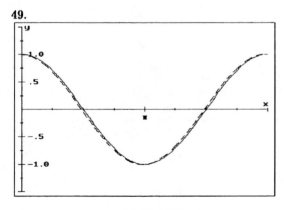

The two graphs almost coincide; that is, for "small" values of Δx, the difference quotient approximates the derivative. The *limit* as Δx approaches 0 of the difference quotient *is* $\cos x$.

50. The two graphs are nearly indistinguishable.

51. $f(\theta) = \tan\theta$; $f(\frac{\pi}{4}) = \tan\frac{\pi}{4} = 1$

$f'(\theta) = \sec^2\theta$; $f'(\frac{\pi}{4}) = \sec^2\frac{\pi}{4} = 2$

The equation of the tangent line is

$$y - 1 = 2(x - \tfrac{\pi}{4})$$

$$4x - 2y - \pi + 2 = 0$$

52. $f(\theta) = \sec\theta$; $f(\frac{\pi}{3}) = \sec\frac{\pi}{3} = 2$

$f'(\theta) = \sec\theta\tan\theta$;

$f'(\frac{\pi}{3}) = \sec\frac{\pi}{3}\tan\frac{\pi}{3} = 2\sqrt{3}$

The equation of the tangent line is

$$y - 2 = 2\sqrt{3}(x - \tfrac{\pi}{3})$$

$$6\sqrt{3}x - 3y + 6 - 2\sqrt{3}\pi = 0$$

53. $f(x) = \sin x$; $f(\frac{\pi}{6}) = \frac{1}{2}$

$f'(x) = \cos x$; $f'(\frac{\pi}{6}) = \frac{\sqrt{3}}{2}$

The equation of the tangent line is

$$y - \tfrac{1}{2} = \tfrac{\sqrt{3}}{2}(x - \tfrac{\pi}{6})$$

$$\sqrt{3}x - 2y + \left(1 - \tfrac{\sqrt{3}\pi}{6}\right) = 0$$

54. $f(x) = \cos x$; $f(\frac{\pi}{3}) = \frac{1}{2}$

$f'(x) = -\sin x$; $f'(\frac{\pi}{3}) = -\frac{\sqrt{3}}{2}$

The equation of the tangent line is

$$y - \tfrac{1}{2} = -\tfrac{\sqrt{3}}{2}(x - \tfrac{\pi}{3})$$

$$3\sqrt{3}x + 6y - 3 - \sqrt{3}\pi = 0$$

55. $y = x + \sin x$; $y(0) = 0$

$y' = 1 + \cos x$; $y'(0) = 2$

The equation of the tangent line is

$$y - 0 = 2(x - 0)$$

$$2x - y = 0$$

56. $x = t\sec t$; $x(0) = 0$

$x' = t\sec t\tan t + \sec t$; $x'(0) = 1$

The equation of the tangent line is

$$x - 0 = 1(t - 0)$$

$$x - t = 0$$

57. $y = x\cos x$, $y(\frac{\pi}{3}) = \frac{\pi}{3}(\frac{1}{2}) = \frac{\pi}{6}$.

$y' = x(-\sin x) + \cos x (1)$

$$= \cos x - x \sin x,$$

$$y'\left(\tfrac{\pi}{3}\right) = \tfrac{1}{2} - \left(\tfrac{\pi}{3}\right)\left(\tfrac{\sqrt{3}}{2}\right) = \frac{3 - \pi\sqrt{3}}{6}$$

The equation of the tangent line is

$$y - \frac{\pi}{6} = \left(\frac{3 - \pi\sqrt{3}}{6}\right)\left(x - \frac{\pi}{3}\right)$$

$$(\pi\sqrt{3} - 3)x + 6y - \pi^2\sqrt{3} = 0$$

58. $y = 2 \sin x + 3 \cos x$

$$\frac{dy}{dx} = 2 \cos x - 3 \sin x$$

$$\frac{d^2y}{dx^2} = -2 \sin x - 3 \cos x$$

$$\frac{d^2y}{dx^2} + y = -2 \sin x - 3 \cos x$$
$$+ (2 \sin x + 3 \cos x) = 0$$

59. $y = 4 \cos x - 2 \sin x$

$$\frac{dy}{dx} = -4 \sin x - 2 \cos x;$$

$$\frac{d^2y}{dx^2} = -4 \cos x + 2 \sin x$$

$$\frac{d^2y}{dx^2} + y = -4 \cos x + 2 \sin x$$
$$+ (4 \cos x - 2 \sin x) = 0$$

60. $y' = -A \sin x + B \cos x$

$$y'' = -A \cos x - B \sin x$$

$$y'' + 2y' + 3y = 2 \sin x$$

$$-A \cos x - B\sin x + 2(-A \sin x + B\cos x)$$
$$+ 3(A \cos x + B \sin x) = 2 \sin x$$

$$(-A + 2B + 3A)\cos x + (-B - 2A + 3B)\sin x$$

$$= 2 \sin x$$

Equating like coefficients leads to:

$$-A + 2B + 3A = 2(A + B) = 0$$

and

$$-B - 2A + 3B = 2(-A + B) = 2$$

Thus, $-A = B$, $B = \tfrac{1}{2}$ and

$$y = -\tfrac{1}{2} \cos x + \tfrac{1}{2} \sin x$$

61. $y' = (Bx + A)\cos x + (B - Ax)\sin x$

$$y'' = (2B - Ax)\cos x - (2A + Bx)\sin x$$

Since $y'' + y = -3 \cos x$ we have

$$(2B - Ax)\cos x - (2A + Bx)\sin x$$

$$+ Ax \cos x + Bx \sin x = -3 \cos x$$

$$2B \cos x - 2A \sin x = -3 \cos x$$

Thus, $2B = -3$ and $-2A = 0$ so that

$$A = 0,\ B = -\tfrac{3}{2}$$

62.   Let $f(x) = \begin{cases} x^2 \sin \tfrac{1}{x} & \text{if } x \neq 0 \\ 0 & \text{if } x = 0 \end{cases}$

Then the derivative is

$$f'(x) = \begin{cases} 2x \sin \tfrac{1}{x} - \cos \tfrac{1}{x} & \text{if } x \neq 0 \\ 0 & \text{if } x = 0 \end{cases}$$

which is not continuous at the origin.

63.   Let $f(x) = \cos x$, then $f'(0) = -\sin 0 = 0$

Also

$$f'(0) = \lim_{h \to 0} \frac{\cos(0 + h) - \cos 0}{h}$$

$$= \lim_{h \to 0} \frac{\cos h - 1}{h}$$

Therefore, $\displaystyle\lim_{h \to 0} \frac{\cos h - 1}{h} = 0$

64. $\dfrac{d}{dx}(\cos x)$

$$= \lim_{\Delta x \to 0} \frac{\cos(x + \Delta x) - \cos x}{\Delta x}$$

$$= \lim_{\Delta x \to 0} \frac{\cos x \cos\Delta x - \sin x \sin\Delta x - \cos x}{\Delta x}$$

$$= \lim_{\Delta x \to 0} \left[\cos x \frac{\cos \Delta x - 1}{\Delta x} - \sin x \frac{\sin \Delta x}{\Delta x}\right]$$

$$= (\cos x)(0) - (\sin x)(1) = -\sin x$$

65. $\dfrac{d}{dx}(\cot x) = \dfrac{d}{dx}\left(\dfrac{\cos x}{\sin x}\right) = \dfrac{-\sin^2 x - \cos^2 x}{\sin^2 x}$

$$= \frac{-1}{\sin^2 x} = -\csc^2 x$$

66. $\dfrac{d}{dx}(\sec x) = \dfrac{d}{dx}\left(\dfrac{1}{\cos x}\right)$

$$= \frac{(\cos x)(0) - (1)(-\sin x)}{\sin^2 x}$$

$$= \frac{\sin x}{\sin^2 x} = \frac{1}{\cos x} \cdot \frac{\sin x}{\cos x} = \sec x \tan x$$

67. $\dfrac{d}{dx}(\csc x) = \dfrac{d}{dx}\left(\dfrac{1}{\sin x}\right) = \dfrac{d}{dx}(\sin x)^{-1}$

$$= (-1)(\sin x)^{-2}(\cos x)$$

$$= \left(\frac{\cos x}{\sin x}\right)\left(\frac{1}{\sin x}\right) = -\cot x \csc x$$

68.   Answers vary.

69.   a. $\triangle AOB$ has height $\sin x$ and base 1, so its area is $\left(\tfrac{1}{2}\right)(1)(\sin x)$. Sector AOB has radius 1 and inscribed angle x (in radians). Its area is $\tfrac{1}{2}x$. The figure shows

that the area of the sector exceeds that of the triangle, so sin x < x.

b. The same reasoning applies if x falls in Quadrant IV.

c. Since sin $0 = 0$ and since both the right and left hand limits of sin x exist (they are 0 as $x \to 0$), sin x is continuous at 0.

d. $\lim\limits_{x \to c} \sin x = \lim\limits_{x \to c} [\sin(x - c) + c]$

$= \lim\limits_{x \to c} [\sin(x - c)\cos c + \cos(x - c)\sin c]$

$= \cos c \lim\limits_{x \to c} (x - c) + \sin c \lim\limits_{x \to c} \cos(x - c)$

$= (\cos c)(0) + (\sin c)(1) = \sin c.$

Therefore sin x is continuous for all real x.

70. a. From Figure 2.12, the area of sector AOB is between that of $\triangle AOD$ and $\triangle BOC$. Thus,

$\frac{1}{2}(\tan x)(1) \geq \frac{1}{2}x(1)^2 \geq \frac{1}{2}\sin x \cos x$

Divide by $\frac{1}{2}\sin x > 0$ (in Quadrant I):

$\frac{1}{\cos x} \geq \frac{x}{\sin x} \geq \cos x$

Since $1 \geq \cos x$, we have

$1 \geq \frac{1}{\cos x} \geq \frac{x}{\sin x}$

$\lim\limits_{x \to 0} 1 \geq \lim\limits_{x \to 0} \frac{1}{\cos x} \geq \lim\limits_{x \to 0} \frac{x}{\sin x} = 1$

Since cos $0 = 1$ and $\lim\limits_{x \to c}(\cos x) = 1$, the cosine is continuous at $x = 0$.

b. $\lim\limits_{x \to c}(\cos x) = \lim\limits_{x \to c}\{\cos[(x - c) + c]\}$

$= \lim\limits_{x \to c} [\cos(x - c)\cos c - \sin(x - c)\sin c]$

$= \cos c \lim\limits_{x \to c} (x - c) + \sin c \lim\limits_{x \to c} \sin(x - c)$

$= (\cos c)(1) + (\sin c)(0) = \cos c.$

71. From Figure 2.12, the area of sector AOB is between that of $\triangle AOD$ and $\triangle BOC$. Thus,

$\frac{1}{2}(\tan x)(1) \geq \frac{1}{2}h(1)^2 \geq \frac{1}{2}\sin h \cos h$

Divide by $\frac{1}{2}\tan h > 0$ (in Quadrant I)

$1 \geq \frac{h}{\tan h} \geq \cos^2 h$

$\lim\limits_{x \to 0} 1 \geq \lim\limits_{x \to 0} \frac{h}{\tan h} \geq \lim\limits_{x \to 0} \cos^2 h$

$1 \geq \lim\limits_{x \to 0} \frac{h}{\tan h} \geq 1$

By the squeeze theorem, $\lim\limits_{x \to 0} \frac{h}{\tan h} = 1.$

72. $\frac{d}{dx}(\tan x) = \lim\limits_{\Delta x \to 0} \frac{\tan(x + \Delta x) - \tan x}{\Delta x}$

$= \lim\limits_{\Delta x \to 0} \frac{1}{\Delta x}\left[\frac{\sin(x + \Delta x)}{\cos(x + \Delta x)} - \frac{\sin x}{\cos x}\right]$

$= \lim\limits_{\Delta x \to 0} \frac{\sin(x+\Delta x)\cos x - \sin x \cos(x+\Delta x)}{\Delta x \cos x \cos(x + \Delta x)}$

$= \lim\limits_{\Delta x \to 0} \frac{\sin \Delta x}{\Delta x} \lim\limits_{\Delta x \to 0} \frac{1}{\cos x \cos(x + \Delta x)}$

$= (1)\left(\frac{1}{\cos^2 x}\right) = \sec^2 x$

73. Answers vary.

74. Yes; consider some examples. For every choice of f and g, then each case we can find a real valued function such that the conditions of the problem are satisfied.

The students do not have sufficient calculus background to prove this result at this time. However, after Section 4.3, we can justify this conclusion using the second fundamental theorem:

If $h(x) = \int_0^1 f'(x)\, g'(t)\, dt,$

then $h'(x) = f'(x)g'(x)$

2.4 Rates of Change: Rectilinear Motion, Page 117

1. The instantaneous rate of change, or rate of change at a point, is given by $f'(x)$.

$f'(x) = 2x - 3, \quad f'(2) = 1$

2. $f'(x) = 1 - 2x; f'(1) = 1 - 2 = -1$

3. $f'(x) = -4x + 1; f'(1) = -4 + 1 = -3$

4. $f'(x) = \frac{(x + 1)(0) - (-2)(1)}{(x + 1)^2} = \frac{2}{(x + 1)^2};$

$f'(1) = \frac{2}{(1 + 1)^2} = \frac{1}{2}$

5. $f'(x) = \frac{(3x + 5)(2) - (2x - 1)(3)}{(3x + 5)^2}$

$= \frac{13}{(3x + 5)^2}; f'(-1) = \frac{13}{4}$

6. $f'(x) = (x^2 + 2)(1 + \frac{1}{2}x^{-1/2}) + (x + x^{1/2})(2x)$

$f'(4) = (4^2 + 2)[1 + \frac{1}{2}(4)^{-1/2}] + (4 + 4^{1/2})(2 \cdot 4)$

$= \frac{141}{2}$

7. $f'(x) = -x \sin x + \cos x; f'(\pi) = -1$

8. $f'(x) = (x + 1)\cos x + \sin x; f'(\frac{\pi}{2}) = 1$

9. $f'(x) = 1 + \frac{(2 - 4x)(0) - 3(-4)}{(2 - 4x)^2}$

$= 1 + \frac{12}{(2 - 4x)^2} = \frac{16x^2 - 16x + 16}{(2 - 4x)^2}$

$$= \frac{4(x^2 - x + 1)}{(1 - 2x)^2}; f'(0) = 4$$

10. $f'(x) = \frac{(x+1)(0) - (1)(1)}{(x+1)^2} - \frac{(x-1)(0) - (1)(1)}{(x-1)^2}$

$f'(3) = \frac{3}{16}$

11. $f'(x) = -\sin^2 x + \cos^2 x; f'(\frac{\pi}{2}) = -1$

12. $f'(x) = \frac{(x^2 + 1)(2x) - x^2(2x)}{(x^2 + 1)^2} = \frac{2x}{(x^2 + 1)^2}$

$f'(1) = \frac{1}{2}$

13. Write $f(x) = \left(x - \frac{2}{x}\right)\left(x - \frac{2}{x}\right)$; then use the product rule:

$f'(x) = \left(x - \frac{2}{x}\right)\left(1 + \frac{2}{x^2}\right) + \left(x - \frac{2}{x}\right)\left(1 + \frac{2}{x^2}\right)$

$= 2\left(\frac{x^2 - 2}{x}\right)\left(\frac{x^2 + 2}{x^2}\right) = \frac{2(x^4 - 4)}{x^3};$

$f'(1) = -6$

14. Treat as a product;

$f'(x) = 2(\sin x)(\cos x) = \sin 2x; f'(\frac{\pi}{4}) = 1$

15. **a.** $s'(t) = v(t) = 2t - 2$.

   b. $s''(t) = a(t) = 2$

   c. Object begins at $s(0) = 6$ and ends at $s(2) = 6$; $s'(t) = 0$ when

   $$2t - 2 = 0$$
   $$t = 1$$

   On $[0, 1)$ object retreats to $s(1) = 5$; on $(1, 2]$ object advances. Distance covered:

   $$\left| s(2) - s(1) \right| + \left| s(1) - s(0) \right| = 2$$

   d. Because $a(t) > 0$, the object is continuously accelerating.

16. **a.** $s'(t) = v(t) = 6t + 2$

   b. $s''(t) = a(t) = 6$

   c. Object begins at $s(0) = -5$ and ends at $s(1) = 0$; $s'(t) = 0$ when

   $$6t + 2 = 0$$
   $$t = -\frac{1}{3}$$

   Since this is not in the interval $[0, 1]$, the object advances on $[0, 1]$. Distance covered: $\left| s(1) - s(0) \right| = \left| 0 - (-5) \right| = 5$

   d. Because $a(t) > 0$, the object is continuously accelerating.

17. **a.** $s'(t) = v(t) = 3t^2 - 18t + 15$

   b. $s''(t) = a(t) = 6t - 18$

   c. Object begins at $s(0) = 25$ and ends at $s(6) = 7$; $s'(t) = 0$ when

   $$3t^2 - 18t + 15 = 0$$
   $$3(t - 5)(t - 1) = 0$$
   $$t = 1, 5$$

   On $[0, 1)$ object advances to $s(1) = 32$; on $(1, 5)$ object retreats to $s(5) = 0$; and on $(5, 6]$ the object advances. Distance covered:

   $$\left| s(1) - s(0) \right| + \left| s(5) - s(1) \right|$$
   $$+ \left| s(6) - s(1) \right| = \left| 32 - 25 \right| + \left| 0 - 32 \right|$$
   $$+ \left| 7 - 0 \right| = 7 + 32 + 7 = 46$$

   d. $s''(t) = 0$ when $\quad\quad 6t - 18 = 0$
   $$t = 3$$

   On $[0, 3)$ the object is decelerating, and on $(3, 6]$ the object is accelerating.

18. **a.** $s'(t) = v(t) = 4t^3 - 12t^2 + 8$

   b. $s''(t) = a(t) = 12t^2 - 24t$

   c. Object begins at $s(0) = 0$ and ends at $s(4) = 32$; $s'(t) = 0$ when

   $$4t^3 - 12t^2 + 8 = 0$$
   $$4(t - 1)(t^2 - 2t - 2) = 0$$
   $$t = 1, 1 \pm \sqrt{3}$$

   Note: $1 - \sqrt{3}$ is not on the interval $[0, 4]$. On $[0, 1)$ object advances to $s(1) = 5$; on $(1, 1 + \sqrt{3})$ object retreats to $s(1 + \sqrt{3}) = -4$; on $(1 + \sqrt{3}, 4]$ object advances to 32. Distance covered:

   $$\left| s(1) - s(0) \right| + \left| s(1 + \sqrt{3}) - s(1) \right|$$
   $$+ \left| s(4) - s(1 + \sqrt{3}) \right| = \left| 5 - 0 \right| + \left| -4 - 5 \right|$$
   $$+ \left| 32 - (-4) \right| = 5 + 9 + 36 = 50$$

   d. $s''(t) = 0$ when $\quad 12t^2 - 24t = 0$
   $$12t(t - 2) = 0$$
   $$t = 0, 2$$

   On $[0, 2)$ the object is decelerating, and on $(2, 4]$ the object is accelerating.

19. **a.** $s'(t) = v(t) = -2t^{-2} - 2t^{-3}$

   b. $s''(t) = a(t) = 4t^{-3} + 6t^{-4}$

   c. Object begins at $s(1) = 3$ and ends at

$s(3) = \frac{7}{9}$; $s'(t) = 0$ when

$$-2t^{-2} - 2t^{-3} = 0$$
$$-2t - 2 = 0$$
$$t = -1$$

-1 is not on the interval $[1, 3]$. On $[1, 3]$ object retreats.

Distance covered:

$$\left| s(3) - s(1) \right| = \left| \frac{7}{9} - 3 \right| = \frac{20}{9}$$

d. $s''(t) = 0$ when

$$4t^{-3} + 6t^{-4} = 0$$
$$4t + 6 = 0$$
$$t = -\frac{3}{2}$$

On $[1, 3]$ the object is accelerating.

20. a. $s'(t) = v(t) = -2t^{-3}$

b. $s''(t) = a(t) = 6t^{-4}$

c. Object begins at $s(1) = 2$ and ends at $s(2) = \frac{5}{4}$;

$s'(t) \neq 0$ so the object retreats on $[1, 2]$.

Distance covered:

$$\left| s(2) - s(1) \right| = \left| \frac{5}{4} - 2 \right| = \frac{3}{4}$$

d. $s''(t) \neq 0$

On $[1, 2]$ the object is accelerating.

21. a. $s'(t) = v(t) = -3 \sin t$

b. $s''(t) = a(t) = -3 \cos t$

c. Object begins at $s(0) = 3$ and ends at $s(2\pi) = 3$; $s'(t) = 0$ when

$$-3 \sin t = 0$$

On $[0, 2\pi]$: $t = 0, \pi, 2\pi$

On $[0, \pi)$ object retreats to $s(\pi) = -3$; on $(\pi, 2\pi]$ object advances to $s(2\pi) = 3$.

Distance covered:

$$\left| s(\pi) - s(0) \right| + \left| s(2\pi) - s(\pi) \right|$$
$$= \left| -3 - 3 \right| + \left| 3 - (-3) \right| = 6 + 6 = 12$$

d. $s''(t) = 0$ when $-3 \cos t = 0$

On $[0, 2\pi]$: $t = \frac{\pi}{2}, \frac{3\pi}{2}$

On $[0, \frac{\pi}{2})$ the object is decelerating, on $(\frac{\pi}{2}, \frac{3\pi}{2})$ the object is accelerating, and

on $(\frac{3\pi}{2}, 2\pi]$ the object is decelerating.

22. a. $s'(t) = v(t) = \sec t \tan t$

b. $s''(t) = a(t) = \sec t \sec^2 t + \sec t \tan^2 t$
$$= \sec^3 t + \sec t \tan^2 t$$

c. Object begins at $s(0) = 2$ and ends at $s(\frac{\pi}{4}) = 1 + \sqrt{2}$.

$s'(t) \neq 0$ on $[0, \frac{\pi}{4}]$

On $[0, \frac{\pi}{4}]$ object advances.

Distance covered:

$$\left| s(\tfrac{\pi}{4}) - s(0) \right| = \left| 1 + \sqrt{2} - 0 \right| = 1 + \sqrt{2}$$

d. $s''(t) \neq 0$ on $[0, \frac{\pi}{4}]$

On $[0, \frac{\pi}{4}]$ the object is accelerating.

23. a. $f'(x) = -6$

b. The rate of change is negative, so the scores are declining. The rate of change is constant, so the drop will not vary from year to year.

24. a. $v(t) = x'(t) = 6t^2 + 6t - 36$

b. $a(t) = x''(t) = 12t + 6$

c. $v(t) = 0$ when $6t^2 + 6t - 36 = 0$
$$6(t - 2)(t + 3) = 2, -3$$

-3 is not on $[0, 3]$

Distance covered:

$$\left| x(2) - x(0) \right| + \left| x(3) - x(2) \right|$$
$$= \left| -4 - 40 \right| + \left| 13 - (-4) \right| = 44 + 17$$
$$= 61$$

25. $v(t) = x'(t) = 3t^2 - 18t + 24 = 0$ at $t = 2, 4$. $x(t)$ is advancing on $[0, 2)$ and $(4, 8]$ and retreating on $(2, 4)$. So the total distance traveled is

$$\left| x(2) - x(0) \right| + \left| x(8) - x(4) \right| + \left| x(4) - x(2) \right|$$
$$= \left| 40 - 20 \right| + \left| 148 - 36 \right| + \left| 36 - 40 \right|$$
$$= 20 + 112 + 4 = 136 \text{ units}$$

26. $v(t) = s'(t) = \dfrac{90t}{3t + 12} = \dfrac{30t}{t + 4}$

$a(t) = v'(t) = \dfrac{30[(t+4)(1) - t(1)]}{(t + 4)^2} = \dfrac{120}{(t + 4)^2}$

$a(10) = \dfrac{120}{14^2} \approx 0.61$

27. a. $s(t) = 10t + \dfrac{5}{t+1}$,

$$v(t) = s'(t) = 10 - \dfrac{5}{(t+1)^2}$$

$$v(4) = 10 - \tfrac{5}{25} \approx 9.8 \text{ m/s}$$

b. $v(t) > 0$ so the distance traveled during the 5th minute is $s(5) - s(4)$,

$$D = \left(50 + \tfrac{5}{6}\right) - \left(40 + \tfrac{5}{5}\right) \approx 9.8 \text{ m}$$

28. a. $Q(t) = 5(1 - 0.04t)(1 - 0.04t)$

$$Q'(t) = 5(1 - 0.04t)(-0.04)$$

$$+ 5(-0.04)(1 - 0.04t)$$

$$= -0.4(1 - 0.04t)$$

$$Q'(2) = -0.4[1 - 0.04(2)]$$

$$= -0.368 \text{ gal/sec}$$

b. $-0.4(1 - 0.04t) = 0$

$$t = \dfrac{1}{0.04} = 25 \text{ sec}$$

c. $Q'(25) = -0.4[1 - 0.04(25)] = 0$

29. a. The rock is stationary at $t = 2$ (on the way up), so $-32(2) + s'(0) = 0$, the initial velocity is 64 ft/sec.

b. $s(t) = -16t^2 + 64t + s(0)$
$s(7) = 0$, so

$$-16(7)^2 + 64(7) + s(0) = 0$$

$$s(0) = 16(7)^2 - 64(7) = 336$$

The cliff is 336 ft high.

c. $s'(t) = -32t + 64$ ft/sec

d. $s'(7) = -32(7) + 64 = -160$ ft/sec

The negative sign indicated downward motion, since upward is positive.

30. a. $v(5) = s'(5) = -32(5) = 320 = 160$

b. $a(3) = s''(3) = -32$

31. The equation guiding the path of the first rock is

$$s_1(t) = -16t^2 + s_1'(0) + 90$$

and the second rock is

$$s_2(t) = -16(t-1)^2 + s'(t-1) + H$$

Time t is 0 when the first rock starts its motion; $s'(0) = 0$ since the rocks are dropped. The first rock hits the ground at time

$$-16t^2 + 90 = 0 \text{ or } t = \pm\dfrac{3\sqrt{10}}{4}$$

Reject the negative value, so $t \approx 2.3717$ sec.

The two rocks hit the ground simultaneously, so $-16\left(\dfrac{3\sqrt{10}}{4} - 1\right)^2 + H = 0$ or $H \approx 30$ ft

32. a. $s(t) = -16t^2 + s'(0)t + s(0)$ with
$s(0) = 0$ and $s'(0) = 160$:

$$s(t) = -16t^2 + 160t; \; s(t) = 0 \text{ when}$$

$$-16t^2 + 160t = 0$$

$$-16t(t - 10) = 0$$

$$t = 0, 10$$

The ball will hit the ground in 10 sec.

b. $s'(t) = -32t + 160; \; s'(10) = -160$ ft/sec

c. $s'(t) = 0$ when $-32t + 160 = 0$ or when

$$t = 5 \text{ sec.}$$

33. $s(t) = -16t^2 + s(0)$

The pavement, that is ground level, is reached in 3 seconds, so $0 = -16(3^2) + s(0)$ or $s(0) = 144$.

34. On the moon, $a_m = -5.5$ ft/s²;
$s'(t) = -5.5t + s'(0)$. It takes the rock 2 sec to reach its maximum height, at which time the velocity is 0. Thus, $s'(0) = (5.5)(2) = 11$ ft/s. The ground is hit in 7 seconds, so

$$0 = s(0) + 11t - \tfrac{1}{2}(5.5)(7)^2$$

$$s(0) = 57.75 \text{ ft}$$

35. On Mars $g = 12$ ft/s², so

$$s(t) = -6t^2 + v_0 t + s_0,$$

$$v(t) = -12t + v_0.$$

Since the rock goes up and back to ground level in 4 s, it reaches its maximum height in 2 s. At that time $v = 0$. So

$$0 = -12(2) + v_0 \text{ and } v_0 = 24 \text{ ft/s}$$

Therefore, the rock passes her on the way down with $v = -24$ ft/s. The equation for the rest of the rock's trip will be:

$$s(t) = -6t^2 + (-24)t + 0$$

$$s(3) = -6(3)^2 - 24(3) = -126 \text{ ft}$$

The cliff is 126 ft. high.

36. $s(t) = 88t - 8t^2; \; s'(t) = 88 - 16t = 0$ at the instant t_1 the car stops. Thus,

$$8(11 - 2t_1) = 0 \text{ or } t_1 = 5.5 \text{ sec}$$

The distance required to stop is

$s(5.5) = 8[(11)(5.5) - (5.5)^2] = 22(11) = 242$ ft

37. **a.** Because $C(t) = 100t^2 + 400t + 5{,}000$ is the circulation t years from now, the rate of change of circulation t years from now is $C'(t) = 200t + 400$ newspapers per year.

  b. The rate of change of circulation 5 years from now is $C'(5) = 200(5) + 400$ $= 1{,}400$ newspapers per year.

  c. The actual change in the circulation during the sixth year is

$$C(6) - C(5) = [100(6^2) + 400(6) + 5{,}000]$$
$$- [100(5^2) + 400(5) + 5{,}000]$$
$$= 1{,}500 \text{ newspapers}$$

38. **a.** Because $f(x) = -\frac{1}{3}x^3 + \frac{1}{2}x^2 + 50x$ is the number of units assembled x hours after 8:00 AM. The rate at which the units are being assembled x hours after 8:00 AM is

$$f'(x) = -x^2 + x + 50 \text{ units per hour.}$$

  b. The rate of assembly at 9:00 AM ($x = 1$) is

$$f'(1) = -(1^2) + 1 + 50 = 50$$

units per hour.

  c. The actual number of units assembled between 9:00 AM (when $x = 1$) and 10:00 AM (when $x = 2$) is

$$f(2) - f(1) = [-\frac{1}{3}(2^3) + \frac{1}{2}(2^2) + 50(2)]$$
$$- [-\frac{1}{3}(1^3) + \frac{1}{2}(1^2) + 50(1)]$$
$$\approx 49 \text{ units}$$

39. $q(t) = 0.05t^2 + 0.1t + 3.4$

  a. $q'(t) = 0.1t + 0.1$; $q'(1) = 0.2$ ppm/yr

  b. Change in first year:

$$q(1) - q(0) = (0.05 + 0.1 + 3.4) - (3.4)$$
$$= 0.15 \text{ ppm}$$

  c. Change in second year:

$$q(2) - q(1) = (0.2 + 0.2 + 3.4) - (3.55)$$
$$= 0.25 \text{ ppm}$$

40. $F = GmMr^{-2}$; $\dfrac{dF}{dr} = (GmM)(-2)r^{-3}$
$= kr^{-3}$ where $k = -2GmM$.

41. $P(t) = P_0 + 61t + 3t^2$; $P'(t) = 61 + 6t$;

$P'(5) = 61 + 30 = 91$ thousand/hr

42. **a.** $g'(t) = 2t + 5$; in 1992 ($t = 2$),

$$g'(2) = 2(2) + 5 = 9 \text{ billion dollars/yr}$$

  b. $g(2) = 4 + 10 + 106 = 120$

The percentage rate of change is $900/120$ $= 7.5\%$ per year

43. $P(x) = 2x + 4x^{3/2} + 5000$
  a. $P'(x) = 2 + 6x^{1/2}$
$P'(9) = 2 + 18 = 20$ persons per mo

  b. $\dfrac{P'(9)}{P(9)}(100) = \dfrac{20}{5126}(100) \approx 0.39\%$ per mo

44. **a.** Because your starting salary is \$30,000 and you obtain a raise of \$3,000 per yr, your salary x yr from now will be $S(x) = 30{,}000 + 3{,}000x$ dollars. The percentage rate of change of this salary x years from now is

$$100\left[\frac{S'(x)}{S(x)}\right] = 100\left[\frac{3{,}000}{30{,}000 + 3{,}000x}\right]$$
$$= \frac{100}{10 + x}$$

percent per year.

  b. The percentage rate of change after one year is $100/11 = 9.08\%$ per year.

  c. In the long run, $100/(10 + x) \to 0$. That is, the percentage rate of change of your salary approaches zero (even though your salary will continue to increase at a constant rate).

45. Let $t = 0$ for 1990 and $G(t)$ the GDP in billions of dollars. $G(0) = 125$, $G(2) = 155$, so the slope of the line is

$$m = \frac{155 - 125}{2} = 15 = G'(t)$$

$G(t) = 125 + 15t$; $G(5) = 125 + 75 = 200$.

The percentage rate of change is
$$\frac{100(15)}{200} = 7.5\% \text{ per year}$$

46. $y = mx + b$, $y' = m$, and the percentage rate of change R is $100m/(mx + b)$.

$$\lim_{x \to \infty} R = 0$$

A "small" (constant) variation in the rate of change (y') will have less and less effect on a larger and larger number (y).

47. $N(t) = 5 - t^2(t - 6) = -t^3 + 6t^2 + 5$
  a. $N'(t) = -3t^2 + 12t$

Percentage change,

$$C(t) = 100\left(\frac{-3t^2 + 12t}{-t^3 + 6t^2 + 5}\right)$$

b. We want to find values of $C(t)$ such that

$$(100)\,\frac{-3t^2 + 12t}{-t^3 + 6t^2 + 5} > 30,$$

$$\frac{3t^2 - 12t}{t^3 - 6t^2 - 5} > 0.3$$

Evaluating $C(t)$ for various weeks:

$C(0) = 0;\ \ C(1) = 0.90;\ C(2) \approx 0.57;$

$C(3) \approx 0.28;\ C(4) = 0;\ C(5) = -0.5;$

$C(6) = -7.2$

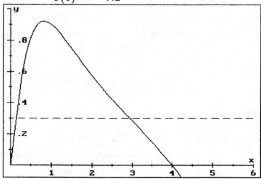

From early in the first week until almost the third week the disease has epidemic level. Epidemic will be declared for the second and third weeks.

48. a. $s(t) = 7\cos t;\ v(t) = s'(t) = -7\sin t$

$a(t) = s''(t) = -7\cos t$

b. $s(0) = 7$, $s(2\pi) = 7$; the period (one revolution) is 2π.

c. The highest point is reached at $t = \pi$ (downward is positive). $s(\pi) = -7$. The amplitude is 7.

49. Let r be the speed of the first car and $2r$ the speed of the second car. The distances traveled by the cars are rt and $2rt$, respectively. The distance between them is

$$D^2(t) = (rt)^2 + (2rt)^2 - 2(rt)(2rt)\cos\frac{\pi}{3}$$
$$= r^2t^2 + 4r^2t^2 - 2r^2t^2$$
$$= 3r^2t^2$$
$$D(t) = \sqrt{3}\,rt$$

The rate of increase of distance is

$D'(t) = \sqrt{3}\ r = 45$, so $r = \dfrac{45}{\sqrt{3}} = 15\sqrt{3}$

This is approximately 26 mi/h.

50. Let the gravity be denoted by g. Then the acceleration is $-g$, and the velocity is $v = -gt + v$, where v_0 is the initial velocity. The rock reaches its maximum height in 2.5 seconds. The velocity is 0 at this instant and $v_0 = 5g/2$. The distance is

$$s(t) = -\frac{gt^2}{2} + v_0 t = -\frac{gt^2}{2} + \frac{5gt}{2}$$
$$= \left[\frac{gt}{2}(5 - t)\right]$$

$s(2.5) = 2.5g(5 - 2.5) = 3.125g$

Since $3.125g = 37.5$, $g = 12$.

Our friendly spy finds himself on Mars.

51. Let x be one edge of the cube. Then $V(x) = x^3;\ V'(x) = 3x^2 = 0.5S(x)$, so they are equal.

52. Let x be the radius of the sphere. Then $V(x) = \frac{4}{3}\pi r^3;\ V'(r) = 4\pi r^2 = S(r)$, so they are equal.

53.
$$\left(P + \frac{A}{V^2}\right)(V - B) = kT$$

$$\left(P + \frac{A}{V^2}\right)(V - B)' + \left(P + \frac{A}{V^2}\right)'(V - B) = 0$$

Remember T is fixed.

$$\left(P + \frac{A}{V^2}\right)(1) + \left(P' - \frac{2A}{V^3}\right)(V - B) = 0$$

$$PV^3 + AV + (V^3 P' - 2A)(V - B) = 0$$

$$V^3 P' - 2A = \frac{PV^3 + AV}{B - V}$$

$$V^3 P' = \frac{V}{B - V}(PV^2 + A) + 2A$$

$$V^3 P' = \frac{V^3}{B - V}\left(P + \frac{A}{V^2}\right) + 2A$$

$$V^3 P' = \frac{-V^3}{(V - B)^2}\left[\left(P + \frac{A}{V^2}\right)(V - B)\right] + 2A$$

$$P' = \frac{-1}{(V - B)^2}\left[\left(P + \frac{A}{V^3}\right)(V - B)\right] + \frac{2A}{V^3}$$

$$P' = \frac{-kT}{(V - B)^2} + \frac{2A}{V^3}$$

54. $P = \frac{4}{3}\pi N\left(\dfrac{\mu^2}{3kT}\right) = \dfrac{4\pi\mu^2 N}{9k}T^{-1}$

$\dfrac{dP}{dT} = -\dfrac{4\pi\mu^2 N}{9k}T^{-2}$

2.5 The Chain Rule, Page 126

1. The chain rule is the differentiation of a function of a function. If $y = f(u)$ and $u = g(x)$, then
$$\frac{dy}{dx} = \frac{dy}{du} \cdot \frac{du}{dx}$$

2. Use the chain rule to differentiate a function of a function.

3. $\frac{dy}{dx} = \frac{dy}{du}\frac{du}{dx} = \frac{d}{du}(u^2 + 1)\frac{d}{dx}(3x - 2)$
$= 2u(3) = 6(3x - 2)$

4. $\frac{dy}{dx} = \frac{dy}{du}\frac{du}{dx} = \frac{d}{du}(2u^2 - u + 5)\frac{d}{dx}(1 - x^2)$
$= (4u - 1)(-2x) = (4 - 4x^2 - 1)(-2x)$
$= 8x^3 - 6x$

5. $\frac{dy}{dx} = \frac{dy}{du}\frac{du}{dx} = \left(\frac{-4}{u^3}\right)(2x) = \frac{-8x}{(x^2 - 9)^3}$

6. $\frac{dy}{du} = \frac{(2u+5)(1) - 2(u + 3)}{(2u + 5)^2} = \frac{-1}{(2u + 5)^2}$
$\frac{du}{dx} = \frac{5(1 - 2x) - (-2)(5x - 3)}{(1 - 2x)^2} = \frac{-1}{(1 - 2x)^2}$
$\frac{dy}{dx} = \frac{dy}{du}\frac{du}{dx} = \frac{1}{(2u + 5)(1 - 2x)^2}$
$= \frac{1}{\left(\frac{10x - 6 + 5 - 10x}{1 - 2x}\right)^2(1 - 2x)^2}$
$= \frac{1}{\left(\frac{-1}{1 - 2x}\right)^2(1 - 2x)^2} = 1$

7. $\frac{dy}{dx} = \frac{dy}{du}\frac{du}{dx} = (-\sin u)(2x)$
$= -2x\sin(x^2 + 7)$

8. $\frac{dy}{dx} = \frac{dy}{du}\frac{du}{dx} = \left\{\left(3x + \frac{6}{x}\right)\sec^2\left(3x + \frac{6}{x}\right)\right.$
$\left. + \tan\left(3x + \frac{6}{x}\right)\right\}(3 - 6x^{-2})$

9. **a.** $g'(u) = 5u^4$ **b.** $u'(x) = 3$
   c. $f'(x) = 5(3x - 1)^4(3) = 15(3x - 1)^4$

10. **a.** $g'(u) = 3u^2$ **b.** $u'(x) = 2x$
    c. $f'(x) = 3(x^2 + 1)^2(2x) = 6x(x^2 + 1)^2$

11. **a.** $g'(u) = 15u^{14}$ **b.** $u'(x) = 6x + 5$
    c. $f'(x) = 15(6x + 5)(3x^2 + 5x - 7)^{14}$

12. **a.** $g'(u) = 7u^6$ **b.** $u'(x) = -8 - 24x$
    c. $f'(x) = 7(5 - 8x - 12x^2)^6(-8 - 24x)$
    $= (-56 - 168x)(5 - 8x - 12x^2)^6$
    $= -7(24x + 8)(12x^2 + 8x - 5)^6$

13. **a.** $f'(x) = 2\cos(2x)$
    b. $g'(x) = \cos(2\cos x)(-2\sin x)$
    $= -2\sin x\cos(2\cos x)$

14. **a.** $f'(x) = \cos(\sin\theta)(\cos\theta) = (\cos\theta)\cos(\sin\theta)$
    b. $f'(\theta) = \cos(\cos\theta)(-\sin\theta)$
    $= (-\sin\theta)\cos(\cos\theta)$

15. $s'(\theta) = [\cos(4\theta + 2)](4) = 4\cos(4\theta + 2)$

16. $c'(\theta) = 3\sin(5 - 3\theta)$

17. $f'(x) = x^2\sec^2 x^2(2x) + 2x\tan x^2$
    $= 2x^3\sec^2 x^2 + 2x\tan x^2$

18. $h'(x) = x^2(3)(2x - 5)^2(2) + (2x)(2x - 5)^3$
    $= (2x - 5)^2(6x^2 + 4x^2 - 10x)$
    $= 10x(x - 1)(2x - 5)^2$

19. $p'(x) = (\sin x^2)\frac{d}{dx}(\cos x^2) + (\cos x^2)\frac{d}{dx}(\sin x^2)$
    $= (\sin x^2)(-\sin x^2)(2x) + (\cos x^2)(\cos x^2)(2x)$
    $= 2x(\cos^2 x^2 - \sin^2 x^2)$ or $2x\cos 2x^2$

    If we applied the double angle identity to the original function our work would have been easier:
    $p(x) = \frac{1}{2}(2\sin x^2\cos x^2) = \frac{1}{2}\sin 2x^2$
    $p'(x) = \frac{1}{2}(\cos 2x^2)(4x) = 2x\cos 2x^2$

20. $f'(x) = 2\csc(\sqrt{x})[-\csc(\sqrt{x})\cot(\sqrt{x})](\frac{1}{2}x^{-1/2})$
    $= -x^{-1/2}\csc^2(\sqrt{x})\cot(\sqrt{x})$

21. $f'(x) = (2x^2 + 1)^4(5)(x^2 - 2)^4(2x)$
    $+ (x^2 - 2)^5(4)(2x^2 + 1)^3(4x)$
    $= x(2x^2+1)^3(x^2 - 2)^4[10(2x^2+1) + 16(x^2 - 2)]$
    $= 2x(2x^2 + 1)^3(x^2 - 2)^4(18x^2 - 11)$

22. $f'(x) = (x^3 + 1)^5(6)(2x^3 - 1)^5(6x^2)$
    $+ 5(x^3 + 1)^4(3x^2)(2x^3 - 1)^6$
    $= 3x^2(x^3 + 1)^4(2x^3 - 1)^5[12x^3 + 12 + 10x^3 - 5]$
    $= 3x^2(x^3 + 1)^4(2x^3 - 1)^5(22x^3 + 7)$

23. $f'(t) = (1 - t^2)(\cos t^2)(2t) + (\sin t^2)(-2t)$
    $= 2t(\cos t^2 - t^2\cos t^2 - \sin t^2)$

24. $f'(x) = 2\sin(\sqrt{x+3})\cos(\sqrt{x+3})[\frac{1}{2}(x+3)^{-1/2}]$
    $= \frac{\sin\sqrt{x+3}\cos\sqrt{x+3}}{\sqrt{x+3}}$

25. $f'(t) = (t^2 - 1)^3[-\sin(3t+2)](3)$

$\qquad + 3(t^2 - 1)^2(2t)[\cos(3t + 2)]$

$\qquad = 3(t^2 - 1)^2[-(t^2 - 1)\sin(3x + 2)]$

$\qquad + 2t \cos(3t + 2)]$

$\qquad = 3(t^2 - 1)^2[2t \cos(3t+2) - t^2\sin(3t+2)$

$\qquad + \sin(3t+2)]$

26. $f'(x) = \frac{1}{2}\left(\frac{x^2-5}{x^2+3}\right)^{1/2}\left[\frac{(x^2-5)(2x)-(x^2+3)(2x)}{(x^2-5)^2}\right]$

$\qquad = \frac{1}{2}\left(\frac{x^2-5}{x^2+3}\right)^{1/2}\left[\frac{-16x}{(x^2-5)^2}\right]$

$\qquad = -8x(x^2+3)^{-1/2}(x^2-5)^{-3/2}$

27. $f'(x) = \frac{1}{2}\left(\frac{2x^2-1}{3x^2+2}\right)^{-1/2}\frac{d}{dx}\left(\frac{2x^2-1}{3x^2+2}\right)$

$\qquad = \frac{1}{2}\left(\frac{2x^2-1}{3x^2+2}\right)^{-1/2}\left[\frac{(3x^2+2)(4x)-(2x^2-1)(6x)}{(3x^2+2)^2}\right]$

$\qquad = \frac{1}{2}\left(\frac{2x^2-1}{3x^2+2}\right)^{-1/2}\left[\frac{14x}{(3x^2+2)^2}\right]$

$\qquad = \frac{14x}{2(2x^2-1)^{1/2}(3x^2+2)^{3/2}}$

$\qquad = \frac{7x}{(2x^2-1)^{1/2}(3x^2+2)^{3/2}}$

28.   **a.** $f'(x) = \frac{1}{2}(\sin x^2)^{-1/2}(2x)\cos x^2 = \frac{x\cos x^2}{\sqrt{\sin x^2}}$

     b. $f'(x) = 2\sin(\sqrt{x})\cos(\sqrt{x})(\frac{1}{2}x^{-1/2})$

$\qquad = \frac{\sin(\sqrt{x})\cos(\sqrt{x})}{\sqrt{x}} = \frac{\sin(2\sqrt{x})}{2\sqrt{x}}$

29.   **a.** $f'(x) = -\frac{1}{2}(\cos x^2)^{-3/2}(-\sin x^2)(2x)$

$\qquad = x\sin x^2(\cos x^2)^{-3/2}$

     b. $f(x) = \sqrt{\frac{1}{\cos^2 x}} = |\cos x|^{-1}$

$\qquad f'(x) = |\sec x|\tan x$ or $\frac{\tan x}{|\cos x|}$

30. $f'(x) = \frac{1}{2}(x + \sqrt{x})^{-1/2}(1 + \frac{1}{2}x^{-1/2})$

$\qquad = \frac{1}{2}\left(1 + \frac{1}{2\sqrt{x}}\right)(x + \sqrt{x})^{-1/2}$

$\qquad = \frac{2\sqrt{x} + 1}{4x(1 + \sqrt{x})}$

31. $f'(x) = \frac{1}{3}(x^2 + 2\sqrt{x})^{-2/3}\left(2x + \frac{1}{\sqrt{x}}\right)$

$\qquad = \frac{2x\sqrt{x} + 1}{3\sqrt{x}(x^2 + 2\sqrt{x})^{2/3}}$

32. $f'(x) = \cos[\sin(\sin x)][\sin(\sin x)]'$

$\qquad = \cos[\sin(\sin x)][\cos(\sin x)](\cos x)$

$\qquad = \cos x \cos(\sin x) \cos[\sin(\sin x)]$

33. $f'(x) = \frac{1}{2}(x^2 + 5)^{-1/2}(2x)$

$\qquad f'(2) = \frac{2}{\sqrt{9}} = \frac{2}{3}$

Using the point-slope formula:

$\qquad y - 3 = \frac{2}{3}(x - 2)$

$\qquad 2x - 3y + 5 = 0$

34. $f'(x) = 3(5x + 4)^2(5)$

$\qquad f'(-1) = 15(-1)^2 = 15$

Using the point-slope formula:

$\qquad y + 1 = 15(x + 1)$

$\qquad 15x - y + 14 = 0$

35. $f'(x) = x^2(2)(x - 1)(1) + 2x(x - 1)^2$

$\qquad = 2x(x - 1)(x + x - 1)$

$\qquad = 2x(x - 1)(2x - 1)$

$\qquad f'(\frac{1}{2}) = 0; f(\frac{1}{2}) = \frac{1}{16}$

Using the point-slope formula:

$\qquad y - \frac{1}{16} = 0$

36. $f'(x) = 3\cos(3x - \pi)$

$\qquad f'(\frac{\pi}{2}) = 3\cos\frac{\pi}{2} = 0; f(\frac{\pi}{2}) = \sin(\frac{3\pi}{2} - \pi) = 1$

Using the point-slope formula: $y - 1 = 0$

37. $f'(x) = x^2[-3\sec^2(4 - 3x)] + 2x\tan(4 - 3x)$

$\qquad = -3x^2\sec^2(4 - 3x) + 2x\tan(4 - 3x)$

$\qquad f'(0) = 0; f(0) = 0$

Using the point-slope formula: $y = 0$

38.

$f'(x) = 2\left(\frac{5-x^2}{3x-1}\right)\left[\frac{(3x-1)(-2x)-(5-x^2)(3)}{(3x-1)^2}\right]$

$\qquad = 2\left(\frac{5-x^2}{3x-1}\right)\left[\frac{-3x^2 + 2x - 15}{(3x-1)^2}\right]$

$\qquad = \frac{2(x^2-5)(3x^2 - 2x + 15)}{(3x-1)^3}$

$\qquad f'(0) = 150; f(0) = 25$

Using the point-slope formula:

$\qquad y - 25 = 150(x - 0)$

$\qquad 150x - y + 25 = 0$

39. $f'(x) = x(\frac{1}{2})(1 - 3x)^{-1/2}(-3) + \sqrt{1 - 3x}$

$= \dfrac{-3x + 2(1 - 3x)}{2(1 - 3x)^{1/2}} = \dfrac{-9x + 2}{2(1 - 3x)^{1/2}}$

$f'(x) = 0$ if $-9x + 2 = 0$ or $x = \frac{2}{9}$

40. $g'(x) = x^2[2(2x + 3)](2) + (2x)(2x + 3)^2$

$= 2x(2x + 3)(4x + 3)$

$f'(x) = 0$ if $x = 0, -\frac{3}{2}, -\frac{3}{4}$

41. $q'(x) = \dfrac{(x + 2)^3 2(x - 1) - (x - 1)^2 3(x + 2)^2(1)}{(x + 2)^6}$

$= \dfrac{(x - 1)(x + 2)^2[2(x + 2) - 3(x - 1)]}{(x + 2)^6}$

$= \dfrac{(x - 1)(x + 2)^2(-x + 7)}{(x + 2)^6}$

$= \dfrac{(x - 1)(-x + 7)}{(x + 2)^4}$

$q'(x) = 0$ and therefore has a horizontal tangent when $x = 1, 7$

42. $f'(x) = 3(2x^2 - 7)^2(4x) = 12x(2x^2 - 7)^2$

$f'(x) = 0$ if $x = 0, \pm\dfrac{\sqrt{14}}{2}$

43. **a.** $f'(x) = (x + 3)^2(2)(x - 2)(1)$

$\qquad\qquad + 2(x + 3)(1)(x - 2)^2$

$= 2(x + 3)(x - 2)(2x + 1)$

$f'(x) = 0$ if $x = -3, 2, -\frac{1}{2}$

b. $f''(x) = 2[(x - 2)(2x + 1)$

$\qquad + (x + 3)(2x + 1) + 2(x + 3)(x - 2)]$

$= 2(6x^2 + 6x - 11)$

$f''(x) = 0$ if $x = \dfrac{-3 \pm 5\sqrt{3}}{6}$

44. **a.** The graph indicates that $u \approx 5$ when $x = 2$. The slope of the tangent line to the curve $u = g(x)$ at $x = 2$ is about 1.

b. The graph indicates that $y \approx 3$ when $u = 5$. The slope of the tangent line is about $\frac{3}{2}$.

c. The slope of $y = f[g(x)]$ is about $(1)(\frac{3}{2}) = 1.5$

45. **a.** Assume the snowball is a sphere;

$V = \frac{4}{3}\pi r^3; \dfrac{dr}{dt} = -\dfrac{10}{2} = -5 \text{ cm/hr}$

$\dfrac{dV}{dt} = 4\pi r^2 \dfrac{dr}{dt} = 4\pi(5^2)(-5)$

$\qquad\qquad = -500\pi \text{ cm}^3/\text{hr}$

b. Since the surface area is $S = 4\pi r^2$,

$\dfrac{dS}{dt} = 8\pi r \dfrac{dr}{dt} = (8\pi)(5)(-5)$

$\qquad\qquad = -200\pi \text{ cm}^2/\text{hr}$

46. Here we have a function of a function, so the chain rule is required.

$\dfrac{dL}{dt} = \dfrac{dL}{dp} \cdot \dfrac{dp}{dt}$

$= (0.05)(\frac{1}{2})(p^2 + p + 58)^{-1/2}(2p + 1)\left[\dfrac{6}{(t + 1)^2}\right]$

$= \dfrac{6(2p + 1)}{4(p^2 + p + 58)^{1/2}(t + 1)^2}$ parts/million

Two years from now, $t = 2$, $p(2) = 18$, and the rate of change is

$\dfrac{dL}{dt} = \dfrac{6[2(18) + 1]}{4(18^2 + 18 + 58)^{1/2}(2 + 1)^2} = \dfrac{37}{120}$

≈ 0.31 parts/million

47. $C(q) = 0.2q^2 + q + 900;$

$q(t) = t^2 + 100t$ and $q(1) = 101$

$C'(q) = (0.4q + 1)\big|_{q=101}(2t + 100)\big|_{t=1}$

$= 41.4(102) = \$4,222.80 \text{ per hr}$

48. $\dfrac{dD}{dt} = \dfrac{dD}{dp} \cdot \dfrac{dp}{dt} = 4,374(-2p^{-3})(0.04t + 0.1)$

$= -8,748p^{-3}(0.04t + 0.1) = -6 \text{ lb/wk}$

49. $D(p) = 8,000p^{-1}$

$p(t) = 0.04t^{3/2} + 15;$

$p(25) = (0.04)(125) + 15 = 02$

$D'(p) = (-8,000p^{-2})\big|_{q=20}(0.06t^{1/2})\big|_{t=25}$

$= -8,000(20)^{-2}(0.3)$

$= -6 \text{ blenders per month.}$

The demand is decreasing.

50. $s(t) = A \sin 2t; \dfrac{ds}{dt} = 2A \cos 2t;$

$\dfrac{d^2s}{dt^2} = -4A \sin 2t; \dfrac{d^2s}{dt^2} + 4s = 0$

51. **a.** $f(x) = L(x^2),$

$f'(x) = L'(x^2)(2x) = \dfrac{1}{x^2}(2x) = \dfrac{2}{x}$

b. $f(x) = L\left(\dfrac{1}{x}\right);$

$f'(x) = L'\left(\dfrac{1}{x}\right)\left(-\dfrac{1}{x^2}\right) = x\left(-\dfrac{1}{x^2}\right)$

$= -\dfrac{1}{x}$

c. $f(x) = L\left(\dfrac{2}{3\sqrt{x}}\right),$

$f'(x) = L'\left(\dfrac{2}{3\sqrt{x}}\right)\left(-\dfrac{1}{3x^{3/2}}\right)$

$$= \left(\frac{3\sqrt{x}}{2}\right)\left(-\frac{1}{3x^{3/2}}\right) = -\frac{1}{2x}$$

d. $f(x) = L\left(\frac{2x + 1}{1 - x}\right),$

$$f'(x) = L'\left(\frac{2x+1}{1-x}\right)$$

$$= \left(\frac{(1-x)2 - (2x+1)(-1)}{(1-x)^2}\right)$$

$$= \left(\frac{1-x}{2x+1}\right)\left(\frac{3}{(1-x)^2}\right)$$

$$= \frac{3}{(2x+1)(1-x)}$$

52. a. $I = 20s^{-2}; \frac{dI}{dt} = -(20)s^{-3}\frac{ds}{dt};$

$\frac{ds}{dt} = -2t$

When $s(t) = 28 - t^2 = 19$, $t = 3$

$\frac{dI}{dt} = -2(20)(19^{-3})(-2)(3) \approx 0.035$

b. When $s(t) = 28 - t^2 = 3$, $t = 5$

$\frac{dI}{dt} = -2(20)(3^{-3})(-2)(5)$

≈ 14.81 lux/sec

53. $I = 40s^{-2}; s = 20 - 15 = 5; \frac{ds}{dt} = -2$ m/sec;

$\frac{dI}{dt} = -80s^{-3}\frac{ds}{dt} = \frac{160}{5^3} = 1.28$ lux/sec

54. $\theta = \theta_m\sin kt; \frac{d\theta}{dt} = k\theta_m\cos kt;$

$\frac{d^2\theta}{dt^2} = -k^2\theta_m\sin kt$

$\frac{d^2\theta}{dt^2} + k^2\theta = -k^2\theta_m\sin kt + k^2(\theta_m\sin kt) = 0$

55. a. From Figure 2.22, $\tan\theta = \frac{s(t)}{2}$, but

$\theta = 6\pi t$, so $s(t) = 2\tan 6\pi t$.

b. $s'(t) = 2(\sec^2 6\pi t)(6\pi) = 12\pi\sec^2 6\pi t$.
Although we are not given t, we are given the distance from the lighthouse is 4. So
$s = 2\sqrt{3}$ and $2\sqrt{3} = 2\tan 6\pi t$,

$\tan 6\pi t = \sqrt{3}$,

$6\pi t = \arctan\sqrt{3}$, $t = \frac{\arctan\sqrt{3}}{6\pi}$,

$t = \frac{1}{18}$ min. We can now find

$s'(t) = 12\pi\sec^2 6\pi t = 12\pi\sec^2\frac{\pi}{3}$

$= 12\pi(4) = 48\pi \approx 150.8$ km/min

56. $\frac{d}{dx}\sin(\frac{\pi}{2} - x) = \cos(\frac{\pi}{2} - x)(-1)$

This can be rewritten as

$\frac{d}{dx}\cos x = -\sin x$

57. $g(x) = f[u(x)], g'(x) = f'[u(x)][u'(x)]$

$g'(-3) = f'[u(-3)][u'(-3)]$

$= [f'(5)](2) = -6$

$g(-3) = f[u(-3)] = f(5) = 3$

The equation of the tangent line is:

$$y - 3 = -6(x + 3)$$

$$6x + y + 15 = 0$$

58. a. $f'(u) = \frac{1}{u^2 + 1};$ let $g(x) = f[u(x)]$ so

$g'(x) = \frac{d}{du}f(u)\frac{du}{dx}$

$u(x) = 3x - 1;$

$g'(x) = \frac{3}{u^2 + 1} = \frac{3}{(3x - 1)^2 + 1}$

b. Let $h(x) = f[u(x)]; u(x) = x^{-1}$

$h'(x) = \frac{x^2}{x^{-2} + 1}(-x^{-2}) = \frac{-1}{x^2 + 1}$

59. $f'(x) = \sqrt{x^2 + 5}; g(x) = x^2 f(\frac{x}{x-1})$

$g'(x) = x^2 f'(\frac{x}{x-1})\left[\frac{-1}{(x-1)^2}\right]$

$\qquad + 2xf(\frac{x}{x-1})$

$= x^2 f'(\frac{x}{x-1})\left[\frac{-1}{(x-1)^2}\right]$

$\qquad + 2x f(\frac{x}{x-1})$

$g'(2) = -2^2 f'(2) + 4f(2)$

$= -4\sqrt{4 + 5} + 4(-3)$

$= -24$

60. $u(x) = \cot x; \frac{df}{dx} = \frac{\sin x}{x}$

$\frac{df}{du}\cdot\frac{du}{dx} = \frac{df}{dx}$

$\frac{df}{du}(-\csc^2 x) = \frac{\sin x}{x}$

Thus, $\frac{df}{du} = -\frac{\sin^3 x}{x}$

61. $\frac{d}{dx}f'[f(x)] = f''[f(x)]f'(x)$

and

$\frac{d}{dx}f[f'(x)] = f'[f'(x)]f''(x)$

62. $a(t) = \frac{dv}{dt} = \frac{dv}{ds}\cdot\frac{ds}{dt} = v\frac{dv}{ds}$

Now with

$$s(t) = -2t^3 + 4t^2 + t - 3$$

$$v(t) = -6t^2 + 8t + 1$$

$$a(t) = -12t + 8 = -4(3t - 2)$$

Thus,

$$\frac{dv}{ds} = \frac{a(t)}{v(t)} = \frac{4(3t - 2)}{6t^2 - 8t - 1}$$

2.6 Implicit Differentiation, Page 134

1. $2x + 2y\dfrac{dy}{dx} = 0$

    $\dfrac{dy}{dx} = -\dfrac{x}{y}$

2. $(2x)(1) + \dfrac{dy}{dx} = (3x^2)(1) + 3y^2\dfrac{dy}{dx}$

    $(1 - 3y^2)\dfrac{dy}{dx} = 3x^2 - 2x$

    $\dfrac{dy}{dx} = \dfrac{3x^2 - 2x}{1 - 3y^2}$

3. $xy' + y = 0$

    $y' = -\dfrac{y}{x}$

4. $2x^2y' + 2(2x)y + 3x(2y)y' + 3y^2 = 0$

    $(2x^2 + 6xy)y' = -(4xy + 3y^2)$

    $y' = \dfrac{4xy + 3y^2}{2x^2 + 6xy}$

5. $2x + 3(x\dfrac{dy}{dx} + y) + 2y\dfrac{dy}{dx} = 0$

    $(3x + 2y)\dfrac{dy}{dx} = -2x - 3y$

    $\dfrac{dy}{dx} = -\dfrac{2x + 3y}{3x + 2y}$

6. $3x^2 + 3y^2y' = 1 + y'$

    $y' = \dfrac{1 - 3x^2}{3y^2 - 1}$

7. $3(x + y)^2(1 + y') + 3y' = 0$

    $3(x + y)^2 + 3(x + y)^2y' + 3y' = 0$

    $[(x + y)^2 + 1]y' = -(x + y)^2$

    $y' = -\dfrac{(x + y)^2}{(x + y)^2 + 1}$

8. $3x^2 + xy' + y + 3y^2y' = 1$

    $y'(x + 3y^2) = 1 - 3x^2 - y$

    $y' = \dfrac{1 - 3x^2 - y}{x + 3y^2}$

9. $-\dfrac{1}{y^2}\dfrac{dy}{dx} - \dfrac{1}{x^2} = 0$

    $\dfrac{dy}{dx} = -\dfrac{y^2}{x^2}$

10. $2(2x + 3y)(2 + 3y') = 0$

    $2 + 3y' = 0$

    $y' = -\dfrac{2}{3}$

11. $\cos(x + y)(1 + y') = 1 - y'$

    $[\cos(x + y) + 1]y' = 1 - \cos(x + y)$

    $y' = \dfrac{1 - \cos(x + y)}{\cos(x + y) + 1}$

12. $\sec^2\dfrac{x}{y}\left(\dfrac{-xy' + y}{y^2}\right) = y'$

    $-xy'\sec^2\dfrac{x}{y} + y\sec^2\dfrac{x}{y} = y^2y'$

    $(-x\sec^2\dfrac{x}{y} - y^2)y' = -y\sec^2\dfrac{x}{y}$

    $y' = \dfrac{y\sec^2\dfrac{x}{y}}{x\sec^2\dfrac{x}{y} + y^2}$

13. $(-\sin xy)\dfrac{d}{dx}(xy) = -2x$

    $(-\sin xy)(x\dfrac{dy}{dx} + y) = -2x$

    $(-x\sin xy)\dfrac{dy}{dx} = y\sin xy - 2x$

    $\dfrac{dy}{dx} = \dfrac{y\sin xy - 2x}{-x\sin xy}$

    $= \dfrac{2x - y\sin xy}{x\sin xy}$

14. $5(x^2 + 3y^2)^4\dfrac{d}{dx}(x^2 + 3y^2) = 2x\dfrac{dy}{dx} + 2y$

    $5(x^2 + 3y^2)^4(2x + 6y\dfrac{dy}{dx}) = 2x\dfrac{dy}{dx} + 2y$

    $10x(x^2 + 3y^2)^4 + 30y(x^2 + 3y^2)^4\dfrac{dy}{dx} = 2x\dfrac{dy}{dx} + 2y$

    $[30y(x^2 + 3y^2)^4 - 2x]\dfrac{dy}{dx} = 2y - 10x(x^2 + 3y^2)^4$

    $\dfrac{dy}{dx} = \dfrac{2y - 10x(x^2 + 3y^2)^4}{30y(x^2 + 3y^2)^4 - 2x}$

15. **a.** $2x + 3y^2\dfrac{dy}{dx} = 0$

    $\dfrac{dy}{dx} = -\dfrac{2x}{3y^2}$

    b. $x^2 + y^3 = 12$

    $y = (12 - x^2)^{1/3}$

    $\dfrac{dy}{dx} = \dfrac{1}{3}(12 - x^2)^{-2/3}(-2x)$

    $= -\dfrac{2x}{3(12 - x^2)^{2/3}}$

16. **a.** $x\dfrac{dy}{dx} + y + 2\dfrac{dy}{dx} = 2x$

$\qquad (x+2)\dfrac{dy}{dx} = 2x - y$

$\qquad\qquad \dfrac{dy}{dx} = \dfrac{2x-y}{x+2}$

b. $xy + 2y = x^2$

$\qquad y = \dfrac{x^2}{x+2}$

$\qquad \dfrac{dy}{dx} = \dfrac{(x+2)(2x) - x^2(1)}{(x+2)^2}$

$\qquad\quad = \dfrac{x^2 + 4x}{(x+2)}$

$\qquad\quad = \dfrac{x(x+4)}{(x+2)^2}$

17. **a.** $1 - \dfrac{1}{y^2}\dfrac{dy}{dx} = 0$

$\qquad\quad \dfrac{dy}{dx} = y^2$

b. $xy + 1 = 5y$

$\qquad (x-5)y = -1$

$\qquad\quad y = \dfrac{-1}{x-5}$

$\qquad \dfrac{dy}{dx} = -\dfrac{-1}{(x-5)^2}$

$\qquad\quad = \dfrac{1}{(x-5)^2}$

18. **a.** $x\dfrac{dy}{dx} + y - 1 = \dfrac{dy}{dx}$

$\qquad (x-1)\dfrac{dy}{dx} = 1 - y$

$\qquad\qquad \dfrac{dy}{dx} = \dfrac{1-y}{x-1}$

b. $xy - x = y + 2$

$\qquad y = \dfrac{x+2}{x-1}$

$\qquad \dfrac{dy}{dx} = \dfrac{(x-1)(1) - (x+2)(1)}{(x-1)^2}$

$\qquad\quad = \dfrac{-3}{(x-1)^2}$

19. $2x + 2yy' = 0$

$\qquad y' = -\dfrac{x}{y};\ y_P{}' = \dfrac{2}{3}$

The equation of the tangent line is

$\qquad y - 3 = \dfrac{2}{3}(x+2)$

$\qquad 3y - 9 = 2x + 4$

$\qquad 2x - 3y + 13 = 0$

20. $3x^2 + 3y^2y' = y'$

$\qquad 3x^2 = y'(1 - 3y^2)$

$y' = \dfrac{3x^2}{1 - 3y^2};\ y_P{}' = -\dfrac{27}{11}$

The equation of the tangent line is

$\qquad y + 2 = -\dfrac{27}{11}(x - 3)$

$\qquad 11y + 22 = -27x + 81$

$\qquad 27x + 11y - 59 = 0$

21. $\dfrac{d}{dx}[\sin(x - y)] = \dfrac{d}{dx}(xy)$

$[\cos(x-y)]\dfrac{d}{dx}(x - y) = x\dfrac{dy}{dx} + y$

$[\cos(x-y)](1 - \dfrac{dy}{dx}) = x\dfrac{dy}{dx} + y$

$\cos(x-y) - [\cos(x-y)]\dfrac{dy}{dx} = x\dfrac{dy}{dx} + y$

$[-x - \cos(x-y)]\dfrac{dy}{dx} = y - \cos(x-y)$

$\qquad\qquad \dfrac{dy}{dx} = \dfrac{y - \cos(x-y)}{-x - \cos(x-y)}$

$\qquad\qquad\quad = \dfrac{\cos(x-y) - y}{\cos(x-y) + x}$

$\dfrac{dy}{dx}(0, \pi) = \dfrac{\cos(-\pi) - \pi}{\cos(-\pi)}$

$\qquad\qquad = \dfrac{-1 - \pi}{-1} = \pi + 1.$

The equation of the tangent line is

$\qquad y - \pi = (\pi + 1)(x - 0)$

$\qquad (\pi + 1)x - y + \pi = 0$

22. $\tan\dfrac{y}{x}\sec\dfrac{y}{x}\left(\dfrac{xy' - y}{x^2}\right) - \dfrac{1}{2}(x+1)^{-1/2} = 0$

$\quad 2\sqrt{x+1}\tan\dfrac{y}{x}\sec\dfrac{y}{x}(xy' - y) - x^2 = 0$

At $P(3, \pi)$,

$6y'\sqrt{4}\tan\dfrac{\pi}{3}\sec\dfrac{\pi}{3} - 2\pi\sqrt{4}\tan\dfrac{\pi}{3}\sec\dfrac{\pi}{3} = 9$

$\qquad y' = \dfrac{8\pi\sqrt{3} + 9}{24\sqrt{3}}$

The equation of the tangent line is

$\qquad y - \pi = \dfrac{8\pi\sqrt{3} + 9}{24\sqrt{3}}(x - 3)$

$(8\pi\sqrt{3} + 9)x - 24\sqrt{3}\,y - \pi = 0$

23. $2(x^2 + y^2)(2x + 2yy') = 4(2xy + x^2y')$

At $(1, 1)$: $\qquad 2 + 2y' = 2 + y'$

$\qquad\qquad y' = 0$

24. $2(x^2 + y^2)(2x + 2yy') = \dfrac{25}{3}(2x - 2yy')$

At $(2, 1)$: $\qquad 4 + 4y' = \dfrac{5}{3}(4 - 2y')$

$\qquad 12 + 12y' = 20 - 10y'$

$\qquad\qquad y' = -\dfrac{1}{11}$

25. $3x^2 + 3y^2\dfrac{dy}{dx} - \dfrac{9}{2}\left(x\dfrac{dy}{dx} + y\right) = 0$

$\qquad\qquad \left(3y^2 - \dfrac{9}{2}x\right)\dfrac{dy}{dx} = \dfrac{9}{2}y - 3x^2$

$\dfrac{dy}{dx} = \dfrac{9y - 6x^2}{6y^2 - 9x} = \dfrac{3y - 2x^2}{2y^2 - 3x}$

At $(2, 1)$: $\quad y' = \dfrac{3-8}{2-6} = \dfrac{5}{4}$

26. $\qquad y^2(-1) + 2yy'(6 - x) = 3x^2$

$\qquad\qquad -y^2 + 2(6 - x)yy' = 3x^2$

At $(3, 3)$: $\qquad -9 + 18y' = 27$

$\qquad\qquad\qquad y' = 2$

27. $\qquad 2x + 2xy' + 2y - 3y'y^2 = 0$

$\qquad\qquad 2 + 2y' - 2 - 3y' = 0$

$\qquad\qquad\qquad y' = 0$

The tangent line is horizontal so the normal line is a vertical line with equation $x - 1 = 0$.

28. $x^2(\frac{1}{2})(y - 2)^{-1/2}y' + 2x\sqrt{y - 2} = 2yy' - 3$

Multiply both sides by $2(y - 2)^{-1/2}$:

$x^2y' + 4x(y - 2) = 2(2yy' - 3)(y - 2)^{-1/2}$

At $(1, 2)$:

$\quad 1^2y' + 2(1)\sqrt{2 - 2} = 2(4y' - 3)(2 - 2)^{-1/2}$

$\qquad\qquad\qquad y' = 0$

The normal line, then, is vertical, namely $x = 1$.

29. $\qquad 7x + 5y^2 = 1$

$\qquad\qquad 7 + 10yy' = 0$

$\qquad\qquad\qquad y' = -\dfrac{7}{10y}.$

For y'' we again differentiate:

$\qquad \dfrac{d}{dx}(y') = \dfrac{d}{dx}\left(-\dfrac{7}{10y}\right)$

$\qquad\qquad y'' = \dfrac{7}{10y^2}\,y'$

$\qquad\qquad\quad = \left(\dfrac{7}{10y^2}\right)\left(-\dfrac{7}{10y}\right)$

$\qquad\qquad\quad = -\dfrac{49}{100y^3}$

30. $\qquad x + 3y^2y' = 0$

$\qquad\qquad y' = -\dfrac{x}{3y^2}$

$\qquad\qquad y'' = -\dfrac{1}{3}\dfrac{y^2 - 2xyy'}{y^4}$

$\qquad\qquad\quad = -\dfrac{1}{3y^3}\left(y + \dfrac{2x^2}{3y^2}\right)$

$\qquad\quad = -\dfrac{2x^2 + 3y^3}{9y^5}$

31. $8x - 6yy' = 0$

$\qquad y' = \dfrac{4x}{3y}$

$\qquad y'' = \dfrac{4}{3}\dfrac{y - xy'}{y^2}$

$\qquad\quad = \dfrac{4}{3y^2}\left(y - \dfrac{4x^2}{3y}\right)$

$\qquad\quad = \dfrac{4}{9y^3}(3y^2 - 4x^2)$

$\qquad\quad = \dfrac{-36}{9y^3} = -\dfrac{4}{y^3}$

32. $2x + 5xy' + 5y - 2yy' = 0$

$\qquad\qquad y' = \dfrac{2x + 5y}{2y - 5x}$

$y'' = \dfrac{(2y - 5x)(2 + 5y') - (2x + 5y)(2y' - 5)}{(2y - 5x)^2}$

$\quad = \dfrac{1}{(2y - 5x)^2}\left\{(2y - 5x)\left(2 + \dfrac{10x + 25y}{2y - 5x}\right)\right.$

$\qquad\qquad \left. - (2x + 5y)\left(\dfrac{4x + 10y}{2y - 5x} - 5\right)\right\}$

$\quad = \dfrac{1}{(2y - 5x)^3}[(2y - 5x)(29y) - (2x + 5y)(29x)]$

$\quad = \dfrac{29}{(2y - 5x)^3}[2y^2 - 5xy - 2x^2 - 5xy]$

$\quad = \dfrac{58}{(2y - 5x)^3}[x^2 + 5xy - y^2]$

$\quad = -\dfrac{58}{(2y - 5x)^3}$

33. **a.** $\qquad b^2y^2 + a^2v^2 = a^2b^2$

$\qquad\qquad b^2uu' + a^2v = 0$

$\qquad\qquad\qquad u' = -\dfrac{a^2v}{b^2u}$

$\quad$ **b.** $\qquad b^2u + a^2vv' = 0$

$\qquad\qquad\qquad v' = -\dfrac{b^2u}{a^2v}$

34. **a.** $(a - b)u^3 - (a + b)v^2 = c$

$\quad 3(a - b)u^2u' - 2(a + b)v = 0$

$\qquad\qquad u' = \dfrac{2(a + b)v}{3(a - b)u^2}$

$\quad$ **b.** $\quad 3(a - b)u^2 - 2(a + b)vv' = 0$

$\qquad\qquad v' = \dfrac{3(a - b)u^2}{2(a + b)v}$

35. $4x + 3xy' + 3y + 2yy' = 0$

$y' = 0$ at $P(a, b)$ if $4a + 3b = 0$;

Substitute $b = -\dfrac{4a}{3}$ into

$$2a^2 + 3ab + b^2 = -2$$

$$2a^2 + 3a\left(-\dfrac{4a}{3}\right) + \left(-\dfrac{4a}{3}\right)^2 = -2$$

$$18a^2 - 36a^2 + 16a^2 = -18$$

$$2a^2 = 18$$

$$a = \pm 3$$

Then $b = \mp 4$. Two such points are $(3, -4)$ and $(-3, 4)$.

36. $2x - 3xy' - 3y + 4yy' = 0$

$$y' = \dfrac{3y - 2x}{4y - 3x}$$

The denominator is not defined if $y = 3x/4$. Substituting in the given equation leads to

$$x^2 - \dfrac{9x^2}{4} + \dfrac{9x^2}{8} = -2$$

$$x = \pm 4$$

Then $y = \pm 3$. Two such points are $(4, 3)$ and $(-4, -3)$.

37. **a.** $2x + 2g(x)[g'(x)] = 0$

$$g'(x) = -\dfrac{x}{g(x)}$$

b. $g(x) = -\sqrt{10 - x^2}$ is a differentiable function of x for which $g(x) < 0$. Using the chain rule:

$$g'(x) = -\dfrac{1}{2\sqrt{10 - x^2}}\dfrac{d}{dx}(10 - x^2)$$

$$= \dfrac{x}{\sqrt{10 - x^2}}$$

Verifying the result of (a):

$$\dfrac{x}{\sqrt{10 - x^2}} = \dfrac{-x}{-\sqrt{10 - x^2}} = \dfrac{-x}{g(x)}$$

38. $3x^2 + 3y^2\dfrac{dy}{dx} = 2Ax\dfrac{dy}{dx} + 2Ay$

$$(3y^2 - 2Ax)\dfrac{dy}{dx} = 2Ay - 3x^2$$

$$\dfrac{dy}{dx} = \dfrac{2Ay - 3x^2}{3y^2 - 2Ax}$$

At (A, A), $dy/dx = -A^2/A^2 = -1$.

The equation of the tangent line is

$$y - A = -(x - A)$$

$$x + y - 2A = 0$$

For the normal, the slope is 1, and the equation of the normal is

$$y - A = x - A$$

$$x - y = 0$$

39. **a.** $2x + 2y\dfrac{dy}{dx} = 6\dfrac{dy}{dx}$

$$(2y - 6)\dfrac{dy}{dx} = -2x$$

$$\dfrac{dy}{dx} = \dfrac{-2x}{2y - 6} = \dfrac{x}{3 - y}$$

b. $x^2 + y^2 = 6y - 10$

$$x^2 + y^2 + 9 = -10 + 9$$

$$x^2 + (y - 3)^2 = -1$$

This is impossible since the sum of two squares is not negative (in the real number system).

c. The derivative does not exist.

40. $2(x^2 + y^2)(2x + 2yy') = 4(2x - 2yy')$

The tangent line is horizontal if $y' = 0$. Thus,

$$x(x^2 + y^2) = 2x$$

$$x(x^2 + y^2) - 2x = 0$$

$$x[(x^2 + y^2) - 2] = 0$$

$$x = 0 \text{ or } x^2 + y^2 = 2$$

Substituting in the original equation leads to

$$x^2 - y^2 = 1$$

Adding equations reveals $2x^2 = 3$ or $x = \pm\dfrac{\sqrt{6}}{2}$ and $y^2 = \dfrac{1}{2}$ or $y = \pm\dfrac{\sqrt{2}}{2}$.

The points at which the tangent lines are horizontal are:

$$\left(\dfrac{\sqrt{6}}{2}, \dfrac{\sqrt{2}}{2}\right), \left(-\dfrac{\sqrt{6}}{2}, \dfrac{\sqrt{2}}{2}\right), \left(\dfrac{\sqrt{6}}{2}, -\dfrac{\sqrt{2}}{2}\right),$$
$$\left(-\dfrac{\sqrt{6}}{2}, -\dfrac{\sqrt{2}}{2}\right)$$

41. $\dfrac{3}{2}(x^2 + y^2)^{1/2}(2x + 2yy')$

$$= \dfrac{1}{2}(x^2 + y^2)^{-1/2}(2x + 2yy')$$

$$\left[3y(x^2 + y^2)^{1/2} - y(x^2 + y^2)^{-1/2}\right]y'$$

$$= x(x^2 + y^2)^{-1/2} - 3x(x^2 + y^2)^{1/2} + 1$$

$$y' = \dfrac{x(x^2 + y^2)^{-1/2} - 3x(x^2 + y^2)^{1/2} + 1}{3y(x^2 + y^2)^{1/2} - y(x^2 + y^2)^{-1/2}}.$$

Multiplying numerator and denominator by $(x^2 + y^2)^{1/2}$ we obtain:

$$y' = \frac{x - 3x(x^2 + y^2) + (x^2 + y^2)^{1/2}}{3y(x^2 + y^2) - y}$$

$$= \frac{x(1 - 3x^2 - 3y^2) + (x^2 + y^2)^{1/2}}{y(3x^2 + 3y^2 - 1)}$$

Now the denominator, $y(3x^2 + 3y^2 - 1) = 0$, when $y = 0$, or $3x^2 + 3y^2 - 1 = 0$.
To find the x-coordinate of the point let $y = 0$ in the equation of the cardioid:

At $y = 0$, $\qquad x^3 = |x| + x$.

For $x < 0$, $\quad x^3 = 0$, $\quad x = 0$.

For $x > 0$. $\quad x^3 = 2x$, $\quad x = 0, \pm\sqrt{2}$

At $(0, 0)$ the derivative is undefined. At $(-\sqrt{2}, 0)$ the function is undefined. So this factor of the denominator gives one vertical tangent at $(\sqrt{2}, 0)$. Now consider the other factor:

$$3x^2 + 3y^2 - 1 = 0$$
$$x^2 + y^2 = \tfrac{1}{3}$$

is a circle with radius of $\sqrt{3}/3$. We need to find the points where it intersects the cardioid. Substituting $x^2 + y^2 = 1/3$ into the cardioid equation:

$$\left(\tfrac{1}{3}\right)^{3/2} = \left(\tfrac{1}{3}\right)^{1/2} + x$$
$$x = -\frac{2\sqrt{3}}{9}$$

Using the circle to find y at this point:

$$\left(-\frac{2\sqrt{3}}{9}\right)^2 + y^2 = \tfrac{1}{3}$$
$$y^2 = \tfrac{5}{27}$$
$$y = \pm\frac{\sqrt{15}}{9}$$

There are two additional vertical tangents, at

$$\left(-\frac{2\sqrt{3}}{9}, \frac{\sqrt{15}}{9}\right) \text{ and } \left(-\frac{2\sqrt{3}}{9}, -\frac{\sqrt{15}}{9}\right).$$

42. $\frac{2}{3}x^{-1/3} + \frac{2}{3}y^{-1/3}y' = 0$

$$y' = -\frac{y^{1/3}}{x^{1/3}}$$

At $(8, 8)$, $y' = -\frac{2}{2} = -1$
The equation of the tangent line is

$$y - 8 = -(x - 8)$$
$$x + y - 16 = 0$$

The x- and y- intercepts are $(0, 16)$ and $(16, 0)$, and the area of the triangle is 128 square units.

43. $\quad x - [f(x)]^2 = 9$

$$[f(x)]^2 = x - 9$$
$$f(x) = \pm\sqrt{x - 9}$$

Neither function is differentiable at $x = 9$. We are dealing with the positive and negative branches of the parabola $y = x - 9$.

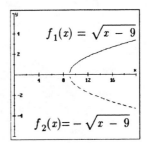

44. $\qquad \frac{x}{a^2} + \frac{yy'}{b^2} = 0$

$$y' = -\frac{b^2 x}{a^2 y}$$

At (x_0, y_0); $m = -b^2 x_0/a^2 y_0$, so the equation of the tangent line is

$$y - y_0 = -\frac{b^2 x_0}{a^2 y_0}(x - x_0)$$
$$a^2 y y_0 - a^2 y_0^2 + b^2 x_0 x - b^2 x_0^2 = 0$$

or $\qquad \frac{x_0^2}{a^2} + \frac{y_0^2}{b^2} = 1$

since (x_0, y_0) lies on the curve and thus satisfies the equation of the curve.

45. $\qquad \frac{x^2}{a^2} - \frac{y^2}{b^2} = 1$

$$\frac{2x}{a^2} - \frac{2yy'}{b^2} = 0$$
$$y' = \frac{b^2 x}{a^2 y}$$

Thus, the slope at (x_0, y_0) is $m = b^2 x_0/a^2 y_0$ and the equation of the line becomes

$$y - y_0 = \frac{b^2 x_0}{a^2 y_0}(x - x_0)$$
$$\frac{y_0 y}{b^2} - \frac{y_0^2}{b^2} = \frac{x_0 x}{a^2} - \frac{x_0^2}{a^2}$$

or $\qquad \frac{x_0^2}{a^2} - \frac{y_0^2}{b^2} = 1$

since (x_0, y_0) lies on the curve, and thus satisfies the equation of the curve.

46.

$$ax^2 + by^2 = c$$

$$2ax + 2by\frac{dy}{dx} = 0$$

$$\frac{dy}{dx} = -\frac{2ax}{2by} = -\frac{ax}{by}$$

$$\frac{d^2y}{dx^2} = \frac{by(a) - ax\left(b\dfrac{dy}{dx}\right)}{(by)^2} = -\frac{aby - abx\dfrac{dy}{dx}}{b^2y^2}$$

Since $\dfrac{dy}{dx} = -\dfrac{ax}{by}$

$$\frac{d^2y}{dx^2} = -\frac{aby - abx\left(-\dfrac{ax}{by}\right)}{b^2y^2} = -\frac{ab^2y^2 + a^2bx^2}{b^3y^3}$$

$$= -\frac{ab(by^2 + ax^2)}{b^3y^3} = -\frac{a(by^2 + ax^2)}{b^2y^3}$$

From the original equation $by^2 + ax^2 = c$ so

$$\frac{d^2y}{dx^2} = -\frac{ac}{b^2y^3}$$

47.

$$x^{1/2} + y^{1/2} = C$$

$$\tfrac{1}{2}x^{-1/2} + \tfrac{1}{2}y^{-1/2}yy' = 0$$

$$y' = -\frac{y^{1/2}}{x^{1/2}}$$

The equation of the tangent line at (x_0, y_0) is

$$y - y_0 = -\frac{y_0^{1/2}}{x_0^{1/2}}(x - x_0)$$

The x-intercept is found by setting $y = 0$:

$$-y_0 = -\frac{y_0^{1/2}}{x_0^{1/2}}(x - x_0)^2$$

$$x = x_0 + \sqrt{y_0x_0}$$

The y-intercept is found by setting $x = 0$:

$$y - y_0 = \frac{y_0^{1/2}}{x_0^{1/2}}(-x_0)$$

$$y = y_0 + \sqrt{y_0x_0}$$

The sum of these intercepts is

$$x_0 + \sqrt{y_0x_0} + y_0 + \sqrt{x_0y_0} = x + y$$

$$x_0 + 2\sqrt{x_0y_0} + y_0 = x + y$$

$$(\sqrt{x_0} + \sqrt{y_0})^2 = x + y$$

$$C^2 = x + y$$

48. Let α be the angle between the tangent to C and the positive x-axis. Let β be the angle between the tangent to C_2 and the positive x-axis. Then

$$\theta = \pi - [\alpha + (\pi - \beta)] = \beta - \alpha$$

$$\tan \theta = \tan(\beta - \alpha) = \frac{\tan \beta - \tan \alpha}{1 + \tan \alpha \tan \beta}$$

$$= \frac{m_2 - m_1}{1 + m_2m_1} \text{ if } m_2 > m_1$$

Similarly, $\tan \theta = \dfrac{m_1 - m_2}{1 + m_1m_2}$ if $m_1 > m_2$;

Thus, $\tan \theta = \dfrac{|m_1 - m_2|}{1 + m_1m_2}$

49. The circles $x^2 + y^2 = 1$ and $x^2 + (y-1)^2 = 1$ intersect at

$$y^2 - 1 = (y - 1)^2 - 1$$

$$-1 = -2y + 1 - 1$$

$$y = \tfrac{1}{2}$$

If $y = \tfrac{1}{2}$, then $x = \pm\sqrt{3}/2$.

If $x = \dfrac{\sqrt{3}}{2}$, then slope of the first circle at $\left(\dfrac{\sqrt{3}}{2}, \dfrac{1}{2}\right)$ is $2x + 2yy' = 0$ or $y' = \sqrt{3}$.

The slope of the second circle at $\left(\dfrac{\sqrt{3}}{2}, \dfrac{1}{2}\right)$ is

$$2x + 2(y-1)y' = 0 \text{ or } y' = \frac{\sqrt{3}/2}{1/2 - 1} = -\sqrt{3}$$

From Problem 48,

$$\tan \theta = \frac{|m_1 - m_2|}{1 + m_1m_2} = \frac{\sqrt{3} + \sqrt{3}}{1 - 3} = -\sqrt{3}$$

which leads to $\theta = -\dfrac{\pi}{3}$.

If $x = -\dfrac{\sqrt{3}}{2}$, then the slope of the first circle at $\left(-\dfrac{\sqrt{3}}{2}, \dfrac{1}{2}\right)$ is $2x + 2y' = 0$ or $y' = \sqrt{3}$.

The slope of the second circle at $\left(-\dfrac{\sqrt{3}}{2}, \dfrac{1}{2}\right)$ is

$$2x + 2(y - 1)y' = 0 \text{ or } y' = \frac{\sqrt{3}/2}{1/2 - 1}$$

$$= -\sqrt{3}. \text{ Thus, } \tan \theta = -\frac{2\sqrt{3}}{1 - 3} = \sqrt{3},$$

which leads to $\theta = \dfrac{\pi}{3}$.

50. By the chain rule,

$$\frac{d(y^2)}{d(x^2)} = \frac{d(y^2)}{d(x)}\frac{dx}{dx^2} = 2yy'\frac{1}{2x} = \frac{dy}{dx}\cdot\frac{y}{x}$$

2.7 Related Rates, Page 141

1. $\qquad 2x\dfrac{dx}{dt} + 2y\dfrac{dy}{dt} = 0$

When $x = 3$, $y = 4$, and $\dfrac{dx}{dt} = 4$

$$3(4) + 4\dfrac{dy}{dt} = 0$$
$$\dfrac{dy}{dt} = -3$$

2. $\qquad 2x\dfrac{dx}{dt} + 2y\dfrac{dy}{dt} = 0$

When $x = 4$, $y = 3$, and $\dfrac{dy}{dt} = 2$

$$4\dfrac{dx}{dt} + 3(2) = 0$$
$$\dfrac{dx}{dt} = -\dfrac{3}{2}$$

3. $\qquad 10x\dfrac{dx}{dt} - \dfrac{dy}{dt} = 0$

When $x = 10$, $y = -400$, and $\dfrac{dx}{dt} = 10$

$$10(10)(10) - \dfrac{dy}{dt} = 0$$
$$\dfrac{dy}{dt} = 1{,}000$$

4. $\qquad 8x\dfrac{dx}{dt} - \dfrac{dy}{dt} = 0$

When $x = 1$, $y = -96$, and $\dfrac{dx}{dt} = -6$

$$8(1)\dfrac{dx}{dt} + 6 = 0$$
$$\dfrac{dx}{dt} = -\dfrac{3}{4}$$

5. $\qquad \dfrac{dy}{dt} = 2(\tfrac{1}{2})x^{-1/2}\dfrac{dx}{dt}$

When $x = 9$, $y = 2\sqrt{9} - 9 = -3$, and $\dfrac{dy}{dt} = 5$

$$5 = (9)^{-1/2}\dfrac{dx}{dt}$$
$$\dfrac{dx}{dt} = 15$$

6. $\qquad \dfrac{dy}{dt} = 5(\tfrac{1}{2})(x+9)^{-1/2}\dfrac{dx}{dt}$

When $x = 7$, $y = 5\sqrt{7+9} = 20$, and $\dfrac{dx}{dt} = 2$

$$\dfrac{dy}{dt} = \dfrac{5}{2}(16)^{-1/2}(2)$$
$$\dfrac{dy}{dt} = \dfrac{5}{4}$$

7. $\qquad x\dfrac{dy}{dt} + y\dfrac{dx}{dt} = 0$

When $x = 5$, $y = 2$, and $\dfrac{dx}{dt} = -2$

$$5\left(\dfrac{dy}{dt}\right) + 2(-2) = 0$$
$$\dfrac{dy}{dt} = \dfrac{4}{5}$$

8. $\qquad x\dfrac{dy}{dt} + y\dfrac{dx}{dy} = 0$

When $x = 1$, $y = 2$, and $\dfrac{dx}{dt} = -2$

$$(1)\dfrac{dy}{dt} + (2)(-2) = 0$$
$$\dfrac{dy}{dx} = 4$$

9. $\qquad 2x\dfrac{dx}{dt} + x\dfrac{dy}{dt} + \dfrac{dx}{dt}y - 2y\dfrac{dy}{dt} = 0$

When $x = 4$,

$$16 + 4y - y^2 = 11$$
$$y^2 - 4y - 5 = 0$$
$$(y+1)(y-5) = 0$$
$$y = -1 \text{ (discard)}, \ 5$$

When $\dfrac{dy}{dt} = 5$,

$$2(4)\dfrac{dx}{dt} + (4)(5) + \dfrac{dx}{dt}(5) - 2(5)(5) = 0$$
$$13\dfrac{dx}{dt} = 30$$
$$\dfrac{dx}{dt} = \dfrac{30}{13}$$

10. $\dfrac{d}{dt}F(x) = -12\dfrac{dx}{dt} = -12(\tfrac{1}{4}) = -3$

11. $\dfrac{d}{dt}F(x) = -12\dfrac{dx}{dt} = -12(\tfrac{1}{4}) = -3$

Notice that since F is a linear function of x, the change in F is a constant, and does not depend upon the value of x. (See Problems 10 and 11.)

12. $y^2 = 4x$ **13.** $4x^2 + y^2 = 4$

$2y\dfrac{dy}{dt} = 4\dfrac{dx}{dt}$ $4x\dfrac{dx}{dt} + y\dfrac{dy}{dt} = 0$

$-4\dfrac{dy}{dt} = 4(3)$ $(2\sqrt{3})(5) + (1)\dfrac{dy}{dx} = 0$

$\dfrac{dy}{dt} = -3$ ft/s $\dfrac{dy}{dt} = -10\sqrt{3}$ units/s

14. The area of the ripple is $A = \pi r^2$ in.^{2}

$$\dfrac{dA}{dt} = 2\pi r\dfrac{dr}{dt} = 2\pi(8)(3) = 48\pi \text{ in.}^2/\text{s}$$

15. The area of the ripple is $A = \pi r^2$ in.^{2}

$$\dfrac{dA}{dt} = 2\pi r\dfrac{dr}{dt} \text{ or } 4 = 2\pi(1)\dfrac{dr}{dt} \text{ so that}$$
$$\dfrac{dr}{dt} = \dfrac{2}{\pi} \approx 0.637 \text{ ft/s}$$

16. $\dfrac{dQ}{dt} = 2p\dfrac{dp}{dt} + 3\dfrac{dp}{dt} = 2(30)(2) + 3(2) = 126$ units/yr

17. $\dfrac{dR}{dt} = x\dfrac{dx}{dt} + 3\dfrac{dx}{dt} = 14(2) + 3(2) = 34$

In 2 years, $x = 14$, so the revenue will be rising at \$26,000/yr.

18. $\dfrac{dN}{dt} = 2p\dfrac{dp}{dt} + 5\dfrac{dp}{dt} = 40(1.2) + 5(1.2) = 54$

The number of patients is increasing by 54,000/yr.

19. Given $PV = C$, we are asked to find $\dfrac{dP}{dt}$ for the specific values of $V = 30$,

$$P = 90 \text{ , and } \frac{dV}{dt} = 10.$$

Differentiating Boyle's law with respect to t:

$$P\frac{dV}{dt} + V\frac{dP}{dt} = 0 .$$

Using our specified values:

$$90(10) + 30\frac{dP}{dt} = 0:$$

$$\frac{dP}{dt} = -30 \text{ lb/in.}^2/\text{s.} \quad \text{The}$$

negative indicates the pressure is decreasing.

20. A related rate problem is a problem involving one or more equations relating to changes with respect to time.

21. 1. Draw a figure and assign variables.
2. Relate these variables through one or more equations (or formulas).
3. Differentiate.
4. Substitute numbers.

22. $V = \frac{4}{3}\pi r^3$

$$\frac{dV}{dt} = 4\pi r^2 \frac{dr}{dt}$$

$$3 = 4\pi(2)^2 \frac{dr}{dt}$$

$$\frac{dr}{dt} = \frac{3}{16\pi} \approx 0.05968 \text{ in./s}$$

23. $V = \frac{4}{3}\pi r^3$

$$\frac{dV}{dt} = 4\pi r^2 \frac{dr}{dt}$$

$$-5 = 4\pi(4)^2 \frac{dr}{dt}$$

$$\frac{dr}{dt} = -\frac{5}{64\pi} \approx -0.025 \text{ in./min}$$

$$S = 4\pi r^2$$

$$\frac{dS}{dt} = 8\pi r \frac{dr}{dt}$$

$$\frac{dS}{dt} = 8\pi(4)\left(-\frac{5}{64\pi}\right) = -2.5 \text{ in.}^2/\text{min.}$$

24. $S = 4\pi r^2$

$$\frac{dS}{dt} = 8\pi r \frac{dr}{dt}$$

$$-3\pi = 8\pi(2)\frac{dr}{dt}$$

$$\frac{dr}{dt} = -\frac{3}{16} = -0.1675 \text{ cm/s}$$

Also $V = \frac{4}{3}\pi r^3$

$$\frac{dV}{dt} = 4\pi r^2 \frac{dr}{dt}$$

$$= 4\pi(2^2)(-\tfrac{3}{16}) = -3\pi \text{ cm}^3/\text{s}$$

25. $V = \frac{4}{3}\pi r^2$

$$\frac{dV}{dt} = 4\pi r^2 \frac{dr}{dt}$$

$$= 4\pi(4^2)(0.3) \approx 60.3 \text{ cm}^3/\text{min}$$

26. Let x be the length of the shadow and y be the distance of the person from the street light. Given $dy/dt = 5$ ft/s, and we wish to find dx/dt. Because of similar triangles,

$$\frac{x}{6} = \frac{x+y}{18}$$

$$y = 2x$$

$$\frac{dy}{dt} = 2\frac{dx}{dt}$$

$$5 = 2\frac{dx}{dt}$$

$$\frac{dx}{dt} = 2.5$$

The shadow is lengthening at 2.5 ft/s.

27. Let x be the distance from the foot of the ladder from the wall, y the distance to the top of the ladder from the ground. We are asked to find dx/dt at the instant when $dy/dt = -3$ and $x = 5$. The Pythagorean theorem is the general relationship here: $x^2 + y^2 = 13^2$. So

$$2x\frac{dx}{dt} + 2y\frac{dy}{dt} = 0$$

Now we need to have a value for y. Find y at this instant by $5^2 + y^2 = 13^2$; $y = 12$. Now find dx/dt:

$$2(5)\frac{dx}{dt} + 2(12)(-3) = 0; \quad \frac{dx}{dt} = 7.2 \text{ ft/s}$$

28. $v_c = 40$ mi/h; $v_t = 30$ mi/h; because the speed is constant, $s_c = 40t$ mi, $s_t = 30t$ mi. The distance is the hypotenuse of a right triangle,

$$D = 10t\sqrt{3^2 + 4^2} = 50t; \frac{dD}{dt} = 50 \text{ mi/h}$$

29. Let x be the radius of the top circle of the body of water and y its height. The radius of the bottom circle is 20, the height of the cone is 40 ft. The unfilled height of the cone is $40 - y$. By similar right triangles,

$$\frac{40 - y}{x} = \frac{40}{20}$$

$$2x = 40 - y$$

$$x = \frac{40 - y}{2}$$

If $y = 12$, $x = 14$. The volume of the body of water is

$$V = \frac{\pi}{3}(20^2)(40) - \frac{\pi}{3}(x^2)(40 - y)$$
$$= \frac{\pi}{3}(20^2)(40) - \frac{\pi}{12}(40 - y)^3$$
$$\frac{dV}{dt} = 0 - \frac{\pi}{4}(40 - y)^2(-\frac{dy}{dt})$$
$$80 = \frac{\pi}{4}(40 - 12)^2 \frac{dy}{dt}$$
$$\frac{dy}{dt} = \frac{80(4)}{\pi(28)^2} \approx 0.13 \text{ ft/min}$$

30. In Problem 29, the change of volume of the body of water is

$$\frac{dV}{dt} = \frac{\pi}{4}(40 - y)^2(\frac{dy}{dt})$$
$$80 - V_{out} = \frac{\pi}{4}(40 - 12)^2(0.05)$$
$$V_{out} = 80 - \frac{\pi}{4}(28)^2(0.05)$$
$$\approx 49.2 \text{ ft}^3/\text{min}$$

31. Let x be the horizontal distance from the boat to the pier and D the inclined distance from the boat to the rope puller. D is the hypotenuse of a right triangle with legs 12 and x.

$$12^2 + x^2 = D^2$$

$$0 + 2x \frac{dx}{dt} = (2D) \frac{dD}{dt}$$

$$2(16)\frac{dx}{dt} = 2D(-6)$$

At the instant when $x = 16$, $D = 20$:

$$\frac{dx}{dt} = -\frac{240}{32} = -7.5$$

The negative value means the distance is decreasing at the rate of 7.5 ft/min.

32. Let x be the horizontal distance from the boat to the pier and D the inclined distance from the boat to windlass. D is the hypotenuse of a right triangle with legs 4 and x.

$$x^2 + 4^2 = D^2$$
$$2x \frac{dx}{dt} = 2D \frac{dD}{dt}$$
$$\frac{dD}{dt} = \frac{x}{D}(2)$$

At the instant when $D = 5$, $x = 3$:
$$\frac{dD}{dt} = \frac{2(3)}{5} = 1.2$$

The rope is unwinding at the rate of 1.2 m/min.

33. $H(t) = -16t^2 + 160$ is the height of the ball at time t. The distance from the shadow to a point directly under the ball on the ground is x. $H(t)$ and x form a right triangle. The lamp is at 160 ft above the ground and at a horizontal distance $x + 10$ from the shadow of the ball. The line through the lamp, ball, and shadow is the hypotenuse of a right triangle similar to the one discussed above. Thus,

$$\frac{x + 10}{160} = \frac{x}{H}$$
$$xH + 10H - 160x = 0$$

If $t = 1$, $H(1) = 160 - 16 = 144$ so that $140x + 1,440 - 160x = 0$ which implies that $x = 90$:

$$x(-16t^2 + 160) + 10(-16t^2 + 160) - 160x = 0$$
$$-16xt^2 - 160t^2 + 1,600 = 0$$
$$-32xt - 16\frac{dx}{dt}t^2 - 320t = 0$$

With $t = 1$, $x = 90$,

$$-32(90)(1) - 16\frac{dx}{dt}(1)^2 - 320(1) = 0$$
$$\frac{dx}{dt} = -200 \text{ ft/s}$$

34. $$I = 40s^{-2}$$
$$\frac{dI}{dt} = -80s^{-3}\frac{ds}{dt}$$
$$s = 20 - 8 = 12, \frac{ds}{dt} = -4 \text{ m/s}$$
$$\frac{dI}{dt} = -80(12)^{-3}(-4) \approx 0.185 \text{ lux/s}$$

35. Use similar triangles:
$$\frac{s}{10 + L} = \frac{6}{L}$$
$$L = \frac{60}{s - 6}$$
$$\frac{dL}{dt} = -\frac{60}{(s - 6)^2}\frac{ds}{dt}$$

Since $s = 30 - 16t^2$ and $ds/dt = -32t$ so when $t = 1$ we know $s = 14$, and $ds/dt = -32$.
$$\frac{dL}{dt} = -\frac{60}{64}(-32) = 30 \text{ ft/s}$$

36. Let x be the distance between the race car and the finish line, then $dx/dt = 200$ km/h.

$$x = 20 \tan \theta$$
$$\frac{dx}{dt} = 20 \sec^2\theta \frac{d\theta}{dt}$$
$$200 = 20 \sec^2 0 \frac{d\theta}{dt}$$
$$10 = \frac{d\theta}{dt}$$

The angle is of the line of sight is changing at the rate of 10 km/hr.

37. At noon, the car is at the origin, while the truck is at $(250, 0)$. At time t, the truck is at position $(250 - x, 0)$, while the car is at $(0, y)$. Let H be the distance between them. Also, we are given $dx/dt = 25$, $dy/dt = 50$,

$x = 25t$ and $y = 50t$.

$H^2 = (250 - x)^2 + y^2$

$H^2 = (250 - 25t)^2 + 50^2 t^2$

$H^2 = 3,125t^2 - 12,500t + 62,500$

a. $2H \dfrac{dH}{dt} = 6,250t - 12,500$

$$\frac{dH}{dt} = \frac{3,125t - 6,250}{\sqrt{3,125t^2 - 12,500t + 62,500}}$$

b. $dH/dt = 0$ when $3,125t - 6,250 = 0$ or when $t = 2$.

c. $H^2 = 3,125(2)^2 - 12,500(2) + 62,500$

$\quad = 50,000$

$H = 100\sqrt{5} \approx 224$ mi   (reject negative)

38. Let y be the height of the balloon and s the distance between the balloon and the observer. When $y = 400$, $s = 500$, and $dy/dt = 10$ ft/s.

$$s^2 = 300^2 + y^2$$

$$2s \frac{ds}{dt} = 2y \frac{dy}{dt}$$

$$2(500) \frac{ds}{dt} = 3(400)(10)$$

$$\frac{ds}{dt} = 8 \text{ ft/s}$$

39. Let x denote the horizontal distance (in miles) between the plane and the observer. Let t denote the time (in hours), and draw a diagram representing the situation so that the angle is labeled θ and the distance is labeled D. It is given $dx/dt = -500$ mi/h. We are asked to find $d\theta/dt$ at the instant when $x = 4$ and $dx/dt = -500$.

$$\tan \theta = \frac{3}{x}.$$

$$(\sec^2 \theta) \frac{d\theta}{dt} = -\frac{3}{x^2} \frac{dx}{dt}$$

We note that $D = 5$ when $x = 4$ so we have

$$\left(\frac{5}{4}\right)^2 \frac{d\theta}{dt} = -\frac{3}{4^2}(-500)$$

$$\frac{d\theta}{dt} = 60 \text{ rad/hr} = 1 \text{ rad/min}$$

40. The velocity of the man walking is $dx/dt = -5$ ft/s.

$$x = (18 - 6)\cot \theta$$

$$\frac{dx}{dt} = -12 \csc^2 \theta \frac{d\theta}{dt}$$

If $x = 9$, $\csc \theta = 15/12 = 5/4$ so that

$$-5 = -12\left(\tfrac{5}{4}\right)^2 \frac{d\theta}{dt}$$

$$\frac{d\theta}{dt} = \frac{4}{15} \approx 0.27 \text{ rad/s}$$

41. $d\theta/dt = .25$ rev/h $= \dfrac{\pi}{2}$ rad/h

$$x = 2 \tan \theta$$

$$\frac{dx}{dt} = 2 \sec^2 \theta \frac{d\theta}{dt}$$

If $x = 1$, the hypotenuse is $\sqrt{2^2 + 1} = \sqrt{5}$ and $\sec \theta = \sqrt{5}/2$.

$$\frac{dx}{dt} = 2\left(\frac{\sqrt{5}}{2}\right)^2 \left(\frac{\pi}{2}\right) = \frac{5\pi}{4} \approx 3.927 \text{ mi/h}$$

42.   a.   Let y be the depth of the water and 10 by $(2 + 2x)$ the dimensions the horizontal surface of the water. Form a right triangle in the trapezoidal cross-section by drawing a line through an end of the 2 ft base perpendicular to the 5 ft base. Now draw in the height y for the (vertical) water level, and the horizontal dimension x for the water level portion in this triangle. These two right triangles reveal that

$$\frac{x}{y} = \frac{\dfrac{5 - 2}{2}}{2}$$

$$x = \frac{3}{4}y$$

The volume of water in the trough is

$$V = \frac{2 + 2x + 2}{2}(y)(10)$$

$$= 10(2 + x)y$$

$$= 10(2 + \tfrac{3}{4}y)y$$

$$= 7.5y^2 + 20y$$

b.

$$V = 7.5y^2 + 20y$$

$$\frac{dV}{dt} = 15y \frac{dy}{dt} + 20 \frac{dy}{dt}$$

$$10 = [15(1) + 20] \frac{dy}{dt}$$

$$\frac{dy}{dt} = \frac{2}{7} \approx 0.28 \text{ ft/m}$$

$$\approx 3.5 \text{ in./m}$$

43. Draw a figure using the variables of A for the distance traveled by the first ship, B for the distance traveled by the second ship, D for the distance between them, and the constant angle of 60°. We are asked to find dD/dt at $t = 2$ and $t = 5$. As the problem suggests, these variables are all generally related by the law of cosines:

$$D^2 = A^2 + B^2 - 2AB \cos \theta.$$

$$2D \frac{dD}{dt} = 2A \frac{dA}{dt} + 2B \frac{dB}{dt} - \left(A \frac{dB}{dt} + B \frac{dA}{dt}\right)$$

(Note that $\cos \theta = \tfrac{1}{2}$ is a constant.)

At $t = 2$ $A = 16$, $B = 12$, $\frac{dA}{dt} = 8$, $\frac{dB}{dt} = 12$,

$$D^2 = 16^2 + 12^2 - 2(16)(12)(\tfrac{1}{2})$$

$$D = \sqrt{208}$$

$$2\sqrt{208}\,\frac{dD}{dt} = 2(16)(8) + 2(12)(12)$$

$$- [16(12) + 12(8)]$$

$$\frac{dD}{dt} = \frac{128}{\sqrt{208}} = \frac{32\sqrt{13}}{13} \approx 8.875 \text{ knots}$$

At $t = 5$, $A = 40$, $B = 48$, $\frac{dA}{dt} = 8$, $\frac{dB}{dt} = 12$,

$$D^2 = 40^2 + 48^2 - 2(40)(48)(\tfrac{1}{2})$$

$$D = \sqrt{1984}$$

$$2\sqrt{1984}\,\frac{dD}{dt} = 2(40)(8) + 2(48)(12)$$

$$- [40(12) + 48(8)]$$

$$\frac{dD}{dt} = \frac{464}{\sqrt{1984}} = \frac{58\sqrt{31}}{31} \approx 10.417 \text{ knots}$$

44. Let y be the depth of the water at the deep end and 25 by x the dimensions of the horizontal rectangular water surface. The volume of the water is (for $y \le 12$)

$$V = \frac{25xy}{2}$$

In a vertical cross-section, the right triangle with legs x and y is similar to the right triangle with legs 60 and 12. Thus,

$$\frac{x}{60} = \frac{y}{12}$$

$$x = 5y$$

By substitution, we have

$$V = \frac{125y^2}{2}$$

Now $y = 5$, $dV/dt = 800$ ft^3/min, we have

$$\frac{dV}{dt} = 125y\,\frac{dy}{dt}$$

$$800 = 125(5)\,\frac{dy}{dt}$$

$$\frac{dy}{dt} = \frac{32}{25} = 1.28 \text{ ft/min}$$

45. Consider the figure at the top of the next column. From similar triangles $\triangle AFG$ and $\triangle ABC$:

$$\frac{h_1}{1} = \frac{h_1 - 1}{\frac{3}{4}}$$

$$h_1 = 4$$

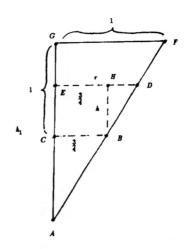

From $\triangle ADE$ and $\triangle AFG$, we have

$$r = \tfrac{1}{4}(3 + h)$$

The volume of water is that of a frustum with height h and radii r and 0.75

$$V = \frac{\pi h}{3}\big[r^2 + 0.75r + 0.75^2\big]$$

$$= \frac{\pi}{48}(h^3 + 9h^2 + 27h)$$

$$\frac{dV}{dt} = \frac{\pi}{48}(3h^2 + 18h + 2)\frac{dh}{dt}$$

When $h = \frac{1}{2}$, $dV/dt = 8$ in.^{3}/min or 1/216 ft^3/min. Now,

$$\frac{1}{216} = \frac{\pi}{48}[3(\tfrac{1}{2})^2 + 18(\tfrac{1}{2}) + 27]\frac{dh}{dt}$$

$$\frac{dh}{dt} = -\frac{8}{1{,}323\pi} \approx 0.001924 \text{ ft/min}$$

or 0.0233 in.min

46. **a.** Consider the right triangle with legs 2 and $s(t)$. The acute angle at the vertex with the lighthouse is θ, and $\theta = 6\pi t$.

$$\frac{s(t)}{2} = \tan\theta$$

$$s(t) = 2\tan\theta$$

$$s(t) = 2\tan 6\pi t$$

b. When the point on the cliff is 4 km from the lighthouse

$$s = \sqrt{16 - 4} = 2\sqrt{3}$$

and the corresponding value of θ is

$$\theta = \cos^{-1}(\tfrac{2}{4}) = \frac{\pi}{3}$$

Therefore,
$$\frac{ds}{dt} = 2(6\pi)\sec^2 6\pi t = 12\pi(4) = 48\pi$$
This is about 150.8 km/h.

47. a.
$$\frac{x}{150} = \cot\theta$$
$$x = 150\cot\theta$$
$$\frac{dx}{dt} = -150\csc^2\theta\,\frac{d\theta}{dt}$$
$$\frac{d\theta}{dt} = -\frac{1}{150}\sin^2\theta\,\frac{dx}{dt} \approx \frac{4}{15}\sin^2\theta$$

b. If $x \to 0$, $\theta \to \frac{\pi}{2}$, and $\sin\theta \to 1$; thus,
$$\lim_{\theta\to 0}\left|\frac{d\theta}{dt}\right| = 0.27$$
$d\theta/dt$ approaches 0.27 rad/s

c. $\frac{d\theta}{dt} = \frac{v}{150}$; as v increases so will $d\theta/dt$ and it becomes more difficult to see the seals.

2.8 Differentials and Tangent Line Approximations, Page 152

1. a. $\sqrt{17} \approx 4.123105626$

b. $y' = \frac{1}{2\sqrt{x}}$; $x_0 = 16$; $y_0 = 4$; $\Delta x = 1$
$$\Delta y = \frac{1}{2\sqrt{16}}(1) = 0.125$$
$$\sqrt{17} \approx 4.00 + 0.125 = 4.125$$

2. a. $\sqrt[3]{28} \approx 3.036588968$

b. $y' = \frac{1}{\sqrt[3]{x^2}}$; $x_0 = 27$; $y_0 = 3$; $\Delta x = 1$
$$\Delta y = \frac{1}{3\sqrt[3]{27^2}}(1) = 0.037$$
$$\sqrt[3]{28} \approx 3.00 + 0.037 = 3.037$$

3. a. $\sqrt[4]{15} \approx 1.967989671$

b. $y' = \frac{1}{4x^{3/4}}$; $x_0 = 16$; $y_0 = 2$; $\Delta x = -1$
$$\Delta y = \frac{1}{(4)(8)}(-1) = -0.03$$
$$\sqrt[4]{15} \approx 2.00 - 0.03 = 1.97$$

4. a. $\sqrt[4]{26} \approx 2.962496065$

b. $y' = \frac{1}{3x^{2/3}}$; $x_0 = 27$; $y_0 = 3$; $\Delta x = -1$
$$\Delta y = \frac{1}{(3)(9)}(-1) = -0.037$$
$$\sqrt[3]{26} \approx 3.00 - 0.037 = 2.963$$

5. a. $\frac{1}{\sqrt{4.5}} \approx 0.47140452098$

b. $y' = \frac{-1}{2x^{3/2}}$; $x_0 = 4$; $y_0 = 0.5$; $\Delta x = 0.5$
$$\Delta y = \frac{-0.5}{(2)(8)}(0.5) = -0.03$$
$$\frac{1}{\sqrt{4.5}} \approx 0.5 - 0.03 = 0.47$$

6. a. $\frac{1}{\sqrt[3]{7.8}} \approx 0.50423749$

b. $y' = \frac{-1}{3x^{4/3}}$; $x_0 = 8$; $y_0 = 0.5$; $\Delta x = -0.2$
$$\Delta y = \frac{-1}{(3)(16)}(-0.2) = 0.0042$$
$$\frac{1}{\sqrt[3]{7.8}} \approx 0.5 + 0.0042 = 0.5042$$

7. a. $(1.1)^4 \approx 1.4641$

b. $y' = 4x^3$; $x_0 = 1.0$; $y_0 = 1.0$; $\Delta x = 0.1$
$$\Delta y = 4(1)(0.1) = 0.4$$
$$(1.1)^4 \approx 1.00 + 0.4 = 1.4$$

8. a. $(1.9)^5 \approx 24.76099$

b. $y' = 5x^4$; $x_0 = 2.0$; $y_0 = 32$; $\Delta x = -0.1$
$$\Delta y = 5(16)(-0.1) = -8$$
$$(1.9)^4 \approx 32 - 8 = 24$$

9. a. $\sqrt{0.99} \approx 0.9949874371$

b. $y' = \frac{1}{2\sqrt{x}}$; $x_0 = 1.0$; $y_0 = 1.0$; $\Delta x = -0.01$
$$\Delta y = \frac{1}{2\sqrt{1}}(-0.01) = -0.05$$
$$\sqrt{0.99} \approx 1.0 - 0.005 = 0.995$$

10. a. $0.99^{100} = 0.366032341$

b. $y' = 100x^{99}$; $x_0 = 1.0$; $y_0 = 1.0$;
$$\Delta x = -0.01; \Delta y = 100(1^{99})(-0.01) = -1$$
$$\sqrt{0.99} \approx 1.0 - 1.0 = 0$$

11. a. $\cos(\frac{\pi}{2} + 0.01) \approx -0.00999983333$

b. $y' = -\sin x$; $x_0 = \frac{\pi}{2}$; $y_0 = 0$; $\Delta x = 0.01$
$$\Delta y = (-\sin\frac{\pi}{2})(0.01) = 0.01$$
$$\cos(\frac{\pi}{2} + 0.01) \approx 0 - 0.01 = -0.01$$

12. a. $\sin(\frac{\pi}{2} - 0.01) \approx -0.99999995$

b. $y' = \cos x$; $x_0 = \frac{\pi}{2}$; $y_0 = 1$; $\Delta x = -0.001$
$$\Delta y = (\cos\frac{\pi}{2})(-0.01) = 0$$
$$\sin(\frac{\pi}{2} - 0.01) \approx 1 - 0 = 1.0$$

13. a. $\sin(\frac{\pi}{4} - 0.001) \approx 0.706399321$

b. $y' = \cos x$; $x_0 = \frac{\pi}{4}$; $y_0 = \frac{\sqrt{2}}{2}$

$\Delta x = -0.001$

$\Delta y = (\sin \frac{\pi}{4})(-0.001) = \frac{\sqrt{2}}{2}(-0.001)$

$\sin(\frac{\pi}{4} - 0.001) \approx \frac{\sqrt{2}}{2}(1 - 0.001)$

≈ 0.7063

14. a. $\cos(\frac{\pi}{3} + 0.001) \approx 0.49913372433$

b. $y' = -\sin x$; $x_0 = \frac{\pi}{3}$; $y_0 = 0.5$;

$\Delta x = 0.001$

$\Delta y = (-\sin \frac{\pi}{3})(0.001) = -0.000866025$

$\cos(\frac{\pi}{3} + 0.0001) \approx 0.5 - 0.000866025$

$= 0.49913397$

15. $f(x) = x^2 + x - 2$; $f'(x) = 2x + 1$

$f'(1) = 2 + 1 = 3$; $\Delta x = 0.001$

$\Delta f = f(x_0 + \Delta x) - f(x_0)$

$\approx f'(x_0)\Delta x = f'(1)(0.001)$

$= 3(0.001) = 0.003$

16. $f(x) = 3x^2 - x + 2$; $f'(x) = 6x - 1$

$f'(2) = 12 - 1 = 11$; $\Delta x = 0.01$

$\Delta f = f(x_0 + \Delta x) - f(x_0)$

$\approx f'(x_0)\Delta x = f'(2)(0.01)$

$= 11(0.01) = 0.11$

17. $f(x) = x^{10} + x^5$; $f'(x) = 10x^9 + 5x^4$

$f'(1) = 10 + 5 = 15$; $\Delta x = -0.01$

$\Delta f = f(x_0 + \Delta x) - f(x_0)$

$\approx f'(x_0)\Delta x = f'(1)(-0.01)$

$= 15(-0.01) = -0.15$

18. $f(x) = 3x^7 + x^4$; $f'(x) = 21x^6 + 4x^3$

$f'(1) = 21 + 4 = 25$; $\Delta x = -0.001$

$\Delta f = f(x_0 + \Delta x) - f(x_0)$

$\approx f'(x_0)\Delta x = f'(1)(-0.001)$

$= 25(-0.001) = -0.025$

19. $f(x) = \sqrt{x} + x^{3/2}$; $f'(x) = \frac{1}{2\sqrt{x}} + \frac{3}{2}\sqrt{x}$

$f'(4) = \frac{1}{4} + 3 = 3.25$; $\Delta x = 0.001$

$\Delta f = f(x_0 + \Delta x) - f(x_0)$

$\approx f'(x_0)\Delta x = f'(4)(0.001)$

$= 3.25(0.001) = 0.00325$

20. $f(x) = x^{1/3} + x^2$; $f'(x) = \frac{1}{3}x^{-2/3} + 2x$

$f'(8) = \frac{1}{12} + 16 = 16.083333$; $\Delta x = -0.01$

$\Delta f = f(x_0 + \Delta x) - f(x_0)$

$\approx f'(x_0)\Delta x = f'(8)(-0.01)$

$= 16.083333(-0.01) = -0.1608333$

21. $d(2x^3) = 6x^2 \, dx$

22. $d(3 - 5x^2) = -10x \, dx$

23. $d(2\sqrt{x}) = x^{-1/2} \, dx$

24. $d(x^5 + \sqrt{x^2 + 5}) = 5x^4 + \frac{1}{2}(x^2 + 5)^{-1/2}(2x) \, dx$

$= [x^4 + x(x^2 + 5)^{-1/2}] \, dx$

25. $d(x \cos x) = (\cos x - x \sin x) dx$

26. $d(x \sin 2x) = [x(\cos 2x)(2) + \sin 2x] \, dx$

$= (2x \cos 2x + \sin 2x) \, dx$

27. $d\left(\frac{\tan 3x}{2x}\right) = \frac{x(3 \sec^2 3x) - (\tan 3x)(1)}{2x^2}$

$= \frac{3x \sec^2 3x - \tan 3x}{2x^2} \, dx$

28. $d\left(\frac{x^2 \sec x}{x - 3}\right)$

$= \frac{(x - 3)(x^2 \sec x \tan x + 2x \sec x) - (x^2 \sec x)(1)}{(x - 3)^2} \, dx$

29. $d\left(x\sqrt{x^2 - 1}\right) = \left[\frac{x^2}{(x^2 - 1)^{1/2}} + (x^2 - 1)^{1/2}\right] dx$

$= \frac{2x^2 - 1}{(x^2 - 1)^{1/2}} \, dx$

30. $d\left(\frac{x - 5}{\sqrt{x + 4}}\right)$

$= \frac{\sqrt{x + 4}(1) - (x - 5)(\frac{1}{2})(x + 4)^{-1/2}}{x + 4} \, dx$

$= \frac{2x + 8 - x + 5}{2(x + 4)^{3/2}} \, dx = \frac{x + 3}{2(x + 4)^{3/2}} \, dx$

31. $dx = \Delta x$; $dy = f'(x) dx$

32. Propagation of error describes the accumulation of errors due to successive approximations. Relative error is df/f. Percentage error is relative error times 100.

33. Let $f(x) = x^5 - 2x^3 + 3x^2 - 2$;

$f'(x) = 5x^4 - 6x^2 + 6x$;

$x_0 = 3$ and $\Delta x = dx = 0.01$.

Now, $f(x_0 + \Delta x) \approx f(x_0) + f'(x_0) dx$, so

$f(3.01) \approx f(3) + f'(3) dx$

$$= [(3)^5 - 2(3)^3 + 3(3)^2 - 2]$$
$$+ [5(3)^4 - 6(3)^2 + 6(3)](0.01)$$
$$= 214 + 3.69 = 217.69$$

Comparing this to a calculator value of $f(3.01) = 217.7155882$ we see an error of approximately 0.0255882.

34. Let $f(x) = x^{1/4} + x^{1/3} + 3x^{1/2}$

$f'(x) = \frac{1}{4}x^{-3/4} + \frac{1}{3}x^{-2/3} + \frac{3}{2}x^{-1/2}$

$x_0 = 4,096$ and $\Delta x = dx = 4$

Now, $f(x_0 + \Delta x) \approx f(x_0) + f'(x_0)dx$, so

$f(4,100) \approx f(4,096) + f'(4,096)dx$

$= [(4,096)^{1/4} + (4,096)^{1/3} + 3(4,096)^{1/2}]$

$+ [\frac{1}{4}(4,096)^{-3/4} - \frac{1}{3}(4,096)^{-2/3} + \frac{3}{2}(4,096)^{1/2}](4)$

$= [8 + 16 + 3(64)] + [\frac{3}{128} + \frac{1}{768} + \frac{1}{2,048}](4)$

$= 216 + \frac{155}{6,144}(4) \approx 216.1009$

Comparing this to a calculator value of $f(4,100) = 216.100886171$ we see an error of less than 0.001.

35. $A(r) = \pi r^2$; $A'(r) = 2\pi r$;

$\left|\frac{dA}{A}\right| = \left|\frac{2\pi r \, dr}{\pi r^2}\right| \approx 0.06$ or 6%

36. $V(r) = 4.5\pi r^2$; $V' = 4\pi r$;

$\left|\frac{dV}{V}\right| = \left|\frac{9\pi r \, dr}{4.5\pi r^2}\right| \approx 0.02$ or 2%

37. $V = \frac{4}{3}\pi r^3$; $V' = 9\pi r^2$;

$\left|\frac{dV}{V}\right| = \left|\frac{3(4\pi r^2) \, dr}{4\pi r^3}\right| = 3\left|\frac{dr}{r}\right| = 0.03$ or 3%

38. Since the circulation will be

$$C(t) = 100t^2 + 400t + 5,000$$

papers t years from now, the increase in circulation during the next 6 months $(\Delta t = 0.5)$ will be

$$\Delta C = C(0.5) - C(0) \approx C'(0)(0.5)$$

Because $C'(t) = 200t + 40$ and $C'(0) = 400$, it follows that

$$\Delta C \approx 400(0.5) = 200 \text{ newspapers}$$

39. Because the average level of carbon monoxide in the air t years from now will be

$$Q(t) = 0.05t^2 + 0.1t + 3.4$$

parts per million, the change in the carbon monoxide level during the next six months

$(\Delta t = 0.5)$ will be

$$\Delta Q = Q(0.5) - Q(0) \approx Q'(0)(0.5)$$

Since $Q'(t) = 0.1t + 0.1$ and $Q'(0) = 01$, it follows that

$$\Delta Q \approx 0.1(0.5) = 0.05 \text{ parts per million}$$

40. Because the cost is

$$C(q) = 0.1q^3 - 0.5q^2 + 500q + 200$$

the change in cost resulting from a decrease in production from 4 units to 3.9 units $(\Delta q = -0.1)$ is

$$\Delta C = C(3.9) - C(4) \approx C'(4)(-0.1)$$

Since $C'(q) = 0.3q^2 - q + 500$ and $C'(4) = 500.80$, it follows that

$$\Delta Q = 500.80(-0.1) = -50.08$$

That is, the cost will decrease by approximately $50.08.

41. $Q(L) = 60,000L^{1/3}$; $Q'(L) = 20,000L^{-2/3}$;

$L_0 = 1,000$; $\Delta L = -60$

$\Delta Q \approx Q'(L_0)\Delta L$

$= 20,000(1000)^{-2/3}(-60)$

$= -12,000 \text{ units}$

42. $S(R) = 1.8 \times 10^5 R^2$, $R = 1.2 \times 10^{-2}$, and

$\Delta R = \pm 5 \times 10^{-4}$.

$\Delta S = S(1.2 \times 10^{-2} \pm 5 \times 10^{-4})$

$- S(1.2 \times 10^{-2})$

$\approx S'(1.2 \times 10^{-2})(\pm 5 \times 10^{-4})$

$S'(R) = 3.6 \times 10^5 R$

$S'(1.2 \times 10^{-2}) = (3.6 \times 10^5)(1.2 \times 10^{-2})$

$= 4.32 \times 10^3$

Thus,

$\Delta'S = (4.3 \times 10^3)(\pm 5 \times 10^{-4})$

$= \pm 2.15 \text{ cm/s}$

43. Because $V = \frac{4}{3}\pi r^3$ and $dV = 4\pi r^2 dR$ when $R = 8.5/2$ and $\Delta R = dR = 1/8$, we have

$dV = 4\pi(4.25)^2(\frac{1}{8}) \approx 28.37 \text{ in.}^3$

44. Let x be the length of an edge of the cube. The top and bottom of the box are, respectively, $T = x^2$, $B = x^2$. The sides are $S = 4x^2$. The cost is 2¢/in.^{2} for the sides, 3¢/in.^{2} for the bottom, and 4¢/in.^{2} for the lid.

Section 2.8, Differentials and Tangent Line Approximations
Page 75

With $\Delta x = 1$,

$\Delta C = (2x)(4 + 3 + 8)\big|_{x=3} = 40(15) = 600$

cent; that is, $\Delta C = \$6.00$. The actual cost
for 20 in. sides is $400(4 + 3 + 8) = 6,000$, the
cost for 21 in. sides is $441(15)= 6,615$. The
actual increase is 615 cents $= \$6.15$.

45. Let $P(x)$ be the pulse in beats/min.

$P(x) = \dfrac{596}{\sqrt{x}}; \ P'(x) = -\dfrac{298}{x^{3/2}}; \ x_0 = 59;$

$\Delta x = 1$, so that

$\Delta P \approx P'(x_0)\,\Delta x = -\dfrac{298}{59^{3/2}}(1) \approx -0.658$

beats/min; or about 2 beats every 3 minutes.

46. Let $C(t)$ be the concentration of the drug.

$C(t) = \dfrac{0.12t}{t^2 + t + 1}$

$C'(t) = (0.12)\dfrac{(t^2 + t + 1)(1) - 2t^2 - t}{(t^2 + t + 1)^2}$

$\quad = (0.12)\dfrac{-t^2 + 1}{(t^2 + t + 1)^2}$

$C'(0.5) = (0.12)\dfrac{0.75}{(1.75)^2} \approx 0.02939$

$\Delta t = \dfrac{5}{60} = \dfrac{1}{12}; \ \Delta C = C'(0.5)\Delta t \approx 0.002$

47. Let $S(R)$ be the speed of the blood.

$S(R) = cR^2$

$S'(R) = 2cR$

Now $\Delta R = \pm 0.01R$ and the percentage rate
change is

$\Delta S = \dfrac{100S'(R)\Delta R}{S(R)} = \dfrac{100[2cR(\pm 0.01R)]}{cR^2}$

$\quad = \pm\dfrac{2cR^2}{cR^2} = \pm 2\%$

48. Let T be the period of the pendulum.
$\Delta L = 0.4\%$ of L or $0.004L$.

$\Delta T = 2\pi(\tfrac{1}{2})\sqrt{\dfrac{g}{L}}\,\Delta L$

The percentage rate of change in T is

$\dfrac{100\Delta T}{T} = \dfrac{(100)(2)\pi(0.5)g^{1/2}L^{-1/2}\Delta L}{2\pi g^{-1/2}L^{1/2}}$

$\quad = \dfrac{50g\Delta L}{L} = 6.4\%$

49. Let N be the number of alpha particles falling
on a unit area of the screen.

$N(\theta) = \dfrac{1}{\sin^4(\frac{\theta}{2})}$

$N'(\theta) = -4[\sin^{-5}(\tfrac{\theta}{2})\cos(\tfrac{\theta}{2})]\tfrac{1}{2}$

$\quad = -2\sin^{-5}(\tfrac{\theta}{2})\cos(\tfrac{\theta}{2});$

With $\theta = 1$ and $\Delta\theta = 0.1$

$\Delta N \approx N'(\theta)\Delta\theta$

$\quad = -2[\sin^{-5}(0.5)\cos(0.5)](0.1)$

$\quad \approx -6.93$ (or about 7) particles/unit area.

50. a. Let $C(q)$ be the cost of manufacturing q
units.

$C(q) = 3q^2 + q + 500$

$C'(q) = 6q + 1; \ C'(40) = 241$

b. $C(41) - C(40)$

$= [3(41)^2 + 41 + 500] - [3(40)^2 + 40 + 500]$

$= 244$

51. a. Since the cost is

$C(q) = 0.1q^3 - 5q^2 + 500q + 200$

$C'(q) = 0.3q^2 - 10q + 500$

The cost of producing the fourth unit is

$C'(3) = 0.3(3^2) - 30 + 500 = 472.7$

b. The actual cost of manufacturing the
fourth unit is

$C(4) - C(3)$

$= 0.1(4^3 - 3^3) - 5(4^2 - 3^2) + 500(4 - 3)$

$= 468.70$

52. a. Let $C(x)$ be the total cost of producing x
units and $p(x)$ the selling price. The
marginal cost is

$C'(x) = \tfrac{2}{7}x + 4$

b. If $10 = \tfrac{2}{7}x + 4$, then $x = 21$ and

$P(21) = \tfrac{1}{4}(80 - 21) = 14.75$

c. The cost of producing the eleventh unit is
approximately

$C'(10) = \dfrac{20}{7} + 4 = 6.86$

d. The actual cost of producing the 11th
unit is
$C(11) - C(10) = \dfrac{11^2 - 10^2}{7} + 4(11 - 10) = 7$

53. a. Marginal cost is: $C'(x) = \tfrac{4}{5}x + 3$

b. $\tfrac{4}{5}x + 3 = 23; \ x = 25;$

$p(25) = \tfrac{1}{5}(20) = \$4.00$

c. The cost of producing the 11th unit is

approximately

$$C'(10) = \tfrac{4}{5}(10) + 3 = \$11.00$$

d. The actual cost of producing the 11th unit is

$$C(11) - C(10) = [\tfrac{2}{5}(11)^2 + 3(11) + 10]$$
$$- [\tfrac{2}{5}(10)^2 + 3(10) + 10]$$
$$= 91.4 - 80 = \$11.40$$

54. Let $Q(L)$ be the daily output.

$$Q'(L) = 120L^{-2/3}$$
$$Q'(1,000) = \frac{120}{1,000^{2/3}} = 1.2$$
$$\Delta Q \approx 1.2(1) = 1.2 \text{ units}$$

55. a. Let $P(x)$ be the profit when producing x units at a total cost of $C(x)$ and revenue of $R(x)$.

$$P(x) = R(x) - C(x)$$
$$P'(x) = R'(x) - C'(x)$$
$$P'(x) = 0 \text{ if } C'(x) = R'(x).$$

b. $A'(x) = \dfrac{xC'(x) - C(x)}{x^2} = \dfrac{1}{x}\left[C'(x) - \dfrac{C(x)}{x}\right]$

$A'(x) = 0$ if $C'(x) = A(x)$.

56. a. $V = \tfrac{4}{3}\pi r^3$; $\Delta r = \pm 0.125$. The propagated error (for $+0.125$) is

$$\Delta V = V(14.125) - V(14)$$
$$= \frac{4\pi(14.125^2 - 14^3)}{3} \approx 310.6 \text{ in.}^3$$

b. $\Delta V = V(14.125) - V(14)$

$$\approx V'(14)(\pm 0.1255)$$

$V'(r) = 4\pi r^2$, so

$$V'(14) = 4\pi 14^2 \approx 2,463 \text{ and}$$

$$\Delta V \approx 2,463(\pm 0.125) \approx \pm 307.9 \text{ in.}^3$$
In order for the result to be 2 in.,

$$\Delta V = 4\pi r^2 \Delta r = 2 \text{ or } \Delta r = \frac{2}{2\pi(14^2)}$$
$$\approx 0.000812. \text{ Thus,}$$

$$13.99919 < r < 14.00081$$

57. a. $f(x) = x^{1/2}$; $f'(x) = \tfrac{1}{2}x^{-1/2}$

With $\Delta x = h$, $f'(1) = \tfrac{1}{2}$ so

$$f(1 + h) \approx f(1) + f'(1)h = 1 + \frac{h}{2}$$

b. Let $f(x) = \tfrac{1}{x} = x^{-1}$; $f'(x) = -x^{-2}$

$$f(1 + h) \approx f(1) + f'(1)h = 1 - h$$

58. $f(x) = \sqrt[n]{x^n}$ so $f'(x) = \dfrac{1}{nx^{1-1/n}}$

$$f'(A^n) = \frac{1}{n(A^n)^{1-1/n}} = \frac{1}{nA^{n-1}}$$
$$f(A^n + h) \approx f(A^n) + f'(A^n)h$$
$$= A + \frac{h}{nA^{n-1}}$$

59. $f(x) = \sqrt{x}$ and $f'(x) = \dfrac{1}{2x^{1/2}}$; $\Delta x = -3$

$$f(97) = f(100 - 3) = f(100) + f'(100)(-3)$$
$$= 10 - \frac{3}{2\sqrt{100}} = 9.85$$

Calculator check: $\sqrt{97} \approx 9.848857802$

If $\Delta x = 16$,

$$f(97) = f(81 + 16)$$
$$= f(81) + f'(81)(16)$$
$$= 9 + \frac{16}{2\sqrt{81}} \approx 9.89$$

2.9 The Newton-Raphson Method for Approximating Roots, Page 158

1. $f(x) = x^3 - 29$; $f'(x) = 3x^2$;

$$x_{n+1} = x_n - \frac{f(x_n)}{f'(x_n)} = x_n - \frac{x_n^3 - 29}{3x_n^2}$$

So $x_{n+1} = \dfrac{2x_n^3 + 29}{3x_n^2}$.

Letting $x_0 = 3$, $x_1 \approx 3.074\,074\,074$, $x_2 \approx 3.072\,317\,830$, $x_3 \approx 3.072\,316\,825$. At this point we can be confident of the value to the nearest 0.00001 as 3.072 31. Compare this to the exact root of $\sqrt[3]{29}$.

For most of the problems in this section, we used computer software. Even though such software is, of course, optional, it can help us through much of the tedium of some of these problems. The output shown is from *Converge* 4.0 available from JEMware, The Kawaiahao Plaza Executive Center, 567 South King Street, Suite 178, Honolulu, Hawaii 96813. Phone: 808-523-9911.

$x_3 = 3.07232$

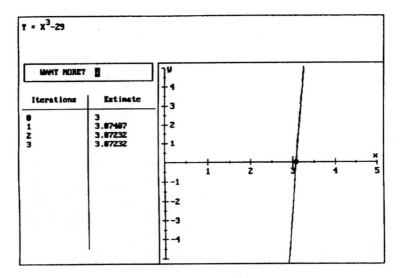

2. $x_3 = 3.07232$

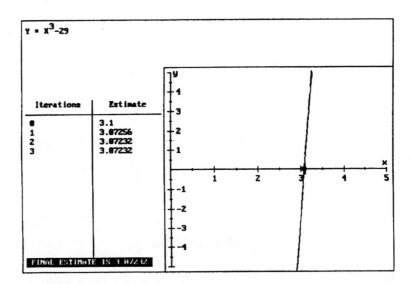

3. $x_3 = 1.44281$

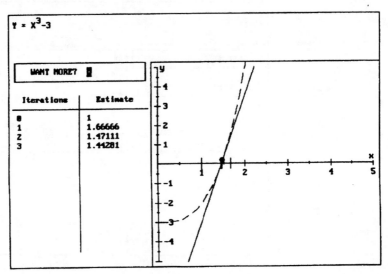

4. $x_3 = 2.01235$

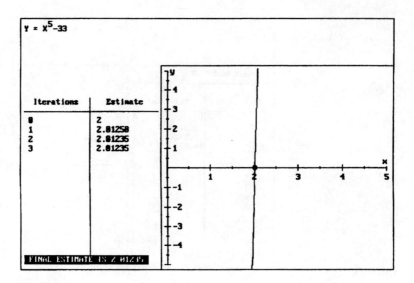

5. $x_3 = 2.00622$

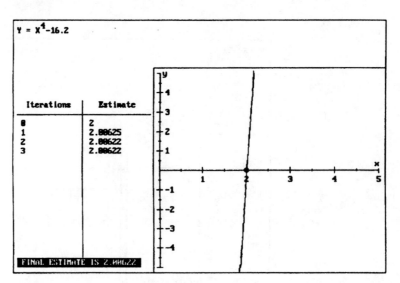

6. $x_3 = 2.57133$

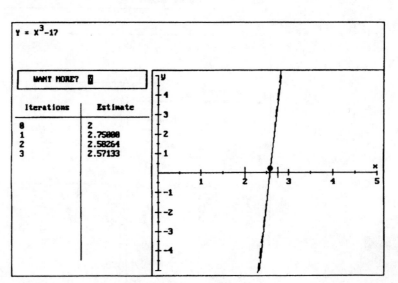

7. $x_3 = -0.000096$

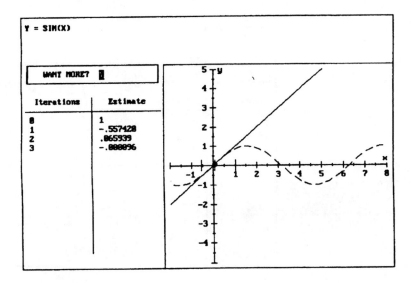

8. $x_3 = 0.000000$

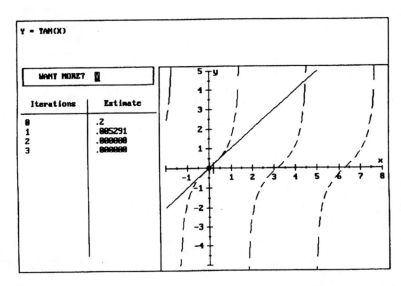

9. $x_3 = 0.453398$

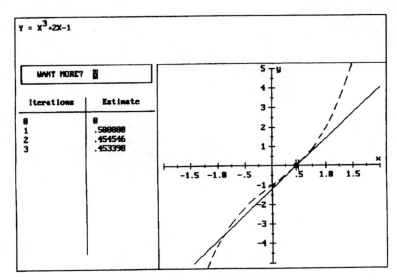

10. $x_3 = -1.45384$

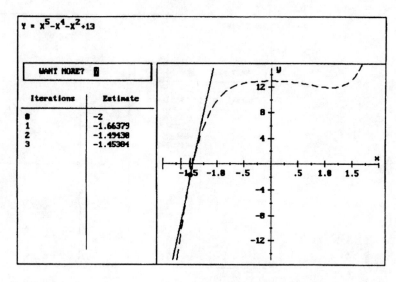

11.   Find where a tangent line close to a root crosses the x-axis. Draw the tangent line at this new x-value and look to see where this tangent line crosses the x-axis. Use this point of intersection as the next estimate of the root to close in on the root. This method may fail if the curve is discontinuous near a root. If the tangent line crosses the x-axis near another root, the process may not converge.

12. $x_4 = 2.53209$

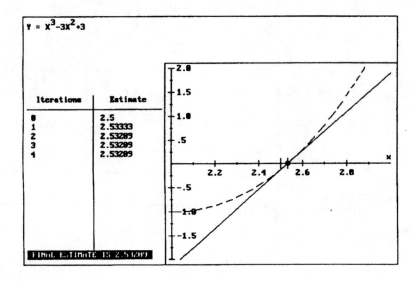

13. $x_3 = 0.322185$

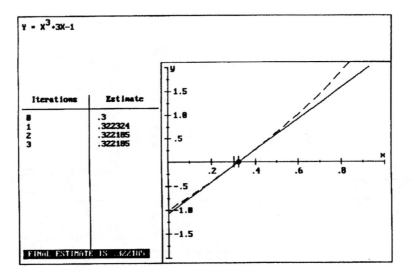

14. $x_4 = 0.453398$

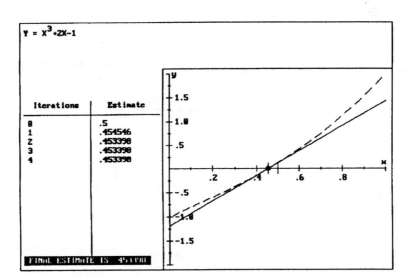

15. $x_4 = -2.54682$

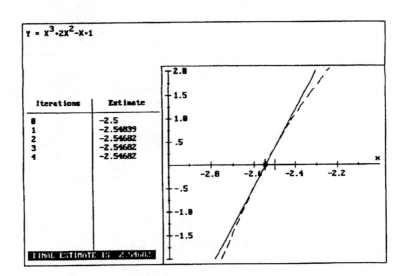

16. $x_3 = 2.67170$

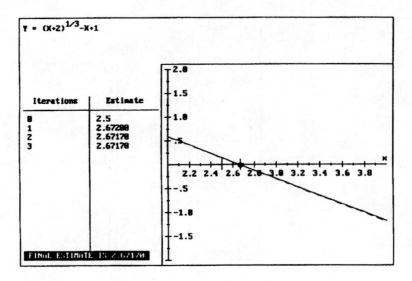

17. $x_3 = -2.79632$

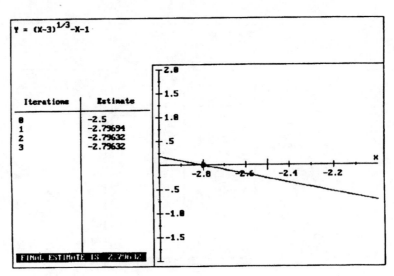

18. $x_4 = 1.28343$

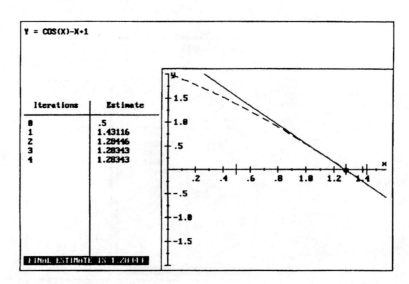

19. $x_3 = 0.876726$

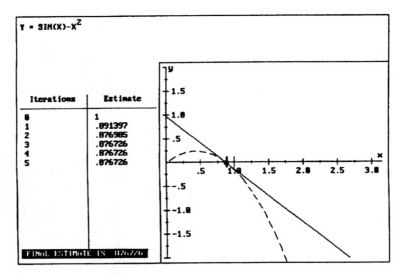

20. $f(x) = 9x^2 + 2x + 3$; $f'(x) = 18x + 2 = 2(9x + 1)$

$f(-\frac{1}{9}) = 0$, so $x_2 = -\frac{1}{9}$ and the tangent line to the graph at $x = -\frac{1}{9}$ is horizontal.

21. $f(x) = 1 - \frac{1}{x}$; $f'(x) = \frac{1}{x^2}$; $x_{n+1} = x_n - \frac{f(x_n)}{f'(x_n)} = x_n - \frac{1 - 1/x_n}{1/x_n^2} = -x_n^2 + 2x_n$

If we let $x_0 = 2$ then $x_1 = 0$ and all other iterations are equal to 0. This would seem to indicate that $x = 0$ is a root, but $f(x)$ is not defined at $x = 0$. More formally,

$$\left| \frac{f(x)f''(x)}{[f'(x)]^2} \right| = \left| \frac{(1 - x^{-1})(-2x^{-3})}{x^{-4}} \right| = \left| \frac{-2x^{-3} + 2x^{-4}}{x^{-4}} \right| = |-2x + 2| = 2|x - 1|$$

When $x_0 = 2$, $2|x - 1|$ is not less than 1, so Newton's method fails.

22.   Let $x_0 = a^6$, $a > 0$. $f(x) = \sqrt[6]{x}$, $f(a^6) = a$, $f'(x) = \frac{1}{6x^{5/6}}$; $f'(a^6) = \frac{1}{6a^5}$

$x_1 = a^6 - \frac{a}{\frac{1}{6}a^5} = a^6 - 6a^6 < 0$. Thus $f(x_2)$ is not defined.

23. $f(x) = 9x^4 - 16x^3 - 36x^2 + 96x - 60$; $f'(x) = 36x^3 - 48x^2 - 72x + 96$

$f'(\frac{4}{3}) = 36(\frac{4}{3})^3 - 48(\frac{4}{3})^2 - 72(\frac{4}{3}) + 96 = 0$, so we cannot form x_{n+1}.

24.   There are 2 roots on the given interval.

$x_{11} = 1.69990$

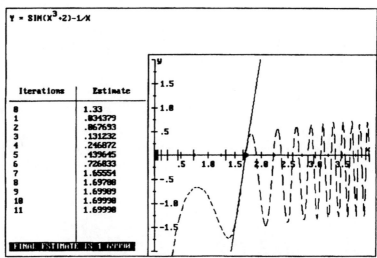

$x_5 = 1.90110$

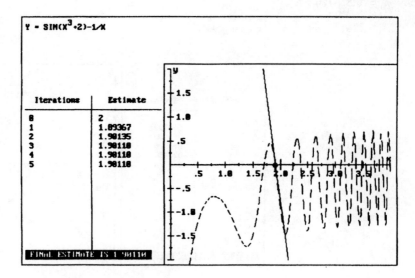

25. $x_3 = 1.53981$

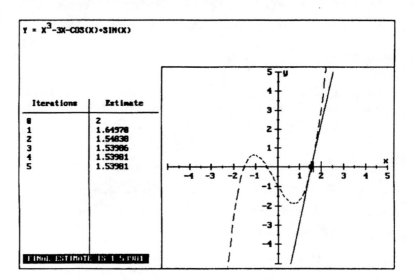

26.   a.

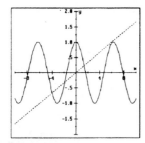

The graphs intersect in three places. You might wish to zoom to see what happens when x is in the neighborhood of 5 to verify that there is no intersection point in that neighborhood.

b. $x_3 = -3.98583$

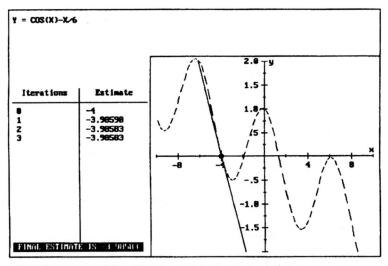

$x_4 = -1.89152$

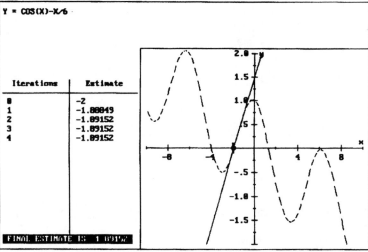

$x_4 = 1.34475$

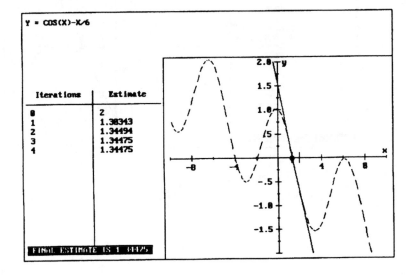

27. $f(x) = x^6 - x^5 + x^3 - 3;$ $f'(x) = 6x^5 - 5x^4 + 3x^2;$ $f(0) = 3$
$f(-1) = -4,$ $f(1) = -2,$ and $f(2) > 0$ so a root exists on $[1, 2]$.

$x_0 = 1.4;$ $f(1.4) \approx 1.895;$ $f'(1.4) \approx 18.9,$ so $x_1 = 1.4 - \dfrac{1.895}{18.9} \approx 1.2997$

$f(1.2997) \approx 0.307;$ $f'(1.2997) \approx 13.1;$ $x_2 = 1.2997 - \dfrac{0.307}{13.1} \approx 1.2763$

$f(1.2763) \approx 0.0147;$ $f'(1.2763) \approx 11.9;$ $x_3 = 1.2763 - \dfrac{0.0147}{11.9} \approx 1.2751$

$x_4 = 1.2751$

Note: there is another real root, as shown below. $x_3 = -1.16137$

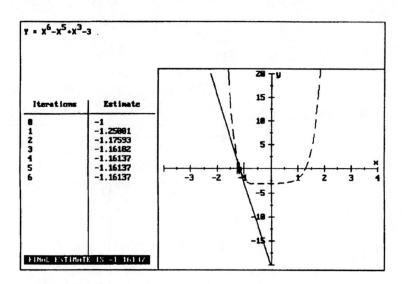

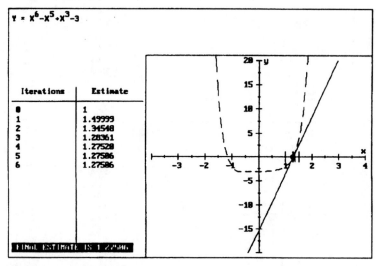

28. Let x = length of side of the square and r = radius of the circle.

$A = x^2$ and $A = \pi r^2$
$P = 4x$ and $C = 2\pi r$

Thus,

$$\begin{cases} 2x^2 = \pi r^2 \\ 4x + 2\pi r = 10 \end{cases}$$

$$r = \frac{10 - 4x}{2\pi}$$

so that

$$2x^2 = \pi\left(\frac{10 - 4x}{2\pi}\right)^2$$

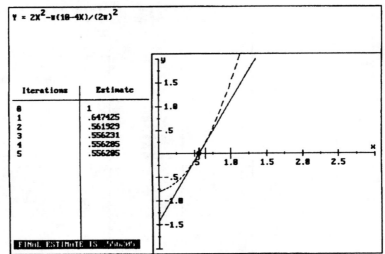

Let $f(x) = 2x^2 - \pi\left(\dfrac{10 - 4x}{2\pi}\right)^2$. From the computer printout we see $x_5 = 0.556285$.

29. $x_5 = -2.34641$

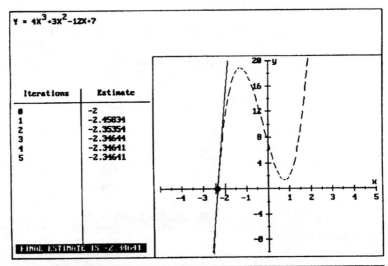

30. $x_3 = 1$

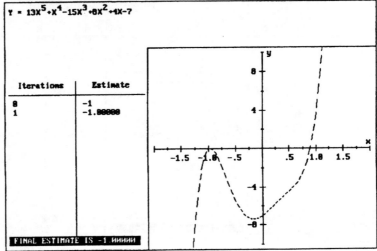

$x_3 = 0.881863$

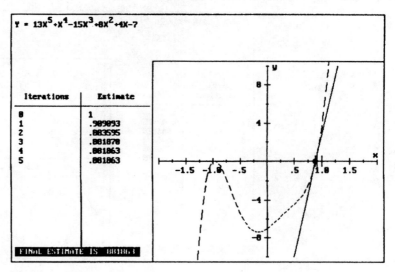

31. The object is stationary when the velocity is zero; that is $v(t) = s'(t) = 12t^3 + 12t^2 - 60t + 1 = 0$

$x_3 = 0.016724;$

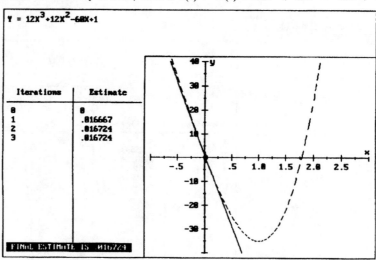

$x_3 = 1.78105$

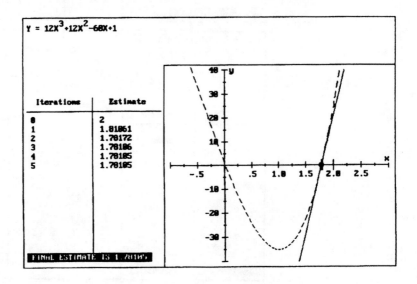

32. The object is stationary when the velocity is zero; that is $v(t) = s'(t) = 3t^2 - 4t - 5 = 0$

$x_3 = -0.786300$

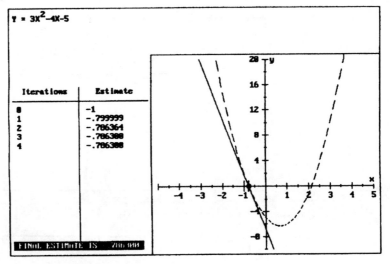

$x_3 = 2.11963$

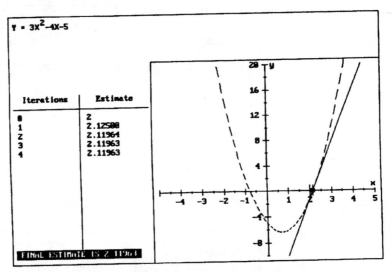

33. $f(x) = -2x^4 + 3x^2 + \frac{11}{8}$ $\qquad f'(x) = -8x^3 + 6x$

a. $f(0) = \frac{11}{8}$ and $f(2) = -\frac{149}{8}$ so there is a root on $[0, 2]$.

Also note that $f(x)$ is an even function and, therefore, symmetric about the y-axis. So it must have at least two roots.

b. $x_{n+1} = x_n - \dfrac{f(x_n)}{f'(x_n)} = x_n - \dfrac{-2x_n^4 + 3x_n^2 + \frac{11}{8}}{-8x_n^3 + 6x_n} = \dfrac{-6x_n^4 + 3x_n^2 - \frac{11}{8}}{-8x_n^3 + 6x_n}$

If $x_0 = \frac{1}{2}$ then $x_1 = -0.5$, $x_2 = 0.5$ and the values will continue to oscillate.

This is due to the symmetry about the y-axis.

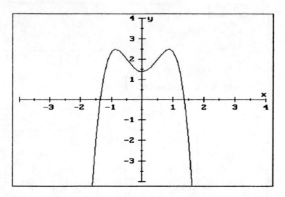

34. **a.** $x_1 = -2.00000$

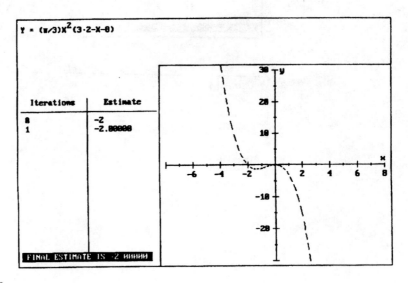

$x_4 = -0.532089$

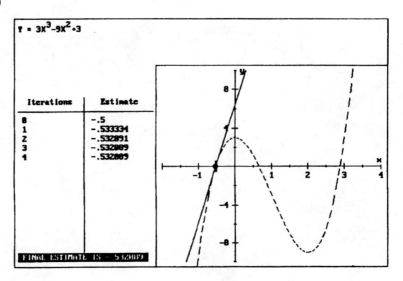

b. $x_4 = 0.652704$

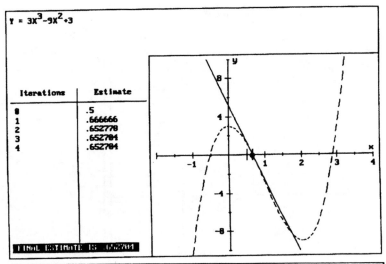

$x_4 = 2.87939$

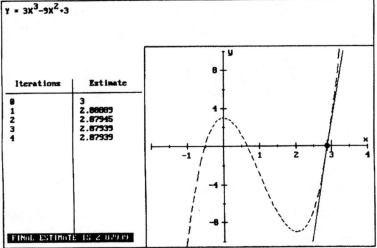

c. If the given equation were $3x_c^3 - 9x_c + 4 = 0$; Newton-Raphson (or a computer program) suggests $x_c = 0.774$.

35. The root is 1.41421, but investigating the errors, we find:

$e_1 \approx 0.0858$, so
$\qquad K \approx .3432$

$e_2 \approx 0.002245$, so
$\qquad K \approx 0.3328$

$e_3 \approx 0.00000179$,
$\qquad$ so $K \approx 0.3$

$K \approx \frac{1}{3}$

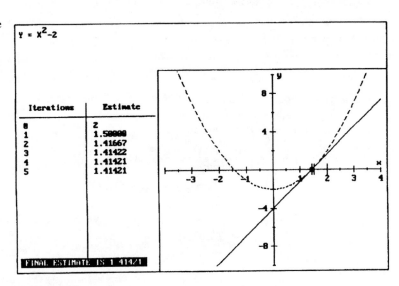

36. $x_4 = 4.93750$

The point, x_n, covers one-half the distance left to reach 5 after each step. The error is halved each time.

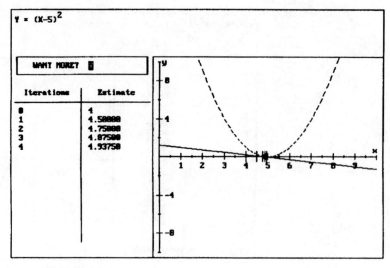

Iterations	Estimate
0	4
1	4.50000
2	4.75000
3	4.87500
4	4.93750

37. $f(x) = -x^4 + x^2 + A$; $f'(x) = -4x^3 + 2x$; $x_0 = \frac{1}{3}$

$f(x) = -\frac{1}{81} + \frac{1}{9} + A = \frac{-1 + 9 + 81A}{81}$; $f'(\frac{1}{3}) = -\frac{4}{27} + \frac{2}{3} = \frac{14}{27}$

$$x_1 = \frac{1}{3} - \frac{\frac{8 + 81A}{81}}{\frac{14}{27}} = \frac{1}{3} - \frac{8 + 81A}{42}$$

Solve

$$\frac{1}{3} - \frac{8 + 81A}{42} = -\frac{1}{3}$$

$$28 = 8 + 81A$$

$$A = \frac{20}{81}$$

38. $f(x) = x^{1/2}$; $f'(x) = \frac{1}{2}x^{-1/2}$; $f(0.05) \approx 0.2236$; $f'(0.05) \approx \frac{1}{0.447} \approx 2.24$

$x_2 = 0.05 - \frac{0.2236}{2.24} < 0$, so $f(x_2)$ is not defined. Note that the tangent line is vertical to the curve at $x = 0$. Another value of x_1 would not have made any difference.

39. If $f(x_n) = 0$, x_n is a root. If $f'(x_n) \neq 0$, the tangent line through that root (x-intercept) is not horizontal. Successive tangent lines in the Newton-Raphson method fall on the same line and all points at which these tangent lines intersect the x-axis are the same as the root. Thus, $x_n = x_{n+1} = \cdots$

40. **a.** $f(x) = x^2 - N$; $f'(x) = 2x$

$$x_{n+1} = x_n - \frac{x_n^2 - N}{2x_n} = \frac{1}{2x_n}(2x_n^2 - x_n^2 + N) = \frac{1}{2}(2x_n - x_n + Nx_n) = \frac{1}{2}\left(x_n + \frac{N}{x_n}\right)$$

b. The square root is $\sqrt{1{,}265}$; $N = 1{,}265$; $x_0 = 34$ since $35^2 = 1{,}225$.

$$x_1 = \frac{1}{2}\left(35 + \frac{1{,}265}{35}\right) \approx 35.57143 \qquad x_2 = \frac{1}{2}\left(35.57143 + \frac{1{,}265}{35.57143}\right) \approx 35.56684$$

$$x_3 = \frac{1}{2}\left(35.56684 + \frac{1{,}265}{35.56684}\right) \approx 35.56684$$

41. $x = \frac{1}{N}$; $f(x) = N - \frac{1}{x} = N - x^{-1}$; $f'(x) = x^{-2}$

$$x_{n+1} = x_n - \frac{N - x^{-1}}{x^{-2}} = 2x_n - Nx_n^2 = x_n(2 - Nx_n); \quad N \approx 2.355673; \quad x_0 = 0.5$$

$x_1 = (0.5)[2 - (2.355673)(0.5)] \approx 0.41108$

$x_2 = (0.41108)[2 - (2.355673)(0.41108)] \approx 0.42408$

$x_3 = (0.42408)[2 - (2.355673)(0.42408)] \approx 0.424512$

CHAPTER 2 REVIEW

Proficiency Examination, Page 160

1. $m_{\text{sec}} = \dfrac{\Delta y}{\Delta x} = \dfrac{f(x + \Delta x) - f(x)}{\Delta x}$

   $m_{\text{tan}} = \lim\limits_{\Delta x \to 0} \dfrac{f(x + \Delta x) - f(x)}{\Delta x}$

2. If $y = f(x)$, then

   $\dfrac{dy}{dx} = \lim\limits_{\Delta x \to 0} \dfrac{f(x + \Delta x) - f(x)}{\Delta x}$,

   provided this limit exists.

3. A normal line is perpendicular to the tangent line at a point on the graph of a function.

4. a. $\dfrac{d}{dx}(k) = 0$ b. $\dfrac{d}{dx}(x^n) = nx^{n-1}$

   c. $(cf)' = cf'$ d. $(f + g)' = f' + g'$

   e. $(f - g)' = f' - g'$

   f. $(af + bg)' = af' + bg'$

   g. $(fg)' = fg' + f'g$

   h. $\left(\dfrac{f}{g}\right)' = \dfrac{gf' - fg'}{g^2}$

   i. $(\cos u)' = (-\sin u)u'$

   j. $(\sin u)' = (\cos u)u'$

   k. $(\tan u)' = (\sec^2 u)u'$

   l. $(\sec u)' = (\sec u \tan u)u'$

   m. $(\csc u)' = (-\csc u \cot u)u'$

   n. $(\cot u)' = (-\csc^2 u)u'$

5. A higher derivative is a derivative of a derivative.

   $y''; \ y'''; \ y^{(4)}; \ \dots; \ \dfrac{d^2y}{dx^2}, \dfrac{d^3y}{dx^3}, \dfrac{d^4y}{dx^4}; \dots$

6. $v(t) = s'(t); \ a(t) = v'(t) = s''(t)$

7. $[f(u(x))]' = f'[u(x)]u'(x)$

8. Apply all the rules of differentiation, treating y as a function of x and remembering the chain rule. For example, $(cxy)' = c(xy' + y)$; now solve for y'.

9. (1) Draw a figure.
   (2) Relate the variables through a formula or equation.
   (3) Differentiate the equations (formulas).
   (4) Substitute numerical values and solve algebraically.

10. $f(b) \approx f(a) + f'(a)(b - a)$

11. Marginal analysis of a function refers to the investigation of the behavior of the first derivative of a function.

12. $dx = \Delta x; \ dy = f'(x)dx$
    See Figure 2.45, page 146.

13. The propagated error is the difference between $f(x + \Delta x)$ and $f(x)$. The relative error is $\Delta f/f$, and the percentage error is $100\left|\Delta f/f\right|$.

14. The Newton-Raphson method approximates a root of a function by locating a point near a root, and then finding where the tangent line at this point intersects the x-axis. That is,

    $$x_{n+1} = x_n - \dfrac{f(x_n)}{f'(x_n)}$$

    Repetition of this technique usually closes in on the root.

15. $y = x^3 + x^{3/2} + \cos 2x$

    $\dfrac{dy}{dx} = 3x^2 + \dfrac{3}{2}x^{1/2} - 2\sin 2x$

16. $\dfrac{dy}{dx} = -4\left(\sqrt{3}x + \dfrac{3}{x^2}\right)^{-5}\left(\sqrt{3} + (-2)\dfrac{3}{x^3}\right)$

    $= -4\left(\sqrt{3}x + \dfrac{3}{x^2}\right)^{-5}\left(\sqrt{3} - \dfrac{6}{x^3}\right)$

    $= \dfrac{-4x^{10}}{(\sqrt{3}x^3 + 3)^5} \cdot \dfrac{\sqrt{3}x^3 - 6}{x^3}$

    $= \dfrac{-4x^7(\sqrt{3}x^3 - 6)}{(\sqrt{3}x^3 + 3)^5}$

17. $\dfrac{dy}{dx} = \dfrac{1}{2}[\sin(3 - x^2)]^{-\frac{1}{2}}[\cos(3 - x^2)](-2x)$

    $= -x[\cos(3 - x^2)][\sin(3 - x^2)]^{-1/2}$

18. $x\dfrac{dy}{dx} + y + 3y^2\dfrac{dy}{dx} = 0$

    $(x + 3y^2)\dfrac{dy}{dx} = -y$

    $\dfrac{dy}{dx} = \dfrac{-y}{x + 3y^2}$

19. $\sin^2 a + \cos^2 a = 1$. So $y = 1$, $\dfrac{dy}{dx} = 0$

20. $2x + xy' + y - 4yy' = 0$

    $(x - 4y)y' = -(2x + y)$

    $y' = \dfrac{2x + y}{4y - x}$

$$y'' = \frac{(4y - x)(2 + y') - (2x + y)(4y' - 1)}{(4y - x)^2}$$

$$= \frac{(4y - x)\left(2 + \frac{2x + y}{4y - x}\right) - (2x + y)\left(4\frac{2x + y}{4y - x} - 1\right)}{(4y - x)^2}$$

$$= \frac{(4y - x)\left(\frac{9y}{4y - x}\right) - (2x + y)\left(\frac{9x}{4y - x}\right)}{(4y - x)^2}$$

$$= \frac{9[(4y^2 - xy) - (2x^2 + xy)]}{(4y - x)^3}$$

$$= \frac{9(4y^2 - 2xy - 2x^2)}{(4y - x)^3}$$

$$= \frac{18(2y^2 - xy - x^2)}{(4y - x)^3}$$

$$= \frac{18(x + 2y)(x - y)}{(x - 4y)^3}$$

21. $f(x) = x - 3x^2;$

$$f(x + \Delta x) = (x + \Delta x) - 3(x + \Delta x)^2$$

$$= x + \Delta x - 3x^2 - 6x\Delta x - 3\Delta x^2$$

$$\frac{dy}{dx} = \lim_{\Delta x \to 0} \frac{f(x + \Delta x) - f(x)}{\Delta x}$$

$$= \lim_{\Delta x \to 0} \frac{(x + \Delta x - 3x^2 - 6x\Delta x - 3\Delta x^2) - (x - 3x^2)}{\Delta x}$$

$$= \lim_{\Delta x \to 0} \frac{(\Delta x - 6x\Delta x - 3\Delta x^2)}{\Delta x}$$

$$= \lim_{\Delta x \to 0} (1 - 6x - 3\Delta x) = 1 - 6x$$

22. $\frac{dy}{dx} = (x^2 + 3x - 2)(-3) + (7 - 3x)(2x + 3)$

At $x = 1$, $\frac{dy}{dx} = 14$

$$y - 8 = 14(x - 1)$$

$$14x - y - 6 = 0$$

23. $y = f(1) = \frac{1}{2}$, so the point is $(1, \frac{1}{2})$

$$\frac{dy}{dx} = 2\sin(\frac{\pi x}{4})\cos(\frac{\pi x}{4})(\frac{\pi}{4})$$

At $x = 1$, $\frac{dy}{dx} = 2\left(\frac{\sqrt{2}}{2}\right)\left(\frac{\sqrt{2}}{2}\right)(\frac{\pi}{4}) = \frac{\pi}{4}$

$$y - \frac{1}{2} = \frac{\pi}{4}(x - 1)$$

The equation of the normal line at that same point must have a slope that is the negative reciprocal of the slope of the tangent, $m = -\frac{4}{\pi}$. The equation of the normal:

$$y - \frac{1}{2} = -\frac{4}{\pi}(x - 1)$$

24. $A = \pi r^2$ so $\frac{dA}{dt} = 2\pi r \frac{dr}{dt}$

$\frac{dr}{dt} = 0.5$ when $r = 2$ so

$$\frac{dA}{dt} = 2\pi(2)(0.5) = 2\pi \text{ ft}^2/\text{s}$$

25. $f(x) = x^3 - x^2 - x - 13;$ $f(0) = -13;$ $f(3) = 2$

$$f'(x) = 3x^2 - 2x - 1$$

The conditions of the intermediate value theorem are met, and we are guaranteed at least one point, c, in the interval such that $f(c) = 0$.   By the Newton-Raphson method,

$$x_{n+1} = x_n - \frac{f(x_n)}{f'(x_n)}$$

$$= x_n - \frac{x_n^3 - x_n^2 - x_n - 13}{3x_n^2 - 2x_n - 1}$$

$$= \frac{2x_n^3 - x_n^2 + 13}{3x_n^2 - 2x_n - 1}$$

$$\frac{x}{3} = \frac{13}{15}, \quad x = \frac{39}{15} \approx 2.6$$

$$x_1 \approx 2.93977$$
$$x_2 \approx 2.89650$$
$$x_3 \approx 2.89571$$
$$x_4 \approx 2.89571$$

Supplementary Problems, Page 161

1. $y' = 4x^3 + 6x - 7$

2. $y' = 5x^4 + 9x^2$

3. $y' = \frac{1}{2}\left(\frac{x^2 - 1}{x^2 - 5}\right)^{-1/2}\left[\frac{(x^2 - 5)(2x) - (x^2 - 1)(2x)}{(x^2 - 1)^{1/2}(x^2 - 5)^{3/2}}\right]$

$$= \frac{-4x}{(x^2 - 1)^{1/2}(x^2 - 5)^{3/2}}$$

4. $y' = \frac{(x + \sin x)(3\sin 3x) - (1 - \cos 3x)(1 + \cos x)}{(x + \sin x)^2}$

$$= \frac{3x\sin 3x + 3\sin x \sin 3x - 1 - \cos x + \cos 3x \cos x}{(x + \sin x)^2}$$

5. $4x - xy' - y + 2y' = 0$

$$y' = \frac{4x - y}{x - 2}$$

6. $y' = \sqrt{\sin(3 - x^2)} - \frac{x^2\cos(3 - x^2)}{\sqrt{\sin(3 - x^2)}}$

7. $y' = -4\left(\sqrt{3x} + \frac{3}{x^2}\right)^{-5}\left(\frac{3}{2\sqrt{3x}} - \frac{6}{x^3}\right)$

8. $y' = 7(x^2 + 3x - 6)^6(2x + 3)$

9. $y' = 10(x^3 + x)^9(3x^2 + 1)$

$$= 10x^9(x^2 + 1)^9(3x^2 + 1)$$

10. $y' = \sqrt{x}[10(x^2+5)^9(2x)] + (x^2+5)^{10}(\frac{1}{2})x^{-3/2}$

$$= \frac{(41x^2+5)(x^2+5)^9}{2\sqrt{x}}$$

11. $y' = \sqrt[3]{x}[5(x^3+1)^4(3x^2)]+(x^3+1)^5(\frac{1}{3})x^{-2/3}$

$$= \frac{1}{3}(x^3+1)^4[45x^2+^{1/3}+(x^3+1)x^{-2/3}]$$

$$= \frac{(x^3+1)^4(46x^3+1)}{3x^{2/3}}$$

12. $y' = (x^3+3)^5[8(x^3-5)^7(3x^2)] + (x^3-5)^8[5(x^3+3)^4(3x^2)]$

$$= 3x^2(x^3-5)^7(x^3+3)^4(13x^3-1)$$

13. $y' = 56x^3(x^4-1)^{10}(2x^4+3)^6 + 40x^3(x^4-1)^9(2x^4+3)^7$

$$= 8x^3(x^4-1)^9(2x^4+3)^6[7(x^4-1)+5(2x^4+3)]$$

$$= 8x^3(x^4-1)^9(2x^4+3)^6(17x^4+8)$$

14. $y' = \frac{1}{2}\sin^{-1/2}5x(\cos 5x)(5) = \frac{5\cos 5x}{2\sqrt{\sin 5x}}$

15. $y' = \frac{1}{2}(\cos x^{1/2})^{-1/2}(\frac{1}{2}x^{-3/2})(-\sin x^{1/2})$

$$= \frac{-\sin\sqrt{x}}{4\sqrt{x}\sqrt{(\cos\sqrt{x})}}$$

16. $y' = 3(\sin x + \cos x)^2(\cos x - \sin x)$

17. $y' = 5(x^{1/2}+x^{1/3})^4(\frac{1}{2}x^{-1/2}+\frac{1}{3}x^{-2/3})$

18. $y' = \frac{1}{2}\left(\frac{x^3+x^2-1}{x^3+2x^2+3}\right)^{-1/2}\left[\frac{(x^3+2x^2+3)(3x^2+x-1)-(x^3+x^2-1)(3x^2+4x)}{(x^3+2x^2+3)^2}\right]$

$$= \frac{1}{2}\left(\frac{x^3+x^2-1}{x^3+2x^2+3}\right)^{-1/2}\left[\frac{3x^5+6x3+9x^2+2x^4+4x^3+6x-3x^5-3x^4+3x^2-4x^4-4x^3+4x}{(x^3+2x^2+3)^2}\right]$$

$$= \frac{1}{2}\left(\frac{x^3+x^2-1}{x^3+2x^2+3}\right)^{-1/2}\left[\frac{-x(5x^3+6x^2+12x+10)}{(x^3+2x^2+3)^2}\right]$$

19. $y' = [\cos(\sin x)]\cos x$ **20.** $y' = -[\sin(\sin x)]\cos x$

21. $y' = \dfrac{2-x^{-1/2}}{y^{-1/2}} = \dfrac{2x^{1/2}y^{1/2}-y^{1/2}}{x^{1/2}}$ **22.** $8x-32yy'=0$

$$y' = \frac{x}{4y}$$

23. $(\cos xy)(xy'+y) = y'+1$

$$y' = \frac{1-y\cos xy}{x\cos xy-1}$$

24. $[\cos(x+y)](1+y') - [\sin(x-y)](1-y') = xy'+y$

$\cos(x+y) + y'(\cos x+y) - \sin(x-y) + y'\sin(x-y) = xy'+y$

$[\cos(x+y)+\sin(x-y)-x]y' = -\cos(x+y)+\sin(x-y)+y$

$$y' = \frac{-\cos(x+y)+\sin(x-y)+y}{\cos(x+y)+\sin(x-y)-x}$$

25. $\dfrac{dy}{dx} = 5x^4-20x^3+21x-6x; \dfrac{d^2y}{dx^2} = 20x^3-60x^2+42x-6$

26. $\dfrac{dy}{dx} = \dfrac{(2x+3)(1)-(x-5)(2)}{(2x+3)^2} + 2(3x-1)(3)$

$= \dfrac{2x+3-2(x-5)}{(2x+3)^2} + 6(3x-1)$

$= \dfrac{13}{(2x+3)^2} + 6(3x-1)$

$= 13(2x+3)^{-2} + 18x - 6$

$\dfrac{d^2y}{dx^2} = (-2)(13)(2x+3)^{-3}(2) + 18 = -52(2x+3)^{-3} + 18$

27. $2x + 3y^2y' = 0 \qquad\qquad y' = -\dfrac{2x}{3y^2}$

$\dfrac{d^2y}{dx^2} = -\dfrac{2}{3}\left[\dfrac{y^2 - x(2yy')}{y^4}\right]$

$= -\dfrac{2(y - 2xy')}{3y^3}$

$= -\dfrac{2\left[y - 2\left(-\dfrac{2x}{3y^2}\right)\right]}{3y^3}$

$= -\dfrac{2(3y^3 + 4x^2)}{9y^5}$

28. $2x + (\cos y)y' = 0 \qquad\qquad y' = -\dfrac{2x}{\cos y}$

$\dfrac{d^2y}{dx^2} = -\dfrac{\cos y - x(-\sin y)y'}{\cos^2 y}$

$= \dfrac{2}{\cos^2 y}\left[\cos y + x \sin y\left(-\dfrac{2x}{\cos y}\right)\right]$

$= \dfrac{4x^2\sin y - 2\cos^2 y}{\cos^3 y}$

29. $y' = 4x^3 - 21x^2 + 2x; \ \text{At } P(0, -3), \ m = y' = 0$

The equation of the (horizontal) tangent line is $y + 3 = 0$

30. $f(x) = (3x^2 + 5x - 7)^3; \ f(1) = 1$

$f'(x) = 3(3x^2 + 5x - 7)^2(6x + 5); \ f'(1) = 33$

The equation of the tangent line is: $\quad y - 1 = 33(x - 1)$

$33x - y - 32 = 0$

31. $f(x) = x \cos x; \ f(\tfrac{\pi}{2}) = 0$

$f'(x) = -x \sin x + \cos x; \ f'(\tfrac{\pi}{2}) = -\tfrac{\pi}{2}$

The equation of the tangent line is: $\quad y = -\tfrac{\pi}{2}(x - \tfrac{\pi}{2})$

$2\pi x + 4y - \pi^2 = 0$

32. $f(x) = \dfrac{\sin x}{\sec x \tan x} = (\cos x)^2; \ f(\pi) = 1$

$f'(x) = -2 \cos x \sin x = -\sin 2x; \ f'(\pi) = 0$

The equation of the (horizontal) tangent line is $y - 1 = 0$.

33. $2xyy' + x^2y' + y^2 + 2xy = 0$

At $(1, 1)$: $2y' + y' + 1 + 2 = 0$, or $y' = -1$

The equation of the tangent line is:

$$y - 1 = -(x - 1)$$

$$x + y - 2 = 0$$

34. $f(x) = (x^3 - 3x^2 + 4)^2$; $f(1) = 4$

$f'(x) = 2(x^3 - 3x^2 + 4)(3x^2 - 6x)$; $f'(1) = -12$

The equation of the tangent line is:

$$y - 4 = -12(x - 1)$$

$$12x + y - 16 = 0$$

35. $f(x) = \dfrac{3x - 4}{3x^2 + x - 5}$; $f(1) = 1$

$f'(x) = \dfrac{(3x^2 + x - 5)(3) - (3x - 4)(6x + 1)}{2}$;

$f'(1) = \dfrac{(-1)(3) - (-1)(7)}{2} = 4$

The equation of the tangent line is:

$$y - 1 = 4(x - 1)$$

$$4x - y - 3 = 0$$

36. $\frac{2}{3}[x^{-1/3} + y^{-1/3}y'] = 0$

At $(1, 1)$, $\frac{2}{3}[1 + y'] = 0$ or $y' = -1$

$$y - 1 = -(x - 1)$$

$$x + y - 2 = 0$$

37. $f(x) = (x^3 - x^2 + 2x - 1)^4$; $f(1) = 1$

$f'(x) = 4(x^3 - x^2 + 2x - 1)^3(3x^2 - 2x + 2)$;

$f'(1) = 12$; thus, $m = 12$ and $m_T = -\frac{1}{12}$

The equation of the tangent line is:

$$y - 1 = 12(x - 1)$$

$$12x - y - 11 = 0$$

The equation of the normal line is:

$$y - 1 = -\frac{1}{12}(x - 2)$$

$$x + 12y - 13 = 0$$

38. $f(x) = (2x + x^{-1})^3$

$f'(x) = 3(2x + x^{-1})^2(2 - x^{-2})$;

$f'(1) = 3(3)^2(1) = 27$; thus, $m = 27$ and $m_T = -\frac{1}{27}$; the equation of the tangent line is:

$$y - 27 = 27(x - 1)$$

$$27x - y = 0$$

The equation of the normal line is:

$$y - 27 = -\frac{1}{27}(x - 1)$$

$$729x + 29y - 1 = 0$$

39. $\dfrac{dy}{dx} = 3x^2 - 7$; $\dfrac{d^2y}{dx^2} = 6x$;

$\dfrac{dx}{dt} = t\cos t + \sin t$; $\dfrac{d^2x}{dt^2} = -t\sin t + \cos t + \cos t$

$\dfrac{d^2y}{dt^2} = (3x^2 - 7)(2\cos t - t\sin t) + 6x(t\cos t + \sin t)^2$

40. $f'(x) = x^2(\cos x^2)(2x) + 2x \sin x^2$

$= 2x^3\cos x^2 + 2x \sin x^2$

$f''(x) = 2x^3(-\sin x^2)(2x) + 2(3x^2)(\cos x^2)$

$\qquad + 2x(\cos x^2)(2x) + 2 \sin x^2$

$= -4x^4(\sin x^2) + 6x^2(\cos x^2)$

$\qquad + 4x^2(\cos x^2) + 2 \sin x^2$

$= 2(1 - 2x^4)(\sin x^2) + 10x^2(\cos x^2)$

41. $f(x) = x^4 - x^{-4}$

$f'(x) = 4x^3 + 4x^{-5}$

$f''(x) = 12x^2 - 20x^{-6}$

$f'''(x) = 24x + 120x^{-7}$

$f^{(4)}(x) = 24 - 840x^{-8}$

42. $f(x) = x(x^2 + 1)^{7/2}$

$f'(x) = x(\frac{7}{2})(x^2 + 1)^{5/2}(2x) + (x^2 + 1)^{7/2}$

$= (x^2 + 1)^{5/2}[7x^2 + x^2 + 1]$

$= (x^2 + 1)^{5/2}(8x^2 + 1)$

$f''(x) = \frac{5}{2}(x^2 + 1)^{3/2}(2x)(8x^2 + 1)$

$\qquad + (x^2 + 1)^{5/2}(16x)$

$= (x^2 + 1)^{3/2}[5x(8x^2 + 1) + (x^2 + 1)(16x)]$

$= x(x^2 + 1)^{3/2}[40x^2 + 5 + 16x^2 + 16]$

$= x(x^2 + 1)^{3/2}(56x^2 + 21)$

$= 7x(x^2 + 1)^{3/2}(8x^2 + 3)$

$f'''(x) = \frac{3}{2}(x^2 + 1)^{1/2}(2x)(56x^3 + 21x)$

$\qquad + (x^2 + 1)^{3/2}(168x^2 + 21)$

$= 3x(x^2 + 1)^{1/2}(7)(8x^3 + 3x)$

$\qquad + 21(x^2 + 1)^{3/2}(8x^2 + 1)$

$= 21(x^2 + 1)^{1/2}[x(8x^3 + 3x)$

$\qquad + (x^2 + 1)(8x^2 + 1)]$

$= 21(x^2 + 1)^{1/2}(16x^4 + 12x^2 + 1)$

43. Let $y = \sqrt[3]{\dfrac{x^4 + 1}{x^4 - 2}}$ so $y^3 = \dfrac{x^4 + 1}{x^4 - 2}$

$$3y^2 y' = \frac{(4x^3)(x^4 - 2 - x^4 - 1)}{(x^4 - 2)^2}$$

$$y' = \frac{(4x^3)(-3)}{3y^2(x^4 - 2)^2}$$

$$= \frac{-4x^3}{(x^4 - 2)^{4/3}(x^4 + 1)^{2/3}}$$

44. $\qquad\qquad x^3 y^3 + x - y = 1$

$$3x^3 y^2 y' + 3x^2 y^3 + 1 - y' = 0$$

$$(3x^3 y^2 - 1)y' = -3x^2 y^3 - 1$$

$$y' = -\frac{3x^2 y^3 + 1}{3x^3 y^2 - 1}$$

$$3x^3 y^2 y'' + 3x^3(2yy')y' + 2(3x^2)y^2 y'$$

$$+ 3x^2(3y^2 y') + 3(2x)y^3 - y'' = 0$$

$$(3x^3 y^2 - 1)y'' = -6x^3 y(y')^2 - 18x^2 y^2 y' - 6xy^3$$

$$y'' = \frac{-6x^3 y(y')^2 - 18x^2 y^2 y' - 6xy^3}{3x^3 y^2 - 1}$$

45. $\qquad\qquad x^2 + 4xy - y^2 = 8$

$$2x + 4xy' + 4y - 2yy' = 0$$

$$y' = \frac{x + 2y}{y - 2x}$$

$$y'' = \frac{(y - 2x)(1 + 2y') - (x + 2y)(y' - 2)}{(y - 2x)^2}$$

$$= \frac{5y - 5xy'}{(y - 2x)^2}$$

$$= \frac{5y - 5x\left(\dfrac{x + 2y}{y - 2x}\right)}{(y - 2x)^2}$$

$$= \frac{5y^2 - 20xy - 5x^2}{(y - 2x)^3}$$

46. $f'(x) = 2x + 5$ if $x \le 0$; $f'(x) = 5$ if $0 < x < 6$; $f'(x) = 2x$ if $x > 6$

There is a corner point at $x = 6$, so the derivative does not exist.

47. $3x^2 - 3y^2 y' = 2(xy' + y)$

At $x = 1$:

$$3 - 3y' = 2(-y' + 1)$$

$$y' = 1$$

The slope of the tangent is 1 and the slope of the normal is -1.

The equation of the tangent line is:

$$y - 1 = x + 1$$

$$x - y + 2 = 0$$

The equation of the normal line is:

$$y - 1 = -(x + 1)$$

$$x + y = 0$$

48. Let $f(x) = x^{3/2} + 2x^{1/2}$; $f(16) = 72$

$f'(x) = \frac{3}{2}x^{1/2} + x^{-1/2}$; $f'(16) = 6.5$

$\Delta x = 0.01$

$$f(16.01) = f(16 + 0.01)$$

$$\approx f(16) + f'(16)(0.01)$$

$$= 72 + 6.5(0.01)$$

$$= 72.065$$

A calculator approximation is 72.06251.

49. $f(x) = \cos x$; $f(\frac{\pi}{6}) = \dfrac{\sqrt{3}}{2}$

$f'(x) = -\sin x$; $f'(\frac{\pi}{6}) = \dfrac{1}{2}$

$\Delta x = \dfrac{\pi}{600}$

$$f(\tfrac{101\pi}{600}) = f(\tfrac{\pi}{6} + \tfrac{\pi}{600})$$

$$\approx f(\tfrac{\pi}{6}) + f'(\tfrac{\pi}{6})(\tfrac{\pi}{600})$$

$$= \frac{\sqrt{3}}{2} + \frac{1}{2}\left(\frac{\pi}{600}\right)$$

$$\approx 0.8686$$

A calculator approximation is 0.8634.

50. $V = \frac{1}{3}\pi r^2 h$; $\dfrac{dV}{dr} = \frac{2}{3}\pi rh$

At $r = 10$, $h = 2$, $V = \frac{1}{3}\pi(100)(2) \approx 209.44$

$\dfrac{dV}{dr} = \frac{1}{3}(40)\pi \approx 41.89$; $\Delta r = 0.01$

$V \approx 209.44 + 0.42 \approx 209.86$

A calculator approximation is 209.8586.

51. An effective marginal tax rate is the derivative of the tax rate.

52. Consider a right triangle with horizontal leg of 600, vertical leg y, and angle of elevation θ.

$$y = 600 \tan \theta$$

$$\frac{dy}{dt} = 600 \sec^2\theta \, \frac{d\theta}{dt}$$

$$20 = 600(\tfrac{5}{3})^2 \frac{d\theta}{dt}$$

$$\frac{d\theta}{dt} = \frac{20(9)}{(600)(25)} = \frac{9}{750} \approx 0.12 \text{ rad/min}$$

53. $\dfrac{d}{dx}f(x^3 - 1) = \dfrac{d}{dx}f(u)\dfrac{du}{dx} = [f'(u)](3x^2)$

$$= 3x^2[2(x^3 - 1)^2 + 3]$$

$$= 6x^2(x^3 - 1)^2 + 9x^2$$

54. $\frac{d}{dx}f(x^2 + x) = \frac{d}{dx}f(u)\frac{du}{dx} = [f'(u)](x + 1)$

$$= (x + 1)[(x^2 + x)^2 + x^2 + x]$$

$$= (x + 1)(x^4 + 2x^3 + 2x^2 + x)$$

55. Let $F(x) = (f \circ f)(x) = f[f(x)] = f(u)$

where $u = f(x)$.

$\frac{dF(x)}{dx} = \frac{df(u)}{du} \cdot \frac{df(x)}{dx} = (6u)(6x) = 6(3x^2 + 1)(6x)$

$$= 36x(3x^2 + 1)$$

56. Let $F(x) = (f \circ g)(x) = f[g(x)] = f(u)$
where $u = g(x)$.

$\frac{dF(x)}{dx} = \frac{df(u)}{du} \cdot \frac{dg(x)}{dx} = (2 \cos 2x^2 - 3 \sin 3x^2)(2x)$

$$= 4x \cos 2x^2 - 6x \sin 3x^2$$

57. $f'(0) = \lim_{\Delta x \to 0} \frac{f(0 + \Delta x) - f(0)}{\Delta x}$

$$= \lim_{\Delta x \to 0} \frac{\Delta x \sin \frac{1}{\Delta x}}{\Delta x}$$

$$= \lim_{\Delta x \to 0} \sin \frac{1}{\Delta x} \text{ is not defined}$$

58. The velocity of the car is 60 mi/h and that of the truck is 45 mi/h. The respective distances covered are $60t$ and $45t$. The distance between the car and the truck is

$$D^2 = 15^2(16t^2 + 9t^2) = 15^2(25t^2)$$

$$D = 75t \text{ (disregard the negative value)}$$

The rate of change of the distance between the car and the truck is 75 mi/h.

59. From Problem 58, the rate of change of the distance between the car and the truck is a constant, so the rate is independent of the particular instant of time. Thus, the answer for this problem is the same as the answer for Problem 58, namely the rate is 75 mi/h.

60. $V = \frac{4}{3}\pi r^3; \frac{dV}{dt} = 4\pi r^2 \frac{dr}{dt}$

$S = 4\pi r^2$ so at the specific instant when $4\pi r^2 = 4\pi$ we have:

$$\frac{dV}{dt} = 4\pi(2)$$

$$\frac{dV}{dt} = 8\pi$$

61. $V = x^3$, when the volume is 27, $x = 3$,

$dx/dt = -1$, the surface area is $S = 6x^2$:

$$\frac{dS}{dt} = 12x \frac{dx}{dt}$$

$$= 12(3)(-1) = -36 \text{ cm}^2/\text{h}$$

62. Let A be the area of a circle of radius r and circumference C.

$$A = \pi r^2 \text{ so } \frac{dA}{dt} = 2\pi r = C$$

63. $v(0) = 1,200$ m/s; $v(2 \times 10^{-3}) = 6,000$ m/s;

$v(t) = at + v(0)$. Thus,

$$(2 \times 10^{-3})a = 6,000 - 1,200$$

$$a = \frac{4,800}{2 \times 10^{-3}} = 2.4 \times 10^6 \text{ m/s}$$

64. Consider a right triangle with horizontal leg of 3,000, vertical leg y, and the angle of elevation θ.

$$y = 3,000 \tan \theta$$

$$\frac{dy}{dx} = 3,000 \sec^2 \theta \frac{d\theta}{dt}$$

$$750 = 3,000 (\frac{5,000}{3,000})^2 \frac{d\theta}{dt}$$

$$\frac{d\theta}{dt} = (\frac{750}{3,000})(\frac{3}{5})^2 = 0.09$$

This is in radians/s; this is about 5°/s.

65. Let A be the cross-sectional area of the artery. $r_0 = 1.2$;

$$A = \pi r^2$$

$$\frac{dA}{dr} = 2\pi r \frac{dr}{dr}$$

$$= 2\pi(1.2 - 0.3)$$

$$= -1.8\pi \approx 5.65 \text{ mm}^2/\text{s}$$

66. **a.** $p'(x) = -2x; \Delta x = 0.05;$

$$\Delta p = -2x \frac{dx}{dt} = -2(2)(0.05) = -0.2$$

b. $\Delta x = 0.1; \Delta p = -2(2)(0.1) = -0.4$

c. $\Delta x = -0.05; \Delta p = -2(2)(-0.05) = 0.2$

67. **a.** Let x be the rate in gallons/mi and t the time in hours. The cost of the driver is:

$C_d = 16t$, where $t = \frac{300}{x}$ hr

The cost of gas is

$C_g = 2(\frac{1}{300})(\frac{1,500}{x} + x)(300) = 2(\frac{1,500}{x} + x)$

The total cost is

$$C(x) = 16(\frac{300}{x}) + 2(\frac{1,500}{x} + x)$$

$$= 7,800x^{-1} + 2x$$

b. $C'(x) = -\frac{4,800}{x^2} + \frac{-3,000}{x^2} + 2$

$$C'(55) \approx -0.5785$$

A change from 55 to 57 mi/h decreases the cost by about
$2(0.5785) \approx \$1.16.$

68. Let y be the height of the object and θ the angle of elevation.

$$y = 30 \tan \theta$$
$$\frac{dy}{dt} = 30 \sec^2\theta \, \frac{d\theta}{dt}$$

At $\theta = \frac{\pi}{4}$,

$$3 = (30)(2) \frac{d\theta}{dt}$$
$$\frac{d\theta}{dt} = 0.05$$

69. $m(v^2 - v_0^2) = k(x_0^2 - x^2)$

$$m(2v) \frac{dv}{dt} = k(-2x) \frac{dx}{dt}$$

Since $\frac{dx}{dt} = v$ and $\frac{dv}{dt} = a$

$$2mva = -2kxv$$
$$ma = -kx$$

Thus, $F = ma = -kx.$

70. $g[f(x)] = x;\ g'[f(x)][f'(x)] = 1;$

Thus, $g'[f(x)] = \dfrac{1}{f'(x)}.$

71. Let H be home base, F first base, P the position of the ball, and R the position of the runner. Let $s(t)$ be the distance between the ball and the runner, s_1 the distance between the runner and home base, $\theta = \angle PHR$, $\theta_1 = \angle RHF$. The runner runs at 30 ft/s and has traveled $90 - 25 = 63$ ft when the catcher throws the ball. Thus, he has been running for 65/30 sec to that time.

$$|\overline{RF}| = \left(\tfrac{65}{30} + t\right)30 = 65 + 30t \text{ ft}$$

where t is the time after the catcher releases the ball. The distance from the base to the runner is

$$s_1 = |\overline{HR}| = \sqrt{|\overline{HF}|^2 + |\overline{RF}|^2}$$
$$= \sqrt{90^2 + (65 + 30t)^2}$$

We also note $|\overline{PH}| = 120t$

Also,

$$\cos\theta = \cos(\tfrac{\pi}{4} - \theta_1)$$
$$= \frac{\sqrt{2}}{2}\cos\theta_1 - \frac{\sqrt{2}}{2}(-\sin\theta_1)$$
$$= \frac{\sqrt{2}}{2}(\cos\theta_1 + \sin\theta_1)$$

$$= \frac{\sqrt{2}}{2}\left(\frac{90}{s_1} + \frac{65 + 30t}{s_1}\right)$$
$$= \frac{\sqrt{2}}{2}\left(\frac{155 + 30t}{s_1}\right)$$

Finally, use the law of cosines

$$s^2 = |\overline{PH}|^2 + s_1^2 - 2|\overline{PH}|s_1\cos\theta$$

$$= (120t)^2 + [90^2 + (65+30t)^2]$$
$$- 2(120t)[90^2 + (65+30t)2]\left(\frac{\sqrt{2}}{2}\right)\left(\frac{155 + 30t}{s_1}\right)$$
$$= (120t)^2 + [90^2 + (65+30t)^2]$$
$$- \sqrt{2}(120t)(155 + 30t)$$

The rate of change of s satisfies

$$2s\frac{ds}{dt} = \frac{1}{2s}[2(120)^2t + 60(65 + 30t)$$
$$- \sqrt{2}(1{,}220)(155 + 60t)]$$

At the time the catcher throws the ball, we have $t = 0$, $s^2 = 0 + [90^2 + 65^2] - 0 \approx 111.02$. Thus,

$$\frac{ds}{dt} \approx \frac{1}{2(111.02)}[0 + 60(65) - \sqrt{2}(120)(155)]$$

$$\approx -100.9 \text{ ft/s}$$

72. Let x be the distance $\overline{OB}$. By the law of cosines

$$25 = 4 + x^2 - 2(5)x\cos\theta$$

Thus,

$$0 = 0 + 2x\frac{dx}{dt} - 10\left(\frac{dx}{dt}\cos\theta + x\sin\theta\frac{d\theta}{dt}\right)$$
$$\frac{dx}{dt}(2x - 109\cos\theta) = x\sin\theta\frac{d\theta}{dt}$$
$$\frac{dx}{dt} = \frac{x\sin\theta}{2x - 109\cos\theta}\frac{d\theta}{dt}$$

Since $\frac{d\theta}{dt} = 3\frac{\text{rad}}{\text{sec}}(2\pi)\frac{\text{rad}}{\text{rev}} = 6\pi\frac{\text{rad}}{\text{sec}}$

$$\frac{d^2x}{dt^2} = \frac{6\pi}{(2x - 109\cos\theta)^2}[(2x - 109\cos\theta)(x\cos\theta\frac{d\theta}{dt}$$
$$+ \frac{dx}{dt}\sin\theta) - (x\sin\theta)(2\frac{dx}{dt} + 109\sin\theta\frac{d\theta}{dt})]$$

$$= \frac{6\pi}{(2x - 109\cos\theta)^2}[(2x - 109\cos\theta)6\pi x\cos\theta$$

$$+ \frac{6\pi x\sin\theta}{2x - 109\cos\theta}\sin\theta$$

$$- (x\sin\theta)\frac{12\pi x\sin\theta}{2x - 109\cos\theta} + 609\pi\sin\theta]$$

73. $PV = kT;\ P'V + P'V = 0;$

$$V' = -\frac{P'V}{P} = -\frac{7(30)}{25} = -8.4 \text{ in.}^3/\text{min}$$

74. **a.** $f(0) = -2 < 0$; $f(10) > 0$, so there is at least one root on $[-10, 10]$.

b.

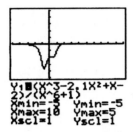

We need to zoom to show a root:

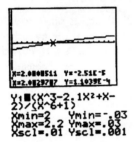

It looks like a root occurs at about 2.08.

c. Values for x_n, $f(x_n)$, and $f'(x_n)$.

n	x_n	$f(x_n)$	$f'(x_n)$
0	2	-0.006154	0.08895
1	2.0692	-0.000789	0.0671
2	2.08095	-0.000018	0.0639
3	2.08012436	0.0	0.0638
4	2.08012438		

We can check these results with further zooms from the calculator:

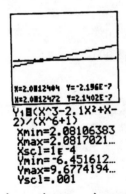

To 6 places, the trace shows $x \approx 2.08124$

d. $x = 3$ is to the right of a maximum of the curve. Tangent lines intersect the x-axis further and further away, to the detriment of convergence.

e. $x = 0.5$ is to the left of a minimum of the curve. Tangent lines intersect the x-axis further and further to the left, to the

detriment of convergence.

f. $x_0 = -1.0$ is to the right of a maximum of the curve. Tangent lines intersect the x-axis further and further away to the left, to the detriment of convergence.

75. **a.** $f(x) = x^3 - x = x(x - 1)(x + 1)$
$f'(x) = 3x^2 - 1$
Note that $f'(s_3) = 0$, which is not good for the Newton-Raphson method. Using $x_0 = s_5$ leads to $x_1 = -s_5$, and vice versa.

b. The concavity of the curve for $x > s_3$ indicates the iterates will monotonically approach 1.0. Likewise for $x < -s_3$ the iterates will approach -1.0.

c. Same argument as shown in part **b.**

d. The numbers verify parts **b** and **c.**

e. The first starting value leads to the zero at -1.0; the second to 1.0, and so on. The point is that this problem is extremely sensitive to the starting value of x.

76. This is Putnam Problem 1, morning session in 1946. Suppose that the function $f(x) = ax^2 + bx + c$, where a, b, c are real constants, satisfies the condition $|f(x)| \leq 1$. Prove $|f'(x)| \leq 4$ for $|x| \leq 1$. If $a \neq 0$, the graph of $y = ax^2 + bx + c$ is a parabola which can be assumed without loss of generality to open upward, *i.e.*, $a > 0$. [We discuss the straight line case later.] By symmetry we may assume that b is nonnegative. Then the vertex falls in the left half-plane and it is clear that $\max_{|x| \leq 1} |f'(x)|$ occurs when $x = 1$, and this maximum value is $2a + b$. It remains to show that $2a + b \leq 4$. Now

$$f(1) = a + b + c \leq 1, \text{ and } f(0) = c \geq -1.$$

Thus, $a + b \leq 2$. Since a and b are both nonnegative, $a \leq 2$ and $2a + b \leq 4$.

For the linear case we have, $a = 0$. If $a = 0$, then

$$f'(x) = b = \frac{f(1) - f(x)}{2}$$

so $$|f'(x)| = \frac{|f(1) + f(-1)|}{2} \leq 1$$

Historical note: The chemist Mendeleev raised the question as to restrictions on $p_n'(x)$ for $-1 \leq x \leq 1$ when $|p_n(x)| \leq 1$ on $-1 \leq x \leq 1$, then $|p_n'(x)| \leq 1$ on $-1 \leq x \leq 1$,

where p_n is a polynomial of degree n. A.A. Markov answered this question in 1890 by proving that if $p_n(x) \leq 1$ on $-1 \leq x \leq 1$, then $\left| p_n'(x) \right| \leq n^2$ on the same interval. The present problem is thus the special case where $n = 2$. It it known that equality occurs if and only if (except for sign) $p_n(x) = 2$. It is known that equality occurs if and only if $p_n(x) = \cos(n \cos^{-1} x)$; i.e. $p_n(x)$ is the polynomial such that $\cos(n \cos^{-1} x)$, i.e. $p_n(x)$ is the polynomial such that $\cos n\theta = p_n(\cos \theta)$. For $n = 2$, $\cos 2\theta (2)\cos^2 \theta = 2x^2 - 1$. The polynomial $p_n(x)$ are called Chebshev polynomials. See John Todd, "A survey of Numerical Analysis," New York, 1962, pp. 138-139. The generalized version appears as Problem 83, in Section 6, Polya and Szego, *Aufgaben und Lehrsatze aus der Analysis*, Vol 2, p. 91 and p. 287.

77. This is Putnam Problem 2, morning session in 1939. Let P have coordinates (x_0, y_0); then the slope at P is $3x_0^2$. The equation of the tangent at P is $y = 3x_0^2(x - x_0) + x_0^3$. The points of intersection of the tangent plane and the original curve are determined by the relation $x^3 = 3x_0^2(x - x_0) + x_0^3$, which is equivalent to $(x - x_0)^2(x + 2x_0) = 0$. Hence the second point of intersection is $(-2x_0, -8x_0^3)$. The slope at this point is $12x_0^2$, which is four times the slope at P, as was to be proved. If $x_0 = 0$, the tangent does not really meet the curve again. However, since the tangent in this case has a triple point of intersection with the curve, instead of the usual double point of intersection, it is reasonable to say that it meets the curve "again" at $(0, 0)$.

78. This is Putnam Problem 6, morning session in 1946. Newton's law of motion for a particle of unit mass takes the form F = force = dv/dt. Since we are given that $x = at + bt^2 + ct^3$, it follows that $v = dx/dt = a + 2bt + 3ct^2$; $dv/dt = 2b + 6ct$. We now express the force in terms of v:

$$F^2 = 4b^2 + 24bct + 36c^2t^2$$
$$= 4b^2 + 12c(2bt + 3ct^2)$$
$$= 4b^2 + 12c(v - a)$$

Hence $F = f(v) = \pm \sqrt{4b^2 - 12ac + 12cv}$.

The radical sign is taken to be the sign of $2b + 6ct$ which, if the hypotheses of the problem are satisfied, cannot change for the interval of time under consideration, since then v would take the same value twice but dv/dt would not.

CHAPTER 3

Additional Applications of the Derivative

3.1 Extreme Values of a Continuous Function, Page 175

1. $f(-3) = -34; f(3) = 26;$
$f'(x) = 10 - 2x$

$$10 - 2x = 0$$
$$x = 5$$

$f(5) = 30$; maximum value is 26 and the minimum value is -34.

2. $f(-4) = -30; f(4) = 18;$

$$f'(x) = 6 - 2x$$
$$6 - 2x = 0$$
$$x = 3$$

$f(3) = -17$; maximum value is 18 and the minimum value is -30.

3. $f(-1) = -4; f(3) = 0;$

$$f'(x) = 3x^2 - 6x$$
$$3x^2 - 6x = 0$$
$$3x(x - 2) = 0$$
$$x = 0, 2$$

$f(0) = 0; f(2) = -4$; maximum value is 0 and the minimum value is -4.

4. $f(-3) = 9; f(3) = 9;$
$f'(x) = 4t^3 - 16t$

$$4t^3 - 16t = 0$$
$$4t(t + 2)(t - 2) = 0$$
$$t = 0, -2, 2$$

$f(0) = 0; f(-2) = -16; f(2) = -16;$
maximum value is 9 and the minimum value is 0.

5. $f(-\frac{1}{2}) = -\frac{1}{8}; f(1) = 1;$

$$f'(x) = 3x^2$$
$$3x^2 = 0$$
$$x = 0$$

$f(0) = 0$; maximum value is 1 and the minimum value is $-\frac{1}{8}$.

6. $g(-2) = -2; g(2) = 2;$
$g'(x) = 3x^2 - 3$

$$3(x - 1)(x + 1) = 0$$
$$x = 1, -1$$

$g(1) = -2; g(-1) = 2$; maximum value is 2 and the minimum value is -2.

7. $f(-1) = -2; f(1) = 0;$
$f'(x) = 5x^4 - 4x^2$

$$5x^4 - 4x^2 = 0$$
$$x^3(5x - 4) = 0$$
$$x = 0, \frac{4}{5}$$

$f(0) = 0; f(\frac{4}{5}) \approx -0.08192$; maximum value is 0 and the minimum value is -2.

8. $g(-1) = 17; g(3) = 189;$
$g'(x) = 15t^4 - 80t^2$

$$15t^4 - 80t^2 = 0$$
$$5t^2(3t^2 - 16) = 0$$
$$t = 0, \pm\frac{4}{\sqrt{3}}$$

$g(0) = 0; g(\pm\frac{4}{3}\sqrt{12}) \approx \pm\frac{256}{3\sqrt{3}}$; maximum value is 189 and the minimum value is

$$-\frac{256}{3\sqrt[3]{3}} \approx -49.3.$$

9. $f(-1) = 1; f(1) = 1;$
$f'(x)$ is not defined at $x = 0; f(0) = 0$

Maximum value is 1 and the minimum value is 0.

10. $f(-4) = 7; f(4) = 1;$
$f'(x)$ is not defined at $x = 3; f(3) = 0.$

Maximum value is 7 and the minimum value is 0.

11. $f(0) = 1; f(2) = \sin^2 2 + \cos 2 \approx 0.41067;$
$f'(u) = 2 \sin u \cos u - \sin u$

$$2 \sin u \cos u - \sin u = 0$$
$$\sin u(2 \cos u - 1) = 0$$
$$u = 0, \frac{\pi}{3}$$

$f(0) = 1; f(\frac{\pi}{3}) = \frac{5}{4}$; maximum value is 1.25

and the minimum value is approximately
0.41067.

12. $g(0) = -1; \ g(\pi) = 1;$

$$g'(u) = \cos u - \sin u$$

$$\cos u - \sin u = 0$$

$$\tan u = 1$$

$$u = \frac{\pi}{4}$$

$g(\frac{\pi}{4}) = 0;$ maximum value is 1 and the

minimum value is $-1.$

13.   **a.**   Find the value of the function at each
endpoint of an interval.
       b.   Find the critical points, that is, points at
which the function is zero or undefined.
       c   Find the value of the function at each
critical point.
       d.   Read the absolute extrema.

14.   The calculator does not seem to take the
derivative at $x = 0$ and $x < 0$ into account.
The derivative is not defined at $x = 0$, but it
certainly is defined for $x < 0$.

15. $f(1) = 0; \ f(-1) = 0;$

$f'(u) = -\frac{2}{3}u^{-1/3}$ which is not defined at

$u = 0.$ $f(0) = 1.$ The maximum value is 1
and the minimum value is 0.

16. $g(-50) = 0; \ g(14) = 16;$

$g'(t) = \frac{2}{3}(50 + t)^{-2/3}$ which is not defined at

$t = -50.$ The maximum value is 16 and the
and the minimum value is 0.

17. $f(-0.1) \approx -3; \ f(\pi + 0.1) \approx -5;$

$$f'(\theta) = 3\cos^2\theta(-\sin \theta) - 8 \cos \theta(-\sin \theta)$$

$$3\cos^2\theta(-\sin \theta) - 8 \cos \theta(-\sin \theta) = 0$$

$$(\cos \theta \sin \theta)(-3 \cos \theta + 8) = 0$$

$\cos \theta = 0$ $\sin \theta = 0$ $-3 \cos \theta + 8 = 0$

$\theta = \frac{\pi}{2}$ $\theta = \pi$ $\cos \theta = \frac{8}{3}$

No values of θ cause $\cos \theta$ to be equal to 8/3.

$f(\frac{\pi}{2}) = 0; \ f(\pi) = -5.$ The maximum value
is 0 and the minimum value is $-5.$

18. $g(-2) = -2 \sin(-2) \approx 1.819;$

$g(2) = 2 \sin 2 \approx 1.819;$

$$g'(\theta) = \theta \cos \theta + \sin \theta$$

$$\theta \cos \theta + \sin \theta = 0$$

$$\tan \theta = -\theta$$

The only solution on $[-2, 2]$ is $\theta = 0.$ The
maximum value is approximately 1.819 and
the minimum value is 0.

19. $f(-1) = 0; \ f(1) = 0;$

$$f'(x) = 20(378\pi)\cos(378\pi x)$$

$$20(378\pi) \cos(378\pi x) = 0$$

$$378\pi x = \frac{n\pi}{2}$$

$$x = \frac{n}{756}$$

for some integer n (so that $n \leq 756$).

$f(\frac{n}{756}) = \pm 20;$ the maximum value is 20

and the minimum value is $-20.$ There will
be 378 points with a maximum of 20, and 378
points with a minimum of $-20.$ Your
graphing calculator will most likely not be
able to show this on $[-1, 1].$ A change of the
horizontal scale gives the following graph:

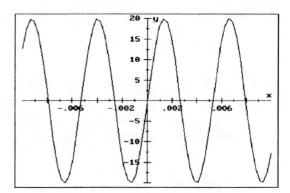

20. $g(-4) = -28; \ g(4) = -60$
$g'(x) = 6x^2 - 6x - 36.$
      $6x^2 - 6x - 36 = 0$
      $6(x - 3)(x + 2) = 0$
              $x = 3, \ -2$
$g(3) = -77; \ g(-2) = 48$
The maximum value is 48 and the minimum
value is $-77.$

21. $g(-4) = 76; \ g(4) = 12$

$$g'(x) = 3x^2 + 6x - 24$$

$$3x^2 + 6x - 24 = 0$$

$$3(x + 4)(x - 2) = 0$$

$$x = -4, 2$$

$$g(2) = -32$$

The maximum value is 76 and the minimum value is -32.

22. $f(-4) = -287\frac{2}{3}; \; f(4) = 117\frac{2}{3}$

$f'(x) = 8x^2 - 10x + 8$

$8x^2 - 10x + 8 = 0$

$4x^2 - 5x + 4 = 0$

$$x = \frac{5 \pm \sqrt{25 - 64}}{8}$$

There are no real roots. The maximum value is approximately 117.7 and the minimum value is approximately -287.7.

23. $f(0) = \frac{1}{6}; \; f(2) = \frac{1}{2}$

$f'(x) = \frac{1}{6}(3x^2 - 12x + 9)$

$3x^2 - 12x + 9 = 0$

$3(x - 1)(x - 3) = 0$

$x = 1, 3$

$f(1) = \frac{5}{6}; \; f(3) = \frac{1}{6}$

The maximum value is $\frac{5}{6}$ and the minimum value is $\frac{1}{6}$.

24. $g(0) = 0; \; g(4) = 6{,}496;$

$g'(u) = 294u^2 - 8u + 72$

$294u^2 - 8u + 72 = 0$

$147u^2 - 4u + 36 = 0$

There are no real roots for this equation. The maximum value is 6,496 and the minimum value is 0.

25. $f(0) = 0; \; f(4) = -2$

$f'(w) = \frac{1}{2}w^{-1/2}(w-5)^{1/3} + \frac{1}{3}w^{1/2}(w-5)^{-2/3}$

$\qquad = \frac{1}{6}w^{-1/2}(w-5)^{-2/3}[3(w-5) + 2w]$

$\qquad = \frac{1}{6}w^{-1/2}(w-5)^{-2/3}(5w - 15)$

$\qquad = \frac{5}{6}w^{-1/2}(w-5)^{-2/3}(w-3)$

$f'(w) = 0$ when $w = 3$

$f(3) = \sqrt{3}(3-5)^{1/3} = -\sqrt[6]{108} \approx -2.1822$

The maximum value is 0 and the minimum value is $-\sqrt[6]{108} \approx -2.1822$.

26. $h(-1) = -9^{1/3} \approx -2.08;$

$h(4) = 16^{1/3} \approx 2.52$

$h'(x) = \frac{1}{3}x^{-2/3}(x-3)^{2/3} + \frac{2}{3}x^{1/3}(x-3)^{-1/3}$

$\qquad = \frac{1}{3}x^{-2/3}(x-3)^{-1/3}[(x-3) + 2x]$

$\qquad = \frac{1}{3}x^{-2/3}(x-3)^{-1/3}(3x-3)$

$\qquad = x^{-2/3}(x-3)^{-1/3}(x-1)$

$h'(x) = 0$ when $x = 1$

$h(1) = 2^{2/3} \approx 1.58$

The maximum value is 2.52 and the minimum value is -2.08.

27. $h(0) = 1$, $h(2\pi) = 0$ but h is not continuous. There is no maximum and no minimum value.

28. $s(0) = 0; \; s(2\pi) = 2\pi$

$s'(t) = t(-\sin t) + \cos t - \cos t = -t \sin t$

$s'(t) = 0$ when $t = 0, \pi$

The maximum value is 2π and the minimum value is $-\pi$.

29. $f(1) = 5; \; f(4) = 8$

$f'(x) = -4$ if $x < 1$; $f'(x) = -2x + 6$ if $x \ge 1$

$f'(x) = 0$ when $x = 3$; $f(3) = 8$

The maximum value is 13 and the minimum value is 5.

30. $f(-1) = 11; \; f(4) = -4; \; f(2) = 2$

$f'(x) = -3$ if $x < 2$; $f'(x) = -2x + 3$

for $x \ge 2$; $f'(x) = 0$ when $x = \frac{3}{2}$.

$f(\frac{3}{2}) = \frac{7}{2}$

The maximum value is 11 and the minimum value is -4.

31. $f(-1) = f(1) = 1$; $f'(x) = 2x = 0$ at $x = 0$; $f(0) = 0$. The smallest value is 0.

32. $f(0)$ is not defined; there is no largest value.

33. $g(1) = 7$; $g(9) = 7$; $g'(x) = -9x^{-2} + 1$ which is equal to 0 when $x = 3$ (-3 is not in the domain). $g(3) = 3$. The smallest value is 3.

34. $g(\pm 1) = 0;$

$g'(x) = \dfrac{(x^2+1)(2x) - (x^2-1)(2x)}{(x^2+1)^2} = \dfrac{4x}{(x^2+1)^2}$

$g'(x) = 0$ when $x = 0$; $g(0) = -1$

The smallest value is -1.

35. $f(-2) = 0; \; f(3) = 0; \; f(1) = 2$

The function f is not continuous at $t = 1$.

$f'(t) = -2t - 1$ if $t < 1$;

$f'(t) = -1$ if $t \ge 1$

$f'(t) = 0$ when $t = -\frac{1}{2}$; $f\left(-\frac{1}{2}\right) = \frac{9}{4}$

The maximum value is $\frac{9}{4}$.

Answers to Problems 36-41 may vary.

36.   a. $f(x) = \begin{cases} x^2 \text{ for } -1 < x < 1 \\ x + 2 \text{ for } 1 \leq x < 2 \end{cases}$

         The minimum is 0, but there is no maximum.

     b. $f(x) = \begin{cases} -x^2 \text{ for } -1 < x < 1 \\ -x + 2 \text{ for } 1 \leq x < 2 \end{cases}$

         The maximum is 0, but there is no minimum.

     c. $f(x) = \begin{cases} \sin x \text{ for } -\pi < x \leq \frac{\pi}{2} \\ 0.5 \text{ for } \frac{\pi}{2} < x < 3 \end{cases}$

         The maximum is 1 and the minimum is -1.

     d. $f(x) = \begin{cases} \sin x \text{ for } 0 < x < \frac{\pi}{2} \\ 0.5 \text{ for } \frac{\pi}{2} < x < 3 \end{cases}$

         There is no maximum and there is no minimum.

37.   a. $f(x) = \begin{cases} x^2 \text{ for } -0.5 \leq x < 1 \\ -x + 2 \text{ for } 1 \leq x < 1.5 \end{cases}$

         The minimum is 0, but there is no maximum.

     b. $f(x) = \begin{cases} -x^2 \text{ for } -1 < x < 1 \\ x - 2 \text{ for } 1 < x < 1.52 \end{cases}$

         The maximum is 0, but there is no minimum.

     c. $f(x) = \begin{cases} \sin x \text{ for } -\pi \leq x \leq \frac{\pi}{2} \\ 0.5 \text{ for } \frac{\pi}{2} < x \leq 3 \end{cases}$

         The maximum is 1 and the minimum is -1.

     d. $f(x) = \begin{cases} -0.5 \text{ for } x = -\frac{\pi}{2} \\ 1 + \sin x \text{ for } x \text{ rational} \\ \sin x \text{ for } x \text{ irrational} \\ 0.5 \text{ for } x = \frac{\pi}{2} \end{cases}$

         There is no maximum and there is no minimum.

38.   a. $f(x) = \sin x$ on $(-\pi, 1)$
         The minimum is 0, but there is no maximum.

     b. $f(x) = \sin x$ on $(0, 4)$.
         The maximum is 1, but there is no minimum.

     c. $f(x) = \sin x$ on $(0, 2\pi)$
         The maximum is 1 and the minimum is -1.

     d. $f(x) = \sin x$ on $(-1, 1)$.
         There is no maximum and no minimum.

39.   a.   No such function can be found because of of the extreme value theorem (theorem 3.1).

     b.   No such function can be found because of the extreme value theorem.

     c. $f(x) = \sin x$ for $[0, 2\pi]$.

     d.   No such function can be found because of the extreme value theorem.

40. $f(x) = x^{-2}$ on $[-1, 1]$ has no maximum.

41. $f(x) = x^{-1}$ on $(0, 1)$ has no extremum.

42. $v(t) = s'(t) = 3t^2 - 12t - 15$

       $v(0) = -15$; $v(4) = -15$

         $v'(t) = 6t - 12 = 0$ when $t = 2$

       $v(2) = -18$

       The maximum value for the velocity is -15.

43. $v(t) = s'(t) = 4t^3 - 6t^2 - 24t + 60$

       $v(0) = 60$; $v(3) = 42$

         $v'(t) = 12t^2 - 12t - 24$

             $= 12(t - 2)(t + 1)$

       $v'(t) = 0$ when $t = -1, 2$

       Reject $t = -1$ since it is not in the domain.

       $v(2) = 20$

       The maximum velocity is 60 when $t = 0$.

44.   Let x and y be the numbers we are seeking on $[0, 12]$. Then $x + y = 8$ and $P = x^2(y - x)^2$.

       $P'(x) = x^2(2)(8 - x)(-1) + 2x(8 - x)^2$

            $= 2x(8 - x)(8 - 2x)$

       $P'(x) = 0$ when $x = 0, 8$, and 4.

       $P(0) = P(8) = 0$; $P(4) = 256$.

       The largest product occurs when $x = y = 4$.

45. Let x and y be the numbers we are seeking on $[0, 6]$. Then $2x + y = 12$ and $P = x(12 - 2x)$.

$P'(x) = 12 - 4x = 0$ when $x = 3$

$P(0) = P(6) = 0$; $P(3) = 18$.

The largest product occurs when $x = 3$ and $y = 6$.

46. $P = xy^3 = x(80 - 3x)^3$ on $[0, \frac{80}{3}]$.
$P(0) = P(\frac{80}{3}) = 0$

$P'(x) = x(3)(80 - 3x)^2(-3) + (80 - 3x)^3$

$\qquad = (80 - 3x)^2(80 - 3x - 9x)$

$\qquad = (80 - 3x)^2(80 - 12x)$

$P'(x) = 0$ when $x = \frac{80}{3}$ and $x = \frac{20}{3}$

$P(\frac{20}{3}) = \frac{20}{3}\left[80 - 3(\frac{20}{3})\right]^3 = 1,200^2$

The largest product occurs when $x = \frac{20}{3}$ and $y = 60$.

47. $P = xy = x(126 - 3x)$ on $[0, 42]$.

$P(0) = P(42) = 0$

$P'(x) = -6x + 126 = 0$ when $x = 21$

$P(21) = 1,323$

The largest product occurs when $x = 21$ and $y = 63$.

48. $P = x^2y = x^2\left(\frac{-18 + 2x}{5}\right) = \frac{2}{5}(x^3 - 9x^2)$ on

$[0, 9]$; $P(0) = P(9) = 0$

$P'(x) = \frac{2}{5}(3x^2 - 18x)$

$\qquad = \frac{2}{5}(3x)(x - 6)$

$P'(x) = 0$ when $x = 0$ and $x = 6$

$P(6) = \frac{2}{5}(6^3 - 9 \cdot 6^2) = -\frac{216}{5}$

The largest product occurs when $x = 6$ and $y = -\frac{6}{5}$.

49. Let x and y be the sides of a rectangle. Then the perimeter is $P = 2x + 2y$ or

$y = \frac{P - 2x}{2}$.

$A(x) = x\left(\frac{P - 2x}{2}\right) = \frac{1}{2}(xP - 2x^2)$

$A'(x) = \frac{1}{2}(P - 4x) = 0$ when $x = \frac{P}{4}$

$A(0) = A(\frac{P}{2}) = 0$; $A(\frac{P}{4}) = \frac{P^2}{16}$

The largest area occurs when $x = P/4$.

$y = \frac{P - 2(P/4)}{2} = \frac{P}{4}$. Thus, $x = y$.

50. Let $P = xy$ be the product; since $x^2 + y^2 = a^2$

we have

$P = x\sqrt{x^2 - a^2}$

$P' = (x^2 - a^2)^{1/2} + \frac{1}{2}x(x^2 - a^2)^{-1/2}(2x)$

$\quad = \frac{2x^2 - a^2}{(x^2 - a^2)^{1/2}}$

Now $P' = 0$ when $x = \frac{a}{\sqrt{2}}$

$P(0) = P(a) = 0$;

$P(\frac{a}{\sqrt{2}}) = \frac{a}{\sqrt{2}}\sqrt{\frac{a^2}{2} - a^2} = \frac{a^2}{2}$

The points are $\left(\frac{a}{\sqrt{2}}, \frac{a}{\sqrt{2}}\right), \left(\frac{a}{\sqrt{2}}, -\frac{a}{\sqrt{2}}\right)$.

51. $A(x) = \frac{C(x)}{x} = 0.125x + \frac{20,000}{x}$

$A'(x) = 0.125 - \frac{20\,000}{x^2} = 0$ when

$x^2 = 160,000$, $x = 400$.

Now consider $A(x) = C'(x)$:

$0.125x + \frac{20\,000}{x} = 0.25x$

$0.125x^2 = 20,000$

$x = 400$

52. a. $\qquad x > x^2$

$x - x^2 > 0$

$x(1 - x) > 0$

Solution is $(0, 1)$

Let $P(x) = x - x^2$; $P'(x) = 1 - 2x = 0$ at $x = \frac{1}{2}$. $P(0) = P(1) = 0$; $P(\frac{1}{2}) = \frac{1}{4}$; The greatest difference occurs at $x = \frac{1}{2}$.

b. $\qquad x > x^3$

$x - x^3 > 0$

$x(1 - x^2) > 0$

Solution for nonnegative x is $(0, 1)$

Let $P(x) = x - x^3$; $P'(x) = 1 - 3x^2 = 0$ at $x = \frac{1}{\sqrt{3}}$ (disregard negative value). $P(0) = P(1) = 0$; $P(\frac{1}{\sqrt{3}}) = \frac{2}{\sqrt{3}}$; The greatest difference occurs at $x = \frac{1}{\sqrt{3}}$.

c. $\qquad x > x^n$

$x - x^n > 0$

$x(1 - x^n) > 0$

Solution for nonnegative x is $(0, 1)$

Let $P(x) = x - x^n$; $P'(x) = 1 - nx^{n-1} = 0$

at $x = (\frac{1}{n})^{1/(n-1)}$.

$P(0)$ $P(1) = 0$; $P[(\frac{1}{n})^{1/(n-1)}] =$

$n^{-1/(n-1)} - n^{-n/(n-1)}$

The greatest difference occurs at

$x = (\frac{1}{n})^{1/(n-1)}$.

53. Let

$S(x) = (a_1 - x)^2 + (a_2 - x)^2 + \cdots + (a_n - x)^2$

$S'(x) = -2(a_1 - x + a_2 - x + \cdots + a_n - x)$

$S'(x) = 0$ if $x = \dfrac{a_1 + a_2 + \cdots + a_n}{n} = \bar{x}$

54. Let $f(x) = mx - 1 + \frac{1}{x}$ be the vertical

distance from the tangent line to the curve.

$f'(x) = m - x^{-2} = 0$ when $x = m^{-1/2}$.

$f(m^{-1/2}) = mm^{-1/2} + m^{1/2} = 2m^{1/2}$

The minimum distance occurs when

$x = m^{-1/2}$

55. $f'(\theta) = -8 \csc \theta \cot \theta + 27 \sec \theta \tan \theta$

$\quad = \dfrac{-8 \cos \theta}{\sin \theta \sin \theta} + \dfrac{27 \sin \theta}{\cos \theta \cos \theta}$

$\quad = \dfrac{-8 \cos^3\theta + 27 \sin^3\theta}{\sin^2\theta \cos^2\theta}$

$f'(\theta) = 0$ when

$\qquad -8 \cos^3\theta + 27 \sin^3\theta = 0$

$\qquad\qquad \tan^3\theta = \dfrac{8}{27}$

$\qquad\qquad \tan \theta = \dfrac{2}{3}$

When $\theta \to 0^+$ or $\theta \to (\frac{\pi}{2})^-$, $f(\theta) \to +\infty$

so a the critical value $\theta = \tan^{-1}(\frac{2}{3})$ is a

minimum.

56. Since $f(x)$ has a relative minimum at $x = c$,

$\qquad\qquad f(c) \leq f(c + \Delta x)$

$\qquad f(c) - f(c + \Delta x) \leq 0$

$\qquad\qquad f(c + \Delta x) \geq 0$

If $\Delta x > 0$, then $\dfrac{f(c + \Delta x) - f(c)}{\Delta x} \geq 0$

$\qquad \lim_{\Delta x \to 0} \dfrac{f(c + \Delta x) - f(c)}{\Delta x} \geq 0$

$\qquad\qquad f'(c) \geq 0$

If $\Delta x < 0$, then $\dfrac{f(c + \Delta x) - f(c)}{\Delta x} \leq 0$

$\qquad \lim_{\Delta x \to 0} \dfrac{f(c + \Delta x) - f(c)}{\Delta x} \leq 0$

$\qquad\qquad f'(c) \leq 0$

Thus, c is a critical point.

3.2 The Mean Value Theorem, Page 181

1. There is at least one number c on (a, b) such that $f'(c) = 0$ if f is continuous on $[a, b]$ and $f'(x)$ exists on (a, b). The tangent line is horizontal at $(c, f(c))$. The importance of this theorem is in its use in proving the mean value theorem.

2. If $f(x)$ is continuous and differentiable on a closed interval, then there is a point at which the tangent line to the curve represented by this function is parallel to the line segment joining the endpoints of the curve. These hypotheses are used in the proof so that the conditions of Rolle's theorem are satisfied. Yes, the conclusion can be true even if all of the hypotheses are not satisfied. For example, if $f(x) = x^{-2}$ on $[-2, 2]$, the tangent line from $(-2, 0.25)$ to $(2, 0.25)$ is horizontal, but there is no horizontal tangent line to the curve. If $f(x) = |x|$ on $[-2, 2]$, the tangent line from $(-2, 2)$ to $(2, 2)$ is horizontal, but there is no horizontal tangent line to the curve.

3. Polynomials are everywhere continuous and differentiable, so the hypotheses of MVT are met. $f'(x) = 4x$, so there exists a c on the interval $[0, 2]$ such that

$$f'(c) = \frac{f(2) - f(0)}{2 - 0}$$

$$4c = \frac{9 - 1}{2}$$

$$8c = 8$$

$$c = 1$$

4. Polynomials are everywhere continuous and differentiable, so the hypotheses of MVT are met. $f'(x) = -2x$, so there exists a c on the interval $[-1, 0]$ such that

$$f'(c) = \frac{f(0) - f(-1)}{0 - (-1)}$$

$$-2c = \frac{4 - 3}{1} \text{ or } c = -\frac{1}{2}$$

5. Polynomials are everywhere continuous and differentiable, so the hypotheses of MVT are met. $f'(x) = 3x^2 + 1$, so there exists a c on the interval $[1, 2]$ such that

$$f'(c) = \frac{f(2) - f(1)}{2 - 1}$$

$$3c^2 + 1 = \frac{10 - 2}{1}$$

$$3c^2 = 7$$

$$c = \pm\sqrt{\frac{7}{3}}$$

The number $\sqrt{\frac{7}{3}} \approx 1.5275$ is on the interval $[1, 2]$.

6. Polynomials are everywhere continuous and differentiable, so the hypotheses of MVT are met. $f'(x) = 6x^2 - 2x$, so there exists a c on the interval $[0, 2]$ such that

$$f'(c) = \frac{f(2) - f(0)}{2 - 0}$$

$$6c^2 - 2c = \frac{12 - 0}{2}$$

$$3c^2 - c - 3 = 0$$

$$c = \frac{1 \pm \sqrt{1 + 36}}{6} \approx 1.1805$$

7. Polynomials are everywhere continuous and differentiable, so the hypotheses of MVT are met. $f'(x) = 4x^3$, so there exists a c on the interval $[-1, 2]$ such that

$$f'(c) = \frac{f(2) - f(-1)}{2 - (-1)}$$

$$4c^3 = \frac{18 - 3}{3}$$

$$c^3 = 15$$

$$c = \sqrt[3]{5} \approx 1.0772$$

8. Polynomials are everywhere continuous and differentiable, so the hypotheses of MVT are met. $f'(x) = 5x^4$, so there exists a c on the interval $[2, 4]$ such that

$$f'(c) = \frac{f(4) - f(2)}{4 - 2}$$

$$5c^4 = \frac{1,027 - 80}{2}$$

$$c^4 = 94.7$$

$$c = \sqrt[4]{94.7} \approx 3.12$$

9. $f(x)$ is continuous and differentiable everywhere on $[1, 4]$, so the hypotheses of MVT are met. $f'(x) = \frac{1}{2\sqrt{x}}$, so there exists a c on the interval $[1, 4]$ such that

$$f'(c) = \frac{f(4) - f(1)}{4 - 1}$$

$$\frac{1}{2\sqrt{c}} = \frac{2 - 1}{3}$$

$$\sqrt{c} = \frac{3}{2}$$

$$c = \frac{9}{4}$$

10. $f(x)$ is continuous and differentiable everywhere on $[1, 4]$, so the hypotheses of MVT are met. $f'(x) = -\frac{1}{2}x^{-3/2}$ so there exists a c on the interval $[1, 4]$ such that

$$f'(c) = \frac{f(4) - f(1)}{4 - 1}$$

$$-\frac{1}{2}c^{-3/2} = \frac{\frac{1}{2} - 1}{2}$$

$$c^{-3/2} = \frac{1}{3}$$

$$c = 3^{2/3} \approx 2.08$$

11. $f(x)$ is continuous and differentiable everywhere on $[0, 2]$, so the hypotheses of MVT are met. $f'(x) = -\frac{1}{(x + 1)^2}$, so there exists a c on the interval $[0, 2]$ such that

$$f'(c) = \frac{f(2) - f(0)}{2 - 0}$$

$$-\frac{1}{(c + 1)^2} = \frac{\frac{1}{3} - 1}{2},$$

$$\frac{2}{3}(c + 1)^2 = 2$$

$$c + 1 = \pm\sqrt{3}$$

$$c = -1 \pm \sqrt{3}$$

Note that only $c = -1 + \sqrt{3} \approx 0.73$ lies in the specified interval.

12. $f(x)$ is continuous and differentiable everywhere on $[1, 4]$, so the hypotheses of MVT are met. $f'(x) = -\frac{1}{x^2}$, so there exists a c on the interval $[1, 4]$ such that

$$f'(c) = \frac{f(4) - f(1)}{4 - 1}$$

$$-\frac{1}{c^2} = \frac{\frac{5}{4} - 2}{3}$$

$$c^2 = 4$$

$$c = \pm 2$$

The desired solution $c = 2$ is on the interval.

13. $f(x)$ is continuous and differentiable everywhere on $[0, \frac{\pi}{2}]$, so the hypotheses of MVT are met. $f'(x) = -\sin x$, so there exists a c on the interval $[0, \frac{\pi}{2}]$ such that

$$f'(c) = \frac{f(\frac{\pi}{2}) - f(0)}{\frac{\pi}{2} - 0}$$

$$-\sin c = \frac{0 - 1}{\frac{\pi}{2} - 0}$$

$$\sin c = \frac{2}{\pi}$$

The desired solution $c = \sin^{-1}(\frac{2}{\pi}) \approx 0.6901$.

14. $f(x)$ is continuous and differentiable everywhere on $[0, 2\pi]$, so the hypotheses of MVT are met. $f'(x) = \cos x - \sin x$ so there exists a c on the interval $[0, 2\pi]$ such that

$$f'(c) = \frac{f(2\pi) - f(0)}{2\pi - 0}$$

$$\cos c - \sin c = \frac{1 - 1}{2\pi}$$

$$\cos c = \sin c$$

$$c = \frac{\pi}{4}, \frac{5\pi}{4} \text{ on } [0, 2\pi]$$

15. $f(x)$ is continuous on the closed interval, but is not differentiable on the open interval as there is a cusp at $(2,0)$. The theorem does not apply.

16. Rolle's theorem is not applicable since $\tan x$ is not continuous at $\pi/2$ and $3\pi/2$.

17. Rolle's theorem is applicable since $f(x) = \sin x$ is continuous on $[0, 2\pi]$ and differentiable on $(0, 2\pi)$.

18. Rolle's theorem is applicable since $f(x) = |x| - 2$ is continuous on $[0, 4]$ and differentiable on $(0, 4)$. It is not differentiable at 0, but that is not part of the open interval.

19. Rolle's theorem is not applicable since $f(x) = \sqrt[3]{x}$ is continuous on $[-8, 8]$ but $f'(0) = \frac{1}{3}0^{-2/3}$ does not exist.

20. Rolle's theorem is applicable since $f(x) = \frac{1}{x - 2}$ is continuous on $[-1, 1]$ and differentiable on $(-1, 1)$. It is not differentiable at $x = 2$, but that is not part of the open interval.

21. Rolle's theorem is not applicable on $[1, 2]$ since $f(x) = \frac{1}{x - 2}$ is not continuous at $x = 2$.

22. Rolle's theorem is not applicable on $[-\pi, \pi]$ since $f(x) = 3x + \sec x$ is not continuous at $x = \pi/2$ or $3\pi/2$.

23. $f(x)$ is continuous everywhere, and

$$f'(x) = 2 \sin x \cos x = \sin 2x \text{ so}$$

$f(x)$ is differentiable everywhere.

$f(a) = f(b) = 1$. Rolle's theorem applies.

24. By the MVT there exists a number c on (a, b) such that

$$\frac{f(b) - f(a)}{b - a} = f'(c)$$

$$f(b) - f(a) = (b - a)f'(c)$$

$$f(b) = f(a) + (b - a)f'(c)$$

25. $f(x) = \tan x$ so that

$$\left|\frac{\tan u - \tan v}{u - v}\right| = \sec^2 c \geq 1$$

$$|\tan u - \tan v| = |u - v|$$

26. The MVT is not applicable on $[-1, 1]$ since $f(x) = 1/x$ is not continuous at $x = 0$.

27. $g(x) = |x|$ is continuous everywhere, but has a cusp at $x = 0$, so is not differentiable there. The theorem does not apply.

28. $f(x) = \cos x$ is continuous on $[x, y]$ and differentiable on (x, y). Also, $f'(x) = \sin x$, so by the MVT we have

$$\frac{\cos x - \cos y}{x - y} = \sin c \leq 1$$

$$\cos x - \cos y \leq x - y \leq |x - y|$$

The slope of the tangent line is at most -1 at $c_1 = 3\pi/2$. If $\cos x$ and $\cos y$ were on that line, $\cos x - \cos y = -1$, but because the cosine curve is not linear $|\cos x - \cos y| < 1$ (even close to c_1). Thus,

$$|\cos x - \cos y| \leq |x - y|$$

29. The zero derivative theorem does not apply because $f(x)$ is not continuous at $x = 0$.

30. **a.** $f(x) = (1 + x)^n$; $f'(x) = n(1 + x)^{n-1}$;

$$f(0) = 1; \frac{f(x) - f(0)}{x - 0} = f'(c)$$

b. $\frac{(1 + x)^n - 1}{x} = n(1 + c)^{n-1}$;

$$\lim_{x \to 0} \frac{(1+x)^n - 1}{x} = \lim_{x \to 0} n(1 + c)^{n-1} = n$$

31. **a.** Let $f(x) = \cos x - 1$ on $[0, x]$. The hypotheses of the MVT apply, so there

exists a w on the interval such that

$$f'(w) = \frac{f(0) - f(x)}{0 - x}$$

$$\sin w = \frac{0 - \cos x + 1}{-x}$$

$$(-\sin w)x = \cos x - 1$$

b. Since w is on the interval $[0, x]$, as x approaches 0, w must approach 0 also. So

$$\lim_{x \to 0} \frac{\cos x - 1}{x} = \lim_{w \to 0} \frac{\cos w - 1}{w}$$

$$= \lim_{w \to 0} (-\sin x) = 0$$

32. $f(x) = \cos x$; $f(\pi) = -1$; $f'(x) = -\sin x$ so

$$\frac{\cos x - \cos \pi}{x - \pi} = -\sin c \text{ for } 0 < c < x$$

$$\lim_{x \to \pi^+} \frac{\cos x + 1}{x - \pi} = -\lim_{c \to 0} \sin c = 0$$

33. $f(x) = 1 + \frac{1}{x}$; $f'(x) = -\frac{1}{x^2} < 0$ for all x

$f(b) > 1$ and $f(a) < 1$. Then, $f(b) - f(a) > 0$

and $\dfrac{f(b) - f(a)}{b - a} > 0$. As a result,

$\dfrac{f(b) - f(a)}{b - a} = f'(w)$ is impossible if

$a < w < b$. The MVT does not apply because $f(x)$ is not continuous at $x = 0$.

34. $f(x) = \sqrt{x}$; $f'(x) = \frac{1}{2\sqrt{x}}$; $f(4) = 2$

By the MVT,

$$\frac{\sqrt{xs} - 2}{x - 4} = \frac{1}{2\sqrt{w}} \quad \text{for } x < w < 4$$

$$\sqrt{x} - 2 = \frac{x - 4}{2\sqrt{w}}$$

If $x > 4$, $\sqrt{w} > 2$, $\frac{1}{\sqrt{w}} < \frac{1}{2}$, and

$$\sqrt{x} < 2 + \frac{x - 4}{2(2)} = 2 + \frac{cx - 4}{4}$$

35. A straight line will be continuous and differentiable, and $v(t_1) = v(t_2)$, so the hypotheses of Rolle's theorem are met. This guarantees that there exists some value c in the time interval for which $v'(c) = 0$. Since the acceleration is $v'(t)$, this is what was to have been shown.

36. Let $s(t)$ be the distance traveled during time t. Then

$$\frac{s(5) - s(0)}{5 - 0} = s'(t) = v(t)$$

Now, $v(t) = \dfrac{6 \text{ mi}}{\text{min}} \cdot \dfrac{60 \text{ min}}{\text{hr}} = 72$ mi/h

37. $\dfrac{P(b) - P(a)}{b - a} = P'(c)$ for $a < c < b$

$P'(x) = R'(x) - C'(x)$. If $P(b) = P(a)$, then $P'(c) = 0$ and $R'(c) = C'(c)$.

38. Let $f(x) = (x - 1)\sin x$, then

$$f'(x) = (x - 1)\cos x + \sin x$$

$f(0) = 0$ and $f(1) = 0$; for $0 < c < 1$,

$$f'(c) = (c - 1)\cos c + \sin c = 0 \text{ or}$$

$$1 - c = \tan c.$$

39. If $x > 15$ then $f(x) = \sqrt{1 + x}$ is continuous and differentiable on $[15, x]$ and the hypotheses of the MVT are met. Therefore, there exists a c on $[15, x]$ such that

$$f'(c) = \frac{1}{2\sqrt{1 + c}}$$

By the MVT,

$$\frac{f(x) - f(15)}{x - 15} = \frac{1}{2\sqrt{1 + c}}$$

$$\frac{\sqrt{1 + x} - 4}{x - 15} = \frac{1}{2\sqrt{1 + c}}$$

$$\sqrt{1 + x} = 4 + \frac{x - 15}{2\sqrt{1 + c}}$$

but c is on $[15, x]$ and so $c > 15$, making the denominator of the fraction greater then 8, and

$$4 + \frac{x - 15}{2\sqrt{1 + c}} < 4 + \frac{x - 15}{8},$$

so by transitivity

$$\sqrt{1 + x} < 4 + \frac{x - 15}{8}$$

40. $f(x) = \dfrac{1}{2x + 1}$; $f'(x) = -\dfrac{2}{(2x + 1)^2}$ for $[0, 2]$

$f(2) = \frac{1}{5}$; With $x < c < 2$, we have from the MVT,

$$\frac{\frac{1}{2x + 1} - \frac{1}{5}}{x - 2} = -\frac{1}{(2x + 1)^2}$$

$$\frac{1}{2x + 1} - \frac{1}{5} = -\frac{2(x - 2)}{(2c + 1)^2}$$

Since $\dfrac{1}{(2c + 1)^2} > \dfrac{1}{25}$,

$$\frac{1}{2x + 1} > \frac{1}{5} + \frac{2(2 - x)}{25}$$

41. The hypotheses of the MVT are not satisfied. $f(x) = \tan x$ is not continuous at $x = \pi/2$.

42. Let $f(x) = x^3 - 3x + a$ and $0 < x < 1$

$f(1) = a - 2$; $f(0) = a$, and

$$\frac{f(1) - f(0)}{1 - 0} = f'(c)$$

$$a - (a - 2) = 3(c^2 - 1) \text{ for } 0 < c < 1$$

This leads to $c = \pm\dfrac{\sqrt{3}}{3}$; only one value c is positive.

43. $f(x) = x^3 + ax - 1 = 0$; $f'(x) = 3x^2 + a$. Let $0 < x < N$ (with $N \to \infty$), then

$$\frac{x^3 + ax - 1 - N^3 - aN + 1}{x - N} = 3c^2 + a$$

$f(0) = -1; f(2) = 7 + 2a > 0$. By the intermediate value theorem, there is at least one real root. Assume that there are two roots, x_1 and x_2. Then $f(x_1) = f(x_2)$ and by Rolle's theorem $3c^2 + a = 0$, which is impossible since $a > 0$.

44. $f'(x) = (2n + 1)x^{2n} + a \geq 0$ because the even exponent makes x^{2n} nonnegative and $a > 0$ by hypothesis. This means the function is strictly increasing or strictly decreasing. Assume that there are two roots, x_1 and x_2. Then $f(x_1) = f(x_2)$ and by Rolle's theorem $(2n + 1)x^{2n} + a = 0$, which is impossible since $a > 0$.

45. Let $F(x) = f(x) - g(x)$. Since, by hypothesis, $f(a) = g(a)$ and $f(b) = g(b)$; $F(a) = F(b) = 0$. By Rolle's theorem, there is a number $a \leq c \leq b$ such that $F'(c) = 0 = f'(c) - g'(c)$. Thus, $f'(c) = g'(c)$.

46. $\dfrac{f'(x_2) - f'(x_1)}{x_2 - x_1} = f''(x) = 0$ with $x_2 \neq x_1$, so $f'(x_2) = f'(x_1)$ for all x. This means that $f(x)$ is a constant function and $f(x) = Ax + B$.

47. $\dfrac{f(x) - f(x_1)}{x - x_1} = f'(x) = Ax_3 + B$ with $x < x_3 < x_1$ for all x (and x_3). Then

$f(x) = f(x_1) + (Ax + B)(x - x_1)$

$= f(x_1) - B(x - x_1) + Ax^2 + Bx$

$= ax^2 + bx + c$, a quadratic

3.3 First Derivative Test, Page 190

1. The first derivative test is a test to determine where a function is increasing or decreasing, leading to relative extrema. Specifically, the first-derivative test is used to (1) find all critical values of f. (2) Classify each critical point $(c, f(c))$ as follows: **a.** relative maximum if the curve is rising to the left of c and falling to the right of c; **b.** relative minimum if the curve is falling to the left of c and rising to the right of c; **c.** not an extremum if the derivative has the same sign on both sides of c.

2. The derivative indicates the rate of change (positive, zero, or negative) of a function with respect to x.

3. Notice that each function has a value of 0 when the other has a horizontal tangent. So this does not tell us which is which. Next consider that when f is increasing f' must be positive. This identifies which graph is f and which is f'. The black curve is the function and the blue one is the derivative.

4. The black curve is the function and the blue curve is the derivative.

5.   **6.**

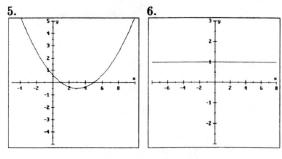

7.   **8.**

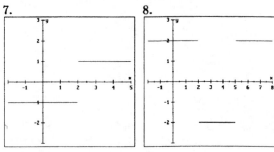

9.   **10.**

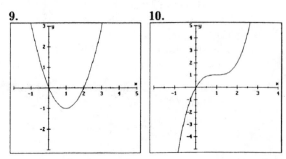

11.

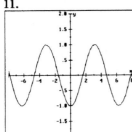

12.

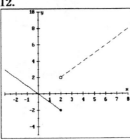

d.

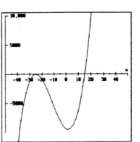

13. **a.** $f'(x) = 3x^2 + 6x = 3x(x + 2)$
critical values: $x = 0,\ x = -2$

b. increasing on $(-\infty, -2) \cup (0, +\infty)$
decreasing on $(-2, 0)$

c. critical points: $(0, 1)$, relative minimum;
$(-2, 5)$, relative maximum

d.

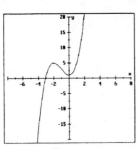

16. **a.** $f'(x) = 5x^4 + 20x^3 - 1,650x^2 - 4,000x + 60,000$; approximate critical values (use computer software or a graphing calculator): $x = -17.862,\ x = -7.79,$
$x = 5.265,\ x = 16.3885$

b. increasing on $(-\infty, -17.862) \cup (-7.79, 5.265) \cup (16.3885, +\infty)$; decreasing on $(-17.862, -7.79) \cup (5.265, 16.3885)$

c. critical points: $(-17.862, 115,300)$
relative maximum; $(-7.79, -339,000)$
relative minimum; $(5.265, 188,100)$
relative maximum; $(16.3885, -431,900)$
relative minimum

d.

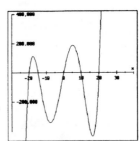

14. **a.** $f'(x) = x^2 - 9 = (x + 3)(x - 3)$
critical values: $x = \pm 3$

b. increasing on $(-\infty, -3)$ and $(3, +\infty)$
decreasing on $(-3, 3)$

c. critical points $(3, -16)$, relative
minimum; $(-3, 20)$, relative maximum

d.

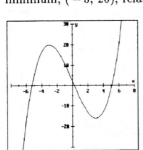

17. **a.** $f'(x) = 5x^4 - 5(4x^3) = 5x^3(x - 4)$
critical values: $x = 0,\ x = 4$

b. increasing on $(-\infty, 0) \cup (4, +\infty)$
decreasing on $(0, 4)$

c. critical points: $(4, -156)$, relative
minimum; $(0, 100)$, relative maximum

d.

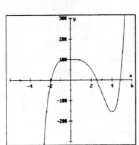

15. **a.** $f'(x) = 3x^2 + 70x - 125$

critical values: $x = \frac{5}{3},\ x = -25$

b. increasing on $(-\infty, -25) \cup (\frac{5}{3}, +\infty)$

decreasing on $(-25, \frac{5}{3})$

c. critical points: $(\frac{5}{3}, -9,481)$, relative

minimum; $(-25, 0)$, relative maximum

18. **a.** $f'(x) = 1 - x^{-2}$
critical values $x = 0,\ x = 4$

b. increasing on $(-\infty, 0)$ and $(4, +\infty)$
decreasing on $(0, 4)$

c. critical points: $(-1, -2)$, relative
maximum; $(1, 2)$, relative minimum

d.

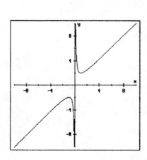

19. a. $f'(x) = \dfrac{(x^2+3)(1) - (x-1)(2x)}{(x^2+3)^2}$

$\qquad = \dfrac{-x^2 + 2x + 3}{(x^2+3)^2}$

critical values: $x = -1$, $x = 3$

b. increasing on $(-1, 3)$
decreasing on $(-\infty, -1) \cup (3, +\infty)$

c. critical points: $(-1, -\frac{1}{2})$, relative
minimum; $(3, \frac{1}{6})$, relative maximum

d.

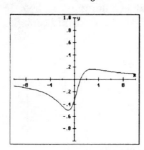

20. a. $f'(x) = \dfrac{(x^2+1)(1) - (x+1)(2x)}{(x^2+1)^2}$

$\qquad = \dfrac{-x^2 - 2x + 1}{(x^2+1)^2}$

critical values: $x = -1 \pm \sqrt{2}$;
$x \approx -2.414, 0.414$

b. increasing on $(-\infty, -2.414) \cup (0.414, +\infty)$; decreasing on $(-2.414, 0.414)$

c. critical points: $(-2.414, -0.2071)$,
relative minimum; $(0.414, 1.2071)$,
relative maximum

d.

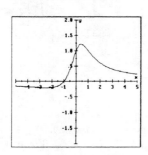

21. a. $f'(t) = (t+1)^2 + 2(t+1)(t-5)$

$\qquad = (t+1)(t+1+2t-10)$

$\qquad = (t+1)(3t - 9)$

$\qquad = 3(t+1)(t-3)$

critical values: $t = -1$, $t = 3$

b. increasing on $(-\infty, -1) \cup (3, +\infty)$
decreasing on $(-1, 3)$

c. critical points: $(3, -32)$, relative
minimum; $(-1, 0)$, relative maximum

d.

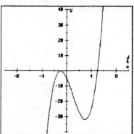

22. a. $f'(t) = (2t-1)^2(2t) + 2(2t-1)(2)(t^2-9)$

$\qquad = 2(2t-1)[t(2t-1) + 2t^2 - 18]$

$\qquad = 2(2t-1)(4t^2 - t - 18)$

$\qquad = 2(2t-1)(t+2)(4t - 9)$

critical values: $t = \frac{1}{2}, -2, \frac{9}{4}$

b. increasing on $(-2, \frac{1}{2}) \cup (\frac{9}{4}, +\infty)$;

decreasing on $(-\infty, -2) \cup (\frac{1}{2}, \frac{9}{4})$

c. critical points: $(-2, -125)$, relative
minimum; $(\frac{1}{2}, 0)$ relative minimum;
$(\frac{9}{4}, -48.5)$, relative minimum

d.

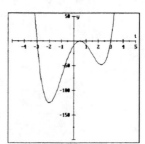

23. a. $f(x) = (x-3)(x-7)(2x+1)$
$\qquad = 2x^3 - 19x^2 + 32x + 21$
$f'(x) = 6x^2 - 38x + 32$
$\qquad = 2(x-1)(3x - 16)$
critical values: $x = 1, \frac{16}{3}$

b. increasing on $(-\infty, 1) \cup (\frac{16}{3}, +\infty)$
decreasing on $(1, \frac{16}{3})$

c. critical points: $(\frac{16}{3}, -\frac{1,225}{27})$, relative
minimum; $(1, 36)$, relative maximum

d.

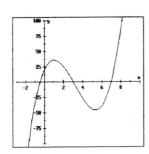

24. a. $f'(x) = 16(2)(x^2 - 10x + 21)(2)(x - 10)$
$= 64(x - 3)(x - 7)(x - 5)$
critical values: $x = 3, 5, 7$

b. increasing on $(3, 5) \cup (7, +\infty)$; decreasing on $(-\infty, 3) \cup (5, 7)$

c. critical points: $(3, 0)$, relative minimum; $(5, 256)$, relative maximum; $(7, 0)$, relative minimum

d.

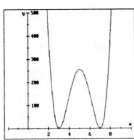

25. a. $f'(x) = \dfrac{x}{\sqrt{x + 1}}$; critical value $x = 0$

b. increasing on $(0, +\infty)$; decreasing on $(-\infty, 0)$

c. critical point: $(0, 1)$, relative minimum

d.

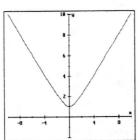

26. a. $f'(x) = \dfrac{x - 1}{\sqrt{x^2 - 2x + 2}}$; critical value $x = 0$

b. increasing on $(1, +\infty)$; decreasing on $(-\infty, 1)$

c. critical point: $(1, 1)$, relative minimum

d.

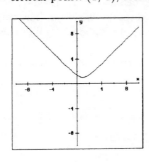

27. a. $g'(x) = x^{2/3}(2) + \frac{2}{3}x^{-1/3}(2x - 5)$
$= \frac{10}{3}(x - 1)x^{-1/3}$

critical values: $x = 1$ and $f'(x)$ is undefined at $x = 0$

b. increasing on $(-\infty, 0) \cup (1, +\infty)$ decreasing on $(0, 1)$

c. critical points: $(1, -3)$, relative minimum; $(0, 0)$, relative maximum

d.

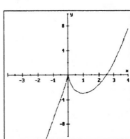

28. a. $g'(x) = x^{1/3}(\frac{1}{2})(x + 15)^{-1/2}(1)$
$+ \frac{1}{3}x^{-2/3}(x + 15)^{1/2}$

$= \frac{1}{6}x^{-2/3}(x + 15)^{-1/2}[3x + 2(x + 15)]$

$= \frac{1}{6}x^{-2/3}(x + 15)^{-1/2}(5x + 30)$

$= \frac{5}{6}x^{-2/3}(x + 15)^{-1/2}(x + 6)$

critical values: $x = -6$, and $f'(x)$ is undefined if $x = 0$, $x = -15$

b. increasing on $(-6, 0) \cup (0, +\infty)$ decreasing on $(-15, -6)$

c. critical points: $(-6, -5.4514)$, relative minimum; $(6, 0)$ neither

d.

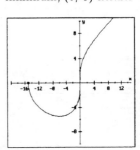

29. a. $f'(\theta) = -2 \sin \theta - 1$

critical values: $x = \frac{7\pi}{6}$, $x = \frac{11\pi}{6}$

b. increasing on $\left(\frac{7\pi}{6}, \frac{11\pi}{6}\right)$

decreasing on $\left(0, \frac{7\pi}{6}\right) \cup \left(\frac{11\pi}{6}, 2\pi\right)$

c. critical points: $\left(\frac{7\pi}{6}, -5.3972\right)$, relative minimum; $\left(\frac{11\pi}{6}, -4.0275\right)$, relative maximum; $(2\pi, -4.28)$, neither

d.

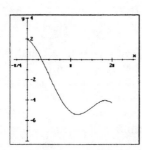

30. a. $c'(\theta) = 1 - 2\sin 2\theta$

critical values: $\theta = \frac{\pi}{12}$, $\theta = \frac{5\pi}{12}$

b. increasing on $(0, \frac{5\pi}{6}) \cup (\frac{5\pi}{12}, \pi)$

decreasing on $(\frac{\pi}{12}, \frac{5\pi}{12})$

c. $(0, 1)$, neither; $(\frac{\pi}{12}, 1.1278)$, neither;

$(\frac{5\pi}{12}, 4.14)$, relative minimum

d.

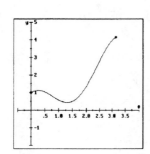

31. a. $t'(x) = 2\tan x \sec^2 x = \dfrac{2\sin x}{\cos^3 x}$

critical values: $x = 0$

b. increasing on $(0, \frac{\pi}{4})$

decreasing on $(-\frac{\pi}{4}, 0)$

c. critical point: $(0, 0)$, relative minimum;

$(-\frac{\pi}{4}, 1)$, neither; $(\frac{\pi}{4}, 1)$, neither

d.

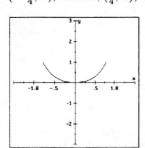

32. a. $f'(x) = -9\sin x - 8\cos x \sin x$

$\qquad = (-\sin x)(9 + 8\cos x)$

critical values: $x = 0$, $x = \pi$; note: no

solution for $\cos x = -\frac{9}{8}$

b. increasing on $(0, \pi)$

c. $(0, 5)$, relative maximum; $(\pi, -13)$,

relative minimum; $(\frac{\pi}{4}, 1)$, neither

d.

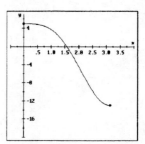

33. a. $f'(x) = \dfrac{1}{50\pi}\cos(\dfrac{x}{50\pi})$

critical values: $x = 25\pi^2$, $x = 75\pi^2$

b. increasing on $(0, 25\pi^2) \cup (75\pi^2, 100\pi^2)$

decreasing on $(25\pi^2, 75\pi^2)$

c. critical points: $(75\pi^2, -1)$, relative

minimum; $(25\pi^2, 1)$, relative maximum

d.

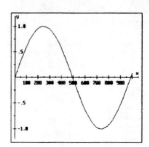

34. a. $f'(x) = 3(x - 2)^2$

critical value: $x = 2$

b. increasing on $(-\infty, 2) \cup (2, +\infty)$

c. critical point: $(2, 0)$, neither

d.

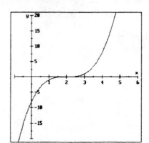

35. If $n > 0$ and n is even, the curve has a

minimum at $x = 2$.

If $n > 0$ and n is odd, the curve has a point

of inflection at $x = 2$.

If $n = 0$, then $f(x) = 1$.

If $n < 0$ and n is even, the curve is not

defined at $x = 2$. It has a vertical

asymptote at $x = 2$. The curve is

symmetric about the line $x = 2$.

If $n < 0$ and n is odd, the curve is not

defined at $x = 2$. It also has a vertical

asymptote at $x = 2$. The curve is

symmetric about the point $(2, 0)$.

36. $f'(x) = 21(x^2 - 1)(x^3 - 3x + 1)^6$

$f'(1^-) < 0$ and $f'(1^+) > 0$; similarly for $x = -1$; relative minimum at $x = \pm 1$.

37. $f'(x) = 20(x^3 - 1)(x^4 - 4x + 2)^4 = 0$
$f'(1^-) < 0$ and $f'(1^+) > 0$; relative minimum at $x = 1$.

38. $f'(x) = (x^2 - 4)^4(3)(x^2 - 1)^2(2x)$
$\qquad\qquad + (x^2 - 1)^3(4)(x^2 - 1)^3(2x)$
$\quad = (x^2 - 4)^3(x^2 - 1)^2(2x)(3x^2 - 12 + 4x^2 - 4)$
$\quad = (x^2 - 4)^3(x^2 - 1)^2(2x)(7x^2 - 16)$

$f'(1^-) > 0$, $f'(1^+) > 0$, $x = 1$ is neither a maximum nor a minimum;
$f'(2^-) > 0$, $f'(2^+) < 0$, $x = 2$ is a maximum.

39. $f'(x) = \frac{1}{3}(x^3 - 48x)^{-2/3}(3x^2 - 48)$

$f'(4^-) < 0$, $f'(4^+) > 0$, $x = 4$ is a relative minimum

40. $f'(x) = (x - 1)^2(x - 2)(x - 4)(x + 5)$ has critical values of $x = 1, 2, 4,$ and -5

$f'(1^-) > 0$, $f'(1^+) > 0$, $x = 2$ is neither a relative maximum nor a relative minimum;
$f'(2^-) > 0$, $f'(2^+) < 0$, $x = 2$ is a maximum;
$f'(4^-) < 0$, $f'(4^+) > 0$, $x = 4$ is a relative minimum
$f'(-5^-) > 0$, $f'(-5^+) > 0$, $x = -5$ is neither a maximum nor a relative minimum

41. Critical values are $x = \frac{1}{2}$, $x = -3$, and $x = 1$

$f'(\frac{1}{2}^-) < 0$, $f'(\frac{1}{2}^+) > 0$, $x = \frac{1}{2}$ is a relative minimum
$f'(-3^-) > 0$, $f'(-3^+) < 0$, $x = -3$ is a relative maximum
$f'(1^-) > 0$, $f'(1^+) > 0$, $x = 1$ is neither a maximum nor a minimum

42.

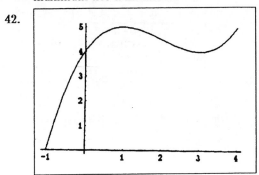

43.

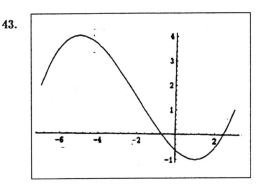

44.

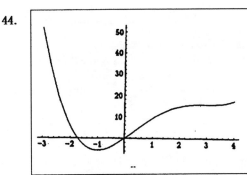

45. This is a system of three variables for which we have three pieces of information. $(5, 12)$ is on the curve so: $12 = 25a + 5b + c$. $(0, 3)$ is on the curve so $3 = c$.

$f'(x) = 2ax + b = 0$ when x is 5, so:

$0 = 10a + b$. Substituting the second and third equations into the first:

$12 = 25a - 50a + 3$, $a = -\frac{9}{25}$, $b = \frac{18}{5}$, $c = 3$

Therefore, $f(x) = -\frac{9}{25}x^2 + \frac{18}{5}x + 3$.

46. $f(x) = ax^2 + bx + c$; $f'(x) = 2ax + b = 0$ when $x_1 = -\frac{b}{2a}$; $f'(x)$ changes sign at x_1 so there is an extremum at this critical value.

47. $f(x) = (x - p)(x - q) = x^2 - (p + q)x + pq$
$f'(x) = 2x - (p + q) = 0$, so $x_1 = \frac{p + q}{2}$
which is midway between $(0, p)$ and $(0, q)$

48. $I = I_0\left(\frac{\sin\theta}{\theta}\right)^2$; $I' = 2I_0\left(\frac{\sin\theta}{\theta}\right)$

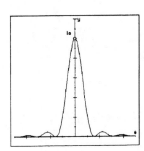

49. $v'(T) = v_0 \frac{1}{2}\left(1 + \frac{1}{273}T\right)^{-1/2}\left(\frac{1}{273}\right).$

For $T > 0$ this is always positive, so the function is monotonic increasing.

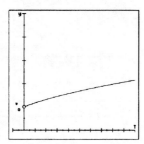

50. $f'(x) = 3x^2 + 2ax + b;$
$f'(-3) = 27 - 6a + b = 0;$
$f'(1) = 3 + 2a + b = 0;$
Solving these equations simultaneously, we find $a = 3$, $b = -9$.
$f(1) = -14 = 1 + 3 - 9 + c$, or $c = -9$;
Thus, $f(x) = x^3 + x^2 - 9x - 9$

51. $f'(x) = 12x^3 + 3Ax^2 + 2Bx + C$
$f(0) = -15$, so $D = -15$;
$f(2) = 1$, so $48 + 8A + 4B + 2C - 15 = 1$
$f'(5) = 0$, so $C = 0$;
$f'(2) = 0$, so $96 + 12A + 4B = 0$
Solving these equations simultaneously, we find $A = -16$, and $B = 24$.
$f(x) = 3x^4 - 16x^3 + 24x^2 - 15$
$(0, -15)$ is a minimum and $(2, 1)$ is neither

52. $f'(x) = (x - A)^{m-1}(x - B)^{n-1}[(n + m)x$
$$- (nA + mB)] = 0$$
if $x = A$, $x = B$, or $x = \dfrac{nA + mB}{n + m}$

If m is odd, $m - 1$ is even and $(x - A)^{m-1}$ is even regardless of whether $x - A$ is positive or negative. In that case, $x = A$ leads to neither a maximum nor a minimum.

But if m is even, $(x - A)^{m-1} < 0$ for $x < A$ and $(x - A)^{m-1} > 0$ for $x > A$. Thus, $x = A$ leads to minimum. The same holds true for $x = B$ and the value of m. Now, for
$$x_1 = \frac{nA + mB}{n + m}$$
$f'(x) < 0$ for $x < x_1$ and $f'(x) > 0$ for $x > x_1$.

Thus, x_1 is a minimum.

53. $f(x) = \left(Ax^{5/3} + Bx^{2/3}\right)^{1/2}$

$f'(x) = \frac{1}{2}\left(Ax^{5/3} + Bx^{2/3}\right)^{-1/2}\left(\frac{5}{3}Ax^{2/3} + \frac{2}{3}Bx^{-1/3}\right)$

$= \frac{1}{6}\left(Ax^{5/3} + Bx^{2/3}\right)^{-1/2}\left(5Ax^{2/3} + 2Bx^{-1/3}\right)$

$= \frac{1}{6}x^{-1/3}\left(Ax^{5/3} + Bx^{2/3}\right)^{-1/2}\left(5Ax + 2B\right)$

$f'(x) = 0$ when $5Ax + 2B = 0$; that is, $x = -2B/5A$; that is, a relative minimum. We also have a critical value when $x = 0$, as the denominator will be 0. The square root is also 0 when $x = -B/A$. So there are two points at which $f'(x)$ is undefined: $x = -B/A$ and $x = 0$.

54. $f'(x) = 3x^2 + \sin x = 0$ when $x = 0$,
$x = -0.33$; $f(0 - 0.33) \approx -0.982$

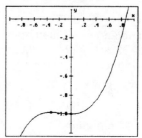

55. $f'(x) = 5x^4 + 2x \cos x^2$
$= x(5x^3 + 2 \cos x^2)$
$f'(x) = 0$ when $x = 0$ or $x \approx -0.71$

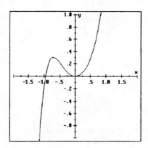

56. $f'(c) = g'(x) = 0$ because c is a critical number. Now let $F(x) = f(x)g(x)$.
$F'(x) = f(x)g'(x) + f'(x)g(x)$
$F'(c) = f(c)g'(c) + f'(c)g(c) = 0$
If a relative maximum occurs at $x = c$, it is not always true that a relative maximum must occur at $x = c$. Let $f(x) = -x^2$, $g(x) = -x^4$, and $F(x) = x^6$. The values $f(0)$ and $g(0)$ are maxima, while $F(0)$ is a minimum.

57. $f'(x) = \lim\limits_{\Delta x \to 0} \dfrac{f(x + \Delta x) - f(x)}{\Delta x} \geq 0$
since $\Delta x > 0$ and $f(x + \Delta x) - f(x) > 0$ because $f(x + \Delta x) > f(x)$ by hypotheses.

58. **a.** Let $f(x) > 0$ and $g(x) > 0$ on $[a, b]$ and $F(x) = f(x)g(x)$. Then,
$$F'(x) = f(x)g'(x) + f'(x)g(x) > 0$$
by hypotheses and by Problem 58.

b. If $f(x) < 0$ and $g(x) < 0$ on $[a, b]$ then

$$F'(x) = f(x)g'(x) + f'(x)g(x) < 0$$

so the function is strictly decreasing.

59. **a.** Let $g(x) = x^n f(x)$, then

$$g'(x) = x^n f'(x) + nx^{n-1}f(x) > 0$$

because $n > 0$, $x > 0$, $f(x) > 0$, and by Problem 57, $f'(x) > 0$.

b. $g(x) = x^n \sin x$, then

$$g'(x) = x^n \cos x + nx^{n-1}\sin x > 0$$

because $n > 0$, $x > 0$, $\cos x > 0$, $\sin x > 0$ on $(0, \frac{\pi}{2})$.

60. Pick x_1 and x_2 on (a, b).

$$\frac{f(x_2) - f(x_1)}{x_2 - x_1} = f(c)$$

$$f(x_2) - f(x_1) = f(c)(x_2 - x_1) < 0$$

Thus, $f(x_2) < f(x_1)$ as $x_2 > x_1$ so that f is strictly decreasing.

61. True; let $F(x) = f(x) + g(x)$, then

$$F(x) = f(x) + g(x)$$

$$F'(a) = f'(a) + g'(a)$$

$f'(x) > 0$ and $g'(x) > 0$ for $x < a$ imply $F'(x) > 0$ for $x < 0$. Similarly, $F'(x) < 0$ when $x > a$. This shows that $f(a)$ is a relative maximum.

62. True; $f(x)$ and $g(x)$ are both differentiable at $x = c$ and $f'(c) = 0$. $g'(c) = 0$ by hypotheses (relative extremum).

63. False; $f(x) = -|x - c|$ increases for $x > c$ and decreases for $x < c$, but the derivative does not exist at $x = c$.

64. False; $(0, f(0))$ could be a point of inflection (a turning point).

65. False; $f(x) = 7$ is not strictly increasing and $f'(x)$ is nonnegative.

66. True; by hypotheses $f(x_2) < f(x_1)$ for all $x_1 > x_2$, but $-f(x_2) > -f(x_1)$, which means that f is strictly increasing.

67. False; let $f(x) = x$ and $g(x) = 2x$, both of which are strictly increasing. Then $F(x) = f(x) - g(x) = -x$ which is strictly decreasing.

68. Let $y = x - \sin x$; then $y' = 1 - \cos x > 0$ for $0 < x < \pi/2$. Thus, y is strictly increasing

and $x > \sin x$ or

$$\frac{1}{x} < \frac{1}{\sin x}$$

3.4 Concavity and the Second-Derivative Test, Page 201

1. The second derivative test is used to determine concavity, which in turn may give information about the relative maximum and relative minimum. Let f be a function such that $f'(c) = 0$ and the second derivative exists on an open interval containing c. If $f''(c) > 0$, there is a relative minimum at $x = c$. If $f''(x) < 0$, there is a relative maximum at $x = c$. If $f''(c) = 0$, then the second-derivative test fails and gives no information.

2. $f''(x) > 0$ implies that the concavity of the graph of a f is up. $f''(x) <$ implies that the concavity of the graph of a f is down. $f''(x) = 0$ denotes a point of inflection.

3. Let $P(t)$ be the price function in terms of time. Then $dP/dt \geq 0$ and $d^2 P/dt < 0$.

4. $f'(x) = 2x + 5$; $f''(x) = 2$

critical point: $(-\frac{5}{2}, -\frac{37}{4})$; relative minimum;

increasing on $(-2.5, +\infty)$;

decreasing on $(-\infty, -2.5)$

concave up on $(-\infty, +\infty)$

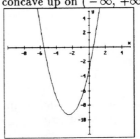

5. $f'(x) = 4(x + 20) - 8$; $f''(x) = 4$

critical point: $(-18, -1)$, relative minimum;

increasing on $(-18, +\infty)$;

decreasing on $(-\infty, -18)$;

concave up on $(-\infty, +\infty)$

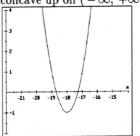

6. $f'(x) = 3x^2 - 3$; $f''(x) = 6x$
critical points: $(1, -6)$, relative minimum;
$\quad (-1, -2)$, relative minimum;
increasing on $(-\infty, -1) \cup (1, +\infty)$;
decreasing on $(-1, 1)$;
concave up on $(0, +\infty)$;
concave down on $(-\infty, 0)$

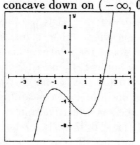

7. $f'(x) = x^2 - 9$; $f''(x) = 2x$
critical points: $(-3, 20)$, relative maximum;
$\quad (3, -16)$, relative minimum;
increasing on $(-\infty, -3) \cup (3, +\infty)$;
decreasing on $(-3, 3)$; inflection point $(0, 2)$
concave up on $(0, +\infty)$;
concave down on $(-\infty, 0)$

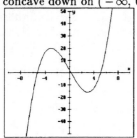

8. $f'(x) = 4(x - 12)^3 - 6(x - 12)^2$
$\quad = 2(x - 12)^2(2x - 27)$
$f''(x) = 12(x - 12)^2 - 12(x - 12)$
$\quad = 12(x - 12)(x - 13)$
critical points: $(13.5, -1.69)$, relative
$\quad$ minimum; $(12, 0)$, point of inflection;
$\quad (13, 1)$, point of inflection
increasing on $(13.5, +\infty)$
decreasing on $(-\infty, 13.5)$
concave up on $(-\infty, 12) \cup (13.5, +\infty)$
concave down on $(12, 13.5)$

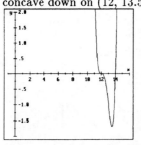

9. $f'(x) = 2 - 18x^{-2}$; $f''(x) = 36x^{-3}$
critical points: $(-3, -11)$, relative
$\quad$ maximum; $(3, 13)$, relative minimum;

increasing on $(-\infty, -3) \cup (3, +\infty)$
decreasing on $(-3, 0) \cup (0, 3)$;
concave up on $(0, +\infty)$;
concave down on $(-\infty, 0)$

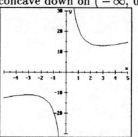

10. $f'(x) = 12u^3 - 6u^2 - 24u + 18$
$\quad = 6(u - 1)^2(2u + 3)$
$f''(x) = 36u^2 - 12u - 24$
$\quad = 12(u - 1)(3u + 2)$
critical points: $(-1.5, -37.06)$ is a relative
$\quad$ minimum; $(1, 2)$ is a point of inflection
increasing on $(-1.5, +\infty)$;
decreasing on $(-\infty, -1.5)$;
concave up on $(-\infty, -1.5) \cup (1, +\infty)$;
concave down on $(-1.5, 1)$

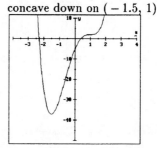

11. $g'(u) = 4u^3 + 18u^2 - 48u = 2u(2u^2 + 9u - 24)$
$g''(u) = 12u^2 + 36u = 12(u + 4)(u - 1)$
critical points: $(0, 26)$, relative maximum;
$\quad (-6.38, -852.22)$, relative minimum;
$\quad (1.88, -6.47)$, relative minimum;
$\quad$ inflection points $(-4, -486)$, $(1, 9)$;
increasing on $(-6.38, 0) \cup (1.88, +\infty)$;
decreasing on $(-\infty, -6.38) \cup (0, 1.88)$
concave up on $(-\infty, -4) \cup (1, +\infty)$;
concave down on $(-4, 1)$

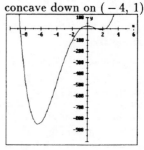

12. $f'(x) = \dfrac{(x-3)(2x) - x^2}{(x-3)^2} = \dfrac{x^2 - 6x}{(x-3)^2}$

$f''(x) = \dfrac{(x-3)^2(2x-6) - (x^2-6x)(2)(x-3)}{(x-3)^4}$

$\qquad = \dfrac{18}{(x-3)^3}$

derivative is not defined at $x = 3$, which is
not in the domain; critical points: $(0, 0)$,
    relative maximum; $(6, 12)$, is a relative
    minimum;
increasing on $(-\infty, 0) \cup (6, +\infty)$
decreasing on $(0, 3) \cup (3, 6)$
concave up on $(3, +\infty)$
concave down on $(-\infty, 3)$

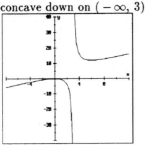

13. $f'(x) = \dfrac{(x+1)(2x-3) - (x^2-3x)(1)}{(x+1)^2}$

$\qquad = \dfrac{(x+3)(x-1)}{(x+1)^2}$

$f''(x) = \dfrac{(x+1)^2(2x+2) - (x^2+2x-3)(2x+2)}{(x+1)^4}$

$\qquad = \dfrac{8}{(x+1)^3}$

critical points: $(-3, -9)$, relative
    maximum; $(1, -1)$, relative minimum;
increasing on $(-\infty, -3) \cup (1, +\infty)$;
decreasing on $(-3, -1) \cup (-1, 1)$;
concave up on $(-1, +\infty)$;
concave down on $(-\infty, -1)$

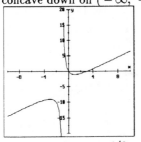

14. $f'(x) = x(x^2 + 1)^{-1/2}$

$f''(x) = x(-\tfrac{1}{2})(x^2+1)^{-3/2}(2x) + (x^2+1)^{-1/2}$

$\qquad = -\tfrac{1}{2}(x^2+1)^{-3/2}[2x^2 - 2(x^2+1)]$

$\qquad = -\tfrac{1}{2}(x^2+1)^{-3/2}[-2]$

$\qquad = (x^2+1)^{-3/2}$

critical point: $(0, 2)$, relative minimum;

increasing on $(0, +\infty)$
decreasing on $(-\infty, 0)$
concave up on $(-\infty, +\infty)$

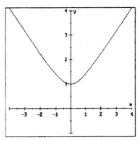

15. $g'(x) = 2(t^3 + t)(3t^2 + 1) = 2t(t^2 + 1)(3t^2 + 1)$

$g''(x) = 2(15t^4 + 12t^2 + 1)$

critical point: $(0, 0)$, relative minimum;
increasing on $(0, +\infty)$; decreasing on
$(-\infty, 0)$; concave up on $(-\infty, +\infty)$

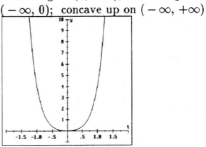

16. $f'(t) = 3(t^3 + 3t^2)^2(3t^2 + 6t)$

$\qquad = 9t(t^3 + 3t^2)(t + 2)$

$\qquad = 9t^5(t + 3)^2(t + 2)$

$f''(t) = 18t^4(2t + 3)(2t + 5)$

critical points: $(0, 0)$, relative minimum;
    $(-2, 64)$, relative maximum; points of
    inflection: $(3, 0), (-2.5, 30.5), (-1.5, 38.44)$
increasing on $(-\infty, -2) \cup (0, +\infty)$
decreasing on $(-2, 0)$
concave up on $(-3, -2.5) \cup (-1.5, +\infty)$
concave down on $(-\infty, -3) \cup (-2.5, -1.5)$

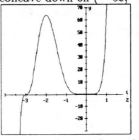

17. $f'(x) = 3t^2 - 12t^3 = 3t^2(1 - 4t^3)$

$f''(x) = 6t - 36t^2 = 6t(1 - 6t)$

critical points: $(0, 0)$, point of inflection;
    $(\tfrac{1}{4}, \tfrac{1}{256})$, relative maximum;
increasing on $(-\infty, \tfrac{1}{4})$;
decreasing on $(\tfrac{1}{4}, +\infty)$;

points of inflection: $(0, 0)$, $(\frac{1}{6}, \frac{1}{432})$;
concave up on $(0, \frac{1}{6})$;
concave down on $(-\infty, 0) \cup (\frac{1}{6}, +\infty)$

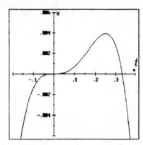

18. $f'(x) = 20t^4 - 20t^3 = 20t^3(t - 1)$
    $f''(x) = 80t^3 - 60t^2 = 20t^2(4t - 3)$
    critical points: $(0, 0)$, relative maximum;
        $(1, -1)$, relative minimum
        $(0.75, -0.63)$, point of inflection;
    increasing on $(-\infty, 0) \cup (1, +\infty)$
    decreasing on $(0, 1)$
    concave up on $(0.75, +\infty)$
    concave down on $(-\infty, 0.75)$

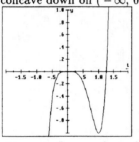

19. $g'(t) = 30t^5 - 18t^2 = 6t^2(5t^3 - 3)$
    $g''(t) = 150t^4 - 36t = 6t(25t^3 - 6)$
    critical points: $0.6^{1/3} \approx 0.84$, so
        $(0.84, -1.8)$ is a relative minimum;
        $0.24^{1/3} \approx 0.62$, so $(0.62, -1.15)$, point
        of inflection; $(0, 0)$, point of inflection;
        $(0.62, 0)$, points of inflection;
    increasing on $(0.84, +\infty)$;
    decreasing on $(-\infty, 0.84)$;
    concave up on $(-\infty, 0) \cup (0.62, +\infty)$;
    concave down on $(0, 0.62)$

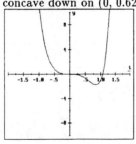

20. $f'(x) = -2x(x^2 + 3)^{-2}$

    $f''(x) = -2(x^2 + 3)^{-2} - 2x(-2)(x^2 + 3)^{-3}(2x)$

$= -2(x^2 + 3)^{-3}[(x^2 + 3) - 4x^2]$

$= -2(x^2 + 3)^{-3}(3 - 3x^2)$

$= 6(x^2 + 3)^{-3}(x - 1)(x + 1)$

critical points: $(0, \frac{1}{3})$ is a relative maximum;
    $(1, \frac{1}{4})$ and $(-1, \frac{1}{4})$ are points of inflection
increasing on $(-\infty, 0)$
decreasing on $(0, +\infty)$
concave up on $(-\infty, -1) \cup (1, +\infty)$
concave down on $(-1, 1)$

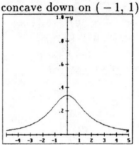

21. $f'(x) = \dfrac{x^2 + 1 - x(2x)}{(x^2 + 1)^2} = \dfrac{1 - x^2}{(x^2 + 1)^2}$

    $f''(x) = \dfrac{(x^2+1)^2(-2x) - (1 - x^2)(2)(x^2+1)(2x)}{(x^2 + 1)^4}$

critical points: $(-1, -\frac{1}{2})$, relative
    minimum; $(1, \frac{1}{2})$, relative maximum;
    $(0, 0)$ is a point of inflection
increasing on $(-1, 1)$;
decreasing on $(-\infty, -1) \cup (1, +\infty)$;
points of inflection: $(\sqrt{3}, \frac{\sqrt{3}}{4})$, $(-\sqrt{3}, \frac{\sqrt{3}}{4})$;
concave up on $(-\sqrt{3}, 0) \cup (\sqrt{3}, +\infty)$;
concave down on $(-\infty, -\sqrt{3}) \cup (0, \sqrt{3})$

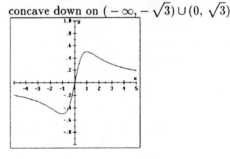

22. $f'(t) = \frac{9}{3}t^{1/3} + 9$
    $f''(t) = \frac{8}{9}t^{-2/3}$

critical point: $(-38.44, -85.48)$, relative
minimum
increasing on $(-38.44, +\infty)$
decreasing on $(-\infty, -38.44)$
concave up on $(-\infty, +\infty)$

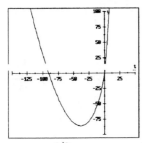

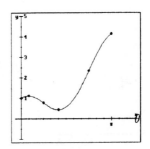

23. $f(x) = x^{4/3}(x - 27) = x^{7/3} - 27x^{4/3}$

     $f'(x) = \frac{7}{3}x^{4/3} - 36x^{1/3} = \frac{1}{3}x^{1/3}(7x - 108)$

     $f''(x) = \frac{28}{9}x^{1/3} - 12x^{-2/3} = \frac{4}{9}x^{-2/3}(7x - 27)$

critical points: $(0, 0)$, relative maximum;
     $(15.4, -444.4)$, relative minimum;
increasing on $(-\infty, 0) \cup (15.4, +\infty)$;
decreasing on $(0, 15.4)$; inflection point is
$(3.9, -140)$; concave up on $(3.9, +\infty)$;
concave down on $(-\infty, 3.9)$

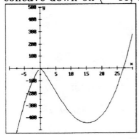

24. $t'(\theta) = \cos\theta + 2\sin\theta$; $t''(\theta) = -\sin\theta + 2\cos\theta$
critical points: $(2.68, 2.24)$, relative minimum;
     $(5.82, -2.24)$, relative minimum
increasing on $(0, 2.68) \cup (5.82, 2\pi)$
decreasing on $(2.68, 5.82)$
concave up on $(0, 1.11) \cup (4.25, 2\pi)$
concave down on $(1.11, 4.25)$

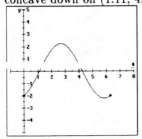

25. $t'(\theta) = 1 - 2\sin 2\theta$; $t''(\theta) = -4\cos 2\theta$
critical points: $(\frac{\pi}{12}, 1.13)$, relative maximum;

     $(\frac{5\pi}{12}, 0.44)$, relative minimum; inflection

     points: $(\frac{\pi}{4}, 0.79)$, $(\frac{3\pi}{4}, 2.4)$;

increasing on $(0, \frac{\pi}{12}) \cup (\frac{5\pi}{12}, \pi)$;

decreasing on $(\frac{\pi}{12}, \frac{5\pi}{12})$; concave up on $(\frac{\pi}{4}, \frac{3\pi}{4})$;

concave down on $(0, \frac{\pi}{4}) \cup (\frac{3\pi}{4}, \pi)$

26. $h'(u) = \dfrac{(2+\cos u)\cos u - \sin u(-\sin u)}{(2 + \cos u)^2}$

         $= \dfrac{2\cos u + 1}{(2 + \cos u)^2}$

   $h''(u) = \dfrac{(2+\cos u)(-2\sin u) + 4\sin u\cos u + 2\sin u}{(2 + \cos u)^3}$

         $= \dfrac{-4\sin u - 2\sin u\cos u + 4\sin u\cos u + 2\sin u}{(2 + \cos u)^3}$

critical points: $(2.09, 0.58)$, relative
     maximum; $(4.19, -0.58)$, relative
     minimum; points of inflection: $(0, 0)$,
     $(\pi, 0)$, $(2\pi, 0)$
increasing on $(0, 2.09) \cup (4.19, 2\pi)$
decreasing on $(2.09, 4.19)$
concave up on $(\pi, 2\pi)$
concave down on $(0, \pi)$

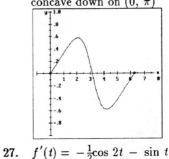

27. $f'(t) = -\frac{1}{2}\cos 2t - \sin t$

     $f''(x) = \sin 2t - \cos t$

no critical points; inflection points:
     $(0.52, 0.65)$, $(1.57, 0)$, $(2.62, -0.65)$;
decreasing on $(0, \pi)$;
concave up on $(0.52, 1.57) \cup (2.62, 2\pi)$;
concave down on $(0, 0.52) \cup (1.57, 2.62)$

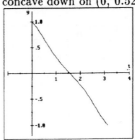

28.

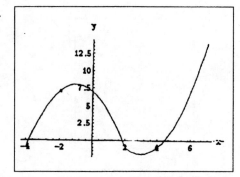

29.

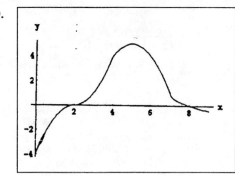

30.

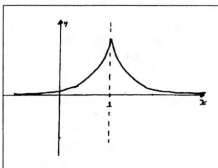

The derivative does not exist at $x = 1$.

31.

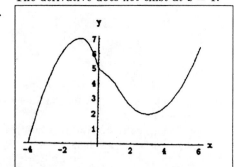

32. **a.**   **b.**

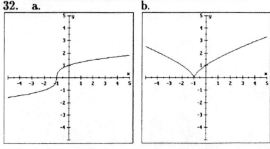

c.   **d.**

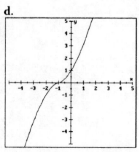

With n an even positive integer, $(-1, 0)$ is a corner point; it has a vertical tangent line and the graph is symmetric with respect to $x = -1$. With n an odd positive integer $(n \neq 3)$, $(-1, 0)$ is a point of inflection; it has a vertical tangent line and the graph is symmetric with respect to $(-1, 0)$

33. $y' = 2Ax + B$ and $y'' = 2A$;
concave up if $y'' > 0$ or $2A > 0$ or $A > 0$;
concave down if $y'' < 0$ or $2A < 0$ or $A < 0$

34. $f(x) = Ax^3 + Bx^2 + C$

$f'(x) = 3Ax^2 + 2Bx$

$f''(x) = 6Ax + 2B$

$E(2, 11)$ is an extremum, so

$f'(2) = 12A + 4B = 0$ or $B = -3A$

$I(1, 5)$ is a point of inflection, so

$f''(1) = 6A + 2B = 0$ or $B = -3A$

Since the points E and I are on the curve, their coordinates must satisfy its equation. Thus,

$f(2) = 8A + 4B + C = 8A - 12A + C = 11$

$f(1) = A + B + C = A - 3A + C = 5$

Solving these equations simultaneously, we find $A = -3$, $B = 9$, and $C = -1$. Thus,

$f(x) = -3x^3 + 9x^2 - 1$

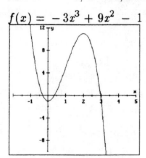

35. $y' = 3x^2 + 2bx + c = 0$ when
$$x = \frac{-b \pm \sqrt{b^2 - 3x}}{3}$$
If $b^2 - 3c > 0$, there are two distinct

extrema. The distance between the extrema changes as b varies.

36. Let $t = 0$ at noon, and let $f(t)$ be the number of sets assembled per hour as a function of time. $f'(t) > 0$ when $0 < t < 8$; $f(4) = 3$ $f''(t) < 0$ when $4 < t < 8$, $f(8) = 5$

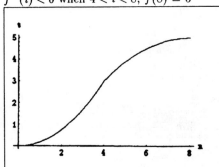

37. $D(x) = \frac{9}{4}x^4 - 7\ell x^3 + 5\ell^2 x^2$ on $[0, \ell]$

$D'(x) = 9x^3 - 21\ell x^2 + 10\ell^2 x$

$\qquad = x(9x^2 - 21\ell x + 10\ell^2)$

$\qquad = x(3x - 5\ell)(3x - 2\ell)$

$D'(x) = 0$ when $x = 0, \frac{2}{3}\ell$

$\qquad$ ($\frac{5}{3}\ell$ is not in domain)

$D''(x) = 27x^2 - 42\ell x + 10\ell^2$

$D''(\frac{2}{3}\ell) < 0$, so $\frac{2}{3}\ell$ is a maximum.

Maximum deflection at $x = \frac{2\ell}{3}$

38. The marginal cost is

$MC = C'(x) = 3x^2 - 48x + 350$

$C''(x) = 6(x - 8)$

Thus, $x = 8$ is a critical point and a relative minimum since $C''(x) > 0$.

39. $C'(x) = 8x^3 - 18x^2 - 24x - 2$

$C'(x) = 24x^2 - 36x - 24 = 12(x - 2)(2x + 1)$

$C''(x) = 48x - 36 = 12(4x - 3)$

$C''(x) = 0$ when $x = 2$, (Note: $-\frac{1}{2}$ is not in

$\qquad$ domain.)

$C'''(2) > 0$, $x = 2$ is a relative minimum

$C'(0) = -2$; $C'(2) = -58$, $C'(3) = -20$

Maximum marginal cost is -2 at $x = 0$

minimum marginal cost is -58 at $x = 2$

40. **a.** At $(0, 0)$, $\frac{dy}{dx} = 0$, so the tangent line is

horizontal with equation $y = 0$

b. $x = 0$ is a secondary critical point; $(0, 0)$ is a point of inflection.

41. **a.** $-(2)^2 + 6(2)^2 + 13(2) = 42$

$\qquad -\frac{1}{3}(2)^3 + \frac{1}{2}(2)^2 + 25(2) = 49\frac{1}{3}$

b. $N(x) = -x^3 + 6x^2 + 13x - \frac{1}{3}(4 - x)^3$

$\qquad\qquad + \frac{1}{2}(4 - x)^2 + 25(4 - x)$

c. $N'(x) = -2x^2 + 5x = 0$ when $x = 2.5$, so the optimum time for the break is 10:30 A.M. The optimum time is half the value for the diminishing return.

42. $P' = \frac{w^2}{2\rho S}\left(-v^{-2}\right) + \frac{3}{2}\rho A v^2 = 0$ when

$v^4 = \frac{w^2}{3\rho^2 SA}$ or $v = \left(\frac{w^2}{3\rho^2 SA}\right)^{1/4}$

$P'' > 0$, so v is a minimum.

43. $\Delta y = f'(c)(x - c)$, so

$\qquad\qquad f(x) - f(c) > \Delta y$

$\qquad\qquad\qquad f(x) > f(c) + f'(c)(x - c)$

The existence of c is guaranteed by the MVT.

44. Take two points, say P and Q, on the curve which represents the graph of a function f which is concave up. There is at least one point R on the curve such that all the points on the curve are above the tangent line at R.

45. $f'(x) = 4x^3 - 9x^2 + 1$

$f''(x) = 12x - 18x = 6x(2 - 3x)$

critical points: $(-0.31, -5.21)$, relative minimum; $(2.20, -11.32)$, relative maximum; $(0.36, -4.76)$, relative minimum;

inflection points: $(0, -5)$, $(\frac{3}{2}, -\frac{137}{16})$

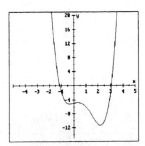

46. $f'(x) = 3x^2 + \cos x > 0$ for all x

$f''(x) = 6x - \sin x = 0$ when $x = 0$

There are no critical points; the function is increasing for all x.

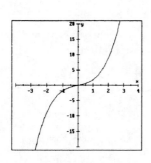

47. $f'(x) = \sec^2 x - 2x$
$f''(x) = 2 \sec^2 x \tan x - 2$
critical points: $(2.08, -3.12)$, $(4.37, -13.29)$,
  $(5.03, -25.34)$;
inflection points: $(-5.68, -28.57)$,
  $(-2.54, -2.77)$, $(0.60, 3.32)$,
  $(3.74, -10.31)$

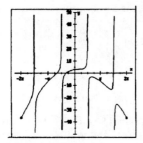

48. $f'(x) = 5x^4 + 6x^2 - 2x = x(5x^3 + 6x - 2)$
$f''(x) = 20x^3 + 12x - 2$
critical points: $(0, 11)$, $(0.31, 10.97)$
inflection point: $(0.16, 10.98)$

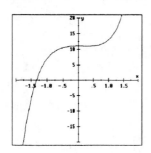

3.5 Infinite Limits and Asymptotes, Page 211

1. $\displaystyle\lim_{x \to +\infty} \frac{2,000}{x+1} = 0$ **2.** $\displaystyle\lim_{x \to +\infty} \frac{7,000}{\sqrt{x}+1} = 0$

3. $\displaystyle\lim_{x \to +\infty} \frac{3 + \frac{5}{x}}{1 - \frac{2}{x}} = \frac{3+0}{1-0} = 3$

4. $\displaystyle\lim_{x \to +\infty} \frac{1 + \frac{2}{x}}{3 - \frac{5}{x}} = \frac{1}{3}$

5. $\displaystyle\lim_{t \to +\infty} \frac{\frac{5}{t}+1}{\frac{7}{t}-1} = -1$

6. $\displaystyle\lim_{s \to +\infty} \frac{3 + \frac{4}{s}}{\frac{7}{s}-2} = -\frac{3}{2}$

7. $\displaystyle\lim_{x \to +\infty} \frac{3 - \frac{7}{x} + \frac{5}{x^2}}{-2 + \frac{1}{x} - \frac{9}{x^2}} = -\frac{3}{2}$

8. $\displaystyle\lim_{t \to +\infty} \frac{9 + \frac{50}{t^3} + \frac{800}{t^5}}{1 - \frac{1,000}{t^5}} = 9$

9. $\displaystyle\lim_{t \to +\infty} \frac{\frac{17}{t} + \frac{800}{t^4} + \frac{1,000}{t^5}}{1 - \frac{1}{t^6}} = 0$

10. $\displaystyle\lim_{x \to -\infty} \frac{\frac{100}{x} + \frac{1}{x^5} - \frac{1}{x^6} + \frac{500}{x^7}}{1 + \frac{2}{x^7}} = 0$

11. $\displaystyle\lim_{x \to +\infty} \frac{8x - \frac{9}{x} + \frac{5}{x^2}}{1 + \frac{300}{x}} = +\infty$

12. $\displaystyle\lim_{x \to +\infty} \frac{\frac{5,000}{x} + \frac{300}{x^2} + \frac{1,000}{x^5}}{1 + \frac{3}{x^5}} = 0$

13. $\displaystyle\lim_{x \to -\infty} \frac{(2 + \frac{5}{x})(1 - \frac{3}{x})}{(7 - \frac{2}{x})(4 + \frac{1}{x})} = \frac{(2)(1)}{(7)(4)} = \frac{1}{14}$

14. $\displaystyle\lim_{x \to +\infty} \frac{(3 - \frac{10}{x^2})(1 + \frac{1}{x})}{(4 - \frac{1}{x} + \frac{1}{x^2})(1 - \frac{7}{x})} = \frac{(3)(1)}{(4)(1)} = \frac{3}{4}$

15. $\displaystyle\lim_{x \to -\infty} \frac{(2 + \frac{1}{x})(1 + \frac{5}{x})(1 - \frac{3}{x})}{x + \frac{9}{x^3}} = 0$

16. $\displaystyle\lim_{x \to +\infty} \frac{(1)(1 + \frac{1}{x})(1 + \frac{2}{x})}{x - \frac{4}{x^3}} = 0$

17. $\displaystyle\lim_{t \to +\infty} \frac{8 + \frac{1}{t}}{7} - \lim_{x \to +\infty} \frac{\frac{1}{t}-1}{1 + \frac{5}{t}} = \frac{8}{7} - (-1) = \frac{15}{7}$

18. $\displaystyle\lim_{x \to +\infty} \frac{1}{1 + \frac{1}{x}} - \lim_{x \to +\infty} \frac{2}{1 - \frac{1}{x}} = 1 - 2 = -1$

19. $\displaystyle\lim_{t \to +\infty} \left(\frac{8 + \frac{5}{t}}{\frac{3}{t} - 2} \right)^3 = (-4)^3 = -64$

20. $\lim\limits_{t\to+\infty} \sqrt{\dfrac{18t^2+t-4}{3t+2t^2}} = \sqrt{\lim\limits_{t\to+\infty} \dfrac{18+\frac{1}{t}-\frac{4}{t^2}}{\frac{3}{t}+2}}$

$= \sqrt{9} = 3$

21. $\lim\limits_{x\to+\infty} \dfrac{x}{\sqrt{x^2+1{,}000}} = \lim\limits_{x\to+\infty} \dfrac{1}{\sqrt{1+\frac{1{,}000}{x^2}}}$

$= \dfrac{1}{\sqrt{1+0}} = 1$

22. $\lim\limits_{x\to-\infty} \dfrac{3x}{\sqrt{4x^2+10}} = \lim\limits_{x\to-\infty} \dfrac{-3}{\sqrt{4+\frac{10}{x^2}}} = -\dfrac{3}{2}$

23. $\sqrt{35} > 5.916$, so the denominator is growing faster than the numerator, and $\lim\limits_{x\to+\infty} f(x) = 0$

24. $\sqrt{37} > 6.083$, so the numerator is growing faster than the denominator, and
$\lim\limits_{x\to+\infty} f(x) = +\infty$

25. $\sqrt{34} > 5.831$, so the numerator is growing faster than the denominator, and
$\lim\limits_{x\to+\infty} f(x) = +\infty$

26. $\sqrt{33} > 5.744$, so the denominator is growing faster than the numerator, and $\lim\limits_{x\to+\infty} f(x) = 0$

27. $\lim\limits_{x\to-2^-} \dfrac{x^2-3x+4}{x+2} = -\infty$

28. $\lim\limits_{x\to1^-} \dfrac{x-1}{|x^2+1|} = \lim\limits_{x\to1^-} \dfrac{x-1}{|x+1||x-1|}$

$= \lim\limits_{x\to1^-} \dfrac{-1}{x+1} = -\dfrac{1}{2}$

29. $\lim\limits_{x\to3^+} \dfrac{(x-3)(x-1)}{(x-3)(x-3)} = \lim\limits_{x\to3^+} \dfrac{x-1}{x-3} = +\infty$

30. $\lim\limits_{x\to3^+} \left(\dfrac{1}{x-7} - \dfrac{1}{x-3}\right)$

$= \lim\limits_{x\to3^+} \dfrac{1}{x-7} - \lim\limits_{x\to3^+} \dfrac{1}{x-3} = -\infty$

31. For small values of x, the $\sin x < x$, so the denominator will approach 0 through positive values. Meanwhile the numerator is approaching 1. The quotient $\frac{1}{0} \to \infty$.

32. $\lim\limits_{x\to\pi/4} \dfrac{\sec x}{\tan x - 1} = \dfrac{\sqrt{2}}{\lim\limits_{x\to\pi/4}(\tan x - 1)} = +\infty$

33. $\lim\limits_{x\to+\infty} x\sin\frac{1}{x} = \lim\limits_{u\to0^+} \dfrac{\sin u}{u} = 1$

34. $\lim\limits_{x\to0^+} \dfrac{x^2-x}{x-\sin x} = \lim\limits_{x\to0^+} \dfrac{x-1}{1-\frac{\sin x}{x}}$

$= \lim\limits_{x\to0^+} \dfrac{-1}{1-1} = -\infty$

35. $\dfrac{x^2\csc x(1+\cos x)}{1-\cos^2 x} = \dfrac{x^2(1+\cos x)}{\sin^3 x}$

$= \dfrac{1+\cos x}{x\,\frac{\sin^3 x}{x^3}} = \dfrac{1+\cos x}{x\left(\frac{\sin x}{x}\right)^3}.$

$\lim\limits_{x\to0} \dfrac{1+\cos x}{x\left(\frac{\sin x}{x}\right)^3} = \dfrac{1+1}{0(1)^3} = \dfrac{2}{0} = +\infty$

36. $\lim\limits_{x\to0^+} \dfrac{x-\sin x}{\sqrt{\sin x}}$

$= \lim\limits_{x\to0^+} \dfrac{x}{\sqrt{\sin x}} - \lim\limits_{x\to0^+} \sqrt{\sin x}$

$= \lim\limits_{x\to0^+} \sqrt{x}\left(\sqrt{\dfrac{x}{\sin x}}\right) = \lim\limits_{x\to0^+} \sqrt{x}(1) = 0$

37. $f'(x) = \dfrac{26}{(7-x)^2};\ f''(x) = \dfrac{52}{(7-x)^3}$

asymptotes: $x = 7,\ y = -3$;
graph rising on $(-\infty, 7)\cup(7, +\infty)$;
concave up on $(-\infty, 7)$;
concave down on $(7, +\infty)$;
no critical points;
no points of inflection

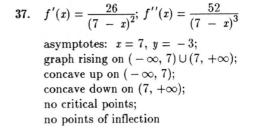

38. $g'(x) = -\dfrac{15}{(t+4)^2};\ g''(x) = \dfrac{30}{(t+4)^3}$

asymptotes: $t = 0,\ y = 0$
graph falling on $(-\infty, -4)\cup(4, +\infty)$
concave up on $(-4, +\infty)$
concave down on $(-\infty, -4)$
no critical points;
no points of inflection

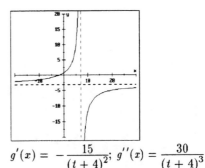

39. $f'(x) = -\dfrac{6}{(t-3)^2}$; $f''(x) = \dfrac{12}{(t-3)^2}$

asymptotes: $x = 3$, $y = 6$;
graph falling on $(-\infty, 3) \cup (3, +\infty)$;
concave up on $(3, +\infty)$;
concave down on $(-\infty, 3)$;
no critical points;
no points of inflection

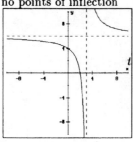

40. $g'(x) = 1 - \dfrac{4}{(4-x)^2}$; $g''(x) = -\dfrac{8}{(4-x)^3}$

asymptotes: $x = 4$; $y = x$
graph rising on $(-\infty, 2) \cup (6, +\infty)$
graph falling on $(2, 4) \cup (4, 6)$
concave up on $(4, +\infty)$
concave down on $(-\infty, 4)$
no critical points;
no points of inflection

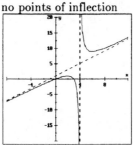

41. $h'(x) = \dfrac{2(x^3 - 1)}{x^2}$; $h''(x) = \dfrac{2(x^3 + 2)}{x^3}$

asymptotes: $x = 0$;
graph rising on $(1, +\infty)$
graph falling on $(-\infty, 0) \cup (0, +\infty)$;
concave up on $(-\infty, -\sqrt[3]{2})$ and $(0, +\infty)$;
concave down on $(-\sqrt[3]{2}, 0)$;
critical point $(1, 3)$ is a relative minimum;
point of inflection is $(-\sqrt[3]{2}, -2\sqrt[3]{4})$

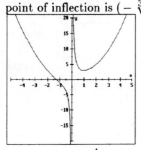

42. $f'(x) = 1 + \dfrac{4}{(x+1)^2}$; $f''(x) = -\dfrac{8}{(x+1)^3}$

asymptotes; $x = -1$; $y = x + 1$
graph rising on $(-\infty, 1) \cup (1, +\infty)$
concave up on $(-\infty, -1)$
concave down on $(-1, +\infty)$

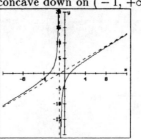

43. $f'(s) = -\dfrac{27s^2}{(s^3 - 8)^2}$; $f''(s) = \dfrac{108s(s^2 + 4)}{(s^3 - 8)^3}$

asymptotes: $x = 2$, $y = 1$;
graph falling on $(-\infty, 2) \cup (2, +\infty)$;
concave up on $(-\sqrt[3]{4}, 0)$ and $(2, +\infty)$;
concave down on $(-\infty, -\sqrt[3]{4})$, and $(2, +\infty)$;
critical point is $(0, -\frac{1}{3})$;
points of inflection $(0, -\frac{1}{8})$, $(-\sqrt[3]{4}, \frac{1}{4})$

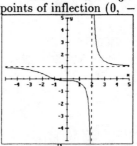

44. $f'(x) = \dfrac{5x^2 - 50x + 45}{(x^2 - 9)^2}$;

$f''(x) = -\dfrac{10}{(x^2 - 9)^3}(-x^3 + 15x^2 - 27x^2 + 45)$

asymptotes: $x = 3$, $x = -3$, $y = 2$
graph rising on $(-\infty, -3) \cup (-3, 1) \cup (9, +\infty)$
graph falling on $(1, 3) \cup (3, 9)$
concave up on $(-\infty, -3)$ and $(3, +\infty)$
concave down on $(-3, 3)$

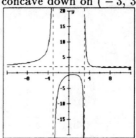

45. $g'(x) = -\dfrac{8(x+4)^2 + 27(x-1)^2}{(x-1)^2(x+4)^2}$

$g''(x) = \dfrac{10(x+1)(7x^2 - 4x + 97)}{(x-1)^3(x+4)^3}$

asymptotes: $x = -4$, $x = 1$, $y = 0$;

graph falling on $(-\infty, -4) \cup (-4, 1)$
    $\cup (1, +\infty)$;
concave up on $(-4, -1), (1, +\infty)$;
concave down on $(-\infty, -4), (-1, 1)$;
no critical points; point of inflection is
    $(-1, 5)$

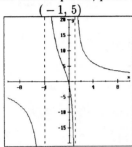

46. $f'(x) = -\dfrac{2(x^2 + 1)}{(x^2 - 1)^2}$; $f''(x) = \dfrac{4x(x^2 + 3)}{(x^2 - 1)^3}$

asymptotes: $x = -1$, $x = 2$, $y = 0$
graph falling on $(-\infty, -1) \cup (-1, 1) \cup (1, +\infty)$
concave down on $(-\infty, -1) \cup (0, 1) \cup (1, +\infty)$
concave up on $(-1, 0)$
$(0, 0)$ is a point of inflection

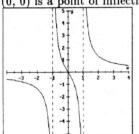

47.

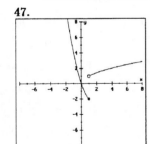

48.

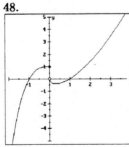

49.

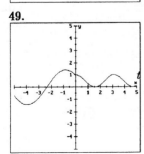

50.

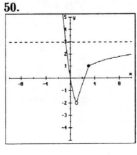

51.

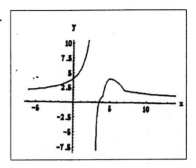

52.

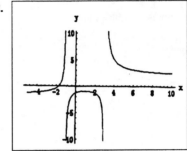

53. $\displaystyle\lim_{x \to 0^+} \left(\dfrac{1}{x^2} - \dfrac{1}{x} \right) = \lim_{x \to 0^+} \dfrac{1 - x}{x^2} = +\infty$

Wake up, Kornerkutter; read the story of
the hare and the tortoise.

54. In order to have a vertical asymptote, it is
necessary that $3 - 5b = 0$, or $b = 3/5$.
Thus,

$$\lim_{x \to +\infty} \dfrac{ax + \dfrac{3}{5}}{3 - \dfrac{3x}{5}} = \lim_{x \to +\infty} \dfrac{5ax + 3}{15 - 3x} = \dfrac{5a}{3}$$

Since this limit is equal to -3, we see
$a = 9/5$.

55. Solution 2 is incorrect.

56. $y' = -\dfrac{2x}{(x^2 - 1)^2}$; $y'' = -\dfrac{2(3x^2 + 1)}{(x^2 - 1)^3}$

The curve has a horizontal asymptote at
$y = 0$ and vertical asymptotes at $x = 1$,
$x = -1$. There is a relative maximum at
$(0, 0)$.

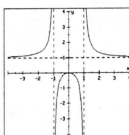

57. **a.** $+\infty$ **b.** $+\infty$ **c.** $-\infty$
    d. $-\infty$ **e.** $+\infty$

58. **a.** $P(0) = \dfrac{57(0) + 8}{3(0) + 4} = 2$; 2,000 people.

b.

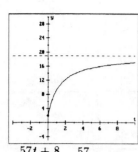

$$\lim_{t \to +\infty} \frac{57t + 8}{3t + 4} = \frac{57}{3} = 19$$

The limit is 19,000 people.

59. a. $\lim_{x \to +\infty} f(x) = \frac{a}{r}$; $f\left(\frac{at - cr}{br - as}\right) = \frac{a}{r}$

b. vertical asymptote

c. graph of g

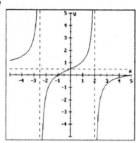

graph of h

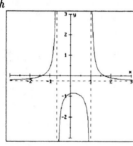

60. If $p(c) \neq 0$ and $d(c) = 0$; $f(c)$ grows beyond all bounds as $|x| \to c$ because the numerator is finite and does not vanish.

61. a. If $m > n$

$$\lim_{x \to +\infty} \frac{a_n x^n + \cdots + a_1 x + a_0}{b_m x^m + \cdots + b_1 x + b_0}$$

$$= \lim_{x \to +\infty} \frac{a_n + \cdots + a_1 x^{1-n} + a_0 x^{-n}}{b_m x^{m-n} + \cdots + b_1 x^{1-n} + b_0 x^{-n}}$$

$$= 0$$

b. If $m = n$, $\lim_{x \to +\infty} f(x) = \frac{a_n}{b_n}$

c. If $m < n$, $\lim_{x \to +\infty} f(x)$ will not have a finite limit L.

62. a. $\lim_{x \to c^+} f(x) = -\infty$

Given $M < 0$, there exists a $\delta > 0$ such that $f(x) < M$ for all x in $(0, \delta)$.

b. $\lim_{x \to c^-} f(x) = +\infty$

Given $M > 0$, there exists a $\delta > 0$ such that $f(x) > M$ for all x in $(0, \delta)$

63. By hypotheses, $\lim_{x \to +\infty} f(x) = L$ and $\lim_{x \to +\infty} g(x) = M$. Thus for each $\epsilon > 0$ there exists $N_1 > 0$ and $N_2 > 0$ such that

$$\left| f(x) - L \right| < \frac{\epsilon}{2} \text{ and } \left| g(x) - M \right| < \frac{\epsilon}{2}$$

Then,

$$\left| [f(x) + g(x)] - (L + M) \right|$$

$$= \left| [f(x) - L] + [g(x) - M] \right|$$

$$\leq \left| f(x) - L \right| + \left| g(x) - M \right|$$

$$< \frac{\epsilon}{2} + \frac{\epsilon}{2} = \epsilon \text{ for } x > N \text{ where } N = \max(N_1, N_2)$$

This means that

$$\lim_{x \to +\infty} [f(x) + g(x)] = M + L$$

$$= \lim_{x \to +\infty} f(x) + \lim_{x \to +\infty} g(x)$$

64. $\lim_{x \to c} f(x) = L$; $\lim_{x \to c} g(x) = A > 0$; because $\lim_{x \to c} g(x) = A$ exists, for each $\epsilon > 0$ (with $\epsilon = \min(\epsilon, 1)$), we have

$$\left| g(x) - A \right| < \epsilon \text{ for } |x - c| < \delta$$

Thus, $g(x) < A + \epsilon$. Now,

$$\left| f(x)g(x) - LA \right| = \left| f(x)g(x) - Lg(x) + Lg(x) - LA \right|$$

$$= \left| [f(x) - L]g(x) + L[g(x) - A] \right|$$

$$\geq \left| \left| f(x) - L \right| \left| g(x) \right| - \left| L \right| \left| g(x) - A \right| \right|$$

$$= \left| f(x) - L \right| (A - \epsilon) + |L|\epsilon$$

$$= \left| \left| f(x) - .L \right| (A - \epsilon) + L\epsilon \right| \to +\infty$$

$$\text{if } L \to +\infty$$

65. $\lim_{x \to c} f(x) = +\infty$; $\lim_{x \to +\infty} g(x) = A < 0$ (and assumed finite). $f(x) > L$ for $x > N$.

$$A - \epsilon < g(x) < A + \epsilon \text{ where } \epsilon = \min(\epsilon, 1) > 0$$

$$\frac{f(x)}{g(x)} \geq \frac{f(x)}{A - \epsilon}$$

$$\lim_{x \to c} \left| \frac{f(x) - L}{g(x)} \right| = \frac{\left| f(x) - L \right|}{g(x)}$$

$$\geq \frac{\left| \frac{1}{2} - La \right|}{A - 1} = -\infty$$

3.6 Summary of Curve Sketching, Page 218

1. In general, to sketch a curve, find intercepts, look for symmetry, set $f'(x) = 0$, rough out intervals of curve increase and decrease, set $f''(x) = 0$, rough out intervals of curve concavity, look for horizontal and vertical asymptotes. That is, use Table 3.2 on page 215.

2. Critical values are values at which extrema occur or the function is not defined. They can lead to relative maxima and/or minima.

3. Concavity indicates the shape of the curve as cupped up or cupped down; a point of inflection indicates where the curve changes concavity.

4. Asymptotes are lines that are approached by a curve the further out (toward $\pm\infty$) the curve is plotted.

5. $f'(x) = 12x(x-1)^2$;
$f''(x) = 12(3x^2 - 4x + 1)$

6. $f'(x) = 12(x^2 - 12x + 27)$
$f''(x) = 12(2x - 12)$

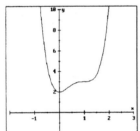

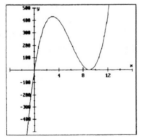

7. $f'(x) = 6(x+1)^2$
$f''(x) = 12(x+1)$

8. $f'(x) = 60x^3(x+1)^2$
$f''(x) = 60x^2(5x^2 + 8x + 3)$

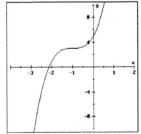

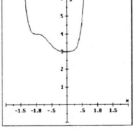

9. $f'(t) = 3(t+1)(t-3)$
$f''(t) = 6(t-1)$

10. $f'(z) = (1 - z^2)$
$f''(z) = 6z$

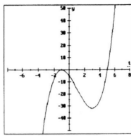

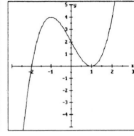

11. $g'(x) = 6(x-2)(x+1)$
$g''(x) = 6(2x-1)$

12. $h'(x) = 3(x^2 + 2x - 1) = 0$
$h''(x) = 6(x+1)$

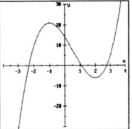

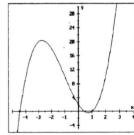

13. $g'(t) = 2t(t^2 + 1)(3t + 1)$

$g''(t) = 2(4t^3 + 3t^2 + 6t + 1)$

14. $f'(x) = \dfrac{16 - 2x^2}{\sqrt{16 - x^2}}$

$f''(x) = \dfrac{2x(x^2 - 24)}{(16 - x^2)^{3/2}}$

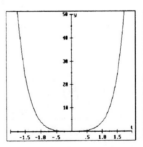

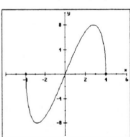

15. $f'(x) = 12(3x^3 - 10x^2 + 12x - 3)$
$f''(x) = 12(9x^2 - 20x + 12)$

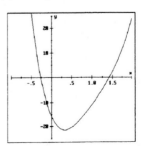

16. $g'(u) = 2(2u^2 + 9u^2 - 24u - 31)$
$g''(u) = 12(u + 4)(u - 1)$

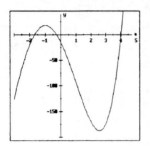

17.

$f'(t) = 20t^3(1 + t)$

$f''(t) = 20t^2(3 + 4t)$

18.

$g'(t) = \dfrac{3t^2 - 1}{2t^{3/2}2}$

$g''(t) = \dfrac{3(t^2 + 1)}{4t^{5/2}}$

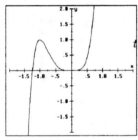

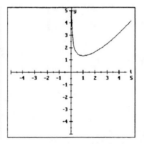

19. $f'(x) = \dfrac{-2x}{(x^2 + 12)^2}$

$f''(x) = \dfrac{6(x - 2)(x + 2)}{(x^2 + 12)^3}$

20. $f'(x) = \dfrac{2x(x + 4)}{(x + 2)^2}$

$f''(x) = \dfrac{16}{(x + 2)^3}$

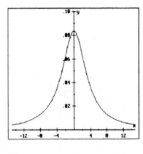

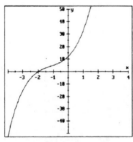

21.

$c'(x) = -25(450\pi)\sin(450\pi x)$
$c''(x) = -25(450\pi)^2\cos(450\pi x)$

22. $f'(x) = 3x(x + 2)$
$f''(x) = 6(x + 1)$

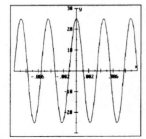

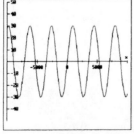

For Problems 23-43, in addition to showing the requested information, as an aid to graphing, we show the critical points and inflection points with coordinates rounded to the the nearest tenth unit.

23. $f'(x) = 3x(x + 2)$
  critical values: $x = 0$, $x = -2$
$f''(x) = 6(x + 1)$
  critical value: $x = -1$
relative maximum at $x = -2$;
relative minimum at $x = 0$;
inflection point at $x = -1$;
critical points $(-2, 5)$, $(0, 1)$;
inflection point: $(-1, 3)$

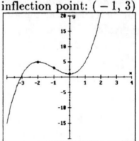

24. $f'(x) = 4x(x + 1)(x - 1)$
  critical values: $x = 0$, -1, 1
$f''(x) = 4(3x^2 - 1)$
  critical values: $x = \pm\sqrt{3}/3$
relative maximum at $x = 0$;
relative minimums at $x = \pm 0.6$
inflection points: $(\pm 0.6, 2.7)$

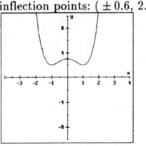

25. $f'(x) = 4x(x - 3)(x + 3)$
  critical values: $x = 0$, -3, 3
$f''(x) = 12(x^2 - 3)$
  critical values: $x = \pm\sqrt{3}$
relative maximum at $x = 0$;
relative minimums at $x = -3$, 3;
inflection points at $x = \pm\sqrt{3}$;
critical points $(-3, 0)$, $(0, 81)$, $(3, 0)$;
inflection points $(-1.7, 36)$, $(1.7, 36)$

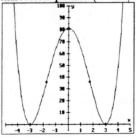

26. $g'(x) = 3(x-2)(x+2)$
             critical values: $x = 2$, $x = -2$
      $g''(x) = 6x$
             critical value: $x = 0$
      relative maximum at $x = -2$
      relative minimum at $x = 2$
      inflection point at $x = 0$

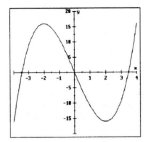

27. $f'(x) = 2 - 18x^{-2}$
             critical values: $x = 3$, $x = -3$
      $f''(x) = -36x^{-3}$
             no critical values
      relative maximum at $x = -3$;
      relative minimum at $x = 3$;
      critical value at $x = 0$ (neither, it is a
             vertical asymptote);
      critical points $(-3, -11)$, $(3, 13)$;
      no inflection points

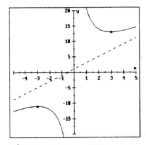

28. $f'(x) = 6(x-2)(x+1)$
             critical values: $x = 2$, $x = -1$
      $f''(x) = 6(2x - 1)$
             critical value: $x = \frac{1}{2}$
      relative maximum at $x = -1$
      relative minimum at $x = 2$
      point of inflection is $(0.5, 6.5)$
      critical points $(2, -7)$, $(-1, 20)$

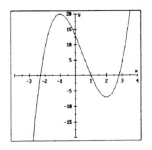

29. $f'(x) = -3(x+1)(x-3)$
             critical values: $x = -1$, $x = 3$
      $f''(x) = -6(x - 1)$
             critical value: $x = 1$
      relative maximum at $x = 3$;
      relative minimum at $x = -1$;
      inflection point at $x = 1$;
      critical points $(-1, 0)$, $(3, 32)$;
      inflection point $(1, 16)$

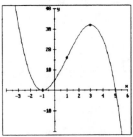

30. $f'(t) = 2(t - 3)(t - 4)(2t - 7)$
             critical values: $t = 3, 4, \frac{7}{2}$
      $f''(t) = 2(6t^2 - 42t + 73)$
             critical values: $t \approx 3.8, 3.2$
      relative maximum at $x = 3.5$
      relative minimums at $x = 3$, $x = 4$
      inflection points at $x = 3.2$, $x = 3.8$

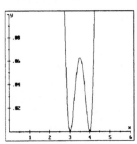

31. $f'(x) = 6(x - 1)^2(2x + 3)$
             critical values: $x = 1$, $-\frac{3}{2}$
      $f''(x) = 12(x - 1)(3x + 2)$
             critical values: $x = 1$, $-\frac{2}{3}$
      relative minimum at $x = -3/2$;
      inflection point at $x = -2/3, 1$;
      critical points $(-1.5, -32.1)$, $(1, 7)$;
      inflection points $(-0.67, -16.1)$

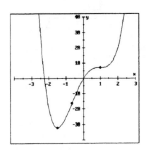

32. $f'(x) = \dfrac{4(x-1)}{3x^{2/3}}$

critical value: $x = 1$;
not defined at $x = 0$

$f''(x) = \dfrac{4(x+2)}{9x^{5/3}}$

critical value: $x = -2$;
not defined at $x = 0$
relative minimum at $x = 1$
point of inflection is $(-2, 7.6)$

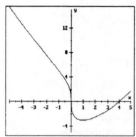

33. $f'(u) = \dfrac{5u - 14}{39u^{1/3}}$

critical value: $x = \frac{14}{5}$
not defined at $x = 0$

$f''(u) = \dfrac{2(5u + 7)}{9u^{4/3}}$

critical value: $x = -\frac{7}{5}$
not defined at $x = 0$
relative maximum at $x = 0$;
relative minimum at $x = 14/5$;
inflection point at $x = -7/5$;
critical points $(0, 0)$, $(2.8, -8.3)$;
inflection point $(-1.4, 10.5)$

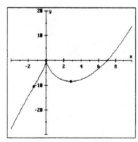

34. $g'(u) = -2(u^2 + u + 1)^{-3}(2u + 1)$

critical value: $u = -\frac{1}{2}$

$g''(u) = 2(u^2 + u + 1)^{-4}(10u^2 + 10u + 1)$

critical values: $u = \dfrac{-5 \pm \sqrt{15}}{10}$

$\approx -0.1, -0.9$

relative maximum $(-0.5, 1.8)$
points of inflection $(-0.9, 1.2)$, $(-0.1, 1.2)$

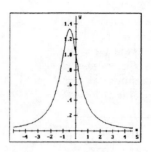

35. $f'(x) = \dfrac{x(x-4)}{(x-2)^2}$

critical values $x = 0, 4$;
not defined at $x = 2$

$f''(x) = \dfrac{8}{(x-2)^3}$

not defined at $x = 2$
relative maximum at $x = 0$;
relative minimum at $x = 4$;
critical value at $x = 2$ (neither; it is a
vertical asymptote);
critical points $(0, 0)$, $(4, 8)$;
no inflection points

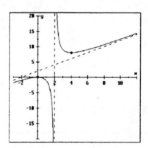

36. $t'(\theta) = \cos\theta + \sin\theta$
critical values at $\theta = \frac{3\pi}{4}, \frac{7\pi}{4}$
$t''(\theta) = -\sin\theta + \cos\theta$
critical values at $\theta = \frac{\pi}{4}, \frac{5\pi}{4}$
relative maximum at $\theta = 3\pi/4$
relative minimum at $\theta = 7\pi/4$
points of inflection at $\theta = \pi/4$, $\theta = 5\pi/4$
critical points $(2.4, 1.4)$, $(5.5, -1.4)$
inflection points $(0.8, 0)$, $(3.9, 0)$

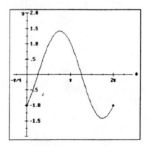

37. $f'(x) = 1 - 2 \cos 2x$
          critical values at $x = \frac{\pi}{6}, \frac{5\pi}{6}$
$f''(x) = 4 \sin 2x$
          critical values at $x = 0, \frac{\pi}{2}$
relative maximum at $x = 5\pi/6$;
relative minimum at $x = \pi/6$;
inflection point at $x = \pi/2$;
critical points $(0.5, -0.3), (2.6, 3.5)$;
inflection point $(1.6, 1.6)$

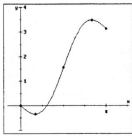

38. $g'(t) = \sec^2 t - \sec t \tan t$
$= \dfrac{1 - \sin t}{\cos^2 t}$
          critical values at $t = \frac{\pi}{2}$
          (which is not in the domain)
$g''(t) = \sec t \left(\dfrac{2 \sin t - 1 - \sin^2 t}{\cos^2 t} \right)$
          critical values $t = \frac{\pi}{2}, -\frac{\pi}{6}$
          (which are not in the domain)
no extrema or points of inflection on $[0, \frac{\pi}{3}]$

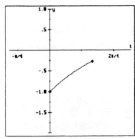

39. $f'(x) = 2 \cos x(\sin x - 1)$
          critical value at $x = \frac{\pi}{2}$
$f''(x) = -4 \sin^2 x + 2 \sin x + 2$
$= -2(\sin x - 1)(2 \sin x + 1)$
          critical value at $x = \frac{\pi}{2}$
relative minimum at $x = \pi/2$;
critical point $(1.6, 0)$

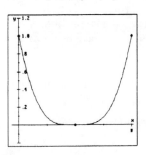

40. $g'(t) = \frac{1}{2}t^{-3/2}(t - 4)$
          critical value at $t = 4$
          not defined at $t = 0$
$g''(t) = \frac{1}{4}t^{-5/2}(12 - t)$
          critical value at $t = 12$
          not defined at $t = 0$
relative minimum at $(4, 4)$
point of inflection at $(12, 4.6)$

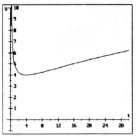

41. $g'(t) = (t^2 + 1)^{-3/2}$
          no critical values
$g''(t) = -3t(t^2 + 1)^{-5/2}$
          critical value at $t = 0$
inflection point $(0, 0)$

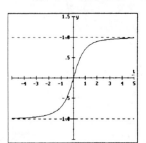

42. $g'(x) = \sin x \left(\dfrac{1 + \cos^2 x}{\cos^2 x} \right)$
          critical values: $x = -\pi, 0, \pi$
          not defined at $x = \frac{\pi}{2}$
$g''(x) = \dfrac{8 - \sin^2 2x}{4 \cos^3 x}$
          no critical values; not defined at $x = \frac{\pi}{2}$
relative minimum at $(0, 0)$

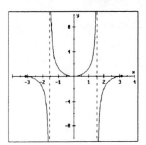

43. $g'(x) = \dfrac{2}{(1 + \cos x)^2}$

critical values $x = \pm \pi$ (neither; vertical asymptotes);

$g''(x) = \dfrac{4 \sin x}{(1 + \cos x)^3}$
critical value at $x = 0$
inflection points at $x = 0$; $(0, 0)$

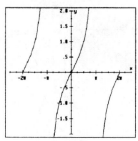

44. $v = \sqrt{gr} \, \tan^{1/2}\theta$

$v' = \dfrac{\sqrt{gr}}{2}\left(\dfrac{\sec^2\theta}{\sqrt{\tan \theta}}\right)$

This is not defined at $\theta = 0$ and $\frac{\pi}{2}$

$v'' = \dfrac{\sqrt{gr}}{2}\left(\dfrac{\sqrt{\tan \theta}(2)\sec^2\theta \tan \theta}{\tan \theta}\right)$

$\qquad - \dfrac{\sqrt{gr}}{2}\left(\dfrac{\sec^2\theta(\frac{1}{2})\tan^{-1/2}\theta \sec^2\theta}{\tan \theta}\right)$

$\qquad = \dfrac{\sqrt{gr}}{2}\left[\dfrac{\sqrt{\sec^2\theta(4\tan^2\theta - \sec^2\theta)}}{(\tan \theta)^{3/2}}\right]$

$v'' = 0$ if

$\sqrt{\sec^2\theta(4 \tan^2\theta - \sec^2\theta)} = 0$

$\sec^2\theta(4 \tan^2\theta - \sec^2\theta) = 0$

$4 \tan^2\theta - \sec^2\theta = 0$

$(2 \tan \theta - \sec \theta)(2 \tan \theta + \sec \theta) = 0$

$2 \tan \theta = \pm \sec \theta$

$\dfrac{2 \sin \theta}{\cos \theta} = \pm \dfrac{1}{\cos \theta}$

$\sin \theta = \pm\dfrac{1}{2}$

On $[0, \frac{\pi}{2}]$ the solution is $\theta = \frac{\pi}{6}$

Point of inflection at $(0.5, .8\sqrt{gr})$

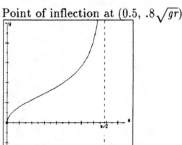

45. $m = m_0(1 - v^2/c^2)^{-1/2}$

$m' = m_0(1 - v^2/c^2)^{-3/2}(v/c^2)$

Since $v > 0$, m' is not defined at $v = c$

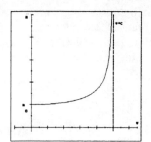

46.   **47.**

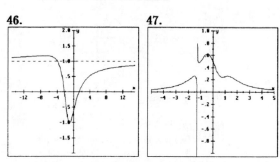

48. True; f and g are concave up by hypotheses (so $f'' > 0$ and $g'' > 0$). Let
$F(x) = f(x) + g(x)$; then
$F''(x) = f''(x) + g''(x) > 0$
so F is concave up.

49. False; let $f(x) = x^6$ is concave up and
$g(x) = -x^2$ is concave down so the
hypotheses are satisfied. Let $F(x) = f(x)g(x)$
$= -x^8$. $F''(x) < 0$ and F is concave down.

50. False; let $f(x) = -20 + x^4$ on $I = [-2, -1]$
and $g(x) = [f(x)]^2 = 400 - 40x^4 + x^8$ so that
the hypotheses are satisfied on I.
$g'(x) = -160x^3 + 8x^7$,
$g''(x) = -480x^2 + 56x^6$
$g''(-1.1) < 0$ and the curve is concave down
at $x = -1.1$.

51. False; let $f(x) = x^2 > 0$ and
$g(x) = (x + 3)^2 > 0$ on $I = [-2, -1]$
Let $F(x) = f(x)/g(x) = x^4 + 6x^3 + 9x^2$
$F'(x) = 4x^3 + 18x^2 + 18x$
$F''(x) = 12x^2 + 364x + 18$ and $F''(-1.9) < 0$
so the curve is concave down at $x = -1.9$.

3.7 Optimization in the Physical Sciences and Engineering, Page 226

1. (1) Understand the question;

(2) Choose the variables; let Q be the quantity to be maximized or minimized.

(3) Express Q in terms of the defined variables.

(4) Determine the domain;

(5) Find the extrema;

(6) Answer the question that was asked.

2. The extrema may not be applicable to a particular problem, or even though may not answer the question asked may lead to the answer.

3. Let x and y be the dimensions of the rectangular plot. The fencing (perimeter) is $P = 2(x + y)$ and $A = xy = 64$. Minimize P, so write

    $$P = 2\left(x + \frac{64}{x}\right)$$
    $$P' = 2(1 - 64x^{-2})$$

    $P' = 0$ if $x = 8$ and $y = 8$. Since $P''(8) > 0$, $x = 8$ is a minimum. The dimensions of the garden should be 8 ft by 8 ft.

4. Let x and y be the dimensions of the picnic area. The fencing (perimeter) is $F = x + 2y$ and $A = xy = 5,000$. Minimize P, so write

    $$P = x + 2\left(\frac{5,000}{x}\right)$$
    $$P' = 1 - 10,000x^{-2}$$

    $P' = 0$ if $x = 100$. Since $P''(100) > 0$, $x = 100$ is a minimum. The dimensions of the picnic area should be 100 yd by 50 yd.

5. Let x and y be the dimensions of the rectangular plot with $A = xy$. We wish to maximize A.

    a. The fencing (perimeter) is $F = 2(x + y) = 320$.

    $$A = x(160 - x) = 160x - x^2$$
    $$A' = 160 - 2x$$

    $A' = 0$ if $x = 80$. Since $A''(80) < 0$, $x = 80$ is a maximum. Since this is a continuous function on a closed interval, $[0, 160]$, we are guaranteed an absolute maximum and an absolute minimum by the extreme value theorem. The endpoints are obvious minimums for the area, so our one remaining candidate must be the maximum. The use of the extreme value theorem often eliminates the need to use a second derivative test. Be certain that the hypotheses are met in

each case. The dimensions that give the maximum area are 80 ft by 80 ft.

b. The fencing is now $F = x + 2y = 320$, so $y = \frac{1}{2}(320 - x)$ on $[0, 320]$.

    $$A = xy = \frac{1}{2}(320x - x^2)$$
    $$A' = \frac{1}{2}(320 - 2x) = 160 - x$$

    $A' = 0$ when $x = 160$. Thus, the dimensions that give the maximum area are 160 ft by 80 ft.

6. The size of the sheet (measured to the nearest centimeter) is 28 cm by 22 cm (found by measuring). Let x be the side of the square that is cut out ($0 \leq x \leq 11$ cm). We want to maximize the volume, V,

    $$V = \ell w h$$
    $$= (28 - 2x)(22 - 2x)x$$
    $$= 4x^3 - 100x^2 + 616x$$
    $$V' = 12x^2 - 200x + 616$$

    $V' = 0$ when

    $$x = \frac{200 \pm \sqrt{(-200)^2 - 4(12)(616)}}{2(12)}$$
    $$\approx 12.59 \text{ (not in domain)}, 4.08$$

    $$V \approx (28 - 2 \cdot 4.08)(22 - 2 \cdot 4.08)(4.08)$$
    $$\approx 1,120 \text{ cm}^3$$

    Since a liter is $1,000$ cm^3, we see that the maximum volume is about 1 liter.

7. Let x be the side $\overline{CD}$ and y the other dimension of the rectangle. The price of the fence is

    $$P = 3\left(2y + \frac{3x}{2}\right)$$

    Since the area is $1,200$ m^2, $xy = 1,200$ or $y = \frac{1,200}{x}$. Thus,

    $$P(x) = 3\left(\frac{2,400}{x} + \frac{3x}{2}\right) = 7,200x^{-1} + 4.5x$$
    $$P'(x) = -7,200x^{-2} + 4.5$$

    $P'(x) = 0$ if $x = 40$. $P(4) = 360$ is a minimum since $P''(40) > 0$. The maximum amount that Jones must pay is $360.

8. Let x and y be the lengths of the sides of the rectangle and R be the radius of the semicircle. Then $y = \sqrt{R^2 - x^2}$ for $0 \leq x \leq R$. Then

    $$A = 2xy = 2x\sqrt{R^2 - x^2}$$

$$A' = 2(R^2 - x^2)^{1/2} + 2x(\tfrac{1}{2})(R^2 - x^2)^{-1/2}(-2x)$$

$$= \frac{2(-2x^2 + R^2)}{\sqrt{R^2 - x^2}}$$

$A' = 0$ if $x_c = \dfrac{R}{\sqrt{2}}$ and $y_c = \sqrt{R^2 - \dfrac{R^2}{2}} = \dfrac{R}{\sqrt{2}}$

$A' > 0$ for $x < x_c$ and $A' < 0$ for $x > x_c$.
$A(0) = A(R) = 0$.

9.   The volume of the box will be given by

$$V = x(24 - 2x)(45 - 2x)$$

$$= 4x^3 - 138x^2 + 1080x$$

$$V' = 12x^2 - 276x + 1080$$

$$= 12(x - 5)(x - 18)$$

$V' = 0$ if $x = 0, 5, 18$

0 is obviously a minimum, 18 is not in the domain, so the maximum volume occurs when squares of 5 in. are cut from the corners. The dimensions of the box will be 5 by 14 by 35 in. and the volume will be 2,450 in.[3]

If the hypotheses of the extreme value theorem apply to a particular problem, we can often avoid testing a candidate to see if it is a maximum or minimum. Suppose we have three candidates and two are obviously minimums - the third must be a maximum.

10. Let R be the radius of the sphere and r the radius of the cylinder with height h; $0 \le h \le R$. Since $r^2 = R^2 - h^2$,

$$V = 2\pi r^2 h$$

$$= 2\pi h(R^2 - h^2)$$

$$= 2\pi(R^2 h - h^3)$$

$$V' = 2\pi(R^2 - 3h^2)$$

$V' = 0$ when $h = R/\sqrt{3}$. The dimensions are $h = \dfrac{R}{3}\sqrt{3}$ and $r = \dfrac{R}{3}\sqrt{6}$.

11. Draw a figure and note $0 \le r \le R$ and $r^2 = R^2 - h^2$.

$$S = 2\pi r h$$

$$= 2\pi r \sqrt{R^2 - r^2}$$

$$S' = 2\pi(R^2 - r^2)^{1/2} + 2\pi r(\tfrac{1}{2})(R^2 - r^2)^{-1/2}(2r)$$

$$= \frac{2\pi(R^2 - r^2)}{\sqrt{R^2 - r^2}}$$

$S' = 0$, then $r = \dfrac{R}{\sqrt{2}}$. The dimensions are

$$h = r = \frac{R}{\sqrt{2}}.$$

12. A vertex of the inner square subdivides a side of the outer square into line segments x and y ($0 \le x \le L$). The given outer square has side $x + y$. Thus, $x^2 + y^2 = L^2$. The area of the outer square is

$$A(x) = (x + y)^2 = (x + \sqrt{L^2 - x^2})^2$$

$$A'(x) = 2(x + \sqrt{L^2 - x^2})\left(1 + \frac{-2x}{2\sqrt{L^2 - x^2}}\right)$$

$A'(x) = 0$ if $x = \sqrt{L^2 - x^2}$ or $x_c = \dfrac{L}{\sqrt{2}}$

which yields a maximum area.

13. Draw the figure and label the dimensions of the cylinder r and h. Maximize the volume of the cylinder: $V = \pi r^2 h$. By similar triangles:

$$\frac{h}{R - r} = \frac{H}{R}$$

Substituting:

$$V = \pi r^2 \left[\frac{H}{R}(R - r)\right]$$

$$= \pi\left(Hr^2 - \frac{H}{R}r^3\right).$$

$$V' = \pi\left(2Hr - \frac{3H}{R}r^2\right)$$

$V' = 0$ when

$$2Hr - \frac{3H}{R}r^2 = 0$$

$$\frac{H}{R}r(2R - 3r) = 0$$

$$r = 0 \text{ or } \frac{2}{3}R$$

The extreme value theorem applies, so the maximum volume of the cylinder occurs when its radius is $\frac{2}{3}$ of the radius of the cone. Its height will be

$$\frac{H}{R}(R - \tfrac{2}{3}R) = \tfrac{1}{3}H$$

that is; $\frac{1}{3}$ the height of the cone.

14. Draw a figure with the car at the origin of a Cartesian coordinate system and the truck at $(250, 0)$. At time t, the truck is at position $(250 - x, 0)$, while the car is at $(0, y)$. Let D be the distance that separates them.

$$\frac{dx}{dt} = 60 \text{ and } \frac{dy}{dt} = 80,$$

so that $x = 60t$ and $y = 80t$.

$$D^2 = (250 - x)^2 + y^2$$

$$D^2 = (250 - 60t)^2 + (80t)^2$$

$$= 2,500(4t^2 - 12t + 25)$$

$$(D^2)' = 110,000(2t - 3)$$

The derivative of the distance squared is 0

when $t = 1.5$ hr. Substituting this into the equation for D^2 produces the shortest distance: $x = 60(1.5) = 90$, $y = 80(1.4) = 120$

$$D^2 = (250 - 90)^2 + 120^2 = 1,600(25)$$

$$D = 40(5) = 200$$

The minimum distance is 200 mi.

15. Let x be the length of a rectangle and y be the width with $2P$ the fixed perimeter. Then,

$$2P = 2x + 2y \text{ or } y = P - x$$

The area is to be maximized:

$$A = xy = x(P - x) = Px - x^2$$

$$A' = P - 2x$$

$A' = 0$ when $x = \frac{P}{2}$. $A'' < 0$, which means that $x = y = P/2$ maximizes the enclosed area. This rectangle is a square.

16. Let x and y denote the dimensions of a rectangle with fixed area A and perimeter P. $A = xy$ or $y = A/x$. The perimeter is to be minimized:

$$P = 2x + 2y = 2x + \frac{2A}{x}$$

$$P' = 2 - \frac{2A}{x^2}$$

$P' = 0$ when $x = \pm\sqrt{A}$. For $x > 0$ and with $P'' = 4A/x^3 > 0$, the minimal perimeter is attained when $x = \sqrt{A}$ and $y = A/x = \sqrt{A}$. This rectangle is a square.

17. We need to maximize $V = \ell wh$, given that $\ell = w$, and $5(\ell w) + 1(\ell w + 2\ell h + 2wh) = 72$. Solving the cost relationship for h:

$$6\ell w + h(2\ell + 2w) = 72$$

$$h = \frac{72 - 6\ell w}{2\ell + 2w}$$

$$= \frac{36 - 3\ell w}{\ell + w}$$

$$V = \ell(\ell)\left(\frac{36 - 3\ell(\ell)}{\ell + \ell}\right)$$

$$= \frac{36\ell^2 - 3\ell^4}{2\ell}$$

$$= \tfrac{3}{2}(12\ell - \ell^3)$$

$$V' = 18 - \tfrac{9}{2}\ell^2$$

$V' = 0$ when $\ell = \pm2$. The extreme value theorem applies, the end points and $\ell = -2$ are not reasonable, so the maximum volume with the given restrictions occurs when

$\ell = w = 2$ ft. and $h = 6$ ft.

18. Let x be the dimension of one side of the square base and y the length. The volume is $V = x^2 y$. The cross sectional perimeter plus the length is $4x + y = 108$ (maximum). Thus, $y = 4(27 - x)$ and

$$V = 4x^2(27 - x)$$

$$= 4(27x^2 - x^3)$$

$$V' = 4(54x - 3x^2) = 12x(18 - x)$$

$V' = 0$ when $x = 18$ and $y = 4(27 - 18) = 36$.

19. Draw a figure; pick a point x km down the road (toward the power plant) from the nearest point on the paved road and have the jeep head toward that point. The distance traveled by the jeep on the sand is $\sqrt{x^2 + 32^2}$ and the distance on the paved road is $16 - x$ (assuming $x < 16$). The total time traveled (in hours) is

$$t = \frac{\sqrt{32^2 + x^2}}{48} + \frac{16 - x}{80}$$

$$t' = \frac{1}{48}(\tfrac{1}{2})(32^2 + x^2)^{-1/2}(2x) - \frac{1}{80}$$

$t' = 0$ when

$$\frac{1}{48}(\tfrac{1}{2})(32^2 + x^2)^{-1/2}(2x) - \frac{1}{80} = 0$$

$$\frac{x}{48(32^2 + x^2)^{1/2}} = \frac{1}{80}$$

$$80x = 48(32^2 + x^2)^{1/2}$$

$$5x = 3(32^2 + x^2)^{1/2}$$

$$25x^2 = 9(32^2 + x^2)$$

$$16x^2 = 9 \cdot 32^2$$

$$x = \frac{3 \cdot 32}{4} = 24$$

This distance is further than the power plant. The minimum time corresponds to heading for the power plant on the sand. The time is

$$t = \frac{\sqrt{32^2 + 16^2}}{48} + 0 = \frac{\sqrt{5}}{3} \approx 0.745$$

This is 44 minutes 43 seconds; so he has about 5 minutes 17 seconds to diffuse the bomb.

20. Refer to the figure; Missy rows to a point P which is x mi from point B. The rowing distance is $H_r = \sqrt{36 + x^2}$ and the corresponding time is $t_r = \frac{\sqrt{36 + x^2}}{6}$. The time taken to walk the $H_w = s - x$ distance is $t_w = \frac{s - x}{10}$. The

total time, t, is

$$t = \frac{\sqrt{36 + x^2}}{6} + \frac{s - x}{10}$$

$$t' = \frac{2x}{12\sqrt{36 + x^2}} - \frac{1}{10}$$

$t' = 0$ when $x = 4.5$ km.

a. When $s = 4$, Missy should row all the way.

b. When $s = 6$, she should land at a point 4.5 km from point B and run the rest of the way (1.5 km).

21. Refer to the figure. The distance from A to B, when projected onto the road is

$$\sqrt{12^2 - (5 - 3)^2} = \sqrt{140}$$

Let $C(x, 0)$ be the desired point on the road and D the sum of the distances.

$$D = \sqrt{x^2 + 25} + \sqrt{(\sqrt{140} - x)^2 + 9}$$

$$D' = x(x^2 + 25)^{-1/2} + \frac{1}{2}[(\sqrt{140} - x)^2 + 9]^{-1/2}(2)(\sqrt{140} - x)(-1)$$

Solve for x when $D' = 0$:

$$\frac{x}{\sqrt{x^2 + 25}} + \frac{x - \sqrt{140}}{\sqrt{(\sqrt{140} - x)^2 + 9}} = 0$$

Solving leads to $x \approx 7.395$ mi. An alternate method consists of recognizing that when BC is reflected about the road, $\overline{A}\,\overline{C}\overline{B}$ lies in a straight line.

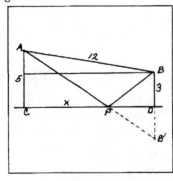

We want to minimize the $L = \overline{AP} + \overline{PB}$. Drawing a line parallel and equal to $\overline{CD}$ through B, we have a right triangle with hypotenuse 12 and leg 2, so $\overline{CD} = \sqrt{140}$. Let $\overline{CP} = x$. Now imagine that B is on the opposite side of the road at B'. The shortest distance will be $\overline{AB'}$, which will be equal to $\overline{AP} + \overline{PB}$.

$$\overline{AB'} = \sqrt{\sqrt{140}^2 + 8^2} = 2\sqrt{51} \approx 14.28 \text{ mi.}$$

To find the desired point P on the highway, use similar triangles:

$$\frac{x}{5} = \frac{\sqrt{140} - x}{3}, \quad x \approx 7.395 \text{ mi from } C$$

22. Draw a figure. Let x and y be the dimensions of the rectangular printed matter. Then, $xy = 108$ or $y = 108/x$. The area of the poster is to be minimized:

$$A = (x + 4)(y + 12) = xy + 4y + 12x + 48$$
$$= 108 + 432x^{-1} + 12x + 48$$
$$A' = 12 - 432x^{-2}$$

$A' = 0$ when $x^2 = 36$ or $x = 6$. Then, $y = 108/6 = 18$. The cost of the poster is

$$C(x) = 20(6 + 4)(18 + 12) = 6,000$$

cents or \$60.00.

23. Refer to the figure. Draw vertical segments of length h from either end of the 14 cm segment so the distance on either side of the endpoint of this segment is labeled s; thus, $x = 2s$ and $h = \sqrt{36 - s^2}$. We wish to maximize the area

$$A = \frac{1}{2}[14 + (14 + 2s)]\sqrt{36 - s^2}$$
$$= (14 + s)\sqrt{36 - s^2}$$
$$A' = \sqrt{36 - s^2} + (14 + s)\left[\frac{1}{2} \cdot \frac{-2s}{\sqrt{36 - s^2}}\right]$$
$$= \sqrt{36 - s^2} + (14 + s)\left[\frac{-s}{\sqrt{36 - s^2}}\right]$$

$A' = 0$ when

$$\sqrt{36 - s^2} = \frac{s(14 + s)}{\sqrt{36 - s^2}}$$
$$36 - s^2 = 14s + s^2$$
$$s^2 + 7s - 18 = 0$$
$$(s - 2)(s + 9) = 0$$
$$s = 2, -9 \text{ (negative not in domain)}$$
$$A_{\max} = (14 + 2)\sqrt{36 - 2^2} = 16\sqrt{32} = 64\sqrt{2}$$

24. The amount of material is $M = 2(\pi r^2) + 2\pi rh$ and the volume is $V = \pi r^2 h = 6.89\pi$. Solving for h we obtain $h = 6.89r^{-2}$ so that

$$M = 2\pi r^2 + 2\pi r(6.89r^{-2})$$
$$= \pi[2r^2 + 13.78r^{-1}]$$
$$M' = \pi[4r - 13.78r^{-2}]$$

$M' = 0$ when $r \approx 1.51$ and $h \approx 3.02$ in.

The actual dimensions of Coke can are $h \approx 2.25$ in. and $r \approx 1.125$. The actual dimensions of a can of cola differ for historical reasons and marketing reasons.

25. We are asked to minimize the cost of material to construct the cylinder.

$$C = 0.5(\pi r^2) + 0.3(2\pi rh)$$

The volume is not specified, but is a constant, $V = \pi r^2 h$. To have the cost as a function of a single variable, solve for $h = V/(\pi r^2)$.

$$C = 0.5\pi r^2 + 0.6\pi r\left(\frac{V}{\pi r^2}\right)$$
$$= 0.5\pi r^2 + \frac{0.6V}{r}$$
$$C' = \pi r - \frac{0.6V}{r^2}$$

$C' = 0$ when

$$r^3 = \frac{0.6V}{\pi}$$
$$r = \sqrt[3]{\frac{0.6V}{\pi}} \text{ or } \sqrt[3]{\frac{3V}{5\pi}}$$

Since the extreme value theorem applies, and the endpoints are obviously maximums, this must be the radius that gives the minimum cost.

26. Draw a figure by completing Figure 3.61 as follows: From point A draw a horizontal segment of length d to the point of intersection of this segment and the ray formed by β. The perpendicular distance from the common point of intersection of the rays to the horizontal segment is h. Let x be the distance from A to this point of intersection. Then, $x = h\tan\alpha$, $d - x = h\tan\beta$, so that the total distance is

$$\ell = \sqrt{x^2 + h^2} + \sqrt{(d - x)^2 + h^2}$$
$$\frac{d\ell}{dx} = \frac{x}{\sqrt{x^2 + h^2}} + \frac{(d - x)(-1)}{\sqrt{(d - x)^2 + h^2}}$$

$d\ell/dx = 0$ if

$$\frac{x}{\sqrt{x^2 + h}} = \frac{d - x}{\sqrt{(d - x)^2 + h^2}}$$
$$\sin\alpha = \sin\beta$$
$$\alpha = \beta$$

27. Let x be the dimension of one side of the equilateral triangle, which is also the dimension of the base of the rectangle. Let y be the height of the rectangle. The perimeter of the window is $3x + 2y = 20$ giving

$$y = \frac{20 - 3x}{2} \text{ with } 0 \le x < \frac{20}{3}$$

We need to find h, the height of the isosceles triangle. With x as the hypotenuse of one-half the isosceles triangle,

$$h = x\sin\frac{\pi}{3} = \frac{\sqrt{3}}{2}x$$

The base of the right triangle is $x/2$. Thus, the area of the entire isosceles triangle is

$$A_t = \frac{\sqrt{3}}{4}x^2$$

The area of the rectangle is

$$A_r = xy = \frac{x(20 - 3x)}{2}$$

Since twice as much light passes through the rectangle as through the stained glass isosceles triangle, then

$$L = k\left[\frac{\sqrt{3}}{4}x^2 + (2)\frac{x(20 - 3x)}{2}\right]$$
$$= k\left[\left(\frac{\sqrt{3}}{4} - 3\right)x^2 + 20x\right]$$
$$L' = k\left[\left(\frac{\sqrt{3}}{2} - 6\right)x + 20\right]$$

$L' = 0$ when $x = \dfrac{20}{6 - \dfrac{\sqrt{3}}{2}} \approx 3.8956 \approx 4$ ft

Then $y = (\frac{1}{2})(20 - 11.6868) \approx 4.1566 \approx 4$ ft;

$L'' < 0$ which signifies that the window that admits the most light is when $x = 4$ and $y = 4$.

28. a. Using similar triangles, $\triangle BAO$ and $\triangle ORS$, since $\alpha = \beta$,

$$\frac{RS}{AB} = \frac{q}{p}$$

From $\triangle COF$ and $\triangle FRS$, we have since $\gamma = \delta$

$$\frac{RS}{OC} = \frac{q - f}{f}$$

Thus,

$$\frac{q}{p} = \frac{q - f}{f}$$
$$fq = pq - pf$$
$$fq + pf = pq$$
$$\frac{fq}{fqp} + \frac{pf}{fqp} = \frac{pq}{fqp}$$
$$\frac{1}{p} + \frac{1}{q} = \frac{1}{f}$$

b. $q = 24 - p$ so that

$$\frac{p(24 - p)}{24} = f$$
$$\frac{df}{dp} = \frac{24 - 2p}{24}$$

$df/dp = 0$ if $p = 12$.

29. $E' = -Mg + \dfrac{2mgx}{\sqrt{x^2 + d^2}} = 0$ when

$$Mg\sqrt{x^2 + d^2} = 2mgx,$$

$$\sqrt{x^2 + d^2} = \frac{2mx}{M}$$

$$x^2 + d^2 = \frac{4m^2x^2}{M^2}$$

$$\left(\frac{4m^2}{M^2} - 1\right)x^2 = d^2$$

$$x^2 = \frac{d^2 M^2}{4m^2 - M^2}$$

$$x = \frac{Md}{\sqrt{4m^2 - M^2}}.$$

30. $D(x) = k(2x^4 - 5Lx^3 + 3L^2x^2)$ on $[0, L]$

$D'(x) = k(8x^3 - 15Lx^2 + 6L^2x)$

$\qquad = kx(8x^2 - 15Lx + 6L^2)$

$D'(x) = 0$ when $x = 0$ or when

$$x = \frac{15L \pm \sqrt{225L^2 - 192L^2}}{16}$$

$$= \frac{15 - \sqrt{33}}{19} L \approx 0.58L$$

31. $P = I^2 R = \left(\dfrac{E}{R + r}\right)^2 R = \dfrac{E^2 R}{(R + r)^2}$

$\dfrac{dP}{dR} = \dfrac{E^2(r - R)}{(R + r)^3} = 0$ if $R = r$.

32. a. $I(\theta) = \dfrac{k}{16} \cos\theta \sin^2\theta$ on $[0, \frac{\pi}{2}]$

$I'(\theta) = \dfrac{k}{16}[\cos\theta(2\sin\theta\cos\theta + \sin^2\theta)(-\sin\theta)]$

$\qquad = \dfrac{k}{16}(2\cos^2\theta\sin\theta - \sin^3\theta)$

$\qquad = \dfrac{k}{16}\sin\theta(2\cos^2\theta - \sin^2\theta)$

b. If $\tan\theta = \sqrt{2}$ then $\dfrac{\sin^2\theta}{\cos^2\theta} = 2$ so that

$2\cos^2\theta - \sin^2\theta = 0$, which from part **a** implies $I'(\theta) = 0$. Also, if $\tan\theta = \sqrt{2}$, then

$$\sin\theta = \sqrt{\tfrac{2}{3}} \text{ and } \cos\theta = \frac{1}{\sqrt{3}}$$

c. Substituting the results of part **b** into part **a**, we have

$$I(\theta) = \frac{k}{16}\cos\theta\sin^2\theta = \frac{k}{16}\left(\frac{1}{\sqrt{3}}\right)\left(\sqrt{\tfrac{2}{3}}\right) = \frac{k}{24\sqrt{3}}$$

33. Since we used trigonometric functions in our solution for Problem 32, we will now do the

solution without them this time.

$I(h) = \dfrac{k\sin\phi}{d^2}$ where $\sin\phi = \dfrac{h}{d}$ and

$d = \sqrt{h^2 + 16}$.

$I(h) = \dfrac{kh}{d^3}$

$I'(h) = \dfrac{kd^3 - 3khd^2 d'}{d^6}$

$\qquad = \dfrac{kd - 3khd'}{d^4}$

$I'(h) = 0$ when

$$k(d - 3hd') = 0$$

Substitute:

$$\sqrt{h^2 + 16} = 3h\left(\frac{h}{\sqrt{h^2 + 16}}\right)$$

$$h^2 + 16 = 3h^2$$

$$h^2 = 8$$

$$h = 2\sqrt{2} \approx 3 \text{ ft.}$$

And, of course, this is the same answer as the previous problem because sine and cosine are cofunctions. That is, $\theta + \phi = 90°$, and $\cos\theta = \dfrac{h}{d} = \sin\phi$.

34. Refer to Figure 3.66. Note: $\cos\theta = \dfrac{x}{L}$ and $\cos 2\theta = -(20 - x)/x$

$$\frac{x - 20}{x} = \cos 2\theta$$

$$= 2\cos^2\theta - 1$$

$$= 2\left(\frac{x}{L}\right)^2 - 1$$

$$L^2(x - 20) = 2x^3 - xL^2$$

$$L^2(x - 20) + xL^2 = 2x^3$$

$$L^2 = \frac{2x^3}{2x - 20}$$

$$L = \frac{x^{3/2}}{(x - 10)^{1/2}}$$

$$L' = \frac{(x - 10)^{1/2}(\frac{3}{2})x^{1/2} - x^{3/2}(\frac{1}{2})(x - 10)^{-1/2}}{x - 10}$$

$$= \frac{3(x - 1)x^{1/2} - x^{3/2}}{2(x - 10)^{3/2}}$$

$$= \frac{x^{1/2}(2x - 30)}{2(x - 10)^{3/2}}$$

$L' = 0$ if $x = 0$ or if $x = 15$ (undefined if $x = 10$)

$$L(15) = \sqrt{\frac{15^3}{5}} = 15\sqrt{3} \approx 26 \text{ cm is a}$$
minimum.

35. Refer to Figure 3.67. Let C denote the horizontal clearance at the corner, as shown in the figure. Consider the pipe as composed of two pieces, $L = x + y$. Parallel lines cut by a transversal form equal corresponding angles, θ. In one triangle

$$\sin \theta = \frac{2\sqrt{2}}{x}$$

and in the other triangle

$$\cos \theta = \frac{2\sqrt{2}}{y}$$

Solve for x and y:

$x = \dfrac{2\sqrt{2}}{\sin \theta}$ and $y = \dfrac{2\sqrt{2}}{\cos \theta}$. Then L, as a function of θ is

$$L(\theta) = \frac{2\sqrt{2}}{\sin \theta} + \frac{2\sqrt{2}}{\cos \theta}$$
$$= 2\sqrt{2}(\sin \theta)^{-1} + 2\sqrt{2}(\cos \theta)^{-1}$$

where θ is restricted to $[0, \frac{\pi}{2}]$. Since $L(\theta)$ is undefined at the endpoints, the minimum value of L occurs when the $L' = 0$.

$$L'(\theta) = -2\sqrt{2}(\sin \theta)^{-2}(\cos \theta)$$
$$\qquad - 2\sqrt{2}(\cos \theta)^{-2}(-\sin \theta)$$
$$= 2\sqrt{2}\left(\frac{\sin \theta}{\cos^2 \theta} - \frac{\cos \theta}{\sin^2 \theta}\right)$$
$$= 2\sqrt{2}\left(\frac{\sin^3 \theta - \cos^3 \theta}{\cos^2 \theta \sin^2 \theta}\right)$$

$L'(\theta) = 0$ when $\cos^3 \theta = \sin^3 \theta$ or when $\theta = \frac{\pi}{4}$. Thus, the minimal clearance is

$$L(\tfrac{\pi}{4}) = \frac{2\sqrt{2}}{\sin \frac{\pi}{4}} + \frac{2\sqrt{2}}{\cos \frac{\pi}{4}} = 4 + 4 = 8$$

The longest pipe that can clear the corner is 8 ft long.

36. Refer to Figure 3.68. Let s be the side of the square base with $0 \le s \le 20\sqrt{2}$. Consider a plane that passes through the vertex of the pyramid (is perpendicular to the base) and is parallel to two sides in the base. The slanted sides of the pyramid intersect this plane in a length ℓ. If one-half of the pyramid is unfolded, we obtain half of our original sheet of paper. The hypotenuse is

$$20\sqrt{2} = 2\ell + s$$
$$\ell = \frac{20\sqrt{2} - s}{2}$$

Return to the pyramid and the vertical cutting plane. Half of this intersection consists of a right plane with altitude h (the height of the pyramid), base $s/2$, and

hypotenuse ℓ.

$$h = \sqrt{\ell^2 - \frac{s^2}{4}}$$
$$= \sqrt{\frac{(20\sqrt{2} - s)^2 - s^2}{4}}$$
$$= \frac{1}{2}\sqrt{800 - 40\sqrt{2}\, s}$$
$$= \sqrt{200 - 10\sqrt{2}\, s}$$

The volume of the pyramid is

$$V = s^2 h = s^2 \sqrt{200 - 10\sqrt{2}\, s}$$

with $0 \le s \le 10\sqrt{2}$.

$$V' = s^2 \frac{-10\sqrt{2}}{2\sqrt{200 - 10\sqrt{2}\, s}} + 2s\sqrt{200 - 10\sqrt{2}\, s}$$
$$= \frac{-5\sqrt{2}\, s^2 + 2s(200 - 10\sqrt{2}\, s)}{\sqrt{200 - 10\sqrt{2}\, s}}$$
$$= \frac{-25\sqrt{2}\, s^2 + 400s}{\sqrt{200 - 10\sqrt{2}\, s}}$$

$V' = 0$ if

$$-25\sqrt{2}\, s^2 + 400s = 0$$
$$-25s(\sqrt{2}\, s - 16) = 0$$
$$s = 0, 8\sqrt{2}$$

If $s = 0$ we obtain a minimum, so if $s = 8\sqrt{2}$

$$V = s^2 \sqrt{200 - 10\sqrt{2}\, s}$$
$$= 128\sqrt{200 - 10\sqrt{2}(8\sqrt{2})}$$
$$= 128\sqrt{200 - 160} = 256\sqrt{10} \approx 809 \text{ in.}^3$$

37. $V'(T) = 1 - 6.42(10^{-5}) + 17.02(10^{-6})T$
$$\qquad - 20.37(10^{-8})T^2$$

$V'(T) = 0$ when

$$T = \frac{0.851(10^{-5}) \pm \sqrt{0.724[10^{-10} - 0.1308(10^{-10})]}}{2.037(10^{-7})}$$

$$\approx 80 \text{ °C}, 4 \text{ °C}$$

Reject 80 °C since the mathematical model is valid around the freezing temperature. $T = 4°$ is a minimum since V' is falling on $(0, 4)$ and rising on $(4, 80)$. Liquid water and solid ice can coexist at 0 °C. The water below the surface is denser since the trapped temperature is a little higher than 0. As the temperature above the surface drops, the

change in temperature is passed to the next higher level of water and transforms to ice. The graph of this volume function is interesting:

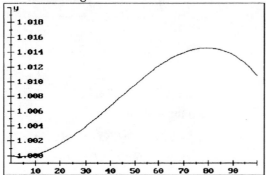

The maximum value occurs when $T \approx 80°$. The minimum value occurs when $T \approx 4°$.

3.8 Optimization in Business, Economics, and the Life Sciences, Page 236

1. The profit is maximized when the marginal revenue equals the marginal cost.

$$R(x) = xp = \tfrac{1}{2}x(75 - x) = \tfrac{1}{2}(75x - x^2)$$
$$R'(x) = \tfrac{1}{2}(75 - 2x)$$
$$R'(x) = C'(x)$$
$$\tfrac{1}{2}(75 - 2x) = \tfrac{1}{4}x + 5$$
$$150 - 4x = x + 20$$
$$5x = 130$$
$$x = 26$$

2. The profit is maximized when the marginal revenue equals the marginal cost.

$$R(x) = xp = \tfrac{1}{5}x(45 - x) = \tfrac{1}{5}(45x - x^2)$$
$$R'(x) = \tfrac{1}{5}(45 - 2x)$$
$$R'(x) = C'(x)$$
$$\tfrac{1}{5}(45 - 2x) = \tfrac{4}{5}x + 3$$
$$x = 5$$

3. The profit is maximized when the marginal revenue equals the marginal cost.

$$R(x) = xp = \frac{70x - x^2}{x + 30}$$
$$R'(x) = \frac{(x + 30)(70 - 2x) - (70x - x^2)}{(x + 30)^2}$$
$$R'(x) = C'(x)$$

$$\frac{(x + 30)(70 - 2x) - (70x - x^2)}{(x + 30)^2} = \frac{1}{5}$$
$$(x + 30)(70 - 2x) - (70x - x^2) = \tfrac{1}{5}(x + 30)^2$$
$$5(-x^2 - 60x + 2{,}100) = x^2 + 60x + 900$$
$$6x^2 + 360x - 9{,}600 = 0$$
$$x^2 + 60x - 1{,}600 = 0$$
$$(x + 80)(x - 20) = 0$$

$x = 20$ (-80 is not in the domain). To maximize profit produce 20 items.

4. $$C'(x) = 8x^3 - 30x^2 - 36x + 1$$
$$C''(x) = 24x^2 - 60x - 36$$
$$= 12(2x^2 - 5x - 3)$$
$$= 12(2x + 1)(x - 3)$$
$$C'(0) = 1; \quad C'(3) = -161; \quad C'(5) = 71$$

The maximum is 71 and the minimum is -161.

5. $$A(x) = \frac{C(x)}{x} = 3x + 1 + 48x^{-1}$$
$$A'(x) = 3 - 48x^{-2} = 0 \text{ at } x = 4$$

$A''(4) > 0$, so $x = 4$ determines the minimum average cost of $A(4) = 12 + 1 + 12 = 25$.

6. The relationship between the number of Mopsy dolls and Flopsy dolls is given by

$$y = \frac{82 - 10x}{10 - x}$$

on $[0, 8]$. Suppose Mopsy sells for \$2 and Flopsy sells for \$1 each. The revenue is

$$R(x) = x + \frac{2(82 - 10x)}{10 - x}$$
$$= \frac{164 - 20x + 10x - x^2}{10 - x}$$
$$= \frac{x^2 + 10x - 164}{x - 10}$$
$$R'(x) = \frac{(x - 10)(2x + 10) - (x^2 + 10x - 164)(1)}{(x - 10)^2}$$
$$= \frac{x^2 - 20x + 64}{(x - 10)^2}$$

$R'(x) = 0$ when

$$x^2 - 20x + 64 = 0$$
$$(x - 16)(x - 4) = 0$$

$x = 16$ (not in domain), 4

$R(0) = 16.4; \quad R(4) = 18; \quad R(8) = 10;$ if $x = 4$,

$y = \dfrac{82 - 40}{10 - 4} = 7$

Revenue is maximized when 400 Flopsys and 700 Mopsys are sold.

7. **a.** Total cost, $C(x)$, is the average cost, $A(x)$, times the number of items produced, x.

$C(x) = 5x + \dfrac{x^2}{50}$.

$R(x) = x\left(\dfrac{380 - x}{20}\right)$.

$\begin{aligned}
P(x) &= R(x) - C(x) \\
&= \dfrac{380x - x^2}{20} - \dfrac{250x + x^2}{50} \\
&= \dfrac{-7x^2 + 1,400x}{100} \\
&= -0.07x^2 + 14x
\end{aligned}$

b. The maximum profit occurs when

$R'(x) = C'(x)$.

$19 - \dfrac{x}{10} = 5 + \dfrac{x}{25}$

$14 = \dfrac{7}{50}x$

$x = 100$

This gives a price per item,

$p = \dfrac{380 - x}{20} = \$140$ per item.

The maximum profit is

$\begin{aligned}
P(x) &= \dfrac{(-7x + 1,400)x}{100} \\
&= \dfrac{(-700 + 1,400)100}{100} \\
&= \$700
\end{aligned}$

8. Let x denote the number of bottles in each shipment. The number of shipments is 600 and the ordering cost is $30(600/x) = 1,800/x$. The purchase cost is $600(4) = 2,400$. The storage cost is $(x/2)(0.9) = 0.45x$. As a result the total cost is

$C(x) = 0.45x + 1,800x^{-1} + 2,400$

$C'(x) = 0.45 - 1,800x^{-2} = 0$

when

$x^2 = \dfrac{1,800}{0.45} = 40,000$

Thus, $x = 200$ which is a minimum for the cost since $C'' > 0$. The number of shipments is $600/200 = 3$. The order should be placed 3 times per year.

9. Let x denote the number of cases of connectors in each shipment, and $C(x)$ the corresponding (variable) cost. Then,

$C(x) = \text{STORAGE COST} + \text{ORDERING COST}$

$= \dfrac{4.5x}{2} + 20\left(\dfrac{18,000}{x}\right)$

$= 2.25x + 360,000x^{-1}$ on $[0, 18,000]$

$C'(x) = 2.25 - 360,000x^{-2}$

$C'(x) = 0$ when $x = 400$. Since this is the only critical point in the interval, and since $C''(x) = 720,000x^{-3} > 0$ when $x = 400$, C is minimized when $x = 400$. The number of shipments should be $18,000/400 = 45$ times per year.

10. **a.** $C(x) = 3x^2 + 5x + 75$ and

$A(x) = \dfrac{C(x)}{x} = 3x + 5 + 75x^{-1}$

$A'(x) = 3 - 75x^{-2} = 0$ when $x = 5$

on the interval $(0, +\infty)$.

$A''(x) = 150x^{-3} > 0$ when $x = 5$.

Average costs are minimized when 5 units are produced.

b. The marginal cost is $C'(x) = 6x + 5$ and equals the average cost when

$6x + 5 = 3x + 5 + \dfrac{75}{x}$

$3x^2 - 75 = 0$

$x^2 = 25$

$x = \pm 5$ (reject -5)

The average cost per unit equals the marginal costs when $x = 5$, which is the same level of production as that of part **a**.

c.

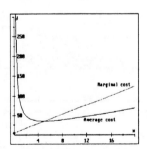

11. **a.** Average revenue per unit is total revenue divided by number of units:

$A(x) = -2x + 68 - \dfrac{128}{x}$

Marginal revenue is $R'(x) = -4x + 68$. They are equal when:

$-2x + 68 - \dfrac{128}{x} = -4x + 68$

$2x = \dfrac{128}{x}$

$$x^2 = 64$$

$$x = 8 \quad (\text{reject } x = -8)$$

b. $A'(x) = -2 + \dfrac{128}{x^2}$;

$A'(8) = 0; \ A'(8^-) > 0, \ A'(8^+) < 0$

So there is a relative maximum at $(8, 36)$. The function is increasing on $[0, 8)$ and decreasing on $(8, +\infty)$.

c.

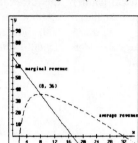

12. The consumer expenditure is $E(p) = px(p)$ $= 73p^{1/2}$. This function is monotonically increasing. The economic expenditure increases when the price is increased.

13. COST = STORAGE COST + ORDERING COST

$$C(x) = \frac{tN}{2} + \frac{QS}{N}$$

With $N = x$

$$C(x) = \frac{tx}{2} + \frac{QS}{x}$$

$$C'(x) = \frac{t}{2} - \frac{QS}{x^2}$$

$C'(x) = 0$ if $x^2 = \dfrac{2QS}{t}$ or $x = \sqrt{\dfrac{2QS}{t}}$

We now substitute this value into the cost equations:

The STORAGE COST $= \dfrac{tx}{2} = \sqrt{\dfrac{2QS}{t}} = \sqrt{\dfrac{tQS}{2}}$

The ORDERING COST $= \dfrac{QS}{x} = \dfrac{QS}{\sqrt{2QS}}$

Thus, the total cost is minimized when the storage cost is equal to the ordering cost.

14. Let x be the number of groups of 50 people above the 600 passenger level. Then the number of passengers is $600 + 50x$. The fare is $500 - 25x$ (in cents). The revenue is

$$R(x) = (600 + 50x)(500 - 25x)$$

$$= 1,250(12 + x)(20 - x)$$

$$= 1,250(240 + 8x - x^2)$$

The domain is $0 \le x \le 20$.

$R'(x) = 1,250(8 - 2x) = 0$ if $x = 4$

$R(0) = 3,000; \ R(4) = 3,200; \ R(20) = 0$
Revenue is maximized at the 450-passenger group level and the fare is
$5.00 - 1.00 = \$4.00$.

15. $\qquad P(x) = R(x) - C(x)$

Let $x =$ the increase in price,
$40 + x =$ cost per board,
$50 - 3x =$ number of boards sold.

$$P(x) = (40 + x)(50 - 3x) - 25(50 - 3x)$$

$$= -3x^2 + 5x - 750$$

$$P'(x) = -6x + 5 = 0 \text{ when } x = \tfrac{5}{6}.$$

Since this is not an integer, and our function has discreet rather than continuous values, we will test the nearest integer values.

$$P(0) = 750; \ P(1) = 752;$$

therefore, raise the price \$1, sell the boards at a price of \$41, sell 47 per month, and have the maximum profit of \$752.

16. Let x be the distance from the 60 ppm plant. Then emission at this distance is $60/x$. For the 240 ppm plant, the emission is $240/(10 - x)$. The total pollution is

$$p = \frac{60}{x} + \frac{240}{10 - x}$$

$$p' = 60\left[-\frac{1}{x^2} + \frac{4}{(10 - x)^2}\right]$$

$$= \frac{60}{x^2(10 - x)^2}[(10 - x)^2 - 4]$$

$p' = 0$ if $x = 4$ or $x = 6$ ($x = 14$ is not in the domain). For $x = 6$, $p'' > 0$, a minimum is reached.

17. Let x be the number of people above the 100 level ($0 \le x \le 50$). Then, the number of travelers is $100 + x$. The fare per traveler is $2,000 - 5x$. The revenue is

$$R(x) = (100 + x)(2,000 - 5x)$$

and the profit is

$$P(x) = (100 + x)(2,000 - 5x)$$

$$- 125,000 - 500(x + 100)$$

$$= -5x^2 + 1,000x + 25,000$$

$$P'(x) = -10x + 1,000 = 0 \text{ at } x = 100$$

But $x = 50$ fills all the available seats;
$P(0) = 25,000; \ P(50) = 37,500$. Thus, lower the fare by $5(50) = \$250$.

18. Let x denote the number of \$2 reductions in the price and $P(x)$ the corresponding profit function. For each \$2 reduction in price, 20 more books than the current 200 will be sold, and so the total number sold is $200 + 2x$. Each book sells for $30 - 2x$ dollars (the current price minus the number of \$2 reductions), and the cost of each book is \$6. Hence, the profit per book is $(30 - 2x) - 6 = 2(12 - x)$ dollars. Then,

$P(x) =$ (NUMBER OF BOOKS SOLD)(PROFIT PER BOOK)

$= (200 + 20x)(2)(12 - x)$ on $[0, 12]$

$P'(x) = 80(1 - x) = 0$ when $x = 1$.
$P(1) = 4,840$; $P(0) = 4,800$; $P(12) = 0$; the greatest profit corresponds to a \$2 reduction so the book sells for \$28.

19. We want to maximize the yield, $Y(x)$. Let x be the additional number of trees to be planted.

NUMBER OF TREES $= 60 + x$

AVERAGE YIELD $= 400 - 4x$

$Y(x) = (60 + x)(400 - 4x)$

$= -4x^2 + 160x + 24,000$

$Y'(x) = -8x + 160 = 0$ when $x = 20$

Plant 80 total trees, have an average yield per tree of 320 oranges, and a maximum total crop of 25,600 oranges.

20. Let x be the number of days following July 1 where the farmer should harvest. The sales price per bushel is then $2 - 0.02x$ and the number of bushels of potatoes is $80 + x$. The revenue, when selling on day x, is

$R(x) = (80 + x)(2 - 0.02x)$

$= 160 + 0.40x - 0.02x^2$

$R'(x) = 0.40 - 0.04x = 0$ when $x = 10$.

$R''(10) < 0$, so $x = 10$ maximizes R. The date of harvest is July 1 plus 10 days, or July 11.

21. Let x denote the number of additional grapevines to be planted and $N(x)$ the corresponding yield. Since there are 50 grapevines to begin with and x additional grapevines are planted, the total number is $50 + x$. For each additional grapevine, the average yield is decreasing by 2 lb, so that the yield is $150 - 2x$ lb. Thus,

$N(x) =$ (NO. OF GRAPES PER VINE)(NO. OF GRAPEVINES)

$= (150 - 2x)(50 + x)$

$= 2(75 - x)(50 + x)$

$= 2(3,750 + 25x - x^2)$ on $[0, 20]$

$N'(x) = 2(25 - 2x) = 0$ when $x = 12.5$

$N(0) = 7,500$; $N(12) = 7,812$; $N(13) = 7,812$; $N(20) = 7,700$. The greatest possible yield is 7,812 pounds of grapes, which is generated by planting 12 or 13 (choose 12 for practical reasons) additional vines; that is, 62 vines are planted.

22. Let x be the number of days and $R(x)$ the corresponding revenue. Over the period of 80 days, 24,000 pounds of glass have been collected at a rate of 300 pounds per day, and so for each day over 80, an additional 300 pounds will be collected. Thus, the total number of pounds collected and sold is $24,000 + 300x$. Currently the recycling center pays 1¢ per pound. For each additional day, it reduces the price it pays by 1¢ per 100 pounds; that is, by 0.01¢/lb. Hence, after x additional days, the price per pound will be $1 - 0.01x$ cents.

$R(x) =$ (NO. OF GRAPES/VINE)(NO. OF VINES)

$= (24,000 + 300x)(1 - 0.01x)$

$= 24,000 + 60x - 3x^2$ on $[0, 100]$

$R'(x) = 60 - 6x = 0$ when $x = 10$

$R(0) = 24,000$; $R(100) = 0$; $R(10) = 24,300$

The most profitable time to conclude the project is 10 days from now.

23. The demand function is the number of items, x, that can be sold at a given price, p.

a. The revenue function, $R(x)$, is the number of items sold times the price per item.

$R(x) = (120 - 0.1x^2)x = -0.1x^3 + 120x$

$R'(x) = -0.3x^2 + 120 = 0$ when

$x^2 = 400$, $x = 20$ (-20 is not in domain)

$R''(x) = -0.6x$, which is negative for all values of x in the domain, so our candidate is a maximum. Revenue is increasing for prices less than 20, and decreasing for prices greater than 20. $p(20) = 80$, so the revenue is maximized when $x = 20$.

b.

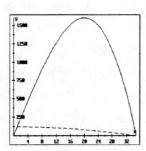

The demand function:
$$p(x) = 120 - 0.1x^2$$
The revenue function:
$$R(x) = -0.1x^3 + 120x, \text{ on } [0, 1200]$$

24. a. $R(x) = xp(x) = \dfrac{bx - x^2}{a}$ on $[0, b]$

$R'(x) = \dfrac{1}{a}(b - 2x) = 0$ if $x = \dfrac{b}{2}$

R is increasing on $(0, \dfrac{b}{2})$ and deceasing on $(\dfrac{b}{2}, b)$.

b.

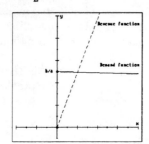

25. Let E be the amount of energy. Since it requires twice as much energy over water than over land, let us double the distance ℓ over the water. Let x be the distance along the shore where the pigeon is crossing the shoreline.

$$E = 2\ell + 10 - x$$
$$= 2\sqrt{9 + x^2} + 10 - x$$
$$\frac{dE}{dx} = \frac{2x}{\sqrt{9 + x^2}} - 1 = 0 \text{ if } x = \sqrt{3}$$

If $\tan\theta = \dfrac{\sqrt{3}}{3}$, then $\theta = \dfrac{\pi}{6}$.

26. $t = \dfrac{s}{v - v_1}$;

$$E = cv^k t = cs\left(\frac{v^k}{v - v_1}\right)$$

$$\frac{dE}{dv} = cs\left[\frac{(v - v_1)kv^{k-1} - v^k}{(v - v_1)^2}\right]$$

$dE/dv = 0$ if $\dfrac{csv^{k-1}}{(v - v_1)^2}(kv - kv_1 - v) = 0$

$$v = \frac{kv_1}{k - 1}$$

27. We wish to minimize the surface area given a

fixed volume: $v_0 = \pi r^2 h$.

$$S(r) = 2\pi r(h + r)$$
$$= 2\pi r\left(\frac{v_0}{\pi r^2} + r\right)$$
$$= \frac{2 v_0}{r} + 2\pi r^2$$
$$S'(r) = -\frac{2 v_0}{r^2} + 4\pi r$$

$S'(r) = 0$ when $4\pi r^3 = 2v_0$, $r = \sqrt[3]{\dfrac{v_0}{2\pi}}$.

$S''(r) = \dfrac{4 v_0}{r^3} + 4\pi > 0$, so the previous value of r gives the minimum.

28. $F'(p) = p^n(M - n)(1 - p)^{M-n-1}(-1)$
$$+ np^{n-1}(1 - p)^{M-n}$$
$$= p^{n-1}(1 - p)^{M-n-1}(-Mp + np + n - np)$$
$= 0$ if $p = \dfrac{n}{M}$; $F'(p) > 0$ for $p < \dfrac{n}{M}$ and
$F'(p) < 0$ for $p > \dfrac{n}{M}$ so $p = \dfrac{n}{M}$ leads to a maximum.

29. $v' = A(2r_0 r - 3r^2) = Ar(2r_0 - 3r) = 0$
when $r = 0$ and $r = \dfrac{2r_0}{3}$. $v(0) = 0$;

$$v\left(\frac{2r_0}{3}\right) = A\left(\frac{4r_0^2}{9}\right)\left(r_0 - \frac{2r_0}{3}\right) = A\left(\frac{4r_0^3}{27}\right)$$

Thus, $r = \dfrac{2r_0}{3}$ leads to a maximum velocity.

30. $E' = \dfrac{1}{v}[2a(v - b)] - \dfrac{1}{v^2}[a(v - b)^2 + c]$
$$= \frac{1}{v^2}(av^2 - ab^2 - c) = 0 \text{ when}$$
$$v^2 = \frac{ab^2 + c}{a} = \frac{(0.04)(36)^2 + 9}{0.04} = 1,521$$

Thus, $v = 39$ units of length per unit of time.

31. The total cost of production is the cost of setting up the machines plus the cost of supervising machine operation. Let $x = $ the number of machines used. The number of hours the machines must run is 8,000 divided by 50 times the number of machines working. Supervision is \$35 for each of these hours.

a. $\quad C(x) = 800x + \dfrac{8,000}{50x}(35)$

$$= 800x + \frac{5,600}{x}$$

$$C'(x) = 800 - \frac{5,600}{x^2} = 0$$

when $x^2 = 7$, $x \approx 2.65$.

Since this is a discrete rather than a continuous function, we need to test the

nearest integer values. $C(2) = 4,400$, $C(3) \approx 4,267$. So if we set up and run 3 machines, the number of operational hours will be $\frac{8000}{3(50)} \approx 53.3$ hrs.

b. The supervisor will earn approximately $(53.3)(35) \approx \$1866.67$

32. a. Let x be number of machines to be set up. The set-up cost is Sx, the number of units produced per hour is nx, the number of hours of operation is $pQ/(nx)$, and the total cost is

$$C = Sx + \frac{pQ}{nx}$$
$$C' = S - \frac{pQ}{nx^2}$$
$$C' = 0 \text{ when } x^2 = \frac{pQ}{nS} \text{ or } x = \sqrt{\frac{pQ}{nS}}$$

b. The set-up cost equals the operating cost if

$$Sx = \frac{pQ}{nx} \text{ or } x^2 = \frac{pQ}{nS}, \ x = \sqrt{\frac{pQ}{nS}}$$

33. The revenue is $R(x) = px$ and its derivative

$$\frac{dR}{dx} = p + x\frac{dp}{dx}$$
$$= p\left[1 + \frac{x}{p}\frac{dp}{dx}\right]$$
$$= \frac{xp}{x}\left[1 + \frac{x}{p}\frac{dp}{dx}\right]$$
$$= \frac{R(x)}{x}\left[1 + \frac{1}{E(x)}\right]$$

3.9 l'Hôpital's Rule, Page 245

1. a. l'Hôpital's rule does not apply.

$\lim_{x \to \pi}(1 - \cos x) \neq 0$ and is defined.
Similarly $\lim_{x \to \pi} x \neq 0$ and is defined.

b. l'Hôpital's rule does not apply;

$$\lim_{x \to \pi/2}\frac{\sin x}{x} = \frac{\pi}{2}$$

2. $\lim_{x \to +\infty}\frac{x}{\sqrt{x^2 - 1}} = \frac{+\infty}{+\infty}$

l'Hôpital's rule applies but $\lim_{x \to +\infty}\frac{\frac{1}{x}}{\frac{x}{\sqrt{x^2 - 1}}}$

$= \lim_{x \to +\infty}\frac{\sqrt{x^2 - 1}}{x}$ poses the same dilemma as

the original problem. Another approach is dividing both numerator and denominator by x (which is finite before we take the limit).

3. $\lim_{x \to 2}\frac{x^4 - 16}{x^2 - 4} = \lim_{x \to 2}\frac{4x^3}{2x} = \lim_{x \to 2} 2x^2 = 8$

4. $\lim_{x \to 1}\frac{x^4 - 1}{x^2 - 1} = \lim_{x \to 1}\frac{4x^3}{2x} = \lim_{x \to 1} 2x^2 = 2$

5. l'Hôpital's rule does not apply because

$$\lim_{x \to 2}\frac{x^3 - 27}{x^2 - 9} \neq \frac{0}{0}$$
$$\lim_{x \to 2}\frac{x^3 - 27}{x^2 - 9} = \frac{8 - 27}{4 - 9} = \frac{19}{5}$$

6. $\lim_{x \to 1}\frac{x^3 - 1}{x^2 - 1} = \lim_{x \to 1}\frac{3x^2}{2x} = \frac{3}{2}$

7. $\lim_{x \to 1}\frac{x^{10} - 1}{x - 1} = \lim_{x \to 1}\frac{10x^9}{1} = 10$

8. $\lim_{x \to -1}\frac{x^{10} - 1}{x + 1} = \lim_{x \to -1} 10x^9 = -10$

9. $\lim_{x \to 0}\frac{1 - \cos^2 x}{\sin^3 x} = \lim_{x \to 0}\frac{-2\cos x(-\sin x)}{3\sin^2 x \cos x}$

$= \frac{2}{3}\lim_{x \to 0}\frac{1}{\sin x}$ which is not defined.

10. $\lim_{x \to 0}\frac{1 - \cos^2 x}{3\sin x} = \lim_{x \to 0}\frac{-2\cos x(-\sin x)}{3\cos x}$

$= \frac{2}{3}\lim_{x \to 0}\sin x = 0$

11. $\lim_{x \to \pi}\frac{\cos\frac{x}{2}}{\pi - x} = \lim_{x \to \pi}\frac{-\frac{1}{2}\sin\frac{x}{2}}{-1} = \frac{1}{2}$

12. $\lim_{x \to 0}\frac{1 - \cos x}{x^2} = \lim_{x \to 0}\frac{\sin x}{2x} = \frac{1}{2}\lim_{x \to 0}\cos x = \frac{1}{2}$

13. l'Hôpital's rule does not apply;

$$\lim_{x \to 0}\frac{\sin ax}{\cos bx} = \frac{\sin 0}{\cos 0} = 0$$

14. $\lim_{x \to 0}\frac{\tan 3x}{\sin 5x} = \lim_{x \to 0}\frac{3\sec^2 3x}{5\cos 5x} = \frac{3}{5}$

15. $\lim_{x \to 0}\frac{x - \sin x}{\tan x - x} = \lim_{x \to 0}\frac{1 - \cos x}{\sec^2 x - 1}$

$= \lim_{x \to 0}\frac{\sin x}{2\sec^2 x \tan x} = \lim_{x \to 0}\frac{1}{2\sec^3 x} = \frac{1}{2}$

16. $\lim_{x \to 0}\frac{1 - \cos^2 x}{x \tan x} = \lim_{x \to 0}\frac{-2\cos x(-\sin x)}{x\sec^2 x + \tan x}$

$= \lim_{x \to 0}\frac{\sin 2x}{x\sec^2 x + \tan x}$

$= \lim_{x \to 0}\frac{2\cos 2x}{x(2\sec^2 x \tan x) + 2\sec^2 x} = \frac{2}{2} = 1$

17. $\lim_{x \to \pi/2}\frac{3\sec t}{2 + \tan t} = 3\lim_{x \to \pi/2}\frac{\sec t \tan t}{\sec^2 t}$

$= 3\lim_{x \to \pi/2}\frac{\tan t}{\sec t} = 3\lim_{x \to \pi/2}\sin t = 3$

18. $\lim_{x \to 0}\frac{x + \sin^3 x}{x^2 + 2x} = \lim_{x \to 0}\frac{1 + 3\sin^2 x \cos x}{2x + 2} = \frac{1}{2}$

19. $\lim\limits_{x\to 0} \dfrac{3x + \sin^3 x}{x \cos x} = \lim\limits_{x\to 0} \dfrac{3 + 3\sin^2 x \cos x}{\cos x - x \sin x}$

$\qquad = \dfrac{3}{1} = 3$

20. $\lim\limits_{x\to 0} \dfrac{x - \sin ax}{x + \sin bx} = \lim\limits_{x\to 0} \dfrac{1 - a\cos ax}{1 + b\cos bx} = \dfrac{1 - a}{1 + b}$

21. $\lim\limits_{x\to 0} \dfrac{x^2 + \sin x^2}{x^2 + x^3} = \lim\limits_{x\to 0} \dfrac{2x + 2x \cos x^2}{2x + 3x^2}$

$\qquad = \lim\limits_{x\to 0} \dfrac{2 + 2\cos^2 x}{2 + 3x} = 2$

22. $\lim\limits_{x\to +\infty} x^2 \sin \frac{1}{x};\quad$ Let $u = \frac{1}{x}$;

$\qquad \lim\limits_{u\to 0^+} \dfrac{\sin u}{u^2} = \lim\limits_{u\to 0^+} \dfrac{\cos u}{2u} = \dfrac{1}{2}\lim\limits_{u\to 0^+} \dfrac{\cos u}{u} = +\infty$

23. $\lim\limits_{x\to +\infty} x^{3/2} \sin \frac{1}{x};\quad$ Let $u = \frac{1}{x}$;

$\qquad \lim\limits_{u\to 0^+} \dfrac{\sin u}{u}\cdot u^{-1/2} = \lim\limits_{u\to 0^+} \dfrac{\sin u}{u}\lim\limits_{u\to 0^+} \dfrac{1}{\sqrt{u}}$

$\qquad = (1)(+\infty) = +\infty$

24. l'Hôpital's rule does not apply

$\qquad \lim\limits_{x\to 1}(1 - \cos x)\cot x = (0)(1) = 0$

25. $\lim\limits_{x\to 1} \dfrac{(x - 1)\sin(x - 1)}{1 - \cos(x - 1)};\quad$ Let $u = x - 1$

$\qquad \lim\limits_{u\to 0} \dfrac{u \sin u}{1 - \cos u} = \lim\limits_{u\to 0} \dfrac{u \cos u + \sin u}{\sin u}$

$\qquad = \lim\limits_{u\to 0} \dfrac{-u\sin u + 2\cos u}{\cos u} = 2$

26. $\lim\limits_{\theta\to 0} \dfrac{\theta - 1 + \cos^2\theta}{\theta^2 + 5\theta} = \lim\limits_{\theta\to 0} \dfrac{1 + 2u\cos\theta(-\sin\theta)}{2\theta + 5}$

$\qquad = \dfrac{1}{5}$

27. $\lim\limits_{x\to 0} \dfrac{x + \sin(x^2 + x)}{3x + \sin x}$

$\qquad = \lim\limits_{x\to 0} \dfrac{1 + (2x + 1)\cos(x^2 + x)}{3 + \cos x}$

$\qquad = \dfrac{1 + 1}{3 + 1} = \dfrac{1}{2}$

28. $\lim\limits_{x\to(\pi/2)^-} \sec 3x \cos 9x$ is not in the proper

form to apply l'Hôpital's rule;

$\qquad \lim\limits_{x\to(\pi/2)^-} \dfrac{\cos 9x}{\cos 3x} = \lim\limits_{x\to(\pi/2)^-} \dfrac{-9\sin 9x}{-3\sin 3x}$

$\qquad = \dfrac{-3}{9} = -3$

29. $\lim\limits_{x\to 0}\left(\dfrac{1}{x} - \dfrac{1}{x\sin x}\right)$ is not in the proper form

to apply l'Hôpital's rule; we do not need
l'Hôpital's rule for this problem:

$\lim\limits_{x\to 0} \dfrac{\sin x - 1}{x\sin x} = \dfrac{-1}{0} = -\infty$

30. $\lim\limits_{x\to 0}\left(\dfrac{1}{\sin 2x} - \dfrac{1}{2x}\right)$ is not in the proper form

to apply l'Hôpital's rule;

$\lim\limits_{x\to 0} \dfrac{2x - \sin 2x}{2x\sin 2x} = \lim\limits_{x\to 0} \dfrac{2 - 2\cos 2x}{2[2x\cos 2x + \sin 2x]}$

is not defined

31. $\lim\limits_{x\to 0}\left(\dfrac{1}{\sin^2 x} - \dfrac{1}{x}\right)$ is not in the proper form to

apply l'Hôpital's rule;

$\lim\limits_{x\to 0} \dfrac{x - \sin^2 x}{x\sin^2 x} = \lim\limits_{x\to 0} \dfrac{1 - 2\sin x\cos x}{2x\sin x\cos x + \sin^2 x}$

$\qquad = \dfrac{1}{0} = +\infty$

32. $\lim\limits_{x\to 0}\left(\cot x - \dfrac{1}{x}\right)$ is not in the proper form to

apply l'Hôpital's rule;

$\lim\limits_{x\to 0}\left(\dfrac{\cos x}{\sin x} - \dfrac{1}{x}\right) = \lim\limits_{x\to 0} \dfrac{x\cos x - \sin x}{x\sin x}$

$\qquad = \lim\limits_{x\to 0} \dfrac{-x\sin x\cos x + \cos x - \cos x}{x\cos x + \sin x}$

$\qquad = \lim\limits_{x\to 0} \dfrac{-x\cos x - \sin x}{-x\sin x + x + \cos x} = \dfrac{0}{1} = 0$

33. $\lim\limits_{x\to 0^+}\left(\dfrac{2\cos x}{\sin 2x} - \dfrac{1}{x}\right)$ is not in the proper form

to apply l'Hôpital's rule;

$\lim\limits_{x\to 0^+} \dfrac{2x\cos x - \sin 2x}{x\sin 2x}$

$\qquad = \lim\limits_{x\to 0^+} \dfrac{-2x\sin x + 2\cos x - 2\cos 2x}{2x\cos 2x + \sin 2x}$

$\qquad = \lim\limits_{x\to 0^+} \dfrac{-2x\cos x - 2\sin x - 2\sin x + 4\sin 2x}{-4x\sin 2x + 2\cos 2x + 2\cos 2x}$

$\qquad = 0$

34. $\lim\limits_{x\to +\infty}\left(\dfrac{x^2}{x - 1} - \dfrac{x^2}{x + 1}\right)$ is not in the proper

form to apply l'Hôpital's rule;

$\lim\limits_{x\to +\infty} \dfrac{x^2(x + 1 - x + 1)}{x^2 - 1} = \lim\limits_{x\to +\infty} \dfrac{2x^2}{x^2 - 1}$

$\qquad = \lim\limits_{x\to +\infty} \dfrac{4x}{2x} = 2$

35. $\lim\limits_{x\to +\infty}\left(\dfrac{x^3}{x^2 - x + 1} - \dfrac{x^3}{x^2 + x - 1}\right)$ is not in

the proper form to apply l'Hôpital's rule

$\lim\limits_{x\to +\infty} \dfrac{2(x^4 - x^3)}{x^4 - x^2 + 2x - 1}$

Dividing by x^4, or several applications of

l'Hôpital's rule, gives a limit of 2.

36. $\lim\limits_{x \to 0} \dfrac{\sin 3x \sin 2x}{x \sin 4x}$

$= \lim\limits_{x \to 0} \dfrac{\sin 3x(2 \cos 2x + 3 \cos 3x \sin 2x)}{4x \cos 4x + \sin 4x}$

$= \lim\limits_{x \to 0} \left(\dfrac{2 \sin 3x(-2\sin 2x)+2(3\cos 3x)\cos 2x}{-16x \sin\ 4x + 4 \cos 4x + 4 \cos 4x} \right.$

$\left. + \dfrac{3 \cos 3x(2 \cos 2x) + 3(-3 \sin 3x) \sin 2x}{-16x \sin\ 4x + 4 \cos 4x + 4 \cos 4x} \right)$

$= \dfrac{12}{8} = 1.5$

37. $\lim\limits_{x \to \pi/2} \dfrac{\sin 2x \cos x}{x \sin 4x}$

$= \lim\limits_{x \to \pi/2} \dfrac{2 \cos 2x \cos x - \sin 2x \sin x}{4x \cos 4x + \sin 4x}$

$= \dfrac{0}{2\pi} = 0$

38. $\lim\limits_{x \to 0} \left(\dfrac{\sin 2x}{x^3} + \dfrac{a}{x^2} + b \right) = \lim\limits_{x \to 0} \dfrac{\sin 2x + ax + bx^3}{x^3}$

$= \lim\limits_{x \to 0} \dfrac{2 \cos\ 2x + a + 3bx^2}{3x^2} = \dfrac{0}{0}$ if $a = -2$;

$\lim\limits_{x \to 0} \dfrac{-4 \sin 2x + 6bx}{6x} = \lim\limits_{x \to 0} \dfrac{-8 \cos 2x + 6b}{6}$

$= \dfrac{-8 + 6b}{6} = 1$ if $-8 + 6b = 6$ or $b = \dfrac{7}{3}$

39. $\lim\limits_{\alpha \to \beta} \left[\dfrac{C}{\alpha^2 - \beta^2}(\sin \alpha t - \sin\beta t) \right]$

$= C \lim\limits_{\alpha \to \beta} \dfrac{\sin \alpha t - \sin \beta t}{\alpha^2 - \beta^2}$

$= C \lim\limits_{\alpha \to \beta} \dfrac{t \cos \alpha t - 0}{2\alpha - 0}$ Note: treat α as variable

$= \dfrac{Ct}{2\beta} \cos \beta t$

3.10 Antiderivatives, Page 251

1. $\int 2\ dx\ = 2x + C$ **2.** $\int -4\ dx = -4x + C$

3. $\int (2x + 1)\ dx = x^2 + x + C$

4. $\int (4 - 5x)\ dx = 4x - \dfrac{5}{2}x^2 + C$

5. $\int (4t^3 + 3t^2)\ dt\ = t^4 + t^3 + C$

6. $\int (-8t^3 + 15t^5)\ dt = -2t^4 + \dfrac{15}{6}t^6 + C$

7. $\int (6u^2 - 3 \cos u)\ du = 2u^3 - 3 \sin u + C$

8. $\int (5t^3 - \sqrt{t})\ dt = \dfrac{5}{4}t^4 - \dfrac{2}{3}t^{3/2} + C$

9. $\int \sec^2\theta\ d\theta\ =\ \tan \theta\ +\ C$

10. $\int \sec \theta \tan \theta\ d\theta = \sec \theta + C$

11. $\int 2 \sin \theta\ d\theta\ =\ -2 \cos \theta + C$

12. $\int 3 \cos \theta\ d\theta = 3 \sin \theta + C$

13. $\int (u^{3/2} - u^{1/2} + u^{-10})\ du$

$= \dfrac{u^{5/2}}{\frac{5}{2}} - \dfrac{u^{3/2}}{\frac{3}{2}} + \dfrac{u^{-9}}{-9} + C$

$= \dfrac{2}{5} u^{5/2} - \dfrac{2}{3} u^{3/2} - \dfrac{1}{9} u^{-9}\ + C$

14. $\int (x^3 - 3x + \sqrt[4]{x} - 5)\ dx$

$= \dfrac{1}{4}x^4 - \dfrac{3}{2}x^2 + \dfrac{4}{5}x^{5/4} - 5x + C$

15. $\int x(x + \sqrt{x})\ dx = \int (x^2 + x^{3/2})\ dx$

$= \dfrac{1}{3}x^3 + \dfrac{2}{5}x^{5/2} + C$

16. $\int y(y^2 - 3y)\ dy = \int (y^3 - 3y^2)\ dy$

$= \dfrac{1}{4}y^4 - y^3 + C$

17. $\int (t^{-2} - t^{-3} + t^{-4})\ dt$

$= -t^{-1} + \dfrac{1}{2}t^{-2} - \dfrac{1}{3}t^{-3} + C$

18. $\int \dfrac{1}{t}\left(\dfrac{2}{t^2} - \dfrac{3}{t^3} \right) dt = \int (2t^{-3} - 3t^{-4})\ dt$

$= -t^{-2} + t^{-3} + C$

19. $\int (2x^2 + 5)^2\ dx = \int (4x^4 + 20x^2 + 25)\ dx$

$= \dfrac{4}{5}x^5 + \dfrac{20}{3}x^3 + 25x + C$

20. $\int (3 - 4x^3)^2\ dx = \int (9 - 24x^3 + 16x^6)\ dx$

$= 9x - 6x^4 + \dfrac{16}{7}x^7 + C$

21. $\int \left(\dfrac{x^2 + 3x - 1}{x^4} \right) dx$

$= \int (x^{-2} + 3x^{-3} - x^{-4})\ dx$

$= -x^{-1} - \dfrac{3}{2}x^{-2} + \dfrac{1}{3}x^{-3} + C$

22. $\int \dfrac{x^2 + \sqrt{x} + 1}{x^2}\ dx = \int (1 + x^{-3/2} + x^{-2})\ dx$

$= x - 2x^{-1/2} - x^{-1} + C$

23. $F(x) = \int (x^2 + 3x)\ dx = \dfrac{1}{3}x^3 + \dfrac{3}{2}x^2 + C$

$F(0) = \dfrac{1}{3}(0) + \dfrac{3}{2}(0) + C = 0$ so $C = 0$;

$F(x) = \frac{1}{3}x^3 + \frac{3}{2}x^2$

24. $F(x) = \int (\sqrt{x} + 3)^2 \, dx = \int (x + 6x^{1/2} + 9) \, dx$

$= \frac{1}{2}x^2 + 4x^{3/2} + 9x + C;$

$F(4) = \frac{1}{2}(4)^2 + 4(4)^{3/2} + 9(4) + C = 36$ so

$C = -40;\ F(x) = \frac{1}{2}x^2 + 4x^{3/2} + 9x - 40$

25. $F(x) = \int (2x - 1)^2 \, dx = \int (4x^2 - 4x + 1) \, dx$

$= \frac{4}{3}x^3 - 2x^2 + x + C;$

$F(1) = \frac{4}{3}(1)^3 - 2(1)^2 + 1 + C = 3$, so

$C = \frac{8}{3};\ F(x) = \frac{4}{3}x^3 - 2x^2 + x + \frac{8}{3}$

26. $F(x) = \int (3 - 2 \sin x) \, dx = 3x - 2(-\cos x) + C$

$F(0) = 0 + 2 + C = 3$, so $C = -1;$

$F(x) = 3x + 2 \cos x - 1$

27. **a.** $F(x) = \int (x^{-1/2} - 4) \, dx = 2x^{1/2} - 4x + C$

$F(1) = 2 - 4 + C = 0$, so $C = 4;$

$F(x) = 2\sqrt{x} - 4x + 4$

b.

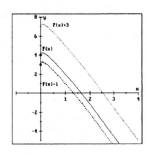

c. $G(x) = F(x) + C_0;$

$G'(x) = F'(x) = \frac{1}{\sqrt{x}} - 4 = 0$

At $x = \frac{1}{16}$ (note the original integrand),

$G(\frac{1}{16}) = 2(\frac{1}{4}) - 4(\frac{1}{16}) + 2 + C_0 = 0$, so

$C_0 = -\frac{9}{4}$

28. $a = -32$ ft/s^2, so $v(t) = \int -32 \, dt$

$= -16t + v_0 = -16t + 96$

The maximum height is reached when $v(t) = 0$
$v(t) = 0$ or $t = 3$ sec. Now,

$s(t) = \int v(t) \, dt = \int (-16t + 96) \, dt$

$= -16t^2 + 96t + s_0$

For convenience, let $s_0 = 0$, so

$s(3) = -16(9) + 96(3) = 144$

The time to reach the ground is the same as
the time it takes to reach the maximum
height.

29. $C(x) = \int C'(x) \, dx = \int (6x^2 - 2x + 5) \, dx$

$= 2x^3 - x^2 + 5x + C$

$C(1) = 2(1)^3 - (1)^2 + 5(1) + C = 5$, so

$C = -1.$

$C(x) = 2x^3 - x^2 + 5x - 1$, and

$C(5) = 250 - 25 + 25 - 1 = \$249.$

30. **a.** $R(x) = \int (-3x^2 + 4x + 32) \, dx$

$= -x^3 + 2x^2 + 32x + C$

$R(0) = 0$, so $C = 0$ and

$R(x) = -x^3 + 2x^2 + 32x = xp(x)$. Thus,

$p(x) = -x^2 + 2x + 32$

b. $R'(x) = -3x^2 + 4x + 32$

$= -(x - 4)(3x + 8)$

$x = 4$ $(-\frac{8}{3}$ is not in the domain$)$

When $x = 4$ the market price is $p(4) = 24$
and the level of production is

$R(4) = -64 + 32 + 128 = 96$ units

31. Let $P(t)$ be the population in the town at
time t in months.

$P'(t) = 4 + 5t^{2/3}$

$P(t) = \int (4 + 5t^{2/3}) \, dt = 4t + 3t^{5/3} + C$

$P(0) = 10,000$, so

$P(t) = 4t + 3t^{5/3} + 10,000$

The population in 8 months will be

$P(8) = 4(8) + 3(8)^{5/3} + 10,000 = 10,128$

32. $a(t) = \sqrt{t} + t^2 = t^{1/2} + t^2;$

$v(t) = \int (t^{1/2} + t^2) \, dt = \frac{2}{3}t^{3/2} + \frac{1}{3}t^3 + C_1$

Since $v(0) = 0 + 0 + C_1 = 2$, $C_1 = 2$

$v(t) = \frac{2}{3}t^{3/2} + \frac{1}{3}t^3 + 2$

$s(t) = \int (\frac{2}{3}t^{3/2} + \frac{1}{3}t^3 + 2) \, dt$

$= \frac{4}{15}t^{5/2} + \frac{1}{12}t^4 + 2t + C_2$

Since $s(0) = 0$, $C_2 = 0$

$s(4) = \frac{4}{15}(2)^5 + \frac{1}{12}(4)^4 + 2(4) = \frac{568}{15}$

$v(4) = \frac{2}{3}(2)^3 + \frac{1}{3}(4)^3 + 2 = \frac{86}{3}$

33. $a(t) = k$, $v(t) = \int a(t)\, dt = \int k\, dt = kx + C$

But $v(0) = 0$, so $C = 0$

$s(t) = \int v(t)\, dt = \int kx\, dt = \frac{kx^2}{2} + C$

But $s(0) = 0$, so again $C = 0$. Now it is given that $s(6) = 360$, so $\frac{k(6)^2}{2} = 360$,

$k = 20$ ft/s^2

34. Let $P(x)$ be the price of bacon in t months.

$P'(t) = 0.984 + 0.012\sqrt{t}$;

$P(t) = \int (0.984 + 0.012\sqrt{t})\, dt$

$= 0.984t + 0.008t^{3/2} + C$

Since $P(0) = 1.80$,

$P(4) = 0.984(4) + 0.0008(8) + 1.80 + 1.80$

$= 5.80$

The price of bacon in 4 mo will be \$5.80/lb.

35. With $a(t) = k$, $v(t) = kt + C_1$; $C_1 = 0$ because the plane starts from rest ($v_0 = 0$) and the distance $s(t) = kt^2/2 + C_2$. For convenience, $s(0) = 0$, so $C_2 = 0$. Let t_1 be the time required for liftoff.

$s(t_1) = \frac{kt_1^2}{2} = 1.2$ and $V(t_1) = kt_1 = 100$

Dividing these equations leads to

$\frac{kt_1^2}{kt_1} = \frac{2.4}{100} = 0.024$ hr

From the velocity equation,

$k = \frac{100}{t_1} = \frac{100}{0.024} \approx 4{,}167$ mi/h^2

36. With $a(t) = k$, $v(t) = kt + C_1$; the initial speed of the car is $v_0 = 88$ so $v(0) = C_1 = 88$ and the distance is $s(t) = kt^2/2 + 88t + C_2$. For convenience, $s(0) = 0$ so $C_2 = 0$. Let t_1 be the time required for stopping.

$s(t_1) = \frac{kt_1^2}{2} + 88t_1 = 121$ and $V(t_1) = 0$

or $t_1 = -\frac{88}{k}$. Substituting into the distance formula leads to

$\frac{88^2 k}{2k} - \frac{88^2}{k} - 121 = 0$, or

$k = -\frac{88^2}{242} = -32$ ft/s

37. $\int [af(x) + bg(x) + ch(x)]$

$= \int \{[af(x) + bg(x)] + ch(x)\}\, dx$

$= \int [af(x) + bg(bx)]\, dx + \int ch(x)\, dx$

$= \int af(x)\, dx + \int bg(x)\, dx + \int ch(x)\, dx$

$= a\int f(x)\, dx + b\int g(x)\, dx + c\int h(x)\, dx$

Chapter 3 Review

Proficiency Examination, Page 252

1. No maximum is larger than the absolute maximum and no minimum is smaller than the absolute minimum. A relative extremum is an extremum only in the neighborhood of the point of interest.

2. A continuous function f on a closed interval $[a, b]$ has an absolute maximum and an absolute minimum.

3. A critical value of a function is a value of the independent variable at which the derivative of the function is zero or is not defined. A critical point has both coordinates.

4. Find critical values, evaluate the function at these values and at the endpoints of the closed interval. Finally, determine the absolute extrema by selecting the largest and smallest of the evaluated functional values.

5. If a function is differentiable on (a, b) and continuous on $[a, b]$ such that $f(a) = f(b)$ for Rolle's theorem, then there exists a value c on (a, b) — with $f'(c) = 0$ for Rolle's theorem, such that
$$f'(c) = \frac{f(b) - f(a)}{b - a} \quad (a \neq b)$$
Rolle's theorem is a special case of the MVT.

6. Given $f(x)$, find c such that $f'(c) = 0$ or $f'(c)$ is not defined. Investigate $f'(x)$ on both sides of $x = c$.

7. Given $f(x)$, find c such that $f''(c) = 0$ or $f''(c)$ is not defined. If $f''(c) < 0$, then f is concave down at $x = c$.

8. For a plane curve, an asymptote is a line which has the property that the distance from a point P on the curve to the line approaches zero as the distance from P to the origin

increases without bound and P is on a suitable portion of the curve.

9. $\lim\limits_{x \to +\infty} f(x) = L$, given an $\epsilon > 0$, there exists a number N_1 such that $\left| f(x) - L \right| < \epsilon$ whenever $x > N_1$ for x in the domain of f.

$\lim\limits_{x \to c} f(x) = +\infty$ if for any number $N > 0$, it is possible to find a number $\delta > 0$ such that $f(x) > N$ whenever $0 < |x - c| < \delta$.

10.   Find the domain and range of a function, locate intercepts, if any. Investigate symmetry, asymptotes, and find extrema and/or points of inflection, if any.

11.   An optimization problem involves finding the largest, best, or smallest value of a function or situation. First, find the domain of the function and find the extrema.

12.   Let f and g be functions that are differentiable on an open interval containing c (except possibly at c itself). If

$\lim\limits_{x \to c} \dfrac{f(x)}{g(x)}$ produces an indeterminate form $\dfrac{0}{0}$

or $\dfrac{\infty}{\infty}$, then $\lim\limits_{x \to c} \dfrac{f(x)}{g(x)} = \lim\limits_{x \to c} \dfrac{f'(x)}{g'(x)}$.

13.   Find a function whose derivative is the given function.

14. $\lim\limits_{x \to \pi/2} \dfrac{\sin 2x}{\cos x} = \lim\limits_{x \to \pi/2} \dfrac{2 \cos 2x}{-\sin x} = \dfrac{-2}{-1} = 2$

15. $\lim\limits_{x \to 1} \dfrac{1 - \sqrt{x}}{x - 1} = \lim\limits_{x \to 1} \dfrac{-\frac{1}{2} x^{-1/2}}{1} = -\dfrac{1}{2}$

16. $\lim\limits_{x \to +\infty} \left(\dfrac{1}{x} - \dfrac{1}{\sqrt{x}} \right) = \lim\limits_{x \to +\infty} \dfrac{1 - \sqrt{x}}{x}$

$= \lim\limits_{x \to +\infty} \dfrac{-\frac{1}{2} x^{-1/2}}{1} = \dfrac{1}{\infty} = 0$

17. $\lim\limits_{x \to +\infty} \left(\dfrac{x^3}{x^2 - 2} - \dfrac{x^3}{x^2 + 2} \right) = \lim\limits_{x \to +\infty} \dfrac{4x^3}{x^4 - 4}$

$= \lim\limits_{x \to +\infty} \dfrac{\frac{4x^3}{x^4}}{\frac{x^4}{x^4} - \frac{4}{x^4}} = \lim\limits_{x \to +\infty} \dfrac{\frac{4}{x}}{1 - \frac{4}{x^4}} = 0$

18. $f(x) = x^3 + 3x^2 - 9x + 2$,

$f'(x) = 3x^2 + 6x - 9$

$= 3(x+3)(x - 1) = 0$ when $x = -3, 1$

$f''(x) = 6x + 6 = 0$ when $x = -1$

Relative maximum at $(-3, 29)$;
relative minimum at $(1, -3)$;
inflection point at $(-1, 13)$

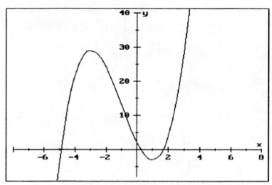

19. $f(x) = 27x^{1/3} - x^{4/3}$

$f'(x) = 9x^{-2/3} - \frac{4}{3}x^{1/3} = 0$ when

$\dfrac{9}{x^{2/3}} = \dfrac{4x^{1/3}}{3}$

$x = \dfrac{27}{4}$ (not defined at $x = 0$)

$f''(x) = -6x^{-5/3} - \frac{4}{9}x^{-2/3} = 0$ when

$x = -\frac{27}{2}$. $f''(-\frac{27}{2}^{-}) > 0$, $f''(-\frac{27}{2}^{+}) < 0$,

so there are inflection points at approximately $(-\frac{27}{2}, -96.43)$ and $(0, 0)$.
Relative maximum at $(\frac{27}{4}, 38.27)$

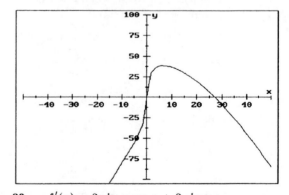

20. $f'(x) = 2 \sin x \cos x + 2 \sin x$
       $= 2 \sin x(\cos x + 1) = 0$ when
$x = 0, \pi, 2\pi$ on $[0, 2\pi]$. The endpoints are 0 and 2π: $f(0) = -2$, $f(2\pi) = -2$. All other values of $f(x)$ are greater than -2, so these are absolute minimums.
$f''(x) = -2 \sin^2 x + 2 \cos^2 x + 2 \cos x$
$f''(\pi) = 0 + 2 - 2 = 0$, so the test fails.
Use the first derivative test to see if there is a change of slope: $f'(\pi^{-}) > 0$,

$f'(\pi^+) < 0$, so there is a relative maximum at $(\pi, 2)$.

To search for inflection points change $f''(x)$ to a function of cosines:

$$f''(x) = -2(1 - \cos^2 x) + 2\cos^2 x + 2\cos x$$

$$= 4\cos^2 x + 2\cos x - 2$$

$$= 2(2\cos^2 x + \cos x - 1)$$

$$= 2(2\cos x - 1)(\cos x + 1) = 0 \text{ when}$$

$\cos x = \frac{1}{2}, -1; x = \frac{\pi}{3}, \pi, \frac{5\pi}{3}$. Checking $f''(x)$ for change of concavity at these points: $f''(\frac{\pi}{3}^-) > 0$, $f''(\frac{\pi}{3}^+) < 0$. So there is an inflection point at $(\frac{\pi}{3}, -\frac{1}{4})$. $f''(\frac{5\pi}{3}^-) < 0$, $f''(\frac{5\pi}{3}^+) > 0$. So there is an inflection point at $(\frac{5\pi}{3}, -\frac{1}{4})$. $f''(\pi^-) < 0$, $f''(\pi^+) < 0$. So there is no inflection point at $x = \pi$; the function is concave up on $(0, \frac{\pi}{3})$ and on $(\frac{5\pi}{3}, 2\pi)$ and is concave down on $(\frac{\pi}{3}, \frac{5\pi}{3})$.

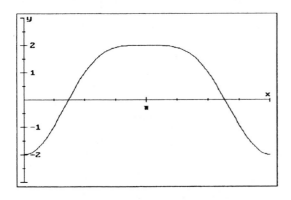

21. Checking for horizontal asymptotes:

$$\lim_{x \to +\infty} f(x) = 1; \lim_{x \to -\infty} f(x) = 1$$

So $y = 1$ is a horizontal asymptote. Factoring the denominator we see that there are two candidates for vertical asymptotes; 2 and -2. Testing these:

$$\lim_{x \to -2^-} f(x) = +\infty; \lim_{x \to -2^+} f(x) = -\infty$$
$$\lim_{x \to 2^-} f(x) = -\infty; \lim_{x \to 2^+} f(x) = +\infty$$

So there are two vertical asymptotes.

$$f'(x) = \frac{(x^2 - 4)2x - (x^2 - 1)2x}{(x^2 - 4)^2} = \frac{-6x}{(x^2 - 4)^2}$$

This is equal to 0 when $x = 0$, and it is easy to see that the slope is positive to the left of 0, and negative to the right of 0. So there is

a relative maximum at $(0, \frac{1}{4})$. This should be sufficient information to sketch the graph:

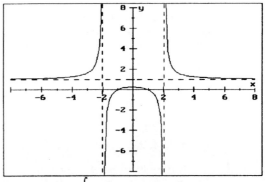

22. $f(x) = \int (3x^2 + 1)\,dx = x^3 + x + C$

If $f(0) = 10$ then $C = 10$ and $f(x) = x^3 + x + 10$

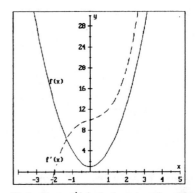

Note that $f'(x)$ designates the slope of $f(x)$.

23. This is a continuous function on a closed interval, so the extreme value theorem guarantees an absolute maximum and an absolute minimum. The candidates are the critical points and the endpoints. $f'(x) = 4x^3 - 10x^4 = x^3(4 - 10x)$ which is 0 at $x = 0, \frac{2}{5}$. Testing our candidates: $f(0) = 5$, $f(\frac{2}{5}) = 5\frac{16}{3125}$, $f(1) = 4$. The absolute maximum is at $(0.4, 5.00512)$ and the absolute minimum at $(1, 4)$.

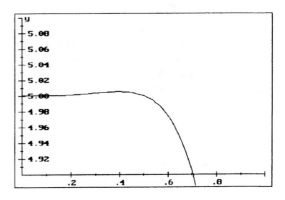

The vertical scale had to be changed to show the critical maximum, so the minimum is off the bottom of the graph at $(1, 4)$.

24. We are asked to minimize the amount of material, $S = x^2 + 4xh$, with the restriction that $V = 2$. We can use the volume formula to find a substitute value for h, and then have S as a function of x.

$$lwh = 2,\; x^2h = 2,\; h = 2/x^2$$

$$S = x^2 + 4x\left(\frac{2}{x^2}\right) = x^2 + \frac{8}{x};\; S' = 2x - \frac{8}{x^2} = 0$$

when $x^3 = 4,\; x = \sqrt[3]{4};\; h = \frac{2}{x^2} = \frac{2}{\sqrt[3]{4^2}}$

The dimensions of the box are approximately 1.587 by 1.587 by .794 ft, or to the nearest inch: 19 in. $\times$ 19 in. $\times$ 10 in.

25. Using increments $\Delta P = \frac{dP}{dt}\Delta t$ where

$$\frac{dP}{dt} = \frac{6}{(t+1)^2};\; \text{at } t = 0,\; \frac{dP}{dt} = 6 \text{ so}$$

$$\Delta P = 6(0.25) = 1.5 \text{ or } 1,500 \text{ people}$$

Check by finding the exact change:

$P(0.25) - P(0)$

$= \left(20 - \frac{6}{0.25+1}\right) - \left(20 - \frac{6}{0+1}\right)$

$= 15.2 - 14 = 1.2 \text{ or } 1,200 \text{ additional}$ people.

Supplementary Problems, Page 252

1. $f'(x) = 3x^2 + 12x + 9 = 3(x+1)(x+3)$
   critical values at $x = -1$ (relative minimum), $x = -3$ (relative maximum)
   $f''(x) = 6x + 12 = 6(x+2)$; critical values at $x = -2$ (point of inflection)

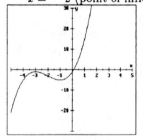

2. $f'(x) = 4x^3 + 12x^2 + 8x = 4x(x+2)(x+1)$
   critical values at $x = -2$ (relative minimum), $x = -1$ (relative maximum), $x = 0$ (relative minimum)
   $f''(x) = 12x^2 + 24x + 8 = 4(3x^2 + 6x + 2)$
   critical values at $x = \dfrac{-3 \pm \sqrt{3}}{3}$
   $\approx -1.5774,\; -0.4226$ (points of inflection)

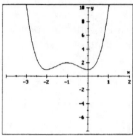

3. $f'(x) = 12x^3 - 12x^2 = 12x^2(x-1)$
   critical values at $x = 0$ (point of inflection), $x = 1$ (relative minimum)
   $f''(x) = 36x^2 - 24x = 12x(3x - 2)$
   critical values at $x = 0$, $x = \frac{2}{3}$ (point of inflection)

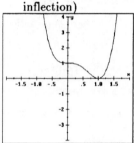

4. $f'(x) = 12x^3 - 8x = 4x(3x^2 - 2)$
   critical values at $x = 0$ (relative maximum), $x = \pm\sqrt{2/3} \approx \pm 0.82$ (relative minima)
   $f''(x) = 36x^2 - 8$;
   critical values $x = \pm\sqrt{2/9} \approx \pm 0.4714$ (points of inflection)

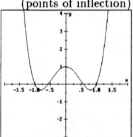

5. $f'(x) = 30x^4 - 60x^3 + 30x^2 = 30x^2(x-1)^2$
   critical values at $x = 0$ (point of inflection), $x = 1$ (point of inflection)
   $f''(x) = 120x^3 - 180x^2 + 60x$
   $= 60x(x-1)(2x-1)$
   critical values $x = 0$, $x = 1$, $x = \frac{1}{2}$ (point of inflection)

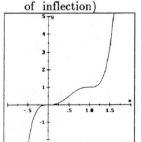

6. $f'(x) = 15x^4 - 30x^2 + 15$
          $= 15(x - 1)^2(x + 1)^2$
     critical values at $x = 1$ (point of
     inflection), $x = -1$ (point of inflection)
     $f''(x) = 60x^3 - 60x = 60x(x - 1)(x + 1)$
     critical values at $x = 0$, $x = 1$, $x = -1$

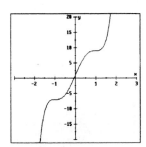

7. $f(x) = \dfrac{9 - x^2}{3 + x^2} = -1 + \dfrac{12}{3 + x^2}$

     $f'(x) = \dfrac{12(-1)(2x)}{(3 + x^2)^2} = \dfrac{-24x}{(x^2 + 3)^2}$

     critical value $x = 0$ (relative maximum)

     $f''(x) = \dfrac{-24[(3 + x^2)^2(1) - x(2)(3 + x^2)(2x)]}{(3 + x^2)^4}$

           $= \dfrac{72(x^2 - 1)}{(x^2 + 3)^3}$

     critical values $x = 1$ (point of inflection),
     $x = -1$ (point of inflection)

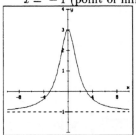

8. $f'(x) = \dfrac{(1 - x) - x(-1)}{(1 - x)^2} = \dfrac{1}{(1 - x)^2}$

     critical values $x = 1$ (function not defined
     at $x = 1$)

     $f''(x) = 2(1 - x)^{-3} \neq 0$

     horizontal asymptote $y = -1$
     vertical asymptote $x = 1$

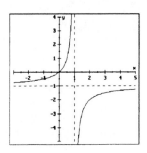

9. $f'(x) = 2\cos 2x - \cos x = 4\cos^2 x - \cos x - 2$
     critical values $x \approx 0.6$ (relative
     maximum), $x \approx 2.2$ (relative minimum),
     $x \approx 4.1$ (relative maximum), $x \approx 5.7$
     (relative minimum); other values $\pm 2\pi$
     $f''(x) = -4\sin 2x + \sin x$
     critical values $x = 0$, π and $x \approx 1.4$, 4.8;
     other values $\pm 2\pi$ (points of inflection)

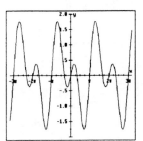

10. $f(x) = \sin x \sin 2x = 2\sin^2 x \cos x$
     $f'(x) = 2[-\sin^3 x + \cos^2 x(2\sin x)]$
         $= 2\sin x(\cos^2 x - \sin^2 x)$
     critical values $x = 0$, $\frac{\pi}{4}$, $\frac{3\pi}{4}$, π, $\frac{5\pi}{4}$, $\frac{7\pi}{4}$, 2π
     (other values $\pm 2\pi$)
     $f''(x) = 2[\cos x(\cos^2 x - \sin^2 x)$
         $+ \sin x(2\cos x \sin x - 2\sin x \cos x)]$
         $= 2\cos x(\cos^2 x - 5\sin^2 x)$
     critical values $x = \frac{\pi}{2}$, $\frac{3\pi}{2}$ and also where

$$\cos^2 x - 5\sin^2 x = 0$$
$$\frac{\cos^2 x}{\sin^2 x} = 5$$
$$\tan x = \pm\sqrt{5}$$
$$x \approx 1.15, \ 1.99, \ 4.29$$

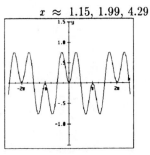

11. $f(x) = \dfrac{x^2 - 4}{x^2} = 1 - \dfrac{4}{x^2}$
     $f'(x) = 8x^{-3} \neq 0$
     $f''(x) = -24x^{-4} \neq 0$

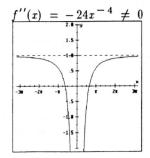

12. $f(x) = \dfrac{(x-1)(x+3)}{(x-1)(x-2)}$

y-intercept $(0, -1.5)$

$y = 1$ is a horizontal asymptote

$x = 2$ is a vertical asymptote

$x = 1$ is a deleted point

$f'(x) = \dfrac{x - 2 - (x+3)}{(x-2)^2} = \dfrac{-5}{(x-2)^2} \neq 0$

$f''(x) = \dfrac{-10}{(x-2)^3} \neq 0$

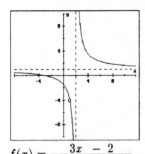

13. $f(x) = \dfrac{3x - 2}{(x+1)^2(x-2)}$

y-intercept $(0, 1)$

$y = 0$ is a horizontal asymptote

$x = -1$ and $x = 2$ are vertical asymptotes

$f'(x) = \dfrac{-6(x^2 - 2x + 2)}{(x-2)^2(x+1)^3}$

f is not defined at $x = 2$, $x = -1$

$f''(x) = \dfrac{6(3x^3 - 10x^2 + 20x - 12)}{(x+1)^4(x-2)^3}$

critical value at $x = 0.89$

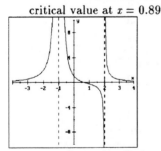

14. $f(x) = \dfrac{x^3 - 2}{x(x+1)(x+2)}$

$y = 1$ is a horizontal asymptote

$x = -2$, $x = -1$, $x = 0$ are vertical asymptotes

$f'(x) = \dfrac{3x^4 + 4x^3 - 9x^2 - 18x - 6}{x^2(x+1)^2(x+2)^2}$

critical values at $x \approx -0.45$, (relative maximum), $x = 2$ (relative minimum)

$f''(x) = \dfrac{2(-x^6 - 6x^5 + 18x^4 + 76x^3 + 99x^2 + 54x + 12)}{x^3(x+1)^3(x+2)^3}$

critical values at $x \approx -1.48$ (point of inflection), $x \approx 3.28$ (point of inflection)

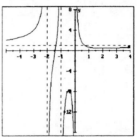

15. $f'(x) = 4x^3 - 16x = 4x(x+2)(x-2)$

critical values $x = 0, -2, 2$

$f(-1) = 5$; $f(2) = -4$; $f(0) = 12$;

maximum $f(0) = 12$; minimum $f(2) = -4$

16. $f'(x) = x^{1/2}(\tfrac{1}{3})(x-5)^{-2/3}(1)$

$\qquad + \tfrac{1}{2}x^{-1/2}(x-5)^{1/3}$

$\quad = \tfrac{1}{6}x^{-1/2}(x-5)^{-2/3}[2x + 3(x-5)]$

$\quad = \dfrac{5(x-3)}{6\sqrt{x}(x-5)^{2/3}}$

critical value at $x = 3$, $x = 5$

$f(0) = 0$; $f(6) \approx 2.45$; $f(3) \approx -2.18$;

$f(5) = 0$

maximum $f(6) \approx 2.45$;

minimum $f(3) \approx -2.18$

17. $\displaystyle\lim_{x \to +\infty} \dfrac{x\sin^2 x}{x^2 + 1} \leq \lim_{x \to +\infty} \dfrac{x}{x^2 + 1} = 0$

18. $\displaystyle\lim_{x \to +\infty} \dfrac{2x^4 - 7}{6x^4 + 7} = \lim_{x \to +\infty} \dfrac{2 - \dfrac{7}{x^4}}{6 + \dfrac{7}{x^4}} = \dfrac{1}{3}$

19. $\displaystyle\lim_{x \to +\infty}(\sqrt{x^2 - x} - x)$

$\displaystyle = \lim_{x \to +\infty} \dfrac{(\sqrt{x^2 - x} - x)(\sqrt{x^2 - x} + x)}{\sqrt{x^2 - x} + x}$

$\displaystyle = \lim_{x \to +\infty} \dfrac{x^2 - x - x^2}{\sqrt{x^2 - x} + x}$

$\displaystyle = \lim_{x \to +\infty} \dfrac{-1}{\sqrt{1 - \dfrac{1}{x}} + 1} = -\dfrac{1}{2}$

20. $\displaystyle\lim_{x \to +\infty}(\sqrt{x(x+b)} - x)$

$\displaystyle = \lim_{x \to +\infty} \dfrac{(\sqrt{x(x+b)} - x)(\sqrt{x(x+b)} + x)}{\sqrt{x(x+b)} + x}$

$\displaystyle = \lim_{x \to +\infty} \dfrac{x^2 + bx - x^2}{\sqrt{x^2 + bx} + x}$

$$= \lim_{x \to +\infty} \frac{b}{\sqrt{1 + \frac{b}{x}} + 1} = \frac{b}{2}$$

21. $\displaystyle\lim_{x \to 0} \frac{x \sin x}{x + \sin^3 x} = \lim_{x \to 0} \frac{x \cos x + \sin x}{1 + 3 \sin^2 x \cos x} = 0$

22. $\displaystyle\lim_{x \to 0} \frac{x \sin x}{x^2 - \sin^3 x} = \lim_{x \to 0} \frac{x \cos x + \sin x}{2x - 2 \sin^2 x \cos x}$

$$= \lim_{x \to 0} \frac{-x \sin x + \cos x + \cos x}{2 - 2 \cos 2x}$$

This limit does not exist.

23. $\displaystyle\lim_{x \to 0} \frac{x \sin^2 x}{x^2 - \sin^2 x}$

$$= \lim_{x \to 0} \frac{x(2 \sin x \cos x) + \sin^2 x}{2x - 2 \sin x \cos x}$$

$$= \lim_{x \to 0} \frac{x(2 \cos 2x) + \sin 2x + 2 \sin x \cos x}{2 - 2 \cos 2x}$$

$$= \lim_{x \to 0} \frac{-2x \sin 2x + \cos 2x + \cos 2x}{-2 \sin 2x}$$

This limit does not exist.

24. $\displaystyle\lim_{x \to 0} \frac{x - \sin x}{\tan^3 x} = \lim_{x \to 0} \frac{1 - \cos x}{3 \tan^2 x \sec^2 x}$

$$= \frac{1}{3} \lim_{x \to 0} \frac{\sin x}{\tan^2 x(2 \sec^2 x) \tan x + 2 \tan x \sec^4 x}$$

$$= \frac{1}{6} \lim_{x \to 0} \frac{\sin x}{\dfrac{\sin^3 x}{\cos^5 x} + \dfrac{\sin x}{\cos^5 x}}$$

$$= \frac{1}{6} \lim_{x \to 0} \frac{\cos^5 x}{\sin^2 x + 1} = \frac{1}{6}$$

25. $\displaystyle\lim_{x \to 0} \frac{\sin^2 x}{\sin x^2} = \lim_{x \to 0} \frac{2 \sin x \cos x}{2x \cos x^2}$

$$= \left[\lim_{x \to 0} \frac{\sin x}{x} \right] \left[\lim_{x \to 0} \frac{1}{\cos x} \right] = (1)(1) = 1$$

26. $\displaystyle\lim_{x \to 0} \left(\frac{1}{x^2} - \frac{1}{x^2 \sec x} \right) = \lim_{x \to 0} \frac{\sec x - 1}{x^2 \sec x}$

$$= \lim_{x \to 0} \frac{\sec x \tan x}{x^2 \sec x \tan x + 2x \sec x}$$

$$= \lim_{x \to 0} \frac{\tan x}{x^2 \tan x + 2x}$$

$$= \lim_{x \to 0} \frac{\sec^2 x}{x(\cdots) + 2} = \frac{1}{2}$$

27. $\displaystyle\lim_{x \to \pi/2} -\frac{\sec^2 x}{\sec^2 3x} = (-1) \lim_{x \to \pi/2} \frac{\cos^2 3x}{\cos^2 x}$

$$= (-1) \left[\lim_{x \to \pi/2} \frac{\cos 3x}{\cos x} \right]^2$$

$$= (-1) \left[\lim_{x \to \pi/2} \frac{-3 \sin 3x}{-\sin x} \right]^2$$

$$= (-1) \left[\frac{-3(-1)}{-1} \right]^2 = -9$$

28. $\displaystyle\lim_{x \to \pi/2} -(1 - \sin x) \tan x$

$$= (-1) \lim_{x \to \pi/2} \frac{1 - \sin x}{\cot x}$$

$$= (-1) \lim_{x \to \pi/2} \frac{-\cos x}{-\csc^2 x}$$

$$= (-1) \lim_{x \to \pi/2} \cos x \sin^2 x = 0$$

29. $\displaystyle\lim_{x \to \pi/2} -(\sec x - \tan x) = \lim_{x \to \pi/2} \frac{\sin x - 1}{\cos x}$

$$= \lim_{x \to \pi/2} \frac{\cos x}{-\sin x} = 0$$

30. $\displaystyle\lim_{x \to +\infty} (\sqrt{x^2 + 4} - \sqrt{x^2 - 4})$

$$= \lim_{x \to +\infty} \frac{(\sqrt{x^2 + 4} - \sqrt{x^2 - 4})(\sqrt{x^2 + 4} + \sqrt{x^2 - 4})}{2}$$

$$= \lim_{x \to +\infty} \frac{x^2 + 4 - x^2 + 4}{\sqrt{x^2 + 4} + \sqrt{x^2 - 4}} = 0$$

31. $\displaystyle\int (5x - 6) \, dx = \frac{5}{2}x^2 - 6x + C$

32. $\displaystyle\int (t - 1)(t + 2) \, dt = \int (t^2 + t - 2) \, dt$

$$= \frac{1}{3}t^3 + \frac{1}{2}t^2 - 2t + C$$

33. $\displaystyle\int x^{-3} \, dx = -\frac{1}{2}x^{-2} + C$

34. $\displaystyle\int \frac{5x^2 - 2x + 1}{\sqrt{x}} \, dx$

$$= \int (5x^{3/2} - 2x^{1/2} + x^{-1/2}) \, dx$$

$$= 2x^{5/2} - \frac{4}{3}x^{3/2} + 2x^{1/2} + C$$

35. The blue curve (g) is the function and the red curve (f) is the derivative. Note: the apparent discontinuity at $x = -4$; it is not clear if g is discontinuous or if it is just off the scale. (We assume that $f = g'$ corresponds to the high point off the top of the page.) The low points of $f = g'$ corresponds to the inflection points of g.

36. $f(x) = ax^3 + bx^2 + C$
$f'(x) = 3ax^2 + 2bx; \ f'(-1) = 3a - 2b = 1$
$f''(x) = 6ax + 2b; \ f''(-1) = -6a + 2b = 0$
Solving this system of equations to find
$a = -\frac{1}{3}$ and $b = -1$. Thus,

$$f(x) = -\tfrac{1}{3}x^3 - x^2 + C$$

$$f(-1) = -\tfrac{1}{3}(-1)^3 - (-1)^2 + C = 2$$

which implies that $C = 8$. This means the desired equation is

$$f(x) = -\tfrac{1}{3}x^3 - x^2 + 8$$

37. The graph of a quadratic polynomial must have an equation of the form
$f(x) = ax^2 + bx + c$;
$f'(x) = 2ax + b$; $f''(x) = 2a \neq 0$
The graph of the quadratic polynomial has no point of inflection because its curvature does not change.
The graph of a cubic polynomial must have an equation of the form
$f(x) = ax^3 + bx^2 + cx + d$
$f'(x) = 3ax^2 + 2bx + c$;
$f''(x) = 6ax + 2b = 0$ at $x = -\dfrac{b}{3a}$
Thus, the graph of a cubic polynomial has one point of inflection.

38. Find the minimum distance for D, the square of the distance from $P(0, 1)$ to a point on the hyperbola.

$$D = (x - 0)^2 + (y - 1)^2 = 4 + y^2 + (y - 1)^2$$
$$\text{since } x^2 = 4 + y^2$$

$$D' = 2y + 2(y - 1) = 4y - 2 = 0 \text{ if } y = \tfrac{1}{2}.$$

Now, $x^2 = 4 + \tfrac{1}{4} = \tfrac{17}{4}$ or $x = \pm\dfrac{\sqrt{17}}{2} = \pm 2.06$

The points closest points to P are
$(-2.06, 0.5)$ and $(2.06, 0.5)$

39. Let $2x$ be the length of the base of the window and y the length of the height of the rectangular vertical side. Then

$$P_0 = 2x + 2y + \pi x$$

$$= (2 + \pi)x + 2y \text{ or } y = \tfrac{1}{2}[P_0 - (2 + \pi)x]$$

The area of the window is

$$A = 2xy + \frac{\pi x^2}{2} = x[P_0 - (2 + \pi)x] + \frac{\pi x^2}{2}$$

$$= xP_0 - 2x^2 - \pi x^2 + \frac{\pi x^2}{2}$$

$$A' = P_0 - 4x - \pi x = P_0 - (4 + \pi)x$$

$$A' = 0 \text{ when } x = \frac{P_0}{4 + \pi}. \text{ Now}$$

$$y = \tfrac{1}{2}\left[P_0 - \frac{2 + \pi}{4 + \pi}P_0\right] = \frac{P_0}{2}\left[\frac{4 + \pi - 2 - \pi}{4 + \pi}P_0\right]$$

$$= \frac{P_0}{4 + \pi}$$

The width of the window is $\dfrac{2P_0}{4 + \pi}$ and the radius is $P_0(\pi + 4)^{-1}$.

40. $f(x) = Ax^3 + Bx^2 + Cx + D$
$f'(x) = 3Ax^2 + 2Bx + C$
Because $x = \pm 1$ are critical points we need
$f'(-1) = 3A - 2B + C = 0$ and
$f'(1) = 3A + 2B + C = 0$
These equations imply $B = 0$ and $C = -3A$.
Thus,

$$f(x) = Ax^3 - 3Ax + D$$

$f(-1) = -A + 3A + D = 1$
$f(1) = A - 3A + D = -1$

These equations imply $D = 0$, $A = \tfrac{1}{2}$, and
$C = -\tfrac{3}{2}$. Thus, $f(x) = \tfrac{1}{2}(x^3 - 3x)$.

41. Let (x, y) be the point on the semicircle that is also a vertex of a rectangle. With $0 \leq x \leq a$, the area is $2xy$. Since $y^2 = a^2 - x^2$,

$$A(x) = 2x\sqrt{a^2 - x^2},$$

$$A'(x) = 2\left(\frac{-x^2}{\sqrt{a^2 - x^2}} + \sqrt{a^2 - x^2}\right)$$

$$= \frac{2(-x^2 + a^2 - x^2)}{\sqrt{a^2 - x^2}} = 0 \text{ if}$$

$x = \dfrac{\sqrt{2}}{2}a$; $A(0) = A(a) = 0$, the maximum
area of the rectangle enclosed in the semicircle
is $A\left(\dfrac{\sqrt{2}a}{2}\right) = \sqrt{2a}\left(\sqrt{a^2 - \dfrac{a^2}{2}}\right) = a^2$ square
units.

42. Let x be the number of $\$20$ increases. The number of units rented is $200 - 5x$
$(0 \leq x \leq 40)$ and the amount of each rent payment is $600 + 2x$. The profit is

$$P(x) = (200 - 5x)(600 + 20x) - (200 - 5x)(80)$$

$$= (200 - 5x)(520 + 20x)$$

$$P'(x) = (200 - 5x)(20) + (-5)(520 + 20x)$$

$$= 4,000 - 100x - 260 + 100x$$

$$= 200(7 - x)$$

$$P'(x) = 0 \text{ if } x = 7;$$

$$P(0) = 104,000; \; P(40) = 0; \; P(7) = 108,000$$

The maximum profit of $\$108,900$ is reached when 165 units are rented at $\$720$ each.

43. Yield $= f(x) = (200 - 5x)(30 + x)$
$= -5x^2 + 50x + 6,000$;

$$f'(x) = -10x + 50 = 0 \text{ at } x = 5$$

$f''(x) < 0$, thus $x = 5$ leads to a maximum yield of $f(5) = 6,125$ for 35 trees per acre.

44. Let Q be the point on the shore straight across from the oil rig and P the point on the shore where the pipe starts on land. With $PQ = x$, the distance along the bank is $8 - x$. The distance across the water is given by the Pythagorean theorem to be $\sqrt{9 + x^2}$. The cost C is

$C = $ COST IN THE WATER + COST ON THE SHORE

$$= (1.5)\sqrt{9 + x^2} + (1)(8 - x)$$

$$C'(x) = \frac{(1.5)(2x)}{2\sqrt{9 + x^2}} - 1 = 0 \text{ when}$$

$$2.25x^2 = 9 + x^2$$

$$1.25x^2 = 9$$

$$x = 2.66833$$

(reject negative value)

$C(0) = 12.45$; $C(8) = 12.81$; $C(2.6833) = 11.35$ That is, if $8 - x = 5.3167$ mi or $28,072$ ft of pipe is laid on the shore gives the minimum cost.

45. Let x be the number of weeks of waiting. The number of cases is $100 + 25x$. The profit per case is $10 - x$ on $[0, 10]$. The profit is

$$P(x) = (100 + 25x)(10 - x)$$

$$= 1,000 + 150x - 25x^2$$

$$P'(x) = 150 - 50x = 0 \text{ when } x = 3;$$

$$P(0) = P(10) = 0; \ P(3) = 1,225$$

That is, the maximum profit of $1,225 is realized when the grower waits 3 weeks.

46. Let x be the number of groups of items sold. The sales price at 60 items is $1.20, so the profit is 80¢. The number of buzzers at a low sales price is $60 + 10x$, while the corresponding profit per buzzer is $80 - 10x$. The profit is

$$P(x) = (60 + 10x)(80 - 10x)$$

$$= 100(48 + 2x - x^2)$$

$$P'(x) = -2(x - 1) = 0 \text{ when } x = 1;$$

$$P(0) = P(100) = 0; \ P(1) = 4,900 \text{ cents or } \$49$$

is the maximum profit.

47. Let x be the number of units of telephones sold. The price per telephone is

$$p(x) = 150 - x \text{ on } [0, 50]$$

The total production cost is

$$C(x) = 2,500 + 30x$$

The profit is

$$P(x) = x(150 - x) - 2,500 - 30x$$

$$= 150x - x^2 - 2,500 - 30x$$

$$= -(x^2 - 120x + 2,500)$$

$$P'(x) = -2x + 120 = 0 \text{ when } x = 60$$

The price is $p(60) = 150 - 60 = 90$ and the maximum profit is
$P(60) = -(60^2 - 120(60) + 2,500) = 1,100$

48. The number of books sold is

$$n = \frac{50,000}{x + 20} - 2,000$$

where x is the price per book. The profit is

$$P(x) = (x - 29)n$$

$$= (x - 29)\left(\frac{50,000}{x + 20} - 2,000\right)$$

$$= \frac{50,000(x - 29)}{x + 20} - 2,000(x - 29)$$

$$P'(x) = \frac{50,000(x + 20 - x + 29)}{(x + 20)^2} - 2,000$$

$$P'(x) = 0 \text{ if } \frac{25(49)}{(x + 20)^2} = 1 \text{ when}$$

$$x = 15, \ x = 55$$

The selling price of the book that costs $29 to produce should be $55.

49. Let x be the number of machines. The number of items produced per machine per hour is $25x$. The number of hours needed is $\frac{5,000}{25x} = \frac{200}{x}$; the cost of supervision is $20\left(\frac{200}{x}\right) = \frac{4,000}{x}$; the set-up cost is $50x$.

Thus, the total cost, C, is

$$C(x) = 50x + \frac{4,000}{x}; \ C'(x) = 50 - \frac{4,000}{x^2} = 0$$

if $x^2 = 80$ or $x \approx 9$ (reject the negative value); $C(0)$ is not defined; $C(9) \approx 894.44$; $C(12) \approx 933.33$. The minimum value occurs when 9 machines are used.

50. $R(N) = -3N^4 + 50N^3 - 261N^2 + 540N$
$R'(N) = -12N^3 + 150N^2 - 522N + 540$
Since 2 is a critical value, divide by $N - 2$, and factor out the common factor of -6:

$$R'(N) = -6(N - 2)(2N^2 - 21N + 45)$$

$$= -6(N - 2)(N - 3)(2N - 15)$$

The critical values are $N = 2, 3, 7.5$
$R(0) = 0$; $R(9) = 486$; $R(2) = 388$;
$R(3) = 378$; $R(7.5) = 970.3125$;
The maximum value is 970.3125 which occurs
when $x = 7.5$. Should hire 7 new employees.

51. $f(x) = a_n x^n + a_{n-1} x^{n-1} + a_{n-2} x^{n-2}$

$$+ \cdots + a_2 x_2 + a_1 x + a_0$$

$f'(x) = na_n x^{n-1} + (n-1)a_{n-1} x^{n-2}$

$$+ \cdots + 2a_2 x + a_1$$

$f''(x) = n(n-1)a_n x^{n-2}$

$$+ (n-1)(n-2)a_{n-1} x^{n-3}$$

$$+ \cdots + 2a_2$$

A polynomial of degree $n - 2$ has at most
$n - 2$ roots. This means that there are at
most $n - 2$ points of inflection.

52. $f(x) = x^n$; $f'(x) = nx^{n-1}$;

$f''(x) = n(n-1)x^{n-2} = 0$ at $x = 0$

If n is even $f''(x) \geq 0$ for all x.

If n is odd, $f''(x) \geq 0$ if $x \geq 0$ and $f''(x) \leq 0$

if $x \leq 0$. The graph of the curve will change
direction of curvature only once in this latter
case.

53. Let $P_0(x_0, y_0)$ be the point of contact
between the line $y = mx + b$ and the circle
$x^2 + y^2 = 1$. The slope of the line is
$dy/dx = -x/y$. At P_0, $m = -x_0/y_0$ so the
equation of the tangent line is

$$y - y_0 = -\frac{x_0}{y_0}(x - x_0)$$

$$y_0 y - y_0^2 + x_0 x - x_0^2 = 0$$

$$y_0 y + x_0 x = x_0^2 + y_0^2 = 1$$

If $x = 0$, $y = y_1$, so $y_1 = 1/y_0$. If $y = 0$,

$x = x_1$, so $x_1 = 1/x_0$. The function to be
minimized is

$$\ell = x_1 + y_1$$

$$= \frac{1}{x_0} + \frac{1}{y_0}$$

$$= \frac{1}{x_0} + \frac{1}{\sqrt{1 - x_0^2}}$$

$$\ell' = -\frac{1}{x_0^2} - \frac{-x_0}{(1 - x_0^2)^{3/2}}$$

$\ell' = 0$ if

$$x_0^3 = (1 - x_0)^{3/2}$$

$$x_0^2 = 1 - x_0^2$$

$$x_0^2 = \frac{1}{2}$$

$$x_0 = \frac{\sqrt{2}}{2}$$

Thus, $y_0 = \sqrt{1 - \frac{1}{2}} = \frac{\sqrt{2}}{2}$.

The equation of the line is

$$y - \frac{\sqrt{2}}{2} = -\frac{\sqrt{2}}{2}$$

$$2y - \sqrt{2} = -2x + \sqrt{2}$$

$$2x + 2y - 2\sqrt{2} = 0$$

$$x + y - \sqrt{2} = 0$$

54. $P(\theta) = \dfrac{a \sin \theta}{(1 - b \cos \theta)^5}$

$P'(\theta)$

$$\frac{a[(1 - b\cos\theta)^5(\cos\theta) - (\sin\theta)(5)(1 - b\cos\theta)^4(b\sin\theta)]}{(1 - b\cos\theta)^{10}}$$

$P'(\theta) = 0$ if $1 - b \cos \theta = 0$ or $\cos \theta = \frac{1}{b}$;

and when

$(1 - b \cos \theta)\cos \theta - 5b \sin^2\theta = 0$ or

$$\cos \theta = \frac{-1 \pm \sqrt{1 + 80b^2}}{8b}$$

(reject the negative value). The largest value
of $\cos \theta = 1$ and that occurs when $b = 1$.

55. $\dfrac{f(x) - f(a)}{x - a} = f'(d)$ for d in (a, b) and that

$f'(d) = c$. Then,

$f(x) = f'(d)(x - a) + f(a) = c(x - a) + f(a)$

56. Apply the MVT to $f'(x)$ over $[0, x_0]$ to obtain
$$\frac{f'(x_0) - f'(0)}{x_0 - 0} = f''(w) \text{ for } 0 < w < x_0$$
$$f'(x_0) = f''(w)x_0 + f'(0)$$
$$= -c^2 f(w)x_0 + 1$$

or

$$f'(x_0) + c^2 f(w)x_0 = 1$$

57. $E(x) = \dfrac{Q_0 x}{(x^2 + R^2)^{3/2}}$;

$$E'(x) = \frac{Q_0(R^2 - 2x^2)}{(x^2 + R^2)^{5/2}} = 0$$

if $x = \dfrac{\sqrt{2}R}{2}$; $E(0) = 0$

58. $f'(x) = 2 \cos 2x$
$f''(x) = -4 \sin 2x$

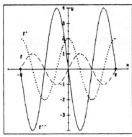

a. $f'(x) > 0$ on $(-\pi, -\frac{3\pi}{4}) \cup (-\frac{\pi}{4}, \frac{\pi}{4}) \cup (\frac{3\pi}{4}, \pi)$

b. $f'(x) < 0$ on $(-\frac{3\pi}{4}, -\frac{\pi}{4}) \cup (\frac{\pi}{4}, \frac{3\pi}{4})$

c. $f''(x) > 0$ on $(-\frac{\pi}{2}, 0) \cup (\frac{\pi}{2}, \pi)$

d. $f''(x) < 0$ on $(-\pi, -\frac{\pi}{2}) \cup (0, \frac{\pi}{2})$

e. $f'(x) = 0$ at $x = \pm\frac{\pi}{4}, \pm\frac{3\pi}{4}$

f. $f'(x)$ exists everywhere

g. $f''(x) = 0$ at $x = 0, \pm\pi, \pm\frac{\pi}{2}$

59. $f'(x) = 3x^2 - 2x - 1$
$f''(x) = 6x - 2$

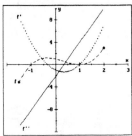

a. $f'(x) > 0$ on $(-\frac{3}{2}, -\frac{1}{3}) \cup (1, 2)$

b. $f'(x) < 0$ on $(-\frac{3}{2}, 1)$

c. $f''(x) > 0$ on $(0, 2)$

d. $f''(x) < 0$ on $(\frac{1}{3}, 2)$

e. $f'(x) = 0$ at $x = -\frac{1}{3}, 1$

f. $f'(x)$ exists everywhere

g. $f''(x) = 0$ at $x = -\frac{1}{3}$

60. $f'(x) = 4x^3 - 4x = 4x(x^2 - 1)$
$f''(x) = 12x^2 - 4 = 4(3x^2 - 1)$

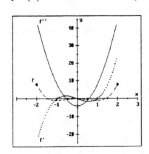

a. $f'(x) > 0$ on $(-1, 0) \cup (1, 2)$

b. $f'(x) < 0$ on $(-2, -1) \cup (0, 1)$

c. $f''(x) > 0$ on $(-2, -\sqrt{3}/3) \cup (\sqrt{3}/3, 2)$

d. $f''(x) < 0$ on $(-\sqrt{3}, \sqrt{3})$

e. $f'(x) = 0$ at $x = \pm 1$

f. $f'(x)$ exists everywhere

g. $f''(x) = 0$ at $x = \pm\sqrt{3}/3$

61. $f'(x) = 3x^2 - 1 - x^{-2}$
$f''(x) = 6x + 2x^{-3}$

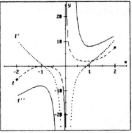

a. $f'(x) > 0$ on $(-2, -0.876) \cup (0.876, 2)$

b. $f'(x) < 0$ on $(-0.876, 0) \cup (0, 0.876)$

c. $f''(x) > 0$ on $(0, 2)$

d. $f''(x) < 0$ on $(-2, 0)$

e. $f'(x) = 0$ at $x = \pm 0.876$

f. $f'(x)$ does not exist at $x = 0$

g. $f''(x) \neq 0$

62. Let x and y be the dimensions of the rectangle.

$$2x + 2y = 60 \text{ or } y = 30 - x$$

$A(x) = 30x - x^2; A'(x) = 2x - 30$

$A'(x) = 0$ at $x = 15$;

$A(0) = 0; A(15) = 225; A(30) = 0;$ the minimum area is 0.

63. With $4x^2 + y^2 = 1$, $y = \sqrt{1 - 4x^2}$ and
$$d = \frac{1}{x} - \sqrt{1 - 4x^2}$$

is the vertical distance between the shore and the road. Using a graphing calculator or computer software we find that this distance is minimized in the neighborhood of $x = 0.460355$.

64.

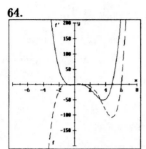

Detail of graph:

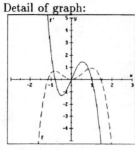

Estimate the relative minimum at about $(4.8, -110)$, and relative maxima at $(1, 1)$ and $(-0.8, 0.7)$.

65. **a.** The cost is to be minimized:

$$C(x) = (1.7)\frac{1}{250}\left(\frac{750}{x} + x\right)(500)$$

$$= \frac{17(x^2 + 750)}{5x}$$

$$C'(x) = \frac{17(x^2 - 750)}{5x^2} = 0 \text{ at}$$

$$x = \pm 5\sqrt{30}$$

(negative value not in the domain); the cost is minimized in the neighborhood of 27 mi/hr

b. The total cost $T(x)$ is found by

$$T(x) = \text{COST OF FUEL} + \text{COST OF DRIVER}$$

$$= \frac{17(x^2 + 750)}{5x} + \left(\frac{500}{x}\right)(28)$$

$$= \frac{17x^2 + 82,750}{5x}$$

$$T'(x) = \frac{17x^2 - 82,750}{5x^2} = 0 \text{ when}$$

$$x = \pm\frac{5\sqrt{56,270}}{17} \approx \pm 69.76$$

Neither of these values are in the domain $[15, 55]$. The total cost is minimized at $x = 55$ mi/hr

66. **a.** The angular velocity is $d\theta/dt = 2$ so $\theta = 2t$. The point (x, y) is a vertex of a right triangle whose hypotenuse is 2. Thus, $x = 2\cos 2t$ and $y = \sin 2t$.

b. The vertical distance q is the sum of y and the vertical leg of the right triangle with hypotenuse from $(0, q)$ to (x, y). The vertical leg of this triangle is $\sqrt{L^2 - x^2}$. Then

$$q(t) = \sin 2t + \sqrt{L^2 - \cos^2 2t}$$

c. $q'(t) = 2\cos 2t + \dfrac{-2\cos 2t(-2\sin 2t)}{2\sqrt{L^2 - \cos^2 2t}}$

d. $q'(t) = 0$ at $(0, 2)$ since the rod has stopped going up and is not yet coming down (at this instant).

e. Answer can be verified using computer software.

f.

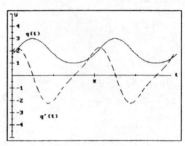

The graphs are as expected; $q(t)$ with a maximum on $[0, 2\pi]$ near $t = 0.8$ and a minimum near 1.4. Also note that $q'(x) > 2$ for small $t > 0$.

g. Strange flat areas in the graph occur and $q'(x) > 3$ in places

h. For $L = 2.02$, at $t = 0.1469$; $q'(t) = 3.634$

i. The value of 3.634 is much larger than the speed of (x, y) which is 2 ft/s

j. For L very close to 1.0, q'_{max} is close is 4.0; we conjecture this as a limit. This is difficult to prove.

67. **a.** $f(r) = \cos\dfrac{25}{r} - \dfrac{r}{r+1}$; from the graph, $f(r) = 0$ at $r \approx 310$.

b. $f'(r) = \dfrac{1}{(r+1)^2} - \dfrac{1}{r+1} + \dfrac{25\sin\frac{25}{r}}{r^2}$

$r_0 = 310$; $r_1 \approx 313.261$; $r_2 \approx 313.331$

68. This is Putnam Problem 11 in the afternoon session of 1938. Any point on the parabola has coordinates of the form $(2mt^2, 2mt)$. Let $\overline{AB}$ be a chord normal to the parabola at A, say $A = (2mt^2, 2mt)$ and $B = (2ms^2, 2ms)$. The slope of $\overline{AB}$ is $1/(s + t)$, and the slope of the tangent at A is $1/(2t)$. Hence

$$s + t = \frac{-1}{2t}$$

The length L of $\overline{AB}$ is given by

$$L^2 = 4m^2[(s^2 - t^2)^2 + (s - t)^2]$$

$$= 4m^2(s - t)^2[(s + t)^2 + 1]$$

Substituting $s = -t - \dfrac{1}{2t}$, we have

$$L^2 = 4m^2\left(\frac{4t^2 + 1}{2t}\right)^2\left(\frac{1 + 4t^2}{4t^2}\right)$$

$$= \frac{m^2}{4}\left[\frac{(4t^2 + 1)^2}{t^4}\right]$$

We seek the value of t which minimizes L, so we may just as well choose t to minimize

$$\frac{4t^2 + 1}{t^{4/3}} = 4t^{2/3} + t^{-4/3}$$

Setting the derivative equal to 0, we find two critical points,

$$t = \pm\frac{\sqrt{2}}{2}$$

Since $L \to +\infty$ as $t \to 0$, these two critical values both given minima. Either of the two shortest chords is of length $3\sqrt{3}\,|m|$.

69. This is Putnam Problem 1 in the morning session of 1941. Make the substitution $x = a(1 - y)$. Then the polynomial becomes

$$a^6\left(y2 - 3y^3 + \tfrac{5}{2}y^2 - \tfrac{1}{2}\right)$$

Since $a^6 y^2$ is surely positive, it suffices to prove that

$$g(y) = y^4 - 3y^3 + \tfrac{5}{2}y^2 - \tfrac{1}{2} < 0 \text{ for } 0 < y < 1$$

Since $g'(y) = 4y^3 - 9y^2 + 5y = y(y - 1)(4y - 5)$,

the critical values for g are 0, 1, $\frac{5}{4}$. Between

consecutive critical values g is strictly monotonic. Therefore, since $g(0) = -\frac{1}{2}$ and $g(1) = 0$, we have

$$-\tfrac{1}{2} < g(y) < 0 \text{ for } 0 < y < 1$$

70. This is Problem 1 of the 1987 Putnam examination.

$x^4 + 36 \leq 13x^2$ is equivalent to

$$(x - 3)(x - 2)(x + 2)(x + 3) \leq 0$$

which is satisfied only if

$$-3 \leq x \leq -2 \text{ or } 2 \leq x \leq 3$$

The function f is increasing on these intervals since $f'(x) = 3(x^2 - 1) > 0$. Thus, the maximum of f over the domain is $\max\{f(-2), f(3)\} = 18$.

71. This is Problem 2 of the 1986 Putnam examination.

$$\frac{A(R) + A(S)}{A(T)} = \frac{ay + bz}{\frac{1}{2}hx}$$

where T is the triangle and the height of the triangle is $h = a + b + c$. By similar triangles,

$$\frac{x}{h} = \frac{y}{b + c} = \frac{z}{c}$$

which implies

$$y = \frac{(b + c)x}{b} \quad \text{and} \quad z = \frac{cx}{h}$$

so

$$\frac{A(R) + A(S)}{A(T)} = \frac{a\left[\frac{(b + c)x}{h}\right] + b\left[\frac{cx}{h}\right]}{\frac{1}{2}hx}$$

$$= \frac{2}{h^2}(ab + ac + bc)$$

We need to maximize $ab + ac + bc$ subject to $a + b + c = h$. To do this, we fix a so that $b + c = b - a$ and

$$ab + ac + b = a(b - a) + bc$$

Then bc is maximized when $b = c$ and $2ab + b^2$ is maximized (subject to $a + 2b = h$) when $a = b = c = h/3$. Thus, the maximum ratio is $2/3$, which is independent of T.

CHAPTER 4

Integration

4.1 Area As the Limit of a Sum; Summation Notation, Page 264

1. $\displaystyle\sum_{k=1}^{\infty} 1 = 6(1) = 6$

2. $\displaystyle\sum_{k=1}^{250} 1 = 250(1) = 250$

3. $\displaystyle\sum_{k=1}^{15} k = \frac{(15)(16)}{2} = 120$

4. $\displaystyle\sum_{k=1}^{10} k = \frac{(10)(11)}{2} = 55$

5. $\displaystyle\sum_{k=1}^{5} k^3 = \frac{(5^2)(6^2)}{2} = 225$

6. $\displaystyle\sum_{k=1}^{7} k^2 = \frac{(7)(8)(15)}{6} = 140$

7. $\displaystyle\sum_{k=1}^{100} (2k - 3) = 2\sum_{k=1}^{100} k - 3\sum_{k=1}^{100} 1 = 2\left[\frac{(100)(101)}{2}\right] - 300 = 9{,}800$

8. $\displaystyle\sum_{k=1}^{100} (k - 1)^2 = 0^2 + 1^2 + 2^2 + \cdots + 99^2 = \sum_{k=1}^{99} k^2 = \frac{(99)(100)(199)}{6} = 328{,}350$

9. $\displaystyle\lim_{n\to+\infty} \sum_{k=1}^{n} \frac{k}{n^2} = \lim_{n\to+\infty} \sum_{k=1}^{n} k = \lim_{n\to+\infty} \frac{1}{n^2}\left[\frac{n(n+1)}{2}\right] = \lim_{n\to+\infty} (1)\left(1 + \frac{1}{n}\right) = \frac{1}{2}$

10. $\displaystyle\lim_{n\to+\infty} \sum_{k=1}^{n} \frac{k^2}{n^3} = \lim_{n\to+\infty} \frac{1}{n^3} \sum_{k=1}^{n} k^2 = \lim_{n\to+\infty} \frac{1}{n^3}\frac{n(n+1)(2n+1)}{6}$

$\displaystyle = \frac{1}{6}\lim_{n\to+\infty} (1)\left(1 + \frac{1}{n}\right)\left(2 + \frac{1}{n}\right) = \frac{1}{3}$

11. $\displaystyle\lim_{n\to+\infty} \sum_{1}^{n}\left(\frac{2}{n} + \frac{2k}{n^2}\right) = \lim_{n\to+\infty}\left[\frac{2}{n}(n) + \frac{2}{n^2}\left(\frac{n(n+1)}{2}\right)\right] = \lim_{n\to+\infty}\left[2 + \left(1 + \frac{1}{n}\right)\right] = 3$

12. $\displaystyle\lim_{n\to+\infty} \sum_{k=1}^{n}\left(1 + \frac{2k}{n}\right)^2\left(\frac{2}{n}\right) = \lim_{n\to+\infty}\left(\frac{2}{n}\right)\left(\sum_{k=1}^{n} 1 + \frac{1}{n}\sum_{k=1}^{n} k + \frac{4}{n^2}\sum_{k=1}^{n} k^2\right)$

$\displaystyle = \lim_{n\to+\infty}\left(\frac{2}{n}\right)\left[n + \frac{4}{n}\frac{n(n+1)}{2} + \frac{4}{n^2}\frac{n(n+1)(2n+1)}{6}\right]$

$\displaystyle = \lim_{n\to+\infty}\left[2 + 4(1)\left(1 + \frac{1}{n}\right) + \left(2 + \frac{1}{n}\right)\right]$

$\displaystyle = 2 + 4 + \frac{8}{3} = \frac{26}{3} \approx 8.67$

13.   **a.** $n = 4$, $\Delta x = 0.25$, $f(a + k\Delta x) = k + 1$; $S = 3.5$

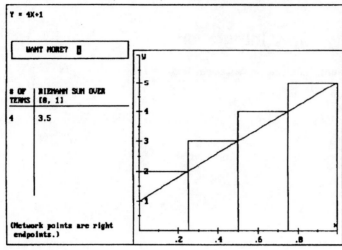

   b. $n = 8$, $\Delta x = 0.125$, $f(a + k\Delta x) = \dfrac{k}{2} + 1$; $S = 3.25$

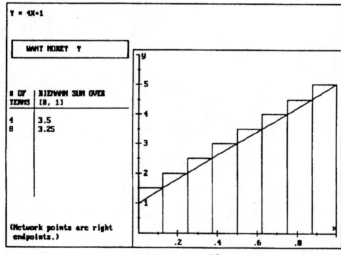

14.   **a.** $n = 3$, $\Delta x = \dfrac{1}{3}$, $f(a + k\Delta x) = 3 - \dfrac{2k}{3}$; $S \approx 1.67$

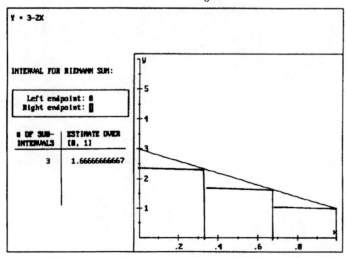

b. $n = 6$, $\Delta x = \frac{1}{6}$, $f(a + k\Delta x) = 3 - \frac{k}{3}$; $S \approx 1.83$

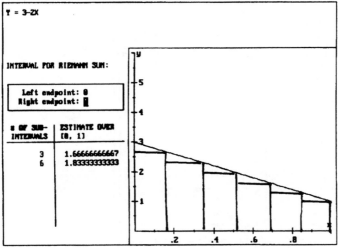

15.   **a.** $n = 4$, $\Delta x = 0.25$, $f(a + k\Delta x) = \left(1 + \frac{k}{4}\right)^2$; $S = 2.71875$

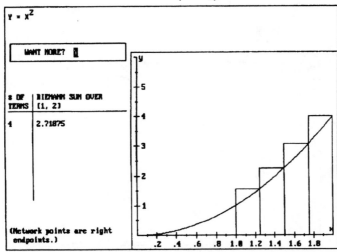

b. $n = 6$, $\Delta x = \frac{1}{6}$, $f(a + k\Delta x) = \left(1 + \frac{k}{6}\right)^2$; $S \approx 2.58796$

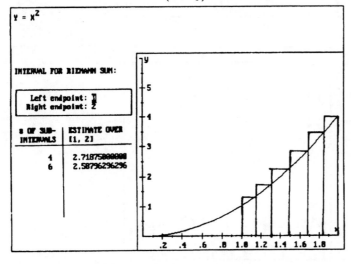

16. $n = 4, \Delta x = \frac{\pi}{8}, f(a + k\Delta x) = \cos(-\frac{\pi}{2} + \frac{k\pi}{8}); S \approx 1.184$

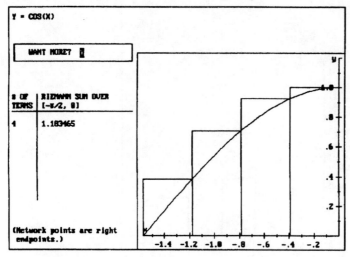

17. $n = 6, \Delta x = \frac{1}{6}, f(a + k\Delta x) = 3 - \frac{k}{3}; S \approx 0.795$

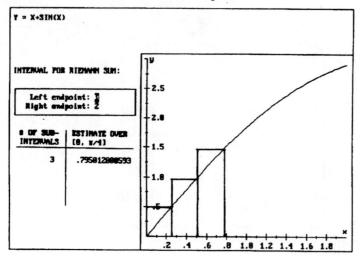

18. $n = 46, \Delta x = 0.25, f(a + k\Delta x) = \left(1 + \frac{k}{4}\right)^{-2}; S \approx 0.42$

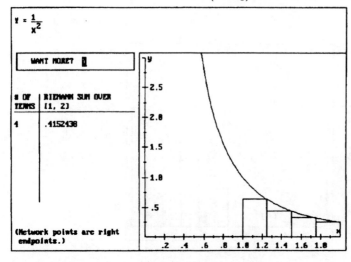

19. $n = 46, \Delta x = 0.25, f(a + k\Delta x) = 2\left(1 + \frac{k}{4}\right); S = 1.269$

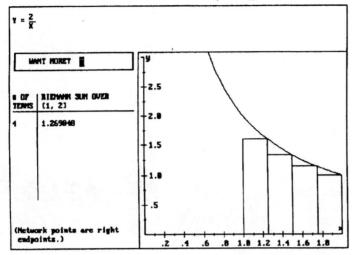

20. $n = 4, \Delta x = 0.25, f(a + k\Delta x) = \sqrt{1 + \frac{3k}{4}}; S \approx 5.03009$

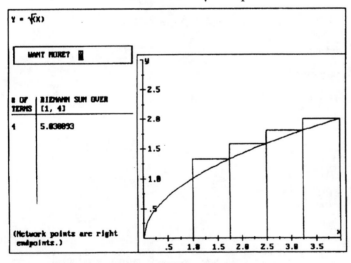

21. $n = 4, \Delta x = 0.25, f(a + k\Delta x) = \frac{3k}{4}; S \approx 1.2033$

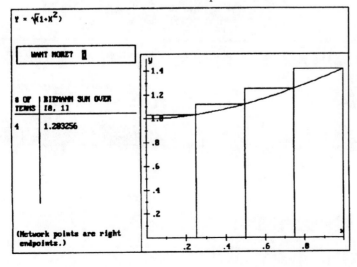

22. $\Delta x = \dfrac{b-a}{n} = \dfrac{2}{n};$

$A = \lim\limits_{n\to+\infty} \sum\limits_{k=1}^{n} f(0 + k\Delta x)\Delta x$

$ = \lim\limits_{n\to+\infty} \sum\limits_{k=1}^{n} f(\tfrac{2k}{n})\Delta x$

$ = \lim\limits_{n\to+\infty} \sum\limits_{k=1}^{n} \left(\dfrac{32k^3}{n^3} + \dfrac{4k}{n}\right)\left(\dfrac{2}{n}\right)$

$ = \lim\limits_{n\to+\infty} \left[\dfrac{32}{n^3}\sum\limits_{k=1}^{n} k^3 + \dfrac{4}{n}\sum\limits_{k=1}^{n} k\right]\left(\dfrac{2}{n}\right)$

$ = \lim\limits_{n\to+\infty}\left[\dfrac{64}{n^4}\dfrac{n^2(n+1)^2}{4} + \dfrac{8n(n+1)}{2n^2}\right]$

$ = 16 + 4 = 20$

23. $\Delta x = \dfrac{b-a}{n} = \dfrac{2-1}{n} = \dfrac{1}{n}$

$A = \lim\limits_{n\to+\infty} \sum\limits_{k=1}^{n} f(1 + k\Delta x)\Delta x$

$ = \lim\limits_{n\to+\infty} \sum\limits_{k=1}^{n} f(1 + \tfrac{k}{n})\Delta x$

$ = \lim\limits_{n\to\infty} \sum\limits_{k=1}^{n} \left[4(1 + \tfrac{k}{n})^3 + 2(1 + \tfrac{k}{n})\right]\left(\tfrac{1}{n}\right)$

$ = \lim\limits_{n\to+\infty} \sum\limits_{k=1}^{n} \left[4\left(1 + \tfrac{3k}{n} + \tfrac{3k^2}{n^2} + \tfrac{k^3}{n^3}\right) + 2 + \tfrac{2k}{n}\right]\left(\tfrac{1}{n}\right)$

$ = \lim\limits_{n\to+\infty} \sum\limits_{k=1}^{n} \left[6 + \tfrac{14k}{n} + \tfrac{12k^2}{n^2} + \tfrac{4k^3}{n^3}\right]\left(\tfrac{1}{n}\right)$

$ = \lim\limits_{n\to+\infty} \left\{6n + \dfrac{14n(n+1)}{2n}\right.$

$ \quad \left. + \dfrac{12n(n+1)(2n+1)}{6n^2} + \dfrac{4n^2(n+1)^2}{4n^3}\right\}\dfrac{1}{n}$

$ = \lim\limits_{n\to+\infty} \left\{6 + \dfrac{14n^2 + 14n}{2n^2} + \dfrac{24n^3 + 36n^2 + 12n}{6n^3}\right.$

$ \quad \left. + \dfrac{4n^4 + 8n^3 + 4n^2}{4n^4}\right\}$

$ = (6 + 7 + 4 + 1) = 18$

24. $\Delta x = \dfrac{b-a}{n} = \dfrac{3-0}{n} = \dfrac{3}{n}$

$A = \lim\limits_{n\to+\infty} \sum\limits_{k=1}^{n} f(\tfrac{3k}{n})\Delta x$

$ = \lim\limits_{n\to+\infty} \sum\limits_{k=1}^{n} \left[6\left(\tfrac{3k}{n}\right)^2 + 2\left(\tfrac{3k}{n}\right) + 4\right]\left(\tfrac{3}{n}\right)$

$ = \lim\limits_{n\to+\infty} \sum\limits_{k=1}^{n} \left(\dfrac{54k^2}{n^2} + \dfrac{6k}{n} + 4\right)\left(\tfrac{3}{n}\right)$

$ = \lim\limits_{n\to+\infty} \left[\dfrac{162}{n^3}\sum\limits_{k=1}^{n} k^2 + \dfrac{18}{n^2}\sum\limits_{k=1}^{n} k + \dfrac{12}{n}\sum\limits_{k=1}^{n} 1\right]$

$ = \lim\limits_{x\to+\infty} \left[\dfrac{162}{n^3}\dfrac{n(n+1)(2n+1)}{6} + \dfrac{18}{n^2}\dfrac{n(n+1)}{2} + \dfrac{12n}{n}\right]$

$ = \dfrac{162}{3} + 9 + 12 = 75$

25. $\Delta x = \dfrac{b-a}{n} = \dfrac{3-1}{n} = \dfrac{2}{n}$

$A = \lim\limits_{n\to+\infty} \sum\limits_{k=1}^{n} f(1 + \tfrac{2k}{n})\Delta x$

$ = \lim\limits_{n\to+\infty} \sum\limits_{k=1}^{n} \left[6\left(1 + \tfrac{2k}{n}\right)^2 + 2\left(1 + \tfrac{2k}{n}\right) + 4\right]\left(\tfrac{2}{n}\right)$

$ = \lim\limits_{n\to+\infty} \sum\limits_{k=1}^{n} \left(12 + \dfrac{28k}{n} + \dfrac{24k^2}{n^2} + 4\right)\left(\tfrac{2}{n}\right)$

$ = \lim\limits_{n\to+\infty} \left[\dfrac{24}{n}\sum\limits_{k=1}^{n} 1 + \dfrac{56}{n^2}\sum\limits_{k=1}^{n} k + \dfrac{48}{n^3}\sum\limits_{k=1}^{n} k^2\right]$

$ = \lim\limits_{n\to+\infty} \left[24 + \dfrac{56}{n^2}\dfrac{n(n+1)}{2} + \dfrac{48}{n^3}\dfrac{n(n+1)(2n+1)}{6}\right]$

$ = 24 + 28 + 16 = 68$

26. $\Delta x = \dfrac{b-a}{n} = \dfrac{1-0}{n} = \dfrac{1}{n}$

$A = \lim\limits_{n\to+\infty} \sum\limits_{k=1}^{n} f(\tfrac{k}{n})\Delta x$

$ = \lim\limits_{n\to+\infty} \sum\limits_{k=1}^{n} \left[\dfrac{3k^2}{n^2} + \dfrac{2k}{n} + 1\right]\left(\tfrac{1}{n}\right)$

$ = \lim\limits_{n\to+\infty} \left[\dfrac{3}{n^3}\sum\limits_{k=1}^{n} k^2 + \dfrac{2}{n}\sum\limits_{k=1}^{n} k + \dfrac{1}{n}\sum\limits_{k=1}^{n} 1\right]$

$ = \lim\limits_{n\to+\infty} \left[\dfrac{3}{n^2}\dfrac{n(n+1)(2n+1)}{6} + \dfrac{2}{n^2}\dfrac{n(n+1)}{2} + \dfrac{n}{n}\right]$

$ = 1 + 1 + 1 = 3$

27. $\Delta x = \dfrac{b-a}{n} = \dfrac{1-0}{n} = \dfrac{1}{n}$

$A = \lim\limits_{n\to+\infty} \sum\limits_{k=1}^{n} f(\tfrac{k}{n})\Delta x$

$ = \lim\limits_{n\to+\infty} \sum\limits_{k=1}^{n} \left[4(\tfrac{k}{n})^3 + 3(\tfrac{k}{n})^2\right]\left(\tfrac{1}{n}\right)$

$ = \lim\limits_{n\to+\infty} \left[\dfrac{4}{n^4}\sum\limits_{k=1}^{n} k^3 + \dfrac{3}{n^3}\sum\limits_{k=1}^{n} k^2\right]$

$ = \lim\limits_{x\to+\infty} \left[\dfrac{4}{n^4}\dfrac{n^2(n+1)^2}{6} + \dfrac{3}{n^3}\dfrac{n(n+1)(2n+1)}{6}\right]$

$ = 1 + 1 = 2$

28. The statement is true. We are dealing with a rectangle of height C and base $b - a$.

29. This statement is false. We are dealing with a trapezoid of height $(C/2)(a + b)$ and base $(b - a)$. The area is
$$A = \frac{C}{2}(b + a)(b - a) = \frac{C(b^2 - a^2)}{2}$$

30. The statement is true. Consider the trapezoid of height $\frac{1}{2}(a^2 + b^2)$ and base $b - a$. The area is
$$A = \frac{1}{2}(b^2 + a^2)(b - a)$$
The area under the parabola is less than the area of the trapezoid.

31. The statement is true. $y = \sqrt{1 - x^2}$ or $x^2 + y^2 = 1$ (for $y \geq 0$) represents the equation of a semicircle. The area is $A = \pi/2$.

32. The statement is false. Let $f(x) = x^2$ on $(0, 1)$, then $[f(x)]^2 = x^4$. Now, $[f(x)]^2 \leq f(x)$ and the area under $[f(x)]^2$ is less than that under $f(x)$.

33. The statement is true. The graph of f is symmetric with respect to the y-axis. The area on the right of the x-axis is equal to that on the left.

34. $f(x) = x^3$ on $[0, 1]$; $\Delta x = \frac{1}{n}$;
$$f(a + kx) = f(\tfrac{k}{n}) = \frac{k^3}{n^3}.$$
$$A = \lim_{n \to +\infty} \frac{1}{n^4} \sum_{k=1}^{n} k^3$$
$$= \lim_{n \to +\infty} \frac{1}{n^4} \frac{n^2(n + 1)^2}{4} = \frac{1}{4}$$

35. $f(x) = C$ on $[a, b]$; $\Delta x = \frac{b - a}{n}$; $f(a + kx) = C$
$$A = \lim_{n \to +\infty} \frac{b - a}{n} \sum_{k=1}^{n} C = \frac{b - a}{n} nC$$
$$= (b - a)C$$
where $b - a$ is the length of the rectangle of width C.

36. $f(x) = \frac{h}{b}x$ where h is the height and b the base of the right triangle; $\Delta x = \frac{b}{n}$;
$$f(a + kx) = \frac{hk}{n}$$
$$A = \lim_{n \to +\infty} \frac{b}{n} \sum_{k=1}^{n} \frac{hk}{n}$$
$$= \lim_{n \to +\infty} \left[\frac{bh}{n^2} \sum_{k=1}^{n} k\right]$$
$$= \lim_{n \to +\infty} \left[\frac{bh}{n^2} \frac{n(n + 1)}{2}\right]$$
$$= \frac{bh}{2}$$

37. a. $f(x) = 2x^2$; $\Delta x = \frac{1}{n}$;
$$f(a + kx) = 2(1 + \tfrac{k}{n})^2$$
$$A = \lim_{x \to +\infty} \left[2(1 + \tfrac{k}{n})^2\right]\left(\tfrac{1}{n}\right)$$
$$= \lim_{x \to +\infty} \left[\frac{2n}{n} + \frac{4}{n^2} \sum_{k=1}^{n} k + \frac{2}{n^3} \sum_{k=1}^{n} k^2\right]$$
$$= \lim_{x \to +\infty} \left[2 + \frac{4n(n + 1)}{2n^2} + \frac{2n(n + 1)(2n + 1)}{6n^3}\right]$$
$$= 2 + 2 + \tfrac{2}{3}$$
$$= \frac{14}{3}$$

b. If $g(x) = \frac{2}{3}x^3$; $g'(x) = 2x^2$;
$$g(2) - g(1) = \frac{16}{3} - \frac{2}{3}$$
$$= \frac{14}{3}$$
$$= A$$

c. If $h(x) = \frac{2}{3}x^3 + C$; $h'(x) = 2x^2$;
$$h(2) - h(1) = \frac{16}{3} + C - \frac{2}{3} - C$$
$$= \frac{14}{3}$$
$$= A$$
regardless of C.

38.   The area seems to be 2 square units.

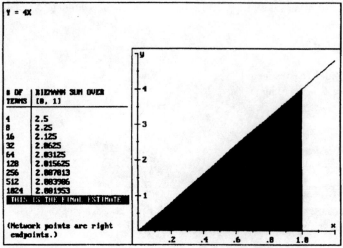

39.   The area seems to be 64/3.

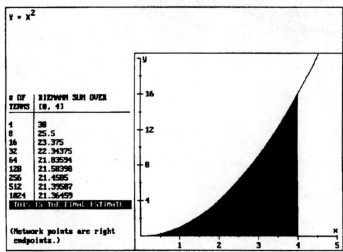

40.   The area seems to be 1.

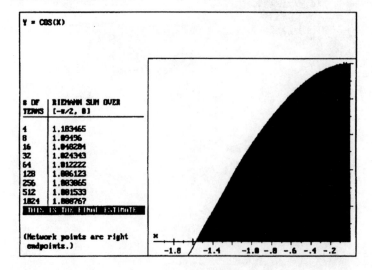

41. The area seems to be 0.6.

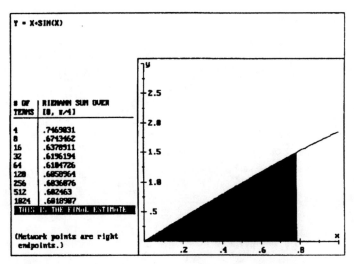

42. a. The area seems to be 2.

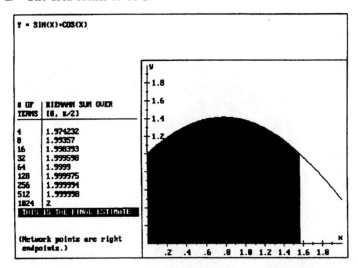

b. With $g(x) = -\cos x + \sin x$; $g'(x) = \sin x + \cos x = f(x)$
$g(\frac{\pi}{2}) = -0 + 1 = 1$; $g(0) = -1 + 0 = -1$, and

$$g(\tfrac{\pi}{2}) - g(0) = 1 - (-1) = 2 = A$$

c. With $h(x) = -\cos x + \sin x + C$; $h'(x) = \sin x + \cos x = f(x)$.
$h(\frac{\pi}{2}) = -0 + 1 + C = 1 + C$; $h(0) = -1 + 0 + C = -1 + C$, and

$$h(\tfrac{\pi}{2}) - h(0) = 1 + C - (-1) - C = 2 = A$$

43. a. $\dfrac{1}{2} \displaystyle\sum_{k=1}^{n} [k^2 - (k^2 - 2k + 1)] + \dfrac{1}{2} \sum_{k=1}^{n} 1 = \dfrac{1}{2} \sum_{k=1}^{n} (2k)$

$$= 1 + 2 + \cdots + n = \sum_{k=1}^{n} k$$

b. $\displaystyle\sum_{k=1}^{n} [k^2 - (k-1)^2] = (1^2 - 0^2) + (2^2 - 1^2) + \cdots + [n^2 - (n-1)^2] = n^2$

c.
$$\sum_{k=1}^{n} k = \frac{1}{2}\sum_{k=1}^{n}[k^2 - (k^2 - 2k + 1)] + \frac{1}{2}n$$
$$\text{From part a.}$$
$$= \frac{1}{2}n^2 + \frac{1}{2}n \qquad \text{From part b.}$$
$$= \frac{1}{2}(n^2 + n) = \frac{n(n+1)}{2}$$

44. a.
$$k^3 = a[k^4 - (k-1)^4] + bk^2 + ck + d$$
$$= a[k^4 - (k^4 - 4k^3 + 6k^2 - 4k + 1)]$$
$$\quad + bk^2 + ck + d$$
$$= 4ak^3 - 6ak^2 + 4ak^2 + 4ak - a$$
$$\quad + bk^2 + ck + d$$
$$= (4a - 1)k^3 + (b - 6a)k^2$$
$$\quad + (4a + c)k + (-a + d)$$

Since $4a - 1 = 1$; $b - 6a = 0$; $4a + c = 0$;
$-a + d = 0$; we find $a = \frac{1}{4}$, $b = \frac{3}{2}$,
$c = -1$, and $d = \frac{1}{4}$.

b.
$$\sum_{k=1}^{n}[k^3 - (k-1)^3] = (1^3 - 0^2)$$
$$+ (2^3 - 1^2) + \cdots + [n^3 - (n-1)^3] = n^3$$

Thus,

$$n^3 = \sum_{k=1}^{n}[k^3 - (k-1)^3]$$
$$= 3\sum_{k=1}^{n}k^2 - 3\sum_{k=1}^{n}k + \sum_{k=1}^{n}1$$
$$= 3\sum_{k=1}^{n}k^2 - 3\frac{n(n+1)}{2} + n$$
$$\text{From Problem 43.}$$
$$= 3\sum_{k=1}^{n}k^2 - \frac{3n^2 + n}{2}$$

Solve for the summation to find:

$$3\sum_{k=1}^{n}k^2 = n^3 + \frac{3n^2 + n}{2}$$
$$= \frac{n(2n^2 + 3n + 1)}{2}$$

Consequently,

$$\sum_{k=1}^{n}k^2 = \frac{n(n+1)(2n+1)}{6}$$

45.
$$\lim_{n\to+\infty}\sum_{k=1}^{n}\left(\frac{k-1}{n}\right)^2\left(\frac{1}{n}\right)$$
$$= \lim_{n\to+\infty}\frac{1}{n^3}\sum_{k=1}^{n}(k^2 - 2k + 1)$$

$$= \lim_{n\to+\infty}\frac{1}{n^3}\left[\frac{n(n+1)(2n+1)}{6} - 2\frac{n(n+1)}{2} + n\right]$$
$$= \frac{2}{6} - 0 + 0 = \frac{1}{3}$$

46.
$$\Delta x = \frac{b-a}{n}; \; x_0 = a + \frac{\Delta x}{2} = a + \frac{b-a}{2n};$$
$$x_1 = a + \frac{3\Delta x}{2} = a + \frac{3(b-a)}{2n};$$
$$x_2 = a + \frac{5\Delta x}{2} = a + \frac{5(b-a)}{2n}; \; \cdots$$
$$x_k = a + \frac{k\Delta x}{2} = a + \frac{k(b-a)}{2n}$$
$$A = \lim_{n\to+\infty}\left[\sum_{k=1}^{n}f\left(a + \frac{k(b-a)}{2n}\right)\left(\frac{b-a}{n}\right)\right]$$

47. a.
$$\sum_{k=1}^{n}c = \sum_{k=1}^{n}ck^0 = c(1^0 + 2^0 + \cdots + n^0) = nx$$

b.
$$\sum_{k=1}^{n}(a_k + b_k) = (a_1 + b_1) + (a_2 + b_2)$$
$$+ \cdots + (a_n + b_n)$$
$$= (a_1 + a_2 + \cdots + a_n)$$
$$+ (b_1 + b_2 + \cdots + b_n)$$
$$= \sum_{k=1}^{n}a_k + \sum_{k=1}^{n}b_k$$

c.
$$\sum_{k=1}^{n}ca_k = ca + ca_2 + \cdots + ca_n$$
$$= c(a_1 + a_2 + \cdots + a_n)$$
$$= c\sum_{k=1}^{n}a_k$$

d. The solution is illustrated in the problem statement.

e.
$$\sum_{k=1}^{n}a_k = (a_1 + a_2 + \cdots + a_m)$$
$$+ (a_{m+1} + a_{m+2} + \cdots + a_n)$$
$$= \sum_{k=1}^{m}a_k + \sum_{k=m+1}^{n}a_k$$

f.
$$\sum_{k=1}^{n}a_k = a_1 + a_2 + \cdots + a_n$$
$$\le b_1 + b_2 + \cdots + b_n = \sum_{k=1}^{n}b_k$$

since the inequalities hold term-by-term.

4.2 Riemann Sums and the Definite Integral, Page 275

1.
$$f(x) = 2x + 1; \; a = 0; \; \Delta x = \frac{1}{4}$$
$$f(a + k\Delta x) = f\left(\frac{k}{4}\right) = \frac{k}{2} + 1$$
$$\int_0^1 (2x + 1)\, dx \approx \sum_{k=1}^{4}\left(\frac{k}{2} + 1\right)\left(\frac{1}{4}\right) = 2.25$$

2. $f(x) = 4x^2 + 2;\ a = 0;\ \Delta x = \frac{1}{4}$

$f(a + k\Delta x) = f(\frac{k}{4}) = 4(\frac{k}{4})^2 + 2$

$\int_0^1 (4x^2 + 2)\ dx \approx \sum_{k=1}^4 (\frac{k^2}{4} + 2)(\frac{1}{4}) = 3.875$

3. $f(x) = x^2;\ a = 1;\ \Delta x = \frac{1}{2}$

$f(a + k\Delta x) = f(1 + \frac{k}{2}) = (1 + \frac{k}{2})^2$

$\int_1^3 x^2\ dx \approx \sum_{k=1}^4 (1 + \frac{k}{2})^2(\frac{1}{2}) = 10.75$

4. $f(x) = x^3;\ a = 0;\ \Delta x = \frac{1}{2}$

$f(a + k\Delta x) = f(\frac{k}{2}) = (\frac{k}{2})^3$

$\int_0^2 x^3\ dx \approx \sum_{k=1}^4 (\frac{k}{2})^3(\frac{1}{2}) = 6.25$

5. $f(x) = 1 - 3x;\ a = 0;\ \Delta x = \frac{1}{4}$

$f(a + k\Delta x) = f(\frac{k}{4}) = 1 - \frac{3k}{4}$

$\int_0^1 (1 - 3x)\ dx \approx \sum_{k=1}^4 (1 - \frac{3k}{4})(\frac{1}{4}) \approx -0.88$

6. $f(x) = x^2 - x^3;\ a = 1;\ \Delta x = \frac{1}{2}$

$f(a + k\Delta x) = f(1 + \frac{k}{2}) = (1 + \frac{k}{2})^2 - (1 + \frac{k}{2})^3$

$\int_1^3 (x^2 - x^3)\ dx \approx \sum_{k=1}^4 (1 + \frac{k}{2})^2 - (1 + \frac{k}{2})^3 = -16.25$

7. $f(x) = \cos x;\ a = -\frac{\pi}{2};\ \Delta x = \frac{\pi}{8}$

$f(a + k\Delta x) = f(-\frac{\pi}{2} + \frac{k\pi}{8}) = \cos(-\frac{\pi}{2} + \frac{k\pi}{8})$

$\int_{-\pi/2}^0 \cos x\ dx \approx \sum_{k=1}^4 \cos(-\frac{\pi}{2} + \frac{k\pi}{8})(\frac{\pi}{8}) \approx 1.18$

8. $f(x) = x + \sin x;\ a = 0;\ \Delta x = \frac{\pi}{16}$

$f(a + k\Delta x) = f(\frac{k\pi}{16}) = \frac{k\pi}{16} + \sin \frac{k\pi}{16}$

$\int_0^{\pi/4} (x + \sin x)\ dx \approx \sum_{k=1}^4 (\frac{k\pi}{16} + \sin \frac{k\pi}{16})(\frac{\pi}{16})$

$= 0.75$

9. $v(t) = 3t + 1;\ a = 1;\ \Delta t = \frac{3}{4}$;

$v(a + k\Delta t) = v(1 + \frac{3k}{4}) = 3(1 + \frac{3k}{4}) + 1$

$S_4 = \sum_{k=1}^4 (4 + \frac{9k}{4})(\frac{3}{4}) = \frac{231}{8} = 28.88$

10. $v(t) = 1 + 2t;\ a = 1;\ \Delta t = \frac{1}{4}$;

$v(a + k\Delta t) = v(1 + \frac{k}{4}) = 3 + \frac{k}{2}$

$S_4 = \sum_{k=1}^4 (3 + \frac{k}{2})(\frac{1}{4}) = \frac{17}{4} = 4.25$

11. $v(t) = \sin t;\ a = 0;\ \Delta t = \frac{\pi}{4}$

$v(a + k\Delta t) = v(\frac{k\pi}{4}) = \sin \frac{k\pi}{4}$

$S_4 = \sum_{k=1}^4 \sin \frac{k\pi}{4}(\frac{\pi}{4}) = \frac{\pi}{4}[\sqrt{2} + 1] \approx 1.896$

12. $v(t) = \cos t;\ a = 0;\ \Delta t = \frac{\pi}{8}$

$v(a + k\Delta t) = v(\frac{k\pi}{8}) = \cos \frac{k\pi}{8}$

$S_4 = \sum_{k=1}^4 \cos \frac{k\pi}{8}(\frac{\pi}{8}) \approx 0.791$

13. $\int_0^{-1} x^2\ dx = -\int_{-1}^0 x^2\ dx = -\frac{1}{3}$

14. $\int_{-1}^2 (x^2 + x)\ dx = \int_{-1}^2 x^2\ dx + \int_{-1}^2 x\ dx = 3 + \frac{3}{2} = \frac{9}{2}$

15. $\int_{-1}^2 (2x^2 - 3x)\ dx = 2\int_{-1}^2 x^2\ dx - 3\int_{-1}^2 x\ dx$

$= 2(3) - 3(\frac{3}{2}) = \frac{3}{2}$

16. $\int_0^2 x^2\ dx = \int_0^{-1} x^2\ dx + \int_{-1}^2 x^2\ dx$

$= -\int_{-1}^0 x^2\ dx + \int_{-1}^2 x^2\ dx = -\frac{1}{3} + 3 = \frac{8}{3}$

17. $\int_{-1}^0 x\ dx = \int_{-1}^2 x\ dx + \int_2^0 x\ dx$

$= \int_{-1}^2 x\ dx - \int_0^2 x\ dx = \frac{3}{2} - 2 = -\frac{1}{2}$

18. $\int_{-1}^0 (3x^2 - 5x)\ dx = 3\int_{-1}^0 x^2\ dx - 5\int_{-1}^0 x\ dx$

$= 3(\frac{1}{3}) - 5(-\frac{1}{2}) = 1 + \frac{5}{2} = \frac{7}{2}$

19. On $[0, 1]$ $x^3 \le x$, so $\displaystyle\int_0^1 x^3 \, dx \le \int_0^1 x \, dx = \frac{1}{2}$

20. $\displaystyle\int_0^\pi \sin x \, dx \le \int_0^1 1 \, dx = \lim_{n \to +\infty} \sum_{k=1}^n \frac{\pi}{n} = \pi$

21. Let $F = \displaystyle\int_{-2}^4 f(x) \, dx$ and $G = \int_{-2}^4 g(x) \, dx$.

Then $\displaystyle\int_{-2}^4 [5f(x) + 2g(x)] \, dx = 5F + 2G = 7.$

$\displaystyle\int_{-2}^4 [3f(x) + g(x)] \, dx = 3F + G = 4$

Subtracting the first from twice the second leads to $F = 1$ and $G = 1$.

22. **a.** Let $F = \displaystyle\int_0^2 f(x)dx = 3$; $G = \int_0^2 g(x)dx = -1$

and $H = \displaystyle\int_0^2 h(x) \, dx = 3.$ Then

$\displaystyle\int_0^2 [2f(x) + 5g(x) - 7h(x)] \, dx = 2F + 5G$

$- \, 7H = 2(3) + 5(-1) - 7(3) = -20$

b. $\displaystyle\int_0^2 [5f(x) + sg(x) - 6h(x)] \, dx$

$= 5F + sG - 6H = 5(3) + (-1)s - 6(3)$

$= -s - 3 = 0$ which implies $s = -3$

23. By the subdivision and opposite properties,

$\displaystyle\int_{-1}^2 f(x)dx = \int_{-1}^1 f(x)dx + \int_1^3 f(x)dx + \int_3^2 f(x) \, dx$

$= \displaystyle\int_{-1}^1 f(x) \, dx + \int_1^3 f(x) \, dx - \int_2^3 f(x) \, dx$

$= 3 + 5 - (-2) = 10$

24.

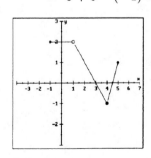

25.

$\displaystyle\int_{-1}^5 dx = \int_{-1}^1 2 \, dx + \int_1^3 (3 - x) \, dx + \int_3^4 (3 - x) \, dx$

$+ \displaystyle\int_4^{9/2} (2x - 9) \, dx + \int_{9/2}^5 (2x - 9) \, dx$

$= 2(2) + \frac{1}{2}(2)(2) - \frac{1}{2}(1)(1) - \frac{1}{2}(0.5)(1)$

$+ \frac{1}{2}(0.5)(1) = \frac{11}{2}$

$\displaystyle\int_{-3}^5 dx = \int_{-3}^{-1} 5 \, dx + \int_{-1}^2 (4 - x)dx + \int_2^5 (2x - 2)dx$

$= 5(2) + 3\left(\frac{5 + 2}{2}\right) + 3\left(\frac{2 + 8}{2}\right)$

$= 10 + \frac{21}{2} + 15 = \frac{71}{2}$

26. $\displaystyle\int_a^b f(x) \, dx = \int_a^c f(x) \, dx + \int_c^b f(x) \, dx$

$= \displaystyle\int_a^c f(x) \, dx + \left[\int_c^d f(x) \, dx + \int_d^b f(x) \, dx\right]$

27. $A = \displaystyle\int_a^b (Cx + D) \, dx$ is the area of a rectangle

of base $(b - a)$ and height $(Ca + D)$, plus that of a trapezoid with base $(b - a)$ and heights $(Ca + D)$ and $(Cb + D)$. Thus,

$A = \frac{1}{2}(b - a)[Ca + (Ca + D + Cb + D)]$

$= (b - a)\left[\frac{C}{2}(b + a) + D\right]$

28. $f(x) = x^2$ and $\Delta x = \dfrac{b - a}{n}$; $S_n = a + \dfrac{b - a}{n}$;

$S_n = \displaystyle\sum_{k=1}^n \left[a + \frac{b - a}{n}k\right]^2 \left(\frac{b - a}{n}\right)$

$$\int_a^b x^2 dx = \lim_{n \to +\infty} \frac{b-a}{n}\left[a^2 + \frac{2(b-a)}{n}ak + \frac{(b-a)^2}{n^2}k^2\right]$$

$$= \lim_{n \to +\infty} \frac{b-a}{n}\left(a^2 n + \frac{2(b-a)}{n}\frac{an(a+1)}{2}\right.$$

$$\left. + \frac{(b-a)^2}{n^2}\frac{n(n+1)(2n+1)}{6}\right)$$

$$= \left(a^2 + ab - a^2 + \frac{(b-a)^2}{3}\right)(b-a)$$

$$= \left(\frac{b-a}{3}\right)(3a^2 + 3ab - 3a^2 + b^2 - 2ab + a^2)$$

$$= \tfrac{1}{3}(b-a)(b^2 + ab + a^2) = \tfrac{1}{3}(b^3 - a^3)$$

29. The respective subintervals are $\Delta x_1 = 0.8$; $\Delta x_2 = 1.1$; $\Delta x_3 = 0.4$; $\Delta x_4 = 0.4$; $\Delta x_5 = 0.3$, The norm is the largest of these deltas, namely, 1.1.

30. $f(x) = 4 - 5x$

k:	1	2	3	4	5
x_k^*:	-0.5	0.8	1	1.3	1.8
$f(x_k^*)$:	6.5	0	-1	-2.5	-5
Δx_k:	0.8	1.1	0.4	0.4	0.3

$R_5 = \sum_1^5 f(x_k)\Delta x_k = 6.5(0.8) + 0 + (-1)(0.4)$
$+ (-2.5)(0.4) + (-5)(0.3) = 2.3$

31. $f(x) = x^3$

k:	1	2	3	4	5
x_k^*:	-1	0	1	$\frac{128}{81}$	$\frac{125}{64}$
$f(x_k^*)$:	-1	0	1	3.94616	7.4506
Δx_k:	0.8	1.1	0.4	0.4	0.3

$R_5 = \sum_1^5 f(x_k)\Delta x_k = -1(0.8) + 0 + 1(0.4) +$
$3.94616(0.4) + 7.4506(0.3) = 3.413644$

32. $\displaystyle\int_a^b f(x)\,dx = \lim_{x \to +\infty}\sum_{k=1}^n f(x_k^*)\Delta x_k$

$\displaystyle\le \sum_{k=1}^n g(x_n^*)\Delta x_k = \int_a^b g(x)\,dx$

because $f(x) \le g(x)$ implies

$f(x_k^*) \le b_k = g(x_k^*)$

33. The statement is false. Let $f(x) = x$ on

$[-1, 1]$. $\displaystyle\int_{-1}^1 x\,dx = 0$

34. The statement is false. Let $f(x) = g(x) = 1$ on $[0, 2]$. $\displaystyle\int_0^2 (1)(1)\,dx = 2$ and $\displaystyle\int_0^2 (1)\,dx = 2$.

$2 = \displaystyle\int_0^2 (1)(1)\,dx$

$\ne \left[\displaystyle\int_0^2 (1)\,dx\right]\left[\displaystyle\int_0^2 (1)\,dx\right]\left[\displaystyle\int_0^2 (1)\,dx\right] = 4$

35. a. Let $g(x) = -f(x) > 0$ on $[a, b]$. The area under $g(x) > 0$ is

$\displaystyle\int_a^b g(x)\,dx = A$ and $\displaystyle\int_a^b [-f(x)]\,dx$

$= \displaystyle\int_a^b (-1)f(x)\,dx = (-1)\int_a^b f(x)dx = -A$

b. $\displaystyle\int_a^b f(x)\,dx = \int_a^b [p(x) + n(x)]\,dx$

$= \displaystyle\int_a^b [p(x) - |n(x)|]\,dx$

$= \displaystyle\int_a^b p(x)\,dx - \int_a^b |n(x)|\,dx = P - N$

Here $p(x)$ is the function that leads to the sum of positive regions and $n(x)$ is the function that generates all the negative regions.

c. $\displaystyle\int_{-20}^1 2x\,dx = -\int_{-2}^0 2x\,dx + \int_0^1 2x\,dx$

$= 4 + 1 = 5$

4.3 The Fundamental Theorem of Calculus; Integration by Substitution, Page 285

1. $\displaystyle\int_{-10}^{10} 7\,dx = 7x\big|_{-10}^{10} = 7[10 - (-10)] = 140$

2. $\displaystyle\int_{-5}^7 (-3)\,dx = -3x\big|_{-5}^7 = -3[7 - (-5)] = -36$

3. $\displaystyle\int_{-3}^{5} (2x + a)\,dx = x^2 + ax\Big|_{-3}^{5}$

$\qquad = (5)^2 + 5a - [(-3)^2 + (-3)a] = 16 + 8a$

4. $\displaystyle\int_{-2}^{2} (b - x)\,dx = (bx - \tfrac{1}{2}x^2)\Big|_{-2}^{2}$

$\qquad = (2b - 2) - (-2b - 2) = 4b$

5. $\displaystyle\int_{-1}^{2} ax^3\,dx = \tfrac{1}{4}ax^4\Big|_{-1}^{2} = \tfrac{1}{4}a[16 - 1] = \tfrac{15}{4}a$

6. $\displaystyle\int_{-1}^{1} (x^3 + bx^2)\,dx = (\tfrac{1}{4}x^4 + \tfrac{1}{3}bx^3)\Big|_{-1}^{1}$

$\qquad = \tfrac{1}{4} + \tfrac{1}{3}b - (\tfrac{1}{4} - \tfrac{1}{3}b) = \tfrac{2}{3}b$

7. $\displaystyle\int_{1}^{2} x^{-3}\,dx = \dfrac{x^{-2}}{-2}\Big|_{1}^{2} = -\tfrac{1}{2}[\tfrac{1}{4} - 1] = \tfrac{3}{8}$

8. $\displaystyle\int_{-2}^{-1} 3x^{-2}\,dx = 3\int_{-2}^{-1} x^{-2}\,dx = -3x^{-1}\Big|_{-2}^{-1}$

$\qquad = -3(-1 - \tfrac{1}{-2}) = \tfrac{3}{2}$

9. a. $\displaystyle\int_{0}^{1} (5u^7 + \pi^2)\,du = 5(\tfrac{1}{8}u^8 + \pi^2 u)\Big|_{0}^{1} = \tfrac{5}{8} + \pi^2$

$\quad$ **b.** $\displaystyle\int_{0}^{1} (8x^7 + \sqrt{\pi}\,)\,dx = 8(\tfrac{1}{8}x^8 + \sqrt{\pi}\,x)\Big|_{0}^{1}$

$\qquad = 1 + \sqrt{\pi}$

10. $\displaystyle\int_{0}^{4} \sqrt{x}(x + 1)\,dx = \int_{0}^{4} (x^{3/2} + x^{1/2})\,dx$

$\qquad = \Big[\tfrac{2}{5}x^{5/2} + \tfrac{2}{3}x^{3/2}\Big]\Big|_{0}^{4} = \tfrac{272}{15}$

11. $\displaystyle\int_{0}^{1} (t^{3/2} - t)\,dt = \Big[\tfrac{2}{5}t^{5/2} - \tfrac{1}{2}t^2\Big]\Big|_{0}^{1} = -\tfrac{1}{10}$

12. $\displaystyle\int_{1}^{2} \dfrac{x^3 + 1}{x^2}\,dx = \int_{1}^{2} (x + x^{-2})\,dx$

$\qquad = \Big[\tfrac{1}{2}x^2 - x^{-1}\Big]\Big|_{1}^{2} = 2$

13. $\displaystyle\int_{1}^{4} \dfrac{x^2 + x + 1}{\sqrt{x}}\,dx = \int_{1}^{4} (x^{3/2} + x^{1/2} - x^{-1/2})\,dx$

$\qquad = \Big[\tfrac{2}{5}x^{5/2} + \tfrac{2}{3}x^{3/2} - 2x^{1/2}\Big]\Big|_{1}^{4} = \tfrac{226}{15}$

14. $\displaystyle\int_{-2}^{3} (\sin^2 x + \cos^2 x)\,dx = \int_{-2}^{3} dx = x\Big|_{-2}^{3} = 5$

15. $\displaystyle\int_{0}^{\pi/4} (\sec^2 x - \tan^2 x)\,dx = \int_{0}^{\pi/4} dx = x\Big|_{0}^{\pi/4} = \tfrac{\pi}{4}$

16. a. $\displaystyle\int_{0}^{4} (2t + 4)\,dt = (t^2 + 4t)\Big|_{0}^{4} = 32$

$\quad$ **b.** $\displaystyle\int_{0}^{4} (2t + 4)^{-1/2}\,dt = \int_{4}^{12} u^{-1/2}(\tfrac{1}{2}\,du)$

$\qquad = u^{1/2}\Big|_{4}^{12} = 2\sqrt{3} - 2$

17. a. $\displaystyle\int_{0}^{\pi/2} \sin\theta\,d\theta = -\cos\theta\Big|_{0}^{\pi/2} = 1$

$\quad$ **b.** $\displaystyle\int_{0}^{\pi/2} \sin 2\theta\,d\theta = \int_{0}^{\pi} \sin u\,(\tfrac{1}{2}\,du)$

$\qquad = -\tfrac{1}{2}\cos u\Big|_{0}^{\pi} = 1$

18. a. $\displaystyle\int_{0}^{\pi} \cos t\,dt = \sin t\Big|_{0}^{\pi} = 0$

$\quad$ **b.** $\displaystyle\int_{0}^{\sqrt{\pi}} t\cos t^2\,dt = \tfrac{1}{2}\int_{0}^{\pi} \cos u\,du = \tfrac{1}{2}\sin u\Big|_{0}^{\pi} = 0$

19. a. $\displaystyle\int_{0}^{4} \sqrt{x}\,dx = \tfrac{2}{3}x^{3/2}\Big|_{0}^{4} = \tfrac{16}{3}$

$\quad$ **b.** $\displaystyle\int_{-4}^{0} \sqrt{-x}\,dx = -\int_{4}^{0} u^{1/2}\,du$

$\qquad = -\tfrac{2}{3}u^{3/2}\Big|_{4}^{0} = \tfrac{16}{3}$

20. a. $\displaystyle\int_{0}^{16} \sqrt[4]{x}\,dx = \tfrac{4}{5}x^{5/4}\Big|_{0}^{16} = \tfrac{128}{5}$

$\quad$ **b.** $\displaystyle\int_{-16}^{0} \sqrt[4]{-x}\,dx = -\int_{16}^{0} u^{1/4}\,du$

$\qquad = -\tfrac{4}{5}u^{5/4}\Big|_{16}^{0} = \tfrac{128}{5}$

21. **a.** $\displaystyle\int_0^3 |5x|\, dx = 5\int_0^3 x\, dx = \frac{5}{2}x^2\Big|_0^3 = \frac{45}{2}$

  b. $\displaystyle\int_{-3}^0 |5x|\, dx = \int_{-3}^0 (-5x)\, dx$

  $\displaystyle = -\frac{5}{2}x^2\Big|_{-3}^0 = \frac{45}{2}$

  c. $\displaystyle\int_{-3}^3 |5x|\, dx = \int_{-3}^0 |5x|\, dx + \int_0^3 |5x| = 45$

22. **a.** $\displaystyle\int_0^2 |x|\, dx = \int_0^2 x\, dx = 2$

  b. $\displaystyle\int_{-2}^0 |x|\, dx = \int_{-2}^0 (-x)\, dx = 2$

  c. $\displaystyle\int_{-2}^2 |x|\, dx = \int_{-2}^0 |x|\, dx + \int_0^2 |x|\, dx = 4$

23. $\displaystyle\int (2x+3)^4\, dx = \frac{1}{2}\int u^4\, du = \frac{1}{2}\cdot\frac{1}{5}u^5 + C$

  $\displaystyle = \frac{1}{10}(2x+3)^5 + C$

24. $\displaystyle\int \sqrt{3t-5}\, dt = \frac{1}{3}\int u^{1/2}\, du = \frac{1}{3}\cdot\frac{2}{3}u^{3/2} + C$

  $\displaystyle = \frac{2}{9}(3t-5)^{3/2} + C$

25. $\displaystyle\int (x-27)^{2/3}\, dx = \frac{3}{5}(x-27)^{5/3} + C$

26. $\displaystyle\int (11-2x)^{-4/5}\, dx = -\frac{1}{2}\int u^{-4/5}\, du$

  $\displaystyle = -\frac{1}{2}(5u^{1/5}) + C = -\frac{5}{2}(11-2x)^{1/5} + C$

27. $\displaystyle\int (x^2 - \cos 3x)\, dx = \frac{1}{3}x^3 - \frac{1}{3}\int \cos u\, du$

  $\displaystyle = \frac{x^3}{3} - \frac{1}{3}\sin 3x + C$

28. $\displaystyle\int \csc^2 5t\, dt = \frac{1}{5}\int \csc^2 u\, du = -\frac{1}{5}\cot 5t + C$

29. $\displaystyle\int \sin(4-x)\, dx = -\int \sin u\, du = \cos(4-x) + C$

30. $\displaystyle\int s\sqrt{s^2+4}\, ds = \frac{1}{2}\int u^{1/2}\, du = \frac{1}{3}(s^2+4)^{3/2} + C$

31. $\displaystyle\int \sqrt{t}(t^{3/2}+5)^3\, dt = \frac{2}{3}\int u^3\, du = \frac{1}{6}(t^{3/2}+5)^4 + C$

32. $\displaystyle\int \frac{(6x-9)\, dx}{(x^2-3x+5)^3} = 3\int u^{-3}\, du$

  $\displaystyle = -\frac{3}{2}(x^2-3x+5)^{-2} + C$

33. **a.** $\displaystyle\int x(3x^2-5)^5\, dx = \frac{1}{6}\int u^5\, du$

  $\displaystyle = \frac{1}{36}(3x^2-5)^6 + C$

  b. $\displaystyle\int x(3x-5)^5\, dx = \int \frac{1}{3}(u+5)u^5\left(\frac{du}{3}\right)$

  Let $u = 3x - 5$; $du = 3\, dx$

  and $x = \frac{1}{3}(u+5)$;

  $\displaystyle = \frac{1}{9}\int (u^6 + 5u^5)\, du = \frac{1}{9}\left(\frac{1}{7}u^7 + \frac{5}{6}u^6\right) + C$

  $\displaystyle = \frac{1}{378}u^6(6u+35) + C$

  $\displaystyle = \frac{1}{378}(3x-5)^6[6(3x-5)+35] + C$

  $\displaystyle = \frac{1}{378}(3x-5)^6(18x+5) + C$

34. **a.** $\displaystyle\int x\sqrt{2x^2-5}\, dx = \frac{1}{4}\int u^{1/2}\, du + C$

  $\displaystyle = \frac{1}{6}(2x^2-5)^{3/2} + C$

  b. $\displaystyle\int x\sqrt{2x-5}\, dx = \int \frac{1}{2}(u+5)(u^{1/2})\left(\frac{du}{2}\right)$

  Let $u = 2x - 5$; $du = 2\, dx$

  and $x = \frac{1}{2}(u+5)$

  $\displaystyle = \frac{1}{4}\int (u^{3/2} + 5u^{1/2})\, du$

  $\displaystyle = \frac{1}{4}\left(\frac{2}{5}u^{5/2} + \frac{10}{3}u^{3/2}\right) + C$

  $\displaystyle = \frac{1}{30}u^{3/2}(3u+25) + C$

  $\displaystyle = \frac{1}{30}(2x-5)^{3/2}[3(2x-5)+25] + C$

  $\displaystyle = \frac{1}{15}(2x-5)^{3/2}(3x+5) + C$

35. **a.** $\displaystyle\int x\sqrt{2x^2+1}\, dx = \frac{1}{4}\int u^{1/2}\, du$

  $\displaystyle = \frac{1}{6}(2x^2+1)^{3/2} + C$

  b. $\displaystyle\int x\sqrt{2x+1}\, dx = \int \frac{1}{2}(u-1)u^{1/2}\left(\frac{du}{2}\right)$

  Let $u = 2x + 1$; $du = 2\, dx$

  and $x = \frac{1}{2}(u-1)$

  $\displaystyle = \frac{1}{4}\int (u^{3/2} - u^{1/2})\, du$

  $\displaystyle = \frac{1}{4}\left(\frac{2}{5}u^{5/2} - \frac{2}{3}u^{3/2}\right) + C$

  $\displaystyle = \frac{1}{30}u^{3/2}(3u-5) + C$

  $\displaystyle = \frac{1}{30}(2x+1)^{3/2}[3(2x+5)-5] + C$

  $\displaystyle = \frac{1}{15}(2x+1)^{3/2}(3x-1) + C$

36. a. $\int x^2(x^3 + 9)^{1/2} dx = \frac{1}{3}\int u^{1/2}\ du$

$\qquad\qquad = \frac{2}{9}(x^3 + 9)^{3/2} + C$

b. $\int x^3(x^2 + 9)^{1/2} dx = \int x(u - 9)u^{1/2}(\frac{du}{2x})$

$\qquad\qquad$ Let $u = x^2 + 9;\ du = 2x\ dx$

$\qquad\qquad$ and $x^2 = u - 9$

$\qquad = \frac{1}{2}\int (u^{3/2} - 9u^{1/2})\ du$

$\qquad = \frac{1}{2}(\frac{2}{5}u^{5/2} - \frac{18}{3}u^{3/2}) + C$

$\qquad = \frac{1}{2}u^{3/2}(u - 15) + C$

$\qquad = \frac{1}{5}(x^2 + 9)(x^2 - 6) + C$

37. a. $\int x(x^2 + 4)^{1/2} dx = \frac{1}{2}\int u^{1/2}\ du$

$\qquad = \frac{1}{2}\cdot\frac{2}{3}\ u^{3/2} + C = \frac{1}{3}(x^2 + 4)^{3/2} + C$

b. $\int x^3(x^2 + 4)^{1/2} dx = \int x(u - 4)u^{1/2}(\frac{du}{2x})$

$\qquad\qquad$ Let $u = x^2 + 4;\ du = 2x\ dx$

$\qquad\qquad$ and $x^2 = u - 4$

$\qquad = \frac{1}{2}\int (u^{3/2} - 4u^{1/2})\ du$

$\qquad = \frac{1}{2}(\frac{2}{5}u^{5/2} - \frac{8}{3}u^{3/2}) + C$

$\qquad = \frac{1}{5}(x^2 + 4)^{5/2} - \frac{4}{3}(x^2 + 4)^{3/2} + C$

$\qquad = \frac{1}{15}(x^2 + 4)^{3/2}(3x^2 - 8) + C$

38. $\int x\sin(3 + x^2)\ dx = \frac{1}{2}\int \sin u\ du$

$\qquad = -\frac{1}{2}\cos(3 + x^2) + C$

39. $\int \sin^3 t\cos t\ dt = \int u^3\ du = \frac{1}{4}u^4 + C$

$\qquad = \frac{1}{4}\sin^4 t + C$

40. $\int_{-1}^{1}(x^2 + 1)\ dx = (\frac{1}{3}x^3 + x)\Big|_{-1}^{1} = \frac{8}{3}$

41. $\int_{0}^{1}\sqrt{t}\ dt = \frac{2}{3}t^{3/2}\Big|_0^1 = \frac{2}{3}$

42. $\int_{-\pi/2}^{\pi/2}\cos x\ dx = \sin x\Big|_{-\pi/2}^{\pi/2} = 2$

43. $\int_{0}^{\pi/6}(\sin x + \cos x)\ dx = (-\cos x + \sin x)\Big|_0^{\pi/6}$

$\qquad = -\frac{\sqrt{3}}{3} + \frac{1}{2} + 1 = \frac{1}{2}(3 - \sqrt{3})$

44. $\int_{0}^{4} t\sqrt{t^2 + 9}\ dt = \frac{1}{2}\int_{9}^{25} u^{1/2}\ du = \frac{1}{3}u^{3/2}\Big|_9^{25} = \frac{98}{3}$

$\qquad\qquad$ Let $u = t^2 + 9;\ du = 2t\ dt;$

$\qquad\qquad$ if $t = 0$, then $u = 0$;

$\qquad\qquad$ if $t = 4$, then $u = 25$

45. $\int_{1/5}^{1}\frac{1}{t^2}\sqrt{5 - \frac{1}{t}}\ dt = \int_{0}^{4} u^{1/2}\ du = \frac{2}{3}u^{3/2}\Big|_0^4 = \frac{16}{3}$

$\qquad\qquad$ Let $u = 5 - t^{-1};\ du = t^{-2};$

$\qquad\qquad$ if $t = 1/5$, then $u = 0$;

$\qquad\qquad$ if $t = 1$, then $u = 4$

46. $\int_{2}^{9} x(x - 1)^{1/3}\ dx = \int_{1}^{8}(u + 1)u^{1/3}\ du$

$\qquad\qquad$ Let $u = x - 1;\ du = dx$

$\qquad\qquad$ if $x = 2$, then $u = 1$;

$\qquad\qquad$ if $x = 9$, then $u = 8$

$\qquad = \int_{1}^{8}(u^{4/3} + u^{1/3})\ du = (\frac{3}{7}u^{7/3} + \frac{3}{4}u^{4/3})\Big|_1^8$

$\qquad = \frac{1,839}{28} \approx 65.6786$

47. $\int_{2}^{3}|x|\ dx = \frac{1}{2}x^2\Big|_2^3 = 2.5$

48. Evaluating an integral can be illustrated and facilitated by taking the area under a curve. The result is a number, which corresponds to that of the area under the curve, but the units associated with the integral are not necessarily square units of area. That is, an area is an integral, but an integral is not necessarily an area.

49. $\int_{-4}^{4} x^{1/2}\ dx$ does not make sense since $x^{1/2}$ is not defined for $x = 0$

50. a. $F(x) = \int\left(\frac{1}{\sqrt{x}} - 4\right)\ dx$

$\qquad = \int (x^{-1/2} - 4)dx = 2\sqrt{x} - 4x + C$

$\qquad F(1) = 2 - 4 + C = 0$, so $C = 2$

$\qquad F(x) = 2\sqrt{x} - 4x + 2$

b.

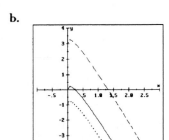

c. $G(x) = F(x) + C = 2\sqrt{x} - 4x + 2 + C$

$G'(x) = x^{-1/2} - 4 = 0$ for $x = \frac{1}{16}$

$G(\frac{1}{16}) = 2\sqrt{\frac{1}{16}} - \frac{4}{16} + 2 + C = 0$, so $C = -\frac{9}{4}$

51. Let $f(x) = \begin{cases} -x + 10 \text{ if } x \le 1 \\ x + 8 \text{ if } 1 < x < 4 \\ -x + 16 \text{ if } x \le 4 \end{cases}$

$f(1) = 9$ and $f(4) = 12$; the function is continuous, but the derivative fails to exist at $x = 1$ and $x = 4$.

52. $\displaystyle\int_0^2 f(x)\,dx = \int_0^1 f(x)\,dx + \int_1^2 f(x)\,dx$

$= \displaystyle\int_0^1 x^3\,dx + \int_1^2 x^4\,dx = \frac{1}{4}x^4\Big|_0^1 + \frac{1}{5}x^5\Big|_1^2$

$= \frac{1}{20}(5 + 128 - 4) = \frac{129}{20}$

53. $\displaystyle\int_0^\pi f(x)\,dx = \int_0^{\pi/2} f(x)\,dx + \int_{\pi/2}^\pi f(x)\,dx$

$= \displaystyle\int_0^{\pi/2} \cos x\,dx + \int_{\pi/2}^\pi x\,dx$

$= \sin x\Big|_0^{\pi/2} + \frac{1}{2}x^2\Big|_{\pi/2}^\pi = 1 - 0 + \frac{\pi^2}{2} - \frac{\pi^2}{8}$

$= \frac{1}{8}(8 + 3\pi^2) \approx 4.7$

54. a. $\displaystyle\int_{-a}^a f(x)\,dx = \int_{-a}^0 f(x)\,dx + \int_0^a f(x)\,dx$

Let $u = -x$, then $du = -dx$, and if $x = -a$, then $u = a$. Thus,

$\displaystyle\int_{-a}^a f(x)\,dx = -\int_a^0 f(-u)\,du + \int_0^a f(x)\,dx$

$= -\displaystyle\int_a^0 [-f(u)]\,du + \int_0^a f(x)\,dx$

$= \displaystyle\int_a^0 f(u)\,du + \int_0^a f(x)\,dx$

$= -\displaystyle\int_0^a f(u)\,du + \int_0^a f(x)\,dx = 0$

b. $\displaystyle\int_{-a}^a f(x)\,dx = \int_{-a}^0 f(x)\,dx + \int_0^a f(x)\,dx$

Let $u = -x$, then $du = -dx$, and if $x = -a$, then $u = a$. Thus,

$\displaystyle\int_{-a}^a f(x)\,dx = -\int_a^0 f(-u)\,du + \int_0^a f(x)\,dx$

$= -\displaystyle\int_a^0 f(u)\,du + \int_0^a f(x)\,dx$

$= \displaystyle\int_0^a f(u)\,du + \int_0^a f(x)\,dx$

$= 2\displaystyle\int_{-a}^0 f(x)\,dx$

55. $\displaystyle\int_{-\pi}^\pi \sin x\,dx = 0$ since $\sin x$ is odd.

56. $\displaystyle\int_{-\pi/2}^{\pi/2} \cos x\,dx = 2$ since $\cos x$ is even.

57. $\displaystyle\int_{-3}^3 x\sqrt{x^4 + 1}\,dx = 0$ since $x\sqrt{x^4 + 1}$ is odd.

58. $\displaystyle\int_{-1}^1 \frac{\sin x\,dx}{x^2 + 1} = 0$ since $\frac{\sin x}{x^2 + 1}$ is odd.

59. a. $\displaystyle\int_{-175}^{175} (7x^{1001} + 14x^{99})\,dx = 0$; true

b. $\displaystyle\int_0^\pi \sin^2 x\,dx = \int_0^\pi \cos^2 x\,dx$; true

c. $\displaystyle\int_{-\pi/2}^{\pi/2} \cos x\,dx = \int_{-\pi}^0 \sin x\,dx$; false

60. $dy/dx = x(x^2 - 1)^{1/3}$;

$F(x) = \displaystyle\int x(x^2 - 1)^{1/3}\,dx = \frac{1}{2}\int u^{1/3}\,du$

$$= \tfrac{1}{2} \cdot \tfrac{3}{4}(x^2 - 1)^{4/3} + C$$

$$F(3) = \tfrac{3}{8}(2)^4 + C \text{ implies } C = -5.$$

Thus,

$$F(x) = \tfrac{3}{8}(x^2 - 1)^{4/3} - 5$$

61. a. $v(t) = t^2(t^3 - 8)^{1/3} = 0$ at $t = 0, 2$;

since $t = 0$ is the starting point, it turns around at $t = 2$.

b. $v(t) = s'(t)$, so

$$s(t) = \int t^2(t^3 - 8)^{1/3}\, dt = \tfrac{1}{3}\int u^{1/3}\, du$$

$$= \tfrac{1}{3} \cdot \tfrac{3}{4}\, u^{4/3} + C = \tfrac{1}{4}\,(t^3 - 8)^{4/3} + C$$

Now when $t = 0$, $s(0) = 1$.

$$s(0) = \frac{(0 - 8)^{4/3}}{4} + C = 1, \text{ so}$$

$$C = -3;$$

$$s(t) = \tfrac{1}{4}\,(t^3 - 8)^{4/3} - 3$$

In part **a** we found that the particle turns when $t = 2$, so

$$s(2) = \tfrac{1}{4}(2^3 - 8)^{4/3} - 3 = -3 \text{ It}$$

turns at $s = -3$.

62. Water flows into the tank at the rate of

$$v'(t) = t(3t^2 + 1)^{-1/2} \text{ ft}^3/\text{s}$$

The volume at time t is

$$v(t) = \tfrac{1}{6}\int (3t^2 + 1)^{-1/2}(6t\, dt)$$

$$= \tfrac{1}{3}\sqrt{3t^2 + 1} + C$$

The tank is empty to start, so

$$v(0) = 0 = \tfrac{1}{3} + C, \text{ so } C = -\tfrac{1}{3}$$

Thus,

$$v(t) = \tfrac{1}{3}(\sqrt{3t^2 + 1} - 1)$$

$$v(4) = \tfrac{1}{3}(\sqrt{49} - 1) = 2 \text{ ft}^3$$

The height h is given by the equation

$$100h = 2, \text{ so } h = \tfrac{1}{50} \text{ ft or } \tfrac{12}{50} = 0.24 \text{ in.}$$

63. $\displaystyle\int [(x^2 - 1)(x + 1)]^{-2/3}\, dx$

$$= \int [(x - 1)(x + 1)^2]^{-2/3}\, dx$$

Let $u = x + 1$; $du = dx$

$$= \int [(u - 2)u^2]^{-2/3}\, du$$

$$= \int [(1 - 2u^{-1})u^3]^{-2/3}\, du$$

$$= \int (1 - 2u^{-1})^{-2/3} u^{-2}\, du$$

Let $t = 1 - 2u^{-1}$; $dt = 2u^{-2}\, du$

$$= \int t^{-2/3}(\tfrac{dt}{2}) = \tfrac{1}{2} 3 t^{1/3} + C$$

$$= \tfrac{3}{2}(1 - 2u^{-1})^{1/3} + C$$

$$= \tfrac{3}{2}(u - 2)^{1/3} u^{-1/3} + C$$

$$= \tfrac{3}{2}[1 - 2(x + 1)^{-1}]^{1/3} + C$$

$$= \tfrac{3}{2}(x - 1)^{1/3}(x + 1)^{-1/3} + C$$

$$= \tfrac{3}{2}\left(\tfrac{x - 1}{x + 1}\right)^{1/3} + C$$

64. a. For our purpose with these counterexamples, we let all the constants of integration be zero.

$$\int x\sqrt{x}\, dx = \int x^{3/2}\, dx = \tfrac{2}{5}x^{5/2}$$

$$\int x\, dx = \tfrac{1}{2}x^2; \quad \int \sqrt{x}\, dx = \tfrac{2}{3}x^{3/2}$$

Since $\tfrac{2}{5}x^{5/2} \neq \left(\tfrac{1}{2}x^2\right)\left(\tfrac{2}{3}x^{3/2}\right)$

the result follows.

b. $\displaystyle\int \frac{\sqrt{x}}{x}\, dx = \int x^{-1/2}\, dx = 2\sqrt{x}$

$$\int \sqrt{x}\, dx = \tfrac{2}{3}x^{3/2}; \quad \int x\, dx = \tfrac{1}{2}x^2$$

Since $2\sqrt{x} \neq \dfrac{\tfrac{2}{3}x^{3/2}}{\tfrac{1}{2}x^2}$, the result follows.

65. a. $\displaystyle\int_a^b f(x)\, dx = \int_a^b f'(x)\, dx = f(x)\Big|_a^b$

$$= f(b) - f(a).$$

b. $\displaystyle\int_a^b [f(x)]^2 dx = \int_a^b f(x)f'(x)\, dx$

$$= \tfrac{1}{2}[f(x)]^2 \Big|_a^b$$

$$= \tfrac{1}{2}\{[f(b)]^2 - [f(a)]^2\}.$$

4.4 Introduction to Differential Equations, Page 293

1. $x^2 + y^2 = 7$

$$2x + 2y \frac{dy}{dx} = 0$$

$$\frac{dy}{dx} = -\frac{x}{y}$$

2. $5x^2 - 2y^2 = 3$

$$10x - 4y \frac{dy}{dx} = 0$$

$$\frac{dy}{dx} = \frac{5x}{2y}$$

3. $xy = C$

$$x \frac{dy}{dx} + y = 0$$

$$\frac{dy}{dx} = -\frac{y}{x}$$

4. $x^2 - 3xy + y^2 = 5$

$$2x\,dx - 3x\,dy - 3y\,dx + 2y\,dy = 0$$

$$(2y - 3x)dy + (2x - 3y)dx = 0$$

5. $\frac{dy}{dx} = A\cos(Ax + B);\ \frac{d^2y}{dx^2} = -A^2\sin(Ax + B)$

Thus,

$$\frac{d^2y}{dx^2} + A^2 y = -A^2\sin(Ax + B) + A^2\sin(Ax + B)$$

$$= 0$$

6. $\frac{dy}{dx} = \frac{x^3}{5} + \frac{A}{x^2};\ \frac{d^2y}{dx^2} = \frac{3x^2}{5} - \frac{2A}{x^3}$

Thus,

$$x \frac{d^2y}{dx^2} + 2\frac{dy}{dx} = x\left(\frac{3x^2}{5} - \frac{2A}{x^3}\right) + 2\left(\frac{x^3}{5} + \frac{A}{x^2}\right)$$

$$= \frac{3x^3}{5} - \frac{2A}{x^2} + \frac{2x^3}{5} + \frac{2A}{x^2} = x^3$$

7. $\frac{dy}{dx} = -\frac{x}{y}$

$$y\,dy = -x\,dx$$

$$\int y\,dy = -\int x\,dx$$

$$\frac{y^2}{2} = -\frac{x^2}{2} + C_1$$

$$x^2 + y^2 = C$$

(Since C_1 is any constant, let $2C_1 = C$.)

8. $\frac{dx}{dt} = x^2 \sqrt{t}$

$$x^{-2}\,dx = t^{1/2}\,dt$$

$$\int x^{-2}\,dx = \int t^{1/2}\,dt$$

$$-x^{-1} = \tfrac{2}{3}t^{3/2} + C_1$$

$$x = -3(2t^{3/2} + C)^{-1}$$

9. $\frac{dy}{dx} = \sqrt{\frac{x}{y}}$

$$y^{1/2}\,dy = x^{1/2}\,dx$$

$$\int y^{1/2}\,dy = \int x^{1/2}\,dx$$

$$\tfrac{2}{3}y^{3/2} = \tfrac{2}{3}x^{3/2} + C_1$$

$$x^{3/2} - y^{3/2} = C$$

10. $\frac{dy}{dx} = \sqrt{\frac{y}{x}}$

$$y^{-1/2}\,dy = x^{-1/2}\,dx$$

$$2y^{1/2} = 2x^{1/2} + C_1$$

$$\sqrt{x} - \sqrt{y} = C$$

11. $\frac{dy}{dx} = \frac{x}{y}\sqrt{1 - x^2}$

$$y\,dy = x\sqrt{1 - x^2}\,dx$$

$$\int y\,dy = \int x\sqrt{1 - x^2}\,dx$$

$$\int y\,dy = \int u^{1/2}\,du$$

Let $u = (1 - x^2);\ du = -2x\,dx$

$$\frac{y^2}{2} = -\tfrac{1}{2}\frac{(1 - x^2)^{3/2}}{\tfrac{3}{2}} + C_1$$

$$\frac{(1 - x^2)^{3/2}}{3} + \frac{y^2}{2} = C_1$$

$$2(1 - x^2)^{3/2} + 3y^2 = C$$

12. $\frac{dy}{dx} = (y - 4)^2$

$$(y - 4)^{-2}\,dy = dx$$

$$\int (y - 4)^{-2}\,dy = \int dx$$

$$(y - 4)^{-1} = x + C_1$$

$$x - (y - 4)^{-1} = C$$

13.
$$xy\ dx + \sqrt{xy}\ dy = 0$$
$$\sqrt{xy}\ dx = -dy \quad (\text{if } xy \neq 0)$$
$$x^{1/2}\ dx = -y^{-1/2}\ dy$$
$$\int x^{1/2}\ dx = -\int y^{-1/2}\ dy$$
$$\tfrac{2}{3}x^{3/2} = -2y^{1/2} + C_1$$
$$x^{3/2} + 3y^{1/2} = C$$

14.
$$4y^3\ dz + 5z^2\ dy = 0$$
$$4z^{-2}dz = -5y^{-3}dy$$
$$\int 4z^{-2}\ dz = -\int 5y^{-3}\ dy$$
$$-4z^{-1} = -\tfrac{5}{-2}y^{-2} + C_1$$
$$5y^{-2} - 8z^{-1} = C$$

15.
$$\frac{dy}{dx} = \frac{\sin x}{\cos y}$$
$$\cos y\ dy = \sin x\ dx$$
$$\int \cos y\ dy = \int \sin x\ dx$$
$$\sin y = -\cos x + C$$
$$\cos x + \sin y = C$$

16.
$$x^2\ dy + \sec y\ dx = 0$$
$$\int \cos y\ dy = \int x^{-2}\ dx$$
$$\sin y = \frac{x^{-1}}{-1} + C$$
$$\sin y = C - \frac{1}{x}$$

17. $x\ dy + y\ dx = 0$; $d(xy) = 0$, so $xy = C$

18. $\dfrac{x\ dy - y\ dx}{x^2} = 0$; $\dfrac{d}{dx}\left(\dfrac{y}{x}\right) = 0$, so $\dfrac{y}{x} = C$ or $y = Cx$

19. Write $y\ dx = x\ dy$ as $\dfrac{x\ dy - y\ dx}{x^2} = 0$;
$$\frac{d}{dx}\left(\frac{y}{x}\right) = 0, \text{ so } \frac{y}{x} = C \text{ or } y = Cx$$

20.
$$x^2y\ dy + xy^2\ dx = 0$$
$$x^2(2y\ dy) + y^2(2x\ dx) = 0$$
$$x^2\ d(y^2) + y^2\ d(x^2) = 0$$

$$d(x^2y^2) = 0$$
$$x^2y^2 = C$$

21. Let Q denote the number of bacteria. Then, dQ/dt is the rate of change of Q, and since this rate of change is proportional to Q, it follows that
$$\frac{dQ}{dt} = kQ$$
where k is a positive constant of proportionality.

22. Let $Q(t)$ be the amount of radium present at time t and k a positive proportionality constant (the negative sign indicates decay). Then,
$$\frac{d}{dt}Q(t) = -kQ(t)$$

23. Let T be temperature, t be time, T_n be temperature of the surrounding medium, and c be the constant of proportionality. Then:
$$\frac{dT}{dt} = c(T - T_n)$$

24. Let Q denote the number of facts recalled and N the total number of relevant facts in the person's memory. Then dQ/dt is the rate of change of Q, and $(N - Q)$ is the number of relevant facts not recalled. Since the rate of change is proportional to $N - Q$, it follows that
$$\frac{dQ}{dt} = k(N - Q)$$
where k is a positive constant of proportionality.

25. Let t denote time, Q the number of residents who have been infected, and B the total number of susceptible residents. The differentiable equation describing the spread of the epidemic is
$$\frac{dQ}{dt} = kQ(B - Q)$$
where k is the positive constant of proportionality.

26. Let Q denote the number of people implicated and N the total number of people not to be implicated. Then dQ/dt is the rate of change of Q, and $N - Q$ is the number involved but not implicated. Since the rate of change is jointly proportional to Q and $N - Q$, it follows that
$$\frac{dQ}{dt} = kQ(N - Q)$$

where k is the positive constant of proportionality.

27. Family of curves: $2x - 3y = C$; differentiating with respect to x leads to the slope of the tangent lines $dy/dx = 2/3$. For the orthogonal trajectories, the slope is the negative reciprocal, or $dY/dX = -3/2$. Integrating leads to the orthogonal trajectories: $2Y + 3X = K$

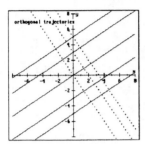

28. Family of curves $y = x + C$; differentiating with respect to x leads to the slope of the tangent lines $dy/dx = 1$. For the orthogonal trajectories, the slope is the negative reciprocal, or $dY/dX = -1$. Integrating leads to $Y = X^{-1} + K_1$ or

$Y = X^{-1} + K.$

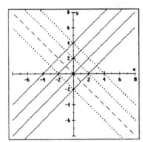

29. Family of curves: $y = x^3 + C$; differentiating with respect to x leads to the slope of the tangent lines $dy/dx = 3x^2$. For the orthogonal trajectories, the slope is the negative reciprocal, or $dY/dX = -X^{-2}/3$. Integrating leads to the orthogonal trajectories: $Y = \frac{1}{3}X^{-1} + K$

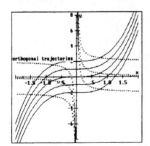

30. Family of curves: $y = x^4 + C$; differentiating with respect to x leads to the slope of the tangent lines $dy/dx = 4x^3$. For the orthogonal trajectories, the slope is the negative reciprocal, or $dY/dX = -X^{-3}/4$.

Integrating leads to $4Y = -\frac{1}{2}X^{-2} + K_1$ or $Y = -\frac{1}{8}X^{-2} + K.$

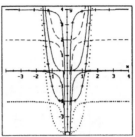

31. Family of curves: $xy^2 = C$; differentiating with respect to x leads to the slope of tangent lines to $dy/dx = -y/2x$. For the orthogonal trajectories, the slope is the negative reciprocal, or $dY/dX = 2X/Y$. Integrating leads to the orthogonal trajectories:

$$Y^2 - 2X^2 = K_1, \text{ or } X^2 - \frac{Y^2}{2} = K$$

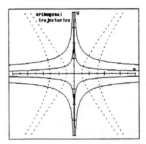

32. Conjecture: the orthogonal trajectories are circles. Family of curves: $x^2 - y^2 = C$; differentiating with respect to x leads to the slope of the tangent lines $dy/dx = x/y$. For the orthogonal trajectories, the slope is the negative reciprocal of $dY/dX = -Y/X$. Integrating leads to $Y^2 + X^2 = K$

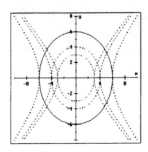

33. a.

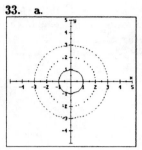

b.

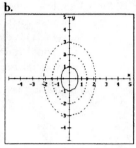

c.

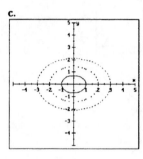

d.

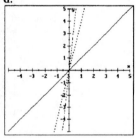

e.

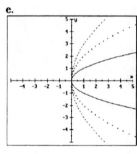

f.

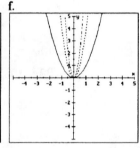

a and *d* are orthogonal trajectories;
b and *e* are orthogonal trajectories;
c and *f* are orthogonal trajectories.

34. Let V be the volume in ft^3 and h the height
at time time t. Then $V = 9\pi h$,

$$\frac{dV}{dt} = 9\pi\frac{dh}{dt}$$

The hole has area $A_0 = \pi\left(\frac{2.0}{12}\right)^2 = \frac{1}{36}\pi$ ft^2
By Toricelli's law

$$\frac{dV}{dt} = -4.8\left(\frac{1}{36}\pi\right)\sqrt{h}$$

Thus,

$$9\pi\frac{dh}{dt} = -4.8\left(\frac{1}{36}\pi\right)$$

$$\frac{dh}{dt} = -\frac{2}{135}\sqrt{h}$$

$$\int h^{-1/2}\,dh = -\int\frac{2}{135}\,dt$$

$$2\sqrt{h} = -\frac{2}{135}t + C$$

When $t = 0$, $h = 5$, so

$$2\sqrt{5} = C$$

Evaluate $2\sqrt{h} = -\frac{2}{135}t + 2\sqrt{5}$

When $h = 0$, $t = 135\sqrt{5} \approx 302$ sec ≈ 5 min.

35. By Toricelli's Law, $\dfrac{dV}{dt} = -4.8A_0\sqrt{h}$.

Given a square hole with $s = 1.5$ in. $= \frac{1}{8}$ ft,
$A_0 = \left(\frac{1}{8}\right)^2 = \frac{1}{64}$. So $\dfrac{dV}{dt} = -4.8\left(\frac{1}{64}\right)\sqrt{h}$.
$\dfrac{dV}{dt} = 9\,\pi\,\dfrac{dh}{dt}$. Substituting, we have:

$$-(4.8)\frac{\sqrt{h}}{64} = 9\pi\frac{dh}{dt}$$

$$\int -\frac{4.8}{9\pi(64)}\,dt = \int h^{-1/2}\,dh$$

$$-\frac{1}{120\pi}\,t = 2h^{1/2} + C$$

$$t = -240\pi\sqrt{h} + C$$

When the tank is full, $t = 0$ and $h = 5$.

$$0 = -240\pi\sqrt{5} + C, \quad C = 240\pi\sqrt{5}$$

We have the formula relating time and the
height of the water:

$$t = -240\pi\sqrt{h} + 240\pi\sqrt{5}$$

The tank will be empty when $h = 0$. The
time required for this is: $t = 240\pi\sqrt{5}$ min
≈ 28 min.

36. a. By Toricelli's law, $dV/dt = -4.8A_0\sqrt{h}$.
$A_0 = \left(\frac{2}{12}\right)^2 = \frac{1}{36}$; $V = 4h$, so

$$\frac{dV}{dt} = 4\left(\frac{dh}{dt}\right)$$

Thus, $4\dfrac{dh}{dt} = -4.8A_0\sqrt{h}$.

b.

$$\frac{dh}{dt} = -\frac{1.2}{36}\sqrt{h}$$

$$\int h^{-1/2}\,dh = -\int\frac{1}{30}\,dt$$

$$2\sqrt{h} = -\frac{1}{30}t + C$$

Since $h(0) = 4$, $C = 4$, so

$$\sqrt{h} = -\frac{1}{60}t + 2$$

so that $h = 0$ when $t = 120$ sec or 2 min.

37. a. $v^2 = \dfrac{2gR^2}{s} + v_0^2 - 2gR$

$$= \frac{2(32)(3{,}956)^2(5{,}280)^2}{(3{,}956)(5{,}280) + 200}$$

$$+ 150^2 - 2(32)(3{,}956)(5{,}280)$$

$$= 9{,}700 \text{ or } v \approx 98.5 \text{ ft/s}$$

b. The velocity at the maximum height is 0. Then, $R/s = 1 - v_0^2/(2gR)$ ≈ 0.9999983168; thus, $s \approx 3,956.067$ and $h \approx 0.067$ for a height of 351.56 ft.

38. a. $v_0 = \sqrt{2gR} = \sqrt{2(5.5)(1,080)(5,280)}$
$\approx 7,920$ ft/s

b. $v_0 = \sqrt{2gR} = \sqrt{2(12)(2,050)(5,280)}$
$\approx 16,118$ ft/s

c. $v_0 = \sqrt{2gR} = \sqrt{2(28)(3,800)(5,280)}$
$\approx 33,520$ ft/s

39. $\dfrac{dP}{dt} = 1500\, t^{-1/2}$ so
$$P(t) = \int 1,500\, t^{-1/2} dt \quad P$$
$$= 3000\, \sqrt{t} + C.$$

1994 is four years after 1990, so

$P(4) = 39,000$, and $39,000 = 3000\sqrt{4} + C$,

or $C = 33,000$.

$P(t) = 3,000\sqrt{t} + 33,000$

a. $P(0) = 33,000$

b. $P(9) = 3,000\sqrt{9} + 33,000 = 42,000$

40. Let $P(t)$ be the price of chicken in cents t weeks from now.

$$\frac{dP}{dt} = (3t + 1)^{1/2}$$

$$P(t) = \int (3t + 1)^{1/2}\, dt = \frac{2}{9}(3t + 1)^{3/2} + C$$

$$P(0) = 300 = \frac{2}{9} + C, \text{ so } C = 300 - \frac{2}{9}$$

In 8 weeks, $P(8) = \frac{2}{9}(25)^{3/2} + 300 - \frac{2}{9}$

≈ 327.56 or \$3.28

41. Let $C(t)$ be the crop in bushels in t years.

$$\frac{dC}{dt} = 0.3t^2 + 0.6t + 1$$

$$C(t) = \int (0.3t^2 + 0.6t + 1)\, dt$$

$$= 0.1t^3 + 0.3t^2 + t + K$$

Today $(t = 0)$, the values of the crop is \$3K. In 20 days,

$3C(20) = 3[.1(20)^3 + 0.3(20)^2 + 20 + K]$

$= 2,840 + 3K$

The increase in value is \$2,820.

42. Let R_x and R_y be the radii of planets X and

Y, respectively. With $g_x = (8/9)g_y$, and $v_{ex} = 6$, we have

$$\frac{v_{ex}^2}{v_{ey}^2} = \frac{2g_x R_x}{2g_y R_y} = \frac{2}{9} \text{ which leads to}$$

$$v_{ey} = 9\sqrt{2} \text{ ft/s.}$$

43.
$$\frac{dP}{dt} = k\sqrt{P}$$

$$\int P^{-1/2}\, dP = \int k\, dt.$$

$$2\sqrt{P} = kt + C$$

$P(0) = 9,000$, so $C = 60\sqrt{10}$, and $P(-10) = 4,000$, so

$$2\sqrt{4,000} = -10k + 60\sqrt{10}$$

$$k = 2\sqrt{10}$$

We now have the equation:

$$2\sqrt{P} = 2\sqrt{10}t + 60\sqrt{10}$$

$$\sqrt{P} = \sqrt{10}t + 30\sqrt{10}$$

To find t for $P = 16,000$,

$$\sqrt{16,000} = \sqrt{10}t + 30\sqrt{10}$$

$$t = 10 \text{ years from now}$$

44. The deceleration is $a(t) = -28$ ft/s^2. The velocity $v(t)$ is an antiderivative of the deceleration, so $v(t) = -28t + C_1$, where C_1 is the initial velocity (when $t = 0$). Now, 60 mi/hr = 88 ft/s. Thus, $v(t) = -28t + 88$. The car will stop when the velocity is 0, or when $t = 88/28 \approx 3.1429$ sec. The distance traveled, $s(t)$, is an antiderivative of the velocity, so

$$s(t) = -\frac{28}{2}t^2 + 88t + C_2$$

Let the point at which the brakes are applied correspond to $s(0) = 0$, so that $C_2 = 0$. Then the braking distance is

$$s(3.1429) = -\frac{28}{2}(3.1429)^2 + 88(3.1429)$$

$$\approx 138.54 \text{ ft}$$

45. $\dfrac{dv}{dt} = -28;\ \dfrac{ds}{dt} = v = -28t + 88$ so that
$$s = -14t^2 + 88t$$

where s is the distance from the point where the brakes are applied. Before the brakes are applied, he travels

$$s_1 = (88)(0.7) = 61.6 \text{ ft}$$

After the brakes are applied, he travels until $v = 0$; that is $t = 88/28 = 22/7$ sec. The

distance traveled after the brakes are applied is

$$s_2 = -14(\tfrac{22}{7})^2 + 88(\tfrac{22}{7}) \approx 138.3$$

Thus, the total stopping distance is

$$s_1 + s_2 = 61.6 + 138.3 = 199.9 \text{ ft}$$

The camel is toast! However, if you want to have some fun with this solution, you might argue that if the camel is standing so that the car is positioned between the camel's front and rear legs, and if the hood of the car is more than 0.89 ft $\approx$ 10.7 in. in length, the camel will escape undamaged. Here, of course, it is assumed that the camel's stomach is above the car's hood.

46. **a.** Let $Q(t)$ be the amount of radioactive substance present at time t. Then,

$$\frac{dQ}{dt} = kQ^2$$

$$\int Q^{-2} dQ = \int k \, dt$$

$$-Q^{-1} = kt + C$$

Since $Q_0 = 100$, $C = -\tfrac{1}{100}$ and

$$-Q^{-1} = kt - \tfrac{1}{100}$$

If $t = 1$, $Q = 80$, and $k = -\tfrac{1}{400}$, so that

$$Q^{-1} = \tfrac{1}{400}t + \tfrac{1}{100}$$

In 6 days,

$$Q^{-1} = \tfrac{1}{400}(6) + \tfrac{1}{100}$$

$$Q = 40$$

that is, 40 g.

b. $\tfrac{1}{10} = \tfrac{1}{400} t + \tfrac{1}{100}$ or $t = 36$ days

47. 50 cm on a side of a square means $A = 2500$, and $k = 0.0025$

$$\frac{dQ}{dt} = -0.0025(2500)\frac{dT}{ds}$$

Since dQ/dt is a constant,

$$\frac{dQ}{dt}\int ds = -6.25 \int dT$$

$$\frac{dQ}{dt}s = -6.25T + C$$

When $s = 0$, $T = 60$,

$$\frac{dQ}{dt}(0) = -6.25(60) + C$$

$C = 375$. When $s = 2$, $T = 5$,

$$\frac{dQ}{dt}(2) = -6.25(5) + 375$$

$$\frac{dQ}{dt} \approx 171.875 \text{ calories/s}$$

48. Differentiating $V = 9\pi h^3$ leads to

$$\frac{dV}{dt} = 9\pi\left(3h^2 \frac{dh}{dt}\right) = -4.8A_0\sqrt{h}$$

from Toricelli's law; converting to ft,

$$27\pi h^2 \frac{dh}{dt} = (-4.8)(\tfrac{\pi}{144})\sqrt{h}$$

$$\int h^{3/2} \, dh = -\int \frac{4.8\pi}{144(27\pi)} \, dt$$

$$\tfrac{2}{5}h^{5/2} = -\tfrac{1}{810}t + C$$

If $t = 0$, then $h = 4$, so that $C = \tfrac{64}{5}$. The height is zero when

$$0 = -\tfrac{1}{810}t + \tfrac{64}{5}$$

$$10,360 = t$$

This is about 173 min or 2 hr and 53 min.

49. **a.** The volume of water left after t seconds, to a depth of h_0 is $V = 16h$ ft^3. According to Torcelli's law,

$$16\frac{dh}{dt} = -\frac{4.8}{144}\sqrt{h}$$

$$\frac{dh}{dt} = -\frac{4.8}{2,304}\sqrt{h}$$

$$= -\frac{1}{480}\sqrt{h}$$

b.

$$\int \frac{dh}{\sqrt{h}} = -\int \frac{dt}{480}$$

$$2\sqrt{h} = -\frac{1}{480}t + C$$

At $h = 0$,

$$\frac{1}{480}t = 2\sqrt{6}$$

$$t = 960\sqrt{6} \approx 2,353 \text{ sec or 39 min}$$

50. **a.** With $v^2 = 2gR\left(\frac{R}{s} - 1\right) + v_0^2$ and (for maximum height) $v = 0$ if

$$\frac{R}{s} - 1 = -\frac{v_0^2}{2gR}$$

$$s = \frac{2gR^2}{2gR - v_0^2}$$

The maximum height above the ground is

$$h = s - R = \frac{v_0^2 R}{2gR - v_0^2}$$

b.
$$R + 450 = 4R\left[(2)\left(\frac{25R}{5,280}\right) - 4\right]$$
$$(R + 450)(\tfrac{5}{528}R - 4) = 4R$$
$$0.00947R^2 - 3.7385R - 1,800 = 0$$

The positive solution of the equation is $R \approx 676$ mi.

51. $F = ma = m\left(\dfrac{k}{s^2}\right)$; if $a = -g$ and $s = R$, then $a = -\dfrac{k}{s^2}$ so that $-g = -\dfrac{k}{R^2}$ or $k = gR^2$. This means that $a = -gR^2/s^2$.

4.5 The Mean Value Theorem for Integrals; Average Value, Page 301

1. $f(x)$ is continuous on $[1, 2]$, so the MVT guarantees the existence of a c such that:
$$\int_1^2 4x^3\, dx = f(c)(2 - 1)$$
$$15 = f(c)$$
$$15 = 4c^3$$
$$c^3 = \frac{15}{4}$$
$$c = \frac{\sqrt[3]{30}}{2} \approx 1.55$$

2. $$\int_0^2 (x^2 + 4x + 1)\, dx = f(c)(2 - 0)$$
$$(\tfrac{1}{3}x^3 + 2x^2 + x)\Big|_0^2 = (c^2 + 4c + 1)(2)$$
$$\frac{38}{6} = c^2 + 4c + 1$$
$$3c^2 + 12c - 16 = 0$$
$$c \approx 1.055, \; -5.055$$

1.055 is in the interval.

3. $$\int_1^5 15x^{-2}\, dx = f(c)(5 - 1)$$
$$-15x^{-1}\Big|_1^5 = 15c^{-2}(4)$$
$$12 = 60c^{-2}$$
$$c^2 = 5$$
$$c = \pm\sqrt{5}$$

$\sqrt{5} \approx 2.24$ is in the interval.

4. The mean value theorem does not apply because $f(x) = 12x^{-3}$ is not continuous at $x = 0$.

5. The mean value theorem does not apply because $f(x) = \csc x$ is discontinuous at $x = 0$.

6. $$\int_{-\pi/2}^{\pi/2} \cos x\, dx = f(c)(\tfrac{\pi}{2} + \tfrac{\pi}{2})$$
$$\sin x\Big|_{-\pi/2}^{\pi/2} = \pi \cos c$$
$$2 = \pi \cos c$$
$$c \approx \pm 0.8807$$

7. By the second FTC, $F'(x) = 4x + 9$;
$$F(x) = \int_0^x (4t + 9)\; dt = (2t^2 + 9t)\Big|_0^x$$
$$= 2x^2 + 9x + C; \; F'(x) = 4x + 9$$

8. By the second FTC, $F'(x) = 3x^2 - 2x + 5$
$$F(x) = \int_0^x (3t^2 - 2t + 5)\; dt$$
$$= (t^3 - t^2 + 5t)\Big|_0^x$$
$$= x^3 - x^2 + 5x + C$$
$$F'(x) = 3x^2 - 2x + 5$$

9. By the second FTC, $F'(x) = x^{1/4} - 4$
$$F(x) = \int_4^x (t^{1/4} - 4)\; dt$$
$$= \frac{4t^{5/4}}{5} - 4t\Big|_4^x$$
$$= \frac{4x^{5/4}}{5} - 4x - \frac{4^{5/4}}{5} + 16$$
$$= \tfrac{4}{5}x^{5/4} - 4x - \frac{16\sqrt{2}}{5} + 16$$
$$F'(x) = x^{1/4} - 4$$

10. By the second FTC, $F'(x) = \dfrac{x}{\sqrt{1 + 3x^2}}$
$$F(x) = \int_1^x \frac{x}{\sqrt{1 + 3x^2}}\; dx$$
$$= \tfrac{1}{6}\int_1^x (1 + 3t^2)^{-1/2}(6t\; dt)$$
$$= \tfrac{1}{3}\sqrt{1 + 3t^2}\Big|_1^x$$
$$= \tfrac{1}{3}(\sqrt{1 + 3x^2} - 2)$$
$$F'(x) = \tfrac{1}{3} \cdot \tfrac{1}{2} \frac{6x}{\sqrt{1 + 3x^2}} = \frac{x}{\sqrt{1 + 3x^2}}$$

11. By the second FTC, $F'(x) = \dfrac{1}{(1 - 3x)^3}$

$$F(x) = \int_1^x \frac{1}{(1 - 3t)^3}\, dt$$

$$= -\frac{1}{3}\int_1^x (1 - 3t)^{-3}(-3\; dt)$$

$$= -\frac{1}{3}\left(-\frac{1}{2}\right)(1 - 3t)^{-2}\bigg|_1^x$$

$$= \frac{1}{6}(1 - 3x)^{-2} - \frac{1}{24}$$

$$\mathrm{F}'(x) = \frac{1}{6}(-2)(1 - 3x)^{-3}(-3) + 0$$

$$= \frac{1}{(1 - 3x)^3}$$

12. By the second FTC, $F'(x) = \sec x \tan x$

$$F(x) = \int_{\pi/3}^x \sec t \tan t\, dt = \sec t \big|_{\pi/3}^x$$

$$= \sec x - 2;\ \mathrm{F}'(x) = \sec x \tan x$$

13. $A = \displaystyle\int_0^{10} \frac{x}{2}\, dx = \frac{x^2}{4}\bigg|_0^{10} = 25$

$$f(c)(b - a) = \left(\frac{c}{2}\right)(10) = 25$$

so $c = 5$ and $f(5) = 2.5$

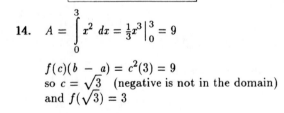

14. $A = \displaystyle\int_0^3 x^2\, dx = \frac{1}{3}x^3\bigg|_0^3 = 9$

$$f(c)(b - a) = c^2(3) = 9$$
so $c = \sqrt{3}$ (negative is not in the domain)
and $f(\sqrt{3}) = 3$

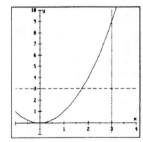

15. $\displaystyle\int_0^2 (x^2 + 2x + 3)\, dx = \left(\frac{1}{3}x^3 + x^2 + 3x\right)\bigg|_0^2 = \frac{38}{3}$

$$f(c)(b - a) = (c^2 + 2c + 3)(2) = \frac{38}{3}$$
so $3c^2 + 6c - 10 = 0$;
 $c \approx 1.08$ (negative is not in the domain)
$f(1.08) \approx 6.33$

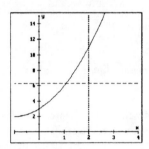

16. $A = \displaystyle\int_{0.5}^2 x^{-2}\, dx = -x^{-1}\bigg|_{0.5}^2 = 1.5$

$$f(c)(b - a) = c^{-2}(1.5) = 1.5$$
so $c = 1$ (negative is not in the domain)
and $f(1) = 1$

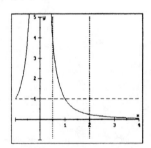

17. $A = \displaystyle\int_{-1}^{1.5} \cos x\, dx = \sin x \bigg|_{-1}^{1.5}$

$$= \sin 1.5 - \sin(-1) \approx 1.839$$

$$f(c)(b - a) = \cos c(2.5) \approx 1.83897$$
so $c \approx 0.744264$ or -0.744264 on $[-1, 1.5]$.
$f(c) \approx 0.735586$

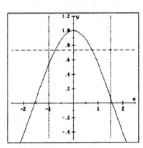

18. $\displaystyle\int_{0.5}^2 (x + \sin x)\, dx = \left(\frac{1}{2}x^2 - \cos x\right)\bigg|_{0.5}^2 \approx 3.1687$

$f(c)(b - a) = (c + \sin c)(1.5) = 3.1687$
so $c \approx 1.1855$; $f(1.1855) \approx 2.1122$

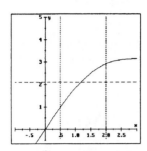

19. $\dfrac{1}{2 - (-1)} \displaystyle\int_{-1}^{2} (x^2 - x + 1)\,dx$

$= \dfrac{1}{3}(\dfrac{1}{3}x^3 - \dfrac{1}{2}x^2 + x)\big|_{-1}^{2} = \dfrac{3}{2}$

20. $\dfrac{1}{1 - (-2)} \displaystyle\int_{-2}^{1} (x^3 - 3x^2)\,dx$

$= \dfrac{1}{3}(\dfrac{1}{4}x^4 - x^3)\big|_{-2}^{1} = \dfrac{13}{12}$

21. $\dfrac{1}{\frac{\pi}{4} - 0} \displaystyle\int_{0}^{\pi/4} \sin x\,dx = -\dfrac{4}{\pi}\cos x\big|_{0}^{\pi/4}$

$= -\dfrac{4}{\pi}(\cos \dfrac{\pi}{4} - 1) \approx 0.3729$

22. $\dfrac{1}{\frac{\pi}{2} - 0} \displaystyle\int_{0}^{\pi/2} (2 \sin x - \cos x)\,dx$

$= \dfrac{2}{\pi}(-2\cos x - \sin x)\big|_{0}^{\pi/2} = \dfrac{2}{\pi} \approx 0.6366$

23. $\dfrac{1}{4 - 0} \displaystyle\int_{0}^{4} \sqrt{4 - x}\,dx = \dfrac{1}{4}(-\dfrac{2}{3})(4 - x)^{3/2}\big|_{0}^{4} = \dfrac{4}{3}$

24. $\dfrac{1}{-7 - 0} \displaystyle\int_{-7}^{0} (1 - x)^{1/3}\,dx$

$= \dfrac{1}{7}\cdot\dfrac{3}{4}(1 - x)^{4/3}\big|_{-7}^{0} = \dfrac{45}{28} \approx 1.6071$

25. $\dfrac{1}{1 - 0} \displaystyle\int_{0}^{1} (2x - 3)^3\,dx = \dfrac{1}{2}\dfrac{(2x - 3)^4}{4}\big|_{0}^{1} = -10$

26. $\dfrac{1}{1 - 0} \displaystyle\int_{0}^{1} x(2x^2 + 7)^{1/2}\,dx = \dfrac{1}{4}\dfrac{2(2x^2 + 7)^{3/2}}{3}\big|_{0}^{1}$

$= \dfrac{1}{6}(27 - 7^{3/2}) \approx 1.4133$

27. $\dfrac{1}{1 - (-2)} \displaystyle\int_{0}^{1} (x^2 + 1)^3\,dx = \dfrac{1}{6}\dfrac{(x^2 + 1)^4}{4}\big|_{-2}^{1}$

$= -\dfrac{609}{24} = -25.375$

28. $\dfrac{1}{3 - 0} \displaystyle\int_{0}^{3} \dfrac{x}{\sqrt{x^2 + 1}}\,dx = \dfrac{1}{6}\displaystyle\int_{0}^{3} (x^2 + 1)^{-1/2}(2x\,dx)$

$= \dfrac{2}{6}(x^2 + 1)^{1/2}\big|_{0}^{3} = \dfrac{1}{3}(\sqrt{10} - 1) \approx 0.7208$

29. $\dfrac{1}{3 - (-3)} \displaystyle\int_{-3}^{3} \sqrt{9 - x^2}\,dx$

$= \dfrac{1}{6}(\text{area of a semi-circle with } r = 3)$

$= \dfrac{1}{6}(\dfrac{9\pi}{2}) = \dfrac{3\pi}{4} \approx 2.36$

30. $\dfrac{1}{4 - 0} \displaystyle\int_{0}^{4} \sqrt{16 - x^2}\,dx$

$= \dfrac{1}{4}(\text{area of a quarter-circle with } r = 4)$

$= \dfrac{1}{4}(\dfrac{16\pi}{4}) = \pi$

31. $F'(x) = \cos x^2$

32. $G(x) = \displaystyle\int_{x}^{-1} \dfrac{t^2 + 3}{t - 1}\,dt = -\displaystyle\int_{-1}^{x} \dfrac{t^2 + 3}{t - 1}\,dt$

$G'(x) = -\dfrac{x^2 + 3}{x - 1}$

33. $H'(t) = \left[\dfrac{\sin(1 - t)}{(1 - t) + 1}\right](-1) - \left[\dfrac{\sin 3t}{3t + 1}\right](3)$

$= \dfrac{\sin (1 - t)}{t - 2} - \dfrac{3 \sin 3t}{3t + 1}$

34. $G'(y) = \dfrac{1}{\sqrt{y} + 1}\left(\dfrac{1}{2\sqrt{y}}\right) - \dfrac{1}{1 - y^2 + 1}(-2y)$

$= \dfrac{1}{2y + 2\sqrt{y}} + \dfrac{2y}{2 - y^2}$

If you use a computer to evaluate this integral, the answer might look like:

$$\dfrac{3y^2 + 4y^{3/2} + 2}{2\sqrt{y}(2 - y^2)(\sqrt{y} + 1)}$$

35. Average height $= \dfrac{1}{t_1 - t_0} \displaystyle\int_{t_0}^{t_1} \left(-\dfrac{1}{2} gt^2 + v_0 t\right) dt$

$= \dfrac{1}{t_1 - t_0}\left(-\dfrac{1}{6} gt^3 + v_0\dfrac{t^2}{2}\right)\bigg|_{t_0}^{t_1}$

$= \dfrac{1}{t_1 - t_0}\left[-\dfrac{1}{6} gt_1^3 + v_0\dfrac{t_1^2}{2} - \left(-\dfrac{1}{6} gt_0^3 + v_0\dfrac{t_0^2}{2}\right)\right]$

$= -\dfrac{g}{6}(t_1^2 + t_1 t_0 + t_0^2) + \dfrac{1}{2}v_0(t_1 + t_0)$

36. $\dfrac{ds}{dt} = v = -gt + v_0$

Average velocity $= \dfrac{1}{t_1 - t_0} \displaystyle\int_{t_0}^{t_1} \left(-gt + v_0\right) dt$

$\qquad = \dfrac{1}{t_1 - t_0} \left(-\tfrac{1}{2}gt^2 + v_0 t\right)\Big|_{t_0}^{t_1}$

$\qquad = -\dfrac{g}{2}(t_1 + t_0) + v_0$

37. Avg temp $= \dfrac{1}{12 - 9} \displaystyle\int_{9}^{12} (-0.3t^2 + 4t + 10)\, dt$

$\qquad = \tfrac{1}{3}[-0.1t^3 + 2t^2 + 10t]\Big|_{9}^{12} = \tfrac{1}{3}(235.2 - 179.1)$

$\qquad = 18.7\,^{\circ}C$

38. a. $\dfrac{1}{3 - 0}\displaystyle\int_{0}^{3}(t + 1)^2\, dt = \tfrac{1}{3}\cdot\tfrac{1}{3}(t + 1)^3\Big|_{0}^{3}$

$\qquad = \dfrac{64}{9}$ ppm

b. Solve $(t + 12)^2 = \dfrac{64}{9}$ to find $t = \dfrac{5}{3}$;

that is, 1 year, 8 mo

39. a. The area under the curve is

$\displaystyle\int_{0}^{x} f(t)\, dt = A(x) = \tan x$ on $[0, \tfrac{\pi}{2}]$

b. $\dfrac{d}{dt}\displaystyle\int_{0}^{x} f(t)\, dt = (\tan x)' = \sec^2 x$

40. $\dfrac{1}{x + 1}\displaystyle\int_{-1}^{x} f(t)\, dt = \sin x;$

$\dfrac{d}{dt}\displaystyle\int_{-1}^{x} f(t)\, dt = [(x - 1)\sin x]\,'$

$\qquad = (x + 1)\cos x + \sin x$

41. $A(x) = \dfrac{1}{x} = \dfrac{1}{x^2 - x}\displaystyle\int_{x}^{x^2} f(t)\, dt$

$x - 1 = \displaystyle\int_{x}^{x^2} f(t)\, dt$

$\dfrac{d}{dx}(x - 1) = \dfrac{d}{dx}\displaystyle\int_{x}^{x^2} f(t)\, dt$

$1 = 2xf(x^2) - f(x)$

At $x = 2$,

$1 = 2(2)f(4) - f(2)$

$1 = 4[2f(2)] - f(2)$

$1 = 7f(2)$

$f(2) = \tfrac{1}{7}$

4.6 Numerical Integration: The Trapezoidal Rule and Simpson's Rule, Page 308

1. $\Delta x = \dfrac{2 - 1}{4} = \dfrac{1}{4}$

$$
\begin{array}{ll}
x_0 = 1 & f(x_0) = 1 \\
x_1 = 1.25 & f(x_1) = 1.5625 \\
x_2 = 1.5 & f(x_2) = 2.25 \\
x_3 = 1.75 & f(x_3) = 3.0625 \\
x_4 = 2 & f(x_4) = 4
\end{array}
$$

Trapezoidal rule:

$A \approx \tfrac{1}{2}[1 + 2(1.5625) + 2(2.25) + 2(3.0625)$

$\qquad\qquad + 4](\tfrac{1}{4})$

$\qquad \approx \tfrac{1}{8}(18.75) \approx 2.34375$

Simpson's rule:

$A \approx \tfrac{1}{3}[1 + 4(1.5625) + 2(2.25) + 4(3.0625)$

$\qquad\qquad + 4](\tfrac{1}{4})$

$\qquad \approx \tfrac{1}{12}(28) \approx 2.33333$

Exact:

$A = \displaystyle\int_{1}^{2} x^2\, dx = \dfrac{x^3}{3}\Big|_{1}^{2} = \dfrac{8}{3} - \dfrac{1}{3} = \dfrac{7}{3}$

2. $\Delta x = \dfrac{4 - 0}{6} = \dfrac{2}{3} \approx 0.6667$

$$
\begin{array}{ll}
x_0 = 1.0000 & f(x_0) = 0.000 \\
x_1 = 0.6667 & f(x_1) = 0.816 \\
x_2 = 1.3334 & f(x_2) = 1.205 \\
x_3 = 2.0000 & f(x_3) = 1.414 \\
x_4 = 2.6667 & f(x_4) = 1.633 \\
x_5 = 3.3333 & f(x_5) = 1.826 \\
x_6 = 4.0000 & f(x_6) = 2.000
\end{array}
$$

Trapezoidal rule:

$A \approx \tfrac{1}{2}[1(0.000) + 2(0.816) + 2(1.205)$

$+ 2(1.414) + 2(1.633) + 2(1.826) + 1(2.000)](\tfrac{2}{3})$

$\qquad \approx \tfrac{1}{3}(15.788) \approx 5.2627$

Simpson's rule:

$A \approx \tfrac{1}{3}[1(0.000) + 4(0.816) + 2(1.205)$

$+ 4(1.414) + 2(1.633) + 4(1.826) + 1(2.000)](\tfrac{2}{3})$

$\qquad \approx \tfrac{2}{9}(23.9) \approx 5.3111$

Exact:

$$A = \int_0^4 \sqrt{x}\, dx = \frac{2}{3} x^{3/2} \Big|_0^4 = \frac{16}{3}$$

3. $\Delta x = \dfrac{1-0}{4} = \dfrac{1}{4}$

$$
\begin{array}{ll}
x_0 = 0 & f(x_0) = 1.00 \\
x_1 = 0.25 & f(x_1) = 0.94 \\
x_2 = 0.5 & f(x_2) = 0.80 \\
x_3 = 0.75 & f(x_3) = 0.64 \\
x_4 = 1 & f(x_4) = 0.50
\end{array}
$$

a. Trapezoidal rule:

$$A \approx \tfrac{1}{2}[1(1)+2(0.94)+2(0.80)+2(0.64)+1(0.50)](\tfrac{1}{4})$$

$$\approx \tfrac{1}{8}(6.26) \approx 0.7825$$

b. Simpson's rule:

$$A \approx \tfrac{1}{3}[1(1)+4(0.94)+2(0.80)+4(0.64)+1(0.50)](\tfrac{1}{4})$$

$$\approx \tfrac{1}{12}(9.24) \approx 0.785$$

4. $\Delta x = \dfrac{0-(-1)}{4} = \dfrac{1}{4}$

$$
\begin{array}{ll}
x_0 = -1 & f(x_0) = 1.414 \\
x_1 = -0.75 & f(x_1) = 1.250 \\
x_2 = -0.5 & f(x_2) = 1.118 \\
x_3 = -0.25 & f(x_3) = 1.031 \\
x_4 = 0 & f(x_4) = 1.000
\end{array}
$$

a. Trapezoidal rule:

$$A \approx \tfrac{1}{2}[1(1.414)+2(1.250)+2(1.118)$$
$$+2(1.031)+1(1.000)](\tfrac{1}{4})$$
$$\approx \tfrac{1}{8}(9.212) \approx 1.1515$$

b. Simpson's rule:

$$A \approx \tfrac{1}{3}[1(1.414)+4(1.250)+2(1.118)$$
$$+4(1.031)+1(1.000)](\tfrac{1}{4})$$
$$\approx \tfrac{1}{12}(13.774) \approx 1.1478$$

5. $\Delta x = \dfrac{4-2}{4} = \dfrac{1}{2}$

$$
\begin{array}{ll}
x_0 = 2 & f(x_0) = 1.381773 \\
x_1 = 2.5 & f(x_1) = 1.264307 \\
x_2 = 3 & f(x_2) = 1.068232 \\
x_3 = 3.5 & f(x_3) = 0.805740 \\
x_4 = 4 & f(x_4) = 0.493151
\end{array}
$$

a. Trapezoidal rule:

$$A \approx \tfrac{1}{2}[1.381773+2(1.264307)+2(1.068232)$$

$$+2(0.805740)+0.493151](\tfrac{1}{2})$$

$$= \tfrac{1}{4}(8.151482) \approx 2.037871$$

b. Simpson's rule:

$$A \approx \tfrac{1}{3}[1.381773+4(1.264307)$$
$$+2(1.068232)+4(0.805740)$$
$$+.493151](\tfrac{1}{2})$$
$$= \tfrac{1}{6}(12.291576) \approx 2.048596$$

6. $\Delta x = \dfrac{2-0}{6} = \dfrac{1}{3} \approx 0.3333$

$$
\begin{array}{ll}
x_0 = 0.0000 & f(x_0) = 0.000 \\
x_1 = 0.3333 & f(x_1) = 0.315 \\
x_2 = 0.6667 & f(x_2) = 0.524 \\
x_3 = 1.0000 & f(x_3) = 0.540 \\
x_4 = 1.3333 & f(x_4) = 0.314 \\
x_5 = 1.6667 & f(x_5) = -0.160 \\
x_6 = 2.0000 & f(x_6) = -0.832
\end{array}
$$

Trapezoidal rule:

$$A \approx \tfrac{1}{2}[1(0.000)+2(0.315)+2(0.524)$$
$$+2(0.540)+2(0.314)+2(-0.160)$$
$$+1(-0.832)](\tfrac{1}{3})$$
$$\approx \tfrac{1}{6}(2.234) \approx 0.3723$$

Simpson's rule:

$$A \approx \tfrac{1}{3}[1(0.000)+4(0.315)+2(0.524)$$
$$+4(0.540)+2(0.314)+4(-0.160)$$
$$+1(-0.832](\tfrac{1}{3})$$
$$\approx \tfrac{1}{9}(3.624) \approx 0.4027$$

7. Approximate the area under a curve (evaluate an integral) by taking the sum of areas of trapezoids whose upper line segment joins two consecutive points on an arc of the curve.

8. Approximate the area under a curve (evaluate an integral) by taking the sum of areas of parabolic sections whose parabola passes through three consecutive points on an arc of the curve.

9. For the trapezoidal rule, $|E_n| \leq \dfrac{(b-a)^3}{12n^2}M$,

where M is the maximum value of $f''(x)$ on $[a, b]$.

$$f(x) = \frac{1}{x^2+1};$$
$$f'(x) = -2x(x^2+1)^{-2} = \frac{-2x}{(x^2+1)^2}$$

$$f''(x) = \frac{-2(x^2+1)^2 - (-2x)(2)(x^2+1)(2x)}{(x^2+1)^4}$$

$$= \frac{6x^2 - 2}{(x^2+1)^3}$$

Candidates for extrema are $\{0, 1\}$.

$f''(0) = -2; \; f''(1) = \frac{1}{2}$

So we need

$$\frac{1}{12n^2}(2) \le 0.05$$

$$\frac{1}{n^2} \le 0.30$$

$$n^2 \ge \frac{10}{3}$$

$$n \ge 2$$

$$A \approx \frac{1}{2}[f(0) + 2f(0.5) + f(1)](\tfrac{1}{2})$$

$$= \frac{1}{4}[1 + 2(0.8) + 0.5] = 0.775;$$

The actual answer is between $0.79 + 0.05$ and $0.79 - 0.05$.

10. $f'(x) = \frac{x}{\sqrt{x^2+1}}; \; f''(x) = (x^2+1)^{-3/2}$

$f'''(x) = -3x(x^2+1)^{-5/2}$

$f^{(4)}(x) = 3(-1 + 4x^2)(x^2+1)^{-7/2}$ whose

maximum value is 3 on $[-1, 2]$. For Simpson's rule

$$\frac{3^5(3)}{180n^4} < 0.05$$

$$n^4 > 81$$

from which we will pick $n = 4$ (n must be even).

$$\Delta x = \frac{2 - (-1)}{4} = \frac{3}{4}$$

$x_0 = -1.00$		$f(x_0) = 1.414$
$x_1 = -0.25$		$f(x_1) = 1.264$
$x_2 = 0.50$		$f(x_2) = 1.118$
$x_3 = 1.25$		$f(x_3) = 1.600$
$x_4 = 2.00$		$f(x_4) = 2.236$

$A \approx 4.335$; the actual answer is between $4.335 - 0.05$ and $4.335 + 0.05$.

11. $f'(x) = -2\sin 2x; \; f''(x) = -4\cos 2x;$

$f'''(x) = 8\sin 2x; \; f^{(4)}(x) = 16\cos 2x$

whose maximum value is 16 on $[0, 1]$. For Simpson's rule

$$\frac{1^5(16)}{180n^4} < 0.0005$$

$$n^4 > 178$$

$$n > 3.65$$

from which we will pick $n = 4$ (n must be even).

$$\Delta x = \frac{1 - 0}{4} = \frac{1}{4}$$

$x_0 = 0.00$		$f(x_0) = 1.000$
$x_1 = 0.25$		$f(x_1) = 0.939$
$x_2 = 0.50$		$f(x_2) = 0.770$
$x_3 = 0.75$		$f(x_3) = 0.535$
$x_4 = 1.00$		$f(x_4) = 0.292$

$A \approx 0.727$; the actual answer is between $0.727 - 0.0005$ and $0.727 + 0.0005$.

12. $f'(x) = -x^{-2}; \; f''(x) = 2x^{-3};$

$f'''(x) = -6x^{-4}; \; f^{(4)}(x) = 24x^{-5}$ whose

maximum value is 24 on $[1, 2]$. For Simpson's rule

$$\frac{1^5(24)}{180n^4} < 0.0005$$

$$n^4 > 267$$

$$n \ge 4.04$$

from which we will pick $n = 6$ (n must be even).

$$\Delta x = \frac{2 - 1}{4} = \frac{1}{4}$$

$x_0 = 1.000$		$f(x_0) = 1.000$
$x_1 = 1.167$		$f(x_1) = 0.857$
$x_2 = 1.333$		$f(x_2) = 0.750$
$x_3 = 1.500$		$f(x_3) = 0.667$
$x_4 = 1.667$		$f(x_4) = 0.600$
$x_5 = 1.833$		$f(x_5) = 0.545$
$x_6 = 2.000$		$f(x_6) = 0.500$

$A \approx 0.693$; the actual answer is between $0.693 - 0.0005$ and $0.693 + 0.0005$.

13. $f(x) = x(4 - x)^{1/2}; \; f'(x) = \frac{8 - 3x}{2(4-x)^{1/2}};$

$$f''(x) = \frac{3x - 16}{4(4-x)^{3/2}}$$

On $[0, 4]$ the candidates for extrema are the endpoints and where $f'(x) = 0; \; x = \frac{8}{3}$.

$f''(4)$ is undefined, $f''(\frac{8}{3}) \approx -1.30$,

$f''(0) = -\frac{1}{2}$, so $M = 1.30$.

$$0.1 \leq \frac{4^3}{12n^2}(1.30)$$

$$n^2 \geq \frac{640}{12}(1.30)$$

$$n \geq 8.32$$

from which we will pick $n = 9$ terms gives $A \approx 8.5$ with an error of less than 0.1.

14. $f'(\theta) = -\theta \sin^2\theta + \cos^2\theta$

$f''(\theta) = -2(\theta \cos^2\theta + \sin 2\theta)$;

$f'''(\theta) = -6 \cos 2\theta + 4\theta \sin 2\theta$;

$f^{(4)}(\theta) = 8(\theta \cos 2\theta + 2 \sin 2\theta)$

whose maximum value is 40 on $[0, \pi]$. For Simpson's rule

$$\frac{\pi^5(40)}{180n^4} < 0.005$$

$$n^4 > 13,566$$

$$n > 10.7$$

from which we will pick $n = 12$ (n must be even). $A \approx 2.47$ with an error of less than 0.01.

15. $f'(x) = -x^{-2}$; $f''(x) = 2x^{-3}$;

$f'''(x) = -6x^{-4}$; $f^{(4)}(x) = 24x^{-5}$

a. $\frac{2^3(2)}{12n^2} \leq 0.00005$ or $n \approx 163.29$;

pick $n = 164$

b. $\frac{2^5(24)}{180n^4} \leq 0.00005$ or $n \approx 17.09$

pick $n = 18$

16. $f'(x) = 3x^2 + 4x$; $f''(x) = 6x + 4$;

$f'''(x) = -6$; $f^{(4)}(x) = 0$

a. $\frac{5^3(28)}{12n^2} \leq 0.00005$ or $n \approx 2,415.23$

pick $n = 2,416$

b. $\frac{2^5(0)}{180n^4} \leq 0.00005$ or $n = 0$

pick $n = 2$ (n must be even for Simpson's rule)

17. $f'(x) = -\frac{1}{2}x^{-3/2}$; $f''(x) = \frac{3}{4}x^{-5/2}$;

$f'''(x) = -\frac{15}{8}x^{-7/2}$; $f^{(4)}(x) = \frac{105}{16}x^{-9/2}$

a. $\frac{27}{12n^2}\left(\frac{3}{4}\right) \leq 0.00005$, or $n \approx 183.71$

pick $n = 184$

b. $\frac{3^5}{180n^4}\left(\frac{105}{16}\right) \leq 0.00005$, or $n \approx 20.52$

pick $n = 22$ (n must be even)

18. $f'(x) = -\sin x$; $f''(x) = -\cos x$;

$f'''(x) = \sin x$; $f^{(4)}(x) = \cos x$

a. $\frac{2^3(1)}{12n^2} \leq 0.00005$ or $n \approx 115.47$;

pick $n = 116$

b. $\frac{2^5(1)}{180n^4} \leq 0.00005$ or $n \approx 7.72$

pick $n = 8$

19. $f(x) = (1 - x^2)^{1/2}$; the second derivative is unbounded on $[0, 1]$, so the number of intervals needed to guarantee a certain accuracy cannot be predicted. If $n = 8$, $T_8 = (0.5)(12.347)(0.125) \approx 0.772$. Thus, $\pi \approx 3.1$.

20. $f(x) = (1 - x^2)^{1/2}$

The second derivative is unbounded on $[0, 1]$, so the number of intervals needed to guarantee a certain accuracy cannot be predicted. If $n = 4$, $S_4 = (0.3333)(9.2508)(0.25) \approx 0.7708$. Thus, $\pi \approx 3.1$.

21. $f'(x) = -x^{-2}$; $f''(x) = 2x^{-3}$; This is a decreasing function, so the maximum occurs at the left endpoint: $f''(1) = 2$.

$\frac{1(2)}{12n^2} \leq 0.0000\ 005$, $n \approx 577.35$;

pick $n = 578$.

22. $\Delta x = \frac{6 - 0}{6} = 1$

$x_0 = 0$	$f(x_0) = 11$
$x_1 = 1$	$f(x_1) = 13$
$x_2 = 2$	$f(x_2) = 14.5$
$x_3 = 3$	$f(x_3) = 10.5$
$x_4 = 4$	$f(x_4) = 11.8$
$x_5 = 5$	$f(x_5) = 11.6$
$x_6 = 6$	$f(x_6) = 9.8$

$A \approx \frac{1}{2}[1(11)+2(13)+2(14.5)+2(10.5)$

$\qquad +2(11.8)+2(11.6)+1(9.8)](1)$

$\qquad \approx \frac{1}{2}(143.6) \approx 71.8$

23. $\Delta x = 5$

$A \approx \frac{1}{3}[(9)+4(15)+2(20)+4(27)+(30)](5)$

$\qquad \approx 412$

24. $\Delta x = \frac{1}{3}$

$A \approx \frac{1}{2}[3.7+2(3.9)+2(4.1)+2(4.1)+2(4.2)$

$\qquad +2(4.4)+2(4.6)+2(4.9)+2(5.2)$

$\qquad +2(5.5)+6](\frac{1}{3}) \approx \frac{1}{6}(91.7) \approx 13.76$

25. $\Delta x = 0.5$

$A \approx \frac{1}{3}[10+4(9.75)+2(10)+4(10.75)+2(12)$

$\qquad +4(13.75)+2(16)+4(18.75)$

$\qquad +2(22)+4(25.75)+30](0.5)$

$\qquad \approx \frac{1}{3}(475)(0.5) = 79$

26. a. $\int_{0}^{\pi} \sin x\, dx = -\cos x\Big|_{0}^{\pi} = 2$

b.

TYPE OF ESTIMATE	# OF SUB-INTERVALS	ESTIMATE OVER [0, π]
Right endpt	10	1.98352353751
Right endpt	20	1.99588597271
Right endpt	40	1.99897181058
Right endpt	80	1.99974297245
Trapezoid	10	1.98352353751
Trapezoid	20	1.99588597271
Trapezoid	40	1.99897181058
Trapezoid	80	1.99974297245
Simpson	10	2.00010951732
Simpson	20	2.00000678444
Simpson	40	2.00000042389
Simpson	80	2.00000002643

From the table, $\quad M_{10} \approx 0.6378$

$\qquad\qquad\qquad M_{20} \approx 0.6363$

$\qquad\qquad\qquad M_{40} \approx 0.6378$

A good indication for M is about 0.64. Also,

$\qquad\qquad\qquad K_{10} \approx 0.644178$

$\qquad\qquad\qquad K_{20} \approx 0.638495$

$\qquad\qquad\qquad K_{40} \approx 0.638495$

$\qquad\qquad\qquad K_{80} \approx 0.636737$

A good indication for K is about 0.64.

27. a. $\int_{0}^{\pi}(9x - x^3)\,dx = \frac{9}{2}x^2 - \frac{1}{4}x^4\Big|_{0}^{\pi}$

$\qquad = \frac{1}{4}(18\pi^2 - \pi^4) \approx 20.06094705$

TYPE OF ESTIMATE	# OF SUB-INTERVALS	ESTIMATE OVER [0, π]
Right endpt	10	19.3882917476
Right endpt	20	19.7855000789
Right endpt	40	19.9384437331
Right endpt	80	20.0035004324
Trapezoid	10	19.8174243188
Trapezoid	20	20.0000663645
Trapezoid	40	20.0457268759
Trapezoid	80	20.0571420038
Simpson	10	20.0609470464
Simpson	20	20.0609470464
Simpson	40	20.0609470464
Simpson	80	20.0609470464

For Simpson's rule the same values occur. The error term involves $f^{(4)}(x)$ which is 0 for cubics; that is, the Simpson error is 0.

b. The problem is that $f^{(4)}(x)$ is unbounded near $x = 2$.

28. $\int_{a}^{b} p(x)\,dx = \int_{a}^{b}(a_3 x^3 + a_2 x^2 + a_1 x + a_0)\,dx$

$\qquad = \left[\frac{a_3}{4}x^4 + \frac{a_2}{3}x^3 + \frac{a_1}{2}x^2 + a_0 x\right]\Big|_{a}^{b}$

$\qquad = \frac{1}{12}(b-a)[3a_3 b^3 + 3a_3 ab^2 + 3a_3 a^2 b$

$\qquad\qquad + 3a_3 a^3 + 4a_2 b^2 + 4a_2 ab$

$\qquad\qquad + 4a_2 a^2 + 12a_1 b + 6a_1 a + 12a_0]$

Now,

$\frac{(b-a)}{6}\left[p(a) + 4p\left(\frac{a+b}{2}\right) + p(b)\right]$

$= \frac{(b-a)}{6}[\frac{3}{2}a_3 b^3 + \frac{3}{2}a_3 ab^2 + \frac{3}{2}a_3 a^2 b$

$\qquad + \frac{3}{2}a_3 a^3 + 2a_2 b^2 + 2a_2 ab + 2a_2 a^2$

$\qquad + 6a_1 b + 3a_1 a + 6a_0 + 2]$

Thus,

$\int_{a}^{b} p(x)\,dx = \frac{(b-a)}{6}\left[p(a) + 4p\left(\frac{a+b}{2}\right) + p(b)\right]$

29. $a = -1,\ b = 2,\ b - a = 3,\ \frac{b-a}{6} = \frac{1}{2},$

and $\frac{a+b}{2} = \frac{1}{2}$

$$\int_{-1}^{2} (x^3 - 3x + 4)\,dx = \{2^3 - 3(2) + 4$$

$$+ 4[(0.5)^3 - 3(0.5) + 4] + (-1)^3$$

$$- 3(-1) + 4\}(0.5) = \frac{45}{4}$$

30. $a = -1$, $b = 3$, $b - a = 4$, $\dfrac{b-a}{6} = \dfrac{2}{3}$,

and $\dfrac{a+b}{2} = 1$

$$\int_{-1}^{3} (x^3 + 2x^2 - 7)\,dx = \frac{32}{3}$$

31.   **a.** $\displaystyle\int_{-1}^{1} p(x)\,dx$

$$= \int_{-1}^{1} (a_3 x^3 + a_2 x^2 + a_1 x + a_0)\,dx$$

$$= \left[\frac{a_3}{4} x^4 + \frac{a_2}{3} x^3 + \frac{a_1}{2} x^2 + a_0 x \right]\Big|_{-1}^{1}$$

$$= \frac{2}{3} a_2 + 2 a_0$$

Now, $p(c) + p(-c) = 2a_2 + 2a_0$

$$\frac{2}{3} a_2 + 2 a_0 = 2 a_2 c^2 + 2 a_0$$

$$c^2 = \frac{1}{3}$$

$$c = \frac{1}{\sqrt{3}}$$

which is on the interval $[0, 1]$.

b. $\displaystyle\int_{-1/2}^{1/2} p(x)\,dx$

$$= \int_{-1/2}^{1/2} (a_3 x^3 + a_2 x^2 + a_1 x + a_0)\,dx$$

$$= \left[\frac{a_3}{4} x^4 + \frac{a_2}{3} x^3 + \frac{a_1}{2} x^2 + a_0 x \right]\Big|_{-1/2}^{1/2}$$

$$= \frac{1}{12} a_2 + a_0$$

Now, $\dfrac{1}{3}[p(-c) + p(0) + p(c)] = \dfrac{2}{3} a_2 c^3 + a_0$

$$\frac{2}{3} a_2 c^3 + a_0 = \frac{1}{12} a_2 + a_0$$

$$c^2 = \frac{1}{8}$$

$$c = \frac{1}{\sqrt{2}}$$

which is on the interval $[0, 1]$.

32. Using the error formula for Simpson's rule, $f^{(4)}(x)$ for a third degree polynomial will be zero, so there has to be a number c in the interval for which the error is zero and the value is exact.

33.   **a.** $P_1(-h, f(-h))$ leads to

$$f(-h) = Ah^2 - Bh + C$$

$P_2(0, f(0))$ leads to $f(0) = C$

$P_3(h, f(h))$ leads to

$$f(h) = Ah^2 + Bh + C$$

Adding to the first equation to the third equation yields

$$2Ah^2 + 2C = f(h) + f(-x)$$

From the second equation we have

$$2Ah^2 + 2f(0) = f(h) + f(-x)$$

Solve this equation for A:

$$A = \frac{1}{2h^2}[f(h) - 2f(0) + f(-h)]$$

Solve the third equation for B:

$$B = \frac{1}{2h}[f(h) - f(-h)]$$

Thus,

$$y = Ax^2 + Bx + C$$

$$= \frac{1}{2h^2}[f(h) - 2f(0) + f(-h)]x^2$$

$$+ \frac{1}{2h}[f(h) - f(-h)]x + f(0)$$

b. $\displaystyle\int_{-h}^{h} p(x)\,dx = \frac{1}{6h^2}[f(h) - 2f(0) + f(-h)]x^3$

$$+ \frac{1}{4h}[f(h) - f(-h)]x^2 + f(0)x\Big|_{-h}^{h}$$

$$= \frac{h}{3}[f(-h) + 4f(0) + f(h)]$$

c.   From part b:

$$\frac{h}{3}[f(-h) + 4f(0) + f(h)] = \int_{-h}^{h} p(x)\,dx$$

Let $u = x + x_2$, then $du = dx$;

if $x = -h$, then $u = x_2 - h$;

if $x = h$, then $u = x_2 + h$

$$\int_{-h}^{h} p(x)\,dx = \int_{x_2 - h}^{x_2 + h} p(x)\,dx$$

Now, write $x_1 = x_2 - h$, $x_2 = x_2 + 0$, $x_3 = x_2 + h$

$$\int_{x_2-h}^{x_2+h} p(x)\ dx = \int_{x_1}^{x_2} p(x)\ dx$$

4.7 Area Between Two Curves, Page 317

1. $-x^2 + 6x - 5 = \frac{3}{2}x - \frac{3}{2}$

$2x^2 - 9x + 7 = 0$

$(2x - 7)(x - 1) = 0$

$x = \frac{7}{2}, 1$

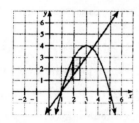

$$\int_{1}^{7/2} [(-x^2 + 6x - 5) - (\tfrac{3}{2}x - \tfrac{3}{2})]\ dx$$

$$= \int_{1}^{7/2} (-x^2 + \tfrac{9}{2}x - \tfrac{7}{2})\ dx$$

$$= (-\tfrac{1}{3}x^3 + \tfrac{9}{4}x^2 - \tfrac{7}{2}x)\Big|_{1}^{7/2} = \frac{125}{48}$$

2. $y^2 - 6y = -y$

$y^2 - 5y = 0$

$y(y - 5) = 0$

$y = 0, 5$

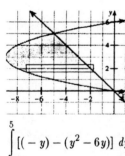

$$\int_{0}^{5} [(-y) - (y^2 - 6y)]\ dy$$

$$= \int_{0}^{5} (5y - y^2)\ dy$$

$$= (\tfrac{5}{2}y^2 - \tfrac{1}{3}y^3)\Big|_{0}^{5} = \frac{125}{6}$$

3. $y^2 - 5y = 0$

$y(y - 5) = 0$

$y = 0, 5$

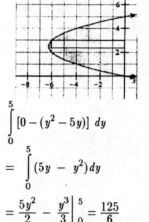

$$\int_{0}^{5} [0 - (y^2 - 5y)]\ dy$$

$$= \int_{0}^{5} (5y - y^2)\ dy$$

$$= \frac{5y^2}{2} - \frac{y^3}{3}\Big|_{0}^{5} = \frac{125}{6}$$

4. $x^2 - 8x = 0$

$x(x - 8) = 0$

$x = 0, 8$

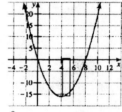

$$\int_{0}^{8} [0 - (x^2 - 8x)]\ dy = \int_{0}^{8} (8x - x^2)\ dx$$

$$= (-\tfrac{1}{3}x^3 + 4x^2)\Big|_{0}^{8} = \frac{256}{3}$$

5. $\sin x = 0$

$x = 0, \pi, 2\pi$

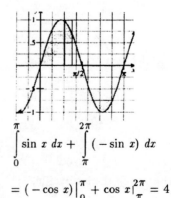

$$\int_{0}^{\pi} \sin x\ dx + \int_{\pi}^{2\pi} (-\sin x)\ dx$$

$$= (-\cos x)\Big|_{0}^{\pi} + \cos x\Big|_{\pi}^{2\pi} = 4$$

6.
$$(x - 1)^3 = x - 1$$
$$(x - 1)^3 - (x - 1) = 0$$
$$(x - 1)[(x - 1)^2 - 1] = 0$$
$$(x - 1)(x - 1 - 1)(x - 1 + 1) = 0$$
$$x = 0, 1, 2$$

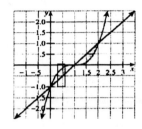

$$\int_0^1 [(x - 1)^3 - (x - 1)]\,dx + \int_1^2 [(x - 1) - (x - 1)^3]\,dx$$

$$[\tfrac{1}{4}(x - 1)^4 - \tfrac{1}{2}(x - 1)^2]\Big|_0^1 + [\tfrac{1}{2}(x - 1)^2 - \tfrac{1}{4}(x - 1)^2]\Big|_1^2 = \tfrac{1}{2}$$

7.

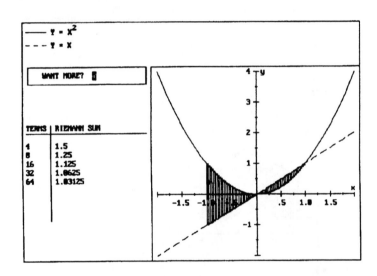

The curves intersect at $(0, 0)$ and $(1, 1)$.

$$A = \int_{-1}^0 (x^2 - x)\,dx$$
$$+ \int_0^1 (x - x^2)\,dx$$
$$= (\tfrac{1}{3}x^3 - \tfrac{1}{2}x^2)\Big|_{-1}^0$$
$$+ (\tfrac{1}{2}x^2 - \tfrac{1}{3}x^3)\Big|_0^1$$
$$= 1$$

8.

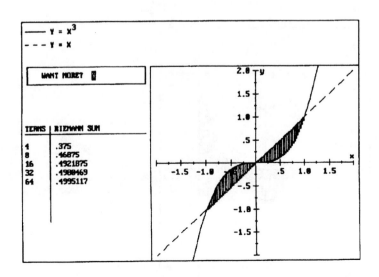

The curves intersect at $(-1, -1)$, $(0, 0)$, and $(1, 1)$

$$A = \int_{-1}^0 (x^3 - x)\,dx$$
$$+ \int_0^1 (x - x^3)\,dx$$
$$= (\tfrac{1}{4}x^4 - \tfrac{1}{2}x^2)\Big|_{-1}^0$$
$$+ (\tfrac{1}{2}x^2 - \tfrac{1}{4}x^4)\Big|_0^1$$
$$= \tfrac{1}{2}$$

9.

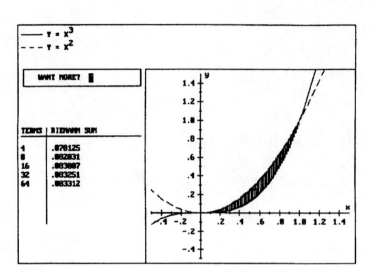

The curves intersect at $(0, 0)$ and $(1, 1)$.

$$A = \int_0^1 (x^2 - x^3)\ dx$$

$$= (\tfrac{1}{3}x^3 - \tfrac{1}{4}x^4)\Big|_0^1$$

$$= \frac{1}{12}$$

10.

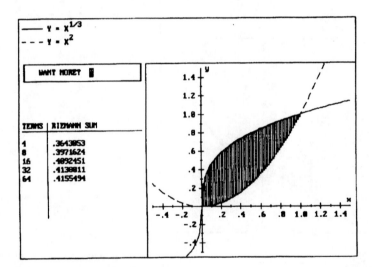

The curves intersect at $(0, 0)$ and $(1, 1)$.

$$A = \int_0^1 (x^{1/3} - x^2)\ dx$$

$$= (\tfrac{3}{4}x^{4/3} - \tfrac{1}{3}x^3)\Big|_0^1$$

$$= \frac{5}{12}$$

11.

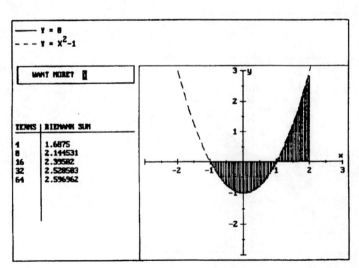

The curves intersect at $(-1, 0)$ and $(1, 0)$.

$$A = \int_{-1}^1 (1 - x^2)\ dx$$

$$+ \int_1^2 (x^2 - 1)\ dx$$

$$= (-\tfrac{1}{3}x^3 + x)\Big|_{-1}^1$$

$$+ (\tfrac{1}{3}x^3 - x)\Big|_1^2$$

$$= \frac{8}{3}$$

12.

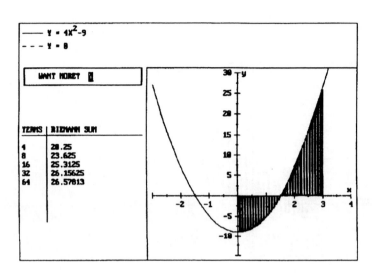

The curves intersect at $(\frac{3}{2}, 0)$.

$$A = \int_0^{3/2} (9 - 4x^2)\, dx$$

$$+ \int_{3/2}^3 (4x^2 - 9)\, dx$$

$$= (9x - \tfrac{4}{3}x^3)\Big|_0^{3/2}$$

$$+ (\tfrac{4}{3}x^3 - 9x)\Big|_{3/2}^3$$

$$= 27$$

13.

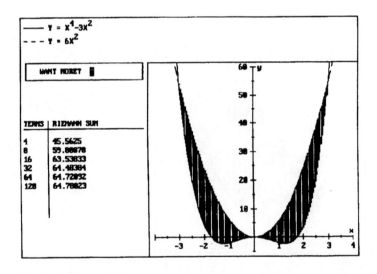

The curves intersect at $(0, 0)$ and $(-3, 54)$ and $(3, 54)$.

$$A = \int_{-3}^0 (9x^2 - x^4)\, dx$$

$$+ \int_0^3 (9x^2 - x^4)\, dx$$

$$= 2(\tfrac{9}{3}x^3 - \tfrac{1}{5}x^5)\Big|_0^3$$

$$= \frac{324}{5}$$

14.

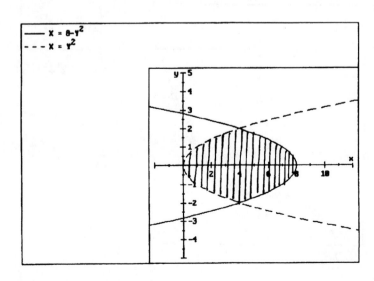

The curves intersect at $(4, 2)$ and $(-4, 2)$.

$$A = \int_{-2}^2 (8 - 2y^2)\, dy$$

$$= 2(8y - \tfrac{2}{3}y^3)\Big|_0^2$$

$$= \frac{64}{3}$$

15.

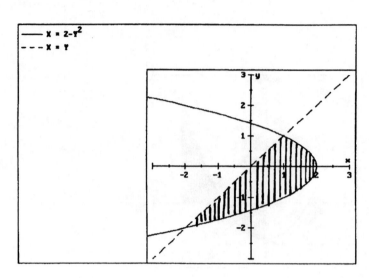

The curves intersect at $(0, 1)$ and $(-4, -2)$.

$$A = \int_{-2}^{1} (2 - y^2 - y)\, dy$$

$$= (2y - \tfrac{1}{3}y^3 - \tfrac{1}{2}y^2)\Big|_{-2}^{0}$$

$$= \frac{9}{2}$$

16.

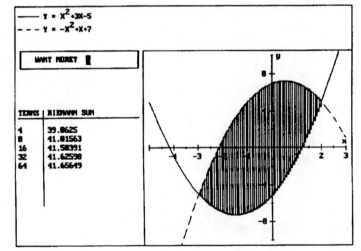

The curves intersect at $(2, 5)$ and $(-3, -5)$.

$$(-x^2 + x + 7) - (x^2 + 3x - 5)$$

$$= -2x^2 - 2x + 12$$

$$A = \int_{-3}^{2} (-2x^2 - 2x + 12)\, dx$$

$$= (-\tfrac{2}{3}x^3 - x^2 + 12x)\Big|_{-3}^{2}$$

$$= \frac{125}{3}$$

17.

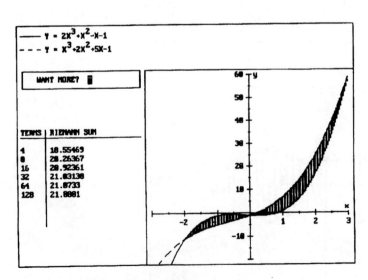

The curves intersect at $(-2, 11)$, and $(3, 59)$.

$$A = \int_{-2}^{0} (x^3 - 2x^2 - 6x)\, dx$$

$$+ \int_{0}^{3} (-x^3 + 2x^2 + 6x)\, dx$$

$$= (\tfrac{1}{4}x^4 - \tfrac{2}{3}x^3 - 3x^2)\Big|_{-2}^{0}$$

$$+ (-\tfrac{1}{4}x^4 + \tfrac{2}{3}x^3 + 3x^2)\Big|_{0}^{3}$$

$$= \frac{253}{12}$$

18.

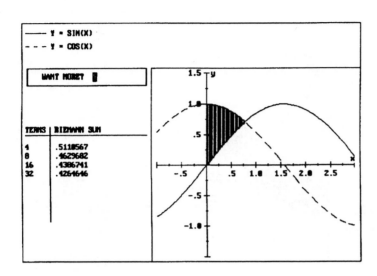

$$A = \int\limits_0^{\pi/4} (\cos x - \sin x)\, dx$$

$$= (\sin x + \cos x)\Big|_0^{\pi/4}$$

$$= \sqrt{2} - 1$$

19.

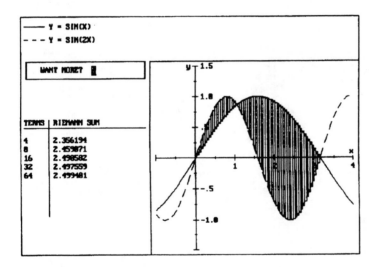

The curves intersect when $x = 0$, $\pi/3$, and π.

$$A = \int\limits_0^{\pi/3} (\sin 2x - \sin x)\, dx$$

$$+ \int\limits_{\pi/3}^{\pi} (\sin x - \sin 2x)\, dx$$

$$= \left(-\tfrac{1}{2}\cos 2x + \cos x\right)\Big|_0^{\pi/3}$$

$$+ \left(-\cos x + \tfrac{1}{2}\cos 2x\right)\Big|_{\pi/3}^{\pi}$$

$$= \tfrac{5}{2}$$

20.

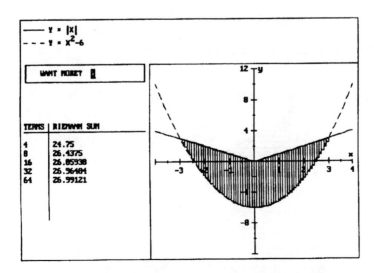

The curves intersect at $(3, 3)$ and $(-3, 3)$. Note the symmetry with respect to the y-axis.

$$A = 2\int\limits_0^3 (x - x^2 + 6)\, dx$$

$$= 2\left(\tfrac{1}{2}x^2 - \tfrac{1}{3}x^3 + 6x\right)\Big|_0^3$$

$$= 2\left(\tfrac{27}{2}\right)$$

$$= 27$$

21.

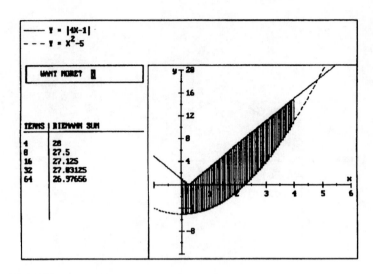

The curves do not intersect on $[0, 4]$, but the absolute value function changes at $x = 1/4$.

$$A = \int_0^{1/4} (-4x + 1 - x^2 + 5)\, dx$$

$$+ \int_{1/4}^4 (4x - 1 - x^2 + 5)\, dx$$

$$= (-2x^2 + x - \tfrac{1}{3}x^3 + 5x)\Big|_0^{1/4}$$

$$+ (2x^2 - x - \tfrac{1}{3}x^3 + 5x)\Big|_{1/4}^4$$

$$= \frac{323}{12}$$

22.

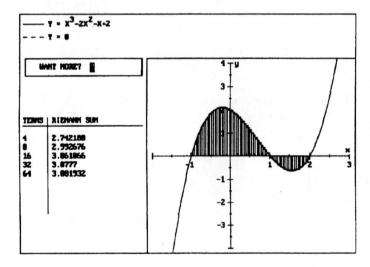

The curves intersect at $(1, 0)$, $(-1, 0)$ and $(2, 0)$

$$A = \int_{-1}^1 (x^3 - 2x^2 - x + 2)\, dx$$

$$+ \int_1^2 (-x^3 + 2x^2 + x - 2)\, dx$$

$$= (\tfrac{1}{4}x^4 - \tfrac{2}{3}x^3 - \tfrac{1}{2}x^2 + 2x)\Big|_{-1}^0$$

$$+ (-\tfrac{1}{4}x^4 + \tfrac{2}{3}x^3 + \tfrac{1}{2}x^2 - 2x)\Big|_1^2$$

$$= \frac{35}{12}$$

23.

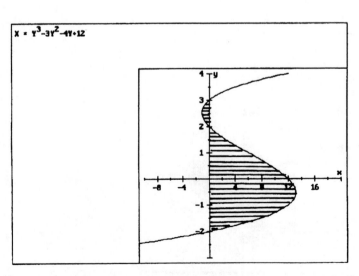

The curves intersect at $(0, 2)$, $(0, -2)$, and $(0, 3)$.

$$A = \int_{-2}^2 (y^3 - 3y^2 - 4y + 12)\, dy$$

$$+ \int_2^3 (-y^3 + 3y^2 + 4y - 12)\, dy$$

$$= (\tfrac{1}{4}y^4 - y^3 - 2y^2 + 12y)\Big|_{-2}^2$$

$$+ (\tfrac{1}{4}y^4 + y^3 + 2y^2 - 12y)\Big|_2^3$$

$$= \frac{131}{4}$$

24. **a.** $q_0 = 1$, $p_0 = 3.5 - (0.5)(1) = 3$;

Consumer's surplus $= \int_0^{q_0} D(q)\ dq - p_0 q_0$

$= \int_0^1 (3.5 - 0.5q)\ dq - 3$

$= (3.5q - \frac{1}{4}q^2)\big|_0^1 - 3$

$= 3.25 - 3 = 0.25$

b. $q_0 = 1.5$, $p_0 = 3.5 - (0.5)(1) = 1.25$;

C.S. $= \int_0^{q_0} D(q)\ dq - p_0 q_0$

$= \int_0^{1.5} (3.5 - 0.5q)\ dq - 1.25(1.5)$

$= (3.5q - \frac{1}{4}q^2)\big|_0^{1.5} - 1.875 = 2.81$

25. **a.** $q_0 = 1$, $p_0 = 2.5 - (1.5)(1) = 1$;

C.S. $= \int_0^{q_0} D(q)\ dq - p_0 q_0$

$= \int_0^1 (2.5 - 1.5q)\ dq - 1$

$= (2.5q - \frac{3}{4}q^2)\big|_0^1 - 1$

$= 2.5 - 0.75 - 1 = 0.75$

b. $q_0 = 0$, $p_0 = 2.5 - 0 = 2.5$;

C.S. $= \int_0^0 (2.5 - 0)\ dq - 0 = 0$

26. **a.** $q_0 = 4$ $p_0 = 100 - 8(4) = 68$;

C.S. $= \int_0^{q_0} D(q)\ dq - p_0 q_0$

$= \int_0^4 (100 - 8q)\ dq - 4(68)$

$= (100q - 4q^2)\big|_0^4 - 272$

$= 400 - 64 - 272 = 56$

b. $q_0 = 10$, $p_0 = 100 - 80 = 20$;

C.S. $= \int_0^{10} (100 - 8q)\ dq - 20(10)$

$= (100q - 4q^2)\big|_0^{10} - 200$

$= 1,000 - 400 - 200 = 400$

27. **a.** $q_0 = 5$, $p_0 = 150 - (6)(5) = 128$;

C.S. $= \int_0^{q_0} D(q)\ dq - p_0 q_0$

$= \int_0^5 (150 - 6q)\ dq - (120)(5)$

$= (150q - 3q^2)\big|_0^5 - 600$

$= 750 - 75 - 600 = 75$

b. $q_0 = 12$, $p_0 = 150 - 6(12) = 78$;

C.S. $= \int_0^{12} (150 - 6q)\ dq - (78)(12)$

$= (150q - 3q^2)\big|_0^{12} - 936$

$= 1,800 - 432 - 936 = 432$

28. If $y = f(x)$ and $y = g(x)$ are given, it is best to use vertical strips because one need not solve for $x = f^{-1}(x)$ and $x = g^{-1}(x)$. There are usually fewer integrals involved. For the same reason, if $x = f(y)$, $x = g(y)$, use vertical strips.

29. The consumer's surplus is the amount the consumer is willing to spend less than the amount actually spent.

30. Equilibrium occurs when $D(q) = S(q)$.

$$14 - q^2 = 2q^2 + 2$$
$$3q^2 = 12$$
$$q = 2 \quad (-2 \text{ is meaningless here.})$$
$$p = D(2) = 10$$

Consumer surplus is:

$\int_0^2 (14 - q^2)\ dq - 2(10)$

$= 14q - \frac{q^3}{3}\big|_0^2 - 20 \approx \$5.33$

31. Equilibrium occurs when $D(q) = S(q)$.

$$25 - q^2 = 5q^2 + 1$$
$$6q^2 = 24$$

$$q = 2 \quad \text{(disregard negative)}$$

$$p = D(2) = 21$$

Consumer surplus is: $\displaystyle\int_0^2 (25 - q^2)\, dq - 2(21) = 25q - \dfrac{q^3}{3}\Big|_0^2 - 42 = 50 - \dfrac{8}{3} - 42 \approx \$5.33$

32. Equilibrium occurs when $D(q) = S(q)$.

$$32 - 2q^2 = \tfrac{1}{3}q^2 + 2q + 5$$

$$7q^2 + 6q - 81 = 0$$

$$(q - 3)(7q + 27) = 0$$

$$q = 3 \quad \text{(disregard negative)}$$

$$p = D(3) = 14$$

Consumer surplus is: $\displaystyle\int_0^3 (32 - 2q^2)\, dq - 3(14) = 32q - \dfrac{2q^3}{3}\Big|_0^3 - 42 = 78 - 42 \approx \$36$

33. Equilibrium occurs when $D(q) = S(q)$.

$$27 - q^2 = \tfrac{1}{4}q^2 + \tfrac{1}{2}q + 5$$

$$5q^2 + 2q - 88 = 0$$

$$(q - 4)(5q + 22) = 0$$

$$q = 4 \quad \text{(disregard negative)}$$

$$p = D(4) = 11$$

Consumer surplus is: $\displaystyle\int_0^4 (27 - q^2)\, dq - 4(11) = 27q - \dfrac{q^3}{3}\Big|_0^4 - 44 = 108 - \dfrac{64}{3} - 44 \approx \$42.67$

34. The lines intersect at $x = -1$; the line $2y = 11 - x$ intersects the parabola at $x = \sqrt{7}$; the line $y = 7x + 13$ intersects the parabola at $x = -2$ and $x = 9$.

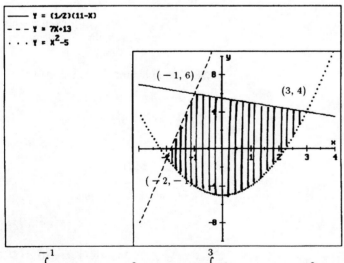

$$A = \int_{-2}^{-1} [7x + 13 - (x^2 - 5)]\, dx + \int_{-1}^{3} [\tfrac{1}{2}(11 - x) - (x^2 - 5)]\, dx$$

$$= [\tfrac{7}{2}x^2 + 13x - \tfrac{1}{3}x^3 + 5x]\Big|_{-2}^{-1} + [\tfrac{11}{2}x - \tfrac{1}{4}x^2 - \tfrac{1}{3}x^3 + 5x]\Big|_{-1}^{3}$$

$$= \frac{31}{6} + \frac{92}{3} = \frac{215}{6} \approx 35.83$$

35.   This is one-eighth of a circle with radius $\sqrt{8}$. $A = \frac{1}{8}\pi(\sqrt{8})^2 = \pi$

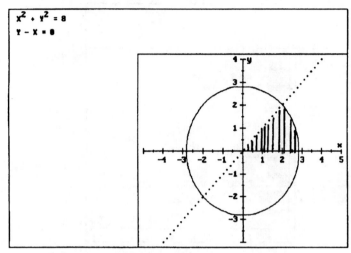

36.   Since $\sqrt{x} + \sqrt{y} = 1$ implies $y = 1 - 2x^{1/2} + x$

$$A = \int_0^1 (1 - 2x^{1/2} + x)\, dx = (x - \tfrac{4}{3}x^{3/2} + \tfrac{1}{2}x^2)\Big|_0^1 = \tfrac{1}{6}$$

37.   **a.**   The use of the machine will be profitable as long as the rate at which revenue is generated is greater than the rate at which costs accumulate, that is, until $R(x) = C(x)$.

$$6{,}025 - 10x^2 = 4{,}000 + 15x^2$$
$$25x^2 = 2{,}025$$
$$x^2 = 81$$
$$x = \pm 9$$

The machine will be profitable for 9 years.

b.   The difference $R(x) - C(x)$ represents the rate of change of the net earnings generated by the machine. Hence, the net earnings over the next 9 years is:

$$\int_0^9 [R(x) - C(x)]\, dx = \int_0^9 [(6{,}025 - 10x^2) - (4{,}000 + 15x^2)]\, dx$$

$$= \int_0^9 (2{,}025 - 25x^2)\, dx = (2{,}025x - \tfrac{25}{3}x^3)\Big|_0^9 = 12{,}150$$

In geometric terms, the net earnings is represented by the area of the region between the curves $y = R(x)$ and $y = C(x)$ from $x = 0$ to $x = 9$.

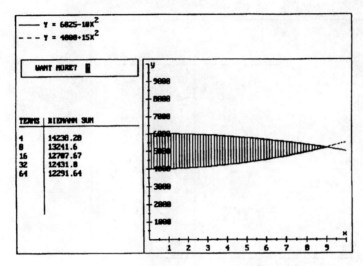

38.   The excess production is the difference between the area under the production curves.

$$\int_0^4 [60 - 2(t - 1)^2 - 50 + 5t] \, dt = \left[10t - \tfrac{2}{3}(t - 1)^3 + \tfrac{5}{2}t^2\right]\Big|_0^4 = \frac{184}{3}$$

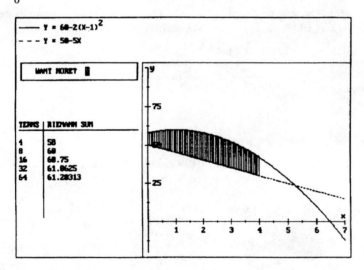

39.   **a.**   Equating the two profit functions:

$$100 + x^2 = 220 + 2x,$$

$$x^2 - 2x - 120 = 0$$

$$(x - 12)(x + 10) = 0, \; 12$$

The second plan is more profitable for 12 years.

   b.   Excess profit $= \displaystyle\int_0^{12} [(220 + 2x) - (100 + x^2)] \, dx$

$$= 120x + x^2 - \frac{x^3}{3}\Big|_0^{12} = 1{,}008$$

In geometric terms, the net excess profit generated by the second plan is the area of the region between the curves $y = R_2(x)$ and $y = R_1(x)$ from $x = 0$ to $x = 12$.

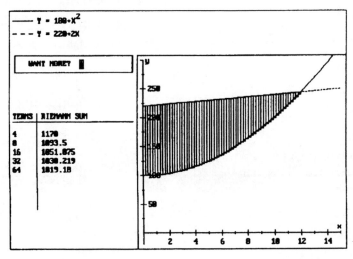

40. **a.** $P(x) = xp(x) - C(x) = (110 - x)x - (x^3 - 25x^2 + 2x + 30) = -x^3 + 24x^2 + 108x - 30$

$P'(x) = -3x^2 + 48x + 108 = -3(x - 18)(x + 2)$

$P'(x) = 0$ when $x = 18$ (disregard negative root);

$P''(x) = -6x + 48 < 0$ when $x = 18$, so $x = 18$ is a maximum.

b. Consumer's surplus $= \displaystyle\int_0^{18} (110 - x)dx - 18(92) = (110x - \tfrac{1}{2}x^2)\Big|_0^{18} - 1{,}656 = 162$

41. **a.** $P(x) = xp(x) - C(x) = (124 - 2x)x - (2x^3 - 59x^2 + 4x + 76) = -2x^3 + 57x^2 + 120x - 76$

$P'(x) = -6x^2 + 114x + 120 = -6(x - 20)(x + 1)$

$P'(x) = 0$ when $x = 20$ (disregard negative root);

$P''(x) = -12x + 114 < 0$ when $x = 20$, so $x = 20$ is a maximum.

b. Consumer's surplus $= \displaystyle\int_0^{20} (124 - 2x)dx - 20(84) = (124x - x^2)\Big|_0^{20} - 1{,}680 = 400$

42. **a.** The cost function is $\displaystyle\int (6 + \tfrac{1}{4}q^2)\, dq = 6q - \tfrac{1}{12}q^3 + K$

Assume $K = 0$ since there is no mention of overhead or other fixed costs. The revenue function is

$$R(q) = qp(q) = 45q - q^3$$

The marginal revenue function is $R'(q) = 45 - 3q^2$

b. The profit function is

$$P(q) = R(q) - C(q)$$
$$= -45q - q^3 - 6q - \tfrac{1}{12}q^3$$
$$= -\tfrac{13}{12}q^3 + 39q$$
$$P'(q) = -\tfrac{13}{4}q^2 + 39 = 0 \text{ when}$$
$$q = \pm\sqrt{12} \text{ (disregard negative value)}$$

c. The consumer's surplus is

$$\int_0^{\sqrt{12}} (45 - q^2)\, dq - \sqrt{12}(33)$$

$$= (45q - \tfrac{1}{3}q^3)\Big|_0^{\sqrt{12}} - 33\sqrt{12}$$

$$= 16\sqrt{3} \approx 27.71$$

The consumer's surplus is $27.71.

43. a. The cost function is

$$\int (\tfrac{3}{4}q^2 + 5)\, dq = \tfrac{1}{4}q^3 + 5q + K$$

Assume $K = 0$ since there is no mention of overhead or other fixed costs. The revenue function is

$$R(q) = qp(q) = \tfrac{1}{4}q(10 - q)^2$$

The marginal revenue function is

$$R'(q) = \tfrac{1}{4}(100 - 40q + 3q^2)$$

$$= \tfrac{1}{4}(10 - q)(10 - 3q)$$

b. The profit function is

$$P(q) = R(q) - C(q)$$
$$= \tfrac{1}{4}(100q - 20q^2 + q^3) - \tfrac{1}{4}q^3 - 5q$$
$$= 20q - 5q^2$$
$$P'(q) = 20 - 10q = 0 \text{ when } q = 2$$

c. The consumer's surplus is

$$\tfrac{1}{4}\int_0^2 (100 - 20q + q^2)\, dq - \tfrac{1}{2}(64)$$

$$= \tfrac{1}{4}(100q - 10q^2 + \tfrac{1}{3}q^3)\Big|_0^2 - 32$$

$$= \tfrac{26}{3} \approx 8.67$$

The consumer's surplus is $8.67.

44. $P(x) = R(x) - C(x) = \sqrt{3}x^{1/2} - 0.2x$

$P'(x) = \dfrac{\sqrt{3}}{2}x^{-1/2} - 0.2 = 0$ when $x = \dfrac{3}{0.16}$

$x \approx 19$ people; the net revenue is

$$\sqrt{3}(19)^{1/2} - 0.2(19) \approx 3.75$$

Thus, the net revenue is $3,750.

45. $R(x) = q(30 - 4q^2) - (q^2 + 6q)$

$$= -4q^3 - q^2 + 14q$$

$R'(x) = -12q^2 - 2q + 14 = -2(q - 1)(6q + 7)$

$R'(x) = 0$ when $q = 1$ (disregard negative)

The consumer's surplus is

$$\int_0^1 (20 - 4q^2)\, dq - (1)(16) = 20 - \tfrac{4}{3} - 16 = \tfrac{8}{3}$$

46. a. A horizontal strip has area

$$dA = 2x\, dy = 2\sqrt{b^2 - y^2}\, dy$$

These slabs lie between $y = -b$ and $y = b$

$$V = 2L\int_{-b}^{b} \sqrt{b^2 - y^2}\, dy$$

b. When the tank is filled to a height of $y = h$

$$V = 2L\int_{-b}^{h} \sqrt{b^2 - y^2}\, dy$$

c. The V values are 75.5299, 196.539, 344.337, 502.655, 660.972, 808.77, 932.78, 1,005.31

47. Let $s(q)$ be the supply function. If the market price is s_0, then q_0 units can be supplied. The cost to the producer will be $s_0 q_0$. The area under the supply curve from $[0, q_0]$ represents the cost to all the supplies for this market. The difference is the producer's surplus.

$$s_0 Q_0 - \int_0^{q_0} s(q)\, dq = \int_0^{q_0} [s_0 - s(q)]\, dq$$

CHAPTER 4 REVIEW

Practice Problems, Page 319

1. Suppose f is continuous and $f(x) \geq 0$ throughout the interval $[a, b]$. Then the **area** of the region under the curve $y = f(x)$ over this interval is given by

$$A = \lim_{n \to +\infty} \sum_{k=1}^{n} f(a + k\Delta x)\Delta x$$

where $\Delta x = \dfrac{b - a}{n}$.

2. a. $\displaystyle\sum_{k=1}^{n} c = \underbrace{c + c + \ldots + c}_{n \text{ terms}} = nc$

b. $\displaystyle\sum_{k=1}^{n} (a_k + b_k) = \sum_{k=1}^{n} a_k + \sum_{k=1}^{n} b_k$

c. $\displaystyle\sum_{k=1}^{n} ca_k = c\sum_{k=1}^{n} a_k = \left(\sum_{k=1}^{n} a_k\right)c$

d. $\displaystyle\sum_{k=1}^{n} (ca_k + db_k) = c\sum_{k=1}^{n} a_k + d\sum_{k=1}^{n} b_k$

e. $\displaystyle\sum_{k=1}^{n} a_k = \sum_{k=1}^{m} a_k + \sum_{k=m+1}^{n} a_k$

f. If $a_k \leq b_k$ for $k = 1, 2, \ldots, n$, then
$$\sum_{k=1}^{n} a_k \leq \sum_{k=1}^{n} b_k$$

g. $\displaystyle\sum_{k=1}^{n} 1 = n$

h. $\displaystyle\sum_{k=1}^{n} k = 1 + 2 + 3 + \ldots + n = \frac{n(n+1)}{2}$

i. $\displaystyle\sum_{k=1}^{n} k^2 = 1^2 + 2^2 + 3^2 + \ldots + n^2$

$$= \frac{n(n+1)(2n+1)}{6}$$

j. $\displaystyle\sum_{k=1}^{n} k^3 = 1^3 + 2^3 + 3^3 + \ldots + n^3$

$$= \frac{n^2(n+1)^2}{4}$$

3. $f(\overset{*}{x}_2)\Delta x_2 + \ldots + f(\overset{*}{x}_n)\Delta x_n = \displaystyle\sum_{k=1}^{n} f(\overset{*}{x}_k)\Delta x_k$

4. If f is defined on the closed interval $[a, b]$ we say f is integrable on $[a, b]$ if

$$I = \lim_{\|P\| \to 0} \sum_{k=1}^{n} f(\overset{*}{x}_k)\Delta x_k$$

exists. This limit is called the definite integral of f from a to b. The definite integral is denoted by

$$I = \int_{a}^{b} f(x)\, dx \qquad \text{or} \qquad I = \int_{x=a}^{x=b} f(x)\, dx$$

5. Suppose f is continuous and $f(x) \geq 0$ on the closed interval $[a, b]$. Then the area under the curve $y = f(x)$ on $[a, b]$ is given by the definite integral of f on $[a, b]$. That is,

$$\text{Area} = \int_{a}^{b} f(x)\, dx$$

6. The distance traveled by an object with continuous velocity $v(t)$ along a straight line from time $t = a$ to $t = b$ is

$$S = \int_{a}^{b} |v(t)|\, dt$$

7. **a.** $\displaystyle\int_{a}^{a} f(x)\, dx = 0$

b. $\displaystyle\int_{a}^{b} f(x)\, dx = -\int_{b}^{a} f(x)\, dx$

8. If f is continuous on the interval $[a, b]$ and F is any function that satisfies $F'(x) = f(x)$ throughout this interval, then

$$\int_{a}^{b} f(x)\, dx = F(b) - F(a)$$

9. Define a new variable of integration, $u = g(x)$. Find dx as a function of du and restate the limits. Make sure that new integrand involves only the new variables, and that it has a form which can be integrated.

10. A differential equation is an equation that contains derivatives. A separable differential equation can be rewritten with one variable in the left side of the equation and the other variable in the right side. Each side of the equation is now integrated (if possible).

11. Two trajectories are orthogonal if the slopes of the tangent lines to both curves are negative reciprocals of each other at each point of intersection of the curves.

12. If f is continuous on the interval $[a, b]$, there is at least one number c between a and b such that

$$\int_{a}^{b} f(x)\, dx = f(c)(b - a)$$

13. If f is continuous on the interval $[a, b]$, the average value of f on this interval is given by the integral

$$\frac{1}{b-a} \int_{a}^{b} f(x)\, dx$$

14. Let $f(t)$ be continuous on the interval $[a, b]$ and define the function G by the integral equation

$$G(x) = \int_{a}^{x} f(t)\, dt$$

for $a \leq x \leq b$. Then G is an antiderivative of f on $[a, b]$; that is,

$$G'(x) = \frac{d}{dx}\left[\int_{a}^{x} f(t)\, dt \right] = f(x)$$

on $[a, b]$.

15. If $u(x)$ and $v(x)$ are differentiable functions of x, then

$$\frac{d}{dx}\left[\int_{v(x)}^{u(x)} f(t)\, dt \right] = f(u)\frac{du}{dx} - f(v)\frac{dv}{dx}$$

16. **a.** Divide the interval $[a, b]$ into n

subintervals, each of width $\Delta x = \frac{b-a}{n}$, and let $\overset{*}{x}_k$ denote the right endpoint of the kth subinterval. The base of the kth rectangle is the kth subinterval, and its height is $f(\overset{*}{x}_k)$. Hence, the area of the kth rectangle is $f(\overset{*}{x}_k)\Delta x$. The sum of the areas of all n rectangles is an approximation for the area under the curve and hence an approximation for the corresponding definite integral. Thus,

$$\int_a^b f(x)\, dx \approx \sum_{k=1}^n f(\overset{*}{x}_k)\Delta x$$

b. Let f be continuous on $[a, b]$. The trapezoidal rule is

$$\int_a^b f(x)\, dx \approx \tfrac{1}{2}[f(x_0) + 2f(x_1) +$$

$$2f(x_2) + \cdots + 2f(x_{n-1}) + f(x_n)]\Delta x$$

where $\Delta x = \frac{b-a}{n}$ and, for the kth subinterval, $x_k = a + k\Delta x$.

c. Let f be continuous on $[a, b]$. Simpson's rule is

$$\int_a^b f(x)\, dx \approx \tfrac{1}{3}[f(x_0) + 4f(x_1) + 2f(x_2)$$

$$+ \cdots + 4f(x_{n-1}) + f(x_n)]\Delta x$$

where $\Delta x = \frac{b-a}{n}$, $x_k = a + k\Delta x$, k an integer and n an even integer. Moreover, the larger the value for n, the better the approximation.

17. If f and g are continuous and satisfy $f(x) \geq g(x)$ on the closed interval $[a, b]$, then the area between the two curves $y = f(x)$ and $y = g(x)$ is given by

$$A = \int_a^b [f(x) - g(x)]\, dx$$

18. If q_0 units of a commodity are sold at a price of p_0 dollars per unit and if $p = D(q)$ is the consumer's demand function for the commodity, then

$$\text{Consumer's surplus} = \int_0^{q_0} D(q)\, dq - p_0 q_0$$

19. $\displaystyle\int_0^1 (2x^4 - 3x^2)\, dx = 2\int_0^1 x^4\, dx - 3\int_0^1 x^2\, dx$

$$= 2(\tfrac{1}{5}) - 3(\tfrac{1}{3}) = -\tfrac{3}{5}$$

20. Using Leibniz's rule:

$$F'(x) = \frac{d}{dx}\int_3^x t^5 \sqrt{\cos(2t+1)}\, dt$$

$$= x^5 \sqrt{\cos(2x+1)}$$

21. $\displaystyle\int_1^4 (x^{1/2} + x^{-3/2})\, dx = \frac{2x^{3/2}}{3} - \frac{2x^{-1/2}}{1}\Big|_1^4$

$$= \tfrac{16}{3} - 1 - \tfrac{2}{3} + 2 = \tfrac{17}{3}$$

22. $\displaystyle\int_0^1 (2x^3 - 18x^2 + 40x - 12)\, dx$

$$= \frac{x^4}{2} - 6x^3 + 20x^2 - 12x\Big|_0^1$$

$$= \tfrac{1}{2} - 6 + 20 - 12 = \tfrac{5}{2}$$

23. Let $u = 1 + \cos x$, $du = -\sin x\, dx$. For the new limits, when $x = 0$, $u = 0$, and when $x = \frac{\pi}{2}$, $u = 1$.

$$\int_0^{\pi/2} (1 + \cos x)^{-2} \sin x\, dx = -\int_0^1 u^{-2}\, du$$

$$= \frac{1}{u}\Big|_2^1 = 1 - \tfrac{1}{2} = \tfrac{1}{2}$$

24. Let $u = 2x^2 + 2x + 5$, $du = (4x + 2)\, dx$. For new limits, when $x = -2$, $u = 9$, and when $x = 1$, $u = 9$.

$$\tfrac{1}{2}\int_9^9 u^{1/2}\, du = 0$$

25. a. $\displaystyle A = \int_{-1}^3 (3x^2 + 2)\, dx = x^3 + 2x\Big|_{-1}^3$

$$= 27 + 6 + 1 + 2 = 36$$

b. The curves intersect when

$$x^6 = x$$

$$x(x^5 - 1) = 0$$

$$x(x - 1)(x^4 + x^3 + x^2 + x + 1) = 0$$

$$x = 0, 1$$

They intersect at $(0, 0)$, $(1, 1)$. On $[0, 1]$ the curve $y^3 = x$ is above $y = x^2$.

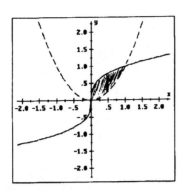

We can use either vertical or horizontal strips. Vertical:

$$A = \int_0^1 (x^{1/3} - x^2)\, dx = \frac{3x^{4/3}}{4} - \frac{x^3}{3}\Big|_0^1$$

$$= \frac{3}{4} - \frac{1}{3} = \frac{5}{12}$$

Horizontal: (The parabola is the leading curve.)

$$A = \int_0^1 (y^{1/2} - y^3)\, dy = \frac{2y^{3/2}}{3} - \frac{y^4}{4}\Big|_0^1$$

$$= \frac{2}{3} - \frac{1}{4} = \frac{5}{12}$$

26. Average value $= \dfrac{1}{b-a}\displaystyle\int_0^{\pi/2} \cos 2x\, dx$

$$= \frac{2}{\pi}\int_0^{\pi/2} \frac{1}{2}\cos 2x\, (2\,dx) = \frac{1}{\pi}(\sin 2x)\Big|_0^{\pi/2}$$

$$= \frac{1}{\pi}(0) = 0$$

27. $a(t) = 2t + 1;\; v(t) = t^2 + t + v_0 = t^2 + t + 2$

$s(t) = \dfrac{t^3}{3} + \dfrac{t^2}{2} + 2t + 4$

28. **a.** The profit ends when $R' = C'$.

$$1575 - 5x^2 = 1200 + 10x^2,$$

$$15x^2 = 375$$

$$x^2 = 25$$

$$x = 5 \text{ years.}$$

b. Earnings, the difference between the revenue and the cost at any given time is

$$\int_0^5 [(1{,}575 - 5x^2) - (1{,}200 + 10x^2)]\, dx$$

$$= \int_0^5 (-15x^2 + 375)\, dx$$

$$= -5x^3 - 375x \Big|_0^5$$

$$= -625 + 1875 = \$1{,}250$$

29. $$\frac{dy}{y^2} = \sin 3x\, dx$$

$$\int y^{-2}\, dy = \frac{1}{3}\int \sin 3x\, (3\,dx)$$

$$-\frac{1}{y} = -\frac{1}{3}\cos 3x + C$$

$$y = \frac{3}{\cos 3x + C}$$

30. **a.** For the trapezoidal rule we need

$$\frac{(b-a)^3}{12n^2} M \le 0.0005 \text{ where } M \text{ is the}$$

maximum of $|f''(x)|$ on $[a, b]$.

$f(x) = \cos x,\; f'(x) = -\sin x,$

$f''(x) = -\cos x,$ so on $[0, \frac{\pi}{2}]$,

$M = |-1| = 1.$

$$\frac{(\frac{\pi}{2})^3}{12n^2}(1) \le 0.0005,\; n^2 \ge 646,\; n \ge 26.$$

b. For Simpson's rule we need $\dfrac{(b-a)^5}{180n^4}$

$K \le 0.0005$ where K is the maximum of $|f^{(4)}(x)|$ on $[0, \frac{\pi}{2}]$. $f'''(x) = \sin x,$

$f^{(4)}(x) = \cos x.$ So on $[0, \frac{\pi}{2}]$, $M = 1.$

$$\frac{(\frac{\pi}{2})^5}{180n^4}(1) \le 0.0005,\; n^4 \ge 106.3,\; n \ge 4$$

Supplementary Problems, Page 320

1. $\displaystyle\int_{-1}^1 [3g(x) + 2f(x)]\, dx = 3\int_{-1}^1 g(x)\, dx\, 1$

$$+ 2\left[\int_{-1}^0 f(x)\, dx + \int_0^1 f(x)\, dx\right]$$

$$= 3(7) + [3 + (-1)] = 25$$

2. $\displaystyle\lim_{n\to+\infty}\left[3\sum_{k=1}^n k^2\left(\frac{1}{n}\right)^2\left(\frac{1}{n}\right) + 2\sum_{k=1}^n \frac{k}{n^2} - \sum_{k=1}^n \frac{1}{n}\right]$

$$= \lim_{n \to +\infty} \left[\frac{3}{n^3} \sum_{k=1}^{n} k^2 + \frac{2}{n^2} \sum_{k=1}^{n} k - \frac{1}{n}(n) \right]$$

$$= \lim_{n \to +\infty} \left[\frac{3}{n^3} \frac{n(n+1)(2n+1)}{6} + \frac{2}{n^2} \frac{n(n+1)}{2} - 1 \right]$$

$$= 1 + 1 - 1 = 1$$

3. $$\lim_{n \to +\infty} \left[4 \sum_{k=1}^{n} k^3 \left(\frac{1}{n}\right)^3 \left(\frac{1}{n}\right) + 6 \sum_{k=1}^{n} \frac{k^2}{n^2} + 3 \sum_{k=1}^{n} \frac{1}{n} \right]$$

$$= \lim_{n \to +\infty} \left[\frac{4}{n^4} \sum_{k=1}^{n} k^3 + \frac{6}{n^3} \sum_{k=1}^{n} k^2 - \frac{3}{n}(n) \right]$$

$$= \lim_{n \to +\infty} \left[\frac{4}{n^4} \frac{n^2(n+1)^2}{4} + \frac{6}{n^3} \frac{n(n+1)(2n+1)}{6} + 3 \right]$$

$$= 1 + 2 + 3 = 6$$

4. $$\int_{0}^{1} (5x^4 - 8x^3 + 1) \, dx = (x^5 - 2x^4 + x)\Big|_{0}^{1} = 0$$

5. $$\int_{-1}^{2} 30(5x - 2)^2 \, dx = 6 \int_{-1}^{2} (5x - 2)^2 \, d(5x - 2)$$

$$= 6(\tfrac{1}{3})(5x - 2)^3 \Big|_{-1}^{2} = 1{,}710$$

6. $$\int_{0}^{1} (x\sqrt{x} + 2)^2 \, dx = \int_{0}^{1} (x^3 + 4x^{3/2} + 4) \, dx$$

$$= [\tfrac{1}{4}x^4 + 4(\tfrac{2}{5})x^{5/2} + 4x]\Big|_{0}^{1} = \frac{117}{20}$$

7. $$\int_{1}^{2} \frac{x^2}{\sqrt{x^3 + 1}} \, dx = \frac{1}{3} \int_{1}^{2} (x^3 + 1)^{-1/2} (3x^2 \, dx)$$

$$= \frac{1}{3}(2\sqrt{x^3 + 1})\Big|_{1}^{2} = \frac{6 - 2\sqrt{2}}{3}$$

8. $$\int_{2}^{2} (x + \sin x)^3 \, dx = 0$$

9. $$\int_{-1}^{0} \frac{dx}{\sqrt{1 - 2x}} = -\frac{1}{2} \int_{-1}^{0} (1 - 2x)^{-1/2}(-2 \, dx)$$

$$= -\frac{1}{2}(2\sqrt{1 - 2x})\Big|_{-1}^{0} = \sqrt{3} - 1$$

10. $$\int_{1}^{2} \frac{dx}{\sqrt{3x - 1}} = \frac{1}{3} \int_{1}^{2} (3x - 1)^{-1/2}(3 \, dx)$$

$$= \frac{2}{3}(3x - 1)^{1/2}\Big|_{1}^{2} = \frac{2}{3}(\sqrt{5} - \sqrt{2})$$

11. $$\int_{-1}^{0} \frac{dx}{\sqrt[3]{1 - 2x}} = -\frac{1}{2} \int_{-1}^{0} (1 - 2x)^{-1/3}(-2 \, dx)$$

$$= -\frac{3}{4}(1 - 2x)^{2/3}\Big|_{-1}^{0} = -\frac{3}{4}(1 - \sqrt[3]{9})$$

$$\approx 0.810063$$

12. $$\int \sqrt{x}(x^2 + \sqrt{x} + 1) \, dx = \int (x^{5/2} + x + x^{1/2}) \, dx$$

$$= \frac{2}{7}x^{7/2} + \frac{1}{2}x^2 + \frac{2}{3}x^{3/2} + C$$

13. $$\int (x - 1)^3 \, dx = \frac{1}{4}(x - 1)^3 + C$$

14. $$\int \frac{x^2 + 1}{x^2} \, dx = \int (1 + x^{-2}) \, dx = x - \frac{1}{x} + C$$

15. $$\int (\sin^2 x + \cos^2 x) \, dx = \int 1 \, dx = x + C$$

16. $$I = \int x(x + 4)\sqrt{x^3 + 6x^2 + 2} \, dx$$

Let $u = x^3 + 6x^2 + 2$; $du = 3x(x + 4) \, dx$

$$I = \frac{1}{3}(x^3 + 6x^2 + 2)^{3/2}(\tfrac{2}{3}) + C$$

$$= \frac{2}{9}(x^3 + 6x^2 + 2)^{3/2} + C$$

17. $$I = \int x(2x^2 + 1)\sqrt{x^4 + x^2} \, dx$$

Let $u = x^4 + x^2$; $du = (4x^3 + 2x) \, dx$

$$I = \frac{1}{4} \int u^{1/2} \, du = \frac{1}{3}u^{3/2} + C$$

$$= \frac{1}{2}(x^4 + x^2)^{3/2} + C$$

$$= \frac{x^3(x^2 + 1)^{3/2}}{3} + C$$

18. $$I = \int \frac{dx}{\sqrt{x}(\sqrt{x} + 1)^2}$$

Let $u = \sqrt{x} + 1$; $du = \frac{dx}{2\sqrt{x}}$

$$I = 2 \int u^{-2} \, du = -2u^{-1} + C$$

$$= -\frac{2}{\sqrt{x} + 1} + C$$

19. $$I = \int x\sqrt{1 - 5x^2} \, dx$$

Let $u = 1 - 5x^2$; $du = -10x$

$$I = -\frac{1}{10}(\tfrac{2}{3})(1 - 5x^2)^{3/2} + C$$

$$= -\frac{1}{15}(1 - 5x^2)^{3/2} + C$$

20. $\displaystyle\int \sqrt{\sin x - \cos x}(\sin x + \cos x)\,dx$

$\displaystyle = \int (\sin x - \cos x)^{1/2} d(\sin x - \cos x)$

$\displaystyle = \tfrac{2}{3}(\sin x - \cos x)^{3/2} + C$

21. $\displaystyle\int_{-10}^{10} [3 + 7x^{73} - 100x^{101}]\,dx$

$\displaystyle = 2\int_{0}^{10} 3\,dx + 0 + 0 = 6x\big|_{0}^{10} = 60$

22. $\displaystyle\int_{-\pi/4}^{\pi/4} [\sin(4x) + 2\cos(4x)]\,dx$

$\displaystyle = 0 + 4\int_{0}^{\pi/4} \cos 4x\,dx = \sin 4x\big|_{0}^{\pi/4} = 0$

23. $\displaystyle F'(x) = \cos^4 t\,\Big|_{x^2}^{x^3} = 3x^8\cos^4 x^3 - 2x^5\cos^4 x^2$

24. $\displaystyle\int_{1}^{4} x^{-2}\,dx = -x^{-1}\big|_{1}^{4} = \tfrac{3}{4}$

25. $\displaystyle\int_{-1}^{1} (2 + x - x^2)\,dx = (2x + \tfrac{1}{2}x^2 - \tfrac{1}{3}x^3)\Big|_{-1}^{1} = \tfrac{10}{3}$

26. $\displaystyle\int_{0}^{2} x^4\,dx = \tfrac{1}{5}x^5\Big|_{0}^{2} = \tfrac{32}{5}$

27. $\displaystyle\int_{-1}^{0} (-x)\sqrt{x^2 + 5}\,dx + \int_{0}^{2} x\sqrt{x^2 + 5}\,dx$

$\displaystyle = \left[-\tfrac{1}{3}(x^2 + 5)^{3/2}\right]\Big|_{-1}^{0} + \left[\tfrac{1}{3}(x^2 + 5)^{3/2}\right]\Big|_{0}^{2}$

$\displaystyle = \tfrac{1}{3}[5^{3/2} - 6^{3/2}] + \tfrac{1}{3}[9^{3/2} - 5^{3/2}] = 9 - 2\sqrt{6}$

28. $\displaystyle\int_{-1/2}^{1} (2y + 2 - 4y^2)\,dy = (y^2 + 2y - \tfrac{4}{3}y^3)\Big|_{1/2}^{1} = \tfrac{17}{6}$

29. $\displaystyle\int_{0}^{1} (x^3 - x^4)\,dx = (\tfrac{1}{4}x^4 - \tfrac{1}{5}x^5)\Big|_{0}^{1} = \tfrac{1}{20}$

30. $\displaystyle\int_{0}^{1} (y^{2/3} - y^2)\,dy = (\tfrac{3}{5}y^{5/3} - \tfrac{1}{3}y^3)\Big|_{0}^{1} = \tfrac{4}{15}$

31. $\displaystyle\int_{0}^{\pi/6} (\cos x - \sqrt{3}\sin x)\,dx - \int_{\pi/6}^{\pi/2} (\cos x - \sqrt{3}\sin x)\,dx$

$\displaystyle = (\sin x + \sqrt{3}\cos x)\Big|_{0}^{\pi/6} - (\sin x + \sqrt{3}\cos x)\Big|_{\pi/6}^{\pi/2}$

$\displaystyle = \tfrac{1}{2} + \sqrt{3}\,\tfrac{\sqrt{3}}{2} - \sqrt{3} - 1 + \tfrac{\sqrt{3}}{2}\sqrt{3} + \tfrac{1}{2}$

$\displaystyle = 3 - \sqrt{3} \approx 1.2679$

32. a. $\displaystyle f'(t) = \int (\sin 4t - \cos 2t)\,dt$

$\displaystyle = -\tfrac{1}{4}\cos 4t - \tfrac{1}{2}\sin 2t + C_1$

Since $f'(\tfrac{\pi}{4}) = 1$, $C_1 = \tfrac{5}{4}$, so

$\displaystyle f'(x) = -\tfrac{1}{4}\cos 4t - \tfrac{1}{2}\sin 2t + \tfrac{5}{4}$

$\displaystyle f(t) = \int (-\tfrac{1}{4}\cos 4t - \tfrac{1}{2}\sin 2t + \tfrac{5}{4})\,dt$

$\displaystyle = -\tfrac{1}{16}\sin 4t + \tfrac{1}{4}\cos 2t + \tfrac{5}{4}t + C_2$

Since $f(\tfrac{\pi}{2}) = 1$, $C_2 = \tfrac{5}{4} - \tfrac{5}{8}\pi$

$\displaystyle f(t) = -\tfrac{1}{16}\sin 4t + \tfrac{1}{4}\cos 2t + \tfrac{5}{4}t + \tfrac{5}{4} - \tfrac{5}{8}\pi$

b. $\displaystyle f''(x) = \tfrac{1}{2}x^4 + \tfrac{1}{3}x^3 + C_1;$

$\displaystyle f''(1) = 2$, so $C_1 = \tfrac{7}{6}$

$\displaystyle f'(x) = \tfrac{1}{10}x^5 + \tfrac{1}{12}x^4 + \tfrac{7}{6}x + C_2;$

$\displaystyle f'(1) = 1$, so $C_2 = -\tfrac{7}{20}$

$\displaystyle f(x) = \tfrac{1}{60}x^6 + \tfrac{1}{60}x^5 + \tfrac{7}{12}x^2 - \tfrac{7}{20}x + C_3;$

$\displaystyle f(1) = 0$, so $C_3 = -\tfrac{4}{15}$

$\displaystyle f(x) = \tfrac{1}{60}(x^6 + x^5 + 35x^2 - 21x - 16)$

33. $\displaystyle\frac{dy}{dx} = (1 - y)^2$

$\displaystyle\int (1 - y)^{-2}\,dy = \int dx$

$\displaystyle -[-(1 - y)^{-1}] = x + C$

$\displaystyle 1 - y = \frac{1}{x - C}$

$\displaystyle y = 1 - \frac{1}{x - C}$

34. $\displaystyle\frac{dy}{dx} = \frac{\cos 4x}{y}$

$\displaystyle\int y\,dy = \int \cos 4x\,dx$

$\displaystyle\tfrac{1}{2}y^2 = \tfrac{1}{4}\sin 4x + C_1$

$$y^2 = \tfrac{1}{2} \sin 4x + C$$

35.
$$\frac{dy}{dx} = \left(\frac{\cos y}{\sin x}\right)^2$$

$$\int \sec^2 y \; dy = \int \csc^2 x \; dx$$

$$\tan y = -\cot x + C$$

$$\tan y + \cot x = C$$

36.
$$\frac{1}{y^2} + \frac{dy}{dx} = \left(\frac{x}{y}\right)^2$$

$$\int y^2 \; dy = \int (x^2 - 1) \; dx$$

$$\tfrac{1}{3} y^3 = \tfrac{1}{3} x^3 - x + C_1$$

$$y^3 = x^3 - 3x + C$$

37.
$$\frac{dy}{dx} = \frac{x-1}{y} + \frac{y - x(y-1)}{y^2}$$

$$\int y^2 \; dy = \int [xy - y + y - x(y-1)] \; dx$$

$$\tfrac{1}{3} y^3 = \tfrac{1}{2} x^2 + C_1$$

$$2y^3 = 3x^2 + C$$

38.
$$\frac{dy}{dx} = \frac{x}{y}\sqrt{\frac{y^2 + 2}{x^2 + 1}}$$

$$\int \frac{y \; dy}{\sqrt{y^2 + 2}} = \int \frac{x \; dx}{\sqrt{x^2 + 1}}$$

$$\sqrt{y^2 + 2} = \sqrt{x^2 + 1} + C$$

39.
$$\frac{dy}{dx} = \frac{x \sin x^2}{y^2 \cos y^3}$$

$$\int y^2 \cos y^3 \; dy = \int x \sin x^2 \; dx$$

$$\tfrac{1}{2} \int \cos y^3 (3y^2 \; dy) = \tfrac{1}{2} \int \sin x^2 (2x \; dx)$$

$$\tfrac{1}{3} \sin y^3 = -\tfrac{1}{2} \cos x^2 + C_1$$

$$\sin y^3 = -\tfrac{3}{2} \cos x^2 + C$$

40.
$$\frac{dy}{dx} = \sqrt{\frac{x}{y}}$$

$$\int y^{1/2} \; dy = \int x^{1/2} \; dx$$

$$\tfrac{2}{3} y^{3/2} = \tfrac{2}{3} x^{3/2} + C_1$$

$$y^{3/2} = x^{3/2} + C$$

41. $\dfrac{4}{\pi} \displaystyle\int_0^{\pi/4} \dfrac{\sin x}{\cos^2 x} \; dx = \dfrac{4}{\pi} \displaystyle\int_0^{\pi/4} \tan x \sec x \; dx$

$$= \tfrac{4}{\pi} \sec x \Big|_0^{\pi/4} = \tfrac{4}{\pi}(\sqrt{2} - 1) \approx 0.5274$$

42. a. $\dfrac{1}{\pi} \displaystyle\int_0^{\pi} \sin x \; dx = -\tfrac{1}{\pi} \cos x \Big|_0^{\pi} = \tfrac{2}{\pi} \approx 0.6366$

b. $\dfrac{1}{2\pi} \displaystyle\int_0^{2\pi} \sin x \; dx = -\dfrac{1}{2\pi} \cos x \Big|_0^{2\pi} = 0$

43. Exact value: $\displaystyle\int_0^{\pi} \sin x \; dx = -\cos x \Big|_0^{\pi} = 2$

$$\Delta x = \frac{\pi - 0}{6} = \frac{\pi}{6} \approx 0.5236$$

$x_0 = 0.0000$	$f(x_0) = 0.0000$
$x_1 = 0.5236$	$f(x_1) = 0.5000$
$x_2 = 1.0472$	$f(x_2) = 0.8660$
$x_3 = 1.5708$	$f(x_3) = 1.0000$
$x_4 = 2.0944$	$f(x_4) = 0.8660$
$x_5 = 2.6180$	$f(x_5) = 0.5000$
$x_6 = 3.1416$	$f(x_6) = 0.0000$

Trapezoidal rule:

$$A \approx \tfrac{1}{2}[1(0) + 2(0.5) + 2(0.8660) + 2(1) + 2(0.8660) + 2(0.5) + 1(0)](\tfrac{\pi}{6})$$

$$\approx \tfrac{\pi}{12}(7.4640) \approx 1.9541$$

44. $\Delta x = \dfrac{1 - 0}{6} = \dfrac{1}{6} \approx 0.1667$

$x_0 = 0.0000$	$f(x_0) = 1.0000$
$x_1 = 0.1667$	$f(x_1) = 1.0023$
$x_2 = 0.3333$	$f(x_2) = 1.0183$
$x_3 = 0.5000$	$f(x_3) = 1.0607$
$x_4 = 0.6667$	$f(x_4) = 1.1367$
$x_5 = 0.8333$	$f(x_5) = 1.2565$
$x_6 = 1.0000$	$f(x_6) = 1.4142$

Trapezoidal rule:

$$A \approx \tfrac{1}{2}[1(1) + 2(1.0023) + 2(1.0183) + 2(1.0607) + 2(1.1367) + 2(1.2565) + 1(1.4142)](\tfrac{1}{6})$$

$$\approx \tfrac{1}{12}(13.2652) \approx 1.1054$$

45. $\Delta x = \dfrac{1 - 0}{8} = \dfrac{1}{8} = 0.125$

$$x_0 = 0.000 \qquad f(x_0) = 1.000$$
$$x_1 = 0.125 \qquad f(x_1) = 0.999$$
$$x_2 = 0.250 \qquad f(x_2) = 0.992$$
$$x_3 = 0.375 \qquad f(x_3) = 0.975$$
$$x_4 = 0.500 \qquad f(x_4) = 0.943$$
$$x_5 = 0.625 \qquad f(x_5) = 0.897$$
$$x_6 = 0.750 \qquad f(x_6) = 0.839$$
$$x_7 = 0.875 \qquad f(x_7) = 0.774$$
$$x_8 = 1.000 \qquad f(x_8) = 0.707$$

Trapezoidal rule:

$$A \approx \tfrac{1}{2}[1(1.000)+2(0.999)+2(0.992)$$
$$+2(0.975)+2(0.943)+2(0.897)$$
$$+2(0.839)+2(0.774)+1(0.707)](\tfrac{1}{8})$$
$$\approx \tfrac{1}{16}(14.5424) \approx 0.9089$$

46. $\Delta x = \dfrac{1-0}{6} = \dfrac{1}{6} \approx 0.1667$

$$x_0 = 0.0000 \qquad f(x_0) = 1.0000$$
$$x_1 = 0.1667 \qquad f(x_1) = 1.0023$$
$$x_2 = 0.3333 \qquad f(x_2) = 1.0183$$
$$x_3 = 0.5000 \qquad f(x_3) = 1.0607$$
$$x_4 = 0.6667 \qquad f(x_4) = 1.1367$$
$$x_5 = 0.8333 \qquad f(x_5) = 1.2565$$
$$x_6 = 1.0000 \qquad f(x_6) = 1.4142$$

Simpson's rule:

$$A \approx \tfrac{1}{3}[1(1.0000)+4(1.0023)+2(1.0183)$$
$$+4(1.0607)+2(1.1367)$$
$$+4(1.2565)+1(1.4142)](\tfrac{1}{6})$$
$$\approx \tfrac{1}{18}(19.8042) \approx 1.1003$$

47. $\Delta x = \dfrac{1-0}{8} = \dfrac{1}{8} = 0.125$

$$x_0 = 0.000 \qquad f(x_0) = 1.000$$
$$x_1 = 0.125 \qquad f(x_1) = 0.999$$
$$x_2 = 0.250 \qquad f(x_2) = 0.992$$
$$x_3 = 0.375 \qquad f(x_3) = 0.975$$
$$x_4 = 0.500 \qquad f(x_4) = 0.943$$
$$x_5 = 0.625 \qquad f(x_5) = 0.897$$
$$x_6 = 0.750 \qquad f(x_6) = 0.839$$
$$x_7 = 0.875 \qquad f(x_7) = 0.774$$
$$x_8 = 1.000 \qquad f(x_8) = 0.707$$

Simpson's rule:

$$A \approx \tfrac{1}{3}[1(1.0000)+4(0.999)+2(0.992)$$
$$+4(0.975)+2(0.943)$$
$$+4(0.897)+2(0.839)$$
$$+4(0.774)+1(0.707)(\tfrac{1}{8})$$
$$\approx \tfrac{1}{24}(21.8902) \approx 0.9096$$

48. Trapezoidal Rule used to calculate estimate.

TYPE OF ESTIMATE	# OF SUB-INTERVALS	ESTIMATE OVER [1, 2]
Trapezoid	4	1.48126048687
Trapezoid	6	1.48058264586
Trapezoid	8	1.48034389072
Trapezoid	18	1.48023311187
Trapezoid	28	1.48008513653
Trapezoid	48	1.48004809471
Trapezoid	88	1.48003883123
Trapezoid	168	1.48003651517
Trapezoid	1888	1.48003576298

The maximum value of $f''(x)$ is about 0.7, so let $M = 1$. $n^2 > [12(0.00005)]^{-1} = 1{,}666.7$ or $n = 41$.

49. Trapezoidal Rule used to calculate estimate.

TYPE OF ESTIMATE	# OF SUB-INTERVALS	ESTIMATE OVER [0, 1]
Trapezoid	4	.217285882353
Trapezoid	6	.215759233382
Trapezoid	8	.215252876377
Trapezoid	18	.215018582773
Trapezoid	28	.214786003261
Trapezoid	48	.214627878269
Trapezoid	88	.214688347819
Trapezoid	168	.214683464287
Trapezoid	1888	.214681878269

The maximum value of $f''(x)$ is about 2, so let $M = 2$. $n^2 > 33.3$ or $n = 6$.

50. Simpson's Rule used to calculate estimate.

TYPE OF ESTIMATE	# OF SUB-INTERVALS	ESTIMATE OVER [1, 2]
Simpson	4	1.48007459248
Simpson	6	1.48004384175
Simpson	8	1.48003835867
Simpson	18	1.48003682586
Simpson	28	1.48003581168
Simpson	48	1.48003574744
Simpson	88	1.48003574341
Simpson	168	1.48003574315
Simpson	1888	1.48003574314

The maximum value of $f''(x)$ is about 7.4, so let $M = 10$. $n^4 > [18(0.00005)]^{-1} = 1{,}111$ or $n = 6$.

51. Trapezoidal Rule used to calculate estimate.

TYPE OF ESTIMATE	# OF SUB-INTERVALS	ESTIMATE OVER [0, 1]
Trapezoid	6	1.01405563943

52. $y = \sin x$ intersects $y = \cos x$ when $\tan x = 1$ at $x = \pi/4$ and $x = 5\pi/4$.

$$A = \int_0^{\pi/4} (\cos x - \sin x)\,dx + \int_{\pi/4}^{5\pi/4} (\sin x - \cos x)\,dx$$

$$+ \int_{5\pi/4}^{2\pi} (\cos x - \sin x)\,dx$$

$$= (\cos x + \sin x)\Big|_0^{\pi/4} + (-\sin x - \cos x)\Big|_{\pi/4}^{5\pi/4}$$

$$+ (\cos x - \sin x)\Big|_{5\pi/4}^{2\pi}$$

$$= \sqrt{2} - 1 - (-2\sqrt{2}) + 1 + \sqrt{2} = 4\sqrt{2}$$

53. The deceleration is $a(t) = -k$ m/s;

$$v(t) = -kt + v_0 = -kt + 25$$

$$s(t) = -\tfrac{1}{2}kt^2 + 25t + 0$$

The car will stop when the velocity is 0 at time t_1 or $t_1 = 25/k$.

$$50 = -\tfrac{1}{2}kt_1^2 + 25t_1$$

$$\tfrac{1}{2}kt_1^2 - 25t_1 + 50 = 0$$

$$\tfrac{1}{2}k(\tfrac{25}{k})^2 - 25(\tfrac{25}{k}) + 50 = 0$$

$$50 - \frac{625}{2k} = 0$$

$$k = 6.25$$

The deceleration is -6.25 m/s.

54. a. $a = \dfrac{dv}{dt} = \dfrac{dv}{ds}\dfrac{ds}{dt} = -4s$

$$v\frac{dv}{ds} = -4s$$

$$\int v\,dv = \int -4s\,ds$$

$$\frac{v^2}{2} = -2s^2 + C$$

Substituting $v = 0$ and $s = 5$ makes the constant $C = 50$ so that $v^2 + 4s^2 = 100$.

b. $v^2 + 4(3^2) = 100$ implies $v = \pm 8$ m/s

55. $R(x) = 150x^{2/3} + C_1$; $C(x) = 0.2x^2 + C_2$;

$$P(x) = 150x^{2/3} - 0.2x^2 + C;$$

$$520 = 150(16^{2/3}) - (0.2)(16^2) + C$$

$$C = -381.24$$

$$P(x) = 150x^{2/3} - 0.2x^2 - 381.24$$

$P(25) \approx 776.24$; the manufacturer's profit is $776.24.

56.
$$\frac{dh}{dt} = 1 + \frac{1}{(t+1)^2}$$

$$\int dh = \int \left[1 + \frac{1}{(t+1)^2}\right] dt$$

$$h = t - \frac{1}{t+1} + C$$

Since the tree was 5 ft tall after 2 years, $C = \tfrac{10}{3}$ and $h(0) = \tfrac{7}{3}$. Thus the tree was 2.33 ft tall when it was transplanted.

57. $R(x) = \displaystyle\int \sqrt{x}(x^{3/2} + 1)^{-1/2}\,dx$

$$= \tfrac{2}{3}\int (x^{3/2} + 1)^{-1/2}(\tfrac{3}{2}x^{1/2}\,dx)$$

$$= 2(x^{3/2} + 1)^{1/2} + C$$

$R(0) = 0 = \tfrac{4}{3} + C$ or $C = -\tfrac{4}{3}$;

$R(4) = \tfrac{4}{3}(8+1)^{1/2} - \tfrac{4}{3} \approx \$2.67$.

58. $\dfrac{dy}{dx} = x\sqrt{x^2 + 5}$

$$y = \tfrac{1}{2}\int (x^2 + 5)^{1/2}(2x\,dx)$$

$$= \tfrac{1}{3}(x^2 + 5)^{3/2} + C$$

Since the curve passes through $(2, 10)$

$$10 = \tfrac{1}{3}(2^2 + 5)^{3/2} + C$$

$$1 = C$$

$$y = \tfrac{1}{3}(x^2 + 5)^{3/2} + 1$$

59. $a(t) = 12(2t + 1)^{-3/2}$; $v(0) = 0$; $x(0) = 3$

$$v(t) = 6\int (2t + 1)^{-3/2}(2\,dt)$$

$$= -12(2t + 1)^{-1/2} + C$$

$$= -12(2t + 1)^{-1/2} + 12$$

$$s(t) = -6\int (2t + 1)^{-1/2}(2\,dt) + \int 12\,dt$$

$$= -6[2(2t + 1)^{1/2} - 2t - 1] + C$$

At $t = 0$, $x = 3$, so $C = 9$;

$$s(t) = -6[2(2t + 1)^{1/2} - 2t - 1] + 9$$

$$s(4) = 27$$

60. $\dfrac{dQ}{dt} = 0.1t + 0.2$

$$Q(t) = 0.05t^2 + 0.2t + C$$

Since the current level of carbon monoxide is 3.4 ppm, $C = 3.4$. In 3 years,

$$Q(3) = 0.05(9) + 0.2(3) + 3.4 = 4.05 \text{ ppm}$$

61. Because of the cow, $a = d$ (the deceleration d is assumed to be a constant).

$$v(t) = dt + v_0$$

Since $3d + v_0 = 0$, $v_0 = 3d$.

$$s(t) = \tfrac{1}{2}dt^2 + v_0 t + 0$$

$$s_0 = \tfrac{9}{2}d + 3v_0 = \tfrac{9}{2}d - 9d = -\tfrac{9}{2}d$$

For the second time, we will assume the same deceleration, d.

$$v(t) = dt + v_0 = 20$$

$v(5) = 0$, so $v_0 = -5d - 20$. Combining the two expressions for v_0 leads to

$$-3d = -5d - 20$$

$$d = -10$$

$v_0 = 30$ ft/s and $s_0 = \tfrac{90}{2} = 45$ ft

$$s(t) = -\tfrac{1}{2}(10)t^2 + (30 + 20)t + 0$$

$$s_1 = -5(5^2) + 50(5) = 125 \text{ ft}$$

62. a.

$$\frac{7{,}105}{T+7} = \int_T^{12} (180 - \tfrac{5}{4}t^2) \, dt$$

$$= 180(12 - T) - \tfrac{5}{12}(12^3 - T^3)$$

$$T = 5 \text{ (using a graphing calculator)}$$

b. The total revenue is

$$R(t) = \int P(t) \, dt = 180t - \tfrac{5}{12}t^3 + C$$

$R(0) = 0$, so $C = 0$ and

$$R(5) = 180(5) - \tfrac{5}{12}(5)^3 \approx \$847.92$$

63. $R'(x) = C'(x)$

$$\frac{1}{\sqrt{2x}} = \frac{1}{3}$$

$$\sqrt{2x} = 3$$

$$x = \tfrac{9}{2}$$

The net revenue is $N(x) = R(x) - C(x)$
$= \sqrt{2x} - \tfrac{1}{3}x$; since x is an integer we try
$x = 4$ and $x = 5$; $N(4) = 1.49509$;
$N(5) = 1.49561$; the greater value is when
$x = 5$ additional people are hired, and the net
revenue is \$1,495.

$$\int_0^5 [(2x)^{1/2} - \tfrac{1}{3}x] \, dx$$

$$= [\tfrac{1}{2} \cdot \tfrac{2}{3}(2x)^{3/2} - \tfrac{1}{6}x^2] \Big|_0^5 \approx \$6{,}374.26$$

64. $P(x) = \int (10 + 2\sqrt{x}) \, dx = 10x + \tfrac{4}{3}x^{3/2} + C$

$P(0) = 10$, so $C = 10$;

$$P(9) - P(0) = 90 + 36 + 10 - 10$$

$$= 126 \text{ people}$$

65.   Let $N(t)$ denote the number of bushels that

that are produced over the next t days. Then
$\frac{dN}{dt} = 0.3t^2 + 0.6t + 1$, and the increase in the
crop over the next six days is

$$N(5) - N(0) = \int_0^6 (0.3t^2 + 0.6t + 1) \, dt$$

$$= (0.1t^3 + 0.3t^2 + t) \Big|_0^6 = 38.4$$

If the price remains fixed at \$2 per bushel, the
corresponding increase in the value of the crop
is \$76.80.

66. Since the price of turkey t months after the
beginning of the year is

$$P(t) = 0.06t^2 - 0.2t + 1.2$$

dollars per pound, the average price during the
first six months is

$$\frac{1}{6-0} \int_0^6 (0.06t^2 - 0.2t + 1.2) \, dx$$

$$= \left[\tfrac{1}{6}(0.02t^3 - 0.1t^2 + 1.2t) \right] \Big|_0^6 \approx 1.32$$

The average price was \$1.32 per pound.

67.

$$R(x) = 50 \int_0^9 (40 + 3\sqrt{x}) \, dx$$

$$= 50(40x + 2x^{3/2}) \Big|_3^2 = \$20{,}700$$

68. Consider a small amount of time. The demand
for beef will be $D(n)$ pounds and the price $P(n)$
dollars/lb. The revenue during that
microsecond is the area of a rectangle, namely
$D(n)P(n)$. The revenue over 9 months is the
sum of the areas of rectangles:

$$A = \int_0^{12} D(x)P(x) \, dx$$

69. $f(x)$ is defined at every point on $[a, b]$. If
$f(x) > 0$, then $-|f(x)| \le f(x) \le |f(x)|$ holds
because $|f(x)| = f(x) > 0 > -|f(x)|$. If
$f(x) < 0$, then $-|f(x)| \le f(x) \le |f(x)|$ holds
because $-|f(x)| = f(x) < |f(x)|$.

70.
$$\left| \int_a^b f(x)\, dx \right| = \left| \lim_{n \to +\infty} \sum_{k=1}^n f(x_k)\Delta x_k \right|$$

$$= \lim_{n \to +\infty} \left| \sum_{k=1}^n f(x_k)\Delta x_k \right|$$

$$\leq \lim_{n \to +\infty} \sum_{k=1}^n |f(x)|\Delta x_k = \int_a^b |f(x)|\, dx$$

The inequality is justified on the basis of the triangle inequality.

71.
$$\left| \int_0^\pi \sin x\, dx \right| \leq \int_0^\pi |\sin x|\, dx \leq \int_0^\pi (1)\, dx = \pi$$

72. Let $X = \dfrac{dx}{dt}$ and $Y = \dfrac{dy}{dt}$; then $2X + 5Y = t$ and $X + 3Y = 7 \cos t$. Solve this system to find $X = -35 \cos t + 3t$ and $Y = 14 \cos t - t$. Thus,

$$y = \int (14 \cos t - t)\, dt = 14 \sin t - \tfrac{1}{2}t^2 + C_1$$

and

$$x = \int (3t - 35 \cos t)\, dt = \tfrac{3}{2}t^2 - 35 \sin t + C_2$$

73. This is a slight modification of Putnam Problem 2 of the afternoon session of 1970.

The equation $\dfrac{1}{T}\displaystyle\int_{-T}^{T} f(t)\, dt = \tfrac{1}{2}[f(t_1) + f(t_2)]$

is satisfied for all values of a, b, c, and d if and only if $t_2 = -t^2 = \pm T/\sqrt{3}$. If $T = 3$ hr, $T/\sqrt{3} \approx 1$ hr, 44 min. Therefore, in the case considered, the critical times are 1 hr 44 min each side of noon.

74. This is Putnam Problem 6 in the morning session of 1951. Choose coordinates so that the equation of the parabola is $4ay = x^2$, $a > 0$. The chord connecting the points $P(2as, as^2)$ to the point $Q(2at, at^2)$ has the equation

$$y = \tfrac{1}{2}(t + s)x - ast$$

and the tangent line has slope t. Hence this line will be normal to the parabola at Q if and only if

$$\tfrac{1}{2}t(t + s) = -1 \text{ or } s = -\tfrac{2}{s} - t$$

We see, therefore, that s and t have opposite signs. Take $s < 0$ and $t > 0$. Then the area cut off by the chord is

$$\int_{2as}^{2at} \left[\tfrac{1}{2}(t + s)x - ast - \frac{x^2}{4a} \right] dx = \frac{a^2}{3}(t - s)^3$$

The area will be minimal when $t - s$ is minimal. But

$$t - s = 2t + \frac{2}{t} = 2\left(\sqrt{t} - \frac{1}{\sqrt{t}}\right)^2 + 4 \geq 4$$

Equality is attained only when $\sqrt{t} = 1$ and hence $t = 1$. Thus, of all the normals to the parabola at points to the right of the axis the normal is $(2a, a)$ cuts off the least area. The area cut off is $64a^2/3$. By symmetry, the normal at $(-2a, a)$ cuts off the lease area among normals at points to the left of the axis. The critical normals can be characterized at those which meet the axis at an angle of $45°$.

75. This is Putnam Problem 1 from the morning session in 1958. If $a_0 + a_1 x + a_2 x^2 + \cdots + a_n x^n = 0$, then

$$\int_0^1 f(x)\, dx = \frac{a_0}{1} + \frac{a_1}{2} + \cdots + \frac{a_n}{n + 1} = 0$$

Hence, by the mean value theorem for integrals, there exists a number c between 0 and 1 such that

$$f(c) = \int_0^1 f(x)\, dx = 0$$

Remark: this problem appears in G. H. Hardy, *A Course in Pure Mathematics*, 7th ed. Cambridge University Press, 1938, p. 243. It is stated that the problem appeared in the *Cambridge Mathematical Tripos* for 1929.

CHAPTER 5

Exponential, Logarithmic, and Inverse Trigonometric Functions

5.1 Exponential Functions; The Number *e*,
Page 332

1.

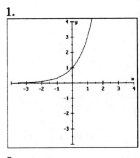

2.

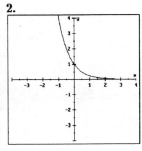

3.

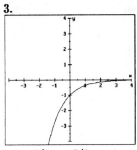

4.

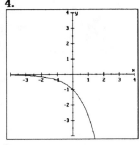

5. $32^{2/5} + 9^{3/2} = 2^2 + 3^3 = 31$

6. $(1 + 4^{3/2})^{-1/2} = (1 + 8)^{-1/2} = 3^{-1} = \frac{1}{3}$

7. 0.7368062997

8. 1.992301858

9. 13.46373803

10. 0.128734903

11. 200.33681

12. 1.156785174

13. 9,783.225896

14. 100,662.188

15. 38,523.62544

16. 569,772.0391

17.
$$3^{x^2 - x} = 9$$
$$3^{x^2 - x} = 3^2$$
$$x^2 - x = 2$$
$$(x - 2)(x + 1) = 0$$
$$x = 2, \ -1$$

18.
$$4^{x^2 + x} = 16$$
$$x^2 + x = 2$$
$$(x + 2)(x - 1) = 0$$
$$x = -2, 1$$

19. $2^x 5^{x+2} = 25,000$
$$2^x 5^x 5^2 = 2^3 5^5$$
$$(2 \cdot 5)^x = 2^3 5^3$$
$$10^x = 10^3$$
$$x = 3$$

20.
$$3^x 4^{x+1/2} = 3,456$$
$$3^x 4^x 4^{1/2} = 3,456$$
$$(3 \cdot 4)^x = 1,728$$
$$12^x = 12^3$$
$$x = 3$$

21.
$$(\sqrt[3]{2})^{x+10} = 2^{x^2}$$
$$2^{x/3 + 10/3} = 2^{x^2}$$
$$\frac{x}{3} + \frac{10}{3} = x^2$$
$$3x^2 - x - 10 = 0$$
$$(x - 2)(3x + 5) = 0$$
$$x = 2, \ -\frac{5}{3}$$

22.
$$(\sqrt[3]{5})^{x+2} = 5^{x^2}$$
$$\frac{x}{3} + \frac{2}{3} = x^2$$
$$3x^2 - x - 2 = 0$$
$$(x - 1)(3x + 2) = 0$$
$$x = 1, \ -\frac{2}{3}$$

23. $e^{2x+3} = 1$
$$e^{2x+3} = e^0$$
$$2x + 3 = 0$$
$$x = -\frac{3}{2}$$

24.
$$\frac{e^{x^2}}{e^{x+6}} = 1$$
$$e^{x^2} = e^{x+6}$$
$$x^2 = x + 6$$
$$(x - 3)(x + 2) = 0$$
$$x = 3, \ -2$$

25. **a.** $y = 1^x = 1$ **b.** $y = 0^x = 0$

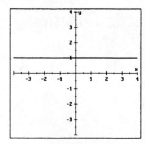 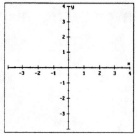

26. In $f(x) = b^x$, let $x = \frac{m}{n}$, then $b^{m/n} = (b^{1/n})^m$

is definitely not defined when n is even and $b < 0$ (in the real number system). In order for $f(x) = b^x$ to be defined when $b < 0$, x needs to be limited to rational numbers with odd denominators when written as a ratio in reduced form. The graph is discontinuous.

27. **a.**

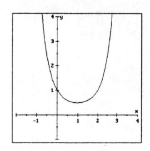

b. $(0, 1)$ is the y-intercept. The exponent exceeds all bounds as $x \to \pm\infty$ so $y \to +\infty$.

c. $(1, 0.5)$ seems to be the lowest point.

28. $A = \$3,600(1 + \frac{0.15}{365})^{7(365)} \approx \$10,285.33$

29. $A = \$9,400(1 + \frac{0.14}{360})^{0.5(360)} \approx \$10,081.44$

30. $A = \$1,000e^{0.06(10)} \approx \$1,822.12$

31. $\$3,000 = Pe^{0.05(4)}$

$P = \frac{\$3,000}{e^{0.2}} \approx \$2,456.19$

32. One year:

$A = 100(1 + \frac{0.06}{12})^{12} \approx 106.17$

$A = 100e^{0.059} \approx 106.08$

For one year the monthly compounding pays more. Five years:

$A = 100(1 + \frac{0.06}{12})^{60} \approx 108.68$

$A = 100e^{5(0.059)} \approx 134.31$

For five years the continuous compounding pays more.

33. **a.** $p(t) = 100e^{-0.03t}$ where $t = 40$:

$p(40) \approx 30.12\%$

b. Failure rate $= 1 - p(50)$

$= 1 - 100e^{-0.03(50)} \approx 77.69\%$

c. $p(40) - p(50) = 30.12 - 22.31 = 7.81\%$

34. $m = \dfrac{10,000(0.01)}{1 - (1 + \frac{0.01}{12})^{-48}} \approx 263.34$

35. $m = \dfrac{110,000(\frac{0.06}{12})}{1 - (1 + \frac{0.06}{12})^{-360}} \approx 659.51$

36. **a.** $F(0) = 70 - A = 35$, so $A = 35$;

$F(30) = 50$, so

$50 = 70 - 36e^{-30k}$

$e^{-30k} = \frac{20}{35} = \frac{4}{7}$

b. $F(6) = 70 - 35e^{-60k}$

$= 70 - 35(e^{-30k})^2$

$= 70 - 35(\frac{4}{7})^2$

≈ 57.57

The temperature is about $58°$

c. As $t \to +\infty$, $F(t) = 70 - 55(0) = 70$

37. $T = A + (B - A)e^{-kt}$ where $A = 75$,

$B = 120$, $t = 30$, $k = 0.01$.

$T = 75 + (120 - 75)e^{-0.3}$

$\approx 75 + (45)(.7408) \approx 108°$

38. Since $N(t) = N_0 2^{kt}$, $N(0) = 2,000$;

$N(10) = 5,000$; $N(0)/N(10) = 2^{10k}/2^0 = \frac{5}{2}$

After 20 minutes the number of bacteria will be

$N(20) = 2,000(2^{20k}) = 2,000(2^{10k})^2$

$= 2,000(\frac{5}{2})^2 = 12,500.$

After one hour the number will be

$N(6) = 2,000(2^{60k}) = 2,000(\frac{5}{2})^6 \approx 488,281$

39. $b^m b^n = \underbrace{(bbb\cdots\cdots b)}_{m} \; \underbrace{(bbb\cdots\cdots b)}_{n}$

$= \underbrace{(bbb\cdots\cdots b)}_{m+n} = b^{m+n}$

40. $\dfrac{b^m}{b^n} = \dfrac{\overbrace{bbb\cdots\cdots b)}^{m}}{\underbrace{bbb\cdots\cdots b}_{n}}$

$= \underbrace{(bbb\cdots\cdots b)}_{m-n} = b^{m-n}$

41. $(b^m)^n = \underbrace{(b^m b^m b^m \cdots\cdots b^m)}_{n}$

$= \underbrace{\overbrace{(bbb\cdots\cdots b)}^{m} \cdots \overbrace{(bbb\cdots\cdots b)}^{m}}_{n} = b^{mn}$

42. $(\sqrt[n]{b})^m = (b^{1/n})^m = b^{m/n} = (b^m)^{1/n} = \sqrt[n]{b^m}$

43. $y = f(x) = f(-x)$ which means $x = f^{-1}(y)$ and $x = -f^{-1}(y)$ or $f^{-1}(y) = -f^{-1}(y)$, which means each y-value corresponds to an $|x|$-value. Thus, $y = f(-x)$ is the reflection of $y = f(x)$ in the y-axis.

44. Consider the graph:

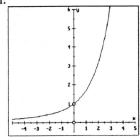

It appears $\displaystyle\lim_{h\to 0} \frac{e^h - 1}{h} = 1$

45. With $x = d$, pick $y = N = e^d$. For $x = d + a > d$, $y = e^x = e^{d+a} = e^d e^a > e^d = N$ Thus, $\displaystyle\lim_{x\to+\infty} e^x = +\infty$.

46. With $x = d$, pick $y = N = e^d$. For $x = d - a < d$, $y = e^x = e^{d-a}$ $= e^d/e^a < e^d = N$. Thus, $\displaystyle\lim_{x\to+\infty} e^x = 0$ Note: $e^x > 0$ for all x.

5.2 Inverse Functions; Logarithms, Page 343

1. $f[g(x)] = 5\left(\dfrac{x-3}{5}\right) + 3 = x$

$g[f(x)] = \dfrac{(5x+3) - 3}{5} = x$

These are inverse functions.

2. $f[g(x)] = \frac{2}{3}(\frac{3}{2}x + 3) = 2 \neq x$

These are not inverse functions.

3. $f[g(x)] = \frac{4}{5}(\frac{5}{4}x + 3) + 4 = x + \frac{12}{5} + 4 \neq x$

These are not inverse functions.

4. $f[g(x)] = \dfrac{1}{(1/x)} = x$

$g[f(x)] = \dfrac{1}{(1/x)} = x$

These are inverse functions.

5. $f[g(x)] = (\sqrt{x})^2 = x$

$g[f(x)] = \sqrt{x^2} = |x| = -x$ (since $x < 0$)

These are not inverse functions.

6. $f[g(x)] = (\sqrt{x})^2 = x$

$g[f(x)] = \sqrt{x^2} = |x| = x$ (since $x \geq 0$)

These are inverse functions.

7. To find the inverse interchange the domain and range values: $\{(5, 4), (3, 6), (1, 7), (4, 2)\}$.

8. Given $y = 2x + 3$; inverse is $x = 2y + 3$ or:

$y = \frac{1}{2}x - \frac{3}{2}$.

9. Given $y = x^2 - 5$, $x \geq 0$; inverse is

$x = y^2 - 5$, $y \geq 0$ or:

$y = \sqrt{x + 5}$ (positive value since $y \geq 0$)

10. Given $y = \sqrt{x} + 5$; inverse is $x = \sqrt{y} + 5$ or:

$y = (x - 5)^2$

11. Given $y = \dfrac{2x - 6}{3x + 3}$; inverse is $x = \dfrac{2y - 6}{3y + 3}$

or:

$3xy + 3x = 2y - 6$

$(3x - 2)y = -3x - 6$

$y = \dfrac{-3(x + 2)}{3x - 2}$ or $\dfrac{3x + 6}{2 - 3x}$

12. Given $y = \dfrac{2x + 1}{x}$; inverse is $x = \dfrac{2y + 1}{y}$

$xy = 2y + 1$

$xy - 2y = 1$

$y = \dfrac{1}{x - 2}$

13. $\log_2 4 + \log_3 \frac{1}{9} = 2 + (-2) = 0$

14. $2^{\log_2 3 - \log_2 5} = 2^{\log_2 3}/2^{\log_2 5} = \frac{3}{5}$

15. $5\log_3 9 - 2\log_2 16 = 5(2) - 2(4) = 2$

16. $(\log_2 \frac{1}{8})(\log_3 27) = (-3)(3) = -9$

17. $\left(3^{\log_7 1}\right)\left(\log_5 0.04\right) = 3^0\left(\log_5 \frac{1}{25}\right) = -2$

18. $e^{5\ln 2} = e^{\ln 2^5} = 2^5 = 32$

19. $\log_3 3^4 - \ln e^{0.5} = 4 - 0.5 = 3.5$

20. $\ln(\log 10^e) = \ln e = 1$

21. $\exp(\ln 3 - \ln 10) = \exp(\ln \frac{3}{10}) = \frac{3}{10}$

22. $\exp(\log_{e^2} 25) = \exp\left(\frac{\ln 25}{\ln e^2}\right) = \exp\left(\frac{\ln 5^2}{2\ln e}\right)$
$= \exp\left(\frac{2\ln 5}{2}\right) = \exp(\ln 5) = 5$

23. $x^2 = 16;\; x = 4$

24. $x = 10^{5.1} \approx 125{,}892.5412$

25. $-3x = \ln 0.5;$
$x = \frac{1}{3}\ln 0.5 \approx 0.23104906$

26. $x^2 = e^9$ so $x \approx 90.0171311$

27. $-x = \log_7 15$ or
$x = -\log_7 15 \approx -1.391662509$

28. $2x = \ln(\ln(4 + e));$
$x = \frac{1}{2}\ln(\ln(4 + e)) \approx 0.322197023$

29. $\frac{1}{2}\log_3 x = \log_2 8$
$\log_3 x = 6$
$x = 3^6 = 729$

30. $\log_2(x^{\log_2 x}) = 4$
$(\log_2 x)(\log_2 x) = 4$
$\log_2 x = \pm 2$
$x = 2^2 = 4$ or $2^{-2} = \frac{1}{4}$

31. $\log_3 x + \log_3(2x + 1) = 1$
$\log_3 x(2x + 1) = 1$
$x(2x + 1) = 3$
$2x^2 + x - 3 = 0$
$(x - 1)(2x + 3) = 0$
$x = 1,\; -\frac{3}{2}$

(Reject the negative value since logarithms of

negative numbers are not defined.)

32. $\ln\left(\frac{x^2}{1 - x}\right) = \ln x + \ln\left(\frac{2x}{1 + x}\right)$
$\ln\left(\frac{x^2}{1 - x}\right) = \ln\left(\frac{2x^2}{1 + x}\right)$
$\frac{x^2}{1 - x} = \frac{2x^2}{1 + x}$
$1 + x = 2(1 - x)$
$x = \frac{1}{3}$

33. $2^{3\log_2 x} = 4\log_3 9$
$2^{\log_2 x^3} = 4(2)$
$x^3 = 8$
$x = 2$

34. If $f(x) = y = 3x + 5$, then the inverse is
$x = 3y + 5$ or $y = \frac{1}{3}(x - 5) = f^{-1}(x)$;
$f^{-1}(2) = -1$
$(f^{-1})'(2) = \frac{1}{f'[f^{-1}(2)]} = \frac{1}{f'(-1)} = \frac{1}{3}$

35. If $f(x) = y = \sin x$, then the inverse is
$x = \sin y$ or $y = \sin^{-1} x = f^{-1}(x)$;
$(f^{-1})'(x) = \frac{1}{f'[f^{-1}(x)]} = \frac{1}{f'(\sin^{-1} x)}$
$= \frac{1}{\cos(\sin^{-1} x)} = \frac{1}{\sqrt{1 - x^2}}$
Let $u = \sin^{-1} x$, then $\sin u = x$ so
$\cos(\sin^{-1} x) = \cos u = \sqrt{1 - \sin u}$
$= \sqrt{1 - x^2}$

36. $\log_b 1{,}296 = 4$, so $b^4 = 1{,}296 = 6^4$, so $b = 4$

37. $\log_{\sqrt{b}} 106 = 2$, so $b = 106$;
$\sqrt{b - 25} = \sqrt{106 - 25} = 9$

38.

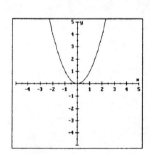

$f(x) = x^2$ for all x does not have an inverse, since the function is not one-to-one.

39.

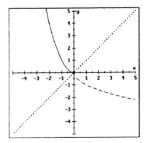

The inverse exists, since it passes the horizontal line test. $f^{-1}(x) = -\sqrt{x}$

40.

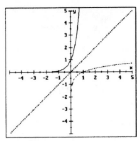

The inverse exists, since it passes the horizontal line test. $f^{-1}(x) = \log x$

41.

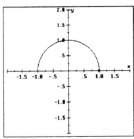

$f(x) = \sqrt{1 - x^2}$ does not have an inverse because it is not a one-to-one function.

42.

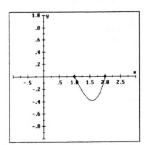

$f(x) = x(x - 1)(x - 2)$ does not have an inverse because it is not a one-to-one function.

43.

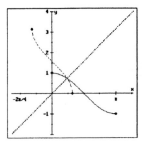

The inverse exists, since it passes the horizontal line test. $f^{-1}(x) = \cos^{-1} x$

44.

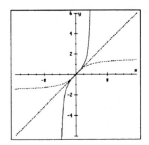

The inverse exists, since it passes the horizontal line test. $f^{-1}(x) = \tan^{-1} x$

45.
$$I(x) = I_0 e^{kx}$$

$I(0) = I_0$; $I(2) = 0.05 I_0$ but also $I(2) = I_0 e^{2k}$ so that

$$0.05 I_0 = I_0 e^{2k}$$
$$e^{2k} = 0.05$$
$$2k = \ln 0.05$$
$$k = -1.497866$$

When does $I(x) = .01 I_0$?

$$0.01 I_0 = I_0 e^{-1.497866x}$$
$$-1.497866x = \ln 0.01$$
$$x \approx 3.07$$

It is approximately 3 meters below the surface.

46. **a.**
$$F(5) = 2F(9)$$
$$e^{-5k} = 2e^{-9k}$$
$$e^{4k} = 2$$
$$4k = \ln 2$$
$$k = 0.25 \ln 2$$

Thus, $F(t) = e^{-0.25t \ln 2} = e^{\ln 2^{-0.25t}}$

$= 2^{-0.25t}$ so $F(7) = 2^{-1.75} \approx 0.2973$

b. $1 - F(10) = 1 - e^{-10} = 1 - 2^{-2.5}$

≈ 0.8232

c. $F(4) - F(5) = 2^{-1} - 2^{-1.25} \approx 0.07955$

47. A is to be $2P$, and $r = .06$;

$2P = Pe^{0.06t}$

$2 = e^{0.06t}$

$\ln 2 = 0.06t$

$t = \dfrac{\ln 2}{0.06} \approx 11.55$ years or 11 yr 202 days

48.
$$2P = Pe^{rt}$$
$$2 = e^{rt}$$
$$\ln 2 = rt$$
$$t = \frac{\ln 2}{r}$$

49. Compare A_1 and A_2 for a fixed P where

$$A_1 = \left(1 + \frac{0.07}{12}\right)^{12t} \text{ and } A_2 = e^{0.0695t}.$$

t	A_1 (1st Nat)	A_2 (World)
0	1	1
1	1.0723	1.072
2	1.1498	1.1491
3	1.2329	1.2318
4	1.3221	1.3205
5	1.4176	1.4155
10	2.0097	2.0037
100	1074.6	1043.1
190	574,559	543,074

First National Bank gives the better deal.

50. $t = \dfrac{\ln 2}{r}$ from Problem 48; $r = \dfrac{\ln 2}{10} \approx 0.0693$

or 6.93%.

51. 1624 to 1990 is 364 years;

$$25.2 \times 10^9 = 24 e^{364r}$$
$$364r = \ln\!\left(\frac{25.2 \times 10^9}{24}\right)$$
$$r = \frac{1}{364}\ln\!\left(\frac{25.2 \times 10^9}{24}\right) \approx 0.0571$$

The interest rate is approximately 5.71%.

52. a. $2P_0 = P_0(2^{60k})$

$60k = 1$

$k = \dfrac{1}{60}$

$P(t) = P_0 2^{t/60}$

Also, $P(20) = 1,000$ so

$1,000 = P_0 2^{1/3}$

$P_0 \approx 793.7$

b. $5,000 = P_0(2^{t/60})$

$\dfrac{t}{60} = \log_2(5,000/P_0)$

$t = 60 \log_2(5,000/P_0)$

$t \approx 159.32$

The time is about 2 hr and 40 min.

53. For a rock concert at 110 decibels,

$I_r = I_0 10^{110/10}$ and for normal conversation,

$I_n = I_0 10^{50/10}$. The difference in loudness

between the concert and normal conversation

is

$$D = 10 \log \frac{10^{11}}{10^5} = 10 \log 10^6 = 60$$

Thus, the concert is 60 times as loud as
normal conversation.

$$\frac{I_r}{I_n} = \frac{10^{11}}{10^5} = 10^6$$

The rock concert is one million times as
intense as normal conversation.

54. a. $1.5M + 11.4 = \log E$

$$E = 10^{1.5M+11.4}$$

b. $\dfrac{E_1}{E_2} = 10^{1.5(M_1 - M_2)} = 10^{1.5(8.5 - 6.5)}$

$= 10^3 = 1,000$

55. $b^{x \log_b x} = b^{\log_b x^x} = x^x$

56. $(\log_a b)(\log_b a) = \left(\dfrac{\ln b}{\ln a}\right)\!\left(\dfrac{\ln a}{\ln b}\right) = 1$

57. $y = \log_b x$; let (u, v) be a point on the
logarithmic curve, then $v = \log_b u$ or $b^v = u$.
Thus, (v, u) is on the curve with equation
$y = b^x$.

58. If $y = f(x)$, then $x = f^{-1}(y)$. Differentiate
with respect to x, and apply the chain rule:

$$1 = \left(\frac{df^{-1}(y)}{dy}\right)\frac{dy}{dx}$$

59. By Problem 57,

$$\frac{df}{dx}\frac{df^{-1}}{dy} = 1 > 0$$

If $\dfrac{df^{-1}}{dy} > 0$, (f^{-1} is increasing), then

$\dfrac{df}{dx} > 0$ (f is also increasing) since the product

of two positive numbers is positive.

Similarly, if $\dfrac{df^{-1}}{dy} < 0$ (f^{-1} is decreasing),

then $\dfrac{df}{dx} < 0$ (f is also decreasing).

60. The slope of the line joining $P(a, b)$ and

$Q(b,\ a)$ is

$$m = \frac{b\ -\ a}{a\ -\ b} = -1$$

which is the negative reciprocal of the slope of the line $y = x$. The midpoint $M(x_m,\ y_m)$ with $x_m = \frac{1}{2}a + b = y_m$ lies on $y = x$, so $y = x$ is the perpendicular bisector of $\overline{AB}$.

61. Suppose $g_1[f(x)] = x = f[g_1(x)]$ and

$g_2[f(x)] = x = f[g_2(x)]$. Thus,

$x = f[g_1(x)] = f[g_2(x)$ and $f^{-1}\{f[g_1(x)]\}$

$= f^{-1}\{f[g_2(x)]\}$ which implies $g_1(x) = g_2(x)$;

that is, the inverse is unique.

5.3 Derivatives Involving e^x and ln x, Page 350

1. $\dfrac{dy}{dx} = 3e^{3x}$　　　　**2.** $\dfrac{dy}{dx} = \dfrac{5}{5x} = \dfrac{1}{x}$

3. $\dfrac{dy}{dx} = \dfrac{1}{3x^4 + 5x}(12x^3 + 5) = \dfrac{12x^3 + 5}{3x^4 + 5x}$

4. $\dfrac{dy}{dx} = 2e^{2x+1}$　　**5.** $\dfrac{dy}{dx} = 2x + 2^x \ln 2$

6. $\dfrac{dy}{dx} = 3^x \ln x + 3x^2$　**7.** $\dfrac{dy}{dx} = e^x + ex^{e-1}$

8. $\dfrac{dy}{dx} = \pi^x \ln \pi + \pi x^{\pi - 1}$

9. $\dfrac{dy}{dx} = 2xe^{x^2}$　　**10.** $\dfrac{dy}{dx} = e^x + ex^{e-1}$

11. $f'(x) = -4e^{-4x} - 2e^{3-2x}$

12. $f'(x) = 3e^{-5x} + 5e^{-5x}$

13. $f'(t) = (2t + 1)\exp(t^2 + t + 5)$

14. $f'(x) = (2x + 3x^{-2})\exp(x^2 - 3x^{-1})$

15. $f'(u) = \frac{1}{2}\left[e^{2u}(2) - e^{-2u}(-2)\right] = e^{2u} + e^{-2u}$

16. $f'(t) = 3e^{3t+1} - 3e^{1-3t}$

17. $f'(x) = -x^2 e^{-x} + 2xe^{-x} = e^{-x}(2x - x^2)$

18. $f'(x) = (x^2 + 3)e^{1+x} + 2xe^{1+x}$

$\qquad = e^{1+x}(x^2 + 2x + 3)$

19. $g'(t) = \dfrac{(3t + 5)(2)e^{2t} - e^{2t}(3)}{(3t + 5)^2} = \dfrac{e^{2t}(6t + 7)}{(3t + 5)^2}$

20. $g'(t) = e^t \sin(2t + 1) + 2e^t \cos(2t + 1)$

21. $g'(t) = 2(e^{\sqrt{t}} - t)\left[e^{\sqrt{t}}(\frac{1}{2}t^{-1/2}) - 1\right]$

$\qquad = 2(e^{\sqrt{t}} - t)\left(\dfrac{e^{\sqrt{t}}}{2\sqrt{t}} - 1\right)$

22. $f'(t) = 2(e^x + \ln 2x)(e^x + \frac{1}{x})$

23. $f'(u) = \dfrac{1}{\ln u}(\frac{1}{u}) = \dfrac{1}{u \ln u}$

24. $f'(x) = e^x(\exp e^x)[\exp(\exp e^x)]$

25. $\dfrac{dy}{dx} = \dfrac{\cos x - \sin x}{\sin x + \cos x}$

26. $\dfrac{dy}{dx} = \dfrac{\sec x \tan x + \sec^2 x}{\sec x + \tan x} = \sec x$

27. $xe^{-x}(-1) + e^{-x} = \dfrac{dy}{dx}$

$\qquad\qquad\qquad \dfrac{dy}{dx} = e^{-x}(1 - x)$

28. $xe^y y' + e^y + ye^x + y'e^x = 2$

$\qquad\qquad (xe^y + e^x)y' = 2 - e^y - ye^x$

$\qquad\qquad\qquad\qquad y' = \dfrac{2 - e^y - ye^x}{xe^y + e^x}$

29. $e^{xy}(xy' + y) = 2x$

$\qquad xe^{xy}y' = 2x - ye^{xy}$

$\qquad\qquad y' = \dfrac{2x - ye^{xy}}{xe^{xy}}$

30. $\quad \ln x + \ln y = e^{2x}$

$\qquad x^{-1} + y^{-1}y' = 2e^{2x}$

$\qquad\qquad y^{-1}y' = 2e^{2x} - x^{-1}$

$\qquad\qquad\qquad y' = \dfrac{y(2xe^{2x} - 1)}{x}$

31. $\quad e^{xy}(xy' + y) + 2y^{-1}y' = 1$

$\qquad (xe^{xy} + 2y^{-1})y' = 1 - ye^{xy}$

$\qquad\qquad\qquad y' = \dfrac{1 - ye^{xy}}{xe^{xy} + 2y^{-1}}$

32. $\qquad\qquad \log_3\left(\dfrac{x^2}{y}\right) = x$

$\qquad \log_3 x^2 - \log_3 y = x$

$\qquad \dfrac{\ln x^2}{\ln 3} - \dfrac{\ln y}{\ln 3} = x$

$\qquad 2\ln x - \ln y = (\ln 3)x$

$\qquad 2x^{-1} - y^{-1}y' = \ln 3$

$\qquad\qquad y^{-1}y' = 2x^{-1} - \ln 3$

$\qquad\qquad\qquad y' = \dfrac{y(2 - x \ln 3)}{x}$

33. a. Differentiate a variable base, constant exponent; $y' = 2x$

b. Differentiate a constant base ($\neq e$), variable exponent; $y' = 2^x \ln 2$

c. Differentiate a base e, variable exponent; $y' = e^x$

d. Differentiate a variable base, constant exponent of e; $y' = ex^{e-1}$

34. Compare the differentiation of common and

natural logarithms.

a. $y' = \dfrac{1}{x \ln 10}$ **b.** $y' = \dfrac{1}{x}$

35. Logarithmic differentiation is advantageous when differentiating a complicated product or quotient. It is a technique for changing the derivative of a product to the derivative of a sum. Along with the product transformation is the changing of the derivative of a power to the differentiation of the base.
Note: $y = uv/w$ becomes

$$\ln y = \ln u + \ln v - \ln w.$$

36. $y = (x^{10} + 1)^{1/6}(x^7 - 3)^{4/9}$ so

$$\ln y = \tfrac{1}{6}\ln(x^{10} + 1) + \tfrac{4}{9}\ln(x^7 - 3)$$

$$\frac{1}{y}\frac{dy}{dx} = \tfrac{1}{6}(x^{10}+1)^{-1}(10x^9) + \tfrac{4}{9}(x^7 - 3)^{-1}(7x^6)$$

$$\frac{dy}{dx} = y\left[\frac{5x^9}{3(x^{10} + 1)} + \frac{28x^6}{9(x^7 - 3)}\right]$$

37. $y = \dfrac{(2x - 1)^5}{(x - 9)^{1/2}(x + 3)^2}$

$$\ln y = 5\ln(2x-1) - \tfrac{1}{2}\ln(x-9) - 2\ln(x+3)$$

$$\frac{1}{y}\frac{dy}{dx} = 10(2x-1)^{-1} - \tfrac{1}{2}(x-9)^{-1} - 2(x+3)^{-1}$$

$$\frac{dy}{dx} = y\left[\frac{10}{2x-1} - \frac{1}{2(x-9)} - \frac{2}{x+3}\right]$$

38. $y = \dfrac{e^{2x}}{(x^2 - 3)(4 - 5x)^{1/2}}$

$$\ln y = 2x - \ln(x^2 - 3) - \tfrac{1}{2}\ln(4 - 5x)$$

$$\frac{1}{y}\frac{dy}{dx} = 2 - 2x(x^2-3)^{-1} + \tfrac{5}{2}(4 - 5x)^{-1}$$

$$\frac{dy}{dx} = y\left[2 - \frac{2x}{x^2 - 3} + \frac{5}{2(4 - 5x)}\right]$$

39. $\ln y = 3x^2 - 2\ln(x^3 + 1) + 2\ln(4x - 7)$

$$\frac{1}{y}\frac{dy}{dx} = 6x - 2(3x^2)(x^3+1)^{-1} + 8(4x-7)^{-1}$$

$$\frac{dy}{dx} = y\left[6x - \frac{6x^2}{x^3 + 1} + \frac{8}{4x - 7}\right]$$

40. $\ln y = x \ln x$

$$\frac{1}{y}\frac{dy}{dx} = xx^{-1} + \ln x$$

$$\frac{dy}{dx} = y(1 + \ln x)$$

41. $\ln y = \ln\sqrt{x}\,\ln x$

$$\ln y = \tfrac{1}{2}\ln x \ln x$$

$$\ln y = \tfrac{1}{2}(\ln x)^2$$

$$\frac{1}{y}\frac{dy}{dx} = (\ln x)(\tfrac{1}{x})$$

$$\frac{dy}{dx} = \frac{y\ln x}{x}$$

42. $\ln y = 2x \ln(x^2 + 3)$

$$\frac{1}{y}\frac{dy}{dx} = 2\ln(x^2 + 3) + 2x(x^2 + 3)^{-1}(2x)$$

$$\frac{dy}{dx} = y\left[2\ln(x^2 + 3) + \frac{4x^2}{x^2 + 3}\right]$$

43. $\ln y = \ln(\sin x)^{\sqrt{x}} = \sqrt{x}\,\ln(\sin x)$

$$\frac{1}{y}\frac{dy}{dx} = \sqrt{x}\,\frac{1}{\sin x}(\cos x) + \ln(\sin x)\frac{1}{2\sqrt{x}}$$

$$\frac{dy}{dx} = y\sqrt{x}\left[\cot x + \frac{\ln(\sin x)}{2x}\right]$$

Note: if $\sin x < 0$, the problem statement is not defined, so absolute values are not necessary.

44. From Problem 40, $(x^x)' = x^x(1 + \ln x)$, so

$$\frac{1}{y}\frac{dy}{dx} = x^x x^{-1} + (\ln x)x^x(1 + \ln x)$$

$$\frac{dy}{dx} = y(x^{x-1} + x^x\ln x + x^x\ln^2 x)$$

45. $\ln y = x \ln x^x = x^2 \ln x$

$$\frac{1}{y}\frac{dy}{dx} = x^2 x^{-1} + 2x\ln x$$

$$\frac{dy}{dx} = y(x + 2x\ln x)$$

46. $\ln y = \sin x \ln x$

$$\frac{1}{y}\frac{dy}{dx} = (\sin x)x^{-1} + (\cos x)(\ln x)$$

$$\frac{dy}{dx} = y[(\tfrac{\sin x}{x}) + (\cos x)(\ln x)]$$

47. $\ln y = x \ln|\sin x|$

$$\frac{1}{y}\frac{dy}{dx} = x(\sin x)^{-1}(\cos x) + \ln|\sin x|$$

$$\frac{dy}{dx} = y[x\cot x + \ln|\sin x|]$$

48. $\ln y = x \ln(\ln x)$

$$\frac{1}{y}y' = x(\ln x)^{-1}(x^{-1}) + \ln(\ln x)$$

At $x = e$, $y = 1$ and $y' = 1$

The equation of the tangent line is

$$y - 1 = (1)(x - e)$$
$$x - y - e + 1 = 0$$

49.   ln $y = (\ln x)(\ln x) = (\ln x)^2$

$$\frac{1}{y} y' = 2(\ln x)(x^{-1})$$

At $x = 1$, $y = 1$ and $y' = 0$

The equation of the tangent line is

$$y - 1 = 0$$

50.   ln $y = (\cos x)[\ln(\sin x)]$

$$\frac{1}{y} y' = (\cos x)(\sin x)^{-1}(\cos x) - (\sin x)[\ln(\sin x)]$$

At $x = \frac{\pi}{2}$, $y = 1$ and $y' = 0$

The equation of the tangent line is

$$y - 1 = 0$$

51.   ln $y = \ln x^{x^x} = x^x \ln x$

$$\frac{1}{y} y' = x^x\left(\frac{1}{x}\right) + (\ln x)x^x(1 + \ln x)$$

At $x = 1$, $y = 1$, and $y' = 1$

The equation of the tangent line is

$$y - 1 = 1(x - 1)$$
$$x - y = 0$$

52. $y' = xe^{-x}(-1) + e^{-x}$

At $x = 0$, $y = 0$ and $y' = 1$

The equation of the tangent line is

$$y - 0 = (1)(x - 0)$$
$$x - y = 0$$

53. $y' = x^{-1}e^x + e^x(-x^{-2})$

At $x = 1$, $y = e$, and $y' = 0$

The equation of the tangent line is

$$y - e = 0$$

54. $y' = e^{-x}(\cos x) - e^{-x}(\sin x)$

At $x = 0$, $y = 0$, and $y' = 1$

The equation of the tangent line is

$$y - 0 = (1)(x - 0)$$
$$x - y = 0$$

55. $2x + 2y' = e^y y'$

At $(1, 0)$, $y' = -2$

The equation of the tangent line is

$$y - 0 = (1)(x - 1)$$
$$2x + y - 2 = 0$$

56. $x\left[\dfrac{y'}{(\ln 2)y}\right] + \dfrac{\ln y}{\ln 2} = 1 - y'$

At $(1, 1)$

$$\frac{y'}{\ln 2} = 1 - y' \text{ so } y' = \frac{\ln 2}{1 + \ln 2}$$

The equation of the tangent line is

$$y - 1 = \left(\frac{\ln 2}{1 + \ln 2}\right)(x - 1)$$
$$y(1 + \ln 2) - 1 - \ln 2 = x(\ln 2) - \ln 2$$
$$(\ln 2)x + (1 + \ln 2)y - 1 = 0$$

57. $y' = x[2^{-3x}(-3)\ln 2] + 2^{-3x}$

At $x = 1$, $y = 2^{-3} = \frac{1}{8}$ and

$$y' = [2^{-3}(-3)\ln 2] + 2^{-3}$$
$$= \tfrac{1}{8}(1 - 3 \ln 2)$$

The equation of the tangent line is

$$y - \tfrac{1}{8} = \tfrac{1}{8}(1 - 3 \ln 2)(x - 1)$$
$$y \approx -0.135x + 0.260$$

58. $y' = 2[xx^{-1} + \ln x]$

At $x = 1$, $y = 0$, and $y' = 2$

The equation of the tangent line is

$$y - 0 = 2(x - 1)$$
$$2x - y - 2 = 0$$

59. $y' = \frac{1}{2} \cdot \dfrac{xx^{-1} - \ln x}{x^2}$

At $x = 4$, $y = \frac{1}{4} \ln 2$, and $y' = \dfrac{1 - \ln 4}{32}$

The equation of the tangent line is

$$y - \frac{\ln 2}{4} = \left(\frac{1 - \ln 4}{32}\right)(x - 1)$$
$$32y - 8 \ln 2 = (1 - 2 \ln 2)(x - 1)$$
$$y \approx -0.121x - 0.004$$

60. $x^{-1} + y^{-1}y' = 2 - 2y'$

At $(1, 1)$, $y' = 1 - 2y'$ or $y' = \frac{1}{3}$

The equation of the tangent line is

$$y - 1 = \tfrac{1}{3}(x - 1)$$
$$x - 3y - 2 = 0$$

61.   ln $y = \ln(2x^2 - 3) + 2\ln(x+2) - 5\ln(x^3 + 3x - 5)$

$\frac{1}{y}\frac{dy}{dx} = (2x^2 - 3)^{-1}(4x) + 2(x + 2)^{-1}(1)$

$\qquad\qquad - 5(x^3 + 3x - 5)^{-1}(3x^2 + 3)$

At $(1, 9)$,

$y' = 9[(-1)^{-1}(4) + 2(3)^{-1} - 5(-1)^{-1}(6)]$

$\quad = 240$

The equation of the tangent line is

$$y - 9 = 240(x - 1)$$

$$240x + y + 231 = 0$$

62. $\ln y = 2x \ln(1 - 3x)$

$\frac{1}{y}\frac{dy}{dx} = 2[x(1 - 3x)^{-1}(-3) + \ln(1 - 3x)]$

At $x = -1$, $y = \frac{1}{16}$, and

$y' = \frac{1}{8}[-1(4)^{-1}(-3) + \ln 4] \approx 0.267$

The equation of the tangent line is

$$y - \frac{1}{16} = 0.267(x + 1)$$

$$y \approx 0.267x + 0.330$$

63. $2 + y' = e^y y'$

At $(1, 0)$, y' is not defined, so the tangent line is vertical and the normal line is horizontal, namely $y = 0$.

64. $y' = x^{1/2}x^{-1} + \frac{1}{2}x^{-1/2}\ln x$

At $x = e^{-2}$, $y = -2e^{-1}$, $y' = 0$

The tangent line is horizontal, so the normal is vertical, namely $x = e^{-2}$.

65. a. If $-\frac{\pi}{2} < x < \frac{\pi}{2}$, $\cos x > 0$, so

$\qquad F'(x) = (\cos x)(-\sin x) = -\tan x$

If $\frac{\pi}{2} < x < \frac{3\pi}{2}$, $\cos x < 0$, so

$\qquad F'(x) = (-\cos x)(\sin x) = -\tan x$

b. $F'(x)$

$\qquad = (\sec x + \tan x)^{-1}(\sec x \tan x + \sec^2 x)$

$\qquad = \sec x$

66. $\lim\limits_{h \to 0} \frac{a^h - 1}{h} = \lim\limits_{x \to 0} \frac{a^h \ln a - 0}{1} = \ln a$

67. If $f(x) = e^x$ then by definition,

$f'(x) = \lim\limits_{h \to 0} \frac{e^{x+h} - e^x}{h}$

$\qquad = \lim\limits_{h \to 0} \frac{e^x e^h - e^x}{h}$

$\qquad = \lim\limits_{h \to 0} \frac{e^x(e^h - 1)}{h} \cdot$

$\qquad = e^x \ln e = e^x \qquad From\ Problem\ 66$

5.4 Applications Involving Derivatives of e^x and $\ln x$, Page 356

1. c   **2.** d   **3.** b   **4.** e

5. $f'(x) = 3^{x-1}(\ln 3)$; $f''(x) = 3^{x-1}(\ln^2 3)$

a. $D = (-\infty, +\infty)$; or the set $\mathbb{R}$

b. None; the graph is always rising.

c. None; intercept $(0, \frac{1}{3})$.

d. Concave up everywhere; no inflection points.

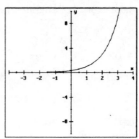

6. $f'(x) = -2^{3-x}(\ln 3)$; $f''(x) = 2^{3-x}(\ln^2 e)$

a. $D = (-\infty, +\infty)$; or the set $\mathbb{R}$

b. None; the graph is always falling.

c. None; intercept $(0, 8)$

d. Concave up everywhere; no inflection points.

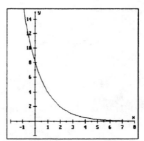

7. $f'(x) = 1 - x^{-1}$; $f''(x) = x^{-2}$

a. $D = (0, +\infty)$

b. None; the graph is rising for $x > 1$ and falling for $0 < x < 1$.

c. Relative minimum at $(1, 1)$; no intercepts.

d. Concave up everywhere; no inflection points.

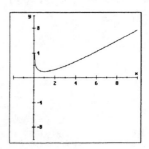

8. $f'(x) = 1 + \ln x;\ f''(x) = x^{-1}$

     a. $D = (0, +\infty)$

     b. $x = e^{-1}$; the graph is rising on $(e^{-1}, +\infty)$; falling on $(0, e^{-1})$

     c.   Relative minimum at $(e^{-1}, -e^{-1})$; intercept at $(1, 0)$

     d.   Concave up on $(0, +\infty)$; no inflection points.

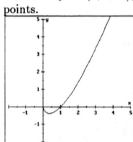

9. $f'(x) = (1 - x)e^{-x};\ f''(x) = (x - 2)e^{-x}$

     a. $D = (-\infty, +\infty)$; or the set $\mathbb{R}$

     b. $x = 1$; the graph is rising on $(-\infty, 1)$; falling on $(1, +\infty)$.

     c.   relative maximum at $(1, e^{-1})$; intercept $(0, 0)$

     d.   concave up on $(2, +\infty)$; concave down on $(-\infty, 2)$; inflection point at $(2, 2e^{-2})$

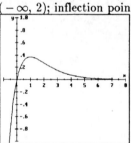

10. $f'(u) = (-2u^2 + 2u)e^{-2u};$
$f''(u) = (4u^2 - 4u)e^{-2u} + (-4u + 2)e^{-2u}$

         $= 2(2u^2 - 4u + 1)e^{-2u}$

     a. $D = (-\infty, +\infty)$; or the set $\mathbb{R}$

     b. $u = 0, 1$; the graph is rising on $(0, 1)$; falling otherwise

     c.   relative maximum at $(1, e^{-2})$; relative minimum at $(0, 0)$; crosses the axis at $(0, 0)$

     d.   concave down on $(0.3, 1.7)$ and concave up otherwise

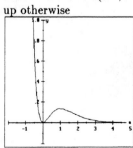

11. $f'(x) = \dfrac{\sqrt{x}(\frac{1}{2})x^{-1} - \frac{1}{2}x^{-1/2}(\ln\sqrt{x})}{x}$

       $= \dfrac{2 - \ln x}{4x^{3/2}}$

   $f''(x) = \dfrac{1}{4}\dfrac{x^{3/2}(-x^{-1}) - \frac{3}{2}x^{1/2}(2 - \ln x)}{x^3}$

       $= \dfrac{-8 + 3\ln x}{8x^{5/2}}$

     a. $D = (0, +\infty)$

     b. $x = e^2$; the graph is rising on $(0, e^2)$; and falling on $(e^2, +\infty)$.

     c.   relative maximum at (e^2, e^{-1}); intercept, $(1, 0)$

     d.   concave down on $(0, e^{8/3})$, concave up on $(e^{8/3}, +\infty)$; inflection point $(e^{8/3}, \frac{4}{3}e^{-4/3})$

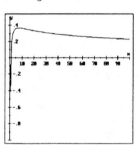

12. $f'(x) = (\ln 3)^{-1}x^{-1};$

   $f''(x) = -(\ln 3)^{-1}x^{-2}$

     a. $D = (0, +\infty)$

     b.   The graph rises for all x

     c.   There are no high or low points. An intercept is $(\frac{1}{2}, 0)$

     d.   The graph is concave down for all x

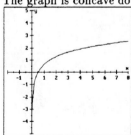

13. $f'(x) = -2xe^{-x^2};$
$f''(x) = -2(-2x^2 + 1)e^{-x^2}$

     a. $D = (-\infty, +\infty)$; or the set $\mathbb{R}$

     b. $x = 0$; the graph is rising on $(-\infty, 0)$; falling on $(0, +\infty)$

     c.   relative maximum at $(0, 1)$; intercept, $(0, 1)$

d. concave up on $(-\infty, -2^{-1/2})$ and on $(2^{-1/2}, +\infty)$; concave down on $(-2^{-1/2}, 2^{-1/2})$; points of inflection $(-2^{-1/2}, e^{-1/2})$, $(2^{-1/2}, e^{-1/2})$

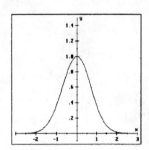

14. $f'(w) = \dfrac{2(2^w)(\ln 2) - 2^w}{w^2} = \dfrac{2^w(\ln 2w - 1)}{w^2}$

$f''(w) = \dfrac{2^w}{w^4}[w^2(\ln 2 + w \ln^2 2 - \ln 2)$

$\qquad\qquad - 2w(w \ln 2 - 1)]$

$\qquad = \dfrac{2^w}{w^3}[w^2(\ln^2 2 - 2w \ln 2 + 2)]$

a. Domain $= (-\infty, 0) \cup (0, +\infty)$
b. $x = 1/\ln 2 \approx 1.44$; the graph is rising on $(1.44, +\infty)$ and is falling on $(-\infty, 1.44)$
c. A relative minimum at $(1.44, 1.88)$; no intercepts
d. concave down on $(-\infty, 0)$; concave up on $(0, +\infty)$

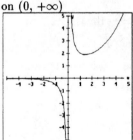

15. $f'(t) = 2t^{-1} + (1 - t)^{-1} = \dfrac{2 - t}{t(1 - t)}$

$f''(t) = -2t^{-2} + (1 - t)^{-2}$

$\qquad = -\dfrac{t^2 - 4t + 2}{t^2(t - 1)^2}$

a. $D = (-\infty, 0) \cup (0, 1)$

b. None; the graph is rising on $(0, 1)$ and falling on $(-\infty, 0)$.

c. None; intercepts, $\left(\dfrac{-1 + \sqrt{5}}{2}, 0\right)$ and

$\left(\dfrac{-1 - \sqrt{5}}{2}, 0\right)$

d. concave down on $(-\infty, 0)$ and on $(0, 2 - \sqrt{2})$; concave up on $(2 - \sqrt{2}, 1)$; point of inflection at $x = 2 - \sqrt{2}$

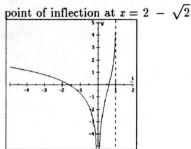

16. $g'(t) = \dfrac{-2t}{4 - t^2}$;

$g''(t) = -2\dfrac{(4 - t^2)(1) - (-2t)t}{(4 - t^2)^2}$

$\qquad = \dfrac{-2(t^2 + 4)}{(t^2 - 4)^2}$

a. Domain $= (-2, 2)$
b. $t = 0$; graph rises on $(-2, 0)$ and falls on $(0, 2)$
c. A relative maximum at $(0, \ln 4)$; intercepts $(\pm\sqrt{3}, 0)$
d. Concave down for all x in the domain.

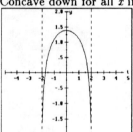

17. $f'(x) = e^x - e^{-x}$; $f''(x) = e^x + e^{-x}$
a. $D = (-\infty, +\infty)$; or the set $\mathbf{R}$
b. The graph is rising on $(0, \infty)$ and falling on $(-\infty, 0)$.
c. relative minimum at $(0, 2)$; intercept, $(0, 2)$
d. concave up everywhere; no inflection points.

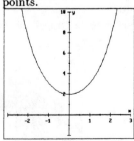

18. $f'(x) = \dfrac{4}{(e^x + e^{-x})}$

$f''(x) = \dfrac{-8(e^x - e^{-x})}{(e^x + e^{-x})^3}$

   a.   Domain $= (-\infty, +\infty)$ or the set **R**

   b. $x = 0$; the graph is rising for all x

   c.   no relative maximums or minimums; intercept at $(0, 0)$

   d.   concave up on $(-\infty, 0)$ and concave down on $(0, +\infty)$; inflection point $(0, 0)$

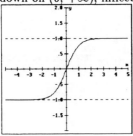

19. $f'(x) = 2(\ln x)x^{-1}$;

$f''(x) = 2[(\ln x)(-x^{-2}) + x^{-2}]$

$\qquad = 2(x^{-2})(1 - \ln x)$

   a. $D = (0, +\infty)$

   b. $x = 1$; graph is falling on $(0, 1)$ and rising on $(1, \infty)$.

   c.   relative minimum at $(1, 0)$; intercept $(1, 0)$

   d.   concave up on $(0, e)$, concave down on (e, ∞); inflection point $(e, 1)$.

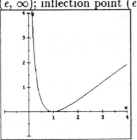

20. $f'(x) = 2x^{-1}$; $f''(x) = -2x^{-2}$

   a.   Domain $(-\infty, 0) \cup (0, +\infty)$

   b. $x = 0$; the graph is rising on $(0, +\infty)$ and falling on $(-\infty, 0)$

   c.   no relative maximums or minimums; intercepts at $(1, 0)$, $(-1, 0)$

   d.   concave down for all values in the domain

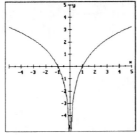

21.   **a.** $S'(x) = \frac{1}{2}(e^x - (-1)e^{-x}) = C(x)$

        $C'(x) = \frac{1}{2}(e^x + (-1)e^{-x}) = S(x)$

   b.

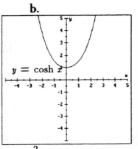 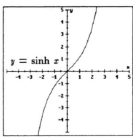

22. $2^{x^2 - 2x} = 4^x$

$x^2 - 2x = 2x$

$x(x - 4) = 0$

$\qquad x = 0, 4$

The points of intersection are $(0, 1)$ and $(4, 256)$.

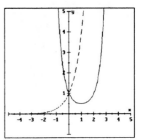

23.

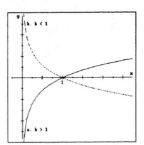

24. $f'(x) = (-2x + 1)e^{-2x} = 0$ at $x = \frac{1}{2}$

$f(-1) = -e^2$; $f(\frac{1}{2}) = \frac{1}{2}e^{-1}$; $f(2) = 2e^{-4}$

The maximum is $(2e)^{-1} \approx 0.184$, and the minimum is $-e^2 \approx -7.389$.

25. $f'(x) = \dfrac{xe^x - e^x}{x^2} = 0$ at $x = 1$

$f(\frac{1}{2}) = 2e^{1/2}$; $f(1) = e$; $f(2) = 2^{-1}e^2$

The maximum is $2^{-1}e^2 \approx 3.695$, and the minimum is $e \approx 2.728$.

26. $f'(t) = \dfrac{t(2)(\ln t)(t^{-1}) - \ln^2 t}{t^2}$

$\qquad = \dfrac{\ln t(2 - \ln t)}{2} = 0$ at $x = 1$, e^2

$f(\frac{1}{2}) = 2\ln^2 2$; $f(1) = 0$; $f(e^2) = 4e^{-2}$; $f(9)$

$= \dfrac{\ln^2 9}{9}$

The maximum is $2\ln^2 2 \approx 0.961$, and the minimum is 0.

27. $f'(t) = t(e^{-t^2})(-2t) + e^{-t^2}(1)$

$\quad = e^{-t^2}(1 - 2t^2) = 0$ at $t = \pm\dfrac{\sqrt{2}}{2}$

$f(\pm 1) = \pm e^{-1}; f(\pm\dfrac{\sqrt{2}}{2}) = \dfrac{\pm 1}{\sqrt{2e}};$

The maximum is $\dfrac{1}{\sqrt{2e}} \approx 0.429$, and the

minimum is $-\dfrac{1}{\sqrt{2e}} \approx -0.429$

28. $\lim\limits_{x\to 0^+} x^x = \lim\limits_{x\to 0^+} \exp(\ln x^x) = \lim\limits_{x\to 0^+} \exp(x \ln x)$

$= \lim\limits_{x\to 0^+} \exp\left(\dfrac{\ln x}{x^{-1}}\right) = \lim\limits_{x\to 0^+} \exp\left(\dfrac{x^{-1}}{-x^{-2}}\right)$

$= -\lim\limits_{x\to 0^+} \exp(-x) = 1$

29. $\lim\limits_{x\to 0^+} x \ln x = \lim\limits_{x\to 0^+} \dfrac{\ln x}{x^{-1}} = \lim\limits_{x\to 0^+} \dfrac{x^{-1}}{-x^{-2}}$

$= -\lim\limits_{x\to 0^+}(-x) = 0$

30. Let $L = \lim\limits_{x\to 0^+} x^{\sin x}$; so

$\ln L = \lim\limits_{x\to 0^+}(\sin x)(\ln x) = \lim\limits_{x\to 0^+}\dfrac{\ln x}{\csc x}$

$= \lim\limits_{x\to 0^+}\dfrac{\frac{1}{x}}{-\csc x \cot x} = \lim\limits_{x\to 0^+}\dfrac{-\sin^2 x}{x \cos x}$

$= \left(\lim\limits_{x\to 0^+}\dfrac{\sin x}{x}\right)\left(\lim\limits_{x\to 0^+}\dfrac{-\sin x}{\cos x}\right) = (1)(0) = 0$

Thus, $L = e^0 = 1$

31. $\ln L = \lim\limits_{x\to 0^+}(\sin x)(\ln x) = \lim\limits_{x\to 0^+}\dfrac{\ln x}{\csc x}$

$= \lim\limits_{x\to 0^+}\dfrac{\frac{1}{x}}{-\csc x \cot x} = \lim\limits_{x\to 0^+}\dfrac{-\sin^2 x}{x \cos x}$

$= \left(\lim\limits_{x\to 0^+}\dfrac{\sin x}{x}\right)\left(\lim\limits_{x\to 0^+}\dfrac{-\sin x}{\cos x}\right) = (1)(0) = 0$

32. $\lim\limits_{x\to +\infty}\dfrac{\ln(\ln x)}{x} = \lim\limits_{x\to +\infty}\dfrac{(\ln x)^{-1}x^{-1}}{1}$

$= \lim\limits_{x\to +\infty}(x \ln x)^{-1} = 0$

33. Let $L = \lim\limits_{x\to +\infty}(1 - x^{-1})^{2x}$;

$\ln L = \lim\limits_{x\to +\infty} 2x \ln(1 - x^{-1})$

$= \lim\limits_{x\to 0}\dfrac{2\ln(1 - x)}{x} = \lim\limits_{x\to 0}\dfrac{\frac{2(-1)}{1-x}}{1} = -2$

Thus, $L = e^{-2}$.

34. Let $L = \lim\limits_{x\to +\infty}(\ln x)^{1/x}$;

$\ln L = \lim\limits_{x\to +\infty}\dfrac{\ln(\ln x)}{x} = 0$ *From Problem 32*

Thus, $L = e^0 = 1$.

35. Let $L = \lim\limits_{x\to 0}(e^x + x)^{1/x}$;

$\ln L = \lim\limits_{x\to 0}\ln(e^x + x)^{1/x}$

$= \lim\limits_{x\to +\infty}\dfrac{1}{x}\ln(e^x + x)$

$= \lim\limits_{x\to 0}\dfrac{e^x + 1}{e^x + x} = 2$

Thus, $L = e^2$.

36. Let $L = \lim\limits_{x\to 1} x^{1/(x-1)}$;

$\ln L = \lim\limits_{x\to 1}\dfrac{\ln x}{x - 1} = \lim\limits_{x\to 1}\dfrac{\frac{1}{x}}{1} = 1$

Thus, $L = e^1 = e$.

37. Let $L = \lim\limits_{x\to 0}(e^x - 1 - x)^x$;

$\ln L = \lim\limits_{x\to 0}\dfrac{\ln(e^x - 1 - x)}{\frac{1}{x}}$

$= \lim\limits_{x\to 0}\dfrac{\frac{e^x - 1}{e^x - 1 - x}}{-\frac{1}{x^2}} = \lim\limits_{x\to 0}\dfrac{x^2(e^x - 1)}{-e^x + 1 + x}$

$= \lim\limits_{x\to 0}\dfrac{x^2 e^x + 2x(e^x - 1)}{1 - e^x}$

$= \lim\limits_{x\to 0}\dfrac{x^2 e^x + 2xe^x + 2(e^x - 1) + 2x}{-e^x}$

$= \lim\limits_{x\to 0}\dfrac{(x^2 + 4x + 2)e^x - 2}{-e^x} = 0$

Thus, $L = e^0 = 1$

38. $f'(x) = e^{-x}(\cos x - \sin x) = 0$ if $\tan x = 1$ or $x = (2n + 1)(\frac{\pi}{4})$. The x-axis is crossed where $\sin x = 0$ at $x = n\pi$ (this is shown in the second figure). This is a damped sine function.

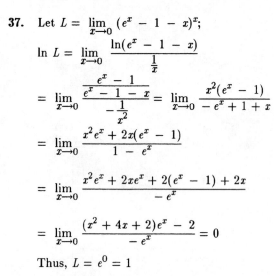

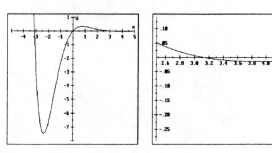

39. Recall that $\lim\limits_{x \to +\infty} \left(1 + \frac{1}{x}\right)^x = e$. Now if

$k > m$, let $m = k - n$.

$$\lim_{x \to +\infty}\left(1 + \frac{1}{x^k}\right)^{x^m} = \lim_{x \to +\infty}\left(1 + \frac{1}{x^k}\right)^{x^{k-n}}$$

$$= \lim_{x \to +\infty}\left(1 + \frac{1}{x^k}\right)^{x^k x^{-n}} = \lim_{x \to +\infty} (e)^{x^{-n}}$$

$$= \lim_{x \to +\infty} (e)^{1/x^n} = e^0 = 1$$

Now if $k < m$, let $m = k + n$.

$$\lim_{x \to +\infty}\left(1 + \frac{1}{x^k}\right)^{x^m} = \lim_{x \to +\infty}\left(1 + \frac{1}{x^k}\right)^{x^{k+n}}$$

$$= \lim_{x \to +\infty}\left(1 + \frac{1}{x^k}\right)^{x^k x^n} = \lim_{x \to +\infty} (e)^{x^n} = e^\infty$$

(increases without limit).

And finally, if $k = m$,

$$\lim_{x \to +\infty}\left(1 + \frac{1}{x^k}\right)^{x^m} = \lim_{x \to +\infty}\left(1 + \frac{1}{x^k}\right)^{x^k} = e$$

40. **a.** $D'(t)$

$$= \frac{1}{\sqrt{2\pi}\sigma}\exp\left[-\frac{1}{2}\left(\frac{t-m}{\sigma}\right)^2\right]\left(-\frac{1}{2\sigma^2}\right)(2)(t-m)$$

$$= 0 \text{ when } t = m; \ D(m) = \frac{1}{\sqrt{2\pi}\sigma}$$

b. $\lim\limits_{t \to \pm\infty} D(t) = 0$

c.

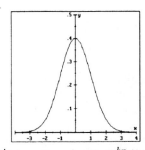

41. $G'(x) = A \exp(-Be^{-kx})(kBe^{-kx})$

$\qquad = kAB\exp(-Be^{-kx})\exp(-kx)$

$\qquad = kAB\exp(-Be^{-kx} - kx)$

$G''(x) = kAB\exp(-Be^{-kx} - kx)(kBe^{-kx} - k)$

Not critical points; $G''(x) = 0$ when $x = \frac{1}{k}\ln B$

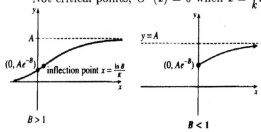

42. $P'(t) = e^{-0.04t}(1) + e^{-0.04t}(t+1)(-0.04)$

$\qquad = 0.04e^{-0.04t}(24 - t) = 0$ if $t = 24$

The population is the largest after 24 minutes.

$$\lim_{x \to +\infty} e^{-0.04t}(t+1) = \lim_{x \to +\infty} \frac{t+1}{e^{0.04t}} = 0$$

$P''(t) = 0.04e^{-0.04t}(-1.96 + 0.04t) = 0$ if

$t = 49$

After 49 minutes, the rate at which the population changes per minute starts to decrease.

43. $g'(t) = t(\frac{1}{t}) + (\ln t)(1) = 1 + \ln t = 0$

$g'(t) = 0$ if $t = 0$; when $\ln t = -1$, $t = \frac{1}{e}$.

Candidates: $\{0, \frac{1}{e}, 4\}$. Using the second derivative test on the critical point:

$g''(t) = \frac{1}{t}$, $g''(\frac{1}{e}) = e > 0$, so there is a relative minimum at $(\frac{1}{e}, 1 - \frac{1}{e})$.

$g(0) = 1$, $g(4) = 4\ln 4 + 1 \approx 6.545$.

So $(\frac{1}{e}, 1 - \frac{1}{e})$ is an absolute minimum and $(4, 6.545)$ is an absolute maximum.

44. **a.** The average cost is:

$$A(x) = 4e^{-x/6} + 30x^{-1}$$

The marginal cost is:

$$C'(x) = 4x^{-x/6}\left(-\frac{1}{6}x + 1\right)$$

The revenue is:

$$R(p) = px = -6p\ln\left(\frac{p}{40}\right)$$

The marginal revenue is:

$$R'(p) = -6\left[1 + \ln\left(\frac{p}{40}\right)\right]$$

b. $P(x) = R(x) - C(x) = px - 4xe^{-x/6} - 30$

Since $-\frac{1}{6}x = \ln\left(\frac{p}{40}\right)$ it follows that

$p = 40e^{-x/6}$, so that

$P'(x) = 36(-\frac{1}{6}x + 1)e^{-x/6} = 0$ when

$x = 6$. Profit is maximized when $x = 6$.

45. **a.** $P(t) = 50e^{0.02t}$; $P'(t) = e^{0.02t}$;

$P'(10) = e^{0.2} \approx 1.22$ million people/yr

b. The percentage rate is $\dfrac{100e^{0.02t}}{50e^{0.02t}} = 2$

that is, 2%.

46. a. $Q(t) = 20,000e^{-0.4t}$;

$Q'(t) = -8,000e^{-0.4t}$;

$Q'(5) = -8,000e^{-2} \approx \$1,082.68/\text{yr}$

b. The percentage rate is $\dfrac{100(-8,000)e^{-0.4t}}{20,000e^{-0.4t}}$

$= 40$; that is 40%.

47. a. $f'(t) = Ae^{-kt}(-k) = -kf(t)$

b. The percentage rate of change $= 100\dfrac{f'(t)}{f(t)}$

$= \dfrac{-100kf(t)}{f(t)} = -100k$ which is a constant.

48. $f'(t) = -Ake^{kt} = kf(t)$

49. a. $f'(x) = 3e^{-0.2x}$; $f'(10)(1) = 3e^{-2}$
≈ 0.406 thousand; that is 406 copies
b. $f(11) - f(10) = 15(e^{-2} - e^{-2.2})$
≈ 0.368 thousand; that is 368 copies

50. The profit is $P(x) = 1,000e^{-0.1x}(x - 5)$;
$P'(x) = 1,000e^{-0.1x}(6 - 0.1x) = 0$ when
$x = 60$. That is, profit is maximized when
$x = 60$.

51. $p(t) = 160(1 + 8e^{-0.01t})^{-1}$

$p'(t) =$

$-160(1+8e^{-0.01t})^{-2}(-0.01)(8e^{-0.01t})$

$p'(t) = 12.8e^{-0.01t}(1 + 8e^{-0.01t})^{-2}$

$p''(t) =$

$12.8\Big(e^{-0.01t}(-2)(1+8e^{-0.01t})^{-3}(8e^{-0.01t})$

$(-0.01)+(1+8e^{-0.01t})^{-2}(e^{-0.01t})(-0.01)\Big)$

$= -0.128(1 + 8e^{-0.01t})^{-2}(e^{-0.01t})$

$\Big(-16e^{-0.01t}(1 + 8e^{-0.01t})^{-1} + 1\Big)$

This expression is 0 when

$(1 + 8e^{-0.01t}) - 16e^{-0.01t} = 0$ or when

$e^{0.01t} = 8$, $t = \dfrac{\ln 8}{0.01} \approx 208$ years

52. The epidemic is spreading most rapidly when
the rate of change $R(t)$ is a maximum, that is,
when $R'(t) = f''(t) = 0$.

$f'(t) = A(-1)(1+Ce^{-kt})^{-2}(-Cke^{-kt})$

$= kACe^{-kt}(1 + Ce^{-kt})^{-2}$

$f''(t) = kAC(1 + Ce^{-kt})^{-4}(1$

$+ Ce^{-kt})^2(-ke^{-kt}) - (e^{-kt})(2)(1$

$+ Ce^{-kt})(-Cke^{-kt})$

$= \dfrac{k^2ACe^{-kt}(Ce^{-kt} - 1)}{(1 + Ce^{-kt})^3} = 0$ if $Ce^{-kt} = 1$

or $t = k^{-1}\ln C$. Substituting in the original
equation leads to

$f(k^{-1}\ln C) = A(1 + Ce^{-\ln C})^{-1} = \tfrac{1}{2}A$

This means half of the total number of
susceptible residents. $R'(t) < 0$ when
$t > (\ln C)k^{-1}$ (the graph of the curve is
decreasing) and $R'(t) > 0$ when $t < (\ln C)k^{-1}$
(the graph of the curve is increasing).

53. a. $v(t) = (-t^{1/2} + \tfrac{1}{2}t^{-1/2})e^{-t}$
$a(t) = \tfrac{1}{4}t^{-3/2}(4t^2 - 4t - 1)e^{-t}$

b. $v(0^+) > 0$; $v(t) = 0$ if $2t - 1 = 0$ or
$t = 0.5$ sec
$v(t) < 0$ for $t > 0.5$
$x(0.5) = \sqrt{0.5}\,e^{-0.5} \approx 0.43$

c. $x(1) = \sqrt{1}\,e^{-1} \approx 0.37$

The total distance traveled is

$(0.43 - 0) + (0.43 - 0.37) \approx 0.49$ units

54. $x(t) = Ae^{kt} + Be - kt$;

$v(t) = kAe^{kt} - kBe - kt$

$a(t) = k^2(Ae^{kt} + Be - kt) = k^2x(t)$

55. Let $g(x) = [f(x)]^2$, then the percentage rate of
change is

$\dfrac{100g'(x)}{g(x)} = \dfrac{100[2f(x)f'(x)]}{[f(x)]^2}$

$= 100\left[\dfrac{2f'(x)}{f(x)}\right] = k\cos x$

From this we see

$\dfrac{f'(x)}{f(x)} = \dfrac{k}{200}\cos x$

Integrate to find

$\ln[f(x)] = \dfrac{k}{200}\sin x + A$

$f(x) = C\exp(k\sin x/200)$

Since $f(0) = 1$, we have

$f(0) = 1 = Ce^0$ implies $C = 1$

$f(x) = \exp(k\sin x/200)$, so $f(\pi) = \exp(0) = 1$

56. $y = f(x) = e^{kx}$; $y = g(x) = e^{-kx}$;
$g(-x) = e^{kx} = f(x)$ for all x so e^{kx} and e^{-kx}

are reflections of each other in the y-axis.

57. $C'(t) = \dfrac{k}{b-a}(ae^{at} - be^{bt});$

$C''(t) = \dfrac{k}{b-a}(a^2 e^{at} - b^2 e^{bt})$

We know that $C'(t_c) = 0$, so we have

$ae^{-at} = b^{-bt_c}$.

$C''(t) = \dfrac{k}{b-a}\left(\dfrac{b}{a}\right)^{a/(b-a)}\left[a^2 - b^2\left(\dfrac{b}{a}\right)^{b/a}\right]$

< 0 since $0 < \dfrac{b}{a}$ and $a^2 < b^2 < \dfrac{b^2 b}{a^{b/a}}$

Thus, the graph of $C'(t)$ is concave downward, which means that t_c is a relative maximum.

5.5 Integrals Involving e^x and In x, Page 362

1. $\displaystyle\int e^{5x}\, dx = \frac{1}{5}\int e^{5x}(5\ dx) = \frac{1}{5}e^{5x} + C$

2. $\displaystyle\int e^{-2x}\, dx = -\frac{1}{2}\int e^{5x}(-2\ dx) = -\frac{1}{2}e^{-2x} + C$

3. $\displaystyle\int \frac{dx}{2x+1} = \frac{1}{2}\int (2x+1)^{-1}(-2\ dx)$

$= \frac{1}{2}\ln|2x+1| + C$

4. $\displaystyle\int \frac{3\ dx}{2x-5} = \frac{3}{2}\int (2x-5)^{-1}(2\ dx)$

$= \frac{3}{2}\ln|2x-5| + C$

5. $\displaystyle\int \ln e^x dx = \int x\ dx = \frac{x^2}{2} + C$

6. $\displaystyle\int \ln e^{x^2} dx = \int x^2\ dx = \frac{x^3}{3} + C$

7. $\displaystyle\int (au^2 - u^{-1} + e^u)\ du = \frac{a}{3}u^3 - \ln|u| + e^u + C$

8. $\displaystyle\int (au^4 + bu^{-1} + ce^u)\ du$

$= \frac{a}{5}u^5 + b\ln|u| + ce^u + C$

9. $\displaystyle\int x\exp(1 - x^2)\ dx = -\frac{1}{2}\int e^{1-x^2}(-2x\ dx)$

$= -\frac{1}{2}e^{1-x^2} + C$

10. $\displaystyle\int (x+1)\exp(x^2 + 2x + 3)\ dx$

$= \frac{1}{2}\int e^{x^2+2x+3}(2x+2)\ dx$

$= \frac{1}{2}e^{x^2+2x+3} + C$

11. $\displaystyle\int\left(\frac{x^2}{1-x^3} + xe^{x^2}\right) dx$

$= \frac{1}{3}\int \frac{-3x^2}{1-x^3}\ dx + \frac{1}{2}\int e^{x^2}(2x\ dx)$

$= -\frac{1}{3}\ln|1 + x^3| + \frac{1}{2}e^{x^2} + C$

12. $\displaystyle\int\left(\frac{4x+6}{\sqrt{x^2+3x}} + xe^{3x^2}\right) dx$

$= 2\int (x^2+3x)^{-1/2}(2x+3)dx + \int e^{3x^2}(6x\ dx)$

$= 4(x^2+3x)^{1/2} + \frac{1}{6}e^{3x^2} + C$

$= 4\sqrt{x^2+3x} + \frac{1}{6}e^{3x^2} + C$

13. $\displaystyle\int \frac{x\ dx}{2x^2+3} = \frac{1}{4}\int \frac{4x\ dx}{2x^2+3} = \frac{1}{4}\ln(2x^2+3) + C$

14. $\displaystyle\int \frac{x^2\ dx}{x^3+1} = \frac{1}{3}\int \frac{3x^2}{x^3+1}\ dx = \frac{1}{3}\ln|x^3+1| + C$

15. $\displaystyle\int \frac{2x+1}{x-5}\ dx = \int \frac{2x-10+11}{x-5}\ dx$

$= 2\int dx + \int \frac{11\ dx}{x-5}$

$= 2x + 11\int \frac{dx}{x-5}$

$= 2x + 11\ln|x-5| + C$

16. $\displaystyle\int \frac{4x\ dx}{2x+1} = 2\int \frac{2x\ dx}{2x+1} = 2\ln|2x+1| + C$

17. $\displaystyle\int \sqrt{x}\, e^{x\sqrt{x}}\ dx = \frac{2}{3}\int e^u\ du = \frac{2}{3}e^{x^{3/2}} + C$

Let $u = x\sqrt{x} = x^{3/2};\ du = \frac{3}{2}\sqrt{x}\ dx.$

18. $\displaystyle\int \frac{e^{\sqrt[3]{x}}}{x^{2/3}}\ dx = 3\int e^u\ du = 3e^{\sqrt[3]{x}} + C$

Let $u = \sqrt[3]{x} = x^{1/3};\ du = \frac{1}{3}x^{-2/3}$

19. $\displaystyle\int 2^{3+u}\ du = \frac{1}{\ln 2}\int 2^{3+u}\,(\ln 2\ du) = \frac{2^{3+u}}{\ln 2} + C$

20. $\displaystyle\int 10^{2t}\ dt = \frac{1}{2\ln 10}\int 10^{2t}(\ln 10)(2\ dt)$

$= \frac{10^{2t}}{2\ln 10} + C$

21. $\displaystyle\int \frac{\ln x}{x}\ dx = \int (\ln x)\frac{dx}{x} = \frac{(\ln x)^2}{2} + C$

22. $\displaystyle\int \frac{\ln(x+1)}{x+1}\ dx = \int \ln(x+1)\,\frac{dx}{x+1}$

$= \frac{1}{2}[\ln(x+1)]^2 + C$

23. $\int \dfrac{dx}{\sqrt{x}(\sqrt{x}+7)} = 2\int u^{-1}\,du$

Let $u = x^{1/2} + 7$; $du = \frac{1}{2}x^{-1/2}\,dx$

$\qquad = 2\ln\left|\sqrt{x}+7\right| + C$

24. $\int \dfrac{dx}{x^{2/3}(\sqrt[3]{x}+1)} = 3\int u^{-1}\,du$

Let $u = x^{1/3} + 1$; $du = \frac{1}{3}x^{-2/3}\,dx$

$\qquad = 3\ln(\sqrt[3]{x}+1) + C$

25. $\int \dfrac{e^t\,dt}{e^t+1} = \ln(e^t+1) + C$

Let $u = e^t + 1$; $du = e^t\,dt$

26. $\int \dfrac{e^{\sqrt{t}}\,dt}{\sqrt{t}(e^{\sqrt{t}}+1)} = 2\int u^{-1}\,du$

Let $u = e^{\sqrt{t}} + 1$; $du = e^{\sqrt{t}}(\frac{1}{2}t^{-1/2}\,dt)$

$\qquad = 2\ln(e^{\sqrt{t}}+1) + C$

27. $\int \dfrac{\cos x\,dx}{5 + 2\sin x} = \frac{1}{2}\int \dfrac{2\cos x\,dx}{5 + 2\sin x}$

$\qquad = \frac{1}{2}\ln(5 + 2\sin x) + C$

Note: $5 + 2\sin x > 0$, so absolute values are superfluous.

28. $\int \dfrac{\sec^2\theta\,d\theta}{1 + \tan\theta} = \int u^{-1}\,du = \ln|1 + \tan\theta| + C$

Let $u = 1 + \tan\theta$; $du = \sec^2\theta\,d\theta$

29. $\int \dfrac{\cot\sqrt{x}}{\sqrt{x}}\,dx = 2\int \cot u\,du = 2\ln|\sin\sqrt{x}| + C$

Let $u = \sqrt{x}$; $du = \dfrac{dx}{2\sqrt{x}}$

30. $\int \dfrac{\tan x\,dx}{\sec x + 1} = \int \dfrac{\sin x\,dx}{1 + \cos x} = -\int \dfrac{-\sin x\,dx}{1 + \cos x}$

$\qquad = -\ln(1 + \cos x) + C$

Note: $1 + \cos x > 0$, so absolute values are superfluous.

31. $\int \dfrac{\sin x - \cos x}{\sin x + \cos x}\,dx$

$\qquad = -\int \dfrac{1}{\sin x + \cos x}(\cos x - \sin x\,dx)$

$\qquad = -\ln|\sin x + \cos x| + C$

32. $\int \dfrac{\sec x\tan x\,dx}{5 - 3\sec x} = -\frac{1}{3}\int \dfrac{-3\sec x\tan x\,dx}{5 - 3\sec x}$

$\qquad = -\frac{1}{3}\ln|5 - 3\sec x| + C$

33. $\displaystyle\int_0^1 \dfrac{5x^2\,dx}{2x^3 + 1} = \frac{5}{6}\int_0^1 \dfrac{6x^2\,dx}{2x^3 + 1}$

Let $u = 2x^3 + 1$; $du = 6x^2\,dx$
$x = 0$, $u = 1$, when $x = 1$, $u = 3$

$\qquad = \frac{5}{6}\displaystyle\int_1^3 \dfrac{du}{u} = \frac{5}{6}(\ln 3 - \ln 1) = \frac{5}{6}\ln 3$

34. $\displaystyle\int_1^4 \dfrac{e^{-\sqrt{x}}}{\sqrt{x}}\,dx = -2\int_1^4 e^{-\sqrt{x}}\,\dfrac{dx}{2\sqrt{x}}$

$\qquad = -2e^{-\sqrt{x}}\Big|_1^4 = -2(e^{-2} - e^{-1})$

$\qquad = 2\left(\dfrac{1}{e} - \dfrac{1}{e^2}\right)$

35. $\displaystyle\int_{-\ln 2}^{\ln 2} \frac{1}{2}(e^x - e^{-x})\,dx$

$\qquad = \frac{1}{2}\displaystyle\int_{-\ln 2}^{\ln 2} e^x\,dx + \frac{1}{2}\int_{-\ln 2}^{\ln 2} e^{-x}(-dx)$

$\qquad = \frac{1}{2}(e^{\ln x} - e^{-\ln x} + e^{-\ln x} - e^{\ln x})\Big|_{-\ln 2}^{\ln 2} = 0$

36. $\displaystyle\int_0^2 (e^x - e^{-x})^2\,dx = \int_0^2 (e^{2x} - 2e^x e^{-x} + e^{-2x})\,dx$

$\qquad = (\frac{1}{2}e^{2x} - 2x - \frac{1}{2}e^{-2x})\Big|_0^2$

$\qquad = \frac{1}{2}(e^4 - e^{-4}) - 4$

37. $\displaystyle\int_1^2 \dfrac{e^{1/x}\,dx}{x^2} = -\int_1^2 e^{1/x}\left(-\dfrac{dx}{x^2}\right) = -\int_1^{1/2} e^u\,du$

Let $u = \frac{1}{x}$; $du = -\dfrac{dx}{x^2}$; if $x = 1$, $u = 1$; if $x = 2$, $u = \frac{1}{2}$.

$\qquad = -e^u\Big|_1^{1/2} = e - e^{1/2}$

38. $\displaystyle\int_1^2 5^{-x}\,dx = -\frac{1}{\ln 5}\int_1^2 5^{-x}(-\ln 5)\,dx$

$\displaystyle\qquad = -\frac{1}{\ln 5}\,5^{-x}\Big|_1^2 = -\frac{1}{\ln 5}(5^{-2}-5^{-1})$

$\displaystyle\qquad = \frac{4}{25\ln 5}$

39. $\displaystyle\int_0^{\pi/6}\tan 2x\,dx = -\frac{1}{2}\int_0^{\pi/6}\frac{-2\sin 2x\,dx}{\cos 2x}$

$\displaystyle\qquad = -\frac{1}{2}\ln|\cos 2x|\Big|_0^{\pi/6}$

$\displaystyle\qquad = -\frac{1}{2}(\ln\cos\frac{\pi}{3} - \ln 1)$

$\displaystyle\qquad = -\frac{1}{2}\ln\frac{1}{2} = \frac{1}{2}\ln 2 \approx 0.3466$

40. $\displaystyle\int_{\pi/4}^{\pi/3}\cot x\csc x\,dx = -\csc x\Big|_{\pi/4}^{\pi/3}$

$\displaystyle\qquad = -\left(\frac{2}{\sqrt{3}} - \frac{2}{\sqrt{2}}\right) = \sqrt{2} - \frac{2}{3}\sqrt{3}$

41. $\displaystyle\int_0^5\frac{0.58}{1+e^{-0.2x}}\,dx = -\frac{0.58}{0.2}\int_0^5\frac{-0.2\,dx}{1+e^{-0.2x}}$

$\displaystyle\qquad = -2.9\int_0^{-1}\frac{du}{1+e^u} = 2.9\int_0^{-1}\frac{-e^{-u}\,du}{e^{-u}+1}$

$\displaystyle\qquad = 2.9\ln|1+e^{-u}|\Big|_0^{-1}$

$\displaystyle\qquad = 2.9[\ln(1+e) - \ln 2]$

$\displaystyle\qquad = 2.9\,\ln\left(\frac{e}{2}+\frac{1}{2}\right) \approx 1.79833$

42. $\displaystyle\int_0^{12}\frac{5{,}000}{1+10e^{-t/5}}\,dt = -25{,}000\int_0^{12}\frac{1}{1+10e^{-t/5}}\cdot\frac{dt}{-5}$

$\displaystyle\qquad = -25{,}000\int_0^{-12/5}\frac{du}{1+10e^u}$

$\displaystyle\qquad = 25{,}000\int_0^{-12/5}\frac{-e^{-u}\,du}{e^{-u}+10}$

$\displaystyle\qquad = 25{,}000\ln|10+e^{-u}|\Big|_0^{-12/5}$

$\displaystyle\qquad = 25{,}000[\ln(10+e^{12/5}) - \ln 11]$

$\displaystyle\qquad \approx 16{,}193.25$

43. $y = \frac{1}{2}(e^x - e^{-x})$ crosses the x-axis at $(0, 0)$.

$\displaystyle A = \int_0^{\ln 2}\frac{1}{2}(e^x - e^{-x})\,dx = \frac{1}{2}(e^x + e^{-x})\Big|_0^{\ln 2} = \frac{1}{4}$

44. $y = \frac{\ln x}{x}$ crosses the x-axis at $(1, 0)$.

$\displaystyle A = \int_1^e\frac{\ln x}{x}\,dx = \frac{1}{2}\ln^2 x\Big|_1^e = \frac{1}{2}$

45. $y = 2^x$ and $y = 2^{2-x}$ cross when $2^x = 2^{1-x}$

or when $x = \frac{1}{2}$

$\displaystyle A = \int_0^{1/2}(2^{1-x} - 2^x)\,dx$

$\displaystyle\qquad = \frac{1}{\ln 2}(-2^{1-x} - 2^x)\Big|_0^{1/2}$

$\displaystyle\qquad = \frac{1}{\ln 2}(3 - 2\sqrt{2}) \approx 0.2475$

46. $y = \cot x$ does not cross the x-axis on $[\frac{\pi}{3}, \frac{\pi}{2}]$.

$\displaystyle A = \int_{\pi/3}^{\pi/2}\cot x\,dx = \int_{\pi/3}^{\pi/2}\frac{\cos x}{\sin x}\,dx = \ln|\sin x|\Big|_{\pi/3}^{\pi/2}$

$\displaystyle\qquad = \ln 1 - \ln\frac{\sqrt{3}}{2} = \ln 2 - \ln\sqrt{3}$

$\displaystyle\qquad \approx 0.1438$

47. $\displaystyle\frac{1}{5}\int_{-1}^4 e^{3x}\,dx = \frac{1}{15}(e^{12} - e^{-3}) \approx 10{,}850$

48. $\displaystyle\frac{1}{2}\int_1^3 xe^{-x^2}\,dx = -\frac{1}{4}e^{-x^2}\Big|_1^3$

$\displaystyle\qquad = \frac{1}{4}(e^{-1} - e^{-9}) \approx 0.09194$

49. $\displaystyle\frac{4}{\pi}\int_0^{\pi/4}\tan x\,dx = \frac{4}{\pi}(-\ln|\cos x|)\Big|_0^{\pi/4}$

$\displaystyle\qquad = \frac{4}{\pi}(-\ln\frac{\sqrt{2}}{2} + \ln 1)$

$\displaystyle\qquad = \frac{4}{\pi}\ln\sqrt{2} = \frac{\ln 4}{\pi} \approx 0.44$

50. $\displaystyle\int\frac{\ln^3 x}{x}\,dx = \int(\ln x)^3\frac{dx}{x} = \frac{\ln^4 x}{4} + C$

51. $\displaystyle\int\frac{dx}{\ln x} = \int\frac{1}{\ln x}\left(\frac{dx}{x}\right) = \ln|\ln x| + C$

52. $\displaystyle\int\frac{dx}{e^x+1} = -\int\frac{1}{1+e^{-x}}(-e^{-x}dx)$

$$= -\ln|1 + e^{-x}| + C = \ln\frac{e^x}{1 + e^x} + C$$

53. Since $v(t) = s'(t)$, $s(t) = \int v(t)\, dt$

$$= \int\left(\frac{1}{t} + t\right) dt = \ln t + \frac{t^2}{2} + C$$

On the given time interval,

$$s = \int_1^{e^2}\left(\frac{1}{t} + t\right) dt = \left(\ln t + \frac{t^2}{2}\right)\Big|_1^{e^2}$$

$$= 2 + \frac{e^4}{2} - \frac{1}{2} = \frac{3 + e^4}{2} \approx 28.8 \text{ ft}$$

54. $C(x) = \int\left(e^{0.01x} + 3\sqrt{x}\right) dx$

$$= 100e^{0.01x} + 2x^{3/2} + K$$

Since $C(4) = 1{,}000$, $K \approx 897.92$

$C(9) = 100e^{0.09} + 2(9^{3/2}) = 897.92 \approx 1{,}061.34$

55. The number of the original 10,000 members present after 10 months is $f(10) = 10{,}000e^{-1}$ $\approx 2{,}943$. Let Δt be a time interval, then $200\Delta t$ will be signed up in that period. The time left (until the 10 months have elapsed) is $10 - t$. The number of these members still around at the 10 month mark will be $200\Delta t f(10 - t)$. The number of members after 10 months will be

$$N = 2{,}943 + \int_0^{10} 200e^{10-t}\, dt$$

$$= 2{,}943 + 200e^{10}\int_0^{10} e^{-t}\, dt$$

$$= 2{,}943 + 200e^{10}(1 - e^{-10})$$

$$\approx 4{,}207 \text{ people}$$

56. The number of the original 200 members still present after 8 months is $f(8) = 200e^{-1.6}$ ≈ 40. Let Δt be a time interval, then $10\Delta t$ will be signed up in that period. The time left (until the 8 months have elapsed) is $8 - t$. The number of these members still around at the 8 month mark will be $10\Delta t f(8 - t)$. The number of members after 8 months will be

$$N = 40 + \int_0^8 10e^{-1.6+0.2t}\, dt$$

$$= 40 + 10e^{-1.6}\frac{1}{0.2}e^{0.2t}\Big|_0^8 = 80 \text{ people}$$

57. Finding the intersections: $\frac{2}{x} = 3 - x$, $x^2 - 3x + 2 = 0$, $x = 1, 2$.

$$A = \int_1^2\left(3 - x - \frac{2}{x}\right) dx = 3x - \frac{x^2}{2} - 2\ln x\Big|_1^2$$

$$= (6 - 2 - \ln 4) - (3 - \frac{1}{2}) = \frac{3}{2} - \ln 4$$

$$\approx 0.11 \text{ square units}$$

58. $f(x) = xe^{-x^2}$ crosses the x-axis at $(0, 0)$. The graph is symmetric with respect to the origin.

$$A = \int_{-3}^3 xe^{-x^2}\, dx = 2\int_0^3 xe^{-x^2}\, dx$$

$$= -\int_0^3 e^{-x^2}(-2x\, dx) = -e^{-x^2}\Big|_0^3$$

$$= 1 - e^{-9} \approx 0.999877$$

59. $A = \int_1^5 2x^{-2}\, dx = 2(-x^{-1})\Big|_1^5 = \frac{8}{5}$

60. **a.** Let 1825 be the base year (that is $t = 0$ is 1825). Then $P(0) = P_0 = 1$.

$$P(t) = e^{kt}$$

Since $1986 - 1825 = 161$

$P(161) = e^{161k} = 5$, so $k \approx 0.0099965088$

The year 0 A.D. is

$$P(-1825) = e^{-1825k} \approx 1.19 \times 10^{-8}$$

which is about 12 people! This puts this mathematical model in doubt when t is small.

b. $\int_{-1825}^{161} e^{kt}\, dt = k^{-1}e^{kt}\Big|_{-1825}^{161}$

$$\approx 500 \text{ billion people}$$

For the year 2000, $t = 175$

$P(175) = e^{175k} \approx 5.751087890$ billion

The percentage of people alive is

$$100\left(\frac{5.751087890}{500}\right) \approx 1.15$$

That is, about 1.15% are alive today.

61. **a.** If $1 < t$, $\frac{1}{t} < 1$, so the inequality holds for the integrals,

$$\int_1^{1+1/n} \frac{1}{t}\, dt \le \int_1^{1+1/n} dt$$

$$\ln|t||_1^{1+1/n} \le t|_1^{1+1/n}$$

$$\ln\left(1 + \tfrac{1}{n}\right) \le \tfrac{1}{n}$$

b.
$$t \le 1 + \tfrac{1}{n}$$

$$\tfrac{1}{t} \ge \frac{1}{1 + \tfrac{1}{n}}$$

$$\tfrac{1}{t} \ge \frac{n}{n+1}$$

$$\int_1^{1+1/n} \tfrac{1}{t}\, dt \ge \int_1^{1+1/n} \frac{n}{n+1}\, dt$$

$$\ln t|_1^{1+1/n} \ge \frac{n}{n+1} t\Big|_1^{1+1/n}$$

$$\ln(1 + \tfrac{1}{n}) \ge \frac{n}{n+1}\left(\tfrac{1}{n}\right)$$

$$\ln\left(1 + \tfrac{1}{n}\right) \ge \frac{1}{n+1}$$

c.
$$\exp[\ln(1 + \tfrac{1}{n})] \ge \exp\left(\frac{1}{n+1}\right)$$

$$1 + \tfrac{1}{n} \ge e^{1/(n+1)}$$

$$(1 + \tfrac{1}{n})^n \ge e$$

d. Raise both sides of the inequality in part **a** to the nth power and combine this result with that from part **c** to obtain the desired result.

62.
$$\int \cos x\, dx = \int \frac{\cos x\, dx}{\sin x} = \int u^{-1}\, du$$

Let $u = \sin x$; $du = \cos x\, dx$

$$= \ln|u| + C = \ln|\sin x| + C$$

5.6 The Inverse Trigonometric Functions, Page 372

1. Trigonometric functions are not one-to-one. Restrictions make them one-to-one so that the inverses can be defined.

2. An inverse trigonometric function represents an angle and defines the ratio of two sides of a right triangle. The Pythagorean theorem determines the third side. All trigonometric functions can be expressed in terms of the given inverse function.

3. **a.** $\frac{\pi}{3}$ **b.** $-\frac{\pi}{3}$ **4.** **a.** $-\frac{\pi}{6}$ **b.** $\frac{2\pi}{3}$

5. **a.** $-\frac{\pi}{4}$ **b.** $\frac{5\pi}{6}$ **6.** **a.** $\frac{3\pi}{4}$ **b.** $\frac{2\pi}{3}$

7. **a.** $-\frac{\pi}{3}$ **b.** π **8.** **a.** $\frac{\pi}{6}$ **b.** $\frac{3\pi}{4}$

9. $\frac{\sqrt{3}}{2}$ **10.** $\frac{\sqrt{2}}{2}$ **11.** 3 **12.** $\frac{\sqrt{2}}{4}$

13. Let $\alpha = \sin^{-1}\tfrac{1}{5}$ and $\beta = \cos^{-1}\tfrac{1}{5}$

Then $\sin\alpha = \tfrac{1}{5}$ and
$$\cos\alpha = \pm\sqrt{1 - \tfrac{1}{25}} = \frac{2\sqrt{6}}{5}$$

Also, $\cos\beta = \tfrac{1}{5}$ and
$$\sin\beta = \pm\sqrt{1 - \tfrac{1}{25}} = \frac{2\sqrt{6}}{5}$$

$$\cos(\sin^{-1}\tfrac{1}{5} + 2\cos^{-1}\tfrac{1}{5}) = \cos(\alpha + 2\beta)$$

$$= \cos\alpha\cos 2\beta - \sin\alpha\sin 2\beta$$

$$= \cos\alpha(\cos^2\beta - \sin^2\beta) - \sin\alpha(2\cos\beta\sin\beta)$$

$$= \frac{2\sqrt{6}}{5}\left[\tfrac{1}{25} - \tfrac{24}{25}\right] - \tfrac{1}{5}\left[2\cdot\tfrac{1}{5}\cdot\frac{2\sqrt{6}}{5}\right]$$

$$= \frac{2\sqrt{6}}{5}\left[-\tfrac{23}{25} - \tfrac{2}{25}\right]$$

$$= -\frac{2\sqrt{6}}{5} \approx -0.9798$$

14. Let $\alpha = \sin^{-1}\tfrac{1}{5}$ and $\beta = \cos^{-1}\tfrac{1}{4}$

Then $\sin\alpha = \tfrac{1}{5}$ and
$$\cos\alpha = \pm\sqrt{1 - \tfrac{1}{25}} = \frac{2\sqrt{6}}{5}$$

Also, $\cos\beta = \tfrac{1}{4}$ and
$$\sin\beta = \pm\sqrt{1 - \tfrac{1}{16}} = \frac{\sqrt{15}}{4}$$

$$\sin(\sin^{-1}\tfrac{1}{5} + \cos^{-1}\tfrac{1}{4}) = \sin(\alpha + \beta)$$

$$= \sin\alpha\cos\beta + \cos\alpha\sin\beta$$

$$= \left(\tfrac{1}{5}\right)\left(\tfrac{1}{4}\right) + \left(\frac{2\sqrt{6}}{5}\right)\left(\frac{\sqrt{15}}{4}\right)$$

$$= \tfrac{1}{20} + \frac{3\sqrt{10}}{10} \approx 0.6825$$

15.
$$\frac{dy}{dx} = \frac{1}{\sqrt{1 - (2x+1)^2}}(2) = \frac{2}{\sqrt{-4x^2 - 4x}}$$

$$= \frac{1}{\sqrt{-x^2 - x}}$$

16.
$$\frac{dy}{dx} = -\frac{1}{\sqrt{1 - (4x+3)^2}}(4)$$

$$= \frac{-2}{\sqrt{-4x^2 - 6x - 2}}$$

17.
$$\frac{dy}{dx} = \frac{1}{1 + (x^2 + 1)} \frac{x}{\sqrt{x^2 + 1}}$$

$$= \frac{x}{(x^2 + 2)\sqrt{x^2 + 1}}$$

18. $\dfrac{dy}{dx} = -\dfrac{1}{1 + (x^2)^2}(2x) = -\dfrac{2x}{1 + x^4}$

19. $\dfrac{dy}{dx} = 3(\sin^{-1}2x)^2 \dfrac{1}{\sqrt{1 - (2x)^2}}(2)$

$\qquad = \dfrac{6(\sin^{-1}2x)^2}{\sqrt{1 - 4x^2}}$

20. $\dfrac{dy}{dx} = 4(\tan^{-1}x^2)^3 \dfrac{1}{1 + (x^2)^2}(2x)$

$\qquad = \dfrac{8x(\tan^{-1}x^2)^3}{1 + x^4}$

21. $\dfrac{dy}{dx} = \dfrac{1}{2\sqrt{\tan^{-1}(2x)}} \dfrac{1}{1 + 4x^2}(2)$

$\qquad = \dfrac{1}{(4x^2 + 1)\sqrt{\tan^{-1}(2x)}}$

22. $\dfrac{dy}{dx} = \dfrac{1}{\sin^{-1}x} \dfrac{1}{\sqrt{1 - x^2}}$

23. $\dfrac{dy}{dx} = -\dfrac{1}{\cos^{-1}x\sqrt{1 - x^2}}$

24. $\dfrac{dy}{dx} = \dfrac{1}{\sqrt{1 - e^{2x}}}(e^x) = \dfrac{e^x}{\sqrt{1 - e^{2x}}}$

25. $\dfrac{dy}{dx} = \dfrac{1}{1 + x^{-2}}(-x^{-2}) = \dfrac{-1}{x^2 + 1}$

26. $\dfrac{dy}{dx} = -\dfrac{1}{\sqrt{1 - (\sin x)^2}}(\cos x)$

$\qquad = -\dfrac{\cos x}{\sqrt{\cos^2 x}} = -\dfrac{\cos x}{|\cos x|}$

27. $\dfrac{dy}{dx} = \dfrac{1}{\sqrt{1 - (\cos x)^2}}(-\sin x)$

$\qquad = -\dfrac{\sin x}{|\sin x|}$

28. $\dfrac{dy}{dx} = \dfrac{1}{\sin^{-1}(e^x)} \dfrac{1}{\sqrt{1 - e^x}}(e^x)$

$\qquad = \dfrac{e^x}{\sin^{-1}(e^x)\sqrt{1 - e^x}}$

29. $x\dfrac{1}{\sqrt{1 - y^2}}y' + \sin^{-1}y + y\dfrac{1}{1 + x^2}$

$\qquad + y'(\tan^{-1}x) = 1$

$y'\left(\dfrac{x}{\sqrt{1 - y^2}} + \tan^{-1}x\right)$

$\qquad = 1 - \sin^{-1}y - \dfrac{y}{1 + x^2}$

30. $y' = \dfrac{1 - \sin^{-1}y - \dfrac{y}{1 + x^2}}{\dfrac{x}{\sqrt{1 - y^2}} + \tan^{-1}x}$

$\qquad \dfrac{1}{\sqrt{1 - y^2}}y' + y' = 2x\,y' + 2y$

$\qquad [(1 - y^2)^{-1/2} + 1 - 2x]y' = 2y$

$\qquad \dfrac{dy}{dx} = \dfrac{2y}{(1 - y^2)^{-1/2} + 1 - 2x}$

31. $\displaystyle\int \dfrac{dx}{x^2 + 16} = \frac{1}{4}\tan^{-1}\frac{x}{4} + C$

32. $\displaystyle\int \dfrac{dx}{x\sqrt{x^2 - 9}} = \frac{1}{3}\sec^{-1}\left|\frac{x}{3}\right| + C$

33. $\displaystyle\int \dfrac{dx}{\sqrt{5 - 2x^2}} = \frac{1}{\sqrt{5}}\int \dfrac{dx}{\sqrt{1 + \frac{2}{5}x^2}}$

Let $u^2 = \frac{2}{5}x^2$; $u = \sqrt{\frac{2}{5}}x$; $du = \sqrt{\frac{2}{5}}\,dx$

$\qquad = \dfrac{1}{\sqrt{2}}\int \dfrac{\sqrt{\frac{2}{5}}\,dx}{\sqrt{1 + \frac{2}{5}x^2}} = \dfrac{1}{\sqrt{2}}\sin^{-1}\sqrt{\frac{2}{5}}\,x + C$

$\qquad = \dfrac{\sqrt{2}}{2}\sin^{-1}\left(\dfrac{x\sqrt{10}}{5}\right) + C$

34. $\displaystyle\int \dfrac{dx}{\sqrt{4 - 7x^2}} = \frac{1}{\sqrt{7}}\int \dfrac{\sqrt{7}\,dx}{\sqrt{2^2 - (\sqrt{7}x)^2}}$

$\qquad = \dfrac{1}{\sqrt{7}}\sin^{-1}\dfrac{\sqrt{7}x}{2} + C$

35. $\displaystyle\int \dfrac{2x + 5}{x^2 + 4x + 5}\,dx = \int \dfrac{2(x + 2) + 1}{(x + 2)^2 + 1}\,dx$

Let $u = x + 2$; $du = dx$

$\qquad = \displaystyle\int \dfrac{2u + 1}{u^2 + 1}\,du = \int \dfrac{2u}{u^2 + 1}\,du + \int \dfrac{1}{u^2 + 1}\,du$

$\qquad = \ln(u^2 + 1) + \tan^{-1}u + C$

$\qquad = \ln(x^2 + 4x + 5) + \tan^{-1}(x + 2) + C$

36. $\displaystyle\int \dfrac{2x - 3}{x^2 + 1}\,dx = \int \dfrac{2x}{x^2 + 1}\,dx - \int \dfrac{3}{x^2 + 1}\,dx$

$\qquad = \ln(x^2 + 1) - 3\tan^{-1}x + C$

37. $\displaystyle\int \dfrac{x\,dx}{x^2 + x + 1} = \frac{1}{2}\int \dfrac{2x + 1 - 1}{x^2 + x + 1}\,dx$

$$= \frac{1}{2}\int \frac{2x+1}{x^2+x+1}\,dx - \frac{1}{2}\int \frac{1}{x^2+x+1}\,dx$$

$$= \frac{1}{2}\ln|x^2+x+1| - \frac{1}{2}\int \frac{dx}{(x+\frac{1}{2})^2+\frac{3}{4}}$$

$$= \frac{1}{2}\ln|x^2+x+1| - \frac{2}{3}\int \frac{dx}{\frac{4}{3}(x+\frac{1}{2})^2+1}$$

$$= \frac{1}{2}\ln|x^2+x+1| - \frac{2}{3}\int \frac{dx}{\left(\frac{2}{\sqrt{3}}\right)^2(x+\frac{1}{2})^2+1}$$

$$= \frac{1}{2}\ln|x^2+x+1| - \frac{2}{3}\frac{\sqrt{3}}{2}\int \frac{\frac{2}{\sqrt{3}}\,dx}{\left(\frac{2}{\sqrt{3}}\right)^2(x+\frac{1}{2})^2+1}$$

$$= \frac{1}{2}\ln|x^2+x+1| - \frac{\sqrt{3}}{3}\tan^{-1}\left(\frac{2}{\sqrt{3}}\right)(x+\frac{1}{2}) + C$$

$$= \frac{1}{2}\ln|x^2+x+1| - \frac{\sqrt{3}}{3}\tan^{-1}\left[\frac{\sqrt{3}}{3}(2x+1)\right] + C$$

38. $\displaystyle\int \frac{2x+1}{\sqrt{1-x^2}}\,dx$

$$= -\int (1-x^2)^{-1/2}(-2x\,dx) + \int \frac{dx}{\sqrt{1-x^2}}$$

$$= -2(1-x^2)^{1/2} + \sin^{-1}x + C$$

$$= -2\sqrt{1-x^2} + \sin^{-1}x + C$$

39. $\displaystyle\int \frac{(2+x)\,dx}{\sqrt{4-2x-x^2}}$

$$= \int \frac{(2+x)\,dx}{\sqrt{5-(1+2x+x^2)}} = \int \frac{(2+x)\,dx}{\sqrt{(\sqrt{5})^2-(x+1)^2}}$$

$$= \int \frac{dx}{\sqrt{(\sqrt{5})^2-(x+1)^2}} - \frac{1}{2}\int \frac{(-2-2x)\,dx}{\sqrt{4-2x-x^2}}$$

$$= \sin^{-1}\left(\frac{x+1}{\sqrt{5}}\right) - \frac{1}{2}\frac{\sqrt{4-2x-x^2}}{\frac{1}{2}} + C$$

$$= \sin^{-1}\left(\frac{x+1}{\sqrt{5}}\right) - \sqrt{4-2x-x^2} + C$$

40. $\displaystyle\int \frac{(8x-1)\,dx}{\sqrt{1-4x^2}} = -\int (1-4x^2)^{-1/2}(-8x\,dx)$

$$-\frac{1}{2}\int \frac{2\,dx}{\sqrt{1-(2x)^2}}$$

$$= -\sqrt{1-4x^2} - \frac{1}{2}\sin^{-1}2x + C$$

41. $\displaystyle\int \frac{dx}{x^2+2x+2} = \int \frac{dx}{(x+1)^2+1}$

$$= \tan^{-1}(x+1) + C$$

42. $\displaystyle\int \frac{dx}{\sqrt{-x^2+3x-2}}$

$$= \int \frac{dx}{\sqrt{-(x^2-3x+\frac{9}{4})-2+\frac{9}{4}}}$$

$$= \int \frac{dx}{\sqrt{(\frac{1}{2})^2-(x-\frac{3}{2})^2}}$$

$$= \sin^{-1}\left(\frac{x-\frac{3}{2}}{\frac{1}{2}}\right) = \sin^{-1}(2x-3)$$

43. $\displaystyle\int_{1/\sqrt{3}}^{1} \frac{dx}{x\sqrt{4x^2-1}} = \int_{1/\sqrt{3}}^{1} \frac{2\,dx}{(2x)\sqrt{(2x)^2-1}}$

$$= \sec^{-1}2x\Big|_{1/\sqrt{3}}^{1} = \frac{\pi}{6}$$

44. $\displaystyle\int_{0}^{\pi/4} \frac{\sin x\,dx}{1+\cos^2 x} = -\int_{0}^{\pi/4} \frac{-\sin x\,dx}{1+(\cos x)^2}$

$$= -\tan^{-1}(\cos x)\Big|_{0}^{\pi/4} = \frac{\pi}{4} - \tan^{-1}\frac{\sqrt{2}}{2}$$

45. $\displaystyle\int_{0}^{\ln 2\sqrt{3}} \frac{e^x\,dx}{e^{2x}+4} = \frac{1}{4}\int_{0}^{\ln 2\sqrt{3}} \frac{e^x\,dx}{\frac{e^{2x}}{4}+1}$

$$= \frac{1}{2}\int_{0}^{\ln 2\sqrt{3}} \frac{\frac{1}{2}e^x\,dx}{\frac{e^{2x}}{4}+1} = \frac{1}{2}\tan^{-1}\frac{1}{2}e^x\Big|_{0}^{\ln 2\sqrt{3}}$$

$$= \frac{1}{2}(\tan^{-1}\sqrt{3} - \tan^{-1}\frac{1}{2})$$

$$= \frac{\pi}{6} - \frac{1}{2}\tan^{-1}\frac{1}{2}$$

46. $\displaystyle\int_{1}^{e^2} \frac{dx}{x[4+(\ln x)^2]} = \int_{1}^{e^2} \frac{1}{2^2+(\ln x)^2}\frac{dx}{x}$

$$= \frac{1}{2}\tan^{-1}(\frac{1}{2}\ln x)\Big|_{1}^{e^2}$$

$$= \frac{1}{2}[\tan^{-1}(\frac{1}{2}\ln e^2) - \tan^{-1}(\frac{1}{2}\ln 1)]$$

$$= \frac{\pi}{8}$$

47. $\displaystyle\int_0^{1/\sqrt{2}} \frac{x\,dx}{\sqrt{1-x^4}} = \frac{1}{2}\int_0^{1/\sqrt{2}} \frac{2x\,dx}{\sqrt{1-(x^2)^2}}$

$\displaystyle = \frac{1}{2}\sin^{-1}x^2\Big|_0^{1/\sqrt{2}} = \frac{\pi}{12}$

48. $\displaystyle\int_0^1 \frac{x^3\,dx}{1+x^8} = \frac{1}{4}\int_0^1 \frac{4x^3\,dx}{1+(x^4)^2}$

$\displaystyle = \frac{1}{4}\tan^{-1}x^4\Big|_0^1 = \frac{\pi}{16}$

49. If $\tan^{-1}x = \theta$, then $\tan\theta = x$, so

$\sec\theta = \sqrt{x^2+1}$ and $\cos\theta = \dfrac{1}{\sqrt{x^2+1}}$

Then $\sin^2\theta = 1 - \dfrac{1}{x^2+1} = \dfrac{x^2}{x^2+1}$

$\sin(2\tan^{-1}x) = \sin 2\theta = 2\sin\theta\cos\theta$

$\displaystyle = 2\left(\frac{x}{\sqrt{x^2+1}}\right)\left(\frac{1}{\sqrt{x^2+1}}\right) = \frac{2x}{x^2+1}$

50. $\tan(2\tan^{-1}x) = \tan 2\theta = \dfrac{2\tan\theta}{1-\tan^2\theta}$

$\displaystyle = \frac{2x}{1-x^2}$

51. If $\cos^{-1}x = \theta$, then $\cos\theta = x$

so $\sec\theta = \dfrac{1}{x}$ and $\tan\theta = \sqrt{\sec^2\theta - 1}$

$\displaystyle = \sqrt{\frac{1}{x^2}-1} = \frac{\sqrt{1-x^2}}{x}$

52. If $\sin^{-1}x = \theta$, then $\sin\theta = x$ and $\sin^2\theta = x^2$,

$\cos^2\theta = 1 - x^2$. Thus,

$\cos(2\sin^{-1}x) = \cos 2\theta = \cos^2\theta - \sin^2\theta$

$= 1 - x^2 - x^2 = 1 - 2x^2$

53. If $\sin^{-1}x = \alpha$, then $\sin\alpha = x$ and if $\cos^{-1}x = \beta$, then $\cos\beta = x$. This means that sine and cosine are equal so that $\alpha = \beta = \dfrac{\pi}{4}$

$\sin(\sin^{-1}x + \cos^{-1}x) = \sin\dfrac{\pi}{2} = 1$

54. $\cos(\sin^{-1}x + \cos^{-1}x) = \cos\dfrac{\pi}{2} = 0$ (See the solution for Problem 53.)

55. a. Let $\theta = \cot^{-1}x$, so that $\cot\theta = x$. Consider a reference triangle with sides 1, x, and $\sqrt{1-x^2}$. Let the other acute

angle be α. Now $\theta = \dfrac{\pi}{2} - \alpha$ and $\tan\alpha = x$. Thus,

$\cot^{-1}x = \theta = \dfrac{\pi}{2} - \tan^{-1}x$

b. Let $\theta = \sec^{-1}x$, so that

$x = \sec\theta$

$x = \dfrac{1}{\cos\theta}$

$\cos\theta = \dfrac{1}{x}$

$\theta = \cos^{-1}\left(\dfrac{1}{x}\right)$

c. Let $\theta = \csc^{-1}x$, so that

$x = \csc\theta$

$x = \dfrac{1}{\sin\theta}$

$\sin\theta = \dfrac{1}{x}$

$\theta = \sin^{-1}\left(\dfrac{1}{x}\right)$

56. a. $f'(x) = 0$ implies f is a constant function. This is not surprising since

$f(x) = \sin^{-1}x + \cos^{-1}x = \dfrac{\pi}{2}$

b. $g(x) = \tan^{-1}x + \cos^{-1}x$

$g'(x) = (1+x^2)^{-1} - (1+x^2)^{-1} = 0$

which implies that g is a constant function. This is not surprising since

$g(x) = \tan^{-1}x + \cot^{-1}x = \dfrac{\pi}{2}$

57. The curve does not intersect the x-axis on $[\sqrt{2}, 2]$.

$\displaystyle A = \int_{\sqrt{2}}^2 \frac{dx}{x\sqrt{x^2-1}} = \sec^{-1}x\Big|_{\sqrt{2}}^2 = \frac{\pi}{12}$

58. $f'(x) = (x+1)(x^2+1)^{-1} + \tan^{-1}x$

$f''(x) = (x^2+1)^{-1} + (x+1)[-(x^2+1)^{-2}(2x)] + (x^2+1)^{-1}$

$= 2(x^2+1)^{-1} + (x+1)(-2x)(x^2+1)^{-2}$

$= (x^2+1)^{-2}(2-2x)$

$= -2(x-1)(x^2+1)^{-2}$

$f''(x) = 0$ when $x = 1$; $f(1) = \dfrac{\pi}{4}$, so the point of inflection is $\left(1, \dfrac{\pi}{2}\right)$.

59. $f'(x) = \dfrac{a^{-1}}{1+a^{-2}x^2} - \dfrac{b^{-1}}{1+b^{-2}x^2}$

$= \dfrac{a}{a^2+x^2} - \dfrac{b}{b^2+x^2}$

$f'(x) = 0$ when

$$a(b^2 + x^2) = b(a^2 + x^2)$$
$$ab^2 + ax^2 = ba^2 + bx^2$$
$$(a - b)x^2 = ab(a - b)$$
$$x = \pm\sqrt{ab}$$
$$f''(x) = \frac{-2ax}{(a^2 + x^2)^2} + \frac{2bx}{(b^2 + x^2)^2}$$

so $f''(\sqrt{ab}) = \dfrac{-2a^{3/2}b^{1/2}}{(a^2 + ab)^2} + \dfrac{2a^{1/2}b^{3/2}}{(b^2 + x^2)^2}$

$$= \frac{2ab}{(a + b)^2}(-a^{-2} + b^{-2}) > 0$$

so $x = \sqrt{ab}$ leads to a relative minimum. Similarly, $x = -\sqrt{ab}$ leads to a relative maximum.

60. $f'(x) = 1 + (1 + x^2)^{-1} > 0$; the curve is rising for all x.

$f''(x) = -2x(1 + x^2)^{-2} = 0$ when $x = 0$; $(0, 0)$ is a point of infection.

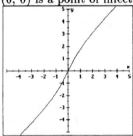

61. $f'(x) = 1 - \dfrac{1}{1 + (2x)^2}(2) = \dfrac{4x^2 - 1}{4x^2 + 1}$

$f'(x) = 0$ when $x = \pm\frac{1}{2}$

$f''(x) = \dfrac{(4x^2 + 1)(8x) - (4x^2 - 1)(8x)}{(4x^2 + 1)^2}$

$= \dfrac{16x}{(4x^2 + 1)^2}$

Relative maximum at $(-\frac{1}{2}, -\frac{1}{2} + \frac{\pi}{4})$ and a relative minimum at $(\frac{1}{2}, \frac{1}{2} - \frac{\pi}{4})$.

There are intercepts when $f(x) = 0$, $\tan^{-1}(2x) = x$, which is not easily solved except for the point $(0, 0)$.

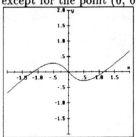

62. $f(x) = \cos^{-1}x + \sin^{-1}x = \frac{\pi}{2}$ on $[-1, 1]$

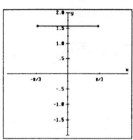

63. $f'(x) = 2 - (1 - x^2)^{-1/2}$
$f'(x) = 0$ when

$$1 - x^2 = \frac{1}{4}$$
$$x = \pm\frac{\sqrt{3}}{2}$$

$f''(x) = -x(1 - x^2)^{-3/2}$

$f''\left(\dfrac{\sqrt{3}}{2}\right) < 0$ so $\dfrac{\sqrt{3}}{2}$ is a maximum; similarly, $-\dfrac{\sqrt{3}}{2}$ is a minimum.

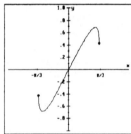

64. $\dfrac{dy}{dx} = 2$ m/s; $\theta = \tan^{-1}\left(\frac{y}{4}\right)$

$\dfrac{d\theta}{dt} = \dfrac{\frac{1}{4}\frac{dy}{dt}}{1 + \frac{y^2}{16}} = \dfrac{4}{16 + y^2}\dfrac{dy}{dt}$

When $y = \frac{3}{2}$

$\dfrac{d\theta}{dt} = \dfrac{4(2)}{16 + \frac{9}{4}} = \dfrac{32}{73} \approx 0.44$ m/s

65. Let the horizontal distance from the camera to the wall be x, and name the resulting angles as indicated:

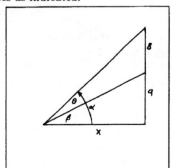

$\theta = \alpha - \beta = \tan^{-1}\frac{17}{x} - \tan^{-1}\frac{9}{x}$.

Maximize with derivatives:

$$\frac{d\theta}{dt} = \frac{1}{1 + \left(\frac{17}{x}\right)^2}\left(-\frac{17}{x^2}\right) - \frac{1}{1 + \left(\frac{9}{x}\right)^2}\left(-\frac{9}{x^2}\right)$$

$$= -\frac{17}{x^2 + 289} + \frac{9}{x^2 + 81}$$

$$= \frac{1224 - 8x^2}{(x^2 + 289)(x^2 + 81)}$$

$\frac{d\theta}{dt} = 0$ when $8(x^2 - 153) = 0$, $x = \sqrt{153}$

Since the endpoint $x = 0$ is obviously a minimum, we have the maximum angle when $x = \sqrt{153} = 3\sqrt{17} \approx 12.4$ ft.

66. $d(\sin^{-1}\frac{u}{a} + C) = \dfrac{1}{\sqrt{1 - \left(\frac{u}{a}\right)^2}}\,d\left(\frac{u}{a}\right)$

$$= \frac{1}{a\sqrt{1 - \left(\frac{u}{a}\right)^2}}\,du = \frac{1}{\sqrt{a^2 - u^2}}\,du$$

Thus, $\displaystyle\int \frac{du}{\sqrt{a^2 - u^2}} = \sin^{-1}\frac{u}{a} + C$

67. $\displaystyle\int \frac{du}{a^2 + u^2} = \frac{1}{a}\int \frac{d\left(\frac{u}{a}\right)}{1 + u^2} = \frac{1}{a}\tan^{-1}\left(\frac{u}{a}\right) + C$

68. $\displaystyle\int \frac{du}{u\sqrt{u^2 - a^2}} = \int \frac{du}{au\sqrt{\left(\frac{u}{a}\right)^2 - 1}}$

$$= \frac{1}{a}\int \frac{d\left(\frac{u}{a}\right)}{u\sqrt{\left(\frac{u}{a}\right)^2 - 1}} = \frac{1}{a}\sec^{-1}\left|\frac{u}{a}\right| + C$$

69. $\dfrac{d}{dx}(\cos^{-1}x) = \dfrac{d}{dx}\left(\dfrac{\pi}{2} - \tan^{-1}x\right) = -\dfrac{1}{1 + x^2}$

70. $\dfrac{d}{dx}(\sec^{-1}x) = \dfrac{d}{dx}[\cos^{-1}(x^{-1})]$

$$= -\frac{1}{\sqrt{1 - x^{-2}}}\left(-\frac{1}{x^2}\right) = \frac{|x|}{x^2\sqrt{x^2 - 1}}$$

$$= \frac{1}{|x|\sqrt{x^2 - 1}}$$

5.7   An Alternative Approach: The Logarithm as an Integral, Page 377

Instructor's Note: As you may know, there are survival hints for the students in the *Student's Solution Manual*. We have included one of these hints here because you might want to share it with your entire class.

SURVIVAL HINT
The material of this section may seem confusing because you already "know" too much about logarithms and exponents. In precalculus you were introduced to $y = b^x$ without any proof that it was continuous for irrational values. The approach here is really better because the area function used *is* continuous to begin with, so the other functions derived from it, e^x, b^x, and $\log_b x$ will also be continuous. Try to read the section as if you were seeing $\ln x$ for the first time.

1. Let $L(x) = \displaystyle\int_1^x \frac{dt}{t}$. Then we have

$L(xy) = L(x) + L(y)$ and $L(x^r) = rL(x)$
in particular,

$$L(2^{-N}) = -NL(2) < 0$$

since

$$L(2) = \int_1^2 \frac{dt}{t} > 0$$

As $N \to +\infty$, $2^{-N} \to 0$ and $L(2^{-N}) \to -\infty$.

Thus,

$$\lim_{x \to 0^+} \ln x = -\infty$$

2. $\Delta x = \dfrac{3 - 1}{8} = \dfrac{1}{4} = 0.25$

$x_0 = 1.00$		$f(x_0) = 1.00$	
$x_1 = 1.25$		$f(x_1) = 0.80$	
$x_2 = 1.50$		$f(x_2) = 0.67$	
$x_3 = 1.75$		$f(x_3) = 0.57$	
$x_4 = 2.00$		$f(x_4) = 0.50$	
$x_5 = 2.25$		$f(x_5) = 0.45$	
$x_6 = 2.50$		$f(x_6) = 0.40$	
$x_7 = 2.75$		$f(x_7) = 0.36$	
$x_8 = 3.00$		$f(x_8) = 0.33$	

Simpson's rule:

$A \approx \frac{1}{3}[1(1.00) + 4(0.80) + 2(0.67)$
$\qquad + 4(0.57) + 2(0.50)$
$\qquad + 4(0.45) + 2(0.40)$
$\qquad + 4(0.36) + 1(0.33)(\frac{1}{4})]$
$\approx \frac{1}{12}(10.19) \approx 1.0992$

The calculator value is 1.098612289.

3. $f'(t) = -t^{-2}$; $f''(t) = 2t^{-3}$;
$f'''(t) = -6t^{-4}$; $f^{(4)}(t) = 24t^{-5}$

Thus, $K = 24$. The error estimate for 8 subintervals is

$$E = \frac{2^5(24)}{180 \cdot 8^4} \approx 0.00104$$

For an error of 0.00005, the number of

subintervals should be

$$0.00005 \geq \frac{2^5(24)}{180n^4}$$

$$n^4 \geq 85{,}333$$

$n \approx 17.09$; the number of subintervals should be 18.

4. From the product rule for logarithms

$$\ln\left[\left(\tfrac{x}{y}\right)y\right] = \ln\left(\tfrac{x}{y}\right) + \ln y$$

$$\ln x - \ln y = \ln\left(\tfrac{x}{y}\right)$$

5. **a.** $F'(x) = \frac{p}{x}$; $G(x) = p\ln x$, so

$$G'(x) = \frac{p}{x} = F'(x).$$

Therefore, $F(x) = G(x) + C$

b. If $x = 1$, then

$$0 = p\cdot 0 + C,\ \text{so}\ C = 0$$

Thus, $F(x) = G(x)$

$$\ln x^p = p\ln x$$

6. Assume $M \neq N$; then $\ln M = \ln N$ implies that there exists a number $M < c < N$ such that $f'(c) = 0$. This is impossible unless $c = 1$ or $M = N$. If $M = N$, then

$$e^M = e^N \text{ or } \ln M = \ln N$$

7. **a.** $f(xy) = f(x) + f(y)$ leads to $f(1) = f(1) + f(1)$ when $x = y = 1$. Thus, $f(1) = 2f(1)$ holds only when $f(1) = 0$.
    b. $f(1) = f(-1) + f(-1)$ when $x = y = -1$. Thus, $0 = f(1) = 2f(-1)$ holds when $f(-1) = 0$
    c. $f(-x) = f(-1) + f(x)$ so $f(-x) = 0 + f(x)$ or $f(-x) = f(x)$
    d. Hold x fixed in the equation

$$f(xy) = f(x) + f(y)$$

and differentiate with respect to y by the chain rule.

$$xf'(xy) = 0 + f'(y)$$

In particular, when $y = 1$:

$$f'(x) = \frac{f'(1)}{x}$$

e. From part **d**, it can be seen that f' is continuous and hence integrable on any

closed interval $[a,\ b]$ not including the origin. By the fundamental theorem of calculus,

$$f(x) - f(c) = \int_c^x f(t)\,dt = f'(1)\int_c^x \frac{dt}{t}$$

for $x > 0$ if $c > 0$ and $x < 0$ if $c < 0$. Since $f(1) = 0$, we can see $c = 1$ to obtain

$$f(x) = f'(1)\int_1^x \frac{dt}{t} \qquad \text{if } x > 0$$

If $x < 0$, then $-x > 0$ and since $f(x) = f(-x)$, we obtain

$$f(x) = f'(1)\int_1^{-x} \frac{dt}{t} \qquad \text{if } x < 0$$

Combining these two formulas:

$$f(x) = f'(1)\int_1^{|x|} \frac{dt}{t} \qquad \text{if } x \neq 0$$

Finally, if $f'(1) \neq 0$ (that is, f is not identically zero), we can let

$$F(x) = \frac{f(x)}{f'(1)} = \int_0^{|x|} \frac{dt}{t}$$

It is easy to show that if

$$f(xy) = f(x) + g(x), \text{ then}$$

$$F(xy) = F(x) + F(y)$$

All solutions of $f(xy) = f(x) + f(y)$ can be obtained by as multiples of $F(x)$

8. Assume $\ln x = \ln y = A$ with $x \neq y$. Then

$$e^{\ln x} = e^{\ln y} = e^A$$

$$x = y = e^A$$

9. It is clear that

$$\ln 2 < \ln e < \ln 3$$

$$e^{\ln 2} < e^{\ln e} < e^{\ln 3}$$

$$2 < e < 3$$

CHAPTER 5 REVIEW

Proficiency Examination, Page 378

1. An exponential function is a function with a constant base and a variable exponent.

2. $e = \lim\limits_{x \to +\infty}\left(1 + \frac{1}{n}\right)^n$; it is the base of natural logarithms.

3. Let r be the rate and t the number of years.

a. $A = (1 + i)^N$ where $i = \frac{r}{n}$; $N = nt$

b. $A = e^{rt}$

4. Let f be a function with domain D and range R. Then the function f^{-1} with domain R and range D is the **inverse of** f if
$f^{-1}[f(x)] = x$ for all x in D
$f[f^{-1}(y)] = y$ for all y in R

5. The horizontal line test for a function (which already meets the vertical line test) checks to see that for a given value of y there is only one value of x. It checks to see if a function is one-to-one.

6. Assume you have the graph for $y = f(x)$. Plot the graph that is symmetric with respect to the line $y = x$ to obtain the graph of the inverse function.

7. $(f^{-1})'(x) = \dfrac{1}{f'[f^{-1}(x)]}$

8. **a.** A logarithm is an exponent. If $y = b^x$ ($b > 0$, $b \neq 1$), then $x = \log_b y$.
   b. A logarithm with base 10.
   c. A logarithm with base e.

9. $\log_a x = \dfrac{\log_b x}{\log_b a}$

10. **a.** $\dfrac{d}{dx}(\ln u) = \dfrac{1}{u}\dfrac{du}{dx}$ **b.** $\dfrac{d}{dx}(\log_b u) = \dfrac{1}{\ln b}\left(\dfrac{1}{u}\right)\dfrac{du}{dx}$

   c. $\dfrac{d}{dx}(e^u) = e^u\dfrac{du}{dx}$ **d.** $\dfrac{d}{dx}(b^u) = b^u \ln b\,\dfrac{du}{dx}$

11. Start with an equation in which both members consist of products or quotients. Take the logarithm of both sides to transform the products into the sum of logarithms and differentiate term-by-term.

12. **a.** $\displaystyle\int e^x\,dx = e^x + C$

   b. $\displaystyle\int x^n\,dx = \dfrac{x^{n+1}}{n+1} + C$

   c. $\displaystyle\int \tan x\,dx = -\ln|\cos x| + C$

   d. $\displaystyle\int \cot x\,dx = \ln|\sin x| + C$

13. $y = \sin^{-1}x$;

$-1 \le x \le 1$;

$-\dfrac{\pi}{2} \le y \le \dfrac{\pi}{2}$;

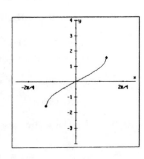

$y = \cos^{-1}x$;

$-1 \le x \le 1$;

$0 \le y \le \pi$;

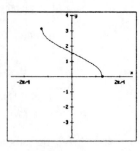

$y = \tan^{-1}x$;

$-\infty < x < +\infty$;

$-\dfrac{\pi}{2} < y < \dfrac{\pi}{2}$

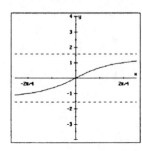

$y = \sec^{-1}x$;

$x \ge 1$ or $x \le -1$;

$0 \le y \le \pi$,

$y \neq \dfrac{\pi}{2}$

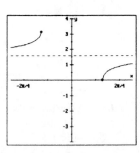

$y = \csc^{-1}x$;

$x \ge 1$ or $x \le -1$;

$-\dfrac{\pi}{2} \le y \le \dfrac{\pi}{2}$,

$y \neq 0$

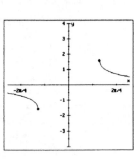

$y = \cot^{-1}x$;

$-\infty < x < +\infty$;

$0 < y < \pi$;

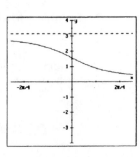

14. $\sin(\sin^{-1}x) = x$ for $-1 \le x \le 1$

$\sin^{-1}(\sin y) = y$ for $-\frac{\pi}{2} \le y \le \frac{\pi}{2}$

$\tan(\tan^{-1}x) = x$ for all x

$\tan^{-1}(\tan y) = y$ for $-\frac{\pi}{2} < y < \frac{\pi}{2}$

15. $\frac{d}{dx}(\sin^{-1}u) = \frac{1}{\sqrt{1-u^2}}\frac{du}{dx}$

$\frac{d}{dx}(\cos^{-1}u) = \frac{-1}{\sqrt{1-u^2}}\frac{du}{dx}$

$\frac{d}{dx}(\tan^{-1}u) = \frac{1}{1+u^2}\frac{du}{dx}$

$\frac{d}{dx}(\sec^{-1}u) = \frac{1}{|u|\sqrt{u^2-1}}\frac{du}{dx}$

$\frac{d}{dx}(\csc^{-1}u) = \frac{-1}{|u|\sqrt{u^2-1}}\frac{du}{dx}$

$\frac{d}{dx}(\cot^{-1}u) = \frac{-1}{1+u^2}\frac{du}{dx}$

16. For a constant $a > 0$:

$\int \frac{du}{\sqrt{a^2-u^2}} = \sin^{-1}\frac{u}{a} + C$

$\int \frac{du}{a^2+u^2} = \frac{1}{a}\tan^{-1}\frac{u}{a} + C$

$\int \frac{du}{|u|\sqrt{u^2-a^2}} = \frac{1}{a}\sec^{-1}\left|\frac{u}{a}\right| + C$

17. **a.** $y' = x^2\left(e^{-\sqrt{x}}\right)\left(-\frac{1}{2\sqrt{x}}\right) + e^{-\sqrt{x}}(2x)$

$= -\frac{1}{2}x^{3/2}e^{-\sqrt{x}} + 2xe^{-\sqrt{x}}$

$= \frac{1}{2}xe^{-\sqrt{x}}\left(4 - \sqrt{x}\right)$

b. $y' = \frac{(\ln 3x)(\frac{1}{x}) - (\ln 2x)(\frac{1}{x})}{(\ln 3x)^2}$

$= \frac{\ln 1.5}{x(\ln 3x)^2}$

18. **a.** $y' = \frac{1}{\sqrt{1-(3x+2)^2}}(3)$

$= \frac{3}{\sqrt{1-(3x+2)^2}}$

b. $y' = \frac{1}{1+(2x)^2}(2) = \frac{2}{1+4x^2}$

19. $\ln y = \ln \frac{\ln(x^2-1)}{\sqrt[3]{x}(1-3x)^3}$

$= \ln[\ln(x^2-1)] - \frac{1}{3}\ln x - 3\ln(1-3x)$

$\frac{1}{y}y' = \frac{1}{\ln(x^2-1)}\frac{1}{x^2-1}(2x) - \frac{1}{3x} - \frac{3}{1-3x}(-3)$

$= \frac{2x}{(x^2-1)\ln(x^2-1)} - \frac{1}{3x} - \frac{9}{3x-1}$

$y' = y\left[\frac{2x}{(x^2-1)\ln(x^2-1)} - \frac{1}{3x} - \frac{9}{3x-1}\right]$

20. $y = (x^2-3)e^{-x} = 0$ when $x = \pm\sqrt{3}$,

When $x = 0$, $y = -3$, which is the y-intercept.

$y' = (x^2-3)e^{-x}(-1) + e^{-x}(2x)$

$= -e^{-x}(x^2-2x-3) = 0$

when $(x-3)(x+1) = 0$, $x = -1, 3$

$y'' = -e^{-x}(2x-2) + (x^2-2x-3)e^{-x}$

$= e^{-x}(x^2-4x-1) = 0$ when

$x = 2\pm\sqrt{5}$

$y''(-1) = 4e > 0$, so there is a relative minimum at $(-1, -2e)$;

$y''(3) = -\frac{4}{e^x} < 0$, so there is a relative maximum at $(3, 6e^{-3})$.

$\lim_{x\to\infty}f(x) = \lim_{x\to\infty}\frac{x^2-3}{e^x} = \lim_{x\to\infty}\frac{2x}{e^x}$

$= \lim_{x\to\infty}\frac{2}{e^x} = 0$

$\lim_{x\to-\infty}f(x) = \lim_{x\to-\infty}\frac{x^2-3}{e^x} = \frac{+\infty}{0} = +\infty$

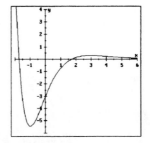

21. **a.** $\frac{1}{2}\int_0^{\ln 2}\left[e^{2x}(2) + e^{-2x}(-2)\right]dx$

$= \frac{1}{2}\left(e^{2x} + e^{-2x}\right)\Big|_0^{\ln 2} = \frac{9}{8}$

b. $\int (\ln x)(\frac{1}{x})\, dx = \frac{(\ln x)^2}{2} + C$

c. $\int \tan 2x\, dx = \int \frac{\sin 2x}{\cos 2x}\, dx$

$= -\frac{1}{2}\int \frac{(-\sin 2x)(2\, dx)}{\cos 2x}$

$$= -\tfrac{1}{2}\ln|\cos 2x| + C$$

22. **a.** $\displaystyle\lim_{x\to 0^+}\left((1+\tfrac{1}{x})^x\right)^4 = \left(\lim_{x\to 0^+}(1+\tfrac{1}{x})^{x4}\right) = e^4$

b. $\displaystyle\lim_{x\to 1^+}\left(\frac{1}{1-x}\right)^x = (-\infty)^x$ so the limit

does not exist.

c. This is of form 0^0, so let $y = x^{\tan x}$,

$\ln y = \ln x^{\tan x} = (\tan x)(\ln x)$.

$\displaystyle\lim_{x\to 0^+}\ln y = \lim_{x\to 0^+}(\tan x)(\ln x)$, which is

of $(0)(\infty)$ form, so write it as:

$\displaystyle\lim_{x\to 0^+}\ln y = \lim_{x\to 0^+}\frac{\ln x}{\cot x}$, which has form

$\tfrac{\infty}{\infty}$, now we can use l'Hôpital's rule:

$$\lim_{x\to 0^+}\ln y = \lim_{x\to 0^+}\frac{\tfrac{1}{x}}{-\csc^2 x}$$

$$= \lim_{x\to 0^+}-\frac{\sin^2 x}{x} = \lim_{x\to 0^+}-\frac{2\sin x\cos x}{1} = 0$$

So now $\displaystyle\lim_{x\to 0^+}\ln y = \ln\left(\lim_{x\to 0^+}y\right) = 0$,

$\displaystyle\lim_{x\to 0^+}\ln y = e^0 = 1$

23. $\displaystyle A = \int_0^1 (e^{2x}-e^x)\,dx = \tfrac{1}{2}e^{2x}-e^x\Big|_0^1$

$$= \tfrac{1}{2}(e^2 - 2e + 1) \approx 1.47624$$

24. **a.** $5 = 2\left(1 + \frac{0.08}{4}\right)^{4t}$

$\ln 2.5 = 4t\ln 1.02$

$t = \frac{\ln 2.5}{4\ln 1.02} \approx 11.57$

or 11 years 208 days

b. $5 = 2\left(1 + \frac{0.08}{12}\right)^{12t}$,

$\ln 2.5 = 12t\ln 1.0066667$

$t = \frac{\ln 2.5}{12\ln 1.00666667} \approx 11.49$

years or 11 years 180 days

c. $2.5 = e^{0.08t}$

$0.08t = \ln 2.5$

$t \approx 11.45$ years or 11 years 166 days

25. Profit = Revenue − Cost

$P = p\left(800e^{-0.01p}\right) - 40\left(800e^{-0.01p}\right)$

$$= \left(800e^{-0.01p}\right)(p - 40)$$

$P' = 800\left[\left(e^{-0.01p}\right)(1)+(p-40)\left(e^{-0.01p}\right)(-0.01)\right]$

$$= 800\left[\left(e^{-0.01p}\right)(1.4 - 0.01p)\right] = 0$$

when $0.01p = 1.4$ or $p = 140$

Sell the cameras for $140.

Supplementary Problems, Page 379

1. **a.** 1.504077   **b.** 16.444647   **c.** 1.107149

2. **a.** 7.4   **b.** 3.7   **c.** 1.25663706

3. **a.** $\frac{3}{5} = 0.6$ **b.** $\frac{63}{65} \approx 0.9692$

4. **a.** $4^{x-1} = 8$ **b.** $2^{x^2+4x} = 16^{-1}$

$2^{2x-2} = 2^3$ $2^{x^2+4x} = 2^{-4}$

$2x - 2 = 3$ $x^2 + 4x = -4$

$x = \frac{5}{2}$ $(x + 2)^2 = 0$

$x = -2$

5. **a.** $\ln(x-1) + \ln(x+1) = 2\ln\sqrt{12}$

$\ln(x-1)(x+1) = \ln 12$

$(x-1)(x+1) = 12$

$x^2 = 13$

$x = \sqrt{13}$

Reject $x = -\sqrt{13}$ since $\ln x$ is not defined at $x = -\sqrt{13}$.

b. $\sqrt{x} = \cos^{-1}0.317 + \sin^{-1}0.317 = \frac{\pi}{2}$

Thus, $x = \frac{\pi^2}{4}$.

6. **a.** $\log_2 2^{x^2} = 4$

$2^{x^2} = 2^4$

$x^2 = 4$

$x = \pm 2$

b. $\log_4\sqrt{x(x-15)} = 1$

$x(x-15) = 4^1$

$x^2 - 15x - 16 = 0$

$(x-16)(x+1) = 0$

$x = 16, -1$

7. **a.** $\log_2 x + \log_2(x-15) = 4$

$\log_2 x(x-15) = 4$

$$2^4 = x(x - 15)$$
$$(x - 16)(x + 1) = 0$$
$$x = 16$$

Reject $x = -1$ since $\log_2 x$ not defined at $x = -1$.

b.
$$3^{2x-1} = 6^x 3^{1-x}$$
$$3^{2x-1} = 2^x 3^x 3^{1-x}$$
$$3^{2x-2} = 2^x$$
$$(2x - 2)\ln 3 = x \ln 2$$
$$2x \ln 3 - x \ln 2 = 2 \ln 3$$
$$x = \frac{2 \ln 3}{2 \ln 3 - \ln 2} \approx 1.4084$$

8. a. $y = 2x^3 - 7$; inverse is $x = 2y^3 - 7$
$$y = f^{-1}(x) = \sqrt[3]{\tfrac{1}{2}(x + 7)}$$

b. $y = \sqrt[7]{2y + 1}$, $x \geq -\frac{1}{2}$; inverse is
$$x = \sqrt[7]{2x + 1},\ y \geq -\tfrac{1}{2}$$
$$y = f^{-1}(x) = \tfrac{1}{2}(x^7 - 1)$$

9. a. $y = \sqrt{e^x - 1}$, $x \geq 0$; inverse is
$$x = \sqrt{e^y - 1},\ y \geq 0$$
$$x^2 = e^y - 1,\ \text{so}$$
$$y = f^{-1}(x) = \ln(x^2 + 1)$$

b. $y = \dfrac{x + 5}{x - 7}$, $x \neq 7$; inverse is
$$x = \frac{y + 5}{y - 7},\ \text{so } x(y - 7) = y + 5$$
$$y = f^{-1}(x) = \frac{7x + 5}{x - 1}$$

10. $y = f(x) = \dfrac{x + a}{x - 1}$; inverse is $x = \dfrac{y + a}{y - 1}$
or $x(y - 1) = y + a$ or $xy - x = y + a$
$$y = f^{-1}(x) = \frac{x + a}{x - 1}$$

11. $y = f(x) = \dfrac{ax + b}{cx + d}$; inverse is $x = \dfrac{ay + b}{cy + d}$
$$xcy + xd = ay + b$$
$$cxy - ay = b - dx$$
$$y = \frac{b - dx}{cx - a}$$
f^{-1} exists if $x \neq a/c$.

12. $y = f(x) = \dfrac{x + 1}{x - 1}$; inverse is $x = \dfrac{y + 1}{y - 1}$

$$xy - x = y + 1$$
$$y = \frac{x + 1}{x - 1}$$
The domain of f^{-1} is all real x, $x \neq -1$.

13. $y' = (4x + 5)\exp(2x^2 + 5x - 3)$

14. $y' = \dfrac{2x}{x^2 - 1}$

15.
$$y' = x(3^{2-x})\ln 3(-1) + 3^{2-x}$$
$$= 3^{2-x}(1 - x \ln 3)$$

16. $y = \log_3(x^2 - 1) = \dfrac{\ln(x^2 - 1)}{\ln 3}$
$$y' = \frac{2x}{(x^2 - 1)\ln 3}$$

17.
$$e^{xy} + 2 = \ln y - \ln x$$
$$e^{xy}(xy' + y) = y^{-1}y' - x^{-1}$$
$$e^{xy}x^2 yy' + e^{xy}xy^2 = xy' - y$$
$$y' = \frac{-(y + xy^2 e^{xy})}{x^2 y e^{xy} - x}$$
$$= \frac{y(1 + xy e^{xy})}{x(1 - xy e^{xy})}$$

18. $\dfrac{dy}{dx} = \dfrac{3\sqrt{x}}{\sqrt{1 - (3x + 2)^3}} + \dfrac{1}{2\sqrt{x}}\sin^{-1}(3x + 2)$

19. $y' = e^{\sin x}(\cos x)$

20. $y = 2^x \log_2 x = \dfrac{2^x \ln x}{\ln 2}$
$$y' = (\ln 2)^{-1}(2^x x^{-1} + 2^x \ln 2 \ln x)$$
$$= (x \ln 2)^{-1} 2^x + 2^x \ln x$$

21. $y = e^{-x}\log_5 3x = (\ln 2)^{-1}(2^x \ln x)$
$$y' = (\ln 5)^{-1}(e^{-x}x^{-1} - e^{-x}\ln 3x)$$
$$= (\ln 5\, e^x)^{-1}(x^{-1} - \ln 3x)$$
$$= \frac{e^{-x}}{x \ln 5}(1 - x \ln 3x)$$

22.
$$x2^y + y2^x = 3$$
$$x(2^y \ln 2)y' + 2^y + y(2^x \ln 2) + 2^x y' = 0$$
$$(x2^y \ln 2 + 2^x)y' = -2^y - y2^x \ln 2$$
$$y' = -\frac{2^y + y2^x \ln 2}{x(2^y \ln 2) + 2^x}$$

23. $\ln(x + y^2) = x^2 + 2y$
$$\frac{1 + 2yy'}{x + y^2} = 2x + 2y'$$
$$1 + 2yy' = 2x(x + y^2) + 2y'(x + y^2)$$
$$2yy' - 2xy' - 2y^2 y' = 2x^2 + 2xy^2 - 1$$

$$y' = \frac{2x^2 + 2xy^2 - 1}{2y - 2x - 2y^2}$$

24. $\dfrac{dy}{dx} = \dfrac{1}{(x + \tan^{-1}x)^2}\left\{(x + \tan^{-1}x)\dfrac{1}{\sqrt{1 - x^2}}\right.$

$$\left. - \sin^{-1}x\left(1 + \frac{1}{1 + x^2}\right)\right\}$$

25. $\dfrac{dy}{dx} = \dfrac{\sin^{-1}x - x(1 - x^2)^{-1/2}}{(\sin^{-1}x)^2}$

$$+ \frac{1}{x^2}\left(\frac{x}{1 + x^2} - \tan^{-1}x\right)$$

26. $\dfrac{dy}{dx} = e^{-x}\dfrac{1}{2x\sqrt{\ln 2x}} - e^{-x}\sqrt{\ln 2x}$

27. $\dfrac{dy}{dx} = (\sin x)(1 - x^2)^{-1/2} + (\cos x)(\sin^{-1}x)$

$$+ x(-1)(1 + x^2)^{-1} + \cot^{-1}x$$

28. $y = x \ln ex = x(\ln e + \ln x) = x + x \ln x$

$y' = 1 + 1 + \ln x = 2 + \ln x$

At $x = 1$, $y = 1$, and $y' = 2$:

$$y - 1 = 2(x - 1)$$

$$2x - y - 1 = 0$$

29. $y' = e^{2x+1}(2x + 1)$

At $x = \frac{1}{2}$, $y = \frac{1}{2}$, and $y' = 2$:

$$y - \frac{1}{2} = 2(x - \frac{1}{2})$$

$$4x - 2y - 1 = 0$$

30. $$e^{xy} = x - y$$

$$e^{xy}(xy' + y) = 1 - y'$$

At $x = 1$, $y = 0$, and $y' = \frac{1}{2}$:

$$y = \frac{1}{2}(x - 1)$$

$$x - 2y + 1 = 0$$

31. $y = (1 - x)^x$ or $\ln y = x\ln(1 - x)$

$$y^{-1}y' = -x(1 - x)^{-1} + \ln(1 - x)$$

At $x = 0$, $y = 1$, and $y' = 0$; so the tangent
line is horizontal with equation $y - 1 = 0$.

32. $y = 2^x - \log_2 x = 2^x - (\ln 2)^{-1}\ln x$

$y' = 2^x \ln 2 - (x \ln 2)^{-1}$

At $x = 1$, $y = 2$, and

$y' = 2\ln 2 - (\ln 2)^{-1} \approx -0.06$

$$y - 2 = -0.06(x - 1)$$

$$0.06x + y - 1.94 = 0$$

33. $\ln y = (2 - x)\ln 5 + 3\ln(x^2 - x) - 4\ln(2x^3 - 3x)$

$$\frac{1}{y}\frac{dy}{dx} = -\ln 5 + \frac{3(2x - 1)}{x^2 - x} - \frac{12(2x^2 - 1)}{2x^3 - 3x}$$

$$\frac{dy}{dx} = y\left[-\ln 5 + \frac{3(2x - 1)}{x^2 - x} - \frac{12(2x^2 - 1)}{2x^3 - 3x}\right]$$

34. $\ln y = \frac{1}{3}\ln(x^2 + 3) + \frac{1}{2}\ln(x^4 + x + 1) - \frac{1}{4}(x^3 + 5)$

$$\frac{1}{y}\frac{dy}{dx} = \frac{2x}{3(x^2 + 3)} + \frac{4x^3 + 1}{2(x^4 + x + 1)} - \frac{3x^2}{4(x^3 + 5)}$$

$$\frac{dy}{dx} = y\left[\frac{2x}{3(x^2 + 3)} + \frac{4x^3 + 1}{2(x^4 + x + 1)} - \frac{3x^2}{4(x^3 + 5)}\right]$$

35. $\ln y = 2x \ln(x^2 + 1)$

$$\frac{1}{y}\frac{dy}{dx} = 2x\frac{2x}{x^2 + 1} + 2\ln(x^2 + 1)$$

$$\frac{dy}{dx} = 2y\left[\frac{2x^2}{x^2 + 1} + \ln(x^2 + 1)\right]$$

36. $\ln y = x \ln x$

$$\frac{1}{y}\frac{dy}{dx} = \ln x + 1$$

$$\frac{dy}{dx} = y(\ln x + 1)$$

37. $\ln y = 2x \ln(3x^2 + 2)$

$$\frac{1}{y}\frac{dy}{dx} = 2[x(6x)(3x^2 + 2)^{-1} + \ln(3x^2 + 2)]$$

$$\frac{dy}{dx} = 2y\left[\frac{6x^2}{3x^2 + 2} + \ln(3x^2 + 2)\right]$$

38. $\ln y = \sin x \ln x$

$$\frac{1}{y}\frac{dy}{dx} = x^{-1}\sin x + \ln x \cos x$$

$$\frac{dy}{dx} = y\left[\frac{\sin x}{x} + \ln x \cos x\right]$$

39. $y = x^2 \ln\sqrt{x} = \frac{1}{2}x^2 \ln x$

Domain $(0, +\infty)$;

$y' = \frac{1}{2}(x^2 x^{-1} + 2x \ln x) = \frac{1}{2}(x + 2x \ln x)$

$y'' = \frac{1}{2}(1 + 2xx^{-1} + 2\ln x) = \frac{1}{2}(3 + 2\ln x)$

relative minimum at $(e^{-1/2}, -0.09)$;
point of inflection at $x = e^{-1.5}$;
intercept $(1, 0)$;

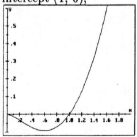

40. $\sin^{-2}x - \cos^{-1}x = 2\sin^{-1}x - \frac{\pi}{2}$
Domain, $-1 \le x \le 1$

$y' = 2(1 - x^2)^{-1/2}; \quad y'' = 2x(1 - x^2)^{-3/2}$

relative maximum at $(1, \frac{\pi}{2})$;
relative minimum at $(-1, -\frac{3\pi}{2})$
point of inflection at $x = 0$
intercept at $(0, \frac{\pi}{2})$

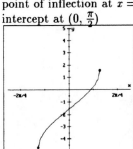

41. $f(x) = \ln\left(\frac{x-1}{x+1}\right) = \ln(x-1) - \ln(x+1)$
domain $(-\infty, -1) \cup (1, +\infty)$;

$f'(x) = (x-1)^{-1} - (x+1)^{-1}$

$\quad = 2(x^2 - 1)^{-1}$

$f''(x) = -4x(x^2 - 1)^{-2}$

no extreme points;
no points of inflection;
no intercepts;
asymptotes $y = 0, \; x = \pm 1$

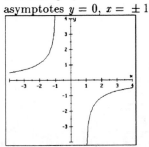

42. $f(x) = xe^{1/x}$
Domain: all real numbers, $x \ne 0$

$f'(x) = e^{1/x}(-x^{-1} + 1)$

$f''(x) = e^{1/x}[x^{-2} + (1 - x^{-1})(-x^{-2})]$

no relative maximum;
relative minimum at $(1, e)$;
there are no intercepts;
$x = 0$ is a vertical asymptotes

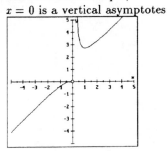

43. $\displaystyle\int \frac{x+1}{x^2 + 2x + 5}\,dx = \frac{1}{2}\int \frac{2(x+1)\,dx}{x^2 + 2x + 5}$

$\quad = \frac{1}{2}\ln(x^2 + 2x + 5) + C$

44. $\displaystyle\int xe^{-x^2}\,dx = -\frac{1}{2}\int e^{-x^2}(-2x\,dx)$

$\quad = -\frac{1}{2}e^{-x^2} + C$

45. $\displaystyle\int \frac{dx}{7 + x^2} = \frac{1}{\sqrt{7}}\tan^{-1}\left(\frac{x}{\sqrt{7}}\right)$

46. $\displaystyle\int \frac{dx}{\sqrt{25 - 9x^2}} = \frac{1}{3}\int \frac{3\,dx}{\sqrt{5^2 - (3x)^2}}$

$\quad = \frac{1}{3}\sin^{-1}\frac{3x}{5} + C$

47. $\displaystyle\int_0^{\pi/2} \frac{\sin x\,dx}{2 + \cos x} = \ln(2 + \cos x)\Big|_0^{\pi/2}$

$\quad = -\ln 2 + \ln 3 = \ln 1.5$

48. $\displaystyle\int_0^{\ln 2}(e^{2x} - e^{-2x})\,dx$

$= \frac{1}{2}\int_0^{\ln 2} e^{2x}(2\,dx) + \frac{1}{2}\int_0^{\ln 2} e^{-2x}(-dx)$

$= \frac{1}{2}(e^{2x} + e^{-2x})\Big|_0^{\ln 2} = \frac{17}{8}$

49. $\displaystyle\int_0^{\ln 2}(e^{2x} + 5)e^x\,dx = \int_0^{\ln 2}(e^{3x} + 5e^x)\,dx$

$= \frac{1}{3}(e^{3x} + 5e^x)\Big|_0^{\ln 2} = \frac{22}{3}$

50. $\displaystyle\int_0^{\pi/6}\sec 2x\,dx = \frac{1}{2}\int_0^{\pi/6}\frac{\sec 2x(\sec 2x + \tan 2x)}{\sec 2x + \tan 2x}(2\,dx)$

$= \frac{1}{2}\ln|\sec 2x + \tan 2x|\Big|_0^{\pi/6} = \frac{1}{2}\ln(2 + \sqrt{3})$

51. $\displaystyle\int \frac{e^{\sqrt{x}}}{\sqrt{x}}\,dx = 2\int e^{\sqrt{x}}\frac{dx}{2\sqrt{x}} = 2e^{\sqrt{x}} + C$

52. $\displaystyle\int \frac{\sin 2x\,dx}{1 + \sin^2 x} = \int \frac{2\sin x\cos x\,dx}{1 + \sin^2 x}$

$= \ln(1 + \sin^2 x) + C$

53. $\displaystyle\int \frac{e^{2x}\,dx}{e^x + 1} = \int \frac{(e^{2x} + e^x - e^x)\,dx}{e^x + 1}$

$$= \int\left(e^x - \frac{e^x}{e^x + 1}\right)dx$$

$$= e^x - \ln(e^x + 1) + C$$

54. $\displaystyle\int \frac{(x^2 + x + 1)\,dx}{x^2 + 1} = \int\left(\frac{x^2 + 1}{x^2 + 1} + \frac{x}{x^2 + 1}\right)dx$

$$= \int dx + \frac{1}{2}\int \frac{2x\,dx}{x^2 + 1} = x + \frac{1}{2}\ln(1 + x^2) + C$$

55. $\displaystyle\int \frac{\cos x - \sin x}{\cos x + \sin x}\,dx = \ln|\cos x + \sin x| + C$

56. $\displaystyle\int \frac{\sec^2 x\,dx}{1 + \tan x} = \ln|1 + \tan x| + C$

57. $\displaystyle\int \frac{1 + \sqrt{1 - x^2}}{\sqrt{1 - x^2}}\,dx = \int \frac{dx}{\sqrt{1 - x^2}} + \int dx$

$$= \sin^{-1}x + x + C$$

58. $\displaystyle\int \frac{(2 + x)\,dx}{1 + x^2} = 2\int \frac{dx}{1 + x^2} + \frac{1}{2}\int \frac{2x\,dx}{1 + x^2}$

$$= \tan^{-1}x + \frac{1}{2}\ln(1 + x^2) + C$$

59. $\displaystyle\int \frac{\cos x\,dx}{1 - \sin^2 x} = \int \frac{\cos x\,dx}{\cos^2 x} = \int \sec x\,dx$

$$= \ln|\sec x + \tan x| + C$$

60. $\displaystyle\int \frac{x\,dx}{x^4 + 1} = \frac{1}{2}\int \frac{2x\,dx}{(x^2)^2 + 1} = \frac{1}{2}\tan^{-1}x^2 + C$

61. $\displaystyle\int \frac{x^2\,dx}{1 + x^6} = \frac{1}{3}\int \frac{3x^2\,dx}{1 + (x^3)^2} = \frac{1}{3}\tan^{-1}x^3 + C$

62. $\displaystyle\int \frac{x^4\,dx}{1 + x^{10}} = \frac{1}{5}\int \frac{5x^4\,dx}{1 + (x^5)^2} = \frac{1}{5}\tan^{-1}x^5 + C$

63. $\displaystyle\int \frac{dx}{x + x(\ln x)^2} = \int \frac{1}{1 + (\ln x)^2}\,\frac{dx}{x}$

$$= \tan^{-1}(\ln x) + C$$

64. $\displaystyle\int \frac{\ln x\,dx}{x + x(\ln x)^4} = \frac{1}{2}\int \frac{2\ln x\,dx}{x}\left(\frac{1}{1 + (\ln^2 x)^2}\right)$

$$= \frac{1}{2}\tan^{-1}(\ln x) + C$$

65. a. $\displaystyle\lim_{x\to 0^+}(1 + x)^{4/x} = \left[\lim_{x\to 0^+}(1 + x)^{1/x}\right]^4 = e^4$

b. $\displaystyle\lim_{x\to 1}\left(\frac{1}{1 - x}\right)^x = +\infty$

66. a. $\displaystyle\lim_{x\to 0^+} x^{\tan x} = \lim_{x\to 0^+}\exp[\ln x^{\tan x}]$

$$= \lim_{x\to 0^+}\exp\left[\frac{\ln x}{\cot x}\right] = \lim_{x\to 0^+}\exp\left[\frac{x^{-1}}{-\csc^2 x}\right]$$

$$= \lim_{x\to 0^+}\exp(-\sin x) = e^0 = 1$$

b. $\displaystyle\lim_{x\to 0}\frac{5^x - 1}{x} = \lim_{x\to 0}\frac{5^x\ln 5}{1} = \ln 5$

67. a. $\displaystyle\lim_{x\to 0}\frac{\ln(x^2 + 1)}{x} = \lim_{x\to 0}\frac{2x}{x^2 + 1} = 0$

b. Let $y = \left(\frac{1}{x}\right)^x$; then $\ln y = x\ln x^{-1}$

$$\lim_{x\to +\infty}\ln y = \lim_{x\to +\infty}[-x\ln x]$$

$$= \lim_{u\to 0^-}[-\frac{1}{u}\ln \frac{1}{u}]$$

$$= \lim_{u\to 0^-}\left[-\frac{\ln\frac{1}{u}}{\frac{1}{u}}\right]$$

$$= \lim_{u\to 0^-}\left[-\frac{-\frac{1}{u}}{1}\right]$$

$$= -\infty$$

$$\lim_{x\to +\infty}\ln y = e^{-\infty} = 0$$

68. a. $\displaystyle\lim_{x\to +\infty}(4 - \frac{1}{x})^x$ This limit tends to 4^x which exceeds all bounds

b. $\displaystyle\lim_{x\to +\infty}\frac{e^x\cos x - 1}{x}$

$$= \lim_{x\to +\infty}\frac{-e^x\sin x + e^x\cos x}{1}$$

$$= \lim_{x\to +\infty}e^x(-\sin x + \cos x) = +\infty$$

69. $\displaystyle A = \int_0^1 2^{-x}\,dx = -\frac{1}{\ln 2}2^{-x}\Big|_0^1 = \frac{1}{2\ln 2}$

$$\approx 0.7213$$

70. $\displaystyle A = \int_1^{e^2}\frac{\ln x}{x}\,dx = \int_1^{e^2}\ln x\left(\frac{dx}{x}\right) = \frac{1}{2}(\ln x)^2\Big|_1^{e^2} = 2$

71. $\displaystyle A = \int_0^{\pi/4}\tan x\,dx = -\int_0^{\pi/4}\frac{-\sin x\,dx}{\cos x}$

$$= -\ln|\cos x|\Big|_0^{\pi/4}$$

$$= -\ln \frac{1}{\sqrt{2}} = \frac{1}{2}\ln 2$$

72. $\displaystyle A = \int_0^1 \frac{x}{1 + e^{-x^2}}\,dx = -\frac{1}{2}\int_0^1 \frac{-2x\,dx}{1 + e^{-x^2}}$

$$= -\frac{1}{2}\int_0^{-1}\frac{du}{1 + e^u} = \frac{1}{2}\int_0^{-1}\frac{-e^{-u}\,du}{e^{-u} + 1}$$

$$= \frac{1}{2}\ln|e^{-u} + 1|\Big|_0^{-1} = \frac{1}{2}\ln(e + 1) - \frac{1}{2}\ln 2$$

73. $f'(x) = (x^2 + bx + b)e^{-x} = 0$ if

$$x_1 = \frac{-b - \sqrt{b^2 - 4b}}{2}; \ x_2 = \frac{-b + \sqrt{b^2 - 4b}}{2}$$

Case 1: $0 < b < 4$; both $x_1 < 0$ and $x_2 < 0$
$f'(0) = b > 0$; the curve rises on
$(-\infty, x_1) \cup (x_2, +\infty)$ and falls on
(x_1, x_2). There is a relative maximum
at x_1 and a relative minimum at x_2.

Case 2: $b = 4$; $x_1 = x_2$, where there is a
relative minimum

Case 3: $b > 4$; there are no relative
extremum. The curve is rising for
all x.

Case 4: $b < 0$; $f'(0) = b < 0$; the curve rises
on $(-\infty, x_1) \cup (x_2, +\infty)$ and falls on
(x_1, x_2). There is a relative maximum
at x_1 and a relative minimum at x_2.

74. $\dfrac{1}{e-1} \displaystyle\int_1^e \ln x \left(\dfrac{dx}{x}\right) = \dfrac{1}{e-1} \dfrac{(\ln x)^2}{2}\Big|_1^e = \dfrac{1}{2(e-1)}$

75. **a.** $2,000 = P\left(1 + \dfrac{0.0625}{4}\right)^{40}; \ P \approx 1,075.71$

b. $2,000 = Pe^{-10(0.0625)}; \ P \approx 1,070.52$

76. Consider a right triangle with the leg along
the shore $\overline{PQ}$ with the length labeled x, and
the leg from the lighthouse to P labeled 4,000
ft. We know that $dx/dt = 3$ ft/s. Let θ be the
acute angle adjacent to $\overline{PQ}$. Then
$x = 4,000 \tan \theta$ and $\theta = \tan^{-1}(x/4,000)$.

$$\frac{d\theta}{dt} = \frac{\frac{1}{4,000}}{1 + \left(\frac{1}{4,000}\right)^2 x^2} \frac{dx}{dt} = \frac{4,000}{4,000^2 + x^2} \frac{dx}{dt}$$

At $x = 1,000$

$$\frac{d\theta}{dt} = \frac{3(4,000)}{4,000^2 + 1,000^2} = \frac{12}{17,000} \text{ radians/s}$$

77. Consider a right triangle with the leg along
the shore $\overline{PQ}$ with the length labeled x, and
the leg from the lighthouse to P labeled 4 mi.
We know that

$$\frac{d\theta}{dt} = \left(2 \frac{\text{rev}}{\text{min}}\right)\left(2\pi \frac{\text{rad}}{\text{rev}}\right) = 4\pi \text{ rad/min}$$

Then

$$\tan \theta = \frac{x}{4}$$

$$\sec^2\theta \frac{d\theta}{dt} = \frac{1}{4} \frac{dx}{dt}$$

At t_0, $x = 2$, $\sec \theta_0 = \dfrac{\sqrt{20}}{4}$ and

$$\frac{dx}{dt} = 4\left(\frac{\sqrt{20}}{4}\right)^2(4\pi) = 20\pi \text{ mi/min}$$

78. Let $\alpha = \tan^{-1}x$ and $\beta = \tan^{-1}y$; then

$$\tan(\alpha + \beta) = \frac{\tan\alpha + \tan\beta}{1 - \tan\alpha\tan\beta} = \frac{x + y}{1 - xy}$$

Thus, $\alpha + \beta = \tan^{-1}\left(\dfrac{x+y}{1-xy}\right)$

a. $\alpha = \tan^{-1}\frac{1}{2}; \ \beta = \tan^{-1}\frac{1}{3};$

$$\alpha + \beta = \tan^{-1}\left(\frac{\frac{1}{2} + \frac{1}{3}}{1 - \frac{1}{2}\cdot\frac{1}{3}}\right) = \frac{\pi}{4}$$

b. $\alpha = \tan^{-1}\frac{1}{3}; \ \beta = \tan^{-1}\frac{1}{7};$

$$\tan 2\alpha = \frac{2\tan\alpha}{1 - \tan^2\alpha} = \frac{\frac{2}{3}}{1 - \frac{1}{9}} = \frac{3}{4}$$

$$2\alpha + \beta = \tan^{-1}\left(\frac{\tan 2\alpha + \tan\beta}{1 - \tan\alpha\tan\beta}\right)$$

$$= \tan^{-1}\left(\frac{\frac{3}{4} + \frac{1}{7}}{1 - \frac{1}{3}\cdot\frac{1}{7}}\right)$$

$$= \tan^{-1}1 = \frac{\pi}{4}$$

c. $\alpha = \tan^{-1}\frac{1}{5}; \ \beta = \tan^{-1}\frac{1}{239};$

$$\tan 2\alpha = \frac{2\tan\alpha}{1 - \tan^2\alpha} = \frac{\frac{2}{5}}{1 - \frac{1}{25}} = \frac{5}{12}$$

$$\tan 4\alpha = \frac{2\tan 2\alpha}{1 - \tan^2 2\alpha} = \frac{\frac{10}{12}}{1 - \frac{25}{144}} = \frac{120}{119}$$

$$4\alpha - \beta = \tan^{-1}\left(\frac{\tan 4\alpha - \tan\beta}{1 - \tan 4\alpha\tan\beta}\right)$$

$$= \tan^{-1}\left(\frac{\frac{120}{119} + \frac{1}{239}}{1 - \frac{120}{119}\cdot\frac{1}{239}}\right)$$

$$= \tan^{-1}\left(\frac{28,561}{28,561}\right) = \frac{\pi}{4}$$

79. $(\cot^{-1}x)' = \left(\frac{\pi}{2} - \tan^{-1}x\right)' = -(1 + x^2)^{-2}$

80. $\dfrac{d}{dx}(\sec^{-1}x) = \dfrac{d}{dx}(\cos^{-1}x^{-1})$

$$= \frac{1}{\sqrt{1 - x^{-2}}}(-x^{-2}) = \frac{1}{x^2\sqrt{\frac{x^2 - 1}{x^2}}}$$

$$= \frac{|x|}{x^2\sqrt{x^2 - 1}} = \frac{1}{|x|\sqrt{x^2 - 1}}$$

81. We are given $T_a = 10$, $T_d = 98.6$, and
$T = 40$. Therefore, we have

$$40 = 10 + (98.6 - 10)(0.97)^t$$

Solving for t, we find

$$(0.97)^t = \frac{30}{88.6} = 0.3386$$

$$t \ln 0.97 = \ln 0.3386$$

$$t = \frac{\ln 0.3386}{\ln 0.97} \approx 35.5536 \text{ hr}$$

Thus, the body had been in the freezer for 35.5536 hr (35 hr and 33 min), so Siggy had been put into the freezer on Wednesday morning at about 1:27 AM (1:30 AM is close enough). André was in the slammer, so Boldfinger must have done it.

82. a. $F(5) = 2F(9)$, so

$$e^{-5k} = 2e^{-9k}$$

$$e^{4k} = 2$$

$$4k = \ln 2$$

$$k = \tfrac{1}{4} \ln 2$$

$$F(t) = e^{(\ln 2)t/4}; \; F(7) \approx 0.2973$$

b. $1 - F(10) = 1 - e^{(\ln 2)(10)/4} \approx 0.8232$

c. $F(4) - F(5) = e^{(\ln 2)(4)/4} - e^{(\ln 2)(5)/4}$

≈ 0.0796

83. First National Bank:

$$A = 100(1+\tfrac{0.07}{12})^{12} \approx 107.23$$

Fells Cargo Bank:

$$A = 100 e^{0.0695} \approx 107.20$$

Thus, First National pays 3¢ more per year per $100 deposit.

84. $A(t) = \displaystyle\int_0^t e^{-kx} \, dx = -\tfrac{1}{k} e^{-kx} \Big|_0^t = \tfrac{1}{k}(1 - e^{-kt})$

$$\lim_{x \to +\infty} A(t) = \tfrac{1}{k}$$

85. $\displaystyle\lim_{x \to +\infty} x(N^{1/x} - 1) = \lim_{x \to +\infty} \frac{N^{1/x} - 1}{x^{-1}}$

$= \displaystyle\lim_{x \to +\infty} \frac{N^{1/x}(\ln N)(-x^{-2})}{-x^{-2}}$

$= (1)\ln N = \ln N$

86. a. $y = e^x$; let the area under the curve be partitioned into n subintervals. A rectangle is mounted on each subinterval. The sum of the area of rectangles is

$$\lim_{n \to +\infty}(e^{1/n} + \cdots + e)(n^{-1}) = \int_0^1 e^x \, dx = e - 1$$

b. $\displaystyle\lim_{n \to +\infty}\left(\frac{1}{1+n} + \frac{1}{2+n} + \cdots + \frac{1}{n+n}\right)$

$= \displaystyle\lim_{n \to +\infty}\left(\frac{1}{1+\frac{1}{n}} + \frac{1}{1+\frac{2}{n}} + \cdots + \frac{1}{1+\frac{1}{n}}\right)\left(\frac{1}{n}\right)$

$= \displaystyle\int_0^1 \frac{1}{1+x} \, dx = \ln|1 + x|\Big|_0^1 = \ln 2$

c. $\displaystyle\lim_{n \to +\infty}\left(\frac{1}{n^2+1} + \frac{2}{n^2+4} + \cdots + \frac{n}{n^2+n^2}\right)$

$= \displaystyle\lim_{n \to +\infty}\left(\frac{\frac{1}{n}}{1+(\frac{1}{n})^1} + \frac{\frac{2}{n}}{1+(\frac{2}{n})^2} + \cdots \right.$

$\left. + \frac{\frac{n}{n}}{1+(\frac{n}{n})^2}\right)\left(\frac{1}{n}\right)$

$= \displaystyle\int_0^1 \frac{x}{1+x^2} \, dx = \tfrac{1}{2}\ln(1+x^2)\Big|_0^1 = \tfrac{1}{2}\ln 2$

87. $\displaystyle\lim_{x \to 0^-} \frac{1}{1+2^{1/x}} = \lim_{x \to 0^+} \frac{1}{1+2^{-1/x}} = 1$

$$\lim_{x \to 0^+} \frac{1}{1+2^{1/x}} = 0$$

88. a. False; $\tan^{-1}1 = \frac{\pi}{4}$, but $\frac{\sin^{-1}1}{\cos^{-1}1}$ is not defined

b. False; $e^{1/2} \approx 1.65$; $e^{-1/2} \approx 0.01$

c. False; $\tan^{-1}1 = \frac{\pi}{4}$;

$(\tan 1)^{-1} = \cot 1 \approx 0.642$

d. True; $\cot^{-1}x = \frac{\pi}{2} - \tan^{-1}x$ because these are complementary angles.

e. True; let $\alpha = \sin^{-1}x$, then $\sin^2\alpha = x$, $\cos^2\alpha = 1 - x^2$; $\cos(\sin \alpha) = \sqrt{1 - x^2}$

f. True; let $\alpha = \cos^{-1}x$, then $\cos \alpha = x$; $\frac{1}{\cos \alpha} = \frac{1}{x} = \sec \alpha$ or $\alpha = \sec^{-1}\frac{1}{x} = \cos^{-1}x$

89. $y = 2^x$ crosses $y = x^{-1}$ near $x = 0.6411$, as indicated in the graph.

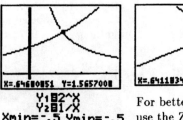

For better accuracy, use the ZOOM.

90. The graphs cross near $x = 1.87$, as indicated in the graph. Using the ZOOM or computer software you can find a better approximation,

say $x \approx 1.88320350591$.

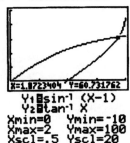

X=1.8723404 Y=60.731762
Y₁◼sin⁻¹ (X-1)
Y₂◼tan⁻¹ X
Xmin=0 Ymin=-10
Xmax=2 Ymax=100
Xscl=.5 Yscl=20

91.

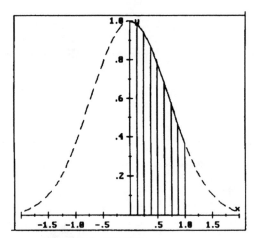

# OF TERMS	TRAPEZOIDAL RULE OVER [0, 1]
4	.7429841
6	.7451194
8	.7458656

92.   Trapezoidal Rule used to calculate estimate.

TYPE OF ESTIMATE	# OF SUB- INTERVALS	ESTIMATE OVER [50000, 60000]
Trapezoid	20	916.388878777

93. This is Putnam Problem 5ii of the morning session of 1939. Let m be the mass of the particle, and let θ be the angular position of the rod, measured from the vertical, at time t. The force of gravity mg can be resolved into two components, $mg \cos \theta$ acting along the rod, and $mg \sin \theta$ acting perpendicular to the rod. The former is counterbalanced by the tension (or compression) in the rod and the latter accelerates the particle along the circle of radius a. By Newton's third law, we have $mg \sin \theta = ma \, d^2\theta/dt^2$. Multiply both sides by $(2/m) \, d\theta/dt$ to obtain

$$2g \sin \theta \frac{d\theta}{dt} = 2a \frac{d\theta}{dt} \frac{d^2\theta}{dt^2}$$

$$\int 2g \sin \theta \frac{d\theta}{dt} = \int 2a \frac{d\theta}{dt} \frac{d^2\theta}{dt^2}$$

$$-2g \cos \theta + k = a\left(\frac{d\theta}{dt}\right)^2$$

From the initial conditions, $\theta = d\theta/dt = 0$ when $t = 0$, we find $k = 2g$. Thus,

$$a\left(\frac{d\theta}{dt}\right)^2 = 2g(1 - \cos \theta) = 4g \sin^2 \frac{\theta}{2}$$

so that

$$\frac{d\theta}{dt} = 2\sqrt{\frac{g}{a}} \sin \frac{\theta}{2}$$

We have chosen the positive square root because $d\theta/dt$ is positive for $0 < \theta \leq \pi$. The time required for the passage from $\theta = \pi/2$ to $\theta = \pi$ is given by

$$\int_{\pi/2}^{\pi} \frac{dt}{d\theta} \, d\theta = \int_{\pi/2}^{\pi} \frac{1}{2}\sqrt{\frac{a}{g}} \csc \frac{\theta}{2} \, d\theta$$

$$= \sqrt{\frac{a}{g}}\left[-\ln \csc \frac{\theta}{2} + \cot \frac{\theta}{2}\right]\Big|_{\pi/2}^{\pi}$$

$$= \sqrt{\frac{a}{g}} \ln(\sqrt{2} + 1)$$

94. This is Putnam Problem 15 of the afternoon session of 1940.

$(\sqrt{n})^{\sqrt{n+1}}$ is greater than $(\sqrt{n+1})^{\sqrt{n}}$ for $n > 8$. Consider the function $f(x) = \frac{\ln x}{x}$ for $x > 0$. Its derivative is $\frac{1 - \ln x}{x^2}$ which is negative for $x > e$. Hence, if $e \leq x < y$ we have $f(x) > f(y)$, and

$$xy\left(\frac{\ln x}{x}\right) > xy\left(\frac{\ln y}{y}\right)$$

Taking exponential we obtain $e^{y \ln x} > e^{x \ln y}$; that is

$$x^y > y^x \text{ provided } e \leq x < y$$

If $n \geq 8$, then $e < \sqrt{n} < \sqrt{n+1}$, so

$$(\sqrt{n})^{\sqrt{n+1}} > (\sqrt{n+1})^{\sqrt{n}}$$

95. This is Putnam Problem 3 of the afternoon session of 1951.

$$\ln\left(1 + \frac{1}{x}\right) = \int_x^{1+x} \frac{dt}{t} > \int_x^{1+x} \frac{dt}{1+x} = \frac{1}{1+x}$$

96. This is Putnam Problem 1 of the morning session of 1961. In the first quadrant the given equation is equivalent to

$$\frac{1}{y} \ln y = \frac{1}{x} \ln x$$

Consider the function given by $f(t) = t^{-1}\ln t$ for $t > 0$. Since $f'(t) = (1 - \ln t)t^{-2}$, it is clear that f is strictly increasing for $t \leq e$, is strictly decreasing for $t \geq e$ and achieves its maximum value for e^{-1} for $t = e$.

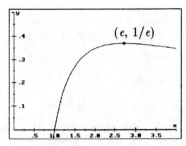

Moreover, $f(t) \to -\infty$ as $t \to 0$ and $f(t) \to 0$ as $t \to +\infty$. It follows for α in $(0, e^{-1})$ the equation $f(t) = \alpha$ has two solutions, one in $(1, e)$, the other in $(e, +\infty)$.

For α near 0, the lower solution is just above 1 and the upper solution is large. As α increases to e^{-1}, the lower solution increases to e and the upper solution decreases to e. Therefore, the set of points satisfying the displayed equation consists of the line $y = x$ and a curve M lying in the quadrant $x > 1$, $y > 1$ and asymptotic to the line $x = 1$ and $y = 1$, as shown in the figure. M is evidently symmetric in the line $y = x$ and crosses that line at (e, e).

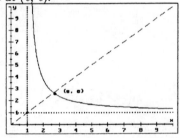

To establish the smoothness of the curve requires material not yet discussed in the book. For reference see "The Real Function Defined by $x^r = y^s$" in *American Mathematical Monthly*, Vol 23 (1916), pp. 233-237. R. Robinson Rowe called the curve "mutuabola" in *Journal of Recreational Mathematics*, Vol. 3 (1970) pp. 176-178.

CHAPTER 6

Additional Applications of the Integral

6.1 Volume: Disks, Washers, and Shells, Page 393

1.

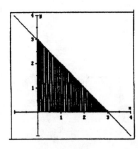

$$V = \int_0^3 (3 - x^2) \, dx$$

$$= -\tfrac{1}{3}(3 - x)^3 \Big|_0^3$$

$$= 9$$

2.

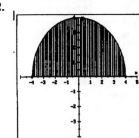

$$V = 2\int_0^4 (16 - x^2) \, dx$$

$$= 2(16x - \tfrac{1}{3}x^3) \Big|_0^4$$

$$= \frac{256}{3}$$

3.

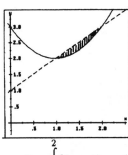

The curves intersect when:
$$x^2 - 2x + 3 = x + 1$$
$$x^2 - 3x + 2 = 0$$
$$(x - 2)(x - 1) = 0$$
$$x = 2, 1$$

$$V = \int_1^2 [(x + 1) - (x^2 - 2x + 3)]^2 \, dx$$

$$= \int_1^2 (x^4 - 6x^3 + 13x^2 - 12x + 4) \, dx$$

$$= (\tfrac{1}{5}x^5 - \tfrac{3}{2}x^4 + \tfrac{13}{3}x^3 - 6x^2 + 4x) \Big|_1^2 = \tfrac{1}{30}$$

4.

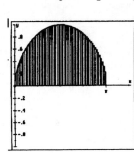

$$V = \int_1^\pi \sin x \, dx$$

$$= -\cos x \Big|_0^\pi$$

$$= 2$$

5.

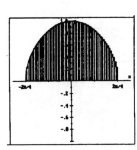

$$V = \int_{-\pi/2}^{\pi/2} \cos x \, dx$$

$$= \sin x \Big|_{-\pi/2}^{\pi/2}$$

$$= 2$$

6.

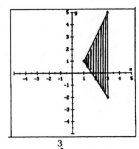

$$m_{AB} = 2;$$
$$y - 1 = 2(x - 1)$$
$$y = 2x - 1$$
$$m_{AC} = -\tfrac{3}{2}$$
$$y - 1 = -\tfrac{2}{3}(x - 1)$$
$$y = -\tfrac{1}{2}(3x - 5)$$

$$V = \int_1^3 [(2x - 1) + \tfrac{1}{2}(3x - 5)]^2 \, dx$$

$$= \frac{49}{4}\int_1^3 (x - 1)^2 \, dx$$

$$= \frac{49}{12}(x - 1)^3 \Big|_1^3 = \frac{98}{3}$$

In Problems 7-12, we note that an equilateral triangle of side a has area $\tfrac{1}{4}\sqrt{3}\, a^2$.

7.

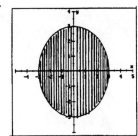

$$V = \tfrac{1}{4}\int_{-3}^3 [\sqrt{3}(2y)^2] \, dx$$

$$= 2\sqrt{3}\int_0^3 (9 - x^2) \, dx$$

$$= 2\sqrt{3}(9x - \tfrac{1}{3}x^3) \Big|_0^3$$

$$= 36\sqrt{3}$$

8.

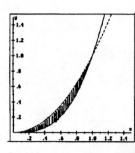

The curves intersect when
$$x^3 = x^2$$
$$x^3 - x^2 = 0$$
$$x(x^2 - 1) = 0$$
$$x(x - 1)(x + 1) = 0$$
$$x = 0, 1, -1$$

We consider twice the area of the region from 0 to 1.

$$V = \frac{\sqrt{3}}{4} \int_0^1 (x^2 - x^3)^2 \, dx = \frac{\sqrt{3}}{4} \int_0^1 (x^4 - 2x^5 + x^6) \, dx$$

$$= \frac{\sqrt{3}}{4} \left(\frac{1}{5}x^5 - \frac{1}{3}x^6 + \frac{1}{7}x^7 \right) \Big|_0^1 = \frac{\sqrt{3}}{420}$$

9.

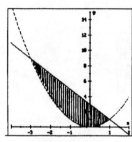

The curves intersect when
$$x^2 = -2x + 3$$
$$x^2 + 2x - 3 = 0$$
$$(x + 3)(x - 1) = 0$$
$$x = -3, 1$$

$$V = \frac{\sqrt{3}}{4} \int_{-3}^1 (x^2 + 2x - 3)^2 \, dx$$

$$= \frac{\sqrt{3}}{4} \int_{-3}^1 (x^4 + 4x^2 + 9 + 4x^3 - 6x^2 - 12x) \, dx$$

$$= \frac{\sqrt{3}}{4} \int_{-3}^1 (x^4 + 4x^3 - 2x^2 - 12x + 9) \, dx$$

$$= \frac{\sqrt{3}}{4} \left(\frac{1}{5}x^5 + x^4 - \frac{2}{3}x^3 - 6x^2 + 9x \right) \Big|_{-3}^1$$

$$= \frac{128\sqrt{3}}{15}$$

10.

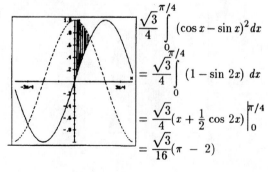

$$\frac{\sqrt{3}}{4} \int_0^{\pi/4} (\cos x - \sin x)^2 \, dx$$

$$= \frac{\sqrt{3}}{4} \int_0^{\pi/4} (1 - \sin 2x) \, dx$$

$$= \frac{\sqrt{3}}{4} \left(x + \frac{1}{2} \cos 2x \right) \Big|_0^{\pi/4}$$

$$= \frac{\sqrt{3}}{16} (\pi - 2)$$

11.

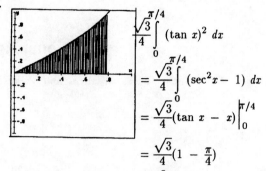

$$\frac{\sqrt{3}}{4} \int_0^{\pi/4} (\tan x)^2 \, dx$$

$$= \frac{\sqrt{3}}{4} \int_0^{\pi/4} (\sec^2 x - 1) \, dx$$

$$= \frac{\sqrt{3}}{4} (\tan x - x) \Big|_0^{\pi/4}$$

$$= \frac{\sqrt{3}}{4} \left(1 - \frac{\pi}{4} \right)$$

12.

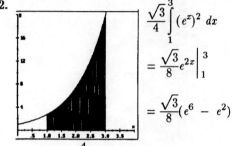

$$\frac{\sqrt{3}}{4} \int_1^3 (e^x)^2 \, dx$$

$$= \frac{\sqrt{3}}{8} e^{2x} \Big|_1^3$$

$$= \frac{\sqrt{3}}{8} (e^6 - e^2)$$

13. **a.** disk: $\pi \int_0^4 (4 - x)^2 \, dx$

   b. shell: $2\pi \int_0^4 x(4 - x) \, dx$

   c. washer: $\pi \int_0^4 [(4 - x)^2 - 1] \, dx$

   or shell: $2\pi \int_0^4 (x + 1)(4 - x) \, dx$

14. **a.** shell: $2\pi \int_0^4 y\sqrt{4 - y} \, dy$

   b. disk: $\pi \int_0^4 y^2 \, dy$

   c. shell: $2\pi \int_0^2 (y + 1)\sqrt{4 - y} \, dy$

15. **a.** shell: $2\pi \int_0^2 y\sqrt{4 - y^2} \, dy$

   b. disk: $\pi \int_0^2 (4 - y^2) \, dy$

c. washer: $\pi \displaystyle\int_0^2 [(1 + \sqrt{4 - y^2})^2 - 1]\, dy$

or shell: $2\pi \displaystyle\int_0^2 (y + 1)\sqrt{4 - y^2}\, dy$

16. a. disk: $\pi \displaystyle\int_{-2}^2 (4 - x)^2\, dx$

b. shell: $2\pi \displaystyle\int_{-2}^2 x\sqrt{4 - x^2}\, dx$

c. washer: $\pi \displaystyle\int_{-2}^2 [(1 + \sqrt{4 - x^2})^2 - 1]\, dx$

17. The curves intersect where

$$x = y^2 = (x^2)^2 = x^4$$
$$x^4 - x = 0$$
$$x(x^3 - 1) = 0$$
$$x = 0, 1$$

a. By shells: $V = 2\pi \displaystyle\int_0^1 (y^{3/2} - y^3)\, dy$

By washers: $V = \pi \displaystyle\int_0^1 (x - x^4)\, dx$

b. By washers: $V = \pi \displaystyle\int_0^1 (y - y^4)\, dy$

By shells: $V = 2\pi \displaystyle\int_0^1 (x^{3/2} - x^3)\, dx$

18. The curves intersect where

$$y = x^2 - 4x + 4 = (y^2)^2 - 4(y^2) + 4$$
$$y^4 - 4y^2 - y - 4 = 0$$

We use computer software or a graphing calculator to find $y = 1$, $y \approx 1.83$ (which is not in the interval of integration)

a. By washers: $V = \pi \displaystyle\int_1^2 x^2 - (x^2 - 4x + 4)^2\, dx$

b. By shells: $V = 2\pi \displaystyle\int_1^2 (x^{3/2} - x^3 + 4x^2 - 4x)\, dx$

19. a. By washers: $V = \pi \displaystyle\int_0^1 [(x^3 + 2x + 1)^2 - 1]\, dx$

b. By shells: $V = 2\pi \displaystyle\int_0^1 (x^4 + 2x^2)\, dx$

20. a. By washers:

$$V = \pi \int_{-2}^0 [(-x^2 - 4x)^2 - (x^2)^2]\, dx$$

b. By shells: $V = 2\pi \displaystyle\int_{-2}^0 (-2x^3 - 4x^2)\, dx$

21. The cross section is a square of side $2y$ and area $4y^2$.

$$V = \int_{-3}^3 4(9 - x^2)\, dx = 8\int_0^3 (9 - x^2)\, dx$$
$$= 8\left(9x - \frac{x^3}{3}\right)\Big|_0^3 = 144 \text{ cubic units}$$

22. The cross section is an equilateral triangle of side $2y$ and area $\frac{1}{4}\sqrt{3}(2y)^2 = \sqrt{3}y^2$.

$$V = \sqrt{3}\int_{-3}^3 (9 - x^2)\, dx$$
$$= 2\sqrt{3}(9x - \tfrac{1}{3}x^3)\Big|_0^3 = 36\sqrt{3} \text{ cubic units}$$

23. The cross section is an isosceles triangle with hypotenuse $2y$ and side $\sqrt{2}y$ with area y^2.

$$V = \int_{-3}^3 \tfrac{1}{2}(9 - x^2)\, dx = \int_0^3 (9 - x^2)\, dx$$
$$= (9x - \tfrac{1}{3}x^3)\Big|_0^3 = 18 \text{ cubic units}$$

24. The cross section is a semicircle with radius y and area $\frac{1}{2}\pi y^2$.

$$V = \tfrac{1}{2}\pi \int_{-3}^3 (9 - x^2)\, dx = 2\pi \int_0^3 (9 - x^2)\, dx$$
$$= 2\pi(9x - \tfrac{1}{3}x^3)\Big|_0^3 = 36\pi \text{ cubic units}$$

25. The curves intersect when

$$x + 1 = x^2 - 1$$
$$x^2 - x - 2 = 0$$
$$x = -1, 2$$

The base of each square is

$(x + 1) - (x^2 - 1) = x + 2 - x^2$ with area

$(x + 2 - x^2)^2 = x^4 - 2x^3 - 3x^2 + 4x + 4.$

$$V = \int\limits_{-1}^{2} (x^4 - 2x^3 - 3x^2 + 4x + 4)\,dx$$

$$= \left(\frac{x^5}{5} - \frac{x^4}{2} - x^3 + 2x^2 + 4x\right)\Big|_{-1}^{2}$$

$$= \frac{81}{10} \text{ cubic units}$$

26. The curves intersect when

$$x^2 - 1 = x + 1$$

$$x^2 - x = 0$$

$$x(x - 1) = 0$$

$$x = 0, 1$$

The cross section is an equilateral triangle of side $x + 1 - (x^2 - 1) = -x^2 + x + 2$ with area $\frac{1}{4}\sqrt{3}(-x^2 + x + 2)^2 = \frac{1}{4}\sqrt{3}(x^4 - 2x^3 - 3x^2 + 3x + 4)$.

$$V = \frac{1}{4}\sqrt{3}\int\limits_{-1}^{2} (x^4 - 2x^3 - 3x^2 + 4x + 4)\,dx$$

$$= \frac{1}{4}\sqrt{3}\left(\frac{1}{5}x^5 - \frac{1}{2}x^4 - x^3 + 2x^2 + 4x\right)\Big|_{-1}^{2}$$

$$= \frac{81\sqrt{3}}{40} \text{ cubic units}$$

27. The curves intersect at $(-1, 0)$ and $(2, 3)$, (see solution for Problem 26 for details). The cross section is a rectangle of side $x + 1 - (x^2 - 1)$ and area $(-x^2 + x + 2)$.

$$V = \int\limits_{-1}^{2} (-x^2 + x + 2)\,dx$$

$$= \left(-\frac{1}{3}x^3 + \frac{1}{2}x^2 + 2x\right)\Big|_{-1}^{2} = \frac{9}{2} \text{ cubic units}$$

28. The curves intersect at $(-1, 0)$ and $(2, 3)$, (see solution for Problem 26 for details). The cross section is a semicircle of diameter $x + 1 - (x^2 - 1)$ and area $\frac{1}{4}\pi(-x^2 + x + 2)^2$

$$V = \frac{1}{4}\pi\int\limits_{-1}^{2} (x^4 - 2x^3 - 3x^2 + 4x + 4)\,dx$$

$$= \frac{1}{4}\pi\left(\frac{1}{5}x^5 - \frac{1}{2}x^4 - x^3 + 2x^2 + 4x\right)\Big|_{-1}^{2}$$

$$= \frac{81}{40}\pi \text{ cubic units}$$

29.

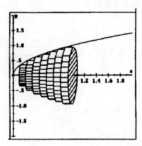

Use disks; $V = \pi\int\limits_{0}^{1} (x^{1/2})^2\,dx = \frac{1}{2}\pi x^2\Big|_{0}^{1} = \frac{\pi}{2}$

30.   Use disks;

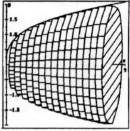

$$V = \pi\int\limits_{0}^{1} (x^{1/3})^2\,dx = \frac{3}{5}\pi x^{5/3}\Big|_{0}^{1} = \frac{96\pi}{5}$$

31.   Use washers;

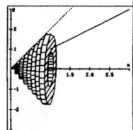

$$V = \pi\int\limits_{0}^{1} [(2x)^2 - x^2]\,dx = \pi$$

32.   Use washers;

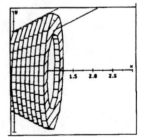

$$V = \pi\int\limits_{0}^{1} [(x + 2)^2 - (x + 1)^2]\,dx$$

$$= \pi(x^2 + 3x)\Big|_{0}^{1} = 4\pi$$

33.   Use washers;

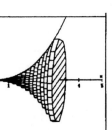

$$V = \pi \int_0^\pi (x^2 + x^3)^2 \, dx$$

$$= \pi \int_0^\pi (x^4 + 2x^5 + x^6) \, dx$$

$$= \pi(\tfrac{1}{5}x^5 + \tfrac{1}{3}x^6 + \tfrac{1}{7}x^7)\Big|_0^\pi$$

$$= \frac{\pi^8}{7} + \frac{\pi^7}{3} + \frac{\pi^6}{5} \approx 2,555$$

34.   Use washers;

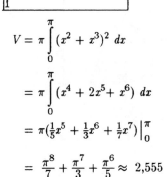

$$V = \pi \int_0^1 [(2 - x^2)^2 - (x^2)^2] \, dx$$

$$= 8\pi \int_0^1 (1 - x^2) \, dx$$

$$= 8\pi(x - \tfrac{1}{3}x^3)\Big|_0^1 = \frac{16\pi}{3}$$

35.   Use washers;

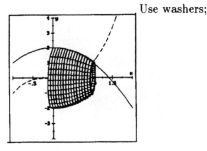

$$V = \pi \int_0^1 [(x^2)^2 - (x^3)^2] \, dx$$

$$= \pi \int_0^1 (x^4 - x^6) \, dx$$

$$= \pi(\tfrac{1}{5}x^5 - \tfrac{1}{7}x^7)\Big|_0^1 = \frac{2\pi}{35}$$

36.   Use disks;

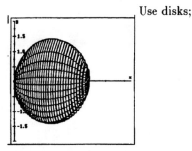

$$V = \pi \int_0^\pi (\sqrt{2 \sin x})^2 \, dx$$

$$= 2\pi \int_0^\pi \sin x \, dx = -2\pi \cos x\Big|_0^\pi = 4\pi$$

37.   Use disks;

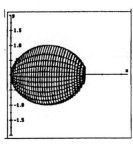

$$V = \pi \int_0^\pi (\sqrt{\sin x})^2 \, dx$$

$$= -\pi \cos x\Big|_0^\pi = 2\pi$$

38.   Use washers;

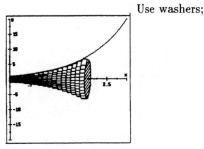

$$V = \pi \int_0^2 [(e^x)^2 - (e^{-x})^2] \, dx$$

$$= \tfrac{1}{2}\pi(e^{2x} + e^{-2x})\Big|_0^2$$

$$= \tfrac{\pi}{2}(e^4 + e^{-4} - 2) \approx 82.6$$

39.

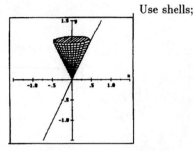

Use shells;

$$V = 4\pi \int\limits_0^1 x^2 \ dx = \frac{4}{3}\pi x^3 \Big|_0^1 = \frac{4\pi}{3}$$

40.

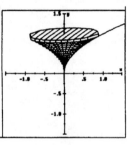

Use shells;

$$V = 2\pi \int\limits_0^1 x(x^{1/2}) \ dx$$

$$= \frac{4}{5}\pi x^{5/2} \Big|_0^1 = \frac{4\pi}{5}$$

41.

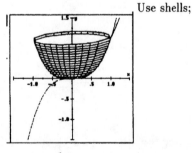

Use shells;

$$V = 2\pi \int\limits_0^1 x(x^2 - x^3) \ dx$$

$$= 2\pi(\tfrac{1}{4}x^4 - \tfrac{1}{5}x^5) \Big|_0^1 = \frac{\pi}{10}$$

42.

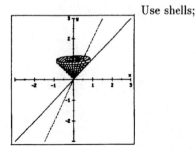

Use shells;

$$V = 2\pi \int\limits_0^1 x(2x - x) \ dx$$

$$= 2\pi \int\limits_0^1 x(2x - x) \ dx = 2\pi(\tfrac{1}{3}x^3) \Big|_0^1 = \frac{2\pi}{3}$$

43.

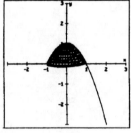

Use shells;

$$V = 2\pi \int\limits_0^1 x(1 - x^2) \ dx$$

$$= 2\pi(\tfrac{1}{2}x^2 - \tfrac{1}{4}x^4) \Big|_0^1 = \frac{\pi}{2}$$

44.

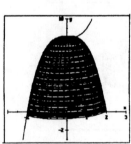

The curve crosses the x-axis at $x = 0$ and $x = 2$. Use shells;

$$V = 2\pi x \int\limits_0^2 x(8 - x^3) \ dx$$

$$= 2\pi(4x^2 - \tfrac{1}{5}x^5) \Big|_0^2 = \frac{96\pi}{5}$$

45.

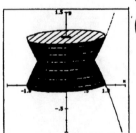

The curves intersect at $\left(\frac{\sqrt{2}}{2}, \frac{1}{2}\right)$. Use shells;

$$V = 2\pi \int\limits_0^{\sqrt{2}/2} x(1 - x^2 - x^2) \ dx$$

$$= 2\pi \int_0^{\sqrt{2}/2} (x - 2x^3)\ dx$$

$$= 2\pi\Big(\frac{x^2}{2} - \frac{x^4}{2}\Big)\Big|_0^{\sqrt{2}/2} = \frac{\pi}{4}$$

46. Use shells;

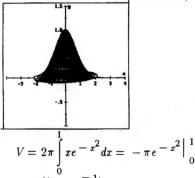

$$V = 2\pi \int_0^1 xe^{-x^2}\ dx = -\pi e^{-x^2}\Big|_0^1$$

$$= (1 - e^{-1})\pi$$

47. The curves intersect at $(0, 0)$ and $(1, 1)$. Use washers;

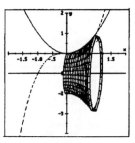

$$V = \pi \int_0^1 [(x^2 + 1)^2 - (x^3 + 1)^2]\ dx$$

$$= \pi \int_0^1 (-x^6 + x^4 - 2x^3 + 2x^2)\ dx$$

$$= \pi\Big(-\frac{1}{7}x^7 + \frac{1}{5}x^5 - \frac{1}{2}x^4 + \frac{2}{3}x^3\Big)\Big|_0^1 = \frac{47\pi}{210}$$

48. Use disks;

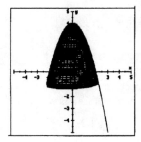

$$V = \pi \int_{-1}^4 (4 - y)\ dy = -\pi(xy - \tfrac{1}{2}y^2)\Big|_{-1}^4 = \frac{25\pi}{2}$$

49. Use shells;

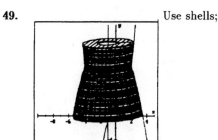

$$V = 2\pi \int_{\frac{1}{2}}^2 (x + 1)(12 - x^2 - x^3)\ dx$$

$$= 2\pi \int_1^2 (-x^4 - 2x^3 - x^2 + 12x + 12)\ dx$$

$$= 2\pi\Big(-\frac{1}{5}x^5 - \frac{1}{2}x^4 - \frac{1}{3}x^3 + 6x^2 + 12x\Big)\Big|_1^2$$

$$= \frac{419\pi}{15}$$

50. Use shells;

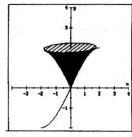

$$V = 2\pi \int_0^2 x[2 - \tfrac{1}{8}(12x - x^3)]\ dx$$

$$= \tfrac{1}{4}\pi \int_0^2 (16x - 12x^2 + x^4)\ dx$$

$$= \tfrac{\pi}{4}(8x^2 - 4x^3 + \tfrac{1}{5}x^5)\Big|_0^2 = \frac{8\pi}{5}$$

51. $m_{AB} = 3$, so $\overline{AB}$ is $y - 1 = 4(x - 1)$

$m_{BC} = -2$, so $\overline{BC}$ is $y - 1 = -2(x - 4)$

The element of area is parallel to the x-axis. Use shells;

$$V = 2\pi \int_1^5 [\tfrac{1}{2}(9 - y) - \tfrac{1}{4}(y + 3)]y\ dy$$

$$= \tfrac{3}{2}\pi \int_1^5 (5y - y^2)\ dy$$

$$= \tfrac{3}{2}\pi(\tfrac{5}{2}y^2 - \tfrac{1}{3}y^3)\Big|_1^5 = 28\pi$$

52. $m_{AB} = \frac{3}{2}$, so $\overline{AB}$ is $y - 1 = \frac{3}{2}(x - 1)$

$m_{BC} = -\frac{3}{2}$, so $\overline{BC}$ is $y - 1 = -\frac{3}{2}(x - 5)$

The element of area is parallel to the x-axis.
Use shells;

$$V = 2\pi \int_1^4 [\tfrac{1}{3}(17 - 2y) - \tfrac{1}{3}(2y + 1)]y \, dy$$

$$= \tfrac{8}{3}\pi \int_1^4 (4y - y^2) \, dy$$

$$= \tfrac{8}{3}\pi (2y^2 - \tfrac{1}{3}y^3)\Big|_1^4 = 24\pi$$

53. Let the fixed diameter be the x-axis. The square cross sections will each have a side of $2y$, and area $(2y)^2 = 4(1 - x^2)$.

$$V = \int_{-1}^1 4(1 - x^2) \, dx = 8\int_0^1 (1 - x^2) \, dx$$

$$= 8\Big(x - \frac{x^3}{3}\Big)\Big|_0^1 = \frac{16}{3}$$

54. The base is an equilateral triangle with one leg $\overline{AC}$ on the y-axis and a vertex B on the positive x-axis. $M_{AB} = -\tan\frac{\pi}{6} = -1/\sqrt{3}$ and the equation of the line $\overline{AB}$ is

$$y - 2 = -\frac{1}{\sqrt{3}}x$$

The cross section is a square with side $2y$, and area $4y^2 = 4(2 - \frac{1}{\sqrt{3}}x)^2$.

$$V = 4\int_0^{2\sqrt{3}} (2 - \frac{1}{\sqrt{3}}x)^2 \, dx$$

$$= -4\sqrt{3}\int_0^{2\sqrt{3}} (2 - \frac{1}{\sqrt{3}}x)^2(-\frac{1}{\sqrt{3}} \, dx)$$

$$= -4\sqrt{3}(\tfrac{1}{3})(2 - \frac{1}{\sqrt{3}}x)^3\Big|_0^{2\sqrt{3}} = \frac{32\sqrt{3}}{3}$$

55. The base of the right triangle has its legs on the coordinate axes with a vertex at the origin. The equation of the line passing through the hypotenuse is $y = -x + 4$. The diameter of the semicircular cross section of the solid is y, and the area is $\frac{1}{2}\pi(\frac{1}{2}y)^2 = \frac{1}{8}\pi y^2$.

$$V = \tfrac{1}{8}\int_0^4 (4 - x)^2 \, dx = -\frac{1}{24}\pi(4 - x)^3\Big|_0^4 = \frac{8\pi}{3}$$

56. a. Use disks; $V = \pi\int_1^4 x^{-1} \, dx = x \ln|x|\Big|_1^4$

$$= \pi \ln 4$$

b. Use shells; $V = 2\pi\int_0^4 x^{1/2} \, dx = 2\pi(\tfrac{2}{3})x^{2/3}\Big|_1^4$

$$= \frac{28\pi}{3}$$

c. $dV = \pi[(y + 2)^2 - 2^2]dx$

$$= \pi[(x^{-1/2} + 2)^2 - 4]dx$$

Use washers;

$$V = \pi\int_1^4 [x^{-1} + 4x^{-1/2} + 4 - 4] \, dx$$

$$= \pi(\ln|x| + 8\sqrt{x})\Big|_1^4 = \pi(8 + \ln 4)$$

$$\approx 29.49$$

57.

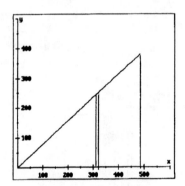

Put the vertex of the pyramid at the origin. Its side will be the the line $y = (375/480)x$. Each rectangular slice perpendicular to the x-axis will have height of y and width $2y$. Twice the sum of all the rectangles from $x = 0$ to $x = 480$ will give the volume.

$$V = 2\int_0^{480} 2y^2 \, dx = 4\int_0^{480} \Big(\frac{375}{480}x\Big)^2 \, dx$$

$$= 4\Big(\frac{375}{480}\Big)^2 \frac{x^3}{3}\Big|_0^{480} = 4(375)^2 \frac{480}{3}$$

$$= 90,000,000 \text{ cubic feet.}$$

58. The cross section is a square with side $2y$. The element of volume is $dV = 4y^2 \, dx$

$$= 1,600\Big(1 - \frac{x^2}{900}\Big) \, dx$$

$$V = 2\int_0^{30} \Big[1,600\Big(1 - \frac{x^2}{900}\Big)\Big] \, dx$$

$$= 3,200\Big(x - \frac{x^3}{2,700}\Big)\Big|_0^{30} = 64,000 \text{ ft}^3$$

59.

Problem 59; Example 1, page 385					
What is a?	0		Number of subdivisions?		12
What is b?	3				
Delta X?	0.25				
X	Delta X	Y1=2X+5	Y2=1-X	Y1-Y2	Volume of Slab
0					
0.25	0.25	5.5	0.75	4.75	5.640625
0.5	0.25	6	0.5	5.5	7.5625
0.75	0.25	6.5	0.25	6.25	9.765625
1	0.25	7	0	7	12.25
1.25	0.25	7.5	-0.25	7.75	15.015625
1.5	0.25	8	-0.5	8.5	18.0625
1.75	0.25	8.5	-0.75	9.25	21.390625
2	0.25	9	-1	10	25
2.25	0.25	9.5	-1.25	10.75	28.890625
2.5	0.25	10	-1.5	11.5	33.0625
2.75	0.25	10.5	-1.75	12.25	37.515625
3	0.25	11	-2	13	42.25
Sum					256.40625

60.

Problem 60; Example 3, page 388					
What is a?	0		Number of subdivisions?		12
What is b?	2				
Delta X?	0.166667				
X	Delta X	R=X^2+1	R^2	Area of base	Volume of disk
0					
0.1666667	0.166667	1.027778	1.056327	3.3185489568	0.5530914928
0.3333333	0.166667	1.111111	1.234568	3.878506642	0.646418107
0.5	0.166667	1.25	1.5625	4.9087375	0.8181229167
0.6666667	0.166667	1.444444	2.08642	6.5548796049	1.0924466008
0.8333333	0.166667	1.694444	2.871142	9.0199566605	1.5033261101
1	0.166667	2	4	12.566368	2.0943946667
1.1666667	0.166667	2.361111	5.574846	17.513890586	2.9189817644
1.3333333	0.166667	2.777778	7.716049	24.240679012	4.0401131687
1.5	0.166667	3.25	10.5625	33.1830655	5.5305109167
1.6666667	0.166667	3.777778	14.2716	44.835559901	7.4725933169
1.8333333	0.166667	4.361111	19.01929	59.750849698	9.9584749496
2	0.166667	5	25	78.5398	13.089966667
Sum					49.718440877

61. $V \approx \frac{1}{2}[1(1.12)+2(1.09)+2(1.05)+2(1.03)$

$\qquad +2(0.99)+2(1.01)+2(0.98)+2(0.99)$

$\qquad +2(0.96)+2(0.93)+1(0.91)](1.0)$

$\qquad = \frac{1}{2}(20.09) \approx 10.05$

62. $V \approx \frac{1}{3}[1(1.12)+4(1.09)+2(1.05)+4(1.03)$

$\qquad +2(0.99)+4(1.01)+2(0.98)+4(0.99)$

$\qquad +2(0.96)+4(0.93)+1(0.91)](1.0)$

$\qquad = \frac{1}{3}(30.19) \approx 10.06$

63. The cross section is a semicircle with radius y, area $\frac{1}{2}\pi y^2$ and volume $dV = \frac{1}{2}\pi y^2\, dx$;

$$V = \frac{1}{2}\pi \int_{-r}^{r} (r^2 - x^2)\, dx = \frac{\pi}{2}(r^2 x - \frac{1}{3}x^3)\Big|_{-r}^{r}$$

$$= \frac{\pi}{6}(6r^3 - 2r^3)$$

Thus, the volume of the entire sphere is

$$V = 2(\frac{\pi}{3}r^3) = \frac{4}{3}\pi r^3.$$

64. Let B be the top vertex of a rectangular tetrahedron of side a. The y-axis is vertical and contains B. The x-axis passes through a vertex A in the base. $|\overline{AB}| = a$. The origin O is the projection of B onto the base. Draw the perpendicular $\overline{OC}$ from O to a side containing A in the base. $|\overline{AC}| = \frac{1}{2}a$. The height of the tetrahedron is H. By the Pythagorean theorem,

$$H^2 + \frac{1}{3}a^2 = a^2 \text{ or } H = \sqrt{\frac{2}{3}}\, a$$

Let (x, y) be a point on $\overline{AB}$. Then, by similar triangles,

$$\frac{\sqrt{\frac{2}{3}}a - y}{\sqrt{\frac{2}{3}}} = \frac{x}{\frac{1}{\sqrt{3}}a}$$

$$x = \frac{1}{\sqrt{2}}\left(\sqrt{\frac{2}{3}}\, a - y\right)$$

Consider a horizontal element of area with cross section an equilateral triangle at altitude y. In this triangle, x corresponds to $\overline{OA} = (1/\sqrt{3})a$ in the base, so the side of the elemental equilateral triangle is $\sqrt{3}x$. The element of volume is

$$dV = \frac{1}{4}\sqrt{3}(\sqrt{3}x)^2\, dy = \frac{3\sqrt{3}}{4}x^2\, dy$$

$$= \frac{3\sqrt{3}}{4}(\frac{1}{2})\left(\sqrt{\frac{2}{3}}\, a - y\right)^2 dy$$

Thus,

$$V = \frac{3\sqrt{3}}{8} \int_{0}^{\sqrt{2/3}\, a} \left(\sqrt{\frac{2}{3}}\, a - y\right)^2 dy$$

$$= -\frac{3\sqrt{3}}{8}(\frac{1}{3})\left(\sqrt{\frac{2}{3}}\, a - y\right)^3\Big|_{0}^{\sqrt{2/3}\, a}$$

$$= \frac{\sqrt{2}}{12}\, a^3 \text{ cubic units}$$

65. First rotate about the x-axis. Use disks with an elemental rectangle perpendicular to the x-axis. The element of volume is $dV = \pi y^2 dx = \pi a^{-2}b^2(a^2 - x^2)\, dx$

$$V_x = 2a^{-2}b^2\pi \int_{0}^{a} (a^2 - x^2)\, dx$$

$$= 2\pi a^{-2}b^2(a^2 x - \tfrac{1}{3}x^3)\Big|_0^a$$

$$= \tfrac{4}{3}\pi ab^2$$

Now rotate about the y-axis. Use disks with an elemental rectangle perpendicular to the y-axis. The element of volume is $dV = \pi x^2 dx = \pi b^{-2}a^2(b^2 - y^2)\,dy$

$$V_y = 2b^{-2}a^2\pi \int_0^b (b^2 - y^2)\,dy$$

$$= 2\pi b^{-2}a^2(b^2 y - \tfrac{1}{3}y^3)\Big|_0^b$$

$$= \tfrac{4}{3}\pi a^2 b$$

66. Use washers with rectangles perpendicular to the y-axis rotated about the y-axis. The element of volume is

$$dV = \pi[(a + \sqrt{b^2 - a^2})^2$$
$$- (a - \sqrt{b^2 - y^2})]\,dy$$

$$V = 2\pi \int_0^b [(a^2 + 2a\sqrt{b^2 - y^2} + b^2 - y^2)$$
$$- (a^2 - 2a\sqrt{b^2 - y^2} + b^2 - y^2)]\,dy$$

$$= 8a\pi \int_0^b \sqrt{b^2 - y^2}\,dy = 8\pi a(\tfrac{1}{4})A$$

$$= 2\pi^2 ab^2$$

where A is the area of a circle of radius b; $A = \pi b^2$

67. $dV = \pi x^2\,dy = \pi(R^2 - y^2)dy$

$$V = \pi \int_h^R (R^2 - y^2)\,dy = \pi(R^2 y - \tfrac{1}{3}y^3)\Big|_h^R$$

$$= \tfrac{1}{3}\pi(2R^3 - 3R^2 h + h^3) \text{ cubic units}$$

68. Consider one-half of a trapezoidal vertical cross section through the axis of symmetry of the frustum of the cone. The coordinates resulting from the upper and lower bases are $A(r_2, h)$ and $(r_1, 0)$. For the line through $\overline{AB}$,

$$m = \frac{h}{r_2 - r_1}, \; y = \frac{h}{r_2 - r_1}(x - r_1), \text{ so}$$

$$dy = \frac{h}{r_2 - r_1}\,dx$$

$$V = \pi \int_{r_1}^{r_2} x^2\Big(\frac{h}{r_2 - r_1}\Big)\,dx = \frac{\pi h}{3(r_2 - r_1)} x^3\Big|_{r_1}^{r_2}$$

$$= \frac{\pi h}{3(r_2 - r_1)}(r_1{}^3 - r_2{}^3)$$

$$= \frac{\pi h(r_2 - r_1)}{3(r_2 - r_1)}(r_2{}^2 + r_1 r_2 + r_1{}^2)$$

$$= \frac{h}{3}(\pi r_2{}^2 + \sqrt{\pi r_1{}^2}\sqrt{\pi r_2{}^2} + \pi r_1{}^2)$$

$$= \frac{h}{3}(A_2 + \sqrt{A_1 A_2} + A_1)$$

69. Let volume V_1 be the base of the gem and V_2 the top of the gem.

$$V_1 = \pi \int_0^{h_1} x^2\,dy = \pi \int_0^{h_1}\left[\frac{\sqrt{R^2 h_1{}^2}\,y}{h_1}\right]^2\,dy$$

$$= \pi\Big(\frac{R^2 - h_1{}^2}{h_1{}^2}\Big)\Big[\frac{y^3}{3}\Big]\Big|_0^{h_1}$$

$$= \frac{\pi}{3}\Big(\frac{R^2 - h_1{}^2}{h_1{}^2}\Big)h_1{}^3 = \frac{\pi}{3}(R^2 - h_1)^2 h_1$$

$$V_2 = \pi \int_{h_1}^{h_2} x^2\,dy = \pi \int_{h_1}^{h_2}(R^2 - y^2)\,dy$$

$$= \pi\Big[R^2 y - \frac{y^3}{3}\Big]\Big|_{h_1}^{h_1}$$

$$= \pi[(R^2 h_2 - \tfrac{1}{3}h_2{}^3) - (R^2 h_1 - \tfrac{1}{3}h_1{}^3)]$$

Total volume

$$V = V_1 + V_2$$

$$= \pi[R^2 h_2 - \tfrac{1}{3}(2h_1{}^3 + h_2{}^3)]$$

6.2 Arc Length and Surface Area, Page 401

1. $\sqrt{1 + [f'(x)]^2} = \sqrt{1 + 3^2} = \sqrt{10}$

$$s = \int_{-1}^{2} \sqrt{10}\,dx = 3\sqrt{10}$$

2. $\sqrt{1 + [f'(x)]^2} = \sqrt{1 + (-4)^2} = \sqrt{17}$

$$s = \int_{-2}^{0} \sqrt{17}\,dx = 2\sqrt{17}$$

3. $\sqrt{1 + [f'(x)]^2}\,dx = \sqrt{1 + (-2)^2} = \sqrt{5}$

$$s = \int_{1}^{3} \sqrt{5}\,dx = 2\sqrt{5}$$

4. $\sqrt{1 + [f'(x)]^2} = \sqrt{1 + \tfrac{9}{4}x}$

$$s = \int_0^4 \sqrt{1 + \tfrac{9}{4}x}\ dx$$

$$= \tfrac{4}{9} \tfrac{2}{3}(1 + \tfrac{9}{4}x)^{3/2}\Big|_0^4 = \tfrac{8}{27}(10\sqrt{10} - 1)$$

5. $\sqrt{1 + [f'(x)]^2} = \sqrt{1 + x}$

$$s = \int_0^4 \sqrt{1 + x}\ dx$$

$$= \tfrac{2}{3}(1 + x)^{3/2}\Big|_0^4 = \frac{10\sqrt{5}}{3} - \frac{2}{3}$$

6. $\sqrt{1 + [f'(x)]^2} = \sqrt{1 + x^2(2 + x^2)} = 1 + x^2$

$$s = \int_0^3 (1 + x^2)\ dx$$

$$= (x + \tfrac{1}{3}x^3)\Big|_0^3 = 12$$

7. $\sqrt{1 + [f'(x)]^2} = \sqrt{1 + (\tfrac{5}{12}x^4 - \tfrac{3}{5}x^{-4})^2}$

$$= \sqrt{(\tfrac{5}{12}x^4)^2 + \tfrac{1}{2} + (\tfrac{3}{5}x^{-4})^2} = \sqrt{(\tfrac{5}{12}x^4 + \tfrac{3}{5}x^{-4})^2}$$

$$s = \int_1^2 (\tfrac{5}{12}x^4 + \tfrac{3}{5}x^{-4})\ dx$$

$$= (\tfrac{1}{12}x^5 - \tfrac{1}{5}x^{-3})\Big|_1^2 = \frac{331}{120}$$

8. $\sqrt{1 + [f'(x)]^2} = \sqrt{1 + (x^2 - \tfrac{1}{4}x^{-2})^2}$

$$= \sqrt{x^4 + \tfrac{1}{2} + \tfrac{1}{16}x^{-4}} = \sqrt{(x^2 + \tfrac{1}{4}x^{-2})^2}$$

$$s = \int_1^4 (x^2 + \tfrac{1}{4}x^{-2})\ dx$$

$$= (\tfrac{1}{3}x^3 - \tfrac{1}{4}x^{-1})\Big|_1^4 = \frac{339}{16}$$

9. $\sqrt{1 + [f'(x)]^2} = \sqrt{1 + (x^3 - \tfrac{1}{4}x^{-3})^2}$

$$= \sqrt{1 + x^6 - \tfrac{1}{2} + \tfrac{1}{16}x^{-6}} = \sqrt{(x^3 + \tfrac{1}{4}x^{-3})^2}$$

$$s = \int_1^2 (x^3 + \tfrac{1}{4}x^{-3})\ dx$$

$$= (\tfrac{1}{4}x^4 - \tfrac{1}{8}x^{-2})\Big|_1^2 = \frac{123}{32}$$

10.

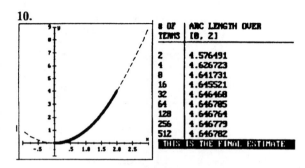

# OF TERMS	ARC LENGTH OVER [0, 2]
2	4.576491
4	4.626723
8	4.641731
16	4.645521
32	4.646468
64	4.646705
128	4.646764
256	4.646779
512	4.646782
	THIS IS THE FINAL ESTIMATE

11.

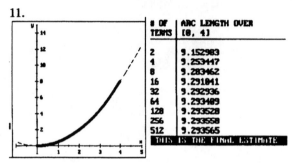

# OF TERMS	ARC LENGTH OVER [0, 4]
2	9.152983
4	9.253447
8	9.283462
16	9.291041
32	9.292936
64	9.293489
128	9.293528
256	9.293558
512	9.293565
	THIS IS THE FINAL ESTIMATE

12.
$$f'(x) = \tfrac{1}{2}(e^{2x} - 1)^{-1/2}e^{2x}(2) - e^{-x}(e^{2x} - 1)^{-1/2}e^x$$

$$= (e^{2x} - 1)(e^{2x} - 1)^{-1/2} = (e^{2x} - 1)^{1/2}$$

$$ds = \sqrt{1 + e^{2x} - 1}\ dx = e^x\ dx$$

$$s = \int_0^{\ln 2} e^x\ dx = e^x\Big|_0^{\ln 2} = 1$$

13. $y = \sqrt[3]{\dfrac{9x^2}{4}}$

Since this does not have a very nice derivative, solve for $x = g(y)$ and use a dy integral.

$$x = \sqrt{\frac{4y^3}{9}} = \frac{2y^{3/2}}{3}, \ g'(y) = y^{1/2}$$

$$\sqrt{1 + g'(y)^2} = \sqrt{1 + y}$$

$$s = \int_0^3 \sqrt{1 + y}\ dy = \tfrac{2}{3}(1 + y)^{3/2}\Big|_0^3$$

$$= \frac{16}{3} - \frac{2}{3} = \frac{14}{3}$$

14. $dy = 3x^{1/2}\ dx; \ ds = \sqrt{1 + 9x}\ dx$

$$s = \int_0^1 (1 + 9x)^{1/2}\ dx = \tfrac{1}{9}\int_0^1 (1 + 9x)^{1/2}(9\ dx)$$

$$= \tfrac{2}{27}(1 + 9x)^{3/2}\Big|_0^1 = \tfrac{2}{27}(10\sqrt{10} - 1)$$

15. $f'(x) = \tfrac{1}{3}(3x^2) + \tfrac{1}{4}(-x^{-2}) = x^2 - \dfrac{1}{4x^2}$

$$1 + [f'(x)]^2 = 1 + x^4 - \frac{1}{2} + \frac{1}{16x^4} = (x^2 + \frac{1}{4x^2})^2$$

a. $S = 2\pi \int_1^3 [\frac{1}{3}x^3 + (4x)^{-1}](x^2 + \frac{1}{4x^2})\, dx$

$$= 2\pi \int_1^3 \left[\frac{1}{3}x^5 + \frac{1}{4}x + \frac{1}{12}x + \frac{1}{16}x^{-3}\right] dx$$

$$= 2\pi\left[\frac{1}{18}x^6 + \frac{1}{6}x^2 - \frac{1}{32}x^{-2}\right]\Big|_1^3 = \frac{1,505\pi}{18}$$

b. $S = 2\pi \int_1^3 x\sqrt{1 + [f'(x)]^2}\, dx$

$$= 2\pi \int_1^3 x(x^2 + \frac{1}{4x^2})\, dx$$

$$= 2\pi\left[\frac{1}{4}x^4 + \frac{1}{4}\ln x\right]\Big|_1^3 = 2\pi[20 + \frac{1}{4}\ln 3]$$

16. $f'(x) = y^{2/3} - \frac{1}{4}y^{-2/3};$

$$ds = \sqrt{1 + (y^{2/3} - \frac{1}{4}y^{-2/3})^2}\, dy$$

$$= \sqrt{1 + (y^{2/3})^2 - \frac{1}{2} + (\frac{1}{4}y^{-2/3})^2}\, dy$$

$$= \sqrt{(y^{2/3} + \frac{1}{2}y^{-2/3})^2}\, dy$$

$$= y^{2/3} + \frac{1}{4}y^{-2/3}\, dy$$

a. About the y-axis, distance is $-f(x)$;

$$S = 2\pi \int_0^1 (-1)\left(\frac{3}{5}y^{5/3} - \frac{3}{4}y^{1/3}\right)\left(y^{2/3} + \frac{1}{4}y^{-2/3}\right) dy$$

$$= -2\pi \int_0^1 (\frac{3}{5}y^{7/3} - \frac{3}{5}y - \frac{3}{16}y^{-1/3}) dy$$

$$= -2\pi(\frac{9}{50}y^{10/3} - \frac{3}{10}y^2 - \frac{9}{32}y^{2/3})\Big|_0^1$$

$$= -2\pi(-\frac{321}{800}) = \frac{321\pi}{400}$$

b. About the x-axis, distance is y;

$$S = 2\pi \int_0^1 y(y^{2/3} + \frac{1}{4}y^{-2/3})\, dy$$

$$= 2\pi \int_0^1 (y^{5/3} + \frac{1}{4}y^{1/3})\, dy$$

$$= 2\pi(\frac{3}{8}y^{8/3} + \frac{3}{16}y^{4/3})\Big|_0^1 = \frac{9\pi}{8}$$

c. About the line $y = -1$, distance is $y + 1$;

$$S = 2\pi \int_0^1 (y + 1)(y^{2/3} + \frac{1}{4}y^{1/3})\, dy$$

$$= 2\pi \int_0^1 [y^{5/3} + y^{2/3} + \frac{1}{4}y^{1/3} + \frac{1}{4}y^{-2/3}]\, dy$$

$$= 2\pi(\frac{3}{8}y^{8/3} + \frac{3}{5}y^{5/3} + \frac{1}{4}\cdot\frac{3}{4}y^{4/3} + \frac{1}{4}\cdot\frac{3}{1}y^{1/3})\Big|_0^1$$

$$= \frac{153\pi}{40}$$

17. $S = 2\pi \int_0^2 (2x + 1)\sqrt{1 + (2)^2}\, dx$

$$= 2\pi\sqrt{5}(x^2 + x)\Big|_0^2 = 12\pi\sqrt{5}$$

18. $S = 2\pi \int_2^6 (\sqrt{x})(1 + \frac{1}{4}x^{-1})^{1/2}\, dx$

$$= 2\pi \int_2^6 (x + \frac{1}{4})^{1/2}\, dx$$

$$= \frac{4}{3}\pi(x + \frac{1}{4})^{3/2}\Big|_2^6 = \frac{49\pi}{3}$$

19. $dy = (x^2 - \frac{1}{4}x^{-2})\, dx$

$$ds = \sqrt{1 + (x^2 - \frac{1}{4}x^{-2})^2}\, dx$$

$$= \sqrt{1 + (x^2)^2 - \frac{1}{2} + (\frac{1}{4}x^{-2})^2}\, dx$$

$$= \sqrt{(x^2)^2 + \frac{1}{2} + (\frac{1}{4}x^{-2})^2}\, dx$$

$$= \sqrt{(x^2 + \frac{1}{4}x^{-2})^2}\, dx = (x^2 + \frac{1}{4}x^{-2})\, dx$$

$$dS = 2\pi y\, ds = 2\pi(\frac{1}{3}x^3 + \frac{1}{4}x^{-1})(x^2 + \frac{1}{4}x^{-2})\, dx$$

$$= 2\pi(\frac{1}{3}x^5 + \frac{1}{12}x + \frac{1}{4}x + \frac{1}{16}x^{-3})\, dx$$

$$S = 2\pi \int_1^2 (\frac{1}{3}x^5 + \frac{1}{3}x + \frac{1}{16}x^{-3})\, dx$$

$$= 2\pi(\frac{1}{18}x^6 + \frac{1}{6}x^2 - \frac{1}{32}x^{-2})\Big|_1^2 = \frac{515\pi}{64}$$

20. $dy = (x^3 - \frac{1}{4}x^{-3})\, dx$

$$ds = 2\pi y\, ds = 2\pi(\frac{1}{4}x^4 + \frac{1}{8}x^{-2})(x^3 - \frac{1}{4}x^{-3})\, dx$$

$$= 2\pi(\tfrac{1}{4}x^7 + \tfrac{1}{16}x + \tfrac{1}{8}x + \tfrac{1}{32}x^{-5})\,dx$$

$$S = 2\pi \int_1^2 (\tfrac{1}{4}x^7 + \tfrac{3}{16}x + \tfrac{1}{32}x^{-5})\,dx$$

$$= 2\pi(\tfrac{1}{32}x^8 + \tfrac{3}{32}x^2 + \tfrac{1}{32}x^{-4})\Big|_1^2 = \frac{16{,}911\pi}{1{,}024}$$

$$\approx 51.88$$

21. $dy = -\tfrac{1}{3}\,dx;\ ds = \sqrt{\tfrac{10}{9}}\,dx = \tfrac{1}{3}\sqrt{10}\,dx$

$$S = \tfrac{2}{3}\sqrt{10}\pi \int_0^3 x\,dx = \tfrac{2}{3}\sqrt{10}\pi\Big(\tfrac{x^2}{2}\Big)\Big|_0^3 = 3\pi\sqrt{10}$$

Check: For the lateral surface of a cone
$S = \pi r l = \pi(3)(\sqrt{10}) = 3\pi\sqrt{10}$

22. $dy = x^{1/2}\,dx;\ ds = \sqrt{1+x}\,dx$

$$dS = 2\pi x\,ds = 2\pi x\sqrt{1+x}\,dx$$

$$S = 2\pi \int_0^3 x\sqrt{1+x}\,dx \quad \text{Integral table \#141}$$

$$= \frac{4\pi(x+1)^{3/2}(3x-2)}{15}\Big|_0^3 = \frac{232\pi}{15}$$

23. $dy = (\tfrac{1}{2}x^{-1/2} - \tfrac{1}{2}x^{1/2})\,dx$

$$ds = \sqrt{1 + (\tfrac{1}{2}x^{-1/2} - \tfrac{1}{2}x^{1/2})^2}\,dx$$

$$= (\tfrac{1}{2}x^{-1/2} + \tfrac{1}{2}x^{1/2})\,dx$$

$$S = 2\pi \int_1^3 x(\tfrac{1}{2}x^{-1/2} + \tfrac{1}{2}x^{1/2})\,dx$$

$$= \pi \int_1^3 (x^{1/2} + x^{3/2})\,dx$$

$$= \pi(\tfrac{2}{3}x^{3/2} + \tfrac{2}{5}x^{5/2})\Big|_1^3 = \pi\Big(\frac{28\sqrt{3}}{5} - \frac{16}{15}\Big)$$

24. $dy = -(4-x)^{-1/2}\,dx$

$$ds = \sqrt{1 + (4-x)^{-1}}\,dx = \sqrt{\frac{x-5}{x-4}}\,dx$$

$$S = 2\pi \int_1^4 x\sqrt{\frac{x-5}{x-4}}\,dx$$

This problem does not lend itself to easy integration. The trapezoidal rule and Simpson's rules fail at $x = 4$. A substitution $u = 1 + (4-x)^{-1}$ will eventually work; however, using computer software we obtained

$$S = \frac{17\pi\ln(\sqrt{3}+2)}{2} + 9\sqrt{3}\,\pi \approx 84.140017$$

25. $f(x) = (1 - x^{2/3})^{3/2};$

$$f'(x) = \tfrac{3}{2}(1 - x^{2/3})^{1/2}(-\tfrac{2}{3}x^{-1/3})$$

$$= -x^{-1/3}(1 - x^{2/3})^{1/2}$$

$$ds = \sqrt{1 + x^{-2/3}(1 - x^{2/3})}\,dx = x^{-1/3}\,dx$$

$$s = 4\int_0^1 x^{-1/3}\,dx = 6x^{2/3}\Big|_0^1 = 6$$

26. See the solution to Problem 25.

$$S = 2\pi \int_{-1}^1 (1 - x^{2/3})^{3/2}x^{-1/3}\,dx$$

$$= 2\pi(2)(\tfrac{3}{2})\int_0^1 (1 - x^{2/3})^{3/2}(\tfrac{2}{3}x^{-1/3}\,dx)$$

$$= -3(\tfrac{2}{5})\pi(1 - x^{2/3})^{5/2}\Big|_0^1 = \frac{12\pi}{5}$$

27. $f(x) = (9 - x^{2/3})^{3/2};$

$$f'(x) = \tfrac{3}{2}(9 - x^{2/3})(-\tfrac{2}{3}x^{-1/3})$$

$$ds = \sqrt{1 + x^{-2/3}(9 - x^{2/3})}\,dx$$

$$= 3x^{-1/3}\,dx$$

$$s = 3\int_1^{27} x^{-1/3}\,dx = \tfrac{9}{2}x^{2/3}\Big|_1^{27} = 36$$

28. See the solution to Problem 26.

$$dS = 2\pi y\,ds = 6\pi(9 - x^{2/3})x^{-1/3}\,dx$$

$$S = 6\pi \int_1^{27} (9 - x^{2/3})x^{-1/3}\,dx$$

$$= -6\pi(\tfrac{3}{2})\int_1^{27} (1 - x^{2/3})^{3/2}(-\tfrac{2}{3}x^{-1/3}\,dx)$$

$$= -6(\tfrac{3}{2})(\tfrac{2}{5}\pi)(9 - x^{3/2})^{5/2}\Big|_1^{27}$$

$$= \frac{2{,}304\sqrt{2}\pi}{5} \approx 2{,}047.28$$

29. $y = \tan x,\ y' = \sec^2 x,$

$$ds = \sqrt{1 + \sec^4 x}\,.$$

$$S = 2\pi \int_0^1 \tan x\sqrt{1 + \sec^4 x}\,dx$$

$$\approx 8.632601 \text{ by Simpson's rule with}$$

$n = 20$. You could also use some computer software to evaluate this integral. The closed

form for this answer is not particularly pleasant:

$$S = \left[\frac{\pi\sqrt{\cos^4 x + 1}}{\cos^2 x} - \pi\ln(\sqrt{\cos^4 x + 1} + \cos^2 x)\right]\Bigg|_0^1$$

30. Let the arc of the graph be subdivided into small elements, each approximated by Δx. Each of these is the hypotenuse of a right triangle with legs Δx and Δy. Then,

$$\Delta s = \sqrt{(\Delta x)^2 + (\Delta y)^2} = \sqrt{1 + \left(\frac{\Delta y}{\Delta x}\right)^2}\,\Delta x$$

and

$$\Delta S = 2\pi x_k^*\sqrt{1 + \left(\frac{\Delta y_k}{\Delta x_k}\right)^2}\,\Delta x_k$$

Thus,

$$S = \lim_{n\to+\infty}\sum_{k=1}^{n} 2\pi x_k^*\sqrt{1 + \left(\frac{\Delta y_k}{\Delta x_k}\right)^2}\,\Delta x_k$$

$$S = 2\pi\lim_{n\to+\infty}\sum_{k=1}^{n} x_k^*\sqrt{1 + \left(\frac{\Delta y_k}{\Delta x_k}\right)^2}\,\Delta x_k$$

$$= 2\pi\int_a^b x\sqrt{1 + \left(\frac{dy}{dx}\right)^2}\,dx$$

$$= 2\pi\int_a^b x\sqrt{1 + [f'(x)]^2}\,dx$$

31. $y = \frac{r}{h}x;\ y' = \frac{r}{h};\ ds = \sqrt{1 + \frac{r^2}{h^2}}\,dx$

$$dS = 2\pi y\,ds = 2\pi(\tfrac{r}{h}x)\frac{\sqrt{h^2 + r^2}}{h}\,dx$$

$$S = 2\pi(\tfrac{r}{h^2})\sqrt{h^2 + r^2}\int_0^h x\,dx$$

$$= \pi r\sqrt{h^2 + r^2}$$

32. a. Each side is a trapezoid with parallel sides s_1 and s_2, and height h_n. The area of a trapezoid is $A = \tfrac{1}{2}h_n(s_1 + s_2)$. The n trapezoids cover an area of

$$A = \tfrac{1}{2}n(h_n)(s_1 + s_2)$$

b. The perimeter of the larger circular base is

$$\lim_{n\to+\infty} ns_1 = 2\pi r_1$$

while the perimeter of the smaller base is

$$\lim_{n\to+\infty} ns_2 = 2\pi r_2$$

When $h_n \to l$,

$$\lim_{n\to+\infty} s_1 = 0 \text{ and } \lim_{n\to+\infty} s_2 = 0,$$

c. $S = \lim_{n\to+\infty} \tfrac{1}{2}nh_n(s_1 + s_2)$

$$= \tfrac{1}{2}\lim_{n\to+\infty} h_n\left[\lim_{n\to+\infty} ns_1 + \lim_{n\to+\infty} ns_2\right]$$

$$= \tfrac{1}{2}l(2\pi r_1 + 2\pi r_2) = \pi(r_1 + r_2)l$$

33. By the MVT there exists an x_k^* such that

$$f'(x_k^*) = \frac{f(x_{k-1}) - f(x_k)}{x_{k-1} - x_k} = \frac{\Delta y_k}{\Delta x_k}$$

Thus,

$$\Delta s = \sqrt{(\Delta x_k)^2 + (\Delta y_k)^2} = \sqrt{1 + \left(\frac{\Delta y_k}{\Delta x_k}\right)^2}\,\Delta x_k$$

$$= \sqrt{1 + [f'(x_k^*)]^2}\,\Delta x_k$$

34. a. From Problem 32, $S_k = \pi(y_{k-1} + y_k)\,l_k$

b. From the MVT,

$$f(c_k) = \tfrac{1}{2}[y_{k-1} + y_k] = \tfrac{1}{2}[f(x_{k-1}) + f(x_k)]$$

since

$$f(x_{k-1}) \le \tfrac{1}{2}[f(x_{k-1}) + f(x_k)] \le f(x_k)$$

$$f(x_{k-1}) \ge \tfrac{1}{2}[f(x_{k-1}) + f(x_k)] \ge f(x_k)$$

depending on whether the curve is falling or rising.

c. When a line segment s_n is rotated about a line at a distance r from its center of mass, the resulting surface area is $2\pi rs_n$. Here the line is the x-axis, and the distance is $y = f(x)$. The element of surface area is (from Problem 33)

$$S = \lim_{n\to+\infty}\sum_{k=1}^{n} 2\pi f(c_k)\sqrt{1 + [f'(x)]^2}\,dx$$

$$= \lim_{\|P\|\to 0}\sum_{k=1}^{n} = 2\pi f(c_k)\sqrt{1 + [f'(x)]^2}\,dx$$

d. $x_{k-1} < c_k \le x_k \to x_k = x$ so $c_k \to x$ and

$$S = \lim_{n\to+\infty}\sum_{k=1}^{n} 2\pi f(c_k)\sqrt{1 + [f'(x)]^2}\,dx$$

$$= 2\pi\int_a^b f(x)\sqrt{1 + [f'(x)]^2}\,dx$$

35. $(x - R)^2 + y^2 = r^2$ or $y = \sqrt{r^2 - (x - R)^2}$

$$y' = \tfrac{1}{2}[r^2 - (x - R)^2]^{-1/2}(2)(x - R)$$

$$= [r^2 - (x - R)^2]^{-1/2}(x - R)$$

$$dS = 2\pi x\sqrt{1 + (y')^2}$$

$$= 2\pi x\sqrt{1 + \left[\frac{(x-R)^2}{r^2 - (x-R)^2}\right]}$$

$$= 2\pi x\sqrt{\frac{r^2 - (x-R)^2 + (x-R)^2}{r^2 - (x-R)^2}}$$

$$= 2\pi x \frac{r}{\sqrt{r^2 - (x-R)^2}}$$

$$S = 2\pi r\int_{R-r}^{R+r} \frac{x}{\sqrt{r^2 - (x-R)^2}}\, dx$$

Let $u = x - R$; $du = dx$; $x = u + R$ if $x = R + r$, and $x = R - r$, then $u = -r$.

$$= 2\pi r\int_{-r}^{r} \frac{u + R}{\sqrt{r^2 - u^2}}\, du$$

$$= 2\pi r\left[\int_{-r}^{r} \frac{u\, du}{\sqrt{r^2 - u^2}} + R\int_{-r}^{r} \frac{du}{\sqrt{r^2 - u^2}}\right]$$

$$= 2\pi r\left[-(r^2 - u^2)^{1/2} + R\sin^{-1}\frac{u}{r}\right]\Big|_{-r}^{r}$$

$$= 2\pi^2 rR$$

This corresponds to rotating the upper semicircle about the y-axis. For the full circle, $S = 4\pi^2 rR$.

6.3 Physical Applications: Work, Liquid Force, and Centroids, Page 413

1. Work is an activity you indulge in to support a chronic habit, an addiction, namely eating. You may find work through an employment agency, or may make your living by doing work. In physics and mathematics, work is a concept that relates the force required to move an object a certain distance. If force and distance are parallel, the formula is $F = Wd$; the work done by the variable force $F(x)$ in moving an object along the x-axis from $x = a$ to $x = b$ is given by

$$W = \int_a^b F(x)\, dx$$

2. Fluid pressure is force per unit area that a fluid exerts on an enclosing body. It depends on the type of liquid (density ρ), the depth y (that is, how many layers of liquid on top weigh on the layer under consideration), and, for the fluid force, the extent (area) of the layer. Find the fluid pressure on an element of the enclosing body and then sum (integrate). The fluid force formula is $F = (\text{pressure})(\text{area}) = \rho hA$. Suppose a flat surface (a plate) is submerged vertically in a fluid of weight density ρ (lb/ft^3) and that the submerged portion of the plate extends from $x = a$ to $x = b$ on a vertical axis. Then the total force, F exerted by the fluid is given by

$$F = \int_a^b \rho\, h(x)L(x)\, dx$$

where $h(x)$ is the depth at x and $L(x)$ is the corresponding length of a typical horizontal approximating strip.

3. Density ρ is a ratio of mass per unit length, or per unit area, or per unit volume. The mass is found by multiplying the units by the density. If the density varies in terms of a variable, find the element of mass and then sum up (integrate). Let f and g be continuous and satisfy $f(x) \geq g(x)$ on the interval $[a, b]$, and consider a thin plate (lamina) of uniform density ρ that covers the region R between the graphs of $y = f(x)$ and $y = g(x)$ on the interval $[a, b]$. Then the centroid of R is the point $(\overline{x}, \overline{y})$ such that

$$\overline{x} = \frac{M_y}{m} = \frac{\rho\displaystyle\int_a^b x[f(x) - g(x)]\, dx}{\rho\displaystyle\int_a^b [f(x) - g(x)]\, dx} \quad\text{and}$$

$$\overline{y} = \frac{M_x}{m} = \frac{\frac{1}{2}\rho\displaystyle\int_a^b \{[f(x)]^2 - [g(x)]^2\}\, dx}{\rho\displaystyle\int_a^b [f(x) - g(x)]\, dx}$$

4. $W = (50)(5) = 250$ ft $\cdot$ lb

5. $W = (850)(15) = 12{,}750$ ft $\cdot$ lb

6. $W = (50)(6) = 300$ ft $\cdot$ lb

7. $F = kx$; if $x = \frac{3}{4}$, $F = 5$, so $k = \frac{20}{3}$

$$W = \frac{20}{3}\int_0^1 x\, dx = \frac{20}{3}\cdot\frac{1}{2}x^2\Big|_0^1 = \frac{10}{3} \text{ ft}\cdot\text{lb}$$

8. $F = kx$; if $x = 100$, $F = 4$, so $k = \frac{1}{50}$

$$W = \frac{1}{50}\int_{10}^{14} x\, dx = \frac{1}{50}\cdot\frac{1}{2}x^2\Big|_{10}^{14} = 0.96 \text{ dyn}\cdot\text{cm}$$

9. The difference in F is 65 lb, the distance is the same. The additional work for the full bucket: $W = (65)(100) = 6,500$ ft-lbs

10. The element of force on an element of rope of length dx is $dF = 0.4\, dx$, $0 \le x \le 30$.

$$W = 0.4 \int_0^{30} x\, dx = 0.2x^2 \Big|_0^{30} = 180 \text{ ft} \cdot \text{lb}$$

11. The cable weighs 20/50 lb/ft. Let x be the distance of cable hanging over the cliff. It is raised a distance x. $dF = \frac{2}{5}x\, dx$

$$W = \frac{2}{5} \int_0^{50} x\, dx = \frac{1}{5}x^2 \Big|_0^{50} = 500$$

The work done by the ball is $W = (30)(50) = 1,500$. Thus, the total work is

$$500 + 1,500 = 2,000 \text{ ft} \cdot \text{lb}$$

12. Label the horizontal 2 ft side as $\overline{AB}$ and extend it by x_1 to O, the origin of a Cartesian coordinate system with positive y-axis pointing downward. The lower vertex C is on the y-axis a distance of h from point O. Let θ be the acute angles of the isosceles triangle (there are two angles labeled θ). Using the law of cosines we find $\cos\theta = \frac{3}{4}$ and from the right $\triangle AOC$ we find $x_1 = \frac{1}{4}$ and $h = 3\sin\theta = \frac{3}{4}\sqrt{7}$.

The desired force is the force by the water on the full $\triangle AOC$ less the force on $\triangle BOC$.

Fluid force on $\triangle AOC$: Consider a layer of vertical rectangular element of thickness dy at a depth of y ft below the surface and with horizontal length x. From similar triangles,

$$\frac{x}{2 + \frac{1}{4}} = \frac{\frac{3}{4}\sqrt{7} - y}{\frac{3}{4}\sqrt{7}}$$

$$x = \frac{9}{4} - \frac{3}{7}\sqrt{7}\, y$$

$$F_1 = \int_0^{3\sqrt{7}/4} 62.4(\tfrac{9}{4} - \tfrac{3}{7}\sqrt{7}y)y\, dy \approx 92.134 \text{ lb}$$

Fluid force on $\triangle BOC$: Consider a layer of vertical rectangular element of thickness dy at a depth of y ft below the surface with horizontal length x. From similar triangles

$$\frac{x}{\frac{1}{4}} = \frac{\frac{3}{4}\sqrt{7} - y}{\frac{3}{4}\sqrt{7}}$$

$$x = \tfrac{1}{4} - \tfrac{1}{21}\sqrt{7}\, y$$

$$F_2 = \int_0^{3\sqrt{7}/4} 62.4(\tfrac{1}{4} - \tfrac{1}{21}\sqrt{7}y)y\, dy \approx 10.2358 \text{ lb}$$

The desired force is $F = F_1 - F_2 \approx 81.898$ lb

Check: Consider a plane surface (the triangle) of area $\frac{3}{4}\sqrt{7}$ with the force of water concentrated at a depth of $\frac{1}{4}\sqrt{7}$, the center of mass for the triangle:

$$62.4(\tfrac{1}{4}\sqrt{7})(\tfrac{3}{4}\sqrt{7}) = 62.4(\tfrac{21}{16}) \approx 81.9$$

13.
$$F = phA = 64.0 \int_0^3 \tfrac{1}{3}(3 - y)y\, dy$$
$$= \tfrac{128}{3}(\tfrac{3}{2}y^2 - \tfrac{1}{3}y^3) \Big|_0^3 = 192 \text{ lb}$$

14.
$$F = 84 \int_0^3 y\sqrt{9 - y^2}\, dy$$
$$= -42(\tfrac{2}{3})(9 - y^3)^{3/2} \Big|_0^3 = 756$$

15.
$$F = 102.4 \int_0^3 y\sqrt{9 - y^2}\, dy$$
$$= -51.2(\tfrac{2}{3})(9 - y^3)^{3/2} \Big|_0^3 = 921.6 \text{ lb}$$

16.
$$F = 62.4 \int_1^2 (4 - y)(y - 1)\, dy$$
$$= 62.4 \int_1^2 (5y - y^2 - 4)\, dy$$
$$= 62.4(\tfrac{5}{2}y^2 - \tfrac{1}{3}y^3 - 4y) \Big|_1^2 = 72.8 \text{ lb}$$

17.
$$F = 849 \int_1^3 y(2)\sqrt{9 - y^2}\, dy$$
$$= 1,698(-\tfrac{1}{3})(9 - y^2)^{3/2} \Big|_1^3 = 9,056\sqrt{2} \text{ lb}$$

18. The curves intersect where
$$x^3 = x^{1/2}$$
$$x^{1/2}(x^{5/2} - 1) = 0$$
$$x = 0, 1$$
$$m = \int_0^1 (x^{1/3} - x^3)\, dx = (\tfrac{3}{4}x^{4/3} - \tfrac{1}{4}x^4) \Big|_0^1 = \tfrac{1}{2}$$
$$M_y = \int_0^1 x(x^{1/3} - x^3)\, dx = \int_0^1 (x^{4/3} - x^4)\, dx$$
$$= (\tfrac{3}{7}x^{7/3} - \tfrac{1}{5}x^5) \Big|_0^1 = \tfrac{8}{35}$$

$$M_x = \int_0^1 \tfrac{1}{2}(x^{1/3} + x^3)(x^{1/3} - x^3) \, dx$$

$$= \tfrac{1}{2}\int_0^1 (x^{2/3} - x^6) \, dx$$

$$= \tfrac{1}{2}(\tfrac{3}{5}x^{5/3} - \tfrac{1}{7}x^7)\Big|_0^1 = \tfrac{8}{35}$$

$$(\overline{x}, \overline{y}) = \left(\frac{\frac{8}{35}}{\frac{1}{2}}, \frac{\frac{8}{35}}{\frac{1}{2}} \right) = (\tfrac{16}{35}, \tfrac{16}{35})$$

19. The curves intersect where

$$x^2 - 9 = 0$$

$$x = -3, 3$$

$$m = 2\int_0^3 0 - (x^2 - 9) \, dx$$

$$= -2(\tfrac{1}{3}x^3 - 9x)\Big|_0^3 = 36$$

$$M_y = 0 \text{ (by symmetry)}$$

$$M_x = 2\int_0^3 \tfrac{1}{2}[0^2 - (x^2 - 9)^2] \, dx$$

$$= \int_0^3 (18x^2 - x^4 - 81) \, dx$$

$$= (6x^3 - \tfrac{1}{5}x^5 - 81x)\Big|_0^3 = -\tfrac{648}{5}$$

$$(\overline{x}, \overline{y}) = \left(0, \frac{-\frac{648}{5}}{36} \right) = (0, -\tfrac{18}{5})$$

20. The curves intersect where

$$4 - x^2 = x + 2$$

$$x^2 + x - 2 = 0$$

$$(x + 2)(x - 1) = 0$$

$$x = -2, 1$$

$$m = \int_{-2}^1 (4 - x^2 - x - 2) \, dx$$

$$= \int_{-2}^1 (2 - x^2 - x) \, dx$$

$$= (2x - \tfrac{1}{3}x^3 - \tfrac{1}{2}x^2)\Big|_{-2}^1 = \tfrac{9}{2}$$

$$M_y = \int_{-2}^1 (2x - x^3 - x) \, dx$$

$$= (x^2 - \tfrac{1}{4}x^4 - \tfrac{1}{2}x^2)\Big|_{-2}^1 = -\tfrac{9}{4}$$

$$M_x = \int_{-2}^1 \tfrac{1}{2}[(4 - x^2)^2 - (x + 2)^2] \, dx$$

$$= \tfrac{1}{2}\int_{-2}^1 (x^4 - 9x^2 - 4x + 12) \, dx$$

$$= \tfrac{1}{2}(\tfrac{1}{5}x^5 - 3x^3 - 2x^2 + 12x)\Big|_{-2}^1 = \tfrac{54}{5}$$

$$(\overline{x}, \overline{y}) = \left(\frac{-\frac{9}{4}}{\frac{9}{2}}, \frac{\frac{54}{5}}{\frac{9}{2}} \right) = (-\tfrac{1}{2}, \tfrac{12}{5})$$

21.
$$m = \int_1^2 x^{-1} dx = \ln x\Big|_1^2 = \ln 2$$

$$M_y = \int_1^2 (x)x^{-1} dx = 2 - 1 = 1$$

$$M_x = \tfrac{1}{2}\int_1^2 (x^{-1})^2 dx = \tfrac{1}{2}(-x^{-1})\Big|_1^2 = \tfrac{1}{4}$$

$$(\overline{x}, \overline{y}) = \left(\frac{1}{\ln 2}, \frac{\frac{1}{4}}{\ln 2} \right) = \left(\frac{1}{\ln 2}, \frac{1}{4 \ln 2} \right)$$

22. The curves intersect where

$$x^{-1} = \tfrac{1}{2}(5 - 2x)$$

$$2 = 5x - 2x^2$$

$$2x^2 - 5x + 2 = 0$$

$$(2x - 1)(x - 2) = 0$$

$$x = \tfrac{1}{2}, 2$$

$$m = \int_{1/2}^2 (\tfrac{5}{2} - x - x^{-1}) \, dx$$

$$= (\tfrac{5}{2}x - \tfrac{1}{2}x^2 - \ln|x|)\Big|_{1/2}^2 = \tfrac{15}{8} - 2\ln 2$$

$$M_y = \int_{1/2}^2 (\tfrac{5}{2}x - x^2 - 1) \, dx$$

$$= (\tfrac{5}{4}x^2 - \tfrac{1}{3}x^3 - x)\Big|_{1/2}^2 = \tfrac{9}{16} \approx 0.5625$$

$$M_x = \int_{1/2}^2 \tfrac{1}{2}[\tfrac{1}{4}(5 - 2x)^2 - x^{-2}] \, dx$$

$$= \tfrac{1}{2}[-\tfrac{1}{24}(5 - 2x)^3 + x^{-1}]\Big|_{1/2}^2 = \tfrac{9}{16}$$

$$(\overline{x}, \overline{y}) \approx \left(\frac{0.5625}{0.4887}, \frac{0.5625}{0.4887}\right) \approx (1.15,\ 1.15)$$

23. $\quad m = \int_1^4 x^{-1/2}\,dx = 2\sqrt{x}\,\Big|_1^4 = 2$

$$M_y = \int_1^4 x^{1/2}\,dx = \tfrac{2}{3}x^{3/2}\,\Big|_1^4 = \tfrac{14}{3}$$

$$M_x = \int_1^4 \tfrac{1}{2}x^{-1}\,dx = \tfrac{1}{2}\ln|x|\,\Big|_1^4 = \ln 2$$

$$(\overline{x}, \overline{y}) \approx \left(\frac{\frac{14}{3}}{2}, \frac{\ln 2}{2}\right) = \left(\frac{7}{3}, \frac{1}{2}\ln 2\right)$$

24. $\quad A = \int_0^4 \sqrt{x}\,dx = \frac{16}{3};\ s = 2\pi\overline{x}$

$$m = \rho\int_0^4 \sqrt{x}\,dx = \frac{16}{3}\rho$$

$$M_y = \rho\int_0^4 x\sqrt{x}\,dx = \frac{64}{5}\rho$$

$$\overline{x} = \frac{12}{5}$$

$$V = As$$

$$= \frac{16}{3}(2\pi)(\tfrac{12}{5}+1) = \frac{544\pi}{15}$$

25. $\quad A = \tfrac{1}{2}(5)(5) = \frac{25}{2};\ \overline{y} = \frac{5}{3}$

$$s = 2\pi(\overline{y}+1)$$

Since the centroid of a triangle is at the intersection of the medians, and that point of concurrency divides the median into segments with ratio $\frac{1}{3}$ to $\frac{2}{3}$. Therefore,

$$V = (\tfrac{25}{2})2\pi\left(\tfrac{5}{3}+1\right) = \frac{200\pi}{3}$$

Using calculus to find $\overline{y}$ is considerably more difficult, as there are two different curves for the upper bound.

$$m = \int_{-3}^0 \tfrac{5}{3}(x+3)\,dx + \int_0^2 -\tfrac{5}{2}(x-2)\,dx$$

$$= \tfrac{5}{3}\left(\tfrac{x^2}{2}+3x\right)\Big|_{-3}^0 - \tfrac{5}{2}\left(\tfrac{x^2}{2}-2x\right)\Big|_0^2 = \frac{25}{2}$$

$$M_x = \tfrac{1}{2}\int_{-3}^0 \left[\tfrac{5}{3}(x+3)\right]^2 dx + \tfrac{1}{2}\int_0^2 \left[-\tfrac{5}{2}(x-2)\right]^2\,dx$$

$$= \tfrac{25}{18}\left(\tfrac{x^3}{3}+3x^2+9x\right)\Big|_{-3}^0 + \tfrac{25}{8}\left(\tfrac{x^3}{3}-2x^2+4x\right)\Big|_0^2$$

$$= \frac{125}{6};\ \overline{y} = \frac{M_x}{M} = \frac{\frac{125}{6}}{\frac{25}{2}} = \frac{5}{3}$$

$$V = 2\pi\left(\tfrac{5}{3}+1\right)\left(\tfrac{25}{2}\right) = \frac{200\pi}{3}\quad\text{(as before)}$$

26. $\quad V = As = \tfrac{\pi}{2}\left[2\pi\left(\tfrac{4}{3\pi}+1\right)\right] = \tfrac{\pi}{3}(4+3\pi)$

27. $\quad A = \tfrac{1}{2}\pi(2)^2 = 2\pi$

$$\overline{x} = \frac{1}{2\pi}\int_{-2}^2 \tfrac{1}{2}\left[\sqrt{4-y^2}\right]^2\,dy$$

$$= \frac{1}{2\pi}\left[\tfrac{1}{2}\left(4y-\tfrac{y^3}{3}\right)\right]\Big|_{-2}^2 = \frac{8}{3\pi}$$

$$V = (2\pi)\left[2\pi\left(\tfrac{8}{3\pi}+2\right)\right] = \frac{8\pi}{3}(4+3\pi)$$

28. $O(0, 0)$, $A(7, 3)$, and $B(7, -2)$ are the vertices of the triangle. The midpoint of AB is $M(7, \frac{1}{2})$. Thus, $\overline{x} = (\frac{2}{3})(7) = \frac{14}{3}$ and $\overline{y} = (\frac{2}{3})(\frac{1}{2}) = \frac{1}{3}$. To check using calculus, we note $y = \frac{3}{7}x$ and $y = -\frac{2}{7}x$ are the equations of OA and OB, respectively.

$$M_x = \int_0^7 x\left(\tfrac{3}{7}x + \tfrac{2}{7}x\right)\,dx = \tfrac{5}{21}x^3\,\Big|_0^7 = \frac{5(49)}{3}$$

$$A = \frac{5(7)}{2}\ \text{and}\ \overline{x} = \frac{5(49)(2)}{3(5)(7)} = \frac{14}{3}$$

$$M_y = \int_0^7 \tfrac{1}{2}\left[\left(\tfrac{3}{7}x\right)^2 - \left(\tfrac{2}{7}x\right)^2\right]\,dx = \tfrac{5}{98}\int_0^7 x^2\,dx$$

$$= \tfrac{5}{98}\cdot\tfrac{1}{3}x^3\,\Big|_0^7 = \frac{35}{6};\ \overline{y} = \frac{5(7)(2)}{6(5)(7)} = \frac{1}{3}$$

$$(\overline{x}, \overline{y}) = (\tfrac{14}{3}, \tfrac{1}{3})$$

29. $A(-2, 0)$, $B(3, -2)$ and $C(3, 5)$ are the vertices of the triangle. The midpoint of $\overline{BC}$ is $M(3, \frac{3}{2})$. Thus,

$$\overline{x} = -2 + \left[\frac{3 - (-2)}{3}\right] = \frac{4}{3}$$

The median from A to M has a vertical change of $\frac{3}{2}$; $\overline{y} = \frac{2}{3}(\frac{3}{2}) = 1$. The centroid of the triangle is at $\left(\frac{4}{3}, 1\right)$.

Now to check by using calculus:

$$m = A = \tfrac{1}{2}(7)(5) = \frac{35}{2}.$$

$$M_x = \tfrac{1}{2}\int_{-2}^3 \left((x+2)^2 - \left\{\tfrac{2}{5}(-x-2)\right\}^2\right)\,dx$$

$$= \tfrac{1}{2}\int_{-2}^3 \left(\tfrac{21}{25}(x+2)^2\right)\,dx$$

$$= \tfrac{21}{50}\left(\tfrac{1}{3}x^3 + 2x^2 + 4x\right)\Big|_{-2}^{3} = \tfrac{35}{2}$$

$$\bar{y} = \frac{M_x}{M} = \frac{\frac{35}{2}}{\frac{35}{2}} = 1$$

$$M_y = \int_{-2}^{3} x\left[(x + 2) - \tfrac{2}{5}(-x - 2)\right] dx$$

$$= \int_{-2}^{3} x\left[\tfrac{7}{5}(x + 2)\right] dx$$

$$= \tfrac{7}{5}\left(\tfrac{1}{3}x^3 + x^2\right)\Big|_{-2}^{3} = \tfrac{70}{3}$$

$$\bar{x} = \frac{M_y}{M} = \frac{\frac{70}{3}}{\frac{35}{2}} = \tfrac{4}{3}; \ (\bar{x}, \bar{y}) = \left(\tfrac{4}{3}, 1\right)$$

30. $W = \int_{1}^{2} (x^4 + 2x^2)\, dx = \left(\tfrac{1}{5}x^5 + \tfrac{2}{3}x^3\right)\Big|_{1}^{2} = \tfrac{23}{15}$ ergs

31. $W = \int_{0}^{\pi} \sin x\, dx - \int_{\pi}^{2\pi} \sin x\, dx = 2\int_{0}^{\pi} \sin x\, dx$

$$= -2\cos x\Big|_{0}^{\pi} = 4 \text{ ergs}$$

32. A representative element (layer, disk) is at a depth of y ft from the bottom (vertex) or $6 - y$ ft from the top. Intersect the cone with a plane through its axis of symmetry with the origin at the vertex. Consider the right triangle in the first quadrant. The horizontal element has radius x. From similar triangles,

$$\tfrac{x}{y} = \tfrac{3}{6} \quad \text{or} \quad x = \tfrac{1}{2}y$$

$$W = 62.4(\tfrac{1}{4})\pi \int_{0}^{6} y^2(6 - y)\, dy$$

$$= 15.6\pi(2y^3 - \tfrac{1}{4}y^4)\Big|_{0}^{6}$$

$$= \frac{8{,}694\pi}{5} \approx 5{,}462.6 \text{ ft}\cdot\text{lb}$$

33. No work will be required, as the tank will empty with the force of gravity.

34. $W = 80(2^2\pi) \int_{0}^{12} y\, dy = 160\pi y^2\Big|_{0}^{12} = 23{,}040\pi$

$$\approx 72{,}382.3 \text{ ft}\cdot\text{lb}$$

35. $W = 62.4\pi \int_{0}^{10} y(100 - y^2)\, dy$

$$= 62.4\pi(50y^2 - \tfrac{1}{4}y^4)\Big|_{0}^{10} \approx 345{,}800 \text{ ft}\cdot\text{lb}$$

36. a. To pump the water to the top of the tank:

$$W = 62.4(600) \int_{0}^{10} (10 - y)\, dy$$

$$= 37{,}440(-\tfrac{1}{2})(10 - y)^2\Big|_{0}^{10}$$

$$= 1{,}872{,}000 \text{ ft}\cdot\text{lb}$$

b. To pump the water to a level 2 ft above the top of the tank:

$$W = 62.4(600) \int_{0}^{10} (12 - y)\, dy$$

$$= 37{,}440(-\tfrac{1}{2})(12 - y^2)\Big|_{0}^{10}$$

$$= 2{,}620{,}800 \text{ ft}\cdot\text{lb}$$

37. $W = 40(3)^2\pi \int_{0}^{2} (12 - y)\, dy$

$$= 360\pi\left(12y - \tfrac{1}{2}y^2\right)\Big|_{0}^{2}$$

$$= 7{,}920\pi \approx 24{,}881 \text{ ft}\cdot\text{lb}$$

38. $F = (0.87)(15) \int_{0}^{20} (20 - y)\, dy$

$$= 2{,}610 \ g \text{ or } 2.61 \text{ km}$$

39. $F = 62.4 \int_{0}^{1} (10 - y)\, dy = 592.8 \text{ lb}$

40. Let a Cartesian coordinate system pass through the center of the circular base of the log at the bottom of the pool. The equation of the circle

is $x^2 + (y - 1)^2 = 1$ or $x = \sqrt{1 - (y - 1)^2}$

$$F = 62.4(2) \int_{0}^{2} (10 - y)\sqrt{1 - (y - 1)^2}\, dy$$

$$= 124.8 \int_{0}^{2} (10 - y)\sqrt{2y - y^2}\, dy$$

$$= \frac{2{,}889\pi}{5} \approx 1{,}815.21 \text{ lb}$$

We use numerical integration (or computer software) to evaluate this integral.

41. Let F_1 be the force on the top, F_2 the force on the larger vertical side, and F_3 be the force on the smaller vertical side, then
$F = F_1 + 2F_2 + 2F_3$.

$$F_1 = 62.4\left(\tfrac{2}{12}\right)\left(\tfrac{3}{12}\right)(9.5) = 24.7 \text{ lb}$$

$$F_2 = \int_{9.5}^{10} 62.4\left(\tfrac{3}{12}\right) x\, dx = 62.4\left(\tfrac{1}{4}\right)\frac{x^2}{2}\Big|_{9.5}^{10} = 76.05 \text{ lb}$$

$$F_3 = \int_{9.5}^{10} 62.4\left(\tfrac{2}{12}\right) x\, dx = 62.4\left(\tfrac{1}{6}\right)\frac{x^2}{2}\Big|_{9.5}^{10} = 50.7 \text{ lb}$$

$$F = 24.7 + 2(76.05) + 2(50.7) = 278.2 \text{ lb}$$

42. On the earth's surface, $F = 800$ lb and $x = 4,000$ mi. Thus,

$$800 = -\frac{1}{4,000^2}k \text{ or } k = -800(4,000)^2$$

$$W = -800(4,000)^2 \int_{4,000}^{4,200} x^{-2}\, dx \approx 152,381 \text{ mi} \cdot \text{lb}$$

43. The right slanted side of the trapezoidal section is on a Cartesian coordinate plane between $(100, 75)$ and $(50, 0)$. Its equation is

$$\frac{y - 0}{x - 50} = \frac{75 - 0}{100 - 50} \text{ or } x = \tfrac{2}{3}y + 50$$

$$F = 62.4 \int_0^{75} (75 - y)(2)(\tfrac{2}{3}y + 50)\, dy$$

$$= 62.4 \int_0^{75} (7{,}500 - \tfrac{4}{3}y^2)\, dy$$

$$= 62.4(7{,}500y - \tfrac{4}{9}y^3)\Big|_0^{75}$$

$$= 23{,}400{,}000 \text{ lb}$$

44. a. $W = 12(625)\displaystyle\int_{10}^{8} x^{-2}\, dx = -12(625)x^{-1}\Big|_{10}^{8}$

$$= -187.5 \text{ ergs}$$

b. $W = \displaystyle\lim_{s \to +\infty}\left[-12(625)x^{-1}\right]\Big|_s^8$

$$= \lim_{s \to +\infty}[12(625)(s^{-1} - \tfrac{1}{8})]$$

$$= -937.5 \text{ ergs}$$

45. $I_x = \displaystyle\int_0^2 y^2(4 - y^2)\,dy = \int_0^2 (4y^2 - y^4)\,dy$

$$= \left(\frac{4y^3}{3} - \frac{y^5}{5}\right)\Big|_0^2 = \frac{64}{15}$$

$$I_y = \int_0^4 x^2(\sqrt{x})\,dx = \frac{x^{7/2}}{\frac{7}{2}}\Big|_0^4 = \frac{256}{7}$$

46. $I_x = \displaystyle\int_0^1 y^2(y)\,dx = \int_0^1 (1 - x^3)\,dx$

$$= (x - \tfrac{1}{4}x^4)\Big|_0^1 = \frac{3}{4}$$

$$I_y = \int_0^1 x^2 y\,dx = \int_0^1 x^2(1 - x^3)^{1/3}dx$$

$$= -\left(\tfrac{1}{3}\right)\left(\tfrac{3}{4}\right)(1 - x^3)^{4/3}\Big|_0^1 = \frac{1}{4}$$

47. $I_y = \displaystyle\int_0^4 x^2 y\,dx = \int_0^4 (4x^2 - x^3)\,dx$

$$= (\tfrac{4}{3}x^3 - \tfrac{1}{4}x^4)\Big|_0^4 = \frac{64}{3}$$

Since $A = 8$ and $I_y = \rho^2 A$ we see

$$\rho = \frac{2\sqrt{2}}{\sqrt{3}} = \tfrac{2}{3}\sqrt{6}$$

48. a. Consider a horizontal rectangular element of area with length $2x$ and width dy. Since $x^2 = a^2 - a^2b^{-2}y^2$;

$$A = \int_{-b}^{b} 2\sqrt{a^2 - a^2 b^{-2}y^2}\, dy$$

$$= \int_{-b}^{b} 2a\sqrt{1 - b^{-2}y^2}\, dy$$

$$= 2a\int_{-b}^{b} \sqrt{1 - \left(\frac{y}{b}\right)^2}\, dy$$

b. $A = 2a\displaystyle\int_{-b}^{b} \sqrt{1 - \left(\frac{y}{b}\right)^2}\, dy$

$$= 2ab^{-1}\int_{-b}^{b} \sqrt{b^2 - y^2}\, dy$$

$$= 2ab^{-1}(\tfrac{1}{2})\pi b^2 = \pi ab$$

c. A horizontal layer of fuel has dimensions $2x$ by L by dy or

$$dV = 2Lx\, dy = 2La\sqrt{1 - b^{-2}y^2}\, dy$$

If the tank is filled to a height h we'll sum up the elements from the bottom $y = -b$ to the fuel level $y = h$, that is

$$V = 2La\int_{-b}^{h} \sqrt{1 - \left(\frac{y}{b}\right)^2}\, dy$$

d. The volumes are 90.6624, 245.674, 430.422, 628.319, 826.215, 1,010.96, 1,165.97, and 1,256.64.

49. a. For $V = 500$, we find $h \approx -0.644391$

b. For $V = 800$, $h = 0.865202$

50. Consider a rectangle with height a and base b. Then $dM_y = ax\, dx$ so $M_y = \tfrac{1}{2}ab^2$. Thus, $\bar{x} = (\tfrac{1}{2}ab^2)(ab)^{-1} = \tfrac{1}{2}b$.

51. Take vertical slices from $x = 0$ to $x = r$; obtain disks, each with a radius of y, where $y = x$.

$$V = \int_0^r \pi x^2 \, dx = \tfrac{1}{3}\pi r^3$$

52. The centroid is at a distance of $s = L/2$ from the axis of rotation. $A = L^2$. By Pappus' theorem,
$$V = 2\pi(s + \tfrac{1}{2}L)L^2 = \pi(2sL^2 + L^3)$$

53. The centroid of a triangle is on a median at a distance $2/3$ from the vertex.

$A = \tfrac{1}{2}L(\tfrac{1}{2}L) = \tfrac{1}{4}L^2$; By Pappus' theorem
$$V = 2\pi\left(\frac{L}{2\sqrt{3}}\right)\left(\frac{\sqrt{3}}{4}L^2\right) = \frac{\pi L^3}{4}$$
Using calculus,
$$V = 2\int_0^{L/2} \pi(\sqrt{3}\, x)^2 \, dx = \frac{\pi L^3}{4}$$

54. Assume the region is under a curve $y = f(x) > 0$ over $[0, a_2]$. Then,
$$\bar{x} = \frac{\int_0^{a_2} xy \, dx}{\int_0^{a_2} y \, dx} = \frac{\int_0^{a_1} xy \, dx + \int_{a_1}^{a_2} xy \, dx}{\int_0^{a_2} y \, dx + \int_{a_1}^{a_2} y \, dx}$$

$$= \frac{1}{A_1 + A_2}\left[A_1 \frac{\int_0^{a_1} xy \, dx}{A_1} + A_2 \frac{\int_{a_1}^{a_2} xy \, dx}{A_2} \right]$$

$$= \frac{1}{A_1 + A_2}\left[A_1 \frac{\int_0^{a_1} xy \, dx}{\int_0^{a_1} y \, dx} + A_2 \frac{\int_{a_1}^{a_2} xy \, dx}{\int_{a_1}^{a_2} y \, dx} \right]$$

$$= \frac{A_1 \bar{x}_1 + A_2 \bar{x}_2}{A_1 + A_2}$$

Similarly, $\bar{y} = \dfrac{A_1\bar{y}_1 + A_2\bar{y}_2}{A_1 + A_2}$. Any region can

be subdivided into subregions whose sum and/or differences are under a curve $y = f(x)$ over some interval $[0, a_2]$.

6.4 Growth, Decay, and First-Order Linear Differential Equations, Page 425

1. $\dfrac{dy}{dx} = 3y$

$$\int y^{-1} \, dy = \int 3 \, dx$$
$$\ln y = 3x + K,$$
$$y = e^{3x+K}, \quad y = Ce^{3x}$$

2. $\dfrac{dy}{dx} = y^2$
$$\int y^{-2} \, dy = \int dx$$
$$-y^{-1} = x + C$$
$$y = \frac{1}{C - x}$$

3. $\displaystyle\int e^{-y} \, dy = \int dx$
$$-e^{-y} = x + K$$
$$e^y = \frac{1}{-K - x}$$
$$y = -\ln|C - x|$$

4. $\dfrac{dy}{dx} = e^{x+y} = e^x e^y$
$$\int e^{-y} \, dy = \int e^{-x} \, dx$$
$$-e^{-y} = e^x + K$$
$$y = -\ln|C - e^x|$$

5. $\displaystyle\int \frac{dy}{y} = \int \frac{dx}{x}$
$$\ln y = \ln x + K$$
$$y = Cx$$

6. $\displaystyle\int y^{-1} \, dy = \int x \, dx$
$$\ln|y| = \tfrac{1}{2}x^2 + C$$
$$y = Ce^{x^2/2}$$

7. $\displaystyle\int \cos y \, dy = \int \cos x \, dx$
$$\sin y = \sin x + C$$
$$y = \sin^{-1}(\sin x + C)$$

8. $\displaystyle\int dy = \int e^x \, dx$
$$y = e^x + C$$

9. $\displaystyle\int \frac{dy}{y + 10} = \int dx$

$$\ln (y + 10) = x + K$$

$$y + 10 = e^{x+K}$$

$$y = Ce^x - 10$$

10. $\int \dfrac{dy}{80 - y} = \int dx$

$$\ln (80 - y) = x + K$$

$$80 - y = e^{x+K}$$

$$y = Ce^x + 80$$

11. **a.** $y' = -2e^{-2x}$; $y'' = 4e^{-2x}$; matches *ii*

 b. $y' = e^x + 4e^{-4x}$; $y'' = e^x - 16e^{-2x}$;

 matches *i*

 c. $y' = e^{-x}(\cos x - \sin x)$

 $y'' = e^x(-\cos x + \sin x - \cos x - \sin x)$

 $= -2e^x \cos x$; matches *v*

 d. $y' = e^{3x}(3x + 1)$; $y'' = e^{3x}(3 + 9x + 3)$

 $= e^{3x}(9x + 6)$; matches *iv*

 e. $y' = e^x - 3e^{-3x}$; $y'' = e^x + 9e^{-3x}$;

 matches *iii*

12. $P(x) = \dfrac{3}{x}$; $Q(x) = x$;

 $I(x) = e^{\int \frac{3}{x} dx} = e^{3 \ln x} = e^{\ln x^3} = x^3$

 $y = \dfrac{1}{x^3}\left(\int x(x^3)\, dx + C \right) = \dfrac{1}{x^3}\left(\dfrac{x^5}{5} + C \right)$

 $= \dfrac{1}{5}x^2 + Cx^{-3}$

13. $P(x) = \dfrac{2}{x}$; $Q(x) = \sqrt{x} + 1$;

 $I(x) = e^{\int \frac{2}{x} dx} = e^{2 \ln x} = e^{\ln x^2} = x^2$

 $y = \dfrac{1}{x^2}\left(\int (\sqrt{x} + 1)x^2\, dx + C \right)$

 $= \dfrac{1}{x^2}\left(\dfrac{2}{7} x^{7/2} + \dfrac{x^3}{3} + C \right)$

 $= \dfrac{2}{7} x^{3/2} + \dfrac{1}{3}x + Cx^{-2}$

14. Divide both sides by x^4: $\dfrac{dy}{dx} + \dfrac{2}{x} y = 5x^{-4}$

 $P(x) = \dfrac{2}{x}$; $Q(x) = 5x^{-4}$;

 $I(x) = e^{\int \frac{2}{x} dx} = e^{2 \ln x} = e^{\ln x^2} = x^2$

 $y = \dfrac{1}{x^2}\left(\int x^2(5x^{-4})\, dx + C \right)$

$$= \dfrac{1}{x^2}(-5x^{-1} + C)$$

$$= -5x^{-3} + Cx^{-2}$$

15. Divide both sides by x^2: $\dfrac{dy}{dx} + \dfrac{1}{x} y = 2x^{-2}$

 $P(x) = \dfrac{1}{x}$; $Q(x) = 2x^{-2}$;

 $I(x) = e^{\int \frac{1}{x} dx} = e^{\ln x} = x$

 $y = \dfrac{1}{x}\left(\int \dfrac{1}{x}(2x^{-2})\, dx + C \right) = \dfrac{1}{x}(2 \ln|x| + C)$

 $= 2x^{-1} \ln|x| + Cx^{-1}$

16. Divide both sides by x: $\dfrac{dy}{dx} + \dfrac{2}{x} y = e^{x^3}$

 $P(x) = \dfrac{2}{x}$; $Q(x) = e^{x^3}$;

 $I(x) = e^{\int \frac{2}{x} dx} = e^{2 \ln x} = x^2$

 $y = \dfrac{1}{x^2}\left(\int \dfrac{2}{x}(e^{x^3})\, dx + C \right) = \dfrac{1}{x^2}\left(\dfrac{1}{3}e^{-x^3} + C \right)$

 $= \dfrac{1}{3}x^{-2}e^{-x^3} + Cx^{-2}$

17. $P(x) = 2 + \dfrac{1}{x}$; $Q(x) = e^{-2x}$; $I(x)$

 $= e^{\int \left(2 + \frac{1}{x}\right) dx} = e^{2x + \ln x} = e^{2x}e^{\ln x} = e^{2x}x$

 $y = \dfrac{1}{e^{2x}x}\left(\int e^{-2x}e^{2x}x\, dx + C \right) = \dfrac{1}{xe^{2x}}\left(\dfrac{x^2}{2} + C \right)$

 $= \dfrac{x}{2e^{2x}} + \dfrac{C}{xe^{2x}}$

18. Divide both sides by x: $\dfrac{dy}{dx} - \dfrac{2}{x} y = 2x^2$

 $P(x) = -\dfrac{2}{x}$; $Q(x) = 2x^2$;

 $I(x) = e^{\int -\frac{2}{x} dx} = e^{-2 \ln x} = x^{-2}$

 $y = \dfrac{1}{x^{-2}}\left(\int (2x^2)(x^{-2})\, dx + C \right)$

 $= x^2(2x + C)$

 $= 2x^3 + Cx^2$

 Since $y = 2$ when $x = 1$, $2 = 2 + C$ or $C = 0$;
 thus, $y = 2x^3$.

19. $$\dfrac{dy}{dx} = -\dfrac{y}{x}$$

 $$\int y^{-1}\, dy = -\int x^{-1} dx$$

 $$\ln|y| = -\ln|x| + K$$

 $$xy = C$$

Since $y = 2$ when $x = 2$, we have $C = 4$; thus, $y = 4x^{-1}$.

20.
$$y^2 = 4kx$$
$$2yy' = 4k$$
$$y' = \frac{2k}{y} = \frac{2\frac{y^2}{4x}}{y} = \frac{y}{2x}$$

The slope of the orthogonal trajectory is the negative reciprocal:
$$\frac{dY}{dX} = -\frac{2X}{Y}$$
$$\int Y\, dY = \int -2X\, dX$$
$$\frac{1}{2}Y^2 = -X^2 + K$$
$$2X^2 + Y^2 = C$$

21.
$$xy = c$$
$$xy' + y = 0$$
$$y' = -\frac{y}{x}$$

The slope of the orthogonal trajectory is the negative reciprocal:
$$\frac{dY}{dX} = \frac{X}{Y}$$
$$\int Y\, dY = \int X\, dX$$
$$\frac{Y^2}{2} = \frac{X^2}{2} + K$$
$$Y^2 = X^2 + C$$

22.
$$x^2 + y^2 = r^2$$
$$2x + 2yy' = 0$$
$$y' = -\frac{x}{y}$$

The slope of the orthogonal trajectory is the negative reciprocal:
$$\frac{dY}{dX} = \frac{Y}{X}$$
$$\int Y^{-1}\, dY = \int X^{-1}\, dX$$
$$\ln Y = \ln X + K$$
$$Y = CX$$

23.
$$y = Ce^{-x}$$
$$y' = -Ce^{-x}$$
$$y' = -y$$

The slope of the orthogonal trajectory is the negative reciprocal:

$$\frac{dY}{dX} = \frac{1}{Y}$$
$$\int Y\, dY = \int dX$$
$$\frac{1}{2}Y^2 = X + K$$
$$Y^2 = 2X + C$$

24. Uninhibited decay model;
$$Q = Q_0 e^{-kt}$$
$$\frac{Q}{Q_0} = e^{-k(2,047)}$$
$$\text{where } t = 1947 + 100$$
$$\approx 0.780655$$
$$\text{where } k = \frac{\ln 2}{5,730}$$

There was still 78% of the ^{14}C still present.

25. Uninhibited decay model:
$$0.28Q_0 = Q_0 e^{-kt}$$
$$-kt = \ln 0.28$$
$$t = -\frac{\ln 0.28}{k} \text{ where } k = \frac{\ln 2}{5,730}$$
$$\approx 10,523 \text{ years or about } 11,000 \text{ yr}$$

26. Uninhibited growth model where t is the time after the year 1989, the growth rate of 5.08% means $k \approx \ln 1.0508$, and $Q_0 = 5,465$ is the initial amount:
$$Q(t) = 5,464e^{t \ln 1.0508}$$
$$Q(11) = 5,464e^{11 \ln 1.0508} \approx 9,423.915773$$

The GDP in the year 2000 will be about \$9,424 billion.

27. Uninhibited growth model where t is the time after the year 1980, a growth rate of 2.5% means $k \approx \ln 1.025$, and $Q_0 = 1,481$ is the initial amount:
$$Q(t) = 1,481e^{t \ln 1.025}$$
$$Q(10) = 1,481e^{10 \ln 1.025} \approx 1,895.00521$$

The predicted GDP is about \$1,895 billion. *The 1992 World Almanac and Book of Facts* reports a GNP of \$5,465.1 billion. This was not an accurate prediction (compare with the data in Problem 26).

28. Uninhibited growth model where t is the time after 1984, a growth rate of 4.9% means $k \approx 1.049$, and $Q_0 = 1,155$ is the initial

amount (in thousands):

$$Q(t) = 1,155e^{t \ln 1.049}$$

$$Q(12) = 1,155e^{12 \ln 1.049} \approx 2,050.63253$$

The predicted number of divorces in 1996 is about 2,051,000.

29. Uninhibited growth model where t is the time after 1984, a growth rate of 10.5% means $k \approx 1.105$, and $Q_0 = 2,487$ is the initial amount (in thousands):

$$Q(t) = 2,487e^{t \ln 1.105}$$

$$Q(12) = 2,487e^{12 \ln 1.105} \approx 8,241.819927$$

The predicted number of marriages in 1996 is about 8,242,000.

30. a. Let $Q(t)$ be the amount of salt in the solution at time t (minutes). Then

$$\frac{dQ}{dt} = \underbrace{(1)(2)}_{\text{inflow}} - \underbrace{\frac{Q}{30}(2)}_{\text{outflow}}$$

$$\frac{dQ}{dt} + \frac{Q}{15} = 2$$

Integrating factor is $\exp\left[\int e^{1/15\,dt}\right] = e^{t/15}$

$$Q(t) = e^{-t/15}\left[\int 2e^{t/15}\,dt + C\right]$$

$$= e^{-t/15}\left[2\,\frac{e^{t/13}}{\frac{1}{15}} + C\right] = 30 + Ce^{-t/15}$$

$$Q(0) = 10 = 30 + Ce^0 \text{ or } C = -20$$

$$Q(t) = 30 - 20e^{-t/15}$$

b. When $Q = 15$, we have

$$15 = 30 - 20e^{-t/15}$$

$$-\frac{t}{15} = \ln\frac{3}{4}$$

$$t = -15 \ln\frac{3}{4} \approx 4.31523 \text{ min}$$

This is about 4 minutes.

31. a.

$$\frac{dQ}{dt} = \underbrace{(1)(2)}_{\text{inflow}} - \underbrace{\frac{Q}{30+t}(1)}_{\text{outflow}}$$

$$\frac{dQ}{dt} + \frac{Q}{30+t} = 2$$

Integrating factor is

$$\exp\left[\int e^{dt/(30+t)}\right] = e^{\ln(30+t)} = 30 + t$$

$$Q(t) = \frac{1}{30+t}\left[\int 2(30+t)\,dt + C\right]$$

$$= \frac{1}{30+t}[(30+t)^2 + C]$$

$$S(0) = 10 = 30 + \frac{C}{30+t} \text{ or } C = -600$$

$$Q(t) = 30 + t - \frac{600}{30+t}$$

The tank is full when $30 + t = 100$ or when $t = 70$ min.

b. $S(70) = 30 + 70 - \dfrac{600}{30+70} = 94$ lb

32.

$$\frac{db}{dt} = \alpha - \beta$$

$$\frac{db}{dt} + \beta = \alpha$$

The integrating factor is $e^{\int \beta\,dt} = e^{\beta t}$

$$b = e^{-\beta t}\left[\int \alpha\,e^{\beta t} + C\right]$$

$$= e^{-\beta t}\left[\frac{\alpha e^{\beta t}}{\beta} + C\right] = \frac{\alpha}{\beta} + Ce^{-\beta t}$$

$b(0) = 0$ implies $0 = \dfrac{\alpha}{\beta} + C$ implies $C = -\dfrac{\alpha}{\beta}$

$$b(t) = \frac{\alpha}{\beta}(1 - e^{-\beta t})$$

In the "long run" (as $t \to +\infty$), the concentration will be

$$\lim_{t \to +\infty} \frac{\alpha}{\beta}(1 - e^{-\beta t}) = \frac{\alpha}{\beta}$$

The "half-way point" is reached where

$$b(t) = \frac{1}{2} \cdot \frac{\alpha}{\beta} = \frac{\alpha}{\beta}(1 - e^{-\beta t})$$

$$e^{-\beta t} = \frac{1}{2}$$

$$-\beta t = \ln\frac{1}{2}$$

$$t = \frac{\ln\frac{1}{2}}{-\beta} = \frac{\ln 2}{\beta}$$

33. $I(t) = \dfrac{E}{R}\left(1 - e^{-Rt/L}\right)$

$$\lim_{t \to \infty} I(t) = \lim_{t \to \infty} \frac{E}{R} - \lim_{t \to \infty} \frac{E}{R} e^{-Rt/L} = \frac{E}{R}$$

If L is doubled but E and R are held constant, there is no change in the long range current in the circuit.

34. Uninhibited growth model where t is the time after 1984 in years, and $Q(t)$ is the number of Hispanics in year t, and $Q_0 = 15,757,000$.

$$Q(6) = 16,098,000 = 15,757,000e^{6k}$$

$$e^{6k} = \frac{16,098}{15,757}$$

$$k = \frac{1}{6}\ln\left(\frac{16,098}{15,757}\right) \approx 0.0035683883$$

Thus,

$$Q(16) = 15,757,000e^{16k} \approx 16,682,811$$

35. Inhibited growth model where t is the time after 1980, and $Q(t)$ is the number of animals in t years.

$$Q(t) = \frac{5,000}{1 + Ae^{-5000kt}}$$

$$Q(0) = 1,800 = \frac{5,000}{1 + A} \text{ or } A = 1.\overline{7}$$

$$Q(2) = 2,000 = 5,000(1 + 1.\overline{7}\,e^{-5,000(6)k})^{-1}$$

$$1 + 1.\overline{7}3e^{-5,000(6)k} = 2.5$$

$$e^{-5,000(6)k} = 0.84375$$

$$-5,000(6)k = \ln 0.84375$$

$$k \approx 5.663301227 \times 10^{-6}$$

Finally,

$$Q(20) = 5,000(1 + 1.\overline{7}\,e^{-5,000(20)k})^{-1}$$

$$\approx 2,489 \text{ or } 2,500$$

36.
$$\frac{dM}{dt} = r\left(\frac{k}{r} - M\right)$$

$$\frac{dM}{\frac{k}{r} - M} = r\,dt$$

$$-\ln\left|\frac{k}{r} - M\right| = rt + K$$

$$\frac{k}{r} - M = e^{-rt}e^{K} = Ce^{rt}$$

$$M(t) = \frac{k}{r} - Ce^{-rt}$$

$$M(0) = 0 = \frac{k}{r} - C, \text{ so } C = \frac{k}{r}$$

Thus, $M(t) = \frac{k}{r}(1 - e^{-rt})$

37. For the first 40 seconds of free fall $k = 0.75$, and $m = \frac{W}{g} = \frac{192}{32} = 6$, $s_0 = 0$, $v_0 = 0$.

Putting these values into the formula for motion of a falling body through a resisting medium:

$$v(t) = \frac{mg}{k} + \left(v_0 - \frac{mg}{k}\right)e^{-kt/m}$$

we have

$$v(40) = \frac{192}{0.75} + \left(0 - \frac{192}{0.75}\right)e^{-0.75(40)/6}$$

$$\approx 254.3$$

as the velocity when the parachute opens. To find the distance he has fallen during the 40 seconds we need the formula for $s(t)$, which can be found by integrating $v(t)$:

$$s(t) = \frac{mgt}{k} + \left(\frac{mv_0}{k} - \frac{gm^2}{k^2}\right)(1 - e^{-kt/m}) + s_0$$

$$s(40) = \frac{192(40)}{0.75}$$

$$+ \left[\frac{6(0)}{0.75} - \frac{32(6^2)}{(0.75)^2}\right](1 - 5\,e^{-0.75(40)/6}) + 0$$

$$s(40) \approx 8,205.8 \text{ ft}$$

The parachute now opens and he is $10,000 - 8,206.8 = 1,793.2$ ft from the ground. We will now find how many seconds later the parachute will have slowed his velocity to 20 ft/s. If the distance traveled in this time is less than 1793.2 he will land safely.

$$20 = \frac{192}{10} + \left(254.3 - \frac{192}{10}\right)e^{-10t/6}$$

$$e^{-10t/6} = 0.00340$$

$$-\frac{5t}{3} = \ln 0.00340$$

$$t \approx 3.41 \text{ sec}$$ Now find the

distance traveled in this time using the parachute.

$$s(3.41) \approx \frac{192(3.41)}{10}$$

$$+ \left[\frac{6(254.3)}{10} - \frac{32(6^2)}{(10)^2}\right][1 - e^{-10(3.41)/6}] + 0$$

$$\approx 206.1 \text{ ft.}$$

He will land safely, but will be floating down with his parachute much too long and may be spotted. He should have spent longer in free fall.

38. The governing differential equation is

$$5\frac{dI}{dt} + 10I = E(t)$$

$$\frac{dI}{dt} + 2I = \frac{1}{5}E(t)$$

The integrating factor is $e^{\int 2\,dt} = e^{2t}$. Then,

$$I(t) = e^{-2t}\left[\int e^{2t}E(t)\,dt + C\right]$$

a. If $E = 15$, then

$$I(t) = e^{-2t}\left[\int 15e^{2t}\,dt + C\right]$$

$$= e^{-2t}\left[\frac{15}{2}e^{2t} + C\right] = \frac{15}{2} + Ce^{-2t}$$

Since $I(0) = 0$, $C = -\frac{15}{2}$, so

$$I(t) = \frac{15}{2}(1 - e^{-2t})$$

b. If $E = 5e^{-2t}\sin t$, then

$$I(t) = e^{-2t}\left[\int e^{2t}(5e^{-2t}\sin t)\ dt + C\right]$$

$$= e^{-2t}[-5\cos t + C]$$

Since $I(0) = 0$, $C = 5$, so

$$I(t) = 5e^{-2t}(1 - \cos t)$$

39. a. If $Q(t) = B$ is a solution of

$$Q'(t) = kQ(B - Q)\ \text{since}$$

$$Q'(t) = 0 = kB(B - B)$$

b. $Q' > 0$, so Q is increasing, but cannot exceed B because of the sign of Q'.

c. $Q' < 0$, so Q is decreasing, but cannot exceed B because of the sign of Q'.

40. a. $\dfrac{Q(t + \Delta t) - Q(t)}{\Delta t} = kQ(B - Q)$

Let $\Delta t = 1$,

$$Q(t + 1) - Q(t) = kQ(B - Q)$$

$$Q(t + 1) = Q(t) + kQ(B - Q)$$

$$= Q(t)[1 + k(B - Q(t)]$$

b. Answers vary; see the *Technology Manuals* for some examples. With $kB > 2$ the solution will monotonically increase with $Q(16) \approx 5.9957$

c. With $kB > 2$, the solution will oscillate around 5, but not converge.

d. Here with $kB > 2$, note $1 - kB < -1$ so the $y(n)$ will, in absolute value, get larger at each step. Note this is the linearized problem, so is only indicative of $Q(n)$ where $Q(n)$ is close to 5.

41. a. If $u = y^{1-n}$, then

$$u'(x) = (1 - n)y^{-n}y'(x),\ \text{and}$$

$$y'(x) = \frac{u'(x)y^n}{(1 - n)}$$

Substituting:

$$\frac{u'(x)y^n}{(1 - n)} + P(x)y = Q(x)y^n$$

Now multiply by $\dfrac{1 - n}{y^n}$ to obtain

$$u' + P(x)(1 - n)y^{1-n} = Q(x)(1 - n)$$

$$u' + (1 - n)P(x)u = (1 - n)Q(x)$$

b. $P(x) = x^{-1}$; $Q(x) = 2$, and $n = 2$; let $u = y^{-1}$, so

$$\frac{du}{dx} = -y^2\frac{dy}{dx}$$

$$\frac{dy}{dx} = -y^2 u'$$

$$-y^2u' + \frac{y}{x} = 2y^2$$

$$-u' + x^{-1}y^{-1} = 2$$

$$-u' + x^{-1}u = 2$$

$$\frac{du}{dx} - x^{-1}u = -2$$

Then the integrating factor is

$$\exp\left[\int\frac{-1}{x}\ dx\right] = e^{-\ln x} = x^{-1}\ \text{and}$$

$$x^{-1}\frac{du}{dx} - x^{-1}x^{-1}u = -2x^{-1}$$

$$d(ux^{-1}) = -2x^{-1}dx$$

$$ux^{-1} = -2\ln|x| + C$$

$$u = x\ln x^{-2} + Cx$$

Thus, $y = [x(C - \ln x^2)]^{-1}$

42. $A_2 = \displaystyle\int_0^y x\ dy$ and $A_1 = \displaystyle\int_0^x y\ dx$;

$$2A_2 = A_1$$

$$2\int_0^y x\ dy = \int_0^x y\ dx$$

$$2x\frac{dy}{dx} = y(x)$$

$$\frac{dy}{dx} = \frac{y}{2x}$$

$$\int y^{-1}\ dy = \int\tfrac{1}{2}x^{-1}\ dx$$

$$\ln|y| = \tfrac{1}{2}\ln|x| + \tfrac{1}{2}\ln|C|$$

$$y = \tfrac{1}{2}Cx$$

43. a. $m\dfrac{dv}{dt} + kv = mg$

Integrating factor is $e^{\int(k/m)dt} = e^{kt/m}$

$$v = e^{-kt/m}\left[\int ge^{kt/m}\ dt + C_1\right]$$

$$= e^{-kt/m}\left[\frac{gm}{k}e^{kt/m} + C_1\right]$$

$$= \frac{gm}{k} + C_1 e^{-kt/m}\qquad v_0 = v(0)$$

$$v_0 = \frac{gm}{k} + C_1\ \text{implies}\ C_1 = v_0 - \frac{gm}{k}$$

$$v = \frac{gm}{k} + (v_0 - \frac{gm}{k})e^{-kt/m}$$

b. $\dfrac{ds}{dt} = \dfrac{gm}{k} + (v_0 - \dfrac{gm}{k})e^{-kt/m}$

$$s = \frac{gm}{k} t + (v_0 - \frac{gm}{k})(-\frac{m}{k}) e^{-kt/m} + C_2$$

$s = s_0$ when $t = 0$, so

$$s_0 = 0 - \frac{m}{k}(v_0 - \frac{gm}{k}) + C^2 \text{ and}$$

$$C_2 = s_0 + \frac{m}{k}(v_0 - \frac{gm}{k})$$

Thus,

$$s = \frac{gm}{k} t - \frac{m}{k}(v_0 - \frac{gm}{k}) e^{-kt/m}$$
$$+ s_0 + \frac{m}{k}(v_0 - \frac{gm}{k})$$
$$= \frac{gm}{k} t + \frac{m}{k}(v_0 - \frac{gm}{k})(1 - e^{-kt/m}) + s_0$$

CHAPTER 6 REVIEW

Proficiency Examination, Page 429

44. Slice a solid-like loaf of bread, approximate each slice by assuming the lateral side(s) perpendicular to the cross sectional surface; sum up each of these slices.

45. Use disks when the solid of revolution is generated by an area that touches the axis of rotation. Use washers when the area does not touch the axis. Use shells when it is convenient to think of the solid sliced by cylindrical tubes stuffed inside each other like a telescope or collapsed TV/car radio antenna.

46. Let f be a function whose derivative f' is continuous on the interval $[a, b]$. Then the arc length, s, of the graph of $y = f(x)$ between $x = a$ to $x = b$ is given by the integral

$$s = \int_a^b \sqrt{1 + [f'(x)]^2} \, dx$$

Similarly, for the graph of $x = g(y)$, where g' is continuous on the interval $[c, d]$, the arc length from $y = c$ to $y = d$ is

$$s = \int_c^d \sqrt{1 + [g'(y)]^2} \, dy$$

47. Suppose f' is continuous on the interval $[a, b]$. Then the surface generated by revolving about the x-axis the arc of the curve $y = f(x)$ on $[a, b]$ has surface area

$$S = 2\pi \int_a^b f(x)\sqrt{1 + [f'(x)]^2} \, dx$$

48. The work done by the variable force $F(x)$ in moving an object along the x-axis from $x = a$ to $x = b$ is given by

$$W = \int_a^b F(x) \, dx$$

49. Suppose a flat surface (a plate) is submerged vertically in a fluid of weight density ρ (lb/ft^3) and that the submerged portion of the plate extends from $x = a$ to $x = b$ on a vertical axis. Then the total force, F exerted by the fluid is given by

$$F = \int_a^b \rho \, h(x) L(x) \, dx$$

where $h(x)$ is the depth at x and $L(x)$ is the corresponding length of a typical horizontal approximating strip.

50. Let f and g be continuous and satisfy $f(x) \geq g(x)$ on the interval $[a, b]$, and consider a thin plate (lamina) of uniform density ρ that covers the region R between the graphs of $y = f(x)$ and $y = g(x)$ on the interval $[a, b]$. Then

The mass of R is: $m = \rho \int_a^b [f(x) - g(x)] \, dx$

The centroid of R is the point $(\overline{x}, \overline{y})$ such that

$$\overline{x} = \frac{M_y}{m} = \frac{\rho \int_a^b x[f(x) - g(x)] \, dx}{\rho \int_a^b [f(x) - g(x)] \, dx} \quad \text{and}$$

$$\overline{y} = \frac{M_x}{m} = \frac{\frac{1}{2}\rho \int_a^b \{[f(x)]^2 - [g(x)]^2\} \, dx}{\rho \int_a^b [f(x) - g(x)] \, dx}$$

51. The solid generated by revolving a region R about a line outside its boundary (but in the same plane) has volume $V = As$, where A is the area of R and s is the distance traveled by the centroid of R.

52. Hooke's law states that when a spring is pulled x units past its equilibrium (rest) position, there is a restoring force $F(x) = kx$ that pulls the spring back toward equilibrium.

10. Pascal's principle states that fluid pressure is the same in all directions.

11. The general solution of the first-order linear differential equation

$$\frac{dy}{dx} + P(x)y = Q(x)$$

is given by

$$y = \frac{1}{I(x)}\left[\int Q(x)I(x)\ dx + C\right]$$

where $I(x) = e^{\int P(x)dx}$.

12. An RL circuit is an electrical system with resistance R and inductance L.

13.-18. The definite integrals could represent the following:
A. Disks revolved about the y-axis.
B. Disks revolved about the x-axis.
C. Slices taken perpendicular to the x-axis.
D. Slices taken perpendicular to the y-axis.
E. Mass of a lamina with density π.
F. Washers taken along the x-axis.
G. Washers taken along the y-axis.

13. All but E are formulas for volumes of solids.
14. A, B, F, G   **15.** F, G   **16.** C, D
17. A, F   **18.** B, G
19. **a.**

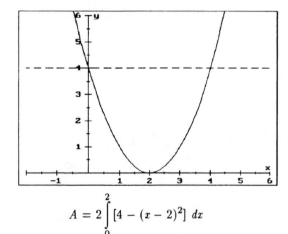

$$A = 2\int_0^2 [4 - (x-2)^2]\ dx$$

$$= 2\int_0^2 (4x - x^2)\ dx$$

$$= 2\left(2x^2 - \frac{x^3}{3}\right)\Big|_0^2 = \frac{32}{3}$$

b. $$V = 2\pi\int_0^2 [R^2 - r^2]\ dx$$

$$= 2\pi\int_0^2 [4^2 - (x-2)^4]\ dx$$

$$= 2\pi\int_0^2 (-x^4 + 8x^3 - 24x^2 + 32x)\ dx$$

$$= 2\pi\left(-\frac{x^5}{5} + 2x^4 - 8x^3 + 16x^2\right)\Big|_0^2$$

$$= \frac{256\pi}{5}$$

c. $$V = \int_0^4 2\pi rh\ dx$$

$$= 2\pi\int_0^4 x[4 - (x-2)^2]\ dx$$

$$= 2\pi\int_0^4 x(4x - x^2)\ dx$$

$$= 2\pi\int_0^4 (4x^2 - x^3)\ dx$$

$$= 2\pi\left(\frac{4x^3}{3} - \frac{x^4}{4}\right)\Big|_0^4 = \frac{128\pi}{3}$$

20. $$V = 2\int_0^2 2\sqrt{4-x^2}\ 4\sqrt{4-x^2}\ dx$$

$$= 16\int_0^2 (4-x^2)\ dx = 16\left(4x - \frac{x^3}{3}\right)\Big|_0^2 = \frac{256}{3}$$

21. $$s = \int_a^b \sqrt{1 + [f'(x)]^2}\ dx$$

$$= \int_0^1 \sqrt{1 + \left(-\frac{3\sqrt{x}}{2}\right)^2}\ dx$$

$$= \int_0^1 \frac{1}{2}\sqrt{4 + 9x}$$

$$= \frac{1}{18}\int_0^1 (4 + 9x)^{1/2}(9x\ dx)$$

$$= \frac{1}{18} \frac{(4 + 9x)^{3/2}}{\frac{3}{2}} \Bigg|_0^1 = \frac{1}{27}(13\sqrt{13} - 8)$$

22. $s = 2\pi \int_0^1 \sqrt{x}\, \sqrt{1 + \left(\frac{1}{2\sqrt{x}}\right)^2}\, dx$

$$= 2\pi \int_0^1 \sqrt{x}\, \sqrt{\frac{4x + 1}{4x}}\, dx$$

$$= \pi \int_0^1 \sqrt{4x + 1}\, dx$$

$$= \frac{\pi}{4} \int_0^1 (4x + 1)^{1/2}(4\, dx)$$

$$= \frac{\pi}{4} \frac{(4x + 1)^{3/2}}{\frac{3}{2}} \Bigg|_0^1 = \frac{\pi}{6}(5\sqrt{5} - 1) \approx 5.33$$

23. $P(x) = \frac{x}{x + 1}, \quad Q(x) = e^{-x}$

$$I(x) = e^{\int \frac{x}{x+1}\, dx} = e^{\int \left(1 - \frac{1}{x+1}\right) dx}$$

$$= e^x e^{-\ln|x + 1|} = e^x (x + 1)^{-1}$$

$$y = \frac{1}{e^x (x + 1)^{-1}} \left[\int e^{-x}\left[e^x(x + 1)^{-1}\right] dx + C \right]$$

$$= \frac{x + 1}{e^x} \left[\int \frac{dx}{x + 1} + C \right]$$

$$= \frac{x + 1}{e^x}\left(\ln|x + 1| + 1\right)$$

24. $\frac{dS}{dt} = \text{Inflow} - \text{Outflow}$

$$= 1.3(5) - \frac{S(t)}{200 + (5 - 3)t}(3)$$

$$= 6.5 - \frac{3S(t)}{200 + 2t}$$

$$\frac{dS}{dt} + \frac{3S}{200 + 2t} = 6.5$$

$$P(t) = \frac{3}{200 + 2t}; \quad Q(t) = 6.5$$

$$I(t) = e^{\int \frac{3}{200 + 2t}\, dt} = e^{(3/2)\, \ln(t + 100)}$$

$$= (t + 100)^{3/2}$$

$$(t + 100)^{3/2}\frac{dS}{dt} + (t + 100)^{3/2}$$

$$\frac{3}{2(t + 100)} S = 6.5(t + 100)^{3/2}$$

$$\frac{d}{dt}\left[(t + 100)^{3/2}S\right] = 6.5(t + 100)^{3/2}$$

$$\int \frac{d}{dt}\left[(t + 100)^{3/2}S\right] dt = \int 6.5(t + 100)^{3/2}\, dt$$

$$(t + 100)^{3/2}S = 6.5(t + 100)^{5/2}\left(\tfrac{2}{5}\right) + C$$

But when $t = 0$, $S = 400$;
$C = 400,000 - 260,000 = 140,000$

$$S(t) = 2.6(t + 100) + \frac{140,000}{(t + 100)^{3/2}}$$

The concentration of salt in the tank at time t is given by $C(t) = S(t)(200 + 2t)$.

$$C(t) = \frac{2.6(t + 100) + \dfrac{140,000}{(t + 100)^{3/2}}}{200 + 2t}$$

$$= 1.3 + \frac{70,000}{(t + 100)^{5/2}}$$

Notice that as we continue to dilute the mixture the minimum concentration will approach the incoming concentration of 1.3, and the maximum concentration occurs when $t = 0$. We wish to find the time required for a concentration of two-thirds the original amount of 2 lb per gallon:

$$\frac{4}{3} = 1.3 + \frac{70,000}{(t + 100)^{5/2}}$$

$$\frac{1}{30} = \frac{70,000}{(t + 100)^{5/2}}$$

$$\frac{1}{30}(t + 100)^{5/2} = 70,000$$

$$(t + 100)^{5/2} = 2,100,000$$

$$t = (2,100,000)^{2/5} - 100$$

$$\approx 238 \text{ min or 3 hr 58 min}$$

25. $M = \int_0^1 [(x - x^3) - (x^2 - x)]\, dx$

$$= \int_0^1 (2x - x^2 - x^3)\, dx$$

$$= x^2 - \frac{x^3}{3} - \frac{x^4}{4}\Big|_0^1 = \frac{5}{12}$$

$$M_y = \int_0^1 x(2x - x^2 - x^3)\, dx$$

$$= \int_0^1 (2x^2 - x^3 - x^4)\, dx$$

$$= \frac{2x^3}{3} - \frac{x^4}{4} - \frac{x^5}{5}\Big|_0^1 = \frac{13}{60}$$

$$M_x = \frac{1}{2}\int_0^1 [(x - x^3)^2 - (x^2 - x)^2]\, dx$$

$$= \frac{1}{2}\int_0^1 (x^6 - 3x^4 + 2x^3)\, dx$$

$$= \frac{1}{2}\left(\frac{x^7}{7} - \frac{3x^5}{5} + \frac{x^4}{2}\right)\Big|_0^1 = \frac{3}{140}$$

$$\bar{x} = \frac{M_y}{M} = \frac{\frac{13}{60}}{\frac{5}{12}} = \frac{13}{25}; \quad \bar{y} = \frac{M_x}{M} = \frac{\frac{3}{140}}{\frac{5}{12}} = \frac{9}{175}$$

The centroid is $(\frac{13}{25}, \frac{9}{175})$.

Supplementary Problems, Page 429

1. $\displaystyle\int_0^9 \pi x\, dx = \frac{1}{2}\pi x^2\Big|_0^9 = \frac{81\pi}{2}$

2. $\displaystyle 2\pi\int_0^{3/\sqrt{2}} (9x - 2x^3)\, dx$
$$= 2\pi(\tfrac{9}{2}x^2 - \tfrac{1}{2}x^4)\Big|_0^{3/\sqrt{2}} = \frac{81\pi}{4}$$

3. $\displaystyle\pi\int_{-1}^1 (x^4 - x^8)\, dx = 2\pi(\tfrac{1}{5}x^5 - \tfrac{1}{9}x^9)\Big|_0^1 = \frac{8\pi}{45}$

4. $\displaystyle\pi\int_{\pi/4}^{\pi/3} \cos x\, dx = \pi\sin x\Big|_{\pi/4}^{\pi/3} = \frac{\pi}{2}(\sqrt{3} - \sqrt{2})$

5. $\displaystyle 8\int_0^4 (4 - y)\, dy = -4(4 - y)^2\Big|_0^4 = 64$

6. $\displaystyle\frac{\pi}{4}\int_0^1 (x - x^2)\, dx = \frac{\pi}{4}(\tfrac{1}{2}x^2 - \tfrac{1}{3}x^3)\Big|_0^1 = \frac{\pi}{24}$

7. $y = 2 - x^2$ intersects $y = x$ at $(-2, 2)$ and $(1, 1)$.
$$A = m = \int_{-2}^1 (2 - x^2 - x)\, dx$$

$$= (2x - \tfrac{1}{3}x^3 - \tfrac{1}{2}x^2)\Big|_{-2}^1 = \frac{9}{2}$$

$$M_x = \frac{1}{2}\int_{-2}^1 (4 - 4x^2 + x^4 - x^2)\, dx$$

$$= [\tfrac{1}{2}(\tfrac{1}{5}x^5) - \tfrac{5}{3}x^3 + 4x]\Big|_{-2}^1 = \frac{9}{5}$$

$$M_y = \int_{-2}^1 (2x - x^3 - x^2)\, dx$$

$$= (x^2 - \tfrac{1}{4}x^4 - \tfrac{1}{3}x^3)\Big|_{-2}^1 = -\frac{9}{4}$$

$$(\bar{x}, \bar{y}) = \left(-\frac{9}{4}\cdot\frac{2}{9}, \frac{9}{5}\cdot\frac{2}{9}\right) = \left(-\frac{1}{2}, \frac{2}{5}\right)$$

8. $-100 = k(8 - 10)$, so $k = 50$ and $F(x) = 50x$
$$W = 50\int_0^{-3} x\, dx = 50(\tfrac{1}{2})x^2\Big|_0^{-3} = 225 \text{ in.}\cdot\text{lb};$$

$$W = 50\int_{-2}^{-3} x\, dx = 125 \text{ in.}\cdot\text{lb}$$

9. $-1{,}000 = k(8 - 10)$, so $k = 500$ and $F(x) = 500x$.
$$W = 500\int_0^1 x\, dx = 500(\tfrac{1}{2}x^2)\Big|_0^1 = 250 \text{ in.}\cdot\text{lb}$$

10. $dV = \pi x^2\, dy = \frac{1}{2}\pi y\, dy$; $dF = 62.4(\tfrac{1}{2})\pi y\, dy$;
at $x = 2$, $y = 8$ so the depth is $d = 8 - y$;
$$dW = 31.2\pi(8y - y^2)\, dy$$
$$W = 31.2\pi\int_0^8 (8y - y^2)\, dy$$
$$= 31.2\pi(4y^2 - \tfrac{1}{3}y^3)\Big|_0^8 \approx 8{,}362 \text{ ft}\cdot\text{lb}$$

11. $dV = \pi x^2\, dy = \frac{1}{2}\pi y\, dy$; $dF = 62.4(\tfrac{1}{2})\pi y\, dy$;
at $x = 2$, $y = 8$ so the depth is $d = 13 - y$;
$$dW = 31.2\pi(13y - y^2)\, dy$$
$$W = 31.2\pi\int_0^8 (13y - y^2)\, dy$$
$$= 31.2\pi(\tfrac{13}{2}y^2 - \tfrac{1}{3}y^3)\Big|_0^8 \approx 24{,}047 \text{ ft}\cdot\text{lb}$$

12. $\displaystyle\int_0^a \pi(x^2 - a^2)\, dx = 2\pi(a^2x - \tfrac{1}{3}x^3)\Big|_0^a = \frac{4}{3}\pi a^3$

13. $\displaystyle\int_0^a \pi(y^2 - a^2)\, dy = 2\pi(a^2y - \tfrac{1}{3}y^3)\Big|_0^a = \frac{4}{3}\pi a^3$

14. $V_1 = 2\int_0^1 (1 - 2x^2 + x^4)\, dx$

$$= 2(x - \tfrac{2}{3}x^3 + \tfrac{1}{5}x^5)\Big|_0^1 = \tfrac{16}{15}$$

$$V_2 = 2\int_0^1 \sqrt{y}\, dy = (\tfrac{4}{3}y^{3/2})\Big|_0^1 = \tfrac{4}{3}$$

15. $S = 2\pi r \int_0^h dy = 2\pi rh$

16. With $y = \tfrac{h}{r}x$, $ds = \sqrt{1 + \tfrac{h^2}{r^2}}\, dx = \tfrac{1}{r}\sqrt{r^2 + h^2}$

$$S = 2\pi \int_0^r x\left(\tfrac{1}{r}\sqrt{r^2 + h^2}\right) dx$$

$$= \tfrac{2\pi}{r}\cdot\tfrac{x^2}{2}\sqrt{r^2 + h^2}\Big|_0^1 = \pi r\sqrt{r^2 + h^2}$$

17. The weight density of the rope is $\tfrac{16}{60} = \tfrac{4}{15}$ lb/ft; the bucket plus the dirt weigh 70 lb.

$$W = \int_0^{60} (70 + \tfrac{4}{15}\, y)\, dy$$

$$= (70y + \tfrac{2}{15}y^2)\Big|_0^{60} = 4{,}680 \text{ ft} \cdot \text{lb}$$

18. Half the vertical cross section of the tank is fitted on a Cartesian coordinate plane with the vertex of the cone at the origin and the 10 ft leg along the y-axis.

$dV = \pi x^2\, dy = \tfrac{1}{25}\pi y^2\, dy$; the depth is $10 - y$;

$$W = \int_0^{10} f(x)\, dx (\tfrac{22}{25})\pi(10y^2 - y^3)\, dy$$

$$= \tfrac{22}{25}\pi(\tfrac{10}{3}y^3 - \tfrac{1}{4}y^4)\Big|_0^{10} = 733 \text{ ft} \cdot \text{lb}$$

19. The side consists of a 30×3 rectangle and a right triangle with legs 30×5; for the rectangle, $dA = 30\, dy$, the depth is y, and $dF = 62.4(30y)\, dy$

$$F = 1{,}872\int_0^3 y\, dy = 936y^2\Big|_0^3 = 8{,}424$$

For the triangle, pick a rectangular element of area $dA = x\, dy$, the depth is $8 - y$, and $dF = 62.4(6)(8y - y^2)\, dy = 374.4(8y - y^2)\, dy$

$$F = 374.4\int_0^5 (8y - y^2)\, dy$$

$$= 374.4(4y^2 - \tfrac{1}{3}y^3)\Big|_0^5 = 21{,}840$$

The total force on the side is $8{,}424 + 21{,}840$

$= 30{,}264 \text{ ft} \cdot \text{lb}$

20. $F = \int_0^{16} 124.8(16y^{1/4} - y^{5/4})\, dy$

$$= 124.8(\tfrac{64}{5}y^{5/4} - \tfrac{4}{9}y^{9/4})\Big|_0^{16} \approx 22{,}719 \text{ lb}$$

21. $F = \int_{-1}^1 (640)(2)(3)\sqrt{1 - y^2}\, dy$

$$= 128\int_{-1}^1 (3)\sqrt{1 - y^2}\, dy - 128\int_{-1}^1 y\sqrt{1 - y^2}\, dy$$

$$= 384(\tfrac{1}{2}\pi) + \tfrac{1}{3}(1 - y^2)^{3/2}\Big|_{-1}^1 = 192\pi$$

22. Half the vertical cross section of the tank is fitted on a Cartesian coordinate plane with the vertex of the cone at the origin and the 12 ft leg along the y-axis.

$x = \tfrac{3}{12}y = \tfrac{1}{4}y$; $dV = \pi x^2\, dy = \tfrac{1}{16}\pi y^2\, dy$; the depth is $14 - y$

$$W = \int_0^6 \tfrac{11}{25}\pi(14y^2 - y^3)\, dy$$

$$= \tfrac{11}{8}\pi(\tfrac{14}{3}y^3 - \tfrac{1}{4}y^4)\Big|_0^6 \approx 2{,}955 \text{ ft} \cdot \text{lb}$$

23. Half the vertical cross section of the tank is fitted on a Cartesian coordinate plane with the lower vertex of the triangle at the origin and the 10 ft altitude along the y-axis.

$x = \tfrac{9}{20}y$; $dA = 2x\, dy = \tfrac{9}{10}y\, dy$; the depth is $10 - y$

$$F = \int_0^{10} (62.4)(\tfrac{9}{10}y)(10 - y)\, dy$$

$$= 56.16\int_0^{10} (10y - y^2)\, dy$$

$$= 56.16(5y^2 - \tfrac{1}{3}y^3)\Big|_0^{10} = 9{,}360 \text{ ft} \cdot \text{lb}$$

24. $W = \int_0^1 \pi y^{2/3}(1 - y)(62.4)\, dy$

$$= 64.2\pi\int_0^1 (y^{2/3} - y^{5/3})\, dy$$

$$= 62.4\pi(\tfrac{3}{5}y^{5/3} - \tfrac{3}{8}y^{8/3})\Big|_0^1 \approx 44.108 \text{ ft} \cdot \text{lb}$$

25.
$$F = \int_{-5}^{-3} (62.4)(-y)(2\sqrt{25 - y^2})\, dy$$

$$= 62.4[\tfrac{2}{3}(25 - y^2)^{3/2}]\Big|_{-5}^{-3} \approx 2{,}662.4 \text{ lb}$$

26. $F = 30x;\ P = -\int_0^s (30x)\, dx = -15s^2$

When $-15s^2 = -20$ ft $\cdot$ lb, $s = \dfrac{2}{\sqrt{3}}$ ft

beyond the spring's equilibrium position.

27. **a.** $P = -\int_{s_0}^{s} (-mg)\, ds = mg(s - s_0)$ ft $\cdot$ lb

    b. $v = -gt + v_0;\ s = -\tfrac{1}{2}gt^2 + v_0 t + s_0;$

$$P = mg(-\tfrac{1}{2}gt^2 + v_0 t);\ K = \tfrac{1}{2}m(v_0 - gt)^2$$

28. Since 1 kg = 1,000 g, $W = r^{-2}$ kg; $m = \dfrac{1}{r^2 g}$;

$F = r^{-2}$;
$$W = \int_{6{,}400}^{10{,}000} r^{-2}\, dr = 5.625 \times 10^{-5} \text{ newtons}$$

29.
$$\frac{dP}{dt} = k(P - A)\cos t$$

$$\int \frac{dP}{P - A} = \int k \cos t\, dt$$

$$\ln|P - A| = k \sin t + C_1$$

$$P - A = e^{k\sin t}e_{}C_1 = Ce^{k\sin t}$$

$$P = A + Ce^{k\sin t}$$

$$P_0 = A + C, \text{ so } C = P_0 - A$$

$$P = A + (P_0 - A)e^{k\sin t}$$

30. $P(x) = R(x) - C(x) = \sqrt{2x} - \tfrac{1}{3}x$

$P'(x) = \dfrac{1}{\sqrt{2}}x^{-1/2} - \tfrac{1}{3}$

$P'(x) = 0$ when $x = \tfrac{9}{2}$ and $P''(\tfrac{9}{2}) < 0$

so a maximum occurs at this critical value.
$P(4) \approx 1.49509$ and $P(5) \approx 1.496109$, so hire 5 new people at an additional revenue of $1,496.11.

31. The curves intersect at $(2, 1)$; assume the density is 1.

$$A = \int_0^1 (5 - 3y - 2y)\, dy = \frac{5}{2}$$

$$M_x = 5\int_0^1 y(1 - y)\, dy = \frac{5}{6}$$

$$M_y = \frac{5}{2}\int_0^1 (5 - y)(1 - y)\, dy$$

$$= \frac{5}{2}\int_0^1 (y^2 - 6y + 5)\, dy = \frac{35}{6}$$

$$(\overline{x}, \overline{y}) = \left(\frac{35/6}{5/2}, \frac{5/6}{5/2}\right) = \left(\frac{7}{3}, \frac{1}{3}\right)$$

32.
$$V = \int_r^a \pi b^2\left(1 - \frac{x^2}{a^2}\right)dx$$

$$= \frac{\pi b^2}{a^2}(a^2 x - \tfrac{1}{3}x^3)\Big|_r^a$$

$$= \frac{\pi b^2}{a^2}\left(\frac{2}{3}a^3 - a^2 r + \frac{1}{3}r^3\right)$$

33.
$$V = \int_a^r \pi a^{-2}b^2(x^2 - a^2)\, dx$$

$$= \pi a^{-2}b^2(\tfrac{1}{3}x^3 - a^2 x)\Big|_a^r$$

$$= \pi b^2\left(\frac{r^3}{3a^2} - r - \frac{2}{3}a\right)$$

34. $A = \tfrac{1}{4}\pi ab;\ V = \tfrac{2}{3}\pi ab^2;\ V = 2\pi \overline{y}A$

$$\overline{y} = \frac{V}{2\pi A} = \frac{2\pi ab^2(4)}{3\pi ab} = \frac{8b}{3};$$

Similarly, $\overline{x} = \dfrac{8a}{b}$

35.
$$W = \int_0^{100} (200 - 0.5x)\, dx$$

$$= (200x - 0.25x^2)\Big|_0^{100} = 17{,}500 \text{ ft} \cdot \text{lb}$$

36. Let t be the time required for the bag to be emptied; $t = 40/0.2 = 200$ s; $F = 40 - 0.2t$ (convert minutes to seconds). The distance covered per second is $y = (31/60)t$

$$W = \int_0^{200} (40 - 0.2t)(\tfrac{31}{60})\, dt$$

$$= \frac{31}{60}(40t - 0.1t^2)\Big|_0^{200} = 1{,}322.7 \text{ ft} \cdot \text{lb}$$

37. Half the vertical side is a right triangle with base $\tfrac{3}{2}$ and height $\tfrac{3}{2}\sqrt{3}$. The vertical leg of this right triangle is along the y-axis, the

lower vertex of the side is at the origin. A rectangular element of area is at a height y with area $dA = 2x\ dy$. By similar triangles,

$$\frac{3}{2}x = \frac{3\sqrt{3}}{2}y$$

$$F = \int_0^{3\sqrt{3}/2} 40\left(\frac{2}{\sqrt{3}}y\right)\left(\frac{3\sqrt{3}}{2} - y\right) dy$$

$$= \frac{80}{\sqrt{3}}\int_0^{3\sqrt{3}}\left(\frac{3\sqrt{3}}{2}y - y^2\right) dy$$

$$= \frac{80}{\sqrt{3}}\left(\frac{3\sqrt{3}}{4}y^2 - \frac{1}{3}y^3\right) dy\ \Big|_0^{3\sqrt{3}/2} = 135\ \text{lb}$$

38. $x^2 + y^2 = 4$; $dA = 2x\ dy = 2\sqrt{4 - y^2}\ dy$; the depth is $-y$;

$$F = \int_{-2}^0 (-62.4)y\sqrt{4 - y^2}\ dy$$

$$= (62.4)(\tfrac{2}{3})(4 - y^2)^{3/2}\Big|_{-2}^0 = 332.8\ \text{lb}$$

39. $W_1 = \int_0^6 (mt + b)\ dt = 18m + 6b = 13$

$$W_2 = \int_0^{12} (mt + b)\ dt = 72m + 12b = 44$$

Solve this system of equations to find $(m, b) = (\frac{1}{2}, \frac{2}{3})$. Thus, $F(t) = \frac{1}{2}t + \frac{2}{3}$.

40. $dV = (10)(6)\ dy$; $dF = (20)(60)\ dy$; $dW = (8 - y)(1,200)\ dy$; the height of the water level is h.

$$W = 1,200\int_0^h (8 - y)\ dy = 1,200(8h - \tfrac{1}{2}h^2)$$

$$W = 2,400\ \text{when}$$

$$8h - \tfrac{1}{2}h^2 = 2$$

$$h^2 - 16h + 4 = 0$$

$$h \approx 0.254\text{ft}$$

41. Use a Cartesian coordinate system with origin at the center of the semicircle. For the hemisphere, $dV = \pi x^2\ dy = \pi(4 - y^2)\ dy$, and the depth is $10 + y$

$$W_h = \int_0^2 24\pi(4 - y^2)(10 + y)\ dy$$

$$= 24\pi\int_0^2 (-y^3 - 10y^2 + 4y + 40)\ dy$$

$$= 24\pi(-\tfrac{1}{4}y^4 - \tfrac{10}{3}y^3 + 2y^2 + 40y)\Big|_0^2$$

$$= 1{,}376\pi$$

For the cylinder, $dV = 4\pi\ dy$; the depth is $10 - y$;

$$W_c = \int_0^1 (10 - y)(96\pi\ dy)$$

$$= 96\pi(10y - \tfrac{1}{2}y^2)\Big|_0^{10} = 4{,}800\pi$$

The total force is $1{,}376\pi + 4{,}800\pi = 6{,}176\pi$.

42. a. $PV^{1.4} = C$ or $P = Cv^{-1.4}$;

$$dW = F\ dx = CV^{-1.4}A\ dx = CV^{-1.4}\ dV$$

$$W = C\int_{V_1}^{V_2} V^{-1.4}\ dV$$

b. $C = (2,500)(0.6)^{1.4} \approx 1{,}222.79$

$$W = 1{,}222.79\int_{0.6}^{0.9} V^{-1.4}\ dV$$

$$= \frac{1{,}222.79}{-0.4}V^{-0.4}\Big|_{0.6}^{0.9} \approx 561.44$$

43. Let the cut-out rectangle have dimensions x by y and its centroid at $(\overline{x}_c, \overline{y}_c)$. The area of the figure remaining after the cutting is $256 - 156 = 100$.

$$100(4.88) + 156\overline{x}_c = 8(256)$$

$$\overline{x}_c = 10$$

This leave $16 - 10 = 6$ cm for $\frac{1}{2}x$ or $x = 12$.

$$y = 256x^{-1} = 13\ \text{so}\ \overline{y}_c = 6.5\ \text{cm}$$

from its (top or bottom) edge.

$$100\overline{y} + 156(\tfrac{13}{2}) = 8(256)$$

$$\overline{y} = 10.34$$

Thus, the desired distances are 4.88 cm (given), $16 - 4.88 = 11.12$ cm, and $16 - 10.34 = 5.66$ cm.

44. The point $(5,15)$ is on the parabola so $y = kx^2$ becomes $15 = 25k$ or $k = 3/5$. Thus,

$$y = \tfrac{3}{5}x^2$$

The rectangular element of area is

$$dA = 2x\ dy = 2\sqrt{\tfrac{5}{3}}\ y^{1/2}dy$$

the depth is $15 - y$, and

$$F = \int_0^{15} (40)(2)\sqrt{\tfrac{5}{3}}(15y^{1/2} - y^{3/2}) \, dy$$

$$\approx (103.28)(10y^{3/2} - \tfrac{2}{5}y^{5/2})\Big|_0^{15}$$

$$\approx 24,000 \text{ lb}$$

45. Reversing the roles of x and y, we obtain

$$\frac{dx}{dy} = \frac{x + e^y y^2}{y}$$

$$\frac{dx}{dy} - \frac{x}{y} = ye^y$$

$$I(y) = e^{\int -1/y \, dy} = e^{-\ln y} = \frac{1}{y}$$

$$x(y) = \frac{1}{1/y}\left[\int \tfrac{1}{y}(ye^y) \, dy + C\right]$$

$$= y\left[\int e^y \, dy + C\right]$$

$$= ye^y + Cy$$

46. Let h be the height of the trapezoid. Then $h^2 + r^2 = \left(\frac{y}{2}\right)^2$ and

$$h = \sqrt{\left(\frac{y}{2}\right)^2 + \left(\frac{y}{4}\right)^2} = \sqrt{\frac{3}{16}}\, y$$

A cross-sectional area is

$$A(x) = \tfrac{1}{2}(y + \tfrac{1}{2}y)\sqrt{\tfrac{3}{16}}\, y$$

$$= \tfrac{3}{4}\sqrt{\tfrac{3}{4}}\, y^2 = \frac{3\sqrt{3}}{16}\, y^2$$

$$V = \int_{-2}^{2} A(x) \, dx = \int_{-2}^{2} \frac{3\sqrt{3}}{16}(4 - x^2) \, dx$$

47. The vertical cross section of the cone is an isosceles triangle with height 1 and base 2 ft. A horizontal element of volume is a cylinder of radius x and height dy. The volume is $dV = \pi x^2 \, dy$. This element is y ft from the base of the cone. From similar right triangles (half of the isosceles triangle), $x = 1 - y$. Thus, $dV = \pi(1 - y)^2 \, dy$, the distance through which the element was lifted is y.

$$W = \int_0^1 140\pi y(1 - y)^2 \, dy$$

$$= 140\pi \int_0^1 (y - 2y^2 + y^3) \, dy$$

$$= 140\pi(\tfrac{1}{2}y - \tfrac{2}{3}y^3 + \tfrac{1}{4}y^4)\Big|_0^1 = \frac{35\pi}{3} \text{ ft} \cdot \text{lb}$$

48. A horizontal element is a cylinder of radius 4 in. or 1/3 ft and height dy. The volume is $dV = (1/9)\pi \, dy$. This element is y ft from the base of the cone, a distance through which the element was lifted.

$$W = \int_0^1 \frac{140}{9}\pi y \, dy = \frac{70}{9}\pi y^2\Big|_0^1 = \frac{70}{9}\pi \text{ ft} \cdot \text{lb}$$

49. Twice the volume on $[0, A]$ equals the volume on $[0, 1]$. Thus,

$$V = \int_0^A \pi(x^2 - x^4) \, dx = 2\pi(\tfrac{1}{3}x^3 - \tfrac{1}{5}x^5)\Big|_0^A$$

$$= \tfrac{2}{3}A^3 - \tfrac{2}{5}A^5$$

$$\tfrac{2}{3}A^3 - \tfrac{2}{5}A^5 = \frac{2}{15}$$

$$6A^5 - 10A^3 + 2 = 0$$

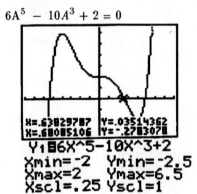

X=.63829787   Y=.03514362
X=.68085106   Y=-.2783078
Y₁◼6X^5-10X^3+2
Xmin=-2   Ymin=-2.5
Xmax=2   Ymax=6.5
Xscl=.25 Yscl=1

We see that on $[0, 1]$ there is one root; by using the ZOOM on this graph we find $A \approx 0.6431$.

50. Consider a Cartesian coordinate system whose origin is in the middle of the base of the isosceles triangle with vertical y-axis. A slant side passes through $(0, 3)$ and $(2.5, 0)$. It equation is

$$y - 3 = -\tfrac{6}{5}x \text{ or } x = \tfrac{5}{6}(3 - y)$$

The height of the triangle is $\frac{3}{2}\sqrt{3}$, and the depth is $4 - y$.

$$F = \int_0^{(3/2)\sqrt{3}} 62.4(\tfrac{5}{3})(3 - y)(4 - y) \, dy$$

$$\approx 104.1667(12y - \tfrac{7}{2}y^2 + \tfrac{1}{3}y^3)\Big|_0^{2.5981}$$

$$\approx 1,395.5 \text{ lb}$$

51. Consider a Cartesian coordinate system whose

origin is in the middle of the base of the isosceles triangle with vertical y-axis; h is the height of the isosceles triangle. One-half of the isosceles triangle is a right triangle with legs h, $\frac{1}{2}B$, and hypotenuse A. Thus,

$$h = \sqrt{A^2 - \tfrac{1}{4}B^2} = \tfrac{1}{2}\sqrt{4A^2 - B^2}$$

Pick a rectangular element of area with length x and height dy at a distance y from the base. From similar triangles,

$$\frac{x}{h - y} = \frac{B}{h} \text{ or } x = \frac{B}{h}(h - y)$$

we see $dA = \frac{B}{h}(h - y)\, dy$ and note that the depth is $d = D + h - y$ so that

$$F = \int_0^h (62.4)\left(\frac{B}{h}\right)(D + h - y)(h - y)\, dy$$

$$= 62.4\left(\frac{B}{h}\right)\int_0^h \left[D(h - y) + (h - y)^2\right]\, dy$$

$$= 62.4\left(\frac{B}{h}\right)\left[-\tfrac{1}{2}(h - y)^2 - \tfrac{1}{3}(h - y)^3\right]\Big|_0^h$$

$$= 62.4B(\tfrac{1}{2}Dh + \tfrac{1}{3}h^2)$$

52. $ds = \sqrt{1 + (3Ax^2 - Bx^{-2})^2}\, dx$

$$= \sqrt{1 + 9A^2x^4 - 6AB + B^2x^{-4}}\, dx$$

$$= \sqrt{9A^2x^4 + 6AB + B^2x^{-4}}\, dx$$

$$= (3Ax^2 + Bx^{-2})\, dx \text{ since } AB = 12$$

$$s = \int_a^b (3Ax^2 + Bx^{-2})\, dx$$

$$= (Ax^3 - Bx^{-1})\Big|_a^b$$

$$= A(b^3 - a^3) - B(b^{-1} - a^{-1})$$

53. $ds = \sqrt{2}\, dx$ and $S = 2\pi\displaystyle\int_a^b (x - 2)\sqrt{2}\, dx$

if $y \geq 0$ on $[a, b]$. Use symmetry to find the actual surface area is twice that on $[2, 3]$:

$$S = 4\pi\int_2^3 (x - 2)\sqrt{2}\, dx$$

$$= 4\pi\sqrt{2}\left[\tfrac{1}{2}(x - 2)^2\right]\Big|_2^3 = 2\pi\sqrt{2}$$

54. $F = -k/r^2$, where k is the proportionality constant and r^2 is the distance squared.

$$\text{Work} = -(4{,}000)^2 W\int_s^{4{,}000} r^{-2}\, dr$$

$$= (4{,}000)^2 W\, \frac{1}{r}\Big|_s^{4{,}000}$$

$$= (4{,}000)^2 W\left(\frac{1}{4{,}000} - \frac{1}{s}\right) \text{mi}\cdot\text{lb}$$

55. $A = \frac{1}{3}[4 + 4\cdot3 + 2\cdot4 + 4\cdot6 + 2\cdot5 + 4\cdot6 + 2\cdot4$

$\qquad + 4\cdot3 + 4](1) = \frac{140}{3}$

$\bar{y} = \frac{52}{11}$; $V = 2\pi(\frac{140}{3})(\frac{52}{11}) \approx 441.21\pi$

56. $V = \displaystyle\int_{-1}^9 \pi\{[f(x) + 1]^2 - 1^2\}\, dx$

$$\approx \pi\left(\frac{10}{2\cdot10}\right)[(5^2 - 1) + 2\{(4^2 - 1)$$

$$+ (5^2 - 1) + (7^2 - 1) + (6^2 - 1) + (7^2 - 1)$$

$$+ (8^2 - 1) + (7^2 - 1) + (5^2 - 1) + (4^2 - 1)$$

$$= \frac{\pi}{2}[24 + 2(15 + 24 + 48 + 35 + 48 + 62 + 48$$

$$+ 24) + 15$$

$$\approx 324.5\pi$$

57. Consider a Cartesian coordinate system with the x-axis through b. $A = \frac{1}{2}ab$, rotate the triangle about the x-axis to obtain a cone of volume $V = \frac{1}{3}\pi ab^2$. Then,

$$\bar{x} = \frac{V}{2\pi A} = \frac{2\pi ab^2}{(2\pi)(3ab)} = \frac{b}{3}$$

Rotate about the y-axis to obtain a cone of volume $V = \frac{1}{3}\pi a^2 b$. Then

$$\bar{y} = \frac{V}{2\pi A} = \frac{2\pi a^2 b}{(2\pi)(3ab)} = \frac{a}{3}$$

The centroid is $\left(\frac{b}{3}, \frac{a}{3}\right)$.

58. Consider a right triangle formed by a vertical cross section with height 480 ft and base 750 ft. Pick a horizontal slab of height dy and base x^2 located at y ft from the top. By similar triangles,

$$\frac{x}{y} = \frac{750}{480} \text{ or } x = \frac{750y}{480}$$

$$W = \int_0^{480} 160y\left(\frac{750}{480}\right)^2 y^2\, dy$$

$$= 160\left(\frac{750}{480}\right)^2\left(\frac{1}{4}y^4\right)\Big|_0^{480}$$

$$\approx 5.184 \times 10^{12} \text{ ft}\cdot\text{lb}$$

59. a. $Q(t) = Q_0 e^{-kt}$ and $\frac{1}{2} = e^{-5.25k}$ or

$k = \frac{1}{5.25} \ln 2$

$Q(5) = Q_0 e^{-5 \ln 2/5.25} \approx 0.5168 Q_0$

The percentage remaining after 5 years is

$\frac{100 Q(5)}{Q_0} \approx 51.7\%$

b. $Q(t_1) = Q_0 e^{-t_1 \ln 2/5.25}$ and also

$Q(t_1) = (1 - 0.9) Q_0$; thus,

$0.1 Q_0 = Q_0 e^{-t_1 \ln 2/5.25}$

$-\frac{t_1 \ln 2}{5.25} = \ln 0.1$

$t_1 = \frac{-5.25 \ln 0.1}{\ln 2} \approx 17.44$

The time for 90% to disintegrate is about $17\frac{1}{2}$ years.

60. Let $Q(t)$ be the amount of undissolved salt in the tank at time t. $Q(0) = 8$; $dQ/dt = -kQ$, so $\ln|Q| = -kt + C$;

$Q(t) = Ce^{-kt}$

$Q_0 = 8$, so $Q(t) = 8e^{-kt}$.

With 2 lb dissolved in 30 minutes,

$8 - 6 = 8e^{-30k}$

$\ln \frac{3}{4} = -20k$

$k = \frac{1}{30} \ln 1.214 \approx 0.009589$

The time to dissolve 1 lb of salt is

$8 - 1 = 8e^{-0.009589 t_1}$

$t_1 \approx 13.93$

The time is about 14 minutes (or 13 min 56 sec).

61. $Q_0 = 10$ million; $Q(1) = 1.02 Q_0$; $1.02 = e^k$ or $k = \ln 1.02$;

$Q(t) = Q_0 e^{(\ln 1.02)t}$

$Q(10) = Q_0 e^{10 \ln 1.02} \approx 12.19$ million

For the population to double,

$2 = e^{t_1 \ln 1.02}$

$t_1 = \frac{\ln 2}{\ln 1.02} \approx 35$ years

62. $y = e^x$; $dy = e^x \, dx$; $ds = \sqrt{1 + e^{2x}} \, dx$

$l = \int_0^1 \sqrt{1 + e^{2x}} \, dx \approx 1.97$

Approximate integral using the trapezoidal rule, or computer software.

63. Use shells;

$V = 2\pi \int_0^2 x(9 - x^2)^{-1/2} \, dx$

$= -2\pi \sqrt{9 - x^2} \Big|_0^2 = 2\pi(3 - \sqrt{5})$

64. Use disks;

$V = \pi \int_1^2 y^2 \, dx = \frac{4}{3}\pi \int_1^2 (3x - 2)^{-1}(3 \, dx)$

$= \frac{4}{3}\pi \ln|3x - 2| \Big|_1^2 = \frac{4}{3}\pi \ln 4 \approx 5.8069$

65. $A = \int_0^{\pi/4} (\sec^2 x - \tan^2 x) \, dx = \int_0^{\pi/4} dx = \frac{\pi}{4}$

66. $y' = e^x + \frac{1}{4}e^{-x}$;

$ds = \sqrt{1 + (e^x - \frac{1}{4}e^{-x})^2} \, dx = (e^x + \frac{1}{4}e^{-x}) \, dx$

$S = 2\pi \int_0^1 (e^x + \frac{1}{4}e^{-x})^2 \, dx$

$= 2\pi(\frac{1}{2}e^{2x} + \frac{1}{2}x - \frac{1}{32}e^{-2x}) \Big|_0^1 \approx 23.44$

67. $V = \frac{\sqrt{3}}{4} \int_0^1 e^{2x} \, dx = \frac{\sqrt{3}}{8} e^{2x} \Big|_0^1 = \frac{\sqrt{3}}{8}(e^2 - 1)$

68. $y = \ln(\sec x) = -\ln(\cos x)$;

$\frac{dy}{dx} = -\frac{\sin x}{\cos x} = \tan x$;

$ds = \sqrt{1 + \tan^2 x} \, dx = \sec x \, dx$

$s = \int_0^{\pi/3} \sec x \, dx = \ln|\sec x + \tan x| \Big|_0^{\pi/3}$

Mathematics Handbook, integration formula 9

$= \ln(\sqrt{3} + 2) \approx 1.3170$

69. $\frac{dx}{dy} = \frac{2ay}{b^2} - \frac{b^2}{8a} \cdot \frac{1}{y}$;

$ds = \sqrt{1 + \frac{4a^2 y^2}{b^4} - \frac{1}{2} + \frac{b^4}{64 a^2 y^2}} \, dy$

$= \left(\frac{2ay}{b^2} + \frac{b^2}{8ay}\right) dy$

$s = \int_b^{2b} \left(\frac{2ay}{b^2} + \frac{b^2}{8a} y^{-1}\right) dy$

$= \left(\frac{a}{b^2} y^2 + \frac{b^2}{8a} \ln|y|\right) \Big|_b^{2b} = 3a + \frac{b^2}{8a} \ln 2$

70. a. The volume of snow cleared is approximately $wh\Delta s = p\Delta t$,

$$wh\frac{\Delta s}{\Delta t} = p$$

$$\lim_{\Delta t \to 0} wh\frac{\Delta s}{\Delta t} = p$$

$$wh\frac{ds}{dt} = p$$

b. The height of the snow equals the rate of the snowfall times the total time it has been snowing. Suppose the rate of snowfall is A (a constant) and that it started snowing t_s hr before noon. Then,

$$h = A(t + t_s) \qquad \text{This is } d = rt.$$

By substitution, we have

$$wA(t + t_s)\frac{ds}{dt} = p$$

$$\int wA\ ds = \int \frac{p}{t + t_s}\ dt$$

$$wAx = p\ln|t + t_s| + C$$

Since $y = 0$ when $t = 0$, it follows that

$$0 = p\ln t_s + C$$

$$C = -p\ln t_s$$

Thus,

$$wAx = p\ln|t + t_s| - p\ln t_s$$

$$x = \frac{p}{wA}\ln\left|\frac{t}{t_s} + 1\right|$$

We also know that $x = 2$ when $t = 1$, which means

$$2 = \frac{p}{wA}\ln\left(\frac{1}{t_s} + 1\right)$$

so that

$$\frac{k}{wA} = \frac{2}{\ln(\frac{1}{t_s} + 1)}$$

and then since $x = 3$ when $t = 2$,

$$3 = \left[\frac{2}{\ln(\frac{1}{t_s} + 1)}\right]\ln\left(\frac{2}{t_s} + 1\right)$$

we have

$$3\ln\left(\frac{1}{t_s} + 1\right) = 2\ln\left(\frac{2}{t_s} + 1\right)$$

$$\ln\left(\frac{1}{t_s} - 1\right)^3 = \ln\left(\frac{2}{t_s} + 1\right)^2$$

$$\left(\frac{1}{t_s} + 1\right)^3 = \left(\frac{2}{t_s} + 1\right)^2$$

$$t_s{}^2 + t_s - 1 = 0$$

$$t = .6180339880$$

(disregard negative value);

$t_s \approx 37$ minutes before noon; that is, it started snowing at approximately 11:23 AM. You might be interested in some variations of this problem: Problem E275 of the Otto Dunkel Memorial Problem Book, *American Mathematical Monthly*, Vol. 64 (1957); p. 54. *Applied Mathematical Notes*, January 1975; pp. 6-11. *American Mathematical Monthly*, Vol. 59 (1952); p. 42 (Problem E963).

71. This is Putnam Examination Problem 1 of the morning session in 1939. The arc in the first quadrant is represented by the equation $y = x^{3/2}$, and its slope is $\frac{3}{2}x^{1/2}$. The point $P(x_0, y_0)$ where the tangent makes an angle of $\pi/4$ is determined from the relation $(3/2)x_0{}^{1/2} = 1$, so that $x_0 = 4/9$. The desired length is therefore

$$\int_0^{4/9}\sqrt{1 + \frac{9x}{4}}\ dx = \frac{8}{25}(1 + \frac{9}{4}x)^{3/2}\Big|_0^{4/9}$$

$$= \frac{8}{27}(2\sqrt{2} - 1)$$

72. This is Putnam Examination Problem 1 of the morning session in 1938.

a.
$$V = \int_{-h/2}^{h/2} (a_0z^3 + a_1z^2 + a_2z + a_3)\ dz$$

$$= \frac{1}{12}a_1h^3 + a_3h$$

The base area and M are given by

$$B_1 = \frac{1}{8}a_0h^3 + \frac{1}{4}a_1h^2 + \frac{1}{2}a_2h + a_3$$

$$B_2 = -\frac{1}{8}a_0h^3 + \frac{1}{4}a_1h^2 - \frac{1}{2}a_2h + a_3$$

$M = a_3$, so the suggested expression

$$\frac{1}{6}h(B_1 + B_2 + 4M)$$

works out to be

$$\frac{1}{6}(\frac{1}{2}a_1h^2 + 65a_3) = \frac{1}{12}a_1h^3 + a_3h = V$$

b. The formula $V = \frac{1}{6}h(B_1 + B_2 + 4M)$

is known in solid geometry as the primoidal formula. It is closely related to Simpson's rule in numerical integration. Indeed, for functions of class C^4 it can be

proved that

$$\int_{-h/2}^{h/2} f(z)\ dz = \tfrac{1}{6}h[f(-\tfrac{h}{2}) + 4f(0) + f(\tfrac{h}{2})]$$

This is known as Simpson's rule. In particular, when f is a polynomial of degree at most 3, $E = 0$, and the result is exact. For the special cases of the cone and the sphere, we proceed as follows. For the cone, let the vertex be in the plane $z = h/2$ and the base in the plane $z = -h/2$. The the area of a cross-section at level z is given by

$$A = (Bh^{-2})(z - \tfrac{h}{2})^2$$

where B is the area of the base. Since the expression for A is a polynomial of degree 2,

$$V = \tfrac{1}{6}h[B + 4(\tfrac{1}{4}B) + 0)] = \tfrac{1}{3}Bh$$

which is a well-known result. For the sphere of radius $r = h/2$, included between two planes $z = -h/2$ and $z = h/2$, the cross-sectional area at level z is given by $A = \pi(r^2 - z^2)$. This expression for A is also a polynomial of degree 2, and we obtain

$$V = \tfrac{1}{6}h(4\pi r^2) = \tfrac{4}{3}\pi r^3$$

For both the sphere and the cone, the coefficient a_0 of z^3 in the cross-sectional area formula is 0.

Cumulative Review for Chapters 1-6, Page 435

1. The limit statement $\lim_{x \to c} f(x) = L$ means that for each $\epsilon > 0$, there corresponds a number $\delta > 0$ with the property that

    $$|f(x) - L| < \epsilon$$
    whenever $0 < |x - c| < \delta$. The notation $\lim_{x \to c} f(x) = L$ is read "the limit of $f(x)$ as x approaches c is L" and means that the functional values $f(x)$ can be made arbitrarily close to L by choosing x sufficiently close to c (but not equal to c).

2. The derivative of f at x is given by

    $$f'(x) = \lim_{\Delta x \to 0} \frac{f(x + \Delta x) - f(x)}{\Delta x}$$

    provided this limit exists. This is the slope of the tangent line to the curve at the point $(x, f(x))$, obtained as the limiting process

from the secant line through $(x, f(x))$ and $(x + h, f(x + h))$.

3. If f is defined on the closed interval $[a, b]$ we say f is integrable on $[a, b]$ if

    $$I = \lim_{\|P\| \to 0} \sum_{k=1}^{n} f(\overset{*}{x}_k)\Delta x_k$$

    exists. This limit is called the definite integral of f from a to b. The definite integral is denoted by

    $$I = \int_{a}^{b} f(x)\ dx$$

4. A differential equation is an equation involving derivatives (differentials). For

    $$\frac{dy}{dx} + P(x)y = Q(x)$$

    multiplication of both sides by

    $$I(x) = e^{\int P(x)\,dx}$$

    transforms the left member into the derivative of a product which can then be integrated.

5. $\lim_{x \to 2} \dfrac{3x^2 - 5x - 2}{3x^2 - 7x + 2} = \lim_{x \to 2} \dfrac{(3x + 1)(x - 2)}{(3x - 1)(x - 2)}$

    $= \lim_{x \to 2} \dfrac{3x + 1}{3x - 1} = \dfrac{7}{5}$

6. $\lim_{x \to +\infty} \dfrac{3x^2 + 7x + 2}{5x^2 - 3x + 3} = \lim_{x \to +\infty} \dfrac{3 + \frac{7}{x} + \frac{2}{x^2}}{5 - \frac{3}{x} + \frac{3}{x^2}} = \dfrac{3}{5}$

7. $\lim_{x \to +\infty} (\sqrt{x^2 + x} - x)$

    $= \lim_{x \to +\infty} \dfrac{(\sqrt{x^2 + x} - x)(\sqrt{x^2 + x} + x)}{\sqrt{x^2 + x} + x}$

    $= \lim_{x \to +\infty} \dfrac{x}{\sqrt{x^2 + x} + x} = \lim_{x \to +\infty} \dfrac{1}{\sqrt{1 + \frac{1}{x}} + 1}$

    $= \dfrac{1}{2}$

8. $\lim_{x \to \pi/2} \dfrac{\cos^2 x}{\cos x^2} = \dfrac{\cos^2 \frac{\pi}{2}}{\cos \frac{\pi^2}{4}} = 0$

9. $\lim_{x \to 0} \dfrac{x \sin x}{x + \sin^2 x} = \lim_{x \to 0} \dfrac{x}{\frac{x}{\sin x} + \sin x} = 0$

10. $\lim_{x \to 0} \dfrac{\sin 3x}{x} = 3$ $\lim_{x \to 0} \dfrac{x \sin 3x}{3x} = 3$

11. $\lim_{x \to +\infty} (1 + x)^{2/x} = \lim_{x \to +\infty} \exp[\ln(1 + x)^{2/x}]$

    $= \lim_{x \to +\infty} \exp\!\left[2\,\dfrac{\ln(1 + x)}{x}\right] = \lim_{x \to +\infty} \exp\!\left[\dfrac{2}{1 + x}\right]$

    $= e^0 = 1$

12. $\lim\limits_{x \to 0} \dfrac{\ln(x^2 + 50)}{2x}$ is not defined

13. $\lim\limits_{x \to 0^+} x^{\sin x} = \lim\limits_{x\to0^+} \exp[\ln x^{\sin x}]$

$= \lim\limits_{x\to0^+} [\sin x \ln x] = \lim\limits_{x\to0^+} \left[\dfrac{\ln x}{\csc x}\right]$

$= \lim\limits_{x\to0^+} \exp\left[\dfrac{1}{x(-\csc x \cot x)}\right]$

$= \lim\limits_{x\to0^+} \exp\left[\dfrac{\sin^2 x}{x(-\cos x)}\right]$

$= \lim\limits_{x\to0^+} \exp\left[\dfrac{\sin x}{-\cos x}\right] = e^0 = 1$

14. $y = 6x^3 - 4x + 2;\ \ y' = 18x - 4$

15. $y = (x^2 + 1)^3(3x - 4)^2$

$y' = (x^2+1)^2(2)(3x-4)(3)$

$\qquad + 3(x^2 + 1)^2(2x)(3x - 4)^2$

$= 6(x^2 + 1)^2(3x - 4)\big[x^2 + 1 + 3x^2 - 4x\big]$

$= 6(2x - 1)^2(x^2 + 1)^2(3x - 4)$

16. $y = \dfrac{x^2 - 4}{3x + 1}$

$y' = \dfrac{(3x + 1)(2x) - (x^2 - 4)(3)}{(3x + 1)^2}$

$= \dfrac{3x^2 + 2x + 12}{(3x + 1)^2}$

17. $y = \dfrac{x}{x + \cos x}$

$y' = \dfrac{x + \cos x - x(1 - \sin x)}{(x + \cos x)^2}$

$= \dfrac{\cos x + x \sin x}{(x + \cos x)^2}$

18. $x^2 + 3xy + y^2 = 0$

$2x + 3xy' + 3y + 2yy' = 0$

$(3x + 2y)y' = -(2x + 3y)$

$y' = -\dfrac{2x + 3y}{3x + 2y}$

19. $y = \csc^2 3x$

$y' = 2 \csc 3x(-\csc 3x \cot 3x)(3)$

$= -6 \csc^2 3x \cot 3x$

20. $y = e^{5x - 4};\ y' = 5e^{5x - 4}$

21. $y = \ln(5x^2 + 3x - 2)$

$y' = \dfrac{10x + 3}{5x^2 + 3x - 2}$

22. $y = \cos^{-1}(x^2 - 3)$

$y' = -\dfrac{2x}{\sqrt{1 - (x^2 - 3)^2}}$

23. $\displaystyle\int_4^9 d\theta = (9 - 4) = 5$

24. $\displaystyle\int_{-1}^{1} 50(2x - 5)^3\, dx = 25\int_{-1}^{1} (2x - 5)^3\, (2\, dx)$

$= \tfrac{25}{4}(2x - 5)^4\Big|_{-1}^{1} = -14{,}500$

25. $\displaystyle\int_0^1 \dfrac{x\, dx}{\sqrt{9 + x^2}} = \tfrac{1}{2}\int_0^1 (9 + x^2)^{-1/2}\, (2\, x\, dx)$

$= 2(\tfrac{1}{2})(9 + x^2)^{1/2}\Big|_0^1 = \sqrt{10} - 3$

26. $\displaystyle\int \csc 3\theta \cot 3\theta\, d\theta = \tfrac{1}{3}\int \csc 3\theta \cot 3\theta\, (3\, d\theta)$

$= -\tfrac{1}{3} \csc \theta + C$

27. $\displaystyle\int \dfrac{e^x\, dx}{e^x + 2} = \ln|e^x + 2| + C = \ln(e^x + 2) + C$

28. $\displaystyle\int \dfrac{x^3 + 2x - 5}{x}\, dx = \int (x^2 + 2 - 5x^{-1})\, dx$

$= \tfrac{1}{3}x^3 + 2x - 5 \ln|x| + C$

29. $\Delta x = \dfrac{4 - 0}{6} = \dfrac{2}{3}$

$x_0 = 0.0000\ \ f(x_0) = 1.0000$
$x_1 = 0.6667\ \ f(x_1) = 0.8783$
$x_2 = 1.3333\ \ f(x_2) = 0.5447$
$x_3 = 2.0000\ \ f(x_3) = 0.3333$
$x_4 = 2.6667\ \ f(x_4) = 0.2238$
$x_5 = 3.3333\ \ f(x_5) = 0.1621$
$x_6 = 4.0000\ \ f(x_6) = 0.1240$

$A \approx \tfrac{1}{3}[1(1)+4(0.8783)+2(0.5447)$

$\qquad +4(0.3333)+2(0.2238)+4(0.1621)$

$\qquad + 1(0.124)](\tfrac{2}{3})$

$\approx \tfrac{1}{3}(8.1561)(\tfrac{2}{3}) \approx 1.812$

30. $y = x^3 - 5x^2 + 2x + 8$

$y' = 3x^2 - 10x + 2$

$y' = 0$ when $x = \dfrac{5 \pm \sqrt{25 + 6}}{3}$

$y'' = 6x - 10$

$y'' = 0$ when $x = \dfrac{5}{3}$

$(-0.19, 8.19)$ is a relative maximum;
$(3.52, -3.30)$ is a relative minimum;
$(1.67, 2.07)$ is a point of inflection

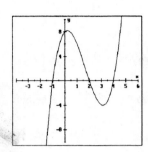

31. $y = \dfrac{4 - x^2}{4 + x^2} = \dfrac{8}{4 + x^2} - 1$

$\displaystyle\lim_{x \to \pm \infty} \dfrac{4 - x^2}{4 + x^2} = -1$ so $y = -1$ is a

horizontal asymptote;

$y' = -\dfrac{8(2x)}{(4 + x^2)^2} = 0$ when $x = 0$

$y'' = -16\left[\dfrac{(4 + x^2)^2 - x(2)(4 + x^2)(2x)}{(4 + x^2)^4}\right]$

$= \dfrac{-16(4 - 3x^2)}{(4 + x^2)^3} = 0$ when $x = \pm 1.15$

$(0, 1)$ is a relative maximum; $(\pm 1.15, -0.5)$ are points of inflection.

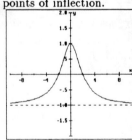

32. $f'(x) = x^2 - 4x + 3 = (x - 1)(x - 3)$
$f'(x) = 0$ when $x = 1, 3$
$f(0) = -10;$ $f(1) = -\frac{26}{3};$
$f(6) = 8;$ $f(3) = -13;$
The maximum value is 8.

33. The curve crosses the x-axis at $(0, 0)$.

$A = \displaystyle\int_{-2}^{0} [0 - x(x^2 - 8)^{1/2}]\, dx$

$\qquad + \displaystyle\int_{0}^{1} [x(x^2 + 8)^{1/2} - 0]\, dx$

$= -\frac{1}{3}(x^2 + 8)^{3/2}\Big|_{-2}^{0} + \frac{1}{3}(x^2 + 8)^{3/2}\Big|_{0}^{1}$

$= \frac{1}{3}[24\sqrt{3} - 32\sqrt{2} + 27] \approx 7.77146$

34. The curves intersect at $x = \frac{\pi}{4}$.

$A = \displaystyle\int_{0}^{\pi/4} (\cos x - \sin x)\, dx + \int_{\pi/4}^{1} (\sin x - \cos x)\, dx$

$= (\cos x + \cos x)\Big|_{0}^{\pi/4} - (\sin x + \cos x)\Big|_{\pi/4}^{1}$

$= 2\sqrt{2} - 1 - \cos 1 - \sin 1 \approx 0.4466$

35. $y' = \dfrac{x \cos x - \sin x}{x^2};$ at $x = \frac{\pi}{4},$

$y' = \dfrac{2\sqrt{2}(\pi - 4)}{\pi^2} \approx -0.246$

36. $\dfrac{dy}{dx} = x^2 y^2 \sqrt{4 - x^3}$

$\displaystyle\int y^{-2}\, dy = \int x^2 (4 - x^3)^{1/2}\, dx$

$-y^{-1} = -\frac{2}{9}[(4 - x^3)^{3/2} + C]$

$y = \dfrac{9}{2(4 - x^3)^{3/2} + C}$

37. $y\, dx - x\, dx = x \ln x\, dx$

$\dfrac{y\, dx - x\, dx}{x^2} = \dfrac{\ln x}{x}\, dx$ *Divide by x^2.*

$\displaystyle\int \dfrac{y\, dx - x\, dx}{x^2} = \int \dfrac{\ln x}{x}\, dx$

$-\dfrac{y}{x} = \frac{1}{2}(\ln x)^2 + C_1$

$y = -\frac{1}{2} x \ln^2 x + Cx$

38. $\dfrac{dy}{dx} = 2(5 - y)$

$\displaystyle\int \dfrac{dy}{5 - y} = \int 2\, dx$

$-\ln|5 - y| = 2x + C$

Since $y = 3$ when $x = 0$, $C = -\ln 2$

$-\ln|5 - y| = 2x - \ln 2$

$\ln \dfrac{2}{|5 - y|} = 2x$

$e^{2x} = \dfrac{2}{|5 - y|}$

$|5 - y| = 2e^{-2x}$

$y = 5 \pm e^{-2x}$

39. $\dfrac{dy}{dx} = e^y \sin x$

$\displaystyle\int e^{-y}\, dy = \int \sin x\, dx$

$-e^{-y} = -\cos x - C$

$$e^{-y} = \cos x + C$$

Since $y = 5$ when $x = 0$, $C = e^{-5} - 1$; Thus,

$$e^{-y} = e^{-5} - 1 + \cos y$$
$$y = -\ln\left|\cos y + e^{-5} - 1\right|$$

40.
$$V = \pi \int_1^2 [2(3x - 2)^{-1/2}]^2 \, dx$$

$$= \pi \int_1^2 4u^{-1} \frac{du}{3} = \frac{4\pi}{3} \ln|u| \Big|_1^2 = \frac{4\pi}{3} \ln 3$$

41. $f'(x) = \frac{1}{2}(e^{2x} - 1)^{-1/2} e^{2x}(2)$

$$ - e^{-x}(e^{2x} - 1)^{-1/2} e^x$$
$$= (e^{2x} - 1)^{-1/2}(e^{2x} - 1)$$
$$= (e^{2x} - 1)^{1/2}$$

$$s = \int_0^1 \sqrt{1 + (e^{2x} - 1)} \, dx = \int_0^1 e^x \, dx$$

$$= e - 1$$

42. The element of rope dx is lifted through a distance $50 - x$, so $dF_r = 0.25 \, dx$ and

$$W = \int_0^{50} 0.25(50 - x) \, dx = -\frac{0.25}{2}(50 - x)^2 \Big|_0^{50}$$

$$= 312.5 \text{ ft} \cdot \text{lb}$$

For the bucket, $W_b = 50(25) = 1{,}250$ ft · lb. Thus, $W = W_b + W_r = 1{,}562.5$ ft · lb.

43. Consider a triangle with legs 5,000 and y, where y is the altitude of the rocket. Use θ for the angle of elevation, opposite the y leg.

$$y = 5{,}000 \tan \theta$$
$$\frac{dy}{dt} = 5{,}000 \sec^2\theta \, \frac{d\theta}{dt}$$
$$\frac{d\theta}{dt} = \frac{1}{5{,}000}\cos^2\theta \, \frac{dy}{dt}$$

Given $dy/dt = 850$; if $y = 4{,}000$, $\cos^2\theta = 25/41$:

$$\frac{d\theta}{dt} = \frac{1}{5{,}000}\left(\frac{25}{41}\right)(850) = \frac{17}{164} \approx 0.1 \text{ rad/s}$$

44. Consider a right triangle with legs 4,000 and x, where x is the beam travels on the shore. Use θ for the angle opposite the y leg. Then

$$y = 4{,}000 \tan \theta$$

$$\frac{dy}{dt} = 4{,}000 \sec^2\theta \, \frac{d\theta}{dt}$$
$$\frac{d\theta}{dt} = \frac{1}{4{,}000} \cos^2\theta \, \frac{dy}{dt}$$

Given $dy/dt = 5$; if $y = 1{,}000$, then $\cos^2\theta = 17/16$:

$$\frac{d\theta}{dt} = \frac{1}{4{,}000}\left(\frac{17}{16}\right)(5) = \frac{17}{64{,}000}$$

$$\approx 0.000265625 \text{ rad/s or}$$

45. Let $S(t)$ be the salt content of the brine at time t. The inflow is $(3)(2) = 6$ lb of salt per minute, while the outflow is

$$\frac{S(t)}{50 + (2 - 1)t}$$

since 2 gallons flow in and 1 flows out.

$$\frac{dS}{dt} = 6 - \frac{S}{50 + t}$$

$$\frac{dS}{dt} + \frac{S}{50 + t} = 6$$

$$I(t) = e^{\int dt/(50+t)} = 50 + t, \text{ so that}$$

$$S = \frac{1}{50 + t}\left[\int 6(50 + t) \, dt + C\right]$$

$$= \frac{1}{50 + t}\left[3t(t + 100) + C\right]$$

Since $S = 0$ when $t = 0$ we see that $C = 0$ so that

$$S = \frac{3t(t + 100)}{50 + t}$$

When $t = 15$, $S = \dfrac{1{,}035}{13} \approx 79.6$ lb

CHAPTER 7

Methods of Integration

7.1 Review of Substitution and Integration by Table, Page 446

1. $\int \dfrac{2x - 1}{(4x^2 - 4x)^2}\, dx$ $\boxed{u = 2x - 1}$

$= \dfrac{1}{16}\int (x^2 - x)^{-2}[(2x - 1)\, dx]$

$= -\dfrac{1}{16}(x^2 - x)^{-1} + C$

2. $\int \dfrac{2x + 5}{\sqrt{x^2 + 5x}}\, dx$ $\boxed{u = x^2 + 5x}$

$= \int (x^2 + 5x)^{-1/2}[(2x + 5)\, dx]$

$= 2(x^2 + 5x)^{1/2} + C$

3. $\int \dfrac{\ln x}{x}\, dx = \dfrac{1}{2}(\ln x)^2 + C$ $\boxed{u = \ln x}$

4. $\int \dfrac{\ln(x + 1)}{x + 1}\, dx$ $\boxed{u = \ln(x + 1)}$

$= \int \ln(x + 1)\, \dfrac{dx}{x + 1}$

$= \dfrac{1}{2}\ln^2(x + 1) + C$

5. $\int \cos x\, e^{\sin x}\, dx$ $\boxed{u = \sin x}$

$= \int e^{\sin x}(\cos x\, dx) = e^{\sin x} + C$

6. $\int \dfrac{x\, dx}{4 + x^4}$ $\boxed{u = x^2}$

$= \dfrac{1}{2}\int \dfrac{2x\, dx}{4 + (x^2)^2} = (\dfrac{1}{2})^2\tan^{-1}\dfrac{1}{2}x^2 + C$

$= \dfrac{1}{4}\tan^{-1}\dfrac{x^2}{2} + C$

7. $\int \dfrac{t^2\, dt}{9 + t^6}$ $\boxed{u = t^3}$

$= \dfrac{1}{3}\int \dfrac{3t^2\, dt}{9 + (t^3)^2} = (\dfrac{1}{3})^2\tan^{-1}\dfrac{1}{3}t^3 + C$

$= \dfrac{1}{9}\tan^{-1}\dfrac{t^3}{3} + C$

8. $\int (1 + \cot x)^4\, \csc^2 x\, dx$ $\boxed{u = 1 + \cot x}$

$= -\int (1 + \cot x)^4(-\csc^2 x\, dx)$

$= -\dfrac{1}{5}(1 + \cot x)^5 + C$

9. $\int \dfrac{4x^3 - 4x}{x^4 - 2x^2 + 3}\, dx$ $\boxed{u = x^4 - 2x^2 + 3}$

$= \int (x^4 - 2x^2 + 3)^{-1}[(4x^3 - 4x)\, dx]$

$= \ln(x^4 - 2x^2 + 3) + C$

10. $\int \dfrac{x^3 - x}{(x^4 - 2x^2 + 3)^2}\, dx$ $\boxed{u = x^4 - 2x^2 + 3}$

$= \int (x^4 - 2x^2 + 3)^{-2}[(4x^3 - 4x)\, dx]$

$= -\dfrac{1}{4}(x^4 - 2x^2 + 3)^{-1} + C$

11. $\int \dfrac{2x + 4}{x^2 + 4x + 3}\, dx$ $\boxed{u = x^2 + 4x + 3}$

$= \int (x^2 + 4x + 3)^{-1}[(2x + 4)\, dx]$

$= \ln|x^2 + 4x + 3| + C$

12. $\int \dfrac{2x + 1}{x^2 + x + 1}\, dx$ $\boxed{u = x^2 + x + 1}$

$= \int (x^2 + 4x + 1)^{-1}[(2x + 1)\, dx]$

$= \ln(x^2 + x + 1) + C$

13. $\int \dfrac{dx}{x^2\sqrt{x^2 - a^2}}$ $\boxed{\text{Formula 201}}$

$= \dfrac{\sqrt{x^2 - a^2}}{a^2 x} + C$

14. $\int \dfrac{dx}{x^2\sqrt{a^2 - x^2}}$ $\boxed{\text{Formula 229}}$

$= -\dfrac{\sqrt{a^2 - x^2}}{a^2 x} + C$

15. $\int x \ln x \, dx$ Formula 502

$= \frac{1}{2}x^2(\ln x - \frac{1}{2}) + C$

16. $\int \ln x \, dx$ Formula 499

$= x \ln x - x + C$

17. $\int xe^{ax} \, dx$ Formula 484

$= a^{-1}e^{ax}(x - a^{-1}) + C$

18. $\int \frac{dx}{a + be^{2x}}$ Formula 489

$= a^{-1}x - (2a)^{-1}\ln|a + be^{2x}| + C$

19. $\int \frac{x^2 \, dx}{\sqrt{x^2 + 1}}$ Formula 174

$= \frac{1}{2}x\sqrt{x^2 + 1} - \frac{1}{2}\ln\left|x + \sqrt{x^2 + 1}\right| + C$

20. $\int \frac{dx}{x^2\sqrt{x^2 + 16}}$ Formula 177

$= -\frac{\sqrt{x^2 + 16}}{16x} + C$

21. $\int \frac{x \, dx}{\sqrt{4x^2 + 1}}$ Formula 173; $u = 2x$

$= \frac{1}{4}\sqrt{4x^2 + 1} + C$

22. $\int \frac{dx}{x\sqrt{1 - 9x^2}}$ Formula 228; $u = 3x$

$= -\ln\left|\frac{1 + \sqrt{1 - 9x^2}}{3x}\right| + C$

23. $\int e^{-4x}\sin 5x \, dx$ Formula 492

$= (16 + 25)^{-1}e^{-4x}(-4\sin 5x - 5\cos 5x) + C$

$= \frac{-4\sin 5x - 5\cos 5x}{41e^{4x}} + C$

24. $\int x \sin^{-1}x \, dx$ Formula 452

$= (\frac{1}{2}x^2 - \frac{1}{4})\sin^{-1}x - \frac{1}{4}x\sqrt{1 - x^2} + C$

25. $\int (1 + bx)^{-1} \, dx$ Formula 24

$= b^{-1}\ln|1 + bx| + C$

26. $\int \frac{x \, dx}{\sqrt{a^2 - x^2}}$ Formula 225

$= -\sqrt{a^2 - x^2} + C$

27. $\int x(1 + x)^3 \, dx$ Formula 31

$= \frac{(x + 1)^5}{5} - \frac{(x + 1)^4}{4} + C$

$= \frac{1}{20}(x + 1)^4(4x - 1) + C$

28. $\int x\sqrt{1 + x} \, dx$ Formula 141

$= \frac{2}{15}(x - 2)\sqrt{(x + 1)^3} + C$

29. $\int xe^{4x} \, dx$ Formula 484

$= \frac{1}{4}e^{4x}(x - \frac{1}{4}) + C$

30. $\int x \ln 2x \, dx$ Formula 502; $u = 2x$

$= \frac{1}{2}(2x)^2\left(\ln|2x| - \frac{1}{2}\right) + C$

$= 2x^2 \ln 2x - x^2 + C$

31. $\int \frac{dx}{\sqrt{5 - 4x - x^2}}$ Formula 257

$= -\sin^{-1}\left(\frac{-2x - 4}{\sqrt{36}} + C\right)$

$= \sin^{-1}\left(\frac{x + 2}{3}\right) + C$

32. $\int \frac{dx}{1 + e^{2x}}$ Formula 489

$= x - \frac{1}{2}\ln\left|1 + e^{2x}\right| + C$

33. $\int \ln^3 x \, dx$ Formula 500

$= x(\ln x)^3 - 3\int (\ln x)^2 \, dx$

                Formula 501

$= x(\ln x)^3 - 3[x(\ln x)^2 - 2x \ln x + 2x] + C$

$= x \ln^3 x - 3x \ln^2 x + 6x \ln x - 6x + C$

34. $\int \frac{x^3 \, dx}{\sqrt{4x^4 + 1}}$ Formula 30; $u = x^4$

$$= \tfrac{1}{12}(4x^4 + 1)^{3/2} + C$$

Note: this can more easily be worked by letting $u = 4x^4 + 1$ and no integral table.

35. $\displaystyle\int \frac{dx}{\sqrt{9 - x^2}}$ $\boxed{\text{Formula 22}}$

$$= \sin^{-1}\tfrac{x}{3} + C$$

By trigonometric substitution: let $x = 3\sin\theta$. Then $dx = 3\cos\theta\,d\theta$, and

$$\sqrt{9 - x^2} = \sqrt{9 - 9\sin^2\theta} = 3\cos\theta;$$

$$\int \frac{dx}{\sqrt{9 - x^2}} = \int \frac{3\cos\theta}{3\cos\theta}\,d\theta$$

$$= \theta + C = \sin^{-1}\left(\tfrac{x}{3}\right) + C$$

36. $\displaystyle\int \frac{x + 1}{\sqrt{4 + x^2}}\,dx$ $\boxed{\text{Formula 172}}$

$$= \tfrac{1}{2}\int \frac{2x\,dx}{\sqrt{4 + x^2}} + \int \frac{dx}{\sqrt{4 + x^2}}$$

$$= \sqrt{4 + x^2} + \ln\left|x + \sqrt{4 + x^2}\right| + C$$

By trigonometric substitution: let $x = 2\tan\theta$. Then $dx = 2\sec^2\theta\,d\theta$, and

$$\sqrt{4 + x^2} = \sqrt{4 + 4\tan^2\theta} = 2\sec\theta;$$

$$\int \frac{x + 1}{\sqrt{4 + x^2}}\,dx = \int \frac{2\tan\theta + 1}{2\sec\theta}(2\sec^2\theta\,d\theta)$$

$$= 2\int \tan\theta\sec\theta\,d\theta + \int \sec\theta\,d\theta$$

$$= 2\sec\theta + \ln|\sec\theta + \tan\theta| + C_1$$

$$= \sqrt{4 + x^2} + \ln\left|\tfrac{1}{2}\sqrt{4 + x^2} + \tfrac{x}{2}\right| + C_1$$

$$= \sqrt{4 + x^2} + \ln\left|\sqrt{4 + x^2} + x\right| - \ln 2 + C_1$$

$$= \sqrt{4 + x^2} + \ln\left|\sqrt{4 + x^2} + x\right| + C$$

37. $\displaystyle\int (9 + x^2)^{1/2}\,dx$ $\boxed{\text{Formula 168}}$

$$= \tfrac{1}{2}[x\sqrt{9 + x^2} + 9\ln(x + \sqrt{9 + x^2})] + C$$

By trigonometric substitution: let $x = 3\tan\theta$. Then $dx = 3\sec^2\theta\,d\theta$, and

$$\sqrt{9 + x^2} = \sqrt{9 + 9\tan^2\theta} = 3\sec\theta;$$

$$\int \sqrt{9 + x^2}\,dx = \int 3\sec\theta(3\sec^2\theta\,d\theta)$$

$$= 9\int \sec^3\theta\,d\theta \quad \boxed{\text{Formula 427}}$$

$$= 9\left[\frac{\sec\theta\tan\theta}{2} + \tfrac{1}{2}\ln|\sec\theta + \tan\theta|\right] + C_1$$

$$= 9\left[\frac{x\sqrt{9 + x^2}}{2(9)}\right] + \tfrac{9}{2}\ln\left|\tfrac{1}{3}\sqrt{9 + x^2} + \tfrac{x}{3}\right| + C_1$$

$$= \tfrac{1}{2}x\sqrt{9 + x^2} + \tfrac{9}{2}\ln(\sqrt{9 + x^2} + x) + \tfrac{9}{2}\ln 3 + C_1$$

$$= \tfrac{1}{2}[x\sqrt{9 + x^2} + 9\ln(x + \sqrt{9 + x^2})] + C$$

38. $\displaystyle\int \frac{dx}{\sqrt{x^2 - 7}}$ $\boxed{\text{Formula 196}}$

$$= \ln\left|x + \sqrt{x^2 - 7}\right| + C$$

By trigonometric substitution: let $x = \sqrt{7}\csc\theta$. Then $dx = -\sqrt{7}\csc\theta\cot\theta\,d\theta$, and

$$\sqrt{x^2 - 7} = \sqrt{7\csc^2\theta - 7} = \sqrt{7}\cot\theta;$$

$$\int \frac{dx}{\sqrt{x^2 - 7}} = \int \frac{-\sqrt{7}\csc\theta\cot\theta\,d\theta}{\sqrt{7}\cot\theta}$$

$$= -\int \csc\theta\,d\theta = \ln|\csc\theta + \cot\theta| + C_1$$

$$= \ln\left|\frac{x}{\sqrt{7}} + \frac{\sqrt{x^2 - 7}}{\sqrt{7}}\right| + C_1$$

$$= \ln\left|x + \sqrt{x^2 - 7}\right| + \tfrac{1}{2}\ln 7 + C_1$$

$$= \ln\left|x + \sqrt{x^2 - 7}\right| + C$$

39. $\displaystyle\int \frac{\sqrt{4x^2 + 1}}{x}\,dx$ $\boxed{\text{Formula 179; } u = 2x}$

$$= \sqrt{4x^2 + 1} - \ln\left|\frac{1 + \sqrt{4x^2 + 1}}{2x}\right| + C$$

40. $\displaystyle\int \sec^3\left(\tfrac{x}{2}\right)\,dx$ $\boxed{\text{Formula 427}}$

$$= \sec\tfrac{x}{2}\tan\tfrac{x}{2} + \ln\left|\sec\tfrac{x}{2} + \tan\tfrac{x}{2}\right| + C$$

41. $\displaystyle\int \sin^6 x\,dx$ $\boxed{\text{Formula 352}}$

$$= -\tfrac{1}{6}\sin^5 x\cos x + \tfrac{5}{6}\int \sin^4 x\,dx$$

$$= -\tfrac{1}{6}\sin^5 x\cos x$$

$$\quad + \tfrac{5}{6}[-\tfrac{1}{4}\sin^3 x\cos x + \tfrac{3}{4}\int \sin^2 x\,dx]$$

$$= -\tfrac{1}{6}\sin^5 x\cos x - \tfrac{5}{24}\sin^3 x\cos x + \tfrac{5}{8}\int \sin^2 x\,dx$$

$$= -\tfrac{1}{6}\sin^5 x\cos x - \tfrac{5}{24}\sin^3 x\cos x + \tfrac{5}{16}x$$

$$\qquad - \tfrac{5}{32}\sin 2x + C$$

42. $\displaystyle\int \frac{dx}{9x^2 + 6x + 1} = \int \frac{dx}{(3x+1)^2}$ $\boxed{\text{Formula 41}}$

$\displaystyle = -\frac{1}{3(3x+1)} + C$

43. $\displaystyle\int (9 - x^2)^{3/2}\, dx$ $\boxed{\text{Formula 245}}$

$\displaystyle = \frac{x(9-x^2)^{3/2}}{4} + \frac{27x(9-x^2)^{1/2}}{8}$

$\displaystyle \qquad + \frac{243}{8} \sin^{-1}\!\left(\frac{x}{3}\right) + C$

44. $\displaystyle\int \frac{\sin^2 x}{\cos x}\, dx$ $\boxed{\text{Formula 382}}$

$\displaystyle = -\sin x + \ln\left|\tan\!\left(\frac{x}{2} + \frac{\pi}{4}\right)\right| + C$

45. $\displaystyle\int \sin^2 x\, dx = \frac{1}{2}\int (1 - \cos 2x)\, dx$

$\displaystyle = \frac{1}{2}\left(x - \frac{1}{2}\sin 2x\right) + C$

$\displaystyle = \frac{x}{2} - \frac{1}{4}\sin 2x + C$

46. $\displaystyle\int \cos^2 x\, dx = \frac{1}{2}\int (1 + \cos 2x)\, dx$

$\displaystyle = \frac{1}{2}\left(x + \frac{1}{2}\sin 2x\right) + C$

$\displaystyle = \frac{x}{2} + \frac{1}{4}\sin 2x + C$

47. Let $u = \sin x$, $du = \cos x\, dx$;

$\displaystyle\int \sin^4 x \cos x\, dx = \frac{u^5}{5} + C = \frac{1}{5}\sin^5 x + C$

48. Let $u = \cos x$, $du = -\sin x\, dx$;

$\displaystyle\int \sin^3 x \cos^4 x\, dx = \int \sin^2 x \cos^4 x\, (\sin x\, dx)$

$\displaystyle = \int (1 - \cos^2 x)\cos^4 x\, (\sin x\, dx)$

$\displaystyle = \int \cos^4 x\, (\sin x\, dx) - \int \cos^6 x\, (\sin x\, dx)$

$\displaystyle = -\int u^4\, du + \int u^6\, du = -\frac{u^5}{5} + \frac{u^7}{7} + C$

$\displaystyle = -\frac{1}{5}\cos^5 x + \frac{1}{7}\cos^7 x + C$

49. $\displaystyle\int \sin^2 x \cos^2 x\, dx = \frac{1}{4}\int (1 - \cos 2x)(1 + \cos 2x)\, dx$

$\displaystyle = \frac{1}{4}\int (1 - \cos^2 2x)\, dx = \frac{1}{4}\int \sin^2 2x\, dx$

$\displaystyle = \frac{1}{4}\left(\frac{1}{2}x - \frac{1}{8}\sin 4x\right) + C$

$\displaystyle = \frac{x}{8} - \frac{1}{32}\sin 4x + C$

50. If m is odd, let $u = \cos x$. If n is odd, let $u = \sin x$. If both m and n are even, use the

identities shown in Problems 45 and 46 until one exponent is odd.

51. Draw a right triangle; if the integral involves $\sqrt{a^2 - u^2}$ let $x = a \sin \theta$.

52. Draw a right triangle; if the integral involves $\sqrt{a^2 + u^2}$ let $x = a \tan \theta$.

53. Draw a right triangle; if the integral involves $\sqrt{u^2 - a^2}$ let $x = a \sec \theta$.

54. $\displaystyle\int \frac{e^x\, dx}{1 + e^{x/2}}$ $\boxed{\text{Let } u = e^{x/2};\ du = \tfrac{1}{2}e^{x/2}\, dx}$

$\displaystyle = 2\int \frac{u\, du}{1 + u} = 2\big[u - \ln|1 + u|\big] + C$
$\boxed{\text{Formula 35}}$

$\displaystyle = 2e^{x/2} - 2\ln(1 + e^{x/2}) + C$

55. $\displaystyle\int \frac{dx}{x^{1/2} + x^{1/4}}$ $\boxed{\text{Let } u = x^{1/4};\ du = \tfrac{1}{4}x^{-3/4}dx}$

$\displaystyle = \int \frac{4u^3\, du}{u^2 + u} = 4\int \frac{u^2\, du}{u + 1}$ $\boxed{\text{Formula 36}}$

$\displaystyle = 4\left[\frac{(u+1)^2}{2} - \frac{2(u+1)}{1} + \ln|u+1|\right] + C_1$

$\displaystyle = 4\left[\tfrac{1}{2}u^2 + u + \tfrac{1}{2} - 2u - 2 + \ln|u+1|\right] + C_1$

$\displaystyle = 4\left[\tfrac{1}{2}u^2 - u + \ln|u+1|\right] + C$

$\displaystyle = 4\left[\frac{x^{1/2}}{2} - x^{1/4} + \ln(x^{1/4} + 1)\right] + C$

56. $\displaystyle\int \frac{18\tan^2 t \sec^2 t}{(2 + \tan^3 t)^2}\, dt$

$\boxed{\text{Let } u = 2 + \tan^3 t;\ du = 3\tan^2 t \sec^2 t\, dt}$

$\displaystyle = \int \frac{6\, du}{u^2} = -6u^{-1} + C$

$\displaystyle = -6(2 + \tan^3 t)^{-1} + C$

57. $\displaystyle\int \frac{4\, dx}{x^{1/3} + 2x^{1/2}}$ $\boxed{\text{Let } x = u^6;\ dx = 6u^5\, du}$

$\displaystyle = \int \frac{4(6u^5\, du)}{u^2 + 2u^3} = 24\int \frac{u^5\, du}{u^2 + 2u^3}$

$\displaystyle = 24\int \frac{u^3\, du}{2u + 1}$

We could use Formula 44, but since the degree of the numerator is greater than the degree of the denominator, do long division to obtain:

$$24 \int \left(\frac{1}{2}u^2 - \frac{1}{4}u + \frac{1}{8} - \frac{\frac{1}{8}}{2u + 1} \right) du$$

$$= 24 \left(\frac{u^3}{6} - \frac{u^2}{8} + \frac{u}{8} - \frac{1}{16} \ln|2u + 1| \right) + C$$

$$= 4x^{1/2} - 3x^{1/3} + 3x^{1/6} - \frac{3}{2} \ln(2x^{1/6} + 1) + C$$

58. $\displaystyle \int \frac{dx}{(x + \frac{1}{2})\sqrt{4x^2 + 4x}} = \frac{1}{2} \int \frac{dx}{(x + \frac{1}{2})\sqrt{x^3 + x}}$

$$= \frac{1}{2} \int \frac{dx}{(x + \frac{1}{2})\sqrt{(x + \frac{1}{2})^2 - \frac{1}{4}}} = \frac{1}{2} \int \frac{du}{u\sqrt{u^2 - (\frac{1}{2})^2}}$$

Let $u = x + \frac{1}{2}$; $du = dx$; then use Formula 200

$$= \frac{1}{\frac{1}{2}} \sec^{-1} \left| \frac{u}{\frac{1}{2}} \right| + C = 2 \sec^{-1}(2x + 1) + C$$

59. $\displaystyle \int \frac{e^{-x} - e^x}{e^{2x} + e^{-2x} + 2} \, dx = \int \frac{e^{-x} - e^x}{(e^x + e^{-x})^2 + 1} \, dx$

Let $u = e^x + e^{-x}$; $du = e^x - e^{-x} dx$

$$= \int \frac{(-1)du}{u^2} = \frac{1}{u} + C = \frac{1}{e^x + e^{-x}} + C$$

$$= -\tan^{-1}(e^x + e^{-x}) + C$$

60. $\displaystyle A = \int_0^4 2x(x^2 + 9)^{-1/2} \, dx$ Let $u = x^2 + 9$

$$= 2(x^2 + 9)^{1/2} \Big|_0^4 = 4$$

61. $\displaystyle A = \int_0^{\pi/4} \cos 2x \, dx = \frac{1}{2} \sin 2x \Big|_0^{\pi/4} = \frac{1}{2}$

62. $\displaystyle V = \pi \int_0^1 x^2(1 - x^2)^{1/2} \, dx$

You could use Formula 152; instead, let $x = \sin\theta$, $dx = \cos\theta \, d\theta$ and $\sqrt{1 - x^2} = \cos\theta$; if $x = 0$, then $\theta = 0$ and if $x = 1$, $\theta = \pi/2$:

$$V = \pi \int_0^{\pi/2} \sin^2\theta \cos\theta(\cos\theta \, d\theta)$$

$$= \pi \int_0^{\pi/2} \sin^2\theta \, 2\theta \, d\theta$$

$$= \frac{\pi}{8} \int_0^{\pi/2} (1 - \cos 4\theta) \, d\theta = \frac{\pi}{8} \theta \Big|_0^{\pi/2} = \frac{\pi^2}{16}$$

63. $\displaystyle V = \pi \int_0^9 \left(\frac{x^{3/2}}{\sqrt{x^2 + 9}} \right)^2 \, dx = \pi \int_0^9 \frac{x^3}{x^2 + 9} \, dx$

Let $u = x^2 + 9$; $du = 2x \, dx$

$$= \frac{\pi}{2} \int_9^{90} \frac{u - 9}{u} \, du = \frac{\pi}{2}[u - 9 \ln u]\Big|_9^{90}$$

$$= \frac{81\pi}{2} - \frac{9\pi}{2} \ln 10 \approx 94.68 \text{ cubic units}$$

64. $\displaystyle V = \pi \int_1^2 (\sqrt[4]{4 - y^2})^2 \, dy = \pi \int_1^2 (4 - y^2)^{1/2} \, dy$

Formula 231

$$= \left[\frac{x(4 - y^2)^{1/2}}{2} + \frac{1}{2} \sin^{-1} y \right]\Big|_1^2$$

$$= \frac{\pi}{6}(4\pi - 3\sqrt{3}) \approx 3.86$$

65. $\displaystyle s = \int_0^1 \sqrt{1 + (2x)^2} \, dx$ Formula 168; $u = 2x$

$$= \frac{1}{2} \left(\frac{2x\sqrt{1 + (2x)^2}}{2} + \frac{1}{2} \ln\left| 2x + \sqrt{1 + (2x)^2} \right| \right)\Big|_0^1$$

$$= \frac{\sqrt{5}}{2} + \frac{1}{4} \ln(2 + \sqrt{5}) \approx 1.4789$$

66. $\displaystyle dy = \frac{\sin x}{\cos x} \, dx = \tan x \, dx$;

$$ds = \sqrt{1 + \tan^2 x} \, dx = \sec x \, dx$$

$$s = \int_0^{\pi/4} \sec x \, dx = \ln|\sec x + \tan x|\Big|_0^{\pi/4}$$

$$= \ln(1 + \sqrt{2})$$

67. $\displaystyle ds = \sqrt{1 + 4x^2} \, dx$; $dS = 2\pi y \, ds$

$$S = 2\pi \int_0^1 x^2\sqrt{1 + 4x^2} \, dx = 2\pi \int_0^2 \frac{u^2}{4}\sqrt{1 + u^2} \, (2 \, dx)$$

Formula 170; $u = 2x$

$$= \pi \left(\frac{u(u^2 + 1)^{3/2}}{4} - \frac{u(u^2 + 1)^{1/2}}{8} \right.$$

$$\left. - \frac{1}{8} \ln\left| u + (u^2 + a^2)^{1/2} \right| \right)\Big|_0^1$$

$$= \frac{9\sqrt{5}\pi}{16} - \frac{\pi \ln(\sqrt{5} + 2)}{32} \approx 3.81$$

68. $\displaystyle dy = 2x \, dx$; $ds = \sqrt{1 + 4x^2} \, dx$; $dS = 2\pi x \, ds$

$$S = 2\pi \int_0^1 x\sqrt{1 + 4x^2} \, dx \qquad \boxed{\text{Let } u = 1 + 4x^2}$$

$$= \frac{\pi}{4}\left(\frac{2}{3}\right)(1 + 4x^2)^{3/2}\Big|_0^1 = \frac{\pi}{6}(5\sqrt{5} - 1)$$

69. $\displaystyle\int \csc x \, dx = -\int \frac{-\csc x(\csc x + \cot x)}{\csc x + \cot x} \, dx$

$$= -\ln|\csc x + \cot x| + C$$

70. a. $\displaystyle\int 2 \sin x \cos x \, dx = -2\int \cos x(-\sin x \, dx)$

$$= -2(\tfrac{1}{2})\cos^2 x + C_1 = -\cos^2 x + C_1$$

b. $\displaystyle\int 2 \sin x \cos x \, dx = 2\int \sin x(\cos x \, dx)$

$$= 2(\tfrac{1}{2})\sin^2 x + C_2 = \sin^2 x + C_2$$

c. $\displaystyle\int 2 \sin x \cos x \, dx = \int \sin 2x \, dx$

$$= -\tfrac{1}{2}\cos 2x + C_3$$

d. All three are the same with $C_1 = C_2 - 1$, $C_2 = C_3 - \frac{1}{2}$, and $C_1 = C_3 - \frac{3}{2}$.

71. Let $u = \pi - x$, $du = -dx$.

$$\int_0^\pi x f(\sin x) \, dx = \int_\pi^0 (\pi - u)f(\sin u)(-du)$$

$$= \int_0^\pi (\pi - u) \, f(\sin u) \, du$$

$$= \int_0^\pi \pi \, f(\sin u) \, du - \int_0^\pi u \, f(\sin u) \, du$$

The value of the integral is independent of the variable used so

$$= \pi \int_0^\pi f(\sin x) \, dx - \int_0^\pi x \, f(\sin x) \, dx$$

Add the second integral to both sides and divide by 2 to obtain the desired result.

72. If $m = n$, $\displaystyle\int_{-\pi}^\pi \sin nx \sin mx \, dx = \int_{-\pi}^\pi \sin^2 nx \, dx$

$$= \frac{1}{2}\int_{-\pi}^\pi (1 - \cos 2nx) \, dx = \pi.$$

If $m \neq n$, then subtract

$$\cos(m + n)x = \cos mx \cos nx - \sin mx \sin nx$$

from

$$\cos(m - n)x = \cos mx \cos nx - \sin mx \sin nx$$

to obtain

$$\sin mx \sin nx = \tfrac{1}{2}[\cos(m - n)x - \cos(m + n)x]$$

Thus,

$$\int_{-\pi}^\pi \sin mx \sin nx \, dx$$

$$= \int_{-\pi}^\pi \tfrac{1}{2}[\cos(m - n)x - \cos(m + n)x] \, dx$$

$$= \tfrac{1}{2}[(m - n)^{-1}\sin(m - n)x$$

$$- (m+n)^{-1}\sin(m+n)x]\Big|_{-\pi}^\pi = 0$$

73. If $m = n$, $\displaystyle\int_{-\pi}^\pi \cos nx \cos mx \, dx = \int_{-\pi}^\pi \cos^2 nx \, dx$

$$= \frac{1}{2}\int_{-\pi}^\pi (1 + \cos 2nx) \, dx = \pi.$$

If $m \neq n$, then add

$$\cos(m + n)x = \cos mx \cos nx - \sin mx \sin nx$$

to

$$\cos(m - n)x = \cos mx \cos nx - \sin mx \sin nx$$

to obtain

$$\cos mx \cos nx = \tfrac{1}{2}[\cos(m - n)x + \cos(m + n)x]$$

Thus,

$$\int_{-\pi}^\pi \cos mx \cos nx \, dx$$

$$= \int_{-\pi}^\pi \tfrac{1}{2}[\cos(m - n)x + \cos(m + n)x] \, dx$$

$$= \tfrac{1}{2}[(m - n)^{-1}\sin(m - n)x$$

$$+ (m+n)^{-1}\sin(m+n)x]\Big|_{-\pi}^\pi = 0$$

74. For the right semicircle

$$x = \sqrt{1 - (y - b)^2} \, dy = -\frac{(y - b) \, dy}{\sqrt{1 - (y - b)^2}}$$

$$ds = \sqrt{1 + \frac{(y - b) \, dy}{\sqrt{1 - (y - b)^2}}} \, dy$$

$$= \frac{dy}{\sqrt{1 - (y - b)^2}}$$

$$dS = 2\pi y \frac{dy}{\sqrt{1 - (y - b)^2}}$$

$$S = 2\pi \int_{b-1}^{b+1} \{(y - b)[1 - (y - b)^2]^{-1/2}$$
$$+ b[1 - (y - b)^2]^{-1/2}\} \, dy$$

7.2 Integration by Parts, Page 453

1. $\int xe^{-2x} \, dx \quad \boxed{u = x; \; dv = e^{-2x} \, dx}$

$= -\frac{1}{2}xe^{-2x} + \frac{1}{2}\int e^{-2x} \, dx$

$= -\frac{1}{2}xe^{-2x} - \frac{1}{4}e^{-2x} + C$

2. $\int x \sin x \, dx \quad \boxed{u = x; \; dv = \sin x \, dx}$

$= -x \cos x + \int \cos x \, dx$

$= -x \cos x + \sin x + C$

3. $\int x \ln x \, dx \quad \boxed{u = \ln x; \; dv = x \, dx}$

$= \frac{1}{2}x^2 \ln x - \frac{1}{2}\int x \, dx$

$= \frac{1}{2}x^2 \ln x - \frac{1}{4}x^2 + C$

4. $\int x \tan^{-1}x \, dx \quad \boxed{u = \tan^{-1}x; \; dv = x \, dx}$

$= \frac{1}{2}x^2 \tan^{-1}x - \frac{1}{2}\int \frac{x^2 \, dx}{x^2 + 1} \quad \boxed{\text{Formula 57}}$

$= \frac{1}{2}x^2 \tan^{-1}x - \frac{1}{2}[x - \tan^{-1}x] + C$

5. $\int \sin^{-1}x \, dx \quad \boxed{u = \sin^{-1}x; \; dv = dx}$

$= x \sin^{-1}x - \int \frac{x \, dx}{\sqrt{1 - x^2}} \quad \boxed{t = 1 - x^2}$

$= x \sin^{-1}x + \sqrt{1 - x^2} + C$

6. $\int x^2 \sin x \, dx \quad \boxed{u = x^2; \; dv = \sin x \, dx}$

$= -x^2 \cos x + 2\int x \cos x \, dx \quad \boxed{u = x; \; dv = \cos x}$

$= -x^2 \cos x + 2\left(x \sin x - \int \sin x \, dx\right)$

$= -x^2 \cos x + 2x \sin x + 2 \cos x + C$

7. $\int \frac{\ln\sqrt{x}}{\sqrt{x}} \, dx \quad \boxed{t = \sqrt{x}; \; dt = dx/(2\sqrt{x})}$

$= 2\int \ln t \, dt = 2(t \ln t - t) + C$

$= \sqrt{x} \ln \sqrt{x} - 2\sqrt{x} + C$

8. $I = \int e^{2x} \sin 3x \, dx \quad \boxed{u = e^{2x}; \; dv = \sin 3x}$

$= -\frac{1}{3}e^{2x} \cos 3x + \frac{2}{3}\int e^{2x} \cos 3x \, dx$

$\boxed{u = e^{2x}; \; dv = \cos 3x}$

$= -\frac{1}{3}e^{2x} \cos 3x + \frac{2}{3}(\frac{1}{3}e^{2x} \sin 3x$

$\qquad - \frac{2}{3}\int e^{2x} \sin 3x \, dx)$

$= -\frac{1}{3}e^{2x} \cos 3x + \frac{2}{9}e^{2x} \sin 3x - \frac{4}{9}I + C_1$

$\frac{13}{9}I = -\frac{1}{3}e^{2x} \cos 3x + \frac{2}{9}e^{2x} \sin 3x + C_1$

$I = -\frac{3}{13}e^{2x} \cos 3x + \frac{2}{13}e^{2x} \sin 3x + C$

9. $\int x \cos^2 x \, dx \quad \boxed{u = x; \; dv = \cos^2 x \, dx}$

$\boxed{du = dx; \; v = \frac{x}{2} + \frac{\sin 2x}{4} \text{ by Formula 317}}$

$= x(\frac{x}{2} + \frac{\sin 2x}{4}) - \frac{1}{2}\int (\frac{x}{2} + \frac{\sin 2x}{4}) \, dx$

$= \frac{1}{2}x^2 + \frac{1}{4}x \sin 2x - \frac{1}{4}x^2 + \frac{1}{8} \cos 2x + C$

$= \frac{1}{4}x^2 + \frac{1}{4}x \sin 2x + \frac{1}{8} \cos 2x + C$

10. $\int \frac{x^3 \, dx}{\sqrt{x^2 + 1}} \quad \boxed{\text{Formula 175}}$

$= \frac{(x^2 + 1)^{3/2}}{3} - \sqrt{x^2 + 1} + C$

11. $\int x^2 \ln x \, dx \quad \boxed{u = x; \; dv = x^2 \, dx}$

$= \frac{1}{3}x^3 \ln x - \frac{1}{3}\int x^2 \, dx = \frac{1}{3}x^3 \ln x - \frac{1}{9}x^3 + C$

12. $\int (x + \sin x)^2 \, dx = \int (x^2 + 2x \sin x + \sin^2 x) \, dx$

$= \frac{1}{3}x^3 + 2\int x \sin x + \int \sin^2 x \, dx$

$= \frac{1}{3}x^3 + 2(\sin x - x \cos x) + (\frac{x}{2} - \frac{\sin 2x}{4}) + C$

$= \frac{1}{3}x^3 + \frac{1}{2}x + 2 \sin x - 2x \cos x - \frac{1}{4}\sin 2x + C$

13. $\int e^{2x}\sqrt{1 - e^x} \, dx \quad \boxed{t = 1 - e^x; \; dt = -e^x \, dt}$

$$= \int (t-1)\sqrt{t}\, dt = \tfrac{2}{5}t^{5/2} - \tfrac{2}{3}t^{3/2} + C$$

$$= \tfrac{2}{5}(1 - e^x)^{5/2} - \tfrac{2}{3}(1 - e^x)^{3/2} + C$$

$$= -\tfrac{2}{15}(3e^x + 2)(1 - e^x)^{3/2} + C$$

14. $\displaystyle\int x \sin x \cos x\, dx = \tfrac{1}{2}\int x \sin 2x\, dx$

$$\boxed{u = x;\ dv = \sin 2x\, dx}$$

$$= \tfrac{1}{2}\left[-\tfrac{1}{2}x \cos 2x - \int -\tfrac{1}{2}\cos 2x\, dx \right]$$

$$= -\tfrac{1}{4}x \cos 2x + \tfrac{1}{4}\left(\tfrac{1}{2}\sin 2x \right) + C$$

$$= -\tfrac{1}{4}x \cos 2x + \tfrac{1}{8}\sin 2x + C$$

15. $\displaystyle\int_1^4 \sqrt{x}\ln x\, dx \qquad \boxed{u = \ln x;\ dv = \sqrt{x}\, dx}$

$$= \tfrac{2}{3}x^{3/2}\ln x\Big|_1^4 - \tfrac{2}{3}\int_1^4 x^{1/2}\, dx$$

$$= \tfrac{16}{3}\ln 4 - \tfrac{4}{9}x^{3/2}\Big|_1^4 = \tfrac{32}{3}\ln 2 - \tfrac{28}{9}$$

16. $\displaystyle\int_1^e x^3 \ln x\, dx \qquad \boxed{u = \ln x;\ dv = x^3\, dx}$

$$= \tfrac{1}{4}x^4 \ln x\Big|_1^e - \tfrac{1}{4}\int_1^e x^3\, dx$$

$$= \tfrac{1}{4}e^4 - \tfrac{1}{16}x^4\Big|_1^e = \tfrac{1}{4}e^4 - \tfrac{1}{16}e^4 + \tfrac{1}{16}$$

$$= \tfrac{1}{16}(3e^4 + 1)$$

17. $\displaystyle\int_1^e (\ln x)^2\, dx \qquad \boxed{u = \ln^2 x;\ dv = dx}$

$$= x(\ln x)^2\Big|_1^e - 2\int_1^e \ln x\, dx$$

$$\boxed{u = \ln x;\ dv = dx}$$

$$= e(\ln e)^2 - 2\left[x \ln x\Big|_1^e - \int_1^e dx \right]$$

$$= e - 2(e - x)\Big|_1^e = e - 2$$

18. $\displaystyle\int_{1/3}^e 3(\ln 3x)^2\, dx \qquad \boxed{t = 3x;\ dt = 3\, dx}$

$$= \int_1^{3e} (\ln t)^2\, dt \qquad \boxed{u = \ln^2 t;\ dv = dt}$$

$$= t(\ln t)^2\Big|_1^{3e} - 2\int_1^{3e} \ln t\, dt$$

$$\boxed{u = \ln t;\ dv = dt}$$

$$= 3e(\ln 3e)^2 - 2\left[t \ln t\Big|_1^{3e} - \int_1^{3e} dt \right]$$

$$= 3e(\ln 3 + 1)^2 - 2[3e(\ln 3e) - t]\Big|_1^{3e}$$

$$= 3e(\ln^2 3 + 2 \ln 3 + 1)$$

$$\qquad - 2[3e(\ln 3 + 1) - 3e + 1]$$

$$= 3e \ln^2 3 + 3e - 2$$

19. $\displaystyle I = \int_0^\pi e^{2x}\cos 2x\, dx \qquad \boxed{u = e^{2x};\ dv = \cos 2x\, dx}$

$$= \tfrac{1}{2}e^{2x}\sin 2x\Big|_0^\pi - \int_0^\pi e^{2x}\sin 2x\, dx$$

$$\boxed{u = e^{2x};\ dv = \sin 2x\, dx}$$

$$= \tfrac{1}{2}e^{2x}\sin 2x\Big|_0^\pi$$

$$\qquad - \left[-\tfrac{1}{2}e^{2x}\cos 2x\Big|_0^\pi + \int_0^\pi e^{2x}\cos 2x\, dx \right]$$

$$= \tfrac{1}{2}e^{2\pi}(0) - (1)(0) + \tfrac{1}{2}e^{2\pi}(1) - \tfrac{1}{2}(1)(1) - I$$

$$2I = \tfrac{1}{2}e^{2\pi} - \tfrac{1}{2}$$

$$I = \tfrac{1}{4}(e^{2\pi} - 1)$$

20. $\displaystyle\int_0^\pi x(\sin x + \cos x)\, dx \quad \boxed{u = x;\ dv = (\sin x + \cos x)dx}$

$$= x(-\cos x + \sin x)\Big|_0^\pi - \int_0^\pi (-\cos x + \sin x)\, dx$$

$$= \pi + (\sin x + \cos x)\Big|_0^\pi = \pi - 2$$

21. Integration by parts is the application of the formula

$$\int u\, dv = uv - \int v\, du$$

The u factor is a part of the integrand that is differentiated and dv is the part that is integrated. Generally, pick dv as complicated as possible yet still integrable, so that the integral on the right is easier to integrate than the original integral.

22. $\displaystyle\int \frac{\ln x \sin(\ln x)}{x}\, dx \qquad \boxed{t = \ln x;\ dt = x^{-1}\, dx}$

$$= \int t \sin t \, dt \quad \boxed{u = t; \; dv = \sin t \, dt}$$

$$= -t \cos t + \int \cos t \, dt = -t\cos t + \sin t + C$$

$$= -\ln x \cos(\ln x) + \sin(\ln x) + C$$

23. $\displaystyle \int [\sin 2x \ln(\cos x)] \, dx = \int 2 \sin x \cos x \ln(\cos x) \, dx$

$$\boxed{t = \cos x; \; dt = -\sin x \, dx}$$

$$= -2 \int t \ln t \, dt \quad \boxed{u = \ln t; \; dv = t \, dt}$$

$$= -2\left[\tfrac{1}{2} \ln t - \tfrac{1}{2} \int t \, dt \right]$$

$$= -t^2 \ln t - \tfrac{1}{2} t^2 + C$$

$$= \tfrac{1}{2} \cos^2 x - (\cos^2 x) \ln(\cos x) + C$$

24. $\displaystyle \int e^{2x} \sin e^x \, dx = \int e^x \sin e^x \, (e^x \, dx)$

$$\boxed{t = e^x; \; dt = e^x \, dx}$$

$$= \int t \sin t \, dt \quad \boxed{u = t; \; dv = \sin t \, dt}$$

$$= -t \cos t + \int \cos t \, dt = -t \cos t + \sin t + C$$

$$= e^x \cos e^x + \sin e^x + C$$

25. $\displaystyle \int [\sin x \ln(2 + \cos x)] \, dx = -\int \ln t \, dt$

$$\boxed{t = 2 + \cos x; \; dt = -\sin x \, dx}$$

$$\boxed{u = \ln t; \; dv = dt}$$

$$= -t \ln t + t + C$$

$$= -(2 + \cos x)\ln(2 + \cos x) + (2 + \cos x) + C$$

26. a. $\displaystyle \int \frac{x^3}{x^2 - 1} \, dx \quad \boxed{u = x^2; \; dv = \frac{x \, dx}{x^2 - 1}}$

$$= \tfrac{1}{2} x^2 \ln |x^2 - 1| - \int x \ln |x^2 - 1| \, dx$$

$$= \tfrac{1}{2} x^2 \ln |x^2 - 1| - \tfrac{1}{2} \int \ln |x^2 - 1| (2x \, dx)$$

$$= \tfrac{1}{2} x^2 \ln |x^2 - 1| - \tfrac{1}{2} (x^2 - 1)\ln |x^2 - 1|$$
$$+ \tfrac{1}{2}(x^2 - 1) + C$$

$$= \tfrac{1}{2}(\ln |x^2 - 1|) + x^2) + C$$

b. $\displaystyle \int \frac{x^3}{x^2 - 1} \, dx = \int [x + x(x^2 - 1)^{-1}] \, dx$

$$= \tfrac{1}{2} x^2 + \tfrac{1}{2} \ln |x^2 - 1| + C$$

$$= \tfrac{1}{2}(\ln |x^2 - 1| + x^2) + C$$

27. $\displaystyle I = \int \sin^2 x \, dx \quad \boxed{u = \sin x; \; dv = \sin x \, dx}$

$$= -\sin x \cos x + \int \cos^2 x \, dx$$

$$= -\sin x \cos x + \int (1 - \sin^2 x) \, dx$$

$$= -\sin x \cos x + x - \int \sin^2 x \, dx$$

$$2I = -\sin x \cos x + x + C$$

$$= \tfrac{1}{2}(x - \sin x \cos x) + C$$

28. $\displaystyle I = \int \cos^2 x \, dx \quad \boxed{u = \cos x; \; dv = \cos x \, dx}$

$$= \sin x \cos x + \int \sin^2 x \, dx$$

$$= \sin x \cos x + \int (1 - \cos^2 x) \, dx$$

$$= \sin x \cos x + x - \int \cos^2 x \, dx$$

$$2I = \sin x \cos x + x + C$$

$$= \tfrac{1}{2}(x + \sin x \cos x) \quad C$$

29. $\displaystyle \int x^n \ln x \, dx \quad \boxed{u = \ln x; \; dv = x^n \, dx}$

$$= \frac{x^{n+1}}{n + 1} \ln x - \frac{1}{n + 1} \int x^n \, dx$$

$$= \frac{x^{n+1}}{n + 1} \ln x - \frac{x^{n+1}}{(n + 1)^2} + C$$

30. Let $x(t)$ be the distance traveled.

$$\frac{dx}{dt} = te^{-t/2}$$

$$x(t) = \int te^{-t/2} \, dt \quad \boxed{u = t; \; dv = e^{-t/2}}$$

$$= -2te^{-t/2} + 2 \int e^{-t/2} dt$$

$$= -2te^{-t/2} - 4e^{-t/2} + C$$

$$= -2e^{-t/2}(2 + t) + C$$

31. Let $Q(t)$ be the number of units produced.

$$\frac{dQ}{dt} = 100te^{-0.05t}$$

$$Q(t) = \int_0^3 100te^{-0.5t} dt \quad \boxed{u = t; \; dv = e^{-0.05t}}$$

$$= -200 \, te^{-0.5t} \Big|_0^3$$

$$-400\int_0^3 e^{-0.5t}(-\tfrac{1}{2})dt$$

$$= [-200\ te^{-0.5t} - 400\ e^{-0.5t}]\Big|_0^3$$

$$= -200\ e^{-0.5t}(t+2)\Big|_0^3$$

$$= -1000\ e^{-1.5} + 400 \approx 177 \text{ units}$$

32. Let $Q(t)$ be the amount of money raised in t weeks.

$$\frac{dQ}{dt} = 2{,}000te^{-0.2t}$$

$$Q(t) = \int_0^5 2{,}000te^{0.2t}dt \quad \boxed{u = t;\ dv = e^{-0.2t}}$$

$$= 2{,}000(-5te^{-0.2t})\Big|_0^5$$

$$-10{,}000\int e^{-0.2t}dt$$

$$= -10{,}000te^{0.2t} - 50{,}000e^{-0.2t}\Big|_0^5$$

$$= -100{,}000e^{-1} + 50{,}000 \approx \$13{,}212$$

33. $$V = 2\pi\int_0^2 xe^{-x}dx \quad \boxed{u = x;\ dv = e^{-x}dx}$$

$$= 2\pi\left(-xe^{-x}\Big|_0^2 + \int_0^2 e^{-x}dx\right)$$

$$= 2\pi(-xe^{-x} - e^{-x})\Big|_0^2$$

$$= 2\pi(1 - 3e^{-2}) \approx 3.73217$$

34. $$V = 2\pi\int_0^{\pi/4} x(\sin x + \cos x)\ dx$$

$$\boxed{u = x;\ dv = (\sin x + \cos x)\ dx}$$

$$= 2\pi(-x\cos x + x\sin x + \sin x + \cos x)\Big|_0^{\pi/4}$$

$$= 2\pi(\tfrac{1}{2}\sqrt{2} - 1] \approx 2.6026$$

35. a. $$V = \pi\int_1^e (\ln x)^2 dx \quad \boxed{u = (\ln x)^2;\ dv = dx}$$

$$= \pi\int_1^e (\ln x)^2 dx$$

$$= \pi x(\ln x)^2\Big|_1^e - 2\pi\int_1^e x\ln x\left(\frac{dx}{x}\right)$$

$$= \pi x(\ln x)^2 - 2\pi(x\ln x - x)\Big|_1^e$$

$$= \pi(e - 2e + 2e - 2) = \pi(e - 2)$$

b. $$V = 2\pi\int_1^e x\ln x\ dx = 2\pi\ [\tfrac{x^2}{4}(2\ln x - 1)]\Big|_1^e$$

$$\text{(See Problem 3)}$$

$$= 2\pi[\tfrac{e^2}{4} + \tfrac{1}{4}] = \tfrac{\pi}{2}(e^2 + 1)$$

36. The curves intersect when $x = \pi/4$; assume the density is 1.

$$A = \int_0^{\pi/4} (\cos x - \sin x)dx = (\sin x + \cos x)\Big|_0^{\pi/4}$$

$$= \sqrt{2} - 1 \approx 0.4142$$

$$M_x = \int_0^{\pi/4} \tfrac{1}{2}(\cos x - \sin x)(\cos x + \sin x)\ dx$$

$$= \tfrac{1}{2}\int_0^{\pi/4} \cos 2x\ dx = \tfrac{1}{4}\sin 2x\Big|_0^{\pi/4} = \tfrac{1}{4} = 0.25$$

$$M_y = \int_0^{\pi/4} x(\cos x - \sin x)\ dx$$

$$\boxed{u = x;\ dv = (\cos x - \sin x)\ dx}$$

$$= x(\sin x + \cos x)\Big|_0^{\pi/4} - \int_0^{\pi/4} (\sin x + \cos x)\ dx$$

$$= \tfrac{1}{4}\pi\sqrt{2} - (-\cos x + \sin x)\Big|_0^{\pi/4}$$

$$= \tfrac{\pi}{4}\sqrt{2} \approx 0.1107$$

$$(\bar{x}, \bar{y}) \approx \left(\frac{0.1107}{0.4142}, \frac{0.25}{0.4142}\right) \approx (0.27, 0.60)$$

37. The curves intersect when $x = 0$. Assume the density is 1.

$$A = \int_0^1 (e^x - e^{-x})\ dx$$

$$= (e^x + e^{-x})\Big|_0^1 \approx 1.0862$$

$$M_x = \tfrac{1}{2}\int_0^1 (e^x - e^{-x})(e^x + e^{-x})\ dx$$

$$= \tfrac{1}{2}\int_0^1 (e^{2x} - e^{-2x})\ dx = \tfrac{1}{4}(e^{2x} + e^{-2x})\Big|_0^1$$

$$\approx 1.3811$$

$$M_y = \int_0^1 x(e^x - e^{-x})\ dx$$

$$\boxed{u = x;\ dv = (e^x - e^{-x})\ dx}$$

$$= x(e^x + e^{-x})\Big|_0^1 - \int_0^1 (e^x + e^{-x})\, dx$$

$$= [x(e^x + e^{-x}) - (e^x - e^{-x})]\Big|_0^1$$

$$= \frac{2}{e} \approx 0.7358$$

$$(\overline{x}, \overline{y}) \approx \left(\frac{0.7358}{1.0862}, \frac{1.3811}{1.0862}\right) \approx (0.68, 1.27)$$

38. $I(x) = \exp\left[\int \frac{x}{1+x}\, dx\right] = \exp\left[\int 1 - \frac{1}{1+x}\right]$

$$= \exp[x - \ln(1+x)] = e^x(x+1)^{-1}$$

$$y = \frac{1}{e^x(x+1)^{-1}}\left[\int x(x+1)\frac{e^x}{(x+1)}dx + C\right]$$

$$= \frac{x+1}{e^x}\left[\int xe^x\, dx + C\right]$$

$$\boxed{u = x;\ dv = e^x\, dx \text{ or Formula 484}}$$

$$= \frac{x+1}{e^x}\left[e^x(x-1) + C\right]$$

$$= x^2 - 1 + Ce^{-x}(x+1)$$

39. $I(x) = \exp\left[\int \frac{2x}{1+x^2}\, dx\right] = x^2 + 1$

$$y = \frac{1}{x^2+1}\int (x^2+1)\sin x\, dx$$

$$= \frac{1}{x^2+1}\left[\int \sin x\, dx + \int x^2 \sin x\, dx\right]$$

The first integral is simple, but the second will require integration by parts twice, or Formula #344.

$$y = \frac{1}{x^2+1}\Big(-\cos x + 2x\sin x$$
$$+ (2 - x^2)\cos x + Cx\Big)$$

But $y = 1$ when $x = 0$ gives $C = 0$;
$$y = \frac{-x^2\cos x + 2x \sin x + \cos x}{x^2+1}$$

40. $\dfrac{dy}{dx} = x^2 + 2y$

$$\frac{dy}{dx} - 2y = x^2$$

$$I(x) = \exp\left[\int -2\, dx\right] = e^{-2x}$$

$$y = \frac{1}{e^{-2x}}\left[\int x^2 e^{-2x}dx + C\right]$$

This integral requires integration by parts twice or Formula 485.

$$= e^{2x}\left[-\tfrac{1}{2}e^{-2x}(x^2 + x + \tfrac{1}{2}) + C\right]$$

$$= -\tfrac{1}{2}x^2 - \tfrac{1}{2}x - \tfrac{1}{4} + Ce^{2x}$$

41. $\displaystyle\int_0^\pi [f(x) + f''(x)]\sin x\, dx$

$$= \int_0^\pi f(x)\sin x\, dx + \int_0^\pi f''(x)\sin x\, dx$$

$$\boxed{u = f(x);\ dv = \sin x\, dx}\quad \boxed{u = \sin x;\ dv = f''(x)\ dx}$$

$$= -\cos x f(x)\Big|_0^\pi + \int_0^\pi f'(x)\cos x\ dx$$

$$\quad + \sin x\, f'(x)\Big|_0^\pi - \int_0^\pi f'(x)\cos x\, dx$$

$$= -\pi \cos \pi\, f(\pi) + \cos 0 f(0)$$

This is 0, so that $f(\pi) = -f(0) = -3$.

42. $v(t) = -r\ln\dfrac{w-kt}{w} - gt$

$$= -8{,}000 \ln \frac{150-t}{150} - 32t$$

$$s(t) = 8{,}000(150)\left[\frac{150-t}{150}\ln\frac{150-t}{150} - \frac{150-t}{150}\right]$$
$$- 16t^2 + C$$

This is true because $\displaystyle\int x\ln x\, dx = x \ln x - x + C$

$s(0) = 0$ so that $C = 1{,}200{,}000$ and

$$s(t) = 8{,}000(150)\left[\frac{150-t}{150}\ln\frac{150-t}{150} - \frac{150-t}{150}\right]$$
$$- 16t^2 + 1{,}200{,}00$$

The distance traveled in 120 sec is

$$s(120) = 1{,}2000{,}000[\tfrac{1}{5}\ln\tfrac{3}{5} - \tfrac{3}{5}] - 16(120)t^2$$
$$+ 1{,}200{,}000;\ s(120) \approx 343{,}335 \text{ ft} \approx 5 \text{ mi}$$

43. Dividing by m gives a first order linear differential equation with

$$P(t) = \frac{k}{m} \text{ and } Q(t) = g$$

$$I(t) = e^{\int \frac{k}{m}\, dt} = e^{kt/m}$$

$$v = e^{-kt/m}\int e^{kt/m}\, g\, dt$$

$$= e^{-kt/m}\frac{mg}{k}\int e^{kt/m}\left(\frac{k}{m}\, dt\right)$$

$$= e^{-kt/m}\left(\frac{mg}{k}e^{kt/m} + C\right)$$

$$= \frac{mg}{k} + e^{-kt/m} + C.$$

But we have initial value of $v = v_0$ when $t = 0$ so $C = v_0 - mg/k$.

$$v = \frac{mg}{k} + \frac{v_0 - \frac{mg}{k}}{e^{kt/m}}$$

44. Let y be the number of pollutants (in millions of cubic feet) in the lake at time t (in days). The number of pollutants flowing into the lake per day is $0.0006(350) = 0.21$ units. The fraction of a lake flowing out per day is 350 million/6 billion, or 0.0583333. Thus, the rate of change is $dy/dt = 0.21 - 0.058333y$;

$$\frac{dy}{dx} + 0.058333y = 0.21$$

$I(x) = e^{0.058333t}$ so that

$$y = \frac{1}{e^{0.058333t}}\left[\int 0.21 e^{0.058333t}\, dt + C\right]$$

$$= 3.6 + Ce^{-0.058333t}$$

Originally there were 0.22% or 0.0022(6,000) $= 13.2$ units of pollutant in the lake. Then,

$$13.2 = 3.6 + 9.6e^{-0.058333t}$$

Then

$$13.20 = 3.6 + 0.6e^{-0.058333t}$$

The concentration of pollutants in the lake is found by solving for t:

$$t \approx -(0.058333)^{-1}\ln 0.5625 \approx 10 \text{ days}$$

45. $\text{AV} = \dfrac{1}{\frac{\pi}{5}}\displaystyle\int_0^{\pi/5} 2.3e^{-0.25t}\cos 5t\, dt$

$$= \frac{5}{\pi}(2.3)\left[\frac{e^{-0.25t}(-0.25\cos 5t + 5\sin 5t)}{(-0.25)^2 + (5)^2}\right]$$

$$\approx 0.0677$$

46. Let θ be the angle subtended by the sign and α the angle subtended by the 1 ft vertical leg of the 4×1 triangle. Then

$$6x^{-1} = \tan(\theta + \alpha) \text{ and } x^{-1} = \tan \alpha$$

$$\tan^{-1}(6x^{-1}) = \theta + \alpha$$

$$\theta = \tan^{-1}(6x^{-1}) - \alpha$$

$$= \tan^{-1}(6x^{-1}) - \tan^{-1}(x^{-1})$$

$$d\theta = [\tan^{-1}(6x^{-1}) - \tan^{-1}(x^{-1})]'\, dt$$

$$\text{AV} = \frac{1}{20-4}\int_4^{20} [\tan^{-1}(6x^{-1}) - \tan^{-1}(x^{-1})]'\, dt$$

$$= \frac{1}{16}[\tan^{-1}(6x^{-1}) - \tan^{-1}(x^{-1})]\Big|_0^1$$

$$\approx -0.31 \text{ rad/ft}$$

47. $\text{AV} = \dfrac{1}{9V_1}\displaystyle\int_{V_1}^{10V_1} nRt \ln \frac{V}{V_1}\, dV$

$$= \frac{nRt}{9V_1}\int_{V_1}^{10V_1} (\ln V - \ln V_1)\, dV.$$

$$\boxed{\text{Formula 499}}$$

$$= \frac{nRt}{9V_1}(V\ln V - V - V\ln V_1)\Big|_{V_1}^{10V_1}$$

$$= \frac{nRt}{9}(10\ln 10 - 9)$$

$$\approx 1.558 nRt$$

48. $\text{AV} = 10(3)(10^5)I_0\displaystyle\int_{10^{-12}}^{3(10^{-5})} \ln \frac{I}{I_0}\frac{dI}{I_0}$

$$= \frac{10^6}{3}I_0\left[\frac{I}{I_0}\ln\frac{I}{I_0} - \frac{I}{I_0}\right]\Big|_{10^{-12}}^{3(10^{-5}}$$

$$= 10(7\ln 3 - 1) \approx 67 \text{ db}$$

49. $\displaystyle\int x^n e^x\, dx \quad \boxed{u = x^n;\ dv = e^x\, dx}$

$$= x(\ln x)^n - n\int (\ln x)^{n-1} dx$$

50. $\displaystyle\int (\ln x)^n\, dx \quad \boxed{u = (\ln x)^n;\ dv = dx}$

$$= x(\ln x)^n - n\int (\ln x)^{n-1} dx$$

51. $I = \displaystyle\int_0^{\pi/2} \sin^2 x\, dx$, n is even;

$$\boxed{u = (\sin x)^{n-1};\ dv = \sin x\, dx}$$

$$= \frac{1}{n}\sin^{n-1} x\cos x\Big|_0^{\pi/2}$$

$$- \frac{n-1}{n}\int_0^{\pi/2} (\sin x)^{n-2}\, dx$$

This recursive formula will be used repeatedly with values of n that decrease by 2.

$$I = 0 + \frac{n-1}{n}\int_0^{\pi/2} \sin^{n-2} x\, dx$$

$$= \frac{n-1}{n}\left[-\frac{\sin^{n-3}x\cos x}{n-2}\right]\Big|_0^{\pi/2}$$

$$+ \frac{n-3}{n-2}\int_0^{\pi/2} \sin^{n-4} x\, dx$$

$$= \frac{(n-1)(n-3)}{n(n-2)} \int_0^{\pi/2} \sin^{n-4}x \, dx$$

$$= \cdots = \frac{(n-1)(n-3)\cdots(3)}{n(n-2)\cdots(4)} \int_0^{\pi/2} \sin^2 x \, dx$$

$$= \frac{(n-1)(n-3)\cdots(3)(1)}{n(n-2)\cdots(4)(2)} \int_0^{\pi/2} (1+\cos 2x) \, dx$$

$$= \frac{1\cdot3\cdot5\cdots(n-3)(n-1)}{2\cdot4\cdot6\cdots(n-2)n} \cdot \frac{\pi}{2}$$

For $\int_0^{\pi/2} \cos^n x \, dx$, develop Formula 321 and then proceed as shown above.

52. $I = \int_0^{\pi/2} \cos^n x \, dx$, n is odd

$$\boxed{u = (\cos x)^{n-1}; \; dv = \cos x \, dx}$$

$$= \frac{1}{n} \cos^{n-1} x \sin x \Big|_0^{\pi/2}$$

$$+ \frac{n-1}{n} \int_0^{\pi/2} (\cos x)^{n-2} dx$$

This recursive formula will be used repeatedly with values of n that decrease by 2.

$$I = 0 + \frac{n-1}{n} \int_0^{\pi/2} \cos^{n-2} x \, dx$$

$$= \frac{n-1}{n} \left[-\frac{\cos^{n-3} x \sin x}{n-2} \right] \Big|_0^{\pi/2}$$

$$+ \frac{n-3}{n-2} \int_0^{\pi/2} \cos^{n-4} x \, dx$$

$$= \frac{(n-1)(n-3)}{n(n-2)} \int_0^{\pi/2} \cos^{n-4} x \, dx$$

$$= \cdots = \frac{(n-1)(n-3)\cdots(2)}{n(n-2)\cdots(3)} \int_0^{\pi/2} \cos x \, dx$$

$$= \frac{2\cdot4\cdot6\cdots(n-3)(n-1)}{1\cdot3\cdot5\cdots(n-2)n}$$

For $\int_0^{\pi/2} \sin^n x \, dx$, develop Formula 352 and then proceed as shown above.

7.3   **The Method of Partial Fractions, Page 463**

1. $\dfrac{1}{x(x-3)} = \dfrac{A_1}{x} + \dfrac{A_2}{x-3}$

$1 = A_1(x-3) + A_2(x)$

If $x = 0$, then $A_1 = -\frac{1}{3}$; if $x = 3$, then $A_2 = \frac{1}{3}$

Thus,

$$\frac{1}{x(x-3)} = \frac{-1}{3x} + \frac{1}{3(x-3)}$$

2. $\dfrac{3x-1}{x^2-1} = \dfrac{A_1}{x+1} + \dfrac{A_2}{x-1}$

$3x - 1 = A_1(x-1) + A_2(x+1)$

If $x = 1$, then $A_2 = 1$; if $x = -1$, then $A_1 = 2$

Thus,

$$\frac{3x-1}{x^2-1} = \frac{2}{x+1} + \frac{1}{x-1}$$

3. $\dfrac{3x^2+2x-1}{x(x+1)} = 3 - \dfrac{x+1}{x(x+1)} = 3 - \dfrac{1}{x}$

4. $\dfrac{2x^2+5x-1}{x(x^2-1)} = \dfrac{A_1}{x} + \dfrac{A_2}{x+1} + \dfrac{A_3}{x-1}$

$2x^2 + 5x - 1$

$= A_1(x^2-1) + A_2 x(x-1) + A_3 x(x+1)$

If $x = 0$, then $A_1 = 1$; $x = 1$, then $A_2 = -2$

$$\frac{2x^2+5x-1}{x(x^2-1)} = \frac{1}{x} + \frac{-2}{x+1} + \frac{3}{x-1}$$

5. $\dfrac{4}{2x^2+x} = \dfrac{4}{x(2x+1)} = \dfrac{A_1}{x} + \dfrac{A_2}{2x+1}$

$4 = A_1(2x+1) + A_2 x$

If $x = 0$, then $A_1 = 4$; if $x = -\frac{1}{2}$, then $A_2 = -8$

$$\frac{4}{2x^2+x} = \frac{4}{x} + \frac{-8}{2x+1}$$

6. $\dfrac{x^2-x+3}{x^2(x-1)} = \dfrac{A_1}{x} + \dfrac{A_2}{x^2} + \dfrac{A_3}{x-1}$

$x^2 - x + 3$

$= A_1 x(x-1) + A_2(x-1) + A_3 x^2$

If $x = 0$, then $A_2 = -3$; if $x = 1$, then $A_3 = 3$; finally equating the coefficients of x^2 we have $1 = A_1 + A_3$, so $A_1 = -2$

$$\frac{x^2-x+3}{x^2(x-1)} = \frac{-2}{x} + \frac{-3}{x^2} + \frac{3}{x-1}$$

7. $F = \dfrac{4x^3+4x^2+x-1}{x^2(x+1)^2}$

$$= \frac{A_1}{x} + \frac{A_2}{x^2} + \frac{A_3}{x+1} + \frac{A_4}{(x+1)^2}$$

$$4x^3 + 4x^2 + x - 1 = A_1 x(x^2 + 2x + 1)$$
$$+ A_2(x^2 + 2x + 1) + A_3(x^3 + x^2) + A_4 x^2$$

If $x = 0$, then $A_2 = 1$; if $x = -1$, then
$A_4 = -2$; finally equate the coefficient of x^3
and those of x to find $4 = A_1 + A_3$ and
$1 = A_1 + 2A_2$, so $A_1 = 3$, and $A_3 = 1$

$$F = \frac{3}{x} + \frac{-1}{x^2} + \frac{1}{x+1} + \frac{-2}{(x+1)^2}$$

8. $\dfrac{x^2 - 5x - 4}{(x^2 + 1)(x - 3)} = \dfrac{A_1 x + B_1}{x^2 + 1} + \dfrac{A_2}{x - 3}$

$$x^2 - 5x - 4 = (A_1 x + B_1)(x - 3) + A_2(x_2 + 1)$$
$$= A_1 x^2 - 3A_1 x + B_1 x - 3B_1 + A_2 x^2 + A_2$$

If $x = 3$, then $A_2 = -1$;

Coefficients of x^2: $1 = A_1 + A_2$, so $A_1 = 2$

Coefficients of x: $-5 = -3A_1 + B_1$, so $B_1 = 1$

$$\frac{x^2 - 5x - 4}{(x^2 + 1)(x - 3)} = \frac{2x + 1}{x^2 + 1} + \frac{-1}{x - 3}$$

9. $F = \dfrac{x^3 + 3x^2 + 3x - 4}{x^2(x + 3)^2}$

$$= \frac{A_1}{x^2} + \frac{A_2}{x} + \frac{A_3}{(x+3)^2} + \frac{A_4}{x+3}$$

$$x^3 + 3x^2 + 3x - 4 = A_1(x + 3)^2$$
$$+ A_2 x(x + 3)^2 + A_3 x^2 + A_4 x(x + 3)$$
$$= (A_2 + A_4)x^3 + (A_1 + 6A_2 + A_3 + 6A_4)x^2$$
$$+ (6A_1 + 9A_2)x + 9A_1$$

If $x = 0$, then $A_1 = -\frac{4}{9}$;

if $x = -3$, then $A_3 = -\frac{13}{9}$

coefficients of x^3: $A_2 + A_4 = 1$

coefficients of x^2: $A_1 + 6A_2 + A_3 + 6A_4 = 3$

coefficients of x: $6A_1 + 9A_2 = 3$

Solving this system of equations, using
previously found values, we find:

$A_2 = \frac{17}{27}$ and $A_4 = \frac{10}{27}$.

$$F = \frac{-\frac{4}{9}}{x^2} + \frac{\frac{17}{27}}{x} + \frac{-\frac{13}{9}}{(x+3)^2} + \frac{\frac{10}{27}}{x+3}$$

$$= \frac{-4}{9x^2} + \frac{17}{27x} - \frac{13}{9(x+3)^2} + \frac{10}{27(x+3)}$$

10. $\dfrac{1}{x^3 - 1} = \dfrac{1}{(x - 1)(x^2 + x + 1)}$

$$= \frac{A_1}{x - 1} + \frac{A_2 x + B_1}{x^2 + x + 1}$$

$$1 = A_1(x^2 + x + 1) + (A_2 x + B_1)(x - 1)$$
$$= A_1 x^2 + A_1 x + A_1 + A_2 x^2 + B_1 x - A_2 x - B_1$$

After solving, we find

$$\frac{1}{x^3 - 1} = \frac{1}{3(x - 1)} + \frac{x + 2}{3(x^2 + x + 1)}$$

11. $\dfrac{1}{1 - x^4} = \dfrac{1}{(1 - x)(1 + x)(1 + x^2)}$

$$= \frac{A_1}{1 - x} + \frac{A_2}{1 + x} + \frac{A_3 x + B_1}{1 + x^2}$$

$$1 = A_1(1 + x)(1 + x^2) + A_2(1 - x)(1 + x^2)$$
$$+ A_3 x(1 - x^2) + B_1(1 - x^2)$$
$$= A_1(x^3 + x^2 + x + 1) + A_2(-x^3 + x^2 - x + 1)$$
$$+ A_3(x - x^3) + B_1 - B_1 x^2$$

After solving, we find

$$\frac{1}{1 - x^4} = \frac{1}{4(1 - x)} + \frac{1}{4(1 + x)} + \frac{1}{2(1 + x^2)}$$

12. $\dfrac{x^4 - x^2 + 2}{x^2(x - 1)} = x + 1 + \dfrac{2}{x^2(x - 1)}$

$$= x + 1 + \frac{A_1}{x} + \frac{A_2}{x^2} + \frac{A_3}{x - 1}$$
$$2 = A_1(x - 1) + A_2(x - 1) + A_3 x^2$$

After solving, we find

$$\frac{x^4 - x^2 + 2}{x^2(x - 1)} = x + 1 + \frac{-2}{x} + \frac{-2}{x^2} + \frac{2}{x - 1}$$

13. $\displaystyle\int \frac{dx}{x(x - 3)} = -\frac{1}{3}\int \frac{dx}{x} + \frac{1}{3}\int \frac{dx}{x - 3}$

$$= -\frac{1}{3}\ln|x| + \frac{1}{3}\ln|x - 3| + C = \frac{1}{3}\ln\left|\frac{x - 3}{x}\right| + C$$

14. $\displaystyle\int \frac{3x - 1}{x^2 - 1}\,dx = \int \frac{2}{x + 1}\,dx + \int \frac{1}{x - 1}\,dx$

$$= 2\ln|x + 1| + \ln|x - 1| + C$$

15. $\displaystyle\int \frac{3x^2 + 2x - 1}{x(x + 1)}\,dx = 3\int dx - \int \frac{1}{x}\,dx$

$$= 3x - \ln|x| + C$$

16. $\displaystyle\int \frac{2x^2 + 5x - 1}{x(x^2 - 1)}\,dx = \int \frac{dx}{x} - \int \frac{2\,dx}{x + 1} + \int \frac{3\,dx}{x - 1}$

$$= \ln|x| - 2\ln|x + 1| + 3\ln|x - 1| + C$$

17. $\displaystyle\int \frac{4\,dx}{2x^2 + x} = \int \frac{4}{x}\,dx - \int \frac{8\,dx}{2x + 1}$

$$= 4\ln|x| - 4\ln|2x + 1| + C$$

18. $\int \dfrac{x^2 - x + 3}{x^2(x - 1)}\, dx$

$$= \int \dfrac{-2}{x}\, dx + \int \dfrac{-3}{x^2}\, dx + \int \dfrac{3\, dx}{x - 1}$$

$$= -2\ln|x| + \dfrac{3}{x} + 3\ln|x - 1| + C$$

19. $\dfrac{2x^3 + 9x - 1}{x^2(x^2 - 1)} = \dfrac{A_1}{x} + \dfrac{A_2}{x^2} + \dfrac{A_3}{x + 1} + \dfrac{A_4}{x - 1}$

Partial fraction decomposition gives $A_1 = 9$, $A_3 = 1$, $A_3 = 6$, $A_4 = 5$.

$$\int \dfrac{2x^3 + 9x - 1}{x^2(x^2 - 1)}\, dx$$

$$= -9\int \dfrac{dx}{x} + \int x^{-2}\, dx + 6\int \dfrac{dx}{x + 1} + 5\int \dfrac{dx}{x - 1}$$

$$= -9\ln|x| - x^{-1} + 6\ln|x + 1| + 5\ln|x - 1| + C$$

20. $\dfrac{x^4 - x^2 + 2}{x^2(x - 1)} = x + 1 + \dfrac{2}{x^2(x - 1)}$

$$= x + 1 + \dfrac{A_1}{x} + \dfrac{A_2}{x^2} + \dfrac{A_3}{x - 1}$$

Partial fraction decomposition gives $A_1 = -2$, $A_2 = -2$, $A_3 = 2$.

$$\int \dfrac{x^4 - x^2 + 2}{x^2(x - 1)}\, dx = \int (x + 1)\, dx$$

$$-2\int \dfrac{dx}{x} - 2\int x^{-2}\, dx + 2\int \dfrac{dx}{x - 1}$$

$$= \tfrac{1}{2}x^2 + x - 2\ln|x| + 2x^{-1} + 2\ln|x - 1| + C$$

21. $\dfrac{x^2 + 1}{x^2 + x - 2} = 1 + \dfrac{-x + 3}{x^2 + x - 2}$

$$= 1 + \dfrac{-x + 3}{(x - 1)(x + 2)}$$

$$= 1 + \dfrac{A_1}{x - 1} + \dfrac{A_2}{x + 2}$$

Partial fraction decomposition gives $A_1 = \tfrac{2}{3}$, $A_2 = -\tfrac{5}{3}$.

$$\int \dfrac{x^2 + 1}{x^2 + x - 2}\, dx$$

$$= \int dx + -\tfrac{5}{3}\int \dfrac{dx}{x + 2} + \tfrac{2}{3}\int \dfrac{dx}{x - 1}$$

$$= x - \tfrac{5}{3}\ln|x + 2| + \tfrac{2}{3}\ln|x - 1| + C$$

22. $\dfrac{1}{x^3 - 8} = \dfrac{1}{(x - 2)(x^2 + 2x + 4)}$

$$= \dfrac{A_1}{x - 2} + \dfrac{A_2 x + B_1}{x^2 + 2x + 4}$$

Partial fraction decomposition gives
$A_1 = \tfrac{1}{12}$, $A_2 = -\tfrac{1}{12}$, $B_1 = -\tfrac{1}{3}$.

$$\int \dfrac{dx}{x^3 - 8}$$

$$= \tfrac{1}{12}\int \dfrac{dx}{x - 2} + \tfrac{1}{12}\int \dfrac{-x - 4}{x^2 + 2x + 4}\, dx$$

$$= \tfrac{1}{12}\int \dfrac{dx}{x - 2} + \tfrac{1}{12}\int \dfrac{-x - 4}{x^2 + 2x + 4}\, dx$$

$$= \tfrac{1}{12}\int \dfrac{dx}{x - 2} - \tfrac{1}{12}\int \dfrac{x + 1 + 3}{x^2 + 2x + 4}\, dx$$

$$= \tfrac{1}{12}\int \dfrac{dx}{x - 2} - \tfrac{1}{12}\int \dfrac{x + 1}{x^2 + 2x + 4}\, dx$$

$$- \tfrac{1}{4}\int \dfrac{dx}{x^2 + 2x + 4}.$$

$$= \tfrac{1}{12}\int \dfrac{dx}{x - 2} - \tfrac{1}{24}\int \dfrac{(2x + 2)\, dx}{x^2 + 2x + 4}$$

$$- \tfrac{1}{4}\int \dfrac{dx}{(x + 1)^2 + 3}$$

$$= \tfrac{1}{12}\ln|x - 2| - \tfrac{1}{24}\ln|x^2 + 2x + 4|$$

$$- \dfrac{\sqrt{3}}{12}\tan^{-1}\left[\dfrac{1}{\sqrt{3}}(x + 1)\right] + C$$

23. Use the results of Problem 11.

$$\int \dfrac{x^4 + 1}{x^4 - 1}\, dx = \int \left[1 + \dfrac{2}{x^4 - 1}\right]\, dx$$

$$= \int dx + 2\left[\tfrac{1}{4}\int \dfrac{dx}{x - 1} - \tfrac{1}{4}\int \dfrac{dx}{x + 1} - \tfrac{1}{2}\int \dfrac{dx}{x^2 + 1}\right]$$

$$= x + \tfrac{1}{2}\ln|x - 1| - \tfrac{1}{2}\ln|x + 1| - \tan^{-1}x + C$$

24. Use the results of Problem 10.

$$\int \dfrac{x^3 + 1}{x^3 - 1}\, dx = \int \left[1 + \dfrac{2}{x^3 - 1}\right]\, dx$$

$$= \int dx + 2\left[\int \dfrac{dx}{3(x - 1)} + \int \dfrac{(x + 2)\, dx}{3(x^2 + x + 1)}\right]$$

$$= x + \tfrac{2}{3}\ln|x - 1| - \tfrac{2}{3}\int \dfrac{x + 2}{x^2 + x + 1}\, dx$$

$$= x + \tfrac{2}{3}\ln|x - 1|$$

$$-\tfrac{2}{3}\left[\int \dfrac{x + \tfrac{1}{2}}{x^2 + x + 1}\, dx + \tfrac{3}{2}\int \dfrac{dx}{(x + \tfrac{1}{2})^2 + \tfrac{3}{4}}\right]$$

$$= x + \tfrac{2}{3}\ln|x - 1| - \tfrac{1}{3}\ln|x^2 + x + 1|$$

$$- \frac{2}{\sqrt{3}}\tan^{-1}\!\left(\frac{2x + 1}{\sqrt{3}}\right) + C$$

25. $\displaystyle\int \frac{x\,dx}{(x + 1)^2} = \int dx - \int \frac{dx}{(x + 1)^2}$

$$= \ln|x + 1| + (x + 1)^{-1} + C$$

26. $\displaystyle\int \frac{2x\,dx}{(x - 2)^2} = 2\int \frac{(x - 2)\,dx}{(x - 2)^2} + \int \frac{2\,dx}{(x - 2)^2}$

$$= 2\ln|x - 2| - 2(x - 2)^{-1} + C$$

27. $\displaystyle\frac{1}{x(x + 1)(x - 2)} = \frac{A_1}{x} + \frac{A_2}{x + 1} + \frac{A_3}{x - 2}$

Partial fraction decomposition gives
$A_1 = -\tfrac{1}{2},\ A_2 = \tfrac{1}{3},\ A_3 = \tfrac{1}{6}.$

$$\int \frac{dx}{x(x + 1)(x - 2)}$$

$$= -\tfrac{1}{2}\int \frac{dx}{x} + \tfrac{1}{3}\int \frac{dx}{x + 1} + \tfrac{1}{6}\int \frac{dx}{x - 2}$$

$$= -\tfrac{1}{2}\ln|x| + \tfrac{1}{3}\ln|x + 1| + \tfrac{1}{6}\ln|x - 2| + C$$

28. $\displaystyle\frac{x + 2}{x(x - 1)^2} = \frac{A_1}{x} + \frac{A_2}{x - 1} + \frac{A_3}{(x - 1)^2}$

Partial fraction decomposition gives
$A_1 = 2,\ A_2 = -2,\ A_3 = 3.$

$$\int \frac{x + 2}{x(x - 1)^2}\,dx$$

$$= 2\int \frac{dx}{x} - 2\int \frac{dx}{x + 1} + 3\int \frac{dx}{(x - 1)^2}$$

$$= 2\ln|x| - 2\ln|x + 1| - 3(x - 1)^{-1} + C$$

29. $\displaystyle\frac{x}{(x + 1)(x + 2)^2} = \frac{A_1}{x + 1} + \frac{A_2}{(x + 2)^2} + \frac{A_3}{x + 2}$

Partial fraction decomposition gives
$A_1 = -1,\ A_2 = 2,\ A_3 = 1.$

$$\int \frac{x\,dx}{(x + 1)(x + 2)^2}$$

$$= -\int \frac{dx}{x + 1} + 2\int \frac{dx}{(x + 2)^2} + \int \frac{dx}{x + 2}$$

$$= -\ln|x + 1| - \frac{2}{x + 2} + \ln|x + 2| + C$$

$$= \ln\left|\frac{x + 2}{x + 1}\right| - \frac{2}{x + 2} + C$$

30. $\displaystyle\frac{x + 1}{x(x^2 + 2)} = \frac{A_1}{x} + \frac{A_2 x + B_1}{x^2 + 2}$

Partial fraction decomposition gives
$A_1 = \tfrac{1}{2},\ A_2 = -\tfrac{1}{2},\ B_1 = 1.$

$$\int \frac{x + 1}{x(x^2 + 2)}\,dx = \tfrac{1}{2}\int \frac{dx}{x} + \tfrac{1}{2}\int \frac{(-x + 2)\,dx}{x^2 + 2}$$

$$= \tfrac{1}{2}\ln|x| - \tfrac{1}{4}\ln|x^2 + 2| + \tfrac{1}{2}\sqrt{2}\,\tan^{-1}\!\left(\frac{x}{\sqrt{2}}\right) + C$$

31. $\displaystyle\frac{5x + 7}{x^2 + 2x - 3} = \frac{5x + 7}{(x - 1)(x + 3)} = \frac{A_1}{x - 1} + \frac{A_2}{x + 3}$

Partial fraction decomposition gives
$A_1 = 3,\ A_2 = 2.$

$$\int \frac{5x + 7}{x^2 + 2x - 3}\,dx = 3\int \frac{dx}{x - 1} + 2\int \frac{dx}{x + 3}$$

$$= 3\ln|x - 1| + 2\ln|x + 3| + C$$

32. $\displaystyle\int \frac{5x\,dx}{x^2 - 6x + 9} = 5\int \frac{(x - 3 + 3)\,dx}{(x - 3)^2}$

$$= 5\int \frac{(x - 3)\,dx}{(x - 3)^2} + 5\int \frac{3\,dx}{(x - 3)^2}$$

$$= 5\int \frac{dx}{x - 3} + 15\int (x - 3)^{-2}\,dx$$

$$= 5\ln|x - 3| - 15(x - 3)^{-1} + C$$

33. $\displaystyle\int \frac{3x^2 - 2x + 4}{x^3 - x^2 + 4x - 4}\,dx$

$$= \ln|x^3 - x^2 + 4x - 4| + C$$

34. $\displaystyle\int \frac{3x^2 + 4x + 1}{x^3 + 2x^2 + x - 2}\,dx$

$$= \ln|x^3 + 2x^2 + x - 2| + C$$

35. Factor the denominator as the product of linear factors with real roots and/or second degree factors with complex roots. The rational function (given fraction) is rewritten as the sum of fractions whose denominators are the factors. The degree of the denominator indicates the number of constants needed.

(1) The factors are linear, there are no multiple roots.

$$F = \frac{A_1}{x - a_1} + \frac{A_2}{x - a_2} + \cdots + \frac{A_n}{x - a_n}$$

(2) The factors are linear, there are multiple roots.

$$F = \frac{A_1}{x - a} + \frac{A_2}{(x - a)^2} + \cdots + \frac{A_n}{(x - a)^n}$$

(3) The factors are quadratic, there are no

multiple roots.

$$F = \frac{A_1 x + B_1}{x^2 + a_1{}^2} + \frac{A_2 x + B_2}{x^2 + a_2{}^2} + \cdots + \frac{A_n x + B_n}{x^2 + a_n{}^2}$$

(4) The factors are quadratic, there are multiple roots.

$$F = \frac{A_1 x + B_1}{x^2 + a^2} + \frac{A_2 x + B_2}{(x^2 + a^2)^2} + \cdots + \frac{A_n x + B_n}{(x^2 + a^2)^n}$$

Usually a combination of these cases applies. Now realize that the left member is identically equal to the decomposed right member. Multiply by the least common denominator. Choose the real roots and evaluate. The values of the constants will fall out one-by-one. In the case of complex roots, equate like coefficients and solve the resulting system of equations. Check that your decomposed fractions add up to the original fraction. Finally, integrate the decomposed fractions.

36. $\displaystyle \int \frac{\cos x \, dx}{\sin^2 x - \sin x - 2}$ $\boxed{t = \sin t; \; dt = \cos t \, dt}$

$$= \int \frac{dt}{t^2 - t - 2} = \int \frac{dt}{(t-2)(t+1)}$$

Now, $\displaystyle \frac{1}{(t-2)(t+1)} = \frac{A_1}{t-2} + \frac{A_2}{t-1}$

Partial fraction decomposition gives $A_1 = 1$, $A_2 = -1$.

$$\int \frac{dt}{(t-2)(t+1)} = \int \frac{dt}{t-2} - \int \frac{dt}{t-1}$$

$$= \ln|t - 2| - \ln|t - 1| + C$$

$$= \ln|\sin x - 2| - \ln|\sin x - 1| + C$$

37. $\displaystyle \int \frac{e^x \, dx}{2e^{2x} - 5e^x - 3}$ $\boxed{t = e^x; \; dt = e^x \, dx}$

$$= \int \frac{dt}{2t^2 - 5t - 3} = \int \frac{dt}{(2t+1)(t-3)}$$

Now, $\displaystyle \frac{1}{(2t+1)(t-3)} = \frac{A_1}{2t+1} + \frac{A_2}{t-3}$

Partial fraction decomposition gives $A_1 = -\frac{2}{7}$, $A_2 = \frac{1}{7}$.

$$\int \frac{dt}{(2t+1)(t-3)} = -\frac{1}{7}\int \frac{2 \, dt}{2t+1} + \frac{1}{7}\int \frac{dt}{t-3}$$

$$= \frac{1}{7}\ln|t - 3| - \frac{1}{7}\ln|2t + 1| + C$$

$$= \frac{1}{7}\ln|e^x - 3| - \frac{1}{7}\ln(2e^x + 1) + C$$

38. $\displaystyle \int \frac{e^x \, dx}{e^{2x} - 1}$ $\boxed{t = e^x; \; dt = e^x \, dx}$

$$= \int \frac{dt}{t^2 - 1} = \int \frac{dt}{(t+1)(t-1)}$$

Now, $\displaystyle \frac{1}{(t+1)(t-1)} = \frac{A_1}{t+1} + \frac{A_2}{t-1}$

Partial fraction decomposition gives $A_1 = -\frac{1}{2}$, $A_2 = \frac{1}{2}$.

$$\int \frac{dt}{(t+1)(t-1)} = -\frac{1}{2}\int \frac{dt}{t+1} + \frac{1}{2}\int \frac{dt}{t-1}$$

$$= \frac{1}{2}\ln|t - 1| - \frac{1}{2}\ln|t + 1| + C$$

$$= \frac{1}{2}\ln|e^x - 1| - \frac{1}{2}\ln|e^x + 1| + C$$

39. $\displaystyle \int \frac{\sin x \, dx}{(1+\cos x)^2} = -\int (1+\cos x)^{-2}(-\sin x \, dx)$

$$= \frac{1}{1 + \cos x} + C$$

40. $\displaystyle \int \frac{\tan x \, dx}{\sec^2 x + 4} = \int \frac{\sin x \cos x \, dx}{1 + 4\cos^2 x}$

$$\boxed{t = \cos x; \; dt = -\sin x \, dx}$$

$$= -\int \frac{t \, dt}{1 + 4t^2} = -\frac{1}{8}\ln(1 + 4t^2) + C$$

$$= -\frac{1}{8}\ln(1 + 4\cos^2 x) + C$$

41. $\displaystyle \int \frac{\sec^2 x \, dx}{\tan x + 4}$ $\boxed{t = \tan x + 4; \; dt = \sec^2 x \, dx}$

$$= \int \frac{dt}{t} = \ln|t| + C = \ln|\tan x + 4| + C$$

42. $\displaystyle \int \frac{dx}{x^{1/4} - x}$ $\boxed{t^4 = x; \; dt = 4t^3 \, dt}$

$$= \int \frac{4t^3 \, dt 1}{t - t^4} = 4\int \frac{t^2 \, dt}{1 - t^3} = -\frac{4}{3}\int \frac{-3t^2 \, dt}{1 - t^3}$$

$$= -\frac{4}{3}\ln|1 - t^3| + C = -\frac{4}{3}\ln|1 - x^{3/4}| + C$$

43. $\displaystyle \int \frac{dx}{x^{1/4} - x^{5/4}}$ $\boxed{t^4 = x; \text{ then } dx = 4t^3 \, dt}$

$$= \int \frac{4t^3 \, dt}{t - t^5} = \int \frac{-4t^2 \, dt}{t^4 - 1}$$

$$= \int \frac{-4t^2 \, dt}{(t^2 + 1)(t + 1)(t - 1)}$$

Now, $\displaystyle \frac{-4t^2}{t^4 - 1} = \frac{A_1 t + B_1}{t^2 + 1} + \frac{A_2}{t + 1} + \frac{A_3}{t - 1}$

Partial fraction decomposition gives $A_1 = 0$, $A_2 = 1$, $A_3 = 1$, $B_1 = -2$.

$$\int \frac{4t^3 \, dt}{t - t^5} = \int \frac{-2 \, dt}{t^2 + 1} + \int \frac{dt}{t + 1} + \int \frac{- \, dt}{t - 1}$$

$$= - \, 2 \tan^{-1} t + \ln|t + 1| - \ln|t - 1| + C$$

$$= - \, 2 \tan^{-1} x^{1/4} + \ln|x^{1/4} + 1|$$
$$- \ln|x^{1/4} - 1| + C$$

44. $\int \dfrac{dx}{\sin x - \cos x}$ $\boxed{\text{Weierstrass substitution}}$

$$= \int \frac{\dfrac{2 \, du}{1 + u^2}}{\dfrac{2u}{1 + u^2} - \dfrac{1 - u^2}{1 + u^2}} = \int \frac{2 \, du}{2u - 1 + u^2}$$

$$= \int \frac{2 \, du}{(u + 1)^2 - 2} \qquad \boxed{t = u + 1; \; dt = du}$$

$$= \int \frac{2 \, dt}{t^2 - 2} \quad \boxed{\text{Use Formula 74 or partial fractions}}$$

$$= \frac{2}{2\sqrt{2}} \ln\left|\frac{t - \sqrt{2}}{t + \sqrt{2}}\right| + C$$

$$= \frac{\sqrt{2}}{2} \ln\left|\frac{u + 1 - \sqrt{2}}{u + 1 + \sqrt{2}}\right| + C$$

$$= \frac{\sqrt{2}}{2} \ln\left|\frac{\tan \frac{x}{2} + 1 - \sqrt{2}}{\tan \frac{x}{2} + 1 + \sqrt{2}}\right| + C$$

45. $\int \dfrac{dx}{3\cos x + 4\sin x}$ $\boxed{\text{Weierstrass substitution}}$

$$= \int \frac{2 \, du}{2u - 1 - u} = 2 \int \frac{du}{3(1 - u^2) + 8u}$$

$$= \int \frac{-2 \, du}{3u^2 - 8u - 3}$$

Now, $\dfrac{-2}{3u^2 - 8u - 3} = \dfrac{-2}{(u - 3)(3u + 1)}$

$$= \frac{A_1}{u - 3} + \frac{A_2}{3u + 1}$$

Partial fraction decomposition gives
$A_1 = -\frac{1}{5}, \; A_2 = \frac{3}{5}.$

$$\int \frac{-2 \, du}{3u^2 - 8u - 3} = -\frac{1}{5} \int \frac{du}{u - 3} + \frac{1}{5} \int \frac{3 \, du}{3u + 1}$$

$$= -\frac{1}{5} \ln|u - 3| + \frac{1}{5} \ln|3u + 1| + C$$

$$= -\frac{1}{5} \ln\left|\tan \frac{x}{2} - 3\right| + \frac{1}{5} \ln\left|3 \tan \frac{x}{2} + 1\right| + C$$

46. $\int \dfrac{dx}{5 \sin x + 4}$ $\boxed{\text{Weierstrass substitution}}$

$$= \int \frac{2 \, du}{10u + 4(1 + u^2)} = \int \frac{du}{2u^2 + 5u + 2}$$

$$= \int \frac{du}{(u + 2)(2u + 1)}$$

Now, $\dfrac{1}{(u + 2)(2u + 1)} = \dfrac{A_1}{u + 2} + \dfrac{A_2}{u + 2}$

Partial fractions decomposition gives
$A_1 = -\frac{1}{3}, \; A_2 = \frac{2}{3}.$

$$\int \frac{du}{(u + 2)(2u + 1)} = -\frac{1}{3} \int \frac{du}{u + 2} + \frac{1}{3} \int \frac{2 \, du}{2u + 1}$$

$$= -\frac{1}{3} \ln|u + 2| + \frac{1}{3} \ln|2u + 1| + C$$

$$= -\frac{1}{3} \ln\left|\tan \frac{x}{2} + 2\right| + \frac{1}{3} \ln\left|2 \tan \frac{x}{2} + 1\right| + C$$

47. $\int \dfrac{\sin x - \cos x}{\sin x + \cos x} \, dx$

$\boxed{t = \sin x + \cos x; \; dx = (\cos x - \sin x) \, dx}$

$$= - \int \frac{dt}{t} = - \ln|t| = - \ln|\sin x + \cos x| + C$$

48. $\int \dfrac{dx}{4 \cos x + 5}$ $\boxed{\text{Weierstrass substitution}}$

$$= \int \frac{2 \, du}{4(1 - u^2) + 5(1 + u^2)} = \int \frac{2 \, du}{u^2 + 9}$$

$$= \frac{2}{3} \tan^{-1} \frac{u}{3} + C = \frac{2}{3} \tan^{-1} \frac{1}{3} \tan \frac{x}{2} + C$$

49. $\int \dfrac{dx}{\sec x - \tan x} = \int \dfrac{(\sec x + \tan x) \, dx}{(\sec x - \tan x)(\sec x + \tan x)}$

$$= \int \frac{(\sec x + \tan x) \, dx}{1} = \int \sec x \, dx + \int \tan x \, dx$$

$$= \ln|\sec x + \tan x| - \ln|\cos x| + C$$

50. $\int \dfrac{dx}{3\sin x + 4\cos x + 5}$ $\boxed{\text{Weierstrass substitution}}$

$$= \int \frac{2 \, du}{6u + 4(1 - u^2) + 5(1 + u^2)}$$

$$= \int \frac{2 \, du}{u^2 + 6u + 9} = \int \frac{2 \, du}{(u + 3)^2} = -\frac{2}{u + 3} + C$$

$$= - \frac{2}{\tan \frac{x}{2} + 3} + C$$

51. $\int \dfrac{dx}{4\sin x - 3\cos x - 5}$ $\boxed{\text{Weierstrass substitution}}$

$$= \int \frac{2 \, du}{8u - 3(1 - u^2) - 5(1 + u^2)}$$

$$= \int \frac{2 \, du}{-2u^2 + 8u - 8} = - \int \frac{du}{u^2 - 4u + 4}$$

$$= - \int \frac{du}{(u - 2)^2} = \frac{1}{u - 2} + C$$

$$= \frac{1}{\tan \frac{x}{2} - 2} + C$$

52. $\displaystyle\int \frac{dx}{2\csc x - \cot x + 2} = \int \frac{\sin x\, dx}{2 - \cos x + 2\sin x}$

$\boxed{\text{Weierstrass substitution}}$

$\displaystyle = \int \frac{4\,u\,du}{2(1+u^2) - (1-u^2) + 4u}$

$\displaystyle = \int \frac{4u\,du}{3u^2 + 4u + 1} = \int \frac{4u\,du}{(u+1)(3u+1)}$

Now, $\displaystyle\frac{4u}{(u+1)(3u+1)} = \frac{A_1}{u+1} + \frac{A_2}{3u+1}$

Partial fraction decomposition gives
$A_1 = 2,\ A_2 = -2.$

$\displaystyle\int \frac{4u\,du}{(u+1)(3u+1)} = 2\int \frac{du}{u+1} - 2\int \frac{du}{3u+1}$

$\displaystyle = 2\ln|u+1| - \tfrac{2}{3}\ln|3u+1| + C$

$\displaystyle = 2\ln\left|\tan\tfrac{x}{2} + 1\right| - \tfrac{2}{3}\ln\left|3\tan\tfrac{x}{2} + 1\right| + C$

53. $\displaystyle\int \frac{dx}{x(3 - \ln x)(1 - \ln x)}$ $\boxed{t = \ln x;\ dt = x^{-1}dx}$

$\displaystyle = \int \frac{dt}{(3 - t)(1 - t)}$

Now, $\displaystyle\frac{1}{(3-t)(1-t)} = \frac{A_1}{t-3} + \frac{A_2}{t-1}$

Partial fraction decomposition gives $A_1 = \tfrac{1}{2}$,
$A_2 = -\tfrac{1}{2}.$

$\displaystyle\int \frac{dt}{(3-t)(1-t)} = \tfrac{1}{2}\int \frac{dt}{t-3} - \tfrac{1}{2}\int \frac{dt}{t-1}$

$\displaystyle = \tfrac{1}{2}\ln|t-3| - \tfrac{1}{2}\ln|t-1| + C$

$\displaystyle = \tfrac{1}{2}\ln|\ln x - 3| - \tfrac{1}{2}\ln|\ln x - 1| + C$

54. $\displaystyle A = \int_0^3 \frac{dx}{x^2 + 5x + 4} = \int_0^3 \frac{dx}{(x+4)(x+1)}$

$\displaystyle = -\tfrac{1}{3}\int_0^3 \frac{dx}{x+4} + \tfrac{1}{3}\int_0^3 \frac{dx}{x+1}$

$\displaystyle = \tfrac{1}{3}\Big[\ln|x+1| - \ln|x+4|\Big]\Big|_0^3$

$\displaystyle = \tfrac{1}{3}\left(\ln\tfrac{4}{7} - \ln\tfrac{1}{4}\right) \approx 0.2756$

55. $\displaystyle A = \int_{4/3}^{7/4} \frac{dx}{x^2 - 5x + 6} = \int_{4/3}^{7/4} \frac{dx}{(x-3)(x-2)}$

$\displaystyle = -\int_{4/3}^{7/4} \frac{dx}{x-2} + \int_{4/3}^{7/4} \frac{1}{x-3}\,dx$

$\displaystyle = \Big[-\ln|x-2| + \ln|x-3|\Big]\Big|_{4/3}^{7/4}$

$\displaystyle = \ln 2 \approx 0.6931$

56. $\displaystyle V = 2\pi \int_0^1 \frac{x\,dx}{(x+4)(x+1)}$

$\displaystyle = 2\pi\left[-\tfrac{1}{3}\int_0^1 \frac{dx}{x+4} + \tfrac{4}{3}\int_0^1 \frac{dx}{x+1}\right]$

$\displaystyle = -\tfrac{2\pi}{3}\ln|x+4| + \tfrac{8\pi}{3}\ln|x+1|\Big|_0^1$

$\displaystyle = \tfrac{2}{3}\pi\left(4\ln\tfrac{5}{4} - \ln 2\right) \approx 0.4177$

57. $\displaystyle V = \pi \int_0^1 \frac{dx}{(x+4)^2(x+1)^2}$

Now, $\displaystyle\frac{1}{(x+4)^2(x+1)^2}$

$\displaystyle = \frac{A_1}{x+1} + \frac{A_2}{(x+1)^2} + \frac{A_3}{x+4} + \frac{A_4}{x+4}$

Partial fraction decomposition gives
$A_1 = -\tfrac{2}{27},\ A_2 = \tfrac{1}{9},\ A_3 = \tfrac{2}{27},\ A_4 = \tfrac{1}{9}$

$\displaystyle\pi \int_0^1 \frac{dx}{(x+4)^2(x+1)^2}$

$\displaystyle = \tfrac{\pi}{27}\int_0^1 \left[\frac{-2}{x+1} + \frac{3}{(x+1)^2} + \frac{2}{x+4} + \frac{3}{(x+4)^2}\right]dx$

$\displaystyle = \tfrac{\pi}{27}\big[-2\ln|x+1| - 3(x-1)^{-1}$

$\displaystyle\qquad\quad + 2\ln|x+4| - 3(x+4)^{-1}\big]\Big|_0^1$

$\displaystyle = \pi\left[\tfrac{11}{180} + \tfrac{2}{27}\ln\tfrac{5}{8}\right] \approx 0.0826$

58. $\displaystyle V = \pi \int_0^3 \frac{dx}{(x+2)^2 - 1}$ $\boxed{t = x+2;\ du = dx}$

$\displaystyle = \pi \int_2^5 \frac{dt}{t^2 - 3}$ $\boxed{\text{Partial fractions or Formula 74}}$

$\displaystyle = \pi\left[\frac{1}{2\sqrt{3}}\ln\left|\frac{t - \sqrt{3}}{t + \sqrt{3}}\right|\right]\Big|_2^5$

$\displaystyle = \frac{\pi}{2\sqrt{3}}\left[\ln\frac{5-\sqrt{3}}{5+\sqrt{3}} - \ln\frac{2-\sqrt{3}}{2+\sqrt{3}}\right] \approx 1.7332$

59. $V = \pi \int_0^4 y^2 \, dx = \pi \int_0^4 \frac{x^2(4-x)}{4+x} \, dx.$

$= \pi \int_0^4 \left(-x^2 + 8x - 32 + \frac{128}{x+4} \right) dx$

$= \pi \left[-\frac{x^3}{3} + 4x^2 - 32x + 128 \ln(x+4) \right]\Big|_0^4$

$= \pi \left(-\frac{256}{3} + 128 \ln 2 \right) \approx 10.648$

60. $V = \pi \int_0^{\pi/4} \frac{1}{(\sin x + \cos x)^2} = \pi \int_0^{\pi/4} \frac{dx}{1 + \sin 2x}$

Formula 365

$= \pi[-\frac{1}{2} \tan(\frac{\pi}{4} - x)]\Big|_0^{\pi/4} = \frac{\pi}{2}$

61. $\frac{dN}{dt} = kN(60 - N)$

$\int \frac{dN}{N(60-N)} = \int k \, dt$

$\frac{1}{60} \int \left[\frac{1}{N} + \frac{1}{60-N} \right] dN = \int k \, dt$

$\frac{1}{60}[\ln N - \ln(60 - N)] = kt + C_1$

$\ln \left| \frac{N}{60 - N} \right| = 60kt + C_2$

$\frac{N}{60 - N} = Ce^{60kt}$

$N = \frac{60}{1 + Ae^{-60kt}}$ where $A = \frac{1}{C}$

$N(0) = 2$ implies $A = 29$ and $N(1) = 3$ implies $e^{-60k} = \frac{57}{87}$; $k \approx 0.007$ (Note: double-oh-seven), How long before $N = 20$ villagers know the secret?

$20 = \frac{60}{1 + 29e^{-60(0.007)t}}$

$e^{-0.42t} = \frac{2}{29} \approx 0.06897$

$t \approx 6.4$ days

The spy is captured.

62. $\frac{1}{a^2 - x^2} = \frac{A_1}{x} + \frac{A_2}{ax + b}$

Partial fraction decomposition gives $A_1 = 1/(2a)$, $A_2 = 1/(2a)$.

$\int \frac{dx}{a^2 - x^2} = \frac{1}{2a} \int \left(\frac{1}{a+x} + \frac{1}{a-x} \right) dx$

$= \frac{1}{2a}[\ln|a + x| - \ln|a - x|] + C$

$= \frac{1}{2a} \ln \left| \frac{a+x}{a-x} \right| + C$

63. $\frac{1}{x(ax + b)} = \frac{A_1}{x} + \frac{A_2}{ax + b}$

Partial fraction decomposition gives $A_1 = 1/b$, $A_2 = -a/b$.

$\int \frac{dx}{a^2 - x^2} = \frac{1}{b} \int \left[\frac{1}{x} - \frac{a}{ax + b} \right] dx$

$= \frac{1}{b}[\ln|x| - \ln|ax + b|] + C$

$= \frac{1}{b} \ln \left| \frac{x}{ax + b} \right| + C$

64. $\int \sec x \, dx = \int \frac{dx}{\cos x}$

Weierstrass substitution

$= \int \frac{2 \, du}{1 - u^2} = \int \left(\frac{1}{1 + u} + \frac{1}{1 - u} \right) du$

$= \ln|1 + u| - \ln|1 - u| + C$

$= \ln \left| \frac{1 + u}{1 - u} \right| + C = \ln \left| \frac{1 + \tan \frac{x}{2}}{1 - \tan \frac{x}{2}} \right| + C$

$= \ln \left| \frac{\cos \frac{x}{2} + \sin \frac{x}{2}}{\cos \frac{x}{2} - \sin \frac{x}{2}} \right| + C$

$= \ln \left| \frac{(\cos \frac{x}{2} + \sin \frac{x}{2})^2}{\cos^2 \frac{x}{2} - \sin^2 \frac{x}{2}} \right| + C$

$= \ln \left| \frac{1 + 2 \sin \frac{x}{2} \cos \frac{x}{2}}{\cos x} \right| + C$

$= \ln \left| \frac{1 + \sin x}{\cos x} \right| + C$

$= \ln|\sec x + \tan x| + C$

65. $\int \csc x \, dx = \int \frac{dx}{\sin x}$

Weierstrass substitution

$= \int \frac{2 \, du}{2u} = \ln|u| + C = \ln \left| \tan \frac{x}{2} \right| + C$

$= \ln \left| \frac{\sin \frac{x}{2}}{\cos \frac{x}{2}} \right| + C$

$= \ln \left| \frac{\cos \frac{x}{2} \sin \frac{x}{2}}{\cos^2 \frac{x}{2}} \right| + C$

$$= \ln \left| \frac{2 \sin \frac{x}{2} \cos \frac{x}{2}}{2 \cos^2 \frac{x}{2}} \right| + C$$

$$= \ln \frac{|\sin x|}{|1 + \cos x|} + C$$

$$= -\ln \left| \frac{1}{\sin x} + \frac{\cos x}{\sin x} \right| + C$$

$$= -\ln|\csc x + \cot x| + C$$

66.
$$\frac{dP}{dt} = k \frac{1}{P(A - P)}$$

$$\int \frac{dP}{P(A - P)} = \int k \, dt$$

Now, $\dfrac{1}{P(A - P)} = \dfrac{A_1}{P} + \dfrac{A_2}{A - P}$

Partial fraction decomposition gives
$A_1 = 1/A$, $A_2 = 1/A$.

$$\int \frac{dP}{P(A - P)} = \frac{1}{A} \int [P^{-1} + (A - P)^{-1}] \, dP$$

$$= A^{-1}(\ln|P| - \ln|A - P| + C_1)$$

$$= \frac{1}{A} \ln \left| \frac{P}{A - P} \right| + C_1$$

Thus,
$$\int \frac{dP}{P(A - P)} = \int k \, dt$$

$$\frac{1}{A} \ln \left| \frac{P}{A - P} \right| + C_1 = kt$$

$$\frac{P}{A - P} = e^{C_3 + kAt} = Ce^{kAt}$$

$$P = ACe^{kAt} - PCe^{kAt}; \text{ at } t = 0, \ C = \frac{P_0}{A - P_0}$$

so
$$P = \frac{AP_0(A - P_0)^{-1}e^{kAt}}{1 + P_0(A - P_0)^{-1}e^{kAt}}$$

$$= \frac{AP_0}{P_0 + (A - P_0)e^{-kAt}}$$

7.4 Summary of Integration Techniques, Page 467

1. $\displaystyle \int \frac{2x - 1}{(x - x^2)^3} \, dx$ $\boxed{u = x - x^2; \ du = (1 - 2x) \, dx}$

$$= -\int u^{-3} \, du = \frac{1}{2} u^{-2} + C = \frac{1}{2(x - x^2)^2} + C$$

$$= \frac{1}{2x^2(x - 1)^2} + C$$

2. $\displaystyle \int \frac{2x + 3}{\sqrt{x^2 + 3x}} \, dx$ $\boxed{u = x^2 + 3x; \ du = (2x+3) \, dx}$

$$= \int u^{-1/2} \, du = 2u^{1/2} + C = 2\sqrt{x^2 + 3x} + C$$

3. $\displaystyle \int (x \sec 2x^2) \, dx$ $\boxed{u = 2x^2; \ du = 4x \, dx}$

$$= \frac{1}{4} \int \sec u \, du = \frac{1}{4} \ln|\sec u + \tan u| + C$$

$$= \frac{1}{4} \ln|\sec 2x^2 + \tan 2x^2| + C$$

4. $\displaystyle \int [x^2 \csc^2(2x^3)] \, dx$ $\boxed{u = 2x^3; \ du = 6x^2 \, dx}$

$$= \frac{1}{6} \int \csc^2 u \, du = -\frac{1}{6} \cot u + C$$

$$= -\frac{1}{6} \cos 2x^3 + C$$

5. $\displaystyle \int (e^x \cot e^x) \, dx$ $\boxed{u = e^x; \ du = e^x \, dx}$

$$= \int \cot u \, du = \ln|\sin e^x| + C$$

6. $\displaystyle \int \frac{\tan \sqrt{x} \, dx}{\sqrt{x}}$ $\boxed{u = x^{1/2}; \ du = \frac{1}{2}x^{-1/2} \, dx}$

$$= 2 \int \tan u \, du = -2 \int \frac{-\sin u \, du}{\cos u}$$

$$= -2 \ln|\cos u| + C = -2 \ln|\cos \sqrt{x}| + C$$

7. $\displaystyle \int \frac{\tan(\ln x) \, dx}{x}$ $\boxed{u = \ln x; \ du = x^{-1} \, dx}$

$$= \int \tan u \, du = -\int \frac{-\sin u \, du}{\cos u} = -\ln|\cos u|$$

$$= -\ln|\cos(\ln x)| + C$$

8. $\displaystyle \int \sqrt{\cot x} \, \csc^2 x \, dx$ $\boxed{u = \cot x; \ du - \csc^2 x \, dx}$

$$= -\int u^{1/2} \, du = -\frac{2}{3} u^{3/2} + C = -\frac{2}{3} \cot^{3/2} x + C$$

9. $\displaystyle \int \frac{(3 + 2 \sin t)}{\cos t} \, dt = \int 3 \sec t \, dt + \int 2 \tan t \, dt$

$$= 3 \ln|\sec t + \tan t| + 2 \ln|\sec t| + C$$

10. $\displaystyle \int \frac{2 + \cos x}{\sin x} \, dx = 2 \int \csc x \, dx + \int \frac{\cos x}{\sin x} \, dx$

$$= -2 \ln|\csc u + \cot u| + \ln|\sin x| + C$$

11. $\displaystyle\int \frac{e^{2t}\,dt}{1 + e^{4t}}$ $\boxed{u = e^{2t};\ du = 2e^{2t}\,dt}$

$\displaystyle = \frac{1}{2}\int \frac{du}{1 + u^2} = \frac{1}{2}\tan^{-1}u + C$

$\displaystyle = \frac{1}{2}\tan^{-1}e^{2t} + C$

12. $\displaystyle\int \frac{\sin 2x\,dx}{1 + \sin^4 x} = \int \frac{2\sin x\cos x\,dx}{1 + (\sin^2 x)^2}$

$\boxed{u = \sin^2 x;\ du = 2\sin x\cos x\,dx}$

$\displaystyle = \int \frac{du}{1 + u^2}$

$\displaystyle = \tan^{-1}u + C = \tan^{-1}(\sin^2 x) + C$

13. $\displaystyle\int \frac{x^2 + x + 1}{x^2 + 9}\,dx = \int \left[1 + \frac{x - 8}{x^2 + 9}\right]dx$

$\displaystyle = \int dx + \int \frac{x\,dx}{x^2 + 9} - 8\int \frac{dx}{x^2 + 9}$

$\displaystyle = x + \frac{1}{2}\ln(x^2 + 9) - \frac{8}{3}\tan^{-1}\frac{x}{3} + C$

14. $\displaystyle\int \frac{3x + 2}{\sqrt{4 - x^2}}\,dx$

$\displaystyle = -\frac{3}{2}\int \frac{-2x\,dx}{\sqrt{4 - x^2}} + 2\int \frac{dx}{\sqrt{4 - x^2}}$

$\displaystyle = -\frac{3}{2}\int (4 - x^2)^{1/2}(-2x)\,dx + 2\int \frac{dx}{\sqrt{4 - x^2}}$

$\displaystyle = -\frac{3}{2}(2)\sqrt{4 - x^2} + 2\sin^{-1}\frac{x}{2} + C$

$\displaystyle = -3\sqrt{4 - x^2} + 2\sin^{-1}\frac{x}{2} + C$

15. $\displaystyle\int \frac{1 + e^x}{1 - e^x}\,dx = \int \frac{(e^{-x/2} + e^{x/2})\,dx}{e^{-x/2} - e^{x/2}}$

$\boxed{u = e^{-x/2} - e^{x/3};\ du = -\frac{1}{2}(e^{-x/2} + e^{x/2})\,dx}$

$\displaystyle = -2\int \frac{du}{u} = -2\ln|u| + C$

$\displaystyle = -2\ln\left|e^{-x/2} - e^{x/2}\right| + C$

$\displaystyle = -2\ln\left|\frac{1 - e^x}{e^{x/2}}\right| + C$

$\displaystyle = -2\ln|1 - e^x| + 2\ln e^{x/2} + C$

$\displaystyle = -2\ln|1 - e^x| + 2\left(\frac{x}{2}\right) + C$

$\displaystyle = x - 2\ln|e^x - 1| + C$

16. $\displaystyle\int \frac{e^{1 - \sqrt{x}}}{\sqrt{x}}\,dx$

$\boxed{u = 1 - \sqrt{x};\ du = -\frac{1}{2}x^{-1/2}\,dx}$

$\displaystyle = -2\int e^u\,du = -2e^u + C = -2e^{1 - \sqrt{x}} + C$

17. $\displaystyle\int \frac{2t^2\,dt}{\sqrt{1 - t^6}}$ $\boxed{u = t^3;\ du = 3t^2\,dt}$

$\displaystyle = \frac{2}{3}\int \frac{du}{\sqrt{1 - u^2}} = \frac{2}{3}\sin^{-1}u + C$

$\displaystyle = \frac{2}{3}\sin^{-1}t^3 + C$

18. $\displaystyle\int \frac{t^3\,dt}{2^8 + t^8}$ $\boxed{u = t^4;\ du = 4t^3\,dt}$

$\displaystyle = \frac{1}{4}\int \frac{du}{(2^4)^2 + u^2} = \frac{1}{4(2^4)}\tan^{-1}u + C$

$\displaystyle = \frac{1}{2^6}\tan^{-1}\frac{t^4}{2^4} + C$

19. $\displaystyle\int \frac{dx}{1 + e^{2x}}$ $\boxed{u = e^{-2x} + 1;\ du = -2e^{-2x}\,dx}$

$\displaystyle = \int \frac{e^{-2x}\,dx}{e^{-2x} + 1} = -\frac{1}{2}\int \frac{du}{u} = -\frac{1}{2}\ln|u| + C$

$\displaystyle = -\frac{1}{2}\ln\left|e^{-2x} + 1\right| + C$

$\displaystyle = x - \frac{1}{2}\ln(e^{2x} + 1) + C$

20. $\displaystyle\int \frac{dx}{4 - e^{-x}} = \frac{1}{4}\int \frac{4e^x\,dx}{4e^x - 1}$

$\boxed{u = 4e^x - 1;\ du = 4e^x\,dx}$

$\displaystyle = \frac{1}{4}\int \frac{du}{u} = \frac{1}{4}\ln|u| + C = \frac{1}{4}\ln|4e^x - 1| + C$

21. $\displaystyle\int \frac{dx}{x^2 + 2x + 2} = \int \frac{dx}{(x + 1)^2 + 1}$

$\displaystyle = \tan^{-1}(x + 1) + C$

22. $\displaystyle\int \frac{dx}{x^2 + x + 4} = \int \frac{dx}{(x + \frac{1}{2})^2 + \frac{15}{4}}$

$\boxed{u = x + \frac{1}{2};\ du = dx}$

$\displaystyle = \int \frac{du}{u^2 + \frac{15}{4}} = \frac{2}{\sqrt{15}}\tan^{-1}\frac{2u}{\sqrt{15}} + C$

$$= \frac{2}{\sqrt{15}} \tan^{-1} \frac{2}{\sqrt{15}}(x + \tfrac{1}{2}) + C$$

23. $\displaystyle\int \frac{dx}{x^2 + x + 1} = \int \frac{dx}{(x + \tfrac{1}{2})^2 + \tfrac{3}{4}}$

$$\boxed{u = x + \tfrac{1}{2};\ du = dx}$$

$$= \int \frac{du}{u^2 + \tfrac{3}{4}} = \frac{2}{\sqrt{3}} \tan^{-1} \frac{2u}{\sqrt{3}} + C$$

$$= \frac{2}{\sqrt{3}} \tan^{-1} \frac{2}{\sqrt{3}}(x + \tfrac{1}{2}) + C$$

24. $\displaystyle\int \frac{dx}{x^2 - x + 1} = \int \frac{dx}{(x - \tfrac{1}{2})^2 + \tfrac{3}{4}}$

$$\boxed{u = x - \tfrac{1}{2};\ du = dx}$$

$$= \int \frac{du}{u^2 + \tfrac{3}{4}} = \frac{2}{\sqrt{3}} \tan^{-1} \frac{2u}{\sqrt{3}} + C$$

$$= \frac{2}{\sqrt{3}} \tan^{-1} \frac{2}{\sqrt{3}}(x - \tfrac{1}{2}) + C$$

25. $\displaystyle\int \tan^{-1}x\ dx \quad \boxed{\text{Formula 457}}$

$$= x \tan^{-1}x - \tfrac{1}{2}\ln(x^2 + 1) + C$$

26. $\displaystyle\int xe^x\sin x\ dx$

$$\boxed{\text{Integrate by parts (4 times) or use Formula 494.}}$$

$$= \frac{x\ e^x(\sin x - \cos x)}{2} - \frac{e^x(-2\cos x)}{4}$$

$$= \tfrac{1}{2}e^x[x\sin x + (1 - x)\cos x] + C$$

27. $\displaystyle\int e^{-x}\cos x\ dx \quad \boxed{\text{Parts or Formula 493}}$

$$= \frac{e^{-x}(-\cos x + \sin x)}{2} + C$$

$$= \tfrac{1}{2}e^{-x}(\sin x - \cos x) + C$$

28. $\displaystyle\int e^{2x}\sin 3x\ dx \quad \boxed{\text{Parts or Formula 492}}$

$$= \frac{e^{2x}(2\sin 3x - 3\cos 3x)}{4 + 9} + C$$

$$= \tfrac{1}{13}e^{2x}(2\sin 3x - 3\cos 3x) + C$$

29. $\displaystyle\int \cos^{-1}(-x)\ dx \quad \boxed{\text{Parts or Formula 445}}$

$$= -\left[(-x)\cos^{-1}(-x) - \sqrt{1 - x^2}\right] + C$$

$$= x\cos^{-1}(-x) + \sqrt{1 - x^2} + C$$

or $x\sin^{-1}x + \dfrac{\pi x}{2} + \sqrt{1 - x^2} + C$

30. $\displaystyle\int \sin^{-1}2x\ dx \quad \boxed{\text{Parts or Formula 451}}$

$$= x\sin^{-1}2x + \sqrt{\tfrac{1}{4} - x^2} + C$$

$$= x\sin^{-1}2x + \tfrac{1}{2}\sqrt{1 - 4x^2} + C$$

31. $\displaystyle\int \sin^3x\ dx = \int \sin^2x \sin x\ dx$

$$= -\int (1 - \cos^2x)(-\sin x)\ dx$$

$$= -\cos x + \tfrac{1}{3}\cos^3x + C$$

32. $\displaystyle\int \cos^5x\ dx = \int (1 - \sin^2x)^2 \cos x\ dx$

$$= \int (1 - 2\sin^2x + \sin^4x)\cos x\ dx$$

$$= \sin x - \tfrac{2}{3}\sin^3x + \tfrac{1}{5}\sin^5x + C$$

33. $\displaystyle\int \sin^3x \cos^2x\ dx$

$$= -\int (1 - \cos^2x)\cos^2x(-\sin x\ dx)$$

$$= \int (\cos^4x - \cos^2x)(-\sin x\ dx)$$

$$= \tfrac{1}{5}\cos^5x - \tfrac{1}{3}\cos^3x + C$$

34. $\displaystyle\int \sin^3x \cos^3x\ dx = \tfrac{1}{8}\int (2\sin x \cos x)^3\ dx$

$$= \tfrac{1}{8}\int \sin^3 2x\ dx = \tfrac{1}{8}\int (1 - \cos^2 2x)\sin 2x\ dx$$

$$= -\tfrac{1}{16}(\cos 2x - \tfrac{1}{3}\cos^3 2x) + C$$

35. $\displaystyle\int \sin^2x \cos^4x\ dx$

$$= \tfrac{1}{8}\int (1 - \cos 2x)(1 + \cos 2x)^2\ dx$$

$$= \tfrac{1}{8}\int (1 - \cos^2 2x)(1 + \cos 2x)\ dx$$

$$= \tfrac{1}{8}\int \sin^2 2x(1 + \cos 2x)\ dx$$

$$= \tfrac{1}{8}\int \sin^2 2x\ dx + \tfrac{1}{8}\int \sin^2 2x(\cos 2x)\ dx$$

$$= \frac{1}{16}\left[x - \frac{1}{4}\sin 4x\right] + \frac{1}{16}\cdot\frac{1}{3}\sin^3 2x + C$$

$$= \frac{1}{16}x - \frac{1}{64}\sin 4x + \frac{1}{48}\sin^3 2x + C$$

36. $\displaystyle\int \sin^2 x \cos^5 x \, dx = \int \sin^2 x (\cos^2 x)^2 \cos x \, dx$

$$= \int \sin^2 x (1 - \sin^2 x)^2 \cos x \, dx$$

$$= \int (\sin^2 x - 2\sin^4 x + \sin^6 x)\cos x \, dx$$

$$= \frac{1}{3}\sin^3 x - \frac{2}{5}\sin^5 x + \frac{1}{7}\sin^7 x + C$$

37. $\displaystyle\int \sin^5 x \cos^4 x \, dx$

$$= -\int (1 - \cos^2 x)^2 \cos^4 x (-\sin x \, dx)$$

$$= -\int (\cos^8 x - 2\cos^6 x + \cos^4 x)(-\sin x \, dx)$$

$$= -\frac{1}{9}\cos^9 x + \frac{2}{7}\cos^7 x - \frac{1}{5}\cos^5 x + C$$

38. $\displaystyle\int \sin^4 x \cos^2 x \, dx$

$$= \frac{1}{8}\int (1 - \cos 2x)^2 (1 + \cos 2x) \, dx$$

$$= \frac{1}{8}\int (1 - \cos 2x)(1 - \cos^2 2x) \, dx$$

$$= \frac{1}{8}\int (1 - \cos 2x)(\sin^2 2x) \, dx$$

$$= \frac{1}{8}\int \sin^2 2x \, dx - \frac{1}{8}\int \sin^2 2x \cos 2x \, dx$$

$$= \frac{1}{16}\int (1 - \cos 4x) \, dx - \frac{1}{48}\int \sin^3 2x \, dx$$

$$= \frac{1}{16}x - \frac{1}{64}\sin 4x - \frac{1}{48}\sin^3 2x + C$$

39. $\displaystyle\int \tan^5 x \sec^4 x \, dx = \int \tan^5 x (\tan^2 x + 1)\sec^2 x \, dx$

$$= \int \tan^7 x \sec^2 x \, dx + \int \tan^5 x \sec^2 x \, dx$$

$$= \frac{1}{8}\tan^8 x + \frac{1}{6}\tan^6 x + C$$

40. $\displaystyle\int \tan^4 x \sec^4 x \, dx = \int \tan^4 x (\tan^2 x + 1)\sec^2 x \, dx$

$$= \int \tan^6 x \sec^2 x \, dx + \int \tan^4 x \sec^2 x \, dx$$

$$= \frac{1}{7}\tan^7 x + \frac{1}{5}\tan^5 x + C$$

41. $\displaystyle\int \frac{\sqrt{1 - x^2}}{x} \, dx$ $\boxed{\text{Formula 235}}$

$$= \sqrt{1 - x^2} - \ln\left|\frac{1 + \sqrt{1 - x^2}}{x}\right| + C$$

42. $\displaystyle\int \frac{dx}{\sqrt{x^2 - 16}}$ $\boxed{\text{Formula 196}}$

$$= \ln\left|x + \sqrt{x^2 - 16}\right| + C$$

43. $\displaystyle\int \frac{2x + 3}{\sqrt{2x^2 - 1}} \, dx = \int \frac{2x \, dx}{\sqrt{2x^2 - 1}} + \int \frac{3 \, dx}{\sqrt{2x^2 - 1}}$

$$= \frac{1}{2}\int \frac{4x \, dx}{\sqrt{2x^2 - 1}} + \frac{3}{\sqrt{2}}\int \frac{dx}{\sqrt{x^2 - \frac{1}{2}}}$$

$\boxed{u = 2x^2 - 1;\ du = 4x \, dx}$ $\boxed{\text{Formula 196}}$

$$= \frac{1}{2}\int u^{-1/2} \, du + \frac{3}{\sqrt{2}}\ln\left|x + \sqrt{x^2 - \frac{1}{2}}\right| + C_1$$

$$= \frac{1}{2}\cdot\frac{2}{1}u^{1/2} + \frac{3}{\sqrt{2}}\ln\left|\sqrt{2}x + \sqrt{2x^2 - 1}\right| + C$$

$$= \sqrt{2x^2 - 1} + \frac{3}{\sqrt{2}}\ln\left|\sqrt{2}x + \sqrt{2x^2 - 1}\right| + C$$

44. $\displaystyle\int \frac{dx}{x\sqrt{x^2 - 1}} = \sec^{-1}|x| + C$

45. $\displaystyle\int \frac{dx}{x\sqrt{x^2 + 1}}$ $\boxed{\text{Formula 176}}$

$$= -\ln\left|\frac{1 + \sqrt{x^2 + 1}}{x}\right| + C$$

$$= \ln|x| - \ln\left|1 + \sqrt{x^2 + 1}\right| + C$$

46. $\displaystyle\int x\sqrt{x^2 + 1} \, dx = \frac{1}{2}\int (x^2 + 1)^{1/2} (2x \, dx)$

$$= \frac{1}{3}(x^2 + 1)^{3/2} + C$$

47. $\displaystyle\int \frac{(2x + 1) \, dx}{\sqrt{4x - x^2 - 2}}$

$$= -\int \frac{(-2x + 4) \, dx}{\sqrt{4x - x^2 - 2}} + 5\int \frac{dx}{\sqrt{2 - (x - 2)^2}}$$

$\boxed{u = 4x - x^2 - 2;\ du = -2x + 4}$

$\boxed{t = x - 2;\ dt = dx}$

$$= -\int u^{-1/2} \, du + 5\int \frac{dt}{\sqrt{2 - t^2}}$$ $\boxed{\text{Formula 224}}$

$$= -2u^{1/2} + 5 \sin^{-1} \frac{t}{\sqrt{2}} + C$$

$$= -2\sqrt{4x - x^2 - 2} + 5 \sin^{-1} \frac{x-2}{\sqrt{2}} + C$$

48. $\displaystyle\int \sqrt{3 + 4x - 4x^2}\, dx$

$$= \int \sqrt{-4(x^2 - x + \tfrac{1}{4}) + 3 + 1}\, dx$$

$$= \int \sqrt{4 - (2x-1)^2}\, dx \quad \boxed{\text{Formula 231}}$$

$$= \frac{(2x-1)\sqrt{4-(2x-1)^2}}{2} + \frac{4}{2} \sin^{-1} \frac{2x-1}{2} + C$$

$$= \tfrac{1}{2}(2x - 1)\sqrt{3 + 4x - 4x^2}$$
$$+ 2 \sin^{-1}\!\left(\frac{2x-1}{2}\right) + C$$

49. $\displaystyle\int \frac{\cos x\, dx}{\sqrt{1 + \sin^2 x}} \quad \boxed{u = \sin x;\ du = \cos x\, dx}$

$$= \int \frac{du}{\sqrt{1 + u^2}} \quad \boxed{\text{Formula 172}}$$

$$= \ln\left| u + \sqrt{1 + u^2} \right| + 1$$

$$= \ln\left(\sin x + \sqrt{1 + \sin^2 x}\right) + C$$

50. $\displaystyle\int \frac{\sec^2 x\, dx}{\sqrt{\sec^2 x - 2}} = \int \frac{\sec^2 x\, dx}{\sqrt{1 + \tan^2 x - 2}}$

$$\boxed{u = \tan x;\ du = \sec^2 x\, dx}$$

$$= \int \frac{du}{\sqrt{u^2 - 1}} \quad \boxed{\text{Formula 196}}$$
$$= \ln\left| u + \sqrt{u^2 - 1} \right| + C$$

$$= \ln\left| \tan x + \sqrt{\tan^2 x - 1} \right| + C$$

51. $\displaystyle\int_0^2 \sqrt{4 - x^2}\, dx \quad \boxed{\text{Formula 231}}$

$$= \frac{x\sqrt{4 - x^2}}{2} + \frac{4}{2} \sin^{-1}\frac{x}{2} \Big|_0^2 = \pi$$

52. $\displaystyle\int_0^1 \frac{dx}{\sqrt{9 - x^2}} = \sin^{-1}\frac{x}{3}\Big|_0^1 = \sin^{-1}\frac{1}{3}$

53. $\displaystyle\int_0^1 \frac{dx}{(x^2 + 2)^{3/2}} \quad \boxed{\text{Formula 182}}$

$$= \frac{x}{2\sqrt{x^2 + 2}}\Big|_0^1 = \frac{1}{2\sqrt{3}} = \frac{\sqrt{3}}{6}$$

54. $\displaystyle\int_0^1 \frac{dt}{4t^2 + 4t + 5} = \int_0^1 \frac{dt}{4(t^2 + 4t + \tfrac{1}{4}) + 4}$

$$= \int_0^1 \frac{dt}{(2t + 1)^2 + 4} = \frac{1}{2}\int_0^1 \frac{2\, dt}{(2t + 1)^2 + 4}$$

$$= \frac{1}{4} \tan^{-1}\!\left(\frac{2t + 1}{2}\right)\Big|_0^1$$

$$= \tfrac{1}{4}(\tan^{-1}\tfrac{3}{2} - \tan^{-1}\tfrac{1}{2})$$

55. $\displaystyle\int_1^2 \frac{dx}{x^4\sqrt{x^2 + 3}} = \int_{x=1}^{x=2} \frac{-\sqrt{3}\,\csc\theta\cot\theta\,d\theta}{9\csc^4\theta\,\sqrt{3}\cot\theta}$

$$\boxed{\begin{array}{l} \text{Trigonometric substitution; } x = \sqrt{3}\csc\theta, \\ dx = -\sqrt{3}\csc\theta\cot\theta\,d\theta; \\ \sqrt{x^2 + 3} = \sqrt{3}\cot\theta. \end{array}}$$

$$= -\frac{1}{9}\int_{x=1}^{x=2} \csc^{-3}\theta\,d\theta = -\frac{1}{9}\int_{x=1}^{x=2} \sin^3\theta\,d\theta$$

$$= -\frac{1}{9}\Big[-\cos\theta + \tfrac{1}{3}\cos^3\theta\Big]\Big|_{x=1}^{x=2}$$

$$= \frac{1}{27}\left(\frac{\sqrt{x^2 + 3}}{x}\right)\left[-3 + \frac{x^2 + 3}{x}\right]\Big|_1^2$$

$$= \tfrac{1}{27}(2 + \tfrac{5}{8}\sqrt{7}) \approx 0.3153$$

56. $\displaystyle\int_0^2 \frac{x^3}{(3 + x^2)^{3/2}}\, dx = \int_{x=0}^{x=2} \frac{3\sqrt{3}\tan^3\theta(-\sqrt{3}\sec^2\theta\,d\theta)}{3\sqrt{3}\sec^3\theta}$

$$\boxed{\begin{array}{l} \text{Trigonometric substitution; } x = \sqrt{3}\tan\theta, \\ dx = -\sqrt{3}\sec^2\theta\,d\theta;\ \sqrt{x^2 + 3} = \sqrt{3}\sec\theta. \end{array}}$$

$$= \sqrt{3}\int_{x=0}^{x=1} \frac{\tan^3\theta}{\sec\theta}\,d\theta = \sqrt{3}\int_{x=0}^{x=1} \frac{\sin^3\theta}{\cos^2\theta}\,d\theta$$

$$= \sqrt{3}\int_{x=0}^{x=1} \frac{1 - \cos^2\theta}{\cos^2\theta}\,(\sin\theta\,d\theta)$$

$$= \sqrt{3}\int_{x=0}^{x=1} \frac{1 - u^2}{u^2}\,du = \sqrt{3}\int_{x=0}^{x=1} (1 - u^{-2})\,du$$

$$= \sqrt{3}(u + u^{-1})\Big|_{x=0}^{x=1}$$

$$= \sqrt{3}\left[\frac{\sqrt{3}}{\sqrt{3 + x^2}} + \frac{\sqrt{3 + x^2}}{\sqrt{3}}\right]\Big|_0^1$$

$$= \tfrac{10}{7}\sqrt{7} - 2\sqrt{3} \approx 0.3155$$

57. $\displaystyle\int_{-2}^{2\sqrt{3}} x^3\sqrt{x^2+4}\;dx$ $\boxed{\text{Formula 171}}$

$$= \frac{(x^2+4)^{5/2}}{5} - \frac{4(x^2+4)^{3/2}}{3}\Bigg|_{-2}^{2\sqrt{3}}$$

$$= \frac{1{,}792 - 64\sqrt{2}}{15} \approx 113.43$$

58. $\displaystyle\int_{0}^{\sqrt{5}} x^2\sqrt{5-x^2}\;dx$ $\boxed{\text{Formula 233}}$

$$= \left(-\frac{x(5-x^2)^{3/2}}{4} + \frac{5x\sqrt{5-x^2}}{8}\right.$$

$$\left. + \frac{25}{8}\sin^{-1}\frac{x}{\sqrt{5}}\right)\Bigg|_{0}^{\sqrt{5}}$$

$$= \frac{25\pi}{16} \approx 4.9087$$

$\boxed{\text{Note: trig substitution also works well.}}$

59. $\displaystyle\int_{0}^{\ln 2} e^t\sqrt{1+e^{2t}}\;dt = \int_{1}^{4} \sqrt{1+u^2}\;du$

$$= \frac{u\sqrt{1+u^2}}{2} + \frac{1}{2}\ln\left|u+\sqrt{1+u^2}\right|\Bigg|_{1}^{4}$$

$\boxed{\text{Formula 168}}$

$$= \tfrac{1}{2}[2\sqrt{5} + \ln(2+\sqrt{5}) - \sqrt{2} - \ln(1+\sqrt{2})]$$

$$\approx 1.8101$$

60. $\displaystyle\int_{1}^{4} \frac{\sqrt{x^2+9}}{x^3}\;dx$ $\boxed{\text{Formula 181}}$

$$= -\frac{\sqrt{x^2+9}}{2x^2} - \frac{1}{2(3)}\ln\left|\frac{3+\sqrt{x^2+9}}{x}\right|\Bigg|_{1}^{4}$$

$$= -\frac{\ln(2\sqrt{10}-6)}{6} + \frac{\sqrt{10}}{2} - \frac{5}{32} \approx 1.6124$$

61. $\displaystyle\int_{0}^{\pi/4} \sin^5 x\;dx = -\int_{0}^{\pi/4} (\sin^4 x)(-\sin x\;dx)$

$$= -\int_{0}^{\pi/4} (1-\cos^2 x)^2(-\sin x\;dx)$$

$$= -\int_{0}^{\pi/4} (\cos^4 x - 2\cos^2 x + 1)(-\sin x\;dx)$$

$$= \left(-\frac{\cos^5}{5} + \frac{2\cos^3 x}{3} - \cos x\right)\Bigg|_{0}^{\pi/4}$$

$$= -\frac{43\sqrt{2}}{120} + \frac{8}{15} \approx 0.0266$$

62. $\displaystyle\int_{0}^{\pi/4} \cos^6 x\;dx = \frac{1}{8}\int_{0}^{\pi/4} (1+\cos 2x)^3\;dx$

$$= \frac{1}{8}\int_{0}^{\pi/4} (1 + 3\cos 2x + 3\cos^2 2x + \cos^3 2x)\;dx$$

$$= \tfrac{1}{8}\left[x + \tfrac{3}{2}\sin 2x + \tfrac{1}{2}\sin 2x - \tfrac{1}{3}\sin^3 2x\right]\Bigg|_{0}^{\pi/4}$$

$$= \frac{5}{192}(3\pi + 8) \approx 0.4538$$

63. $\displaystyle\int_{0}^{\pi/4} \tan^4 x\;dx = \int_{0}^{\pi/4} (\sec^2 x - 1)\tan^2 x\;dx$

$$= \int_{0}^{\pi/4} \tan^2 x\sec^2 x\;dx - \int_{0}^{\pi/4} (\sec^2 x - 1)\;dx$$

$$= \left(\tfrac{1}{3}\tan^3 x - \tan x + x\right)\Bigg|_{0}^{\pi/4} = \frac{\pi}{4} - \frac{2}{3}$$

64. $\displaystyle\int_{0}^{\pi/3} \sec^4 x\;dx = \int_{0}^{\pi/3} \sec^2 x(\tan^2 x + 1)\;dx$

$$= \left(\tfrac{1}{3}\tan^3 x + \tan x\right)\Bigg|_{0}^{\pi/3} = 2\sqrt{3}$$

65. $\displaystyle\int \frac{e^x\;dx}{\sqrt{1+e^{2x}}}$ $\boxed{u = e^x;\; du = e^x dx}$

$$\int \frac{du}{\sqrt{1+u^2}} = \ln\left(\sqrt{1+e^{2x}} + e^x\right) + C$$

66. $\displaystyle\int \frac{dx}{x\sqrt{4x^2+4x+2}}$ $\boxed{u = 2x+1;\; du = 2\;dx}$

$$= \int \frac{2\;dx}{(2x)\sqrt{(2x+1)^2+1}}$$

$$= \int \frac{du}{(u-1)\sqrt{u^2+1}}$$

$\boxed{u = \cot\theta;\; du = -\csc^2\theta\;d\theta;\; \sqrt{u^2+1} = \csc\theta}$

$$= \int \frac{-\csc^2\theta\;d\theta}{(\cot\theta - 1)\csc\theta} = -\int \frac{\csc\theta\;d\theta}{\cot\theta - 1}$$

$$= -\int \frac{d\theta}{\cos\theta - \sin\theta} = -\int \frac{(\cos\theta + \sin\theta)\;d\theta}{\cos^2\theta - \sin^2\theta}$$

$$= -\int \frac{\cos\theta\;d\theta}{1 - 2\sin^2\theta} + \int \frac{\sin\theta\;d\theta}{2\cos^2\theta - 1}$$

$\boxed{s = \sin\theta \text{ in first integral, and } t = \cos\theta \text{ in second}}$

$$= -\int \frac{ds}{1 - 2s^2} + \int \frac{dt}{2t^2 - 1}$$

$\boxed{\text{Formulas 93, 74}}$

$$= \frac{\sqrt{2}}{4} \ln\left|\frac{\sqrt{2}\,s - 1}{\sqrt{2}\,s + 1}\right| + \frac{\sqrt{2}}{4} \ln\left|\frac{\sqrt{2}\,t - 1}{\sqrt{2} + 1}\right| + C$$

$$= \frac{\sqrt{2}}{2} \ln\left|\frac{\sqrt{2x^2 + 2x + 1} - x - 1}{x}\right| + C$$

67. $\int \frac{x^2 + 4x + 3}{x^3 + x^2 + x}\, dx = \int \frac{x^2 + 4x + 3}{x(x^2 + x + 1)}\, dx$

$$\frac{x^2 + 4x + 3}{x(x^2 + x + 1)} = \frac{A_1}{x} + \frac{A_2 x + B_1}{x^2 + x + 1}$$

Partial fraction decomposition gives
$A_1 = 3,\ A_2 = -2,\ B_1 = 1.$

$$\int \frac{x^2 + 4x + 3}{x(x^2 + x + 1)}\, dx = 3\int x^{-1}\,dx + \int \frac{-2x + 1}{x^2 + x + 1}\, dx$$

$$= 3\ln|x| + \int \frac{-2x - 1}{x^2 + x + 1}\, dx + 2\int \frac{dx}{(x + \frac{1}{2})^2 + \frac{3}{4}}$$

$$= 3\ln|x| - \ln|x^2 + x + 1|$$

$$+ 2\left(\frac{2}{\sqrt{3}}\right)\left(\tan^{-1}\frac{2(x + \frac{1}{2})}{\sqrt{3}}\right) + C$$

$$= 3\ln|x| - \ln|x^2 + x + 1|$$

$$+ \frac{4}{\sqrt{3}}\tan^{-1}\frac{\sqrt{3}}{3}(2x + 1) + C$$

68. $\int \frac{6x^2 + 8x}{x^3 + 2x^2}\, dx$

$\boxed{u = x^3 + 2x^2;\ du = 3x^2 + 4x\ dx}$

$$= 2\int \frac{du}{u} = 2\ln|x^3 + 2x^2| + C$$

69. $\int \frac{5x^2 + 18x + 34}{(x - 7)(x + 2)^2}\, dx$

$$\frac{5x^2 + 18x + 34}{(x - 7)(x + 2)^2} = \frac{A_1}{x - 7} + \frac{A_2}{x + 2} + \frac{A_3}{(x + 2)^2}$$

Partial fraction decomposition gives
$A_1 = 5,\ A_2 = 0,\ A_3 = -2$

$$\int \frac{5x^2 + 18x + 34}{(x - 7)(x + 2)^2}\, dx = 5\int \frac{dx}{x - 7} - 2\int \frac{dx}{x + 2}$$

$$= 5\ln|x - 7| + 2(x + 2)^{-1} + C$$

70. $\int \frac{-3x^2 + 9x + 21}{(x + 2)^2(2x + 1)}\, dx$

$$\frac{-3x^2 + 9x + 21}{(x + 2)^2(2x + 1)} = \frac{A_1}{x + 2} + \frac{A_2}{(x + 2)^2} + \frac{A_3}{2x + 1}$$

Partial fraction decomposition gives
$A_1 = -5,\ A_2 = 3,\ A_3 = 7.$

$$\int \frac{-3x^2 + 9x + 21}{(x + 2)^2(2x + 1)}\, dx$$

$$= -5\int \frac{dx}{x + 2} + 3\int \frac{dx}{(x + 2)^2} + 7\int \frac{dx}{2x + 1}$$

$$= 5\ln|x + 2| - 3(x + 2)^{-1} + \tfrac{7}{2}\ln|2x + 1| + C$$

71. $\int \frac{3x + 5}{x^2 + 2x + 1}\, dx = 3\int \frac{x + 1}{(x + 1)^2}\, dx + 2\int \frac{dx}{(x + 1)}$

$$= 3\ln|x + 1| - 2(x + 1)^{-1} + C$$

72. $\int \frac{3x^2 + 2x + 1}{x^3 + x^2 + x}\, dx = \int \frac{du}{u} = \ln|x^3 + x^2 + x| + C$

$\boxed{u = x^3 + x^2 + x;\ du = 3x^2 + 2x + 1}$

73. $\int \frac{5x^2 + 3x - 2}{x^3 + 2x^2}\, dx = \int \frac{5x^2 + 3x - 2}{x^2(x + 2)}\, dx$

$$\frac{5x^2 + 3x - 2}{x^2(x + 1)} = \frac{A_1}{x} + \frac{A_2}{x^2} + \frac{A_3}{x + 2}$$

Partial fraction decomposition gives
$A_1 = 2,\ A_2 = -1,\ A_3 = 3.$

$$\int \frac{5x^2 + 3x - 2}{x^2(x + 2)}\, dx$$

$$= 2\int \frac{dx}{x} - \int \frac{dx}{x^2} + 3\int \frac{1}{x + 2}\, dx$$

$$= 2\ln|x| + \tfrac{1}{x} + 3\ln|x + 2| + C$$

74. $\int \frac{6x + 5}{4x^2 + 4x + 1}\, dx = \int \frac{6x + 3 + 2}{(2x + 1)^2}\, d$

$$= \int \frac{3\,dx}{2x + 1} + \int \frac{2}{(2x + 1)^2}\, dx$$

$$= \tfrac{3}{2}\ln|2x + 1| - (2x + 1)^{-1} + C$$

75. $\int \frac{x\,dx}{(x + 1)(x + 2)(x + 3)}$

$$\frac{x}{(x+1)(x+2)(x+3)}$$

$$= \frac{A_1}{x+1} + \frac{A_2}{x+2} + \frac{A_3}{x+3}$$

Partial fraction decomposition gives
$A_1 = -\frac{1}{2}$, $A_2 = 2$; $A_3 = -\frac{3}{2}$.

$$\int \frac{x\,dx}{(x+1)(x+2)(x+3)}$$

$$= -\frac{1}{2}\int \frac{dx}{x+1} + 2\int \frac{dx}{x+2} - \frac{3}{2}\int \frac{dx}{x+3}$$

$$= -\frac{1}{2}\ln|x+1| + 2\ln|x+2| - \frac{3}{2}\ln|x+3| + C$$

76. $\displaystyle\int \frac{5x^2 - 4x + 9}{x^3 - x^2 + 4x - 4}\,dx = \int \frac{5x^2 - 4x + 9}{(x^2 + 4)(x - 1)}\,dx$

$$\frac{5x^2 - 4x + 9}{(x^2 + 4)(x - 1)} = \frac{A_1 x + B_1}{x^2 + 4} + \frac{A_2}{x - 1}$$

Partial fraction decomposition gives
$A_1 = 3$, $A_2 = 2$, $B_1 = -1$.

$$\int \frac{5x^2 - 4x + 9}{(x^2 + 4)(x - 1)}\,dx$$

$$= 3\int \frac{x\,dx}{x^2 + 4} - \int \frac{dx}{x^2 + 4} + 2\int \frac{dx}{x - 1}$$

$$= \frac{3}{2}\ln(x^2 + 4) - \frac{1}{2}\tan^{-1}\frac{|x|}{2} + 2\ln|x + 1| + C$$

77. $\displaystyle\int \frac{dx}{5\cos x - 12\sin x}$ $\boxed{\text{Weierstrass substitution}}$

$$= \int \frac{2\,du}{5(1 - u^2) - 24u} = \int \frac{-2\,du}{5u^2 + 24u - 5}$$

$$\frac{-2}{(u + 5)(5u - 1)} = \frac{A_1}{u + 5} + \frac{A_2}{5u - 1}$$

Partial fraction decomposition gives
$A_1 = \frac{1}{13}$, $A_2 = -\frac{5}{13}$.

$$\int \frac{-2\,du}{(u+5)(5u-1)} = \frac{1}{13}\int \frac{du}{u + 5} - \frac{5}{13}\int \frac{du}{5u - 1}$$

$$= \frac{1}{13}\bigl(\ln|u + 5| - \ln|5u - 1|\bigr) + C$$

$$= \frac{1}{13}\ln\left|\frac{u + 5}{5u - 1}\right| + C$$

$$= \frac{1}{13}\ln\left|\frac{\tan\frac{x}{2} + 5}{5\tan\frac{x}{2} - 1}\right| + C$$

78. On $[0, 4]$ the graph never drops below the x-axis.

$$A = \int_0^4 (\sin x + \cos 2x)\,dx$$

$$= -\cos x + \sin x \cos x \Big|_0^4$$

$$\approx 2.15$$

79. $\displaystyle\text{AV} = \frac{1}{1 - 0}\int_0^1 x \sin^3 x^2\,dx$

$$= \frac{1}{2}\int_0^1 (1 - \cos^2 x^2)\sin x^2 (2x\,dx)$$

$$= \frac{1}{2}\left[-\cos x^2 + \frac{\cos^3 x^2}{3}\right]\Big|_0^1$$

$$= \frac{1}{6}\cos^3 1 - \frac{1}{2}\cos 1 + \frac{1}{3} \approx 0.09$$

80. $\displaystyle s(t) = \int_0^{\pi/3} (\sin t + \sin^2 t \cos^3 t)\,dt$

$$= \int_0^{\pi/3} \sin t\,dt + \int_0^{\pi/3} \sin^2 t(1 - \sin^2 t)\cos t\,dt$$

$$= \left[\cos t + \frac{1}{3}\sin^3 t - \frac{1}{5}\sin^5 t\right]\Big|_0^{\pi/3}$$

$$= \frac{1}{2} + \frac{11\sqrt{3}}{160} \approx 0.0619078$$

81. $\displaystyle I = \int \cot^m x \csc^n x\,dx$

a. m and n are both odd; separate out $\csc x \cot x\,dx$ and then use the identity $\cot^2 x + 1 = \csc^2 x$ to find the integral of a power of cosecant.

b. n is even; separate out $\csc^2 x\,dx$ and use the identity $\cot^2 x + 1 = \csc^2 x$ to find the integral of a power of cotangent.

c. m is even and n is odd;, use the identity $\cot^2 x + 1 = \csc^2 x$ and integrate by parts.

82. $\displaystyle V = \pi\int_0^{\pi/2} \cos^2 x\,dx = \pi\left[\frac{x}{2} + \frac{1}{4}\sin 2x\right]\Big|_0^{\pi/2} = \frac{\pi^2}{4}$

83. $\displaystyle\text{AV} = \frac{1}{\frac{\pi}{4} - 0}\int_0^{\pi/4} \sec x\,dx$

$$= \frac{4}{\pi}[\ln|\sec x + \tan x|]\Big|_0^{\pi/4} = \frac{4}{\pi}\ln(1 + \sqrt{2})$$

84. $\displaystyle\text{AV} = \int_0^{\pi/4} \csc x\,dx = -\ln|\csc x + \cot x|\Big|_0^{\pi/4}$

$$\approx 1.05$$

85. $\quad A = \displaystyle\int_0^4 \frac{2x \, dx}{\sqrt{x^2+9}} = 2\sqrt{x^2+9}\,\Big|_0^4 = 4$

86. $\quad V = \pi \displaystyle\int_0^2 xe^{-2x^2} \, dx = -\tfrac{1}{4}\pi e^{-2x^2}\,\Big|_0^2$

$\qquad = -\dfrac{\pi}{4}(e^{-8}-1)$

87. $\quad V = 2\pi \displaystyle\int_0^3 \left(x\sqrt{9-x^2}\right)^2 \, dx$

$\qquad = 2\pi \displaystyle\int_0^3 (9x^2 - x^4) \, dx$

$\qquad = 2\pi\left[3x^3 - \dfrac{x^5}{5}\right]\Big|_0^3 = \dfrac{324\pi}{5} \approx 203.5752$

88. $\quad y = \ln(\sec x)$, then $y' = \dfrac{\sec x \tan x}{\sec x}$;

$\qquad ds = \sqrt{1 + \tan^2 x}\, dx = \sec x \, dx$

$\qquad s = \displaystyle\int_0^{\pi/4} \sec x \, dx = \ln|\sec x + \tan x|\,\Big|_0^{\pi/4}$

$\qquad = \ln(1 + \sqrt{2})$

89. $\quad m = \rho \displaystyle\int_0^1 x^2 e^{-x} dx$

$\qquad = \rho\left(-x^2 e^{-x} - 2xe^{-x} - 2e^{-x}\right)\Big|_0^1$

$\qquad = 2 - 5e^{-1}$

$\qquad M_y = \rho \displaystyle\int_0^1 x^3 e^{-x} \, dx = 6 - 16e^{-1}$

$\qquad M_x = \dfrac{\rho}{2}\displaystyle\int_0^1 x^4 e^{-2x} \, dx = \dfrac{3}{8} - \dfrac{21e^{-2}}{8}$

$\qquad (\overline{x}, \overline{y}) \approx (0.709, 0.123)$

90. $\quad I = \displaystyle\int e^{ax} \sin bx \, dx \quad \boxed{u = e^{ax};\; dv = \sin bx \, dx}$

$\qquad = -\dfrac{1}{b} e^{ax} \cos bx + \dfrac{a}{b}\displaystyle\int e^{ax} \cos bx \, dx$

$\qquad \boxed{u = e^{ax};\; dv = \cos bx \, dx}$

$\qquad = -\dfrac{1}{b} e^{ax} \cos bx + \dfrac{a}{b}\left(\dfrac{1}{b} e^{ax} \sin bx\right.$

$\qquad\quad \left. - \dfrac{a}{b}\displaystyle\int e^{ax} \sin bx \, dx\right)$

$\qquad = -\dfrac{1}{b} e^{ax} \cos bx + \dfrac{a}{b}\left(\dfrac{1}{b} e^{ax} \sin bx - \dfrac{a}{b} I\right)$

$\qquad = -\dfrac{1}{b} e^{ax} \cos bx + \dfrac{a}{b^2} e^{ax} \sin bx - \dfrac{a^2}{b^2} I$

$I + \dfrac{a^2}{b^2} I = -\dfrac{1}{b} e^{ax} \cos bx + \dfrac{a}{b^2} e^{ax} \sin bx + C$

$\left(\dfrac{b^2 + a^2}{b^2}\right) I = \dfrac{(a \sin bx - b \cos bx)e^{ax}}{b^2}$

$I = \dfrac{(a \sin bx - b \cos bx)e^{ax}}{a^2 + b^2}$

91. $\quad I = \displaystyle\int e^{ax} \cos bx \, dx \quad \boxed{u = \cos bx;\; dv = e^{ax} \, dx}$

$\qquad = \dfrac{1}{a} e^{ax} \cos bx + \dfrac{b}{a}\displaystyle\int e^{ax} \sin bx \, dx$

$\qquad \boxed{u = \sin bx;\; dv = e^{ax} \, dx}$

$\qquad = \dfrac{1}{a} e^{ax} \cos bx + \dfrac{b}{a}\left(\dfrac{1}{a} e^{ax} \sin bx\right.$

$\qquad\quad \left. - \dfrac{b}{a}\displaystyle\int e^{ax} \cos bx \, dx\right)$

$\qquad = \dfrac{1}{a} e^{ax} \cos bx + \dfrac{b}{a}\left(\dfrac{1}{a} e^{ax} \sin bx - \dfrac{b}{a} I\right)$

$\qquad = \dfrac{1}{a} e^{ax} \cos bx + \dfrac{b}{a^2} e^{ax} \sin bx - \dfrac{b^2}{a^2} I$

$I + \dfrac{b^2}{a^2} I = \dfrac{1}{a} e^{ax} \cos bx + \dfrac{b}{a^2} e^{ax} \sin bx$

$\left(\dfrac{a^2 + b^2}{a^2}\right) I = \dfrac{(a \cos bx + b \sin bx)e^{ax}}{a^2} + C$

$I = \dfrac{(a \cos bx + b \sin bx)e^{ax}}{a^2 + b^2} + C$

92. $\quad \displaystyle\int_a^b xf''(x) \, dx \quad \boxed{u = x;\; dv = f'(x) \, dx}$

$\qquad = xf'(x)\,\Big|_a^b - \displaystyle\int_a^b f'(x) \, dx$

$\qquad = bf'(b) - af'(a) - f(x)\,\Big|_a^b$

$\qquad = bf'(b) - af'(a) - f(b) + f(a)$

93. $\quad \displaystyle\int x^m (\ln x)^n \, dx \quad \boxed{u = (\ln x)^n;\; dv = x^m \, dx}$

$\qquad = \dfrac{x^{m+1}(\ln x)^n}{m+1} - \dfrac{n}{m+1}\displaystyle\int x^m (\ln x)^{n-1} \, dx$

For example,

$\displaystyle\int x^2 (\ln x)^3 \, dx = \dfrac{x^3 (\ln x)^3}{3} - \dfrac{x^3 (\ln x)^2}{3}$

$\qquad\qquad + \dfrac{2}{3}\displaystyle\int x^2 (\ln x) \, dx$

$\qquad = \dfrac{x^3}{3}(\ln x)^3 - \dfrac{x^3}{3}(\ln x)^2$

$$+ \frac{2}{3}\left[\frac{x^3(\ln x)}{3} - \frac{n}{3}\int x^2 \, dx \right]$$

$$= \frac{x^3}{3}(\ln x)^3 - \frac{x^3}{3}(\ln x)^2 + \frac{2x^3}{9}(\ln x) - \frac{2x^3}{27} + C$$

94. $I = \displaystyle\int \sin^n Ax \, dx$

$$\boxed{u = \sin^n Ax; \ dv = \sin Ax \, dx}$$

$$= -\frac{1}{A}\sin^{n-1}ax\cos Ax$$

$$+ (n-1)\int \sin^{n-2}Ax\cos^2 Ax \, dx$$

$$= -\frac{1}{A}\sin^{n-1}ax\cos Ax$$

$$+ (n-1)\int \sin^{n-2}Ax(1 - \sin^2 Ax) \, dx$$

$$= -\frac{1}{A}\sin^{n-1}ax\cos Ax$$

$$+ (n-1)\int \sin^{n-2}Ax \, dx$$

$$- (n-1)\int \sin^n Ax \, dx$$

$$= -\frac{1}{A}\sin^{n-1}ax\cos Ax$$

$$+ (n-1)\int \sin^{n-2}Ax \, dx - (n-1)I$$

$$I + (n-1) = -\frac{1}{A}\sin^{n-1}ax\cos Ax$$

$$+ (n-1)\int \sin^{n-2}Ax \, dx$$

$$I = -\frac{1}{nA}\sin^{n-1}Ax\cos Ax$$

$$+ \frac{n-1}{n}\int \sin^{n-2}Ax \, dx + C$$

For example,

$$\int \sin^3 4x \, dx = -\frac{1}{12}\sin^2 4x \cos 4x + \frac{2}{3}\int \sin 4x \, dx$$

$$= -\frac{1}{12}\sin^2 4x \cos 4x - \frac{1}{6}\cos 4x + C$$

7.5 Improper Integrals, Page 476

1. An improper integral is a definite integral with either the integrand undefined on the interval or a limit to infinity

2. To evaluate $\displaystyle\int_a^{+\infty} f(x) \, dx$ investigate

$I = \displaystyle\int_0^N f(x) \, dx$ and then find $\displaystyle\lim_{N \to +\infty} I.$

To evaluate $\displaystyle\int_{-\infty}^b f(x) \, dx$ investigate

$I = \displaystyle\int_n^b f(x) \, dx$ and then find $\displaystyle\lim_{n \to -\infty} I.$

To evaluate $\displaystyle\int_a^b f(x) \, dx$ with $f(c)$ undefined and $a < c < b$, find $I_1 = \displaystyle\lim_{\epsilon \to 0}\int_a^{c-\epsilon} f(x) \, dx$ and

$I_2 = \displaystyle\lim_{\epsilon \to 0}\int_{b+\epsilon}^b f(x) \, dx.$

3. $\displaystyle\int_1^{+\infty} \frac{dx}{x^3} = \lim_{t \to +\infty}\int_1^t \frac{dx}{x^3} = \lim_{t \to +\infty} -\frac{1}{2x^2}\bigg|_0^t$

$$= \lim_{t \to +\infty}\left(-\frac{1}{2t^2} + \frac{1}{2} \right) = \frac{1}{2}$$

4. $\displaystyle\int_1^{+\infty} \frac{dx}{\sqrt[3]{x}} = \lim_{t \to +\infty}\int_1^t x^{-1/3} \, dx$

$$= \frac{3}{2}\lim_{t \to +\infty}x^{2/3}\bigg|_1^t = +\infty; \text{ diverges}$$

5. $\displaystyle\int_1^{+\infty} \frac{dx}{x^{0.99}} = \lim_{t \to +\infty}\int_1^t \frac{dx}{x^{0.99}}$

$$= 100\lim_{t \to +\infty}x^{0.01}\bigg|_1^t$$

$$= 100\lim_{t \to +\infty}(t^{0.01} - 1) = +\infty; \text{ diverges}$$

6. $\displaystyle\int_1^{+\infty} \frac{dx}{\sqrt{x}} = \lim_{t \to +\infty}\int_0^t x^{-1/2} \, dx$

$$= 2\lim_{t \to +\infty}(t^{1/2} - 1) = +\infty; \text{ diverges}$$

7. $\displaystyle\int_1^{+\infty} \frac{dx}{x^{1.1}} = \lim_{t \to +\infty}\int_1^t \frac{dx}{x^{1.1}}$

$$= \lim_{t \to +\infty}\left[-\frac{1}{0.1x^{0.1}} \right]\bigg|_1^t$$

$$= \lim_{t \to +\infty}\left[-\frac{10}{t^{0.1}} + 10 \right] = 10$$

8. $\displaystyle\int_1^{+\infty} x^{-2/3}\,dx = \lim_{t\to+\infty}\int_1^t x^{-2/3}\,dx$

$\displaystyle = 3\lim_{t\to+\infty} x^{1/3}\Big|_1^t = 3\lim_{t\to+\infty}(t^{1/3}-1)$

$= +\infty;$ diverges

9. $\displaystyle\int_3^{+\infty} \frac{dx}{2x-1} = \frac{1}{2}\lim_{t\to+\infty}\int_3^t \frac{2\,dx}{2x-1}$

$\displaystyle = \frac{1}{2}\lim_{t\to+\infty} \ln|2x-1|\Big|_3^t = +\infty;$ diverges

10. $\displaystyle\int_3^{+\infty} \frac{dx}{\sqrt[3]{2x-1}} = \frac{1}{2}\lim_{t\to+\infty}\int_3^t (2x-1)^{-1/3}(2\,dx)$

$\displaystyle = \frac{3}{4}\lim_{t\to+\infty}(2x-1)^{2/3}\Big|_3^t = +\infty;$ diverges

11. $\displaystyle\int_3^{+\infty} \frac{dx}{(2x-1)^2} = \lim_{t\to+\infty}\int_3^t \frac{dx}{(2x-1)^2}$

$\displaystyle = \frac{1}{2}\int_3^t \frac{2\,dx}{(2x-1)^2} = -\frac{1}{2}\lim_{t\to+\infty}(2x-1)^{-1}\Big|_3^t$

$\displaystyle = \lim_{t\to+\infty}\left[-\frac{1}{4t-2}+\frac{1}{10}\right] = \frac{1}{10}$

12. $\displaystyle\int_0^{+\infty} e^{-x}\,dx = \lim_{t\to+\infty}\int_0^t e^{-x}\,dx$

$\displaystyle = \lim_{t\to+\infty} e^{-x}\Big|_0^1 = 1$

13. $\displaystyle\int_0^{+\infty} 5e^{-2x}\,dx = 5\lim_{t\to+\infty}\int_0^t e^{-2x}\,dx$

$\displaystyle = -\frac{5}{2}\lim_{t\to+\infty} e^{-2x}\Big|_0^t = \frac{5}{2}$

14. $\displaystyle\int_1^{+\infty} e^{1-x}\,dx = \lim_{t\to+\infty}\int_0^t e^{1-x}\,dx$

$\displaystyle = -\lim_{t\to+\infty} e^{1-x}\Big|_1^t = 1$

15. $\displaystyle\int_1^{+\infty} \frac{x^2\,dx}{(x^3+2)^2} = \frac{1}{3}\lim_{t\to+\infty}\int_1^t \frac{3x^2\,dx}{(x^3+2)^2}$

$\displaystyle = \lim_{t\to+\infty}\frac{1}{3}\left(-\frac{1}{x^3+2}\right)\Big|_1^t$

$\displaystyle = \lim_{t\to\infty}\left(-\frac{1}{3t^3+6}+\frac{1}{9}\right) = \frac{1}{9}$

16. $\displaystyle\int_1^{+\infty} \frac{x^2\,dx}{x^3+2} = \frac{1}{3}\lim_{t\to+\infty}\int_1^t \frac{3x^2\,dx}{x^3+2}$

$\displaystyle = \frac{1}{3}\lim_{t\to+\infty} \ln|x^3+2|\Big|_1^t = +\infty;$ diverges

17. $\displaystyle\int_1^{+\infty} \frac{x^2\,dx}{\sqrt{x^3+2}} = \frac{1}{3}\lim_{t\to+\infty}\int_1^t (x^3+2)^{-1/2}(3x^2\,dx)$

$\displaystyle = \frac{2}{3}\lim_{t\to+\infty}\sqrt{x^3+2}\Big|_1^t = +\infty;$ diverges

18. $\displaystyle\int_0^{+\infty} xe^{-x^2}\,dx = -\frac{1}{2}\lim_{t\to+\infty}\int_0^t e^{-x^2}(-2x^2\,dx)$

$\displaystyle = -\frac{1}{2}\lim_{t\to+\infty} e^{-x^2}\Big|_0^t = \frac{1}{2}$

19. $\displaystyle\int_1^{+\infty} \frac{e^{-\sqrt{x}}}{\sqrt{x}}\,dx = \lim_{t\to+\infty}\int_1^t \frac{e^{-\sqrt{x}}}{\sqrt{x}}\,dx$

$\displaystyle = -2\lim_{t\to+\infty}\int_1^t -\frac{e^{-\sqrt{x}}}{2\sqrt{x}}\,dx$

$\displaystyle = -2\lim_{t\to+\infty} e^{-\sqrt{x}}\Big|_1^t$

$\displaystyle = -2\lim_{t\to+\infty}\left(\frac{1}{e^{\sqrt{t}}}-\frac{1}{e}\right) = \frac{2}{e}$

20. $\displaystyle\int_0^{+\infty} xe^{-x}\,dx \quad \boxed{u=x;\ dv=e^{-x}\,dx}$

$\displaystyle = \lim_{t\to+\infty}\left[-xe^{-x}\Big|_0^t + \int_0^t e^{-x}\,dx\right]$

$\displaystyle = -\lim_{t\to+\infty}(-e^{-x})\Big|_1^t = 1$

Note: $\displaystyle\lim_{x\to+\infty}\frac{x}{e^x} = \lim_{x\to+\infty}\frac{1}{e^x} = 0$

by l'Hôpital's rule.

21. $\displaystyle\int_0^{+\infty} 5xe^{10-x}\,dx = 5x \quad \boxed{u = x;\ dv = e^{-x}\,dx}$

$$= \lim_{t\to+\infty}\left[-5e^{10}xe^{-x}\Big|_0^t + \int_0^t e^{-x}\,dx\right]$$

$$= 5e^{10} - \lim_{t\to+\infty}(-e^{-x})\Big|_1^t = 5e^{10}$$

Note: $\displaystyle\lim_{x\to+\infty}\frac{x}{e^x} = \lim_{x\to+\infty}\frac{1}{e^x} = 0$

by l'Hôpital's rule.

22. $\displaystyle\int_1^{+\infty}\frac{\ln x\,dx}{x} = \frac{1}{2}\lim_{t\to+\infty}(\ln x)^2\Big|_1^t = +\infty;$

diverges

23. $\displaystyle\int_2^{+\infty}\frac{dx}{x\ln x} = \lim_{t\to+\infty}\int_2^t(\ln x)^{-1}\frac{dx}{x}$

$$= \lim_{t\to+\infty}\ln(\ln x)\Big|_2^t$$

$$= \lim_{t\to\infty}[\ln(\ln t) - \ln(\ln 2)] = +\infty;\ \text{diverges}$$

24. $\displaystyle\int_2^{+\infty}\frac{dx}{x\sqrt{\ln x}} = \lim_{t\to+\infty}\int_2^t(\ln x)^{-1/2}\frac{dx}{x}$

$$= 2\lim_{t\to+\infty}(\ln x)^{1/2}\Big|_2^t = +\infty;\ \text{diverges}$$

25. $\displaystyle\int_0^{+\infty} x^2 e^{-x}\,dx \quad \boxed{\text{Parts or Formula 485}}$

$$= \lim_{t\to+\infty}\frac{e^{-x}}{-1}\left(x^2 - \frac{2x}{-1} + \frac{2}{1}\right)\Big|_0^t$$

$$= \lim_{t\to+\infty}[2 - e^{-t}(t^2 + 2t + 2)] = 2$$

26. $\displaystyle\int_0^{+\infty} x^3 e^{-x^2}\,dx \quad \boxed{u = x^2;\ du = 2x\,dx}$

$$= \frac{1}{2}\int_0^{+\infty} x^2 e^{-x^2}(2x\,dx) = \frac{1}{2}\int_0^{+\infty} ue^{-u}\,du = \frac{1}{2}$$

(from Problem 20)

27. $\displaystyle\int_{-\infty}^0 \frac{2x\,dx}{x^2+1} = \lim_{t\to-\infty}\int_t^0\frac{2x\,dx}{x^2+1}$

$$= \lim_{t\to-\infty}\ln(x^2+1)\Big|_t^0$$

$$= \lim_{t\to-\infty}-\ln(t^2+1) = -\infty;\ \text{diverges}$$

28. $\displaystyle\int_1^{+\infty}\frac{x\,dx}{(1+x^2)^2} = \frac{1}{2}\lim_{t\to+\infty}\int_0^t(1+x^2)^{-2}(2x\,dx)$

$$= -\frac{1}{2}\lim_{t\to+\infty}\frac{1}{1+x^2}\Big|_0^t = \frac{1}{2}$$

29. $\displaystyle\int_{-\infty}^0\frac{dx}{\sqrt{2-x}} = \lim_{t\to-\infty}\int_t^0(2-x)^{-1/2}\,dx$

$$= -2\lim_{t\to-\infty}2\sqrt{2-x}\Big|_t^0 = +\infty;\ \text{diverges}$$

30. $\displaystyle\int_{-\infty}^4\frac{dx}{(5-x)^2} = \lim_{t\to-\infty}\int_t^4(5-x)^{-2}\,dx$

$$= \lim_{t\to-\infty}\frac{1}{5-x}\Big|_t^4 = 1$$

31. $\displaystyle\int_{-\infty}^{\infty} xe^{-|x|}\,dx = 0$

This function is symmetric about the origin since $f(x) = -f(-x)$, and, therefore, is an odd function, which gives the value zero.

32. $\displaystyle\int_{-\infty}^{\infty}\frac{dx}{x^2+1};\quad f(x) = \frac{1}{x^2+1}$ is an even function,

$$\int_{-\infty}^{\infty}\frac{dx}{x^2+1} = 2\int_0^{+\infty}\frac{dx}{x^2+1} = 2\lim_{t\to+\infty}\tan^{-1}x\Big|_0^t$$

$$= 2\left(\tfrac{1}{2}\right)\pi = \pi$$

33. $\displaystyle\int_0^1\frac{dx}{x^{1/5}} = \lim_{t\to0^+}\int_t^1 x^{-1/5}\,dx$

$$= \frac{5}{4}\lim_{t\to0^+}x^{4/5}\Big|_t^1 = \frac{5}{4}$$

34. $\displaystyle\int_0^4\frac{dx}{x\sqrt{x}} = \lim_{t\to0^+}\int_t^4 x^{-3/2}\,dx$

$$= -2\lim_{t\to0^+}x^{-1/2}\Big|_t^4 = +\infty;\ \text{diverges}$$

35. $\displaystyle\int_0^1\frac{dx}{(1-x)^{1/2}} = \lim_{t\to1^-}\int_0^t(1-x)^{-1/2}\,dx$

$$= -2\lim_{t\to1^-}\sqrt{1-x}\Big|_0^t$$

$$= \lim_{t\to1^-}-2(\sqrt{1-t}-1) = 2$$

36. $\displaystyle\int_{-\infty}^{+\infty}\frac{3x\,dx}{(3x^2+2)^3} = 0$

This function is symmetric about the origin since $f(x) = -f(-x)$, and, therefore, is an

odd function, which gives the value zero.

37. $\displaystyle\int_0^1 \ln x\, dx = \lim_{t\to 0^+}\int_t^1 \ln x\, dx$

$$\boxed{\text{parts or Formula 499}}$$

$$= \lim_{t\to 0^+}(x\ln x - x)\Big|_t^1$$

$$= -1 - \lim_{t\to 0^+}\frac{t^{-1}}{-t^{-2}} = -1$$

38. $\displaystyle\int_1^{+\infty} \ln x\, dx = \lim_{t\to +\infty}\int_1^{+\infty} \ln x\, dx$

$$\boxed{\text{parts or Formula 499}}$$

$$= \lim_{t\to +\infty}(x\ln x - x)\Big|_1^t$$

$$= \lim_{t\to +\infty}\frac{\ln t - 1}{t^{-1}} = +\infty;\ \text{diverges}$$

39. $\displaystyle\int_e^{+\infty}\frac{dx}{x(\ln x)^2} = \lim_{t\to +\infty}\left[-\frac{1}{\ln x}\right]\Big|_e^t$

$$= \lim_{t\to +\infty}\left(-\frac{1}{\ln t}+1\right) = 1$$

40. $\displaystyle\int_0^1 \frac{x\, dx}{1-x^2} = -\frac{1}{2}\lim_{t\to 1^-}\int_0^t (1-x^2)^{-1}(-2x\, dx)$

$$= -\frac{1}{2}\lim_{t\to 1^-}\ln|1-x^2|\Big|_0^t = -\infty;\ \text{diverges}$$

41. $\displaystyle\int_0^1 e^{-(1/2)\ln x}dx = \lim_{t\to 0^+}\int_t^1 x^{-1/2}\, dx$

$$= \lim_{t\to 0^+} 2\sqrt{x}\,\Big|_t^1 = 2$$

42. $\displaystyle\int_0^{+\infty}\frac{dx}{e^x+e^{-x}} = \lim_{t\to +\infty}\int_0^t\frac{e^{-x}\, dx}{1+(e^{-x})^2}$

$$= \lim_{t\to +\infty}\tan^{-1}e^x\Big|_0^t = \frac{\pi}{4}$$

43. $\displaystyle A = \int_0^{+\infty}\frac{2\, dx}{(x-4)^3} = 2\lim_{t\to +\infty}\left(-\frac{1}{2}\right)(x-4)^{-2}\Big|_0^t$

$$= \lim_{t\to +\infty}[-(t-4)^{-2}-(6-4)^{-2}] = \frac{1}{4}$$

44. $\displaystyle y = \frac{2}{(x-4)^3} < 0$ on $(-\infty, 2]$;

$$A = -\int_{-\infty}^2\frac{2\, dx}{(x-4)^3} = 2\lim_{t\to -\infty}\frac{1}{2(x-4)^2}\Big|_t^2$$

$$= \frac{1}{4}$$

45. $\displaystyle A = \lim_{T\to +\infty}\int_0^T 200 e^{-0.002t}\, dt$

$$= -100,000 \lim_{T\to +\infty} e^{-0.002t}\Big|_0^t = 100,000$$

There are 100,000 millirads.

46. $\displaystyle N = 100\int_0^{+\infty}(e^{-0.02t}-e^{-0.1t})\, dt$

$$= 100\lim_{t\to +\infty}(-50e^{0.02t}-100e^{-0.1t})\Big|_0^t$$

$$= 4,000 \text{ thousands of barrels}$$

47. $\displaystyle I = \int_{-\infty}^{+\infty} f(x)\, dx = \lim_{t\to -\infty}\int_t^{-1} e^{x+1}dx + \int_{-1}^1 dx$

$$+ \lim_{t\to +\infty}\int_1^t x^{-2}\, dx$$

$$= \lim_{t\to -\infty} e^{x+1} + 2 - \lim_{t\to +\infty} x^{-1}\Big|_1^t = 4$$

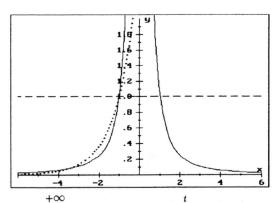

48. $\displaystyle I = \int_2^{+\infty}(\ln x)^{-p}\frac{dx}{x} = \lim_{t\to +\infty}\int_2^t(\ln x)^{-p}\frac{dx}{x}$

$$= \lim_{t\to +\infty}\frac{(\ln x)^{-p+1}}{1-p}\Big|_2^t$$

$$= \lim_{t\to +\infty}\frac{1}{(1-p)(\ln x)^{-1+p}}\Big|_2^t$$

$$= (p-1)(\ln 2)^{1-p} \text{ if } p > 1$$

49. $\displaystyle I = \int_0^1\frac{dx}{x^p} = \lim_{t\to 0^+}\int_t^1 x^{-p}\, dx = \lim_{t\to 0^+}\frac{x^{-p+1}}{1-p}\Big|_t^1$

$$= (1-p)^{-1} \text{ if } p < 1$$

50. $I = \int\limits_{0}^{1/2} \dfrac{dx}{x(\ln x)^p} = \lim\limits_{t \to 0^+} \int\limits_{t}^{1/2} (\ln x)^{-p} \dfrac{dx}{x}$

$= \lim\limits_{t \to 0^+} \dfrac{(\ln x)^{-p+1}}{1 - p} \Big|_{t}^{1/2}$

$= \lim\limits_{t \to 0^+} \dfrac{1}{(1 - p)(\ln x)^{p-1}} \Big|_{t}^{1/2}$

$= (1 - p)^{-1}(-\ln 2)^{p-1}$ if $p > 1$

51. The integration is correct, but has overlooked the fact that the function is undefined (has a vertical asymptote) at $x = 0$. It should be written as:

$$\lim\limits_{t \to 0^-} \int\limits_{-1}^{t} \dfrac{dx}{x^2} + \lim\limits_{t \to 0^+} \int\limits_{t}^{1} \dfrac{dx}{x^2}.$$

Both of these integrals diverge (see Example 3).

52. Although the answer is correct, the student did not work the problem correctly because l'Hôpital's rule was applied to a quotient that had not yet become undetermined.

$I = \int\limits_{1}^{+\infty} (x-1)e^{-x}dx = \lim\limits_{t \to +\infty} \int\limits_{1}^{t} (x-1)e^{-x}dx$

$\boxed{u = x - 1;\ dv = e^{-x}\,dx}$

$= \lim\limits_{t \to +\infty} \left[-(x - 1)e^{-x}\Big|_{1}^{t} + \int\limits_{0}^{t} e^{-x}\,dx \right]$

$= \lim\limits_{t \to +\infty} \left[-e^{t} + e^{-1} \right] = e^{-1}$

53. $I = \int\limits_{0}^{2} f(x)\,dx$

$= 4 \lim\limits_{t \to 0^+} x^{1/4}\Big|_{t}^{1} - 4 \lim\limits_{t \to 2^-} (2-x)^{1/4}\Big|_{1}^{t} = 8$

54. Consider the region with lower bound the sphere with center at the origin and radius 1 (there is no upper bound).

55. a. $\mathcal{L}\{f(x)\} = \lim\limits_{n \to +\infty} \int\limits_{0}^{n} e^{-st}f(t)\,dt$

$\mathcal{L}\{e^{at}\} = \lim\limits_{N \to +\infty} \int\limits_{0}^{N} e^{(a-s)t}e^{at}\,dt$

$= \dfrac{1}{a - s} \lim\limits_{N \to +\infty} e^{(a-s)at}\,dt\Big|_{0}^{N}$

$= -\dfrac{1}{a-s} = \dfrac{1}{s-a}$ since $s - a > 0$

b. $\mathcal{L}\{a\} = \lim\limits_{N \to \infty} -\dfrac{a}{s}\int\limits_{0}^{N} e^{-st}(-s\,dt)$

$= -\dfrac{a}{s}e^{-st}\Big|_{0}^{N} = \dfrac{a}{s}$

c. $\mathcal{L}\{\sin at\} = \int\limits_{0}^{+\infty} e^{-st}\sin at\,dt$

$= \lim\limits_{N \to +\infty} \dfrac{(-s\sin at - a\cos at)e^{-st}}{s^2 + a^2}\Big|_{0}^{N}$

$= \lim\limits_{N \to +\infty} \dfrac{(-s\sin aN - a\cos aN)e^{-sN}}{s^2 + a^2}$

$- \dfrac{-a}{s^2 + a^2} = \dfrac{a}{s^2 + a^2}$

The result of Problem 90, Section 7.4 was used.

d. $\mathcal{L}\{\cos at\} = \int\limits_{0}^{+\infty} e^{-st}\cos at\,dt$

$= \lim\limits_{N \to +\infty} \dfrac{(-s\cos at + a\sin at)e^{-st}}{s^2 + a^2}\Big|_{0}^{N}$

$= \lim\limits_{N \to +\infty} \dfrac{(-s\cos aN - a\sin aN)e^{-sN}}{s^2 + a^2}$

$- \dfrac{-s}{s^2 + a^2} = \dfrac{s}{s^2 + a^2}$

The result of Problem 91, Section 7.4 was used.

56. $\mathcal{L}\{af + bg\} = \int\limits_{0}^{+\infty} e^{-st}(af + bg)$

$= a\int\limits_{0}^{+\infty} e^{-st}f\,dt + b\int\limits_{0}^{+\infty} e^{-st}g\,dt$

$= a\mathcal{L}\{f\} + b\mathcal{L}\{g\}$

57. a. $\mathcal{L}^{-1}\{\tfrac{5}{s}\} = 5\mathcal{L}\{\tfrac{1}{s}\} = 5$

b. $\mathcal{L}^{-1}\left\{\dfrac{s + 2}{s^2 + 4}\right\}$

$= \mathcal{L}^{-1}\left\{\dfrac{s}{s^2 + 4}\right\} + \mathcal{L}^{-1}\left\{\dfrac{2}{s^2 + 4}\right\}$

$= \cos 2t + \sin 2t$

c. $\dfrac{2s^2 - 3s + 3}{s^2(s-1)} = \dfrac{A_1}{s} + \dfrac{A_2}{s^2} + \dfrac{A_3}{s-1}$

Partial fraction decomposition gives
$A_1 = 0$, $A_2 = -3$, $A_3 = 2$.

$\mathcal{L}^{-1}\left\{\dfrac{2s^2 - 3s+3}{s^2(s-1)}\right\} = \mathcal{L}^{-1}\left\{\dfrac{-3}{s^2}\right\} + \mathcal{L}\left\{\dfrac{2}{s-1}\right\}$

To complete this evaluation we need to evaluate

$\mathcal{L}\{t\} = \displaystyle\int_0^{+\infty} te^{-st}\,dt$ [parts or Formula 484]

$= \displaystyle\lim_{N\to+\infty} \dfrac{e^{-st}}{-s}\left(t + \dfrac{1}{s}\right)\Big|_0^N$

$= \displaystyle\lim_{t\to+\infty}\left[\dfrac{1}{s^2} - \dfrac{Ns+1}{e^{Ns}s^2}\right] = \dfrac{1}{s^2}$

$\mathcal{L}^{-1}\left\{\dfrac{-3}{s^2}\right\} + \mathcal{L}\left\{\dfrac{2}{s-1}\right\} = -3t + 2e^t$

d. $\dfrac{3s^3+2s^2 - 3s - 17}{(s^2 + 4)(s^2 - 1)} = \dfrac{A_1 s + B_1}{s^2+4} + \dfrac{A_2}{s-1} + \dfrac{A_3}{s+1}$

Partial fraction decomposition gives
$A_1 = 3$, $A_2 = -\frac{3}{2}$, $A_3 = \frac{3}{2}$, $B_1 = 5$.

$\mathcal{L}^{-1}\left\{\dfrac{3s^3+2s^2 - 3s - 17}{(s^2 + 4)(s^2 - 1)}\right\}$

$= \mathcal{L}^{-1}\left\{\dfrac{3s + 5}{s^2+4}\right\} + \mathcal{L}^{-1}\left\{\dfrac{-\frac{3}{2}}{s-1}\right\} + \mathcal{L}^{-1}\left\{\dfrac{\frac{3}{2}}{s+1}\right\}$

$= 3\mathcal{L}^{-1}\left\{\dfrac{s}{s^2 + 4}\right\} + \dfrac{5}{2}\mathcal{L}^{-1}\left\{\dfrac{2}{s^2 + 4}\right\}$

$\quad - \dfrac{3}{2}\mathcal{L}^{-1}\left\{\dfrac{1}{s - 1}\right\} + \dfrac{3}{2}\mathcal{L}^{-1}\left\{\dfrac{1}{s + 1}\right\}$

(Problem 56)

$= 3\cos 2t + \dfrac{5}{2}\sin 2t - \dfrac{3}{2}\sin e^t + \dfrac{3}{2}e^{-t}$

58. $F(s) = \mathcal{L}\{f(t)\} = \displaystyle\int_0^{+\infty} e^{-st}f(t)\,dt$

$\mathcal{L}\{f(t)e^{at}\} = \displaystyle\int_0^{+\infty} e^{-st}e^{at}f(t)\,dt$

$= \displaystyle\int_0^{+\infty} e^{-(s-a)t}f(t)\,dt$

Let $s_1 = s - a$; then

$\mathcal{L}\{f(t)e^{at}\} = \displaystyle\int_0^{+\infty} e^{-s_1 t}f(t)\,dt = F(s_1)$

$= F(s - a)$

59. a. From the solution of Problem 57c,
$\mathcal{L}\{t\} = \dfrac{1}{s^2}$; similarly, $\mathcal{L}\{t^3\} = \dfrac{3!}{s^4}$

Thus, $\mathcal{L}\{t^3 e^{-2t}\} = \dfrac{3!}{[s-(-2)]^4}$

$\qquad\qquad = \dfrac{6}{(s+2)^4}$

b. $\mathcal{L}\{\cos 2t\} = \dfrac{s}{s^2 + 4}$; then

$\mathcal{L}\{e^{-3t}\cos 2t\} = \dfrac{s+3}{(s+3)^2 + 4}$

c. $\mathcal{L}^{-1}\left\{\dfrac{5}{(s-1)^2}\right\} = e^t \mathcal{L}^{-1}\left\{\dfrac{5}{s^2}\right\} = 5te^t$

d. $\mathcal{L}^{-1}\left\{\dfrac{4s}{s^2+4s+5}\right\} = \mathcal{L}^{-1}\left\{\dfrac{4s}{(s+2)^2 + 1}\right\}$

$= 4e^{-2t}\mathcal{L}^{-1}\left\{\dfrac{1}{s^2 + 1}\right\} = 4e^{-2t}\sin t$

60. $F'(s) = \dfrac{d}{ds}F(s) = \dfrac{d}{ds}\displaystyle\int_0^{+\infty} e^{-st}f(t)\,dt$

$= \displaystyle\int_0^{+\infty} \dfrac{d}{ds}[e^{-st}f(t)]\,dt = \int_0^{+\infty}[-te^{-st}]f(t)\,dt$

$= -\displaystyle\int_0^{+\infty}[te^{-st}]f(t)\,dt = -\mathcal{L}\{tf(t)\} = -F'(s)$

7.6 The Hyperbolic and Inverse Hyperbolic Functions, Page 484

1. 3.6269 **2.** 10.0677 **3.** -0.7616 **4.** 0.0000

5. 1.1995 **6.** 0.0000 **7.** 0.9624 **8.** 0.7500

9. 1.6667 **10.** 2.2924 **11.** 0.6481 **12.** -0.3466

13. $y' = (\cosh 3x)(3) = 3\cosh 3x$

14. $y' = \sinh(1 - 2x^2)(-4x) = -4x\sin(1 - 2x^2)$

15. $y' = \sinh(2x^2 + 3x)(4x + 3)$

$\qquad = (4x+3)\sinh(2x^2 + 3)$

16. $y' = \cosh x^{1/2}(\frac{1}{2}x^{-1/2}) = \frac{1}{2}x^{-1/2}\cosh\sqrt{x}$

17. $y' = (\cosh x^{-1})(-x^{-2}) = -x^{-2}\cosh x^{-1}$

18. $y' = \dfrac{2x}{\sqrt{x^4 - 1}}$ **19.** $y' = \dfrac{3x^2}{\sqrt{1 + x^6}}$

20. $y' = \tanh^{-1}(3x) + x\left(\dfrac{3}{1 - 9x^2}\right)$

$\qquad = \tanh^{-1}(3x) + \dfrac{3x}{1 - 9x^2}$

Alternately, $y = x\tan^{-1}(3x) = \dfrac{x}{2}\ln\dfrac{1 + 3x}{1 - 3x}$

$$y' = \frac{1}{2}\left[\ln\frac{1+3x}{1-3x}\right] + \frac{x}{2}\left[\frac{3}{1+3x} - \frac{-3}{1-3x}\right]$$

$$= \frac{1}{2}\ln\frac{1+3x}{1-3x} + \frac{3x}{1-9x^2}$$

21. $y' = \dfrac{\sec^2 x}{\sqrt{1+\tan^2 x}} = \dfrac{\sec^2 x}{\sec x} = \sec x$

22. $y' = \dfrac{\sec x \tan x}{\sqrt{\sec^2 x - 1}} = \dfrac{\sec x \tan x}{\tan x} = \sec x$

23. $y' = \dfrac{\cos x}{1-\sin^2 x} = \dfrac{\cos x}{\cos^2 x} = \sec x$

24. $y' = \text{sech}\left[\dfrac{1-x}{1+x}\right]\tanh\left[\dfrac{1-x}{1+x}\right]\left(\dfrac{(1+x)-(1-x)}{(1+x)^2}\right)$

$$= \frac{-2}{(1+x)^2}\,\text{sech}\left[\frac{1-x}{1+x}\right]\tanh\left[\frac{1-x}{1+x}\right]$$

25. $y = \dfrac{\sinh^{-1}x}{\cosh^{-1}x} = \dfrac{\ln(x + \sqrt{x^2+1})}{\ln(x + \sqrt{x^2-1})}$

$$y' = \frac{\ln(x+\sqrt{x^2-1})\left(\dfrac{1}{\ln(x+\sqrt{x^2+1})}\right)(1+x)(x^2+1)^{-1/2}}{[\ln(x+\sqrt{x^2-1})]^2}$$

$$- \frac{\ln(x+\sqrt{x^2+1})\left(\dfrac{1}{\ln(x+\sqrt{x^2-1})}\right)(1+x)(x^2-1)^{-1/2}}{[\ln(x+\sqrt{x^2-1})]^2}$$

$$y' = \frac{\dfrac{\ln(x+\sqrt{x^2-1})}{\ln(x+\sqrt{x^2+1})}\dfrac{1+x}{\sqrt{x^2+1}} - \dfrac{\ln(x+\sqrt{x^2+1})}{\ln(x+\sqrt{x^2-1})}\dfrac{1+x}{\sqrt{x^2-1}}}{[\ln(x+\sqrt{x^2-1})]^2}$$

$$= \frac{\cosh^{-1}x\left(\dfrac{1}{\sqrt{x^2+1}}\right) - \sinh^{-1}x\left(\dfrac{1}{\sqrt{x^2-1}}\right)}{(\cosh^{-1}x)^2}$$

$$= \frac{\sqrt{x^2-1}\,\cosh^{-1}x - \sqrt{x^2+1}\,\sinh^{-1}x}{(\sqrt{x^2-1})\sqrt{x^2+1}(\cosh^{-1}x)^2}$$

$$= \frac{\sqrt{x^2-1}\,\cosh^{-1}x - \sqrt{x^2+1}\,\sinh^{-1}x}{\sqrt{x^4-1}(\cosh^{-1}x)^2}$$

26. $y' = \dfrac{1}{\sqrt{1+x^2}} - \dfrac{1}{2}(x^2+1)^{-1/2}(2x) = \dfrac{1-x}{\sqrt{x^2+1}}$

27. $y' = \cosh^{-1}x + \dfrac{x}{\sqrt{x^2-1}} - \dfrac{1}{2}(x^2-1)^{-1/2}(2x)$

$\qquad = \cosh^{-1}x$

28.
$$x\cosh y = y\sinh x + 5$$
$$x(\sinh y)y' + \cosh y = y\cosh x + (\sinh x)y'$$
$$(x\sinh y - \sinh x)y' = y\cosh x - \cosh y$$
$$y' = \frac{y\cosh x - \cosh y}{x\sinh y - \sinh x}$$

29. $e^x\sinh^{-1}x + e^{-x}\cosh^{-1}y = 1$

$$e^x\left(\frac{1}{\sqrt{x^2+1}}\right) + e^x\sinh^{-1}x + e^{-x}\left(\frac{1}{\sqrt{y^2-1}}\right)y'$$
$$- e^{-x}\cosh^{-1}y = 0$$

$$e^{-x}\left(\frac{1}{\sqrt{y^2-1}}\right)y'$$

$$= e^{-x}\cosh^{-1}y - e^x\sinh^{-1}x - \frac{e^x}{\sqrt{x^2+1}}$$

$$y' = \sqrt{y^2-1}\left[\cosh^{-1}y - e^{2x}\sinh^{-1}x - \frac{e^{2x}}{\sqrt{x^2+1}}\right]$$

30. $\displaystyle\int x\cosh(1-x^2)\,dx = -\frac{1}{2}\int\cosh(1-x^2)(-2x\,dx)$

$$= -\frac{1}{2}\sinh(1-x^2) + C$$

31. $\displaystyle\int\frac{\sinh\frac{1}{x}\,dx}{x^2} = -\int\sinh(x^{-1})(-x^{-2}\,dx)$

$$= -\cosh x^{-1} + C$$

32. $\displaystyle\int\frac{\text{sech}^2(\ln x)\,dx}{x} = \tanh(\ln x) + C$

33. $\displaystyle\int\coth x\,dx = \int\frac{\cosh x\,dx}{\sinh x} = \ln|\sinh x| + C$

34. $\displaystyle\int\frac{dx}{\sqrt{4x^2+16}} = \frac{1}{2}\int\frac{\left(\dfrac{dx}{2}\right)}{\sqrt{\left(\dfrac{x}{2}\right)^2+1}} = \frac{1}{2}\sinh^{-1}\frac{x}{2} + C$

35. $\displaystyle\int\frac{dt}{\sqrt{9t^2-16}} = \frac{1}{3}\int\frac{\left(\dfrac{3\,dt}{4}\right)}{\sqrt{\left(\dfrac{3t}{4}\right)^2-1}} = \frac{1}{3}\cosh^{-1}\frac{3t}{4} + C$

36. $\displaystyle\int\frac{dt}{36-16t^2} = \frac{1}{36}\int\frac{dt}{1-\left(\dfrac{4t}{6}\right)^2}$

$$= \frac{3}{2(36)}\int\frac{\left(\dfrac{2\,dt}{3}\right)}{1-\left(\dfrac{2t}{3}\right)^2} = \frac{1}{24}\tanh^{-1}\frac{2t}{3} + C$$

37. $\displaystyle\int\frac{\cos x\,dx}{\sqrt{1+\sin^2 x}} = \sinh^{-1}(\sin x) + C$

38. $\displaystyle\int\frac{x\,dx}{\sqrt{1+x^4}} = \frac{1}{2}\int\frac{2x\,dx}{\sqrt{1+(x^2)^2}} = \frac{1}{2}\sinh^{-1}x^2 + C$

39. $\displaystyle\int\frac{x^2\,dx}{1-x^6} = \frac{1}{3}\int\frac{3x^2\,dx}{1-(x^3)^2} = \frac{1}{3}\tanh^{-1}x^3 + C$

40. $\displaystyle\int_0^{1/2}\frac{dx}{1-x^2} = \tanh^{-1}x^2\Big|_0^{1/2} = \tanh^{-1}\frac{1}{2} = \frac{1}{2}\ln 3$

41. $\displaystyle\int_2^3\frac{dx}{1-x^2} = \coth^{-1}x^2\Big|_2^3 = \coth^{-1}3 - \coth^{-1}2$

$$= \tfrac{1}{2}(\ln \tfrac{4}{2} - \ln 3) = \tfrac{1}{2} \ln \tfrac{2}{3}$$

42. $\displaystyle\int_0^1 \frac{t^5 \, dt}{\sqrt{1 + t^{12}}} = \frac{1}{6}\int_0^1 \frac{6t^5 \, dt}{1 + (t^6)^2} = \frac{1}{6} \sinh^{-1} t^6 \Big|_0^1$

$$= \tfrac{1}{6} \ln(1 + \sqrt{2})$$

43. $\displaystyle\int_1^2 \frac{e^x \, dx}{\sqrt{e^{2x} - 1}} = \cosh^{-1} e^x \Big|_1^2$

$$= \cosh^{-1} e^2 - \cosh^{-1} e$$

$$= \ln\left(e^2 + \sqrt{e^4 - 1}\right) - \ln\left(e + \sqrt{e^2 - 1}\right)$$

44. $\displaystyle\int_0^{\ln 2} \sinh 3x \, dx = \tfrac{1}{3} \cosh 3x \Big|_0^{\ln 2}$

$$= \tfrac{1}{3}[\cosh(3 \ln 2) - \cosh 0]$$

$$= \tfrac{1}{3}[\tfrac{1}{2}(e^{\ln 8} + e^{-\ln 8}) - 1] = \frac{49}{48}$$

45. $\displaystyle\int_0^1 x \operatorname{sech}^2 x^2 \, dx = \frac{1}{2}\int_0^1 \operatorname{sech}^2 x^2 \, (2x \, dx)$

$$= \tfrac{1}{2} \tanh x^2 \Big|_0^1 = \tfrac{1}{2} \tanh 1 = \frac{e^2 - 1}{2(e^2 + 1)}$$

46. $\tanh(x + y) = \dfrac{\sinh(x + y)}{\cosh(x + y)}$

$$= \frac{\sinh x \cosh y + \cosh x \sinh y}{\cosh x \cosh y + \sinh x \sinh y}$$

$$= \frac{\tanh x + \tanh y}{1 + \tanh x \tan y}$$

47. a. $\sinh 2x = \sinh x \cosh x + \cosh x \sinh x$
$$= 2 \sinh x \cosh x$$

b. $\cosh 2x = \cosh x \cosh x + \sinh x \sinh x$
$$= \cosh^2 x + \sinh^2 x$$

48. a. $\tanh x = \dfrac{e^x - e^{-x}}{e^x + e^{-x}} = \dfrac{e^{2x} - 1}{e^{2x} + 1} < 1$

$\tanh(-x) = \dfrac{e^{-x} - e^x}{e^{-x} + e^x} = \dfrac{1 - e^{2x}}{1 + e^{2x}} < -1$

Thus, $-1 < \tanh x < 1$

b. $\displaystyle\lim_{x \to +\infty} \tanh x = \lim_{x \to +\infty} \frac{e^x - e^{-x}}{e^x + e^{-x}}$

$$= \lim_{x \to +\infty} \frac{e^{2x} - 1}{e^{2x} + 1} = \lim_{x \to +\infty} \frac{2e^{2x}}{2e^{2x}} = 1$$

49. $y' = \operatorname{sech}^2 x$

$$= (\cosh^2 x)^{-1}$$

The curve is rising for all x.

$y'' = -2 \operatorname{sech}^2 x \tanh x$

$y'' = 0$ when $x = 0$

The curve is concave up on $(-\infty, 0)$, and concave down on $(0, +\infty)$.

50. $y' = -\cosh^2 x$

$$= -(\sinh^2 x)^{-1}$$

The curve is falling for all x.

$y'' = 2 \cosh^2 x \coth x$

The curve is concave up on $(0, +\infty)$ and concave down on $(-\infty, 0)$.

51. $(\cosh x + \sinh x)^n = [\tfrac{1}{2}(e^x + e^{-x} + e^x - e^{-x})]^n$
$$= (e^x)^n = e^{nx}$$

52. $\cosh^2 t - \sinh^2 t = \left(\frac{x}{a}\right)^2 - \left(\frac{y}{b}\right)^2 = 1$

53. a. $y' = ac \sinh cx + bc \cosh cx$

$y'' = c^2(a \cosh cx + b \sinh cx) = c^2 y$

Thus, $y'' - c^2 y = 0$

b. Let $c = 2$; then $y = a \cosh 2x + b \sinh 2x$ is a solution of $y'' - 4y = 0$. Since $y'(0) = 2$, $2 = 2b$ or $b = 1$. Since $y = 1$ when $x = 0$, $a = 1$ and

$$y = \cosh 2x + \sinh 2x$$

54. $A = \displaystyle\int_0^1 (\cosh x - \sinh x) \, dx = (\sinh x - \cosh x)\Big|_0^1$

$$= \sinh 1 - \cosh 1 + 1 \approx 0.8161$$

55. $s = \displaystyle\int_{-a}^{a} \sqrt{1 + (y')^2} \, dx$

$$= 2\int_0^a \sqrt{1 + \sinh^2 \tfrac{x}{a}} \, dx = 2a \int_0^a \cosh \tfrac{x}{a}\left(\tfrac{1}{a} \, dx\right)$$

$$= 2a \sinh \tfrac{x}{a}\Big|_0^a = 2a(\sinh 1 - \sinh 0)$$

$$= (e - e^{-1})a \approx 2.3504a$$

56. $V = \pi\displaystyle\int_0^1 \tanh^2 x \, dx = \pi\int_0^1 (1 + \operatorname{sech}^2 x) \, dx$

$$= \pi(x + \tanh x)\Big|_0^1 = \pi(1 + \tanh 1) \approx 5.5342$$

57. $y = \cosh x$; $ds = \sqrt{1 + \sinh^2 x} \, dx = \cosh x \, dx$;

$$S = 2\int_0^1 2\pi \cosh^2 x \, dx = 2\pi\int_0^1 (\cosh 2x + 1) \, dx$$

$$= 2\pi(\tfrac{1}{2}\sinh 2x + x)\Big|_0^1 = \pi \sinh 2 + 2\pi$$

$$= \tfrac{\pi}{2}(e^2 - e^{-2} + 4) \approx 17.68$$

58. **a.** $\cosh^2 x - \sinh^2 x = 1$

$$\frac{\cosh^2 x}{\cosh^2 x} - \frac{\sinh^2 x}{\cosh^2 x} = \frac{1}{\cosh^2 x}$$

$$1 - \tanh^2 x = \mathrm{sech}^2 x$$

$$\mathrm{sech}^2 x + \tanh^2 x = 1$$

b. $\cosh^2 x - \sinh^2 x = 1$

$$\frac{\cosh^2 x}{\sinh^2 x} - \frac{\sinh^2 x}{\sinh^2 x} = \frac{1}{\sinh^2 x}$$

$$\coth^2 x - 1 = \mathrm{csch}^2 x$$

$$\coth^2 x - \mathrm{csch}^2 x = 1$$

59. $y = \cosh u = \tfrac{1}{2}(e^u + e^{-u})$

$$\frac{dy}{dx} = \tfrac{1}{2}(e^u - e^{-u})\frac{du}{dx} = \sinh u \frac{du}{dx}$$

$y = \tanh u = \dfrac{\sinh u}{\cosh u}$

$$\frac{dy}{dx} = \frac{\cosh u(\cosh u \frac{du}{dx}) - \sinh u(\sinh^2 u \frac{du}{dx})}{\cosh^2 u}$$

$$= \frac{\cosh^2 u - \sinh^2 u}{\cosh^2 u}\frac{du}{dx}$$

$$= \frac{1}{\cosh^2 u}\frac{du}{dx}$$

$$= \mathrm{sech}^2 u \frac{du}{dx}$$

$y = \mathrm{sech}\, u = (\cosh u)^{-1}$

$$\frac{dy}{dx} = -(\cosh u)^{-2}\sinh u \frac{du}{dx}$$

$$= -\frac{\sinh u}{\cosh u}\frac{1}{\cosh x}\frac{du}{dx}$$

$$= \tanh u\, \mathrm{sech}\, u \frac{du}{dx}$$

60. $y = \cosh^{-1} x$ means $x = \cosh y$

$$1 = \sinh y \frac{dy}{dx}$$

$$\frac{1}{\sinh y} = \frac{dy}{dx}$$

$$\frac{dy}{dx} = \frac{1}{\sqrt{\cosh^2 y - 1}} = \frac{1}{x^2 - 1}$$

$y = \tanh^{-1} x$ means $x = \tanh y$

$$\mathrm{sech}^2 y \frac{dy}{dx} = 1$$

$$\frac{dy}{dx} = \frac{1}{\mathrm{sech}^2 y} = \frac{1}{1 - \tanh^2 y} = \frac{1}{1 - x^2}$$

$y = \mathrm{sech}^{-1} x$ means $x = \mathrm{sech}\, y$

$$1 = -\mathrm{sech}\, y \tanh y \frac{dy}{dx}$$

$$\frac{dy}{dx} = -\frac{1}{\mathrm{sech}\, y \tanh y}$$

$$= -\frac{1}{\mathrm{sech}\, y\sqrt{1 - \mathrm{sech}^2 y}}$$

$$= -\frac{1}{x\sqrt{1 - x^2}}$$

61. **a.** $\cosh[\ln(x + \sqrt{x^2 - 1})]$

$$= \tfrac{1}{2}\left[e^{\ln(x+\ln\sqrt{x^2-1})} + e^{-\ln(x+\ln\sqrt{x^2-1})}\right]$$

$$= \tfrac{1}{2}\left[x + \sqrt{x^2-1} + x - \sqrt{x^2-1}\right]$$

$$= \tfrac{1}{2}[2x] = x$$

Since $\cosh[\ln(x + \sqrt{x^2 - 1})] = x$ we know $\cosh^{-1} x = \ln(x + \sqrt{x^2 - 1})$

b.
$$x = \tanh y$$

$$x = \frac{e^y - e^{-y}}{e^y + e^{-y}}$$

$$xe^y + xe^{-y} = e^y - e^{-y}$$

$$e^y(x - 1) = -e^{-y}(x + 1)$$

$$e^{2y} = \frac{1 - x}{1 + x}$$

$$2y = \ln\frac{1 - x}{1 + x}$$

$$y = \tfrac{1}{2}\ln\frac{1 - x}{1 + x}$$

Thus $\tan^{-1} x = \tfrac{1}{2}\ln\dfrac{1 - x}{1 + x}$

62. $\displaystyle\int \frac{dx}{\sqrt{x^2 + a^2}}$ $\boxed{x = a\sinh u;\ dx = a\cosh u\, du}$

$$= \int \frac{a\cosh u\, du}{a\cosh u} = u + C = \sinh^{-1}\tfrac{x}{a} + C$$

$$\int \frac{dx}{\sqrt{a^2 - x^2}} = \sin^{-1}\tfrac{x}{a} + C$$

Proof: $\dfrac{d}{dx}(\sin^{-1}\tfrac{x}{a}) = \dfrac{a}{\sqrt{1 - (\frac{x}{a})^2}}$

$$= \frac{1}{\sqrt{a^2 - x^2}}$$

Thus, $\displaystyle\int \frac{dx}{\sqrt{a^2 - x^2}} = \sin^{-1}\tfrac{x}{a} + C$

CHAPTER 7 REVIEW

Proficiency Examination, Page 485

1. Let u replace a more complicated symbol or function of the (dummy) variable of

integration, say x. Obtain all forms of x in terms of u. Substitute, integrate, return form of answers from u back to x. In the case of a definite integral transform the limits of x into limits for u and use in the integrated form in terms of u.

2. $\displaystyle \int u\ dv = uv - \int v\ du$

3. A reduction integration formula is an integration formula in which the same general form of the integrand appears in the integrals on both sides of the equal sign. Generally, the integrand on the left is a function raised to a positive integral power, and so is the integrand on the right, but to a lesser power. Thus, the formula can be used repeatedly until the integrand becomes manageable.

4. A trigonometric substitution may be handy when the integrand contains one of the following forms: $\sqrt{x^2 + a^2}$; $\sqrt{x^2 - a^2}$; or $\sqrt{a^2 - x^2}$.

5. The method of partial fractions may be handy when integrating a rational function.

6. The Weierstrass substitutions are:

$u = \tan \frac{x}{2}$, $\sin x = \dfrac{2u}{1 + u^2}$; $\cos u = \dfrac{1 - u^2}{1 + u^2}$, and $du = \dfrac{2\ du}{1 + u^2}$.

7. See Page 465:
Step 1: simplify
Step 2: use basic formulas (check table)
Step 3: substitute
Step 4: classify; parts, trig powers, Weierstrass substitution, trig substitutions, or partial fractions
Step 5: try again

8. An improper integral is one in which a limit of integration is infinite and/or at least one value in the interval of integration leads to an undefined integrand.

9. $\sinh x = \frac{1}{2}(e^x - e^{-x})$; $\cosh x = \frac{1}{2}(e^x + e^{-x})$; $\tanh x = \dfrac{e^x - e^{-x}}{e^x - e^{-x}}$

10. Let u be a differentiable function of x. Then:

$\dfrac{d}{dx}(\sinh u) = \cosh u\ \dfrac{du}{dx}$

$\dfrac{d}{dx}(\cosh u) = \sinh u\ \dfrac{du}{dx}$

$\dfrac{d}{dx}(\tanh u) = \mathrm{sech}^2 u\ \dfrac{du}{dx}$

$\dfrac{d}{dx}(\coth u) = -\mathrm{csch}^2 u\ \dfrac{du}{dx}$

$\dfrac{d}{dx}(\mathrm{sech}\ u) = -\mathrm{sech}\ u\ \tanh u\ \dfrac{du}{dx}$

$\dfrac{d}{dx}(\mathrm{csch}\ u) = -\mathrm{csch}\ u\ \coth u\ \dfrac{du}{dx}$

11. $\displaystyle \int \sinh x\ dx = \cosh x + C$

$\displaystyle \int \cosh x\ dx = \sinh x + C$

$\displaystyle \int \mathrm{sech}^2 x\ dx = \tanh x + C$

$\displaystyle \int \mathrm{csch}^2 x\ dx = -\coth x + C$

$\displaystyle \int \mathrm{sech}\ x\ \tanh x\ dx = -\mathrm{sech}\ x + C$

$\displaystyle \int \mathrm{csch}\ x\ \coth x\ dx = -\mathrm{csch}\ x + C$

12. $\sinh^{-1} x = \ln(x + \sqrt{x^2 + 1})$, all x

$\mathrm{csch}^{-1} x = \ln\left(\dfrac{1}{x} + \dfrac{\sqrt{1 + x^2}}{|x|}\right)$, $x \neq 0$

$\cosh^{-1} x = \ln(x + \sqrt{x^2 - 1})$, $x \geq 1$

$\mathrm{sech}^{-1} x = \ln\left(\dfrac{1 + \sqrt{1 - x^2}}{x}\right)$, $0 < x \leq 1$

$\tanh^{-1} x = \frac{1}{2} \ln \dfrac{1 + x}{1 - x}$, $|x| < 1$

$\coth^{-1} x = \frac{1}{2} \ln \dfrac{x + 1}{x - 1}$, $|x| > 1$

13. $\dfrac{d}{dx}(\sinh^{-1} u) = \dfrac{1}{\sqrt{1 + u^2}}\ \dfrac{du}{dx}$

$\displaystyle \int \dfrac{du}{\sqrt{1 + u^2}} = \sinh^{-1} u + C$

$\dfrac{d}{dx}(\cosh^{-1} u) = \dfrac{1}{\sqrt{u^2 - 1}}\ \dfrac{du}{dx}$, if $|u| > 1$

$\displaystyle \int \dfrac{du}{\sqrt{u^2 - 1}} = \cosh^{-1} u + C$

$\dfrac{d}{dx}(\tanh^{-1} u) = \dfrac{1}{1 - u^2}\ \dfrac{du}{dx}$, if $|u| < 1$

$\displaystyle \int \dfrac{du}{1 - u^2} = \tanh^{-1} u + C$, if $|u| < 1$

$\dfrac{d}{dx}(\mathrm{csch}^{-1} u) = \dfrac{-1}{|u|\sqrt{1 + u^2}}$

$\displaystyle \int \dfrac{du}{u\sqrt{1 + u^2}} = -\mathrm{csch}^{-1} |u| + C$

$\dfrac{d}{dx}(\mathrm{sech}^{-1} u) = \dfrac{-1}{u\sqrt{1 - u^2}}\ \dfrac{du}{dx}$, $0 < u < 1$

$\displaystyle \int \dfrac{du}{u\sqrt{1 - u^2}} = -\mathrm{sech}^{-1} u + C$

$\dfrac{d}{dx}(\coth^{-1} u) = \dfrac{1}{1 - u^2}\ \dfrac{du}{dx}$, if $|u| > 1$

$\displaystyle \int \dfrac{du}{1 - u^2} = \coth^{-1} u + C$, if $|u| > 1$

14. **a.** $\tanh^{-1}(0.5) = \frac{1}{2} \ln \frac{1+0.5}{1-0.5} \approx 0.5493$

b. $\sinh(\ln 3) = \frac{e^{\ln 3} - e^{-\ln 3}}{2} = \frac{4}{3}$

c. $\coth^{-1} 2 = \frac{1}{2} \ln \frac{2+1}{2-1} = \frac{1}{2} \ln 3 \approx 0.5493$

15. $\int \frac{2x+3}{\sqrt{x^2+1}} dx = \int \frac{2x \, dx}{\sqrt{x^2+1}} + \int \frac{3 \, dx}{\sqrt{x^2+1}}$

$\boxed{\text{Formula 172}}$

$= 2\sqrt{x^2+1} + 3 \sinh^{-1} x + C$

16. $\int x \sin 2x \, dx \quad \boxed{u = x; \; dv = \sin 2x \, dx}$

$= -\frac{x}{2} \cos 2x - \int -\frac{1}{2} \cos 2x \, dx$

$= -\frac{x}{2} \cos 2x + \frac{1}{4} \sin 2x + C$

17. $\int \sinh(1-2x) \, dx = \frac{1}{2} \int \sinh(1-2x)(-2 \, dx)$

$= -\frac{1}{2} \cosh(1-2x) + C$

18. $\int \frac{dx}{\sqrt{4-x^2}} = \sin^{-1} \frac{x}{2} + C$

19. $\frac{x^2}{(x^2+1)(x-1)} = \frac{A_1 x + B_1}{x^2+1} + \frac{A_2}{x-1}$

Partial fraction decomposition gives
$A_1 = \frac{1}{2}, \; A_2 = \frac{1}{2}, \; B_1 = \frac{1}{2}.$

$\int \frac{x^2 \, dx}{(x^2+1)(x-1)} = \frac{1}{2} \int \frac{x+1}{x^2+1} dx + \frac{1}{2} \int \frac{dx}{x-1}$

$= \frac{1}{4} \int \frac{2x \, dx}{x^2+1} + \frac{1}{2} \int \frac{dx}{x^2+1} + \frac{1}{2} \int \frac{dx}{x-1}$

$= \frac{1}{4} \ln(x^2+1) + \frac{1}{2} \tan^{-1} x + \frac{1}{2} \ln|x-1| + C$

20. $\int \frac{x^3 \, dx}{x^2-1} = \int \left(x + \frac{x}{x^2-1} \right) dx$

$= \int x \, dx + \frac{1}{2} \int \frac{2x \, dx}{x^2-1}$

$= \frac{x^2}{2} + \frac{1}{2} \ln|x^2-1| + C$

21. $\int_1^2 x \ln x^3 \, dx = 3 \int_1^2 x \ln x \, dx$

$\boxed{\text{Parts or Formula 502}}$

$= 3\left(\frac{x^2}{2} \ln x - \frac{x^2}{4} \right)\Big|_1^2 = 6 \ln 2 - \frac{9}{4} \approx 1.9089$

22. $\frac{1}{(x-1)^2(x+2)} = \frac{A_1}{(x-1)^2} + \frac{A_2}{x-1} + \frac{A_3}{x+2}$

Partial fraction decomposition gives
$A_1 = \frac{1}{3}, \; A_2 = -\frac{1}{9}, \; A_3 = \frac{1}{9}$

$\int_2^3 \frac{dx}{(x-1)^2(x+2)}$

$= \frac{1}{9} \int_2^3 \left(\frac{3}{(x-1)^2} - \frac{1}{x-1} + \frac{1}{x+2} \right) dx$

$= \frac{1}{9} \left[-\frac{3}{x-1} - \ln(x-1) + \ln(x+2) \right]\Big|_2^3$

$= \frac{1}{9}\left(\ln \frac{5}{8} + \frac{3}{2} \right) \approx 0.1144$

23. $\int_3^4 \frac{dx}{2x-x^2} = \int_3^4 \left(\frac{\frac{1}{2}}{x} + \frac{\frac{1}{2}}{2-x} \right) dx$

$= \frac{1}{2} \ln x - \frac{1}{2} \ln|2-x| \Big|_3^4 = \frac{1}{2} \ln \frac{2}{3} = -0.2027$

24. $\int_0^{\pi/4} (\sec^2 x)(\sec x \tan x \, dx) = \frac{\sec^3 x}{3}\Big|_0^{\pi/4}$

$= \frac{2\sqrt{2}-1}{3} \approx 0.6095$

25. $\lim_{t \to +\infty} \int_0^t x e^{-2x} \, dx \quad \boxed{\text{parts or Formula 484}}$

$= \lim_{t \to +\infty} -\frac{x}{2} e^{-2x} - \frac{1}{4} e^{-2x}\Big|_0^t$

$= \lim_{t \to +\infty} \left[-\frac{1}{4} e^{-2t}(2t+1) + \frac{1}{4} \right] = \frac{1}{4}$

26. $\int_0^\pi \frac{\cos x \, dx}{1 - \cos x} = \lim_{t \to 0^+} \int_t^\pi \left(-1 + \frac{1}{1 - \cos x} \right) dx$

$= -x + \lim_{t \to 0^+} \int_t^\pi \frac{1(1+\cos x)}{(1-\cos x)(1+\cos x)} dx$

$= -x + \lim_{t \to 0^+} \left[\int_t^\pi \frac{dx}{\sin^2 x} + \int_t^\pi \frac{\cos x \, dx}{\sin^2 x} \right]$

$= -x - \cot x - \frac{1}{\sin x}\Big|_t^\pi; \quad \text{diverges}$

27. $\frac{2x+3}{x^2(x-2)} = \frac{A_1}{x^2} + \frac{A_2}{x} + \frac{A_3}{x-2}$

Partial fraction decomposition gives
$A_1 = -\frac{3}{2}, \; A_2 = -\frac{7}{4}; \; A_3 = \frac{7}{4}$

$\int_0^1 \frac{2x+3}{x^2(x-2)} dx$

$= \lim_{t \to 0^+} \int_t^1 \left(\frac{-\frac{3}{2}}{x^2} + \frac{-\frac{7}{4}}{x} + \frac{\frac{7}{4}}{x-2} \right) dx$

$= \lim_{t \to 0^+} \left[\frac{3}{2x} + \frac{7}{4} \ln\left| \frac{x-2}{x} \right| \Big|_t^1 \right]$

$$= \lim_{t \to 0^+}\left[\frac{3}{2} + \frac{7}{4}\ln 1 - \frac{3}{2t} - \ln\left|\frac{t-2}{t}\right|\right]$$

$$= +\infty; \text{ diverges}$$

28. $\displaystyle\int_0^{+\infty} e^{-x}\sin x\, dx = \lim_{t \to +\infty}\int_0^t e^{-x}(\sin x)(-dx)$

$$\boxed{\text{Formula 492}}$$

$$= \lim_{t \to +\infty}\left[\frac{e^{-x}(-\sin x - \cos x)}{2}\bigg|_0^t\right]$$

$$= \lim_{t \to +\infty}\left[\frac{e^{-t}(-\sin t - \cos t)}{2} + \frac{1}{2}\right] = \frac{1}{2}$$

29. $\displaystyle y' = \frac{1}{2\sqrt{\tanh^{-1}2x}}\left[\frac{2}{1 - 4x^2}\right]$

$$= \frac{1}{(1 - 4x^2)\sqrt{\tanh^{-1}2x}}$$

30. $\displaystyle 2\pi\int_0^2 x\left(\frac{1}{\sqrt{9 - x^2}}\right)dx = -\pi\int_0^2 \frac{-2x\,dx}{\sqrt{9 - x^2}}$

$$= -2\pi\left(\sqrt{9 - x^2}\right)\bigg|_0^2 = -2\pi\left(\sqrt{5} - 3\right)$$

$$\approx 4.7999$$

Supplementary Problems, Page 485

1. $\displaystyle y' = \frac{1}{1 - (x^{-1})^2}(-x^{-2}) = (1 - x^2)^{-1}$

2. $\displaystyle y' = \cosh^{-1}(3x + 1) + x\left(\frac{1}{\sqrt{(3x + 1)^2 - 1}}\right)(3)$

$$= \cosh^{-1}(3x + 1) + \frac{3x}{\sqrt{(3x + 1)^2 - 1}}$$

3. $\displaystyle y = \frac{\sinh x}{e^x} = \frac{1}{2}e^{-x}(e^x - e^{-x}) = \frac{1}{2}(1 - e^{-2x})$

$$y' = \frac{1}{2}(2e^{-2}) = e^{-2x}$$

4. $\displaystyle y' = \sinh^2 x + \cosh^2 x = \cosh 2x$

5. $\displaystyle y' = x\cosh x + \sinh x$
$$+ (e^x - e^{-x})\cosh(e^x + e^{-x})$$

6. $\displaystyle \int\cos^{-1}x\,dx \ \boxed{\text{Parts or Formula 445}}$

$$= x\cos^{-1}x - \sqrt{1 - x^2} + C$$

7. $\displaystyle \int\frac{x^2\,dx}{\sqrt{4 - x^2}} \quad \boxed{\text{Formula 226}}$

$$= -\frac{x}{2}\sqrt{4 - x^2} + 2\sin^{-1}\frac{x}{2} + C$$

8. $\displaystyle \int\frac{x\,dx}{x^2 - 2x + 5} = \int\frac{(x - 1 + 1)\,dx}{x^2 - 2x + 5}$

$$= \int\frac{(x - 1)\,dx}{x^2 - 2x + 5} + \int\frac{dx}{x^2 - 2x + 5}$$

$$= \frac{1}{2}\int u^{-1}du + \int\frac{dx}{(x - 1)^2 + 4}$$

$$= \frac{1}{2}\ln(x^2 - 2x + 5) + \frac{1}{2}\tan^{-1}\frac{1}{2}(x - 1) + C$$

9. $\displaystyle \frac{3x - 2}{x^3 - 2x^2} = \frac{3x - 2}{x^2(x - 2)} = \frac{A_1}{x} + \frac{A_2}{x^2} + \frac{A_3}{x - 2}$

Partial fraction decomposition gives
$A_1 = -1, A_2 = 1, A_3 = 1$.

$$\int\frac{3x - 2}{x^3 - 2x^2}\,dx = -\int\frac{dx}{x} + \int x^{-2}dx + \int\frac{dx}{x - 2}$$

$$= -\ln|x| - x^{-1} + \ln|x - 2| + C$$

10. Consider a right triangle with hypotenuse $\sqrt{x^2 + x + 1}$, side opposite θ is $x + \frac{1}{2}$, and remaining leg $\frac{1}{2}\sqrt{3}$. Then,

$x + \frac{1}{2} = \frac{1}{2}\sqrt{3}\tan\theta; \ \sqrt{x^2 + x + 1} = \frac{1}{2}\sqrt{3}\sec\theta;$
and $dx = \frac{1}{2}\sqrt{3}\sec^2\theta\,d\theta$

$$\int\frac{dx}{(x^2 + x + 1)^{3/2}} = \int\frac{\frac{1}{2}\sqrt{3}\sec^2\theta\,d\theta}{\frac{3}{8}\sqrt{3}\sec^3\theta}$$

$$= \frac{4}{3}\int\frac{d\theta}{\sec\theta} = \frac{4}{3}\int\cos\theta\,d\theta = \frac{4}{3}\sin\theta + C$$

$$= \frac{2(2x + 1)}{3\sqrt{x^2 + x + 1}} + C$$

11. $\displaystyle \int\frac{dx}{\sqrt{x}(1 + \sqrt[4]{x})} \quad \boxed{x = u^4;\ dx = 4u^3\,du}$

$$= \int\frac{4u^3\,du}{u^2(1 + u)} = 4\int\frac{u\,du}{1 + u} = 4\int du - 4\int\frac{du}{1 + u}$$

$$= 4u - 4\ln|1 + u| + C$$

$$= 4x^{1/4} - 4\ln(1 + x^{1/4}) + C$$

12. $\displaystyle \int\frac{\sqrt{9x^2 - 1}}{x}\,dx = 3\int\frac{\sqrt{x^2 - \frac{1}{9}}}{x}\,dx \ \boxed{\text{Formula 207}}$

$$= 3\sqrt{x^2 - \frac{1}{9}} - 3(\frac{1}{3})\sec^{-1}|3x| + C$$

$$= \sqrt{9x^2 - 1} - \sec^{-1}(3x) + C$$

13. $\displaystyle \int x^2\tan^{-1}x\,dx \quad \boxed{\text{Formula 459}}$

$$= \frac{x^3}{3}\tan^{-1}x - \frac{x^2}{6} + \frac{1}{6}\ln(1 + x^2) + C$$

14. $\displaystyle \int\frac{dx}{\sin x + \tan x} = \int\frac{\cos x\,dx}{\sin x\cos x + \sin x}$

$$\boxed{\text{Weierstrass substitution}}$$

$$= \int \frac{(1 - u^2)(2\,du)}{2u(1 - u^2) + 2u} = \int \frac{(u^2 - 1)\,du}{u(u^2 - 2)}$$

$$\frac{u^2 - 1}{u(u^2 - 2)} = \frac{A_1}{u} + \frac{A_2}{u + \sqrt{2}} + \frac{A_3}{u - \sqrt{2}}$$

Partial fraction decomposition gives
$A_1 = \frac{1}{2}$, $A_2 = \frac{1}{4}$, $A_3 = \frac{1}{4}$.

$$= \frac{1}{2}\int \frac{du}{u} + \frac{1}{4}\int \frac{du}{u + \sqrt{2}} + \frac{1}{4}\int \frac{du}{u - \sqrt{2}}$$

$$= \frac{1}{4}\Big[2\ln|u| + \ln|u + \sqrt{2}| + \ln|u - \sqrt{2}|\Big] + C_1$$

$$= \frac{1}{4}\Big[2\ln\big|u(u^2 - 2)\big|\Big] + C$$

$$= \frac{1}{4}\Big|\tan \frac{x}{2}\Big|\big(\tan^2 \tfrac{x}{2} - 2\big) + C$$

15. $\displaystyle \int e^x \sqrt{4 - e^{2x}}\,dx$ $\boxed{t = e^x;\ dt = e^x\,dx}$

$$= \int \sqrt{4 - t^2}\,dt \quad \boxed{\text{Formula 231}}$$

$$= \frac{x\sqrt{4 - t^2}}{2} + 2\sin^{-1}\tfrac{t}{2} + C$$

$$= \frac{1}{2}e^x\sqrt{4 - e^{2x}} + 2\sin^{-1}\big(\tfrac{e^x}{2}\big) + C$$

16. $\displaystyle \int \cos \tfrac{x}{2} \sin \tfrac{x}{3}\,dx$ $\boxed{t = \tfrac{x}{6};\ dt = \tfrac{1}{6}\,dx}$

$$= 6\int \cos 3t \sin 2t\,dt\ \boxed{\text{Product-to-sum identity}}$$

$$= 3\int [\sin 5t - \sin t]\,dt$$

$$= -\tfrac{3}{5}\cos 5t + 3\sin t + C$$

$$= -\tfrac{3}{5}\cos \tfrac{5}{6}x + 3\sin \tfrac{1}{6}x + C$$

17. $\displaystyle \int \frac{\sqrt{1 + 1/x^2}}{x^5}\,dx = \int \frac{1}{x^2}\sqrt{1 + \frac{1}{x^2}}\,\frac{dx}{x^3}$

$\boxed{u = x^{-2};\ du = -2x^3}$

$$= -\frac{1}{2}\int u\sqrt{1 + u}\,du \quad \boxed{\text{Formula 141}}$$

$$= -\frac{1}{2}\Big[\frac{2(3u - 2)}{15}\sqrt{(1 + u)^3}\Big] + C$$

$$= -\frac{1}{15}(u + 1)^{3/2}[3u - 2] + C$$

$$= \frac{1}{15}(x^{-2} + 1)^{3/2}(2 - 3x^{-2}) + C$$

18. $\displaystyle \int \frac{\sin x\,dx}{\cos^5 x} = -\int \frac{du}{u^5} = \frac{1}{4}u^{-4} + C$

$$= \frac{1}{4}\sec^4 x + C$$

19. $\displaystyle \int \sqrt{1 + \sin x}\,dx = \int \frac{\sqrt{1 - \sin^2 x}\,dx}{\sqrt{1 - \sin x}}$

$$= \int \frac{\cos x\,dx}{\sqrt{1 - \sin x}} = \int \frac{du}{\sqrt{1 - u}}$$

$$= -2\sqrt{1 - u} + C = -2\sqrt{1 - \sin x} + C$$

20. $\displaystyle \int \cos x \ln(\sin x)\,dx = \int \ln u\,du$

$$= u\ln|u| - u + C = \sin x \ln(\sin x) - \sin x + C$$

21. $\displaystyle I = \int \sin(\ln x)\,dx$ $\boxed{u = \sin(\ln x);\ dv = dx}$

$$= x\sin(\ln x) - \int \cos(\ln x)\,dx$$

$$\boxed{u = \cos(\ln x);\ dv = dx}$$

$$= x\sin(\ln x) - x\cos(\ln x) - I + C_1$$

$$2I = x\sin(\ln x) - x\cos(\ln x) + C_1$$

$$I = \tfrac{x}{2}[\sin(\ln x) - x\cos(\ln x)] + C$$

22. $\displaystyle \int e^{2x}\operatorname{sech}(e^{2x})\,dx$ $\boxed{u = e^{2x};\ du = 2e^{2x}\,dx}$

$$= \frac{1}{2}\int \operatorname{sech} u\,du = \int \frac{du}{e^u - e^{-u}} = \int \frac{e^u\,du}{(e^u)^2 + 1}$$

$$= \tan^{-1}e^u + C = \tan^{-1}e^{e^{2x}} + C$$

23. $\displaystyle \int \frac{\sinh x\,dx}{2 + \cosh x} = \int \frac{du}{u} = \ln(2 + \cosh x) + C$

When using computer software, you might
obtain the equivalent form

$$\ln(e^{2x} + 4e^x + 1) - x + C$$

24. $\displaystyle \int \frac{\tanh^{-1}x\,dx}{1 - x^2}$ $\boxed{u = \tanh^{-1}x;\ du = \dfrac{dx}{1 - x^2}}$

$$= \int u\,du = \tfrac{1}{2}(\tanh^{-1} x)^2 + C$$

25. $\displaystyle \int x^2\cot^{-1}x\,dx$ $\boxed{u = \cot^{-1}x;\ dv = x^2\,dx}$

$$= \tfrac{1}{3}x^3\cot^{-1}x + \frac{1}{3}\int \frac{x^3\,dx}{1 + x^2} \boxed{u = 1 + x^2;\ du = x\,dx}$$

$$= \tfrac{1}{3}x^3\cot^{-1}x + \frac{1}{6}\int \frac{u - 1}{u}\,du$$

$$= \tfrac{1}{3}x^3\cot^{-1}x + \tfrac{1}{6}[u - \ln|u|] + C_1$$

$$= \frac{x^3}{3}\cot^{-1}x + \frac{x^2 + 1}{6} - \tfrac{1}{6}\ln(1 + x^2) + C_1$$

$$= \frac{x^3}{3}\cot^{-1}x + \frac{x^2}{6} - \tfrac{1}{6}\ln(1 + x^2) + C$$

26. $\int x(1 + x)^{1/3} \, dx$ $\boxed{u^3 = 1 + x; \; dx = 3u^2 \, du}$

$= \int (u^3 - 1)u(3u^2 \, du) = 3 \int (u^6 - u^3) du$

$= \frac{3}{7} u^7 - \frac{3}{4} u^4 + C$

$= \frac{3}{7}(1 + x)^{7/3} - \frac{3}{4}(1 + x)^{4/3} + C$

27. $\int \dfrac{x^2 + 2}{x^3 + 6x + 1} \, dx$ $\boxed{u = x^3 + 6x + 1}$

$= \frac{1}{3} \int \dfrac{du}{u} = \frac{1}{3} \ln \left| x^3 + 6x + 1 \right| + C$

28. $\int \dfrac{\sin x - \cos x}{(\sin x + \cos x)^{1/4}} \, dx$ $\boxed{u = \sin x + \cos x}$

$= - \int u^{-1/4} \, du = -\frac{4}{3} u^{3/4} + C$

$= -\frac{4}{3}(\sin x + \cos x)^{3/4} + C$

29. $\int \cos(\sqrt{x + 2}) \, dx$ $\boxed{t = \sqrt{x + 2}; \; dx = 2t \, dt}$

$= 2 \int t \cos t \, dt$ $\boxed{\text{parts or Formula 312}}$

$= 2[\cos t + t \sin t] + C$

$= 2 \cos\sqrt{x + 2} + 2\sqrt{x + 2} \, \sin\sqrt{x + 2} + C$

30. $\int \sqrt{5 + 2\sin^2 x} \, \sin 2x \, dx$ $\boxed{u = 5 + 2\sin^2 x;}$

$\boxed{du = 4\sin x \cos x \, dx = 2 \sin 2x \, dx}$

$= \frac{1}{2} \int \sqrt{u} \, du = \frac{1}{3} u^{3/2} + C$

$= \frac{1}{3}(5 + 2\sin^2 x)^{3/2} + C$

31. $\int \dfrac{x^3 + 2x}{x^4 + 4x^2 + 3} \, dx$ $\boxed{u = x^4 + 4x^2 + 3}$

$= \frac{1}{4} \int \dfrac{du}{u} = \frac{1}{4} \ln(x^4 + 4x^2 + 3) + C$

32. $\int \dfrac{x \, dx}{\sqrt{5 - x^2}} = -\frac{1}{2} \int u^{-1/2} \, du = -\sqrt{5 - x^2} + C$

33. $\int \dfrac{\sqrt{5 - x^2}}{x} \, dx$ $\boxed{\text{Formula 235}}$

$= \sqrt{5 - x^2} - \sqrt{5} \ln \left| \dfrac{\sqrt{5} + \sqrt{5 - x^2}}{x} \right| + C$

If you use computer software you might obtain the following equivalent form

$\sqrt{5} \ln \left| \dfrac{\sqrt{5 - x^2} - \sqrt{5}}{x} \right| + \sqrt{5 - x^2}$

34. $\int \dfrac{\sqrt{x^2 + x}}{x} \, dx = \int \dfrac{\sqrt{x + 1} \, dx}{\sqrt{x}}$ $\boxed{t^2 = x + 1}$

$= 2 \int \dfrac{t^2 \, dt}{\sqrt{t^2 - 1}}$ $\boxed{\text{parts or Formula 198}}$

$= 2 \left[\dfrac{t\sqrt{t^2 - 1}}{2} + \frac{1}{2} \ln \left| t + \sqrt{t^2 - 1} \right| \right] + C$

$= \sqrt{x^2 + x} + \ln \left| \sqrt{x} + \sqrt{x + 1} \right| + C$

35. $\int x^3(x^2 + 4)^{-1/2} \, dx$ $\boxed{u = x^2; \; dv = \dfrac{x \, dx}{\sqrt{x^2 + 4}}}$

$= x^2\sqrt{x^2 + 4} - \int \sqrt{x^2 + 4}(2x \, dx)$

$= x^2\sqrt{x^2 + 4} - \frac{2}{3}(x^2 + 4)^{3/2} + C$

$= \frac{1}{3}\sqrt{x^2 + 4}(x^2 - 8) + C$

36. $\displaystyle\int_0^{\pi/2} \dfrac{\cos x \, dx}{\sqrt{\sin x}} = \lim_{t \to 0^+} \int_t^{\pi/2} \dfrac{\cos x \, dx}{\sqrt{\sin x}}$

$= \lim_{t \to 0^+} 2\sqrt{\sin x} \, \Big|_t^{\pi/2} = 2$

37. $\displaystyle\int_{-\infty}^{+\infty} \dfrac{dx}{4 + x^2} = \lim_{s \to -\infty} \int_s^0 \dfrac{dx}{4 + x^2} + \lim_{t \to +\infty} \int_0^t \dfrac{dx}{4 + x^2}$

$= \frac{1}{2} \left[\lim_{s \to -\infty} \tan^{-1}\dfrac{x}{2} \Big|_s^0 + \lim_{t \to +\infty} \tan^{-1}\dfrac{x}{2} \Big|_0^t \right]$

$= \frac{1}{2} \left[\dfrac{\pi}{2} + \dfrac{\pi}{2} \right] = \dfrac{\pi}{2}$

38. $\displaystyle\int_{-\infty}^{+\infty} \dfrac{dx}{x^2 + 4x + 6}$

$= \lim_{s \to -\infty} \int_s^0 \dfrac{dx}{(x + 2)^2 + 2} + \lim_{t \to +\infty} \int_0^t \dfrac{dx}{(x + 2)^2 + 2}$

$= \dfrac{1}{\sqrt{2}} \left[\lim_{s \to -\infty} \tan^{-1}\dfrac{x + 2}{\sqrt{2}} \Big|_s^0 \right]$

$\qquad + \dfrac{1}{\sqrt{2}} \left[\lim_{t \to +\infty} \tan^{-1}\dfrac{x + 2}{\sqrt{2}} \Big|_0^t \right]$

$= \sqrt{2}\left(\dfrac{\pi}{2} - \tan^{-1}\sqrt{2} \right)$

39. $\displaystyle\int_1^{+\infty} \dfrac{dx}{x^4 + x^2} = \lim_{t \to +\infty} \int_1^t \dfrac{dx}{x^2(x^2 + 1)}$

$\dfrac{1}{x^2(x^2 + 1)} = \dfrac{A_1}{x} + \dfrac{A_2}{x^2} + \dfrac{A_3 x + B_1}{x^2 + 1}$

Partial fraction decomposition gives

$$A_1 = 0, \; A_2 = 1, \; A_3 = 0, \; B_1 = -1.$$

$$\lim_{t \to +\infty} \int_1^t \frac{dx}{x^2(x^2+1)}$$

$$= \lim_{t \to +\infty} \left[\int_0^t \frac{dx}{x^2} - \int_0^t \frac{dx}{x^2+1} \right]$$

$$= -\lim_{t \to +\infty} \left[x^{-1} - \tan^{-1} x \right] \Big|_1^t$$

$$= -\left[\frac{\pi}{2} - \left(1 + \frac{\pi}{4}\right)\right] = 1 - \frac{\pi}{4}$$

40. $\displaystyle \int_0^{+\infty} x^n e^{-ax}\,dx = \lim_{t \to +\infty} \int_0^t x^n e^{-an}\,dx$

$$\boxed{\text{Formula 486}}$$

$$= \lim_{t \to +\infty} \left[\frac{e^{-ax}}{a}\left(x^n - \frac{nx^{n-1}}{a} + \cdots + \frac{(-1)^n n!}{a^n}\right)\right]$$

$$= \frac{n!}{a^{n+1}}$$

41. $\displaystyle \int \frac{\sin x \, dx}{1 + \cos^2 x} = -\int \frac{du}{1 + u^2} = -\tan^{-1} u + C$

$$= -\tan^{-1}(\cos x) + C$$

42. $\displaystyle \int_0^1 x^m(1-x)^n dx \quad \boxed{t = 1 - x; \; dx = -dt}$

$$= -\int_1^0 (1-t)^m t^n \, dt = \int_0^1 x^n(1-x)^m \, dx$$

43. **a.** The statement is true;

$$\tanh(\tfrac{1}{2}\ln x) = \tanh \ln\sqrt{x}$$

$$= \frac{e^{\ln\sqrt{x}} - e^{-\ln\sqrt{x}}}{e^{\ln\sqrt{x}} + e^{-\ln\sqrt{x}}}$$

$$= \frac{\sqrt{x} - \sqrt{x^{-1}}}{\sqrt{x} + \sqrt{x^{-1}}} = \frac{x-1}{x+1}$$

b. This statement is true; let
$a = \sinh^{-1}(\tan x)$, then $\sinh a = \tan x$

$$\tan x = \tfrac{1}{2}(e^a - e^{-1}),$$

$$\cos x = \frac{1}{\sqrt{1 + \tan^2 x}}$$

$$= \frac{1}{\sqrt{1 + 0.25(e^a - e^{-a})^2}}$$

$$= \frac{2}{\sqrt{4 + e^{2a} - 2 + e^{-2a}}}$$

$$= \frac{2}{e^a + e^{-a}} = \operatorname{sech} a$$

Let $b = \tanh^{-1}(\sin x)$, then $\tanh b = \sin x$,
$\cos x = \operatorname{sech} a = \operatorname{sech} b$, so $a = b$ is a
solution.

44. $\displaystyle \int_0^1 xf''(3x)\,dx \quad \boxed{t = 3x; \; dt = 3\,dx}$

$$= \frac{1}{9} \int_0^3 tf''(t)\,dt \quad \boxed{u = t; \; dv = f''(x)\,dx}$$

$$= \left[\frac{1}{9} t f'(x)\Big|_0^3 - \int_0^3 f'(t)\,dt \right]$$

$$= \tfrac{1}{9}[3f'(3) - 0 - f(3) + f(0)] = -1$$

45. $\displaystyle I_1 = \int_0^{\pi/2} f(x) \cos x \, dx; \; I_2 = \int_0^{\pi/2} f''(x) \cos x \, dx$

$$I_2 = \int_0^{\pi/2} f''(x)\cos x \, dx \quad \boxed{u = \cos x; \; dv = f''(x)\,dx}$$

$$= f'(x) \cos x \Big|_0^{\pi/2} + \int_0^{\pi/2} f'(x) \sin x \, dx$$

$$= 0 - f'(0) + \int_0^{\pi/2} f'(x) \sin x \, dx$$

$$= -(-1) + \int_0^{\pi/2} f'(x) \sin x \, dx$$

$$\boxed{u = \sin x; \; du = f'(x)\,dx}$$

$$= 1 + \sin x\, f(x)\Big|_0^{\pi/2} - \int_0^{\pi/2} f(x) \cos x\, dx$$

$$= 1 + f(\tfrac{\pi}{2}) - 0 - I_1$$

Thus, $I_1 + I_2 = 1 + 5 = 6$

46. Let P denote the amount of poison gas (in ft^3)
and t the time (in minutes). The concentration
of gas satisfies

$$\frac{dP}{dt} = \underbrace{0.2}_{\text{incoming gas}} - \underbrace{\frac{P}{400}(0.2)}_{\text{outgoing gas}}$$

$$I(x) = e^{\int 0.2\,dt/400} = e^{0.0005t}, \text{ so}$$

$$P = \frac{1}{e^{0.0005t}}\left[\int 0.2e^{0.0005t}\,dt + C \right]$$

$$= 400 + Ce^{-0.0005t}$$

Since there is no poison gas in the closet when
he enters, we have $P(0) = 0$, so $C = -400$ and

$$P = 400(1 - e^{-0.0005t})$$

The air in the closet becomes deadly when
$P = 0.8$. Solving

$$0.8 = 400(1 - e^{-0.0005t})$$

$$-0.0005t = \ln(1 - 0.002)$$

$$t \approx 4.004 \text{ minutes}$$

He survives with about 15 seconds to spare.

47. $f(x) = 1/\ln t$

# OF TERMS	TRAPEZOIDAL RULE OVER [2, 1000]
4	328.9931
8	245.8846
16	208.3376
32	190.8812
64	182.6252
128	178.9596

48. $m = \rho \displaystyle\int_0^{\pi/4} (\sin x + \cos x)\, dx$

$$= -\rho(\cos x - \sin x)\Big|_0^{\pi/4} = \rho$$

$$M_y = \rho \int_0^{\pi/4} x(\sin x + \cos x)\, dx$$

$$\boxed{u = x;\ dv = (\sin x + \cos x)\, dx}$$

$$= \rho\left[x(\sin x - \cos x) + (\sin x + \cos x)\Big|_0^{\pi/4} \right]$$

$$= \rho(\sqrt{2} - 1)$$

$$M_x = \rho \int_0^{\pi/4} \tfrac{1}{2}(\sin x + \cos x)^2\, dx$$

$$= \tfrac{1}{2}\rho \int_0^{\pi/4} (1 + \sin 2x)\, dx$$

$$= \tfrac{1}{2}\rho(x - \tfrac{1}{2}\cos 2x)\Big|_0^{\pi/4} = \tfrac{1}{8}\rho(\pi + 2)$$

$$(\overline{x}, \overline{y}) = \left(\frac{\rho(\sqrt{2} - 1)}{\rho}, \frac{\tfrac{1}{8}\rho(\pi + 2)}{\rho} \right)$$

$$= (\sqrt{2} - 1, \tfrac{1}{8}(\pi + 2)) \approx (0.41, 0.64)$$

49. $m = \rho \displaystyle\int_0^{\pi/3} \sec^2 x\, dx = \rho \tan x\Big|_0^{\pi/4} = \sqrt{3}\rho$

$$M_y = \rho \int_0^{\pi/3} x \sec^2 x\, dx$$

$$\boxed{u = x;\ dv = \sec^2 x\, dx}$$

$$= \rho\left[x \tan x\Big|_0^{\pi/3} - \int_0^{\pi/3} \frac{\sin x\, dx}{\cos x} \right]$$

$$= \rho(\tfrac{1}{3}\sqrt{3}\pi + \ln 0.5)$$

$$M_x = \rho \int_0^{\pi/3} \tfrac{1}{2}\sec^4 x\, dx = \tfrac{1}{2}\rho \int_0^{\pi/3} (1 + \tan^2 x)\sec^2 x\, dx$$

$$= \tfrac{1}{2}\rho(\tan x + \tfrac{1}{3}\tan^3 x)\Big|_0^{\pi/3} = \sqrt{3}\rho$$

$$(\overline{x}, \overline{y}) = \left(\frac{\rho(\tfrac{1}{3}\sqrt{3}\pi + \ln 0.5)}{\sqrt{3}\rho}, \frac{\sqrt{3}\rho}{\sqrt{3}\rho} \right)$$

$$\approx (0.64, 1)$$

50. $V = 2\pi \displaystyle\int_0^1 x \sinh x\, dx$ $\boxed{u = x;\ dv = \sinh x\, dx}$

$$= 2\pi x \cosh x\Big|_0^1 - 2\pi \int_0^1 \cosh x\, dx$$

$$= 2\pi(\cosh 1 - \sinh 1) = 2\pi e^{-1}$$

51. $V = 2\pi \displaystyle\int_0^{\cosh^{-1} 2} x(2 - \cosh x)\, dx$

$$= 2\pi x^2 + 2\pi \int_0^{\cosh^{-1} 2} x \cosh x\, dx$$

$$\boxed{u = x;\ dv = \cosh x\, dx}$$

$$= 2\pi x^2 + 2\pi x \sinh x\Big|_0^{\cosh^{-1} 2}$$

$$- 2\pi \int_0^{\cosh^{-1} 2} \sinh x\, dx$$

$$\approx 2.848$$

52. $V = \pi \displaystyle\int_0^{\pi} \sin^2 x\, dx = \frac{\pi}{2}\int_0^{\pi} (1 - \cos 2x)\, dx$

$$= \frac{\pi}{2}(x - \tfrac{1}{2}\sin 2x)\Big|_0^{\pi} = \frac{\pi^2}{2}$$

$$dy = \cos x\, dx;\ ds = \sqrt{1 + \cos^2 x}\, dx;\ dS = 2\pi y\, ds$$

$$S = 2\pi \int_0^{\pi} \sin x\sqrt{1 + \cos^2 x}\, dx$$

$$= 4\pi \int_0^{\pi/2} \sin x\sqrt{1 + \cos^2 x}\, dx$$

$$\boxed{u = \cos x;\ du = -\sin x\, dx}$$

$$= -4\pi \int_1^0 \sqrt{1 + u^2}\, du \quad \boxed{\text{Formula 168}}$$

$$= -4\pi\left[\frac{u\sqrt{u^2+1}}{2} + \frac{1}{2}\ln\left|x + \sqrt{x^2+1}\right|\right]\Bigg|_1^0$$

$$= -4\pi\left[-\frac{\sqrt{2}}{2} - \frac{1}{2}\ln(\sqrt{2}+1)\right]$$

$$= 2\pi[\sqrt{2} + \ln(1+\sqrt{2})] \approx 7.212$$

53. $ds = \sqrt{1 + x^{1/2}}\, dx;$

$$S = \int_0^1 \sqrt{1 + x^{1/2}}\, dx$$

$$\boxed{u^2 = 1 + \sqrt{x};\ dx = 4(u^3 - u)\, du}$$

$$= 4\int_1^{\sqrt{2}} (u^4 - u^2)\, du = 4\left[\frac{1}{5}u^5 - \frac{1}{3}u^3\right]\Bigg|_1^{\sqrt{2}}$$

$$= \frac{16\sqrt{2}}{15} \approx 1.5085$$

54. $V = \pi\int_0^3 (9 - x^2)^{1/2}\, dx = \pi(\frac{9}{4}\pi) = \frac{9}{4}\pi^2$

55. $V = \pi\int_0^4 \frac{2x\, dx}{1 + (x^2)^2} = \frac{\pi}{2}\tan^{-1}x^2\Big|_0^4$

$$= \frac{\pi^2}{2} - \pi\tan^{-1}\frac{1}{16} \approx 4.739$$

56. $V = \frac{4}{3}\pi\int_1^2 \frac{3\, dx}{3x - 2} = \frac{4}{3}\pi\ln|3x - 2|\Big|_1^2$

$$= \frac{4\pi}{3}\ln 4 \approx 5.807$$

57. $dy = (e^x - \frac{1}{4}e^{-x})\, dx;$

$$ds = \sqrt{1 + (e^x - \frac{1}{4}e^{-x})^2}\, dx$$

$$= (e^x + \frac{1}{4}e^{-x})\, dx$$

$$S = 2\pi\int_0^1 (e^x + \frac{1}{4}e^{-x})^2\, dx$$

$$= 2\pi\left[\frac{1}{2}e^{2x} + \frac{1}{2}x - \frac{1}{32}e^{-2x}\right]\Big|_0^1$$

$$= \pi(e^2 - \frac{1}{16}e^{-2} + \frac{1}{16}) \approx 23.38$$

58. $V = \frac{1}{4}\int_0^1 e^{2x}\, dx = \frac{1}{8}e^{2x}\Big|_0^1 = \frac{1}{8}(e^2 - 1) \approx 0.7986$

59. $A_1 = \int_1^a \frac{dx}{x} = \ln|a|;\quad A_2 = \int_k^{ka} \frac{dx}{x} = \ln|x|\Big|_k^{ka} = \ln|a|$

60. $dy = \tan x\, dx;\quad ds = \sqrt{1 + \tan^2 x}\, dx = \sec x\, dx$

$$s = \int_0^{\pi/3} \sec x\, dx = \ln|\sec x + \tan x|\Big|_0^{\pi/3}$$

$$= \ln(2 + \sqrt{3}) \approx 1.317$$

61. $dy = \frac{2ay}{b^2}\, dy - \frac{b^2}{8ay}\, dy;$

$$ds = \sqrt{1 + \left(\frac{2ay}{b^2} - \frac{b^2}{8ay}\right)^2}\, dy = \left(\frac{2ay}{b^2} + \frac{b^2}{8ay}\right)dy$$

$$s = \int_b^{2b} \left(\frac{2ay}{b^2} + \frac{b^2}{8ay}\right)dy = \left(\frac{ay^2}{b^2} + \frac{b^2}{8a}\ln|y|\right)\Bigg|_b^{2b}$$

$$= 3|a| + \frac{b^2}{8|a|}\ln 2$$

62. a. $\Gamma(s) = \int_0^{+\infty} e^{-t}t^{s-1}\, dt = \lim_{N\to+\infty}\int_0^N e^{-t}t^{s-1}\, dt$

$$\boxed{u = e^{-t};\ dv = t^{s-1}\, dt}$$

$$= \lim_{N\to+\infty}\left[\frac{t^s}{se^t}\Big|_0^N - \frac{1}{s}\int_0^N e^{-t}t^s\, dt\right]$$

$$\le \lim_{N\to+\infty}\frac{1}{s}\int_0^N e^{-t}\, dt = \frac{1}{s}$$

The first fraction vanishes as seen by using l'Hôpital's rule, $\Gamma(s)$ converges.

b. $\Gamma(s - 1)$

$$= \lim_{N\to+\infty}\left[-\frac{t^{s-1}}{e^t}\Big|_0^N + (s-1)\int_0^N e^{-t}t^{s-2}\, dt\right]$$

The first fraction vanishes as seen by using l'Hôpital's rule. The integral can be rewritten as

$$\Gamma(s) = (s - 1)\Gamma(s - 1)$$

which can be restated

$$\Gamma(s + 1) = s\Gamma(s)$$

c. $\Gamma(n + 1) = n\Gamma(n) = n(n-1)\Gamma(n - 1)$

$$= n(n - 1)\cdots(3)(2)\Gamma(1)$$

$$= n!\lim_{N\to+\infty}\int_0^N e^{-t}\, dt = n!$$

63. a. Since $\Gamma(s + 1) = s\Gamma(s)$, $\Gamma[x + (n-1) + 1]$

$$= (x + n - 1)\Gamma(x + n - 2 + 1)$$

$$= (x + n - 1)(n - 2)\Gamma(x + n - 3 + 1)$$

$$\cdots (x + 1)\Gamma(x + 1)$$

$$= (x + n - 1)(n - 2)\Gamma(x + n - 3 + 1)$$
$$\cdots (x + 1)x\Gamma(x)$$

b. $\displaystyle\int_T^{+\infty} e^{-t}t^{x-1}dt \le \lim_{N\to+\infty}\int_T^N e^{-t}\,dt = 0$

$e^{-T} \approx 0.005$, so $T \approx -\ln(0.005) \approx 5.3$; Use 6 for the approximation.

c. $\Gamma(7 + 0.3)$

$$= (6.3)(5.3)(4.3)(3.3)(2.3)(1.3)(0.3)\Gamma(0.3)$$

$$\Gamma(0.3) = \int_0^6 e^{-t}t^{0.3-1}dt$$

$$\le \int_0^6 e^{-t}\,dt \approx 0.9975$$

Thus, $\Gamma(7.3) \approx 1{,}271.42$

d. To integrate $\Gamma(7.3)$ accurately we would need a much larger T, as a look at the integrands shows.

e. $\Gamma(7.3) \approx 1{,}271.83$ to six figures; our procedure gives $1{,}271.42$. The relative error in this is related to the 0.005 error we allowed in $\Gamma(0.3)$.

64. a. According to the plot of the ratio of $C = \Gamma(x + 1)/(x^{x+0.5}e^{-x})$, in the neighborhood of 200,000, $C \approx 2.50663$

b. Note that $\sqrt{2\pi} \approx 2.5066319$ and $\Gamma(x + 1) = \sqrt{2\pi}\,x^{x+0.5}e^{-x}$

c. For $n = 10$, the relative error is about 0.008; for $n = 100$, the relative error is about 0.001.

65. $I = \displaystyle\int (a^2 - x^2)^n \, dx$ $\boxed{u = (a^2 - x^2)^n;\ dv = dx}$

$$= x(a^2 - x^2)^n + 2n\int x^2(a^2 - x^2)^{n-1}dx$$

$$= x(a^2 - x^2)^n + 2n\int (a^2 - x^2)^{n-1}(x^2 - a^2)\,dx$$

$$+ 2a^2n\int(a^2 - x^2)^{n-1}dx$$

$$= x(a^2 - x^2)^n - 2nI + 2a^2n\int(a^2 - x^2)^{n-1}dx$$

$$(2n + 1)I = x(a^2 - x^2)^n + 2a^2n\int(a^2 - x^2)^{n-1}dx$$

$$I = \frac{x(a^2 - x^2)^n}{2n + 1} + \frac{2a^2n}{2n + 1}\int(a^2 - x^2)^{n-1}dx$$

For example

$$\int(3^2 - x^2)^{5/2}dx$$

$$= \frac{x(9 - x^2)^{5/2}}{5 + 1} + \frac{45}{5+1}\int(9 - x^2)^{3/2}dx$$

$$= \frac{1}{6}x(9 - x^2)^{5/2}$$

$$+ \frac{15}{2}\left[\frac{x}{4}(9 - x^2)^{3/2} + \frac{27}{4}\int(9 - x^2)^{1/2}dx\right]$$

$$= \frac{1}{6}x(9 - x^2)^{5/2} + \frac{15}{8}x(9 - x^2)^{3/2} + \frac{405}{8}I_1$$

$$I_1 = \int(9 - x^2)^{1/2}dx = \frac{x\sqrt{9 - x^2}}{2} + \frac{9}{2}\sin^{-1}\frac{x}{3}$$

Thus, $\displaystyle\int(9 - x^2)^{5/2}dx$

$$= \frac{x}{6}(9 - x^2)^{5/2} + \frac{15x}{8}(9 - x^2)^{3/2}$$

$$+ \frac{3{,}645}{16}\sin^{-1}\frac{x}{3} + \frac{405x}{16}(9 - x^2)^{1/2} + C$$

66. $\displaystyle\int\frac{\sin^n x\, dx}{\cos^m x}$ $\boxed{u = \sin^{n-1}x;\ dv = \dfrac{\sin x}{\cos^m x}\,dx}$

$$= \frac{\sin^{n-1}x}{(m-1)\cos^{m-1}x} + \frac{n-1}{1-m}\int\frac{\sin^{n-2}x\, dx}{\cos^{m-1}x}$$

$$= \frac{\sin^{n-1}x}{(m-1)\cos^{m-1}x} - \frac{n-1}{m-1}\int\frac{\sin^{n-2}x\, dx}{\cos^{m-1}x}$$

67. $\displaystyle\int\frac{\cos^n x\, dx}{\sin^m x}$ $\boxed{u = \cos^{n-1}x;\ dv = \dfrac{\cos x}{\sin^m x}\,dx}$

$$= \frac{\cos^{n-1}x}{(m-1)\sin^{m-1}x} + \frac{n-1}{m-1}\int\frac{\cos^{n-2}x\, dx}{\sin^{m-1}x}$$

68. $I = \displaystyle\int x^n(x^2 + a^2)^{-1/2}dx$

$$\boxed{u = x^{n-1};\ dv = x(x^2 + a^2)^{-1/2}\,dx}$$

$$= x^{n-1}(x^2 + a^2)^{1/2}$$

$$- (n-1)\int x^{n-2}(x^2 + a^2)^{1/2}dx$$

$$= x^{n-1}(x^2 + a^2)^{1/2}$$

$$- (n-1)\int x^{n-2}(x^2 + a^2)(x^2 + a^2)^{-1/2}dx$$

$$= x^{n-1}(x^2 + a^2)^{1/2}$$

$$- (n-1)\int x^n(x^2 + a^2)^{-1/2}dx$$

$$- a^2(n-1)\int x^{n-2}(a^2 + x^2)^{-1/2}dx$$

$$= x^{n-1}(x^2 + a^2)^{1/2} - (n-1)I$$

$$- a^2(n-1)\int x^{n-2}(a^2 + x^2)^{-1/2}dx$$

$$nI = x^{n-1}(x^2 + a^2)^{1/2}$$

$$- a^2(n-1)\int x^{n-2}(a^2 + x^2)^{-1/2}dx$$

$$I = \frac{x^{n-1}(x^2 + a^2)^{1/2}}{n}$$

$$- \frac{(n-1)a^2}{n}\int x^{n-2}(a^2 + x^2)^{-1/2}dx$$

69. $I = \displaystyle\int \frac{dx}{x^n\sqrt{ax + b}}$ $\boxed{u = x^{-n};\; dv = (ax+b)^{-1}dx}$

$$= \frac{2\sqrt{ax + b}}{ax^n} + \frac{2n}{a}\int \frac{\sqrt{ax + b}}{x^{n+1}}\,dx$$

$$= \frac{2\sqrt{ax + b}}{ax^n} + \frac{2n}{a}\int \frac{ax}{x^{n+1}\sqrt{ax + b}}\,dx$$

$$+ \frac{2n}{a}\int \frac{b}{x^{n+1}\sqrt{ax + b}}\,dx$$

$$= \frac{2\sqrt{ax + b}}{ax^n} + 2n\int \frac{1}{x^n\sqrt{ax + b}}\,dx$$

$$+ \frac{2nb}{a}\int \frac{1}{x^{n+1}\sqrt{ax + b}}\,dx$$

Since this formula is true for n, it must also be true for $n - 1$:

$$I = \frac{2\sqrt{ax + b}}{ax^{n-1}} + 2(n-1)\int \frac{dx}{x^{n-1}\sqrt{ax + b}}$$

$$+ \frac{2(n-1)b}{a}\int \frac{dx}{x^n\sqrt{ax + b}}$$

$$= \frac{2\sqrt{ax + b}}{ax^{n-1}} + 2(n-1)\int \frac{dx}{x^{n-1}\sqrt{ax + b}}$$

$$+ \frac{2(n-1)b}{a}I$$

$$\frac{2(n-1)b}{a}I = \frac{2\sqrt{ax + b}}{ax^{n-1}}$$

$$+ 2(n-1)\int \frac{dx}{x^{n-1}\sqrt{ax + b}}$$

$$I = \frac{\sqrt{ax + b}}{x^{n-1}(n-1)b} - \frac{(2n-3)a}{(2n-2)b}\int \frac{dx}{x^{n-1}\sqrt{ax + b}}$$

70. $I = \displaystyle\int_1^{+\infty}\left[\frac{2Ax^3}{x^4 + 1} - \frac{1}{x + 1}\right]dx$

$$= \lim_{t \to +\infty}\left[\frac{A}{2}\ln(x^4 + 1) - \ln|x + 1|\right]\Big|_1^t$$

$$= \lim_{t \to +\infty}\left[\ln\frac{(t^4 + 1)^{A/2}}{t + 1} - \ln 2^{A/2 - 1}\right]$$

Consider the first term:

$$\lim_{t \to +\infty}\ln\frac{(t^4 + 1)^{A/2}}{t + 1}$$

$$= \lim_{t \to +\infty}\ln\frac{(t^4)^{A/2} + \frac{1}{2}A(t^4)^{A/2 - 1} + \cdots}{t + 1}$$

The numerator and denominator are the same degree if $A = \frac{1}{2}$.

$$I = \int_1^{+\infty}\left[\frac{x^3}{x^4 + 1} - \frac{1}{x + 1}\right]dx$$

$$= \lim_{t \to +\infty}\left[\ln\frac{(x^4 + 1)^{1/4}}{x + 1}\right]\Big|_1^t$$

$$= \ln \lim_{t \to +\infty}\left[\ln\frac{(t^4 + 1)^{1/4}}{t + 1} - \ln 2^{1/4 - 1}\right]$$

$$= \frac{3}{4}\ln 2 \approx 0.52$$

71. $\displaystyle\int_0^{+\infty}\frac{\sqrt{x}\ln x\,dx}{(x + 1)(x^2 + x + 1)} = \int_0^{+\infty}\frac{\sqrt{x}(x - 1)\ln x\,dx}{(x + 1)(x^3 - 1)}$

$$= \int_0^1\frac{\sqrt{x}\ln x\,dx}{x^3 + \cdots} + \int_1^{+\infty}\frac{\sqrt{x}\ln x\,dx}{x^3 + \cdots}$$

$$\geq \int_0^1\frac{\sqrt{x}\ln x\,dx}{(2x)(3x^2)} + \int_1^{+\infty}\frac{\sqrt{x}\ln x\,dx}{(1)(1)}$$

$$= \lim_{t \to 0^+}\int_t^1\frac{\sqrt{x}\ln x\,dx}{(2x)(3x^2)} + \lim_{N \to +\infty}\int_1^N\sqrt{x}\ln x\,dx$$

$$\boxed{u = \ln x;\; dv = x^{-5/2}\,dx}$$

$$= \lim_{t \to 0^+}\frac{1}{6}\left[-\frac{2}{3}\frac{\ln x}{x^{3/2}}\Big|_t^1 + \frac{2}{3}\int_t^1 x^{-5/2}\,dx\right]$$

$$+ \lim_{N \to +\infty}\int_1^N\sqrt{x}\ln x\,dx$$

This diverges because of the 0 in the denominator of the first fraction after the equal

sign and application of l'Hôpital's rule.

72. This is Putnam examination Problem 1 in the morning session of 1968. The standard approach, from elementary calculus applies. By division, rewrite the integrand as a polynomial plus a rational function with numerator of degree less than 2. The solution follows easily.

73. This is Putnam examination Problem 3 in the morning session of 1980. Let I be the given integral and $\sqrt{2} = r$. We show that $I = \pi/4$. Using $x = \frac{1}{2}\pi - u$, we have

$$I = \int_{\pi/2}^{0} \frac{-du}{1 + \cot^r u} = \int_{0}^{\pi/2} \frac{\tan^r u\, du}{\tan^r u + 1}$$

$$2I = \int_{0}^{\pi/2} \frac{1 + \tan^r x\, dx}{1 + \tan^r x} = \int_{0}^{\pi/2} dx = \frac{\pi}{2}$$

Thus, $I = \frac{\pi}{4}$.

74. This is Putnam examination Problem B5 in the afternoon session of 1985. Let

$$I(x) = \int_{0}^{\infty} t^{-1/2} e^{-at - xt^{-1}}\, dt \text{ where } a = 1985$$

$$I'(x) = \int_{0}^{\infty} t^{-1/2} e^{-at - xt^{-1}}(-t^{-1}\, dt)$$

$$= -\int_{0}^{\infty} t^{-3/2} e^{-at - xt^{-1}}\, dt$$

Let $u = t^{-1}$, $du = -1/t^2\, dt$. Then,

$$I'(x) = -\int_{\infty}^{0} u^{3/2}\, e^{-au^{-1} - xu}\left(\frac{-du}{u2}\right)$$

$$= -\int_{0}^{\infty} u^{-1/2} e^{-au^{-1} - xu}\, du$$

Now let $w = (x/a)u$, $dw = (x/a)\, du$.

$$I'(x) = -\int_{0}^{\infty} \left(\frac{aw}{x}\right)^{-1/2} e^{-x/w - aw}\left(\frac{a}{x}\, dw\right)$$

$$= -\left(\frac{a}{x}\right)^{1/2} \int_{0}^{\infty} w^{-1/2} e^{-x/w - aw}\, dw$$

$$= -\left(\frac{a}{x}\right)^{1/2} I(x)$$

Solving this differential equation, we obtain

$$\int \frac{I'(x)}{I(x)} = \int -\left(\frac{a}{x}\right)^{1/2}\, dx$$

$$I(x) = -2\sqrt{ax} + C$$

$$I(x) = ke^{-2\sqrt{ax}}$$

$$I(0) = ke^0 = k = \int_{0}^{\infty} t^{-1/2} e^{-a/t - 0}\, dt = \sqrt{\frac{\pi}{a}}$$

Thus,

$$I(a) = \sqrt{\frac{\pi}{a}}\, e^{-2\sqrt{a^2}} = \sqrt{\frac{\pi}{a}}\, e^{-2a}$$

In particular, $\displaystyle\int_{0}^{\infty} t^{-1/2} e^{-1985(t + t^{-1})}\, dt$

$$I(1985) = \sqrt{\frac{\pi}{1985}}\, e^{-2(1985)}$$

CHAPTER 8

Infinite Series

8.1 Sequences and Their Limits, Page 499

1. The limit of a sequence is that unique number that the elements of a sequence approaches as more and more numbers in the sequence are considered.

2. A bounded sequence has no element less than a predetermined number M or greater than a predetermined number N. A monotonic sequence is one whose elements do not increase or do not decrease. A sequence converges if it has a limit.

3. $0, 2, 0, 2, 0$

4. $-\frac{1}{8}, \frac{1}{16}, -\frac{1}{32}, \frac{1}{64}, -\frac{1}{128}$

5. $1, 0, \frac{1}{3}, \frac{1}{4}, \frac{1}{5}$

6. $1, 0, -3, 0, 5$

7. $\frac{4}{3}, \frac{7}{4}, 2, \frac{13}{6}, \frac{16}{7}$

8. $0, \frac{1}{3}, \frac{1}{2}, \frac{3}{5}, \frac{2}{3}$

9. $256, 16, 4, 2, \sqrt{2}$

10. $-1, 1, 4, 8, 13$

11. $1, 3, 13, 183, 33{,}673$

12. $\lim\limits_{n \to +\infty} \frac{5n+8}{n} = \lim\limits_{n \to +\infty} \frac{5 + \frac{8}{n}}{1} = 5$

13. $\lim\limits_{n \to +\infty} \frac{5n}{n+7} = \lim\limits_{n \to +\infty} \frac{5}{1 + \frac{7}{n}} = 5$

14. $\lim\limits_{n \to +\infty} \frac{2n+1}{3n-4} = \lim\limits_{n \to +\infty} \frac{2 + \frac{1}{n}}{3 - \frac{4}{n}} = \frac{2}{3}$

15. $\lim\limits_{n \to +\infty} \frac{4 - 7n}{8 + n} = \lim\limits_{n \to +\infty} \frac{\frac{4}{n} - 7}{\frac{8}{n} + 1} = -7$

16. $\lim\limits_{n \to +\infty} \frac{8n^2 + 800n + 5{,}000}{2n^2 - 1{,}000n + 2}$

$= \lim\limits_{n \to +\infty} \frac{8 + \frac{800}{n} + \frac{5{,}000}{n^2}}{2 - \frac{1{,}000}{n} + \frac{2}{n^2}} = \frac{8}{2} = 4$

17. $\lim\limits_{n \to +\infty} \frac{100n + 7{,}000}{n^2 - n - 1} = \lim\limits_{n \to +\infty} \frac{\frac{100}{n} + \frac{7{,}000}{n^2}}{1 - \frac{1}{n} - \frac{1}{n^2}} = 0$

18. $\lim\limits_{n \to +\infty} \frac{8n^2 + 6n + 4{,}000}{n^3 + 1} = 0$

19. $\lim\limits_{n \to +\infty} \frac{n^3 - 6n^2 + 85}{2n^3 - 5n + 170} = \frac{1}{2}$

20. $\lim\limits_{n \to +\infty} \frac{2n}{n + 7\sqrt{n}} = \lim\limits_{n \to +\infty} \frac{2}{1 + \frac{7}{\sqrt{n}}} = 2$

21. $\lim\limits_{n \to +\infty} \frac{8n - 500\sqrt{n}}{2n + 800\sqrt{n}} = \frac{8}{2} = 4$

22. $\lim\limits_{n \to +\infty} \frac{3\sqrt{n}}{5\sqrt{n} + \sqrt[4]{n}} = \lim\limits_{n \to +\infty} \frac{3}{5 + \frac{1}{\sqrt[4]{n}}} = \frac{3}{5}$

23. $\lim\limits_{n \to +\infty} \frac{\ln n}{n^2} = \lim\limits_{n \to +\infty} \frac{\frac{1}{n}}{2n} = \lim\limits_{n \to +\infty} \frac{1}{2n^2} = 0$

24. $\lim\limits_{n \to +\infty} 2^{5/n} = \lim\limits_{n \to +\infty} \exp[\ln(2^{5/n})]$

$= \lim\limits_{n \to +\infty} \exp[5n^{-1}\ln 2] = 1$

25. $\lim\limits_{n \to +\infty} n^{3/n} = \lim\limits_{n \to +\infty} \exp[\ln(n^{3/n})]$

$= \lim\limits_{n \to +\infty} \exp\left(\frac{3\ln n}{n}\right) = 1$

26. $\lim\limits_{n \to +\infty} \left(1 + \frac{3}{n}\right)^n = \left[\lim\limits_{n \to +\infty}\left(1 + \frac{3}{n}\right)^{n/3}\right]^3$

$= \left[\lim\limits_{n \to +\infty}\left(1 + \frac{1}{t}\right)^t\right]^3 = e^3$

27. $\lim\limits_{n \to +\infty} (n+4)^{1/n}$

$= \lim\limits_{n \to +\infty} \exp[\ln(n+4)^{1/(n+2)}]$

$= \lim\limits_{n \to +\infty} \exp\left(\frac{\ln n}{n+2}\right) = \lim\limits_{n \to +\infty} e^{1/n} = 1$

28. $\lim\limits_{n \to +\infty} n^{1/(n+2)} = \lim\limits_{n \to +\infty} \exp[\ln n^{1/(n+2)}]$

$= \lim\limits_{n \to +\infty} \exp\left(\frac{\ln n}{n+2}\right) = \lim\limits_{n \to +\infty} e^{1/n} = 1$

29. $\lim\limits_{n \to +\infty} (\ln n)^{1/n} = \lim\limits_{n \to +\infty} \exp[\ln(\ln n)^{1/n}]$

$= \lim\limits_{n \to +\infty} \exp\left(\frac{\ln(\ln n)}{n}\right)$

$= \lim\limits_{n \to +\infty} \exp\left(\frac{1}{n \ln n}\right) = 1$

30. $\lim\limits_{n \to +\infty} \int_0^\infty e^{-nx}\, dx$

$= \lim\limits_{n \to +\infty} \left[\lim\limits_{t \to +\infty} \int_0^t e^{-nx}\, dx\right]$

$= \lim\limits_{n \to +\infty} \left[-n^{-1} \lim\limits_{t \to +\infty} e^{-nx}\Big|_0^t\right]$

$= \lim\limits_{n \to +\infty} (-n^{-1})\left[\lim\limits_{t \to +\infty}(e^{-nt} - 1)\right] = 0$

31. $\lim_{n \to +\infty} (\sqrt{n^2 + n} - n)$

$= \lim_{n \to +\infty} \dfrac{(\sqrt{n^2 + n} - n)(\sqrt{n^2 + n} + n)}{\sqrt{n^2 + n} + n}$

$= \lim_{n \to +\infty} \dfrac{n^2 + n - n}{\sqrt{n^2 + n} + n}$

$= \lim_{n \to +\infty} \dfrac{n}{\sqrt{n^2 + n} + n}$

$= \lim_{n \to +\infty} \dfrac{1}{\sqrt{1 + \frac{1}{n}} + 1} = \dfrac{1}{2}$

32. $\lim_{n \to +\infty} \sqrt{n + 5\sqrt{n}} - \sqrt{n}$

$= \lim_{n \to +\infty} \dfrac{(\sqrt{n + 5\sqrt{n}} - \sqrt{n})(\sqrt{n + 5\sqrt{n}} + \sqrt{n})}{\sqrt{n + 5\sqrt{n}} + \sqrt{n}}$

$= \lim_{n \to +\infty} \dfrac{n + 5\sqrt{n} - n}{\sqrt{n + 5\sqrt{n}} + \sqrt{n}}$

$= \lim_{n \to +\infty} \dfrac{5}{\sqrt{1 + \frac{5}{\sqrt{n}}} + 1} = \dfrac{5}{2}$

33. $\lim_{n \to +\infty} \sqrt[n]{n} = \lim_{n \to +\infty} \exp[\ln \sqrt[n]{n}]$

$= \lim_{n \to +\infty} \exp[\ln n^{1/n}] = \lim_{n \to +\infty} \exp\left[\dfrac{\ln n}{n}\right]$

$= \lim_{n \to +\infty} e^{1/n} = 1$

34. $\lim_{n \to +\infty} (an + b)^{1/n} = \lim_{n \to +\infty} \exp[\ln(an + b)^{1/n}]$

$= \lim_{n \to +\infty} \exp\left[\dfrac{\ln(an + b)}{n}\right]$

$= \lim_{n \to +\infty} e^{a/(an+b)} = 1$

35. $\lim_{n \to +\infty} [\ln n - \ln(n + 1)]$

$= \lim_{n \to +\infty} \ln \dfrac{n}{n + 1} = \lim_{n \to +\infty} \ln 1 = 0$

36. The elements of $\left\{ \dfrac{n}{2^n} \right\}$ lie on the curve
$f(x) = 2^{-x}x; \; f'(x) = -2^{-x}x\ln 2 + 2^{-x}$
$= (-x\ln 2 + 1)2^{-x} < 0$ whenever
$(\ln 2)^{-1} < x$. Thus, $f(x)$ and $\{n2^{-n}\}$ are
both decreasing. $M = 0$ is a lower bound of
the sequence (the elements are
positive). Thus, $\{n2^{-n}\}$.

37. The elements of $\{a_n\} = \left\{ \ln\left(\dfrac{n + 1}{n}\right) \right\}$
lie on the curve $f(x) = \ln(x + 1) - \ln x$.
$f'(x) = (x + 1)^{-1} - x^{-1} < 0$. Thus, $f(x)$
and $\{a_n\}$ are both decreasing. M is a lower
bound of the sequence (the elements are
positive since $\ln(n + 1) > \ln n$). Thus $\{a_n\}$
converges.

38. The elements of $\{a_n\} = \left\{ \dfrac{3n - 2}{n} \right\}$
lie on the curve $f(x) = 3 - 2n^{-1}$.
$f'(x) = 2n^{-2} > 0$. Thus, $f(x)$ and $\{a_n\}$ are
both increasing. $M = 3$ is an upper bound of
the sequence. Thus $\{a_n\}$
converges.

39. The elements of $\{a_n\} = \left\{ \dfrac{4n + 5}{n} \right\}$
lie on the curve $f(x) = 4 + 5n^{-1}$.
$f'(x) = -5n^{-2} < 0$. Thus, $f(x)$ and $\{a_n\}$
are both decreasing. $M = 4$ is a lower bound
of the sequence. Thus $\{a_n\}$
converges.

40. The elements of $\{a_n\} = \left\{ \dfrac{3n - 7}{2^n} \right\}$
lie on the curve $f(x) = 2^{-x}(3x - 7)$.
$f'(x) = 2^{-x}[3 - (3x - 7)\ln 2] < 0$. Thus,
$f(x)$ and $\{a_n\}$ are both decreasing. $M = 0$ is a
lower bound of the sequence.
Thus $\{a_n\}$ converges.

41. The elements of $\{a_n\} = \{ \sqrt[n]{n} \}$ lie on the
curve
$f(x) = \sqrt[x]{x} = \exp[\ln x^{1/x}]$

$= \exp[x^{-1}\ln x].$

$f'(x) = \exp[\dfrac{\ln x}{x}](x^{-2} - x^{-1}\ln x) < 0$

(when $e < x$). Thus, $f(x)$ and $\{a_n\}$ are both
decreasing. M is a lower bound of the
sequence (the elements are positive since $\ln(n + 1) > \ln n$). Thus $\{a_n\}$
converges.

42. The elements $\{a_n\} = \{1 + (-1)^n\}$
$= \{0, 2, 0, 2, \cdots\}$. There is no unique finite
number that the elements approach. The
sequence diverges by oscillation.

43. The elements $\{a_n\} = \{\cos n\pi\}$
alternate between -1 and 1, and,
therefore, is divergent by oscillation.

44. The elements of $\{a_n\} = \left\{ \dfrac{n^3 - 7n + 5}{100n^2 + 219} \right\}$
diverges because $\lim_{n \to +\infty} a_n = +\infty$. The
sequence is unbounded.

45. The elements of $\{a_n\} = \{ \sqrt{n} \}$ diverges
because $\lim_{n \to +\infty} a_n = +\infty$. The sequence is
unbounded.

46. The elements of $\{E_n\} = \left\{ \dfrac{n^2 h^2}{8ma^2} \right\}$

$$= \left\{ \left[\frac{(6.63)(10^{-27})n^2}{(8)(100)} \right] \right\}^2 \approx 6.87(10^{-57})$$

$E_1 \approx 6.87 \times 10^{-59}$; $E_2 \approx 2.75 \times 10^{-58}$;

$E_3 \approx 6.18 \times 10^{-58}$; $E_4 \approx 1.10 \times 10^{-57}$

47. $a_1 = \frac{1}{2}$, $a_2 = \frac{1}{4}$, $a_3 = \frac{1}{8}$, $a_4 = \frac{1}{16}$, $\cdots$, $a_n = \left(\frac{1}{2}\right)^n$

$a_4 = 6.25\%$, $a_n = 100\left(\frac{1}{2}\right)^n \%$

48. **a.** $a_1 = 1$; $a_2 = 1$; $a_3 = 2$; $a_3 = 3$; $a_4 = 4$; $\cdots$
Each new element is found by adding together the number of pairs of rabbits alive the previous two months; that is

$$a_n = a_{n-1} + a_{n-2}$$

b. $r_1 = 1$, $r_2 = 2$, $r_3 = \frac{3}{2}$, $r_4 = \frac{5}{3} \approx 1.67$,

$r_5 = \frac{8}{5} = 1.6$, $r_6 = \frac{13}{8} \approx 1.63$,

$r_7 = \frac{21}{13} \approx 1.62$, $r_8 = \frac{34}{21} \approx 1.62$,

$r_9 = \frac{55}{34} \approx 1.618$, $r_{10} = \frac{89}{55} \approx 1.618$

c.
$$L = 1 + L^{-1}$$
$$L^2 = L + 1$$
$$L^2 - L - 1 = 0$$
$$L = \frac{1}{2}(1 \pm \sqrt{5}) \approx 1.618034$$

(for the positive root)

49.
$$\left| \frac{n}{n+1} - 1 \right| < 0.01$$
$$\frac{n - n - 1}{n+1} < 0.01$$
$$n + 1 > 100$$
$$n > 99$$

Choose $N = 100$ so that $n > N$.

50.
$$\left| \frac{2n+1}{n+3} - 2 \right| < 0.01$$
$$\frac{2n + 1 - 2n - 6}{n+3} < 0.01$$
$$n + 3 > 500$$
$$n > 497$$

Choose $N = 498$ so that $n > N$.

51.
$$\left| \frac{n^2 + 1}{n^3} - 0 \right| < 0.001$$
$$\frac{n^2 + 1}{n^3} < 0.001$$
$$n^3 - 1{,}000n^2 - 1{,}000 > 0$$
$$n^2(n - 1{,}000) > 1{,}000$$
$$n > 1{,}000$$

Choose $N = 1{,}001$ so that $n > N$.

52. $e^n > 1{,}000$, $n > 3 \ln 10$, so $n > 6.9$

Choose $N = 7$, so that $n > N$.

53. **a.** Recall the Binomial theorem:
$$(a + b)^n = \sum_{k=0}^{n} \binom{n}{k} a^{n-k} b^k$$
Thus,
$$\left(1 + \tfrac{1}{n}\right)^n = \frac{n!}{n!} + \frac{n!}{(n-1)!}\frac{1}{n} + \frac{n!}{(n-2)!2!}\frac{1}{n^2}$$
$$+ \frac{n!}{(n-3)!3!}\frac{1}{n^3} + \cdots$$
$$= 1 + 1 + \frac{1}{2}\left(1 - \frac{1}{n}\right) + \frac{1}{3!}\left(\frac{n-1}{n}\right)\left(\frac{n-2}{n}\right)$$
$$+ \cdots + \frac{1}{n!}\left(\frac{n-1}{n}\right)\left(\frac{n-2}{n}\right) \cdot \cdot \left(\frac{n-(n-1)}{n}\right)$$
$$= 1 + 1 + \frac{1}{2}\left(1 - \frac{1}{n}\right) + \frac{1}{3!}\left(1 - \frac{1}{n}\right)\left(1 - \frac{2}{n}\right)$$
$$+ \cdots + \frac{1}{n!}\left(1 - \frac{1}{n}\right)\left(1 - \frac{2}{n}\right) \cdots \left(1 - \frac{n-1}{n}\right)$$

b. $\left(1 + \frac{1}{n}\right)^n < 1 + 1 + \frac{1}{2} + \frac{1}{2^2} + \frac{1}{2^3} + \cdots$
$$= 2 + \frac{1}{2(1 - 0.5)} = 2 + 1 = 3$$

c. $a_1 = 1$, $a_2 = 2$, $a_3 = a_2 + \frac{1}{2}(1 - \frac{1}{3})$, so $\{a_n\}$ is increasing and by part **a** it is bounded above, so the sequence converges.

54. **a.** $a_n \le a_{n+1} \le M$; let A be the least upper bound. There is a number $0 < c$ such that $a_m < c < A - c < A$. Add $0 < c < A$ and $0 < a_m \le A$ to obtain

$$0 < c + a_m < 2A$$
$$-A < 0 < c + a_m - A$$
$$A - c < a_m$$

Let $c = \epsilon$ and $n = N$, then

$$A - \epsilon < a_N < A$$

(since A is the least upper bound).

b. By hypotheses
$$a_n \le a_n + 1$$
$$A - \epsilon < a_{n+1} < A$$
$$A - \epsilon < a_{n+2} < A$$
$$\vdots$$

c.
$$A - \epsilon < a_n < A$$
$$A - \epsilon < a_n < A + \epsilon$$
$$\left| a_n - A \right| < \epsilon$$

for all $N < n$ implies $\lim\limits_{n \to +\infty} a_n = A$.

8.2 Introduction to Infinite Series: Geometric Series, Page 507

1. A sequence is a collection of elements. A series is a sum of the elements of the sequence.

2. A geometric sequence is a sequence whose successive terms have a common ratio. A geometric series is the sum of the elements of a geometric sequence. The sum of n terms of a geometric series is

   $$S_n = \frac{a(r^n - 1)}{r - 1}$$

   The sum of an infinite geometric series is

   $$S = \frac{a}{1 - r}$$

3. $S = \sum\limits_{k=0}^{\infty} \left(\frac{4}{5}\right)^k = \dfrac{1}{1 - \frac{4}{5}} = 5$

4. $S = \sum\limits_{k=0}^{\infty} \left(-\frac{4}{5}\right)^k = \dfrac{1}{1 + \frac{4}{5}} = \dfrac{5}{9}$

5. $S = \sum\limits_{k=0}^{\infty} \dfrac{2}{3^k} = \dfrac{2}{1 - \frac{1}{3}} = 3$

6. $S = \sum\limits_{k=0}^{\infty} \dfrac{2}{(-3)^k} = \dfrac{2}{1 + \frac{1}{3}} = \dfrac{3}{2}$

7. $S = \sum\limits_{k=1}^{\infty} \left(\frac{3}{2}\right)^k$; this is a geometric series with $r = \frac{3}{2} > 1$, so it diverges.

8. $S = \sum\limits_{k=1}^{\infty} \dfrac{3}{2^k} = \dfrac{3}{2}\left(\dfrac{1}{1 - \frac{1}{2}}\right) = 3$

9. $S = \sum\limits_{k=2}^{\infty} \dfrac{3}{(-4)^k} = \dfrac{3}{16}\left(\dfrac{1}{1 + \frac{1}{4}}\right) = \dfrac{3}{20}$

10. $S = \sum\limits_{k=1}^{\infty} 5(0.9)^k = 4.5\left(\dfrac{1}{1 - 0.9}\right) = 45$

11. $S = \sum\limits_{k=1}^{\infty} e^{-0.2k} = e^{-0.2}(1 - e^{-0.2})^{-1}$
    ≈ 4.5167

12. $S = \sum\limits_{k=1}^{\infty} \dfrac{3^k}{4^{k+2}} = \dfrac{3}{64}\left(\dfrac{1}{1 - \frac{3}{4}}\right) = \dfrac{3}{16}$

13. $S = \sum\limits_{k=2}^{\infty} \dfrac{(-2)^{k-1}}{3^{k+1}} = -\dfrac{2}{27}\left(\dfrac{1}{1 + \frac{2}{3}}\right) = -\dfrac{2}{45}$

14. $S = \sum\limits_{k=2}^{\infty} (-1)^k \dfrac{2^{k+1}}{3^{k-3}} = 24\left(\dfrac{1}{1 + \frac{2}{3}}\right) = \dfrac{72}{5}$

15. $S = \dfrac{1}{2} - \dfrac{1}{2^2} + \dfrac{1}{2^3} - \dfrac{1}{2^4} + \dots = \dfrac{1}{2}\left(\dfrac{1}{1 + \frac{1}{2}}\right) = \dfrac{1}{3}$

16. $S = 1 + \pi + \pi^2 + \pi^3 + \dots$; the series diverges because $r = \pi > 1$.

17. $S = \dfrac{1}{4} + \left(\dfrac{1}{4}\right)^4 + \left(\dfrac{1}{4}\right)^7 + \left(\dfrac{1}{4}\right)^{10} + \dots$
    $= \dfrac{1}{4}\left(\dfrac{1}{1 - \frac{1}{64}}\right) = \dfrac{16}{63}$

18. $S = \dfrac{2}{3} - \left(\dfrac{2}{3}\right)^3 + \left(\dfrac{2}{3}\right)^5 - \left(\dfrac{2}{3}\right)^7 + \dots$
    $= \dfrac{2}{3}\left(\dfrac{1}{1 + \frac{4}{9}}\right) = \dfrac{6}{13}$

19. $S = 2 + \sqrt{2} + 1 + \dots = 2\left(\dfrac{1}{1 - 1/\sqrt{2}}\right) =$
    $= 2(2 + \sqrt{2})$

20. $S = 3 + \sqrt{3} + 1 + \dots = 3\left(\dfrac{1}{1 - 1/\sqrt{3}}\right)$
    $= \dfrac{3}{2}(3 + \sqrt{3})$

21. $S = (1 + \sqrt{2}) + 1 + (-1 + \sqrt{2}) + \dots$
    $= \dfrac{1 + \sqrt{2}}{1 - (1 + \sqrt{2})} = \dfrac{1}{2}(4 + 3\sqrt{2})$

22. $S = (\sqrt{2} - 1) + 1 + (\sqrt{2} + 1) + \dots$
    diverges because $r = (\sqrt{2} - 1)^{-1} > 1$

23. $\sum\limits_{k=1}^{\infty} \left[\dfrac{1}{k^{0.1}} - \dfrac{1}{(k+1)^{0.1}}\right] = \left(1 - \dfrac{1}{2^{0.1}}\right)$
    $+ \left(\dfrac{1}{2^{0.1}} - \dfrac{1}{3^{0.1}}\right) + \dots + \left[\dfrac{1}{n^{0.1}} - \dfrac{1}{(n+1)^{0.1}}\right]$
    $= 1 - \dfrac{1}{(n+1)^{0.1}}; \ \lim\limits_{n \to +\infty}\left[1 - \dfrac{1}{(n+1)^{0.1}}\right] = 1$
    The series converges to 1.

24. $\sum\limits_{k=2}^{\infty} \left[\dfrac{1}{\sqrt{k}} - \dfrac{1}{\sqrt{k+1}}\right] = \left(\dfrac{1}{2^{0.5}} - \dfrac{1}{3^{0.5}}\right)$
    $+ \left(\dfrac{1}{3^{0.5}} - \dfrac{1}{4^{0.5}}\right) + \dots + \left[\dfrac{1}{n^{0.5}} - \dfrac{1}{(n+1)^{0.5}}\right]$
    $= 1 - \dfrac{1}{(n+1)^{0.5}}; \ \lim\limits_{n \to +\infty}\left[\dfrac{1}{2^{0.5}} - \dfrac{1}{(n+1)^{0.5}}\right]$
    $= \dfrac{1}{\sqrt{2}}$ The series converges to $\dfrac{1}{\sqrt{2}}$.

25. $S_n = \sum\limits_{k=0}^{n} \dfrac{1}{(k+1)(k+2)} = \sum\limits_{k=0}^{n}\left[\dfrac{1}{k+1} - \dfrac{1}{k+2}\right]$

    (This is found using a partial fraction decomposition.)

$$S_n = (1 - \tfrac{1}{2}) + (\tfrac{1}{2} - \tfrac{1}{3}) + \cdots$$
$$+ [(N + 1)^{-1} - (N + 2)^{-1}]$$
$$= 1 - (N + 2)^{-1}$$
$$S = \lim_{n \to +\infty} S_n = 1$$

26. $S_n = \sum_{k=0}^{n} \dfrac{1}{(k+2)(k+3)} = \sum_{k=0}^{n} \left[\dfrac{1}{k+2} - \dfrac{1}{k+3}\right]$

(This is found using a partial fraction decomposition.)

$$S_n = (\tfrac{1}{2} - \tfrac{1}{3}) + (\tfrac{1}{3} - \tfrac{1}{4}) + \cdots$$
$$+ [(N + 2)^{-1} - (N + 3)^{-1}]$$
$$= 1 - (N + 3)^{-1}$$
$$S = \lim_{n \to +\infty} S_n = \tfrac{1}{2}$$

27. $S_n = \sum_{k=1}^{n} \ln\left(1 + \dfrac{1}{k}\right) = \sum_{k=1}^{n} [\ln(k+1) - \ln k]$

$$= (\ln 2 - \ln 1) + (\ln 3 - \ln 2)$$
$$+ (\ln 4 - \ln 3) + \cdots$$
$$+ [\ln(N + 1) - \ln N]$$
$$= \ln N$$
$$S = \lim_{n \to +\infty} \ln N; \text{ series diverges}$$

28. $S_n = \sum_{k=1}^{n} \dfrac{1}{(2k-1)(2k+1)} = \sum_{k=1}^{n} \left[\dfrac{1}{2k-1} - \dfrac{1}{2k+1}\right]$

(This is found using a partial fraction decomposition.)

$$S_n = (1 - \tfrac{1}{3}) + (\tfrac{1}{3} - \tfrac{1}{5}) + \cdots$$
$$+ [(2N - 1)^{-1} - (2N + 1)^{-1}]$$
$$= 1 - (2N + 1)^{-1}$$
$$S = \lim_{n \to +\infty} S_n = 1$$

29. $S_n = \sum_{k=1}^{n} \dfrac{2k+1}{k^2(k+1)^2} = \dfrac{1}{2} \sum_{k=1}^{n} \left[\dfrac{1}{k^2} - \dfrac{1}{(k+1)^2}\right]$

(This is found using partial fraction decomposition.)

$$S_n = (1 - \tfrac{1}{4}) + (\tfrac{1}{4} - \tfrac{1}{9}) + \cdots$$
$$+ [N^{-2} - (N + 1)^{-2}]$$
$$= 1 - (N + 1)^{-2}$$
$$S = \lim_{n \to +\infty} S_n = 1$$

30. $S_n = \sum_{k=1}^{n} \dfrac{\sqrt{k+1} - \sqrt{k}}{\sqrt{k^2 + k}}$

$$= \sum_{k=1}^{n} \left[\dfrac{\sqrt{k+1}}{\sqrt{k(k+1)}} - \dfrac{\sqrt{k}}{\sqrt{k(k+1)}}\right]$$

$$= \sum_{k=1}^{n} \left[\dfrac{1}{\sqrt{k}} - \dfrac{1}{\sqrt{k+1}}\right]$$

$$= (1 - \dfrac{1}{\sqrt{2}}) + (\dfrac{1}{\sqrt{2}} - \dfrac{1}{\sqrt{3}}) + \cdots$$
$$+ [N^{-1/2} - (N + 1)^{-1/2}]$$
$$= 1 - (N + 1)^{-1/2}$$
$$S = \lim_{n \to +\infty} S_n = 1$$

31. $0.0101010\cdots = \dfrac{1}{100} + \dfrac{1}{10,000} + \dfrac{1}{1,000,000} + \cdots,$

is a geometric series with $r = \dfrac{1}{100}.$

$$S = \dfrac{\frac{1}{100}}{1 - \frac{1}{100}} = \dfrac{\frac{1}{100}}{\frac{99}{100}} = \dfrac{1}{99}$$

32. $2.2311111\cdots = 2.23 + 0.001[1 + \tfrac{1}{10} + (\tfrac{1}{10})^2 +$

$$\cdots + (\tfrac{1}{10})^k + \cdots]$$

is a geometric series with $r = \dfrac{1}{10}.$

$$S = \dfrac{223}{100} + \dfrac{\frac{1}{1,000}}{1 - \frac{1}{10}} = \dfrac{223}{100} + \dfrac{1}{900} = \dfrac{502}{225}$$

33. $1.4505405\cdots = 1 + 0.405[1 + \tfrac{1}{1,000} + (\tfrac{1}{1,000})^2$

$$+ \cdots + (\tfrac{1}{1,000})^k + \cdots]$$ is a geometric series with

$$r = \dfrac{1}{1,000}.$$

$$S = 1 + \dfrac{\frac{405}{1,000}}{1 - \frac{1}{1,000}} = 1 + \dfrac{405}{999} = \dfrac{52}{37}$$

34. $41.201010010\cdots = 41.2 + 0.0010[1 + \tfrac{1}{1,000} +$

$$(\tfrac{1}{1,000})^2 + \cdots + (\tfrac{1}{1,000})^k + \cdots]$$ is a geometric

series with $r = \dfrac{1}{1,000}.$

$$S = \dfrac{412}{10} + \dfrac{\frac{1}{1,000}}{1 - \frac{1}{1,000}} = \dfrac{206}{5} + \dfrac{1}{999} = \dfrac{205,799}{4,995}$$

35. a. Adding on the right:
$$\dfrac{2Ak - (Bk + 1)}{2^{k+1}}$$

So $k - 1 = 2Ak - Bk - 1$, $1 = 2A - B$, which has numerous solutions. Choose $A = B = 1$.

b. $S_n = \sum_{1}^{n} \dfrac{k}{2^k} - \sum_{1}^{n} \dfrac{k+1}{2^{k+1}}$

$$= \lim_{n\to\infty}\left(\frac{1}{2} - \frac{2}{2^2} + \frac{2}{2^2} - \frac{3}{2^3} + \frac{3}{2^3} - \frac{4}{2^4}\right.$$

$$\left. + \cdots + \frac{n}{2^n} - \frac{n+1}{2^{n+1}}\right)$$

$$= \lim_{n\to\infty}\left(\frac{1}{2}\right) - \lim_{n\to\infty}\left(\frac{n+1}{2^{n+1}}\right)$$

Using l'Hôpital's rule:

$$S = \frac{1}{2} - \lim_{n\to\infty}\frac{1}{2^{n+1}\ln 2} = \frac{1}{2}$$

36. **a.** Adding on the right:

$$\frac{2k-1}{3^{k+1}}\left(\frac{Ak}{3^k}\right) - \frac{Bk+1}{d^{k+1}}$$

So $2k = (3A - B)k$. Choose $A = 1$, $B = 1$.

b. $S_n = \displaystyle\sum_{k=1}^{n}\frac{2k-1}{3^{k+1}} = \sum_{k=1}^{n}\left[\frac{k}{3^k} - \frac{k+1}{3^{k+1}}\right]$

$$= \left(\frac{1}{3} - \frac{1}{3^2}\right) + \left(\frac{2}{3^2} - \frac{3}{3^3}\right) + \cdots$$

$$+ \left(\frac{n}{3^n} - \frac{n+1}{3^{n+1}}\right) + \cdots$$

$$= \frac{1}{3} - \frac{n+1}{3^{n+1}}$$

$$S = \frac{1}{3} - \lim_{n\to+\infty}\frac{n+1}{3^{n+1}}$$

$$= \frac{1}{3} - \lim_{n\to+\infty}\frac{1}{3^{n+1}\ln 3} = \frac{1}{3}$$

37. $\dfrac{\ln\left(\dfrac{n^{n+1}}{(n+1)^n}\right)}{n(n+1)} = \dfrac{1}{n(n+1)}[n\ln n - n\ln(n+1)]$

$$S_n = \sum_{k=1}^{n}\left[\frac{(n+1)\ln n}{n(n+1)} - \frac{n\ln(n+1)}{n(n+1)}\right]$$

$$= \sum_{k=1}^{n}\left[\frac{\ln n}{n} - \frac{\ln(n+1)}{n+1}\right]$$

$$= (0 - \tfrac{1}{2}\ln 2) + (\tfrac{1}{2}\ln 2 - \tfrac{1}{3}\ln 3)$$

$$+ (\tfrac{1}{3}\ln 3 - \tfrac{1}{4}\ln 4) + \cdots$$

$$+ [n^{-1}\ln n - (n+1)^{-1}\ln(n+1)]$$

$$S = \lim_{n\to+\infty}\left[\frac{\ln n}{n} - \frac{\ln(n+1)}{n+1}\right] = 0$$

38. $S_N = \displaystyle\sum_{n=1}^{N}\frac{n}{(n+1)!} = \sum_{n=1}^{N}\left[\frac{1}{n!} - \frac{1}{(n+1)!}\right]$

$$= (1 - \tfrac{1}{2!}) + (\tfrac{1}{2!} - \tfrac{1}{3!}) + \cdots$$

$$+ \left(\frac{1}{N!} - \frac{1}{(N+1)!}\right) = 1 - \frac{1}{(N+1)!}$$

$$S = 1 - \lim_{N\to+\infty}S_N = 1 - \lim_{N\to+\infty}\frac{1}{(N+1)!}$$

$$= 1$$

39. $2\displaystyle\sum_{k=0}^{\infty}a_k + \sum_{k=0}^{\infty}\frac{1}{2^k} = 2(3.57) + 2 = 9.14$

40. $\dfrac{1}{2}\displaystyle\sum_{k=0}^{\infty}b_k - \dfrac{1}{2}\sum_{k=0}^{\infty}\frac{1}{3^k} = \frac{0.54}{2} - \frac{1}{2}\left(\frac{1}{1-\frac{1}{3}}\right)$

$$\approx -2.73$$

41. $\displaystyle\sum_{k=0}^{\infty}(2^{-k} + 3^{-k})^2$

$$= \sum_{k=0}^{\infty}[4^{-k} + 2(6^{-k}) + 9^{-k}]$$

$$= \frac{1}{1-\frac{1}{4}} + 2\left(\frac{2}{1-\frac{1}{6}}\right) + \frac{1}{1-\frac{1}{9}} = \frac{583}{120}$$

42. $\displaystyle\sum_{k=0}^{\infty}\left[\left(\frac{2}{3}\right)^k + \left(\frac{3}{4}\right)^k\right]^2 = \sum_{k=0}^{\infty}\left[\left(\frac{4}{9}\right)^k + 1 + \left(\frac{9}{16}\right)^k\right]$

diverges because $\displaystyle\sum_{k=0}^{\infty}1$ diverges.

43. $S_n = \displaystyle\sum_{k=0}^{n}\frac{1}{(a+k)(a+k+1)}$

$$= \sum_{k=0}^{n}\left(\frac{1}{a+k} - \frac{1}{a+k+1}\right)$$

$$= \left(\frac{1}{a+0} - \frac{1}{a+1}\right) + \left(\frac{1}{a+1} - \frac{1}{a+2}\right)$$

$$+ \cdots + \left(\frac{1}{a+n} - \frac{1}{a+n+1}\right)$$

$$= \frac{1}{a} - \frac{1}{a+n+1}$$

$$S = \lim_{n\to+\infty}S_n = \frac{1}{a}$$

44. $S_n = \displaystyle\sum_{k=0}^{n}\frac{1}{(2+k)^2+2+k} = \sum_{k=0}^{\infty}\frac{1}{(2+k)(2+k+1)}$

This is the same as Problem 41 with $a = 2$; thus $S = 1/2$.

45. $\displaystyle\sum_{k=0}^{\infty}(a_k - b_k)^2 = \sum_{k=0}^{\infty}(a_k^2 - 2a_kb_k + b_k^2)$

$$= \sum_{k=0}^{\infty}a_k^2 - 2\sum_{k=0}^{\infty} + \sum_{k=0}^{\infty}b_k^2$$

$$= 4 - 2(3) + 4 = 2$$

46. $L = 20 + (0.9)(20) + (0.9)^2(20) + \cdots$

$$+ (0.9)^k(20) = 20\left(\frac{1}{1-0.9}\right) = 200$$

47. $N = 500 + \frac{2}{3}(500) + \left(\frac{2}{3}\right)^2(500) + \cdots$

$$+ \left(\frac{2}{3}\right)^k(500) = 500\left(\frac{1}{1-\frac{2}{3}}\right) = 1,500$$

48. $D = 10 + (0.6)(20) + (0.6)^2(20) + \cdots$

$\qquad + (0.6)^k(20) = 10 + 20\left(\dfrac{1}{1-0.6}\right) = 60$

49. $D = h + (0.75)(2h) + (0.75)^2(2h) + \cdots$

$\qquad + (0.75)^k(2h) + \cdots$

$\qquad = h + 0.75(2h)[1 + 0.75 + 0.75^2 + \cdots]$

$\qquad = k + 0.75(2h)\left(\dfrac{1}{1-0.75}\right)$

$\qquad = h + 0.75(2h)(4) = 7h$

If $7h = 21$, then $h = 3$ ft.

50. $S = 0.92(50) + (0.92)^2(50) + \cdots + (0.92)(50)$

$\qquad + \cdots = 50\left(\dfrac{1}{1-0.92}\right) = 575$

51. $D = (10,000)(0.2) + (10,000)(0.2)^2$

$\qquad + (10,000)(.02)^3 + \cdots$

$\qquad = 10,000(0.2)(1 + .02 + 0.2^2 + \cdots)$

$\qquad = 8,000\left(\dfrac{1}{1-0.2}\right) = 10,000$

52. $G = (100) + (100)(0.2) + (100)(.02)^2 + \cdots$

$\qquad = 100(1 + .02 + 0.2^2 + \cdots)$

$\qquad = 100\left(\dfrac{1}{1-0.2}\right) = 125$

The tax is $125 - 100 = 25$.

53. $S = 20e^{-1/2} + 20e^{-1} + \cdots$

$\qquad = 20e^{-1/2}(1 + e^{-1/2} + \cdots)$

$\qquad = 20e^{-1/2}\left(\dfrac{1}{1-e^{-1/2}}\right)$

$\qquad \approx 30.8$

54. $S = 1 + \frac{1}{4} + \left(\frac{1}{4}\right)^2 + \cdots = \dfrac{1}{1-\frac{1}{4}} = \frac{4}{3}$ gm

55. Let a_n be the number of trustees on 12-31 of the nth year.

$N = 6 + 6e^{-0.2} + 6e^{-0.2(2)} + 6e^{-0.2(3)}$

$\qquad + \cdots = \dfrac{6}{1-e^{-0.2}} \approx 33$

56. $P = 2,000e^{-0.15} + 2,000(e^{-0.15})^2$

$\qquad + 2,000(e^{-0.15})^3 + \cdots$

$\qquad = 2,000(e^{-0.15})(1 - e^{-0.15})^{-1}$

$\qquad \approx \$12,358.32$

57. Let T be the time it takes to run the first half of the course. The total time is

$T + \frac{1}{2}T + \frac{1}{4}T + \frac{1}{8}T + \cdots = T(1 + \frac{1}{2} + \frac{1}{4} + \cdots)$

$= T\left(\dfrac{1}{1-\frac{1}{2}}\right) = 2T$

58. The ultimate concentration is the sum R of the geometric series

$\qquad 0.04e^{-0.322} + 0.04e^{-0.322(2)} + \cdots$

That is, if $r = e^{-0.322} \approx 0.725$, then

$\qquad R = 0.04(r + r^2 + r^3 + \cdots) = \dfrac{0.04r}{1-r}$

After n applications of the drug (just before the next interrogation), the drug concentration in the spy's bloodstream is

$R_n = 0.04(r + r^2 + \cdots + r^n) = \dfrac{0.04(r - r^{n+1})}{1-r}$

We want ro find n so that $R_n = 0.8R$; that is,

$0.8\left[\dfrac{0.04r}{1-r}\right] = \dfrac{0.04(r - r^{n+1})}{1-r}$

$\qquad 0.8 = 1 - r^n$

$\qquad r^n = 1 - 0.8 = 0.2$

$\qquad n \ln r = \ln 0.2$

$\qquad n = \dfrac{\ln 0.2}{\ln r} \approx \dfrac{\ln 0.2}{\ln 0.725} \approx 5$

The spy can hold out for 5 doses of the drug. After that ...

59. a. $S_n = \displaystyle\sum_{k=1}^{n} a_k; \ S_{n-1} = \sum_{k=1}^{n-1} a_k$

$\qquad S_n - S_{n-1} = a_n$

b. $a_n = S_n - S_{n-1} = \dfrac{n}{2n+3} - \dfrac{n-1}{2(n-1)+3}$

$\qquad = \dfrac{3}{4x^2 + 8n + 3}$

60. Let $S_a = \Sigma a_k$, then $A = \Sigma a_k$ implies that for $\epsilon_1 > 0$, there exists an N_1 such that $|A - S_a| < \epsilon_1$ for $N_1 < n$. Similarly with S_b Σb_k implies that for $\epsilon_2 > 0$, there exists an N_2 such that $|A - S_b| < \epsilon_2$ for $N_2 < n$. Then

$|(A + B) - (S_a + S_b)| = |[A - S_a] + [B - S_b]|$

$\qquad \le |A - S_a| + |B - S_b| < 2\epsilon$ for $N < n$

where $\epsilon = \min(\epsilon_1, \epsilon_2)$ and $N = \max(N_1, N_2)$. This is restated as $A + B = \displaystyle\sum_{k=1}^{\infty}(a_k + b_k)$.

61. By hypotheses, $\Sigma a_k = A$ and Σb_k diverges.

$b_k = a_k = (a_k - b_k)$, so

$\qquad \Sigma b_k = \Sigma a_k - \Sigma(a_k - b_k)$

Thus, $\Sigma(a_k - b_k) = \Sigma b_k - \Sigma a_k$

which means that $\Sigma(a_k - b_k)$ diverges. It is infinity (or divergence by oscillation) minus a finite number.

62. a. Let $\Sigma a_k = \Sigma(\frac{4}{3})^k$ and $\Sigma b_k = \Sigma(\frac{3}{2})^k$, then

$\Sigma(a_k + b_k)$ also diverges.

b. Let $\Sigma a_k = \Sigma(-\frac{4}{3})^k$ and $\Sigma b_k = \Sigma(\frac{3}{2})^k$, then

$\Sigma(a_k + b_k) = \Sigma(\frac{1}{6})^k$

which converges.

63. If $r > 1$ then $\lim\limits_{n \to \infty} a_n = \infty$ and S_n must

diverge. If $r = 1$, $S_n = an$ and

$\lim\limits_{n \to \infty} an = \infty$.

64. a. $\sum\limits_{k=1}^{n} (a_k - a_{k+2}) = (a_1 - a_3) + (a_2 - a_4)$

$+ (a_3 - a_5) + \cdots + (a_{n-2} - a_n)$

$+ (a_{n-1} - a_{n+1}) + (a_n - a_{n+2})$

$= a_1 - a_3 + a_2 - a_4 + a_3 - a_5 + \cdots$

$+ a_{n-2} - a_n - a_{n+1} - a_n - a_{n+2}$

$= a_1 + a_2 + 0 + \cdots + 0 - a_{n+1} - a_{n+2}$

b. $\sum\limits_{k=1}^{\infty} (a_k - a_{k+2}) = \lim\limits_{n \to \infty} \sum\limits_{k=1}^{n} (a_k - a_{k+2})$

$= a_1 + a_2 - \lim\limits_{n \to \infty} a_{n+1} - \lim\limits_{n \to \infty} a_{n+2}$

$= a_1 + a + 2 - 2A$

c. $\lim\limits_{k \to \infty} k^{1/k} = \lim\limits_{k \to \infty} \exp[\ln k^{1/k}]$

$= \lim\limits_{k \to \infty} \exp[k^{-1} \ln k] = \lim\limits_{k \to \infty} e^{1/k} = 1$

$\sum\limits_{k=1}^{\infty} [k^{1/k} - (k+2)^{k+2}]$

$= 1 + (2)^{1/2} - 1 - 1 = \sqrt{2} - 1$

d. $\lim\limits_{k \to \infty} \sum\limits_{k=2}^{n} \frac{1}{k^2 - 1}$

$= \lim\limits_{k \to \infty} \sum\limits_{k=2}^{n} \left[\frac{\frac{1}{2}}{k-1} + \frac{-\frac{1}{2}}{k+1} \right]$

$= \frac{1}{2} \lim\limits_{k \to \infty} \sum\limits_{k=1}^{n} \left[\frac{1}{k} - \frac{1}{k+2} \right] = \frac{3}{4}$

65. The original square has side, s, of 1. The first inscribed square has

$s_1 = \frac{1}{\sqrt{2}}$ and area $\frac{1}{2^3}$

The nth inscribed square has

$s_n = \left(\frac{1}{\sqrt{2}} \right)^n$ and area $\frac{1}{2^{n+2}}$

Thus the total area is

$T = \frac{1}{2^3}(1 + \frac{1}{2} + \frac{1}{2^2} + \cdots + \frac{1}{2^{n-1}} + \cdots)$

$= \frac{1}{2^3}\left(\frac{1}{1 - \frac{1}{2}} \right) = \frac{1}{4}$

66. The original square has side, s, of a. The first inscribed square has

$s_1 = \frac{a}{\sqrt{2}}$ and area $\frac{1}{2^3}a^2$

The nth inscribed square has

$s_n = \left(\frac{1}{\sqrt{2}} \right)^n a$ and area $\frac{1}{2^{n+2}}a^2$

Thus the total area is

$T = \frac{1}{2^3}(1 + \frac{1}{2} + \frac{1}{2^2} + \cdots + \frac{1}{2^{n-1}} + \cdots)a^2$

$= \frac{1}{2^3}\left(\frac{1}{1 - \frac{1}{2}} \right)a^2 = \frac{1}{4}a^2$

8.3 The Integral Test: *p*-series, Page 515

1. The *p*-series is one of the form $\sum\limits_{k=1}^{\infty} \frac{1}{k^p}$

(with p a constant).

2. If $a_k = f(k)$ for $k = 1, 2, \ldots$, where f is a positive, continuous, and decreasing function of x for $x \geq 1$, then

$\sum\limits_{k=1}^{\infty} a_k \quad$ and $\quad \int\limits_{1}^{\infty} f(x)\, dx$

either both converge or both diverge.

3. $p = 3$; converges   **4.** $p = \frac{1}{2}$; diverges

5. $p = \frac{1}{3}$; diverges   **6.** $p = \frac{3}{2}$; converges

7. $\lim\limits_{k \to \infty} \frac{\ln k}{k^2} = \lim\limits_{k \to \infty} \frac{\frac{1}{k}}{2k} = 0$, so the necessary condition for convergence is met. The hypotheses for the integral test are met so evaluate:

$\lim\limits_{n \to \infty} \int\limits_{1}^{n} \frac{\ln x}{x^2}\, dx \quad \boxed{\text{Formula 505}}$

$= \lim\limits_{n \to \infty} \left[-\frac{1}{x}(\ln x + 1) \right]\Big|_{1}^{n}$

$= \lim\limits_{n \to \infty} \left(-\frac{\ln n}{n} - \frac{1}{n} + 1 \right) = 1$

The series converges.

8. $S = \lim\limits_{n \to \infty} \frac{\ln k}{k} = \lim\limits_{n \to \infty} \frac{\frac{1}{k}}{1} = 0;$

$I = \lim\limits_{n \to \infty} \int\limits_{0}^{b} \frac{\ln x}{x} = \lim\limits_{n \to \infty} \frac{1}{2}(\ln x)^2 \Big|_{0}^{b} = \infty$

I diverges and, therefore, S diverges.

9. $S = \sum_{k=1}^{\infty} \left(2 + \frac{3}{k}\right)^k > 0$; the series diverges.

10. $S = \sum_{k=1}^{\infty} \left(1 + \frac{2}{k}\right)^k > 0$; the series diverges.

11. $S = \sum_{k=1}^{\infty} \frac{1}{k^4}$; this is the p-series where

$p = 4 > 1$, so S converges.

12. $S = \sum_{k=1}^{\infty} \frac{5}{\sqrt{k}}$; this is the p-series where

$p = \frac{1}{2} < 1$, so S diverges.

13. $S = \sum_{k=1}^{\infty} k^{-3/4}$; this is the p-series where

$p = \frac{3}{4} < 1$, so S diverges.

14. $S = \sum_{k=1}^{\infty} k^{-4/3}$; this is a p-series where

$p = \frac{4}{3} > 1$, so S converges.

15. $S = \sum_{k=1}^{\infty} \frac{k}{k^2 + 1}$

$$I = \lim_{n \to \infty} \int_1^n \frac{x\,dx}{x^2 + 1} = \lim_{n \to \infty} \frac{1}{2} \ln(x^2 + 1)\Big|_1^n$$

$$= \lim_{n \to \infty} \frac{1}{2} [\ln(n^2 + 1) - \ln 2] = \infty$$

The series diverges.

16. $S = \sum_{k=1}^{\infty} \frac{k^2}{(k^3 + 2)^2}$

$$I = \lim_{n \to \infty} \int_1^n \frac{x^2\,dx}{(x^3 + 2)^2}$$

$$= \lim_{n \to \infty} -\frac{1}{3}(x^3 + 2)^{-1}\Big|_0^n = \frac{1}{9}$$

The series converges.

17. $S = \sum_{k=1}^{\infty} \frac{k^2}{\sqrt{k^3 + 2}}$

$$I = \lim_{n \to \infty} \int_1^n \frac{x^2\,dx}{\sqrt{x^3 + 2}}$$

$$= \lim_{n \to \infty} \frac{2}{3} \sqrt{x^3 + 2}\Big|_1^b = \infty$$

The series diverges.

18. $S = \sum_{k=1}^{\infty} ke^{-k^2}$

$$I = \lim_{n \to \infty} \int_0^n xe^{-x^2}\,dx$$

$$= \lim_{n \to \infty} -\frac{1}{2}e^{-x^2}\Big|_1^b = \frac{1}{2e}$$

The series converges.

19. $S = \sum_{k=1}^{\infty} \frac{1}{(0.25)^k}$

Note: $\frac{1}{\left(\frac{1}{4}\right)^k} = 4^k$; $\lim_{k \to \infty} 4^k = \infty$

The necessary condition is not met. The series diverges.

20. $S = \sum_{k=1}^{\infty} \frac{1}{4^k}$

This is a convergent geometric series because $r = 1/4 < 1$.

21. $S = \sum_{k=2}^{\infty} \frac{1}{k(\ln k)^2}$

$$I = \lim_{n \to \infty} \int_2^n \frac{dx}{x(\ln x)^2}$$

$$= \lim_{n \to \infty} -(\ln x)^{-1}\Big|_2^n = (\ln 2)^{-1}$$

The series converges.

22. $S = \sum_{k=2}^{\infty} \frac{1}{k\sqrt{\ln k}}$

$$I = \lim_{n \to \infty} \int_2^n (\ln x)^{-1/2} \frac{dx}{x}$$

$$= \lim_{n \to \infty} 2(\ln x)^{1/2}\Big|_2^n = \infty$$

The series diverges.

23. $S = \sum_{k=1}^{\infty} \frac{k}{e^k}$

$$I = \lim_{n \to \infty} \int_1^n \frac{x}{e^x}\,dx \quad \boxed{u = x,\; dv = e^{-x}\,dx}$$

$$= \lim_{n \to \infty}\left[-\frac{x}{e^x} + \int_1^n e^{-x}dx\right]$$

$$= \lim_{n \to \infty} -\frac{1}{e^x}(x + 1)\Big|_1^n$$

$$= \lim_{n \to \infty}\left(-\frac{n}{e^n} - \frac{1}{e^n} + \frac{2}{e}\right) = \frac{2}{e}$$

The series converges.

24. $S = \sum_{k=1}^{\infty} \frac{k^2}{e^k}$

$$I = \lim_{n \to \infty} \int_1^n \frac{x^2}{e^x}\,dx \quad \boxed{u = x^2;\; dv = e^{-x}\,dx}$$

$$= \lim_{n \to \infty}\left[-\frac{x^2}{e^x} + 2\int_1^n \frac{x\,dx}{e^x}\right]$$

$$= \lim_{n\to\infty}\left[-e^{-x}(x^2+2x+2)\right]\Big|_1^n = -\frac{5}{e}$$

The series converges.

25. $S = \sum\limits_{k=1}^{\infty} \frac{\tan^{-1}k}{1+k^2}$

$$I = \lim_{n\to\infty}\int_1^n \frac{\tan^{-1}x\,dx}{1+x^2}$$

$$= \lim_{n\to\infty}\tfrac{1}{2}(\tan^{-1}x)^2\Big|_1^n = \frac{\pi^2}{8} - \frac{\pi^2}{32}$$

The series converges.

26. $S = \sum\limits_{k=1}^{\infty} \cot^{-1}k$

$$I = \lim_{n\to\infty}\int_1^n \cot^{-1}x\,dx$$

$$= \lim_{n\to\infty} \ln|\sin x|\Big|_1^n$$

$$= \lim_{n\to\infty}\left[\ln|\sin n| - \ln|\sin 1|\right] = \infty$$

The series diverges.

27. $S = \sum\limits_{k=1}^{\infty} \frac{1}{e^k + e^{-k}}$

Recall that $\cosh x = \frac{e^x + e^{-x}}{2}$;

$$I = \lim_{n\to\infty}\int_1^n \frac{\frac{1}{2}}{\cosh x}\,dx = \lim_{n\to\infty}\tfrac{1}{2}\int_1^n \operatorname{sech} x\,dx$$

$$= \lim_{n\to\infty}\tfrac{1}{2}\sin^{-1}(\tanh x)\Big|_1^n$$

$$= \tfrac{1}{2}\lim_{n\to\infty}\left[\sin^{-1}(\tanh n) - \sin^{-1}(\tanh 1)\right]$$

$$= \tfrac{1}{2}\left[\frac{\pi}{2} - \sin^{-1}\left(\frac{e^2-1}{e^2+1}\right)\right]$$

The series converges.

28. $S = \sum\limits_{k=1}^{\infty} 3e^{-2k}$

S is a convergent geometric series because $r = 1/e^2 < 1$.

29. $S = \sum\limits_{k=1}^{\infty} \frac{k}{k+1}$

Diverges because $\lim\limits_{k\to\infty}\frac{k}{k+1} = 1 \neq 0$
The necessary condition for convergence is not satisfied.

30. $S = \sum\limits_{k=1}^{\infty} \frac{k^2+1}{k+5}$

Diverges because $\lim\limits_{k\to\infty}\frac{k}{k+1} = 1 \neq 0$
The necessary condition for convergence is not satisfied.

31. $S = \sum\limits_{k=1}^{\infty} \frac{k^2+1}{k^3} = \sum\limits_{k=1}^{\infty}\left(\frac{1}{k} + \frac{1}{k^3}\right)$

The first term is the harmonic divergent series, so S diverges. Never mind that the second term is a convergent p-series.

32. $S = \sum\limits_{k=1}^{\infty} \frac{2k^4+3}{k^5} = \sum\limits_{k=1}^{\infty}\left(\frac{1}{k} + \frac{3}{k^5}\right)$

The first term is the harmonic divergent series, so S diverges. Never mind that the second term is a convergent p-series.

33. $S = \sum\limits_{k=1}^{\infty} \frac{1}{k^2}$

This is a p-series with $p = 2 > 1$, so S converges.

34. $S = \sum\limits_{k=1}^{\infty} \frac{-k^5+k^2+1}{k^5+2}$

$$\lim_{k\to\infty}\left|\frac{-k^5+k^2+1}{k^5+2}\right| = |-1| \neq 0$$

The necessary condition for convergence is not satisfied, so S diverges.

35. $S = \sum\limits_{n=1}^{\infty} \frac{n^{\sqrt{3}}+1}{n^{2.7321}}$

$$= \sum\limits_{n=1}^{\infty}\left[\frac{1}{n^{\sqrt{3}-2.7321}} + \frac{1}{n^{2.7321}}\right]$$

Both of these are p-series with $|p| > 1$, so the series converges.

36. $S = \sum\limits_{n=1}^{\infty} \frac{n^{\sqrt{5}}+1}{n^{2.236}}$

$$= \sum\limits_{n=1}^{\infty}\left[\frac{1}{n^{\sqrt{5}-2.236}} + \frac{1}{n^{2.7321}}\right]$$

The first series is a divergent p-series, so S diverges.

37. $S = \sum\limits_{k=1}^{\infty}\left[\frac{1}{k} + \frac{k+1}{k+2}\right]$

S diverges because $\lim\limits_{k\to\infty}\left[\frac{1}{k} + \frac{k+1}{k+2}\right] = 1 \neq 0$

38. $S = \sum\limits_{k=1}^{\infty}\left[\frac{1}{2k} + \left(\frac{3}{2}\right)^k\right]$

S diverges because $\lim\limits_{k \to \infty} \left[\dfrac{1}{2k} + \left(\dfrac{3}{2}\right)^k\right] = \infty \ne 0$

39. $S = \sum\limits_{k=1}^{\infty} \left[\dfrac{1}{2^k} + \dfrac{2k+3}{3k+4}\right]$

S diverges because $\lim\limits_{k \to \infty} \left(\dfrac{1}{2^k} + \dfrac{2k+3}{3k+4}\right)$

$= \dfrac{2}{3} \ne 0$

40. $S = \sum\limits_{k=1}^{\infty} \left[\dfrac{1}{2^k} - \dfrac{1}{k}\right] = \sum\limits_{k=1}^{\infty} \dfrac{k}{2^k} - \sum\limits_{k=1}^{\infty} \dfrac{1}{k}$

The second series is a divergent p-series (harmonic series), so S diverges.

41. $S = \sum\limits_{k=1}^{\infty} \dfrac{1}{(2+3k)^2}$

$f > 0$, continuous on $[1, \infty)$, and decreasing.

$I = \int\limits_{1}^{\infty} \dfrac{dx}{(2+3x)^2} = \lim\limits_{b \to \infty} \int\limits_{1}^{b} (2+3x)^2\, dx$

$= -\dfrac{1}{3} \lim\limits_{b \to \infty} (2+3x)^{-1} \Big|_{1}^{b} = \dfrac{1}{15}$

I converges, so S converges.

42. $S = \sum\limits_{k=1}^{\infty} (2+k)^{-3/2}$

$f > 0$, continuous on $[1, \infty)$, and decreasing.

$I = \int\limits_{1}^{\infty} (2+x)^{-3/2}\, dx$

$= \lim\limits_{b \to \infty} \int\limits_{1}^{b} (2+x)^{-3/2}\, dx$

$= -2 \lim\limits_{b \to \infty} (2+3x)^{-1/2} \Big|_{1}^{b} = \dfrac{2}{\sqrt{3}}$

I converges, so S converges.

43. $S = \sum\limits_{k=2}^{\infty} \dfrac{\ln k}{k}$

$f > 0$, continuous on $[2, \infty)$, and decreasing.

$I = \int\limits_{2}^{\infty} \dfrac{\ln x}{x}\, dx = \lim\limits_{b \to \infty} \int\limits_{2}^{b} \ln x\, \dfrac{dx}{x}$

$= \dfrac{1}{2} \lim\limits_{b \to \infty} (\ln x)^2 \Big|_{2}^{b} = \infty$

I diverges, so S diverges.

44. $S = \sum\limits_{k=2}^{\infty} \dfrac{1}{k(\ln k)^2}$

$f > 0$, continuous on $[2, \infty)$, and decreasing.

$I = \int\limits_{2}^{\infty} \dfrac{dx}{x(\ln x)^2} = \lim\limits_{b \to \infty} \int\limits_{2}^{b} (\ln x)^{-2}\, \dfrac{dx}{x}$

$= \lim\limits_{b \to \infty} -\dfrac{1}{\ln x} \Big|_{2}^{b} = \dfrac{1}{\ln 2}$

I converges, so S converges.

45. $S = \sum\limits_{k=1}^{\infty} \dfrac{(\tan^{-1} k)^2}{1+k^2}$

$f > 0$, continuous on $[1, \infty)$, and decreasing.

$I = \int\limits_{1}^{\infty} \dfrac{(\tan^{-1} x)^2}{1+x^2}\, dx$

$= \dfrac{1}{3} \lim\limits_{b \to \infty} (\tan^{-1} x)^3 \Big|_{1}^{b} = \dfrac{\pi^3}{24} - \dfrac{\pi^3}{192}$

I converges, so S converges.

46. $S = \sum\limits_{k=1}^{\infty} k e^{-k^2}$

$f > 0$, continuous on $[1, \infty)$, and decreasing.

$I = \int\limits_{1}^{\infty} x e^{-x^2}\, dx = -\dfrac{1}{2} \lim\limits_{b \to \infty} \int\limits_{1}^{b} e^{-x^2}(-2x\, dx)$

$= -\dfrac{1}{2} \lim\limits_{b \to \infty} e^{-x^2} \Big|_{1}^{b} = \dfrac{1}{2e}$

I converges, so S converges.

47. $S = \sum\limits_{k=2}^{\infty} \dfrac{k}{(k^2-1)^p}$

The hypotheses for the integral test are met, so the series will converge if the integral test gives a finite value.

$I = \lim\limits_{n \to \infty} \dfrac{1}{2} \int\limits_{2}^{n} (x^2-1)^{-p}(2x\, dx)$

$= \lim\limits_{n \to \infty} \dfrac{1}{2} \dfrac{(x^2-1)^{-p+1}}{-p+1} \Big|_{2}^{n}$

$= \dfrac{1}{2(1-p)} \lim\limits_{n \to \infty} \left[(n^2-1)^{1-p} - 3^{1-p}\right]$

which converges if $p > 1$.

48. $S = \sum\limits_{k=1}^{\infty} \dfrac{k^2}{(k^2+4)^p}$

$I = \lim\limits_{b \to \infty} \int\limits_{1}^{b} x^2(x^2+4)^{-p}\, dx$

$\boxed{u = x;\; dv = x(x^2+4)^{-p}}$

$$= \lim_{b \to \infty} \left[\frac{x}{2(x^2 + 4)^{p-1}} - \int_1^b \frac{dx}{(x^2 + 4)^{p-1}} \right]$$

$$= \frac{1}{2(5^{p-1})} - \frac{1}{2} I_1$$

With $p = 1$, $I_1 = \frac{1}{2} \tan(-\frac{3}{2}x) \big|_1^b$ always exists. If $p > 1$, the denominator of I_1 increases and the value of I_1 decreases. I converges when $p > 1$, so S converges when $p > 1$.

49. $S = \sum_{k=2}^{\infty} \frac{1}{k^p \ln k}$

$$I = \lim_{b \to \infty} \int_2^b \frac{dx}{x \ln x} = \lim_{b \to \infty} \ln|\ln x| \big|_2^b$$

which diverges. If $p < 1$, the integrand of I increases, so I diverges. If $1 < p$,

$$I = \lim_{b \to \infty} \int_2^b \frac{dx}{x^p \ln x} = \lim_{b \to \infty} \int_2^b \frac{x^{-p}}{\ln x} \, dx$$

$$\leq \lim_{b \to \infty} \int_2^b \frac{x^{-p}}{x} \, dx = \lim_{b \to \infty} \frac{x^{-p+1}}{1 - p} \big|_2^b$$

which converges since the denominator increases beyond all bounds. Thus, S converges if $p > 1$.

50. $S = \sum_{k=2}^{\infty} \frac{\ln k}{k^p}$

Let $p = 1$;

$$I_1 = \lim_{b \to \infty} \int_2^b \frac{\ln x}{x} \, dx = \frac{1}{2} \lim_{b \to \infty} (\ln x)^2 \big|_2^b$$

I_1 diverges. Now let $p < 1$; the integrand of I_1 increases, so I diverges. Finally, let $p > 1$.

$$I_2 = \lim_{b \to \infty} \int_2^b \frac{\ln x}{x^p} \, dx = \lim_{b \to \infty} \int_2^b x^{-p} \ln x \, dx$$

$$\boxed{u = \ln x; \ dv = x^{-p} \, dx}$$

$$= \lim_{b \to \infty} \frac{\ln x}{(1 - p)x^{p-1}} \big|_2^b$$

$$+ (1 - p)^{-1} \lim_{b \to \infty} \int_2^{\infty} x^{-p} dx$$

$$= \lim_{b \to \infty} \left[\frac{\ln x}{(1-p)x^{p-1}} + (1-p)\frac{1}{x^{-p+1}} \right] \Big|_2^b$$

I_2 converges. Thus, S converges if $p > 1$.

51. $S = \sum_{k=3}^{\infty} \frac{1}{k \ln k [\ln(\ln k)]^p}$

$$= \lim_{n \to \infty} \int_3^n [\ln (\ln x)]^{-p} \frac{dx}{x \ln x}$$

$$= \lim_{n \to \infty} \frac{[\ln (\ln x)]^{-p+1}}{1 - p} \Big|_3^p$$

$$= \frac{1}{1 - p} \lim_{n \to \infty} \left([\ln(\ln n)]^{1-p} - [\ln(\ln 3)]^{1-p} \right)$$

which converges if $p > 1$.

52. $S = \sum_{k=2}^{\infty} \frac{1}{k(\ln k)^p}$

$$I = \lim_{b \to \infty} \int_2^b (\ln x)^{-p} \frac{dx}{x}$$

$$= \lim_{b \to \infty} (\ln x)^{-p+1} \big|_2^b$$

I converges when $p > 1$ because $\ln x$ will be in the denominator so S also converges when $p > 1$.

53. a. Partition $[N+1, \infty)$ into segments of unit length. Mount a rectangle on each segment with height $1/x_k^*$ where x_k^* is the right endpoint of the interval. Since $f(x) = 1/x^p$ is decreasing, the area of the first rectangle (base equal to one) is

$$\frac{1}{N^p} \leq \int_N^{N+1} \frac{1}{x^p} \, dx$$

This same inequality holds for the other triangles. Summing up leads to

$$\sum_{k=N+1}^{\infty} \frac{1}{k^p} \leq \int_N^{\infty} \frac{dx}{x^p}$$

b. $\lim_{b \to \infty} \int_N^b \frac{dx}{x^2} = - \lim_{b \to \infty} \frac{1}{x} \big|_N^b = 0.01$ or $\frac{1}{100}$; $N > 100$

c. $\lim_{b \to \infty} \sum_{k=1}^b \frac{1}{k^2} = \frac{\pi}{6} \approx 1.644934067$

Using computer software, we find

$$\sum_{k=1}^{100} \frac{1}{k^2} \approx 1.63498$$

These answers differ by 0.00995, so $N > 100$ is a good choice.

54. a.
$$\lim_{b \to \infty} \int_N^b \frac{1}{x^3} = -\frac{1}{2} \lim_{b \to \infty} \frac{1}{x^2} \Big|_N^b = \frac{1}{2n^2}$$

$$\frac{1}{2N^2} = 0.0005$$

$$N^2 = 1,000$$

$$N \approx 32$$

b. Using computer software,
$$S_{32} = \sum_{k=1}^{32} \frac{1}{k^3} \approx 1.20158 \text{ and}$$

$$S_{64} = \sum_{k=1}^{64} \frac{1}{k^3} \approx 1.20194$$

These answers differ by 0.00035, so $N \geq 32$ appears to be a good choice.

55. For $a_k = \frac{1}{k^{1.1}}$ and $b_k = \frac{1}{\sqrt{k}}$, $S_c = \lim_{k \to \infty} a_k = 0$

$\Sigma a_k b_k = \Sigma k^p$ where $p = 1.1 + 0.5 > 1$, so S_c converges. Σa_k is a convergent p-series with $p = 1.1 > 1$ and Σb_k is a divergent p-series with $p = 0.5 < 1$.

56. For $a_k = 1/k$, and $b_k = 1/\sqrt{k}$, $S_c = \lim_{k \to \infty} a_k = 0$. $\Sigma a_k b_k = \Sigma k^p$ where $p = 1.5 > 1$, so S_c converges. Σa_k is a divergent p-series with $p = 1$ and Σb_k is a divergent p-series with $p = 0.5 < 1$.

57. From the proof of the integral test, we have
$$\int_1^{n+1} f(x)\, dx < S_n < a_1 + \int_1^n f(x)\, dx$$
where $S_n = \sum_{k=1}^{\infty} a_k$ is the nth partial sum of $\sum_{k=1}^{\infty} a_k$. Since $\sum_{k=1}^{n} a_k$ converges, we know that $\lim_{n \to \infty} S_n$ exists (and is finite). Thus, by the squeeze theorem
$$\lim_{n \to \infty} \int_1^n f(x)\, dx \leq \lim_{n \to \infty} S_n \leq a_1 + \lim_{n \to \infty} \int_1^n f(x)\, dx$$
So we have
$$\int_1^{\infty} f(x)\, dx \leq \sum_{k=1}^{\infty} a_k \leq a_1 + \int_1^{\infty} f(x)\, dx$$
as required.

8.4 Comparison Tests, Page 521

1. The geometric series converges when $|r| < 1$.

2. The p-series converges when $p > 1$.

3. $S = \sum_{n=1}^{\infty} \cos^n(\frac{\pi}{6})$ is a convergent geometric series with $r = \cos \frac{\pi}{6} < 1$.

4. $S = \sum_{k=0}^{\infty} 0.5^k$ is a convergent geometric series with $r = 0.5 < 1$.

5. $S = \sum_{k=0}^{\infty} 1.5^k$ is a divergent geometric series with $r = 1.5 > 1$.

6. $S = \sum_{k=0}^{\infty} 2^{k/2} = \sum_{k=0}^{\infty} (2^{1/2})^k$ is a divergent geometric series with $r = \sqrt{2} > 1$.

7. $S = \sum_{k=1}^{\infty} \frac{1}{k}$ is a divergent p-series since $p = 1$.

8. $S = \sum_{k=1}^{\infty} \frac{1}{k^{0.5}}$ is a divergent p-series since $p = 0.5 < 1$.

9. $S = \sum_{k=1}^{\infty} \frac{1}{k^{3/2}}$ is a convergent p-series since $p = 1.5 > 1$.

10. $S = \sum_{k=1}^{\infty} \sqrt{\frac{2}{k}} = \sqrt{2} \sum_{k=1}^{\infty} \frac{1}{k^{1/2}}$ is a divergent p-series since $p = 0.5 < 1$.

11. $S = \sum_{k=0}^{\infty} 1$ is a divergent p-series with $p = 0 < 1$; it is also a divergent geometric series with $r = 1$.

12. $S = \sum_{k=1}^{\infty} e^k$ is a divergent geometric series with $r = e > 1$.

13. $S = \sum_{k=1}^{\infty} \frac{1}{k^2 + k}$ compares with the convergent p-series $\sum_{k=1}^{\infty} \frac{1}{k^2}$.

14. $S = \sum_{k=1}^{\infty} \frac{1}{k^2 + 3k + 2}$ compares with the convergent p-series $\sum_{k=1}^{\infty} \frac{1}{k^2}$.

15. $S = \sum_{k=1}^{\infty} \frac{1}{\sqrt{k}} = \sum_{k=1}^{\infty} \frac{1}{k^{1/2}}$ is the divergent p-series with $p = 1/2$.

16. $S = \sum_{k=1}^{\infty} \frac{1}{k\sqrt{k}} = \sum_{k=1}^{\infty} \frac{1}{k^{3/2}}$ is the convergent p-series with $p = 3/2$.

17. $S = \sum_{k=1}^{\infty} \frac{1}{\sqrt{2k+3}}$ compares with the divergent p-series $\sum_{k=1}^{\infty} \frac{1}{k^{1/2}}$.

18. $S = \sum_{k=1}^{\infty} \frac{1}{\sqrt{k(k+1)}} = \sum_{k=1}^{\infty} \frac{1}{(k^2 + k)^{1/2}}$ compares with the divergent p-series $\sum_{k=1}^{\infty} \frac{1}{k}$.

19. $S = \sum\limits_{k=1}^{\infty} \dfrac{1}{\sqrt{k^3 + 2}}$ compares with the convergent p-series $\sum\limits_{k=1}^{\infty} \dfrac{1}{k^{3/2}}$.

20. $S = \sum\limits_{k=1}^{\infty} \dfrac{1}{\sqrt{k^2 + 1}}$ compares with the divergent p-series $\sum\limits_{k=1}^{\infty} \dfrac{1}{k}$.

21. $S = \sum\limits_{k=1}^{\infty} \dfrac{2k^2}{k^4 - 4}$ compares with the convergent p-series $\sum\limits_{k=1}^{\infty} \dfrac{1}{k^2}$.

22. $S = \sum\limits_{k=1}^{\infty} \dfrac{k+1}{k^2 + 1}$ compares with the divergent p-series $\sum\limits_{k=1}^{\infty} \dfrac{1}{k}$.

23. $S = \sum\limits_{k=1}^{\infty} \dfrac{(k+2)(k+3)}{k^{7/2}}$ compares with the convergent p-series $\sum\limits_{k=1}^{\infty} \dfrac{1}{k^{3/2}}$,

24. $S = \sum\limits_{k=1}^{\infty} \dfrac{(k+1)^3}{k^{9/2}}$ compares with the convergent p-series $\sum\limits_{k=1}^{\infty} \dfrac{1}{k^{3/2}}$

25. $S = \sum\limits_{k=1}^{\infty} \dfrac{2k+3}{k^2 + 3k + 2}$ compares with the divergent p-series $\sum\limits_{k=1}^{\infty} \dfrac{2}{k}$.

26. $S = \sum\limits_{k=1}^{\infty} \dfrac{3k^2 + 2}{k^2 + 3k + 2}$

    $\lim\limits_{k \to \infty} \dfrac{3k^2 + 2}{k^2 + 3k + 2} = 3 \neq 0$, so the

    necessary condition for convergence is not satisfied; S diverges.

27. $S = \sum\limits_{k=1}^{\infty} \dfrac{k}{(k+2)2^k}$ compares with the convergent geometric series $\sum\limits_{k=1}^{\infty} \dfrac{1}{2^k}$.

28. $S = \sum\limits_{k=1}^{\infty} \dfrac{5}{4^k + 3}$ compares with the convergent geometric series $\sum\limits_{k=1}^{\infty} \dfrac{1}{4^k}$.

29. $S = \sum\limits_{k=1}^{\infty} \dfrac{1}{k(k+2)}$ compares with the convergent p-series $\sum\limits_{k=1}^{\infty} \dfrac{1}{k^2}$.

30. $S = \sum\limits_{k=1}^{\infty} \dfrac{1}{(k+2)(k+3)}$ compares with the convergent p-series $\sum\limits_{k=1}^{\infty} \dfrac{1}{k^2}$.

31. $S = \sum\limits_{k=1}^{\infty} \dfrac{1}{\sqrt{k}\, 2^k}$; compares with the convergent geometric series $\sum\limits_{k=1}^{\infty} \dfrac{1}{2^k}$.

32. $S = \sum\limits_{k=1}^{\infty} \dfrac{1,000}{\sqrt{k}\, 3^k}$ compares with the convergent series $\sum\limits_{k=1}^{\infty} \dfrac{1,000}{3^k}$ by using the zero-infinity limit comparison test.

33. $S = \sum\limits_{k=1}^{\infty} \dfrac{|\sin(k!)|}{k^2}$ compares with the convergent p-series $\sum\limits_{k=1}^{\infty} \dfrac{1}{k^3}$.

34. $S = \sum\limits_{k=1}^{\infty} \dfrac{|\cos k^3|}{\sqrt{k}}$ compares with the divergent p-series $\sum\limits_{k=1}^{\infty} \dfrac{1}{\sqrt{k}}$.

35. $\sum\limits_{k=1}^{\infty} \dfrac{2k^3 + k + 1}{k^3 + k^2 + 1}$ diverges since

    $\lim\limits_{k \to \infty} \dfrac{2k^3 + k + 1}{k^3 + k^2 + 1} = 2 \neq 0$; the necessary condition for convergence is not satisfied.

36. $S = \sum\limits_{k=1}^{\infty} \dfrac{6k^3 - k - 4}{k^3 - k^2 - 3}$ diverges since

    $\lim\limits_{k \to \infty} \dfrac{6k^3 - k - 4}{k^3 - k^2 - 3} = 2 \neq 0$; the necessary condition for convergence is not satisfied.

37. $S = \sum\limits_{k=1}^{\infty} \dfrac{k}{4k^3 - 5}$ compares with the convergent p-series $\sum\limits_{k=1}^{\infty} \dfrac{1}{k^2}$.

38. $S = \sum\limits_{k=1}^{\infty} \dfrac{\ln k}{\sqrt{2k + 3}}$ compares with the divergent p-series $\sum\limits_{k=1}^{\infty} \dfrac{1}{\sqrt{k}}$ by using the zero-infinity comparison test.

39. $S = \sum\limits_{k=1}^{\infty} \dfrac{k^2 + 1}{(k^2 + 2)k^2}$ compares with the convergent p-series $\sum\limits_{k=1}^{\infty} \dfrac{1}{k^2}$.

40. $S = \sum\limits_{k=1}^{\infty} \sin \dfrac{1}{k}$ compares with the divergent p-series $\sum\limits_{k=1}^{\infty} \dfrac{1}{k}$.

41. $S = \sum\limits_{k=1}^{\infty} \dfrac{6k^2 + 2k + 1}{k^{1.1}(4k^2 + k + 4)}$ compares with the convergent p-series $\sum\limits_{k=1}^{\infty} \dfrac{1}{k^{1.1}}$.

42. $S = \sum\limits_{k=1}^{\infty} \dfrac{6k^2 + 2k + 1}{k^{0.9}(4k^2 + k + 4)}$ compares with the divergent p-series $\sum\limits_{k=1}^{\infty} \dfrac{1}{k^{0.9}}$.

43. $S = \sum_{k=1}^{\infty} \frac{\sqrt[6]{k}}{\sqrt[4]{k^3+2}\ \sqrt[8]{k}} = \sum_{k=1}^{\infty} \frac{k^{4/24}}{(k^3+2)^{6/24}k^{3/24}}$

$= \sum_{k=1}^{\infty} \frac{k^{1/24}}{(k^3+2)^{6/24}}$ compares with the

divergent p-series $\sum_{k=1}^{\infty} \frac{1}{k^{17/24}}$.

44. $S = \sum_{k=1}^{\infty} \frac{\sqrt{k}}{\sqrt[3]{k^3+1}\ \sqrt[6]{k^5}} = \sum_{k=1}^{\infty} \frac{k^{3/6}}{(k^3+1)^{2/6}k^{5/6}}$

$= \sum_{k=1}^{\infty} \frac{1}{(k^3+2)^{1/3}k^{1/3}}$ compares with the

convergent p-series $\sum_{k=1}^{\infty} \frac{1}{k^{4/3}}$

45. $S = \sum_{k=1}^{\infty} \frac{1}{k^3+4}$ compares with the convergent

p-series $\sum_{k=1}^{\infty} \frac{1}{k^3}$.

46. $S = \sum_{k=2}^{\infty} \frac{\ln k}{k-1}$ compares with the divergent

p-series $\sum_{k=1}^{\infty} \frac{1}{k}$.

47. $S = \sum_{k=1}^{\infty} \frac{\ln(k+1)}{(k+1)^3}$ compares with the log-

power quotient series with $q = 3 > 1$, so S
converges.

48. $S = \sum_{k=1}^{\infty} \frac{\ln k}{k^2}$ compares with the log-power

quotient series with $q = 2 > 1$, so S
converges.

49. $S = \sum_{k=2}^{\infty} \frac{1}{(k+3)(\ln k)^{1.1}}$ compares with

Problem 52 of Section 8.3; $p = 1.1 > 1$;
converges

50. $S = \sum_{k=2}^{\infty} \frac{1}{(k+3)(\ln k)^{0.9}}$ compares with

Problem 52 of Section 8.3; $p = 0.9 < 1$;
diverges

51. $S = \sum_{k=1}^{\infty} k^{(1-k)/k} = \sum_{k=1}^{\infty} \frac{1}{k^{1-1/k}}$ compare
with a divergent p-series $(p = 1 - k^{-1} < 1)$.

52. $S = \sum_{k=1}^{\infty} k^{(1+k)/k} = \sum_{k=1}^{\infty} \frac{1}{k^{1+1/k}}$ compares
with a convergent p-series $(p = 1 + k^{-1} > 1)$.

53. $S = \sum_{k=1}^{\infty} \frac{k+3}{(k+3)!} = \frac{4}{4!} + \frac{5}{5!} + \frac{6}{6!} + \cdots$

First show that $T = \sum_{k=1}^{\infty} \frac{1}{(k+2)!}$ converges.

$T = \sum_{k=3}^{\infty} \frac{1}{k!} < \sum_{k=1}^{\infty} \frac{1}{k^n}$ and $\lim_{n\to\infty} \sum_{k=1}^{\infty} \frac{1}{k^n}$

converges when compared to a p-series with
$p = n > 1$. Now compare the series S with
the convergent series T to conclude that S
converges.

54. $S = 1 + \frac{1}{1\cdot 3} + \frac{1}{1\cdot 3\cdot 5} + \frac{1}{1\cdot 3\cdot 5\cdot 7}$

$+ \cdots + \frac{1}{(2k+1)!}$

$= 1 + \frac{2}{(1)(2)(3)} + \frac{(2)(4)}{5!} + \cdots$

$+ \frac{(2)(4)\cdots(2k)}{(2k+1)!} + \cdots$

$= \lim_{n\to\infty} \sum_{k=1}^{n} \left[1 + \frac{2^k k!}{(2k+1)!} \right]$

Compare with $T = \sum_{k=1}^{\infty} \frac{1}{k!}$ which converges

(see Problem 53). By the zero-infinite limit
comparison test,

$\lim_{k\to\infty} \frac{2^k k!}{(2k+1)!k!} = \lim_{k\to\infty} \frac{2^k}{(2k+1)!} = 0$

Thus, S converges.

55. $S = \sum_{k=2}^{\infty} \frac{1}{(\ln k)^{\ln k}}$

$= \sum_{k=2}^{N} \frac{1}{(\ln k)^{\ln k}} + \sum_{k=N}^{\infty} \frac{1}{(\ln k)^{\ln k}}$

$= a + \sum_{k=N}^{\infty} \frac{1}{(\ln k)^{\ln k}}$ for a finite number a

$= a + \sum_{k=N}^{\infty} \frac{1}{(e^2)^{\ln k}} = a + \sum_{k=N}^{\infty} \frac{1}{k^2}$

Since the last summation converges, we
conclude that S converges.

56. $S = \sum_{k=2}^{\infty} \frac{(\ln k)^p}{k^q}$; use the integral test with

$I = \int_{2}^{\infty} \frac{(\ln x)^p}{x^q}\, dx \quad \boxed{u = (\ln x)^q;\ dv = x^{-n}\, dx}$

$= \lim_{b\to\infty} \left[\frac{(\ln x)^p}{(1-q)x^{q-1}} \Big|_{2}^{b} + \frac{p}{1-q} \int_{2}^{b} \frac{(\ln x)^{p-1}}{x^q}\, dx \right]$

Case 1: If $p \le q$, then the first fraction
converges by l'Hôpital's rule and the integral
converges.
Case 2: If $q < p$, then the first fraction
diverges.
Thus, S converges if $p \le q$.

57. a. $S = \Sigma a_k$ and $T = \Sigma k^{-p}$, a p-series which
converges when $p > 1$. By the limit
comparison test

$0 < \lim_{n\to\infty} \frac{a_k}{k^{-p}} < \infty$

shows that Σa_k converges.

b. $\displaystyle\lim_{n\to\infty} \frac{k^2}{e^{k^2}} = \lim_{n\to\infty} \frac{2k}{2ke^{k^2}} = 0$

Thus, $p = 2$ and $a_k = e^{-k^2}$ in part **a**, we

conclude that Σe^{-k^2} converges.

58. Assume $T = \Sigma a_k b_k$ converges. Then,

$$0 < \lim_{k\to\infty} \frac{a_k}{a_k b_k} < \infty$$

since Σb_k converges. Thus, Σa_k converges.

Now assume $T = \Sigma a_k$ converges. Then,

$$0 < \lim_{k\to\infty} \frac{a_k}{a_k b_k} < \infty$$

since Σb_k converges. Thus, $\Sigma a_k b_k$ converges.

59. $S = \Sigma b_k^{-2} \le \Sigma k^{-2}$ since, by hypotheses,

$b_k \ge k^2$ and so S converges.

$$\sum \frac{A}{b_k} \ge \sum \frac{a_k}{b_k}$$

converges.

60. If Σa_k converges, then $\displaystyle\lim_{k\to\infty} a_k = 0$,

$\displaystyle\lim_{k\to\infty} a_k^{-1} \ne 0$, so Σa_k^{-1} diverges

61. $S = \Sigma a_k^2$ converges. Just compare it with

Σa_k which converges.

$$0 < \lim_{k\to\infty} \frac{a_k^2}{a_k} < \infty$$

62. $\displaystyle\lim_{k\to\infty} \frac{a_k}{b_k} = 0$ implies $a_k < \epsilon b_k$ for any $\epsilon > 0$.

$S = \Sigma a_k$ converges because

$$0 < \frac{a_k}{a_k b_k} < \infty$$

since Σb_k converges.

8.5 The Ratio Test and the Root Test, Page 528

1. In the ratio test, evaluate the limit of the
ratio; it is particularly useful if a_k involves $k!$,
k^p, or a^k, whereas the root test is more useful
if it is easy to find $\sqrt[k]{k}$. These tests are
similar in that if the limit is L, the series
converges if $L < 1$, diverges if $L > 1$, and fails
if $L = 1$.

2. The steps are outlined in Table 8.1, Page 527.

3. $a_k = \frac{1}{k!}$; use the ratio test.

$$\lim_{k\to\infty} \frac{\frac{1}{(k+1)!}}{\frac{1}{k!}} = \lim_{k\to\infty} \frac{k!}{(k+1)!}$$

$$= \lim_{k\to\infty} \frac{1}{k+1} = 0 < 1$$

The series converges.

4. $a_k = \frac{k!}{2^k}$; use the ratio test.

$$\lim_{k\to\infty} \frac{(k+1)!2^k}{2^{k+1}k!} = \frac{1}{2} \lim_{k\to\infty} (k+1) = \infty > 1$$

The series diverges.

5. $a_k = \frac{k!}{2^{3k}}$; use the ratio test.

$$\lim_{k\to\infty} \frac{(k+1)!2^{3k}}{2^{3k+3}} = \frac{1}{8} \lim_{k\to\infty} (k+1) = \infty > 1$$

The series diverges.

6. $a_k = \frac{3^k}{k!}$; use the ratio test.

$$\lim_{k\to\infty} \frac{2^{k+1}k!}{(k+1)!3^k} = 3 \lim_{k\to\infty} \frac{1}{k+1} = 0 < 1$$

The series converges.

7. $a_k = \frac{k}{2^k}$; use the ratio test.

$$\lim_{k\to\infty} \frac{(k+1)2^k}{2^{n+1}k} = \frac{1}{2} < 1$$

The series converges.

8. $a_k = \frac{2^k}{k^2}$; use the ratio test.

$$\lim_{k\to\infty} \frac{a_{k+1}}{a_k} = \lim_{k\to\infty} \frac{k^2 2^{k+1}}{2^k(k+1)^2} = 2 < 1$$

The series converges.

9. $a_k = \frac{k^{100}}{e^k}$; use the ratio test.

$$\lim_{k\to\infty} \frac{(k+1)^{100}e^k}{e^{k+1}k^{100}} = e^{-1} < 1$$

The series converges.

10. $a_k = e^{-k}$; this is a geometric series with
$r = e^{-1} < 1$, so the series converges.
Following the directions, we use the ratio test:

$$\lim_{k\to\infty} \frac{e^{-k-1}}{e^{-k}} = \lim_{k\to\infty} e^{-1} = e^{-1} \le 1$$

The series converges.

11. $a_k = k(\frac{4}{3})^k$; diverges since the necessary

condition for convergence is not satisfied.

That is, $\displaystyle\lim_{k\to\infty} k\frac{4^k}{3^k} = \infty \ne 0$. Following the

directions, we use the ratio test:

$$\lim_{k\to\infty} \frac{(k+1)4^{k+1}3^k}{3^{k+1}k3^k} = \infty$$

The series diverges.

12. $a_k = k(\frac{3}{4})^k$; use the ratio test.

$$\lim_{k \to \infty} \frac{(k + 1)(3/4)^{k+1}}{k(3/4)^k} = \frac{3}{4} < 1$$

The series converges.

13. $a_k = \left(\frac{2}{k}\right)^k$; use the root test.

$$\lim_{k \to \infty} \sqrt[k]{\left(\frac{2}{k}\right)^k} = \lim_{k \to \infty} \frac{2}{k} = 0 < 1$$

The series converges.

14. $a_k = \frac{k^{10}2^k}{k!}$; use the ratio test.

$$\lim_{k \to \infty} \frac{k!(k+1)^{10}2^{k+1}}{k^{10}2^k(k + 1)!} = 2 \lim_{k \to \infty} \frac{(k + 10)^{10}}{k^{10}(k + 1)} = 0 < 1$$

The series converges.

15. $\sum_{k=1}^{\infty} \frac{k^5}{10^k}$; use the ratio test.

$$\lim_{k \to \infty} \frac{10^k(k + 1)^5}{k^5 10^{k+1}} = \frac{1}{10} \lim_{k \to \infty} \frac{(k + 1)^5}{k^5} = \frac{1}{10} < 1$$

The series converges.

16. $a_k = \frac{2^k}{k^2}$; use the ratio test.

$$\lim_{k \to \infty} \frac{2^{k+1}k^2}{(k + 1)^2 2^k} = \lim_{k \to \infty} \frac{k^2}{(k + 1)^2} = 2 > 1$$

The series diverges.

17. $a_k = \left(\frac{k}{3k + 1}\right)^k$; use the root test.

$$\lim_{k \to \infty} \sqrt[k]{\left(\frac{k}{3k + 1}\right)^k} = \lim_{k \to \infty} \frac{k}{3k + 1} = \frac{1}{3} < 1$$

The series converges.

18. $a_k = \frac{3k + 1}{2^k}$; use the ratio test.

$$\lim_{k \to \infty} \frac{2^k(3k+4)}{(3k+1)2^{k+1}} = \frac{1}{2} \lim_{k \to \infty} \frac{3k + 4}{3k + 1} = \frac{1}{2} < 1$$

The series converges.

19. $a_k = \frac{k!}{(k+2)^4}$; use the ratio test.

$$\lim_{k \to \infty} \frac{(k+2)^4(k+1)!}{k!(k + 3)^4} = \frac{1}{2} \lim_{k \to \infty} \frac{(k+2)(k+1)}{(k + 3)^4}$$

$= \infty > 1$; the series diverges.

20. $a_k = \frac{k^5 + 100}{k!}$; use the ratio test.

$$\lim_{k \to \infty} \frac{k![(k + 1)^5 + 100]}{(k^5 + 100)(k + 1)!}$$

$$= \lim_{k \to \infty} \frac{(k + 1)^5 + 100}{(k^5 + 100)(k + 1)} = 0 < 1$$

The series converges.

21. $a_k = \frac{(k!)^2}{(2k)!}$; use the ratio test.

$$\lim_{k \to \infty} \frac{(2k)![(k + 1)!]^2}{(k!)^2(2k + 2)!} = \lim_{k \to \infty} \frac{(k + 1)^2}{(2k+2)(2k+1)}$$

$= \frac{1}{4} < 1$; the series converges.

22. $a_k = k^2 2^{-k}$; use the ratio test.

$$\lim_{k \to \infty} \frac{2^k(k + 1)}{k^2 2^{k+1}} = \frac{1}{2} \lim_{k \to \infty} \frac{(k + 1)^2}{k^2} = \frac{1}{2} < 1$$

The series converges.

23. $a_k = \frac{(k!)^2}{[(2k)!]^2}$; use the ratio test.

$$\lim_{k \to \infty} \frac{[(k + 1)!]^2[(2k)!]^2}{[(2k + 2)!]^2(k!)^2}$$

$$= \lim_{k \to \infty} \frac{(k + 1)^2}{[(2k + 2)(2k + 1)]^2} = \frac{1}{16} < 1$$

The series converges.

24. $a_k = k^4 3^{-k}$; use the ratio test.

$$\lim_{k \to \infty} \frac{3^k(k + 1)^4}{k^4 3^{k+1}} = \frac{1}{3} \lim_{k \to \infty} \frac{(k + 1)^4}{k^4} = \frac{1}{3} < 1$$

The series converges.

25. $a_k = \frac{(k^2)!}{[(2k)^2]!}$; use the ratio test.

$$\lim_{k \to \infty} \frac{[(k + 1)^2]![(2k)^2]!}{[(2k + 2)^2]!(k^2)!}$$

$$= \lim_{k \to \infty} \frac{(k + 1)^2}{[(2k + 2)(2k + 1)]^2} = \frac{1}{16} < 1$$

The series converges.

26. $a_k = \left(\frac{k}{2k+1}\right)^k$; use the root test.

$$\lim_{k \to \infty} \sqrt[k]{\left(\frac{k}{2k + 1}\right)^k} = \lim_{k \to \infty} \frac{k}{2k + 1} = \frac{1}{2} < 1$$

The series converges.

27. $S = \sum_{k=1}^{\infty} \frac{1,000}{k}$; directly compare with the divergent p-series $T = \sum_{k=1}^{\infty} \frac{1}{k}$, so S diverges.

28. $S = \sum_{k=1}^{\infty} \frac{5,000}{k\sqrt{k}}$; directly compare with the convergent p-series $T = \sum_{k=1}^{\infty} \frac{1}{k^{3/2}}$, so S converges.

29. $S = \sum_{k=1}^{\infty} \frac{5k+2}{k2^k}$; use the ratio test.

$$\lim_{k \to \infty} \frac{(5k+7)(k)2^k}{(k+1)2^{k+1}(5k+2)}$$

$$= \lim_{k \to \infty} \frac{(5k+7)(k)2^k}{(k+1)2^{k+1}(5k+2)}$$

$$= \frac{1}{2} < 1; \ S \text{ converges.}$$

30. $S = \sum_{k=1}^{\infty} \frac{(k!)^2}{k^k}$; use the ratio test.

$$\lim_{k \to \infty} \frac{[(k+1)!]^2 k^k}{(k+1)^{k+1}(k!)^2}$$

$$= \lim_{k \to \infty} \frac{(k+1)^2 k^k}{(k+1)^{k+1}}$$

$$= \lim_{k \to \infty} (k+1)\frac{1}{(1+\frac{1}{k})^k}$$

$$= e^{-1} \lim_{k \to \infty} (k+1) = \infty > 1; \ S \text{ diverges}$$

31. $S = \sum_{k=1}^{\infty} \frac{\sqrt{k!}}{2^k}$; use the ratio test

$$\lim_{k \to \infty} \frac{(\sqrt{k+1})!}{2^{k+1}\sqrt{k!}} = \frac{1}{2} \lim_{k \to \infty} \sqrt{k+1} = \infty > 1$$
S diverges

32. $S = \sum_{k=1}^{\infty} \frac{3k+5}{k3^k}$; use the ratio test.

$$\lim_{k \to \infty} = \frac{(3k+8)(k)3^k}{(k+1)3^{k+1}(3k+5)} = \frac{1}{3} < 1$$

S converges

33. $S = \sum_{k=1}^{\infty} \frac{2^k k!}{k^k}$; use the ratio test.

$$\lim_{k \to \infty} \frac{2^{k+1}(k+1)! k^k}{(k+1)^{k+1} 2^k k!} = 2 \lim_{k \to \infty} \frac{k^k}{(k+1)^k}$$

$$= 2 \lim_{k \to \infty} \frac{1}{(1+\frac{1}{k})^k} = \frac{2}{e} < 1; \ S \text{ converges}$$

34. $S = \sum_{k=1}^{\infty} \frac{2^{2k} k!}{k^k}$; use the ratio test.

$$\lim_{k \to \infty} \frac{4^{k+1}(k+1)k^k}{(k+1)^{k+1} 4^k k!} = 4 \lim_{k \to \infty} \frac{k^k}{(k+1)^k}$$

$$= 4 \lim_{k \to \infty} \frac{1}{(1+\frac{1}{k})^k} = \frac{4}{e} > 1; \ S \text{ diverges}$$

35. $S = \sum_{k=1}^{\infty} \frac{\sqrt{k+1}}{k^{k+0.5}}$; use the ratio test.

$$\lim_{k \to \infty} \frac{\sqrt{k+2} \ k^{k+0.5}}{(k+1)^{k+1.5}\sqrt{k+1}}$$

$$= \lim_{k \to \infty} \left(\frac{1}{k}\right) \sqrt{\frac{k+2}{k+1}} = 0; \ S \text{ converges}$$

36. $S = \sum_{k=1}^{\infty} \frac{1}{k^k}$; use the root test.

$$\lim_{k \to \infty} \sqrt[k]{\frac{1}{k^k}} = 0 < 1; \ S \text{ converges}$$

37. $S = \sum_{k=1}^{\infty} \frac{k!}{(k+1)!} = \sum_{k=1}^{\infty} \frac{1}{k+1}$; directly compare with harmonic series to see that S diverges. (Integral test also works.)

38. $S = \sum_{k=1}^{\infty} \frac{2^{1,000k}}{k^{k/2}}$; use the root test.

$$\lim_{k \to \infty} \sqrt[k]{\frac{2^{1,000k}}{k^{k/2}}} = \lim_{k \to \infty} \frac{2^{1,000}}{k^{1/2}} = 0 < 1;$$

S converges

39. $S = \sum_{k=1}^{\infty} \left(1+\frac{1}{k}\right)^{-k^2}$; use the root test.

$$\lim_{k \to \infty} \sqrt[k]{\left(1+\frac{1}{k}\right)^{-k^2}} = \lim_{k \to \infty} \left(1+\frac{1}{k}\right)^{-k}$$

$$= \lim_{k \to \infty} \frac{1}{(1+\frac{1}{k})^k} = \frac{1}{e} < 1; \ S \text{ converges}$$

40. $S = \sum_{k=1}^{\infty} \left(\frac{k+2}{k}\right)^{-k^2}$; use the root test.

$$\lim_{k \to \infty} \sqrt[k]{\left(1+\frac{2}{k}\right)^{-k^2}} = \lim_{k \to \infty} \left(1+\frac{2}{k}\right)^{-k}$$

$$= \lim_{k \to \infty} \frac{1}{[(1+\frac{2}{k})^{k/2}]^2} = \frac{1}{e^2} < 1; \ S \text{ converges}$$

41. $S = \sum_{k=1}^{\infty} \left|\frac{\cos k}{2^k}\right|$; directly compare with the convergent geometric series $T = \sum_{k=1}^{\infty} \frac{1}{2^k}$

$$0 < \lim_{k \to \infty} |\cos k| < \infty$$

S converges; (disregard $\cos k = 0$; these terms do not contribute to these original series).

42. $S = \sum_{k=1}^{\infty} \left|\frac{\sin k}{3^k}\right|$; directly compare with the convergent geometric series $T = \sum_{k=1}^{\infty} \frac{1}{3^k}$

$$0 < \lim_{k \to \infty} |\sin k| < \infty$$

S converges; (disregard $\sin k = 0$; these terms do not contribute to these original series).

43. $S = \sum\limits_{k=2}^{\infty} \left(\frac{\ln k}{k}\right)^k$; use the root test.

$$\lim_{k \to \infty} \sqrt[k]{\frac{(\ln k)^k}{k}} = \lim_{k \to \infty} \frac{\ln k}{k} = \lim_{k \to \infty} \frac{1}{k}$$

$= 0 < 1$; S converges

44. $S = \sum\limits_{k=2}^{\infty} \frac{1}{(\ln k)^k}$; use the root test;

$$\lim_{k \to \infty} \sqrt[k]{\left(\frac{1}{\ln k}\right)^k} = \lim_{k \to \infty} \frac{1}{\ln k} = 0 < 1;$$

S converges

45. $S = \sum\limits_{k=1}^{\infty} k^2 x^k$; use the ratio test.

$$\lim_{k \to \infty} \frac{(k+1)^2 x^{k+1}}{k^2 x^k} = x$$

By the ratio test, S converges when $x < 1$. Since the ratio test fails when the ratio equals 1, investigate

$$S = \sum_{k=1}^{\infty} k^2$$

separately to see that S diverges at $x = 1$. The interval of convergence is $(0, 1)$.

46. $\sum\limits_{k=1}^{\infty} k x^k$; use the ratio test.

$$\lim_{k \to \infty} \frac{(k+1) x^{k+1}}{k x^k} = x$$

By the ratio test, S converges when $x < 1$. Since the ratio test fails when the ratio equal 1, investigate

$$S = \sum_{k=1}^{\infty} k$$

separately to see that S diverges at $x = 1$. The interval of convergence is $(0, 1)$.

47. $S = \sum\limits_{k=1}^{\infty} \frac{(x + 0.5)^k}{k\sqrt{k}}$; use the ratio test.

$$\lim_{k \to \infty} \frac{\dfrac{(x+0.5)^{k+1}}{(k+1)^{3/2}}}{\dfrac{(x+0.5)^k}{k^{3/2}}} = \lim_{k \to \infty} \frac{(x+0.5) k^{3/2}}{(k+1)^{3/2}} = x + 0.5$$

By the ratio test, S converges when $x + 0.5 < 1$ or $x < 0.5$. Since the ratio test fails when the ratio equals 1, investigate

$$S = \sum_{k=1}^{\infty} k^{-3/2}$$

separately to see that S diverges if $x = 0.5$. The interval of convergence is $(0, 0.5)$.

48. $S = \sum\limits_{k=1}^{\infty} \frac{(3x - 0.4)^k}{k^2}$; use the ratio test.

$$\lim_{k \to \infty} \frac{(3x + 0.4)^{k+1} k^2}{(k+1)^2 (3x + 0.4)^k}$$

$$= (3x + 0.4) \lim_{k \to \infty} \frac{k^2}{(k+1)^2} = 3x + 0.4$$

By the ratio test, S converges when $3x + 0.4 < 1$ or $x < \frac{2}{15}$. Since the ratio test fails when the ratio test equals 1, investigate

$$S = \sum_{k=1}^{\infty} \frac{1}{k^2}$$

separately to see that this p-series converges at $x = \frac{2}{15}$. The interval of convergence is $(0, \frac{2}{15}]$.

49. $S = \sum\limits_{k=1}^{\infty} \frac{x^k}{k!}$; use the ratio test.

$$\lim_{k \to \infty} \frac{x^{k+1} k!}{(k+1) x^k} = \lim_{k \to \infty} \frac{x}{k+1} = 0 < 1$$

for all x. The interval of convergence is $(0, \infty)$.

50. $S = \sum\limits_{k=1}^{\infty} \frac{x^{2k}}{k}$; use the ratio test.

$$\lim_{k \to \infty} \frac{x^{2k+2} k}{(k+1) x^{2k}} = \lim_{k \to \infty} x^2 = x^2$$

By the ratio test, S converges when $x < 1$. Since the ratio test fails when the ratio is 1, we test

$$S = \sum_{k=1}^{\infty} \frac{1}{k}$$

separately to see that the p-series diverges when $x = 1$. The interval of convergence for $(0, 1)$.

51. $S = \sum\limits_{k=1}^{\infty} (ax)^k$; use the ratio test.

$$\lim_{k \to \infty} \frac{(ax)^{k+1}}{(ax)^k} = \lim_{k \to \infty} (ax) = ax$$

By the ratio test, S converges when $ax < 1$ or when $x < a^{-1}$. Since the ratio test fails when the ratio is is 1, investigate

$$S = \sum_{k=1}^{\infty} 1^k$$

separately to see that the series S diverges at $x = a^{-1}$. The interval of converges for $(0, a^{-1})$.

52. $S = \sum\limits_{k=1}^{\infty} k x^{2k}$; use the ratio test.

$$\lim_{k \to \infty} \frac{(k+1) x^{2k+2}}{k x^{2k}} = \lim_{k \to \infty} x^2$$

By the ratio test, S converges when $x^2 < 1$ or when $x < 1$. Since the ratio test fails when

the ratio equals 1, investigate

$$S = \sum_{k=1}^{\infty} k$$

separately to see that S diverges at $x = 1$.
The interval of convergence is $(0, 1)$.

53. $S = \sum_{k=1}^{\infty} k^p e^{-k}$; use the root test.

$$\lim_{k \to \infty} \frac{k^{p/k}}{e} = 0$$

By the ratio test, S for all p since $0 < 1$.
converges when $x^2 < 1$, or when $x < 1$.
The integral test shows that

$$\int_{1}^{\infty} x^p e^{-x} dx$$

also converges for all p.

54. a. $S = \sum_{k=1}^{\infty} 2^{-k+(-1)^k}$; use the ratio test.

$$\lim_{k \to \infty} \frac{2^{-k-1+(-1)^{k+1}}}{2^{-k+(-1)^k}}$$

$$= \lim_{k \to \infty} 2^{-k-1+(-1)^{k+1}+k-(-1)^k}$$

$$= \lim_{k \to \infty} 2^{-1} = 2^{-1}$$

S converges since $2^{-1} < 1$.

b. $S = \sum_{k=1}^{\infty} \frac{x^k}{k!}$; use the root test.

$$\lim_{k \to \infty} \sqrt[k]{2^{-k+(-1)^k}}$$

$$= \lim_{k \to \infty} 2^{-1+(-1)^k/k} = 2^{-1}$$

S converges since $2^{-1} < 1$.

55. a. Since $L < 1$ and $\lim_{k \to \infty} \frac{a_{k+1}}{a_k} = L$

the series converges and the necessary condition for convergence is satisfied; that is,

$$\lim_{k \to \infty} a_k = 0$$

b. Since $S = \sum_{k=1}^{\infty} \frac{x^k}{k!}$; use the ratio test.

$$\lim_{k \to \infty} \frac{x^{k+1} k!}{(k+1) x^k} = \lim_{k \to \infty} \frac{x}{k+1} = 0$$

S converges for all x so that

$$\lim_{k \to \infty} \frac{x^k}{k!} = 0$$

56. a. $1 + \frac{1}{2} + \frac{1}{2} + \frac{1}{4} + \frac{1}{4} + \frac{1}{8} + \frac{1}{8} + \frac{1}{16} + \cdots$

The ratios $\frac{a_{k+1}}{a_k}$ are

$\frac{1}{2}, 1, \frac{1}{2}, 1, \frac{1}{2}, \cdots$

Obviously $\lim_{k \to \infty} \frac{a_{k+1}}{a_k}$ does not exist.

b. The sequence of roots $\sqrt[k]{a_k}$ is:

$1, \sqrt{\frac{1}{2}}, \sqrt[3]{\frac{1}{2}}, \sqrt[4]{\frac{1}{4}}, \sqrt[5]{\frac{1}{4}}, \sqrt[6]{\frac{1}{8}}, \cdots$

This sequence approaches $\frac{1}{\sqrt{2}}$

The root test tells us the series converges.

8.6 Alternating Series; Absolute and Conditional Convergence, Page, 537

1. Consider the alternating series $A = \Sigma(-1)^k a_k$ (with positive a_k). If Σa_k converges, then a converges absolutely. If Σa_k diverges, then A may converges conditionally.

2. $\sum_{k=1}^{\infty} \frac{(-1)^k}{k^p}$ converges for $p > 0$

(See Example 4, page 532.)

3. All individual elements of the series need to be finite.

4. $S = \sum_{k=1}^{\infty} \frac{(-1)^{k+1} k}{k^2 + 1}$

To show $\left\{ \frac{k}{k^2 + 1} \right\}$ is decreasing, let

$f(x) = \frac{x}{x^2 + 1}$; then $f'(x) = \frac{x^2 + 1 - x(2x)}{(x^2 + 1)^2}$

$= \frac{-x^2 + 1}{(x^2 + 1)^2} < 0$ if $x > 1$. Thus $\left\{ \frac{k}{k^2 + 1} \right\}$ is

decreasing and $\lim_{k \to \infty} \frac{k}{k^2 + 1} = 0$. Thus, S

converges.

$T = \sum_{k=1}^{\infty} \frac{k}{k^2 + 1}$ diverges because it can be

compared with the divergent series $\sum_{k=1}^{\infty} \frac{1}{k}$.

Thus, S converges conditionally.

5. $S = \sum_{k=1}^{\infty} \frac{(-1)^{k+1} k^2}{k^3 + 1}$

To show $\left\{ \frac{k^2}{k^3 + 1} \right\}$ is decreasing, let

$f(x) = \frac{x^2}{x^3 + 1}$; $f'(x) = \frac{x(2 - x^3)}{(x^3 + 1)^2} < 0$

if $x > 1$. Thus $\left\{ \frac{k^2}{k^3 + 1} \right\}$ is decreasing and

$\lim_{k \to \infty} \frac{k^2}{k^3 + 1} = 0$. Thus, S converges.

$T = \sum_{k=1}^{\infty} \dfrac{k^2}{k^3 + 1}$ diverges because it can be

compared with the divergent series $\sum_{k=1}^{\infty} \dfrac{1}{k}$.

Thus, S converges conditionally.

6. $S = \sum_{k=1}^{\infty} \dfrac{(-1)^{k+1}k}{2k+1}$ diverges because

$$\lim_{k \to \infty} \frac{k+1}{2k+1} = \tfrac{1}{2} \neq 0$$

7. $S = \sum_{k=1}^{\infty} \dfrac{(-1)^{k+1}k^2}{k^2 + 1}$ diverges because

$$\lim_{k \to \infty} \frac{k^2}{k^2 + 1} = 1 \neq 0$$

8. $S = \sum_{k=1}^{\infty} \dfrac{(-1)^{k+1}}{k^{3/2}}$ converges absolutely

because $T = \sum_{k=1}^{\infty} \dfrac{1}{k^{3/2}}$ is a convergent
p-series

9. $S = \sum_{k=1}^{\infty} \dfrac{(-1)^{k+1}k}{2^k}$;

Apply the ratio test to the series of absolute values:

$$\lim_{k \to \infty} \frac{\frac{k+1}{2^{k+1}}}{\frac{k}{2^k}} = \lim_{k \to \infty} \frac{1}{2}\frac{k+1}{k} = \frac{1}{2} < 1$$

S is absolutely convergent.

10. $S = \sum_{k=1}^{\infty} (-1)^{k+1}\dfrac{k^2}{e^k}$

Apply the ratio test to the series of absolute values:

$$\lim_{k \to \infty} = \frac{(k+1)^2 e^k}{2(k+1)k^2} = \frac{1}{e} < 1$$

S is absolutely convergent.

11. $S = \sum_{k=1}^{\infty} \dfrac{(-1)^k}{\sqrt{k}}$

To show $\left\{\dfrac{1}{\sqrt{k}}\right\}$ is decreasing, let

$f(x) = x^{-1/2}; \ f'(x) = -\tfrac{1}{2}x^{-3/2} < 0$, so

$\left\{\dfrac{1}{\sqrt{k}}\right\}$ is decreasing and $\lim_{k \to \infty} \dfrac{1}{\sqrt{k}} = 0$.

Thus, S converges.

$T = \sum_{k=1}^{\infty} \dfrac{1}{k^{1/2}2}$ is a divergent p series $(p = \tfrac{1}{2})$,

so S is converges conditionally.

12. $S = \sum_{k=1}^{\infty} (-1)^k \dfrac{(1 + k^2)}{k^3}$

$$= \sum_{k=1}^{\infty} \frac{(-1)^k}{k^3} + \sum_{k=1}^{\infty} \frac{(-1)^k}{k}$$

The first summation converges absolutely, but the second one converges conditionally (it is the alternating harmonic series). Thus, S converges conditionally.

13. $S = \sum_{k=1}^{\infty} \dfrac{(-1)^{k+1}k!}{k^k}$

Apply the ratio test to the series of absolute values:

$$\lim_{k \to \infty} \frac{(k+1)!k^k}{(k+1)^{k+1}k!} = \lim_{k \to \infty} \frac{1}{(1 + \frac{1}{k})^k} = \frac{1}{e} < 1$$

S is absolutely convergent. (This is similar to example 2 page 615.)

14. $S = \sum_{k=2}^{\infty} (-1)^k \dfrac{k!}{\ln k}$ diverges because

$$\lim_{k \to \infty} \frac{k!}{\ln k} = \infty \neq 0.$$

15. $S = \sum_{k=1}^{\infty} (-1)^k \dfrac{k!}{k^k}$

Apply the ratio test to the series of absolute values:

$$\lim_{k \to \infty} \frac{(k+1)!k^k}{(k+1)^{k+1}k!} = \lim_{k \to \infty} \frac{1}{(1 + \frac{1}{k})^k} = \frac{1}{e}$$

S is absolutely convergent.

16. $S = \sum_{k=1}^{\infty} (-1)^{k+1} \dfrac{2^k}{k!}$

Apply the ratio test to the series of absolute values:

$$\lim_{k \to \infty} \frac{2^{k+1}k!}{(k+1)!2^k} = \lim_{k \to \infty} \frac{2}{k+1} = 0 < 1$$

S is absolutely convergent.

17. $S = \sum_{k=1}^{\infty} \dfrac{(-2)^k}{k!}$

Apply the ratio test to the series of absolute values:

$$\lim_{k \to \infty} \frac{2^{k+1}k!}{(k+1)!2^k} = \lim_{k \to \infty} \frac{2}{k+1} = 0 < 1$$

S is absolutely convergent.

18. $S = \sum_{k=1}^{\infty} (-1)^{k+1} \dfrac{2^{2k+1}}{k!}$

Apply the ratio test to the series of absolute values:

$$\lim_{k\to\infty}\frac{2^{2k+3}k!}{(k+1)!2^{2k+1}}=\lim_{k\to\infty}\frac{4}{k+1}=0<1$$

S is absolutely convergent.

19. $S=\sum_{k=2}^{\infty}\frac{(-1)^{k+1}}{\ln k}$

To show that $\left\{\frac{1}{\ln k}\right\}$ is decreasing, let

$f(x)=\frac{1}{\ln x};\ f'(x)=\frac{-1}{x\ln^2 x}<0$, so $\left\{\frac{1}{\ln k}\right\}$ is

decreasing and $\lim_{k\to\infty}\frac{1}{\ln k}=0$. Thus, S

converges.

$T=\sum_{k=2}^{\infty}\frac{1}{\ln k}$ diverges because it can be

compared with the divergent harmonic series

$\sum_{k=1}^{\infty}\frac{1}{k}$. S converges conditionally.

20. $S=\sum_{k=1}^{\infty}\frac{(-1)^{k+1}k}{(k+1)(k+2)}$

$\left\{\frac{k}{(k+1)(k+2)}\right\}$ is decreasing because

$$\frac{k+1}{(k+2)(k+3)}\le\frac{k}{(k+1)(k+2)}$$

and $\lim_{k\to\infty}\frac{k}{(k+1)(k+2)}=0$; thus S

converges.

$T=\sum_{k=1}^{\infty}\frac{k}{(k+1)(k+2)}$ diverges because

it can be compared with the divergent

harmonic series $\sum_{k=1}^{\infty}\frac{1}{k}$. Thus, S converges

conditionally.

21. $S=\sum_{k=2}^{\infty}\frac{(-1)^{k+1}}{(\ln k)^4}$

This series is alternating, decreasing, and $\lim_{k\to\infty}a_k=0$, so it converges. Considering the series of absolute values:

$$\frac{1}{(\ln k)^4}>\frac{1}{(k\ln k)^4}>\frac{1}{(k\ln k)^k}$$

Using the root test:

$$\lim_{k\to\infty}\left[\frac{1}{(k\ln k)^k}\right]^{1/k}=\lim_{k\to\infty}\frac{1}{(k\ln k)}$$

which diverges by the integral test. S is conditionally convergent.

22. $S=\sum_{k=2}^{\infty}\frac{(-1)^{k+1}}{\ln(\ln k)}$

$\left\{\frac{1}{\ln(\ln k)}\right\}$ is decreasing because

$$\frac{1}{\ln[\ln(k+1)]}\le\frac{1}{\ln(k+1)}$$

and $\lim_{k\to\infty}\frac{1}{\ln(k+1)}=0$; thus S converges.

$T=\sum_{k=1}^{\infty}\frac{1}{\ln(\ln k)}$ diverges by comparison (use

zero-infinity limit comparison) with the

divergent harmonic $\sum_{k=2}^{\infty}\frac{1}{k}$. Thus, S

converges conditionally.

23. $S=\sum_{k=2}^{\infty}\frac{(-1)^{k+1}}{k\ln k}$

$\left\{\frac{1}{k\ln k}\right\}$ is decreasing because

$$\frac{1}{(k+1)\ln(k+1)}\le\frac{1}{k\ln k}$$

and $\lim_{k\to\infty}\frac{1}{k\ln k}=0$; thus, S converges.

$T=\sum_{k=2}^{\infty}\frac{1}{k\ln k}$; use the integral test.

$$I=\int_{2}^{\infty}\ln x\,\frac{dx}{x}=\lim_{b\to\infty}\ln|\ln x|\Big|_{2}^{b}=\infty$$

Thus, S converges conditionally.

24. $S=\sum_{k=1}^{\infty}\frac{(-1)^{k+1}\ln k}{k}$

To show $\left\{\frac{1}{k\ln k}\right\}$ is decreasing, let

$f(x)=\frac{\ln x}{x}$, $f'(x)=\frac{1-\ln x}{x^2}<0$ (if $x>e$),

and $\lim_{k\to\infty}\frac{\ln k}{k}=0$. Thus, S converges.

$T=\sum_{k=1}^{\infty}\frac{\ln k}{k}$; use the integral test.

$$I=\int_{1}^{\infty}\ln x\,\frac{dx}{x}=\frac{1}{2}\lim_{b\to\infty}(\ln x)^2\Big|_{1}^{b}=\infty$$

Thus, S converges conditionally.

25. $S=\sum_{k=2}^{\infty}\frac{(-1)^{k+1}k}{\ln k}$

$$\lim_{k\to\infty}\frac{k}{\ln k}=\lim_{k\to\infty}\frac{1}{\frac{1}{k}}=\infty>0$$

S is divergent.

26. $S=\sum_{k=1}^{\infty}(-1)^{k+1}\frac{\ln k}{k^2}$; use the integral test;

$$I = \int_1^\infty \frac{\ln x}{x^2}\,dx \quad \boxed{\text{parts or Formula 505}}$$

$$= \lim_{k\to\infty}\left[-\frac{\ln x}{x} - \frac{1}{x}\right]\Big|_1^b = 1$$

Thus, S is absolutely convergent.

27. $S = \sum_{k=1}^\infty \frac{(-1)^{k+1}k}{(k+2)^2}$

$\left\{\frac{1}{k\ln k}\right\}$ is decreasing because

$$\frac{k+1}{(k+3)^2} \le \frac{k}{(k+2)^2}$$

and $\lim_{k\to\infty}\frac{k}{(k+2)^2} = 0$; thus, S converges.

$T = \sum_{k=1}^\infty \frac{k}{(k+2)^2}$; use the integral test.

$$I = \int_1^\infty \frac{x\,dx}{(x+2)^2} < \int_1^\infty \frac{x\,dx}{x^2} = \lim_{b\to\infty}\ln|x|\Big|_1^b = \infty$$

Thus, S converges conditionally.

28. $S = \sum_{k=1}^\infty (-1)^{k+1}\left(\frac{k}{k+1}\right)^k$

$$\lim_{k\to\infty}\left(\frac{k}{k+1}\right)^k = \lim_{k\to\infty}\frac{1}{(1+\frac{1}{k})^k} = \frac{1}{e} \ne 0$$

S diverges since the necessary condition for convergence is not satisfied.

29. $S = \sum_{k=2}^\infty (-1)^{k+1}\frac{\ln(\ln k)}{k\ln k}$

To show $\left\{\frac{\ln(\ln k)}{k\ln k}\right\}$ is decreasing, let

$$f(x) = \frac{\ln(\ln k)}{x\ln x};$$

$$f'(x) = \frac{(x\ln x)(x\ln x)^{-1} - \ln(\ln x)(1+\ln x)}{(x\ln x)^2}$$

$$< 0$$

$$\lim_{k\to\infty}\frac{\ln(\ln x)}{x\ln x} = \lim_{k\to\infty}\frac{1}{x\ln x(1+\ln x)}$$

Thus, S converges.

$T = \sum_{k=1}^\infty \frac{\ln(\ln k)}{k\ln k}$; compare with the divergent series $\sum_{k=1}^\infty \frac{1}{k\ln k}$. T diverges, so S is conditionally convergent.

30. $S = \sum_{k=1}^\infty (-1)^{k+1}\left(\frac{1}{k}\right)^{1/k}$

$$\lim_{k\to\infty}\frac{1}{k^{1/k}} = \lim_{k\to\infty}\exp[\ln k^{-1/k}]$$

$$= \lim_{k\to\infty}\exp\left[-\frac{\ln k}{k}\right] = \lim_{k\to\infty}e^{-1/k} = 1 \ne 0$$

S diverges because the necessary condition for convergence is not satisfied.

31. $S = \sum_{k=1}^\infty (-1)^{k+1}\frac{k^5 5^{k+2}}{2^{3k}}$; use the generalized ratio test.

$$\lim_{k\to\infty}\frac{2^{3k}(k+1)5^{k+3}}{k^5 5^{k+2}2^{3k+3}} = \frac{5}{8}\lim_{k\to\infty}\frac{(k+1)^5}{k^5}$$

$$= \frac{5}{8}\lim_{k\to\infty}\frac{(k+1)^5}{k^5} = \frac{5}{8} < 1;$$

S is absolutely convergent.

32. $S = \sum_{k=1}^\infty \frac{(-1)^{k+1}}{2^{2k-2}}$

a. $S_4 = 1 - \frac{1}{4} + \frac{1}{16} - \frac{1}{64} = \frac{51}{64}$

$|S - S_4| < a_5 = \frac{1}{256} \approx 0.0039$

b. $\frac{1}{2^{2n-1}} < 0.0005$

$2^{2n-1} > 2{,}000$

$2n - 1 > \log_2 2{,}000$

$n > 12.97$

Choose $n = 7$; $S_7 = \frac{3{,}277}{4{,}096} \approx 0.800$

33. $S = \sum_{k=1}^\infty \frac{(-1)^{k+1}}{k!}$
a. $S_4 = 1 - \frac{1}{2} + \frac{1}{6} - \frac{1}{24} = \frac{5}{8}$

$|S - S_4| < a_5 = \frac{1}{120} \approx 0.0833$

b. $\frac{1}{n!} < 0.0005$

$n! > 2{,}000$

$n > 6$

Choose $n = 7$; $S_7 = \frac{177}{280} \approx 0.632$

34. $S = \sum_{k=1}^\infty \frac{(-1)^k}{k^2}$

a. $S_4 = -1 + \frac{1}{4} - \frac{1}{9} + \frac{1}{16}$

$= -\frac{115}{144} \approx -0.7986$

$|S - S_4| < a_5 = \frac{1}{25} \approx 0.04$

b. $\frac{1}{n^2} < 0.005$

$n^2 > 2{,}000$

$n > \sqrt{2{,}000}$

Choose $n = 45$; $S_{45} \approx -0.823$ (using software).

35. $S = \sum_{k=1}^{\infty} \frac{(-1)^k}{3^{k+1}}$

a. $S_4 = \frac{1}{9} - \frac{1}{27} + \frac{1}{81} - \frac{1}{243}$

$= -\frac{20}{243} \approx -0.083$

$|S - S_4| < a_5 = \frac{1}{729} \approx 0.001372$

b. $\frac{1}{3^{n+1}} < 0.0005$

$3^{n+1} > 2,000$

$n + 1 > \log_3 2,000$

$n > 5.918$

Choose $n = 6$; $S_6 = -\frac{182}{2,187} \approx -0.083$

36. $S = \sum_{k=1}^{\infty} \frac{(-1)^{k+1}}{k^3}$

a. $S_4 = 1 - \frac{1}{8} + \frac{1}{27} - \frac{1}{64} = \frac{1,559}{1,728} \approx 0.8964$

$|S - S_4| < a_5 = \frac{1}{125} \approx 0.008$

b. $\frac{1}{n^3} < 0.0005$

$n^3 > 2,000$

$n > 12.599$

Choose $n = 13$; $S_{13} \approx 0.902$ (using software)

37. $S = \sum_{k=1}^{\infty} \left(\frac{-1}{5}\right)^k$

a. $S_4 = -\frac{1}{5} + \frac{1}{25} - \frac{1}{125} + \frac{1}{625}$

$= -\frac{104}{625} \approx -0.1664$

$|S - S_4| < a_5 = \frac{1}{3,125} = 0.00032$

b. $\frac{1}{5^n} < 0.0005$

$5^n > 2,000$

$n > 4.72$

Choose $n = 5$; $S_5 = -\frac{521}{3,125} \approx -0.167$

38. $S = \sum_{k=1}^{\infty} \frac{x^k}{k}$; use generalized ratio test.

$\lim_{k \to \infty} \left| \frac{x^{k+1} k}{(k+1)x^k} \right| = |x| < 1$

For $x = 1$, $S = \sum_{k=1}^{\infty} \frac{1}{k}$ diverges.

For $x = -1$, $S = \sum_{k=1}^{\infty} \frac{(-1)^{k+1}}{k}$ converges.
The interval of convergence is $[-1, 1)$.

39. $S = \sum_{k=1}^{\infty} \frac{x^k}{\sqrt{k}}$; use generalized ratio test.

$\lim_{k \to \infty} \left| \frac{x^{k+1}\sqrt{k}}{\sqrt{k+1}x^k} \right| = |x| < 1$

For $x = 1$, $S = \sum_{k=1}^{\infty} \frac{1}{k^{1/2}}$ diverges.

For $x = -1$, $S = \sum_{k=1}^{\infty} \frac{(-1)^{k+1}}{\sqrt{k}}$ converges.
The interval of convergence is $[-1, 1)$.

40. $S = \sum_{k=1}^{\infty} \frac{2^k x^k}{k!}$; use generalized ratio test.

$\lim_{k \to \infty} \left| \frac{2^{k+1} x^{k+1} k!}{(k+1)! 2^k x^k} \right| = \lim_{k \to \infty} \frac{2|x| k!}{(k+1)!} = 0 < 1$

Converges for all x, so the interval of convergence is $(-\infty, \infty)$.

41. $S = \sum_{k=1}^{\infty} \frac{(k+2)x^k}{k^2(k+3)}$; use the generalized ratio test.

$\lim_{k \to \infty} \left| \frac{k^2(k+3)^2 x^{k+1}}{(k+1)^2(k+2)(k+4)x^k} \right| = |x| < 1$

If $x = 1$, $S = \sum_{k=1}^{\infty} \frac{k+2}{k^3 + 3k}$ converges.

If $x = -1$, $S = \sum_{k=1}^{\infty} \frac{(-1)^k(k+2)}{k^2(k+3)}$ converges.

The interval of convergence is $[-1, 1]$.

42. $S = \sum_{k=1}^{\infty} (-1)^{k+1} \left(\frac{x}{k}\right)^k$; use generalized ratio test.

$\lim_{k \to \infty} \left| \frac{x^{k+1} k}{(k+1)x^k} \right| = |x| < 1$

For $x = 1$, $S = (-1)^{k+1} \left(\frac{1}{k}\right)^k$

For $x = -1$, $S = (-1)^{k+1} \left(\frac{-1}{k}\right)^k$

$= (-1)^k \left(\frac{1}{k}\right)^k$

Check absolute convergence using the ratio test.

$\lim_{k \to \infty} \frac{k^k}{(k+1)^{k+1}} = \lim_{k \to \infty} \frac{1}{(k+1)(1+\frac{1}{k})^k} = 0 < 1$

S converges absolutely at $x = \pm 1$. The interval of convergence is $[-1, 1]$.

43. $S = \sum\limits_{k=1}^{\infty} k^p (-1)^k x^k$ for $p > 0$; use generalized ratio test.

$$\lim_{k \to \infty} \left| \frac{(k+1)^p x^{k+1}}{k^p x^k} \right| = |x| < 1$$

For $x = \pm 1$, $S = \sum\limits_{k=1}^{\infty} k^p$ diverges because the

necessary condition for convergence is not satisfied. The interval of convergence is $(-1, 1)$.

44. $|S - S_5| < \frac{1}{6}$ **45.** $|S - S_5| < \frac{1}{6^2}$

46. $|S - S_7| < \frac{1}{\ln 8}$ **47.** $|S - S_7| < \frac{7}{2^7}$

48. $S = \sum\limits_{k=2}^{\infty} \frac{(-1)^{k+1}}{k(\ln k)^p}$; use the integral test.

$$\frac{1}{[\ln(k+1)]^p} < \frac{1}{(\ln k)^p} \text{ and } \lim_{k \to \infty} \frac{1}{k(\ln k)^p} = 0$$

$$I = \int\limits_{2}^{\infty} \frac{dx}{x(\ln x)^p} = \lim_{b \to \infty} \int\limits_{2}^{b} \frac{1}{(1-p)(\ln x)^{p-1}} \bigg|_{2}^{b}$$

which converges when $p > 1$. Thus, S converges when $p > 1$.

49. $S = \sum\limits_{k=1}^{\infty} \frac{\sin \sqrt[4]{2}}{k^2} \leq \sum\limits_{k=1}^{\infty} \frac{1}{k^2}$, a convergent p-

series. Thus, S converges.

50. $S = \sum\limits_{k=1}^{\infty} \frac{x^k}{k!}$; use the ratio test.

$$\lim_{k \to \infty} \frac{x^{k+1} k!}{(k+1)! x^k} = \lim_{k \to \infty} \frac{x}{k+1} = 0 < 1$$

for all x, so S converges for all x. Since it is convergent, the necessary condition must be satisfied. Thus, $S = 0$.

51. Let $f(x) = \frac{1}{x}$; For the kth interval $[k-1, k]$, the area of the rectangle is less than the area under the curve, so

$$\sum\limits_{k=1}^{n} \frac{1}{k} \leq \int\limits_{1}^{n} \frac{dx}{x}$$

$$\sum\limits_{k=1}^{n} \frac{1}{k} \leq \ln n - \ln 1$$

$$\sum\limits_{k=1}^{n} \frac{1}{k} - \ln n \leq 0$$

$$\lim_{n \to \infty} \left| \sum\limits_{k=1}^{n} \frac{1}{k} - \ln n \right| \leq 0$$

This means that

$$\lim_{n \to \infty} (\tfrac{1}{1} + \tfrac{1}{2} + \tfrac{1}{3} + \cdots \tfrac{1}{n} - \ln x) \text{ converges.}$$

52. **a.** $S_{2m} = \sum\limits_{k=1}^{\infty} \frac{(-1)^{k+1}}{k}$

$$= 1 - \tfrac{1}{2} + \tfrac{1}{3} - \tfrac{1}{4} + \cdots + \frac{1}{k^{2m-1}} - \frac{1}{k^{2m}}$$

$$H_{2m} - H_m = 1 + (\tfrac{1}{2} - 1) + \tfrac{1}{3} + (\tfrac{1}{4} - \tfrac{1}{2})$$

$$+ \cdots + \frac{1}{2^{2m-1}} - \frac{1}{k^{2m}}$$

Thus, $S_{2m} = H_{2m} - H_m$.

b. $S_{2m} = H_{2m} - H_m$

$$= [H_{2m} - \ln(2m)] - [H_m - \ln m]$$

$$+ \ln(2m) - \ln m$$

$$S = \lim_{2m \to \infty} S_{2m}$$

$$= \lim_{n \to \infty} \{ [H_{2m} - \ln(2m)] - [H_m - \ln m] $$

$$+ \ln(2m) - m \} = \gamma - \gamma + \ln 2$$

53. **a.** Suppose n is a multiple of 3, say $n = 3k$. Then the partial sum S_{3m} has $2m$ positive terms and m negative terms; we can write

$$S_{3m} = \sum\limits_{k=1}^{2m} \frac{1}{2k-1} - \sum\limits_{k=1}^{m} \frac{1}{2k}$$

For example, with $m = 2$, we have $n = 3m = 6$ and

$$S_6 = 1 + \tfrac{1}{3} - \tfrac{1}{2} + \tfrac{1}{5} + \tfrac{1}{7} - \tfrac{1}{4}$$

$$= (1 + \tfrac{1}{3} + \tfrac{1}{5} + \tfrac{1}{7}) - (\tfrac{1}{2} + \tfrac{1}{4})$$

We find that

$$S_{3m} = \sum\limits_{k=1}^{2m} \frac{1}{2k-1} - \sum\limits_{k=1}^{m} \frac{1}{2k}$$

$$= \left(\sum\limits_{k=1}^{4m} \frac{1}{k} - \sum\limits_{k=1}^{2m} \frac{1}{2k} \right) - \frac{1}{2} \sum\limits_{k=1}^{m} \frac{1}{k}$$

$$= \sum\limits_{k=1}^{4m} \frac{1}{k} - \frac{1}{2} \sum\limits_{k=1}^{2m} \frac{1}{k} - \frac{1}{2} \sum\limits_{k=1}^{m} \frac{1}{k}$$

$$= H_{4m} - \tfrac{1}{2} H_{2m} - \tfrac{1}{2} H_m$$

b. $\lim\limits_{m \to \infty} S_{3m} = \lim\limits_{m \to \infty} (H_{4m} - \ln 4m)$

$$- \tfrac{1}{2} \lim_{m \to \infty} (H_{2m} - \ln 2m) + \ln 4m$$

$$- \tfrac{1}{2} \ln 2m - \tfrac{1}{2} \ln m$$

$$= \gamma - \tfrac{1}{2} \gamma - \tfrac{1}{2} \gamma + \ln 4m - \tfrac{1}{2} \ln 2m - \tfrac{1}{2} \ln m$$

$$= \ln 4 + \ln m - \tfrac{1}{2} \ln 2 - \tfrac{1}{2} \ln m - \tfrac{1}{2} \ln m$$

$$= 2 \ln 2 - \tfrac{1}{2} \ln 2 = \tfrac{3}{2} \ln 2$$

So the rearranged series converges with sum $\frac{3}{2}\ln 2$.

54. Conditionally convergent series should not be rearranged.

55. a. $S = \sum_{k=1}^{\infty} (-1)^{k+1} 2^{1/k}$

$\lim_{k \to \infty} 2^{1/k} = 1 \neq 0$; S diverges

b. $S = \sum_{k=1}^{\infty} (-1)^{k+1}(1 - 2^{1/k})$

This is an alternating series of terms $\{1 - 2^{1/k}\}$; this is a decreasing series and

$$\lim_{k \to \infty}(1 - 2^{1/k}) = 0$$

S is a convergent series. Checking for absolute convergence,

$$\sum_{k=1}^{\infty} 2^{1/k} - 1 > \sum_{k=1}^{\infty} \frac{\ln 2}{k}$$

Use $2^{1/k} - 1 = \frac{\ln 2}{\ln k} + \frac{(\ln 2)^2}{2k^2} + \frac{(\ln 2)^3}{6k^3} + \cdots$

However, it does converge by the alternating series test (conditionally):

$$\lim_{k \to \infty}(1 - 2^{1/k}) = 0$$

and $2^{1/k} - 1 > 2^{1/(n+1)} - 1$

56. $S = \lim_{n \to \infty} \sum_{k=1}^{n} a_k x^k < \lim_{n \to \infty} \sum_{k=1}^{n} A^k x^k$

By the ratio test,

$$\lim_{k \to \infty} \frac{(Ax)^{k+1}}{(Ax)^k} = Ax$$

Now $Ax \leq 1$ since $x \leq A^{-1}$. Thus, S converges absolutely.

57. $S = \sum_{k=1}^{\infty} k^{-2/3}$ is a convergent p-series while

$S = \sum_{k=1}^{\infty} (k^{-2/3})^2 = \sum_{k=1}^{\infty} k^{-4/3}$ diverges.

58. Let $a_k = k^{-4}$; $S = \sum_{k=1}^{\infty} k^{-4}$ is a convergent p-series and so is $S = \sum_{k=1}^{\infty} k^{-2}$.

59. $S = \sum_{k=1}^{\infty} a_k$

By the ratio test, $\lim_{k \to \infty} \frac{a_{k_1}}{a_k} = L < 1$

since the series converges, by hypotheses. Then,

$T = \sum_{k=1}^{\infty} a_k^2$ converges

because $\lim_{k \to \infty} \frac{a_{k_1}^2}{a_k^2} = L^2 < 1$.

60. $\lim_{k \to \infty} \sqrt[k]{a_k} = L$ means that

$\left| \sqrt[k]{a_k} - L \right| < \epsilon$ for $N < k$

Thus, $|a_N| < L^N$,

$S = |a_{N+1}| < L^{N+1}$;

$S = |a_{N+2}| < L^{N+2}; \cdots$

Then, $S = \Sigma a_k \leq \Sigma |a_k|$

Since

$\Sigma |a_k| = |a_1| + |a_2| + \cdots + |a_N| + |a_{N+1}|$

$S = \Sigma a_k \leq L + L^2 + \cdots + L^N + L^{N+1}$

This is a convergent geometric series if $L < 1$. If $L > 1$, then

$a_{N+1} < La_N,\ a_{N+2} < La_{N+1} < L^2 a_N$

The necessary condition for convergence is not satisfied.

If $L = 1$, $\sum \frac{1}{k}$ diverges and $\sum \frac{1}{k^2}$ converges; thus the test fails.

61. Σk^{-1} diverges and Σk^{-2} converges

62. $S_3 = a_1 - a_2 + a_3 = a_1 - (a_2 - a_3) \leq a_1$

$S_5 = a_1 - a_2 + a_3 - a_4 + a_5$

$\quad = a_1 - (a_2 - a_3) - (a_4 - a_5)$

$\quad \leq a_1$

$S_1, S_2, S_5, \cdots$ is a decreasing sequence because

$a_{k-1} - a_k \geq 0$.

8.7 Power series, Page 547

1. $\lim_{k \to \infty} \left| \frac{x^{k+1}(k+1)^2}{(k+2)x^k k} \right| = |x| < 1$ so $R = 1$

At $x = 1$, $S = \sum_{k=1}^{\infty} \frac{k}{k+1}$ diverges because it does not satisfy the necessary condition.

At $x = -1$, $S = \sum_{k=1}^{\infty} \frac{(-1)^k k}{k+1}$ diverges, so the interval of convergence is $(-1, 1)$.

2. $\lim_{k \to \infty} \left| \frac{x^{n+1}(k+1)^3}{(k+2)x^k k^2} \right| = |x| < 1$ so $R = 1$

At $x = 1$, $S = \sum_{k=1}^{\infty} \frac{k^2}{k+1}$ diverges because it does not satisfy the necessary condition, namely $\lim_{k \to \infty} \frac{k^2}{k+1} \neq 0$.

At $x = -1$, $S = \sum_{k=1}^{\infty} \frac{(-1)^k k^2}{k+1}$ diverges, so the interval of convergence is $(-1, 1)$.

3. $\lim_{k \to \infty} \left| \frac{x^{k+1}(k+1)(k+2)^2}{(k+3)(k+1)(k)x^k} \right| = |x| < 1$ so $R = 1$

At $x = 1$, $\sum_{k=1}^{\infty} \frac{k(k+1)}{k+2}$ diverges because it

does not satisfy the necessary condition,

namely $\lim_{k \to \infty} \frac{k(k+1)}{k+2} \neq 0$

At $x = -1$, $S = \sum_{k=1}^{\infty} \frac{(-1)^k k(k+1)}{k+2}$ diverges,

so the interval of convergence is $(-1, 1)$.

4. $\lim_{k \to \infty} \left| \frac{\sqrt{k}x^{k+1}}{\sqrt{k-1}x^k} \right| = |x| < 1$, so $R = 1$

At $x = 1$, $S = \sum_{k=1}^{\infty} \sqrt{k-1}$ diverges, since

$\lim_{k \to \infty} \sqrt{k-1} \neq 0$.

At $x = -1$, $S = \sum_{k=1}^{\infty} (-1)^k \sqrt{k-1}$ diverges

so the interval of convergence is $(-1, 1)$.

5. $\lim_{k \to \infty} \left| \frac{(k+1)^2 3^{k+1}(x-3)^{k+1}}{k^2 3^k (x-3)^k} \right| = 3|x-3| < 1$,

so $R = \frac{1}{3}$. The interval of absolute

convergence is $(\frac{8}{3}, \frac{10}{3})$.

At $x = \frac{8}{3}$, $S = \sum_{k=1}^{\infty} (-1)^k k^2 3^k$ diverges;

at $x = \frac{10}{3}$, $S = \sum_{k=1}^{\infty} k^2 3^k$ diverges. The
interval of convergence is $(\frac{8}{3}, \frac{10}{3})$.

6. $\lim_{k \to \infty} \left| \frac{(k+1)^2 3^k (x-2)^{k+1}}{k^2 3^{k+1}(x-2)^k} \right| = \frac{1}{3}|x-2| < 1$,

so $R = 3$; The interval of absolute
convergence is $(-1, 5)$.

At $x = -1$, $S = \sum_{k=1}^{\infty} (-1)^k k^2$ diverges since

$\lim_{k \to \infty} k^2 \neq 0$.

At $x = 5$, $S = \sum_{k=1}^{\infty} k^2$ diverges. The interval
of convergence is $(-1, 5)$.

7. $\lim_{k \to \infty} \left| \frac{3^{k+1}(x+3)^{k+1}4^k}{4^{k+1}3^k(x+3)^k} \right| = \frac{3}{4}|x+3| < 1$, so

$R = \frac{4}{3}$. Interval of absolute convergence is

$(-\frac{13}{3}, \frac{5}{3})$.

At $x = -\frac{13}{5}$, $S = \sum_{k=1}^{\infty} \frac{3^k(\frac{4}{3})^k(-1)^k}{4^k}$

$= \sum_{k=1}^{\infty} (-1)^k$ which diverges.

At $x = \frac{5}{3}$, $S = \sum_{k=1}^{\infty} 1$ also diverges.

The interval of convergence is $(-\frac{13}{3}, -\frac{5}{3})$.

8. $\lim_{k \to \infty} \left| \frac{4^{k+1}(x+1)^{k+1}3^k}{3^{k+1}4^k(x+1)^k} \right| = \frac{4}{3}|x+1| < 1$, so

$R = \frac{3}{4}$. Interval of absolute convergence is

$(-\frac{7}{4}, -\frac{1}{4})$.

At $x = -\frac{7}{4}$, $S = \sum_{k=1}^{\infty} (-1)^k$ diverges;

at $x = -\frac{1}{4}$, $S = \sum_{k=1}^{\infty} 1$ also diverges.

The interval of convergence is $(-\frac{7}{4}, -\frac{1}{4})$.

9. $\lim_{k \to \infty} \left| \frac{(k+1)!(x-1)^{k+1}5^k}{5^{k+1}k!(x-1)^k} \right| = \infty$, so $R = 0$.

This series converges only at $x = 1$ when all
the terms vanish.

10. $\lim_{k \to \infty} \left| \frac{(x-15)^{k+1}\ln(k+1)}{\ln(k+2)(x-15)^k} \right| = |x-15| < 1$

so $R = 1$.

If $x = 14$, $S = \sum_{k=1}^{\infty} \frac{(-1)^k}{\ln(k+1)}$ converges

because $\lim_{k \to \infty} \frac{1}{\ln(k+1)} = 0$ and

$\frac{1}{\ln(k+2)} \leq \frac{1}{\ln(k+1)}$

If $x = 16$, $S = \sum_{k=1}^{\infty} \frac{1}{\ln(k+1)}$ which diverges

by the zero-infinity limit comparison with

$\sum_{k=1}^{\infty} \frac{1}{k}$. The interval of convergence is $[14, 16)$.

11. $\lim_{k \to \infty} \left| \frac{(k+1)^2(x-1)^{k+1}2^k}{2^{k+1}k^2(x-1)^k} \right| = \frac{1}{2}|x-1|$, so

$R = 2$. The interval of absolute converges is
$(-1, 3)$.

If $x = -1$, $S = \sum_{k=1}^{\infty} (-1)^k k^2$, which

diverges. At $x = 3$, $S = \sum_{1}^{\infty} k^2$, which

diverges. The interval of convergence is
$(-1, 3)$.

12. $\lim\limits_{k \to \infty} \left| \dfrac{2^{k+1}(x-3)^{k+1}(k)(k+1)}{(k+1)(k+2)2^k(x-3)^k} \right|$
$= 2|x-3| < 1$, so $R = \frac{1}{2}$. The interval of
absolute convergence is $(\frac{5}{2}, \frac{7}{2})$.

At $x = \frac{5}{2}$, $S = \sum\limits_{k=1}^{\infty} \dfrac{(-1)^k}{k(k+1)}$ converges by

direct comparison with $\sum\limits_{k=1}^{\infty} k^{-2}$.

At $x = \frac{7}{2}$, $S = \sum\limits_{k=1}^{\infty} \dfrac{1}{k(k+1)}$ also converges.

The interval of convergence $[\frac{5}{2}, \frac{7}{2}]$.

13. $\lim\limits_{k \to \infty} \left| \dfrac{(k+1)^3(3x-4)^{k+1}}{(k+2)^2(k)(3x-4)^k} \right| = |3x-4| < 1$,
so $R = 1$. The interval of absolute

convergence is $(1, \frac{5}{3})$.

At $x = 1$, $S = \sum\limits_{k=1}^{\infty} \dfrac{(-1)^k k}{(k+1)^2}$ converges by

comparison with the negative harmonic series.

At $x = \frac{5}{3}$, $S = \sum\limits_{k=1}^{\infty} \dfrac{k}{(k+1)^2}$ diverges by direct

comparison with the harmonic series. The
interval of convergence is $[1, \frac{5}{3})$.

14. $\lim\limits_{k \to \infty} \left| \dfrac{(2x+3)^{k+1}4^k}{4^{k+1}(2x+3)^k} \right| = \frac{1}{4}|2x+3| < 1$, so

$R = 2$. The interval of absolute convergence

is $\left(-\frac{7}{2}, \frac{1}{2}\right)$.

At $x = -\frac{7}{2}$, $S = \sum\limits_{k=1}^{\infty} (-1)^k$ diverges;

at $x = \frac{1}{2}$, $S = \sum\limits_{k=1}^{\infty} 1$ diverges. The interval of

convergence is $\left(-\frac{7}{2}, \frac{1}{2}\right)$.

15. $\lim\limits_{k \to \infty} \left| \dfrac{(k+1)x^{k+1}7^k}{7^{k+1}kx^k} \right| = \dfrac{|x|}{7}$, so $R = 7$.

Interval of absolute convergence is $(-7, 7)$.

At $x = -7$, $S = \sum\limits_{k=1}^{\infty} k(-1)^k$, which diverges;

at $x = 7$, $S = \sum\limits_{k=1}^{\infty} k$, which diverges. The

interval of convergence is $(-7, 7)$.

16. $\lim\limits_{k \to \infty} \left| \dfrac{(2k+2)! x^{k+1}(3k)!}{(3k+3)!(2k)! x^k} \right|$
$= \lim\limits_{k \to \infty} |x| \left[\dfrac{(2k+2)(2k+1)}{(3k+3)(3k+2)(3k+1)} \right] = 0$,
so the radius of convergence is $R = \infty$ and
the interval of convergence is $(-\infty, \infty)$.

17. $\lim\limits_{k \to \infty} \left| \dfrac{[(k+1)!]^2 x^{k+1} k^k}{(k+1)^{k+1}(k!)^2 x^k} \right| = \lim\limits_{k \to \infty} |x| \left[\dfrac{k+1}{(1+\frac{1}{k})^k} \right]$
$= \infty$, so $R = 0$ and the interval of
convergence is the point $x = 0$

18. $\lim\limits_{k \to \infty} \left| \dfrac{(k+1)x^{k+1}\ln(k+2)}{\ln(k+3)(k)x^k} \right|$
$= \lim\limits_{k \to \infty} |x| \left[\dfrac{(k+1)(k+3)}{(k+2)(k)} \right] = |x| < 1$, so

$R = 1$. The interval of absolute convergence
is $(-1, 1)$.

At $x = -1$, $S = \sum\limits_{k=1}^{\infty} \dfrac{k}{\ln(k+2)}$ diverges by

the zero-infinity comparison with $\sum\limits_{k=1}^{\infty} \dfrac{1}{k+1}$.

At $x = 1$, $S = \sum\limits_{k=1}^{\infty} \dfrac{(-1)^k k}{\ln(k+2)}$ also diverges.

The interval of convergence is $(-1, 1)$.

19. $\lim\limits_{k \to \infty} \left| \dfrac{x^{k+1} k (\ln k)^2}{(k+1)[\ln(k+1)]^2 x^k} \right| = |x|$, so

$R = 1$. Interval of absolute convergence:
$(-1, 1)$.
At $x = -1$, $S = \sum\limits_{k=1}^{\infty} \dfrac{1}{k(\ln k)^2}$ by using the

integral test: $\lim\limits_{b \to \infty} \displaystyle\int_1^b (\ln x)^{-2} \frac{1}{x} \, dx$

$= \lim\limits_{b \to \infty} \left[-(\ln x)^{-1} \right]\Big|_1^n = \infty$, so it diverges.

At $x = 1 = \sum\limits_{1}^{\infty} \dfrac{(-1)^k}{k(\ln k)^2}$ which is alternating,

decreasing, and has $\lim\limits_{k \to \infty} a_k = 0$, so it

converges. The interval of convergence is
$(-1, 1]$.

20. $\lim\limits_{k \to \infty} \left| \dfrac{(3x)^{k+1}2^{k+1}}{2^{k+2}(3x)^k} \right| = \frac{3}{2}|x| < 1$, so $R = \frac{2}{3}$.

The interval of absolute convergence is $\left(-\frac{2}{3}, \frac{2}{3}\right)$.

At $x = -\frac{2}{3}$, $S = \sum\limits_{k=1}^{\infty} \frac{(-1)^k}{2}$ diverges;

at $x = \frac{2}{3}$, $S = \sum\limits_{k=1}^{\infty} \frac{1}{2}$ also diverges.

The interval of convergence is $\left(-\frac{2}{3}, \frac{2}{3}\right)$.

21. $\lim\limits_{k \to \infty} \left| \frac{(2x)^{2k+2}k!}{(k+1)!(2x)^{2k}} \right| = 0 < 1$ for all x, so

$R = \infty$. The interval of convergence is $(-\infty, \infty)$.

22. $\lim\limits_{k \to \infty} \left| \frac{(x+2)^{2k+2}3^k}{2^{k+1}(x+2)^{2k}} \right| = \frac{1}{3}(x+2)^2$

$\frac{1}{3}(x+2)^2 < 1$

$(x+2)^2 < 3$

$-\sqrt{3} < x + 2 < \sqrt{3}$

$-2 - \sqrt{3} < x < -2 + \sqrt{3}$

The radius of convergence is $\sqrt{3}$, and the interval of absolute convergence is $(-2 - \sqrt{3}, -2 + \sqrt{3})$.

At $x = -2 - \sqrt{3}$, $S = \sum\limits_{k=1}^{\infty} (-1)$ diverges;

at $x = -2 + \sqrt{3}$, $S = \sum\limits_{k=1}^{\infty} 1$ also diverges.

The interval of convergence is $(-2 - \sqrt{3}, -2 + \sqrt{3})$.

23. $\lim\limits_{k \to \infty} \left| \frac{(k+1)!(3x)^{3k+3}2^k}{2^{k+1}k!(3x)^{3k}} \right|$

$= \lim\limits_{k \to \infty} \left| \frac{(k+1)(3x)^3}{2} \right| = \infty$, so $R = 0$.

The series converges only in the trivial case when $x = 0$.

24. $\lim\limits_{k \to \infty} \left| \frac{(3x)^{3k+3}x^k}{x^{k+1}(3x)^{3k}} \right| = 3^3 |x|^2$

$3^3 |x|^2 \le 1$

$x^2 \le \frac{1}{27}$

$-\frac{1}{9}\sqrt{3} < x < \frac{1}{9}\sqrt{3}$

The radius of convergence is $R = \frac{\sqrt{3}}{9}$.

At $x = -\frac{1}{9}\sqrt{3}$, $S = \sum\limits_{k=1}^{\infty} (-1)^k$ diverges;

at $x = \frac{1}{9}\sqrt{3}$, $S = \sum\limits_{k=1}^{\infty} 1$ also diverges.

The interval of convergence is $\left(-\frac{1}{9}\sqrt{3}, \frac{1}{9}\sqrt{3}\right)$.

25. $\lim\limits_{k \to \infty} \left| \frac{2^{k+1}(2x-1)^{2k+2}k!}{(k+1)!2^k(2x-1)^{2k}} \right|$

$= 2(2x-1)^2 \lim\limits_{k \to \infty} \frac{1}{k+1} = 0$, so $R = \infty$ and

the interval of convergence is $(-\infty, \infty)$.

26. $\lim\limits_{k \to \infty} \left| \frac{(3x)^{3k+2}2^{k+1}}{2^k(3x)^{3k}} \right| = 54|x|^3$, so

$R = (3\sqrt[3]{2})^{-1}$.

At $x = -(3\sqrt[3]{2})^{-1}$, $S = \sum\limits_{k=1}^{\infty} (-1)^k$ diverges;

At $x = (3\sqrt[3]{2})^{-1}$, $S = \sum\limits_{k=1}^{\infty} 1$ diverges.

The interval of convergence is $\left(\frac{-1}{3\sqrt[3]{2}}, \frac{1}{3\sqrt[3]{2}}\right)$.

27. $\lim\limits_{k \to \infty} \left| \frac{3^{k+1}(5x)^{4k+4}}{3^k(5x)^{4k}} \right| = |3(5x)^4| < 1$,

$|(5x)^4| < \frac{1}{3}$

$|x| < \sqrt[4]{\frac{1}{3(5^4)}} = \frac{\sqrt[4]{27}}{15}$

$R = \frac{1}{15}\sqrt[4]{27}$. It is divergent at both

endpoints, so the interval of convergence

is $\left(-\frac{1}{15}\sqrt[4]{27}, \frac{1}{15}\sqrt[4]{27}\right)$.

28. $\lim\limits_{k \to \infty} \left| \frac{(k+1)!(x+1)^{3k+3}}{k!(x+1)^{3k}} \right|$

$= |x+1|^3 \lim\limits_{k \to \infty} (k+1) = \infty$ so $R = 0$ and

the interval of convergence is reduced to the point $x = -1$.

29. $\lim\limits_{k \to \infty} \left| \frac{(k+1)^2(x+1)^{2k+3}}{k^2(x+1)^{2k+1}} \right| = (x+1)^2 < 1$, so

the interval of absolute convergence is $(-2, 0)$.

At $x = -2$, $S = \sum\limits_{k=1}^{\infty} (-1)^{2k+1}k^2$ diverges;

at $x = 0$, $S = \sum\limits_{k=1}^{\infty} k^2$ diverges so the interval

of convergence is $(-2, 0)$.

30. $\lim\limits_{k \to \infty} \sqrt[k]{k^{-\sqrt{k}}x^k} = |x| \lim\limits_{k \to \infty} k^{-1/\sqrt{k}}$

$= |x| \left[\lim\limits_{k \to \infty} \exp(\ln k^{-1/\sqrt{k}}) \right]^{-1}$

$= |x| \left[\lim\limits_{k \to \infty} \exp\left(\frac{\ln k}{\sqrt{k}}\right) \right]^{-1}$

$= |x| \left[\lim\limits_{k \to \infty} \exp\left(\frac{1/k}{1/(2k)}\right) \right]^{-1} = |x| < 1$

At $x = -1$, $S = \sum\limits_{k=1}^{\infty} (-1)^k k^{-1/2}$ converges

since $\lim\limits_{k \to \infty} k^{-1/2} = 0$ and

$(k + 1)^{-1/2} \leq k^{-1/2}$

At $x = 1$, $S = \sum_{k=1}^{\infty} k^{-1/2}$ diverges because it is a p-series with $p = 0.5$. The interval of convergence is $[-1, 1)$.

31. $\lim_{k \to \infty} \left| 2^{(\sqrt{k+1} - \sqrt{k})}(x - 1) \right| = |x - 1| < 1$

At $x = 0$, $S = \sum_{k=1}^{\infty} (-1)^k 2^{\sqrt{k}}$ diverges;

at $x = 2$, $S = \sum_{k=1}^{\infty} 2^{\sqrt{k}}$ also diverges.
The interval of convergence is $(0, 2)$.

32. $\lim_{k \to \infty} \sqrt[k]{k^{\sqrt{k}}(x-1)^k} = |x| \lim_{k \to \infty} k^{1/\sqrt{k}}$

$= |x| \lim_{k \to \infty} \exp\left(\ln k^{1/\sqrt{k}}\right) = |x| < 1$

At $x = -1$, $S = \sum_{k=1}^{\infty} (-1)^k k^{\sqrt{k}}$ diverges;

at $x = 1$, $S = \sum_{k=1}^{\infty} k^{\sqrt{k}}$ diverges. The interval of convergence is $(-1, 1)$.

33. $\lim_{k \to \infty} \left| \frac{(k+1)(ax)^{k+1}}{k(ax)^k} \right| = a|x| < 1$

At $x = -a^{-1}$, $S = \sum_{k=1}^{\infty} (-1)^k k$ diverges
(does not satisfy the necessary condition);

at $x = a^{-1}$, $S = \sum_{k=1}^{\infty} k$ also diverges (p-series with $p = -1$). The interval of convergence is $\left(-\frac{1}{a}, \frac{1}{a}\right)$.

34. $\lim_{k \to \infty} \left| \frac{(a^2 x)^{k+1}}{(a^2 x)^k} \right| = a^2 |x|$

At $x = -a^{-2}$ and $x = a^{-2}$ the corresponding series both diverge. The interval of convergence is $\left(-\frac{1}{a^2}, \frac{1}{a^2}\right)$.

35. $f(x) = 1 + \frac{x}{2} + \frac{x^2}{4} + \frac{x^3}{8} + \frac{x^4}{16} + \cdots$

$f'(x) = (1)\frac{1}{2} + (2)\frac{x}{4} + (3)\frac{x^2}{8} + (4)\frac{x^3}{16} + \cdots$

$= \sum_{k=0}^{\infty} \frac{k x^{k-1}}{2^k}$

36. $f(x) = x + \frac{x^2}{2} + \frac{x^3}{3} + \frac{x^4}{4} + \cdots$

$f'(x) = 1 + x + x^2 + x^3 + \cdots = \sum_{k=1}^{\infty} x^{k-1}$

37. $f(x) = 2 + 3x + 4x^2 + 5x^3 + \cdots$

$f'(x) = 3 + 8x + 15x^2 + \cdots = \sum_{k=1}^{\infty} k(k+2)x^{k-1}$

38. $f(x) = 1 + x + 2x^2 + 3x^3 + \cdots$

$f'(x) = 1 + 4x + 9x^2 + \cdots = \sum_{k=1}^{\infty} k^2 x^{k-1}$

39. $f(x) = 1 + \frac{x}{2} + \frac{x^2}{4} + \frac{x^3}{8} + \frac{x^4}{16} + \cdots$

$F(x) = \int_0^x f(x) \, dx = x + \frac{x^2}{2(2)} + \frac{x^3}{3(2)^2} + \frac{x^4}{4(2)^3} + \cdots = \sum_{k=1}^{\infty} \frac{x^k}{k(2)^{k-1}}$

Alternatively, we can write $f(x) = \sum_{k=0}^{\infty} \left(\frac{x}{2}\right)^k$

$F(x) = \sum_{k=0}^{\infty} \int_0^x \frac{x^k}{2^k} \, dx = \sum_{k=0}^{\infty} \frac{x^{k+1}}{(k+1)2^k}$

40. $f(x) = \sum_{k=1}^{\infty} \frac{x^k}{k}$;

$F(x) = \sum_{k=1}^{\infty} \int_0^x \frac{x^k}{k} \, dx = \sum_{k=1}^{\infty} \frac{x^{k+1}}{(k+1)2^k}$

41. $f(x) = \sum_{k=0}^{\infty} (k+2)x^k$;

$F(x) = \sum_{k=0}^{\infty} \int_0^x (k+2)x^k \, dx = \sum_{k=0}^{\infty} \frac{(k+2)x^{k+1}}{(k+1)}$

42. $f(x) = \sum_{k=0}^{\infty} k x^k$;

$F(x) = \sum_{k=0}^{\infty} \int_0^x k x^k \, dx = \sum_{k=1}^{\infty} \frac{x^{k+1}}{(k+1)(k+2)}$

43. $S = \sum_{k=1}^{\infty} \frac{\sin(k! x)}{k^2} \leq \sum_{k=1}^{\infty} \frac{1}{n^2}$

S converges (the lower bound is -1) when compared to the convergent p-series with $p = 2$.

$T = S' = \sum_{k=1}^{\infty} \frac{k! \cos(k! x)}{k^2}$

T diverges by the ratio test:

$\lim_{k \to \infty} \left| \frac{(k+1)! \cos[(k+1)! x]}{(k+1)^2 k! \cos(k! x)} \right|$

$= \lim_{k \to \infty} \left| (k+1) \frac{\cos[(k+1)! x]}{\cos(k! x)} \right| = \infty$

Thus, T diverges for all x. Note that Theorem 8.25 (Term-by-term differentiation and integration of a power series) applies to power series, but S is not a power series.

44. Let $S = \sum_{k=1}^{\infty} k a_k x^{k-1}$ and $T = \sum_{k=1}^{\infty} \int_1^x k a_k x^{k-1} dx$

By the root test, $\lim\limits_{k\to\infty} \sqrt[k]{|a_k|x^k|} = \sqrt[k]{|a_k|}|x|$

$= \dfrac{|x|}{R} < 1$ for convergence. This means

$-R < x < R$ is the radius of convergence.

45. Let $S = \sum\limits_{k=1}^{\infty} ka_k x^{kp}$; by the ratio test

$\lim\limits_{k\to\infty} \left| \dfrac{a_{k+1} x^{kp+p}}{a_k x^{kp}} \right| = |x|^p \lim\limits_{k\to\infty} \left| \dfrac{a_{k+1}}{a_k} \right|$

$= |x|^p \lim\limits_{k\to\infty} \dfrac{1}{R} < 1$ because of convergence. Thus, $|x|^p < R$ or $|x| < R^{1/p}$ is the radius of convergence.

46. $\sum\limits_{k=1}^{\infty} \dfrac{x}{k+x}$ behaves like $\sum\limits_{k=1}^{\infty} \dfrac{1}{k}$ for $x > 0$.

Thus, the series diverges (limit comparison test) for all $x > 0$. If $x = 0$, the series obviously converges. If $x < 0$, then eventually $k + x > 0$ and $x/(k + x) < 0$ (for all $k > N$, for some N). The series

$\sum\limits_{k=1}^{\infty} \dfrac{(-x)}{k+x}$ will behave like $\sum\limits_{k=1}^{\infty} \dfrac{1}{k}$, so it will

also diverge. Thus, $\sum\limits_{k=1}^{\infty} \dfrac{x}{k+x}$ converges only for $x = 0$.

47. Let $S = \sum\limits_{k=1}^{\infty} \dfrac{k}{x^k}$; let $t = x^{-1}$ and $S = \sum\limits_{k=1}^{\infty} kt^k$ which converges when $|t| < 1$ (by the ratio test). This means S converges when $|x| > 1$.

48. Let $S = \sum\limits_{k=1}^{\infty} \dfrac{1}{kx^k}$; let $t = x^{-1}$ and

$S = \sum\limits_{k=1}^{\infty} k^{-1} t^k$ which converges when

$|t| < 1$ (by the ratio test). This means S converges when $|x| > 1$.

49. Let $S = \sum\limits_{k=1}^{\infty} \dfrac{(k+3)x^k}{k!(k+4)!}$; use the ratio test.

$\lim\limits_{k\to\infty} |x| \dfrac{[(k+4)!]^2 k!}{(k+1)!(k+3)!(k+5)!}$

$= |x| \lim\limits_{k\to\infty} \dfrac{k+4}{(k+1)(k+5)} = 0 < 1$

for all x. The radius of absolute convergence is $R = \infty$.

50. Let $S = \sum\limits_{k=1}^{\infty} \dfrac{1 \cdot 2 \cdot 3 \dots k(-x)^{2k-1}}{1 \cdot 3 \cdot 5 \dots (2k-1)}$

$= |x^2| \lim\limits_{k\to\infty} \dfrac{k+1}{2k+1} = \tfrac{1}{2}x^2$

The radius of convergence is found by solving $\tfrac{1}{2}x^2 < 1$; thus $-\sqrt{2} < x < \sqrt{2}$. The radius of convergence is $R = \sqrt{2}$.

51. $f(x) = \sum\limits_{k=0}^{\infty} \dfrac{(-1)^k x^{2k+1}}{(2k+1)!}$;

$f'(x) = \sum\limits_{k=0}^{\infty} \dfrac{(-1)^k (2k+1) x^{2k}}{(2k+1)!}$;

$f''(x) = \sum\limits_{k=0}^{\infty} \dfrac{(-1)^k (2k+1)(2k) x^{2(k-1)}}{(2k+1)!}$

$= \sum\limits_{k=0}^{\infty} \dfrac{(-1)^k x^{2(k-1)}}{(2k-1)!}$

$= -\sum\limits_{k=1}^{\infty} \dfrac{(-1)^k x^{2k}}{(2k+1)!} = -f(x)$

52. $f(x) = \sum\limits_{k=0}^{\infty} \dfrac{x^{2k}}{(2k)!}$; $f'(x) = \sum\limits_{k=0}^{\infty} \dfrac{(2k)x^{2k-1}}{(2k)!}$;

$f''(x) = \sum\limits_{k=0}^{\infty} \dfrac{(2k)(2k-1)x^{2(k-1)}}{(2k)!}$

$= \sum\limits_{k=0}^{\infty} \dfrac{x^{2(k-1)}}{2(k-1)!} = \sum\limits_{k=0}^{\infty} \dfrac{x^{2k}}{(2k)!} = f(x)$

8.8 Taylor and Maclaurin Series, Page 562

1. Suppose there is an open interval I containing c throughout which the function f and all its derivatives exist. Then the power series

$$f(c) + \frac{f'(c)}{1!}(x - c) + \frac{f''(c)}{2!}(x - c)^2$$
$$+ \frac{f'''(c)}{3!}(x - c)^3 + \dots$$

is called the Taylor series of f at c. The special case where $c = 0$ is called the Maclaurin series of f:

$$f(0) + \frac{f'(0)}{1!}x + \frac{f''(0)}{2!}x^2 + \frac{f'''(0)}{3!}x^3 + \dots$$

2. The binomial function $(1 + x)^p$ is represented by its Maclaurin series

$$(1 + x)^p = 1 + px + \frac{p(p-1)}{2!}x^2$$
$$+ \frac{p(p-1)(p-2)}{3!}x^3 + \dots$$
$$+ \frac{p(p-1)\cdots(p-k+1)}{k!}x^k + \dots$$

$-1 < x < 1$ if $p \le -1$;
$-1 < x \le 1$ if $-1 < p < 0$;
$-1 \le x \le 1$ if $p > 0$, p not an integer;
all x if p is a nonnegative integer.
for all x if p is a nonnegative integer.

3. $e^{2x} = 1 + 2x + \frac{1}{2!}(2x)^2 + \dots + \frac{1}{k!}(2x)^k + \dots$

$$= \sum_{k=0}^{\infty} \frac{(2x)^k}{k!}$$

4. $e^{-x} = 1 + (-x) + \frac{1}{2!}(-x)^2 + \cdots + \frac{1}{k!}(-x)^k + \cdots$

$$= \sum_{k=0}^{\infty} \frac{(-1)^k x^k}{k!}$$

5. $e^{x^2} = 1 + x^2 + \frac{1}{2!}(x^2)^2 + \cdots + \frac{1}{k!}(x^2)^k + \cdots$

$$= \sum_{k=0}^{\infty} \frac{x^{2k}}{k!}$$

6. $e^{at} = 1 + ax + \frac{1}{2!}(ax)^2 + \cdots + \frac{1}{k!}(ax)^k + \cdots$

$$= \sum_{k=0}^{\infty} \frac{(ax)^k}{k!}$$

7. $\sin x^2 = x^2 - \frac{1}{3!}(x^2)^3 + \frac{1}{5!}(x^2)^5 - \cdots$

$$= \sum_{k=0}^{\infty} \frac{(-1)^k x^{4k+2}}{(2k+1)!}$$

8. $\sin^2 x = \frac{1}{2}(1 - \cos 2x)$

$$= \frac{1}{2}\left[1 - 1 - \frac{1}{2!}(2x)^2 - \frac{1}{4!}(2x)^4 + \cdots \right]$$

$$= \frac{1}{2} \sum_{k=0}^{\infty} \frac{(-1)^{k+1}(2x)^{2k}}{(2k)!}$$

9. $\sin ax = ax - \frac{1}{3!}(ax)^3 + \frac{1}{5!}(ax)^5 - \cdots$

$$= \sum_{k=0}^{\infty} \frac{(-1)^k(ax)^{2k+1}}{(2k+1)!}$$

10. $\cos ax = 1 - \frac{1}{2!}(ax)^2 + \frac{1}{4!}(ax)^4 + \cdots$

$$= \sum_{k=0}^{\infty} \frac{(-1)^k(ax)^{2k}}{(2k)!}$$

11. $\cos 2x^2 = 1 - \frac{1}{2!}(2x^2)^2 + \frac{1}{4!}(2x^2)^4 + \cdots$

$$= \sum_{k=0}^{\infty} \frac{(-1)^k(2x^2)^{2k}}{(2k)!}$$

12. $\cos x^3 = 1 - \frac{1}{2!}(x^3)^2 + \frac{1}{4!}(x^3)^4 + \cdots$

$$= \sum_{k=0}^{\infty} \frac{(-1)^k(x^3)^{2k}}{(2k)!}$$

13. $x^2\cos x = x^2 - \frac{1}{2!}(x)^4 + \frac{1}{4!}(x)^6 + \cdots$

$$= \sum_{k=0}^{\infty} \frac{(-1)^k(x)^{2k+2}}{(2k)!}$$

14. $\sin \frac{x}{4} = \frac{1}{2}x - \frac{1}{2^3(3!)}(x)^3 + \frac{1}{2^5(5!)}(x)^5 + \cdots$

$$= \sum_{k=0}^{\infty} \frac{(-1)^k(x)^{2k+1}}{(2k)!}$$

15. $x^2 + 2x + 1$ is its own Maclaurin series.

16. $x^3 - 2x^2 + x - 5$ is its own Maclaurin series.

17. $xe^x = x + 2x^2 + \frac{1}{2!}(x)^3 + \cdots = \sum_{k=0}^{\infty} \frac{(x)^{k+1}}{k!}$

18. $e^{-x} + e^{2x} = [1+1] + [-x+2x]$

$$+ \frac{1}{2!}[x^2 + (2x)^2] + \frac{1}{3!}[-x^3 + (2x)^3] + \cdots$$

$$= \sum_{k=0}^{\infty} \frac{[(-1)^k + 2^k]x^k}{k!}$$

19. $e^x + \sin x = \sum_{k=0}^{\infty} \frac{x^k}{k!} + \sum_{k=0}^{\infty} \frac{(-1)^k x^{2k+1}}{(2k+1)!}$

$$= \sum_{k=0}^{\infty} \left[\frac{x^k}{k!} + \frac{(-1)^k x^{2k+1}}{(2k+1)!} \right]$$

20. $\sin x + \cos x$

$$= \sum_{k=0}^{\infty} \frac{(-1)^k x^{2k+1}}{(2k+1)!} + \sum_{k=0}^{\infty} \frac{(-1)^k(x)^{2k}}{(2k)!}$$

$$= \sum_{k=0}^{\infty} \left[\frac{(-1)^k x^{2k+1}}{(2k+1)!} + \frac{(-1)^k(x)^{2k}(2k+1)}{(2k+1)(2k)!} \right]$$

$$= \sum_{k=0}^{\infty} \frac{(-1)^k x^{2k}[(2k+1) + x]}{(2k+1)!}$$

21. $\frac{1}{1+4x} = \frac{1}{1-(-4x)}$

$$= 1 + (-4x) + (-4x)^2 + \cdots = \sum_{k=0}^{\infty} (-4x)^k$$

22. $\frac{1}{1-ax} = 1 + (ax) + (ax)^2 + (ax)^4 + \cdots$

$$= \sum_{k=0}^{\infty} (ax)^k$$

23. $\frac{1}{a+x} = \frac{1}{a\left(1 + \frac{x}{a}\right)} = \frac{1}{a\left(1 - \left[-\frac{x}{a}\right]\right)}$

$$= \frac{1}{a} \sum_{k=0}^{\infty} \left(-\frac{x}{a}\right)^k$$

24. $\frac{1}{a^2+x^2} = \frac{1}{a^2\left(1 + \frac{x^2}{a^2}\right)} = \frac{a^{-2}}{1 - (-a^{-2}x^2)}$

$$= a^{-2}[1 + (-a^{-2}x^2) + (-a^{-2}x^2)^2 + \cdots]$$

$$= \sum_{k=0}^{\infty} \frac{(-1)^k x^{2k}}{a^{2(k+1)}} = \frac{1}{a^2} \sum_{k=0}^{\infty} (-1)^k \left(\frac{x}{a}\right)^{2k}$$

25. $\ln(3+x) = \ln 3\left(1 + \frac{x}{3}\right) = \ln 3 + \ln\left(1 + \frac{x}{3}\right)$

$$= \ln 3 + \frac{1}{3}x - \frac{1}{2(3^2)}x^2 + \frac{1}{3(3^3)}x^3 - \frac{1}{4(3^4)}x^4 + \cdots$$

$$= \ln 3 + \sum_{k=0}^{\infty} \frac{(-1)^k x^{k+1}}{(k+1)3^{k+1}}$$

26. $\log(1+x) = (\ln 10)^{-1}\ln(1+x)$

$\quad = (\ln 10)^{-1}[x - \frac{1}{2}x^2 + \frac{1}{3}x^3 - \frac{1}{4}x^4 + \cdots]$

$\quad = \sum\limits_{k=0}^{\infty} \dfrac{(-1)^k x^{k+1}}{(k+1)\ln 10}$

27.   Since $\tan^{-1} u = u - \dfrac{u^3}{3} + \dfrac{u^5}{5} - \dfrac{u^7}{7} + \cdots$

$\quad = \sum\limits_{k=0}^{\infty} \dfrac{(-1)^k x^{2k+1}}{2k+1}$

$\tan^{-1}(2x) = \sum\limits_{k=0}^{\infty} \dfrac{(-1)^k (2x)^{2k+1}}{2k+1}$

28. $\sqrt{x} = x^{1/2} = [1 + (x-1)]^{1/2}$

$\quad = 1 + \frac{1}{2}(x-1)(1^{-1/2})$

$\quad - \dfrac{1}{2^2 2!}(x-1)^2(1^{-3/2}) + \dfrac{(1)(3)}{2^3 3!}(1^{-5/2}) + \cdots$

$\quad = 1 + \frac{1}{2}(x-1)$

$\quad + \sum\limits_{k=2}^{\infty} \dfrac{(-1)^{k+1}(3)(5)\cdots(2k-3)}{2^k k!}(x-1)^k$

29. $e^{-x^2} = 1 - x^2 + \dfrac{1}{2!}(x^2)^2 - \cdots = \sum\limits_{k=0}^{\infty} \dfrac{(-1)^k x^{2k}}{k!}$

30. $\dfrac{\sin x}{x} = x^{-1}[x - \frac{1}{3!}x^3 + \frac{1}{5!}x^5 - \cdots]$

$\quad = \sum\limits_{k=0}^{\infty} \dfrac{(-1)^k x^{2k}}{(2k+1)!}$

31. $e^x \approx e + e(x-1) + \dfrac{1}{2!}e(x-1)^2 + \dfrac{1}{3!}e(x-1)^3$

32. $\ln x \approx \ln 3 + \dfrac{x-3}{3} - \dfrac{(x-3)^2}{18} + \dfrac{(x-3)^3}{81}$

33. $\cos x \approx \cos\frac{\pi}{3} - (x - \frac{\pi}{3})\sin\frac{\pi}{3}$

$\quad - \dfrac{(x-\frac{\pi}{3})^2}{2!}\cos\frac{\pi}{3} + \dfrac{(x-\frac{\pi}{3})^3}{3!}\sin\frac{\pi}{3}$

34. $\sin x \approx \sin\frac{\pi}{4} + (x - \frac{\pi}{4})\cos\frac{\pi}{4}$

$\quad - \dfrac{(x-\frac{\pi}{4})^2}{2!}\sin\frac{\pi}{4} - \dfrac{(x-\frac{\pi}{4})^3}{3!}\cos\frac{\pi}{4}$

35. $\tan x = \dfrac{\sin x}{\cos x} = \dfrac{\sum\limits_{k=0}^{\infty}\dfrac{(-1)^k x^{2k+1}}{(2k+1)!}}{\sum\limits_{k=0}^{\infty}\dfrac{(-1)^k x^{2k}}{(2k)!}}$

$\quad \approx x + \dfrac{x^3}{3} + \dfrac{2x^5}{15} + \dfrac{17x^7}{315}$

Note: you could use the *Mathematics Handbook*, Formula 35, page 149

36. $f(x) = x^2 + 2x + 1$

$\quad = [(x-200) + 200]^2 + 2[(x-200) + 200] + 1$

$\quad = (x-200)^2 + 402(x-200) + 40{,}401$

37. $f(x) = x^3 - 2x^2 + x - 5$

$\quad = [(x-2) + 2]^3 - 2[(x-2) + 2]^2$

$\qquad + [(x-2) + 2] - 5$

$\quad = (x-2)^3 + 4(x-2)^2 + 5(x-2) - 3$

38. $f(x) = x^{1/2};\ f(9) = 3$

$\quad f'(x) = \frac{1}{2}x^{-1/2};\ f'(9) = \frac{1}{6}$

$\quad f''(x) = -\frac{1}{4}x^{-3/2};\ f''(9) = -\frac{1}{108}$

$\quad f'''(x) = \frac{3}{8}x^{-5/2};\ f'''(9) = \frac{1}{648}$

$\quad \sqrt{x} \approx 3 + \frac{1}{6}(x-9) - \frac{1}{216}(x-9)^2 + \frac{1}{3{,}888}(x-9)^3$

39. $f(x) = (2-x)^{-1};\ f(5) = -\frac{1}{3}$

$\quad f'(x) = (2-x)^{-2};\ f'(5) = \frac{1}{9}$

$\quad f''(x) = 2(2-x)^{-3};\ f''(5) = -\frac{2}{27}$

$\quad f'''(x) = 6(2-x)^{-4};\ f'''(5) = \frac{6}{81}$

$\quad f(x) \approx -\frac{1}{3} + \frac{1}{9}(x-5) - \frac{1}{27}(x-5)^2 + \frac{1}{81}(x-5)^3$

40.   We show an alternative method of solution from the method shown in Problem 39.

$\quad f(x) = \dfrac{1}{4-x} = \dfrac{1}{6-(x+2)} = \dfrac{1}{6}\left[\dfrac{1}{1 - \frac{x-2}{6}}\right]$

$\quad = \dfrac{1}{6}\left[1 + \dfrac{x+2}{6} + \left(\dfrac{x+2}{6}\right)^2 + \left(\dfrac{x+2}{6}\right)^3 + \cdots\right]$

$\quad \approx \dfrac{1}{6} + \dfrac{1}{6^2}(x+2) + \dfrac{1}{6^3}(x+2)^2 + \dfrac{1}{6^4}(x+2)^3$

41. $f(x) = \dfrac{3}{2x-1} = \dfrac{3}{2(x-2)+3} = \dfrac{1}{1 + \frac{2}{3}(x-2)}$

$\quad = 1 + [-\frac{2}{3}(x-2)] + [-\frac{2}{3}(x-2)^2]$

$\qquad + [-\frac{2}{3}(x-2)^3] + \cdots$

$\quad \approx 1 - \frac{2}{3}(x-2) - \frac{2}{3}(x-2)^2 - (x-2)^3$

42. $f(x) = \dfrac{5}{3x+2} = \dfrac{5}{3(x-2)+8} = \dfrac{\frac{5}{8}}{1 + \frac{3}{8}(x-2)}$

$\quad = \frac{5}{8}\left[1 - \frac{3}{8}(x-2) - \frac{3}{8}(x-2)^2 - \frac{3}{8}(x-2)^3 + \cdots\right]$

$\quad \approx \frac{5}{8} - \frac{15}{64}(x-2) - \frac{15}{64}(x-2)^2 - \frac{15}{64}(x-2)^3$

43. $f(x) = (1+x)^{1/2}$

$\quad = 1 + \frac{1}{2}x + \dfrac{\frac{1}{2}(\frac{1}{2}-1)x^2}{2!} + \dfrac{\frac{1}{2}(\frac{1}{2}-1)(\frac{1}{2}-2)x^3}{3!}$

$\quad = 1 + \frac{1}{2}x - \frac{1}{8}x^2 + \frac{1}{16}x^3 - \frac{5}{128}x^4 + \cdots$

If p is greater than 0 and not an integer, the interval of absolute convergence is $(-1, 1)$.

44. $f(x) = (1+x^2)^{-1/2}$

$\quad = 1 + \frac{1}{2}x^2 + \dfrac{\frac{1}{2}(-\frac{1}{2})}{2!}(x^2)^2 + \dfrac{\frac{1}{2}(-\frac{1}{2})(-\frac{3}{2})}{3!}(x^3)^2 + \cdots$

$\approx 1 + \frac{1}{2}x^2 - \frac{1}{8}x^4 - \frac{1}{16}x^6;$

Interval of absolute convergence is $(-1, 1)$.

45. $f(x) = (1 + x)^{2/3}$

$= 1 + \frac{2}{3}x + \frac{\frac{2}{3}(-\frac{1}{3})}{2!}x^2 + \frac{\frac{2}{3}(-\frac{1}{3})(-\frac{4}{3})}{3!}x^3 + \cdots$

$\approx 1 + \frac{2}{3}x - \frac{1}{9}x^2 + \frac{4}{81}x^3$

Interval of absolute convergence is $(-1, 1)$.

46. $f(x) = (4 + x)^{-1/3} = 4^{-1/3}(1 + \frac{x}{4})^{-1/3}$

$= 4^{-1/3}\left[1 + (-\frac{1}{3})(\frac{x}{4}) + \frac{(-\frac{1}{3})(-\frac{2}{3})}{2!}(\frac{x}{4})^2\right.$

$\left. + \frac{(-\frac{1}{3})(-\frac{2}{3})(-\frac{5}{3})}{3!}(\frac{x}{4})^3 + \cdots\right]$

$\approx \frac{1}{\sqrt[3]{4}}\left[1 - \frac{1}{12}x + \frac{1}{72}x^2 - \frac{7}{2,592}x^3\right]$

Interval of absolute convergence is $(-4, 4)$.

47. $f(x) = x(1 - x^2)^{-1/2}$

$= x\left[1 - (-\frac{1}{2})x^2 + \frac{(-\frac{1}{2})(-\frac{3}{2})x^4}{2!}\right.$

$\left. - \frac{(-\frac{1}{2})(-\frac{3}{2})(-\frac{5}{2})x^6}{3!} + \cdots\right]$

$\approx x + \frac{1}{2}x^3 + \frac{3}{8}x^5 + \frac{5}{16}x^7$

Interval of absolute convergence is $(-1, 1)$.

48. $f(x) = (2 - x)^{1/4} = 2^{1/4}(1 - \frac{x}{2})^{1/4}$

$= 2^{1/4}$

$= 2^{1/4}\left[1 + (\frac{1}{4})(\frac{x}{2}) + \frac{(\frac{1}{4})(-\frac{3}{4})}{2!}(\frac{x}{2})^2\right.$

$\left. + \frac{(\frac{1}{4})(-\frac{3}{4})(-\frac{7}{4})}{3!}(\frac{x}{2})^3 + \cdots\right]$

$\approx \frac{1}{\sqrt[4]{2}}\left[1 + \frac{1}{8}x - \frac{3}{128}x^2 + \frac{7}{1,024}x^3\right]$

Interval of absolute convergence is $(-2, 2)$.

49. $\ln(1 + x) = \int_0^x \frac{1}{1 + t}\, dt$

$= \int_0^x [1 - t + t^2 - t^3 + \cdots + (-1)^{k-1}t^{k-1} + \cdots]$

$= [t - \frac{1}{2}t^2 + \frac{1}{3}t^3 + \cdots + (-1)^{k-1}t^k + \cdots]\Big|_0^x$

$= \sum_{k=1}^{\infty} \frac{(-1)^{k-1}x^k}{k}$

with interval of convergence $(-1, 1)$.

50. $\cosh x = \frac{1}{2}(e^x + e^{-x})$

$= \frac{1}{2}\Big([1 + x + \frac{1}{2!}x^2 + \frac{1}{3!}x^3 + \cdots]$

$+ [1 - x + \frac{1}{2!}x^2 - \frac{1}{3!}x^3 + \cdots]\Big)$

$= 1 + \frac{1}{2!}x^2 + \frac{1}{4!}x^4 + \cdots = \sum_{k=0}^{\infty} \frac{x^{2k}}{(2k)!}$

51. $\sinh x = \frac{1}{2}(e^x - e^{-x})$

$= \frac{1}{2}\Big([1 + x + \frac{1}{2!}x^2 + \frac{1}{3!}x^3 + \cdots]$

$- [1 - x + \frac{1}{2!}x^2 - \frac{1}{3!}x^3 + \cdots]\Big)$

$= x + \frac{1}{3!}x^3 + \frac{1}{5!}x^5 + \cdots = \sum_{k=0}^{\infty} \frac{x^{2k+1}}{(2k+1)!}$

52. $R_n(\frac{1}{2}) \le \frac{e^{z_n}}{(n+1)!}(\frac{1}{2})^{n+1}$ for $0 < z_{n+1} < \frac{1}{2} < 1$

$\frac{e^{z_n}}{(n+1)!2^{n+1}} < \frac{3}{2^{n+1}(n+1)!} < 0.0005$

Thus, $2^{n+1}(n+1)! > 6,000$ or choose $n = 5$.

53. $R_n(\frac{1}{3}) \le \frac{e^{z_n}}{(n+1)!}(\frac{1}{3})^{n+1}$ for $0 < z_{n+1} < \frac{1}{2} < 1$

$\frac{e^{z_n}}{(n+1)!3^{n+1}} < \frac{3}{3^{n+1}(n+1)!} < 0.0005$

Thus, $3^{n+1}(n+1)! > 6,000$ or choose $n = 4$.

54. $f(x) = \frac{2x}{x^2 - 1} = \frac{1}{x+1} - \frac{1}{1-x}$

$= [1 - x + x^2 - x^3 + \cdots]$

$- [1 + x + x^2 + x^3 + \cdots]$

$= -2x - 2x^3 - \cdots - 2x^{2k+1}$

$= -2x \sum_{k=0}^{\infty} x^{2k}$

55. $f(x) = \frac{6-x}{4-x^2} = \frac{6-x}{(2-x)(2+x)}$

$= \frac{2}{x+2} - \frac{1}{x-2}$ by partial fractions

$= \frac{2}{2(1 + \frac{x}{2})} + \frac{1}{2(1 - \frac{x}{2})}$

$= \sum_{k=0}^{\infty}\left(-\frac{x}{2}\right)^k + \frac{1}{2}\sum_{k=0}^{\infty}\left(\frac{x}{2}\right)^k$

$= \sum_{k=0}^{\infty}\left[(-1)^k + \frac{1}{2}\right]\left(\frac{x}{2}\right)^k$

56. $f(x) = \frac{3(1-x)}{9-x^2} = -\frac{1}{3}\left[\frac{1}{1 - \frac{x}{3}}\right] + \frac{2}{3}\left[\frac{1}{1 + \frac{x}{3}}\right]$

$= -\frac{1}{3}\left[1 + \frac{1}{3}x + (\frac{1}{3}x)^2 + \cdots\right]$

$+ \frac{2}{3}\left[1 - \frac{1}{3}x + (\frac{1}{3}x)^2 + \cdots\right]$

$= \sum_{k=0}^{\infty}\left[-\frac{1}{3^{k+1}} + (-1)^k\frac{2}{3^{k+1}}\right]x^k$

57. $f(x) = \dfrac{1}{x^2 - 3x + 2} = \dfrac{1}{(x-2)(x-1)}$

$= \dfrac{1}{1-x} - \dfrac{\frac{1}{2}}{1 - \frac{x}{2}}$

$= [1 + x + x^2 + \cdots] - \frac{1}{2}[1 + \frac{1}{2}x + (\frac{1}{2}x)^2 + \cdots]$

$= \displaystyle\sum_{k=0}^{\infty} \left[1 - \dfrac{1}{2^{k+1}}\right] x^k$

58. $f(x) = \dfrac{2x - 3}{x^2 - 3x + 2} = -\dfrac{1}{1-x} - \dfrac{\frac{1}{2}}{1 - \frac{x}{2}}$

$= -[1 + x + x^2 + \cdots] - \frac{1}{2}[1 + \frac{1}{2}x + (\frac{1}{2}x)^2 + \cdots]$

$= -\displaystyle\sum_{k=0}^{\infty} \left[1 + \dfrac{1}{2^{k+1}}\right] x^k$

59. $f(x) = \dfrac{-x^2}{(2+x)(1-x^2)}$

$= \dfrac{\frac{4}{3}}{2(1 + \frac{x}{2})} + \dfrac{-\frac{1}{6}}{1-x} + \dfrac{-\frac{1}{2}}{1+x}$

$= \dfrac{2}{3} \displaystyle\sum_{k=0}^{\infty} \left(-\dfrac{x}{2}\right)^k - \dfrac{1}{6} \displaystyle\sum_{k=0}^{\infty} x^k - \dfrac{1}{2} \displaystyle\sum_{k=0}^{\infty} (-x)^k$

$= \displaystyle\sum_{k=0}^{\infty} \left[\dfrac{2}{3}(-\dfrac{1}{2})^k - \dfrac{1}{6} - \dfrac{1}{2}(-1)^k\right] x^k$

60. $f(x) = \dfrac{x^2 - 6x + 7}{(1-x)(2-x)(3-x)}$

$= \dfrac{1}{1-x} + \dfrac{\frac{1}{2}}{1 - \frac{x}{2}} - \dfrac{\frac{1}{3}}{1 - \frac{x}{3}}$

$= \displaystyle\sum_{k=0}^{\infty} x^k + \dfrac{1}{2} \displaystyle\sum_{k=0}^{\infty} (\dfrac{x}{2})^k - \dfrac{1}{3} \displaystyle\sum_{k=0}^{\infty} (\dfrac{x}{3})^k$

$= \displaystyle\sum_{k=0}^{\infty} \left[1 + \dfrac{1}{2^{k+1}} - \dfrac{1}{3^{k+1}}\right] x^k$

61. $f(x) = \sin x \cos x = \frac{1}{2} \sin 2x$

$= \dfrac{1}{2} \displaystyle\sum_{k=0}^{\infty} (-1)^k \dfrac{(2x)^{2k+1}}{(2k+1)!}$

62. a. $f(x) = \cos^3 x = \frac{1}{4}[\cos 3x + 3 \cos x]$

$= \dfrac{1}{4} \displaystyle\sum_{k=0}^{\infty} (-1)^k \dfrac{3^{2k} + 3}{(2k)!} x^{2k}$

b. $f(x) = \sin^3 x = \cos^3(\frac{\pi}{2} - x)$

$= \dfrac{1}{4} \displaystyle\sum_{k=0}^{\infty} (-1)^k \dfrac{3^{2k} + 3}{(2k)!} (\frac{\pi}{2} - x)^{2k}$

63. $f(x) = \left(\cos \dfrac{3x}{2}\right)\left(\cos \dfrac{x}{2}\right)$

$= \frac{1}{2}(\cos 2x + \cos x)$

Mathematics Handbook, Identity 27, Page 43

$= \dfrac{1}{2}\left[\displaystyle\sum_{k=0}^{\infty} (-1)^k \dfrac{(2x)^{2k}}{(2k)!} + \displaystyle\sum_{k=0}^{\infty} (-1)^k \dfrac{x^{2k}}{(2k)!}\right]$

$= \dfrac{1}{2} \displaystyle\sum_{k=0}^{\infty} (-1)^k \left[\dfrac{1 + 2^{2k}}{(2k)!}\right] x^{2k}$

64. $f(x) = \ln[(1 + 2x)(2 + 3x)]$

$= \ln(1 + 2x) + \ln(1 + 3x)$

$= \displaystyle\sum_{k=0}^{\infty} (-1)^k \left[\dfrac{2^{k+1} + 3^{k+1}}{k+1}\right] x^{k+1}$

65. $f(x) = \ln\left[\dfrac{1 + 2x}{1 - 3x + 2x^2}\right]$

$= \ln(1 + 2x) - \ln(1 - 2x) - \ln(1 - x)$

$= \displaystyle\sum_{k=0}^{\infty} (-1)^k \left[\dfrac{2^{k+1} - (-2)^{k+1} - (-1)^{k+1}}{k+1}\right]$

$= \displaystyle\sum_{k=0}^{\infty} [(-1)^k 2^{k+1} + 2^{k+1} + 1] \dfrac{x^{k+1}}{k+1}$

66. $f(x) = \dfrac{x + \sin x}{x}$

$= 1 + x^{-1} \displaystyle\sum_{k=0}^{\infty} (-1)^k \dfrac{x^{2k+1}}{(2k+1)!}$

$\displaystyle\lim_{x \to \infty} f(x) = \lim_{x \to \infty}\left[1 + x^{-1} \displaystyle\sum_{k=0}^{\infty} (-1)^k \dfrac{x^{2k+1}}{(2k+1)!}\right]$

$= \displaystyle\lim_{x \to \infty}[1 + 1] = 2$

67. $g(x) = \dfrac{e^x - 1}{x} = \dfrac{\displaystyle\sum_{k=0}^{\infty} \dfrac{x^k}{k!} - 1}{x}$

$= 1 + \dfrac{x}{2!} + \dfrac{x^2}{3!} + \dfrac{x^3}{4!} + \cdots$

$= \displaystyle\sum_{0}^{\infty} \dfrac{x^k}{(k+1)!}$

$\displaystyle\lim_{x \to 0} g(x) = 1$ (by l'Hôpital's rule)

68. $\left|f^{(n+1)}(z_n)\right| = |\cos z_n| \le 1$ or

$\left|f^{(n+2)}(z_n)\right| = |\sin z_n| \le 1$ with $0 < z_n < x$

Thus,

$|R_{2n}(x)| = \dfrac{\left|f^{(2n+1)}(z_n)\right|}{(2n+1)!} |x - 0|^{2n+1}$

$\le \dfrac{|x|^{2n+1}}{(2n+1)!}$

Note that the coefficients of x^{2k} are 0.

69. $\left|f^{(n+1)}(z_n)\right| = |\cos z_n| \le 1$ or

$\left|f^{(n+2)}(z_n)\right| = |\sin z_n| \le 1$ with $0 < z_n < x$

Thus,

$|R_{2n+1}(x)| = \dfrac{\left|f^{(2n+1)}(z_n)\right|}{(2n+1)!} |x - 0|^{2n+1}$

$$\leq \frac{|x|^{2n+2}}{(2n+2!)}$$

Note that the coefficients of x^{2k+1} are 0.

70. $f'(x) = \sum_{k=0}^{n-1} (-1)^k x^{2k} + R'_{2n}(x)$

$f''(x) = \sum_{k=0}^{n-1} (-1)^k (2k) x^{2k-1} + R''_{2n}(x)$

$f'''(x) = \sum_{k=0}^{n-1} (-1)^k (2k)(2k-1) x^{2k-2}$
$\qquad\qquad + R'''_{2n}(x)$

$f^{(2n+1)}(x) = (-1)^n (2n)! x + R_{2n}^{(2n+1)}(x)$

Thus, with $0 < z_n \leq 1$,

$\left| R_{2n}(x) \right| = \frac{\left| f^{(2n+1)}(z_n) \right|}{(2n+1)!} |x - 0|^{2n+1}$

$\qquad\qquad \leq \frac{|x|^{2n+1}}{2n+1}$

71. a. $f(x) = x \cos 2x$

$= x[1 - \frac{1}{2!}(2^2 x^2) + \frac{1}{4!}(2^4 x^4) - \frac{1}{6!}(2^6 x^6) + \cdots]$

$= x - 2x^3 + \frac{2}{3}x^5 - \frac{4}{45}x^7 + \cdots$

$\qquad P_5(x) = x - 2x^3 + \frac{2}{3}x^5$

b. $f(x) - P_5(x) \approx \frac{4}{45}(\frac{1}{2})^7 \approx 0.00070$

c. $P_5(0.5) - 0.5 \cos 1 \approx 0.00068218$

d.

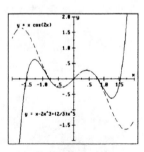

72. a. The software returns the original form, and does not integrate the function.

b. $P_7(x) = x + \frac{5}{6}x^3 + \frac{27}{40}x^5 - \frac{1,303}{5,040}x^7$

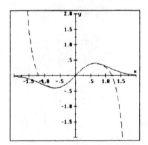

c. $y'(x) = \int P_7(x)\, dx - 0.1$

$= \frac{1}{2}x^2 - \frac{7}{24}x^4 + \frac{9}{80}x^6 - \frac{1,303}{40,320}x^8 - 0.1$

$y(x) = \int y'(x)\, dx - 0.3$

$= \frac{1}{6}x^3 - \frac{7}{120}x^5 + \frac{9}{560}x^7 - \frac{1,303}{362,880}x^9$

$\qquad + 0.1x + 0.3$

73. Let $F(x) = \frac{1}{1-x} = 1 + x + x^2 + x^3 + \cdots$
$\qquad\qquad + x^k + \cdots$

$F'(x) = \frac{1}{(1-x)^2} = 1 + 2x + 3x^2 + 4x^3$
$\qquad\qquad + \cdots$

$F''(x) = \frac{2}{(1-x)^3} = 2 + 2(3)x + (3)(4)x^2$
$\qquad\qquad + \cdots$

Thus, $f(x) = \frac{1}{2} \sum_{k=0}^{\infty} (k+1)(k+2)x^k$.

74. $F(x) = \int_0^x e^{-t^2}\, dt$

$F'(x) = e^{-x^2} = 1 - x^2 + \frac{1}{2!}x^4 - \cdots$
$\qquad\qquad + \frac{(-1)^k x^{2k}}{k!}$

$F(x) = x - \frac{1}{3}x^3 + \frac{1}{(5)2!}x^5 - \frac{1}{(7)3!}x^7 + \cdots$
$\qquad\qquad + \frac{(-1)^k x^{2k+1}}{(2k+1)(k!)} + \cdots$

$= \sum_{k=0}^{\infty} \frac{(-1)^k x^{2k+1}}{(2k+1)(k!)}$

75. Let $F(x) = \int_0^x t^{0.2} e^t\, dt$

$F'(x) = x^{0.2} e^x = x^{0.2} + x^{1.2} + \frac{1}{2!}x^{2.2} + \frac{1}{3!}x^{3.2}$
$\qquad\qquad + \cdots + \frac{x^{k+0.2}}{k!} + \cdots$

$F(x) = \frac{x^{1.2}}{1.2} + \frac{x^{2.2}}{2.2} + \frac{x^{3.2}}{(3.2)2!} + \frac{x^{4.2}}{(4.2)3!} + \cdots$
$\qquad\qquad + \frac{x^{k+1.2}}{(k+1.2)(k!)} + \cdots$

$$F(1) = \frac{1}{1.2} + \frac{1}{2.2} + \frac{1}{(3.2)2!} + \frac{1}{(4.2)3!} + \cdots + \frac{1}{(k+1.2)(k!)} + \cdots$$

$$= \sum_{k=0}^{\infty} \frac{1}{(2k+1)(k!)}$$

76. a. $J_0(x) = \sum_{k=0}^{\infty} \frac{(-1)^k x^{2k}}{(k!)2^{2k}}$; ratio test

$$\lim_{k \to \infty} \frac{x^{2k+2}(k!)2^{2k}}{[(k+1)!]^2 2^{2k+2} x^{2k}} = \frac{1}{2} \lim_{k \to \infty} \frac{1}{(k+1)^2} = 0 < 1$$

Thus, $J_0(x)$ converges for all x.

$$J_1(x) = \sum_{k=0}^{\infty} \frac{(-1)^k x^{2k}}{k!(k+1)!2^{2k}} < J_0(x)$$

so $J_1(x)$ also converges.

b. $J_0(x) = 1 - \frac{x^2}{2^2} + \frac{x^{2(2)}}{(2!)^2 2^{2(2)}} - \frac{x^{2(3)}}{(3!)^2 2^{2(3)}} + \cdots + \frac{(-1)^k x^{2k}}{(k!)^2 2^{2k}}$

$$J'_0(x) = -\frac{2x}{2^2} + \frac{(2)(2)x^{2(2)-1}}{(2!)^2 2^{2(2)}} - \frac{(2)(3)x^{2(3)-1}}{(3!)^2 2^{2(3)}} + \cdots + \frac{(-1)^k(2k)x^{2k-1}}{(k!)^2 2^{2k}}$$

$$= -\left[\frac{2x}{(1!)^2 2^2} - \frac{(2)(2)x^{2(2)-1}}{(2!)^2 2^{2(2)}} - \cdots \right] - \left[-\frac{(-1)^k(2k)x^{2k-1}}{(k!)^2 2^{2k}} + \frac{(-1)^k(2k+2)x^{2k+1}}{2} \right]$$

$$= -\frac{x}{2}\left[1 - \frac{(2)(2)x^{2(2-1)}}{(1)(2!)2^{2(2-1)}} + \frac{x^{2(3-1)}}{(2!)(3!)2^{2(3-1)}} - \cdots \right]$$

$$= -\frac{x}{2} J_1(x)$$

77. $J_0(x) = 1 - \frac{x^2}{2^2} + \frac{x^{2(2)}}{(2!)^2 2^{2(2)}} - \frac{x^{2(3)}}{(3!)^2 2^{2(3)}} + \cdots + \frac{(-1)^k x^{2k}}{(k!)^2 2^{2k}}$

$$J'_0(x) = -\frac{2x}{2^2} + \frac{(2)(2)x^{2(2)-1}}{(2!)^2 2^{2(2)}} - \frac{(2)(3)x^{2(3)-1}}{(3!)^2 2^{2(3)}} + \cdots + \frac{(-1)^k(2k)x^{2k-1}}{(k!)^2 2^{2k}}$$

$$J''_0(x) = -\frac{1}{2} + \frac{(4)(3)x^2}{(2!)^2 2^4} - \frac{(6)(5)x^4}{(3!)^2 2^6} + \cdots + \frac{(-1)^k(2k)(2k-1)x^{2(k-1)}}{(k!)^2 2^{2k}}$$

$$x^2 J''_0(x) + x J'_0(x) + x^2 J_0(x) = x^2\left[1 - \frac{1}{2} - \frac{1}{2} \right] + \frac{x^4}{(2!)^2}\left[-1 + \frac{(4)(4)}{2^4} \right]$$

$$+ \frac{x^6}{(3!)^2 2^4}\left[9 - \frac{(6)(6)}{2^2} \right] - \cdots + \frac{(-1)^k x^{2k+1}}{[(k+1)!]^2 (2^{2k})}\left[(k+1)^2 - \frac{1}{2^2}[2(k+1) + 2(k+1)(2k+1)] \right] + \cdots$$

$$= 0 + \frac{(-1)^k x^{2k+1}(k+1)}{4[(k+1)!]^2 (2^{2k})}[4k^2 + 8k + 4 - 2k - 2 - 4k^2 - 6k - 2] = 0$$

78. $f'(x) = 1 + x + x^2 + \cdots + x^{k-1} + \cdots = \frac{1}{1-x}$;

$$\int_0^x f'(t)\, dt = \int_0^x (1 + t + t^2 + \cdots + t^{k-1} + \cdots)\, dt = -\ln|1-x| \text{ for } x \text{ in } (-1, 1)$$

CHAPTER 8 REVIEW

Proficiency Examination, Page 565

1. A sequence is a succession of numbers that are listed according to a given prescription or rule.

2. If the terms of the sequence approach the number L as n increases without bound, we say that the sequence *converges to the limit L* and write $L = \lim_{n \to \infty} a_n$.

3. A sequence converges if the limit of the nth element is finite (and unique). If not, it diverges.

4. **a.** The elements of a bounded sequence lie within a finite range.
   b. A sequence is monotonic if it is nondecreasing or nonincreasing.
   c. A sequence is strictly monotonic if it is increasing or decreasing.
   d. A sequence $\{a_n\}$ is nonincreasing if
   $$a_1 \geq a_2 \geq \ldots \geq a_{k-1} \geq a_k \geq \ldots$$

5. A monotonic sequence $\{a_n\}$ converges if it is bounded and diverges otherwise.

6. An infinite series is a sum of infinitely many terms.

7. A sequence $\{a_n\}$ converges if $\lim_{n \to \infty} a_n = L$ is finite (and unique). A series converges if
   $$\sum_{k=0}^{\infty} a_k$$
   is finite. For the series to converge, it is necessary that $L = 0$.

8. The middle terms of a telescoping series vanish (by addition and subtraction of the same numbers).

9. The harmonic series is a p-series with $p = 1$. It diverges, but the alternating harmonic series converges.

10. The ratio of consecutive terms of a geometric series is a constant, r. The geometric series converges if $|r| < 1$. With a the first germ of the geometric series, the sum is $S = a/(1-r)$.

11. $S = \sum_{k=0}^{\infty} a_k$ diverges if $\lim_{k \to \infty} a_n \neq 0$.

12. If $a_k = f(k)$ for $k = 1, 2, \ldots$, where f is a positive, continuous, and decreasing function of x for $x \geq 1$, then
    $$\sum_{k=1}^{\infty} a_k \quad \text{and} \quad \int_1^{\infty} f(x)\, dx$$

either both converge or both diverge.

13. $\sum_{k=1}^{\infty} \frac{1}{k^p}$ is a convergent p-series if $p > 1$. It diverges when $p \leq 1$.

14. Let $0 \leq a_k \leq c_k$ for all $k \geq N$ for some N. If $\sum_{k=1}^{\infty} c_k$ converges, then $\sum_{k=1}^{\infty} a_k$ also converges. Let $0 \leq d_k \leq a_k$ for all k. If $\sum_{k=1}^{\infty} d_k$ diverges, then $\sum_{k=1}^{\infty} a_k$ also diverges.

15. Suppose $a_k > 0$ and $b_k > 0$ for all sufficiently large k and that
    $$\lim_{k \to \infty} \frac{a_k}{b_k} = L$$
    where L is finite and positive ($0 < L < \infty$). Then Σa_k and Σb_k either both converge or both diverge.

16. Suppose $a_k > 0$ and $b_k > 0$ for all sufficiently large k. Then,
    If $\lim_{k \to \infty} \frac{a_k}{b_k} = 0$ and Σb_k converges, the series Σa_k converges.
    If $\lim_{k \to \infty} \frac{a_k}{b_k} = \infty$ and Σb_k diverges, the series Σa_k diverges.

17. Given the series Σa_k with $a_k > 0$, suppose that $\lim_{k \to \infty} \frac{a_{k+1}}{a_k} = L$. The ratio test states the following:
    If $L < 1$, then Σa_k converges.
    If $L > 1$ or if L is infinite, then Σa_k diverges.
    If $L = 1$, the test is inconclusive.

18. Given the series Σa_k with $a_k \geq 0$, suppose that $\lim_{k \to \infty} \sqrt[k]{a_k} = L$. The root test states the following:
    If $L < 1$, then Σa_k converges.
    If $L > 1$ or if L is infinite, then Σa_k diverges.
    If $L = 1$, the root test is inconclusive.

19. If $a_k > 0$, then an alternating series
    $$\sum_{k=1}^{\infty} (-1)^k a_k \quad \text{or} \quad \sum_{k=1}^{\infty} (-1)^{k+1} a_k$$
    converges if both of the following two conditions are satisfied:
    1. $\lim_{k \to \infty} a_k = 0$
    2. $\{a_k\}$ is a decreasing sequence; that is,
       $$a_{k+1} < a_k \text{ for all } k.$$

20. Suppose an alternating series
    $$\sum_{k=1}^{\infty} (-1)^k a_k \quad \text{or} \quad \sum_{k=1}^{\infty} (-1)^{k+1} a_k$$
    satisfies the conditions of the alternating

series test; namely, $\lim_{k \to \infty} a_k = 0$ and $\{a_k\}$ is a decreasing sequence $(a_{k+1} < a_k)$. If the series has sum S, then $|S - S_n| < a_{n+1}$, where S_n is the nth partial sum of the series.

21. A series of real numbers Σa_k must converge if the related absolute value series $\Sigma |a_k|$ converges.

22. The series Σa_k is absolutely convergent if the related series $\Sigma |a_k|$ converges. The series Σa_k is conditionally convergent if it converges but $\Sigma |a_k|$ diverges.

23. For the series Σa_k, suppose $a_k \neq 0$ for $k \geq 1$ and that
$$\lim_{k \to \infty} \left| \frac{a_{k+1}}{a_k} \right| = L$$
where L is a real number or ∞. Then:
If $L < 1$, the series Σa_k converges absolutely and hence converges.
If $L > 1$ or if L is infinite, the series Σa_k diverges.
If $L = 1$, the test fails.

24. An infinite series of the form
$$\sum_{k=0}^{\infty} a_k (x - c)^k$$
$$= a_0 + a_1(x - c) + a_2(x - c)^2 + \ldots$$
is called a power series in $(x - c)$.

25. For a power series $\sum_{k=1}^{\infty} a_k x^k$, exactly one of the following is true:

1. The series converges for all x.
2. The series converges only for $x = 0$.
3. The series converges absolutely for all x in an open interval $(-R, R)$ and diverges for $|x| > R$. *Note:* The series should be checked separately at the endpoints, because it could converge absolutely, or converge conditionally, or diverge at $x = R$ and $x = -R$.

26. The interval of convergence of a power series consists of those values of x for which the series converges. Think of that interval as the diameter of a circle. The radius of convergence is the radius of that circle.

27. $P_n(x) = \sum_{k=1}^{\infty} a_k (x - c)^k$ is a Taylor polynomial if $a_k = \dfrac{f^{(k)}(c)}{k!}$.

28. If f and all its derivatives exist in an open interval I containing c, then for each x in I

$$f(x) = f(c) + \frac{f'(c)}{1!}(x - c) + \frac{f''(c)}{2!}(x - c)^2$$
$$+ \ldots + \frac{f^{(n)}(c)}{n!}(x - c)^n + R_n(x)$$

where the remainder function $R_n(x)$ is given by
$$R_n(x) = \frac{f^{(n+1)}(z_n)}{(n + 1)!}(x - c)^{n+1}$$

for some z_n that depends on x and lies between c and x.

29. The Taylor series is $T = \sum_{k=1}^{\infty} a_k(x - c)^k$ and is a Maclaurin series if $c = 0$.

30. The binomial function $(1 + x)^p$ is represented by its Maclaurin series
$$(1 + x)^p = 1 + px + \frac{p(p - 1)}{2!}x^2$$
$$+ \frac{p(p - 1)(p - 2)}{3!}x^3 + \ldots$$
$$+ \frac{p(p - 1)\cdots(p - k + 1)}{k!}x^k + \ldots$$

$-1 < x < 1$ if $p \leq -1$;
$-1 < x \leq 1$ if $-1 < p < 0$;
$-1 \leq x \leq 1$ if $p > 0$, p not an integer;
all x if p is a nonnegative integer.

31. The sequence has an upper bound of 4, a lower bound of 0, and after $n = 4$ is monotonic decreasing:
$$\lim_{n \to \infty} \frac{\dfrac{e^{n+1}}{(n + 1)!}}{\dfrac{e^n}{n!}} = \lim_{n \to \infty} \frac{e}{n + 1} = 0$$

The sequence converges.

32. $\displaystyle \lim_{n \to \infty} \frac{3n^2 - n + 1}{(1 - 2n)n} = \lim_{n \to \infty} \frac{6n - 1}{-4n + 1} = -\frac{3}{2}$

33. This is the definition of e given in Chapter 5.

34. $S = \displaystyle\sum_{k=1}^{\infty} \frac{e^k}{k!}$; use ratio test.
$$\lim_{k \to \infty} \frac{e^{k+1} k!}{(k + 1)! e^k} = \lim_{k \to \infty} \frac{e}{k + 1} = 0 < 1$$
S converges.

35. $S = \displaystyle\sum_{k=1}^{n} \frac{3k^2 - k + 1}{(1 - 2k)k}$; check the necessary condition: $\displaystyle\lim_{k \to \infty} \frac{3k^2 - k + 1}{(1 - 2k)k} = -\frac{3}{2} \neq 0$

S diverges.

36. $S = \sum\limits_{k=2}^{\infty} \dfrac{1}{k \ln k}$; use integral test.

$$\lim_{b \to \infty} \int_{2}^{b} (\ln x)^{-1} \frac{dx}{x} = \lim_{b \to \infty} \ln|\ln x|\Big|_{2}^{b} = \infty$$

S diverges.

37. $S = \sum\limits_{k=0}^{\infty} \dfrac{(-1)^{k+1}}{k^2}$ converges absolutely when

compared with the convergent p-series with $p = 2 > 1$.

38. $S = \sum\limits_{k=0}^{\infty} (-1)^k k x^k$

$$\lim_{k \to \infty} \left| \frac{(k+1)\, x^{k+1}}{k x^k} \right| = |x| < 1$$

The series is absolutely convergent on $(-1, 1)$. At $x = \pm 1$, the series diverges by the divergence test. The interval of convergence is $(-1, 1)$.

39. $\sin x = \sum\limits_{k=0}^{\infty} \dfrac{(-1)^k x^{2k+1}}{(2k+1)!}$

$\sin 2x = \sum\limits_{k=0}^{\infty} \dfrac{(-1)^k (2x)^{2k+1}}{(2k+1)!}$

40. $f(x) = \dfrac{1}{x-3}$ at $c = \dfrac{1}{2}$

$$= \frac{1}{x - \frac{1}{2} - \frac{5}{2}}$$

$$= \frac{-1}{\frac{5}{2} - \left(x - \frac{1}{2}\right)}$$

$$= \frac{-\frac{2}{5}}{1 - \frac{2}{5}\left(x - \frac{1}{2}\right)}$$

$$= -\frac{2}{5}\left[1 + \frac{2}{5}\left(x - \frac{1}{2}\right) + \cdots\right]$$

$$= -\frac{2}{5} \sum\limits_{k=0}^{\infty} \left(\frac{2}{5}\right)^k \left(x - \frac{1}{2}\right)^k$$

Supplementary Problems, Page 566

1. $\lim\limits_{n \to \infty} \dfrac{(-2)^n}{n^2+1} = \lim\limits_{n \to \infty} \dfrac{(-2)^{n-1}}{2} = \infty$;

sequence diverges

2. $\lim\limits_{n \to \infty} \dfrac{(\ln n)^2}{\sqrt{n}} = \lim\limits_{n \to \infty} \dfrac{2^2(\ln n)\sqrt{n}}{n}$

$= 4 \lim\limits_{n \to \infty} \dfrac{\ln n}{\sqrt{n}} = 4 \lim\limits_{n \to \infty} \dfrac{2}{\sqrt{n}} = 0$

sequence converges

3. $\lim\limits_{n \to \infty} \left(1 - \dfrac{2}{n}\right)^n = \lim\limits_{n \to \infty} \left[\left(1 + \dfrac{-2}{n}\right)^{-n/2}\right]^{-2}$

$= e^{-2}$; the sequence converges

4. $\lim\limits_{n \to \infty} \dfrac{e^{0.1n}}{n^5 - 3n + 1} = \lim\limits_{n \to \infty} \dfrac{(0.1)e^{0.1n}}{5n^4 - 3}$

$= \lim\limits_{n \to \infty} \dfrac{(0.1)e^{0.1n}}{5!} = \infty$; sequence diverges

5. $\lim\limits_{n \to \infty} \dfrac{n + (-1)^n}{n} = \lim\limits_{n \to \infty} [1 + (-1)^n n^{-1}]$

$= 1 + 0$; sequence converges

6. $\lim\limits_{n \to \infty} \dfrac{3^n}{3^n + 2^n} = \lim\limits_{n \to \infty} \dfrac{1}{1 + \left(\frac{2}{3}\right)^n} = 1$;

sequence converges

7. $\lim\limits_{n \to \infty} \dfrac{5n^4 - n^2 - 700}{3n^4 - 10n^2 + 1} = \lim\limits_{n \to \infty} \dfrac{5(4!)}{3(4!)} = \dfrac{5}{3}$;

sequence converges

8. $\lim\limits_{n \to \infty} \left(1 + \dfrac{e}{n}\right)^{2n} = e^{2e}$; sequence converges

9. $\lim\limits_{n \to \infty} (\sqrt{n+1} - \sqrt{n})$

$= \lim\limits_{n \to \infty} \dfrac{(\sqrt{n-1} - \sqrt{n})(\sqrt{n-1} + \sqrt{n})}{\sqrt{n-1} + \sqrt{n}}$

$= \lim\limits_{n \to \infty} \dfrac{n - 1 - n}{\sqrt{n-1} + \sqrt{n}}$

$= 0$; sequence converges

10. $\lim\limits_{n \to \infty} (\sqrt{n^4 + 2n^2} - n^2)$

$= \lim\limits_{n \to \infty} \dfrac{(\sqrt{n^4 + 2n^2} - n^2)(\sqrt{n^4 + 2n^2} + n^2)}{\sqrt{n^4 + 2n^2} + n^2}$

$= \lim\limits_{n \to \infty} \dfrac{n^4 + 2n^2 - n^4}{\sqrt{n^4 + 2n^2} + n^2}$

$= \lim\limits_{n \to \infty} \dfrac{2}{\sqrt{1 + 2n^{-2}} + 1}$

$= 1$; sequence converges

11. $\lim\limits_{n \to \infty} \dfrac{\ln n}{n} = \lim\limits_{n \to \infty} \dfrac{1}{n} = 0$; sequence converges

12. $\lim\limits_{n \to \infty} [1 + (-1)^n]$; sequence diverges by

oscillation

13. $\lim\limits_{n \to \infty} \left[\left(1 + \dfrac{4}{n}\right)^n\right] = e^4$; sequence converges

14. $\lim\limits_{n \to \infty} 5^{2/n} = 5^0 = 1$

15. $\lim\limits_{n \to \infty} \dfrac{n^{3/4} \sin n^2}{n + 4} \le \lim\limits_{n \to \infty} \dfrac{n^{3/4}}{n + 4} = 0$

sequence converges

16. $\lim_{n\to\infty} \sum_{k=1}^{n} \frac{n}{n^2 + k^2}$ converges if

$$I = \lim_{b\to\infty} \int_1^b \frac{dx}{2x} = \frac{1}{2} \lim_{n\to\infty} \ln x \Big|_1^b = \infty$$

Note that on the kth interval $x \le k$, so

$$\frac{x}{x^2 + k^2} \le \frac{1}{2x}$$

The sequence diverges.

17. $\sum_{k=-123,456,788}^{123,456,789} \frac{k}{370,370,367}$

$$= \frac{123,456,789}{370,370,367} = \frac{1}{3}$$

A finite series always converges.

18. $\sum_{k=1}^{\infty} 4\left(\frac{2}{3}\right)^k = 4\left(\frac{2}{3}\right)\frac{1}{1-\frac{2}{3}} = 8$

19. $\sum_{k=1}^{\infty} \left(\frac{e}{3}\right)^k = \left(\frac{e}{3}\right)\frac{1}{1-\frac{e}{3}} = \frac{e}{3-e}$

20. $\sum_{k=2}^{\infty} \frac{1}{k^2-1} = \sum_{k=2}^{\infty} \left[\frac{\frac{1}{2}}{k-1} - \frac{\frac{1}{2}}{k+1}\right]$

$$= \frac{1}{2}\lim_{n\to\infty}[1 - \frac{1}{3} + \frac{1}{2} - \frac{1}{4} + \frac{1}{3} - \frac{1}{5} + \cdots$$
$$- \frac{1}{n+2} - \frac{1}{n+3}]$$

$$= \frac{1}{2}(1 + \frac{1}{2}) = \frac{3}{4}$$

21. $\sum_{k=0}^{\infty} \left[\left(\frac{-3}{8}\right)^k + \left(\frac{3}{4}\right)^{2k}\right] = \frac{1}{1+\frac{3}{8}} + \frac{1}{1-\frac{9}{16}} = \frac{232}{77}$

22. $\sum_{k=1}^{\infty} \frac{1}{4k^2-1} = \sum_{k=1}^{\infty} \left[\frac{\frac{1}{2}}{2k-1} - \frac{\frac{1}{2}}{2k+1}\right]$

$$= \frac{1}{2}\lim_{n\to\infty}\left[1 - \frac{1}{3} + \frac{1}{3} - \frac{1}{5} + \frac{1}{5} - \frac{1}{7} + \cdots - \frac{1}{2n+1}\right]$$

$$= \frac{1}{2}$$

23. $\sum_{k=0}^{\infty} \frac{e^k + 3^{k-1}}{6^{k+1}} = \sum_{k=0}^{\infty} \left[\frac{1}{6}\left(\frac{e}{6}\right)^k + \frac{1}{36}\left(\frac{3}{6}\right)^{k-1}\right]$

$$= \frac{\frac{1}{6}}{1-\frac{e}{6}} + \frac{1}{36}\left[2 + 1 + \frac{1}{2} + \cdots + \left(\frac{1}{2}\right)^{k-1} + \cdots\right]$$

$$= \frac{1}{6-e} + \frac{\frac{1}{18}}{1-\frac{1}{2}} = \frac{1}{6-e} + \frac{1}{9} = \frac{15-e}{9(6-e)}$$

24. $\sum_{k=1}^{\infty} (-1)^{k+1}\left(\frac{1}{k} + \frac{1}{k+1}\right)$

$$= \lim_{n\to\infty}\left[(1+\frac{1}{2}) + (-\frac{1}{2}-\frac{1}{3}) + \cdots \pm \left(\frac{1}{n} + \frac{1}{n+1}\right) + \cdots\right]$$

$$= 1$$

25. $\sum_{k=2}^{\infty} \left[\frac{1}{\ln(k+1)} - \frac{1}{\ln k}\right] = \lim_{n\to\infty}\left(\frac{1}{\ln 3} - \frac{1}{\ln 2}\right)$

$$+ \left(\frac{1}{\ln 4} - \frac{1}{\ln 3}\right) + \cdots + \left[\frac{1}{\ln(n+1)} - \frac{1}{\ln n}\right] + \cdots$$

$$= -\frac{1}{\ln 2}$$

26. $\sum_{k=0}^{\infty} \left[3\left(\frac{2}{3}\right)^{2k} - \left(\frac{-1}{3}\right)^{4k}\right] = \frac{3}{1-\frac{4}{9}} - \frac{1}{1+\frac{1}{81}}$

$$= \frac{1,809}{410}$$

27. $\sum_{k=1}^{\infty} \frac{k}{(k+1)(k+2)(k+3)}$

$$= \frac{1}{2}\sum_{k=0}^{\infty} \left(\frac{-1}{k+1} + \frac{4}{k+2} + \frac{-3}{k+3}\right)$$

$$= \frac{1}{2}\left(-\frac{1}{2} + \frac{4}{3} - \frac{1}{3}\right) = \frac{1}{4}$$

28. $S = \sum_{k=1}^{\infty} \frac{5^k k!}{k^k}$; use ratio test.

$$\lim_{n\to\infty} \frac{5^{k+1}(k+1)!k^k}{(k+1)^{k+1}5^k k!} = 5\lim_{k\to\infty}\left[\left(1+\frac{1}{k}\right)^k\right]^{-1}$$

$$= \frac{5}{e} > 1, \text{ so } S \text{ diverges}$$

29. $S = \sum_{k=1}^{\infty} \frac{1}{\sqrt{k^2+4}}$; use limit comparison with

$T = \sum_{k=1}^{\infty} \frac{1}{k}$ which diverges.

30. $S = \sum_{k=0}^{\infty} \frac{k!}{2^k}$; use ratio test.

$$\lim_{k\to\infty} \frac{(k+1)!2^k}{2^{k+1}k!} = \frac{1}{2} < 1, \text{ so } S \text{ converges}$$

31. $S = \sum_{k=1}^{\infty} \frac{1}{2k-1}$; use limit comparison with

$T = \sum_{k=1}^{\infty} \frac{1}{k}$ which diverges.

32. $S = \sum_{k=0}^{\infty} ke^{-k}$; use ratio test.

$$\lim_{k\to\infty} \frac{(k+1)e^{-(k+1)}}{ke^{-k}} = \frac{1}{e} < 1, \text{ so } S \text{ converges.}$$

33. $S = \sum_{k=0}^{\infty} \frac{k^3}{k!}$; use ratio test.

$$\lim_{k\to\infty} \frac{(k+1)k!}{(k+1)!k^3} = \lim_{k\to\infty} \frac{1}{k+1} = 0 < 1, \text{ so } S$$

converges.

34. $S = \sum\limits_{k=1}^{\infty} \dfrac{7^k}{k^2}$; use ratio test.

$$\lim_{k\to\infty} \frac{7^{k+1}k^2}{(k+1)^2 7^k} = 7 > 1, \text{ so } S \text{ diverges}$$

35. $S = \sum\limits_{k=1}^{\infty} \dfrac{k}{(2k-1)!}$; use ratio test.

$$\lim_{k\to\infty} \frac{(k+1)k!}{(2k+1)!k} = \lim_{k\to\infty} \frac{1}{2k+1} = 0 < 1,$$

so S converges.

36. $S = \sum\limits_{k=0}^{\infty} \dfrac{k^2}{(k^3+1)^2}$; use limit comparison with

$T = \sum\limits_{k=1}^{\infty} \dfrac{1}{k^4}$ which converges.

37. $S = \sum\limits_{k=0}^{\infty} \dfrac{k^2}{3^k}$; use ratio test.

$$\lim_{k\to\infty} \frac{(k+1)^2 3^k}{3^{k+1}k^2} = \frac{1}{3} < 1, \text{ so } S \text{ converges.}$$

38. $S = \sum\limits_{k=2}^{\infty} \dfrac{1}{k(\ln k)^{1.1}}$; use integral test.

$$I = \lim_{b\to\infty} \int_2^b (\ln x)^{-1.1}\, \frac{dx}{x} = \lim_{b\to\infty} \frac{-0.1}{(\ln x)^{0.1}}\Big|_2^b$$

$$= \frac{0.1}{\ln 2}, \text{ which converges, so } S \text{ converges.}$$

39. $S = \sum\limits_{k=2}^{\infty} \dfrac{1}{k(\ln k)^2}$; use integral test.

$$I = \lim_{b\to\infty} \int_2^b (\ln x)^{-2}\, \frac{dx}{x} = \lim_{b\to\infty} \frac{-1}{\ln x}\Big|_2^b = \frac{1}{\ln 2}$$

which converges, so S converges.

40. $S = \sum\limits_{k=1}^{\infty} \dfrac{3^k k!}{k^k}$; use the ratio test.

$$\lim_{k\to\infty} \frac{3^{k+1}(k+1)3^k}{(k+1)^{k+1}3^k k!} = 3 \lim_{k\to\infty} \frac{1}{\left(1+\frac{1}{k}\right)^k}$$

$$= \frac{3}{e} > 1, \text{ so } S \text{ diverges.}$$

41. $S = \sum\limits_{k=2}^{\infty} \dfrac{1}{(\ln k)^{1/k}}$; use divergence test.

$$\lim_{k\to\infty} (\ln k)^{-1/k} = \lim_{k\to\infty} \exp[\ln(\ln k)^{-1/k}]$$

$$= \lim_{k\to\infty} e^{-1/(k \ln k)} = 1 \neq 0;\ S \text{ diverges}$$

42. $S = \sum\limits_{k=1}^{\infty} e^{-k^2}$; use root test.

$$\lim_{k\to\infty} \sqrt[k]{e^{-k^2}} = \lim_{k\to\infty} e^{-k} = 0 < 1, \text{ so } S$$

diverges

43. $S = \sum\limits_{k=0}^{\infty} (\sqrt{k^3+1} - \sqrt{k^3})$

Note that $\sqrt{k^3+1} - \sqrt{k^3}$

$$= \frac{(\sqrt{k^3+1} - \sqrt{k^3})(\sqrt{k^3+1} + \sqrt{k^3})}{\sqrt{k^3+1} + \sqrt{k^3}}$$

$$= \frac{1}{\sqrt{k^3+1} + \sqrt{k^3}} \leq \frac{1}{2\sqrt{k^3}} = \frac{1}{2k^{3/2}}$$

Directly compare S with the convergent p-series with $p = 3/2 > 1$; thus S converges.

44. $S = \sum\limits_{k=1}^{\infty} \dfrac{1}{\sqrt{k(k+1)}}$; use limit comparison with

the harmonic series to conclude that S diverges.

45. $S = \sum\limits_{k=1}^{\infty} \dfrac{k-1}{k2^k}$; use a limit comparison with

the convergent geometric series $\sum\limits_{k=1}^{\infty} \dfrac{1}{2^k}$, so S

converges.

46. $S = \sum\limits_{k=0}^{\infty} \dfrac{k!}{k^2(k+1)^2}$; use the ratio test.

$$\lim_{k\to\infty} \frac{(k+1)!k^2(k+1)^2}{(k+1)^2(k+2)^2 k!} = \lim_{k\to\infty} (k+1)$$

$$= \infty > 1, \text{ so } S \text{ diverges.}$$

47. $S = \sum\limits_{k=0}^{\infty} \dfrac{1}{1+\sqrt{k}}$; use limit comparison with

the divergent p-series $T = \sum\limits_{k=0}^{\infty} \dfrac{1}{k^{1/2}}$, so S

diverges.

48. $S = \sum\limits_{k=1}^{\infty} \dfrac{k^2 3^k}{k!}$; use the ratio test.

$$\lim_{k\to\infty} \frac{(k+1)^2 3^{k+1} k!}{(k+1)!k^2 3^k} = 3 \lim_{k\to\infty} \frac{1}{k+1}$$

$$= 3 > 1, \text{ so } S \text{ diverges.}$$

49. $S = \dfrac{1}{1\cdot 2} - \dfrac{1}{3\cdot 2} + \dfrac{1}{5\cdot 2} - \dfrac{1}{7\cdot 2} + \cdots$

$$= \frac{1}{2}\sum_{k=1}^{\infty} \frac{(-1)^k}{(2k-1)}$$

S is decreasing (because $\dfrac{1}{2k+1} \leq \dfrac{1}{2k-1}$ and

$\lim\limits_{k\to\infty} \dfrac{1}{(2k-1)} = 0$), so S converges.

Since $T = \sum\limits_{k=1}^{\infty} \dfrac{1}{(2k-1)}$ diverges (compare

with the harmonic series), S converges conditionally.

50. $S = \dfrac{1}{1 \cdot 2} - \dfrac{1}{2 \cdot 3} + \dfrac{1}{3 \cdot 4} - \dfrac{1}{4 \cdot 5} + \ldots$

$= \sum\limits_{k=1}^{\infty} \dfrac{(-1)^{k+1}}{k(k+1)}$

S is decreasing (because $\dfrac{1}{(k+1)(k+2)} \leq \dfrac{1}{k(k+1)}$

and $\lim\limits_{k \to \infty} \dfrac{1}{k(k+1)} = 0$), so S converges.

Since $T = \sum\limits_{k=0}^{\infty} \dfrac{1}{k(k+1)}$ converges (compare

with the p-series with $p = 2$), S converges absolutely.

51. $S = \dfrac{3}{2} - \dfrac{4}{3} + \dfrac{5}{4} - \dfrac{6}{5} + \ldots$ diverges by the

divergence test since $\lim\limits_{k \to \infty} \dfrac{k+1}{k} \neq 0$; that is

the necessary condition for convergence is not satisfied.

52. $S = 1 - \dfrac{1}{3} + \dfrac{1}{9} - \dfrac{1}{27} + \dfrac{1}{81} - \ldots$

The series $T = \sum\limits_{k=0}^{\infty} \dfrac{1}{3^k}$ converges because it is

a geometric series with ratio $r = \dfrac{1}{3} < 1$, we conclude S converges absolutely.

53. $S = 1 - \dfrac{1}{2} + \dfrac{1}{3} - \dfrac{1}{4} + \ldots = \sum\limits_{k=0}^{\infty} \dfrac{1}{k+1}$

S is decreasing because $\dfrac{1}{k+1} \leq \dfrac{1}{k}$ and

$\lim\limits_{k \to \infty} \dfrac{1}{k+1} = 0$, so S converges.

Since $T = \sum\limits_{k=1}^{\infty} \dfrac{1}{k}$ diverges (harmonic series),

S converges conditionally.

54. $S = -1 + \dfrac{1}{\sqrt{2}} - \dfrac{1}{\sqrt[3]{3}} + \dfrac{1}{\sqrt[4]{4}} - \ldots$

diverges by the divergence test.

55. $S = \sum\limits_{k=1}^{\infty} \dfrac{(-1)^k}{k \, [\ln(k+1)]^2}$; let

$T = \sum\limits_{k=1}^{\infty} \dfrac{1}{k[\ln(k+1)]^2}$; use the integral test.

$I = \lim\limits_{b \to \infty} \int\limits_{1}^{b} \dfrac{dx}{x[\ln(x+1)]^2}$

$= -\lim\limits_{b \to \infty} \dfrac{1}{\ln(x+1)} \Big|_{1}^{b} = \dfrac{1}{\ln 2}$

Then I and T converges, and therefore S

converges absolutely.

56. $S = \sum\limits_{k=1}^{\infty} (-1)^k \left(\dfrac{3k+85}{4k+1} \right)^k$; let

$T = \sum\limits_{k=1}^{\infty} \left(\dfrac{3k+85}{4k+1} \right)^k$; use limit comparison

test with $\sum\limits_{k=1}^{\infty} \dfrac{3^k}{4^k}$ to see T converges so that S

converges absolutely.

57. $S = \sum\limits_{k=1}^{\infty} (-1)^k \tan^{-1} \left(\dfrac{1}{2k+1} \right)$

S is decreasing because

$\tan^{-1} \left(\dfrac{1}{2k+3} \right) \leq \tan^{-1} \left(\dfrac{1}{2k+1} \right)$

and $\lim\limits_{k \to \infty} \tan^{-1} \left(\dfrac{1}{2k+1} \right) = 0$, so S

converges. Since $T = \sum\limits_{k=1}^{\infty} \tan^{-1} \left(\dfrac{1}{2k+1} \right)$

diverges by the limit comparison test (with the harmonic series), we see that S converges conditionally.

58. $S = \sum\limits_{k=1}^{\infty} \dfrac{(-1)^{k(k+1)/2}}{2^k}$ converges absolutely

since $\sum\limits_{k=1}^{\infty} \dfrac{1}{2^k}$ is a convergent geometric series.

59. $\lim\limits_{k \to \infty} \left| \dfrac{(k+1)x^{k+1} 3^k}{3^{k+1} k x^k} \right| = \dfrac{1}{3}|x| < 1$ if

$-3 < x < 3$. At $x = \pm 3$, S diverges by the divergence test. The interval of convergence is $(-3, 3)$.

60. $\lim\limits_{k \to \infty} \left| \dfrac{(k+1)(x-1)^{k+1}}{k(x-1)^k} \right| = |x-1| < 1$

if $0 < x < 2$. At $x = 0$ and at $x = 2$, the series diverges by the divergence test. The interval of convergence is $(0, 2)$.

61. $\lim\limits_{k \to \infty} \left| \dfrac{x^{k+1} k(k+1)}{(k+1)(k+2)x^k} \right| = |x| < 1$ is

$-1 < x < 1$. At $x = -1$, the series converges absolutely (directly compare with the p-series with $p = 2$). At $x = 1$, the series converges. The interval of convergence is $[-1, 1]$.

62. $\lim\limits_{k \to \infty} \left| \dfrac{\ln(k+1)x^{2k+2}\sqrt{k}}{\sqrt{k+1} \, \ln k x^{2k}} \right| = x^2 < 1$ if

$-1 < x < 1$. At $x = -1$, the series converges because if

$f(x) = \dfrac{\ln x}{\sqrt{x}};$

$$f'(x) = \frac{\sqrt{x}\,x^{-1} - (\ln x)\left(\frac{1}{2}\right)x^{-1/2}}{x} < 0$$

Thus $\left\{\dfrac{\ln k}{\sqrt{k}}\right\}$ is decreasing and $\displaystyle\lim_{k \to \infty} \dfrac{\ln k}{\sqrt{k}} = 0$.

At $x = 1$, the series diverges because

$$\sum_{k=0}^{\infty} \frac{\ln k(x^2)^k}{\sqrt{k}} \geq \sum_{k=0}^{\infty} \frac{1}{k^{1/2}}.$$ The interval of

convergence is $[-1, 1)$.

63. $\displaystyle\lim_{k \to \infty}\left|\frac{(k+1)^2(x+1)^{2k+2}2^k}{2^{k+1}k^2(x+1)^{2k}}\right| = \frac{1}{2}(x+1)^2 < 1$

if $-1 - \sqrt{2} < x < -1 + \sqrt{2}$.
At $x = -1 \pm \sqrt{2}$, the series converges
because by the root test,

$$\lim_{k \to \infty}\frac{k^{1/k}}{2} = \frac{1}{2}\lim_{k \to \infty}\exp[\ln k^{2/k}] = \frac{1}{2} < 1$$

The interval of convergence is

$[-1 - \sqrt{2},\ -1 + \sqrt{2}]$.

64. $\displaystyle\lim_{k \to \infty}\left|\frac{x^{k+1}k^k}{(k+1)^{k+1}x^k}\right|$

$= |x|\displaystyle\lim_{k \to \infty}\left[\frac{k}{k+1}(x+1)^2\right]^k\frac{1}{k+1}$

$= |x|\displaystyle\lim_{k \to \infty}\left[\left(1 + \frac{1}{k}\right)^k\right]^{-1} = e|x|(0) < 1$ for all

x. The interval of convergence is $(-\infty, \infty)$.

65. $\displaystyle\lim_{k \to \infty}\left|\frac{(2x-1)^{k+1}k^2}{(k+1)^2(2x-1)^k}\right| = |2x-1| < 1$ if

$0 < x < 1$. At $x = 0$, the series converges
absolutely because $\displaystyle\sum_{k=0}^{\infty}\frac{1}{k^2}$ is a convergent p-
series, so the series also converges at $x = 1$.
The interval of convergence is $[0, 1]$.

66. $\displaystyle\lim_{k \to \infty}\left|\frac{(x+2)^{k+1}k\ln(k+1)}{(k+1)\ln(k+2)(x+2)^k}\right| = |x+2| < 1$

if $-3 < x < -1$. At $x = -1$, the series
diverges by the integral test:

$$I = \int_{1}^{\infty}[\ln(x+1)]^{-1}\frac{dx}{x} = \infty$$

At $x = -3$, the series converges because

$$\frac{1}{(k+1)\ln(k+2)} \leq \frac{1}{k\ln(k+1)}$$

and $\displaystyle\lim_{k \to \infty}\frac{1}{k\ln(k+1)} = 0$. The interval of

convergence is $[-3, -1]$.

67. $\displaystyle\lim_{k \to \infty}\left|\frac{(k+1)(x-3)^{k+1}(k+3)!}{(k+4)!k(x-3)^k}\right|$

$= |x-3|\left(\displaystyle\lim_{k \to \infty}\frac{1}{k+4}\right) = 0 < 1$ for all x.

The interval of convergence is $(-\infty, \infty)$.

68. $\displaystyle\lim_{n \to \infty}\left|\frac{(3n+3)!x^{n+1}n!}{(n+1)!(3n)!x^n}\right|$

$= |x|\left(\displaystyle\lim_{n \to \infty}[3(3n+2)(3n+1)]\right) = \infty$

The interval of convergence is reduced to a
single point, $x = 0$.

69. $\displaystyle\lim_{n \to \infty}\left|\frac{(3n+1)(3n-1)x^{n+1}n!}{(3n+4)(3n+2)x^n}\right| = |x| < 1$ if

$-1 < x < 1$. At $x = -1$, the series converges
(use the limit comparison with the convergent
p-series with $p = 2$). At $x = 1$, the series also
converges, so that the interval of convergence
is $[-1, 1]$.

70. $x - \dfrac{x^3}{3!} + \dfrac{x^5}{5!} - \dfrac{x^7}{7!} + \ldots + \dfrac{(-1)^{k+1}x^{2k-1}}{(2k-1)!}$

$+ \ldots = \displaystyle\sum_{k=0}^{\infty}\frac{(-1)^k x^{2k+1}}{(2k+1)!} = \sin x$, which

converges for all x. Thus, the interval of
convergence is $(-\infty, \infty)$.

71. $f(x) = x^2 e^{-3x} = x^2\displaystyle\sum_{k=0}^{\infty}\frac{(-1)^k(3x)^k}{k!}$

$= \displaystyle\sum_{k=0}^{\infty}(-1)^k\frac{3^k x^{k+2}}{k!}$

72. $f(x) = x^3\cos x = x^3\displaystyle\sum_{k=0}^{\infty}\frac{(-1)^k(x)^{2k+1}}{(2k+1)!}$

$= \displaystyle\sum_{k=0}^{\infty}(-1)^k\frac{x^{2k+4}}{(2k+1)!}$

73. $f(x) = 3^x = 1 + x\ln 3 + \dfrac{1}{2!}(x\ln 3)^2 + \dfrac{1}{3!}(x\ln 3)^3 + \cdots$

Series Formula 28, p. 150 of *Mathematics Handbook*.
    Alternately, $f'(x) = 3^x\ln 3$; $f''(x) = 3^x(\ln 3)^2$,
$\cdots f^{(n)}(x) = 3^x(\ln 3)^n$

74. $f(x) = \cos^3 x = \dfrac{1}{4}[\cos 3x + 3\cos x]$

$= \dfrac{1}{4}\displaystyle\sum_{k=0}^{\infty}(-1)^k\frac{3^{2k} + 3}{(2k)!}x^{2k}$

75. $f(x) = \dfrac{5 + 7x}{1 + 2x - 3x^2} = -\dfrac{7x - 5}{(3x+1)(x-1)}$

$= \dfrac{2}{3x+1} + \dfrac{3}{1-x}$

$= 2\displaystyle\sum_{k=0}^{\infty}(-1)^k(3x)^k + 3\displaystyle\sum_{k=0}^{\infty}x^k$

$$= \sum_{k=0}^{\infty} [2(-3)^k + 3]x^k$$

76. $f(x) = \dfrac{11x - 1}{2 + x - 3x^2} = \dfrac{11x - 1}{(3x + 2)(x - 1)}$

$$= \dfrac{-5}{3x + 2} + \dfrac{2}{1 - x}$$

$$= -\dfrac{5}{2}\sum_{k=0}^{\infty}(-1)^k(\tfrac{3}{2}x)^k + 2\sum_{k=0}^{\infty} x^k$$

$$= \sum_{k=0}^{\infty}[2 + 5(-1)^{k-1}(\tfrac{3}{2})^k]x^k$$

77. $1 - \dfrac{1}{2} + \dfrac{1}{2!}\left(\dfrac{1}{2}\right)^2 - \dfrac{1}{3!}\left(\dfrac{1}{2}\right)^3 + \ldots = e^{-1/2}$

≈ 0.607

78. $|S - S_8| \le a_9 = \dfrac{8}{9!} \approx 0.000025$

79. $12.342\overline{132} = 12.234 + 0.000132 +$

$(132)10^{-9} + (132)10^{-12} + \cdots$

$$= \dfrac{12{,}342}{1{,}000} + \dfrac{132}{1{,}000{,}000}[1 + 10^{-3} + (10^{-3})^2$$

$$+ \cdots + (10^{-3})^k + \cdots]$$

$$= \dfrac{12{,}329{,}790}{990{,}000} = \dfrac{410{,}993}{33{,}300}$$

80. $20 = A + 2A[0.8 + (0.8)^2 + \cdots + (0.8)^{k+1} + \cdots]$

$$= A\left(1 + 1.6[1 + 0.8 + (0.8)^2 + \cdots]\right)$$

$$= A\left(1 + \dfrac{16}{1 - 0.8}\right) = 9A$$

Thus, $A = \dfrac{20}{9}$ ft.

81. $\tan x = x + \dfrac{x^3}{3} + \dfrac{2x^5}{15} + \dfrac{17x^7}{315} + \cdots$

$$\int \tan x \, dx = \int \left[x + \dfrac{x^3}{3} + \dfrac{2x^5}{15} + \dfrac{17x^7}{315} + \cdots\right]dx$$

$-\ln|\cos x| = \dfrac{1}{2}x^2 + \dfrac{1}{12}x^4 + \dfrac{1}{45}x^6 + \cdots$

$\ln|\cos x| = -\dfrac{1}{2}x^2 - \dfrac{1}{12}x^4 - \dfrac{1}{45}x^6 - \cdots$

82. $f(x) = \cos x; \ f'(x) = -\sin x; \ f''(x) = -\cos x;$

$f'''(x) = \sin x, \cdots$

$\cos x = \cos c - (x - c)\sin c$

$\quad - \dfrac{1}{2!}(\cos c)(x - c)^2 + \dfrac{1}{3!}(\sin c)(x - c)^3 + \cdots$

83. $f(x) = e^x; \ f'(x) = e^x; \ f''(x) = e^x; \cdots$

$e^x = e^c + e^c(x - c) + \dfrac{1}{2!}e^c(x - c)^2 + \cdots$

84. $\sqrt{x^3 + 1} = (x^3 + 1)^{1/2} = 1 + \dfrac{1}{2}x^3$

$$+ \dfrac{\frac{1}{2}(-\frac{1}{2})}{2!}(x^3)^2 + \dfrac{\frac{1}{2}(-\frac{1}{2})(-\frac{3}{2})}{3!}(x^3)^3 + \cdots$$

$$= 1 + \dfrac{1}{2}x^3 - \dfrac{1}{8}x^6 + \dfrac{5}{16}x^9 + \cdots$$

$$\int \sqrt{x^3 + 1} \, dx = \int\left[1 + \dfrac{1}{2}x^3 - \dfrac{1}{8}x^6 + \dfrac{5}{16}x^9 + \cdots\right]dx$$

$$= x + \dfrac{1}{8}x^4 - \dfrac{1}{2}x^7 + \dfrac{5}{160}x^{10} + \cdots$$

85. $\displaystyle\int_0^{0.4} \dfrac{dx}{\sqrt{1 + x^3}}$

$$= \int_0^{0.4}\left[1 - \dfrac{1}{2}x^3 + \dfrac{(1)(3)}{2^2(2!)}x^6 - \dfrac{(1)(3)(5)}{2^3(3!)}x^9 + \cdots\right]dx$$

$$= x - \dfrac{1}{8}x^4 + \dfrac{3}{56}x^7 - \dfrac{1}{32}x^{10} + \cdots\Big|_0^{0.4}$$

≈ 0.396884

86. $f(x) = a^x, \ f(0) = 1; \ f'(x) = a^x\ln a, \ f'(0) =$

$\ln a; \ f''(x) = a^x(\ln a)^2; \ f''(0) = (\ln a)^2; \cdots$

$a^x = 1 + (\ln a)x + \dfrac{1}{2!}(\ln a)^1 x^2 + \dfrac{1}{3!}(\ln a)^3 x^3 +$

$\cdots = \displaystyle\sum_{k=0}^{\infty}\dfrac{1}{k!}(\ln a)^k x^k$

87. $f(x) = x^{-1}, \ f(c) = c^{-1}; \ f'(x) = -x^{-2},$

$f'(c) = -c^{-2}; \ f''(x) = 2x^{-3}, f''(c) = 2c^{-3};$

$\dfrac{1}{x} = \dfrac{1}{c} - \dfrac{1}{c^2}(x - c) + \dfrac{1}{c^3}(x - c)^2 + \dfrac{1}{c^4}(x - c)^3 + \cdots$

88. $\sin u = u - \dfrac{1}{3!}u^3 + \dfrac{1}{5!}u^5 - \cdots$

$\sin 0.2 \approx 0.2 - \dfrac{1}{3!}(0.2)^3 + \dfrac{1}{5!}(0.2)^5 + \cdots$

≈ 0.198670

89. $\ln u = (u - 1) - \dfrac{1}{2}(u - 1)^2 + \dfrac{1}{3}(u - 1)^3 - \cdots$

$\ln 1.05 \approx 0.05 - \dfrac{1}{2}(0.05)^2 + \dfrac{1}{3}(0.05)^3 - \cdots$

≈ 0.048790

90. $e^{2x} \approx 1 + 2x + \dfrac{1}{2!}x^2 + \dfrac{1}{3!}x^3$

$$R_3(x) = \dfrac{2^4 e^{2z_c}z_c^4}{4!} \le \dfrac{2}{3}(10)^{-4}$$

$$= 6.6(10^{-13}) < 10^{-12}$$

91. $R_n(x) = \dfrac{f^{(n+3)}(z_n)}{(n + 1)!}(x - c)^{n+1};$

if $0 < x < 1, |1 - x| < 1$

$\left|f^{(k)}(x)\right| = \left|k!(1 - x)^{-(k+1)}\right| < \left|(k + 1)!\right|$

because $0 < x < 1$ and

$\left|R_n(x)\right| < \dfrac{n!}{(n + 1)!}x^{n+1} = \dfrac{x^{n+1}}{n + 1}$

If $-1 \le x < 0, \ 1 < |1 - x|$ but

$\dfrac{1}{|1 - x|} < 1$ and

$$\left| R_n(x) \right| < \frac{n!}{(n+1)!(1-x)^{n+1}} x^{n+1} = \frac{x^{n+1}}{n+1}$$

For ln 1.2,

$$\left| R_n \right|(0.0005) < \frac{(0.2)^{n+1}}{n+1}$$

$n > 8$ to guarantee 6 place accuracy, so for three terms the error is at most

$$\frac{(0.2)^4}{4} = 0.0004$$

92. $\sin x = x - \frac{1}{3!}x^3 + R_4(x)$

$\left| R_4(x) \right| \leq \frac{1}{5!}x^5 = 0.0005$, so $|x| < 0.5797$

93. $\cos x = 1 - \frac{1}{2!}x^2 + R_3(x)$

$\left| R_3(x) \right| \leq \frac{1}{4!}x^4 = 000005$, so $|x| < 0.187$

94. $\dfrac{2\tan x}{1 + \tan^2 x} = \dfrac{2\tan x}{\sec^2 x} = \dfrac{2\sin x \cos^2 x}{\cos x} = \sin 2x$

$$= \sum_{k=0}^{\infty} (-1)^k \frac{(2x)^{2k+1}}{(2k+1)!}$$

95. **a.** $S_n = \frac{n}{2}(a + a_n)$ or

$$S_n = \frac{n}{2}\left[2a + (n-1)d\right]$$

b. $S_n = \dfrac{a}{1-r} - \left(\dfrac{a}{1-r}\right)r^n$

$$= \frac{a(1-r^n)}{1-r}$$

c. If $|r| < 1$, $\lim\limits_{k \to \infty} |r^n| = 0$, so

$$\lim_{k \to \infty} S_n = \frac{a}{1-r}$$

96. This is Putnam Problem A2 in the morning session of 1984. Let S_n denote the nth partial sum of the series. Then

$$S_n = \sum_{k=1}^{n} \frac{6^k}{(3^{k+1} - 2^{k+1})}$$

$$= \sum_{k=1}^{n} \left[\frac{3^k}{3^k - 2^k} - \frac{3^{k+1}}{3^{k+1} - 2^{k+1}}\right]$$

$$= 3 - \frac{3^{n+1}}{3^{n+1} - 2^{n-1}}$$

and the series converges to

$$\lim_{n \to \infty} S_n = 3 - 1 = 2$$

97. This is Putnam Problem A-4 in the morning session of 1977.

$$\sum_{k=0}^{N} \frac{x^{2^n}}{1 - x^{2^{n+1}}}$$

$$= \sum_{k=0}^{\infty} \left[\frac{1}{1 - x^{2^n}} - \frac{1}{1 - x^{2^{n+1}}}\right]$$

$$= \lim_{N \to \infty} \left[\frac{1}{1 - x} - \frac{1}{1 - x^{2^{n+1}}}\right]$$

$$= \frac{1}{1-x} - 1 = \frac{x}{1-x} \text{ since } |x| < 1$$

98. This is Putnam Problem A-2 in the morning session of 1982. Since $x = \log_n 2 > 0$,

$$B_n(x) = 1^x + 2^x + 3^x + \cdots + n^x \leq n(n^x)$$

and

$$0 \leq \frac{B_n(\log_n 2)}{(n \log_2 n)^2} \leq \frac{n(n^{\log_n 2})}{(n \log_2 n)^2}$$

As

$$\sum_{n=2}^{\infty} \frac{2}{n(\log_2 n)^2}$$

converges by the integral test, the given series converges by the comparison test.

CHAPTER 9

Polar Coordinates and the Conic Sections

9.1 The Polar Coordinate System, Page 575

1. To plot (r, θ) first use a moving ray with fixed point at the pole and make an angle θ with the positive x-axis. The distance r along the ray is the point that is plotted.

2. In the right triangle, r is the hypotenuse, x and y are the horizontal and vertical legs, respectively. and the angle θ at the origin. Then from the definitions of the trigonometric functions we have $x = r \cos \theta$ and $y = r \sin \theta$.

   Also, $\tan \theta = y/x$ and $r = \sqrt{x^2 + y^2}$.

3. polar: $(4, \frac{\pi}{4})$, $(-4, \frac{5\pi}{4})$; rect.: $(2\sqrt{2}, 2\sqrt{2})$

4. polar: $(6, \frac{\pi}{3})$, $(-6, \frac{4\pi}{3})$; rect.: $(3, 3\sqrt{3})$

5. polar: $(5, \frac{2\pi}{3})$, $(-5, \frac{5\pi}{3})$; rect.: $(-\frac{5}{2}, \frac{5}{2}\sqrt{3})$

6. polar: $(3, \frac{11\pi}{6})$, $(-3, \frac{5\pi}{6})$; rect.: $(\frac{3}{2}\sqrt{3}, -\frac{3}{2})$

7. polar: $(\frac{3}{2}, \frac{7\pi}{6})$, $(-\frac{3}{2}, \frac{\pi}{6})$; rect.: $(-\frac{3}{4}\sqrt{3}, -\frac{3}{4})$

8. polar: $(-4, 4)$, $(4, 0.86)$; rect.: $(2.61, 3.03)$

9. polar: $(1, \pi)$, $(-1, 0)$; rect.: $(-1, 0)$

10. polar: $(-2, -\frac{3\pi}{2})$, $(2, \frac{\pi}{2})$; rect.: $(0, 2)$

11. polar: $(0, \pi - 3)$ and $(0, \theta)$ for any number θ; rect.: $(0, 0)$

Plotted points for Problems 3-11.

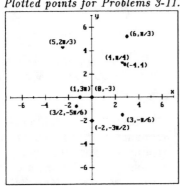

12. $(5\sqrt{2}, \frac{\pi}{4})$, $(-5\sqrt{2}, \frac{5\pi}{4})$

13. $(2, \frac{2\pi}{3})$, $(-2, \frac{5\pi}{3})$ 14. $(4, \frac{5\pi}{3})$, $(-4, \frac{2\pi}{3})$

15. $(2\sqrt{2}, \frac{5\pi}{4})$, $(-2\sqrt{2}, \frac{\pi}{4})$

16. $(3\sqrt{2}, \frac{7\pi}{4})$, $(-3\sqrt{2}, \frac{3\pi}{4})$

17. $(\sqrt{58}, \tan^{-1}\frac{17}{3}) \approx (7.6, 1.2)$,
    $(-\sqrt{58}, \pi + \tan^{-1}\frac{17}{3}) \approx (-7.6, 4.3)$

18. $(6, \frac{5\pi}{3})$, $(-6, \frac{2\pi}{3})$ 19. $(2, \frac{11\pi}{3})$, $(-2, \frac{5\pi}{6})$

20. $(-3, 0)$, $(3, \pi)$

Plotted points for Points 12-20.

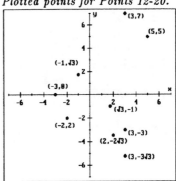

21. **a.** United States   **b.** India
    c. Greenland   **d.** Canada

22. **a.** $(5, 280°)$ **b.** $(5, 240°)$ **c.** $(6, 260°)$ **d.** $(3.5, 40°)$

23.
$$r = 4 \sin \theta$$
$$r^2 = 4r \sin \theta$$
$$x^2 + y^2 = 4y$$
$$x^2 + (y - 2)^2 = 4$$

24.
$$r = 16$$
$$r^2 = 256$$
$$x^2 + y^2 = 256$$

25.
$$r = 1 - \sin \theta$$
$$r^2 = r - r \sin \theta$$
$$x^2 + y^2 = \sqrt{x^2 + y^2} - y$$
$$x^2 + y^2 + y = \sqrt{x^2 + y^2}$$
$$x^2 + y^2 = (x^2 + y^2 + y)^2$$

26.
$$r = 2 \cos \theta$$
$$r^2 = 2r \cos \theta$$
$$x^2 + y^2 = 2x$$
$$(x - 1)^2 + y^2 = 1$$

27.
$$r = \sec \theta$$
$$r \cos \theta = 1$$
$$x = 1$$

28.
$$r = 4 \tan \theta$$
$$r \cos \theta = 4 \sin \theta$$
$$r \cos \theta = \frac{4r \sin \theta}{r}$$
$$x = \frac{4y}{\sqrt{x^2 + y^2}}$$
$$x\sqrt{x^2 + y^2} = 4y$$
$$x^2(x^2 + y^2) = 16y^2$$

29.
$$r^2 = \frac{2}{1 + \sin^2\theta}$$
$$r^2 + r^2 \sin^2\theta = 2$$
$$x^2 + y^2 + y^2 = 2$$
$$x^2 + 2y^2 = 2$$

30.
$$r^2 = \frac{2}{3 \cos^2\theta - 1}$$
$$3r^2 \cos^2\theta - r^2 = 2$$
$$3x^2 - x^2 - y^2 = 2$$
$$2x^2 - y^2 = 2$$

31.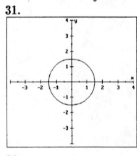

32.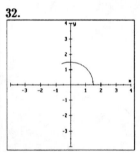

33.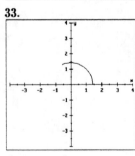

34.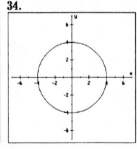

35.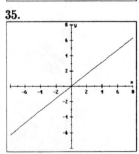

36.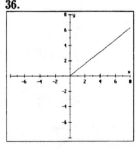

37. 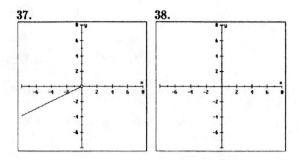   **38.**

39. yes   **40.** no   **41.** no   **42.** no   **43.** yes

44. no   **45.** yes   **46.** yes   **47.** no   **48.** yes

49. no   **50.** yes   **51.** no   **52.** yes

Answers to Problems 53-60 may vary.

53. $(3, 0)$ and $(-3, \pi)$; $(3, \pi)$ and $(-3, 0)$;
$(3, 2)$ and $(-3, 2 + \pi)$

54. $(3, 0)$ and $(-3, \pi)$; $(3, \pi)$ and $(-3, 0)$;
$(3, 2)$ and $(-3, 2 + \pi)$

55. $(1, \frac{1}{3})$ and $(-1, \frac{1}{3} + \pi)$; $(-1, -\frac{1}{3})$ and
$(1, \pi - \frac{1}{3})$; $(0, 2)$ and $(0, 2 + \pi)$

56. $(1, \frac{1}{5})$, $(-1, \frac{1}{5} + \pi)$; $(-1, -\frac{1}{5})$
and $(1, \pi - \frac{1}{5})$; $(0, 1)$ and $(0, 1 + \pi)$

57. $(2, 0)$ and $(-2, \pi)$; $(\frac{1}{2}, \frac{\pi}{6})$ and $(-\frac{1}{2}, \frac{7\pi}{6})$;
$(\frac{1}{2}, \frac{5\pi}{6})$ and $(-\frac{1}{2}, \frac{11\pi}{6})$

58. $(4, 0)$ and $(-4, \pi)$; $(2, \frac{\pi}{2})$ and $(-2, \frac{3\pi}{2})$;
$(0, 3)$ and $(0, \pi + 3)$

59. $(2, \pi)$ and $(-2, 0)$; $(2, 0)$ and $(-2, \pi)$;
$(4, \frac{3\pi}{2})$ and $(-4, \frac{\pi}{2})$

60. $(-8, 0)$ and $(8, \pi)$; $(8, \frac{\pi}{2})$ and $(-8, \frac{3\pi}{2})$;
$(\frac{8}{3}, \pi)$ and $(-\frac{8}{3}, 0)$

61. a. The distance formula is for rectangular coordinates. When given polar coordinates, you need to convert to rectangular coordinates or use the law of cosines: $c^2 = a^2 + b^2 - 2ab \cos \gamma$
$$d = \sqrt{3^2 + 7^2 - 2(3)(7)\cos(\frac{\pi}{3} - \frac{\pi}{4})}$$
$$= \sqrt{58 - 42 \cos \frac{\pi}{12}} \approx 4.175$$
b. $d = \sqrt{r_1^2 + r_2^2 - 2r_1 r_2\cos(\theta_2 - \theta_1)}$

62. $a^2 = r^2 + R^2 - rR\cos(\theta - \alpha)$

63.
$$r = a \sin \theta + b \cos \theta$$
$$r^2 = ar \sin \theta + br \cos \theta$$
$$x^2 + y^2 = ay + bx$$
$$x^2 - bx + y^2 - ay = 0$$
$$x^2 - bx + \frac{b^2}{4} + y^2 - ay + \frac{a^2}{4} = \frac{a^2}{4} + \frac{b^2}{4}$$
$$\left(x - \frac{b}{2}\right)^2 + \left(y - \frac{a}{2}\right)^2 = \frac{a^2 + b^2}{4}$$

We recognize this as a circle with center $\left(\frac{b}{2}, \frac{a}{2}\right)$ and radius $\dfrac{\sqrt{a^2 + b^2}}{2}$.

9.2 Graphing in Polar Coordinates, Page 584

1. If polar-form equation has a form which is shown in Table 9.1, Page 583, plot the critical points and sketch the curve. If the equation has a form you do recognize, then plot points.

2. Let $P(r, \theta)$ satisfy $r = f(\theta)$;
If $(r, -\theta)$ also satisfies $r = f(\theta)$, then the curve is symmetric with respect to the x-axis.
If $(r, \pi - \theta)$ also satisfies $r = f(\theta)$, then the curve is symmetric with respect to the y-axis.
If $(-r, \theta)$ also satisfies $r = f(\theta)$, then the curve is symmetric with respect to the origin.
If any two of these symmetries hold, then the third one necessarily holds.

3. The limaçon $r = a(1 - b \cos \theta)$ has an inner loop ($b > 1$). The cardioid is a special case of the limaçon ($b = 1$). The rose $r = \cos n\theta$, $n > 1$, has n petals if n is odd, and $2n$ petals if n is even. The lemniscate $r^2 = a \cos 2\theta$ is like a rose with two leaves, but covering only half the xy-plane since $r^2 \geq 0$.

4.
 a. lemniscate   **b.** circle
 c. rose (3 petals)   **d.** none (spiral)
 e. cardioid   **f.** line
 g. lemniscate   **h.** limaçon

5.
 a. rose (4 petals)   **b.** lemniscate
 c. circle   **d.** rose (16 petals)
 e. none   **f.** lemniscate
 g. rose (3 petals)   **h.** cardioid

6.
 a. rose (4 petals)   **b.** circle
 c. limaçon   **d.** cardioid
 e. line   **f.** line
 g. rose (5 petals)   **h.** line

7.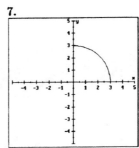

8.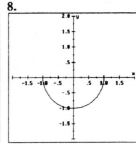

9.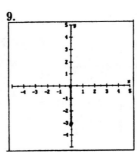

10.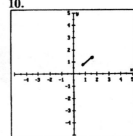

11.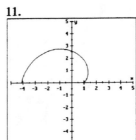

12.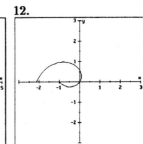

13.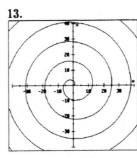

14.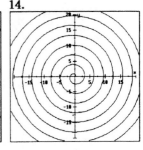

15.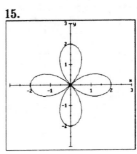

16.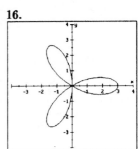

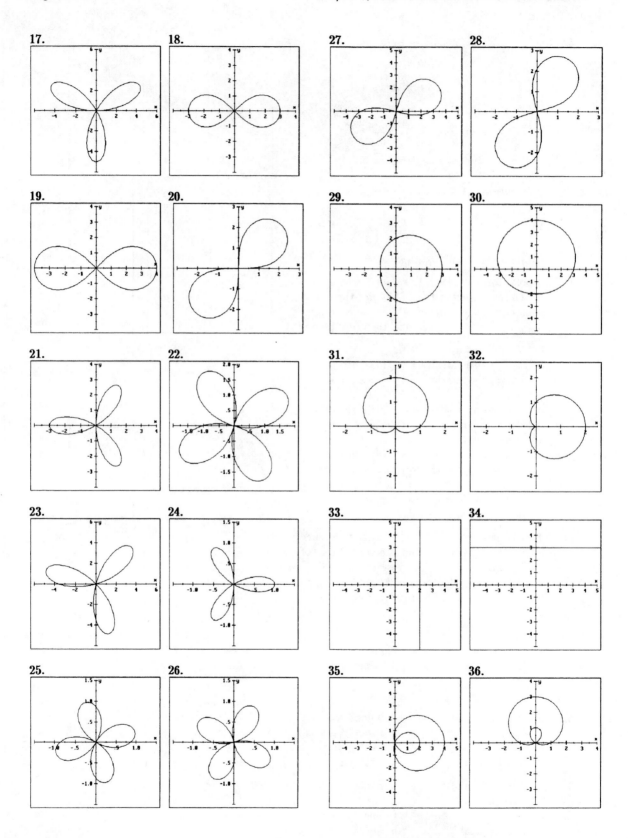

37. **38.** **47.** **48.**

39. **40.** **49.** **50.**

41. **42.** **51.** **52.**

43. **44.** **53.** **54.**

45. **46.** **55.**

Let $P(r, \theta)$ be a
point on $r = \cos\theta + 1$.
Then $(-r, \theta + \pi)$ is
the same point:
$-r = \cos(\theta + \pi) + 1$
$-r = \cos\theta\cos\pi$
$\qquad - \sin\theta\sin\pi + 1$
So
$r = \cos\theta - 1$

The graph shows that both graphs are the
same.

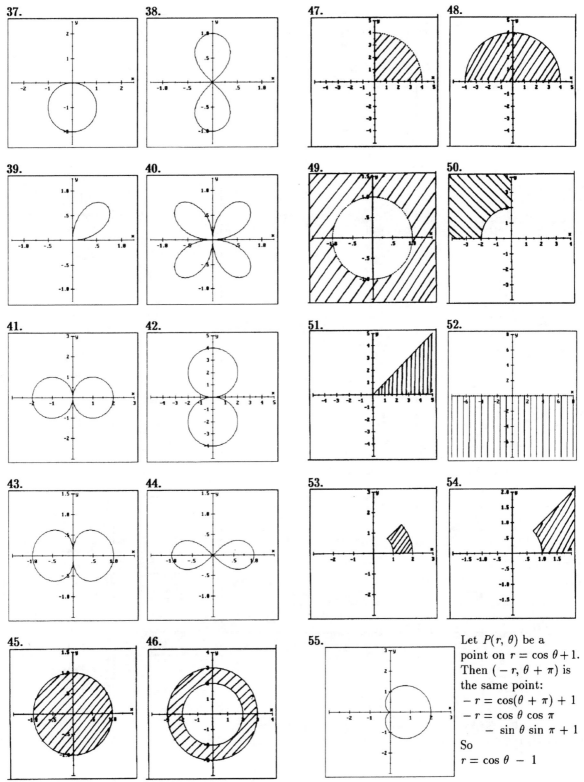

56. a.

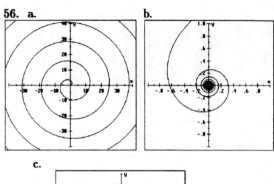

b.

c.

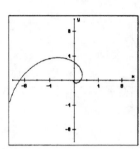

57.

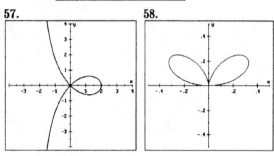

58.

59.

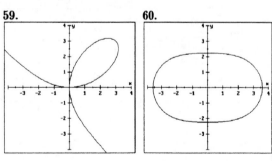

60.

61.

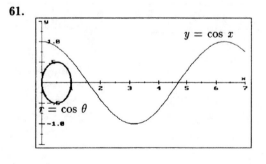

62.

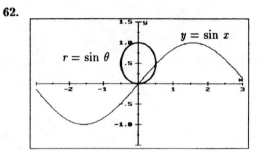

63.

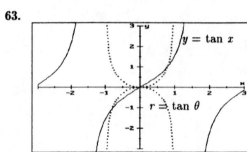

64.

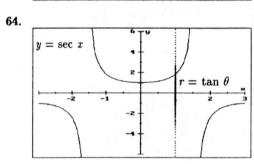

65.

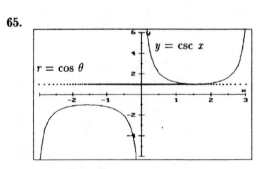

66.

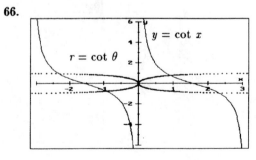

67. a. Graph for $0 < \theta < \pi$

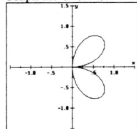

Graph for $0 < \theta < 2\pi$

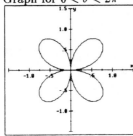

b. When θ is in the first or third quadrants, $r > 0$ and the portions of the curve are drawn in these quadrants. When θ is in the second quadrant, $r < 0$ and the portion of the curve is drawn in the opposite (fourth) quadrant. When θ is in the fourth quadrant, $r < 0$ and the portion of the curve is drawn in the opposite (second) quadrant.

c. $r = \sin m\theta$ has m petals when m is odd (the same curve is traced out twice). $r = \sin m\theta$ has $2m$ petals when m is even.

68. a. If n is odd, $r = \sin \frac{\theta}{n}$, the period, P, is $2n\pi$. Half the curve is traced out when $0 \le \theta \le n\pi$, the other half, which is symmetric with respect to the origin, when $n\pi \le \theta \le 2n\pi$.

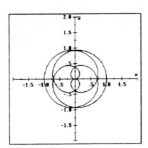

This graph is for $n = 4$.

b. If n is even and m odd, there are $2m$ loops. If n is odd and m is even, there are $2m$ loops. If n is odd and m is odd, there are 2 loops. The graph shown is for $m = 10$ and $n = 21$.

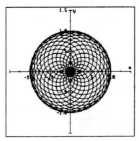

c. If m and n are both odd:
When $m\theta/n$ is in quadrants III or IV, $\sin(m\theta/n) < 0$, $r < 0$, the points are plotted on the extension of the ray, so they have already been plotted.

If m or n is even:
$2m\pi/n$ reduces to a multiple of π (not 2π). Thus, twice the angle needs to be chosen. For instance, if $m = 2$ and $n = 4$, $\alpha = m\theta/n = \theta/2$. If α is in quadrant I, $0 < \theta < \pi$. If α is in quadrant II, $\pi < \theta < 2\pi$. If α is in quadrant III, $2\pi < \theta < 3\pi$, which means that $\sin 2(\theta/2)$ is positive, and the points are plotted on the ray, rather than on the extension of the ray. The whole period is needed to draw the entire curve.

69. Consider a circle with radius r with $P_1(r, \theta)$ in the first quadrant. $P_2(r, \pi - \theta)$ in the second quadrant, and $P_3(r, -\theta)$ in the third quadrant. Label the origin O, the projection of P_2 onto the x-axis A, and the projection of P_1 or P_3 onto the x-axis B.

$$\triangle OAP_2 \sim \triangle OBP_1; \ \triangle OBP_1 \sim \triangle OBP_3$$

OP_2 makes an angle $\pi - \theta$ with respect to the positive x-axis. The point P_3 lies on its extension. Thus, $P_3(-r, \pi - \theta)$ is the image of P_1 reflected in the x-axis. The other primary form of the point is $P_3{}^*(-r, \pi - \theta)$.

70. Refer to Problem 69; $\overline{AP_2} = \overline{BP_1}$ and $\overline{OA} = \overline{OB}$, so $P_1(r, \theta)$ and $P_2(r, \pi - \theta)$ are images of each other reflected in the y-axis. The other primary form of the point is $P_2{}^*(-r, -\theta)$.

71. Consider a right triangle with vertex O at the origin, $A(r, \theta)$, B on the line $\theta = \pi/4$, and acute angle θ_2 opposite let $\overline{AB}$. Right $\triangle OBC$ is a reflection of $\triangle OBA$ in the line OB. Acute angle θ_1 is opposite $\overline{BC}$, $C(r, \pi/2 - \theta)$. The angle α is such that

$$\alpha + \theta_1 + \theta_2 + \theta = \frac{\pi}{2}$$

By hypotheses, $\theta_1 + \theta_2 + \theta = \frac{\pi}{2} - \theta$

By construction, $\theta_2 + \theta = \frac{\pi}{4}$. Thus,

$\theta_1 + \theta = \frac{\pi}{4}$ and $\theta_2 = \theta_1$. This means

$\triangle OBC \sim \triangle OBA$

Therefore, $\overline{CB} = \overline{AB}$ and the points C and A are reflections of each other in the line $y = x$.

72. Let $\theta^* = \theta - \theta_0$ and let $r^* = f(\theta - \theta^*)$. Now $r = r^*$, so $(r^*, \theta - \theta_0)$ and $(r, \theta B)$ represent the same point on the curve. This means that $r = f(\theta - \theta_0)$ is the curve $r = f(x)$ rotated θ_0.

73. **a.** Point $P(r, \theta)$ is at a distance $r = r^*$ from the pole. OP is the angle θ with respect to the original polar axis. Let that axis be rotated through an angle α. Then $\theta - \alpha$ is the angle between the rotated axis and $\overline{OP^*}$. Since $r = r^*$, $f(\theta) = f(\theta - \alpha)$.

   b. $r = 2\sec(\theta - \pi/3)$ becomes $r^* = 2\sec\theta^*$ when the polar axis is rotated $\pi/3$. This becomes $x^* = r^*\cos\theta^* = 2$, a line perpendicular to the (rotated) polar axis.

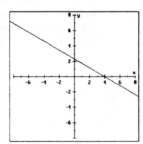

74.

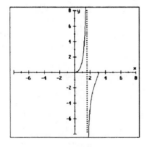

9.3 Area and Tangent Lines in Polar Coordinates, Page 593

1. First sketch the curves to estimate the number of points of intersection. Always check to see if the pole lies on both curves. Then set $r_1 = r_2$; if you suspect that they are still points whose coordinates were not found, set $-r_1^* = f(\theta + \pi)$ and $r_1^* = r_2$. If you suspect that they are still points whose

coordinates were not found, set $-r_2^* = f(\theta + \pi)$ and $r_2^* = r_1$. As a last resort, solve $r_1^* = r_2^*$.

2. Approximate the area by circular sectors $dA = \frac{1}{2}r^2 \, d\theta$. Sum up (integrate) from θ_1 to θ_2.

3. Slope is defined as the ratio of the rise Δy over the run Δx in Cartesian coordinates. In polar coordinates $f'(\theta)$ is the change of f with respect to the angle between the polar axis (x-axis) and the ray from the origin through the point of contact.

4. $P_1(0, 0)$ is a point of intersection.

   $r_1 = r_2$

   $4\cos\theta = 4\sin\theta$

   $\tan\theta = 1$

   $\theta = \frac{\pi}{4}$

   $P_2(2\sqrt{2}, \frac{\pi}{4})$

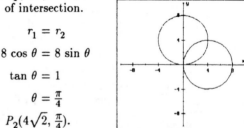

5. $P_1(0, 0)$ is a point of intersection.

   $r_1 = r_2$

   $8\cos\theta = 8\sin\theta$

   $\tan\theta = 1$

   $\theta = \frac{\pi}{4}$

   $P_2(4\sqrt{2}, \frac{\pi}{4})$.

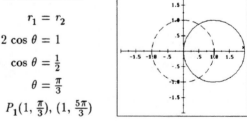

6. The pole is not a point of intersection.

   $r_1 = r_2$

   $2\cos\theta = 1$

   $\cos\theta = \frac{1}{2}$

   $\theta = \frac{\pi}{3}$

   $P_1(1, \frac{\pi}{3}), (1, \frac{5\pi}{3})$

7. The pole is not a point of intersection.

   $r_1 = r_2$

   $4\sin\theta = 2$

   $\sin\theta = \frac{1}{2}$

   $\theta = \frac{\pi}{6}$

   $P_1(2, \frac{\pi}{6})$ and $P_2(2, \frac{5\pi}{6})$

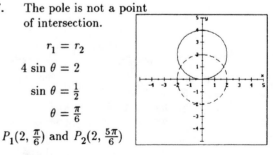

8. The pole is not a point
of intersection.

$$r_1{}^2 = r_2{}^2$$

$$9 \cos 2\theta = 9$$

$$\cos 2\theta = 1$$

$$2\theta = 0$$

$$\theta = 0, 2\pi$$

$$P_1(3, 0) \text{ and } P_2(3, \pi)$$

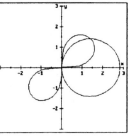

9. The pole is not a point
of intersection.

$$r_1{}^2 = r_2{}^2$$

$$4 = 4 \sin^2 2\theta$$

$$\sin^2 2\theta = 1$$

$$2\theta = \frac{\pi}{2}, \frac{5\pi}{2}$$

$$\theta = \frac{\pi}{4}, \frac{5\pi}{4}$$

$$P_1(2, \tfrac{\pi}{4}), \quad P_2(2, \tfrac{5\pi}{4})$$

10. $P_1(0, 0)$ is a point
of intersection.

$$r_1 = r_2$$

$$1 + \cos \theta = 1 - \cos \theta$$

$$\cos \theta = 0$$

$$\theta = \frac{\pi}{2}, \frac{3\pi}{2}$$

$$P_2(2, \tfrac{\pi}{2}), \ P_3(2, \tfrac{3\pi}{2})$$

11. $P_1(0, 0)$ is a point
of intersection.

$$r_1 = r_2$$

$$1 + \sin \theta = 1 - \sin \theta$$

$$\sin \theta = 0$$

$$\theta = 0, \pi$$

$$P_2(2, 0), \ P_3(2, \pi)$$

12. The pole is not a point
of intersection.

$$r_1{}^2 = r_2{}^2$$

$$9 \sin 2\theta = 9$$

$$\sin 2\theta = 1$$

$$2\theta = \frac{\pi}{2}, \frac{5\pi}{2}$$

$$\theta = \frac{\pi}{4}, \frac{5\pi}{4}$$

$$P_1(3, \tfrac{\pi}{4}), \ P_2(3, \tfrac{5\pi}{4})$$

13. $P_1(0, 0)$ is a point
of intersection.

$$r_1{}^2 = r_2{}^2$$

$$4 \sin 2\theta = 8 \cos^2\theta$$

$$8 \sin \theta \cos \theta = 8 \cos^2\theta$$

$$\cos \theta(\cos \theta - \sin \theta) = 0$$

$$\cos \theta = 0 \text{ and } \cos \theta = \sin \theta$$

$$\theta = \frac{\pi}{2}, \frac{3\pi}{2} \qquad \theta = \frac{\pi}{4}, \frac{5\pi}{4}$$

When $\theta = \frac{\pi}{2}, \frac{3\pi}{2}$, $r = 0$ which is the pole.

When $\theta = \frac{5\pi}{4}$, r is negative and leads to an

alternate primary representation.

$$P_2(2, \tfrac{\pi}{4})$$

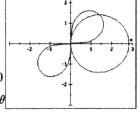

14. $P_1(0, 0)$ is a point
of intersection.

$$r_1{}^2 = r_2{}^2$$

$$9 \cos 2\theta = 18 \sin^2\theta$$

$$9(1 - 2\sin^2\theta) = 18 \sin^2\theta$$

$$\sin^2\theta = \frac{1}{4}$$

$$\sin \theta = \pm\frac{1}{2}$$

$$\theta = \frac{\pi}{6}, \frac{5\pi}{6}$$

$$P_2\!\left(\frac{3\sqrt{2}}{2}, \frac{\pi}{6}\right), \ P_3\!\left(\frac{3\sqrt{2}}{2}, \frac{5\pi}{6}\right)$$

15. The pole is not a point
of intersection.

$$r_1{}^2 = r_2{}^2$$

$$4 \cos 2\theta = 4$$

$$\cos 2\theta = 1$$

$$2\theta = 0, 2\pi$$

$$\theta = 0, \pi$$

$$P_1(2, 0), \ P_2(2, \pi)$$

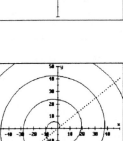

16. $P_1(0, 0)$ is a point
of intersection.

For $\theta = \frac{\pi}{3} + 2n\pi$,

$$r = 3\theta \text{ and}$$

$$P_n(\pi \pm 6n\pi, \tfrac{\pi}{3})$$

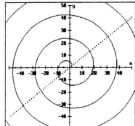

17. $P_1(0, 0)$ is a point of intersection.

$$r_1{}^2 = r_2{}^2$$

$$2 \sin^2\theta = \cos 2\theta$$

$$2 \sin^2\theta = 1 - 2\sin^2\theta$$

$$4 \sin^2\theta = 1$$

$$\sin^2\theta = \tfrac{1}{4}$$

$$\sin \theta = \pm\tfrac{1}{2}$$

$$\theta = \tfrac{\pi}{6},\ \tfrac{5\pi}{6}$$

$$P_2\!\left(\tfrac{\sqrt{2}}{2},\ \tfrac{\pi}{6}\right),\ P_3 = \left(\tfrac{\sqrt{2}}{2},\ \tfrac{5\pi}{6}\right)$$

18. $P_1(0, 0)$ is a point of intersection.

$$r_1 = r_2$$

$$2(1 - \cos \theta) = 4 \sin \theta$$

$$1 - \cos \theta = 2 \sin \theta$$

$$1 - 2 \cos \theta + \cos^2\theta = 4 \sin^2\theta$$

$$1 - 2 \cos \theta + \cos^2\theta = 4(1 - \cos^2\theta)$$

$$5 \cos^2\theta - 2 \cos \theta - 3 = 0$$

$$(5 \cos \theta + 3)(\cos \theta - 1) = 0$$

$$\cos \theta = 1,\ -\tfrac{3}{5}$$

$$\cos \theta = 1 \qquad \cos \theta = -\tfrac{3}{5}$$

$$\theta = 0,\ \pi \text{ (pole)} \qquad \theta = \cos^{-1}\!\left(-\tfrac{3}{5}\right)$$

$$\approx 2.214$$

$$P_2(3.2,\ 2.2)$$

19. $P_1(0, 0)$ is a point of intersection.

$$r_1 = r_2$$

$$2(1 + \cos \theta) = -4 \sin \theta$$

$$1 + \cos \theta = -2 \sin \theta$$

$$1 + 2 \cos \theta + \cos^2\theta = 4 \sin^2\theta$$

$$1 + 2 \cos \theta + \cos^2\theta = 4(1 - \cos^2\theta)$$

$$5 \cos^2\theta + 2 \cos \theta - 3 = 0$$

$$(5 \cos \theta - 3)(\cos \theta + 1) = 0$$

$$\cos \theta = -1 \text{ (pole)} \qquad \cos \theta = \tfrac{3}{5}$$

$$\theta = \cos^{-1}\tfrac{3}{5}$$

$$\approx 0.927$$

In quadrant IV, $\theta \approx 5.356$

$$P_2(3.2,\ 5.4)$$

20. $P_1(0, 0)$ is a point of intersection.

$$r_1 = r_2$$

$$2(1 - \sin \theta) = 4 \cos \theta$$

$$1 - \sin \theta = 2 \cos \theta$$

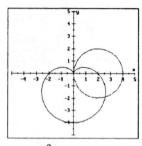

$$1 - 2 \sin \theta + \sin^2\theta = 4 \cos^2\theta$$

$$1 - 2 \sin \theta + \sin^2\theta = 4(1 - \sin^2\theta)$$

$$5 \sin^2\theta - 2 \sin \theta - 3 = 0$$

$$(5 \sin \theta + 3)(\sin \theta - 1) = 0$$

$$\sin \theta = 1 \text{ (pole)} \qquad \sin \theta = -\tfrac{3}{5}$$

$$\theta = \sin^{-1}\!\left(-\tfrac{3}{5}\right)$$

$$\approx -0.6435$$

$$P_2(3.2,\ 5.6)$$

21. $P_1(0, 0)$ is a point of intersection.

$$r_1 = r_2$$

$$2 \cos \theta + 1 = \sin \theta$$

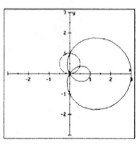

$$\sin^2\theta = 4 \cos^2\theta + 4 \cos \theta + 1$$

$$1 - \cos^2\theta = 4 \cos^2\theta + 4 \cos \theta + 1$$

$$5 \cos^2\theta + 4 \cos \theta = 0$$

$$\cos \theta(5 \cos \theta + 4) = 0$$

$$\cos \theta = 0 \qquad \cos \theta = -\tfrac{4}{5}$$

$$\theta = \tfrac{\pi}{2} \qquad \theta = \cos^{-1}\!\left(-\tfrac{4}{5}\right)$$

$$\theta^* \approx 0.6435 \quad \text{(ref angle)}$$

$$P_2(1,\ \tfrac{\pi}{2}),\ P_3(0.6,\ 0.6)$$

22. $P_1(0, 0)$ is a point of intersection.

$$r_1 = r_2$$
$$2 \sin \theta + 1 = \cos \theta$$

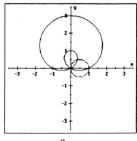

$$4 \sin^2\theta + 4 \sin \theta + 1 = 1 - \sin^2\theta$$
$$5 \sin^2\theta + 4 \sin \theta = 0$$
$$\sin \theta(5 \sin \theta + 4) = 0$$

$\sin \theta = 0$	$\sin \theta = -\frac{4}{5}$
$\theta = 0, \pi$	$\theta = \sin^{-1}(-\frac{4}{5})$
	$\theta^* = 0.92$ (ref angle)

$$P_2(1, 0),\ P_3(0.6, 0.9)$$

23. The pole is not a point of intersection.

$$r_1 = r_2$$
$$\frac{5}{3 - \cos \theta} = 2$$
$$6 - 2 \cos \theta = 5$$
$$2 \cos \theta = 1$$
$$\cos \theta = \frac{1}{2}$$
$$\theta = \frac{\pi}{3}, \frac{5\pi}{3}$$
$$P_1(2, \frac{\pi}{3}),\ P_2(2, \frac{5\pi}{3})$$

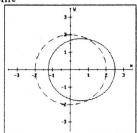

24. The pole is not a point of intersection.

$$r_1 = r_2$$
$$\frac{2}{1 + \cos \theta} = 2$$
$$\cos \theta = 0$$
$$\theta = \frac{\pi}{2}, \frac{3\pi}{2}$$
$$P_1(2, \frac{\pi}{2}),\ P_2(2, \frac{3\pi}{2})$$

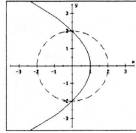

25. The pole is not is a point of intersection.

$$r_1 = r_2$$
$$2 \cos \theta = \frac{1}{1 - \cos \theta}$$
$$2\cos \theta - 2\cos^2\theta = 1$$
$$2\cos^2\theta - 2\cos \theta + 1 = 0$$

There are no intersection points.

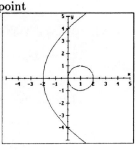

26. The pole is not a point of intersection.

$$r_1 = r_2$$
$$\frac{1}{1 + \cos \theta} = 2(2 - \cos \theta)$$
$$2(1 - \cos^2\theta) = 1$$
$$2 \sin^2\theta = 1$$
$$\sin \theta = \pm\frac{1}{\sqrt{2}}$$
$$\theta = \frac{\pi}{4}, \frac{3\pi}{4}, \frac{5\pi}{4}, \frac{7\pi}{4}$$

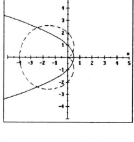

The points of intersection are when $\theta = \frac{3\pi}{4}, \frac{5\pi}{4}$

For these values, $r = 2 \pm \sqrt{2} \approx 0.5858, 3.4142$

$P_1(0.59, \frac{\pi}{4})$, $P_2(0.59, \frac{7\pi}{4})$, $P_3(3.43, \frac{5\pi}{4})$,

$P_4(3.43, \frac{7\pi}{4})$

27. The pole is not a point of intersection.

$$r_1 = r_2$$
$$2 \cos \theta = \sec \theta$$
$$\cos^2\theta = \frac{1}{2}$$
$$\theta = \frac{\pi}{4}, \frac{3\pi}{4}, \frac{5\pi}{4}, \frac{7\pi}{4}$$

The points of

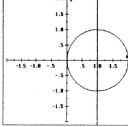

intersection are when $\theta = \frac{\pi}{4}, \frac{7\pi}{4}$.

$$P_1(\sqrt{2}, \frac{\pi}{4}),\ P_2(\sqrt{2}, \frac{7\pi}{4})$$

28. The pole is not a point of intersection.

$$r_1 = r_2$$
$$2 \sin \theta = 2 \csc \theta$$
$$\sin^2\theta = 1$$
$$\theta = \frac{\pi}{2}, \frac{3\pi}{2}$$

The point of

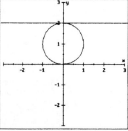

intersection is when $\theta = \frac{\pi}{2}$; $P_1(2, \frac{\pi}{2})$

29. The pole is not a point of intersection.

$$r_1 = r_2$$
$$4 \sin^2 \theta = 1$$
$$\sin \theta = \pm\frac{1}{2}$$
$$\theta = \frac{\pi}{6}, \frac{5\pi}{6}$$
$$(2, \frac{\pi}{6}),\ (2, \frac{5\pi}{6})$$

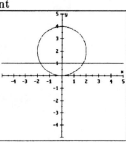

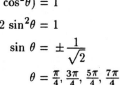

30. $A = \frac{1}{2} \int_{0}^{\pi/6} \sin^2\theta \; d\theta = \frac{1}{4}(\theta - \frac{1}{2}\sin 2\theta)\Big|_{0}^{\pi/6} = \frac{\pi}{24} - \frac{\sqrt{3}}{16}$

31. $A = \frac{1}{2} \int_{0}^{\pi/6} \cos^2\theta \; d\theta = \frac{1}{4}(\theta + \frac{1}{2}\sin 2\theta)\Big|_{0}^{\pi/6} = \frac{\pi}{24} + \frac{\sqrt{3}}{16}$

32. $A = \frac{1}{2} \int_{-\pi/4}^{\pi/4} \sec^2\theta \; d\theta = \tan 2\theta\Big|_{0}^{\pi/4} = 1$

33. $A = \frac{1}{2} \int_{\pi/6}^{\pi/2} \sin\theta \; d\theta = -\frac{1}{2}\cos\theta\Big|_{\pi/6}^{\pi/2} = \frac{\sqrt{3}}{4}$

34. $A = \frac{1}{2} \int_{0}^{2\pi} (e^{\theta/2})^2 \; d\theta = \frac{1}{2}e^{\theta}\Big|_{0}^{2\pi} = \frac{1}{2}(e^{2\pi} - 1)$

35. $A = \frac{1}{2} \int_{0}^{\pi/4} (\sin\theta + \cos\theta)^2 \; d\theta = \frac{1}{2} \int_{0}^{\pi/4} (1 + \sin 2\theta) \; d\theta$

$= \frac{1}{2}(\theta - \frac{1}{2}\cos 2\theta)\Big|_{0}^{\pi/4} = \frac{1}{8}(\pi + 2)$

36. $A = \frac{1}{2} \int_{0}^{2\pi} \frac{\theta^2}{\pi^2} \; d\theta = \frac{1}{2\pi^2} \frac{\theta^3}{3}\Big|_{0}^{2\pi} = \frac{4\pi}{3}$

37. $A = \frac{1}{2} \int_{0}^{2\pi} \left(\frac{\theta^2}{\pi}\right)^2 d\theta = \frac{1}{2\pi^2} \frac{\theta^5}{5}\Big|_{0}^{2\pi} = \frac{16\pi^3}{5}$

38. $f'(\theta) = \sin\theta; \; f'(\frac{\pi}{4}) = \frac{\sqrt{2}}{2}$

$m = \dfrac{\frac{2 - \sqrt{2}}{2}\frac{\sqrt{2}}{2} + \frac{\sqrt{2}}{2}\frac{\sqrt{2}}{2}}{\frac{2 - \sqrt{2}}{2}\frac{\sqrt{2}}{2} + \frac{\sqrt{2}}{2}\frac{\sqrt{2}}{2}} = 1 + \sqrt{2}$

39. At the pole, $0 = 4\cos\theta + 2$ or

$\theta = \cos^{-1}(-\frac{1}{2}) = \frac{2\pi}{3}, \frac{4\pi}{3}$

$f(\frac{2\pi}{3}) = f(\frac{4\pi}{3}) = 0$

$f'(\theta) = -4\sin\theta; \; f'(\frac{2\pi}{3}) = -2\sqrt{3}$

$f'(\frac{4\pi}{3}) = 2\sqrt{3}$

$m = \dfrac{0 - 2\sqrt{3}\sin\frac{2\pi}{3}}{0 - 2\sqrt{3}\cos\frac{2\pi}{3}} = -\sqrt{3}$

$m = \dfrac{0 + 2\sqrt{3}\sin\frac{4\pi}{3}}{0 + 2\sqrt{3}\cos\frac{4\pi}{3}} = \sqrt{3}$

40. At the pole, $0 = \sqrt{\cos 2\theta}$ or $\theta = \frac{\pi}{4}, \frac{3\pi}{4}$

$f(\frac{\pi}{4}) = f(\frac{3\pi}{4}) = 0$

As noted in the text, the formula used in the solution to Problems 38 and 39 is not easy to remember. Instead, you might find it easier to use $dy/dx = (dy/d\theta)/(dx/d\theta)$

$x = \sqrt{\cos 2\theta}\cos\theta$

$\dfrac{dx}{d\theta} = \sqrt{\cos 2\theta}(-\sin\theta) - \dfrac{\cos\theta\sin 2\theta}{\sqrt{\cos 2\theta}}$

$= \dfrac{-\cos 2\theta\sin\theta - \cos\theta\sin 2\theta}{\sqrt{\cos 2\theta}}$

$= \dfrac{-\sin 3\theta}{\sqrt{\cos 2\theta}};$

$y = \sqrt{\cos 2\theta}\sin\theta$

$\dfrac{dy}{d\theta} = \sqrt{\cos 2\theta}\cos\theta - \dfrac{\sin\theta\sin 2\theta}{\sqrt{\cos 2\theta}}$

$= \dfrac{\cos 2\theta\cos\theta - \sin\theta\sin 2\theta}{\sqrt{\cos 2\theta}}$

$= \dfrac{\cos 3\theta}{\sqrt{\cos 2\theta}}$

$m = -\dfrac{\sin 3\theta}{\cos 3\theta} = -\tan 3\theta$

At $\theta = \frac{\pi}{4}, \frac{3\pi}{4}, \; m = \pm 1$

41. $x = 2\cos\theta, \dfrac{dx}{d\theta} = -2\sin\theta,$

$y = 2\sin\theta, \dfrac{dy}{d\theta} = 2\cos\theta$

$m = \dfrac{dy/d\theta}{dx/d\theta} = \dfrac{2\cos\theta}{-2\sin\theta} = -\cot\theta$

$m(\frac{\pi}{3}) = -\cot\frac{\pi}{3} = -\dfrac{\sqrt{3}}{3}$

42. $f'(\theta) = 2\sec\theta\tan\theta \; d\theta$

$m = \dfrac{2\sec\frac{\pi}{4} + 2\sec\frac{\pi}{4}\tan\frac{\pi}{4}}{-2\sec\frac{\pi}{4} + 2\sec\frac{\pi}{4}\tan\frac{\pi}{4}}$ is not defined

43. $f'(\theta) = d\theta; \; m = \dfrac{\frac{\pi}{2}(0) + 1}{-\frac{\pi}{2} + 0} = -\dfrac{2}{\pi}$

44. $f(\pi) = \dfrac{4}{3\sin\pi - 2\cos\pi} = 2$

$f'(\theta) = -\dfrac{4(3\cos\theta + 2\sin\theta)}{(3\sin\theta - 2\cos\theta)^2}$

$f'(\pi) = -\dfrac{4(3\cos\pi + 2\sin\pi)}{(3\cos\pi - 2\cos\pi)^2} = 3$

$m = \dfrac{2\cos\pi + 3\sin\pi}{-2\sin\pi + 3\cos\pi} = \dfrac{2}{3}$

45. $2r\cos\theta + 3r\sin\theta = 3$

$2x + 3y = 3, \; m = -\dfrac{2}{3}$ for all values of θ

46. $f(\frac{\pi}{3}) = 4\sin\frac{\pi}{3}\cos^2\frac{\pi}{3} = \sqrt{3}/2$

$f'(\theta) = 4[\sin\theta(2\cos\theta)(-\cos\theta) + \cos^3\theta]$

$f'(\frac{\pi}{3}) = 4[\sin\frac{\pi}{3}(2\cos\frac{\pi}{3})(-\cos\frac{\pi}{3}) + \cos^3\frac{\pi}{3}]$

$= -\dfrac{5}{2}$

$$m = \frac{\frac{\sqrt{3}}{2}\frac{1}{2} + \frac{-5}{2}\frac{\sqrt{3}}{2}}{-\left(\frac{\sqrt{3}}{2}\right)^2 + \frac{-5}{2}\frac{1}{2}} = \frac{\sqrt{3}}{2}$$

47. $f(\frac{\pi}{6}) = 2 \cos \frac{\pi}{6} \sin^2 \frac{\pi}{6} = \frac{\sqrt{3}}{4}$

$f'(\theta) = 2[\cos \theta(2 \sin \theta)(\cos \theta) - \sin^3\theta]$

$f'(\frac{\pi}{6}) = 2[\cos \frac{\pi}{6}(2 \sin \frac{\pi}{6})(\cos \frac{\pi}{6}) - \sin^3\frac{\pi}{6}] = \frac{5}{4}$

$$m = \frac{\frac{\sqrt{3}}{4}\frac{\sqrt{3}}{2} + \frac{5}{4}\frac{1}{2}}{-\frac{\sqrt{3}}{4}\frac{1}{2} + \frac{5}{4}\frac{\sqrt{3}}{2}} = \frac{2}{\sqrt{3}}$$

48. $y = \sin \theta + \sin^2\theta; \; dy/d\theta = \cos \theta + 2 \sin \theta \cos \theta$

$\frac{dy}{d\theta} = 0$ if $\cos \theta(1 + 2 \sin \theta) = 0$ or when

$\theta = \frac{\pi}{2}, \frac{3\pi}{2}, \frac{7\pi}{6}, \frac{11\pi}{6}$

$f(\frac{\pi}{2}) = 1 + 1 = 2;$

$f(\frac{3\pi}{2}) = 1 - 1 = 0$ (vertical tangent line)

$f(\frac{7\pi}{6}) = f(\frac{11\pi}{6}) = \frac{1}{2}$

$P_1(2, \frac{\pi}{2}), \; P_2(\frac{1}{2}, \frac{7\pi}{6}), \; P_3(\frac{1}{2}, \frac{11\pi}{6})$

49. $y = a \sin \theta + a \sin \theta \cos \theta.$

$\frac{dy}{d\theta} = a \cos \theta - a \sin^2\theta + a \cos^2 \theta$

$= a(2 \cos^2 \theta + \cos \theta - 1)$

$= a(2\cos \theta - 1)(\cos \theta + 1)$

$\frac{dy}{d\theta} = 0$ if $\cos \theta = \frac{1}{2}, -1$ or when

$\theta = \frac{\pi}{3}, \frac{5\pi}{3}, \pi$. $(0, \pi)$ is not a solution because

$\frac{dx}{d\theta} = 0$ there and the slope is undefined.

$P_1(\frac{3a}{2}, \frac{\pi}{3}), \; P_2(\frac{3a}{2}, \frac{5\pi}{3})$

50. $x = a(\cos \theta - \cos^2\theta);$

$\frac{dx}{d\theta} = a(-\sin \theta + 2 \cos \theta \sin \theta)$

$= a \sin \theta(-1 + 2 \cos \theta)$

$\frac{dx}{d\theta} = 0$ if $\sin \theta = 0$ or $\cos \theta = \frac{1}{2}$; that is,

when $\theta = 0, \pi, \frac{\pi}{3}, \frac{5\pi}{3}$

$f(0) = 0$ (horizontal tangent line)

$f(\pi) = 2a; \; f(\frac{\pi}{3}) = \frac{a}{2}; \; f(\frac{5\pi}{3}) = \frac{a}{2}$

$P_1(2a, \pi), \; P_2(\frac{a}{2}, \frac{\pi}{3}), \; P_3(\frac{a}{2}, \frac{5\pi}{3})$

51. $y = 2 \sin^2\theta; \frac{dy}{dx} = 4 \sin \theta \cos \theta = 2 \sin 2\theta$

$x = 2 \cos \theta \sin \theta = \sin 2\theta; \frac{dx}{d\theta} = 2 \cos 2\theta$

$m = \frac{\sin 2\theta}{\cos \theta} = \tan 2\theta; \; m = 1$ when the tangent

line is parallel to the ray $\theta = \frac{\pi}{4}$, so $\theta = \frac{\pi}{8}, \frac{5\pi}{8}$

$P_1(2 \sin \frac{\pi}{8}, \frac{\pi}{8}) \approx (0.77, 0.39);$

$P_2(2 \sin \frac{5\pi}{8}, \frac{5\pi}{8}) \approx (1.85, 1.96)$

52. $A = 2(\frac{1}{2})\int_0^{\pi/4} 4 \cos^2\theta \; d\theta = 2[\theta + \frac{1}{4}\sin \theta]\Big|_0^{\pi/4} = \frac{\pi}{2}$

53. $r = a(\sin 3\theta)$ has a tip of a leaf when $\sin 3\theta = 1, 3\theta = \frac{\pi}{2}, \theta = \frac{\pi}{6}$. We will find the area from $\theta = 0$ to $\theta = \frac{\pi}{6}$ and multiply by 6.

$A = (6)\frac{1}{2}\int_0^{\pi/6} a^2 \sin^2 3\theta \; d\theta = 3a^2\int_0^{\pi/6} \left(\frac{1 - \cos 6\theta}{2}\right) d\theta$

$= \frac{3a^2}{2}(\theta - \frac{1}{6}\sin 6\theta)\Big|_0^{\pi/6} = \frac{\pi a^2}{4}$

54. The curves intersect when $4 \cos \theta = 2$ or when $\theta = \pi/3$.

$A = 2\left(\frac{1}{2}\right)\int_1^{\pi/3} (16 \cos^2 \theta - 4) \; d\theta$

$= (8\theta + 4 \sin 2\theta - 4\theta)\Big|_0^{\pi/3} = \frac{4\pi}{3} + 2\sqrt{3}$

55. The curves intersect when $\cos \theta = 0$ or when $\theta = \pi/2$.

$A = 2\left(\frac{a^2}{2}\right)\int_0^{\pi/2} [1 - (1 - \cos \theta)^2] \; d\theta$

$= a^2[2 - \frac{1}{2}(\theta + \frac{1}{2}\sin 2\theta)]\Big|_0^{\pi/2} = a^2(2 - \frac{\pi}{4})$

56. The curves intersect when $r = 0$ or when $1 - \cos \theta = \sin \theta$ or when $\theta = \pi/2$.

$A = \frac{1}{2}\int_0^{\pi/2} [\sin^2\theta - (1 - \cos \theta)^2] \; d\theta$

$= \frac{1}{2}\int_0^{\pi/2} (-1 + 2 \cos \theta - \cos 2\theta) \; d\theta$

$= \frac{1}{2}[-\theta + 2 \sin \theta - \frac{1}{2}\sin 2\theta]\Big|_0^{\pi/2} = 1 - \frac{\pi}{4}$

57. Solving simultaneously:

$6 \cos \theta = 2 + 2 \cos \theta$

$4 \cos \theta = 2$

$\cos \theta = \frac{1}{2}$

$\theta = \frac{\pi}{3}, \frac{5\pi}{3}$

$$A = (2)\frac{1}{2}\int_0^{\pi/3} [36\cos^2\theta - 4(1 + \cos\theta)^2]\, d\theta$$

$$= 4\int_0^{\pi/3} [9\cos^2\theta - 1 - 2\cos\theta - \cos^2\theta]\, d\theta$$

$$= 4\int_0^{\pi/3} [8\cos^2\theta - 2\cos\theta - 1]\, d\theta$$

$$= 4\left(3\theta + 2\sin 2\theta - 2\sin\theta\right)\Big|_0^{\pi/3} = 4\pi$$

58. The curves intersect when $4 = 8\cos\theta$ or when $\theta = \pi/3$.

$$A = 4\left(\frac{1}{2}\right)\int_1^{\pi/3} [8\cos\theta - 4]\, d\theta$$

$$= 3(8\sin\theta - 4\theta)\Big|_0^{\pi/3} = 8\left(\sqrt{3} - \frac{\pi}{3}\right)$$

59. The regions in the fourth and first quadrants represent half the area of the limaçon without subtracting the inner loop. The inner loop corresponds to $\pi/6 \le \pi/2$.

$$A_o = \int_{-\pi/2}^{\pi/6} 4(1 - 2\sin\theta)^2\, d\theta$$

$$= 4\int_{-\pi/2}^{\pi/6} (1 - 4\sin\theta + 4\sin^2\theta)\, d\theta$$

$$= 4[3\theta + 4\cos\theta - \sin 2\theta]\Big|_{-\pi/2}^{\pi/6}$$

$$= 8\pi + 6\sqrt{3}$$

$$A_i = \int_{\pi/6}^{\pi/2} 4(1 - 2\sin\theta)^2\, d\theta$$

$$= 4[3\theta + 4\cos\theta - \sin 2\theta]\Big|_{\pi/6}^{\pi/2}$$

$$= 4\pi - 6\sqrt{3}$$

$$A = A_o - A_i = 4\pi + 12\sqrt{3}$$

60. $y = \theta^2\sin\theta$; $\dfrac{dy}{d\theta} = \theta^2\cos\theta + 2\theta\sin\theta$

$dy/d\theta = 0$ if $\theta\cos\theta = -2\sin\theta$ or when $\tan\theta = -2/\theta$

$x = \theta^2\cos\theta$; $\dfrac{dx}{d\theta} = -\theta^2\sin\theta + 2\theta\cos\theta$

The slope is $m = \dfrac{\theta\cos\theta + 2\sin\theta}{-\theta\sin\theta + 2\cos\theta}$.

At the pole, $\theta = 0$, so $m = 0$.

61. $r_1 = a(1 + \sin\theta)$

$y_1 = a(\sin\theta + \sin^2\theta)$;

$$\frac{dy_1}{d\theta} = a(\cos\theta + 2\sin\theta\cos\theta)$$

$x_1 = a(\cos\theta + \frac{1}{2}\sin 2\theta)$

$$\frac{dx_1}{d\theta} = a(-\sin\theta + \cos 2\theta)$$

$$m_1 = \frac{\cos\theta + \sin 2\theta}{-\sin\theta + \cos 2\theta}$$

Similarly, $r_2 = b(1 - \sin\theta)$

$y_2 = b(\sin\theta - \sin^2\theta)$;

$$\frac{dy_2}{d\theta} = b(\cos\theta - 2\sin\theta\cos\theta)$$

$x_2 = b(\cos\theta - \frac{1}{2}\sin 2\theta)$

$$\frac{dx_2}{d\theta} = b(-\sin\theta - \cos 2\theta)\, d\theta$$

$$m_2 = \frac{\cos\theta - \sin 2\theta}{-\sin\theta + \cos 2\theta}$$

$$m_1 m_2 = \left(\frac{\cos\theta + \sin 2\theta}{-\sin\theta + \cos 2\theta}\right)\left(\frac{\cos\theta - \sin 2\theta}{-\sin\theta + \cos 2\theta}\right)$$

$$= \frac{\cos^2\theta - \sin^2 2\theta}{\sin^2\theta - 2\sin\theta\cos 2\theta + \cos^2 2\theta}$$

$$= \frac{\cos^2\theta(1 - 4\sin^2\theta)}{\sin^2\theta - (\cos^2\theta - \sin^2\theta)^2}$$

$$= \frac{\cos^2\theta(1 - 4\sin^2\theta)}{\sin^2\theta - (\cos^4\theta - 2\sin^2\theta\cos^2\theta + \sin^4\theta)}$$

$$= \frac{\cos^2\theta - 4\sin^2\theta\cos^2\theta}{1 - \sin^2\theta - 4\sin^2\theta\cos^2\theta} = -1$$

Therefore, the tangent lines are perpendicular.

62. Let ϕ be the angle between the tangent line through P and the positive x-axis. Since $y = f(x)\sin\theta$ and $x = f(\theta)\cos\theta$.

$$m = \tan\phi = \frac{dy}{dx} = \frac{f(\theta)\cos\theta + f'(\theta)\sin\theta}{-f(\theta)\sin\theta + f'(\theta)\cos\theta}$$

$\phi = \theta + \alpha$, so $\tan\alpha = \dfrac{\tan\phi - \tan\theta}{1 + \tan\phi\tan\theta}$

$$= \frac{\dfrac{f(\theta)\cos\theta + f'(\theta)\sin\theta}{-f(\theta)\sin\theta + f'(\theta)\cos\theta} - \dfrac{\sin\theta}{\cos\theta}}{1 + \dfrac{f(\theta)\cos\theta + f'(\theta)\sin\theta}{-f(\theta)\sin\theta + f'(\theta)\cos\theta}\dfrac{\sin\theta}{\cos\theta}}$$

$$= \frac{f(\theta)\cos^2\theta + f'(\theta)\sin\theta\cos\theta + f(\theta)\sin^2\theta - f'(\theta)\sin\theta\cos\theta}{-f(\theta)\sin\theta\cos\theta + f'(\theta)\cos^2\theta + f(\theta)\sin\theta\cos\theta + f'(\theta)\sin^2\theta}$$

$$= \frac{f(\theta)}{f'(\theta)}$$

63. **a.** $\tan \alpha = \dfrac{a \cos \theta}{-a \sin \theta} = -\cot \theta$

b. $\dfrac{dr}{d\theta} = 2 \sin \theta$; $\tan \alpha = \dfrac{1 - \cos \theta}{\sin \theta}$

c. $\dfrac{dr}{d\theta} = 6e^{3\theta}$; $\tan \alpha = \dfrac{2e^{3\theta}}{6e^{3\theta}} = \dfrac{1}{3}$

9.4 Parametric Representation of Curves, Page 602

1. A parameter is an arbitrary constant or a variable in a mathematical expression, which distinguishes various specific cases. In particular, it refers to a variable other than those referring to the coordinate variables.

2. If $x = x(t)$ and $y = y(t)$, then $\dfrac{dy}{dx} = \dfrac{dx/dt}{dy/dt}$

3. $t = x - 1$, so
$y = (x - 1) - 1$
$\quad = x - 2;$
$1 \le x \le 3$

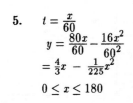

4. $t = -x$
$y = 3 - 2(-x)$
$\quad = 2x + 3$
$-1 \le x \le 0$

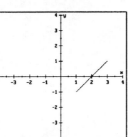

5. $t = \dfrac{x}{60}$
$y = \dfrac{80x}{60} - \dfrac{16x^2}{60^2}$
$\quad = \dfrac{4}{3}x - \dfrac{1}{225}x^2$
$0 \le x \le 180$

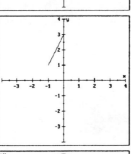

6. $t = \dfrac{x}{30}$
$y = \dfrac{60x}{30} - \dfrac{9x^2}{900}$
$\quad = 2x - \dfrac{1}{100}x^2$
$-30 \le x \le 60$

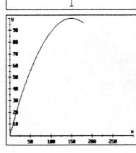
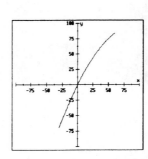

7. $t = x$
$y = 2 + \dfrac{2}{3}(x - 1)$
$\quad = \dfrac{2}{3}x + \dfrac{4}{3}$
$2 \le x \le 5$

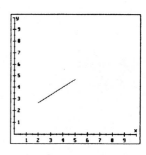

8. $t = x$
$y = \dfrac{3}{5}(5 - t - 2)$
$\quad = \dfrac{9}{5} - \dfrac{3}{5}x$
$-1 \le x \le 3$

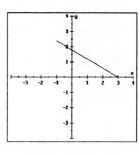

9. $t^2 = x - 1$
$y = t^2 - 2$
$y = (x - 1) - 1$
$\quad = x - 2$
$1 \le x \le 3$

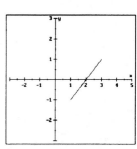

10. $t^2 = \dfrac{x}{2}$
$y = \dfrac{x}{2} + 2$
$0 \le x \le 4$

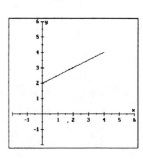

11. $t = x^{1/3}$
$x^{1/3} = y^{1/2}$
$y = x^{2/3}, \ x \ge 0$

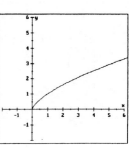

12. $x = y^2$
$0 \leq x \leq 4$

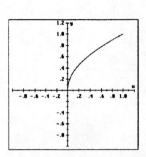

13. $x^2 + y^2 = 9$
$-3 \leq x \leq 3$

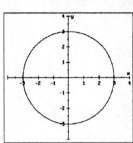

14. $x^2 + y^2 = 4$
$-2 \leq x \leq 2$

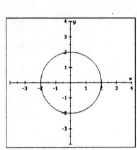

15. $\sin t = x - 1$
$\cos t = y + 2$
$(x-1)^2 + (y+2)^2 = 1$
$0 \leq x \leq 2$

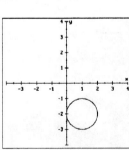

16. $x - 1 = \sin^2 t$
$(y + 2)^2 = \cos^2 t$
$(x-1) + (y+2)^2 = 1$
$x = -(y+2)^2 + 2$
$1 \leq x < 2$ or
$-1 \leq y < -3$

17. $1 + \dfrac{x^2}{16} = \dfrac{y^2}{9}$

$\dfrac{y^2}{9} - \dfrac{x^2}{16} = 1$

$-\infty \leq x \leq \infty$

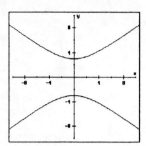

18. $1 + \dfrac{y^2}{4} = \dfrac{x^2}{16}$

$\dfrac{x^2}{16} - \dfrac{y^2}{4} = 1$

$x \geq 4$ or $x \leq -4$

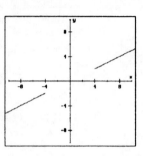

19. $3(3^t) = y$
$3^t = x; \quad y = 3x$
$x \geq 1$

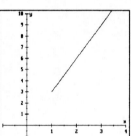

20. $y = \dfrac{2}{x}$
$x \geq 1$

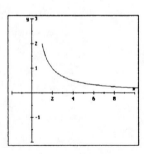

21. $y = ex$
$x \geq 1$

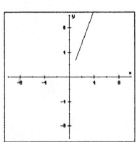

22. $y = \frac{e}{x}$

　　$x \geq 1$

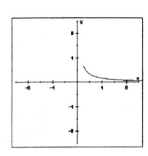

23. $y = \ln x$

　　$x \geq 0$

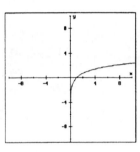

24. $y = \frac{1}{x}$

　　$x > 0$

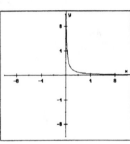

25. $\frac{dy}{dt} = 4t^3$; $\frac{dx}{dt} = 2t$; $\frac{dy}{dx} = \frac{4t^3}{2t} = 2t^2$;

　　$\frac{d^2y}{dx} = \frac{4t}{2t} = 2$

26. $\frac{dy}{dt} = 4t^3$; $\frac{dx}{dt} = 2t$; $\frac{dy}{dx} = \frac{4t^3}{2t} = 2t^2$;

　　$\frac{d^2y}{dx} = \frac{4t}{2t} = 2$

27. $\frac{dy}{dt} = 2 \cos 2t$; $\frac{dx}{dt} = 4e^{4t}$; $\frac{dy}{dx} = \frac{\cos 2t}{2e^{4t}}$;

　　$\frac{d^2y}{dt} = \frac{e^{4t}(-2 \sin 2t) - 4e^{4t}(\cos 2t)}{2e^{4t}}$

　　$\frac{d^2y}{dx} = -\frac{\sin 2t + 2 \cos 2t}{4e^{8t}}$

28. $\frac{dy}{dt} = e^t(t^2 + 2t)$; $\frac{dx}{dt} = e^t$; $\frac{dy}{dx} = t^2 + 2t$;

　　$\frac{d^2y}{dt} = 2t + 2$, so $\frac{d^2y}{dx} = \frac{2t + 2}{e^t} = 2e^{-1}(t + 1)$

29. $\frac{dy}{dt} = b \cos t$; $\frac{dx}{dt} = -a \sin t$; $\frac{dy}{dx} = -\frac{b}{a} \cot t$

　　$\frac{d^2y}{dt} = \frac{b}{a} \csc^2 t$, so

　　$\frac{d^2y}{dx} = -\frac{b}{a^2} \frac{\csc^2 t}{\sin t} = -\frac{b}{a^2 \sin^3 t}$

30. $\frac{dy}{dt} = b \cosh t$; $\frac{dx}{dt} = a \sinh t$; $\frac{dy}{dx} = -\frac{b}{a} \coth t$

　　$\frac{d^2y}{dt} = \frac{b}{a} \operatorname{csch}^2 t$, so $\frac{d^2y}{dx} = -\frac{b}{a^2} \frac{\operatorname{csch}^3 t}{\sin t}$

31. $\frac{dx}{dt} = 2t$; $\frac{dy}{dt} = 4t^3$;

　　$\frac{dy}{dx} = \frac{4t^3}{2t} = 2t^2 = 2(x - 1)$

32. $\frac{dx}{dt} = -2a \sin t$; $\frac{dy}{dt} = 2a \sin t \cos t$;

　　$\frac{dy}{dx} = \frac{2a \sin t \cos t}{-2a \sin t} = -\cos t = -\frac{x}{2a}$

33. $\frac{dx}{dt} = e^{-t}$; $\frac{dy}{dt} = e^t$;

　　$\frac{dy}{dx} = e^{2t} = (y - 1)^2$

34. $\frac{dx}{dt} = 2t$; $\frac{dy}{dt} = \frac{1}{t}$; $\frac{dy}{dx} = \frac{1}{2t^2} = \frac{1}{2x}$

35. $A = \int_0^1 y \frac{dx}{dt} dt = \int_0^1 t^2(4t^3) \, dt = \frac{4t^6}{6}\Big|_0^1 = \frac{2}{3}$

36. $A = \int_0^1 (t + 2)^2(4t^3 + 2t) \, dt$

　　$= \int_0^1 (2t^5 + 8t^4 + 9t^3 + 4t^2 + 4t) \, dt = \frac{451}{30}$

37. $A = \int_0^{\pi/4} (1 + \cos \theta)(1 - \cos \theta) \, d\theta$

　　$= \int_0^{\pi/4} (1 - \cos^2\theta) \, d\theta = \frac{1}{2}\int_0^{\pi/4} (1 - \cos 2\theta) \, d\theta$

　　$= \frac{1}{2}\Big(\frac{\pi}{4} - \frac{1}{2} \sin 2\theta\Big)\Big|_0^{\pi/4} = \frac{\pi}{8} - \frac{1}{4}$

38. $A = \int_0^{\pi/6} \sec^2\theta \sec^2\theta \, d\theta = \int_0^{\pi/6} (1 + \tan^2\theta)\sec^2\theta \, d\theta$

　　$= \Big(\tan \theta + \frac{1}{3} \tan^3\theta\Big)\Big|_0^{\pi/6} = \frac{10\sqrt{3}}{27}$

39. $A = \int_0^1 u^3\Big(\frac{du}{1 + u^2}\Big) = \int_0^1 \Big(u - \frac{u}{u^2 + 1}\Big)du$

　　$= \Big(\frac{u^2}{2} - \frac{1}{2} \ln(u^2 + 1)\Big)\Big|_0^1 = \frac{1}{2}(1 - \ln 2)$

40. $A = \int_0^{1/2} u\Big(\frac{du}{\sqrt{1 - u^2}}\Big) = -\frac{1}{2}\Big[\frac{\sqrt{1 - u^2}}{1/2}\Big]\Big|_0^{1/2}$

　　$= \frac{2 - \sqrt{3}}{2}$

41. $\frac{dy}{dt} = \frac{3}{4}t^{-1/4}; \frac{dx}{dt} = \frac{1}{2}t^{-1/2}$

$ds = \sqrt{\left(\frac{dx}{dt}\right)^2 + \left(\frac{dy}{dt}\right)^2} = \frac{1}{4}\sqrt{9t^{-1/2} + 4t^{-1}}$

$ = \frac{1}{4}\sqrt{\frac{9 + 4t^{-1/2}}{t^{1/2}}} = \frac{1}{4}(9t^{1/2} + 4)^{1/2}t^{-1/2}$

$s = \frac{1}{4}\int_0^4 (9t^{1/2} + 4)^{1/2}t^{-1/2}\, dt$

$ = \frac{1}{27}(9t^{1/2} + 4)^{1/2}\Big|_0^4 = \frac{1}{27}(22^{3/2} - 8)$

42. $\frac{dy}{dt} = 3t^2; \frac{dx}{dt} = 4t;$

$ds = \sqrt{\left(\frac{dx}{dt}\right)^2 + \left(\frac{dy}{dt}\right)^2} = t\sqrt{16t + 9t^2}$

$s = \int_1^2 (16 + 9t^2)^{1/2}t\, dt$

$ = \frac{1}{18}(16 + 9t^2)^{3/2}(\frac{2}{3})\Big|_1^2$

$ = \frac{1}{27}(52^{3/2} - 125)$

43. $\frac{dx}{dt} = \frac{t}{t^2 - 1}; \frac{dy}{dt} = \frac{t}{\sqrt{t^2 - 1}}$

$ds = \sqrt{\left(\frac{dx}{dt}\right)^2 + \left(\frac{dy}{dt}\right)^2}$

$ = \sqrt{\left(\frac{t}{t^2 - 1}\right)^2 + \left(\frac{t}{\sqrt{t^2 - 1}}\right)^2}$

$ = \sqrt{\frac{t^4}{(t^2 - 1)^2}} = \frac{t^2}{t^2 - 1}$

$ = 1 + \frac{1}{t^2 - 1}$

$s = \int_3^7 \left(1 + \frac{1}{t^2 - 1}\right) dt$

$ = \left(t + \frac{1}{2}\ln\frac{t-1}{t+1}\right)\Big|_3^7 = 4 + \ln\sqrt{\frac{3}{2}}$

$ = 4 + \frac{1}{2}\ln\frac{3}{2}$

44. $\frac{dx}{dt} = t\cos t + \sin t - \sin t;$

$\frac{dy}{dt} = \cos t + t\sin t - \cos t$

$s = \int_1^{\pi/2} t\, dt = \frac{t^2}{2}\Big|_0^{\pi/2} = \frac{\pi^2}{8}$

45. $\frac{dx}{dt} = \frac{2}{\sqrt{1 - t^2}}; \frac{dy}{dt} = \frac{-2t}{1 - t^2}$

$ds = \sqrt{\left(\frac{dx}{dt}\right)^2 + \left(\frac{dy}{dt}\right)^2}$

$ = \frac{2}{1 - t^2}\sqrt{1 + \frac{t^2}{1 - t^2}} = \frac{2}{1 - t^2}$

$s = \int_0^{\sqrt{3}/2} \frac{2\, dx}{1 - t^2} = \int_0^{\sqrt{3}/2} \left[\frac{1}{1 + t} + \frac{1}{1 - t}\right] dt$

$ = (\ln|1 + t| - \ln|1 - t|)\Big|_0^{\sqrt{3}/2}$

$ = \ln\frac{1 + \sqrt{3}/2}{1 - \sqrt{3}/2} = \ln(7 + 4\sqrt{3})$

46. Since $\cos 2u = 1 - \sin^2 u$, $y = 1 - 2x^2$ where $-1 \le x \le 1$; $-1 \le y \le 1$; this is a parabolic arc.

47. Since $\cos^2 t + \sin^2 t = 1$,

$\frac{y}{b} + \frac{x^2}{16a^2} = 1$

$y = \frac{b}{16a^2}(16a^2 - x^2)$

48. $A = \int_0^{2\pi} 2(1 - \cos\theta)(2)(1 - \cos\theta)\, d\theta$

$ = 4\int_0^{2\pi} \left[1 - 2\cos\theta + \frac{1}{2}(1 + \cos 2\theta)\right] d\theta$

$ = 4\theta - 8\sin\theta + 2\theta + \sin 2\theta)\Big|_0^{2\pi} = 12\pi$

49. $\frac{dx}{dt} = -6\cos^2 t\sin t; \frac{dy}{dt} = 6\sin^2 t\cos t$

$ds = \sqrt{\left(\frac{dx}{dt}\right)^2 + \left(\frac{dy}{dt}\right)^2} = 6\sin t\cos t = 3\sin 2t$

$s = 4\int_0^{\pi/4} (3\sin 2t)\, dt = 6(-\cos 2t)\Big|_0^{\pi/2} = 12$

50. $ds = \sqrt{[f(\theta)]^2 + [f'(\theta)]^2}\, d\theta = \sqrt{e^{2\theta} + e^{2\theta}}\, d\theta$

$s = \int_0^{\pi/2} \sqrt{e^{2\theta} + e^{2\theta}}\, d\theta = \sqrt{2}\int_0^{\pi/2} e^{\theta}\, d\theta$

$ = \sqrt{2}(e^{\pi/2} - 1)$

51. $r = f(\theta) = 2\cos\theta, f'(\theta) = -2\sin\theta$

$ds = \sqrt{[f(\theta)]^2 + [f'(\theta)]^2}\, d\theta$

$ = \sqrt{\cos^2\theta + \sin^2\theta}\, d\theta$

$s = \int_0^{\pi/3} d\theta = \frac{2\pi}{3}$

52. $r = f(\theta) = 2(1 - \cos\theta)$, $f'(\theta) = 2\sin\theta$

$$ds = \sqrt{[f(\theta)]^2 + [f'(\theta)]^2}\ d\theta$$

$$= \sqrt{(1 - \cos\theta)^2 + \sin^2\theta}\ d\theta$$

$$= \sqrt{2 - 2\cos^2\theta}\ d\theta$$

$$s = 2\sqrt{2}\int_0^\pi \sqrt{1 - \cos^2\theta}\ d\theta$$

$$= 2\sqrt{2}\int_0^\pi \sin\frac{\theta}{2}\ d\theta = 4\sqrt{2}[-\cos\frac{\theta}{2}]\Big|_0^\pi$$

$$= 4\sqrt{2}$$

53. $r = f(\theta) = e^{-\theta}$, $f'(\theta) = -e^{-\theta}$

$$ds = \sqrt{[f(\theta)]^2 + [f'(\theta)]^2}\ d\theta$$

$$= \sqrt{e^{-2\theta} + e^{-2\theta}}\ d\theta$$

$$= \sqrt{2}\,e^{-\theta}\ d\theta$$

$$s = \int_0^\infty \sqrt{2}\,e^{-\theta}\ d\theta = \sqrt{2}$$

54. Answers vary; see the computational window on Page 599 for some ideas.

55. Answers vary; see the computational window on Page 599 for some ideas.

56. Answers vary; see the computational window on Page 599 for some ideas.

57. Answers vary; see the computational window on Page 599 for some ideas.

58. **a.** Consider a right triangle with hypotenuse ds, horizontal leg dx, and vertical leg dy. A similar triangle is obtained by dividing all three sides by dt. From the Pythagorean theorem,

$$\frac{ds}{dt} = \sqrt{\left(\frac{dx}{dt}\right)^2 + \left(\frac{dy}{dt}\right)^2} = \sqrt{(x')^2 + (y')^2}$$

Then

$$dW = F\frac{ds}{dt}\ dt = F(t)\sqrt{(x')^2 + (y')^2}$$

b. $W = \lim\limits_{\|P\|\to\infty} \sum\limits_{k=1}^n F(t)\sqrt{(x')^2 + (y')^2}\ \Delta t$

$$= \int_a^b F(t)\sqrt{(x')^2 + (y')^2}\ dt$$

c. $W = \displaystyle\int_0^1 (2e^t)\sqrt{6^2(\sin^2 t + 1)}\ dt$

$$= 12\int_0^1 e^t\sqrt{1 + \sinh^2 t}\ dt$$

$$= 12\int_0^1 e^t \cosh t\ dt = 3e^2 + 3$$

59. **a.** $\dfrac{dx}{dt} = -2\sin t$; $\dfrac{dy}{dt} = 3\cos t$

$$\frac{ds}{dt} = \sqrt{4\sin^2 t + 9\cos^2 t} = \sqrt{4 + 5\cos^2 t}$$

$$s = 4\int_0^{\pi/2} \sqrt{4 + 5\cos^2 t} \approx 15.8654$$
(using computer software)

$\frac{s}{4}$ is a ballpark approximation since the hypotenuse of a right triangle with legs 1 and 3 is $\sqrt{10} \approx 3.162$ and $\frac{s}{4} \approx 3.9664$. Since the elliptic arc is larger than the hypotenuse, so this is a reasonable answer.

b. $W = \displaystyle\int_0^{2\pi} 5|3\cos t|\sqrt{1 + 8\cos^2 t}\ dt$

$$\approx 147.507 \text{ (using computer software)}$$

c. Working with the circle first we obtain (by hand) $W \approx 125.623$. Also, we calculate that a numerical integration of $e^{-t/10}$ would need to be over the interval $[0, 76]$ to obtain the requested accuracy. On the real spiral, we numerically compute $W \approx 126.645$.

60. Suppose that the spy and Boldfinger travel on a polar plane and that Boldfinger's submarine dives at the pole O when the spy is 3 miles away on the polar axis. After spotting his prey, the spy stems directly toward the origin for 2 miles to cover the possibility that Boldfinger is traveling directly toward him. When he reaches point $S_1(1, 0)$, he begins to travel on a polar path $r = f(\theta)$. The spy wants his path to intersect Boldfinger's escape route at exactly the same time the submarine is there. Suppose the paths intersects at point P. If the distance traveled by the submarine in the time since the destroyer began its search is $f(\theta) - 1$, then in that same time, the destroyer travels twice as far. Using the formula for polar arc length, the spy finds that the distance traveled by the

destroyer along $r = f(\theta)$ is

$$s = \int_0^\theta \sqrt{[f'(u)]^2 + [f(u)]^2}\, du$$

so the paths intersect when

$$2[f(\theta) - 1] = \int_0^\theta \sqrt{[f'(\theta)]^2 + [f(\theta)]^2}\, du$$

By first differentiating with respect to θ and then squaring both sides and combining terms, we obtain

$$2f'(\theta) = \sqrt{[f'(\theta)]^2 + [f(\theta)]^2}$$
$$4[f'(\theta)]^2 = [f'(\theta)]^2 + [f(\theta)]^2$$
$$3[f'(\theta)]^2 = [f(\theta)]^2$$

So that $f'(\theta) = \dfrac{1}{\sqrt{3}} f(\theta)$

By separating variables, we find

$$\frac{df}{f} = \frac{d\theta}{\sqrt{3}}$$

so that

$$\ln|f| = \frac{1}{\sqrt{3}}\theta + C_1$$
$$f(\theta) = Ce^{\theta/\sqrt{3}}$$

Since $f(0) = 1$ — the path begins at $S_1(1, 0)$ — we have $C = 1$, and the destroyer should follow the spiral path $r = e^{\theta/\sqrt{3}}$ once it reaches point S_1.

61. $r = f(\theta)$; $x = f(\theta)\cos\theta$; $y = f(\theta)\sin\theta$

62. Consider a fixed circle of radius a with center at the origin O. Let $A(a, 0)$. A ray makes an angle θ with the positive x-axis and contains the center D of a moving circle of radius R which makes contact with the fixed circle at B. A ray from D to a point $P(x, y)$ — drawn to the right and below D for convenience — on this moving circle also makes an angle θ with respect to OD. The point P was originally at A, before the second circle started moving. Let ϕ be the angle between DB and DP. Arcs AB and BP are the same length. We have

$$a\theta = \phi R \text{ or } \phi = \frac{a\theta}{R}$$

Consider a right triangle with DP as hypotenuse and third vertex above P and below D (for convenience). Label the acute angle at D, α, and the label the legs x_1 and y_1, respectively. Then,

$$\alpha = \pi - \phi - \theta = \pi - \frac{a\theta}{R} - \theta = \pi - \frac{a + R}{R}\theta$$

$$x_1 = R\cos\alpha = R\cos\left[\pi - \frac{(a + R)\theta}{R}\right]$$
$$= R\left[\cos\pi\cos\frac{(a + R)\theta}{R} + \sin\pi\sin\frac{(a + R)\theta}{R}\right]$$
$$= -R\cos\frac{(a + R)\theta}{R}$$

$$y_1 = R\sin\alpha = R\sin\left[\pi - \frac{(a + R)\theta}{R}\right]$$
$$= R\left[\sin\pi\cos\frac{(a + R)\theta}{R} - \cos\pi\sin\frac{(a + R)\theta}{R}\right]$$
$$= R\sin\frac{(a + R)\theta}{R}$$

The distance x_2 from O to the projection of D on the x-axis is $x_2 = (a + R)\cos\theta$. The vertical distance y_1 from D to the x-axis is $y_2 = (a + R)\sin\theta$. Putting all this together, we have

$$x = x_2 + x_1 = (a + R)\cos\theta - R\cos\frac{(a + R)\theta}{R}$$
$$y = y_2 + y_1 = (a + R)\sin\theta - R\sin\frac{(a + R)\theta}{R}$$

63. Consider a fixed circle of radius a with center at the origin O. Let $E(a, 0)$. A ray makes an angle θ with the positive x-axis and contains the center A of a moving circle of radius R which makes contact with the fixed circle at D. A ray from A to a point $P(x, y)$ — drawn to the right and below A for convenience — on this moving circle also makes an angle ϕ with respect to OD. This point P was originally at E, before the second circle started moving. Arcs DP and DE are the same length. We have

$$a\theta = \phi R \text{ or } \phi = \frac{a\theta}{R}$$

Consider a right triangle with AP as hypotenuse and third vertex above P and below A (for convenience). Label the acute angle at A, α, and the label the legs x_1 and y_1, respectively. Then,

$$\alpha = \phi - \theta = \frac{\alpha - R}{R}$$
$$x_1 = R\cos\alpha = R\cos\frac{(a + R)\theta}{R}$$
$$y_1 = R\sin\alpha = R\sin\frac{(a + R)\theta}{R}$$

The distance x_2 from O to the projection of A on the x-axis is $x_2 = (a + R)\cos\theta$. The vertical distance y_1 from A to the x-axis is $y_2 = (a + R)\sin\theta$. Putting all this together, we have

$$x = x_2 + x_1 = (a - R)\cos\theta + R\cos\frac{(a - R)\theta}{R}$$

$$y = y_2 + y_1 = (a - R)\sin\theta - R\sin\frac{(a - R)\theta}{R}$$

64. Consider an arc which is approximated by a line segment Δs. Use this Δs as the hypotenuse of a right triangle with legs Δx and Δy, respectively. Now,

$$\Delta s = \sqrt{(\Delta x)^2 + (\Delta y)^2}$$

A similar right triangle is obtained if the three sides of the triangle are divided by Δt.

$$\frac{\Delta s}{\Delta t} = \sqrt{\left(\frac{\Delta x}{\Delta t}\right)^2 + \left(\frac{\Delta y}{\Delta t}\right)^2}$$

Then,

$$L = \lim_{n\to\infty} \sum_{k=1}^{n} \Delta s = \int_a^b \sqrt{\left(\frac{dx}{dt}\right)^2 + \left(\frac{dy}{dt}\right)^2}\, dt$$

65. $x = f(\theta)\cos\theta,\ y = f(\theta)\sin\theta;$

$$dx = [-f(\theta)\sin\theta + f'(\theta)\cos\theta]\,d\theta$$

$$dy = [f(\theta)\cos\theta + f'(\theta)\sin\theta]\,d\theta$$

$$\left(\frac{dx}{d\theta}\right)^2 = f^2\sin^2\theta - 2\sin\theta\cos\theta ff' + (f')^2\cos^2\theta$$

$$\left(\frac{dy}{d\theta}\right)^2 = f^2\cos^2\theta + 2\sin\theta\cos\theta ff' + (f')^2\sin^2\theta$$

$$\left(\frac{dx}{d\theta}\right)^2 + \left(\frac{dy}{d\theta}\right)^2 = [f(\theta)]^2 + [f'(\theta)]^2$$

$$L = \lim_{n\to\infty} \sum_{k=1}^{n} \Delta s = \int_a^b \sqrt{[f(\theta)]^2 + [f'(\theta)]^2}\, d\theta$$

9.5 Conic Section: the Parabola, Page 611

1. Write the given equation to match one of the four standard-form equations for a parabola (see Page 606). Plot the vertex (h, k), plot a couple of points, and use symmetry for a couple of additional points.

2. Write the given equation to match one of the four standard polar equations for parabolas (see Page 609). The vertex is at the pole; assume $p > 0$. Plot points for $\theta = \pi/2,\ \pi,$ and $-\pi/2$. A couple of additional points may be necessary.

3.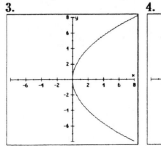

4.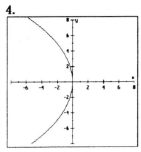

5.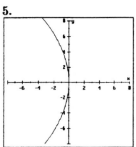

6.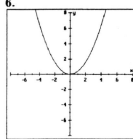

7.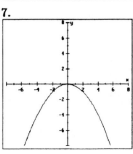

8.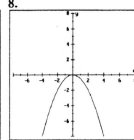

9.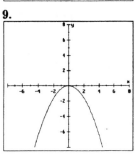

10.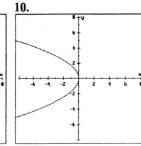

11.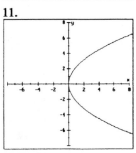

12.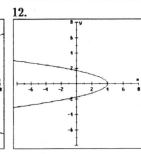

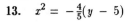
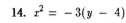

13. $x^2 = -\frac{4}{5}(y - 5)$ **14.** $x^2 = -3(y - 4)$

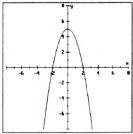
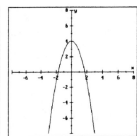

15.

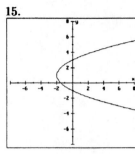

16.

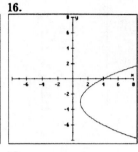

17.

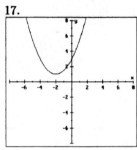

18.

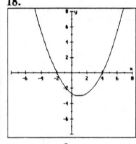

19. $(y - \frac{3}{2})^2 = -4(x - \frac{5}{16})$ **20.** $(y + 5)^2 = 4(x + 3)$

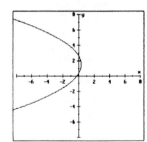

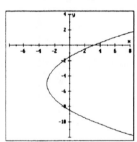

21. $(y + 2)^2 = 10(x - 7)$ **22.** $(x - 3)^2 = -9(y - 1)$

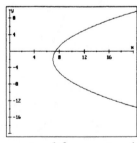

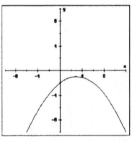

23. $(x + \frac{1}{3})^2 = -2(y - \frac{4}{3})$ **24.** $(x + 1)^2 = -\frac{2}{3}(y - \frac{16}{3})$

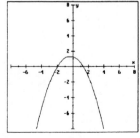

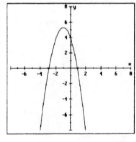

25. $V(\frac{5}{2}, 0)$; $4c = 10$; $y^2 = 10(x - \frac{5}{2})$

26. $V(0, -\frac{3}{2})$; $4c = 6$; $x^2 = -6(y - \frac{3}{2})$

27. $c = 4$, so $4c = 16$; $(y - 2)^2 = -16(x + 1)$

28. $c = 3$, so $4c = 12$; $(x - 4)^2 = 12(y + 1)$

29. $c = 6$, so $4c = 24$; $(x + 2)^2 = 24(y + 3)$

30. $c = 4$, so $4c = 16$; $(y - 4)^2 = 16(x + 3)$

31. The form of the equation is
$$(x + 3)^2 = -4c(y - 2)$$
Since the curve passes through $(-2, 1)$,
$$(-2 + 3)^2 = -4c(1 - 2)$$
$$c = \frac{1}{12}$$
Thus, $(x + 3)^2 = -\frac{1}{3}(y - 2)$

32. The form of the equation is
$$(y - 2)^2 = -4c(x - 4)$$
Since the curve passes through $(-3, -4)$,
$$(-4 + 2)^2 = -4c(-3 - 4)$$
$$c = \frac{9}{7}$$
Thus, $(y - 2) = -\frac{36}{7}(x - 4)$

33. $y^2 = -12(x - 3)$ **34.** $x^2 = 8(y + 2)$

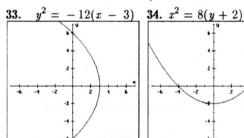

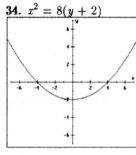

35. $x^2 = 18(y + 9/2)$ **36.** $y^2 = -4(x - 1)$

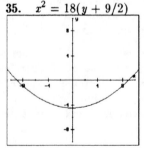

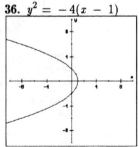

37. $y^2 = 8(x + 2)$ **38.** $y^2 = -6(x - 3/2)$

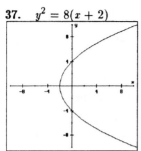 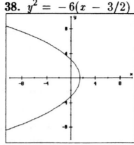

39. $V(4, 0)$; $p = 8$; $r = \dfrac{8}{1 + \cos\theta}$

40. $V(2, \pi)$; $p = 4$; $r = \dfrac{4}{1 - \cos\theta}$

41. $p = 4$; $r = \dfrac{4}{1 - \sin\theta}$

42. $p = 3$; $r = \dfrac{3}{1 + \cos\theta}$

43. $y^2 = 4x$ **44.** $x^2 = -2y$

   $(r\sin\theta)^2 = 4r\cos\theta$ $(r\cos\theta)^2 = -2(r\sin\theta)$

    $r\sin^2\theta = 4\cos\theta$ $r\cos^2\theta = 2\sin\theta$

45. $4x^2 = y - 3$

    $4r^2\cos^2\theta = r\sin\theta - 3$

46. $x + 1 = 2(y - 3)^2$

        $x + 1 = 2y^2 - 12y + 18$

    $r\cos\theta + 1 = 2(r\sin\theta - 3)^2$

47. Minimize the function giving the distance from the point (x_1, y_1) on the parabola to the point $(2, 0)$.

$$d^2 = (x_1 - 2)^2 + y_1{}^2$$

Minimizing d^2 also minimizes d, so substitute $y_1{}^2$ and set the derivative equal to 0:

$$2(x_1 - 2) + 4 = 0$$

$x_1 = 0$, $y_1 = 0$; the origin is the closest point.

48. The focus is $F(0, c)$;

$$d^2 = x_1{}^2 + (y_1 - c)^2$$
$$= 4cy_1 + y_1{}^2 - 2cy_1 + c^2$$

Set the derivative equal to 0:

$$4c + 2y_1 - 2c = 0$$
$$y_1 = -c$$

Since $c \geq 0$, the closest point is when $y_1 = 0$; the origin is the closest point.

49. $y^2 = 4x$

         $2y\dfrac{dy}{dx} = 4$

   $m = \dfrac{dy}{dx} = 2y^{-1}$; at $(1, -2)$, $m = -1$

The slope of the normal is 1; the desired equations are given.

tangent line: $y + 2 = -(x - 1)$ or
       $x + y + 1 = 0$

normal line: $y + 2 = x - 1$ or
       $x - y - 3 = 0$

50. $\sqrt{(x-0)^2 + (y-3)^2} = \sqrt{(x-4)^2 + (y-3)^2}$

   $x^2 + (y - 3)^2 = (x - 4)^2 + (y - 3)^2$

         $x^2 = x^2 - 8x + 16$

            $x = 2$

51. $\sqrt{(x-2)^2 + (y-1)^2} = \sqrt{(x-4)^2 + (y-3)^2}$

   $4x + 4 - 2y + 1 = -8x + 16 - 6y + 9$

       $12x + 4y = 20$

      $3x + y - 5 = 0$

52. $4c = 6$, so $c = \frac{3}{2}$; $V = (\frac{11}{2}, -2)$ so

   $(y + 2)^2 = 6(x - \frac{11}{2})$

53. $V(0, -3)$, so

$$x^2 = -4c(y + 3)$$

Parabola passes through $(6, -9)$ so

$$36 = 4c(-9 + 3)$$
$$c = -\frac{3}{2}$$

The desired equation is $x^2 = -6(y + 3)$.

54. $V(0, 0)$, so

$$x^2 = 4cy$$

Parabola passes through $(4, 6)$ so

$$16 = 4c(6)$$
$$c = \frac{3}{2}$$

The desired equation is $x^2 = \frac{4}{3}y$.

55. $V(0, 0)$, so

$$y^2 = 4cx$$

Parabola passes through $(4, 6)$ so

$$36 = 4c(4)$$
$$c = \frac{9}{4}$$

The focus is $F(\frac{9}{4}, 0)$.

56. $V(0, 0)$; $F(0, -c)$ or
$$x^2 + (y + c)^2 = 0 + (y - c)^2$$
$$x^2 + 2c = -2cy$$
$$x^2 = -4cy$$

57. $V(0, 0)$; $F(c, 0)$ or
$$(x - c)^2 + y^2 = (x + c)^2 + 0$$
$$-2cx + y^2 = 2cx$$
$$y^2 = 4cx$$

58. $V(0, 0)$; $F(-c, 0)$ or
$$(x + c)^2 + y^2 = (x - c)^2 + 0$$
$$2cx + y^2 = -2cx$$
$$y^2 = -4cx$$

59. $V(0, 0)$; $F(c, 0)$ or $y^2 = 4cx$; $P(c, y_0)$ so
$y^2 = 4c^2$ which implies $y = \pm 2c$.
$$d = \sqrt{(c - c)^2 + (2c - 0)^2} = 2c$$
The focal chord has length $2d$ or $4c$.

60. Let $P(r, \theta)$ be a point on $r = \dfrac{p}{1 - \cos \theta}$. The alternate primary form is $P(-r, \theta + \pi)$. Thus,
$$-r = \frac{p}{1 - \cos(\theta + \pi)}$$
$$r = \frac{-p}{1 - \cos \theta \cos \pi - \sin \theta \sin \pi}$$
$$= \frac{-p}{1 + \cos \theta}$$

61. $y^2 = 4cx$, $x \geq 0$ and $c > 0$. The distance from the vertex to $P(x, y)$ is
$$(x - c)^2 + y^2 = d^2$$
$$(x - c)^2 + 4cx = d^2$$
To minimize d, we minimize d^2, so we take the derivative on the left and set it equal to 0:
$$2x - 2c + 4c = 0$$
$$x = -2c$$
The minimum value occurs when $x = 0$.

62. $y^2 = 4cx$, $x \geq 0$ and $c > 0$. The derivative is $2yy' = 4c$; at $P_1(c, 2c)$ we have
$$m_1 = y' = 2c/y; \text{ at } (c, 2c), \; m_1 = -1$$
The tangent line is T_1: $y - 2c = x - c$ or $y = x + c$. At $P_2(c, -2c)$,
$$m_2 = y' = 2c/y; \text{ at } (c, 2c), \; m = -1$$
The tangent line is T_2: $y + 2c = -x + c$ or

$y = -x + c$. These tangents intersect when
$$x + c = -x - c$$
$$x = -c$$
If $x = -c$, then $y = 0$ and the ends of the focal chord intersect on the directrix.

63. $x^2 = 4cy$; by Problem 62, the two tangent lines at the ends of the focal chord intersect on the directrix, so the altitude of the triangle is the distance from the focus to this point of intersection; namely $2c$. The length of the focal chord is $4c$, so the area is found by
$$A = \tfrac{1}{2}(4c)(2c) = 4c^2$$

64. Let $(x - h)^2 + (y - k)^2 = R^2$ be the equation for the circle and $x^2 = 4cy$ the parabola. The abscissas of the common points satisfy
$$(x - h)^2 + \left(\frac{x^2}{4c} - k\right)^2 = R^2$$
$$x^2 - 2hx + h^2 + \frac{x^4}{16c^2} - \frac{kx^2}{2c} + k^2 = R^2$$
Note the coefficient of x^3 is 0. Suppose the polynomial in the left member has roots x_1, x_2, x_3, and x_4. Then
$$(x - x_1)(x - x_2)(x - x_3)(x - x_4) = 0$$
If we expand the left member the coefficient of x^3 is $-(x_1 + x_2 + x_3 + x_4)$ which is 0 as found above.

65. **a.** $y = \dfrac{x^2}{4c}$; $y' = \dfrac{2x}{4c} = \dfrac{x}{2c}$

At $P(x_0, y_0)$, $m_T = x_0/(2c)$. The equation of the tangent line is
$$y - y_0 = \frac{x_0}{2c}(x - x_0)$$

b. For $Q(0, y)$, $y = y_0 - \dfrac{x_0^2}{2c}$

c. For $|\overline{FP}|^2$ we have
$$x_0^2 + (y_0 - c)^2 = x_0^2 + \left(\frac{x_0^2}{4c} - c\right)^2$$
$$= x_0^2 + \frac{x_0^4}{16c^2} - \frac{x_0^2}{2} - c^2$$
$$= \frac{x_0^4}{16c^2} + \frac{x_0^2}{2} + c^2$$
$$= \left(\frac{x_0^2}{4c} + c\right)^2$$

For $|\overline{FQ}|^2$ we have
$$\left(y_0 - \frac{x_0^2}{2c} - c\right)^2 = \left(-\frac{x_0^2}{4c} - c\right)^2$$

$$= (-1)^2 \left(\frac{x_0^2}{4c} + c \right)^2$$

$$= \left(\frac{x_0^2}{4c} + c \right)^2$$

Thus, $|\overline{FP}| = |\overline{FQ}|$ so $\triangle QFP$ is isosceles.

d. $\phi = \angle FQP = \theta = \angle FPQ$; since L is parallel to $\overline{FQ}$, $\angle LPT = \theta$ also.

9.6　Conic Sections: The Ellipse and the Hyperbola, Page 623

1.　Write the equation in standard form and then plot the intercepts. Draw the ellipse passes through the four plotted points.

2.　Write the equation in standard form and then plot the intercepts and the pseudo-intercepts. Draw the central rectangle, then the asymptotes. Draw the hyperbola using the two intercepts and the asymptotes.

3.　If $Ax^2 + Bxy + Cy^2 + Ex + Ey + F = 0$. If $B^2 - 4AC < 0$, the equation is that of an ellipse. If $B^2 - 4AC = 0$, the equation represents a parabola. If $B^2 - 4AC > 0$, the equation is that of a hyperbola.

4.　Write the equation in the forms
$$r = \frac{\epsilon p}{1 \pm \epsilon \cos \theta} \text{ or } r = \frac{\epsilon p}{1 \pm \epsilon \sin \theta}$$
If $\epsilon < 1$, the equation represents an ellipse;

if $\epsilon = 1$, the equation represents a parabola;

if $\epsilon > 1$, the equation represents a hyperbola.

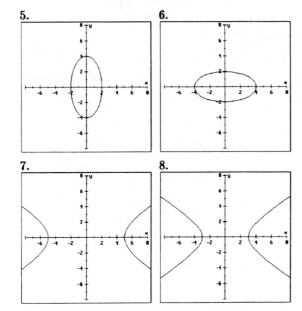

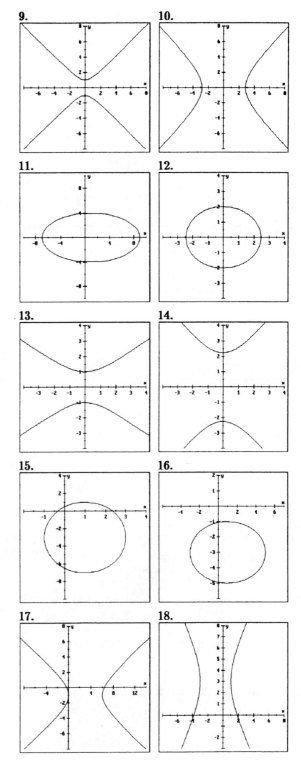

19. $\dfrac{(y-3)^2}{4} - \dfrac{(x-1)^2}{9} = 1$ **20.** $\dfrac{(x+1)^2}{4} - \dfrac{(y-1)^2}{1} = 1$

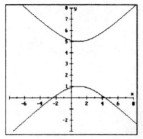 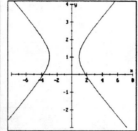

21. $\dfrac{(x+1)^2}{1/4} - \dfrac{(y-1)^2}{1} = 1$ **22.** $\dfrac{(x+1)^2}{4} + \dfrac{(y-1)^2}{1} = 1$

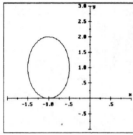 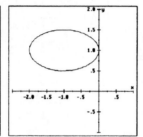

23. $\dfrac{x^2}{4} + \dfrac{(y-1)^2}{9} = 1$ **24.** $\dfrac{(x-1)^2}{9} + \dfrac{y^2}{4} = 1$

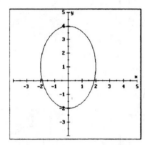 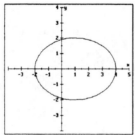

25. $C(0, 5)$; $2a = 8 - 2 = 6$; $b^2 = 9 - 5 = 4$
$$\dfrac{x^2}{4} + \dfrac{(y-5)^2}{9} = 1$$

26. $C(0, 0)$; $a = 4$, $c = 3$; $b^2 = 16 - 9 = 7$
$$\dfrac{y^2}{16} - \dfrac{x^2}{7} = 1$$

27. $C(0, 0)$; $b = 2$, $c = 2$; $a^2 = 4 + 4 = 8$;
$$\dfrac{x^2}{8} + \dfrac{y^2}{4} = 1$$

28. $C(0, 0)$; $c = 3$, $a^2 = 4$; $b^2 = 9 - 4 = 5$
$$\dfrac{y^2}{4} - \dfrac{x^2}{5} = 1$$

29. $C(0, 0)$; $c = \sqrt{2}$, $a = 1$; $b^2 = 2 - 1 = 1$
$$x^2 - y^2 = 1$$

30. $C(0, 0)$; $a = 5$, $c = 7$; $b^2 = 49 - 25 = 24$
$$\dfrac{x^2}{25} - \dfrac{y^2}{24} = 1$$

31. Ellipse with center $(0, 0)$ and $a = 4$, $b = 3$
$$\dfrac{x^2}{16} + \dfrac{y^2}{9} = 1$$

32. Hyperbola with center $(0, 0)$ and $a = 3$, $b = 4$
$$\dfrac{x^2}{9} - \dfrac{y^2}{16} = 1$$

33. $C(2, 1)$; $a = (6+4)/2 = 5$, $b = 3$
$$\dfrac{(x-2)^2}{9} + \dfrac{(y-1)^2}{25} = 1$$

34. Ellipse with center $(0, 0)$ and $a = 6$, $c = 1$
so that $b^2 = 36 - 1 = 35$
$$\dfrac{x^2}{36} + \dfrac{y^2}{35} = 1$$

35. Hyperbola with center $(0, 0)$ and $a = 2$, $c = 6$,
so that $b^2 = 36 - 4 = 32$.
$$\dfrac{y^2}{4} - \dfrac{x^2}{32} = 1$$

36. Ellipse with center $(0, -3)$, $a = 6$, $c = 4$, and
$b^2 = 36 - 16 = 20$.
$$\dfrac{x^2}{36} + \dfrac{y^2}{20} = 1$$

37. Hyperbola with center $(0, -3)$, $a = 3$, $c = 4$,
and $b^2 = 16 - 9 = 7$.
$$\dfrac{x^2}{9} - \dfrac{(y+3)^2}{7} = 1$$

38. Since $(y + 3x)(y - 3x) = 0$, the equation of
hyperbola is $y^2 - 9x^2 = K$. With $P(3, 0)$,
$$0^2 - 81 = K \text{ so that } y^2 - 9x^2 = -81$$

The desired equation is
$$\dfrac{x^2}{9} - \dfrac{y^2}{81} = 1$$

39.   **40.**

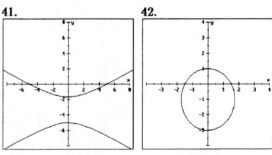

41.   **42.**

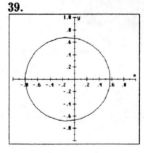 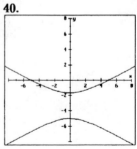

43.
$$5x^2 + 4y^2 = 56$$
$$10x + 8yy' = 0$$
$$y' = -\frac{5x}{4y}$$
At $P(-2, 3)$, $m = \frac{10}{12} = \frac{5}{6}$. The tangent line is
$$y - 3 = \tfrac{5}{6}(x + 2) \text{ or } 5x - 6y + 28 = 0$$

44. There are two cases. Either $2a = 6$ or $2b = 6$. If $a = 3$, $b < 1$ varies. The center is at $(b, 3)$. The equation of the ellipse is
$$\frac{(x - b)^2}{9} + \frac{(y - 3)^2}{b^2} = 1$$
The roles of a and b are interchanged if $b > 1$. Then
$$\frac{(x - a)^2}{a^2} + \frac{(y - 3)^2}{9} = 1$$

45. The equation of the ellipse can be written
$$y^2 = 1 - \frac{x^2}{4}$$
The equation of the tangent line through $P(0, -2)$ is $y + 2 = m(x - 0)$ or $y = mx - 2$. The tangent line intersects the ellipse when
$$(mx - 2)^2 = 1 - \frac{x^2}{4}$$
$$x^2(4m^2 + 1) - 16mx + 12 = 0$$
$$x = \frac{8m \pm \sqrt{64m^2 - 48m^2 - 12}}{4m^2 + 12}$$
For these points to become a single point, the discriminant must be zero. That is,
$$16m^2 - 12 = 0 \text{ or } m = \pm\frac{\sqrt{3}}{2}$$
Thus, $x = \frac{8(\sqrt{3}/2)}{3 + 1} = \sqrt{3}$. The equation of the tangent line is $y = \frac{\sqrt{3}}{2}x - 2$, and the y-value at the point of contact is
$$y = \frac{\sqrt{3}}{2}\sqrt{3} - 2 = -\frac{1}{2}$$
The points of contact are $P(\pm\sqrt{3}, -\frac{1}{2})$.

46. The equation of the ellipse can be written
$$y^2 = -\tfrac{1}{2}(3x^2 + 6x - 3)$$
The square of the distance from $P(2, 0)$ to a point $Q(x, y)$ on the ellipse is
$$d^2 = D = (x - 2)^2 + y^2$$
$$= (x - 2)^2 - \tfrac{1}{2}(3x^2 + 6x - 3)$$
$$D' = 2(x - 2) - \tfrac{1}{2}(6x + 6)$$
$$= 2x - 4 - 3x - 3$$
$$= -x - 7$$
$D' = 0$ when $x = -7$; but this is impossible

since
$$-[3(49) + 6(-7) - 3] < 0 \neq y^2$$
Note that the equation of the ellipse can be rewritten
$$3(x^2 + 2x + 1) + 2y^2 = 3 + 3$$
$$\frac{(x + 1)^2}{2} + \frac{y^2}{3} = 6$$
Thus,
$$-1 - \sqrt{2} \le x \le -1 + \sqrt{2}$$
At $x = -1 - 2$, $y = 0$ and
$$d = |x - 2| = 2 - (-1 - \sqrt{2}) = 3 + \sqrt{2}$$
At $x = -1 + \sqrt{2}$, $y = 0$ and
$$d = |x - 2| = 2 - (-1 + \sqrt{2}) = 3 - \sqrt{2}$$

47. The equation of the ellipse can be written as
$$y^2 = \frac{b^2}{a^2}(a^2 - x^2)$$
The volume is
$$V = 2\pi \frac{b^2}{a^2} \int_0^a (a^2 - x^2)\, dx$$
$$= \frac{2\pi b^2}{a^2}\left(a^2 x - \frac{x^3}{3}\right)\Big|_0^a = \frac{4\pi a b^2}{3}$$

48. The equation of the ellipse can be written as
$$x^2 = \frac{a^2}{b^2}(b^2 - y^2)$$
The volume is
$$V = 2\pi \frac{a^2}{b^2} \int_0^b (b^2 - y^2)\, dy$$
$$= \frac{2\pi a^2}{b^2}\left(b^2 y - \frac{y^3}{3}\right)\Big|_0^b = \frac{4\pi a^2 b}{3}$$

49. The equation of the ellipse can be written as
$$x^2 = 4 - 4y^2 \text{ or } x = 2\sqrt{1 - y^2}$$
Use shells; the volume is
$$V = 2(2\pi) \int_{-1}^1 x(y + 2)\, dy$$
$$= 4\pi \int_{-1}^1 2\sqrt{1 - y^2}\,(y + 2)\, dy$$
$$= 8\pi \int_{-1}^1 y\sqrt{1 - y^2}\, dy + 16\pi \int_{-1}^1 \sqrt{1 - y^2}\, dy$$

$$= 8\pi(0) + 16(\text{area of semicircle})$$
$$= 16\pi(\tfrac{1}{2}\pi) = 8\pi^2$$

50. The asymptotes intersect at

$$\tfrac{9}{4}x - \tfrac{13}{4} = -\tfrac{9}{4}x + \tfrac{5}{4}$$
$$x = 1$$

If $x = 1$, then $y = -1$. The equations of the asymptotes can be rewritten

$$y + 1 = \tfrac{9}{4}(x - 1) \text{ and } y + 1 = -\tfrac{9}{4}(x - 1)$$

Thus, the equation of the hyperbola is

$$(y + 1)^2 - \tfrac{81}{16}(x - 1)^2 = K$$

The point $P(3, -1)$ is on the hyperbola so

$$0 - \tfrac{81}{16}(4) = K$$

and

$$(y + 1)^2 - \tfrac{81}{16}(x - 1)^2 = -\tfrac{81}{4}$$
$$\frac{(x - 1)^2}{4} - \frac{4(y + 1)^2}{81} = 1$$

51. The asymptotes are the lines $y = \pm x$. Thus, the equation of the hyperbola is $x^2 - y^2 = K$. The point $P(9, 0)$ is on the hyperbola, so $81 = K$ and

$$\frac{x^2}{81} - \frac{y^2}{81} = 1$$

52. $a = 2$ and

$$\frac{x^2}{4} - \frac{y^2}{b^2} = 1$$

Using $P(3, \sqrt{5}/2)$ we find

$$\frac{9}{4} - \frac{5}{4b^2} = 1$$
$$b^2 = 1$$

The equation of the hyperbola is

$$\frac{x^2}{4} - \frac{y^2}{1} = 1$$

53. $\cosh^2 t = \dfrac{x - x_0}{a}$ and $\sinh t = \dfrac{y - y_0}{b}$

Since $\cosh^2 t - \sinh^2 t = 1$, we have

$$\frac{(x - x_0)^2}{a^2} - \frac{(y - y_0)^2}{b^2} = 1$$

provided that $x \geq x_0 + a$ because $\cosh t \geq 1$.

54. $\sinh^2 t = \dfrac{x - x_0}{a}$ and $\cosh t = \dfrac{y - y_0}{b}$

Since $\cosh^2 t - \sinh^2 t = 1$, we have

$$\frac{(y - y_0)^2}{b^2} - \frac{(x - x_0)^2}{a^2} = 1$$

provided that $y \geq y_0 + b$ because $\cosh t \geq 1$.

55. $\dfrac{x^2}{4} - \dfrac{y^2}{9} = 1$. The vertex is at $(2, 0)$.

$$y^2 = \tfrac{9}{4}(x^2 - 4), \quad y = \tfrac{3}{2}\sqrt{x^2 - 4}.$$

Finding the first quadrant area and doubling:

$$A = 2\int_2^4 y\, dx = 3\int_2^4 \sqrt{x^2 - 2^2}\, dx$$

(Formula 203 or trigonometric substitution)

$$= 3\left[\frac{x\sqrt{x^2 - 4}}{2} - 2\ln\left(x + \sqrt{x^2 - 4}\right)\right]\Bigg|_2^4$$
$$= 6[2\sqrt{3} - \ln(2 + \sqrt{3})] \approx 12.88$$

56. $y = \pm x$ are the asymptotes, which are perpendicular to each other (their slopes are negative reciprocals). If the asymptotes are perpendicular to each other, their slopes are negative reciprocals and $m_1 = -1/m_2 = m$. Since the asymptotes make an angle of $\pm \pi/4$ with the positive x-axis, $m = 1$. The equations of the asymptotes then are $y = x + b$ and $y = -x + b$, with $b = 0$ because of standard position. Then $(y + x)(y - x) = K$ is the equation of the hyperbola $y^2 - x^2 = K$. The point $(a, 0)$ is a vertex, so $K = -a^2$ and $x^2 - y^2 = a^2$.

57. Write the equation of the hyperbola as

$$y^2 = \frac{9x^2 - 36}{4}$$

a. Use the method of disks.

$$V = \pi\int_2^4 y^2\, dx = \pi\int_2^4 \tfrac{9}{4}(x^2 - 4)\, dx$$
$$= \tfrac{9}{4}\pi\left(\frac{x^3}{3} - 4x\right)\Bigg|_2^4 = 24\pi$$

b. Use the method of shells.

$$V = 2\pi\int_2^4 xy\, dx = 2\pi\int_2^4 \tfrac{3}{2}\sqrt{x^2 - 4}\; x\, dx$$
$$= 3\pi\int_2^4 x\sqrt{x^2 - 4}\, dx$$
$$= \frac{3\pi}{2}\left[\frac{(x^2 - 4)^{3/2}}{3/2}\right]\Bigg|_2^4 = 24\sqrt{3}\pi$$

58. $a = 50$, $b = \dfrac{81}{2}$; $\dfrac{x^2}{2,500} + \dfrac{4y^2}{6,561} = 1$

$$c = \sqrt{2,500 - \frac{6,561}{4}} \approx 29.3215$$

The foci are $2c \approx 59.643$ million miles apart.

59. Put the sun (focus) at the origin. The general equation is:

$$r = \frac{\epsilon p}{1 - \epsilon \cos \theta}.$$

The least distance, l, is the intercept on the negative x-axis, $r(\pi)$, and the greatest distance, g, is the intercept on the positive x-axis, $r(0)$. The major axis, $2a$, is equal to $r(0) + r(\pi)$.

$$2a = \frac{\epsilon d}{1 + \epsilon} + \frac{\epsilon d}{1 - \epsilon} = \frac{2\epsilon d}{1 - \epsilon^2}$$

$$\epsilon d = a(1 - \epsilon^2) \approx 9.3 \times 10^7 (1 - 0.017^2)$$

$$\approx 9.2973 \times 10^7$$

$$l = r(\pi) = \frac{\epsilon d}{1 + \epsilon} \approx 9.1419 \times 10^7 \text{mi}$$

$$g = r(0) = \frac{\epsilon d}{1 - \epsilon} \approx 9.4581 \times 10^7 \text{mi}.$$

60. Let d_1 be the distance that sound travels from the gun at A to the person at P. Let d_2 be the distance that sound travels from the gong to B to the person at P. Let x be the distance that it takes the bullet to reach the gong at B. Then, $d_1 - d_2 = x$ and the person should stand on a branch of the hyperbola with foci at A and B.

61. Let the polar form coordinates of the first station be $O(0,0)$, those of the second $A(4, 0)$, those of the third $B(4, \frac{\pi}{4})$, and $P(r, \theta)$. Also let t_{op} be the time for the wave to travel from O to P, t_{ap} be the time to travel from A to P, and t_{bp} be the time for the wave to travel from B to P.

$$t_{op} = \frac{2}{c} + t_{ap} \text{ or } \frac{r}{c} = \frac{2}{c} + \frac{x}{c}$$

from which we find $x = ri - 2$. Also,

$$t_{bp} = t_{ap}$$

so $\triangle OAP \sim \triangle OPB$. Thus,

$$\theta = \angle AOP = \angle POB = \frac{\pi}{8}$$

$$x^2 = (r - 2)^2$$

$$r^2 + 16 - 8r \cos \theta = r^2 - 4r + 4$$

$$4 - 2r \cos \theta = -r + 1$$

$$r = \frac{3}{2 \cos \theta - 1}$$

Thus, when $\theta = \frac{\pi}{8}$, $r \approx 3.5387$, so the airplane is located at point $(3.5387, \frac{\pi}{8})$.

62. Let d_1 be the distance from $P(x, y)$ to $F_1(-c, 0)$ and d_2 be the distance from $P(x, y)$ to $F_2(c, 0)$. By definition,

$$d_1 + d_2 = 2a$$

$$d_1 = 2a - d_2$$

$$d_1^2 = 4a^2 - 4ad_2 + d_2^2$$

$$(x + c)^2 + y^2 = 4a^2 - 4ad_2 + 4(x - c)^2 + y^2$$

$$2cx = 4a^2 - 4ad_2 - 2cx$$

$$cx = a^2 - ad_2$$

$$ad_2 = a_2 - cx$$

$$a^2[(x - c)^2 + y^2] = a^4 - 2a^2cx + c^2x^2$$

$$a^2x^2 - 2a^2cx + a^2c^2 + a^2y^2 = a^4 - 2a^2cx + c^2x^2$$

$$(a^2 - c^2)x^2 + a^2y^2 = a^2(a^2 - c^2)$$

$$b^2x^2 + a^2y^2 = a^2b^2$$

$$\frac{x^2}{a^2} + \frac{y^2}{b^2} = 1$$

63. Let d_1 be the distance from $P(x, y)$ to $F_1(-c, 0)$ and d_2 be the distance from $P(x, y)$ to $F_2(c, 0)$. By definition,

$$d_1 - d_2 = 2a$$

$$d_1 = 2a + d_2$$

$$d_1^2 = 4a^2 + 4ad_2 + d_2^2$$

$$(x + c)^2 + y^2 = (x - c)^2 + y^2 + 4a^2 + 4ad_2$$

$$2cx = 4a^2 + 4ad_2 - 2cx$$

$$cx = a^2 + ad_2$$

$$ad_2 = -a_2 + cx$$

$$a^2[(x - c)^2 + y^2] = a^4 - 2a^2cx + c^2x^2$$

$$a^2x^2 - 2a^2cx + a^2c^2 = a^4 - 2a^2cx + c^2x^2$$

$$(a^2 - c^2)x^2 + a^2y^2 = a^2(a^2 - c^2)$$

$$-b^2x^2 + a^2y^2 = -a^2b^2$$

$$\frac{x^2}{a^2} - \frac{y^2}{b^2} = 1$$

64. **a.** $A = C = 0$
   b. A and C have the same signs
   c. $A = C$
   d. $A = C$ and $D = E = 0$
   e. A and C have opposite signs
   f. $A > 0$ and

$$\frac{D^2}{4A} + \frac{E^2}{4C} < F$$

   or

   $A < 0$ and

$$\frac{D^2}{4A} + \frac{E^2}{4C} > F$$

65. **a.** $\frac{x^2}{a^2} + \frac{y^2}{b^2} = 1$

$$\frac{2x}{a^2} + \frac{2yy'}{b^2} = 0$$

$$y' = -\frac{b^2 x}{a^2 y}$$

The slope at $P_0(x_0, y_0)$ is

$$m = y' = -\frac{b^2 x_0}{a^2 y_0}$$

The equation of the tangent line is

$$y - y_0 = -\frac{b^2 x_0}{a^2 y_0}(x - x_0)$$

$$\frac{y_0 y}{b^2} - \frac{y_0^2}{b^2} + \frac{x_0 x}{a^2} - \frac{x_0^2}{a^2} = 0$$

$$\frac{x_0 x}{a^2} + \frac{y_0 y}{b^2} = \frac{x_0^2}{a^2} + \frac{y_0^2}{b^2}$$

$$\frac{x_0 x}{a^2} + \frac{y_0 y}{b^2} = 1$$

(Since $P_0(x_0, y_0)$ lies on the ellipse.)

b. At $P_0(\pm a, 0)$,

$$\frac{\pm ax}{a^2} = 1$$

or $x = \pm a$, a vertical tangent line.

At $P_0(0, \pm b)$

$$\frac{\pm by}{b^2} = 1$$

or $y = b$,, a horizontal tangent line.

66. From the ellipse

$$y = \pm \frac{b}{a}\sqrt{x^2 - a^2}$$

and from the asymptote $y = \pm \frac{b}{a}x$. Let d be the vertical distance between points on the ellipse, and on the asymptote x. Then,

$$\lim_{x \to \infty} \frac{b}{a}(\sqrt{x^2 - a^2} - x)$$

$$= \lim_{x \to \infty} \frac{b}{a}\left[\frac{(\sqrt{x^2 - a^2} - x)(\sqrt{x^2 - a^2} + x)}{\sqrt{x^2 - a^2} + x}x\right]$$

$$= \lim_{x \to \infty} \frac{b}{a}\left[\frac{x^2 - a^2 - x^2}{\sqrt{x^2 - a^2} + x}\right]$$

$$= -ab\lim_{x \to \infty} \frac{1}{\sqrt{x^2 - a^2} + x} = 0$$

67.
$$\frac{x^2}{a^2} - \frac{y^2}{b^2} = 1$$

$$\frac{2x}{a^2} - \frac{2yy'}{b^2} = 0$$

$$y' = \frac{b}{a}x$$

An equation of a line parallel to an asymptote is $y = b/a + d$ where $d \neq 0$. Squaring both sides leads to

$$y^2 = \frac{b^2 x^2}{a^2} + \frac{2bdx}{a} + d^2$$

From the equation of the hyperbola we have $y^2 = \frac{b^2}{a^2}(x^2 - a^2)$ so we can now see

$$\frac{b^2 x^2}{a^2} + \frac{2bdx}{a} + d^2 = \frac{b^2}{a^2}(x^2 - a^2)$$

$$-\frac{a^2}{b^2} = \frac{2bdx}{a} + d^2$$

$$x = -\frac{a}{2bd}\left(\frac{a^2}{b^2}\right) + d^2$$

Since this is only one value, we see that there must be a single point of intersection.

68. a. For the ellipse $\frac{x^2}{a^2} + \frac{y^2}{b^2} = 1$, $c^2 = a^2 - b^2$, At $x = \pm c$,

$$\frac{c^2}{a^2} + \frac{y^2}{b^2} = 1 \text{ or } y^2 = \frac{b^2}{a^2}b^2 \text{ or } y = \pm\frac{b^2}{a}$$

The length of the focal chord is $2y = \frac{2b^2}{a}$.

For hyperbola $\frac{x^2}{a^2} - \frac{y^2}{b^2} = 1$, $c^2 = a^2 + b^2$, At $x = \pm c$,

$$\frac{c^2}{a^2} - \frac{y^2}{b^2} = 1 \text{ or } y^2 = \frac{b^2}{a^2}b^2 \text{ or } y = \pm\frac{b^2}{a}$$

The length of the focal chord is $2y = \frac{2b^2}{a}$.

b. For the directrix ,

$$\epsilon = \frac{|\overline{PF}|}{|\overline{PL}|} = \frac{a - c}{x - a}$$

$$\epsilon x - \epsilon a = a - c$$

$$x = \frac{a}{\epsilon} = \frac{a^2}{c}$$

From Problem 65, the equation of the tangent line to the ellipse is

$$\frac{x_0 x}{a^2} + \frac{y_0 y}{b^2} = 1$$

At the focal points $P(\pm c, \frac{a^2}{b})$, it becomes

$$\frac{\pm cx}{a^2} + \frac{a^2 y}{b^3} = 1$$

If $y = 0$,

$$\frac{\pm cx}{a^2} = 1 \text{ or } x = \pm\frac{a^2}{c}$$

The x-values are the first components for the directrix.

For the hyperbola, it can be shown (just as in Problem 65 for the ellipse), that the equation of the tangent line is

$$\frac{x_0 x}{a^2} - \frac{y_0 y}{b^2} = 1$$

At the focal point $P(\pm c, \frac{a^2}{b})$ it becomes

$$\frac{\pm cx}{a^2} - \frac{a^2 y}{b^3} = 1$$

If $y = 0$,

$$\frac{\pm cx}{a^2} = 1 \text{ or } x = \pm\frac{a^2}{c}$$

The x-values are the first components for

the directrix.

c. At the pole, (a focus is at the origin), $r = \epsilon p$ and the length of the focal chord is $\ell = 2\epsilon p$.

69. We know (from Problem 65) that the tangent line at point $P_0(x_0, y_0)$ on the ellipse has the equation

$$\frac{xx_0}{a^2} + \frac{yy_0}{b^2} = 1$$

It is easily seen that this line has slope

$$m_0 = -\frac{x_0 b^2}{y_0 a^2}$$

We find that the slope of the line through points P_0 and F_2 is

$$m_1 = \frac{y_0 - 0}{x_0 - c}$$

and by using the formula for the angle between two lines given in Problem 40 of Section 1.3,

$$\tan\theta_2 = \frac{m_0 - m_1}{1 + m_0 m_1} = \frac{\left(\dfrac{-x_0 b^2}{y_0 a^2}\right) - \left(\dfrac{y_0}{x_0 - c}\right)}{1 + \left(\dfrac{-x_0 b^2}{y_0 a^2}\right)\left(\dfrac{y_0}{x_0 - c}\right)}$$

$$= \frac{-x_0 b^2(x_0 - c) - y_0^2 a^2}{y_0 a^2(x_0 - c) - x_0 y_0 b^2}$$

$$= \frac{-(b^2 x_0^2 + a^2 y_0^2) + x_0 b^2 c}{x_0 y_0(a^2 - b^2) - c y_0 a^2}$$

$$= \frac{-(a^2 b^2) + x_0 b^2 c}{xyc^2 - cy_0 a^2} \quad \text{since } \frac{x_0^2}{a^2} + \frac{y_0^2}{b^2} = 1 \\ \text{and } a^2 - b^2 = c^2$$

$$= \frac{b^2(-a^2 + x_0 c)}{y_0 c(x_0 c - a^2)} = \frac{b^2}{y_0 c}$$

Similarly, the slope of $\overline{P_0 F_1}$ is $m_2 = \dfrac{y_0}{x_0 + c}$, and

$$\tan\theta_1 = \frac{m_2 - m_0}{1 + m_1 m_0} = \frac{\left(\dfrac{y_0}{x_0 + c}\right) - \left(\dfrac{-x_0 b^2}{y_0 a^2}\right)}{1 + \left(\dfrac{y_0}{x_0 + c}\right)\left(\dfrac{-x_0 b^2}{y_0 a^2}\right)}$$

$$= \frac{y_0^2 a^2 - (-x_0 b^2)(x_0 + c)}{(x_0 + c)(y_0 a^2) - x_0 y_0 b^2}$$

$$= \frac{(y_0^2 a^2 + x_0^2 b^2) + x_0 b^2 c}{x_0 y_0(a^2 - b^2) + c y_0 a^2}$$

$$= \frac{a^2 b^2 + x_0 b^2 c}{x_0 y_0 c^2 + c y_0 a^2}$$

$$= \frac{b^2(a^2 + x_0 c)}{y_0 c(x_0 c + a^2)} = \frac{b^2}{y_0 c}$$

Thus, $\tan\theta_1 = \tan\theta_2$, and we see that $\theta_1 = \theta_2$.

CHAPTER 9 REVIEW

Proficiency Examination, Page 626

1. The primary representation of a point in polar form are (r, θ) and $(-r, \theta + \pi)$

2. $r = \sqrt{x^2 + y^2}; \; \overline{\theta} = \tan^{-1}\left|\dfrac{y}{x}\right|$

3. $x = r\cos\theta, \; y = r\sin\theta$

4. A cardioid is heart-shaped (see Table 9.1, Page 583); $r = a(1 \pm \sin\theta), \; r = a(1 \pm \cos\theta)$

5. $r = f(\theta - \theta_0)$, where the graph is rotated through an angle θ_0.

6. Let $P(r, \theta)$ be a point on the graph. If the following forms of the point also satisfy the graph, then the graph has the indicated symmetry.
   Symmetry with respect to the x-axis:
      $P(r, -\theta)$ or $P(-r, \pi - \theta)$
   Symmetry with respect to the y-axis:
      $P(-r, -\theta)$ or $P(r, \pi - \theta)$
   Symmetry with respect to the origin:
      $P(-r, \theta)$ or $P(r, -\theta)$

7. A limaçon (which means snail in French) looks like a heart with a loop or dent (see Table 9.1, Page 583); $r = b \pm a\cos\theta$ or $r = b \pm a\sin\theta$. If $a = b$, it is a cardioid; if $|a| > |b|$, an inner loop exists.

8. A rose curve looks like a flower (see Table 9.1, Page 583); $r = a\cos n\theta$ or $r = a\sin n\theta$, where n is an integer. If n is odd, the curve has n leaves; if n is even, the curve has $2n$ leaves.

9. A lemniscate is a figure-eight curve (see Table 9.1, Page 583); $r^2 = a\cos 2\theta$ or $r^2 = a\sin 2\theta$.

10. If $f(\theta)$ is a differentiable function of θ, then the slope of the tangent line to the polar curve $r = f(\theta)$ at the point $P(r_0, \theta_0)$ is given by

$$m = \frac{f(\theta_0)\cos\theta_0 + f'(\theta_0)\sin\theta_0}{-f(\theta_0)\sin\theta_0 + f'(\theta_0)\cos\theta_0}$$

whenever the denominator is not zero.

11. Step 1. Find the simultaneous solution of the given system of equations. Step 2. Determine whether the pole lies on the two graphs. Step 3. Graph the curves to look for other

points of intersection.

12. $dA = \frac{1}{2}r^2\,d\theta$

13. $A = \int_{\theta_1}^{\theta_2} r^2\,d\theta$

14. Parametric equations of a curve relate the dependent and independent variable through a third variable called a parameter.

15. A trochoid is the path described by a point attached to a fixed radius on a moving wheel. The cycloid is the special case of the point being on the rim of the wheel.

16. $\dfrac{dy}{dx} = \dfrac{dy/dt}{dx/dt}$

17. $s = \int_a^b \sqrt{\left(\dfrac{dx}{dt}\right)^2 + \left(\dfrac{dy}{dt}\right)^2}$

18. $A = \int_{t_1}^{t_2} y(t)\,\dfrac{dx}{dt}\,dt$

19. A parabola is the set of all points in the plane that are equidistant from a fixed point (called the focus) and a fixed line (called the directrix). The standard forms are $y^2 = \pm 4cx$ or $x^2 = \pm 4cy$.

20. Let P be the point of contact of a tangent line to a parabola. The line from P to the focus makes the same angle with the tangent line as the line through P parallel to the axis of symmetry of the parabola.

21. An ellipse is the set of all points in the plane the sum of whose distances from two fixed points is a constant. The standard forms are
$$\frac{x^2}{a^2} + \frac{y^2}{b^2} = 1 \text{ or } \frac{y^2}{a^2} + \frac{x^2}{b^2} = 1$$

22. A hyperbola is the set of all points in the plane such that, for each point on the hyperbola, the difference of its distances from two fixed points is a constant. The standard forms are
$$\frac{x^2}{a^2} - \frac{y^2}{b^2} = 1 \text{ or } \frac{y^2}{a^2} - \frac{x^2}{b^2} = 1$$

23. A ray emanating from one focus bounces off the ellipse and passes through the other focus.

24. $c^2 = a^2 - b^2$ for an ellipse;
$c^2 = a^2 + b^2$ for a hyperbola

25. A conic section is the intersection of a plane with a cone; this intersection is a parabola, ellipse, hyperbola, intersection lines, circle, or a point. Let F be a point in the plane and let L be a line in the same plane. Then the set of

all points P in the plane that satisfy
$$\frac{\text{DISTANCE FROM P TO F}}{\text{DISTANCE FROM P TO L}} = \epsilon$$
is a conic section, and ϵ is a fixed number for each conic called the eccentricity of the conic. The conic is an ellipse if $\epsilon < 1$; a parabola if $\epsilon = 1$; a hyperbola if $\epsilon > 1$. The principal axis of a hyperbola or ellipse contains the foci. It is also referred to as the major axis of the ellipse, or the transverse axis of the hyperbola. The major axis of an ellipse is the longer axis, which contains the foci. The minor axis of an ellipse is the shorter axis, which is perpendicular to the minor axis. The transverse axis of a hyperbola contains the foci. The conjugate axis of a hyperbola passes through the center and is perpendicular to the transverse axis.

26. $r = \dfrac{\epsilon p}{1 + \epsilon \sin \theta}$ or $r = \dfrac{\epsilon p}{1 - \epsilon \sin \theta}$

In this form, the coefficient of the cosine or sine is the eccentricity ($\epsilon < 1$ for an ellipse, $\epsilon = 1$ for a parabola, $\epsilon > 1$ for a hyperbola).

27. **a.** parabola   **b.** hyperbola
     c. circle (ellipse)   **d.** circle
     e. line   **f.** parabola
     g. 4-leafed rose   **h** cardioid

28. parabola

29. circle

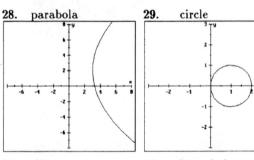

30. ellipse

$$\frac{(x-1)^2}{3} + \frac{(y+2)^2}{3} = 1$$

31. hyperbola

$$\frac{(x-1)^2}{1} - \frac{(y-6)^2}{9} = 1$$

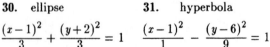

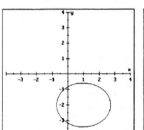

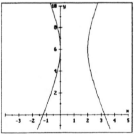

32.   lemniscate

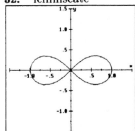

33.   limacon

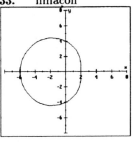

7.   parabola

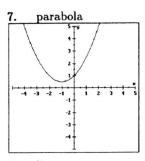

8.   ellipse

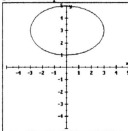

34.   Solving simultaneously, or by symmetry, we see that the intersection is at $(\sqrt{2}, \frac{\pi}{4})$.

$$A = 2a^2(\tfrac{1}{2})\int_0^{\pi/4} r^2 \, d\theta = \int_0^{\pi/4} 4a^2\sin^2\theta \, d\theta$$

$$= 2a^2\left(\theta - \frac{\sin 2\theta}{2}\right)\Big|_0^{\pi/4} = \frac{a^2}{2}(\pi - 2)$$

9.   line

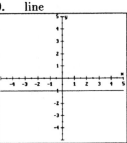

10.   rose curve

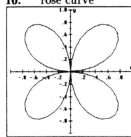

Supplementary Problems, Page 627

1.   circle

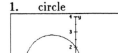

2.   spiral

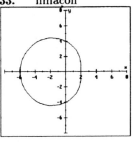

11.   circle

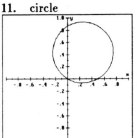

12.   circle

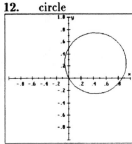

3.   rose curve

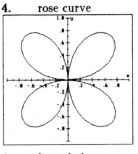

4.   rose curve

13.   rose curve

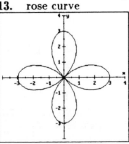

14. figure 8 (not lemniscate)

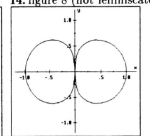

5.   parabola

6.   hyperbola

15.   parabola

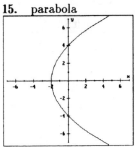

16.   parabola

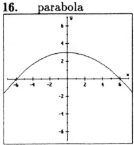

17. hyperbola

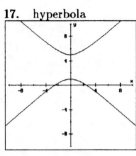

18. ellipse

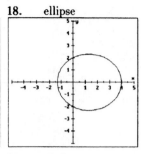

19.

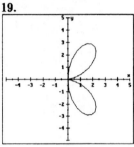

20. circle

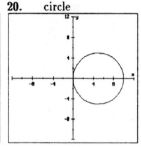

21.

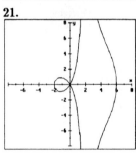

22. rose curve

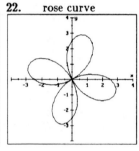

23.

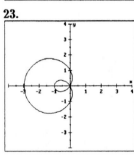

$y = r \sin \theta$
$\quad = \sin \theta - 2 \cos \theta \sin \theta$
$dy/dx = \cos \theta$
$\quad\quad - 2(\cos^2\theta - \sin^2\theta)$
Tangent is horizontal
when $dy/dx = 0$ or
$\cos\theta - 2(\cos^2\theta - \sin^2\theta) = 0$

$$\cos \theta = \frac{1 \pm \sqrt{1+32}}{8}$$

$$\approx = 0.84307, \; -0.5931$$

$$\theta \approx 0.6, \, 2.2$$

The points are $(2.2, 2.2)$, $(2.2, -2.2)$, $(-0.6, 0.6)$, $(-0.6, -0.6)$.

24.

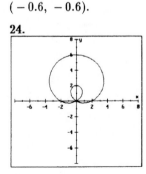

$y = r \sin \theta$
$\quad = 2 \sin \theta + 4 \sin^2\theta$
$dy/dx = 2 \cos \theta$
$\quad\quad + 8 \sin \theta \cos \theta$
Tangent is horizontal
when $dy/dx = 0$ or

$2 \cos \theta + 8 \sin \theta \cos \theta = 0$

$2 \cos \theta(1 + 4 \sin \theta) = 0$

$\cos \theta = 0 \quad$ or $\quad \sin \theta = -\frac{1}{4}$

$\theta = \frac{\pi}{2}, \frac{3\pi}{2}$ and $\theta \approx 2.88, 6.03$

The points are $(6, \frac{\pi}{2})$, $(-2, \frac{3\pi}{2})$, $(1, 6)$, $(-1, 3)$.

25.
$$y = r \sin \theta = \frac{\sin \theta}{1 - \sin \theta}$$

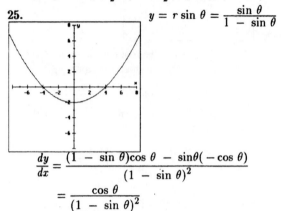

$$\frac{dy}{dx} = \frac{(1 - \sin \theta)\cos \theta - \sin\theta(-\cos \theta)}{(1 - \sin \theta)^2}$$

$$= \frac{\cos \theta}{(1 - \sin \theta)^2}$$

The tangent line is horizontal when $\cos \theta = 0$ or when $\theta = \pi/2, 3\pi/2$. At $\theta = \pi/2$, r is not defined. At $\theta = 3\pi/2$, $r = 2$ so the desired point is $(2, \frac{3\pi}{2})$.

26.
$$y = r \sin \theta$$
$$= \frac{3 \sin \theta}{4 - 3 \cos \theta}$$

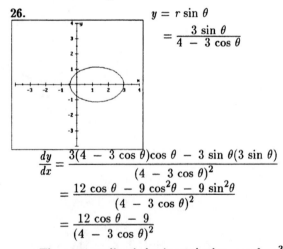

$$\frac{dy}{dx} = \frac{3(4 - 3 \cos \theta)\cos \theta - 3 \sin \theta(3 \sin \theta)}{(4 - 3 \cos \theta)^2}$$

$$= \frac{12 \cos \theta - 9 \cos^2\theta - 9 \sin^2\theta}{(4 - 3 \cos \theta)^2}$$

$$= \frac{12 \cos \theta - 9}{(4 - 3 \cos \theta)^2}$$

The tangent line is horizontal when $\cos \theta = \frac{3}{4}$ or when $\theta \approx \pm 0.72$. At $\theta = \pm 0.72$, $r \approx 1.71$. The desired points are $(1.71, \pm 0.72)$.

27.
$$2x + 3y = 4$$
$$2r \cos \theta + 3r \sin \theta = 4$$
$$r = \frac{4}{2 \cos \theta + 3 \sin \theta}$$

28.
$$r = -2 \sin \theta$$
$$r^2 = -2r \sin \theta$$
$$x^2 + y^2 = 2y$$

29. $x^2 + y^2 - 3x = 5$

$\quad r^2 - 3r\cos\theta = 5$

30. $\qquad\qquad r = \cos 2\theta$

$\qquad\qquad r = \cos^2\theta - \sin^2\theta$

$\qquad\qquad r^3 = r^2\cos^2\theta - r^2\sin^2\theta$

$\quad (x^2 + y^2)^{3/2} = x^2 - y^2$

31. $x^2 + y^2 = \tan^{-1}\frac{y}{x}$

$\qquad\qquad r^2 = \tan^{-1}\left(\dfrac{r\sin\theta}{r\cos\theta}\right)$

$\qquad\qquad r^2 = \theta$

32. $\qquad\qquad r^2 = 1 + \cot\theta$

$\qquad\qquad r^2 = \dfrac{\sin\theta + \cos\theta}{\sin\theta}$

$\qquad r^2\sin\theta = r\sin\theta + r\cos\theta$

$\qquad y\sqrt{x^2 + y^2} = x + y$

33. $\qquad\quad y = mx$

$\quad r\sin\theta = mr\cos\theta$

$\quad \tan\theta = m$

34. $x^2 + y^2 + ax = a\sqrt{x^2 + y^2}$

$\qquad\quad r^2 + ar\cos\theta = ar$

$\qquad\quad r + a\cos\theta = a$

$\qquad\qquad r = a(1 - \cos\theta)$

35. $\qquad (x - a)^2 + y^2 = a^2$

$\quad x^2 + y^2 - 2ax + a^2 = a^2$

$\qquad\quad r^2 - 2ar\cos\theta = 0$

$\qquad\qquad r = 2a\cos\theta$

36. $\quad (x^2 + y^2)^2 = a^2(x^2 - y^2)$

$\qquad\qquad r^4 = a^2 r^2(\cos^2\theta - \sin^2\theta)$

$\qquad\qquad r^2 = a^2\cos 2\theta$

37. $\qquad\qquad r^2 = \dfrac{36}{13\cos^2\theta - 4}$

$\quad 13r^2\cos^2\theta - 4r^2 = 36$

$\quad 13x^2 - 4(x^2 + y^2) = 36$

$\qquad\qquad \dfrac{x^2}{4} - \dfrac{y^2}{9} = 1$

38. $\qquad\qquad r^2 = \dfrac{36}{5\cos^2\theta + 4}$

$\quad 5r^2\cos^2\theta + 4r^2 = 36$

$\quad 5x^2 + 4(x^2 + y^2) = 36$

$\qquad\qquad \dfrac{x^2}{4} + \dfrac{y^2}{9} = 1$

39. $c = 5;\ x^2 = -20y$

40. $a = 8,\ c = 4;\ b^2 = 64 - 16 = 48;$

$\qquad \dfrac{x^2}{48} + \dfrac{y^2}{64} = 1$

41. $a = 4,\ c = 8;\ b^2 = 64 - 16 = 48;$

$\qquad \dfrac{y^2}{16} - \dfrac{x^2}{48} = 1$

42. $a = 2,\ c = 1;\ b^2 = 4 - 1 = 3;$

$\qquad \dfrac{(x - 2)^2}{4} + \dfrac{(y - 1)^2}{3} = 1$

43. $a = 2,\ c = 3;\ b^2 = 9 - 4 = 5;$

$\qquad \dfrac{(x - 6)^2}{4} - \dfrac{y^2}{5} = 1$

44. $c = 3;\ (y - 3)^2 = 12(x - 4)$

45. $a = 1,\ c = 2;\ b^2 = 4 - 1 = 3$

$\qquad \dfrac{(x + 5)^2}{1} - \dfrac{(y - 4)^2}{3} = 1$

46. $a = 6,\ c = 2;\ b^2 = 36 - 4 = 32$

$\qquad \dfrac{(x + 5)^2}{36} + \dfrac{(y - 4)^2}{32} = 1$

47. $a = \dfrac{c}{\epsilon} = \dfrac{5}{2};\ b^2 = \dfrac{25}{4} - \dfrac{9}{4} = 4$

$\qquad \dfrac{4\left(x - \frac{1}{2}\right)^2}{25} + \dfrac{(y - 3)^2}{4} = 1$

48. $a = 3;\ c = \epsilon a = \dfrac{5(3)}{3} = 5;\ b^2 = 25 - 9 = 16$

$\qquad \dfrac{y^2}{9} - \dfrac{x^2}{16} = 1$

49.

The pole is a point of intersection.

$\qquad\qquad r_1 = r_2$

$2\cos\theta = 1 + \cos\theta$

$\qquad \cos\theta = 1$

$\qquad\qquad \theta = 0$

The points of intersection are $P_1(0,\ 0)$ and $P_2(2,\ 0)$.

50.

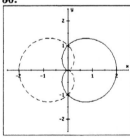

The pole is a point of intersection.

$\qquad\qquad r_1 = r_2$

$1 - \cos\theta = 1 + \cos\theta$

$\qquad \cos\theta = 0$

$\qquad\qquad \theta = \dfrac{\pi}{2},\ \dfrac{3\pi}{2}$

The points of intersection are $P_1(0,\ 0)$, $P_2(1,\ \frac{\pi}{2})$, and $P_3(1,\ \frac{3\pi}{2})$.

51.

The pole is a point of intersection.

$\qquad\qquad r_1 = r_2$

$1 + \sin\theta = 1 + \cos\theta$

$\qquad \tan\theta = 1$

$\qquad\qquad \theta = \dfrac{\pi}{4},\ \dfrac{5\pi}{4}$

The points of intersection are $P_1(0, 0)$, $P_2\left(\frac{2+\sqrt{2}}{2}, \frac{\pi}{4}\right)$, and $P_2\left(\frac{2-\sqrt{2}}{2}, \frac{5\pi}{4}\right)$.

52.

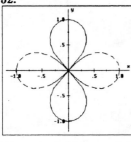

The pole is a point of intersection.
$$r_1^2 = r_2^2$$
$$\cos 2\left(\theta - \frac{\pi}{2}\right) = \cos 2\theta$$
$$-\cos 2\theta = \cos 2\theta$$
$$2\cos 2\theta = 0$$
$$\theta = \frac{\pi}{4}$$

The point of intersection is $P(0, 0)$.

53.

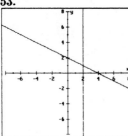

Recognize that these equations have simple rectangular form.
$$\begin{cases} x + 2y = 4 \\ x = 2 \end{cases}$$
Point of intersection is in rectangular form: $(2, 1)$.
In polar form: $(2.24, 0.464)$

54.

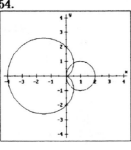

The pole is a point of intersection.
$$r_1 = r_2$$
$$2 - 2\cos\theta = 2\cos\theta$$
$$\cos\theta = \frac{1}{2}$$
$$\theta = \frac{\pi}{3}, \frac{5\pi}{3}$$

The points of intersection are $P_1(0, 0)$, $P_2\left(1, \frac{\pi}{3}\right)$, and $P_3\left(1, \frac{5\pi}{3}\right)$.

55.

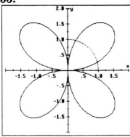

The pole is not a point of intersection.
$$r_1 = r_2$$
$$2\sin 2\theta = 1$$
$$\sin 2\theta = \frac{1}{2}$$
$$\theta = \frac{5\pi}{6}, \frac{\pi}{12},$$

The points of intersection are $P_1\left(1, \frac{\pi}{12}\right)$, $P_2\left(1, \frac{5\pi}{12}\right)$.

56.

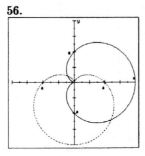

The pole is a point of intersection.
$$r_1 = r_2$$
$$a(1 + \cos\theta) = a(1 - \sin\theta)$$
$$\tan\theta = -1$$
$$\theta = \frac{3\pi}{4}, \frac{7\pi}{4}$$

The points of intersection are $P_1(0, 0)$,

$P_2\left(\frac{a(2-\sqrt{2})}{2}, \frac{3\pi}{4}\right)$, $P_3\left(\frac{a(2+\sqrt{2})}{2}, \frac{7\pi}{4}\right)$

57.

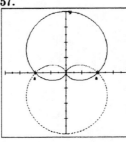

The pole is a point of intersection.
$$r_1 = r_2$$
$$a(1 + \sin\theta) = a(1 - \sin\theta)$$
$$\sin\theta = 0$$
$$\theta = 0, \pi$$
The points of intersection are $P_1(0, 0)$, $P_2(a, 0)$, and $P_3(a, \pi)$.

58.

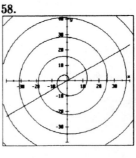

The pole is a point of intersection.
$$\theta = \frac{\pi}{3} + n\pi$$
(Half-measure because the point can be one the ray or on its extension.)
Points of intersection are $P_1(0, 0)$,
$$P_n\left(\frac{2\pi}{3} + 2n\pi, \frac{\pi}{3} + n\pi\right)$$

59. $A = \dfrac{8}{2}\displaystyle\int_0^{\pi/4} (16\cos^2 2\theta)\, d\theta$

$= 32\left(\theta + \frac{1}{4}\sin 4\theta\right)\Big|_0^{\pi/4} = 32\left(\frac{\pi}{4}\right) = 8\pi$

60. $A = \dfrac{2}{2}\displaystyle\int_0^{\pi/6} (4\sin^2\theta)\, d\theta + \dfrac{2}{2}\displaystyle\int_{\pi/6}^{\pi/2} 1\, d\theta$

$= (2\theta - \sin 2\theta)\Big|_0^{\pi/6} + \theta\Big|_{\pi/6}^{\pi/2} = \dfrac{2\pi}{3} - \dfrac{\sqrt{3}}{2}$

61. $A = \dfrac{2}{2}\displaystyle\int_0^{\pi/4} (\cos 2\theta)\, d\theta = \frac{1}{2}\sin 2\theta\Big|_0^{\pi/4} = \dfrac{1}{2}$

62. $A = \dfrac{2}{2}\displaystyle\int_0^{\pi/4} \sin^2 2\theta\, d\theta = \left(\frac{1}{2}\theta - \frac{1}{8}\sin 4\theta\right)\Big|_0^{\pi/4} = \dfrac{\pi}{8}$

63. $A = \dfrac{2a^2}{2}\displaystyle\int_0^{\pi/6} 4\sin^2\theta\, d\theta + \dfrac{2a^2}{2}\displaystyle\int_{\pi/6}^{\pi/2} 1\, d\theta$

$= (2a^2\theta - a^2\sin 2\theta)\Big|_0^{\pi/6} + \theta\Big|_{\pi/6}^{\pi/2}$

$= \dfrac{2a^2\pi}{3} - \dfrac{\sqrt{3}a^2}{2}$

64. $A = \displaystyle\int_0^{\pi/4} (\sin\theta)(4\cos^3\theta\sin\theta)\, d\theta$

$$= 4 \int_0^{\pi/4} \sin^2\theta (1 - \sin^2\theta)\cos\theta \; d\theta$$

$$= 4 \int_0^{\sqrt{2}/2} (u^2 - u^4) \; du = 4\left(\frac{u^3}{3} - \frac{u^5}{5}\right)\Bigg|_0^{\sqrt{2}/2}$$

$$= \frac{7\sqrt{2}}{30}$$

65. $A = \frac{1}{2}\int_0^{\pi} (e^{2\theta})^2 \; d\theta = \frac{1}{8}e^{4\theta}\Big|_0^{\pi} = \frac{1}{8}(e^{4\pi} - 1)$

66. $x = \frac{2\pi}{3}\cos\frac{\pi}{3} = \frac{\pi}{3}$; $y = \frac{2\pi}{3}\sin\frac{\pi}{3} = \frac{\pi\sqrt{3}}{3}$

$$m = \frac{2\theta\cos\theta + 2\sin\theta}{-2\theta\sin\theta + 2\cos\theta}; \text{ at } \theta = \frac{\pi}{3},$$

$$m = \frac{2\pi + 6\sqrt{3}}{-2\sqrt{3} + 6} \approx -3.4151$$

The equation of the tangent line is approximately

$$y - 1.8 = -3.4(x - 1.0)$$

67. $\frac{dx}{dt} = -2\sin wt$; $\frac{dy}{dt} = 2\cos 2t$; $ds = 2\sqrt{1}\; dt$

$$s = 2\int_{-\pi/4}^{\pi/4} dt = \pi$$

68. $\frac{dx}{dt} = \frac{1}{2}(-\sin 2t)$; $\frac{dy}{dt} = \cos t$;

$$\frac{ds}{dt} = \sqrt{\left(\frac{dx}{dt}\right)^2 + \left(\frac{dy}{dt}\right)^2}$$

$$= \sqrt{\frac{\sin^2 2t}{4} + \cos^2 2t}$$

$$= \cos t\sqrt{1 + \sin^2 t}$$

$$s = \int_0^{\pi/2} \sqrt{1 + \sin^2 t}\; (\cos t \; dt)$$

$$= \int_0^1 \sqrt{1 + u^2}\; du \quad \boxed{\text{Formula 168}}$$

$$= \frac{u\sqrt{u^2+1}}{2} + \frac{1}{2}\ln\left|u + \sqrt{u^2+1}\right|\Bigg|_0^1$$

$$= \frac{1}{2}\ln\sqrt{1 + \sqrt{2}} + \frac{\sqrt{2}}{2}$$

69. $\frac{dx}{dt} = -t^{-2}$; $\frac{dy}{dt} = -2t^{-3}$;

$$\frac{ds}{dt} = t^{-3}\sqrt{4 + t^2}$$

$$s = \int_1^2 t^{-3}\sqrt{4 + t^2}\; dt$$

$$\boxed{t = 2\tan\theta; \; dt = 2\sec^2\theta\; d\theta}$$

$$= \int_{t=1}^{t=2} \frac{4\sec^3\theta}{8\tan^3\theta}\; d\theta = \frac{1}{2}\int_{t=1}^{t=2} \csc^3\theta\; d\theta$$

$$= \frac{1}{4}\left[-2t^{-2}\sqrt{4 + t^2} - \ln\left(\frac{\sqrt{8}}{2} + 1\right) - \frac{1}{4}\right]\Bigg|_1^2$$

$$\approx 0.905$$

70. Write the equation of the hyperbola as $y^2 = 9x^2 - 36$, so that $y' = 9x/y$. The point(s) of contact are $m = 9x_0/y_0$. The equation of the tangent line is

$$y - 6 = \frac{9x_0}{y_0}(x - 0) \text{ or } y = \frac{9x_0 x}{y_0} + 6$$

Substituting this into the equation of the hyperbola, we have

$$\left(\frac{9x_0^2}{y_0} + 6\right)^2 = 9x_0^2 - 36$$

Substitute $x_0^2 = \dfrac{36 + y_0^2}{9}$ reveals

$$\left[\frac{(36 + y_0)^2}{y_0} + 6\right]^2 = y_0^2$$

From the square root property,

$$\frac{36 + y_0^2}{y_0} + 6 = \pm y_0$$

For the minus sign, $2y_0^2 + 6y_0 + 36 = 0$ which has no real solutions. For the plus sign, $6y_0 = -36$ or $y_0 = -6$ and $9x_0 = 72$ implies $x_0 = \pm 2\sqrt{2}$. The points of intersection are $(2\sqrt{2}, -6)$ and $(-2\sqrt{2}, -6)$

71. Since $x = a\cot\theta$ and $y = a\sin^2\theta$, we see $\cot\theta = \frac{x}{a}$ and $\csc^2\theta = \frac{a}{y}$.

$$1 + \cot^2\theta = \csc^2\theta$$

$$1 + \frac{x^2}{y^2} = \frac{a}{y}$$

$$a^2 + x^2 = \frac{a^3}{y}$$

$$y = \frac{a^3}{a^2 + x^2}$$

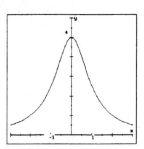

72. **a.** $\frac{dy}{dt} = 3\cos t$; $\frac{dx}{dt} = -4\sin t$; at $t = \frac{\pi}{6}$, $x = 2\sqrt{3}$ and $y = -\frac{3}{2}$ so $\frac{dy}{dt} = \frac{3}{2}\sqrt{3}$, $\frac{dx}{dt} = -2$

$$m = \frac{dy}{dx} = -\frac{\frac{3}{2}\sqrt{3}}{2} = -\frac{3}{4}\sqrt{3}$$

The equation of the tangent line is

$$y - \frac{3}{2} = -\frac{3}{4}\sqrt{3}(x - 2\sqrt{3})$$

b. At $t = \frac{\pi}{4}$, $x = 2\sqrt{2}$, $y = \frac{3}{2}\sqrt{2}$;

$$\frac{dy}{dt} = \frac{3}{2}\sqrt{2}, \frac{dx}{dt} = -2\sqrt{2}.$$

For the tangent and normal lines,

$$m_T = -\frac{\frac{3}{2}\sqrt{2}}{2\sqrt{2}} = -\frac{3}{4} \text{ and } m_N = \frac{4}{3}$$

Thus, the equation of the normal is

$$y - \frac{3}{2}\sqrt{2} = \frac{4}{3}(x - 2\sqrt{2})$$

73. $\frac{dy}{d\theta} = a\sin\theta; \frac{dx}{d\theta} = a(1 - \cos\theta)$

$$m = \frac{dy}{dx} = \frac{\sin\theta}{1 - \cos\theta}$$

a. At $\theta = \frac{\pi}{2}$, $x = \frac{a\pi}{2}$, $y = 0$, and $m = 1$;

the equation of the tangent is

$$y = x - a(\frac{\pi}{2} - 2)$$

b. At $\theta = \pi$, $x = a\pi$, $y = 2a$, and $m = 0$; the equation of the tangent is $y = 2a$.

c. At $\theta = 2\pi$, $x = 2\pi a$, $y = 0$, m_T is undefined. The equation of the tangent is $x = 2\pi a$.

74. a. $\frac{dy}{d\theta} = a\sin\theta; \frac{dx}{d\theta} = a(1 - \cos\theta)$;

$$m = \frac{dy}{dx} = \frac{\sin\theta}{1 - \cos\theta} = \cot\frac{\theta}{2}$$

At $\theta = \theta_0$, $m = \cot\frac{\theta_0}{2}$.

b. The tangent line is horizontal if $\theta = \pi$ or $\theta = 3\pi$ (that is, when $\cot\theta/2 = 0$).

c. The tangent line is vertical if $\theta = 0$ or $\theta = 2\pi$ (that is, $\cot\theta/2$ is not defined).

75. $\frac{dy}{dt} = 1; \frac{dx}{dt} = 4t; m = \frac{1}{4t}$ and the equation of the tangent line is

$$y - t + 1 = \frac{1}{4t}(7 - 2t^2 - 1)$$

At $P(7, 1)$:

$$1 - t + 1 = \frac{1}{4t}(7 - 2t^2 - 1)$$

$$8t - 4t^2 = 6 - 2t^2$$

$$t^2 - 4t + 3 = 0$$

$$(t - 3)(t - 1) = 0$$

$$t = 3, 1$$

At $t = 1$, the point is $(3, 0)$ and at $t = 3$, the point is $(19, 2)$.

76. $x_1 = r_1\cos\theta_1; y_1 = r_1\sin\theta_1; x_2 = r_2\cos\theta_2; y_2 = r_2\sin\theta_2$. The slope is

$$m = \frac{r_2\sin\theta_2 - r_1\sin\theta_1}{r^2\cos\theta_2 - r_1\cos\theta_1}$$

and the equation of the tangent line is

$$y - r_1\sin\theta_1 = m(x - r_1\cos\theta_1)$$

77. $A = \int_{x_1}^{x_2} y\, dx = \int_0^{2\pi} y\frac{dx}{d\theta}\, d\theta$

$$= a^2\int_0^{2\pi}(1 - \cos\theta)^2\, d\theta$$

$$= a^2\int_0^{2\pi}(1 - 2\cos\theta + \cos^2\theta)\, d\theta = 3\pi a^2$$

78. $\frac{dy}{d\theta} = 9\sin\theta; \frac{dx}{d\theta} = 9(1 - \cos\theta)$

$$\frac{ds}{d\theta} = 9\sqrt{\sin^2\theta + 1 - 2\cos\theta + \cos^2\theta}$$

$$= 9\sqrt{2}(\sqrt{1 - \cos\theta}) = 18\sin\frac{\theta}{2}$$

$$s = \int_0^{2\pi} 18\sin\frac{\theta}{2}\, d\theta = 72$$

79. $x^2 + y^2 = \left(\frac{1 - t^2}{1 + t^2}\right)^2 + \left(\frac{2t}{1 + t^2}\right)^2$

$$= \frac{(1 - t^2)^2 + 4t^2}{(1 + t^2)^2}$$

$$= \frac{1 + 2t^2 + t^4}{1 + 2t^2 + t^4} = 1$$

80. $\cos^2 t + \sin^2 t = \frac{(x - 3)^2}{16} + \frac{(y - 4)^2}{9} = 1$

The tangent line is horizontal at the upper vertex $V_1(3, 7)$ and the lower vertex $V_2(3, 7)$.

81. a. $\frac{dy}{dt} = -\sin t + \cos t; \frac{dx}{dt} = \cos t - \sin t$

$$m = \frac{dy}{dx} = \frac{\sin t + \cos t}{\sin t - \cos t} = -\frac{x}{y}$$

b. $-y\, dy = x\, dx$

$$\int -y\, dy = \int x\, dx$$

$$-\frac{y^2}{2} = \frac{x^2}{2} + \frac{C^2}{2}$$

$$x^2 + y^2 = C^2$$

82. $\frac{dy}{dx} = y' = \frac{dy/dt}{dx/dt}$

$$\frac{dy'}{dt} = \frac{\frac{dx}{dt}\frac{d^2y}{dt^2} - \frac{dy}{dt}\frac{d^2x}{dt^2}}{\left(\frac{dx}{dt}\right)^2}$$

$$\frac{d^2y}{dx^2} = \frac{dy'}{dx} = \frac{dy'/dt}{dx/dt}$$

$$= \frac{\frac{dx}{dt}\frac{d^2y}{dt^2} - \frac{dy}{dt}\frac{d^2x}{dt^2}}{\left(\frac{dx}{dt}\right)^3}$$

83. $f'(\theta) = \frac{a}{2}\sec^2\frac{\theta}{2}$; at $\theta = \frac{\pi}{2}$,

$$r = f\left(\frac{\pi}{2}\right) = a;\ f'\left(\frac{\pi}{2}\right) = a$$

$$m = \frac{f'\left(\frac{\pi}{2}\right)}{-f\left(\frac{\pi}{2}\right)} = \frac{a}{-a} = -1$$

84. $\dfrac{dr}{dt} = \dfrac{dr}{d\theta}\dfrac{d\theta}{dt} = -\dfrac{4(2)\sqrt{2}}{2} = -4\sqrt{2}$

85. $\dfrac{dr}{d\theta} = -4(1 + \cos\theta)^{-2}(-\sin\theta)$

$$\frac{dr}{dt} = -3;\ \frac{dr}{dt} = \frac{dr}{d\theta}\frac{d\theta}{dt}$$

$$\frac{d\theta}{dt} = \frac{dr/dt}{dr/d\theta} = \frac{-3}{\dfrac{2\sqrt{2}}{\left(1+\frac{\sqrt{2}}{2}\right)^2}} = \frac{-9 - 6\sqrt{2}}{4\sqrt{2}}$$

86.

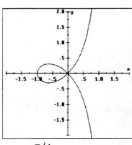

$$A = \int_0^{\pi/4} (\sec\theta - 2\cos\theta)^2\ d\theta$$

$$= \int_0^{\pi/4} (\sec^2\theta - 4 + 2 + 2\cos 2\theta)^2\ d\theta$$

$$= (\tan\theta - 2\theta + \sin 2\theta)\Big|_0^{\pi/4} = 2 - \tfrac{1}{2}\pi$$

87. a. Let $P_1(x_1, y_1)$ be a point on the cardioid
$r = 1 + \sin\theta$. Then $x_1 = (1 + \sin\theta)\cos\theta$
and $y_1 = (1 + \sin\theta)\sin\theta$. The image of
P reflected in the origin is $P_2(x_2, y_2)$ with
$x_2 = [1 + \sin(\theta + \pi)]\cos(\theta + \pi)$ and
$y_1 = [1 + \sin(\theta + \pi)]\sin(\theta + \pi)$.
The square of the distance is

$$d^2 = [(1 + \sin\theta)\cos\theta + (1 - \sin\theta)\cos\theta]^2$$

$$= 4\cos^2\theta + 4\sin^2\theta = 4$$

Thus, the "diameter" of the cardioid is 2,
just like the diameter of the circle.

b. The area of the circle is π square units.
The area of the cardioid is

$$A = \int_{-\pi/2}^{\pi/2} (1 + \sin\theta)^2\ d\theta = \frac{3\pi}{2}$$

88. Write the equation of the ellipse as

$$x^2 = \frac{a^2}{b^2}(b^2 - y^2)$$

$$V = 2\pi \int_0^b \frac{a^2}{b^2}(b^2 - y^2)\ dy$$

$$= \frac{2\pi a^2}{b^2}\left(b^2 y - \frac{y^3}{3}\right)\Big|_0^b = \frac{4\pi a^2 b}{3}$$

With $a = 5$, $V = \dfrac{250\pi}{3}$.

89. Consider a right triangle with hypotenuse L
in the first quadrant and legs on the
coordinate axes. θ is the acute angle between
the horizontal and L. The vertical leg is
$y = L\cos\theta$ and horizontal leg is $x = L\cos\theta$.
For $P(x_p, y_p)$, $y_p = \frac{3}{4}L\sin\theta$ and $x_p\ \frac{1}{4}L\cos\theta$.
Since $\cos^2\theta + \sin^2\theta = 1$,

$$\frac{(4x_p)^2}{L^2} + \frac{(4y_p)^2}{(3L)^2} = 1$$

$$\frac{x^2}{1} + \frac{y^2}{9} = \frac{L^2}{16}$$

90.

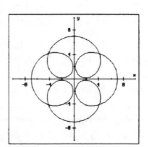

91. $\dfrac{dx}{d\theta} = (a - b)(-\sin\theta) - \dfrac{b(a - b)}{b}\sin\dfrac{(a-b)\theta}{b}$

$\dfrac{dy}{d\theta} = (a - b)(\cos\theta) - \dfrac{b(a - b)}{b}\cos\dfrac{(a - b)\theta}{b}$

$\left(\dfrac{dy}{dt}\right)^2 + \left(\dfrac{dx}{dt}\right)^2 = (a - b)^2 + (a - b)^2$

$\qquad + 2(a - b)^2\sin\theta\sin\left[\dfrac{(a - b)\theta}{b}\right]$

$$- 2(a - b)^2 \cos \theta \cos\left[\frac{(a - b)\theta}{b}\right]$$

$$= 2(a - b)^2\left[1 - \cos\left(1 + \frac{a - b}{b}\theta\right)\right]$$

$$\frac{ds}{d\theta} = \sqrt{2}(a - b)\sqrt{1 - \frac{\cos a\theta}{b}}$$

$$= 2(a - b)\sin\left(\frac{a\theta}{2b}\right)$$

$$s = \int_0^{\pi b/a} 2(a - b)\sin\left(\frac{a\theta}{2b}\right)$$

$$= 2(a - b)\left(\frac{2b}{a}\right)(-1)\frac{a\theta}{2b}\Big|_0^{\pi b/a}$$

$$= \frac{4b(a - b)}{a}$$

92. $\frac{ds}{dt} = 0.187$ mi/μs; $s = 0.187t$ mi;

let $A(0, 0)$, $B(100, 0)$ be the LORAN stations, and $P(x, y)$ the position of the airplane. Also, d_1 and d_2 are the distances of the airplane from A and B, respectively, so that $d_1 - d_2 = (0.187)(4) = 74.8$ mi. The center is $C(50, 0)$ so $c = 50$, $a = 74.8/2 = 37.2$ mi, and $b^2 = c^2 - a^2 = 1{,}116.16$.

$$\frac{(x - 50)^2}{1{,}384} + \frac{y^2}{1{,}116} = 1$$

$$\frac{(x - 50)^2}{1{,}284} - \frac{4{,}900}{1{,}116} = 1$$

$$(x - 50)^2 = 7{,}460.7$$

$$x \approx 136.37, \ -36.37$$

Reject the positive value (not in the domain), so the airplane must travel about 36 miles before reaching the point closest to A. The airplane's position is at $(-36, 70)$.

93. Let $A(0, 0)$ and $B(10, 0)$ be the locations of the two stores and $P(x, y)$ be the position in the city which gives equal cost. Let d_1 and d_2 be the distance of the shopper from A and B, respectively. The cost of traveling is $30d_1$ and $30d_2$ cents, respectively. The center is at $C(5, 0)$, so $c = 5$. The cost at A is $300 + 30d_1$, while the cost at B is $275 + 30d_2$. Equating cost leads to

$$300 + 30d_1 = 275 + 30d_2$$

$$30d_2 - 30d_1 = 25$$

$$d_2 - d_1 = \tfrac{5}{6}$$

Thus, $a = \frac{5}{6}$; $b^2 = 25 - \left(\frac{5}{6}\right)^2 \approx 24.83$

Thus, the equation of the hyperbola is

$$\frac{(x - 5)^2}{0.1736} - \frac{y^2}{24.83} = 1$$

94. This is Putnam Problem 5 of the morning session of 1974. Let F be the fixed focus, M the moving focus, and T the (varying) point of mutual tangency. The reflecting property of parabolas tells us that the tangent line at T makes equal angles with $\overline{FT}$ and with a vertical line. This and congruence of the two parabolas imply that $\overline{MT}$ is vertical and that the segments $\overline{FT}$ and $\overline{MT}$ are equal. Now M must be on the horizontal fixed directrix, $y = 1/4$ by the focus-directrix definition of a parabola.

95. This is Putnam Problem 4 of the afternoon session of 1976. We let $P = (x, y)$ and the ellipse have the equation $b^2x^2 + a^2y^2 = a^2b^2$, with $a > b > 0$. Then $F_1 = (-c, 0)$ and $F_2(c, 0)$ with $c^2 = a^2 - b^2$. Let $r_1 = \overline{PF_1}$ and $r_2 = \overline{PF_2}$. Then $r_1 + r_2 = 2a$ and

$$r_1 r_2 = \tfrac{1}{2}[(r_1 + r_2)^2 - r_1^2 - r_2^2]$$

$$= \tfrac{1}{2}[4a - (x + c)^2 - y^2 - (x - c)^2 - y^2]$$

$$= 2a^2 - x^2 - y^2 - c^2$$

$$= a^2 + b^2 - x^2 - y2$$

A point (u, v) on the tangent to the ellipse at P satisfies

$$\frac{xu}{a^2} + \frac{yv}{b^2} = 1$$

Putting this in the form $u \cos \theta + v \sin \theta = d$ we find that

$$d^2 = \frac{1}{\left(\frac{x}{a^2}\right)^2 + \left(\frac{y}{b^2}\right)^2} = \frac{a^4 b^4}{b^4 x^2 + a^4 y^2}$$

But

$$b^2 x^4 + a^4 y^2 = b^2(a^2 b^2 - a^2 y^2) + a^2(a^2 b^2 - b^2 x^2)$$

$$= a^2 b^2(a^2 + b^2 - x^2 - y^2)$$

$$= a^2 b^2 r_1 r_2$$

Therefore,

$$d^2 r_1 r_2 = \frac{a^4 b^4 r_1 r_2}{a^2 b^2 r_1 r_2} = a^2 b^2$$

which is a constant.

96. This is Putnam Problem 2 of the afternoon session of 1984. The problem asks for the minimum distance between the quarter circles $x^2 + y^2 = 2$ in the open first quarter and the half of the hyperbola $xy = 9$ in that quadrant. Since the tangents to the respective curves at $(1, 1)$ and $(3, 3)$ separate the curves and are both perpendicular to $x = y$, the minimum distance is 8.

CHAPTER 10

Vectors in the Plane and in Space

10.1 Introduction to Vectors, Page 637

1.

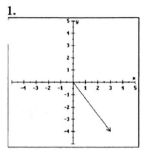

2.

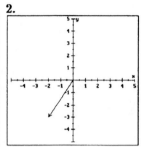

3.

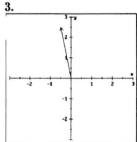

4.

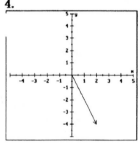

5. $4\mathbf{i} + 3\mathbf{j}$

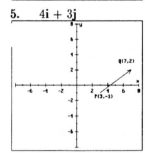

6. $10\mathbf{j}$

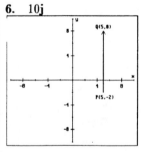

7. $-5\mathbf{i}$

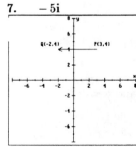

8. $(-7/2)\mathbf{i} - 8\mathbf{j}$

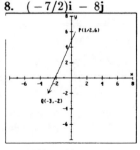

9. $\mathbf{PQ} = (1 + 1)\mathbf{i} + (-2 + 2)\mathbf{j} = 2\mathbf{i}$
$\|\mathbf{PQ}\| = \sqrt{(1 + 1)^2 + (-2 + 2)^2} = 2$

10. $\mathbf{PQ} = (6 - 5)\mathbf{i} + (8 - 7)\mathbf{j} = \mathbf{i} + \mathbf{j}$
$\|\mathbf{PQ}\| = \sqrt{(6 - 5)^2 + (8 - 7)^2} = \sqrt{2}$

11. $\mathbf{PQ} = (0 + 4)\mathbf{i} + (-1 + 3)\mathbf{j} = 4\mathbf{i} + 2\mathbf{j}$
$\|\mathbf{PQ}\| = \sqrt{(0 + 4)^2 + (-1 + 3)^2} = 2\sqrt{5}$

12. $\mathbf{PQ} = (2 - 3)\mathbf{i} + (8 + 5)\mathbf{j} = -\mathbf{i} + 13\mathbf{j}$
$\|\mathbf{PQ}\| = \sqrt{(2 - 3)^2 + (8 + 5)^2} = \sqrt{170}$

13. Let $\mathbf{v} = \mathbf{i} + \mathbf{j}$; $\|\mathbf{v}\| = \sqrt{1^2 + 1^2} = \sqrt{2}$
$$\mathbf{u} = \frac{\mathbf{v}}{\|\mathbf{v}\|} = \frac{1}{\sqrt{2}}(\mathbf{i} + \mathbf{j})$$

14. Let $\mathbf{v} = \frac{1}{2}\mathbf{i} + \frac{1}{4}\mathbf{j}$; $\|\mathbf{v}\| = \sqrt{(\frac{1}{2})^2 + (\frac{1}{4})^2} = \frac{1}{4}\sqrt{5}$
$$\mathbf{u} = \frac{\mathbf{v}}{\|\mathbf{v}\|} = \frac{4}{\sqrt{5}}(\frac{1}{2}\mathbf{i} + \frac{1}{4}\mathbf{j}) = \frac{2}{\sqrt{5}}\mathbf{i} + \frac{1}{\sqrt{5}}\mathbf{j}$$

15. Let $\mathbf{v} = 3\mathbf{i} - 4\mathbf{j}$; $\|\mathbf{v}\| = \sqrt{3^2 + (-4)^2} = 5$
$$\mathbf{u} = \frac{\mathbf{v}}{\|\mathbf{v}\|} = \frac{1}{5}(3\mathbf{i} - 4\mathbf{j}) = \frac{3}{5}\mathbf{i} - \frac{4}{5}\mathbf{j}$$

16. Let $\mathbf{v} = -4\mathbf{i} + 7\mathbf{j}$; $\|\mathbf{v}\| = \sqrt{(-4)^2 + 7^2} = \sqrt{65}$
$$\mathbf{u} = \frac{\mathbf{v}}{\|\mathbf{v}\|} = \frac{1}{\sqrt{65}}(-4\mathbf{i} + 7\mathbf{j}) = -\frac{4}{\sqrt{65}}\mathbf{i} + \frac{7}{\sqrt{65}}\mathbf{j}$$

17. $s\mathbf{u} + t\mathbf{v} = \langle -3s + t, 4s - t \rangle$ so that
$-3s + t = 6$ and $4s - t = 0$; solving
simultaneously, we have $s = 6$ and $t = 24$.

18. $-3s + t = 0$ and $4s - t = -3$; solving
simultaneously, we have $s = -3$ and $t = -9$.

19. $-3s + t = -2$ and $4s - t = 1$; solving
simultaneously, we have $s = -1$ and $t = -5$.

20. $-3s + t = 8$ and $4s - t = 11$; solving
simultaneously, we have $s = 19$ and $t = 65$.

21. $2\mathbf{u} + 3\mathbf{v} - \mathbf{w} = (6 + 12 - 1)\mathbf{i} + (-8 - 9 - 1)\mathbf{j}$
$$= 17\mathbf{i} - 18\mathbf{j}$$

22. $\frac{1}{2}(\mathbf{u} + \mathbf{v}) - \frac{1}{4}\mathbf{w}$
$= \frac{1}{2}(3\mathbf{i} - 4\mathbf{j}) + \frac{1}{2}(4\mathbf{i} - 3\mathbf{j}) - \frac{1}{4}(\mathbf{i} + \mathbf{j})$
$= \frac{13}{4}\mathbf{i} - \frac{15}{4}\mathbf{j}$

23. $\|\mathbf{v}\| = \sqrt{4^2 + (-3)^2} = 5$; $\|\mathbf{u}\| = \sqrt{3^2 + (-4)^2} = 5$
$\|\mathbf{v}\|\mathbf{u} + \|\mathbf{u}\|\mathbf{v} = 5(3\mathbf{i} - 4\mathbf{j}) + 5(4\mathbf{i} - 3\mathbf{j})$
$= 35\mathbf{i} - 35\mathbf{j}$

24. $\|\mathbf{u}\|\|\mathbf{v}\|\mathbf{w} = 5(5)(\mathbf{i} + \mathbf{j}) = 25\mathbf{i} + 25\mathbf{j}$

25. $x - y - 1 = 0$ and $2x + 3y - 12 = 0$;
    solving simultaneously, we have $x = 3$, $y = 2$?

26. $x = 5 - 3y$ and $-4y^2 = 10 - 7x$; solving
    simultaneously, we have $(2, 1)$ and $(\frac{95}{4}, -\frac{25}{4})$

27. $x^2 + y^2 = 16$ and $y = x + 2$; solving
    simultaneously, we have $(-1 + \sqrt{7}, 1 + \sqrt{7})$
    and $(-1 - \sqrt{7}, 1 - \sqrt{7})$

28. $y - 1 = \log x$ and $y = \log 2 + \log(x + 4)$;
    By substitution, $\log x + 1 = \log 2 + \log(x+4)$
    Solving for x, we find $x = 8/(e - 2)$ so that
    $y = \log[8e/(e - 2)]$.

29. $\mathbf{i} = \cos 30°$, $\mathbf{j} = \sin 30°$; $\frac{\sqrt{3}}{2}\mathbf{i} + \frac{1}{2}\mathbf{j}$

30. Let $\mathbf{v} = 2\mathbf{i} - 3\mathbf{j}$ so $\|\mathbf{v}\| = \sqrt{13}$; then
    $\mathbf{u} = \frac{1}{\sqrt{13}}(2\mathbf{i} - 3\mathbf{j}) = \frac{2}{\sqrt{13}}\mathbf{i} - \frac{3}{\sqrt{13}}\mathbf{j}$

31. Let $\mathbf{v} = -4\mathbf{i} + \mathbf{j}$ so $\|\mathbf{v}\| = \sqrt{17}$; then
    $\mathbf{u} = -\frac{1}{\sqrt{17}}(-4\mathbf{i} + \mathbf{j}) = \frac{4}{\sqrt{17}}\mathbf{i} - \frac{1}{\sqrt{17}}\mathbf{j}$

32. Let $\mathbf{v} = (7 + 1)\mathbf{i} + (-3 - 5)\mathbf{j} = 8\mathbf{i} - 8\mathbf{j}$
    $\mathbf{u} = \frac{1}{8\sqrt{2}}(8\mathbf{i} - 8\mathbf{j}) = \frac{1}{\sqrt{2}}\mathbf{i} - \frac{1}{\sqrt{2}}\mathbf{j}$

33. $\mathbf{u} + \mathbf{v} = 5\mathbf{i} + \mathbf{j}$; $\|\mathbf{u} + \mathbf{v}\| = \sqrt{26}$;
    The desired unit vector is $\frac{5}{\sqrt{26}}\mathbf{i} + \frac{1}{\sqrt{26}}\mathbf{j}$

34. $\mathbf{u} - 2\mathbf{v} + 2\mathbf{w} = -4\mathbf{i} + 3\mathbf{j}$; this vector has
    The desired vector is $-\frac{12}{5}\mathbf{i} + \frac{9}{5}\mathbf{j}$.

35. $(3, 10)$ 36. $(0, -4)$

37. **a.** The midpoint of $\overline{PQ} = \frac{1}{2}\langle 12, 6\rangle = \langle 6, 3\rangle$.
    If the tail of this vector is at P, its head
    will be at: $(3, -5)$.
    b. $\frac{5}{6}\langle 12, 6\rangle = \langle 10, 5\rangle$. This vector, with tail
    at P, will end at $(7, -3)$.

38. $-9 \le \eta\|\mathbf{v}\| \le 3$

39. $\|\mathbf{v}\| = \sqrt{\cos^2\theta + \sin^2\theta} = 1$

40. $\|r\mathbf{v}\| = \|\mathbf{u}\|$

41. Not necessarily equal. Equal magnitudes say
    nothing about their direction.

42. $\|\mathbf{v} - \mathbf{u}\| = \|(x - 2)\mathbf{i} + (y + 3)\mathbf{j}\|$
    $= \sqrt{(x - 2)^2 + (y + 3)^2} \le 2$

This is the set of points on or interior to the
circle with center $(2, -3)$ and radius 2.

43. **a.** $\|\mathbf{u} - \mathbf{u}_0\| = \sqrt{(x - x_0)^2 + (y - y_0)^2} = 1$
    This is the set of points on the circle with
    center (x_0, y_0) and radius 1.
    b. $\|\mathbf{u} - \mathbf{u}_0\| \le 2$ is the set of points on or
    interior to the circle with center (x_0, y_0)
    and radius 2.

44. $a\mathbf{u} + b\mathbf{v} = (3a - 6b)\mathbf{i} + (-a + 2b)\mathbf{j}$
    Thus, $3a - 6b = 2$ and $-a + 2b = 5$;
    solving simultaneously gives no solution.

45. Since we do not know the components of $\mathbf{u}$
    and $\mathbf{v}$, we can guarantee the sum is 0 if each
    of the vectors is 0.
    $$a\mathbf{u} + b\mathbf{u} - b\mathbf{v} + c\mathbf{u} + c\mathbf{v} = \mathbf{0}$$
    $$(a + b + c)\mathbf{u} = \mathbf{0} \text{ and } (-b + c)\mathbf{v} = \mathbf{0}$$
    So $b = c$, and $a + 2b = 0$, $a = -2b$.
    Thus, $a = -2t$, $b = t$, $c = t$.

46. Let $\mathbf{w} = c\mathbf{u} + (1 - c)\mathbf{v}$;
    a. $c = 0$; $\mathbf{w} = \mathbf{v}$
    b. $c = \frac{1}{4}$;
    $\mathbf{w} = \frac{1}{4}(\mathbf{u} + 3\mathbf{v})$
    c. $c = \frac{1}{2}$;
    $\mathbf{w} = \frac{1}{2}(\mathbf{u} + \mathbf{v})$
    d. $c = \frac{3}{4}$;

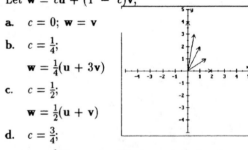

    $\mathbf{w} = \frac{1}{4}(3\mathbf{u} + \mathbf{v})$ All the terminal points
    on the line $2x + y = 4$.
    e. $c = 1$; $\mathbf{w} = \mathbf{u}$

47. Let $\mathbf{F}_3 = a\mathbf{i} + b\mathbf{j}$. Thus,
    $3\mathbf{i} + 4\mathbf{j} + 3\mathbf{i} - 7\mathbf{j} + a\mathbf{i} + b\mathbf{j} = 0\mathbf{i} + 0\mathbf{j}$
    so that $a = -6$ and $b = 3$; $\mathbf{F}_3 = -6\mathbf{i} + 3\mathbf{j}$

48. Let $\mathbf{F}_4 = a\mathbf{i} + b\mathbf{j}$. Thus,
    $(\mathbf{i} - 2\mathbf{j}) + (3\mathbf{i} - 7\mathbf{j}) + (\mathbf{i} + \mathbf{j}) + (a\mathbf{i} + b\mathbf{j}) = 0\mathbf{i} + 0\mathbf{j}$
    so that $a = -5$ and $b = 8$; $\mathbf{F}_4 = -5\mathbf{i} + 8\mathbf{j}$

49. $x = 60 \cos 40° \approx 45.95$ ft/s
    $y = 60 \sin 40° = 38.57$ ft/s

50. Let $\mathbf{m}$ be the direction of the boat with
    respect to the ground, and let $\mathbf{v}$ be the
    component perpendicular to the flow of the
    river; also let $\mathbf{R}$ be the component along the
    river. $\|\mathbf{R}\| = 3.1$; $t = 1/2$ hr, and the distance
    across is $d = 2.1$ mi; $\|\mathbf{v}\| = 2.1/1/2 = 4.2$
    mi/h. Since $\mathbf{m} = \mathbf{v} + \mathbf{R}$,

$$\|\mathbf{m}\| = \sqrt{4.2^2 + 3.1^2} \approx 5.22 \text{ mi/h}$$

Let θ be the angle between the direction of the boat and $\mathbf{v}$. Then

$$\theta = \tan^{-1}\frac{3.1}{4.2} \approx 36.4°$$

51. $\mathbf{F}_1 = \langle 10\cos\frac{\pi}{6}, 10\sin\frac{\pi}{6}\rangle = \langle 5\sqrt{3}, 5\rangle$

$\mathbf{F}_2 = \langle 0, 8\rangle$;

$\mathbf{F}_3 = \langle 5\cos\frac{4\pi}{3}, 10\sin\frac{4\pi}{3}\rangle = \langle -\frac{5}{2}, -5\sqrt{3}\rangle$

$\mathbf{F}_4 = -(\mathbf{F}_1 + \mathbf{F}_2 + \mathbf{F}_3)$

$= -[(5\sqrt{3} + 0 - \frac{5}{2})\mathbf{i} + (5 + 8 - 5\sqrt{3})\mathbf{j}]$

$= (\frac{5}{2} - 5\sqrt{3})\mathbf{i} - (13 - 5\sqrt{3})\mathbf{j}$

52. Let P, Q, R, and S, be consecutive vertices in counterclockwise order of a parallelogram. The point T is where the diagonals intersect. Then,

$$\mathbf{Pt} = a\mathbf{PR}; \quad \mathbf{QT} = b\mathbf{QS}; \quad \mathbf{PT} + \mathbf{TQ} = \mathbf{PQ}$$

so

$$a\mathbf{PR} - b\mathbf{QT} = \mathbf{PQ}$$

$$\mathbf{RT} = (1 - a)\mathbf{RP}; \quad \mathbf{TS} = (1 - b)\mathbf{QS}$$

$$\mathbf{RS} = \mathbf{RT} + \mathbf{TS}$$

$$= (1 - a)\mathbf{RP} + (1 - b)\mathbf{QS}$$

It follows that $a = (1 - a)$ or $a = \frac{1}{2}$ and $b = (1 - b)$ or $b = \frac{1}{2}$

53. Draw a triangle and label the vertices P, Q, and R. Let $\mathbf{u}$ be the vector from P to the midpoint of $\overline{QR}$, $\mathbf{v}$ from Q to the midpoint of $\overline{PR}$, and $\mathbf{w}$ from R to the midpoint of $\overline{PQ}$. Let $P = (x_1, y_1)$, $Q = (x_2, y_2)$ and $R = (x_3, y_3)$. Write each of the three vectors in standard form and add them. Note that $(x_1 + x_2 + x_3)\mathbf{i} = 0$, since you return to the starting point.

54. a. Given $a\mathbf{u} = b\mathbf{v}$; assume $ab \neq 0$, so that $ab = k$ or $b = k/a$ and $a\mathbf{u} = (a/k)\mathbf{v}$ or $k\mathbf{u} = \mathbf{v}$,m which means that $\mathbf{u}$ and $\mathbf{v}$ are orthogonal, contrary to the assumption. We must conclude that $a = b = 0$.

b. $a_1\mathbf{i} + b_1\mathbf{j} = a_2\mathbf{i} + b_2\mathbf{j}$

$(a_1 - a_2)\mathbf{i} = (b_1 - b_2)\mathbf{j}$

were $\mathbf{i}$ and $\mathbf{j}$ are linearly independent; thus, $a_1 = a_2$ and $b_1 = b_2$.

55. a. $\mathbf{CN} = \mathbf{CA} + \mathbf{AN}$

$= -\mathbf{AC} + \frac{1}{2}\mathbf{AB}$

$= \frac{1}{2}\mathbf{AB} - \mathbf{AC}$

$\mathbf{BM} = \mathbf{BA} + \mathbf{AM}$

$= -\mathbf{AB} + \frac{1}{2}\mathbf{AC}$

$= \frac{1}{2}\mathbf{AC} - \mathbf{AB}$

b. $\mathbf{CP} = r\mathbf{CN} = r\left[\frac{1}{2}\mathbf{AB} - \mathbf{AC}\right]$

$\mathbf{BP} = s\mathbf{BM} = s\left[\frac{1}{2}\mathbf{AC} - \mathbf{AB}\right]$

$\mathbf{CB} = \mathbf{CP} + \mathbf{PB} = \mathbf{CP} - \mathbf{BP}$

$= \frac{r}{2}\mathbf{AB} - r\mathbf{AC} - \frac{s}{2}\mathbf{AC} + s\mathbf{AB}$

$= \left(\frac{r}{2} + s\right)\mathbf{AB} + \left(-r - \frac{s}{2}\right)\mathbf{AC}$

Since $\mathbf{CB} = \mathbf{CA} + \mathbf{AB}$ we see

$$\left(\frac{r}{2} - 1 + s\right)\mathbf{AB} = \left(1 - \frac{s}{2} - r\right)\mathbf{CA}$$

where $\mathbf{AB}$ and $\mathbf{CA}$ are linearly independent. From Problem 54,

$\frac{r}{2} - 1 + s = 0$ and $1 - \frac{s}{2} - r = 0$;

solving simultaneously, $r = \frac{2}{3}$, $s = \frac{2}{3}$.

The same procedure applies to the other two medians.

c. $A = (x_1, y_1)$, $B = (x_2, y_2)$, $C = (x_3, y_3)$. let $N(x_N, y_N)$ be the midpoint of $\overline{AB}$; then

$$x_N = \frac{x_1 + x_2}{2} \text{ and } y_N = \frac{y_1 + y_2}{2}$$

and $x_P = x_3 + \frac{2}{3}\left[\frac{x_1 + x_2}{2} - x_3\right]$

$= \frac{x^3}{3} + \frac{x^2 + x_2}{3} = \frac{x_1 + x_2 + x_3}{3}$

Similarly, $y_P = \frac{y_1 + y_2 + y_3}{3}$

56. Plot the points P and Q on an xy-coordinate plane. Let $R(a, d)$ be the intersection of a horizontal through P and a vertical through Q. Then $\mathbf{PQ} = (a - c)\mathbf{i} + (b - d)\mathbf{j}$ $= \langle a - c, b - d\rangle$.

57. a. $\mathbf{u} + \mathbf{v} = \langle u_1, u_2\rangle + \langle v_1, v_2\rangle$

$= \langle u_1 + v_1, u_2 + v_2\rangle$

$= \langle v_1 + u_1, v_2 + u_2\rangle$

$= \mathbf{v} + \mathbf{u}$

b. $\mathbf{u} + \mathbf{0} = \langle u_1, u_2\rangle + \langle 0, 0\rangle$

$= \langle u_1 + 0, u_2 + 0\rangle$

$= \langle u_1, u_2\rangle$

$= \mathbf{u}$

c. $(s + t)\mathbf{u} = (s + t)\langle u_1, u_2\rangle$

$$= \langle (s+t)u_1, (s+t)u_2 \rangle$$
$$= \langle su_1 + tu_1, su_2 + tu_2 \rangle$$
$$= \langle su_1, su_2 \rangle + \langle tu_1. tu_2 \rangle$$
$$= s\langle u_1, u_2 \rangle + t\langle u_1, u_2 \rangle$$
$$= s\mathbf{u} + t\mathbf{u}$$

d. $\quad (st)\mathbf{u} = (st)\langle u_1, u_2 \rangle$
$$= \langle stu_1, stu_2 \rangle$$
$$= s\langle tu_1, tu_2 \rangle$$
$$= s(t\mathbf{u})$$

e. $\mathbf{u} + (-\mathbf{u}) = \langle u_1, u_2 \rangle + \langle -u_1, u_2 \rangle$
$$= \langle u_1 - u_1, u_2 - u_2 \rangle$$
$$= \langle 0, 0 \rangle$$
$$= \mathbf{0}$$

f. $\quad s(\mathbf{u} + \mathbf{v}) = s[\langle u_1, u_2 \rangle + \langle v_1, v_2 \rangle]$
$$= s\langle u_1 + v_1, u_2 + v_2 \rangle$$
$$= \langle s(u_1 + v_1), s(u_2 + v_2) \rangle$$
$$= \langle su_1 + sv_1, su_2 + su_2 \rangle$$
$$= \langle su_1, su_2 \rangle + \langle sv_1, sv_2 \rangle$$
$$= s\langle u_1, u_2 \rangle + s\langle v_1, v_2 \rangle$$
$$= s\mathbf{u} + s\mathbf{v}$$

10.2 Quadric Surfaces and Graphing in Three Dimensions, Page 647

1. $\quad d = \sqrt{(1-3)^2 + (5+4)^2 + (-3-5)^2}$
$$= \sqrt{149}$$

2. $\quad d = \sqrt{(-2-0)^2 + (5-3)^2 + (-7-0)^2}$
$$= \sqrt{57}$$

3. $\quad d = \sqrt{6^2 + 11^2 + (-15)^2} = \sqrt{382}$

4. $\quad d = \sqrt{2^2 + (-6)^2 + 3^2} = 7$

5. $(x - 0)^2 + (y - 0)^2 + (z - 0)^2 = 1$
$x^2 + y^2 + z^2 = 1$

6. $(x + 3)^2 + (y - 5)^2 + (z - 7)^2 = 4$

7. $(x - 0)^2 + (y - 4)^2 + (z + 5)^2 = 9$
$x^2 + (y - 4)^2 + (z + 5)^2 = 9$

8. $(x + 2)^2 + (y - 3)^2 + (z + 1)^2 = 5$

9. $\quad x^2 + y^2 + z^2 - 2y + 2z - 2 = 0$
$x^2 + (y^2 - 2y + 1^2) + (z^2 + 2z + 1^2) = 2 + 1 + 1$

$x^2 + (y - 1)^2 + (z + 1)^2 = 4$
$C(0, 1, -1); r = 2$

10. $\quad x^2 + y^2 + z^2 + 4x - 2z - 8 = 0$
$(x^2 + 4x + 2^2) + y^2 + (z^2 - 2z + 1^2) = 8 + 4 + 1$
$(x + 2)^2 + y^2 + (z - 1)^2 = 13$
$C(-2, 0, 1); r = \sqrt{13}$

11. $\quad x^2 + y^2 + z^2 - 6x + 2y - 2z + 10 = 0$
$(x^2 - 6x + 3^2) + (y^2 + 2y + 1^2) + (z^2 - 2z + 1^2)$
$\quad = -10 + 9 + 1 + 1$
$(x - 3)^2 + (y + 1)^2 + (z - 1)^2 = 1$
$C(3, -1, 1); r = 1$

12. $\quad x^2 + y^2 + z^2 - 2x - 4y + 8z + 17 = 0$
$(x^2 - 2x + 1^2) + (y^2 - 4y + 2^2) + (z^2 + 8z + 4^2)$
$\quad = -17 + 1 + 4 + 16$
$(x - 1)^2 + (y - 2)^2 + (z + 4)^2 = 4$
$C(1, 2, -4); r = 2$

13. circular cone; B **14.** elliptic cone; J

15. hyperboloid of one sheet; E

16. hyperboloid of one sheet; F

17. sphere; A **18.** paraboloid; H

19. paraboloid, G **20.** hyperbolic hyperboloid; D

21. hyperboloid of two sheets; I

22. hyperbolic hyperboloid; L

23. $\|\mathbf{AB}\|^2 = 16 + 4 + 16 = 36$
$\|\mathbf{AC}\|^2 = 4 + 16 + 16 = 36$
$\|\mathbf{BC}\|^2 = 36 + 4 + 0 = 40$
$\triangle ABC$ is isosceles, but not right

24. $\|\mathbf{AB}\|^2 = 4 + 4 + 1 = 9$
$\|\mathbf{AC}\|^2 = 4 + 16 + 16 = 36$
$\|\mathbf{BC}\|^2 = 0 + 36 + 9 = 45$
$\triangle ABC$ is right, but not isosceles

25. $\|\mathbf{AB}\|^2 = 16 + 0 + 1 = 17$
$\|\mathbf{AC}\|^2 = 0 + 36 + 0 = 36$
$\|\mathbf{BC}\|^2 = 16 + 36 + 1 = 53$
$\triangle ABC$ is neither right nor isosceles

26. $\|\mathbf{AB}\|^2 = 25 + 4 + 49 = 78$

$\| \mathbf{AC} \|^2 = 64 + 16 + 169 = 249$

$\| \mathbf{BC} \|^2 = 9 + 36 + 36 = 81$

$\triangle ABC$ is neither right nor isosceles

27. First draw the coordinate axes. Sketch the traces (intersections) of the surface with the coordinate planes (or planes parallel to these). Use the given equation with $x = 0$, then $y = 0$, and finally $z = 0$. Infer the shape in three dimensions − see drawing lesson 3 on Page 644.

28. In identifying quadrics, let each variable, one at a time, equal zero and identify the resulting second degree conic in the coordinate plane. Two or more conics of the same type give the quadric the "oid" name and the remaining conic describes the type. For instance, if two coordinate planes have a parabola, and the third an ellipse, the surface is an elliptic paraboloid.

29. Plane; intercepts are (3, 0, 0), (0, 6, 0), and (0, 0, 2)

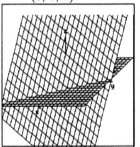

30. Plane; intercept (4, 0, 0); it is perpendicular to the x-axis

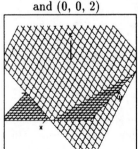

31. Plane; intercepts: (10, 0, 0), (0, 5, 0) and (0, 0, 2)

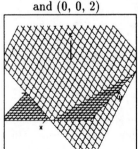

32. Plane; intercepts are (1, 0, 0), (0, 1, 0) and (0, 0, 1)

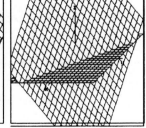

33. Plane; intercepts: (4, 0, 0), (0, −6, 0) and (0, 0, −12)

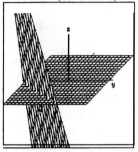

34. Parabolic cylinder

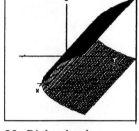

35. Plane; intercept is (−1, 0, 0); it is perpendicular to the x-axis

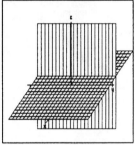

36. Right circular cylinder

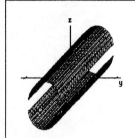

37. Cylinder cross-sections are exponential curves

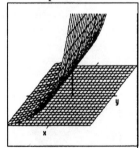

38. Cylinder cross-sections are logarithmic curves

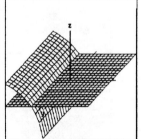

39. Plane

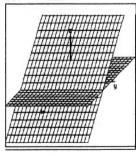

40. Hyperbolic cylinder

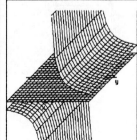

41. Ellipsoid

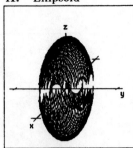

42. Ellipsoid

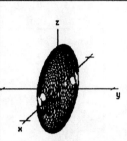

43. Hyperboloid of one sheet

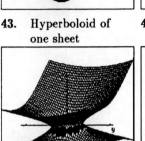

44. Hyperboloid of of two sheets

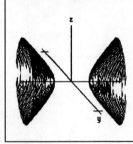

45. Elliptic paraboloid

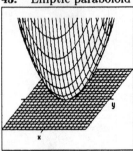

46. Hyperbolic paraboloid

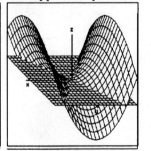

47. Elliptic cone

48. Hyperboloid of two sheets

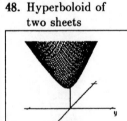

49. $M(-\frac{1}{2}, \frac{5}{2}, 0)$; the diameter is

$$d = \sqrt{(-3)^2 + 1^2 + 6^2} = \sqrt{46}$$

$r^2 = \frac{46}{4} = \frac{23}{2}$; the equation of the sphere is

$$(x + \frac{1}{2})^2 + (y - \frac{5}{2})^2 + z^2 = \frac{23}{2} \text{ or}$$

$$(2x + 1)^2 + (2y - 5)^2 + 4z^2 = 46$$

50. $\|\mathbf{PQ}\|^2 = 25 + 1 + (c + 1)$

$\|\mathbf{PR}\|^2 = (c - 3)^2 + 1 + 1$

$\|\mathbf{QR}\|^2 = (c + 2)^2 + c^2$

There are three possibilities:

a. Let $\overline{PQ}$ be the hypotenuse; then, by the Pythagorean theorem

$26 + c^2 + 2c + 1 = c^2 - 6c + 9 + 2 + c^2 + 2c + 4 + c^2$

$2c^2 - 6c - 12 = 0$

$$c = \frac{3 \pm \sqrt{9 + 24}}{2} = \frac{3 \pm \sqrt{33}}{2}$$

b. Let $\overline{PR}$ be the hypotenuse;

$c^2 - 6c + 9 + 2 = 26 + c^2 + 2c + 1 + c^2 + 4c + 4 + c^2$

$c^2 + 6c + 10 = 0$

This has no real solution since the discriminant is negative.

c. Let $\overline{QR}$ be the hypotenuse;

$c^2 + 4c + 4 + c^2 = 26 + c^2 + 2c + 1 - 6c + 9 + 2$

$c^2 - 8c + 34 = 0$

This has no real solution since the discriminant is negative.

51. The midpoint M of $\overline{BC}$ is $M(1, 2, 2)$. Let $P(x, y, z)$ so that

$\mathbf{AP} = (x + 1)\mathbf{i} + (y - 3)\mathbf{j} + (z - 9)\mathbf{k}$

$\mathbf{AM} = 2\mathbf{i} - \mathbf{j} - 7\mathbf{k}$

$\mathbf{AP} = \frac{2}{3}\mathbf{AM}$ so $3\mathbf{AP} = 2\mathbf{AM}$ or

$3(x + 1) = 4$ or $x = \frac{1}{3}$;

$3(y - 3) = -2$ or $y = \frac{7}{3}$

$3(z - 9) = -14$, so $z = \frac{13}{3}$

The desired point is $P(\frac{1}{3}, \frac{7}{3}, \frac{13}{3})$

10.3 The Dot Product, Page 655

1. If two vectors are given in their component form, the dot product is the sum of the product of their x, y, z components, respectively. Work is the dot product of force and distance vectors.

2. Let $\mathbf{u}$ be the vector that results from projecting $\mathbf{v}$ onto $\mathbf{w}$. The scalar projection is $\|\mathbf{u}\|$ while the vector projection is $\mathbf{u}$.

3. $\mathbf{PQ} = (-1 - 1)\mathbf{i} + [1 - (-1)]\mathbf{j} + (4 - 3)\mathbf{k}$

$= -2\mathbf{i} + 2\mathbf{j} + \mathbf{k}$

$\|\mathbf{PQ}\| = \sqrt{(-2)^2 + (2)^2 + (1)^2} = 3$

4. $\mathbf{PQ} = 2\mathbf{i} + \mathbf{j} - 3\mathbf{k}; \|\mathbf{PQ}\| = \sqrt{4+1+9} = \sqrt{14}$

5. $\mathbf{PQ} = -4\mathbf{i} - 4\mathbf{j} - 4\mathbf{k};$

   $\|\mathbf{PQ}\| = 4\sqrt{1+1+1} = 4\sqrt{3}$

6. $\mathbf{PQ} = -3\mathbf{i} - 4\mathbf{j} + 7\mathbf{k};$

   $\|\mathbf{PQ}\| = \sqrt{9+16+49} = \sqrt{74}$

7. $\mathbf{v} \cdot \mathbf{w} = 3(2) + (-2)(-1) + 4(-6) = -16$

8. $\mathbf{v} \cdot \mathbf{w} = 2(0) + (-6)(-3) + 0(7) = 18$

9. $\mathbf{v} \cdot \mathbf{w} = 2(-3) + (3)(5) + (-1)(4) = 5$

10. $\mathbf{v} \cdot \mathbf{w} = 3(2) + (-1)(5) = 1$

11. $\mathbf{v} \cdot \mathbf{w} = 0;$ orthogonal

12. $\mathbf{v} \cdot \mathbf{w} = 0;$ orthogonal

13. $\mathbf{v} \cdot \mathbf{w} = 3(6) + (-2)(9) = 0;$ orthogonal

14. $\mathbf{v} \cdot \mathbf{w} = 4(8) + (-5)(10) + (1)(-2) = -20;$ not orthogonal

15. $\|\mathbf{i} + \mathbf{j} + \mathbf{k}\| = \sqrt{1+1+1} = \sqrt{3}$

16. $\|\mathbf{i} - \mathbf{j} + \mathbf{k}\| = \sqrt{1+1+1} = \sqrt{3}$

17. $\|2\mathbf{i} + \mathbf{j} - 3\mathbf{k}\|^2 = 4+1+9 = 14$

18. $\|2(\mathbf{i} - \mathbf{j} + \mathbf{k}) - 3(2\mathbf{i} + \mathbf{j} - \mathbf{k})\|^2$

   $= \|-4\mathbf{i} - 5\mathbf{j} + 5\mathbf{k}\|^2 = 16 + 25 + 25 = 66$

19. $2\mathbf{v} - 3\mathbf{w} = [2(1) - 3(2)]\mathbf{i} + [2(-2) - 3(4)]\mathbf{j}$

   $+ [2(2) - 3(-1)]\mathbf{k} = -4\mathbf{i} - 16\mathbf{j} + 7\mathbf{k}$

20. $\|\mathbf{v}\|\mathbf{w} = \sqrt{1+4+4}(2\mathbf{i} + 4\mathbf{j} - \mathbf{k})$

   $= 6\mathbf{i} + 12\mathbf{j} - 3\mathbf{k}$

21. $\|2\mathbf{v} - 3\mathbf{w}\| = \|2\mathbf{i} - 4\mathbf{j} + 4\mathbf{k} - 6\mathbf{i} - 12\mathbf{j} + 3\mathbf{k}\|$

   $= \sqrt{16 + 256 + 49} = \sqrt{321}$

22. $\|\mathbf{v} - \mathbf{w}\|(\mathbf{v} + \mathbf{w}) = \|-\mathbf{i} - 6\mathbf{j} + 3\mathbf{k}\|(3\mathbf{i} + 2\mathbf{j} + \mathbf{k})$

   $= \sqrt{46}(3\mathbf{i} + 2\mathbf{j} + \mathbf{k})$

23. Let $\mathbf{T} = 2\mathbf{i} - 3\mathbf{j} + 5\mathbf{k}; \mathbf{u} \neq s\mathbf{T}$, so $\mathbf{T}$ is not parallel to $\mathbf{u}$.

24. $\mathbf{v} = \langle -4, 6, -10 \rangle = 2\langle 2, -3, 5 \rangle$, so $\mathbf{T}$ is parallel to $\mathbf{v}$.

25. $\mathbf{w} = \langle 1, -\frac{3}{2}, 2 \rangle = \frac{1}{2}\langle 2, -3, 4 \rangle$, so $\mathbf{T} \neq s\mathbf{w}$, so $\mathbf{T}$ is not parallel to $\mathbf{w}$.

26. $\mathbf{v} = \langle -\frac{1}{2}, \frac{3}{2}, 2 \rangle = -\frac{1}{2}\langle 1, -3, 4 \rangle$, $\mathbf{T} \neq s\mathbf{v}$, so $\mathbf{T}$ is not parallel to $\mathbf{w}$.

27. $(\mathbf{v} + \mathbf{w}) \cdot (\mathbf{v} - \mathbf{w})$

   $= 4(2) + (-1)(-3) + (0)(2) = 11$

28. $(\mathbf{v} \cdot \mathbf{w})\mathbf{w} = (3 - 2 - 1)(\mathbf{i} + \mathbf{j} - \mathbf{k}) = 0$

29. $(\|\mathbf{v}\|\mathbf{w}) \cdot (\|\mathbf{w}\|\mathbf{v})$

   $= [(\sqrt{9+4+1})(\mathbf{i}+\mathbf{j} - \mathbf{k})] \cdot [\sqrt{1+1-1}(3\mathbf{i} - 2\mathbf{j} + \mathbf{k})]$

   $= \sqrt{14}\sqrt{30}(3 - 2 + 1) = 0$

30. $\dfrac{2\mathbf{v} + 3\mathbf{w}}{\|3\mathbf{v} + 2\mathbf{w}\|} = \dfrac{9\mathbf{i} - \mathbf{j} - \mathbf{k}}{\|11\mathbf{i} - 4\mathbf{j} + \mathbf{k}\|}$

   $= \dfrac{9}{\sqrt{138}}\mathbf{i} - \dfrac{1}{\sqrt{138}}\mathbf{j} - \dfrac{1}{\sqrt{138}}\mathbf{k}$

31. $\cos\theta = \dfrac{\mathbf{v} \cdot \mathbf{w}}{\|\mathbf{v}\|\|\mathbf{w}\|} = \dfrac{1 - 1 + 1}{3} = \dfrac{1}{3};$

   $\theta \approx 71°$

32. $\cos\theta = \dfrac{\mathbf{v} \cdot \mathbf{w}}{\|\mathbf{v}\|\|\mathbf{w}\|} = \dfrac{-3}{\sqrt{5}\sqrt{10}} = -\dfrac{3}{5\sqrt{2}};$

   $\theta \approx 115°$

33. $\cos\theta = \dfrac{\mathbf{v} \cdot \mathbf{w}}{\|\mathbf{v}\|\|\mathbf{w}\|} = \dfrac{-2}{\sqrt{5}\sqrt{5}} = -\dfrac{2}{5};$

   $\theta \approx 114°$

34. $\cos\theta = \dfrac{\mathbf{v} \cdot \mathbf{w}}{\|\mathbf{v}\|\|\mathbf{w}\|} = \dfrac{8 - 6 + 5}{\sqrt{18}\sqrt{4+9+25}}$

   $= \dfrac{7}{\sqrt{18}\sqrt{38}}; \theta \approx 75°$

35. The scalar projection: $\left|\dfrac{\mathbf{v} \cdot \mathbf{w}}{\|\mathbf{w}\|}\right| = \dfrac{2}{2} = 1$

   The vector projection: $\left(\dfrac{\mathbf{v} \cdot \mathbf{w}}{\mathbf{w} \cdot \mathbf{w}}\right)\mathbf{w} = \dfrac{2}{4}\mathbf{w}$

   $= \dfrac{1}{2}(2)\mathbf{k} = \mathbf{k}$

36. The scalar projection: $\left|\dfrac{\mathbf{v} \cdot \mathbf{w}}{\|\mathbf{w}\|}\right| = 0$

   The vector projection: $\left(\dfrac{\mathbf{v} \cdot \mathbf{w}}{\mathbf{w} \cdot \mathbf{w}}\right)\mathbf{w} = \mathbf{0}$

37. The scalar projection: $\left|\dfrac{\mathbf{v} \cdot \mathbf{w}}{\|\mathbf{w}\|}\right| = \dfrac{6}{\sqrt{13}}$

   The vector projection: $\left(\dfrac{\mathbf{v} \cdot \mathbf{w}}{\mathbf{w} \cdot \mathbf{w}}\right)\mathbf{w}$

   $= -\dfrac{6}{13}(2\mathbf{j} - 3\mathbf{k}) = -\dfrac{12}{13}\mathbf{j} + \dfrac{18}{13}\mathbf{k}$

38. The scalar projection: $\left|\dfrac{\mathbf{v} \cdot \mathbf{w}}{\|\mathbf{w}\|}\right| = 0$

   The vector projection: $\left(\dfrac{\mathbf{v} \cdot \mathbf{w}}{\mathbf{w} \cdot \mathbf{w}}\right)\mathbf{w} = \mathbf{0}$

39. For $\mathbf{u}$ to be orthogonal to both $\mathbf{v}$ and $\mathbf{w}$, we need $\mathbf{u} \cdot \mathbf{v} = 0$ and also $\mathbf{u} \cdot \mathbf{w} = 0$. This gives $1u_1 + 0u_2 + 1u_3 = 0$, and $1u_1 + 0u_2 - 2u_3 = 0$. Solving this simultaneous system we find $u_1 = 0$, $u_3 = 0$ Since u_2 has a coefficient of 0, it could be any value. Let $u_2 = \pm 1$, so $\mathbf{u} = \pm\mathbf{j}$.

40. For $\mathbf{u}$ to be orthogonal to both $\mathbf{v}$ and $\mathbf{w}$, we need $\mathbf{u} \cdot \mathbf{v} = 0$ and also $\mathbf{u} \cdot \mathbf{w} = 0$. This gives $u_1 + u_2 - u_3 = 0$, and $-u_1 + u_2 + u_3 = 0$. Solving this

simultaneous system we find $u_1 = \sqrt{2}/2$, $u_2 = 0$, $u_3 = \sqrt{2}/2$, so $u = \pm \frac{\sqrt{2}}{2}i \pm \frac{\sqrt{2}}{2}k$

41. For u to be orthogonal to both v and w, we need $u \cdot v = 0$ and also $u \cdot w = 0$. This gives $2u_1 + u_2 + 2u_3 = 0$, and $-u_1 + 2u_2 - u_3 = 0$. Solving this simultaneous system we find $u_1 = \sqrt{2}/2$, $u_2 = 0$, $u_3 = -\sqrt{2}/2$, so $u = \pm\frac{\sqrt{2}}{2}(i - k)$

42. For u to be orthogonal to both v and w, we need $u \cdot v = 0$ and also $u \cdot w = 0$. This gives $u_1 + u_2 + u_3 = 0$, and $3u_1 + u_2 = 0$. Solving this simultaneous system we find $u_1 = -a$, $u_2 = 3a$, $u_3 = -2a$ for an arbitrary constant a. For a unit vector, $a = 1/14$ so
$$u = \pm \frac{\sqrt{14}}{14}(i - 3j - 2k)$$

43. $w = -v = s(-2i - 3j + 2k)$;

The value of s is arbitrary, for a unit vector,
$$u = -\frac{1}{\sqrt{17}}(2k + 3j - 2k)$$

44. $u = \frac{1}{\sqrt{6}}(i + 2j - k)$; the vector $\frac{1}{3}$ of the original vector is $w = \frac{\sqrt{6}}{18}(i + 2j - k)$

45. $x(i+j+k) + y(i-j+2k) + z(i+k) = 2i + k$ so $x + y + z = 2$, $x - y = 0$, and $x + 2y + z = 1$; solving simultaneously, we find $x = -1$, $y = -1$, and $z = 4$.

46. $x(i - k) + y(j + k) + z(i - j) = 5i - k$ so $x + z = 5$, $y - z = 0$, and $-x + y = -1$; solving simultaneously, we find $x = 3$, $y = 2$, and $z = 2$.

47. The two vectors, call them u and v, must have a dot product of 0.
$u \cdot v = 6 - 2a - 2a = 0$ or $a = \frac{3}{2}$

48. $v \cdot w = 0$, so $3x - x - = 4 = 0$ or $x = 2$

49. $\cos \alpha = \frac{u \cdot i}{\|u\|} = \frac{2}{\sqrt{6}}$; $\alpha \approx 35°$
$\cos \beta = \frac{u \cdot j}{\|u\|} = \frac{1}{\sqrt{6}}$; $\beta \approx 66°$
$\cos \gamma = \frac{u \cdot k}{\|u\|} = \frac{-1}{\sqrt{6}}$; $\gamma \approx 114°$

50. $\cos \theta = \frac{v \cdot w}{\|v\|\|w\|} = \frac{2 - 1 - 2}{6} = -\frac{1}{6}$
The vector that is the projection of v onto w is $u = -\frac{1}{6}w = -\frac{1}{3}i - \frac{1}{6}j + \frac{1}{6}k$

51. a. $8 - 3 - 1 = 4$

b. $\cos \theta = \frac{v \cdot w}{\|v\| \|w\|} = \frac{4}{\sqrt{18}\sqrt{14}} = \frac{2\sqrt{7}}{21}$

c. $v \cdot (v - sw) = 0$
$(4i-j + k) \cdot [(4 - 2s)i + (-1 - 3s)j + (1+s)k] = 0$
so $s = \frac{9}{2}$

d. $(sv + w) \cdot w = 0$
$[(4s+2)i + (-s+3)j + (s-1)k] \cdot (2i+3j - k) = 0$,
$8s + 4 - 3s + 9 - s + 1 = 0$, so $s = -\frac{7}{2}$

52. a. $v \cdot w = 8 + 18 = 26$

b. $\cos \theta = \frac{v \cdot w}{\|v\| \|w\|} = \frac{26}{(7)(5)} = \frac{26}{35}$

c. $v \cdot (v - sw) = 0$
$$s = \frac{\|v\|^2}{v \cdot w} = \frac{49}{26}$$

d. $(v + tw) \cdot w = v \cdot w + t\|w\|^2 = 0$
$$t = -\frac{25}{26}$$

53. $F_v = \frac{F \cdot v}{\|v\|} = \frac{4 + 2 + 6}{\sqrt{6}} = \frac{12}{\sqrt{6}} = 2\sqrt{6}$

54. $PQ = 2i+j+3k$; $W = F \cdot PQ = 4+3+3 = 10$

55. $PQ = [4 - (-3)]i + [9 - (-5)]j + (11 - 4)k$
$= 7i + 14j + 7k$
$W = F \cdot PQ = \frac{6}{7}(7) + (-\frac{2}{7})(14) + \frac{6}{7}(7) = 8$

56. Consider a right-handed Cartesian coordinate system with the log being pulled along the positive x-axis. Fred's rope is directed above the fourth quadrant and Sam's above the first. The position vector along Fred's rope is
$L_f = x_f i - j + 2k$; $\|L_f\| = x_f^2 + 1 + 4 = 64$,
$x_f = \frac{\sqrt{59}}{8}$; a unit vector along Fred's rope is
$$L_f = \frac{\sqrt{59}}{8}i - \frac{1}{8}j + \frac{2}{8}k$$
The direction cosine for the x-component is
$\frac{\sqrt{59}}{8}$ and the component of force is
$\frac{20\sqrt{59}}{8} \approx 19.2$ lb.
The position vector along Sam's rope is
$L_S = x_s i + j + k$; $\|L_s\| = x_s^2 + 1 + 1 = 62$,
$x_s = \frac{\sqrt{62}}{8}$; a unit vector along Sam's rope is

$$\mathbf{L}_2 = \frac{\sqrt{62}}{8}\mathbf{i} + \frac{1}{8}\mathbf{j} + \frac{1}{8}\mathbf{k}$$

The direction cosine for the x-component is

$\frac{\sqrt{62}}{8}$ and the component of force is

$\frac{30\sqrt{62}}{8} \approx 29.5$ lb.

The total force along the x-axis is

$$F_x = 29.5 + 19.2 = 48.7 \text{ lb}$$

57. A unit vector along the road is
$\mathbf{T} = \cos 10°\mathbf{i} + \sin 10°\mathbf{j}$ and a unit normal
vector is $\mathbf{N} = \sin 10°\mathbf{i} - \cos 10°\mathbf{j}$. A 5,000 lb
force along the road becomes
$\mathbf{F} = 868\mathbf{i} - 4,924\mathbf{j}$. The desired force
required to keep the truck from moving along
the road is 868 lb.

58. a. The horizontal vector is
$\mathbf{F} = 50[(\cos \frac{\pi}{3})\mathbf{i} + (\sin \frac{\pi}{3})\mathbf{j}]$ and the
"drag" vector is $\mathbf{D} = 20\mathbf{i}$. Work is
$W = 50[(\cos \frac{\pi}{3})\mathbf{i} + (\sin \frac{\pi}{3})\mathbf{j}] \cdot 20\mathbf{i} = 500$ ft $\cdot$ lb
b. The horizontal vector is
$\mathbf{F} = 50[(\cos \frac{\pi}{4})\mathbf{i} + (\sin \frac{\pi}{4})\mathbf{j}]$ and the
"drag" vector is $\mathbf{D} = 20\mathbf{i}$. Work is
$W = 50[(\cos \frac{\pi}{4})\mathbf{i} + (\sin \frac{\pi}{4})\mathbf{j}] \cdot 20\mathbf{i}$
$= 500\sqrt{2}$ ft $\cdot$ lb

59. The wind's component on the path of the boat is
$1000 \cos 60° = 500$. $W = \mathbf{F} \cdot \mathbf{B} = 500(50)$
$= 25,000$ ft $\cdot$ lb

60. a. $\int_{-b}^{b} x^2(x^3 - 5x)\, dx = \left(\frac{x^6}{6} - \frac{5x^4}{4}\right)\Big|_{-b}^{b} = 0$

b. $\int_{-b}^{b} \sin kx \sin nx\, dx = \frac{1}{2}\int_{-\pi}^{\pi} [\cos(k - n)x]\, dx$

$- \frac{1}{2}\int_{-\pi}^{\pi} [\cos(k + n)x]\, dx$

$= \frac{1}{2}\left[\frac{\sin(k - n)x}{k - n} - \frac{\sin(k + n)x}{k + n}\right]\Big|_{-\pi}^{\pi} = 0$

c. The region between the two sine curves is
symmetric with respect to the point $P(\pi, 0)$.
This means that for every region with an
area that is positive ($y > 0$), there is another
region that is negative ($y < 0$).

61. a. Fourier series is extremely important in
engineering.

$\int_{-\pi}^{\pi} \sin^2 kx\, dx = \frac{1}{2}\int_{-\pi}^{\pi} (1 - \cos 2kx)\, dx$

$= \frac{x}{2} - \frac{\sin 2kx}{4k}\Big|_{-\pi}^{\pi} = \pi$

Also, $\int_{-\pi}^{\pi} f(x) \sin kx\, dx = b_1\int_{-\pi}^{\pi} \sin x \sin kx\, dx$

$+ b_2\int_{-\pi}^{\pi} \sin 2x \sin kx\, dx + \cdots + b_k\int_{-\pi}^{\pi} \sin^2 kx\, dx + \cdots$

$= 0 + 0 + \cdots + \pi b_k + 0 + \cdots$

Thus, $b_k = \frac{1}{\pi}\int_{-\pi}^{\pi} f(x) \sin kx\, dx$

b. $\sin kx \sin x$ is an even function, so

$\int_{-\pi}^{\pi} \sin^2 kx\, dx = 2\int_{0}^{\pi} \sin^2 kx\, dx$

c. $b_1 = \frac{2}{\pi}\int_{0}^{\pi} x \sin x\, dx = \frac{2}{\pi}(\sin x - x \cos x)\Big|_{0}^{\pi}$

$= \frac{2}{\pi}(0 + \pi) = 2$ $\boxed{\text{Formula 343}}$

$b_2 = \frac{2}{\pi}\int_{0}^{\pi} x \sin 2x\, dx = \frac{2}{\pi}\left[\frac{\sin 2x}{4} - \frac{x \cos 2x}{2}\right]\Big|_{0}^{\pi}$

$= -\frac{2}{\pi}\left[\frac{\pi - 0}{2}\right] = -1$

$b_3 = \frac{2}{\pi}\int_{0}^{\pi} x \sin 3x\, dx = \frac{2}{\pi}\left[\frac{\sin 3x}{3^2} - \frac{x \cos 3x}{3}\right]\Big|_{0}^{\pi}$

$= \frac{2}{\pi}\left[\frac{\pi}{3}\right] = \frac{2}{3}$

$b_4 = \frac{2}{\pi}\int_{0}^{\pi} x \sin 4x\, dx = \frac{2}{\pi}\left[\frac{\sin 4x}{4^2} - \frac{x \cos 4x}{4}\right]\Big|_{0}^{\pi}$

$= -\frac{2}{\pi}\left[\frac{\pi}{4}\right] = -\frac{2}{4}$

$b_n = \frac{2}{\pi}\int_{0}^{\pi} x \sin nx\, dx = \frac{2}{\pi}\left[\frac{\sin nx}{n^2} - \frac{x \cos nx}{n}\right]\Big|_{0}^{\pi}$

$= (-1)^n \frac{2}{\pi}\left[\frac{\pi}{n}\right] = (-1)^n \frac{2}{n}$

d. $f(x) = x(\pi - x)(\pi + x) = x(\pi^2 - x^2)$

$= \pi^2 x - x^3$

$b_n = \frac{2}{\pi}\int_{0}^{\pi} (\pi^2 x - x^3) \sin nx\, dx$ $\boxed{\text{Formula 345}}$

$= -\frac{2}{\pi}\left[\left(\frac{3x^2}{n^2} - \frac{6}{n^4}\right)\sin nx + \left(\frac{6x}{n^3} - \frac{x^3}{n}\right)\cos nx\right]\Big|_{0}^{\pi}$

$= (-1)^{n+1}\frac{2\pi^2}{n} + \frac{2}{\pi}(-1)^n\left(\frac{6\pi}{n^3} - \frac{\pi^3}{n}\right)$

$= (-1)^{n+1}\frac{12}{n^3}$

Thus, $b_1 = 12$, $b_2 = -1.5$, $b_3 = 0.44$, $b_4 = -0.1875$, $b_5 = 0.096$, $b_6 = -0.056$, $b_7 = 0.035$. We have,

$$f(x) = 12 \sin x - 1.5 \sin 2x + 0.44 \sin 3x$$
$$- 0.1875 \sin 4x + 0.96 \sin 5x$$
$$- 0.56 \sin 6x + 0.035 \sin 7x - \cdots$$

62. $\|\mathbf{u}_0 - \mathbf{u}\|^2 = (x-a)^2 + (y-b)^2 + (z-c)^2 < r^2$

The region inside a sphere with center (a, b, c) and radius r.

63. The area of an equilateral triangle with side s, is given by $\dfrac{s^2\sqrt{3}}{4}$. The angle between the two sides is $\dfrac{\pi}{3}$. If one vector is $\langle s, 0 \rangle$, then the other is $\langle \frac{1}{2}, \frac{\sqrt{3}}{2} \rangle$. If $A = 25\sqrt{3}$, then $A = xy = \dfrac{\sqrt{3}s^2}{4}$ so $s^2 = 100$, and $s = 10$. Let $\mathbf{u} = s\mathbf{i} = 10\mathbf{i}$ be the vector along the base and

$$\mathbf{w} = s[(\cos \tfrac{\pi}{3})\mathbf{i} + (\sin \tfrac{\pi}{3})\mathbf{j}] = 5\mathbf{i} + 5\sqrt{3}\mathbf{j}$$

be the vector along the inclined side. Then

$$\mathbf{v} \cdot \mathbf{w} = 50$$

64. Let $\mathbf{OP} = \mathbf{i} + \mathbf{j} + \mathbf{k}$ be the vector along a diagonal of the cube and $\mathbf{OQ} = \mathbf{i} + \mathbf{j}$ be the vector along the diagonal of one of its faces. If α is the angle between $\mathbf{OP}$ and $\mathbf{OQ}$,

$$\cos \alpha = \frac{\mathbf{OP} \cdot \mathbf{OQ}}{\|\mathbf{OP}\|\|\mathbf{OQ}\|} = \frac{2}{\sqrt{3}\sqrt{2}} \text{ or } \alpha = 35°$$

65. $\mathbf{v} \cdot \mathbf{v} = \|\mathbf{v}\|^2 = 0$ means that $\mathbf{v} = 0$; $\mathbf{w}$ can be any vector if $\mathbf{v} \cdot \mathbf{w} = 0$.

66. **a.** $(\mathbf{v} + \mathbf{w}) \cdot (\mathbf{v} + \mathbf{w}) = \mathbf{v} \cdot \mathbf{v} + \mathbf{v} \cdot \mathbf{w} + \mathbf{w} \cdot \mathbf{v} + \mathbf{w} \cdot \mathbf{w}$
$$= \|\mathbf{v}\|^2 + 2\mathbf{v} \cdot \mathbf{w} + \|\mathbf{w}\|^2$$

b. $\|\mathbf{v} + \mathbf{w}\|^2 = (\mathbf{v} + \mathbf{w}) \cdot (\mathbf{v} + \mathbf{w})$
$$= \|\mathbf{v}\|^2 + \|\mathbf{w}\|^2 + 2\|\mathbf{v}\|\|\mathbf{w}\|\cos\theta$$
$$\leq \|\mathbf{v}\|^2 + \|\mathbf{w}\|^2 + 2\|\mathbf{v}\|\|\mathbf{w}\| + 2\|\mathbf{v}\|\|\mathbf{w}\|$$
$$= (\|\mathbf{v}\| + \|\mathbf{w}\|)^2$$

Thus, $\|\mathbf{v} + \mathbf{w}\| \leq \|\mathbf{v}\| + \|\mathbf{w}\|$

67. **a.** $\|\mathbf{v} \cdot \mathbf{w}\| = \|\mathbf{v}\|\|\mathbf{w}\| |\cos\theta| \leq \|\mathbf{v}\|\|\mathbf{w}\|$

b. The equality occurs when $|\cos\theta| = 1$; that is, either when $\theta = 0$ or when $\theta = \pi$. This occurs only when $\mathbf{v}$ and $\mathbf{w}$ are parallel; that is, when $\mathbf{v} = t\mathbf{w}$.

c. $\|\mathbf{v} + \mathbf{w}\|^2 = \|\mathbf{v}\|^2 + \|\mathbf{w}\|^2 + 2(\mathbf{v} \cdot \mathbf{w})$
$$\leq \|\mathbf{v}\|^2 + \|\mathbf{w}\|^2 + 2\|\mathbf{v}\| \cdot \|\mathbf{w}\|$$
$$= (\|\mathbf{v}\| + \|\mathbf{w}\|)^2$$

Thus, $\|\mathbf{v} + \mathbf{w}\| \leq \|\mathbf{v}\| + \|\mathbf{w}\|$.

10.4 The Cross Product, Page 663

1. $\mathbf{v} \times \mathbf{w} = \mathbf{i} \times \mathbf{j} = \mathbf{k}$

2. $\mathbf{v} \times \mathbf{w} = \mathbf{k} \times \mathbf{k} = 0$

3. $\mathbf{v} \times \mathbf{w} = \begin{vmatrix} \mathbf{i} & \mathbf{j} & \mathbf{k} \\ 3 & 0 & 2 \\ 2 & 1 & 0 \end{vmatrix} = -2\mathbf{i} + 4\mathbf{j} + 3\mathbf{k}$

4. $\mathbf{v} \times \mathbf{w} = \begin{vmatrix} \mathbf{i} & \mathbf{j} & \mathbf{k} \\ 1 & -3 & 0 \\ 1 & 0 & 5 \end{vmatrix} = -15\mathbf{i} - 5\mathbf{j} + 3\mathbf{k}$

5. $\mathbf{v} \times \mathbf{w} = \begin{vmatrix} \mathbf{i} & \mathbf{j} & \mathbf{k} \\ 3 & -2 & 4 \\ 1 & 4 & -7 \end{vmatrix} = -2\mathbf{i} + 25\mathbf{j} + 14\mathbf{k}$

6. $\mathbf{v} \times \mathbf{w} = \begin{vmatrix} \mathbf{i} & \mathbf{j} & \mathbf{k} \\ 5 & -1 & 2 \\ 2 & 1 & -3 \end{vmatrix} = \mathbf{i} + 19\mathbf{j} + 7\mathbf{k}$

7. $\mathbf{v} \times \mathbf{w} = \begin{vmatrix} \mathbf{i} & \mathbf{j} & \mathbf{k} \\ 3 & -1 & 2 \\ 2 & 3 & -4 \end{vmatrix} = -2\mathbf{i} + 16\mathbf{j} + 11\mathbf{k}$

8. $\mathbf{v} \times \mathbf{w} = \begin{vmatrix} \mathbf{i} & \mathbf{j} & \mathbf{k} \\ 0 & -1 & 4 \\ 5 & 0 & 6 \end{vmatrix} = -6\mathbf{i} + 20\mathbf{j} + 5\mathbf{k}$

9. $\mathbf{v} \times \mathbf{w} = \begin{vmatrix} \mathbf{i} & \mathbf{j} & \mathbf{k} \\ 1 & -6 & 10 \\ -1 & 5 & -6 \end{vmatrix} = -14\mathbf{i} - 4\mathbf{j} - \mathbf{k}$

10. $\mathbf{v} \times \mathbf{w} = \begin{vmatrix} \mathbf{i} & \mathbf{j} & \mathbf{k} \\ \cos\theta & \sin\theta & 0 \\ -\sin\theta & \cos\theta & 0 \end{vmatrix} = \mathbf{k}$

11. $\sin\theta = \dfrac{\|\mathbf{v} \times \mathbf{w}\|}{\|\mathbf{v}\|\,\|\mathbf{w}\|} = \dfrac{\sqrt{3}}{\sqrt{2}\sqrt{2}} = \dfrac{\sqrt{3}}{2}$

12. $\sin\theta = \dfrac{\|\mathbf{v} \times \mathbf{w}\|}{\|\mathbf{v}\|\,\|\mathbf{w}\|} = \dfrac{\sqrt{2}}{\sqrt{2}\sqrt{3}} = \dfrac{\sqrt{3}}{3}$

13. $\sin\theta = \dfrac{\|\mathbf{v} \times \mathbf{w}\|}{\|\mathbf{v}\|\,\|\mathbf{w}\|} = \dfrac{\sqrt{3}}{\sqrt{2}\sqrt{2}} = \dfrac{\sqrt{3}}{2}$

14. $\sin\theta = \dfrac{\|\mathbf{v} \times \mathbf{w}\|}{\|\mathbf{v}\|\,\|\mathbf{w}\|} = \dfrac{\sqrt{3}}{\sqrt{2}\sqrt{2}} = \dfrac{\sqrt{3}}{2}$

15. $\sin \theta = \dfrac{\|\mathbf{v} \times \mathbf{w}\|}{\|\mathbf{v}\| \, \|\mathbf{w}\|} = \dfrac{3\sqrt{6}}{\sqrt{14} \, \sqrt{77}} = \dfrac{3\sqrt{33}}{77}$

16. $\sin \theta = \dfrac{\|\mathbf{v} \times \mathbf{w}\|}{\|\mathbf{v}\| \, \|\mathbf{w}\|} = \dfrac{(-\cos^2\theta + \sin^2\theta)}{\sqrt{1} \, \sqrt{1}}$

$\qquad = \cos 2\theta$

17. $\mathbf{v} \times \mathbf{w} = \begin{vmatrix} \mathbf{i} & \mathbf{j} & \mathbf{k} \\ 2 & 0 & 1 \\ 1 & -1 & -1 \end{vmatrix} = \mathbf{i} + 3\mathbf{j} - 2\mathbf{k}$

The unit normal is $\dfrac{\mathbf{i} + 3\mathbf{j} - 2\mathbf{k}}{\sqrt{1^2 + 3^2 + (-2)^2}}$

$= \dfrac{1}{\sqrt{14}}\mathbf{i} + \dfrac{3}{\sqrt{14}}\mathbf{j} - \dfrac{2}{\sqrt{14}}\mathbf{k}$

18. $\mathbf{v} \times \mathbf{w} = \begin{vmatrix} \mathbf{i} & \mathbf{j} & \mathbf{k} \\ 0 & 1 & -3 \\ -1 & 1 & 1 \end{vmatrix} = 4\mathbf{i} + 3\mathbf{j} + \mathbf{k}$

The unit normal is $\dfrac{4\mathbf{i} + 3\mathbf{j} + \mathbf{k}}{\sqrt{4^2 + 3^2 + 1^2}}$

$= \dfrac{4}{\sqrt{26}}\mathbf{i} + \dfrac{3}{\sqrt{26}}\mathbf{j} + \dfrac{1}{\sqrt{26}}\mathbf{k}$

19. $\mathbf{v} \times \mathbf{w} = \begin{vmatrix} \mathbf{i} & \mathbf{j} & \mathbf{k} \\ 1 & 1 & 1 \\ 3 & 12 & -4 \end{vmatrix} = -16\mathbf{i} + 7\mathbf{j} + 9\mathbf{k}$

The unit normal is $\dfrac{-16\mathbf{i} + 7\mathbf{j} + 9\mathbf{k}}{\sqrt{(-16)^2 + 7^2 + 9^2}}$

$= \dfrac{-16}{\sqrt{386}}\mathbf{i} + \dfrac{7}{\sqrt{386}}\mathbf{j} + \dfrac{9}{\sqrt{386}}\mathbf{k}$

20. $\mathbf{v} \times \mathbf{w} = \begin{vmatrix} \mathbf{i} & \mathbf{j} & \mathbf{k} \\ 2 & -2 & 1 \\ 4 & 2 & -3 \end{vmatrix} = 4\mathbf{i} + 10\mathbf{j} + 12\mathbf{k}$

The unit normal is $\dfrac{4\mathbf{i} + 10\mathbf{j} + 12\mathbf{k}}{\sqrt{4^2 + (10)^2 + (12)^2}}$

$= \dfrac{2}{\sqrt{65}}\mathbf{i} + \dfrac{5}{\sqrt{65}}\mathbf{j} + \dfrac{6}{\sqrt{65}}\mathbf{k}$

21. $\mathbf{v} \times \mathbf{w} = \begin{vmatrix} \mathbf{i} & \mathbf{j} & \mathbf{k} \\ 3 & 4 & 0 \\ 1 & 1 & -1 \end{vmatrix} = -4\mathbf{i} + 3\mathbf{j} - \mathbf{k}$

$A = \| -4\mathbf{i} + 3\mathbf{j} - \mathbf{k} \| = \sqrt{16 + 9 + 1} = \sqrt{26}$

22. $\mathbf{v} \times \mathbf{w} = \begin{vmatrix} \mathbf{i} & \mathbf{j} & \mathbf{k} \\ 2 & -1 & 2 \\ 4 & -3 & 0 \end{vmatrix} = 6\mathbf{i} + 8\mathbf{j} - 2\mathbf{k}$

$A = \| 6\mathbf{i} + 8\mathbf{j} - 2\mathbf{k} \| = \sqrt{36 + 64 + 4} = 2\sqrt{26}$

23. $\mathbf{v} \times \mathbf{w} = \begin{vmatrix} \mathbf{i} & \mathbf{j} & \mathbf{k} \\ 4 & -1 & 1 \\ 2 & 3 & -1 \end{vmatrix} = -2\mathbf{i} + 6\mathbf{j} + 14\mathbf{k}$

$A = \| -2\mathbf{i} + 6\mathbf{j} + 14\mathbf{k} \| = \sqrt{4 + 36 + 196} = 2\sqrt{59}$

24. $\mathbf{v} \times \mathbf{w} = \begin{vmatrix} \mathbf{i} & \mathbf{j} & \mathbf{k} \\ 2 & 0 & 3 \\ 0 & 2 & -3 \end{vmatrix} = -6\mathbf{i} + 6\mathbf{j} + 4\mathbf{k}$

$A = \| -6\mathbf{i} + 6\mathbf{j} + 4\mathbf{k} \| = \sqrt{36 + 36 + 16} = 2\sqrt{22}$

25. $\mathbf{v} \times \mathbf{w} = \begin{vmatrix} \mathbf{i} & \mathbf{j} & \mathbf{k} \\ 1 & 0 & -1 \\ 1 & -1 & 0 \end{vmatrix} = -\mathbf{i} - \mathbf{j} - \mathbf{k}$

$A = \tfrac{1}{2}\| -\mathbf{i} - \mathbf{j} - \mathbf{k} \| = \tfrac{1}{2}\sqrt{1 + 1 + 1} = \tfrac{1}{2}\sqrt{3}$

26. $\mathbf{v} \times \mathbf{w} = \begin{vmatrix} \mathbf{i} & \mathbf{j} & \mathbf{k} \\ 1 & 1 & -1 \\ -1 & 1 & 2 \end{vmatrix} = 3\mathbf{i} - \mathbf{j} + 2\mathbf{k}$

$A = \tfrac{1}{2}\| 3\mathbf{i} - \mathbf{j} + 2\mathbf{k} \| = \tfrac{1}{2}\sqrt{9 + 1 + 4} = \tfrac{1}{2}\sqrt{14}$

27. $\mathbf{v} \times \mathbf{w} = \begin{vmatrix} \mathbf{i} & \mathbf{j} & \mathbf{k} \\ 1 & 1 & -2 \\ 2 & -1 & -1 \end{vmatrix} = -3\mathbf{i} - 3\mathbf{j} - 3\mathbf{k}$

$A = \tfrac{1}{2}\| -3\mathbf{i} - 3\mathbf{j} - 3\mathbf{k} \| = \tfrac{3}{2}\sqrt{1 + 1 + 1} = \tfrac{3}{2}\sqrt{3}$

28. $\mathbf{v} \times \mathbf{w} = \begin{vmatrix} \mathbf{i} & \mathbf{j} & \mathbf{k} \\ 2 & 0 & 0 \\ 0 & 2 & 0 \end{vmatrix} = 4\mathbf{k}; \; A = \tfrac{1}{2}\| 4\mathbf{k} \| = 2$

29. **a.** does not exist   **b.** scalar

30. **a.** vector   **b.** does not exist

31. **a.** scalar   **b.** vector

32. $V = \begin{vmatrix} 1 & 1 & 0 \\ 1 & 2 & 0 \\ 0 & 0 & 3 \end{vmatrix} = 3$

33. $V = \begin{vmatrix} 0 & 1 & 1 \\ 2 & 1 & 2 \\ 5 & 0 & 0 \end{vmatrix} = 5$

34. $\begin{vmatrix} 1 & 1 & 1 \\ 1 & -1 & -1 \\ 2 & 0 & 3 \end{vmatrix} = -6; \; V = 6$

35. $\begin{vmatrix} 2 & 1 & -1 \\ 3 & 0 & 1 \\ 0 & 1 & 1 \end{vmatrix} = -8; \; V = 8$

36. The dot product of two vectors is a scalar. The dot product of two vectors is 0 if and

only if the vectors are orthogonal. It can also be used to find work and the projection of a vector onto another, including finding direction cosines. The cross product of two vectors is a vector normal to the plane formed by the given vectors. Its magnitude can be used to find the area of the parallelogram formed by these vectors.

37. The right hand rule allows finding the proper direction of the unit vector used in the cross product $\mathbf{u} \times \mathbf{v}$. Place the fingers of your right hand along $\mathbf{u}$, turn your hand toward $\mathbf{v}$ (use the smaller angle), then the direction of your thumb is the direction of the unit vector.

38. $\left|(\mathbf{u} \times \mathbf{v}) \cdot \mathbf{w}\right|$ is the area of the parallelogram formed by $\mathbf{u} \times \mathbf{v}$ (the base) times the height of a parallelepiped (the third side $\mathbf{w}$ is projected onto the normal to the base).

39. a. No; if $\mathbf{u} \times \mathbf{v} = \mathbf{u} \times \mathbf{w}$,
$\|\mathbf{u}\| \|\mathbf{v}\| \sin \theta_1 = \|\mathbf{u}\| \|\mathbf{w}\| \sin \theta_2$
$\|\mathbf{u}\| \sin \theta_1 = \|\mathbf{v}\| \sin \theta_2$ is possible with
$\mathbf{u} \neq \mathbf{v}$
b. No; if $\mathbf{u} \cdot \mathbf{v} = \mathbf{u} \cdot \mathbf{w}$,
$\|\mathbf{u}\| \|\mathbf{v}\| \cos \theta_1 = \|\mathbf{u}\| \|\mathbf{w}\| \cos \theta_2$
$\|\mathbf{u}\| \cos \theta_1 = \|\mathbf{v}\| \cos \theta_2$ is possible with
$\mathbf{u} \neq \mathbf{v}$
c. Yes;
$\|\mathbf{u}\| \|\mathbf{v}\| \sin \theta_1 = \|\mathbf{u}\| \|\mathbf{w}\| \sin \theta_2$
$\|\mathbf{u}\| \cos \theta_1 = \|\mathbf{v}\| \cos \theta_2$ means (after division) $\tan \theta_1 = \tan \theta_2$, so $\theta_1 = \theta_2$,
$\|\mathbf{u}\| = \|\mathbf{v}\|$

40. $\mathbf{u} = \mathbf{i}, \mathbf{v} = \mathbf{i} + \mathbf{j} + \mathbf{k}, \mathbf{w} = \mathbf{i} + 2\mathbf{j} + s\mathbf{k}$;
$\mathbf{n}$ is normal to the plane of $\mathbf{u}$ and $\mathbf{v}$ if
$$\mathbf{n} = \mathbf{u} \times \mathbf{v} = \begin{vmatrix} \mathbf{i} & \mathbf{j} & \mathbf{k} \\ 1 & 0 & 0 \\ 1 & 1 & 1 \end{vmatrix} = -\mathbf{j} + \mathbf{k}$$
$\mathbf{n}$ must be orthogonal to every vector in the plane, so $\mathbf{n} \cdot \mathbf{w} = 0$, so $0 - 2 + s = 0$ or $s = 2$.

41. $\mathbf{u} = \mathbf{i} + \mathbf{j}, \mathbf{v} = 2\mathbf{i} - \mathbf{j} + \mathbf{k}, \mathbf{w} = \mathbf{i} + \mathbf{j} + t\mathbf{k}$;
$\mathbf{n}$ is normal to the plane of $\mathbf{u}$ and $\mathbf{v}$ if
$$\mathbf{n} = \mathbf{u} \times \mathbf{v} = \begin{vmatrix} \mathbf{i} & \mathbf{j} & \mathbf{k} \\ 1 & 1 & 0 \\ 2 & -1 & 1 \end{vmatrix} = \mathbf{i} - \mathbf{j} - 3\mathbf{k}$$
$\mathbf{n}$ must be orthogonal to every vector in the plane, so $\mathbf{n} \cdot \mathbf{w} = 0$, so $1 - 1 - 3t = 0$ or $t = 0$.

42. Let $\mathbf{v} = 2\mathbf{i} - \mathbf{j} + \mathbf{k}; \mathbf{u} = \mathbf{PQ} = -2\mathbf{i} + 4\mathbf{j}$;
$\mathbf{w} = \mathbf{PR} = 4\mathbf{j} - 6\mathbf{k}$;
$\mathbf{n}$ is normal to the plane of $\mathbf{u}$ and $\mathbf{w}$ if
$$\mathbf{n} = \mathbf{u} \times \mathbf{v} = \begin{vmatrix} \mathbf{i} & \mathbf{j} & \mathbf{k} \\ -2 & 4 & 0 \\ 0 & 4 & -6 \end{vmatrix} = -24\mathbf{i} - 12\mathbf{j} - 8\mathbf{k}$$

Now, $\cos \theta = \dfrac{\mathbf{n} \cdot \mathbf{v}}{\|\mathbf{n}\| \|\mathbf{v}\|} = \dfrac{-48 + 12 - 8}{\sqrt{784}\sqrt{6}} = \dfrac{-11}{7\sqrt{6}}$
so that $\theta \approx 130°$.

43. $\mathbf{n}_1 = \mathbf{u} \times \mathbf{v} = \begin{vmatrix} \mathbf{i} & \mathbf{j} & \mathbf{k} \\ 1 & 1 & 0 \\ 2 & -1 & 1 \end{vmatrix} = \mathbf{i} - \mathbf{j} - 3\mathbf{k}$

$\mathbf{n}_1 \times \mathbf{w} = \begin{vmatrix} \mathbf{i} & \mathbf{j} & \mathbf{k} \\ 1 & -1 & -3 \\ 3 & 0 & 0 \end{vmatrix} = -9\mathbf{i} + 3\mathbf{k}$

Also, $\mathbf{n}_2 = \mathbf{v} \times \mathbf{w} = \begin{vmatrix} \mathbf{i} & \mathbf{j} & \mathbf{k} \\ 2 & -1 & 1 \\ 0 & 0 & 3 \end{vmatrix} = -3\mathbf{i} - 6\mathbf{k}$

$\mathbf{u} \times \mathbf{n}_2 = \begin{vmatrix} \mathbf{i} & \mathbf{j} & \mathbf{k} \\ 1 & 1 & 0 \\ -3 & -6 & 0 \end{vmatrix} = -3\mathbf{k}$

Cross product is not an associative operation.

44. $a\mathbf{u} \times b\mathbf{v} = \begin{vmatrix} \mathbf{i} & \mathbf{j} & \mathbf{k} \\ au_1 & au_2 & au_3 \\ bv_1 & bv_2 & bv_3 \end{vmatrix}$

$= ab \begin{vmatrix} \mathbf{i} & \mathbf{j} & \mathbf{k} \\ u_1 & u_2 & u_3 \\ v_1 & v_2 & v_3 \end{vmatrix} = ab(\mathbf{u} \times \mathbf{v})$

45. Given $\mathbf{v} \times \mathbf{w} = \mathbf{w}$, let $\mathbf{n} = \mathbf{v} \times \mathbf{w}$ and $t\mathbf{n} = \mathbf{w}$, thus, $\mathbf{w} = 0$. Alternately, we can note that for $\mathbf{v} \times \mathbf{w} = \mathbf{w}$,

$\begin{vmatrix} \mathbf{i} & \mathbf{j} & \mathbf{k} \\ v_1 & v_2 & v_3 \\ w_1 & w_2 & w_3 \end{vmatrix} = \langle w_1, w_2, w_3 \rangle$,

$\langle v_2 w_3 - v_3 w_2, v_3 w_1 - v_1 w_3, v_1 w_2 - v_2 w_1 \rangle$
$= \langle w_1, w_2, w_3 \rangle$.

$v_2 w_3 - v_3 w_2 = w_1, \quad v_3 w_1 - v_1 w_3 = w_2,$
$v_1 w_2 - v_2 w_1 = w_3.$

Solving this system simultaneously we find
$0 = w_1^2 + w_2^2 + w_3^2$. This can be true only if $\mathbf{w} = \mathbf{0}$.

46. $\mathbf{v} \cdot \mathbf{w} = 0$ implies that $\mathbf{v}$ is normal to $\mathbf{w}$ or $\mathbf{v} = 0$ and/or $\mathbf{w} = 0$. $\mathbf{v} \times \mathbf{w} = \mathbf{0}$ implies that $\mathbf{v}$ is parallel to $\mathbf{w}$ or $\mathbf{v} = 0$ and/or $\mathbf{w} = 0$

47. Let $\mathbf{F} = \mathbf{u} = -40\mathbf{k}$ and

$\mathbf{PQ} = 2\left[(\cos \frac{\pi}{6})\mathbf{j} + (\sin \frac{\pi}{6})\mathbf{k}\right] = \sqrt{3}\mathbf{j} + \mathbf{k}$

$\mathbf{T} = \mathbf{PQ} \times \mathbf{F} = \begin{vmatrix} \mathbf{i} & \mathbf{j} & \mathbf{k} \\ 0 & \sqrt{3} & 1 \\ 0 & 0 & -40 \end{vmatrix} = -40\sqrt{3}\mathbf{i}$

48. Let $\mathbf{F} = \mathbf{u} = -3\mathbf{k}$ and

$$\mathbf{PQ} = 5\left[(\cos\tfrac{\pi}{3})\mathbf{j} + (\sin\tfrac{\pi}{3})\mathbf{k}\right] = \tfrac{5}{2}(\mathbf{j} + \sqrt{3}\mathbf{k})$$

$$\mathbf{T} = \mathbf{PQ} \times \mathbf{F} = -\frac{15}{2}\begin{vmatrix} \mathbf{i} & \mathbf{j} & \mathbf{k} \\ 0 & 1 & \sqrt{3} \\ 0 & 0 & -3 \end{vmatrix} = \frac{45}{2}\mathbf{i}$$

49.
$$(\mathbf{u}\times\mathbf{v})\cdot\mathbf{w} = \begin{vmatrix} \mathbf{i} & \mathbf{j} & \mathbf{k} \\ u_1 & u_2 & u_3 \\ v_1 & v_2 & v_3 \end{vmatrix}\cdot(w_1\mathbf{i} + w_2\mathbf{j} + w_3\mathbf{k})$$

$$= \left(\begin{vmatrix} u_2 & u_3 \\ v_2 & v_3 \end{vmatrix}\mathbf{i} - \begin{vmatrix} u_1 & u_3 \\ v_1 & v_3 \end{vmatrix}\mathbf{j} + \begin{vmatrix} u_1 & u_2 \\ v_1 & v_2 \end{vmatrix}\mathbf{k}\right)$$
$$\cdot(w_1\mathbf{i} + w_2\mathbf{j} + w_3\mathbf{k})$$

$$= \begin{vmatrix} u_2 & u_3 \\ v_2 & v_3 \end{vmatrix}w_1 - \begin{vmatrix} u_1 & u_3 \\ v_1 & v_3 \end{vmatrix}w_2 + \begin{vmatrix} u_1 & u_2 \\ v_1 & v_2 \end{vmatrix}w_3$$

$$= \begin{vmatrix} u_1 & u_2 & u_3 \\ v_1 & v_2 & v_3 \\ w_1 & w_2 & w_3 \end{vmatrix}$$

50. If $\mathbf{u}\cdot(\mathbf{v}\times\mathbf{w}) = 0$ or $(\mathbf{u}\times\mathbf{v})\cdot\mathbf{2} = 0$, then the volume of the parallelepiped formed by $\mathbf{u}$, $\mathbf{v}$, and $\mathbf{w}$ is 0, which means the height is 0 and the endpoints of all three vectors lie in the same plane (assuming they all start at the same point).

51. Let $P(x_1, x_2)$, $Q(x_2, y_2)$, and $R(x_3, y_3)$ be the vertices of the triangle. Also, let

$$\mathbf{v} = \mathbf{PR} = \langle x_3 - x_1, y_3 - y_1, 0\rangle,$$

$$\mathbf{u} = \mathbf{PQ} = \langle x_2 - x_1, y_2 - y_1, 0\rangle$$

$$A = \tfrac{1}{2}\|\mathbf{u}\times\mathbf{v}\| = \frac{1}{2}\begin{vmatrix} \mathbf{i} & \mathbf{j} & \mathbf{k} \\ x_2 - x_1 & y_2 - y_1 & 0 \\ x_3 - x_1 & y_3 - y_1 & 0 \end{vmatrix}$$

$$= \tfrac{1}{2}(x_2 y_3 - x_2 y_1 - x_1 y_3 + x_1 y_1 - x_3 y_2 + x_3 y_1 + x_1 y_2 - x_1 y_1)$$

$$= \tfrac{1}{2}[(x_2 y_3 - x_3 y_2) - (x_1 y_3 - x_3 y_1) + (x_1 y_2 - x_2 y_1)]$$

$$= \frac{1}{2}\begin{vmatrix} x_1 & y_1 & 1 \\ x_2 & y_2 & 1 \\ x_3 & y_3 & 1 \end{vmatrix}$$

52.
$$\mathbf{u}\cdot(\mathbf{v}\times\mathbf{w}) = \begin{vmatrix} u_1 & u_2 & u_3 \\ v_1 & v_2 & v_3 \\ w_1 & w_2 & w_3 \end{vmatrix} \text{ and}$$

$$(\mathbf{u}\times\mathbf{v})\cdot\mathbf{w} = \mathbf{w}\cdot(\mathbf{u}\times\mathbf{v}) = \begin{vmatrix} w_1 & w_2 & w_3 \\ u_1 & u_2 & u_3 \\ v_1 & v_2 & v_3 \end{vmatrix}$$

Because the value of a determinant changes algebraic sign when two rows (or columns) are interchanged. In the above, $\mathbf{u}$ and $\mathbf{v}$ were interchanged and then $\mathbf{v}$ and $\mathbf{w}$.

53. The area of the base is $\tfrac{1}{2}\|\mathbf{AB}\times\mathbf{AC}\|$; then

$$V = \tfrac{1}{3}\left(\tfrac{1}{2}\right)\|\mathbf{AB}\times\mathbf{AC}\|\,\|\mathbf{AD}\|\cos\theta$$
$$= \tfrac{1}{6}(\mathbf{AB}\times\mathbf{AC})\cdot\mathbf{AD}$$

54. $\mathbf{u} + \mathbf{v} + \mathbf{w} = 0$; cross with $\mathbf{v}$ to find

$$\mathbf{u}\times\mathbf{v} + \mathbf{v}\times\mathbf{v} + \mathbf{w}\times\mathbf{v} = 0 \text{ or}$$

$$\mathbf{u}\times\mathbf{v} = -\mathbf{w}\times\mathbf{v} = \mathbf{v}\times\mathbf{w} \text{ since } \mathbf{v}\times\mathbf{v} = \mathbf{0}$$

Similarly, if we cross with $\mathbf{w}$, then

$$\mathbf{u}\times\mathbf{w} + \mathbf{v}\times\mathbf{w} + \mathbf{w}\times\mathbf{w} = 0 \text{ or}$$

$$\mathbf{v}\times\mathbf{w} = -\mathbf{u}\times\mathbf{w} = \mathbf{w}\times\mathbf{u} \text{ since } \mathbf{w}\times\mathbf{w} = \mathbf{0}$$

Thus, $\mathbf{u}\times\mathbf{v} = \mathbf{v}\times\mathbf{w} = \mathbf{w}\times\mathbf{u}$.

55. $\mathbf{u}\times\mathbf{v} = \mathbf{w}$ which means that $\mathbf{w}$ and $\mathbf{u}$ are orthogonal and $\mathbf{w}$ and $\mathbf{v}$ are also orthogonal. Since $\mathbf{u}\cdot\mathbf{v} = 0$, $\mathbf{v}$ and $\mathbf{u}$ are orthogonal. Thus, the three vectors are mutually orthogonal. Consequently, for scalars s and t, we have

$$\mathbf{v} = s(\mathbf{w}\times\mathbf{u}) \text{ and } \mathbf{u} = t(\mathbf{v}\times\mathbf{w})$$

56.
$$c\mathbf{u}\times d\mathbf{v} = \begin{vmatrix} \mathbf{i} & \mathbf{j} & \mathbf{k} \\ cu_1 & cu_2 & cu_3 \\ dv_1 & dv_2 & dv_3 \end{vmatrix}$$

$$= cd\begin{vmatrix} \mathbf{i} & \mathbf{j} & \mathbf{k} \\ u_1 & u_2 & u_3 \\ v_1 & v_2 & v_3 \end{vmatrix} = cd(\mathbf{u}\times\mathbf{v})$$

57. Recall that $\cos\theta = \dfrac{\mathbf{v}\cdot\mathbf{w}}{\|\mathbf{v}\|\,\|\mathbf{w}\|}$, and

$\sin\theta = \dfrac{\|\mathbf{v}\times\mathbf{w}\|}{\|\mathbf{v}\|\,\|\mathbf{w}\|}$ so $\tan\theta = \dfrac{\|\mathbf{v}\times\mathbf{w}\|}{\mathbf{v}\cdot\mathbf{w}}$.

58. Let the vertices of the triangles be at $O(0, 0)$, $A(a, 0)$, and $B(b, c)$. The area of the triangle is

$$A = \tfrac{1}{2}\|\mathbf{u}\times\mathbf{v}\| = \frac{1}{2}\begin{vmatrix} \mathbf{i} & \mathbf{j} & \mathbf{k} \\ a & 0 & 0 \\ b & c & 0 \end{vmatrix} = \frac{|ac|}{2}$$

If $|a|, |b|$, and $|c|$ are integers, then $|ac|\geq 1$ and $A\geq 0.5$. Note: $abc\neq 0$, otherwise $A = 0$.

59.
$$\mathbf{u}\cdot(\mathbf{v}\times\mathbf{w}) = \begin{vmatrix} u_1 & u_2 & u_3 \\ v_1 & v_2 & v_3 \\ w_1 & w_2 & w_3 \end{vmatrix} \text{ and}$$

$$\mathbf{v}\cdot(\mathbf{u}\times\mathbf{w}) = \begin{vmatrix} v_1 & v_2 & v_3 \\ u_1 & u_2 & u_3 \\ w_1 & w_2 & w_3 \end{vmatrix}$$

The two determinants have the same absolute

value since interchanging two rows merely reverses the algebraic sign.

$$|\mathbf{w} \cdot (\mathbf{u} \times \mathbf{v})| = |\mathbf{v} \cdot (\mathbf{u} \times \mathbf{w})| = |\mathbf{u} \cdot (\mathbf{v} \times \mathbf{w})|$$

60. Let c be any vector in $\mathbf{R}^3$, but pass the x-axis through c. Then $\mathbf{c} = c_1\langle 1, 0, 0\rangle$. Let $\mathbf{b}$ be any vector in $\mathbf{R}^3$, but pass the y-axis through the plane of $\mathbf{c}$ and $\mathbf{b}$. Then $\mathbf{b} = \langle 1, b_2, 0\rangle$. Let $\mathbf{a} = a_1\langle 1, a_2, a_3\rangle$ be any vector in $\mathbf{R}^3$. Then

$$\mathbf{n}_1 = b_1 c_1 \begin{vmatrix} \mathbf{i} & \mathbf{j} & \mathbf{k} \\ 1 & b_2 & 0 \\ 1 & 0 & 0 \end{vmatrix} = b_1 b_2 c_1 \langle 0, 0, 1\rangle$$

Now,

$$\mathbf{a} \times \mathbf{n}_1 = a_1 b_1 b_2 c_1 \begin{vmatrix} \mathbf{i} & \mathbf{j} & \mathbf{k} \\ 1 & a_2 & a_3 \\ 0 & 0 & -1 \end{vmatrix}$$

$$= a_1 b_1 b_2 c_1 \langle -a_2, 1, 1\rangle$$

On the other hand,

$$(\mathbf{c} \cdot \mathbf{a})\mathbf{b} - (\mathbf{b} \cdot \mathbf{a})\mathbf{c} = a_1 b_1 c_1 \langle 1, b_2, 0\rangle$$
$$= -a_1 b_1 c_1 \langle 1 + a_1 b_2, 0, 0\rangle$$
$$= a_1 b_1 b_2 c_1 \langle -a_2, 1, 0\rangle$$

61. Let $\mathbf{n} = \mathbf{w} \times \mathbf{z}$; then

$$(\mathbf{u} \times \mathbf{v}) \times (\mathbf{w} \times \mathbf{z}) = (\mathbf{u} \times \mathbf{v}) \times \mathbf{n}$$
$$= (\mathbf{u} \cdot \mathbf{n})\mathbf{v} - (\mathbf{v} \cdot \mathbf{n})\mathbf{u}$$
$$= [\mathbf{u} \cdot (\mathbf{w} \times \mathbf{z})]\mathbf{v} - [\mathbf{v} \cdot (\mathbf{w} \times \mathbf{z})]\mathbf{u}$$

62. $\mathbf{u} \times (\mathbf{v} \times \mathbf{w}) + \mathbf{v} \times (\mathbf{w} \times \mathbf{u}) + \mathbf{w} \times (\mathbf{u} \times \mathbf{v})$

$$= (\mathbf{u} \cdot \mathbf{w})\mathbf{v} - (\mathbf{u} \cdot \mathbf{v})\mathbf{w} + (\mathbf{v} \cdot \mathbf{u})\mathbf{w} - (\mathbf{v} \cdot \mathbf{w})\mathbf{u}$$
$$- (\mathbf{w} \cdot \mathbf{v})\mathbf{u} - (\mathbf{w} \cdot \mathbf{u})\mathbf{v}$$
$$= (\mathbf{v} \cdot \mathbf{u})\mathbf{w} - (\mathbf{u} \cdot \mathbf{v})\mathbf{w} + (\mathbf{v} \cdot \mathbf{w})\mathbf{u} - (\mathbf{w} \cdot \mathbf{v})\mathbf{u}$$
$$+ (\mathbf{u} \cdot \mathbf{w})\mathbf{v} - (\mathbf{w} \cdot \mathbf{u})\mathbf{v}$$
$$= 0$$

63. Let $\mathbf{u} = u_1\langle 1, u_2, u_3\rangle$ and $\mathbf{v} = v_1\langle 1, v_2, v_3\rangle$, then

$$\mathbf{u} \times \mathbf{v} = u_1 v_1 \begin{vmatrix} \mathbf{i} & \mathbf{j} & \mathbf{k} \\ 1 & u_2 & u_3 \\ 1 & v_2 & v_3 \end{vmatrix}$$

$$= u_1 v_1 \langle u_2 v_3 - u_3 v_2, u_3 - v_3, v_2 - u_2\rangle$$

Now $\mathbf{u} \cdot \mathbf{v} \times \mathbf{i} = u_1 v_1 \begin{vmatrix} 1 & u_2 & u_3 \\ 1 & v_2 & v_3 \\ 1 & 0 & 0 \end{vmatrix}$

$$= u_1 v_1 \langle 0, u_3 - v_3, 0\rangle.$$

Finally, $\mathbf{u} \cdot \mathbf{v} \times \mathbf{k} = u_1 v_1 \begin{vmatrix} 1 & u_2 & u_3 \\ 1 & v_2 & v_3 \\ 0 & 0 & 1 \end{vmatrix}$

$$= u_1 v_1 \langle 0, 0, v_2 - u_2\rangle.$$

Thus, $u_1 v_1 \langle u_2 v_3 - u_3 v_2, u_3 - v_3, v_2 - u_2\rangle$

$$= u_1 v_1 \langle u_2 v_3 - u_3 v_2, u_3 - v_3, v_2 - u_2\rangle$$
$$= u_1 v_1 \langle 0, u_3 - v_3, v_1\rangle$$
$$= u_1 v_1 \langle 0, 0, v_2 - u_2\rangle$$

64. $\mathbf{u} \times (\mathbf{u} \times \mathbf{v}) = (\mathbf{v} \cdot \mathbf{u})\mathbf{u} - (\mathbf{u} \cdot \mathbf{u})\mathbf{v}$ and

$$\mathbf{u} \times [\mathbf{u} \times (\mathbf{u} \times \mathbf{v})] = \mathbf{u} \times (\mathbf{v} \cdot \mathbf{u})\mathbf{u} - \mathbf{u} \times (\mathbf{u} \cdot \mathbf{u})\mathbf{v}$$
$$= 0 - (\mathbf{u} \cdot \mathbf{u})(\mathbf{u} \times \mathbf{v})$$
$$= -\|\mathbf{u}\|^2 (\mathbf{u} \times \mathbf{v})$$
$$= -\|\mathbf{u}\|^2 (\mathbf{u} \times \mathbf{v}) \cdot \mathbf{u}$$
$$= -\|\mathbf{u}\|^2 \mathbf{u} \cdot \mathbf{v} \times \mathbf{w}$$

10.5 Lines and Planes in Space, Page 673

1. $\dfrac{x - x_0}{A} = \dfrac{y - y_0}{B} = \dfrac{z - z_0}{C}$ are the symmetric equations of a line; setting each fraction equal to t and solving leads to the parametric equations: $x = x_0 + tA$, $y = y_0 + tB$, $z = z_0 + tC$. There is only one equation of a plane. Two parameters are needed to describe a surface.

2. If $Ax + By + Cz + D = 0$ is the equation of a plane, a vector normal to the plane is given by $\mathbf{N} = A\mathbf{i} + B\mathbf{j} + C\mathbf{k}$. Note that the coefficients of the variables are the direction numbers of the normal.

3. $4(x + 1) - 2(y + 1) + 6(z - 2) = 0$
$4x + 4 - 2y - 2 + 6z - 12 = 0$
$4x - 2y + 6z = 10$
$2x - y + 3z = 5$

4. $5(x - 2) - 3(y + 2) + 4(z + 3) = 0$
$5x - 10 - 3y - 6 + 4z + 12 = 0$
$5x - 3y + 4z - 4 = 0$

5. $-3(x - 4) + 2(y + 1) - 2(z + 1) = 0$
$-3x + 12 + 2y + 2 - 2z - 2 = 0$
$3x - 2y + 2z - 12 = 0$

6. $-2(x + 1) + 4(y - 3) - 8z = 0$
$-2x - 2 + 4y - 12 - 8z = 0$
$-2x + 4y - 8z - 14 = 0$
$x - 2y + 4z + 7 = 0$

7. $x = 1 + 3t$, $y = -1 - 2t$, $z = -2 + 5t$
$\dfrac{x - 1}{3} = \dfrac{y + 1}{-2} = \dfrac{z + 2}{5}$

8. $\dfrac{x - 1}{3} = \dfrac{y}{4}$, $z = -1$ and
$x = 1 + 3t$, $y = 4t$, $z = -1$

9. $\frac{x-1}{1} = \frac{y+1}{2} = \frac{z-2}{1}$

$x = 1 + t, \; y = -1 + 2t, \; z = 2 + t$

10. $\mathbf{v} = (1-2)\mathbf{i} + (3-2)\mathbf{j} + (-1-3)\mathbf{k}$

$\quad = -\mathbf{i} + \mathbf{j} - 4\mathbf{k}$

$\frac{x-2}{1} = \frac{y-2}{-1} = \frac{z-3}{4}$

$x = 2 + t, \; y = 2 - t, \; z = 3 + 4t$

11. $\frac{x-1}{1} = \frac{y+3}{-3} = \frac{z-6}{-5}$

$x = 1 + t, \; y = -3 - 3t, \; z = 6 - 5t$

12. $\frac{x-1}{4} = \frac{y+1}{5} = \frac{z-2}{1}$

$x = 1 + 4t, \; y = -1 + 5t, \; z = 2 + t$

13. $\frac{x}{11} = \frac{y-4}{-6} = \frac{z+3}{10}$

$x = 11t, \; y = 4 - 6t, \; z = -3 + 10t$

14. $\frac{x-1}{3} = \frac{y}{1} = \frac{z+4}{2};$

$x = 1 + 3t, \; y = t, \; z = -4 + 2t$

15. $x = 3$ and $z = 0; \; x = 3, \; y = -1 + t, \; z = 0$

16. $\frac{x+1}{3} = \frac{y-1}{1} = \frac{z-6}{-2}$

$x = -1 + 3t, \; y = 1 + t, \; z = 6 - 2t$

17. $(0, -6, -3), \; (8, 0, -1), \; (12, 3, 0)$

18. $(0, 0, 9), \; (8, 0, 9), \; (3, -6, 0)$

19. $(0, 4, 9), \; (8, 0, -3), \; (6, 1, 0)$

20. $(0, 4, -4), \; (12, 0, 4), \; (6, 2, 0)$

21. A vector parallel to the first line is is $\mathbf{v}_1 = 2\mathbf{i} - 3\mathbf{j} + 5\mathbf{k}$. A vector parallel to the second line is $\mathbf{v}_2 = 4\mathbf{i} - 6\mathbf{j} + 10\mathbf{k}$ $= 2(2\mathbf{i} - 3\mathbf{j} + 5\mathbf{k})$; since these vectors have the same direction numbers the lines are either coincident or parallel; since $P(4, 6, -2)$ is a point of the first line, but not on the second line $\mathbf{v}_1$ is parallel to $\mathbf{v}_2$ and the first line is parallel to the second.

22. A vector parallel to the first line is is $\mathbf{v}_1 = -2\mathbf{i} + 6\mathbf{j} - 4\mathbf{k} = -2(\mathbf{i} - 3\mathbf{j} + 2\mathbf{k})$. A vector parallel to the second line is $\mathbf{v}_2 = \mathbf{i} - 3\mathbf{j} + 2\mathbf{k}$; since these vectors have the same direction numbers the lines are either coincident or parallel; since $P(4, 0, 7)$ is a point of the first line, but not on the second line, $\mathbf{v}_1$ is parallel to $\mathbf{v}_2$ and the first line is parallel to the second.

23. A vector parallel to the first line is is $\mathbf{v}_1 = 3\mathbf{i} - 3\mathbf{j} - 7\mathbf{k}$. A vector parallel to the second line is $\mathbf{v}_2 = 3\mathbf{i} - 3\mathbf{j} - 7\mathbf{k}$. Since these lines have the same direction numbers the lines are either coincident or parallel; since $P(3, 1, -4)$ is a point of the first line, but not on the second line, $\mathbf{v}_1$ is parallel to $\mathbf{v}_2$ and the first line is parallel to the second.

24. A vector parallel to the first line is is $\mathbf{v}_1 = -4\mathbf{i} + \mathbf{j} + 5\mathbf{k}$. A vector parallel to the second line is $\mathbf{v}_2 = 3\mathbf{i} - \mathbf{j} - 2\mathbf{k}$. Since these lines do not have proportional direction numbers, the lines are not parallel or coincident. Now we try to find a point of intersection; for x, $2 - 4t_1 = 3t_2$ or $4t_1 + 3t_2 = 2$; for y, $1 + t_1 = -2 - t_2$ or $t_1 + t_2 = -3$. Solving this system simultaneously we find $t_1 = 11, \; t_2 = -14$. These values do not yield a single value for z, so the lines are skew.

25. A vector parallel to the first line is is $\mathbf{v}_1 = 2\mathbf{i} - \mathbf{j} + \mathbf{k}$. A vector parallel to the second line is $\mathbf{v}_2 = 3\mathbf{i} - \mathbf{j} + \mathbf{k}$. Since these lines do not have proportional direction numbers, the lines are not parallel or coincident. Now we try to find a point of intersection; for x, $3 + 2t_1 = -2 + 3t_2$ or $2t_1 - 3t_2 = -5$; for y, $1 - t_1 = 3 - t_2$ or $t_1 - t_2 = -2$. Solving this system simultaneously we find $t_1 = -1, \; t_2 = 1$. For z, the first line gives $z = 4 - 1 = 3$, while the second line gives $z = 2 + 1 = 3$. Thus, $x = 3 - 2 = 1$ and $y = 1 - (-1) = 2$. The point of intersection is $(1, 2, 3)$.

26. A vector parallel to the first line is is $\mathbf{v}_1 = 2\mathbf{i} - \mathbf{j} + \mathbf{k}$. A vector parallel to the second line is $\mathbf{v}_2 = 2\mathbf{i} + 3\mathbf{j} - 4\mathbf{k}$. Since these lines do not have proportional direction numbers, the lines are not parallel or coincident. Now we try to find a point of intersection; for x, $-1 + 2t_1 = -1 + 2t_2$ or $t_1 = t_2$; for y, $3 - t_1 = -1 + 3t_2$ or $t_1 + 3t_2 = 4$. Solving this system simultaneously we find $t_1 = t_2 = 1$. These values yield a single value for z, so the lines are skew.

27. $\|\mathbf{v}\| = \sqrt{2^2 + (-3)^2 + (-5)^2} = \sqrt{38}$

$a_1 = 2, \; a_2 = -3, \; a_3 = -5$

$\cos \alpha = \frac{2}{\sqrt{38}}, \; \alpha \approx 1.24 \text{ or } 71°;$

$\cos \beta = \frac{-3}{\sqrt{38}}, \; \beta \approx 2.08 \text{ or } 119°;$

$\cos \gamma = \frac{-5}{\sqrt{38}}, \; \gamma \approx 2.52 \text{ or } 144°$

28. $\|\mathbf{v}\| = \sqrt{3^2 + (2)^2} = \sqrt{13}$

$a_1 = 3, \ a_2 = 0, \ a_3 = -2$

$\cos \alpha = \dfrac{3}{\sqrt{13}}, \ \alpha \approx 0.59 \text{ or } 34°;$

$\cos \beta = 0, \ \beta \approx 1.57 \text{ or } 90°;$

$\cos \gamma = \dfrac{-2}{\sqrt{13}}, \ \gamma \approx 2.15 \text{ or } 124°$

29. $\|\mathbf{v}\| = \sqrt{5^2 + (-4)^2 + (3)^2} = 5\sqrt{2}$

$a_1 = 5, \ a_2 = -4, \ a_3 = 3$

$\cos \alpha = \dfrac{5}{5\sqrt{2}}, \ \alpha \approx 0.79 \text{ or } 45°;$

$\cos \beta = \dfrac{-4}{5\sqrt{2}}, \ \beta \approx 2.17 \text{ or } 124°;$

$\cos \gamma = \dfrac{3}{5\sqrt{2}}, \ \gamma \approx 1.13 \text{ or } 65°$

30. $\|\mathbf{v}\| = \sqrt{(1)^2 + (-5)^2} = \sqrt{26}$

$a_1 = 0, \ a_2 = 1, \ a_3 = -5$

$\cos \alpha = 0, \ \alpha \approx 1.57 \text{ or } 90°;$

$\cos \beta = \dfrac{1}{\sqrt{26}}, \ \beta \approx 1.37 \text{ or } 79°;$

$\cos \gamma = \dfrac{-5}{\sqrt{26}}, \ \gamma \approx 2.94 \text{ or } 169°$

31. $\|\mathbf{v}\| = \sqrt{1^2 + (-3)^2 + (9)^2} = \sqrt{91}$

$a_1 = 1, \ a_2 = -3, \ a_3 = 9$

$\cos \alpha = \dfrac{1}{\sqrt{91}}, \ \alpha \approx 1.47 \text{ or } 84°;$

$\cos \beta = \dfrac{-3}{\sqrt{91}}, \ \beta \approx 1.89 \text{ or } 108°;$

$\cos \gamma = \dfrac{9}{\sqrt{91}}, \ \gamma \approx 0.34 \text{ or } 19°$

32. $\|\mathbf{v}\| = \sqrt{1^2 + (-1)^2 + (3)^2} = \sqrt{11}$

$a_1 = 1, \ a_2 = -1, \ a_3 = 3$

$\cos \alpha = \dfrac{1}{\sqrt{11}}, \ \alpha \approx 1.26 \text{ or } 72°;$

$\cos \beta = \dfrac{-1}{\sqrt{11}}, \ \beta \approx 1.88 \text{ or } 108°;$

$\cos \gamma = \dfrac{3}{\sqrt{11}}, \ \gamma \approx 0.44 \text{ or } 25°$

33. $2(x + 1) + 4(y - 3) - 3(z - 5) = 0$

$\qquad\qquad 2x + 4y - 3z + 5 = 0$

34. $-(x - 0) + 0(y + 7) + (z - 1) = 0$

$\qquad\qquad\qquad x - z + 1 = 0$

35. $(x - 0) - 2(y + 3) + 3(z - 0) = 0$

$\qquad\qquad\qquad 2y - 3z + 6 = 0$

36. $-(x - 1) - 2(y - 1) + 3(z + 1) = 0$

$\qquad\qquad\qquad x + 2y - 3z - 6 = 0$

37. $z = 0$ **38.** $x = 0$

39. $\pm \dfrac{1}{\sqrt{21}}(4\mathbf{i} + 2\mathbf{j} + \mathbf{k})$

40. $\pm \dfrac{1}{\sqrt{21}}(2\mathbf{i} + 4\mathbf{j} + \mathbf{k})$

41. $\pm \dfrac{1}{\sqrt{29}}(2\mathbf{i} + 4\mathbf{j} - 3\mathbf{k})$

42. $\pm \dfrac{1}{\sqrt{38}}(5\mathbf{i} - 3\mathbf{j} + 2\mathbf{k})$

43. $\mathbf{PQ} = 2\mathbf{i} + \mathbf{j} - 2\mathbf{k}; \ \mathbf{v} \cdot \mathbf{PQ} = 6 - 4 - 2 = 0,$ so $\mathbf{v}$ and $\mathbf{PQ}$ are orthogonal.

44. $\mathbf{PQ} = 5\mathbf{i} - 5\mathbf{j} - 5\mathbf{k}; \ \mathbf{v} \cdot \mathbf{PQ} = 35 - 20 - 15 = 0,$ so $\mathbf{v}$ and $\mathbf{PQ}$ are orthogonal

45. A vector normal to the first plane is $\mathbf{N}_1 = \mathbf{i} + \mathbf{j}$ and a vector normal to the second plane is $\mathbf{N}_2 = \mathbf{i} - 2\mathbf{k}$. Then

$$\mathbf{N}_1 \times \mathbf{N}_2 = \begin{vmatrix} \mathbf{i} & \mathbf{j} & \mathbf{k} \\ 1 & 1 & 0 \\ 1 & 0 & -2 \end{vmatrix} = -2\mathbf{i} + 2\mathbf{j} - \mathbf{k}$$

The unit vectors are $\mathbf{N} = \pm \frac{1}{3}(2\mathbf{i} - 2\mathbf{j} + \mathbf{k})$.

46. A vector normal to the first plane is $\mathbf{N}_1 = \mathbf{i} + \mathbf{j} + \mathbf{k}$ and a vector normal to the second plane is $\mathbf{N}_2 = \mathbf{i} - \mathbf{j} + \mathbf{k}$. Then

$$\mathbf{N}_1 \times \mathbf{N}_2 = \begin{vmatrix} \mathbf{i} & \mathbf{j} & \mathbf{k} \\ 1 & 1 & 1 \\ 1 & -1 & 1 \end{vmatrix} = 2\mathbf{i} - 2\mathbf{k}$$

The unit vectors are $\mathbf{N} = \pm \dfrac{1}{\sqrt{2}}(\mathbf{i} - \mathbf{k})$.

47. A vector normal to the first plane is $\mathbf{N}_1 = \mathbf{i} + \mathbf{j} - \mathbf{k}$ and a vector normal to the second plane is $\mathbf{N}_2 = 2\mathbf{i} - \mathbf{j} + 3\mathbf{k}$. Then

$$\mathbf{N}_1 \times \mathbf{N}_2 = \begin{vmatrix} \mathbf{i} & \mathbf{j} & \mathbf{k} \\ 1 & 1 & -1 \\ 2 & -1 & 3 \end{vmatrix} = 2\mathbf{i} - 5\mathbf{j} - 3\mathbf{k}$$

Find any point in the intersection of the planes, say $P(1, 4, 1)$. Then an equation of the line is

$$\frac{x - 1}{2} = \frac{y - 4}{-5} = \frac{z}{-3}$$

48. A vector normal to the first plane is $\mathbf{N}_1 = \mathbf{i} + 2\mathbf{j} + \mathbf{k}$ and a vector normal to the second plane is $\mathbf{N}_2 = \mathbf{i} + \mathbf{j} - 2\mathbf{k}$. Then

$$\mathbf{N}_1 \times \mathbf{N}_2 = \begin{vmatrix} \mathbf{i} & \mathbf{j} & \mathbf{k} \\ 1 & 2 & 1 \\ 1 & 1 & -2 \end{vmatrix} = -5\mathbf{i} + 3\mathbf{j} - \mathbf{k}$$

Find any point in the intersection of the planes, say $P(5, -1, 0)$. Then an equation of the line is

$$\frac{x-5}{5} = \frac{y+1}{-3} = \frac{z}{1}$$

49. A vector parallel to the line through
$P(1, -1, 2)$ and $Q(2, 1, 3)$ is $\mathbf{v} = \mathbf{i} + 2\mathbf{j} + \mathbf{k}$,
and this vector is normal to the desired plane.
Thus,
$$(x - 1) + 2(y + 1) + (z - 2) = 0$$
$$x + 2y + z - 1 = 0$$

50. A vector normal to the given plane is
$\mathbf{N} = 2\mathbf{i} - 3\mathbf{j} + \mathbf{k}$, and this vector is parallel to
the desired line. Thus,
$$\frac{x-1}{2} = \frac{y+5}{-3} = \frac{z-3}{1}$$

51. An orthogonal to the given line will have
attitude numbers $[3, 5, 2]$. It must also pass
through $(2, 1, -1)$. It will have equation:
$$3(x - 2) + 5(y - 1) + 2(z + 1) = 0$$
$$3x + 5y + 2z = 9$$

52. Two parallel planes have the same normal
vectors, so the desired equation is
$$2x - y + 3z + D = 0$$
The plane passes through $P(1, 2, -1)$, so
$$2 - 2 - 3 + D = 0 \text{ or } D = 3$$
Thus, $2x - y + 3z + 3 = 0$.

53. A vector parallel to the given line is
$\mathbf{v} = 2\mathbf{i} + 3\mathbf{j} + 4\mathbf{k}$. $\mathbf{N} = \mathbf{i} - 2\mathbf{j} + \mathbf{k}$ is a
normal to the given plane. The line is
parallel to the plane since
$$\mathbf{v} \cdot \mathbf{N} = 2 - 6 + 4 = 0$$

54. The parametric equation of the given line is
$x = 1 + 2t, y = -1 - t, z = 3t$. This line
intersection the plane when
$$3(1 + 2t) + 2(-1 - t) - 3t = 5$$
$$6t + 3 - 2t - 2 - 3t = 5$$
$$t = 4$$
Thus, the given point is $P(9, -5, 12)$.

55. $\mathbf{N}_1 = 2\mathbf{i} + \mathbf{j} - 4\mathbf{k}$, $\mathbf{N}_2 = \mathbf{i} - \mathbf{j} + \mathbf{k}$.
Let θ be the angle between the normals.
$$\cos\theta = \frac{|\mathbf{N}_1 \cdot \mathbf{N}_2|}{||\mathbf{N}_1|| \; ||\mathbf{N}_2||} = \frac{|2 - 1 - 4|}{\sqrt{21}\sqrt{3}} = \frac{\sqrt{7}}{7}$$
$\theta \approx 1.18$ or $68°$.

56. $\mathbf{N}_1 = \mathbf{i} + 2\mathbf{j} - 3\mathbf{k}$, $\mathbf{N}_2 = \mathbf{i} - 2\mathbf{j} + \mathbf{k}$
$$\mathbf{N}_1 \times \mathbf{N}_2 = \begin{vmatrix} \mathbf{i} & \mathbf{j} & \mathbf{k} \\ 1 & 2 & -3 \\ 1 & -2 & 1 \end{vmatrix} = -4\mathbf{i} - 4\mathbf{j} - 4\mathbf{k}$$

The desired line is
$$\frac{x-2}{1} = \frac{y-3}{1} = \frac{z-1}{1}$$

57. $\mathbf{N}_1 = 2\mathbf{i} + \mathbf{j} - 2\mathbf{k}$, $\mathbf{N}_2 = 3\mathbf{i} - 6\mathbf{j} - 2\mathbf{k}$
$$\mathbf{N}_1 \times \mathbf{N}_2 = \begin{vmatrix} \mathbf{i} & \mathbf{j} & \mathbf{k} \\ 2 & 1 & -2 \\ 3 & -6 & -2 \end{vmatrix} = -14\mathbf{i} - 2\mathbf{j} - 15\mathbf{k}$$
The desired line is
$$\frac{x}{14} = \frac{y-1}{2} = \frac{z+1}{15}$$

58. $\mathbf{N}_1 = 2\mathbf{i} + 3\mathbf{j}$, $\mathbf{N}_2 = 3\mathbf{i} - \mathbf{j} + \mathbf{k}$
$$\mathbf{N}_1 \times \mathbf{N}_2 = \begin{vmatrix} \mathbf{i} & \mathbf{j} & \mathbf{k} \\ 2 & 3 & 0 \\ 3 & -1 & 1 \end{vmatrix} = 3\mathbf{i} - 2\mathbf{j} - 11\mathbf{k}$$

59. $\mathbf{N}_1 = 3\mathbf{i} + \mathbf{j} - \mathbf{k}$, $\mathbf{N}_2 = \mathbf{i} - 6\mathbf{j} - 2\mathbf{k}$
$$\mathbf{N}_1 \times \mathbf{N}_2 = \begin{vmatrix} \mathbf{i} & \mathbf{j} & \mathbf{k} \\ 3 & 1 & -1 \\ 1 & -6 & -2 \end{vmatrix} = -8\mathbf{i} + 5\mathbf{j} - 19\mathbf{k}$$
Now to find a point; if $x = 0$, $y - z = 5$ and
$-6y - 2z = 10$. Solving simultaneously, we
find the point $(0, 0, -5)$. The line of
intersection is:
$$\frac{x}{8} = \frac{y}{-5} = \frac{z+5}{19}$$

60. $\mathbf{N}_1 = 2\mathbf{i} - \mathbf{j} + \mathbf{k}$, $\mathbf{N}_2 = \mathbf{i} + \mathbf{j} - \mathbf{k}$
$$\mathbf{N}_1 \times \mathbf{N}_2 = \begin{vmatrix} \mathbf{i} & \mathbf{j} & \mathbf{k} \\ 2 & -1 & 1 \\ 1 & 1 & -1 \end{vmatrix} = 3\mathbf{j} + 3\mathbf{k}$$
Now to find a point; if $z = 0$, $2x - y = 8$ and
$x + y = 5$. Solving simultaneously, we find
the point $(0, 0, -5)$. The line of
intersection is:
$$\frac{y - \frac{2}{3}}{1} = \frac{z}{-1} \text{ and } x = \frac{13}{3}$$

61. $\mathbf{v} = 2\mathbf{i} + \mathbf{j}$, $\mathbf{w} = 2\mathbf{i} - \mathbf{j} + 3\mathbf{k}$
$$\mathbf{v} \times \mathbf{w} = \begin{vmatrix} \mathbf{i} & \mathbf{j} & \mathbf{k} \\ 2 & 1 & 0 \\ 2 & -1 & 3 \end{vmatrix} = 3\mathbf{i} - 6\mathbf{j} - 4\mathbf{k}$$
$||\mathbf{v}||^2 = 9 + 36 + 16 = 61$; $\cos\alpha = \frac{3}{\sqrt{61}}$ so
$\alpha \approx 1.17$ or $67°$; $\cos\beta = \frac{-6}{\sqrt{61}}$ so $\beta \approx 2.45$
or $140°$; $\cos\gamma = \frac{-4}{\sqrt{61}}$ so $\gamma \approx 2.11$ or $121°$

62. $\mathbf{N}_1 = \mathbf{i} + \mathbf{j} + \mathbf{k}$, $\mathbf{N}_2 = 2\mathbf{i} + 3\mathbf{j} - \mathbf{k}$
$$\mathbf{N}_1 \times \mathbf{N}_2 = \begin{vmatrix} \mathbf{i} & \mathbf{j} & \mathbf{k} \\ 1 & 1 & 1 \\ 2 & 3 & -1 \end{vmatrix} = -4\mathbf{i} + 3\mathbf{j} + \mathbf{k}$$

$\|\mathbf{v}\|^2 = 16 + 9 + 1 = 26$; $\cos \alpha = \dfrac{-4}{\sqrt{26}}$ so
$\alpha \approx 2.74$ or $142°$; $\cos \beta = \dfrac{3}{\sqrt{26}}$ so $\beta \approx 0.94$
or $54°$; $\cos \gamma = \dfrac{1}{\sqrt{26}}$ so $\gamma \approx 1.37$ or $79°$

63. Vectors parallel to the given lines are
$\mathbf{v}_1 = \langle a_1,\ b_1,\ c_1 \rangle$, and $\mathbf{v}_2 = \langle a_2,\ b_2,\ c_2 \rangle$.
If $\mathbf{v}_1 \cdot \mathbf{v}_2 = a_1 a_2 + b_1 b_2 + c_1 c_2 = 0$, then the
vectors (and the given lines) are perpendicular
to each other and intersect at $P(x_0,\ y_0,\ z_0)$.

64. **a.** $\mathbf{N} = a\mathbf{i} + b\mathbf{j} + c\mathbf{k}$ is normal to the plane.
$\mathbf{v} = A\mathbf{i} + B\mathbf{j} + C\mathbf{k}$ is parallel to the line.
The cosine of the acute angle between the
vectors is

$$\cos \theta = \frac{|\mathbf{v} \cdot \mathbf{N}|}{\|\mathbf{v}\| \|\mathbf{N}\|} = \frac{aA + bB + cC}{\sqrt{a^2 + b^2 + c^2}\sqrt{A^2 + B^2 + C^2}}$$

b. $\cos \theta = \dfrac{2 + 3 - 1}{\sqrt{3}\sqrt{14}}$ so $\theta \approx 52°$

65. $\dfrac{a}{a} + \dfrac{0}{b} + \dfrac{0}{c} = 1$, so $P(a, 0, 0)$ is the x-intercept
of the plane; $\dfrac{0}{a} + \dfrac{b}{b} + \dfrac{0}{c} = 1$, so $Q(0,\ b,\ 0)$ is
the y-intercept of the plane; $\dfrac{0}{a} + \dfrac{0}{b} + \dfrac{c}{c} = 1$, so
$R(0,\ 0,\ c)$ is the z-intercept of the plane.

66. $\mathbf{N}_1 = \mathbf{v}_1 \times \mathbf{w}_1$ is normal to p_1 and
$\mathbf{N}_2 = \mathbf{v}_2 \times \mathbf{w}_2$ is normal to p_2. Then,
$\mathbf{v} = \mathbf{N}_1 \times \mathbf{N}_2$ is parallel to both planes, or
aligned with their line of intersection.

10.6 Vector Methods for Measuring Distance in $\mathbb{R}^3$, Page 677

1. $d = \dfrac{|12 - 20 + 8|}{5} = 0$

2. $d = \dfrac{|27 + 12 + 8|}{5} = \dfrac{47}{5}$

3. $d = \dfrac{|48 - 15 - 2|}{13} = \dfrac{31}{13}$

4. $d = \dfrac{|1 + 18 + 15|}{\sqrt{10}} = \dfrac{34}{\sqrt{10}}$

5. $d = \dfrac{|8 - 42 + 15|}{\sqrt{10}} = \dfrac{19}{\sqrt{10}}$

6. $d = \dfrac{|8 - 25|}{\sqrt{29}} = \dfrac{17}{\sqrt{29}}$

7. $d = \dfrac{|1 + 0 + 1 - 1|}{\sqrt{3}} = \dfrac{\sqrt{3}}{3}$

8. $d = \dfrac{|-10|}{\sqrt{4 + 9 + 25}} = \dfrac{10}{\sqrt{38}}$

9. $d = \dfrac{|1 - 1 - 2 - 4|}{\sqrt{6}} = \sqrt{6}$

10. $d = \dfrac{|6 - 4 - 2 + 1|}{\sqrt{26}} = \dfrac{\sqrt{26}}{26}$

11. $d = \dfrac{|2a^2 + a + 2a^2 - 4a|}{\sqrt{4a^2 + 1 + a^2}} = \dfrac{|4a^2 - 3a|}{\sqrt{5a^2 + 1}}$

12. $d = \dfrac{|3a - 4a + 3a + \frac{1}{a}|}{\sqrt{9 + 4 + 1}} = \dfrac{|2a + \frac{1}{a}|}{\sqrt{14}} = \dfrac{2a^2 + 1}{|a|\sqrt{14}}$

13. $\mathbf{N} = \mathbf{AC} \times \mathbf{AC} = \begin{vmatrix} \mathbf{i} & \mathbf{j} & \mathbf{k} \\ 1 & 2 & 4 \\ -2 & -1 & 1 \end{vmatrix} = 6\mathbf{i} - 9\mathbf{j} + 3\mathbf{k}$

The equation of the plane containing the
given points is $2x - 3y + z = 0$, so

$$d = \frac{|-2 - 6 + 1|}{\sqrt{4 + 9 + 1}} = \frac{7}{\sqrt{14}}$$

14. The equation of the plane normal to the given
vector is $2x - y + 2z - 4 = 0$, so

$$d = \frac{|-2 - 2 + 2 - 4|}{\sqrt{4 + 1 + 4}} = \frac{6}{3} = 2$$

15. The equation of the plane normal to the given
vector is $3x + y + 5z - (-9 + 5 + 5) = 0$, so

$$d = \frac{|-3 + 2 + 5 - 1|}{\sqrt{9 + 1 + 25}} = \frac{3}{\sqrt{35}}$$

16. $\mathbf{N} = \mathbf{AB} \times \mathbf{AC} = \begin{vmatrix} \mathbf{i} & \mathbf{j} & \mathbf{k} \\ 5 & 2 & 6 \\ 4 & -2 & -1 \end{vmatrix}$
$= 10\mathbf{i} + 29\mathbf{j} - 18\mathbf{k}$

The equation of the plane containing the
given points is $10x + 29y - 18z - 1 = 0$, so

$$d = \frac{|-10 + 58 - 18 - 1|}{\sqrt{100 + 29^2 + 18^2}} = \frac{29}{\sqrt{1{,}265}}$$

17. Let $Q(2, -1, 1)$ be a point on L; then

$\mathbf{PQ} \times \mathbf{v} = \begin{vmatrix} \mathbf{i} & \mathbf{j} & \mathbf{k} \\ 1 & -1 & 2 \\ 3 & 1 & 2 \end{vmatrix} = -4\mathbf{i} + 4\mathbf{j} + 4\mathbf{k}$

$$d = \frac{\|\mathbf{PQ} \times \mathbf{v}\|}{\|\mathbf{v}\|} = \frac{\sqrt{(-4)^2 + 4^2 + 4^2}}{\sqrt{9 + 1 + 4}} = \frac{4\sqrt{3}}{\sqrt{14}}$$

18. Let $Q(0, 1, 0)$ be a point on L; then

$\mathbf{PQ} \times \mathbf{v} = \begin{vmatrix} \mathbf{i} & \mathbf{j} & \mathbf{k} \\ -1 & 1 & -1 \\ 3 & 2 & 1 \end{vmatrix} = 3\mathbf{i} - 2\mathbf{j} - 5\mathbf{k}$

$$d = \frac{\|\mathbf{PQ} \times \mathbf{v}\|}{\|\mathbf{v}\|} = \frac{\sqrt{3^2 + (-2)^2 + (-5)^2}}{\sqrt{9 + 4 + 1}} = \frac{\sqrt{38}}{\sqrt{14}}$$

19. Let $Q(0, 0, 0)$ be a point on L; then

$$\mathbf{PQ} \times \mathbf{v} = \begin{vmatrix} \mathbf{i} & \mathbf{j} & \mathbf{k} \\ -1 & 2 & -2 \\ 1 & -\frac{1}{2} & -1 \end{vmatrix} = -\mathbf{i} - 3\mathbf{j} - \frac{5}{2}\mathbf{k}$$

$$d = \frac{\|\mathbf{PQ} \times \mathbf{v}\|}{\|\mathbf{v}\|} = \frac{\sqrt{(-1)^2 + (-3)^2 + (-\frac{5}{2})^2}}{\sqrt{1 + \frac{1}{4} + 1}}$$

$$= \frac{\sqrt{65}}{3}$$

20. Let $Q(\frac{1}{2}, 0, 0)$ be a point on L; then

$$\mathbf{PQ} = \frac{1}{2}\mathbf{i} - \mathbf{j} + \mathbf{k} = \frac{1}{2}(\mathbf{i} - 2\mathbf{j} + 2\mathbf{k});$$

$$\mathbf{PQ} \times \mathbf{v} = \frac{1}{2} \begin{vmatrix} \mathbf{i} & \mathbf{j} & \mathbf{k} \\ 1 & -2 & 2 \\ 1 & -2 & -1 \end{vmatrix} = \frac{1}{2}(-2\mathbf{i} + 3\mathbf{j} + 4\mathbf{k})$$

$$d = \frac{\|\mathbf{PQ} \times \mathbf{v}\|}{\|\mathbf{v}\|} = \frac{\sqrt{4 + 9 + 16}}{2\sqrt{6}} = \frac{\sqrt{29}}{2\sqrt{6}}$$

21. Let $Q(-a, a, a)$ be a point on L; then

$$\mathbf{PQ} = -2a\mathbf{i} + a\mathbf{j} + 2a\mathbf{k} = a(-2\mathbf{i} + \mathbf{j} + 2\mathbf{k})$$

$$\mathbf{PQ} \times \mathbf{v} = a \begin{vmatrix} \mathbf{i} & \mathbf{j} & \mathbf{k} \\ -2 & 1 & 2 \\ 2 & 1 & 2 \end{vmatrix} = 4a(2\mathbf{j} - \mathbf{k})$$

$$d = \frac{\|\mathbf{PQ} \times \mathbf{v}\|}{\|\mathbf{v}\|} = \frac{4|a|\sqrt{5}}{3}$$

22. Let $Q(a, 0, -4a)$ be a point on L; then

$$\mathbf{PQ} = \frac{a}{2}(2\mathbf{i} - 2\mathbf{j} - 9\mathbf{k})$$

$$\mathbf{PQ} \times \mathbf{v} = a \begin{vmatrix} \mathbf{i} & \mathbf{j} & \mathbf{k} \\ 2 & -2 & -9 \\ 1 & 1 & 1 \end{vmatrix} = \frac{a}{2}(7\mathbf{i} - 11\mathbf{j} + 4\mathbf{k})$$

$$d = \frac{\|\mathbf{PQ} \times \mathbf{v}\|}{\|\mathbf{v}\|} = \frac{a\sqrt{26}}{2\sqrt{3}}$$

23. $d = \dfrac{|2(-2) + 3(3) + (-6)(7) - 5|}{\sqrt{2^2 + 3^2 + (-6)^2}} = 6$

$$(x + 2)^2 + (y - 3)^2 + (z - 7)^2 = 36$$

24. a. A vector parallel to the given line is $\mathbf{v} = 3\mathbf{i} - 2\mathbf{j} + \mathbf{k}$ and the normal to the given plane is $\mathbf{N} = \mathbf{i} + 2\mathbf{j} + \mathbf{k}$.

$\mathbf{N} \cdot \mathbf{v} = 3 - 4 + 1 = 0$, so the vectors are orthogonal which means that the line is parallel to the given plane.

b. $d = \dfrac{|1 + 0 - 1 - 1|}{\sqrt{6}} = \dfrac{1}{\sqrt{6}}$

25. $\mathbf{QR} = 7\mathbf{i} - 6\mathbf{j} - 3\mathbf{k}$; $\mathbf{QS} = 3\mathbf{i} + \mathbf{j} - 8\mathbf{k}$

$$\mathbf{QR} \times \mathbf{QS} = \begin{vmatrix} \mathbf{i} & \mathbf{j} & \mathbf{k} \\ 7 & -6 & -3 \\ 3 & 1 & -8 \end{vmatrix} = 51\mathbf{i} + 47\mathbf{j} + 25\mathbf{k}$$

$$\|\mathbf{QR} \times \mathbf{QS}\| = \sqrt{51^2 + 47^2 + 25^2} = \sqrt{5{,}435}$$

26. $\|\mathbf{P_0P}\|^2 = (x + 1)^2 + (y - 2)^2 + (z - 4)^2$

$$d^2 = \frac{|2x - 5y + 3z - 7|^2}{4 + 25 + 9}$$

Thus,

$$(x + 1)^2 + (y - 2)^2 + (z - 4)^2$$
$$= \frac{1}{38}(2x - 5y + 3z - 7)^2$$

27. We need the distance from the point to the line to equal 5.

$$d = \frac{\|\mathbf{v} \times \mathbf{QP}\|}{\|\mathbf{v}\|} = \left| \frac{Ax_0 + By_0 + Cz_0 + D}{\sqrt{A^2 + B^2 + C^2}} \right| = 5$$

$\mathbf{v} = \langle 4, -1, 3 \rangle$, and $Q(1, -1, 0)$ is on the line,
$\mathbf{QP} = \langle x - 1, y + 1, z \rangle$

$$\mathbf{v} \times \mathbf{QP} = \begin{vmatrix} \mathbf{i} & \mathbf{j} & \mathbf{k} \\ 4 & -1 & 3 \\ x-1 & y+1 & z \end{vmatrix}$$

$$= \langle -3(y+1) - z, \ 3(x-1) - 4z, \ 4(y+1) + (x-1) \rangle$$

Now, $\dfrac{\|\mathbf{v} \times \mathbf{QP}\|}{\|\mathbf{v}\|}$

$$= \frac{\sqrt{(3y + z + 3)^2 + (3x - 4z - 3)^2 + (x + 4y + 3)^2}}{\sqrt{26}}$$

$= 5$; thus,

$$(3y + z + 3)^2 + (3x - 4z - 3)^2 + (x + 4y + 3)^2 = 650$$

28. Let $\mathbf{v}_1 = 3\mathbf{i} - 2\mathbf{j} + \mathbf{k}$ and $\mathbf{v}_2 = 5\mathbf{i} + \mathbf{j} + 3\mathbf{k}$ be vectors associated with the given lines, and let $P_1(-1, 2, 1)$ and $P_2(2, -1, 0)$ be points on the two given lines, respectively.
$\mathbf{P_1P_2} = 3\mathbf{i} - 3\mathbf{j} - \mathbf{k}$;

$$\mathbf{N} = \mathbf{v}_1 \times \mathbf{v}_2 = \begin{vmatrix} \mathbf{i} & \mathbf{j} & \mathbf{k} \\ 3 & -2 & 1 \\ 5 & 1 & 3 \end{vmatrix} = -7\mathbf{i} - 4\mathbf{j} + 13\mathbf{k}$$

$$d = \frac{|\mathbf{P_1P_2} \cdot \mathbf{N}|}{\|\mathbf{N}\|} = \frac{|21 - 12 + 13|}{\sqrt{49 + 16 + 169}} = \frac{22}{\sqrt{234}}$$

29. Let $\mathbf{v}_1 = -\mathbf{i} + 2\mathbf{j} + 3\mathbf{k}$ and $\mathbf{v}_2 = 2\mathbf{i} - \mathbf{j} + 2\mathbf{k}$ be vectors associated with the given lines, and let $P_1(2, 5, 0)$ and $P_2(0, -1, 1)$ be points on the two given lines, respectively.
$\mathbf{P_1P_2} = -2\mathbf{i} - 6\mathbf{j} + \mathbf{k}$;

$$\mathbf{N} = \mathbf{v}_1 \times \mathbf{v}_2 = \begin{vmatrix} \mathbf{i} & \mathbf{j} & \mathbf{k} \\ -1 & 2 & 3 \\ 2 & -1 & 2 \end{vmatrix} = 7\mathbf{i} + 8\mathbf{j} - 3\mathbf{k}$$

$$d = \frac{|\mathbf{P_1P_2} \cdot \mathbf{N}|}{\|\mathbf{N}\|} = \frac{|-14 - 48 - 3|}{\sqrt{49 + 64 + 9}} = \frac{65}{\sqrt{122}}$$

30. Let $v_1 = i + 2j + 3k$ be associated with the given line and let $v_2 = -i - 2j + k$ be the vector determined by the points $A(0, -2, 1)$ and $B(1, -2, 3)$. Let $P(-1, 3, -2)$ be a point on the first line. Then
$BP = -2i + 5j - k$;
$$N = v_1 \times v_2 = \begin{vmatrix} i & j & k \\ 1 & 2 & 3 \\ -1 & 2 & -1 \end{vmatrix} = -4i - 2j$$
$$d = \frac{|BP \cdot N|}{\|N\|} = \frac{|8 - 10|}{\sqrt{16 + 4}} = \frac{2}{\sqrt{20}} = \frac{1}{\sqrt{5}}$$

31. Let $v_1 = i - 2j$ be associated with the given line and let $v_2 = i - j + k$ be the vector determined by the points $A(0, -1, 2)$ and $B(1, -2, 3)$. Let $P(-1, 0, 3)$ be a point on the first line. Then $BP = -2i + 2j$;
$$N = v_1 \times v_2 = \begin{vmatrix} i & j & k \\ 1 & -2 & 0 \\ 1 & -1 & 1 \end{vmatrix} = -2i - j + k$$
$$d = \frac{|BP \cdot N|}{\|N\|} = \frac{|4 - 2|}{\sqrt{4 + 1 + 1}} = \frac{2}{\sqrt{6}} = \tfrac{1}{3}\sqrt{6}$$

32. $N = v_1 \times v_2$ defines a direction normal to both lines.
$$d = \|P_1 P_2\| \cos \theta$$
$$= \|P_1 P_2\| \frac{|P_1 P_2 \cdot N|}{\|P_1 P_2\| \|N\|} = \frac{|P_1 P_2 \cdot N|}{\|N\|}$$

33. a. The distance from the origin to the first plane is
$$\frac{|D_1|}{\sqrt{A^2 + B^2 + C^2}}$$
and the distance form the second plane is
$$\frac{|D_2|}{\sqrt{A^2 + B^2 + C^2}}$$
Thus, the distance between the two planes is
$$d = \frac{|D_1 - D_2|}{\sqrt{A^2 + B^2 + C^2}}$$

b. $d = \dfrac{|4 - 2|}{\sqrt{1 + 1 + 4}} = \dfrac{2}{\sqrt{6}} = \tfrac{1}{3}\sqrt{6}$

CHAPTER 10 REVIEW

Proficiency Examination, Page 679

1. A vector is a directed line segment and a scalar is a real number.

2. If $v = \langle a_1, a_2, a_3 \rangle$, then $sv = \langle sa_1, sa_2, sa_3 \rangle$; geometrically, av is a $a\|v\|$ units long and points along v if $a > 0$ or in the opposite

direction if $a < 0$.

3. The triangular rule says that the sum of two vectors is a vector that extends from the initial point of the first vector to the terminal point of the second vector when the two given vectors are placed so that the initial point of the second vector coincides with the terminal point of the first vector. See Figure 10.3a, Page 632.

4. If two vectors are arranged to that their initial points coincide, then the sum of the vectors is the diagonal of the parallelogram formed by the two vectors. See Figure 10.3b, Page 632.

5. a. $u + v = v + u$
   b. $(u + v) + w = u + (v + w)$
   c. $(st)u = s(tu)$
   d. $u + 0 = u$
   e. $u + (-u) = 0$
   f. $(s + t)u = su + tu$
   g. $s(u + v) = su + sv$
   h. $v \cdot v = \|v\|^2$
   i. u is parallel to v if $u = tv$
   j. $v \cdot w = w \cdot v$
   k. $c(v \cdot w) = (cv) \cdot w = u \cdot (cw)$
   l. $u \cdot (v + w) = u \cdot v + u \cdot w$
   m. $(sv) \times (tw) = st(v \times w)$
   n. $v \times 0 = 0 \times v = 0$
   o. $v \times w = -(w \times v)$
   p. $u \times (v + w) = (u \times v) + (u \times w)$ or $(u + v) \times w = (u \times w) + (v \times w)$
   q. $v \times v = 0$; if $w = sv$, then $v \times w = 0$.

6. $\|u\| = \|ai + bj + ck\| = \sqrt{a^2 + b^2 + c^2}$

7. $\|u + v\| \le \|u\| + \|v\|$

8. i, j, k

9. $(x - a)^2 + (y - b)^2 + (z - c)^2 = r^2$

10. Draw a planar curve. Define a direction (generatrix). Move a line along the curve parallel to the generatrix. The resulting surface is a cylinder.

11. The distance $|P_1 P_2|$ between $P_1(x_1, y_1, z_1)$ and $P_2(x_2, y_2, z_2)$ is
$$|P_1 P_2| = \sqrt{(x_2 - x_1)^2 + (y_2 - y_1)^2 + (z_2 - z_1)^2}$$

12. $v = \frac{u}{\|u\|}$

13. u and v are parallel if $u = sv$ or $u \times v = 0$

14. The dot product of vectors $v = a_1 i + a_2 j + a_3 k$ and $w = b_1 i + b_2 j + b_3 k$ is the scalar denoted by $v \cdot w$ and given by

$$\mathbf{v} \cdot \mathbf{w} = a_1 b_1 + a_2 b_2 + a_3 b_3$$

Alternately, we can say, $\mathbf{u} \cdot \mathbf{v} = \|\mathbf{u}\| \|\mathbf{v}\| \cos \theta$ where θ is the angle between the vectors $\mathbf{u}$ and $\mathbf{v}$.

15. If θ is the angle between the nonzero vectors $\mathbf{v}$ and $\mathbf{w}$, then

$$\cos \theta = \frac{\mathbf{v} \cdot \mathbf{w}}{\|\mathbf{v}\| \|\mathbf{w}\|}$$

16. $\mathbf{u}$ and $\mathbf{v}$ are orthogonal vectors if the lines determined by those vectors are perpendicular.

17. The vector projection of $\mathbf{AB}$ onto $\mathbf{AC}$ is the vector from $\mathbf{A}$ to the projection of B onto a line through $\overline{AC}$. The formula for the vector projection of $\mathbf{v}$ in the direction of $\mathbf{w}$ is (a vector): $\left(\frac{\mathbf{v} \cdot \mathbf{w}}{\mathbf{w} \cdot \mathbf{w}}\right)\mathbf{w}$

18. The scalar projection of $\mathbf{AB}$ onto $\mathbf{AC}$ is the length of the vector from A to the projection of B onto a line through AC. The formula for the scalar projection of $\mathbf{v}$ onto $\mathbf{w}$ is (a number): $\left|\frac{\mathbf{v} \cdot \mathbf{w}}{\|\mathbf{w}\|}\right|$

19. $W = \mathbf{F} \cdot \mathbf{R}$

20. If $\mathbf{v} = a_1\mathbf{i} + a_2\mathbf{j} + a_3\mathbf{k}$ and $\mathbf{w} = b_1\mathbf{i} + b_2\mathbf{j} + b_3\mathbf{k}$, the cross product, written $\mathbf{v} \times \mathbf{w}$, is the vector

$$\mathbf{v} \times \mathbf{w} = (a_2 b_3 - a_3 b_2)\mathbf{i} + (a_3 b_1 - a_1 b_3)\mathbf{j}$$
$$+ (a_1 b_2 - a_2 b_1)\mathbf{k}$$

These terms can be obtained by using a determinant

$$\mathbf{v} \times \mathbf{w} = \begin{vmatrix} \mathbf{i} & \mathbf{j} & \mathbf{k} \\ a_1 & a_2 & a_3 \\ b_1 & b_2 & b_3 \end{vmatrix}$$

21. To find the unit vector of $\mathbf{u} \times \mathbf{v}$ place the fingers of the right hand along vector $\mathbf{u}$, turn to $\mathbf{v}$ along the acute angle, then the direction of the thumb is the direction of $\mathbf{n}$.

22. If $\mathbf{v}$ and $\mathbf{w}$ are nonzero vectors in $\mathbb{R}^3$, that are not multiples of one another, then $\mathbf{v} \times \mathbf{w}$ is orthogonal to both $\mathbf{v}$ and $\mathbf{w}$.

23. If $\mathbf{v}$ and $\mathbf{w}$ are nonzero vectors in $\mathbb{R}^3$ with θ the angle between $\mathbf{v}$ and $\mathbf{w}$ ($0 \leq \theta \leq \pi$), then $\|\mathbf{v} \times \mathbf{w}\| = \|\mathbf{v}\| \|\mathbf{w}\| \sin \theta$

24. If $\mathbf{u} = a_1\mathbf{i} + a_2\mathbf{j} + a_3\mathbf{k}$, $\mathbf{v} = b_1\mathbf{i} + b_2\mathbf{j} + b_3\mathbf{k}$, and $\mathbf{w} = c_1\mathbf{i} + c_2\mathbf{j} + c_3\mathbf{k}$, then the triple scalar product can be found by evaluating the determinant

$$(\mathbf{u} \times \mathbf{v}) \cdot \mathbf{w} = \begin{vmatrix} a_1 & a_2 & a_3 \\ b_1 & b_2 & b_3 \\ c_1 & c_2 & c_3 \end{vmatrix}$$

25. The volume of a parallelepiped formed by $\mathbf{u}$, $\mathbf{v}$, and $\mathbf{w}$ is the absolute value of the triple scalar product, $(\mathbf{u} \times \mathbf{v}) \cdot \mathbf{w}$.

26. $x = x_0 + tA,\ y = y_0 + tB,\ z = z_0 + tC$

27. $\dfrac{x - x_0}{A} = \dfrac{y - y_0}{B} = \dfrac{z - z_0}{C}$

28. If $\mathbf{v} = a_1\mathbf{i} + a_2\mathbf{j} + a_3\mathbf{k}$ is a nonzero vector, then the direction cosines of $\mathbf{v}$ are

$$\cos \alpha = \frac{a_1}{\|\mathbf{v}\|}; \qquad \cos \beta = \frac{a_2}{\|\mathbf{v}\|}; \qquad \cos \gamma = \frac{a_3}{\|\mathbf{v}\|}$$

29. The plane $Ax + By + Cz + D = 0$ has normal vector $\mathbf{N} = A\mathbf{i} + B\mathbf{j} + C\mathbf{k}$

30. $A(x - x_0) + B(y - y_0) + C(z - z_0) = 0$

31. $Ax + By + Cz + D = 0$

32. The distance from the point (x_0, y_0, z_0) to the plane $Ax + By + Cz + D = 0$ is given by

$$d = \left| \frac{Ax_0 + By_0 + Cz_0 + D}{\sqrt{A^2 + B^2 + C^2}} \right|$$

33. The shortest distance from the point P to the line L is given by the formula

$$d = \frac{\|\mathbf{v} \times \mathbf{QP}\|}{\|\mathbf{v}\|}$$

where $\mathbf{v}$ is a vector aligned with L and Q is any point on L.

34. a.
$$2\mathbf{v} + 3\mathbf{w} = [2(2) + 3(3)]\mathbf{i}$$
$$+ [2(-3) + 3(-2)]\mathbf{j}$$
$$+ [2(1) + 3(0)]\mathbf{k}$$
$$= 13\mathbf{i} - 12\mathbf{j} + 2\mathbf{k}$$

b.
$$\|\mathbf{v}\|^2 - \|\mathbf{w}\|^2 = [2^2 + (-3)^2 + 1^2]$$
$$- [3^2 + (-2)^2] = 14 - 13 = 1$$

c. $\left(\dfrac{\mathbf{v} \cdot \mathbf{w}}{\mathbf{w} \cdot \mathbf{w}}\right)\mathbf{w} = \dfrac{12}{13}(3\mathbf{i} - 2\mathbf{j})$

d. $\left|\dfrac{\mathbf{w} \cdot \mathbf{v}}{\|\mathbf{v}\|}\right| = \dfrac{12}{\sqrt{14}} = \dfrac{6\sqrt{14}}{7}$

35. a. $\mathbf{v} \cdot \mathbf{w} = 2(0) + (-5)(1) + 1(-3) = -8$

b. $\begin{vmatrix} \mathbf{i} & \mathbf{j} & \mathbf{k} \\ 2 & -5 & 1 \\ 0 & 1 & -3 \end{vmatrix} = 14\mathbf{i} + 6\mathbf{j} + 2\mathbf{k}$

36. a. $\begin{vmatrix} \mathbf{i} & \mathbf{j} & \mathbf{k} \\ 2 & -3 & 1 \\ 1 & 1 & -2 \end{vmatrix} \cdot (3\mathbf{i} + 5\mathbf{j})$

$$= (5\mathbf{i} + 5\mathbf{j} + 5\mathbf{k}) \cdot (3\mathbf{i} + 5\mathbf{k}) = 40$$

b. Not possible to take the cross product of a scalar and a vector.

c. $\begin{vmatrix} \mathbf{i} & \mathbf{j} & \mathbf{k} \\ 2 & -3 & 1 \\ 1 & 1 & -2 \end{vmatrix} \times (3\mathbf{i} + 5\mathbf{j})$

$$= (5\mathbf{i} + 5\mathbf{j} + 5\mathbf{k}) \times (3\mathbf{i} + 5\mathbf{k}) = 40$$

$$= \begin{vmatrix} \mathbf{i} & \mathbf{j} & \mathbf{k} \\ 5 & 5 & 5 \\ 3 & 0 & 5 \end{vmatrix} = 25\mathbf{i} - 10\mathbf{j} - 15\mathbf{k}$$

d. Not possible to take the dot product of a scalar and a vector.

37. Using Q and the vector $\mathbf{PQ} = \langle 1, -6, 4 \rangle$:
$$\frac{x}{1} = \frac{y+2}{-6} = \frac{z-1}{4}$$

38. Using point P and the direction numbers [2, 0, 3]:
$$2(x - 1) + 0(y - 1) + 3(z - 3) = 0$$
$$2x + 3z - 11 = 0$$

39. The direction numbers for each plane give a normal vector. The cross product of these vectors give the direction of the line of intersection. We can find a point by looking at intercepts.

$$\mathbf{N_1} \times \mathbf{N_2} = \begin{vmatrix} \mathbf{i} & \mathbf{j} & \mathbf{k} \\ 2 & 3 & 1 \\ 0 & 1 & -3 \end{vmatrix} = \langle -10, 6, 2 \rangle$$

If $z = 0$ in the second plane, $y = 5$. Use these values in the first plane to find $x = -\frac{13}{2}$. We now have a point on the line of intersection and its direction. The line is:

$$\frac{x + \frac{13}{2}}{-10} = \frac{y - 5}{6} = \frac{z}{2}$$

$$\frac{x + \frac{13}{2}}{5} = \frac{y - 5}{-3} = \frac{z}{-1}$$

40. Use P as the point, and find the direction numbers by finding $\mathbf{PQ} \times \mathbf{PR}$.
$$\mathbf{PQ} \times \mathbf{PR} = \begin{vmatrix} \mathbf{i} & \mathbf{j} & \mathbf{k} \\ 1 & -5 & 6 \\ 3 & -2 & -1 \end{vmatrix} = \langle 17, 19, 13 \rangle$$
The plane is:
$$17x + 19(y - 2) + 13(z + 1) = 0$$
$$17x + 19y + 13z - 25 = 0$$

41. The direction numbers for the line: [2, -3, 3]. Using the point $(1, 2, -1)$:
$$\frac{x-1}{2} = \frac{y-2}{-3} = \frac{z+1}{3}$$

42. $\cos \alpha = \dfrac{-2}{\sqrt{(-2)^2 + 3^2 + 1^2}} = \dfrac{-2}{\sqrt{14}};$ so
$$\alpha \approx 2.13 \text{ or } 122°$$

$\cos \beta = \dfrac{3}{\sqrt{14}};$ so $\beta \approx 0.64$ or $37°$

$\cos \gamma = \dfrac{1}{\sqrt{14}};$ so $\gamma \approx 1.30$ or $74°$

43. a. The direction numbers are not scalar multiples, so they are not parallel. If $z = 3$, $t = \frac{3}{2}$, $x = 0 \neq -2$. They are skew.

b. The direction numbers are not scalar multiples, so they are not parallel. To determine if they intersect we need to solve a system of equations.
1st line:
$$x = 7 + 5t_1, \; y = 6 + 4t_1, \; z = 8 + 5t_1$$
2nd line:
$$x = 8 + 6t_2, \; y = 6 + 4t_2, \; z = 9 + 6t_2$$
Solving these equations simultaneous, we find $t_1 = t_2 = -1$. Both equations contain the point $(2, 2, 3)$.

44. a. $V = \| (\mathbf{u} \times \mathbf{v}) \cdot \mathbf{w} \|$

$$= \begin{vmatrix} \mathbf{i} & \mathbf{j} & \mathbf{k} \\ 2 & 1 & 0 \\ 1 & -1 & -1 \end{vmatrix} \cdot \langle 3, 0, 1 \rangle$$
$$= |\langle -1, 2, -3 \rangle \cdot \langle 3, 0, 1 \rangle| = 6$$

b. Volume of a tetrahedron is $\frac{1}{6} |(\mathbf{u} \times \mathbf{v}) \cdot \mathbf{w}|$
$$2 = \frac{1}{6} A^2 |(\mathbf{u} \times \mathbf{v}) \cdot \mathbf{w}|$$
$$12 = A^2 \begin{vmatrix} \mathbf{i} & \mathbf{j} & \mathbf{k} \\ 2 & 1 & 0 \\ 1 & -1 & -1 \end{vmatrix} \cdot (3\mathbf{i} + \mathbf{k})$$
$$12 = A^2(6)$$
$$A^2 = 2$$
$$A = \sqrt{2}$$

45. The distance from a point to a plane is given by
$$d = \frac{\| \mathbf{PQ} \times \mathbf{N} \|}{\| \mathbf{v} \|}$$
$$= \left| \frac{2(-1) + 5(1) + (-1)(4) - 3}{\sqrt{2^2 + 5^2 + (-1)^2}} \right|$$
$$= \frac{4}{\sqrt{30}} = \frac{2\sqrt{30}}{15}$$

46. The distance must be measured perpendicular to both lines. That will be in the direction of their cross product.

$$\mathbf{v_1} \times \mathbf{v_2} = \begin{vmatrix} \mathbf{i} & \mathbf{j} & \mathbf{k} \\ 1 & 2 & 3 \\ -1 & 1 & 1 \end{vmatrix} = -\mathbf{i} - 4\mathbf{j} + 3\mathbf{k}$$

Now if $\mathbf{P_1 P_2}$ is a vector joining a point on the

first line to a point on the second line, then the length of the common normal is the component of P_1P_2 in the direction of the normal.

$$d = \frac{|n \cdot P_1P_2|}{\|n\|}$$

$(0, 0, -1)$ is a point on the first line and $(1, 2, 0)$ is a point on the second line, so $P_1P_2 = i + 2j + k$.

$$d = \frac{|\langle -1, -4, 3 \rangle \cdot \langle 1, 2, 1 \rangle|}{\sqrt{(-1)^2 + (-4)^2 + 3^2}} = \frac{6}{\sqrt{26}} = \frac{3\sqrt{26}}{13}$$

47. The distance from a point to a line is given by:

$$d = \frac{\|v \times QP\|}{\|v\|},$$ where v is a vector in the direction of the line, P is the given point, and Q is any point on the line.

$$v \times QP = \begin{vmatrix} i & j & k \\ 3 & 5 & -1 \\ -2 & -5 & -1 \end{vmatrix} = -10i + 5j - 5k$$

$$\|v \times QP\| = \sqrt{(-10)^2 + 5^2 + (-5)^2} = 5\sqrt{6}$$

$$\|v\| = \sqrt{3^2 + 5^2 + (-1)^2} = \sqrt{35}$$

$$d = \frac{5\sqrt{6}}{\sqrt{35}} = \frac{\sqrt{210}}{7}$$

48. The path of the plane is represented by the sum of the vector of the plane, $\langle 0, -200 \rangle$, and the vector of the wind, $\langle 25\sqrt{2}, 25\sqrt{2}, \rangle$. Path: $\langle 25\sqrt{2}, 25\sqrt{2} - 200 \rangle$. The magnitude of this vector is the ground speed:

$$\sqrt{(25\sqrt{2})^2 + (25\sqrt{2} - 200)^2} \approx 168.4 \text{ mph}$$

49. The work performed on the sled is the horizontal component of the force times the displacement. In vector terms:

$$W = F \cdot PQ = \left\langle \frac{3\sqrt{3}}{2}, \frac{3}{2} \right\rangle \cdot \langle 50, 0 \rangle$$
$$= 75\sqrt{3} \approx 130 \text{ ft} \cdot \text{lbs}$$

Supplementary Problems, Page 680

1. $\|AB\| + \|AC\| + \|BC\| = \sqrt{18} + \sqrt{14} + \sqrt{50}$
$$= 8\sqrt{2} + \sqrt{14}$$

2. $AB = i - j + 4k,\ AC = i - 2j - 3k;$
$$A = \tfrac{1}{2}\|AB \times AC\|$$
$$= \frac{1}{2}\begin{vmatrix} i & j & k \\ 1 & -1 & 4 \\ 1 & -2 & -3 \end{vmatrix} = \tfrac{1}{2}\|11i + 7j - k\|$$
$$= \tfrac{1}{2}\sqrt{121 + 49 + 1} = \tfrac{1}{2}\sqrt{171}$$

3. $AB = i - j + 4k,\ AC = i - 2j - 3k;$
$BC = -j - 7k$

$$\cos A = \frac{AB \cdot AC}{\|AB\|\|AC\|} = \frac{1 + 2 - 12}{\sqrt{18}\sqrt{14}}; A \approx 125°$$

$$\cos B = \frac{BA \cdot BC}{\|BA\|\|BC\|} = \frac{-1 + 28}{\sqrt{18}\sqrt{50}}; B \approx 26°$$

$$\cos C = \frac{CA \cdot CB}{\|CA\|\|CB\|} = \frac{2 + 21}{\sqrt{14}\sqrt{50}}; C \approx 30°$$

4. $AD = pi + (p-2)j + k;$
$BD = (p-1)i + (p-1)j - 3k$
$CD = (p-1)i + pj + 4k$

Volume of a tetrahedron is $\tfrac{1}{6}|(AB \times BD) \cdot CD|$

$$= \frac{1}{6}\begin{vmatrix} p & p-2 & 1 \\ p-1 & p-1 & -3 \\ p-1 & p & 4 \end{vmatrix} = \tfrac{1}{6}(18p + 15)$$

Since the volume is 100 cubic units, we have

$$3p + \frac{15}{6} = 100$$
$$p = \frac{615}{18} = \frac{205}{6}$$

5. $\dfrac{x - 1}{3} = \dfrac{y + 2}{1} = \dfrac{z - 3}{-1};$

$x = 1 + 3t,\ y = -2 + t,\ z = 3 - t$

If $t = -1$, $P_1(-2, -3, 4)$, and if $t = 2$, $P_2(7, 0, 1)$

6. $\dfrac{x - 1}{-5} = \dfrac{y}{3} = \dfrac{z + 3}{1};$

$x = 1 - 5t,\ y = 3t,\ z = -3 + t$

If $t = -1$, $P_1(6, -3, -4)$, and if $t = 2$, $P_2(-9, 6, -1)$

7. $\dfrac{x - 1}{2} = \dfrac{z}{1}$ and $y = 4$

$x = 1 + 2t,\ y = 4,\ z = t$

If $t = 1$, $P_1(3, 4, 1)$, and if $t = -1$, $P_2(-1, 4, -1)$

8. $\dfrac{x - 1}{1} = \dfrac{y - 2}{5} = \dfrac{z + 5}{-7};$

$x = 1 + t,\ y = 2 + 5t,\ z = -5 - 7t$

If $t = 1$, $P_1(2, 7, -12)$ and if $t = -1$, $P_2(0, -3, 2)$

9. $v_1 = i + 2j + 2k;\ v_2 = 2i + j - 3k;$

$$\mathbf{v} = \begin{vmatrix} \mathbf{i} & \mathbf{j} & \mathbf{k} \\ 1 & 2 & -3 \\ 2 & 1 & -3 \end{vmatrix} = -8\mathbf{i} + 7\mathbf{j} - 3\mathbf{k}$$

$$L: \frac{x-3}{8} = \frac{y-4}{-7} = \frac{z+1}{3}$$

10. $\mathbf{v}_1 = \mathbf{i} + 3\mathbf{j} - \mathbf{k}$; $\mathbf{v}_2 = 3\mathbf{i} - \mathbf{j} - \mathbf{k}$;

$$\mathbf{v} = \begin{vmatrix} \mathbf{i} & \mathbf{j} & \mathbf{k} \\ 1 & 3 & -1 \\ 3 & -1 & -1 \end{vmatrix} = -4\mathbf{i} - 2\mathbf{j} - 10\mathbf{k}$$

$$L: \frac{x+1}{2} = \frac{y}{1} = \frac{z+2}{5}$$

11. $\mathbf{v}_1 = 2\mathbf{i} + 3\mathbf{j} - 4\mathbf{k}$; $\mathbf{v}_2 = \mathbf{i} - 5\mathbf{j} + 2\mathbf{k}$;

$$\mathbf{v} = \begin{vmatrix} \mathbf{i} & \mathbf{j} & \mathbf{k} \\ 2 & 3 & -4 \\ 1 & -5 & 2 \end{vmatrix} = -14\mathbf{i} - 8\mathbf{j} - 13\mathbf{k}$$

$$L: \frac{x-2}{14} = \frac{y-3}{8} = \frac{z+1}{13}$$

12. $\mathbf{v}_1 = \mathbf{i} - \mathbf{j} - \mathbf{k}$; $\mathbf{v}_2 = 2\mathbf{i} + 3\mathbf{j} - \mathbf{k}$;

$$\mathbf{v} = \begin{vmatrix} \mathbf{i} & \mathbf{j} & \mathbf{k} \\ 2 & 3 & -1 \\ 2 & 3 & -1 \end{vmatrix} = 4\mathbf{i} - \mathbf{j} + 5\mathbf{k}$$

$$L: \frac{x-5}{4} = \frac{y-1}{-1} = \frac{z+3}{5}$$

13. $z = 0$ **14.** $z = 7$

15. $3(x-1) + 4(y+3) - (z-4) = 0$

$$3x + 4y - z + 13 = 0$$

16. $4(x+1) + 4(y-4) - 3(z-5) = 0$

$$4x + 4y - 3z + 3 = 0$$

17. $5(x-4) - 2(y+3) + 3(z-2) = 0$

$$5x - 2y + 3z - 32 = 0$$

18. $\mathbf{v}_1 = 4\mathbf{i} + 2\mathbf{k}$; $\mathbf{v}_2 = 3\mathbf{i} + \mathbf{j} - 2\mathbf{k}$;

$$\mathbf{N} = \begin{vmatrix} \mathbf{i} & \mathbf{j} & \mathbf{k} \\ 4 & 0 & 2 \\ 3 & 1 & -2 \end{vmatrix} = 2(-\mathbf{i} + 7\mathbf{j} + 2\mathbf{k})$$

$$-(x-3) + 7(y+2) + 2(z-1) = 0$$

$$x - 7y - 2z - 15 = 0$$

19. Let $P(4, 1, 3)$, $Q(-4, 2, 1)$, $R(1, 0, 2)$
$\mathbf{PQ} = -8\mathbf{i} + \mathbf{j} - 2\mathbf{k}$; $\mathbf{PR} = -3\mathbf{i} - \mathbf{j} - \mathbf{k}$;

$$\mathbf{N} = \begin{vmatrix} \mathbf{i} & \mathbf{j} & \mathbf{k} \\ 8 & -1 & 2 \\ 3 & 1 & 1 \end{vmatrix} = -3\mathbf{i} - 2\mathbf{j} + 11\mathbf{k}$$

$$-3(x-1) - 2(y-0) + 11(z-2) = 0$$

$$3x + 2y - 11z + 19 = 0$$

20. $\mathbf{v}_1 = 3\mathbf{i} - \mathbf{j} + 2\mathbf{k}$; $\mathbf{v}_2 = \mathbf{i} + 2\mathbf{j} + 3\mathbf{k}$;

$$\mathbf{N} = \begin{vmatrix} \mathbf{i} & \mathbf{j} & \mathbf{k} \\ 3 & -1 & 2 \\ 1 & 2 & 3 \end{vmatrix} = -7\mathbf{i} - 7\mathbf{j} + 7\mathbf{k}$$

$$-7(x-4) - 7(y+1) + 7(z+2) = 0$$

$$x + y - z - 1 = 0$$

21. $\|\mathbf{v}\| = \sqrt{9 + 4 + 1} = \sqrt{14}$

$$\mathbf{v} - \mathbf{w} = (3\mathbf{i} - 2\mathbf{j} + \mathbf{k}) - (4\mathbf{i} + \mathbf{j} - 3\mathbf{k})$$

$$= -\mathbf{i} - 3\mathbf{j} + 4\mathbf{k}$$

$$2\mathbf{v} + 3\mathbf{w} = 2(3\mathbf{i} - 2\mathbf{j} + \mathbf{k}) + 3(4\mathbf{i} + \mathbf{j} - 3\mathbf{k})$$

$$= 18\mathbf{i} - \mathbf{j} - 7\mathbf{k}$$

22. $\|\mathbf{v}\| = 1$; $\mathbf{v} - \mathbf{w} = \mathbf{i} - \mathbf{j}$

$$2\mathbf{v} + 3\mathbf{w} = 2\mathbf{i} + 3\mathbf{j}$$

23. $\|\mathbf{v}\| = \sqrt{25 + 9 + 4} = \sqrt{38}$

$$\mathbf{v} - \mathbf{w} = (5\mathbf{i} - 3\mathbf{j} + 2\mathbf{k}) - (-\mathbf{i} + 2\mathbf{j} - 3\mathbf{k})$$

$$= 6\mathbf{i} - 5\mathbf{j} + 5\mathbf{k}$$

$$2\mathbf{v} + 3\mathbf{w} = 2(5\mathbf{i} - 3\mathbf{j} + 2\mathbf{k}) + 3(-\mathbf{i} + 2\mathbf{j} - 3\mathbf{k})$$

$$= 7\mathbf{i} - 5\mathbf{k}$$

24. $\|\mathbf{v}\| = \sqrt{25 + 16} = \sqrt{41}$

$$\mathbf{v} - \mathbf{w} = (5\mathbf{i} + 4\mathbf{k}) - (\mathbf{j} + 3\mathbf{k}) = 5\mathbf{i} - \mathbf{j} + \mathbf{k}$$

$$2\mathbf{v} + 3\mathbf{w} = 2(5\mathbf{i} + 4\mathbf{k}) + 3(\mathbf{j} + 3\mathbf{k})$$

$$= 10\mathbf{i} + 3\mathbf{j} + 17\mathbf{k}$$

25. $\mathbf{v} \cdot \mathbf{w} = 0 + 0 + 0 = 0$

$$\mathbf{v} \times \mathbf{w} = \begin{vmatrix} \mathbf{i} & \mathbf{j} & \mathbf{k} \\ 1 & 0 & 0 \\ 0 & 1 & 0 \end{vmatrix} = \mathbf{k}$$

26. $\mathbf{v} \cdot \mathbf{w} = 0 + 0 + 1 = 1$

$$\mathbf{v} \times \mathbf{w} = \begin{vmatrix} \mathbf{i} & \mathbf{j} & \mathbf{k} \\ 0 & 0 & 1 \\ 0 & 0 & 1 \end{vmatrix} = \mathbf{0}$$

27. $\mathbf{v} \cdot \mathbf{w} = 6 + 0 + 0 = 6$

$$\mathbf{v} \times \mathbf{w} = \begin{vmatrix} \mathbf{i} & \mathbf{j} & \mathbf{k} \\ 3 & 0 & 2 \\ 2 & 1 & 0 \end{vmatrix} = -2\mathbf{i} + 4\mathbf{j} + 3\mathbf{k}$$

28. $\mathbf{v} \cdot \mathbf{w} = 1 + 0 + 0 = 1$

$$\mathbf{v} \times \mathbf{w} = \begin{vmatrix} \mathbf{i} & \mathbf{j} & \mathbf{k} \\ 1 & -3 & 0 \\ 1 & 0 & 5 \end{vmatrix} = -15\mathbf{i} - 5\mathbf{j} + 3\mathbf{k}$$

29. $\mathbf{v} \cdot \mathbf{w} = 2 + 0 + 0 = 2$

$$\mathbf{v} \times \mathbf{w} = \begin{vmatrix} \mathbf{i} & \mathbf{j} & \mathbf{k} \\ 2 & 0 & 3 \\ 1 & -2 & 0 \end{vmatrix} = 6\mathbf{i} + 3\mathbf{j} - 4\mathbf{k}$$

30. $\mathbf{v} \cdot \mathbf{w} = 1 - 1 - 1 = -1$

$$\mathbf{v} \times \mathbf{w} = \begin{vmatrix} \mathbf{i} & \mathbf{j} & \mathbf{k} \\ 1 & -1 & 1 \\ 1 & 1 & -1 \end{vmatrix} = 2\mathbf{j} + 2\mathbf{k}$$

31. **a.** $5\mathbf{v} - 3\mathbf{w} = 20\mathbf{i} + 10\mathbf{j} + 5\mathbf{k} - 6\mathbf{i} - 3\mathbf{j} + 15\mathbf{k}$

$\qquad = 14\mathbf{i} + 7\mathbf{j} + 20\mathbf{k}$

b. $\|2\mathbf{v} - \mathbf{w}\| = \|8\mathbf{i} + 4\mathbf{j} + 2\mathbf{k} - 2\mathbf{i} - \mathbf{j} + 5\mathbf{k}\|$

$\qquad = \|6\mathbf{i} + 3\mathbf{j} + 7\mathbf{k}\| = \sqrt{36 + 9 + 49} = \sqrt{94}$

c. $\dfrac{(\mathbf{v} \cdot \mathbf{w})\mathbf{w}}{\mathbf{w} \cdot \mathbf{w}} = \dfrac{8 + 2 - 5}{4 + 1 + 25}(2\mathbf{i} + \mathbf{j} - 5\mathbf{k})$

$\qquad = \frac{1}{3}\mathbf{i} + \frac{1}{6}\mathbf{j} - \frac{5}{6}\mathbf{k}$

d. $\dfrac{\mathbf{v} \cdot \mathbf{w}}{\|\mathbf{v}\|} = \dfrac{5}{\sqrt{21}}$

32. $\mathbf{v} \cdot \mathbf{w} = 6 + 10 + 0 = 16$

$$\mathbf{v} \times \mathbf{w} = \begin{vmatrix} \mathbf{i} & \mathbf{j} & \mathbf{k} \\ 2 & 5 & 0 \\ 3 & 2 & -4 \end{vmatrix} = 20\mathbf{i} + 8\mathbf{j} - 11\mathbf{k}$$

33. $\mathbf{v} = \begin{vmatrix} \mathbf{i} & \mathbf{j} & \mathbf{k} \\ 2 & 1 & 0 \\ 1 & 1 & -3 \end{vmatrix} = -3\mathbf{i} + 6\mathbf{j} + \mathbf{k}$

$\cos \alpha = \dfrac{-3}{\sqrt{46}}; \ \cos \beta = \dfrac{6}{\sqrt{46}}; \ \cos \gamma = \dfrac{1}{\sqrt{46}}$

34. $\|\mathbf{w}\| = \sqrt{25 + 16 + 9} = \sqrt{50} = 5\sqrt{2}$

$\cos \alpha = \dfrac{-5}{5\sqrt{2}}; \ \alpha \approx 135°$

$\cos \beta = \dfrac{4}{5\sqrt{2}}; \ \beta \approx 56°$

$\cos \gamma = \dfrac{-3}{5\sqrt{2}}; \ \gamma \approx 115°$

35. $\mathbf{v} = \pm\left(\dfrac{3}{\sqrt{19}}\mathbf{i} + \dfrac{1}{\sqrt{19}}\mathbf{j} + \dfrac{3}{\sqrt{19}}\mathbf{k} \right)$

36. $\mathbf{v}_1 = 3\mathbf{i} - \mathbf{j} + 2\mathbf{k}; \ \mathbf{v}_2 = 2\mathbf{i} - \mathbf{j} - 2\mathbf{k}$

$\mathbf{v}_1 \cdot \mathbf{v}_2 = 6 + 1 - 4 = 3;$

$\cos \theta = \dfrac{|\mathbf{v}_1 \cdot \mathbf{v}_2|}{\|\mathbf{v}_1\| \|\mathbf{v}_2\|} = \dfrac{3}{\sqrt{14}\sqrt{9}}; \ \theta = 74°$

37. Let $\mathbf{v} = 3\mathbf{i} - 4\mathbf{j}$ and $\mathbf{w} = -\mathbf{i} - \mathbf{j} + \mathbf{k}$

$$A = \|\mathbf{v} \times \mathbf{w}\| = \left\|\begin{vmatrix} \mathbf{i} & \mathbf{j} & \mathbf{k} \\ 3 & -4 & 0 \\ -1 & -1 & 1 \end{vmatrix}\right\|$$

$\qquad = \|-4\mathbf{i} - 3\mathbf{j} - 7\mathbf{k}\|$

$\qquad = \sqrt{16\ 9 + 49} = \sqrt{74}$

38. $4x^2 + 4y^2 + 4z^2 + 12y - 4z + 1 = 0$

$4x^2 + 4(y^2 + 3y + \frac{9}{4}) + 4(z^2 - z + \frac{1}{4}) = -1 + 9 + 1$

$x^2 + (y + \frac{3}{2})^2 + (z - \frac{1}{2})^2 = \frac{9}{4}$

The center is $(0, -\frac{3}{2}, \frac{1}{2})$ and the radius is $\frac{3}{2}$.

39. If $x = 0$, $t = -2$ so $y = 14$ and $z = -10$; the desired point is $P(0, 14, -10)$. If $y = 0$, $t = 5$ so $x = 21$ and $z = 25$; the desired point is $Q(21, 0, 25)$. If $z = 0$, $t = 0$ so $x = 6$, $y = 10$; the desired point is $R(6, 10, 0)$.

40. If $x = 0$, $t = \frac{1}{3}$ so $y = \frac{10}{3}$ and $z = -\frac{4}{3}$; the desired point is $P(0, \frac{10}{3}, -\frac{4}{3})$. If $y = 0$, $t = -\frac{1}{2}$ so $x = -\frac{5}{2}$ and $z = -8$; the desired point is $Q(-\frac{5}{2}, 0, -8)$. If $z = 0$, $t = \frac{1}{2}$ so $x = \frac{1}{2}$, $y = 4$; the desired point is $R(\frac{1}{2}, 4, 0)$.

41. Simultaneously solve the system of equations representing the planes to find $P(3, -2, 1)$.

42. Simultaneously solve the system of equations representing the planes to find $P(-2, 3, 4)$.

43. $\mathbf{v}_1 = 3\mathbf{i} - 2\mathbf{j} + 6\mathbf{k}; \ \mathbf{v}_2 = \mathbf{i} - 3\mathbf{j} + 2\mathbf{k};$ these lines intersect because the distance between them is 0, as shown below. Let $P_1(-3, 0, 7)$ be a point on the first line and $Q(-6, -5, 1)$ be a point on the second line.

$$\mathbf{N} = \begin{vmatrix} \mathbf{i} & \mathbf{j} & \mathbf{k} \\ 3 & -2 & 6 \\ 1 & -3 & 2 \end{vmatrix} = 14\mathbf{i} - 7\mathbf{k}$$

$\mathbf{PQ} = -3\mathbf{i} - 5\mathbf{j} - 6\mathbf{k}; \ \mathbf{PQ} \cdot \mathbf{N} = 0$, so the equation of the plane is

$\qquad 2(x + 3) - (z - 7) = 0$

$\qquad\qquad 2x - z + 13 = 0$

44. $d = \dfrac{|2 - 1 - 2 - 4|}{\sqrt{4 + 1 + 1}} = \dfrac{5}{\sqrt{6}} = \dfrac{5\sqrt{6}}{6}$

45. $\mathbf{PQ} = -3\mathbf{i} + 2\mathbf{j} - 2\mathbf{k}; \ \mathbf{v} = 2\mathbf{i} + \mathbf{j} + 4\mathbf{k}$

$$\mathbf{v} \times \mathbf{PQ} = \begin{vmatrix} \mathbf{i} & \mathbf{j} & \mathbf{k} \\ 2 & 1 & 4 \\ -3 & 2 & -2 \end{vmatrix} = -10\mathbf{i} - 8\mathbf{j} + 7\mathbf{k}$$

$d = \dfrac{|\mathbf{v} \times \mathbf{PQ}|}{\|\mathbf{v}\|} = \dfrac{\sqrt{100 + 64 + 49}}{\sqrt{21}} = \dfrac{213}{\sqrt{21}} = \dfrac{\sqrt{497}}{7}$

46. $\mathbf{PQ} = 2\mathbf{i} + 3\mathbf{k}; \ W = |\mathbf{F} \cdot \mathbf{PQ}| = 10 + 3 = 13$

47. $W = |\mathbf{F} \cdot \mathbf{PQ}| = 100(25)\cos \frac{\pi}{6}$

$\qquad = \dfrac{2,500\sqrt{3}}{2} = 1,250\sqrt{3}$

48. Suppose the helicopter is at point $P_1(a, b, c)$ when it is first sighted. Then we have

$$\mathbf{AP_1} = (a - 7)\mathbf{i} + b\mathbf{j} + c\mathbf{k} = (-4\mathbf{i} + 2\mathbf{j} + 5\mathbf{k})s_1$$

$$\mathbf{BP_1} = a\mathbf{i} + (b - 4)\mathbf{j} + c\mathbf{k} = (3\mathbf{i} - 2\mathbf{j} + 5\mathbf{k})s_2$$

for constants s_1 and s_2, so that

From the first equation ($\mathbf{AP_1}$),

$$a - 7 = -4s_1; \ b = 2s_1; \ \text{and} \ c = 5s_1$$

From the second equation ($\mathbf{BP_1}$)

$$a = 3s_2; \ b - 4 = -2s_2; \ \text{and} \ c = 5s_2$$

Solving simultaneously, we find $s_1 = s_2 = 1$ and $a = 3$, $b = 2$, $c = 5$, so the first point of sighting is $P_1(3, 2, 5)$. Similarly, we find the second point of sighting to be $P_2(13, 7, 5.005)$. Boldfinger's helicopter travels

$$\| \mathbf{P_1 P_2} \| = \sqrt{(13 - 3)^2 + (7 - 2)^2 + (5.005 - 5)^2}$$

$$\approx 11.18 \ \text{thousand ft/min}$$

Suppose the spy intercepts Boldfinger at point $P(d, e, f)$. Then, since P lies on the line through $P_1(3, 2, 5)$ and $P_2(13, 7, 5.005)$, we have

$$d - 3 = (13 - 3)t; \ e - 2 = (7 - 2)t; \ \text{and}$$

$$f - 5 = (5.005 - 5)t \ \text{so that}$$

$$d = 3 + 10t, \ e = 2 + 5t, \ f = 5 + 5.005t$$

Since Boldfinger travels on the line through P_1 and P_2, we must have $d = 3 + 10s$, $e = 2 + 5s$, $f = 5 + 0.005s$ for some s. We have seen that Boldfinger travels at about 11.176 thousand ft/min and the spy travels at

$$\frac{150 \ \text{mi}}{\text{hr}} \ \frac{5,280 \ \text{ft}}{\text{mi}} \ \frac{1 \ \text{hr}}{60 \ \text{min}} = 13.2 \ \text{thousand ft/min}$$

Suppose Boldfinger reaches the intercept point P at t minutes after noon. Then, the distance from P_1 to P_2 is

$$\sqrt{(d - 3)^2 + (e - 2)^2 + (f - 5)^2} \approx 11.176t$$

$$\sqrt{(10s)^2 + (5s)^2 + (0.005)^2} \approx 11.176t$$

$$s \approx \frac{11.176t}{11.180} \approx 0.9996t$$

Since the spy travels $t - 10$ minutes to reach the same point, using the distance from Q to P, we have

$$\sqrt{(d - 0)^2 + (e - 0)^2 + (f - 1)^2} \approx 13.2(t - 10)$$

and since $d = 3 + 10s$, $e = 2 + 5s$, $f = 4 + 0.005s$

$$\sqrt{(3 + 10s)^2 + (2 + 5s)^2 + (4 + 0.005s)^2}$$

$$\approx 13.2t \approx 13.2\left(\frac{s}{0.9996} - 10\right)$$

Solving for s, we obtain $s \approx 67$, so

$$t \approx \frac{11.180s}{11.176} \approx 66.98$$

and the intercept point is at $(673, 337, 5335)$. Thus, the spy should travel from $Q(0, 0, 1)$ in the direction of the vector

$$\mathbf{v} = 673\mathbf{i} + 337\mathbf{j} + 5,335\mathbf{k}$$

49. $(\| \mathbf{v} \|\mathbf{w} + \| \mathbf{w} \|\mathbf{v}) \cdot (\| \mathbf{v} \|\mathbf{w} - \| \mathbf{w} \|\mathbf{v})$

$$= \| \mathbf{v} \|^2\mathbf{w} \cdot \mathbf{w} - \| \mathbf{v} \| \| \mathbf{w} \| \mathbf{w} \cdot \mathbf{v}$$

$$+ \| \mathbf{w} \| \| \mathbf{v} \|\mathbf{v} \cdot \mathbf{w} - \| \mathbf{w} \|^2\mathbf{v} \cdot \mathbf{v} = 0$$

Thus, $\| \mathbf{v} \|\mathbf{w} + \| \mathbf{w} \|\mathbf{v}$ and $\| \mathbf{v} \|\mathbf{w} - \| \mathbf{w} \|\mathbf{v}$ are perpendicular.

50. $\mathbf{v} = (\cos\theta)\mathbf{i} + (\sin\theta)\mathbf{j}$ and $\mathbf{w} = (\cos\phi)\mathbf{i} + (\sin\phi)\mathbf{j}$

$$\mathbf{v} \times \mathbf{w} = \begin{vmatrix} \mathbf{i} & \mathbf{j} & \mathbf{k} \\ \cos\theta & \sin\theta & 0 \\ \cos\phi & \sin\phi & 0 \end{vmatrix}$$

$$= (\cos\theta \sin\phi - \sin\theta \cos\phi)\mathbf{k}$$

The absolute value is the area of the parallelogram formed by $\mathbf{v}$ and $\mathbf{w}$. The same area is

$$A = \| \mathbf{v} \| \| \mathbf{w} \| = (1)(1) \sin(\phi - \theta)$$

Thus,

$$\cos\theta \sin\phi - \sin\theta \cos\phi = \sin(\phi - \theta)$$

51. Let $A(1, -2, 3)$, $B(-1, 2, -3)$, and $C(2, 1, -3)$; then $\mathbf{AB} = -2\mathbf{i} + 4\mathbf{j} - 6\mathbf{k}$, $\mathbf{AC} = \mathbf{i} + 3\mathbf{j} - 6\mathbf{k}$, and $\mathbf{BC} = 3\mathbf{i} - \mathbf{j}$

$$\cos\alpha = \frac{\mathbf{AB} \cdot \mathbf{AC}}{\| \mathbf{AB} \| \| \mathbf{AC} \|} = \frac{-1 + 6 + 18}{\sqrt{14}\sqrt{46}}; \ \alpha \approx 25°$$

$$\cos\beta = \frac{\mathbf{BA} \cdot \mathbf{BC}}{\| \mathbf{BA} \| \| \mathbf{BC} \|} = \frac{3 + 2}{\sqrt{14}\sqrt{10}}; \ \beta \approx 65°$$

$$\cos\gamma = \frac{\mathbf{CA} \cdot \mathbf{CB}}{\| \mathbf{CA} \| \| \mathbf{CB} \|} = \frac{3 - 3}{\sqrt{46}\sqrt{10}}; \ \gamma \approx 90°$$

52. $\mathbf{v} = a_1\mathbf{i} + b_1\mathbf{j} + c_1\mathbf{k}$; $\mathbf{w} = \mathbf{i} - 2\mathbf{j}$; $\mathbf{u} = 2\mathbf{i} + \mathbf{k}$

$$\cos\alpha = \frac{\mathbf{u} \cdot \mathbf{w}}{\| \mathbf{v} \| \| \mathbf{w} \|} = \frac{a_1 - 2b_1}{\sqrt{5}\sqrt{a_1^2 + b_1^2 + c_1^2}}$$

$$\cos\beta = \frac{\mathbf{u} \cdot \mathbf{v}}{\| \mathbf{u} \| \| \mathbf{w} \|} = \frac{2a_1 + c_1}{\sqrt{5}\sqrt{a_1^2 + a_2^2 + c_1^2}}$$

Since the angles are the same, we see

$$a_1 - 2b_1 = 2a_1 + c_1$$

$$a_1 + 2b_1 + c_1 = 0$$

53. Let $P(a, 0, 0)$, $Q(0, a, 0)$, $R(a, 0, a)$; then $\mathbf{PQ} = -a\mathbf{i} + a\mathbf{j}$; $\mathbf{PR} = a\mathbf{k}$

$$N = \begin{vmatrix} i & j & k \\ -a & a & 0 \\ 0 & 0 & a \end{vmatrix} = a^2(i + j)$$

The equation of the plane is

$$(x - a) + y = 0 \text{ or } x + y - a = 0$$

54. Let $P(b, -b, 0)$, $Q(0, b, -b)$, $R(-b, 0, b)$;
then $\mathbf{PQ} = -bi + 2bj - bk$;
$\mathbf{PR} = -2bi + bj + bk$

$$N = \begin{vmatrix} i & j & k \\ -b & 2b & -b \\ -2b & b & b \end{vmatrix} = 3b^2(i + j + k)$$

The equation of the plane is

$$(x - b) + (y + b) + z = 0 \text{ or } x + y + z = 0$$

55. The vertices of the triangle $A(0, -2, 1)$,
$B(1, 2, -1)$, and $C(3, -1, 2)$.
$\mathbf{AB} = i + 3j - 3k$, and $\mathbf{AC} = 3i$.
$$A = \tfrac{1}{2}\sqrt{(1 + 9 + 9)(9) - 3^2} = \frac{9\sqrt{2}}{2}$$

56. Let $P(x, y, z)$, $Q(0, 0, 6)$, and $R(x, y, 0)$; then

$$\|\mathbf{PQ}\|^2 = \|\mathbf{PR}\|^2 \text{ or}$$
$$x^2 + y^2 + (z - 6)^2 = z^2$$
$$x^2 + y^2 - 12z + 36 = 0$$

57. $\mathbf{v} = i - j + k$ and $\mathbf{w} = 2i + j - 2k$;
$$A = \tfrac{1}{2}\sqrt{(3)(9) - (2 - 1 - 2)^2} = \frac{\sqrt{26}}{2}$$

58. $\mathbf{v} = i - 2j - 3k$ and $\mathbf{w} = -i + j + 2k$

$$N = \mathbf{v} \times \mathbf{w} = \begin{vmatrix} i & j & k \\ 1 & -2 & -3 \\ -1 & 1 & 2 \end{vmatrix} = -i + j - k$$

The equation of the plane is $x - y + z = 0$

59. $p_1: 2x - y + z = 4$; $x + 3y - z = 2$;

$$N_1 = 2i - j + k; \ N_2 = i + 3j - k$$

$$N_3 = N_1 \times N_2 = \begin{vmatrix} i & j & k \\ 2 & -1 & 1 \\ 1 & 3 & -1 \end{vmatrix} = -2i + 3j + 7k$$

The equation of the plane is $2x - 3y - 7z = 0$.

60. $p_1: 2Ax + 3y + z = 1$; $p_2: x - Ay + 3z = 5$;

$$N_1 = 2Ai + 3j + k; \ N_2 = i - Aj + 3k;$$

$N_1 \cdot N_2 = 0$, so $2A - 3A + 3 = 0$ or $A = 3$

61. $\mathbf{v} = \mathbf{v_1} \times \mathbf{v_2} = \begin{vmatrix} i & j & k \\ 1 & -1 & 0 \\ 1 & -1 & 2 \end{vmatrix} = -2i - 2j$

Thus, the equation of L is

$$\frac{x + 1}{1} = \frac{y - 2}{1} \text{ and } z = 0$$

62. $\mathbf{u} = i - j + k$, $\mathbf{v} = i + 2j - k$; $\mathbf{w} = 2i + 2j + k$

$$N = \mathbf{u} \times \mathbf{v} = \begin{vmatrix} i & j & k \\ 1 & -1 & 1 \\ 1 & 2 & -1 \end{vmatrix} = -i + 2j + 3k$$

Thus, the altitude h is the scalar projection of
$\mathbf{w}$ onto N:
$$h = \frac{|\mathbf{w} \cdot N|}{\|N\|} = \frac{3}{\sqrt{14}} = \frac{3\sqrt{14}}{14}$$

63. $y = x^2$; $\frac{dy}{dx} = 2x$; the slope at $P(3, 9)$ is $m = 6$.
Thus, $\mathbf{T} = i + 6j$ is the tangent vector. The
unit tangent vectors are

$$\mathbf{T} = \pm\frac{1}{\sqrt{37}}(i + 6j); \quad N = \pm\frac{1}{\sqrt{37}}(6i - j)$$

64. $N_1 = \mathbf{u} \times \mathbf{v}$ is orthogonal to $\mathbf{u}$ and
$N_2 = \mathbf{u} \times \mathbf{w}$ is orthogonal to $\mathbf{u}$; thus,
$N_1 \times N_2$ is parallel to $\mathbf{u}$.

65. Let $P(a, b, c)$; then $\mathbf{r} = \mathbf{v} + \mathbf{w}t$ is the set of
position vectors in the line $x = a + At$,
$y = b + Bt$, and $z = c + Ct$.

66. Let $O(0, 0)$, $A(a, 0)$, $B(a, b)$ and $C(0, b)$ be
the vertices of a rectangle; $\mathbf{OB} = ai + bj$
By hypotheses, $\mathbf{OB} \cdot \mathbf{AC} = 0$, so $-a^2 + b^2 = 0$
or $|a| = |b|$ which means that the rectangle is
a square.

67. Let $A(0, 0)$, $B(b, 0)$, $C(b + c, d)$, and $D(c, d)$
be the vertices of a parallelogram. The sum
of the squares of the sides is

$$b^2 + (b + c - b)^2 + d^2 + b^2 + c^2 + d^2 = 2(b^2 + c^2 + d^2)$$

The sum of the squares of the diagonals is

$$(b + c)^2 + d^2 + (b - c)^2 + d^2 = 2(b^2 + c^2 + d^2)$$

68. Let $O(0, 0)$ be the center of the circle and
$A(-a, 0)$, $B(a, 0)$ be the endpoints of the
diameter of the semicircle; $C(x, \sqrt{a^2 - x^2})$
is a point of the semicircle.

$$\mathbf{CA} = (-a - x)i - (\sqrt{a^2 - x^2})j$$
$$\mathbf{CB} = (a - x)i - (\sqrt{a^2 - x^2})j$$
$$\mathbf{CA} \cdot \mathbf{CB} = x^2 - a^2 + a^2 - x^2 = 0$$

which shows that the angle C is $\pi/2$.

69. $\cos \alpha = \frac{\mathbf{B} \cdot \mathbf{v}}{\|\mathbf{B}\| \|\mathbf{v}\|} = \frac{\|\mathbf{v}\| \|\mathbf{w} \cdot \mathbf{v} + \|\mathbf{w}\| \|\mathbf{v}\|^2}{\|\mathbf{B}\| \|\mathbf{v}\|}$

$$= \frac{\mathbf{w} \cdot \mathbf{v} + \|\mathbf{w}\| \|\mathbf{v}\|}{\|\mathbf{B}\|}$$

$\cos \beta = \frac{\mathbf{B} \cdot \mathbf{w}}{\|\mathbf{B}\| \|\mathbf{w}\|} = \frac{\|\mathbf{v}\| \|\mathbf{w}\|^2 + \|\mathbf{w}\| \mathbf{v} \cdot \mathbf{w}}{\|\mathbf{B}\| \|\mathbf{w}\|}$

$$= \frac{\|\mathbf{w}\| \|\mathbf{v}\| + \mathbf{v} \cdot \mathbf{w}}{\|\mathbf{B}\|}$$

Thus, $\cos \alpha = \cos \beta$, or $\alpha = \beta$ is a solution.

70.
$$a\mathbf{u} + b\mathbf{v} + c\mathbf{w} = 0$$

$$a(-\mathbf{i} + 2\mathbf{k}) + b(2\mathbf{i} - \mathbf{j} + 3\mathbf{k}) + c(\mathbf{i} + 3\mathbf{j} - 2\mathbf{k}) = 0$$

$$(-a + 2b + c)\mathbf{i} + (-b + 3c)\mathbf{j} + (2a + 3b - 2c)\mathbf{k} = 0$$

From the coefficient of $\mathbf{j}$ we see that $b = 3c$; the coefficient of $\mathbf{i}$ gives $-a + 6c + c = 0$ or $a = 7c$; finally, the coefficient of $\mathbf{k}$ reveals that $-2a + 9c - 2a = 0$ or $2a = 7c$. Thus, $a = b = c = 0$ and $\mathbf{u}$, $\mathbf{v}$, and $\mathbf{w}$ are linearly independent.

71. a.
$$\mathbf{BP} = \mathbf{BM} + \mathbf{MP}$$

$$b\mathbf{BD} = \tfrac{1}{2}\mathbf{BA} + a\mathbf{MC}$$

$$b(\mathbf{BA} + \mathbf{AD}) = \tfrac{1}{2}\mathbf{BA} + a(\mathbf{MB} + \mathbf{BC})$$

$$\tfrac{1}{2}\mathbf{AB} + b(\mathbf{AD} - \mathbf{AB}) = a(\tfrac{1}{2}\mathbf{AB} + \mathbf{AD})$$

(We are dealing with a parallelogram, so $\mathbf{BC} = \mathbf{AD}$.)

b. Because of the property of vector equality for $\mathbf{AB}$:

$$\tfrac{1}{2} - b = \frac{a}{2}$$

Similarly for $\mathbf{AD}$: $b = a$

Thus, $\tfrac{1}{2} = \frac{3a}{2}$ or $a = \tfrac{1}{3}$.

72. If $\mathbf{u}$, $\mathbf{v}$, and $\mathbf{w}$ are linearly independent, then

$$a(a_1\mathbf{i} + a_2\mathbf{j} + a_3\mathbf{k}) + b(b_1\mathbf{i} + b_2\mathbf{j} + c_2\mathbf{k})$$
$$+ c(c_1\mathbf{i} + c_2\mathbf{j} + c_3\mathbf{k}) = 0$$

This implies

$$aa_1 + bb_1 + cc_1 = 0; \; aa_2 + bb_2 + cc_2 = 0;$$

$$aa_3 + bb_3 + cc_3 = 0.$$

This system of three equations in three unknowns has nontrivial solution only if

$$\begin{vmatrix} a_1 & b_1 & c_1 \\ a_2 & b_2 & c_2 \\ a_3 & b_3 & c_3 \end{vmatrix} = 0$$

This can be rewritten

$$\begin{vmatrix} a_1 & a_2 & a_3 \\ b_1 & b_2 & b_3 \\ c_1 & c_2 & c_3 \end{vmatrix} = 0$$

On the other hand, if

$$\begin{vmatrix} a_1 & a_2 & a_3 \\ b_1 & b_2 & b_3 \\ c_1 & c_2 & c_3 \end{vmatrix} = 0$$

The volume of the parallelepiped formed by the three vectors is 0, the vectors are coplanar, which means that one vector, say $\mathbf{w}$, can be written as a linear combination of $\mathbf{u}$ and $\mathbf{v}$. Then, $\mathbf{w} = r\mathbf{u} + s\mathbf{v}$.

$$a\mathbf{u} + b\mathbf{v} + c\mathbf{w} = a\mathbf{u} + b\mathbf{v} + cr\mathbf{u} + cs\mathbf{v} = 0$$

if $a = -cr$ and $b = -cs$. This means $as = br$. Then, $a = r \ne 0$ and $b = s \ne 0$, so the vectors are linearly dependent.

73. If $\mathbf{u}$, $\mathbf{v}$, and $\mathbf{w}$ are coplanar, then the parallelepiped formed by the three vectors has 0 volume. Thus, $\mathbf{u} \times \mathbf{v} \cdot \mathbf{w} = 0$. The converse also holds. If the volume of the parallelpiped is 0, then the height is 0, and the three vectors are coplanar.

74. Vector $\mathbf{u}_1$ is arbitrary, with length u_{11}. Force the x-axis through $\mathbf{u}_1$. Then, $u_1 = u_{11}\mathbf{i}$. Vector u_2 is arbitrary, with length u_{22}. Force the y-axis through the plane of $\mathbf{u}_1$ and $\mathbf{u}_2$. Then, $u_2 = u_{21}\mathbf{i} + u_{22}\mathbf{j}$. Vectors $\mathbf{v}_1$ and $\mathbf{v}_2$ are also arbitrary and may have components along all three axes (which are in place after $\mathbf{v}_1$ and $\mathbf{v}_2$ have been consider). Thus,

$$\mathbf{v}_1 = v_{11}\mathbf{i} + v_{12}\mathbf{j} + v_{13}\mathbf{k};$$

$$\mathbf{v}_2 = v_{21}\mathbf{i} + v_{22}\mathbf{j} + v_{23}\mathbf{k}'$$

$$\begin{bmatrix} \mathbf{u}_1 \cdot \mathbf{v}_1 & \mathbf{u}_1 \cdot \mathbf{v}_2 \\ \mathbf{u}_2 \cdot \mathbf{v}_1 & \mathbf{u}_2 \cdot \mathbf{v}_2 \end{bmatrix}$$

$$= \begin{bmatrix} u_{11}v_{11} & u_{11}v_{21} \\ u_{21}v_{11} + u_{22}v_{12} & u_{21}v_{21} + u_{22}v_{22} \end{bmatrix}$$

$$= u_{11}(u_{21}v_{11}v_{21} + u_{22}v_{11}v_{22} - u_{21}v_{11}v_{21} - u_{22}v_{11}v_{21})$$

$$= u_{11}u_{22}(v_{11}v_{22} - v_{12}v_{21})$$

We now show that the right-hand side is the same.

$$\mathbf{u}_1 \times \mathbf{u}_2 = \begin{vmatrix} \mathbf{i} & \mathbf{j} & \mathbf{k} \\ u_1 & 0 & 0 \\ u_2 & u_2 & 0 \end{vmatrix} = u_{11}u_{22}\mathbf{k}$$

$$\mathbf{v}_1 \times \mathbf{v}_2 = \begin{vmatrix} \mathbf{i} & \mathbf{j} & \mathbf{k} \\ v_{11} & v_{12} & v_{13} \\ v_{21} & v_{22} & v_{23} \end{vmatrix}$$

$$= (\cdots)\mathbf{i} - (\cdots)\mathbf{j} + (v_{11}v_{22} - v_{12}v_{22})\mathbf{k}$$

Thus,

$$(\mathbf{u}_1 \times \mathbf{u}_2) \cdot (\mathbf{v}_1 \times \mathbf{v}_2) = u_{11}u_{22}(v_{11}v_{22} - v_{12}v_{21})$$

The result follows since this result agrees with the one shown above.

75.
$$\|\mathbf{u} \times \mathbf{v}\|^2 = \|\mathbf{u}\|^2 \|\mathbf{v}\|^2 \sin^2\theta$$
$$= \|\mathbf{u}\|^2 \|\mathbf{v}\|^2 (1 - \cos^2\theta)$$
$$= \|\mathbf{u}\|^2 \|\mathbf{v}\|^2 - (\|\mathbf{u}\| \|\mathbf{v}\| \cos\theta)^2$$
$$= (\mathbf{u} \cdot \mathbf{u})(\mathbf{v} \cdot \mathbf{v}) - (\mathbf{u} \cdot \mathbf{v})^2$$

76. The midpoint of **AB** is $D(-\frac{1}{2}, 4, 2)$ and $\mathbf{CD} = -\frac{5}{2}\mathbf{i} - 4\mathbf{j} + 3\mathbf{k}$. For the intersection $M(x, y, z)$ of the medians, $\mathbf{CM} = (2/3)\mathbf{CD}$ or $3\mathbf{CM} = 2\mathbf{CD}$. Thus,

$$3[(x-2)\mathbf{i} + (y-8)\mathbf{j} + (z+1)\mathbf{k}] = 2(-\tfrac{5}{2}\mathbf{i} - 4\mathbf{j} + 3\mathbf{k})$$

From this equation we find $x = \frac{1}{3}$, $y = \frac{16}{3}$, and $z = 1$. Thus, the centroid is $M(\frac{1}{3}, \frac{16}{3}, 1)$.

77. Let O, A, B, and C be the vertices of the tetrahedron. The midpoint of **AB** is D, the centroid for $\triangle OAB$ is at N, the centroid for $\triangle ABC$ is at M. Suppose **CN**, **CD**, and **CN** intersect at P. Then,

$$\mathbf{OP} = \mathbf{ON} + \mathbf{NP}$$
$$\mathbf{OM} = \tfrac{2}{3}\mathbf{OD} = (1 - a)\mathbf{NC}$$
$$b\mathbf{OM} = \tfrac{2}{3}\mathbf{OD} + (1 - a)(\mathbf{ND} + \mathbf{DC})$$
$$b\mathbf{OM} = \tfrac{2}{3}\mathbf{OD} + (1 - a)[\tfrac{1}{3}\mathbf{OD} - \mathbf{CD}]$$
$$b[\mathbf{OD} + \tfrac{1}{3}\mathbf{DC}] = \tfrac{2}{3}\mathbf{OD} + (1 - a)[\tfrac{1}{3}\mathbf{OD} - \mathbf{CD}]$$
$$b\mathbf{OD} - \tfrac{b}{3}\mathbf{CD} = \tfrac{2}{3}\mathbf{OD} + \tfrac{1 - a}{3}\mathbf{OD} - (1 - a)\mathbf{CD}$$

By the property of equality of vectors, for the coefficients of **OD**:

$$b = \tfrac{2}{3} + \tfrac{1 - a}{3}$$

and for **CD**:

$$-\tfrac{b}{3} = a - 1$$

Solving these equations simultaneously, we find $a = 3/4$, $b = 3/4$.

78.
$$2A^2 = \|\mathbf{u} \times \mathbf{v}\|^2$$
$$= \|\mathbf{u}\|^2 \|\mathbf{v}\|^2 (1 - \cos^2\theta)$$
$$= \|\mathbf{u}\|^2 \|\mathbf{v}\|^2 - (\|\mathbf{u}\| \|\mathbf{v}\| \cos\theta)^2$$
$$= (\mathbf{u} \cdot \mathbf{u})(\mathbf{v} \cdot \mathbf{v}) - (\mathbf{u} \cdot \mathbf{v})^2$$

Thus, $A = \frac{1}{2}\sqrt{(\mathbf{u} \cdot \mathbf{u})(\mathbf{v} \cdot \mathbf{v}) - (\mathbf{u} \cdot \mathbf{v})^2}$

79. **a.** Let s and t be chosen so that $\mathbf{PQ} = s\mathbf{AN}$ and $\mathbf{QN} = t\mathbf{AN}$. By symmetry, $\mathbf{RP} = s\mathbf{CM}$ and $\mathbf{PM} = t\mathbf{CM}$. We have

$$\mathbf{AP} + \mathbf{PM} = \mathbf{AM}$$
$$\underbrace{(1 - s - t)\mathbf{AN}}_{\mathbf{AP}} + \underbrace{t\mathbf{PM}}_{\mathbf{PM}} = \mathbf{AM}$$

Since
$$\mathbf{AN} = \mathbf{AB} + \tfrac{1}{3}\mathbf{BC} = \mathbf{AB} + \tfrac{1}{3}(\mathbf{AC} - \mathbf{AB})$$
$$\mathbf{CM} = \mathbf{AM} - \mathbf{AC} = \tfrac{1}{3}\mathbf{AB} - \mathbf{AC}$$

if follows that

$$(1 - s - t)[\mathbf{AB} + \tfrac{1}{3}(\mathbf{AC} - \mathbf{AB})] + t[\tfrac{1}{3}\mathbf{AB} - \mathbf{AC}] = \tfrac{1}{3}\mathbf{AB}$$

Gathering terms we find

$$[\tfrac{2}{3}(1 - s - t) + \tfrac{1}{3}t - \tfrac{1}{3}]\mathbf{AB} + [\tfrac{1}{3}(1 - s - t) - t]\mathbf{AC} = 0$$

so that

$$\tfrac{2}{3}s + \tfrac{1}{3}t = \tfrac{1}{3}$$
$$\tfrac{1}{3}s + \tfrac{4}{3}t = \tfrac{1}{3}$$

since **AB** and **AC** are linearly independent (are not parallel). Thus, we have $s = 3/7$ and $t = 1/7$. Thus, $\mathbf{PQ} = \frac{3}{7}\mathbf{AN}$.

Next, note that

$$\mathbf{AN} = \mathbf{AB} + \tfrac{1}{3}(\mathbf{AC} - \mathbf{AB}) = \tfrac{2}{3}\mathbf{AB} + \tfrac{1}{3}\mathbf{AC}$$

By the law of cosines,

$$\|\mathbf{AN}\|^2 = (\tfrac{1}{3}\|\mathbf{AB}\|)^2 + \|\mathbf{AB}\|^2 - 2(\tfrac{1}{3}\|\mathbf{AB}\|)(\|\mathbf{AB}\|)\cos 60°$$
$$= \tfrac{1}{9}\|\mathbf{AB}\|^2 + \|\mathbf{AB}\|^2 - 2[\tfrac{1}{3}\|\mathbf{AB}\|^2(\tfrac{1}{2})]$$
$$= \tfrac{7}{9}\|\mathbf{AB}\|^2$$

b. The area of an equilateral triangle of side s is $A = (\sqrt{3}/4)s^2$. If the given triangle $\triangle ABC$ has side with length L, then the inner triangle $\triangle PQR$ (which is also equilateral, by symmetry) has side

$$\|\mathbf{PQ}\| = \left\|\tfrac{3}{7}\mathbf{AN}\right\| = \tfrac{3}{7}\sqrt{\tfrac{7}{9}}\|\mathbf{AB}\| = \tfrac{1}{\sqrt{7}}\|\mathbf{AB}\| = \tfrac{1}{\sqrt{7}}L$$

Thus,

Area of $\triangle ABC = \dfrac{\sqrt{3}}{4}L^2$

Area of $\triangle PQR = \dfrac{\sqrt{3}}{4}\left(\dfrac{1}{\sqrt{7}}L\right)^2$
$$= \tfrac{1}{7}\left(\dfrac{\sqrt{3}}{4}L^2\right)$$
$$= \tfrac{1}{7}[\text{Area of } \triangle ABC]$$

c. The conjecture would not hold because the relationships would be different. For instance, the statement "The area is $(\sqrt{3}/4)s^2$" for a side of length s does not hold.

80. **a.**
$$\alpha \doteq \mathbf{v} - \dfrac{(\mathbf{v} \cdot \mathbf{u})\mathbf{u}}{\|\mathbf{u}\|^2}$$
$$\beta = \mathbf{w} - \dfrac{(\mathbf{w} \cdot \mathbf{u})\mathbf{u}}{\|\mathbf{u}\|^2} - \dfrac{(\mathbf{w} \cdot \alpha)\alpha}{\|\alpha\|^2}$$

$$\boldsymbol{\alpha} \cdot \mathbf{u} = \mathbf{v} \cdot \mathbf{u} - \frac{(\mathbf{v} \cdot \mathbf{u})\mathbf{u} \cdot \mathbf{u}}{\|\mathbf{u}\|^2} = 0$$

$$\boldsymbol{\beta} \cdot \mathbf{u} = \mathbf{w} \cdot \mathbf{u} - \frac{(\mathbf{w} \cdot \mathbf{u})\mathbf{u} \cdot \mathbf{u}}{\|\mathbf{u}\|^2} - \frac{(\mathbf{w} \cdot \boldsymbol{\alpha})\boldsymbol{\alpha} \cdot \mathbf{u}}{\|\boldsymbol{\alpha}\|^2}$$

$$= \mathbf{w} \cdot \mathbf{u} - \frac{(\mathbf{w} \cdot \mathbf{u})\|\mathbf{u}\|^2}{\|\mathbf{u}\|^2} - 0 = 0$$

$$\boldsymbol{\alpha} \cdot \boldsymbol{\beta} = \boldsymbol{\alpha} \cdot \mathbf{w} - \frac{(\mathbf{w} \cdot \mathbf{u})\boldsymbol{\alpha} \cdot \mathbf{u}}{\|\mathbf{u}\|^2} - \frac{(\mathbf{w} \cdot \boldsymbol{\alpha})\boldsymbol{\alpha} \cdot \boldsymbol{\alpha}}{\|\boldsymbol{\alpha}\|^2}$$

$$= \boldsymbol{\alpha} \cdot \mathbf{w} - \frac{(\mathbf{w} \cdot \mathbf{u})(0)}{\|\mathbf{u}\|^2} - \frac{(\mathbf{w} \cdot \boldsymbol{\alpha})\|\boldsymbol{\alpha}\|^2}{\|\boldsymbol{\alpha}\|^2} = 0$$

b. Let a, b, c satisfy $\boldsymbol{\gamma} = a\mathbf{u} + b\boldsymbol{\alpha} + c\boldsymbol{\beta}$; then

$$\boldsymbol{\gamma} \cdot \mathbf{u} = a(\mathbf{u} \cdot \mathbf{u}) + b(\boldsymbol{\alpha} \cdot \mathbf{u}) + c(\boldsymbol{\beta} \cdot \mathbf{u})$$

$$= a(\mathbf{u} \cdot \mathbf{u}) + 0 + 0 \text{ so that}$$

$$a = \frac{\boldsymbol{\gamma} \cdot \mathbf{u}}{\mathbf{u} \cdot \mathbf{u}} = \frac{\boldsymbol{\gamma} \cdot \mathbf{u}}{\|\mathbf{u}\|^2}$$

Similarly,

$$\boldsymbol{\gamma} \cdot \boldsymbol{\alpha} = b(\boldsymbol{\alpha} \cdot \boldsymbol{\alpha}) \text{ so that } b = \frac{\boldsymbol{\gamma} \cdot \boldsymbol{\alpha}}{\|\boldsymbol{\alpha}\|^2}$$

$$\boldsymbol{\gamma} \cdot \boldsymbol{\beta} = c(\boldsymbol{\beta} \cdot \boldsymbol{\beta}) \text{ so that } c = \frac{\boldsymbol{\gamma} \cdot \boldsymbol{\beta}}{\|\boldsymbol{\beta}\|^2}$$

Thus,

$$\boldsymbol{\gamma} = \frac{(\boldsymbol{\gamma} \cdot \mathbf{u})\mathbf{u}}{\|\mathbf{u}\|^2} + \frac{(\boldsymbol{\gamma} \cdot \boldsymbol{\alpha})\boldsymbol{\alpha}}{\|\boldsymbol{\alpha}\|^2} + \frac{(\boldsymbol{\gamma} \cdot \boldsymbol{\beta})\boldsymbol{\beta}}{\|\boldsymbol{\beta}\|^2}$$

81. This is Putnam Problem 4 of the morning session of 1939. Suppose the required line L meets the given lines at points A, B, C, and D, respectively. Then $A(1, 0, a)$, $B(b, 1, 0)$, $C(0, c, 1)$, $D(6d, 6d, -d)$ for some a, b, c, and d. Treat A, B, C, and D as vectors. The condition that they be collinear is that the vectors
$\mathbf{B} - \mathbf{A} = \langle b - 1, 1, a \rangle$,
$\mathbf{C} - \mathbf{A} = \langle -1, c, 1 - a \rangle$,
$\mathbf{D} - \mathbf{A} = \langle 6d - 1, 6d, -d - a \rangle$
The proportionality of the first two tells us that

$$c = \frac{1}{1 - b} = \frac{a - 1}{a}$$

while the first and third give

$$6d = \frac{1 - 6d}{1 - b} = \frac{a + d}{a}$$

Rewrite the middle member here using the first string of equations

$$6d = (1 - 6d)\left(\frac{a - 1}{a}\right) = \frac{a + d}{a}$$

Clearing fractions $6ad = a + d$,

$$a + 6d - 1 - 6ad = a + d$$

Adding these equations, we find $4d = a + 1$,

so

$$6a(a + 1) = 24ad = 4(a + d) = 5a + 1$$

This quadratic equation has roots $a = \frac{1}{3}$, $-\frac{1}{2}$ and the corresponding values of the other unknowns are

$$b = \frac{3}{2}, \frac{2}{3}; \quad c = -2, 3; \quad d = \frac{1}{3}, \frac{1}{8}$$

The direction vectors of the lines (proportional to $\mathbf{B} - \mathbf{A}$, $\mathbf{C} - \mathbf{A}$, and $\mathbf{D} - \mathbf{A}$) in the two cases are $(3, 6, -2)$ and $(-2, 6, 3)$, respectively. The two lines are given parametrically by

L_1: $s \to (1, 0, \frac{1}{3}) + s(3, 6, -2)$

L_2: $t \to (1, 0, -\frac{1}{2}) + t(-2, 6, 3)$.

These lines cross the given lines (in order) for

$s = 0, \frac{1}{6}, -\frac{1}{3}, \frac{1}{3}$ and $t = 0, \frac{1}{6}, \frac{1}{2}, \frac{1}{8}$

In nonparametric form, L_1 is given by

$$y = 2(x - 1) = 1 - 3z$$

and L_2 is given by

$$y = 3(1 - x) = 2z + 1.$$

For a general treatment of this problem, see D. M. Y. Sommerville, *Analytic Geometry of Three Dimensions*, Cambridge, 1934, p. 184.

82. This is Putnam Problem 5 of the afternoon session of 1959. Let the given lines be ℓ and m. Then there is a unique segment $\overline{PQ}$ perpendicular to both lines with P on ℓ and Q on m. The required sphere has $\overline{PQ}$ as its diameter. Suppose ℓ and m are given in terms of a parameter t by $\mathbf{a} + t\mathbf{v}$ and $\mathbf{b} + t\mathbf{w}$, respectively, where $\mathbf{a}$, $\mathbf{b}$, $\mathbf{v}$, and $\mathbf{w}$ are vectors. If the lines are not parallel, $\mathbf{v}$ and $\mathbf{w}$ are linearly independent. Since $\mathbf{PQ}$ is perpendicular to both lines, it has the direction of $\mathbf{v} \times \mathbf{w}$, say $\mathbf{PQ} = \rho\mathbf{v} \times \mathbf{w}$, where ρ is a scalar. Let P and Q be the points $\mathbf{a} + \sigma\mathbf{v}$ and $\mathbf{b} + \tau\mathbf{w}$, respectively. Then

$$\mathbf{PQ} = \mathbf{b} - \mathbf{a} - \sigma\mathbf{v} + \tau\mathbf{w} \text{ and}$$

$$\mathbf{a} - \mathbf{b} = -\rho(\mathbf{v} \times \mathbf{w}) - \sigma\mathbf{v} + \tau\mathbf{w}$$

Hence we can calculate ρ, σ, and τ by expression $\mathbf{a} - \mathbf{b}$ in terms of the independent vectors $\mathbf{v} \times \mathbf{w}$, $\mathbf{v}$ and $\mathbf{w}$. Then the center of the required sphere is at

$$\mathbf{a} + \sigma\mathbf{v} + \tfrac{1}{2}\rho(\mathbf{v} \times \mathbf{w})$$

and its radius is $\frac{1}{2}|\rho|\|\mathbf{v} \times \mathbf{w}\|$. For the example in question, $\mathbf{a} = \langle 1, 4, 5 \rangle$, $\mathbf{b} = \langle -12, 8, 17 \rangle$,

$\mathbf{v} = \langle 1, 2, -3 \rangle$, $\mathbf{w} = \langle 4, -1, 1 \rangle$, and $\mathbf{v} \times \mathbf{w} = \langle -1, -13, -9 \rangle$. Then ρ, σ, and τ, are found from the equations

$$13 = \rho - \sigma + 4\tau$$

$$-4 = 13\rho - 2\sigma - \tau$$

$$-12 = 9\rho + 3\sigma + \tau$$

Solving simultaneously, we find $\rho = \frac{-147}{251}$, $\sigma = \frac{-782}{251}$, and $\tau = \frac{657}{251}$. The center of the sphere is, therefore, at $\left(\frac{-147}{251}, \frac{-782}{251}, \frac{657}{251} \right)$. The square of the radius is

$$\tfrac{1}{4}\rho^2 \| \mathbf{v} \times \mathbf{w} \|^2 = \left(\tfrac{147}{502} \right)^2 (251) = \frac{147^2}{1{,}004}$$

The equation of the sphere is

$$(502x + 915)^2 + (502y - 791)^2 + (502z - 8525)^2$$

$$= 251(147)^2$$

83. This is Putnam Problem 2 of the morning session of 1983. Let **OA** be the long hand and **OB** be the short hand. We can think of **OA** as fixed and **OB** as rotating at constant speed. Let **v** be the vector of point **B** under this assumption. The rate of change of the distance between A and B is the component of **v** in the direction of **AB**. Since **v** is orthogonal to **OB** and the magnitude of **v** is constant, this component is maximal when $\angle OAB$ has is closed to form a triangle; that is, when the distance **AB** is $\sqrt{4^2 - 3^2} = \sqrt{7}$. Alternately, let x be the distance **AB** and $\theta = \angle AOB$. By the law of cosines,

$$x^2 = 3^2 + 4^2 - 2(3)(4) \cos \theta = 25 - 24 \cos \theta$$

Now,

$$2x \frac{dx}{dt} = 24 \sin \theta$$

$$\frac{dx}{d\theta} = \frac{12 \sin \theta}{\sqrt{25 - 24 \cos \theta}}$$

Since dx/dt is an odd function of θ, $\left| dx/ds \right|$ is a minimum when $dx/d\theta$ is a maximum or a minimum. Since dx/ds is a periodic differentiable function of θ, $d^2x/ds^2 = 0$ at the extremes for dx/ds. For such θ,

$$12 \cos \theta = x \frac{d^2x}{d\theta^2} + \left(\frac{dx}{d\theta} \right)^2 = \left(\frac{dx}{d\theta} \right)^2 = \frac{144 \sin^2\theta}{x^2}$$

$$x^2 = \frac{12 \sin^2\theta}{\cos \theta} = \frac{12 - 12 \cos^2\theta}{\cos \theta} = 25 - 24 \cos \theta$$

and it follows that

$$12 \cos^2\theta - 25 \cos \theta + 12 = 0$$

The only allowable solution for $\cos \theta$ is $\cos \theta = \frac{3}{4}$ and hence

$$x = \sqrt{25 - 24 \cos \theta} = \sqrt{25 - 18} = \sqrt{7}$$

CHAPTER 11
Vector Calculus

11.1 Introduction to Vector Functions, Page 690

1. $t \neq 0$

2. $t \geq 0$, $t \neq 2$

3. $t \neq \dfrac{(2n+1)\pi}{2}$, n an integer

4. $t \neq n\pi$, n an integer

5. $t \neq \dfrac{n\pi}{2}$, n an integer

6. $t \neq 0$, $t \leq 10$

7. $t > 0$

8. $t \neq -2$, $t \leq 0$

9.
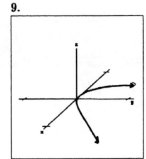

10.
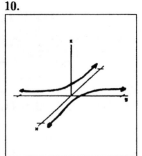

11.
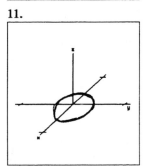

12.
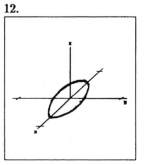

13.
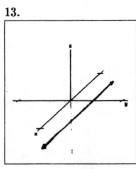

14.
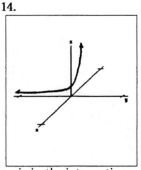

15. circular helix
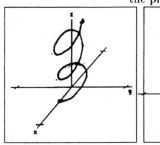

16. curve is in the intersection of the cylinder $y = 1/z$ with the plane $x = y$.
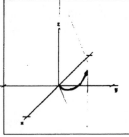

17. curve is in the intersection of the parabolic cylinder $y = (1-x)^2$ with the plane $x + z = 1$

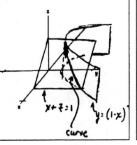

18. the graph is the intersection of the paraboloid $z = 1 - x^2 - y^2$ with the plane $x = y$.

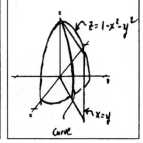

19. curve is the intersection of the cylinder $y = x^2 + 1$ and the plane $y = z + 1$

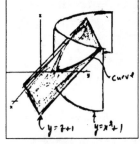

20. curve is a circle in the plane $z = 3$

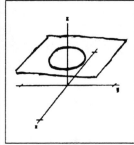

21. A vector-valued function $\mathbf{F}$ with domain D assigns a unique vector $\mathbf{F}(t)$ to each scalar t in the set D. The set of all vectors $\mathbf{v}$ of the form $\mathbf{v} = \mathbf{F}(t)$ for t in D is the range of $\mathbf{F}$. In this text we consider vector functions whose range is in $\mathbb{R}^2$ or $\mathbb{R}^3$. That is, $\mathbf{F}(t) = f_1(t)\mathbf{i} + f_2(t)\mathbf{j}$ in $\mathbb{R}^2$ (plane), or $\mathbf{F}(t) = f_1(t)\mathbf{i} + f_2(t)\mathbf{j} + f_3(t)\mathbf{k}$ in $\mathbb{R}^3$ (space) where f_1, f_2, and f_3 are real-valued (scalar-valued) functions of the real number t defined on the domain set D. In this context f_1, f_2, and f_3 are called the components of $\mathbf{F}$.

22. Suppose the components f_1, f_2, f_3 of the vector function $\mathbf{F}(t) = f_1(t)\mathbf{i} + f_2(t)\mathbf{j} + f_3(t)\mathbf{k}$ all have finite limits as $t \to t_0$, where t_0 is any number or $\pm\infty$. Then the limit of $\mathbf{F}(t)$ as $t \to t_0$ is the vector $\lim\limits_{t \to t_0} \mathbf{F}(t)$
$$= [\lim\limits_{t \to t_0} f_1(t)]\mathbf{i} + [\lim\limits_{t \to t_0} f_2(t)]\mathbf{j} + [\lim\limits_{t \to t_0} f_3(t)]\mathbf{k}$$

23. $x = 2\sin t$, $y = 2\sin t$, $z = \sqrt{8}\cos t$.
$$x^2 + y^2 + z^2 = 4\sin^2 t + 4\sin^2 t + 8\cos^2 t$$

$= 8(\sin^2 t + \cos^2 t) = 8$

Which is a sphere with center at the origin and a radius of $2\sqrt{2}$.

24. z is increasing with $t < 0$ and decreasing with $t > 0$; it is an elliptic spiral.

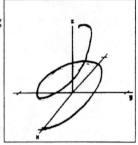

25. $\mathbf{F}(t) = t\mathbf{i} + t^2\mathbf{j} + 2\mathbf{k}$

26. $\mathbf{F}(t) = 2(\cos t)\mathbf{i} + 2(\sin t)\mathbf{j} - \mathbf{k}$

27. $\mathbf{F}(t) = 2t\mathbf{i} + (1 - t)\mathbf{j} + (\sin t)\mathbf{k}$

28. $\mathbf{F}(t) = (2 + 3t)\mathbf{i} + (1 + 2t)\mathbf{j} + 4t\mathbf{k}$

29. $\mathbf{F}(t) = t^2\mathbf{i} + t\mathbf{j} + \sqrt{9 - t^2 - t^4}\,\mathbf{k}$

30. $\mathbf{N}_1 = \mathbf{i} + \mathbf{j} + 3\mathbf{k}; \mathbf{N}_2 = \mathbf{i} - \mathbf{j} - \mathbf{k}$

$$\mathbf{v} = \mathbf{N}_1 \times \mathbf{N}_2 = \begin{vmatrix} \mathbf{i} & \mathbf{j} & \mathbf{k} \\ 2 & 1 & 3 \\ 1 & -1 & -1 \end{vmatrix} = 2\mathbf{i} + 5\mathbf{j} - 3\mathbf{k}$$

For a point of intersection, let $z = 0$, so that $x = 7/3$, $y = 4/3$:

$x = \frac{7}{3} + 2t;\ y = \frac{4}{3} + 5t;\ z = -3t;$ thus,

$\mathbf{F}(t) = (\frac{7}{3} + 2t)\mathbf{i} + (\frac{4}{3} + 5t)\mathbf{j} - 3t\mathbf{k}$

31. $2\mathbf{F}(t) - 3\mathbf{G}(t) = [2(2t) - 3(1 - t)]\mathbf{i}$

$+ [2(-5) - 3(0)]\mathbf{j} + [2(t^2) - 3(\frac{1}{t})]\mathbf{k}$

$= (7t - 3)\mathbf{i} - 10\mathbf{j} + (2t^2 - \frac{3}{t})\mathbf{k}$

32. $t^2\mathbf{F}(t) - 3\mathbf{H}(t) = t^2(2t\mathbf{i} - 5\mathbf{j} + t^2)\mathbf{k}$

$- 3(\sin t\,\mathbf{i} + e^t)\mathbf{j}$

$= (2t^3 - 3\sin t)\mathbf{i} - (5t^2 + 3e^t)\mathbf{j} + t^4\mathbf{k}$

33. $\mathbf{F}(t) \cdot \mathbf{G}(t) = (2t\mathbf{i} - 5\mathbf{j} + t^2\mathbf{k}) \cdot [(1 - t)\mathbf{i} + (\frac{1}{t})\mathbf{k}]$

$= (2t - 2t^2) + t = 3t - 2t^2$

34. $\mathbf{F}(t) \cdot \mathbf{H}(t) = (2t\mathbf{i} - 5\mathbf{j} + t^2\mathbf{k}) \cdot (\sin t\,\mathbf{i} + e^t\mathbf{j})$

$= 2t\sin t - 5e^t$

35. $\mathbf{G}(t) \cdot \mathbf{H}(t) = (1 - t)\sin t$

36. $\mathbf{F}(t) \times \mathbf{G}(t) = \begin{vmatrix} \mathbf{i} & \mathbf{j} & \mathbf{k} \\ 2t & -5 & t^2 \\ 1 - t & 0 & \frac{1}{t} \end{vmatrix}$

$= -\frac{5}{t}\mathbf{i} - (2 + t^3 - t^2)\mathbf{j} + 5(1 - t)\mathbf{k}$

37. $\mathbf{F}(t) \times \mathbf{H}(t) = \begin{vmatrix} \mathbf{i} & \mathbf{j} & \mathbf{k} \\ 2t & -5 & t^2 \\ \sin t & e^t & 0 \end{vmatrix}$

$= -t^2 e^t\mathbf{i} + t^2\sin t\mathbf{j} + (2te^t + \sin t)\mathbf{k}$

38. $\mathbf{G}(t) \times \mathbf{H}(t) = \begin{vmatrix} \mathbf{i} & \mathbf{j} & \mathbf{k} \\ 1 - t & 0 & \frac{1}{t} \\ \sin t & e^t & 0 \end{vmatrix}$

$= -\frac{e^t}{t}\mathbf{i} + \frac{\sin t}{t}\mathbf{j} + (1 - t)e^t\mathbf{k}$

39. $2e^t\mathbf{F}(t) + t\mathbf{G}(t) + 10\mathbf{H}(t)$

$= [2e^t(2t) + t(1 - t) + 10\sin t]\mathbf{i}$

$+ [2e^t(-5) + 10e^t]\mathbf{j} + [2e^t t^2 + 1]\mathbf{k}$

$= (4te^t - t^2 + t + 10\sin t)\mathbf{i} + (2e^t t^2 + 1)\mathbf{k}$

40. $\mathbf{F}(t) \cdot [\mathbf{H}(t) \times \mathbf{G}(t)] = \begin{vmatrix} 2t & -5 & t^2 \\ \sin t & e^t & 0 \\ 1 - t & 0 & \frac{1}{t} \end{vmatrix}$

$= 2e^t + \frac{5}{t}\sin t - t^3 + t^2$

41. $\mathbf{G}(t) \cdot [\mathbf{H}(t) \times \mathbf{F}(t)] = \begin{vmatrix} 1 - t & 0 & \frac{1}{t} \\ \sin t & e^t & 0 \\ 2t & -5 & t^2 \end{vmatrix}$

$= t^2 e^t - t^3 e^t - 2e^t - \frac{5}{t}\sin t$

42. $\mathbf{H}(t) \cdot [\mathbf{G}(t) \times \mathbf{F}(t)] = \begin{vmatrix} \sin t & e^t & 0 \\ 1 - t & 0 & \frac{1}{t} \\ 2t & -5 & t^2 \end{vmatrix}$

$= \frac{5}{t}\sin t + 2e^t + t^3 e^t - t^2 e^t$

43. $\lim_{t \to 1}[3t\mathbf{i} + e^{2t}\mathbf{j} + (\sin\pi t)\mathbf{k}] = 3\mathbf{i} + e^2\mathbf{j}$

44. $\lim_{t \to 0}\left[\frac{(\sin t)\mathbf{i} - t\mathbf{k}}{t^2 + t - 1}\right] = \mathbf{0}$

45. $\lim_{t \to 1}\left[\frac{t^3 - 1}{t - 1}\mathbf{i} + \frac{t^2 - 3t + 2}{t^2 + t - 2}\mathbf{j} + (t^2 + 1)e^{t-1}\mathbf{k}\right]$

$= \lim_{t \to 1}\left[(t^2 + t + 1)\mathbf{i} + \frac{(t - 2)(t - 1)}{(t + 2)(t - 1)}\mathbf{j} + (t^2 + 1)e^{t-1}\mathbf{k}\right]$

$= -3\mathbf{i} - \frac{1}{3}\mathbf{j} + 2\mathbf{k}$

46. $\lim_{t \to 0}\left[\frac{te^t}{1 - e^t}\mathbf{i} + \frac{e^t - 1}{\cos t}\mathbf{j}\right]$

$= \lim_{t \to 0}\left[\frac{te^t + e^t}{-e^t}\mathbf{i} + \frac{e^t - 1}{\cos t}\mathbf{j}\right] = -\mathbf{i} + e^{-1}\mathbf{j}$

47. $\lim_{t \to 0}\left[\frac{\sin t}{t}\mathbf{i} + \frac{1 - \cos t}{t}\mathbf{j} + e^{1-t}\mathbf{k}\right] = \mathbf{i} + e\mathbf{k}$

48. $\lim\limits_{t\to 0^+}\left[\dfrac{\sin 3t}{\sin 2t}\mathbf{i} + \dfrac{\ln(\sin t)}{\ln(\tan t)}\mathbf{j} + t^t\mathbf{k}\right]$

$= \lim\limits_{t\to 0^+}\left[\dfrac{3}{2}\dfrac{\sin 3t}{3t}\dfrac{2t}{\sin 2t}\,\mathbf{i} + \dfrac{\cos t\tan t}{\sin t\sec^2 t}\mathbf{j} + e^{t\ln t}\mathbf{k}\right]$

$= \dfrac{3}{2}\mathbf{i} + \mathbf{j} + \mathbf{k}$

49. $\lim\limits_{t\to 1}\left[2t\mathbf{i} - 3\mathbf{j} + e^t\mathbf{k}\right] = 2\mathbf{i} - 3\mathbf{j} + e\mathbf{k}$

50. $\lim\limits_{t\to 2}\left[(2\mathbf{i} - t\mathbf{j} + e^t\mathbf{k}) \times (t^2\mathbf{i} + 4\sin t\,\mathbf{j})\right]$

$= (2\mathbf{i} - 2\mathbf{j} + e^2\mathbf{k}) \times (4\mathbf{i} + 4\sin 2\,\mathbf{j})$

$= \begin{vmatrix} \mathbf{i} & \mathbf{j} & \mathbf{k} \\ 2 & -2 & e^2 \\ 41 & 4\sin 2 & 0 \end{vmatrix}$

$= (-4e^2\sin 2)\mathbf{i} + 4e^2\mathbf{j} + (8\sin 2 + 8)\mathbf{k}$

51. continuous for all t

52. continuous for $t \neq 0$

53. continuous for $t \neq 0$, $t \neq -1$

54. continuous for all t

55. continuous for all $t \neq 0$

56. continuous for $t \neq 0$ and for $t > 0$

57. **a.** $\lim\limits_{t\to 0} e^t\mathbf{F}(t) = \lim\limits_{t\to 0} e^t(t\mathbf{i} + t^2\mathbf{j} + t^3\mathbf{k}) = \mathbf{0}$

$\left[\lim\limits_{t\to 0} e^t\right]\left[\lim\limits_{t\to 0} \mathbf{F}(t)\right] = (1)(0\mathbf{i} + 0\mathbf{j} + 0\mathbf{k}) = \mathbf{0}$

b. $\lim\limits_{t\to 1} \mathbf{F}(t)\cdot \mathbf{G}(t)$

$= \lim\limits_{t\to 1}\left[(t\mathbf{i} + t^2\mathbf{j} + t^3\mathbf{k})\cdot(\tfrac{1}{t}\mathbf{i} - e^t\mathbf{j})\right]$

$= 1 - e$

$\left[\lim\limits_{t\to 1}\mathbf{F}(t)\right]\cdot\left[\lim\limits_{t\to 1}\mathbf{G}(t)\right]$
$= \left[\lim\limits_{t\to 1}(t\mathbf{i} + t^2\mathbf{j} + t^3\mathbf{k})\right]\cdot\left[\lim\limits_{t\to 1}(\tfrac{1}{t}\mathbf{i} - e^t\mathbf{j})\right]$
$= (\mathbf{i} + \mathbf{j} + \mathbf{k})\cdot(\mathbf{i} - e\mathbf{j})$
$= 1 - e$

c. $\lim\limits_{t\to 1}\left[\mathbf{F}(t) \times \mathbf{G}(t)\right]$

$= \lim\limits_{t\to 1}\begin{vmatrix} \mathbf{i} & \mathbf{j} & \mathbf{k} \\ t & t^2 & t^3 \\ \frac{1}{t} & e^t & 0 \end{vmatrix}$

$= \lim\limits_{t\to 1}\left[-t^3 e^t\mathbf{i} + t^2\mathbf{j} + (te^t - t)\mathbf{k}\right]$

$= -e\mathbf{i} + \mathbf{j} + (e - 1)\mathbf{k}$

$\left[\lim\limits_{t\to 1}\mathbf{F}(t)\right]\times\left[\lim\limits_{t\to 1}\mathbf{G}(t)\right]$

$= \begin{vmatrix} \mathbf{i} & \mathbf{j} & \mathbf{k} \\ 1 & 1 & 1 \\ 1 & e & 0 \end{vmatrix} = -e\mathbf{i} + \mathbf{j} + (e - 1)\mathbf{k}$

58. $(\mathbf{F} \times \mathbf{H})(t) = \mathbf{F}(t + \Delta t) \times \mathbf{G}(t + \Delta t) - \mathbf{F}(t) \times \mathbf{G}(t)$

$= \mathbf{F}(t + \Delta t) \times \mathbf{G}(t + \Delta t) - \mathbf{F}(t + \Delta t) \times \mathbf{G}(t)$
$\quad + \mathbf{F}(t + \Delta t) \times \mathbf{G}(t) - \mathbf{F}(t) \times \mathbf{G}(t)$

$= \mathbf{F}(t + \Delta t) \times \mathbf{G}(t + \Delta t) - \mathbf{G}(t)$
$\quad + [\mathbf{F}(t + \Delta t) - \mathbf{F}(t)] \times \mathbf{G}(t)$

$= \mathbf{F}(t + \Delta t) \times \Delta\mathbf{G}(t) + \Delta\mathbf{F}(t) \times \mathbf{G}(t)$

59. $\lim\limits_{t\to t_0}\left[\mathbf{F}(t) + \mathbf{G}(t)\right]$

$= \lim\limits_{t\to t_0}[f_1(t)\mathbf{i} + f_2(t)\mathbf{j} + f_3(t)\mathbf{k}$
$\quad + g_1(t)\mathbf{i} + g_2(t)\mathbf{j} + g_3(t)\mathbf{k}]$

$= \lim\limits_{t\to t_0}[f_1(t)\mathbf{i} + f_2(t)\mathbf{j} + f_3(t)\mathbf{k}]$
$\quad + \lim\limits_{t\to t_0}[g_1(t)\mathbf{i} + g_2(t)\mathbf{j} + g_3(t)\mathbf{k}]$

$= \lim\limits_{t\to t_0}\mathbf{F}(t) + \lim\limits_{t\to t_0}\mathbf{G}(t)$

60. $\lim\limits_{t\to t_0}\left[\mathbf{F}(t) - \mathbf{G}(t)\right]$

$= \lim\limits_{t\to t_0}[f_1(t)\mathbf{i} + f_2(t)\mathbf{j} + f_3(t)\mathbf{k}$
$\quad - g_1(t)\mathbf{i} - g_2(t)\mathbf{j} - g_3(t)\mathbf{k}]$

$= \lim\limits_{t\to t_0}[f_1(t)\mathbf{i} + f_2(t)\mathbf{j} + f_3(t)\mathbf{k}]$
$\quad - \lim\limits_{t\to t_0}[g_1(t)\mathbf{i} + g_2(t)\mathbf{j} + g_3(t)\mathbf{k}]$

$= \lim\limits_{t\to t_0}\mathbf{F}(t) - \lim\limits_{t\to t_0}\mathbf{G}(t)$

61. $\lim\limits_{t\to t_0}\left[h(t)\mathbf{F}(t)\right]$

$= \lim\limits_{t\to t_0}[h(t)f_1(t)\mathbf{i} + h(t)f_2(t)\mathbf{j} + h(t)f_3(t)\mathbf{k}]$

$= \lim\limits_{t\to t_0}h(t)[f_1(t)\mathbf{i} + f_2(t)\mathbf{j} + f_3(t)\mathbf{k}]$

$= h(t)\lim\limits_{t\to t_0}\mathbf{F}(t)$

62. $\lim\limits_{t\to t_0}\left[\mathbf{F}(t) \times \mathbf{G}(t)\right]$

$= \lim\limits_{t\to t_0}\begin{vmatrix} \mathbf{i} & \mathbf{j} & \mathbf{k} \\ f_1(t) & f_2(t) & f_3(t) \\ g_1(t) & g_2(t) & g_3(t) \end{vmatrix}$

$= \lim\limits_{t\to t_0}[(f_2 g_3 - f_3 g_2)(t)\mathbf{i} - (f_1 g_3 - f_3 g_1)(t)\mathbf{j}$
$\quad + (f_1 g_2 - f_2 g_1)(t)\mathbf{k}]$

$= \left[\left(\lim\limits_{t\to t_0}f_2(t)\right)\left(\lim\limits_{t\to t_0}g_3(t)\right) - \left(\lim\limits_{t\to t_0}f_3(t)\right)\left(\lim\limits_{t\to t_0}g_2(t)\right)\right]\mathbf{i}$

$$+ \left[\left(\lim_{t \to t_0} f_1(t) \right)\left(\lim_{t \to t_0} g_3(t) \right) - \left(\lim_{t \to t_0} f_3(t) \right)\left(\lim_{t \to t_0} g_1(t) \right) \right]\mathbf{j}$$

$$+ \left[\left(\lim_{t \to t_0} f_1(t) \right)\left(\lim_{t \to t_0} g_2(t) \right) - \left(\lim_{t \to t_0} f_2(t) \right)\left(\lim_{t \to t_0} g_1(t) \right) \right]\mathbf{k}$$

$$= \begin{vmatrix} \mathbf{i} & \mathbf{j} & \mathbf{k} \\ \lim_{t \to t_0} f_1(t) & \lim_{t \to t_0} f_2(t) & \lim_{t \to t_0} f_3(t) \\ \lim_{t \to t_0} g_1(t) & \lim_{t \to t_0} g_2(t) & \lim_{t \to t_0} g_3(t) \end{vmatrix}$$

$$= \left[\lim_{t \to t_0} \mathbf{F}(t) \right] \times \left[\lim_{t \to t_0} \mathbf{G}(t) \right]$$

63. a. $3\mathbf{F}(t_0) + 5\mathbf{G}(t_0)$ exists and

$$\lim_{t \to t_0} [3\mathbf{F}(t) + 5\mathbf{G}(t)]$$
$$= 3 \lim_{t \to t_0} \mathbf{F}(t) + 5 \lim_{t \to t_0} \mathbf{G}(t)$$
$$= 3\mathbf{F}(t_0) + 5\mathbf{G}(t_0)$$

The function is continuous at t_0.

b. $\mathbf{F}(t_0) \cdot \mathbf{G}(t_0)$ exists and

$$\lim_{t \to t_0} [\mathbf{F}(t_0) \cdot \mathbf{G}(t_0)]$$

$$= \left[\lim_{t \to t_0} \mathbf{F}(t) \right] \cdot \left[\lim_{t \to t_0} \mathbf{G}(t_0) \right]$$

$$= \mathbf{F}(t_0) \cdot \mathbf{G}(t_0)$$

The function is continuous at t_0.

c. $h(t_0)\mathbf{F}(t_0)$ exists and

$$\lim_{t \to t_0} [h(t)\mathbf{F}(t)]$$

$$= \left[\lim_{t \to t_0} h(t) \right]\left[\lim_{t \to t_0} \mathbf{F}(t) \right]$$

$$= h(t_0)\mathbf{F}(t_0)$$

The function is continuous at t_0.

d. $\mathbf{F}(t_0) \times \mathbf{G}(t_0)$ exists and

$$\lim_{t \to t_0} [\mathbf{F}(t) \times \mathbf{G}(t)]$$

$$= \left[\lim_{t \to t_0} \mathbf{F}(t) \right] \times \left[\lim_{t \to t_0} \mathbf{G}(t) \right]$$

$$= \mathbf{F}(t_0) \times \mathbf{G}(t_0)$$

The function is continuous at t_0.

11.2 Differentiation and Integration of Vector Functions, Page 700

1. $\mathbf{F}'(t) = \mathbf{i} + 2t\mathbf{j} + (1 + 3t^2)\mathbf{k}$

2. $\mathbf{F}'(s) = (1 + 4s)\mathbf{i} + (2s - 1)\mathbf{j} + 2s\mathbf{k}$

3. $\mathbf{F}'(s) = \frac{1}{s}(s\mathbf{i} + 5\mathbf{j} - e^s\mathbf{k}) + (\ln s)(\mathbf{i} - e^s\mathbf{k})$
$= (1 + \ln s)\mathbf{i} + 5s^{-1}\mathbf{j} - e^s(\ln s + s^{-1})\mathbf{k}$

4. $\mathbf{F}'(\theta) = \cos\theta \sec^2\theta\,\mathbf{j} - \sin\theta(\mathbf{i} + \tan\theta)\mathbf{j}$
$$- (3\sin\theta)\mathbf{k}$$
$= (-\sin\theta)\mathbf{i} + (\sec\theta - \sin\theta\tan\theta)\mathbf{j} - 3\sin\theta\mathbf{k}$

5. $\mathbf{F}'(t) = 2t\mathbf{i} - t^{-2}\mathbf{j} + 2e^{2t}\mathbf{k}$
$\mathbf{F}''(t) = 2\mathbf{i} + 2t^{-3}\mathbf{j} + 4e^{2t}\mathbf{k}$

6. $\mathbf{F}'(s) = -4s\mathbf{i} + (\cos s - s\sin s)\mathbf{j} - \mathbf{k}$
$\mathbf{F}''(s) = -4\mathbf{i} + (-2\sin s - s\cos s)\mathbf{j}$

7. $\mathbf{F}'(s) = (\cos s)\mathbf{i} - (\sin s)\mathbf{j} + 2s\mathbf{k}$
$\mathbf{F}''(s) = (-\sin s)\mathbf{i} - (\cos s)\mathbf{j} + 2\mathbf{k}$

8. $\mathbf{F}'(\theta) = 2\sin\theta\cos\theta\,\mathbf{i} - 2\sin 2\theta\,\mathbf{j} + 2\theta\,\mathbf{k}$
$= \sin 2\theta\,\mathbf{i} - 2\sin 2\theta\,\mathbf{j} + 2\theta\,\mathbf{k}$
$\mathbf{F}''(\theta) = 2\cos 2\theta\,\mathbf{i} - 4\cos 2\theta\,\mathbf{j} + 2\mathbf{k}$

9. $f(x) = 2x^2 - 3x^3 - 3x^2 = -3x^3 - x^2$
$f'(x) = -9x^2 - 2x$

10. $f(x) = 1 - x^3 - 2x^2$
$f'(x) = -3x^2 - 4x$

11. $g(x) = \sqrt{1 + 4x^2}$; $g'(x) = \dfrac{4x}{\sqrt{1 + 4x^2}}$

12. $f(x) = \left\| \mathbf{i} + (x^2 - e^x)\mathbf{j} - 2\mathbf{k} \right\|$
$= \sqrt{5 + (x^2 - e^x)^2}$
$f'(x) = \dfrac{(x^2 - e^x)(2x - e^x)}{\sqrt{5 + (x^2 - e^x)^2}}$

13. $\mathbf{V}(t) = \mathbf{R}'(t) = \mathbf{i} + 2t\mathbf{j} + 2\mathbf{k}$
$\mathbf{A}(t) = \mathbf{V}'(t) = \mathbf{R}''(t) = 2\mathbf{j}$; $\mathbf{A}(1) = 2\mathbf{j}$
speed $= \|\mathbf{V}(1)\| = \sqrt{1^2 + 2^2 + 2^2} = 3$
Direction of motion is that of the unit vector
$\dfrac{\mathbf{V}}{\|\mathbf{V}\|} = \frac{1}{3}\mathbf{i} + \frac{2}{3}\mathbf{j} + \frac{2}{3}\mathbf{k}$

14. $\mathbf{V}(t) = -2\mathbf{i} - 2t\mathbf{j} + e^t\mathbf{k}$; $\mathbf{V}(0) = -2\mathbf{i} + \mathbf{k}$
$\mathbf{A}(t) = -2\mathbf{j} + e^t\mathbf{k}$; $\mathbf{A}(0) = -2\mathbf{j} + \mathbf{k}$
$\|\mathbf{V}(0)\| = \sqrt{5}$
Direction of motion is that of the unit vector
$\dfrac{\mathbf{V}}{\|\mathbf{V}\|} = -\frac{2}{\sqrt{5}}\mathbf{j} + \frac{1}{\sqrt{5}}\mathbf{k}$

15. $\mathbf{V}(t) = -\sin t\,\mathbf{i} + \cos t\,\mathbf{j} + 3t\mathbf{k}$;

$$\mathbf{V}(\tfrac{\pi}{4}) = -\frac{\sqrt{2}}{2}\mathbf{i} + \frac{\sqrt{2}}{2}\mathbf{j} + 3\mathbf{k}$$

$$\mathbf{A}(t) = -\cos t\mathbf{i} - \sin t\mathbf{j};$$

$$\mathbf{A}(\tfrac{\pi}{4}) = -\frac{\sqrt{2}}{2}\mathbf{i} - \frac{\sqrt{2}}{2}\mathbf{j}$$

$$\|\mathbf{V}(\tfrac{\pi}{4})\| = \sqrt{\tfrac{2}{4} + \tfrac{2}{4} + 9} = \sqrt{10}$$

Direction of motion is that of the unit vector

$$\frac{\mathbf{V}}{\|\mathbf{V}\|} = -\frac{1}{2\sqrt{5}}\mathbf{i} + \frac{1}{2\sqrt{5}}\mathbf{j} + \frac{3}{\sqrt{10}}\mathbf{k}$$

16. $\mathbf{V}(t) = -2\sin t\mathbf{i} + 2t\mathbf{j} + 2\sin t\mathbf{k};$

$$\mathbf{V}(\tfrac{\pi}{2}) = -2\mathbf{i} + \pi\mathbf{j}$$

$$\mathbf{A}(t) = -2\cos t\mathbf{i} + 2\mathbf{j} - 2\sin t\mathbf{k};$$

$$\mathbf{A}(\tfrac{\pi}{2}) = 2\mathbf{j} - 2\mathbf{k}$$

$$\|\mathbf{V}(\tfrac{\pi}{2})\| = \sqrt{4 + \pi^2}$$

Direction of motion is that of the unit vector

$$\frac{\mathbf{V}}{\|\mathbf{V}\|} = -\frac{2}{\sqrt{4 + \pi^2}}\mathbf{i} + \frac{\pi}{\sqrt{4 + \pi^2}}\mathbf{j}$$

17. $\mathbf{V}(t) = e^t\mathbf{i} - e^{-t}\mathbf{j} + 2e^{2t}\mathbf{k}$

$$\mathbf{V}(\ln 2) = e^{\ln 2}\mathbf{i} - \exp[\ln 2^{-1}]\mathbf{j} + 2e^{2\ln 2}\mathbf{k}$$

$$= 2\mathbf{i} - \tfrac{1}{2}\mathbf{j} + 8\mathbf{k}$$

$$\mathbf{A}(t) = e^t\mathbf{i} + e^{-t}\mathbf{j} + 4e^{2t}\mathbf{k}$$

$$\mathbf{A}(\ln 2) = 2\mathbf{i} + \tfrac{1}{2}\mathbf{j} + 16\mathbf{k}$$

$$\|\mathbf{V}(\ln 2)\| = \sqrt{2^2 + (\tfrac{1}{2})^2 + 8^2} = \frac{\sqrt{273}}{2}$$

Direction of motion is that of the unit vector

$$\frac{\mathbf{V}}{\|\mathbf{V}\|} = \frac{1}{\sqrt{273}}(4\mathbf{i} - \mathbf{j} + 16\mathbf{k})$$

18. $\mathbf{V}(t) = \tfrac{1}{t}\mathbf{i} + \tfrac{3t^2}{2}\mathbf{j} - \mathbf{k};\ \mathbf{V}(1) = \mathbf{i} + \tfrac{3}{2}\mathbf{j} - \mathbf{k}$

$$\mathbf{A}(t) = -\tfrac{1}{t^2}\mathbf{i} + 3t\mathbf{j};\ \mathbf{A}(1) = -\mathbf{i} + 3\mathbf{j}$$

$$\|\mathbf{V}(1)\| = \tfrac{1}{2}\sqrt{4 + 9 + 4} = \frac{17}{2}$$

Direction of motion is that of the unit vector

$$\frac{\mathbf{V}}{\|\mathbf{V}\|} = \frac{1}{\sqrt{17}}(2\mathbf{i} + 3\mathbf{j} - 2\mathbf{k})$$

19. $\mathbf{F}'(t) = 2t\mathbf{i} + 2\mathbf{j} + (3t^2 + 2t)\mathbf{k};$
$\mathbf{F}'(0) = 2\mathbf{j};$
$\mathbf{F}'(1) = 2\mathbf{i} + 2\mathbf{j} + 5\mathbf{k};$
$\mathbf{F}'(-1) = -2\mathbf{i} + 2\mathbf{j} + \mathbf{k}$

20. $\mathbf{F}'(t) = -3t^{-4}\mathbf{i} - 2t^{-3}\mathbf{j} - t^{-2}\mathbf{k};$
$\mathbf{F}'(1) = -3\mathbf{i} - 2\mathbf{j} - \mathbf{k};\ \mathbf{F}'(-1) = -3\mathbf{i} + 2\mathbf{j} - \mathbf{k}$

21. $\mathbf{F}'(t) = \dfrac{1}{(1 + 2t)^2}[(1 + 2t - 2t)\mathbf{i}$

$$+ (2t + 4t^2 - 2t^2)\mathbf{j} + (3t^2 + 6t^3 - 2t^3)\mathbf{k}];$$

$$= \frac{1}{(1 + 2t)^2}[\mathbf{i} + (2t + 2t^2)\mathbf{j} + (3t^2 + 4t^3)\mathbf{k}]$$

$\mathbf{F}'(0) = \mathbf{i};$

$\mathbf{F}'(2) = \tfrac{1}{25}(\mathbf{i} + 12\mathbf{j} + 44\mathbf{k});$

22. $\mathbf{F}'(t) = 2t\mathbf{i} - \sin t\mathbf{j} + (-t^2\sin t + 2t\cos t)\mathbf{k};$

$\mathbf{F}'(0) = \mathbf{0};$

$\mathbf{F}'(\tfrac{\pi}{2}) = \pi\mathbf{i} - \mathbf{j} - \dfrac{\pi^2}{4}\mathbf{k};$

23. $\mathbf{F}'(t) = \cos t\mathbf{i} - \sin t\mathbf{j} + a\mathbf{k};$

$\mathbf{F}'(\tfrac{\pi}{2}) = -\mathbf{j} + a\mathbf{k}$

$\mathbf{F}'(\pi) = -\mathbf{i} + a\mathbf{k};$

24. $\mathbf{F}'(t) = e^t(\pi\cos\pi t + \sin\pi t)\mathbf{i}$

$$+ e^t(-\pi\sin\pi t + \cos\pi t)\mathbf{j}$$

$$+ \pi(\cos\pi t - \sin\pi t)\mathbf{k};$$

$\mathbf{F}'(0) = \pi\mathbf{i} + \mathbf{j} + \pi\mathbf{k};$

$\mathbf{F}'(1) = -\pi e\mathbf{i} - e\mathbf{j} - \pi\mathbf{k};$

$\mathbf{F}'(2) = \pi e^2\mathbf{i} + e^2\mathbf{j} + \pi\mathbf{k}$

25. $\displaystyle\int (t\mathbf{i} - e^{3t}\mathbf{j} + 3\mathbf{k})\,dt = \frac{t^2}{2}\mathbf{i} - \frac{e^{3t}}{3}\mathbf{j} + 3t\mathbf{k} + \mathbf{C}$

26. $\displaystyle\int (\cos t\,\mathbf{i} + \sin t\,\mathbf{j} - 2t\,\mathbf{k})\,dt$

$$= \sin t\mathbf{i} - \cos t\mathbf{j} - t^2\mathbf{k} + \mathbf{C}$$

27. $\displaystyle\int (\ln t\,\mathbf{i} - t\ln t\,\mathbf{j} + 3\ln 3\,\mathbf{k})\,dt$

$$= (t\ln t - t)\mathbf{i} - \tfrac{1}{4}t^2(2\ln t - 1)\mathbf{j}$$

$$+ (3t\ln t - 3t)\mathbf{k} + \mathbf{C}$$

28. $\displaystyle\int (3e^{-t}\mathbf{i} + te^{-t}\mathbf{j} + e^{-t}\sin t\,\mathbf{k})\,dt$

$$= -3e^{-t}\mathbf{i} - (te^{-t} + e^{-t})\mathbf{j}$$

$$- \tfrac{1}{2}e^{-t}(\cos t + \sin t)\mathbf{k} + \mathbf{C}$$

29. $\displaystyle\int [t\ln t\,\mathbf{i} - \sin(1 - t)\mathbf{j} + t\mathbf{k}]\,dt$

$$= \frac{t^2}{2}(\ln t - \tfrac{1}{2})\mathbf{i} - \cos(1 - t)\mathbf{j} + \frac{t^2}{2}\mathbf{k} + \mathbf{C}$$

30. $\displaystyle\int (\sinh t\,\mathbf{i} - 3\mathbf{j} + \cosh t\,\mathbf{k})\,dt$

$$= \cosh t\mathbf{i} - 3t\mathbf{j} + \sinh t\mathbf{k} + \mathbf{C}$$

31. A smooth curve has no corners; the first derivative is continuous.

32. The speed indicates how far an object moves along a tangent to the path of the motion. It is a scalar. The acceleration vector indicates how fast the velocity vector changes. The acceleration vector is not usually along the tangent to the path of the motion.

33. Use the linearity rule. Since all the coefficients are constants, their derivatives

are 0. $\dfrac{d}{dt}(\mathbf{v} + t\mathbf{w}) = \mathbf{w} = \mathbf{i} + 2\mathbf{j} - 3\mathbf{k}$

34. Let $f(t) = \mathbf{v} \cdot t^4\mathbf{w} = 2t^4 - 2t^4 - 15t^4$

$\dfrac{d}{dt}f(t) = -60t^3; \dfrac{d^2}{dt^2}f(t) = -180t^2$

35. Let $f(t) = t\|\mathbf{v}\| + t^2\|\mathbf{w}\|$ where $\|\mathbf{v}\|$ and $\|\mathbf{w}\|$

are constants; $\dfrac{d}{dt}f(t) = \|\mathbf{v}\| + 2t\|\mathbf{w}\|$

$\dfrac{d^2}{dt^2}f(t) = 2\|\mathbf{w}\| = 2\sqrt{1 + 4 + 9} = 2\sqrt{14}$

36. $\mathbf{v} \times \mathbf{w} = \begin{vmatrix} \mathbf{i} & \mathbf{j} & \mathbf{k} \\ 2 & -1 & 5 \\ 1 & 2 & -3 \end{vmatrix} = -7\mathbf{i} + 11\mathbf{j} + 5\mathbf{k}$

Let $f(t) = t\mathbf{v} \times t^2\mathbf{w} = t^3\mathbf{v} \times \mathbf{w}$, so

$\dfrac{d}{dt}f(t) = 3t^2(-7\mathbf{i} + 11\mathbf{j} + 5\mathbf{k})$

37. $3\mathbf{F} - 2\mathbf{G} = [9 + 3t^2 - 2\sin(2 - t)]\mathbf{i}$

$\qquad - 3\cos 3t\,\mathbf{j} + (3t^{-1} + 2e^{2t})\mathbf{k}$

$(3\mathbf{F} - 2\mathbf{G})'(t) = [6t + 2\cos(2 - t)]\mathbf{i}$

$\qquad + 9\sin 3t\,\mathbf{j} - (3t^{-2} - 4e^{2t})\mathbf{k}$

Also,

$3\mathbf{F}'(t) = 6t\mathbf{i} + 9\sin 3t\mathbf{j} - t^{-2}\mathbf{k}$

$-2\mathbf{G}'(t) = 4\cos(2 - t)\mathbf{i} + 4e^{2t}\mathbf{k}$

and,

$3\mathbf{F}'(t) - 2\mathbf{G}'(t) = [6t + 2\cos(2 - t)]\mathbf{i}$

$\qquad + 9\sin 3t\,\mathbf{j} - (3t^{-2} - 4e^{2t})\mathbf{k}$

Therefore, $(3\mathbf{F} - 2\mathbf{G})'(t) = 3\mathbf{F}'(t) - 2\mathbf{G}'(t)$

38. $(\mathbf{F} \cdot \mathbf{G})'(t) = [(3 + t^2)\sin(2 - t) - t^{-1}e^{2t}]'$

$= -(3 + t^2)\cos(2 - t)$

$\qquad + 2t\sin(2 - t) - 2t^{-1}e^{2t} + t^{-2}e^{2t}$

Also,

$\mathbf{F}'(t) = 2t\mathbf{i} + 3\sin 3t\mathbf{j} - t^{-2}\mathbf{k};$

$\mathbf{G}'(t) = -\cos(2 - t)\mathbf{i} - 2e^{2t}\mathbf{k}$

and

$(\mathbf{F}' \cdot \mathbf{G})(t) + (\mathbf{F} \cdot \mathbf{G}')(t)$

$= [2t\mathbf{i} + 3\sin 3t\mathbf{j} - t^{-2}\mathbf{k}][\sin(2 - t)(\mathbf{i} - 2e^{2t}\mathbf{k})]$

$\qquad + [(3 + t^2)\mathbf{i} - \cos 3t\mathbf{j}$

$\qquad + t^{-1}\mathbf{k}] \cdot [-\cos(2 - t)\mathbf{i} - 2e^{2t}\mathbf{k}]$

$= 2t\sin(2 - t) + t^{-2}e^{2t}$

$\qquad - (3 + t^2)\cos(2 - t) - 2t^{-1}e^{2t}$

Therefore, $(\mathbf{F} \cdot \mathbf{G})'(t) = (\mathbf{F}' \cdot \mathbf{G})(t) + (\mathbf{F} \cdot \mathbf{G}')(t)$

39. $\mathbf{V}(t) = \dfrac{d\mathbf{R}}{dt} = e^t\mathbf{i} + t^2\mathbf{j};$

$\mathbf{R}(t) = (e^t + C_1)\mathbf{i} + (\tfrac{t^3}{3} + C_2)\mathbf{j}; \mathbf{R}(0) = \mathbf{i} - \mathbf{j}$

Thus, $C_1 = 0$ and $C_2 = -1$, so

$\mathbf{R}(t) = e^t\mathbf{i} + (\tfrac{1}{3}t^3 - 1)\mathbf{j}$

40. $\dfrac{d}{dt}\mathbf{v}(t) = \dfrac{d^2}{dt^2}\mathbf{R}(t) = 24t^2\mathbf{i} + 4\mathbf{j}$

$\mathbf{V}(t) = (8t^3 + C_1)\mathbf{i} + (4t + C_2)\mathbf{j}; \mathbf{V}(0) = \mathbf{0}$

Thus, $C_1 = C_2 = 0$, so

$\mathbf{R}(t) = (2t^4 + 1)\mathbf{i} + (2t^2 + 2)\mathbf{j}$

41. $\displaystyle\int_0^1 [(t\sqrt{1 + t^2})\mathbf{i} + (1 + t^2)^{-1}\mathbf{j}]\,dt$

$= \tfrac{1}{3}(1 + t^2)^{3/2}\mathbf{i} + \tan^{-1}t\Big|_0^1$

$= \tfrac{1}{3}(2\sqrt{2} - 1)\mathbf{i} + \tfrac{\pi}{4}\mathbf{j}$

42. $\displaystyle\int_0^{\pi/2} [\cos t\,\mathbf{i} + \sin t\,\mathbf{j} + \sin t\cos t\,\mathbf{k}]\,dt$

$= \sin t\,\mathbf{i} - \cos t\,\mathbf{j} + \tfrac{1}{2}\sin^2 t\Big|_0^{\pi/2}$

$= \mathbf{i} + \mathbf{j} + \tfrac{1}{2}\mathbf{k}$

43. $\mathbf{F}'(t) = k(e^{kt}\mathbf{i} - e^{-kt}\mathbf{j})$

$\mathbf{F}''(t) = k^2(e^{kt}\mathbf{i} + e^{-kt}\mathbf{j}) = k^2\mathbf{F}''(t)$

Thus, $\mathbf{F}''(t)$ is parallel to $\mathbf{F}(t)$.

44. $\mathbf{F}'(t) = k(-\sin(kt)\mathbf{i} + \cos(kt)\mathbf{j})$

$\mathbf{F}''(t) = -k^2[\cos(kt)\mathbf{i} + \sin(kt)\mathbf{j}] = -k^2\mathbf{F}''(t)$

Thus, $\mathbf{F}''(t)$ is parallel to $\mathbf{F}(t)$.

45. $\mathbf{F}'(t) = u'(t)\mathbf{i} + v'(t)\mathbf{j} + w'(t)\mathbf{k}$

$\mathbf{F} \cdot \mathbf{F}'(t) = u(t)u'(t) + v(t)v'(t) + w(t)w'(t)$

$\|\mathbf{F}(t)\| = \sqrt{u^2 + v^2 + w^2}$

$\mathbf{F}'(t) = \dfrac{u(t)u'(t) + v(t)v'(t) + w(t)w'(t)}{\sqrt{u^2 + v^2 + w^2}}$

$\|\mathbf{F}'(t)\|\|\mathbf{F}(t)\|' = u(t)u'(t) + v(t)v'(t) + w(t)w'(t)$

46. $[h(t)u(t)]' = [h(t)u(t)]'\mathbf{i} + [h(t)v(t)]'\mathbf{j} + [h(t)w(t)]'\mathbf{k}$

$= [h(t)u'(t) + h'(t)u(t)]\mathbf{i} + [h(t)v'(t) + h'(t)v(t)]\mathbf{j}$

$\qquad + [h'(t)w(t) + h(t)w'(t)]\mathbf{k}$

$= [h(t)u'(t)\mathbf{i} + h(t)v'(t)\mathbf{j} + h(t)w'(t)\mathbf{k}]$

$\qquad + [h'(t)u(t)\mathbf{i} + h'(t)v(t)\mathbf{j} + h'(t)w(t)\mathbf{k}]$

$= h(t)\mathbf{F}'(t) + h'(t)\mathbf{F}$

47. Let $\mathbf{F}(t) = u_1(t)\mathbf{i} + v_1(t)\mathbf{j} + w_1(t)\mathbf{k}$

and $\mathbf{G}(t) = u_2(t)\mathbf{i} + v_2(t)\mathbf{j} + w_2(t)\mathbf{k}$

$[\mathbf{F} \cdot \mathbf{G}]'(t) = [u_1(t)v_1(t)]' + [u_2(t)v_2(t)]' + [w_1(t)w_2(t)]'$

$= u_1(t)u_2'(t) + u_1'(t)u_2(t) + v_1(t)v_2'(t)$

$$+ v_1{}'(t)v_2(t) + w_1(t)w_2{}'(t) + w_1{}'(t)w_2(t)$$

$$= [u_1(t)\mathbf{i} + v_1(t)\mathbf{j} + w_1(t)\mathbf{k}] \cdot [u_2{}'(t)\mathbf{i} + v_2{}'(t)\mathbf{j} + w_2{}'(t)\mathbf{k}]$$

$$= (\mathbf{F}' \cdot \mathbf{G})(t) + (\mathbf{F} \cdot \mathbf{G}')(t)$$

48. Let $\mathbf{F}(t) = u_1(t)\mathbf{i} + v_1(t)\mathbf{j} + w_1(t)\mathbf{k}$

and $\mathbf{G}(t) = u_2(t)\mathbf{i} + v_2(t)\mathbf{j} + w_2(t)\mathbf{k}$

$$[\mathbf{F} \times \mathbf{G}]'(t) = \frac{d}{dt}\begin{vmatrix} \mathbf{i} & \mathbf{j} & \mathbf{k} \\ u_1(t) & v_1(t) & w_1(t) \\ u_2(t) & v_2(t) & w_2(t) \end{vmatrix}$$

$$= [v_1(t)w_2(t) - v_2(t)w_1(t)]'\mathbf{i}$$

$$- [u_1(t)w_2(t) - u_2(t)w_1(t)]'\mathbf{j}$$

$$+ [u_1(t)v_2(t) - u_2(t)v_1(t)]'\mathbf{k}$$

$$= [v_1{}'(t)w_2(t) + v_1(t)w_2{}'(t)$$

$$- v_2(t)w_1{}'(t) - v_2{}'(t)w_1(t)]\mathbf{i}$$

$$+ [u_1{}'(t)w_2(t) + u_1(t)w_2{}'(t)$$

$$- u_2(t)w_1{}'(t) - u_2{}'(t)w_1(t)]\mathbf{j}$$

$$+ [u_1{}'(t)v_2(t) + u_1(t)v_2{}'(t)$$

$$- u_1(t)v_1{}'(t) - u_2{}'(t)v_1(t)]\mathbf{k}$$

Also,

$$(\mathbf{F}' \times \mathbf{G})(t) + (\mathbf{F} \times \mathbf{G}')(t)$$

$$= \begin{vmatrix} \mathbf{i} & \mathbf{j} & \mathbf{k} \\ u_1{}'(t) & v_1{}'(t) & w_1{}'(t) \\ u_2(t) & v_2(t) & w_2(t) \end{vmatrix}$$

$$+ \begin{vmatrix} \mathbf{i} & \mathbf{j} & \mathbf{k} \\ u_1(t) & v_1(t) & w_1(t) \\ u_2{}'(t) & v_2{}'(t) & w_2{}'(t) \end{vmatrix}$$

$$= [v'(t)w_2(t) - v_2{}'(t)w_1(t)\mathbf{i}$$

$$- [u_1(t)w_3(t) - u_3(t)w_1(t)]\mathbf{j}$$

$$+ [u_1{}'(t)v_2(t) - u_2(t)v_1(t)]\mathbf{k}$$

$$+ [v_1(t)w_2{}'(t) - v_2{}'(t)w_1(t)]\mathbf{i}$$

$$- [u_1(t)w_2{}'(t) - u_2{}'(t)w_1(t)]\mathbf{j}$$

$$+ [u_1(t)v_2{}'(t) - v_1(t)u_2(t)]\mathbf{k}$$

Thus,

$$[\mathbf{F} \times \mathbf{G}]'(t) = (\mathbf{F}' \times \mathbf{G})(t) + (\mathbf{F} \times \mathbf{G}')(t)$$

49. $\dfrac{\mathbf{F}}{\|\mathbf{F}\|} = \dfrac{\|\mathbf{F}\|\mathbf{F}' - \mathbf{F}\|\mathbf{F}\|'}{\|\mathbf{F}\|^2} = \dfrac{\mathbf{F}'}{\|\mathbf{F}\|} - \dfrac{\|\mathbf{F}\|'}{\|\mathbf{F}\|^2}\mathbf{F}$

We need to show that

$$\frac{\|\mathbf{F}\|'}{\|\mathbf{F}\|^2} = \frac{\mathbf{F} \cdot \mathbf{F}'}{\|\mathbf{F}\|^3} \quad \text{or} \quad \|\mathbf{F}\|' = \frac{\mathbf{F} \cdot \mathbf{F}'}{\|\mathbf{F}\|}$$

Since $\mathbf{F} = \langle u, v, w \rangle$ and $\mathbf{F}' = \langle u', v', w' \rangle$

$$\mathbf{F} \cdot \mathbf{F}' = uu' + vv' + ww';$$

$$\|\mathbf{F}\| = \sqrt{u^2 + v^2 + w^2}$$

$$\|\mathbf{F}\|' = \frac{uu' + vv' + ww'}{\sqrt{u^2 + v^2 + w^2}} = \frac{\mathbf{F} \cdot \mathbf{F}'}{\|\mathbf{F}\|}$$

11.3 Modeling Ballistics and Planetary Motion, Page 708

1. $T_f = \frac{2}{g}v_0\sin\alpha = \frac{2}{32}(128)\sin 35° \approx 4.6$ sec

$R = \frac{v_0{}^2}{g}\sin 2\alpha = \frac{128^2}{32}\sin 2(35°) \approx 481$ ft

2. $T_f = \frac{2}{g}v_0\sin\alpha = \frac{2}{32}(80)\sin 45° \approx 3.5$ sec

$R = \frac{v_0{}^2}{g}\sin 2\alpha = \frac{80^2}{32}\sin 2(45°) \approx 200$ ft

3. $T_f = \frac{2}{g}v_0\sin\alpha = \frac{2}{9.8}(850)\sin 48.5° \approx 129.9$ sec

$R = \frac{v_0{}^2}{g}\sin 2\alpha = \frac{850^2}{9.8}\sin 2(48.5°) \approx 73{,}175$ m

4. $T_f = \frac{2}{g}v_0\sin\alpha = \frac{2}{9.8}(185)\sin 43.5° \approx 26.0$ sec

$R = \frac{v_0{}^2}{g}\sin 2\alpha = \frac{185^2}{9.8}\sin 2(43.5°) \approx 3{,}488$ m

5. $T_f = \frac{2}{g}v_0\sin\alpha = \frac{2}{9.8}(23.3)\sin 23.74° \approx 1.9$ sec

$R = \frac{v_0{}^2}{g}\sin 2\alpha = \frac{23.3^2}{9.8}\sin 47.48 \approx 41$ m

6. $T_f = \frac{2}{g}v_0\sin\alpha = \frac{2}{9.8}(38.14)\sin 31.04° \approx 4.0$ sec

$R = \frac{v_0{}^2}{g}\sin 2\alpha = \frac{38.14^2}{9.8}\sin 62.08° \approx 131$ m

7. $T_f = \frac{2}{g}v_0\sin\alpha = \frac{2}{32}(100)\sin 14.11° \approx 1.5$ sec

$R = \frac{v_0{}^2}{g}\sin 2\alpha = \frac{100^2}{32}\sin 28.22° \approx 148$ ft

8. $T_f = \frac{2}{g}v_0\sin\alpha = \frac{2}{32}(88)\sin 78.09° \approx 5.4$ sec

$R = \frac{v_0{}^2}{g}\sin 2\alpha = \frac{88^2}{32}\sin 156.18° \approx 98$ m

9. $r = \|\mathbf{R}(t)\| = \sqrt{(2t)^2 + t^2} = t\sqrt{5}$

$\dfrac{dr}{dt} = \sqrt{5}; \dfrac{d^2r}{dt^2} = 0;$

$\theta = \tan^{-1}\dfrac{t}{2t} = \tan^{-1}\frac{1}{2}; \dfrac{d\theta}{dt} = \dfrac{d^2\theta}{dt^2} = 0$

$\mathbf{V}(t) = \sqrt{5}\,\mathbf{u}_r; \mathbf{A}(t) = \mathbf{0}$

10. $r = \|\mathbf{R}(t)\| = \sqrt{(\cos t)^2 + (\sin t)^2} = 1$

$\dfrac{dr}{dt} = \dfrac{d^2r}{dt^2} = 0;$

$\theta = \tan^{-1}\dfrac{\cos t}{\sin t} = t;\ \dfrac{d\theta}{dt} = 1;\ \dfrac{d^2\theta}{dt^2} = 0$

$\mathbf{V}(t) = (0)\mathbf{u}_r + (1)(-1)\mathbf{u}_\theta = -\mathbf{u}_\theta$

$\mathbf{A}(t) = [0 - (1)(-1)^2]\mathbf{u}_r + [1(0)+2(0)(-1)]\mathbf{u}_\theta$

$\quad = -\mathbf{u}_\theta$

11. $\mathbf{V} = \dfrac{dr}{dt}\mathbf{u}_r + r\dfrac{d\theta}{dt}\mathbf{u}_\theta = (2\cos 2t)\mathbf{u}_r + (2\sin 2t)\mathbf{u}_\theta$

$\mathbf{A} = \left[\dfrac{d^2r}{dt^2} - r\left(\dfrac{d\theta}{dt}\right)^2\right]\mathbf{u}_r + \left[r\dfrac{d^2\theta}{dt^2} + 2\dfrac{dr}{dt}\dfrac{d\theta}{dt}\right]\mathbf{u}_\theta$

$\quad = [-4\sin 2t - (\sin 2t)(4)]\mathbf{u}_r$

$\qquad + [(\sin 2t)(0) + 2(2)(2\cos 2t)]\mathbf{u}_\theta$

$\quad = (-8\sin 2t)\mathbf{u}_r + (8\cos 2t)\mathbf{u}_\theta$

12. $\mathbf{V} = \dfrac{dr}{dt}\mathbf{u}_r + r\dfrac{d\theta}{dt}\mathbf{u}_\theta = e^{t-1}\mathbf{u}_r + e^{t-1}(-1)\mathbf{u}_\theta$

$\quad = e^{t-1}(\mathbf{u}_r - \mathbf{u}_\theta)$

$\mathbf{A} = \left[\dfrac{d^2r}{dt^2} - r\left(\dfrac{d\theta}{dt}\right)^2\right]\mathbf{u}_r + \left[r\dfrac{d^2\theta}{dt^2} + 2\dfrac{dr}{dt}\dfrac{d\theta}{dt}\right]\mathbf{u}_\theta$

$\quad = e^{t-1}[(1-1)\mathbf{u}_r + (-2)\mathbf{u}_\theta] = -2e^{t-1}\mathbf{u}_\theta$

13. $r = 5 + 5[\cos(2t+1)];\ \dfrac{dr}{dt} = 10\sin(2t+1)$

$\dfrac{d^2r}{dt^2} = -20\cos(2t+1);\ \dfrac{d\theta}{dt} = 2;\ \dfrac{d^2\theta}{dt^2} = 0$

$\mathbf{V} = \dfrac{dr}{dt}\mathbf{u}_r + r\dfrac{d\theta}{dt}\mathbf{u}_\theta$

$\quad = -10\sin(2t+1)\mathbf{u}_r + 10[1 + \cos(2t+1)]\mathbf{u}_\theta$

$\mathbf{A} = \left[\dfrac{d^2r}{dt^2} - r\left(\dfrac{d\theta}{dt}\right)^2\right]\mathbf{u}_r + \left[r\dfrac{d^2\theta}{dt^2} + 2\dfrac{dr}{dt}\dfrac{d\theta}{dt}\right]\mathbf{u}_\theta$

$\quad = [-20\cos(2t+1) - 20[1 + \cos(2t+1)]]\mathbf{u}_r$

$\qquad + -4[-10\sin(2t+1)]\mathbf{u}_\theta$

$\quad = [-40\cos(2t+1) - 20]\mathbf{u}_r - 40\sin(2t+1)\mathbf{u}_\theta$

14. $r = \dfrac{1}{1 - \cos t};\ \dfrac{dr}{dt} = -\dfrac{\sin t}{(1 - \cos t)^2}$

$\dfrac{d^2r}{dt^2} = -\dfrac{(1 - \cos t)^2(\cos t) - (\sin t)(2)(1 - \cos\theta)(\sin t)}{(1 - \cos t)^4}$

$\quad = -\dfrac{\cos t - \cos^2 t - 2\sin^2 t}{(1 - \cos t)^3}$

$\dfrac{d\theta}{dt} = 1;\ \dfrac{d^2\theta}{dt^2} = 0$

$\mathbf{V} = \dfrac{dr}{dt}\mathbf{u}_r + r\dfrac{d\theta}{dt}\mathbf{u}_\theta$

$\quad = -\dfrac{\sin t}{(1 - \cos t)^2}\mathbf{u}_r + \dfrac{1}{1 - \cos t}\mathbf{u}_\theta$

$\mathbf{A} = \left[\dfrac{d^2r}{dt^2} - r\left(\dfrac{d\theta}{dt}\right)^2\right]\mathbf{u}_r + \left[r\dfrac{d^2\theta}{dt^2} + 2\dfrac{dr}{dt}\dfrac{d\theta}{dt}\right]\mathbf{u}_\theta$

$\quad = \left[-\dfrac{\cos t - \sin^2 t - 1}{(1 - \cos t)^3} - \dfrac{1}{1 - \cos t}\right]\mathbf{u}_r$

$\qquad + \left[0 - \dfrac{2\sin t}{(1 - \cos t)^2}\right]\mathbf{u}_\theta$

$\quad = \left[\dfrac{\cos^2 t - \cos t - \sin^2 t}{(1 - \cos t)^3}\right]\mathbf{u}_r - \left[\dfrac{2\sin t}{(1 - \cos t)^2}\right]\mathbf{u}_\theta$

15. $\alpha = 45°$, so we can use the formula for maximum range: $R_m = v_0^2/g$.

$2000 = \dfrac{v_0^2}{9.8},\ v_0 \approx 140$ m/sec

16. $\sin 2\alpha = \dfrac{Rg}{v_0^2} = \dfrac{1{,}500(32)}{(300)^2} \approx 0.5333;\ \alpha \approx 16°$

17. $\sin 2\alpha = \dfrac{Rg}{v_0^2} = \dfrac{600(32)}{(167.1)^2} \approx 0.6876;\ \alpha \approx 21.7°$

18. a. $y' = v_0\sin\alpha - gt$ if

$\quad t = \dfrac{v_0\sin\alpha}{g} = \dfrac{280}{32\sqrt{2}} = \dfrac{35}{4\sqrt{2}}$

For the maximum height,

$y = \dfrac{280}{\sqrt{2}}\dfrac{35}{4\sqrt{2}} - \dfrac{35^2(32)}{2(16)(2)} \approx 1{,}207$ ft

b. $T_f = \dfrac{2v_0\sin\alpha}{g} = \dfrac{2(280)\sqrt{2}}{32(2)} \approx 12.4$ sec

$R = \dfrac{v_0^2}{g} = \dfrac{280^2\sin 90°}{32} = 2{,}450$ ft

c. $\mathbf{V}(t) = \dfrac{280}{\sqrt{2}}\mathbf{i} + [280\sqrt{2} - 32(12.37)]\mathbf{j}$

$\|\mathbf{V}\| = \sqrt{39{,}2{,}00 + 39{,}144.6} \approx 280$ ft/s

19. The maximum height is reached when $y'(t) = 0$. We can use this equation to find the time at which this occurs, then use that time in $y(t)$ to find the height.

$y(t) = -16t^2 + (V_0\sin\alpha)t + s_0$

$\quad = -16t^2 + 45t + 4$

$y'(t) = -32t + 45 = 0$ when $t = \frac{45}{32}$ sec

$y(\frac{45}{32}) = -16(\frac{45}{32})^2 + 45(\frac{45}{32}) + 4 \approx 35.64$ ft

The ball will land when $y = 0$. We can use $y(t) = 0$ to find the time of flight, and then use that time to find $x(t)$.

$-16t^2 + 45t + 4 = 0$ when $t \approx 2.8987$ sec

$x(t) = (v_0\cos\alpha)t = \dfrac{\sqrt{3}}{2}(90)t = 45\sqrt{3}\,t$

$x(2.8987) \approx 225.93$ ft

To find the distance to the fence we will find t for $y(t) = 5$, then use that time in $x(t)$.

$-16t^2 + 45t + 4 = 5$ when $t \approx 2.79$ sec.

$x(2.79) = 45\sqrt{3}(2.79) \approx 217$ ft

20. 400 ft $= v_0 \cos 24°t$, so $t \approx \dfrac{400}{v_0 \cos 24°}$

$9 = (v_0 \sin 24°)t - 16t^2$

$\quad = v_0 \sin 24° \left[\dfrac{400}{v_0 \cos 24°} \right] - 16 \left[\dfrac{400^2}{v_0{}^2 \cos^2 24°} \right]$

$v_0 = \pm \dfrac{1,600}{\sqrt{\cos 24°(400 \sin 24° - 9 \cos 24°)}}$

$\quad \approx 134.69$ ft/s

Thus, $t \approx \dfrac{400}{134.69 \cos 24°} \approx 3.25$ sec

21. $312 = v_0 \cos 32°t$, so $v_0 \approx \dfrac{312}{t \cos 32°}$

$8 = (v_0 \sin 32°)t - 16t^2 + 3.5$

$\quad = -16t^2 + \left(\dfrac{312}{t \cos 32°} \right)(\sin 32°)t + 3.5$

$t^2 = \dfrac{4.5 - 312 \tan 32°}{-16} \approx 3.45$

Thus, $t \approx \dfrac{312}{107.6 \cos 32°} \approx 3.43$ sec

22. Solve $-\tfrac{1}{2}(32)t^2 + 50(\tfrac{\sqrt{2}}{2})t + 6.5 = 6$

The positive solution is $t \approx 2.22376$

$s = (50 \cos 45°)t \approx 78.62$ ft

Jerry is 78.62 ft from Steve when he catches the pass. Since Jerry runs at 32 ft/sec the *most* he can be is 71.168 ft from the line of scrimmage when he catches the pass. Therefore, Steve must be $78.62 - 71.168 \approx 7.452$ ft (or less) behind the line of scrimmage when he releases the ball. Therefore, Steve fades back about 7.452 ft before releasing the ball.

23. $T_f = \dfrac{2v_0 \sin \alpha}{g} = \dfrac{2(125)\sqrt{2}}{2(32)} \approx 5.5$ sec

24. The time to reach the maximum height is $t = v_0 \sin \alpha / 32$

$12 = \dfrac{v_0{}^2 \sin^2 \alpha^2}{32} - \dfrac{v_0{}^2 \sin^2 \alpha^2}{64}$

$v_0 \sin \alpha = 16\sqrt{3}$

At the basket,

$10 = (v_0 \sin \alpha)t - 16t^2 + 7$

$10 = 16\sqrt{3}\,t - 16t^2 + 7$

$t = \dfrac{2\sqrt{3} \pm 3}{4} \approx 1.616,\ 0.116$

(disregard the second because it is not plausible)

Now, $\sin \alpha = \dfrac{16\sqrt{3}}{v_0}$; $\cos \alpha = \dfrac{20}{v_0 t} \approx \dfrac{12.3762}{v_0}$

so that

$\cos^2 \alpha + \sin^2 \alpha = \left(\dfrac{12.3762}{v_0} \right)^2 + \left(\dfrac{16\sqrt{3}}{v_0} \right)^2 = 1$

Solving for v_0 we find $v_0 \approx 30.35$ ft.

25. $s_0 = 5$ ft; $\alpha = 46°$; $v_0 = 25$ ft/s

$0 = (25 \sin 46°)t - 16t^2 + 5$

$t = 1.3547$ sec (disregard negative solution)

$x = (v_0 \cos \alpha)t = 25(\cos 46°)(1.3547) \approx 23.53$ ft

26. $R = 4,700$ ft; $\alpha = 45°$; $R = \dfrac{v_0{}^2 \sin 2\alpha}{g}$

$4,700 = \dfrac{v_0{}^2}{32}$ so that $v_0 \approx 387.8$ ft/s

27. $\mathbf{V} = \dfrac{dr}{dt}\mathbf{u}_r + r\dfrac{d\theta}{dt}\mathbf{u}_\theta$

$\quad = (2 \cos t)\mathbf{u}_r + (3 + 2 \sin t)3t^2 \mathbf{u}_\theta.$

$\mathbf{A} = \left[\dfrac{d^2 r}{dt^2} - r \left(\dfrac{d\theta}{dt} \right)^2 \right]\mathbf{u}_r + \left[r\dfrac{d^2\theta}{dt^2} + 2\dfrac{dr}{dt}\dfrac{d\theta}{dt} \right]\mathbf{u}_\theta$

$\quad = \left[-2 \sin t - (3 + 2 \sin t)(3t^2)^2 \right]\mathbf{u}_r$

$\qquad + \left[(3 + 2 \sin t)(6t) + 2(2 \cos t)(3t^2) \right]\mathbf{u}_\theta$

$\quad = \left[-2 \sin t - 27t^4 - 18t^4 \sin t \right]\mathbf{u}_r$

$\qquad + \left[18t + 12t \sin t + 12t^2 \cos t \right]\mathbf{u}_\theta$

28. $\dfrac{d\mathbf{r}}{dt} = (-a\omega \sin \omega t)\mathbf{i} + (a\omega \cos \omega)\mathbf{j}$

$\mathbf{r} \cdot \dfrac{d\mathbf{r}}{dt} = -a^2 \omega \sin \omega t \cos \omega t + a^2 \omega \sin \omega t \cos \omega t$

$\quad = 0$

29. $\mathbf{r} = (a \cos \omega t)\mathbf{i} + (a \sin \omega t)\mathbf{j}$; $\left\| \dfrac{d\mathbf{r}}{dt} \right\| = a\omega$

30. $\mathbf{r} = (a \cos \omega t)\mathbf{i} + (a \sin \omega t)\mathbf{j}$;

$\mathbf{v} = \dfrac{d\mathbf{r}}{dt} = (-a\omega \sin \omega t)\mathbf{i} + (a\omega \cos \omega t)\mathbf{j}$;

$\mathbf{a} = (-a\omega^2 \cos \omega t)\mathbf{i} + (-a\omega^2 \sin \omega t)\mathbf{j} = -a\omega^2 \mathbf{r}$

31. $\mathbf{r}(t) = (a \cos \omega t)\mathbf{i} + (a \sin \omega t)\mathbf{j}$

$\mathbf{v}(t) = \mathbf{r}'(t) = (-\omega a \sin \omega t)\mathbf{i} + (\omega a \cos \omega t)\mathbf{j}$

$\mathbf{a}(t) = \mathbf{v}'(t) = \mathbf{r}''(t)$

$\quad = (-\omega^2 a \cos \omega t)\mathbf{i} + (-\omega^2 a \sin \omega t)\mathbf{j}.$

$\| \mathbf{a}(t) \| = \sqrt{(-\omega^2 a \cos \omega t)^2 + (-\omega^2 a \sin \omega t)^2}$

$\quad = a\omega^2$

32. $\| \mathbf{F} \| = mA_N = 2$ lb;

$$2 = \frac{3/16}{32}\kappa\left(\frac{ds}{dt}\right)^2$$

When $\kappa = 1$, $\frac{ds}{dt} \approx 18.5$ ft/s

33. The initial value of the rock is

$$\mathbf{V}_r = 25\left(\frac{\sqrt{3}}{2}\mathbf{i} + \frac{1}{2}\mathbf{j}\right)$$

The effective velocity is

$$\mathbf{V}_0 = \left(\frac{25\sqrt{3}}{2} + 15\right)\mathbf{i} + \frac{25}{2}\mathbf{j}; \; s_0 = -10\mathbf{i} + 30\mathbf{j}$$

$$\mathbf{A} = -32\mathbf{j}$$

$$\mathbf{V} = (-32t)\mathbf{j} + \mathbf{V}_0$$

$$= -32\mathbf{j} + \left(\frac{25\sqrt{3}}{2} + 15\right)\mathbf{i} + \frac{25}{2}\mathbf{j};$$

$$= \left(\frac{25\sqrt{3}}{2} + 15\right)\mathbf{i} + \left(\frac{25}{2} - 32t\right)\mathbf{j}$$

$$\mathbf{s} = \left[\left(\frac{25\sqrt{3}}{2} + 15\right)t - 10\right]\mathbf{i} + \left[-16t^2 + \frac{25t}{2} + 30\right]\mathbf{j}$$

The rock hits the bottom when

$$-16t^2 + \frac{25t}{2} + 30 = 0$$

$$t = \frac{25 \pm \sqrt{8,305}}{64}$$

If we disregard the negative value, we find $t \approx 1.82$ sec. The distance from the base of the cliff is

$$x = \left(\frac{25\sqrt{3}}{2} + 15\right)\left(\frac{25 + \sqrt{8,305}}{64}\right) - 10 \approx 56.7 \text{ ft}$$

34. $\mathbf{A}(t) = \frac{d}{dt}\mathbf{V}$

$$= \frac{d}{dt}\left[\frac{dr}{dt}\mathbf{u}_r + r\frac{d\theta}{dt}\mathbf{u}_\theta\right]$$

$$= \frac{dr}{dt}\frac{d\mathbf{u}_r}{dt} + \mathbf{u}_r\frac{d^2r}{dt^2}$$

$$+ r\frac{d\theta}{dt}\frac{d\mathbf{u}_\theta}{dt} + r\frac{d^2\theta}{dt^2}\mathbf{u}_\theta + \frac{dr}{dt}\frac{d\theta}{dt}\mathbf{u}_\theta$$

$$= \frac{dr}{dt}\mathbf{u}_\theta\frac{d\theta}{dt} + \mathbf{u}_r\frac{d^2r}{dt^2} + r\left(\frac{d\theta}{dt}\right)^2(-1)\mathbf{u}_r$$

$$+ r\frac{d^2\theta}{dt^2}\mathbf{u}_\theta + \frac{dr}{dt}\frac{d\theta}{dt}\mathbf{u}_\theta$$

$$= \left[\frac{d^2r}{dt^2} - r\left(\frac{d\theta}{dt}\right)^2\right]\mathbf{u}_r + \left[2\frac{dr}{dt}\frac{d\theta}{dt} + r\frac{d^2\theta}{dt^2}\right]\mathbf{u}_\theta$$

35. From Problem 34

$$\mathbf{A}(t) = \left[\frac{d^2r}{dt^2} - r\left(\frac{d\theta}{dt}\right)^2\right]\mathbf{u}_r + \left[2\frac{dr}{dt}\frac{d\theta}{dt} + r\frac{d^2\theta}{dt^2}\right]\mathbf{u}_\theta$$

Since $\mathbf{F} = m\mathbf{A}$, $\mathbf{FR} = F_r\mathbf{u}_r + F_\theta\mathbf{u}_\theta$ where

$$F_r = m\left[\frac{d^2r}{dt^2} - r\left(\frac{d\theta}{dt}\right)^2\right] \text{ and}$$

$$F_\theta = m\left[2\frac{dr}{dt}\frac{d\theta}{dt} + r\frac{d^2\theta}{dt^2}\right]$$

36. From Problem 35,

$$rF_\theta = 2mr\frac{dr}{dt}\frac{d\theta}{dt} + mr^2\frac{d^2\theta}{dt^2}$$

Also, $\frac{d}{dt}\left(mr^2\frac{d\theta}{dt}\right) = mr^2\frac{d^2\theta}{dt^2} + 2mr\frac{dr}{dt}\frac{d\theta}{dt}$

Since the two results are the same the desired conclusion is immediate.

37. $\frac{dA}{dt} = \frac{1}{2}r^2\frac{d\theta}{dt} = $ a constant (by Problem 38)

11.4 Unit Tangent and Normal Vectors; Curvature, Page 719

1. $\mathbf{R}(t) = t^2\mathbf{i} + t^3\mathbf{j}$, $t \neq 0$

$$\mathbf{R}'(t) = 2t\mathbf{i} + 3t^2\mathbf{j}; \|\mathbf{R}'(t)\| = t\sqrt{4 + 9t^2}$$

$$\mathbf{T}(t) = \frac{\mathbf{R}'(t)}{\|\mathbf{R}'(t)\|} = \frac{2}{\sqrt{4 + 9t^2}}\mathbf{i} + \frac{3t}{\sqrt{4 + 9t^2}}\mathbf{j}$$

$$\mathbf{T}'(t) = -\frac{18t}{(4 + 9t^2)^{3/2}}\mathbf{i} + \frac{12}{(4 + 9t^2)^{3/2}}\mathbf{j}$$

$$\|\mathbf{T}'(t)\| = \frac{6}{9t^2 + 4}$$

$$\mathbf{N}(t) = \frac{\mathbf{T}'(t)}{\|\mathbf{T}'(t)\|} = \frac{-3t}{\sqrt{4 + 9t^2}}\mathbf{i} + \frac{2}{\sqrt{4 + 9t^2}}\mathbf{j}$$

2. $\mathbf{R}(t) = t^2\mathbf{i} + \sqrt{t}\mathbf{j}$, $t > 0$

$$\mathbf{R}'(t) = 2t\mathbf{i} + \frac{1}{2\sqrt{t}}\mathbf{j}; \|\mathbf{R}'(t)\| = \frac{\sqrt{16t^3 + 1}}{2\sqrt{t}}$$

$$\mathbf{T}(t) = \frac{4t^{3/2}}{\sqrt{16t^3 + 1}}\mathbf{i} + \frac{1}{\sqrt{16t^3 + 1}}\mathbf{j}$$

$$\mathbf{N}(t) = \frac{-1}{\sqrt{16t^3 + 1}}\mathbf{i} + \frac{4t^{3/2}}{\sqrt{16t^3 + 1}}\mathbf{j}$$

3. $\mathbf{R}(t) = e^t\cos t\,\mathbf{i} + e^t\sin t\,\mathbf{j}$

$$\mathbf{R}'(t) = e^t(-\sin t + \cos t)\mathbf{i}$$
$$+ e^t(\cos t + \sin t)\mathbf{j}$$

$$\|\mathbf{R}'(t)\| = \sqrt{2}\,e^t$$

$$\mathbf{T}(t) = \frac{\sqrt{2}}{2}[(\cos t - \sin t)\mathbf{i} + (\cos t + \sin t)\mathbf{j}]$$

$$\mathbf{N}(t) = -\frac{\sqrt{2}}{2}[(\sin t + \cos t)\mathbf{i} + (\sin t - \cos t)\mathbf{j}]$$

4. $\mathbf{R}(t) = t\cos t\,\mathbf{i} + t\sin t\,\mathbf{j}$

$$\mathbf{R}'(t) = (-t\sin t + \cos t)\mathbf{i} + (t\cos t + \sin t)\mathbf{j}$$

$$\|\mathbf{R}'(t)\| = \sqrt{t^2 + 1}$$

$$\mathbf{T}(t) = \frac{1}{\sqrt{t^2+1}}[(-t\sin t + \cos t)\mathbf{i}$$
$$+ (t\cos t + \sin t)\mathbf{j}]$$

$$\mathbf{N}(t) = \frac{1}{\sqrt{t^2+1}}[(\sin t + t\cos t)\mathbf{i}$$
$$+ (t\sin t - \cos t)\mathbf{j}]$$

5. $\mathbf{R}(t) = \cos t\,\mathbf{i} + \sin t\,\mathbf{j} + t\mathbf{k}$

$\mathbf{R}'(t) = (-\sin t)\mathbf{i} + (\cos t)\mathbf{j} + \mathbf{k}$

$\|\mathbf{R}'(t)\| = \sqrt{2}$

$\mathbf{T}(t) = \frac{1}{\sqrt{2}}(-\sin t\,\mathbf{i} + \cos t\,\mathbf{j} + \mathbf{k})$

$\mathbf{T}'(t) = \frac{1}{\sqrt{2}}(-\cos t\,\mathbf{i} - \sin t\,\mathbf{j})$

$\mathbf{N}(t) = -\cos t\,\mathbf{i} - \sin t\,\mathbf{j}$

6. $\mathbf{R}(t) = \sin t\,\mathbf{i} + \cos t\,\mathbf{j} + t\mathbf{k}$

$\mathbf{R}'(t) = (\cos t)\mathbf{i} + (-\sin t)\mathbf{j} + \mathbf{k}$

$\|\mathbf{R}'(t)\| = \sqrt{2}$

$\mathbf{T}(t) = \frac{1}{\sqrt{2}}(\cos t\,\mathbf{i} - \sin t\mathbf{j} + \mathbf{k})$

$\mathbf{T}(t) = \frac{1}{\sqrt{2}}(-\sin t\,\mathbf{i} - \cos t\,\mathbf{j})$

$\mathbf{N}(t) = \cos t\,\mathbf{i} + \sin t\,\mathbf{j}$

7. $\mathbf{R}(t) = \ln t\,\mathbf{i} + t^2\mathbf{k}$

$\mathbf{R}'(t) = \frac{1}{t}\mathbf{i} + 2t\mathbf{k}$

$\|\mathbf{R}'(t)\| = \frac{1}{t}\sqrt{1+4t^4}$

$\mathbf{T}(t) = \frac{1}{\sqrt{1+4t^4}}(t\mathbf{i} + 2t^2\mathbf{k})$

$\mathbf{T}'(t) = \frac{-8t^3\mathbf{i} + 4t\mathbf{k}}{(1+4t^4)^{3/2}}$

$\|\mathbf{T}'(t)\| = \sqrt{\frac{(-8t^3)^2 + (4t)^2}{(1+4t^4)^3}}$

$= \frac{4t\sqrt{4t^4+1}}{(1+4t^4)^{3/2}} = \frac{4t}{1+4t^4}$

$\mathbf{N}(t) = \frac{\mathbf{T}'(t)}{\|\mathbf{T}'(t)\|} = \frac{1}{\sqrt{1+4t^4}}(-2t^2\mathbf{i} + t\mathbf{k})$

8. $\mathbf{R}(t) = e^{-t}\sin t\,\mathbf{i} + e^{-t}\mathbf{j} + e^{-t}\cos t\mathbf{k}$

$\mathbf{R}'(t) = e^{-t}(\cos t - \sin t)\mathbf{i} + (-e^{-t})\mathbf{j}$

$+ e^{-t}(-\sin t - \cos t)\mathbf{k}$

$\|\mathbf{R}'(t)\| = \sqrt{3}e^{-t}$

$\mathbf{T}(t) = \frac{\mathbf{R}'(t)}{\|\mathbf{R}'(t)\|} = \frac{1}{\sqrt{3}}[(\cos t - \sin t)\mathbf{i}$

$- \mathbf{j} - (\cos t + \sin t)\mathbf{k}]$

$\mathbf{T}'(t) = \frac{1}{\sqrt{3}}[(-\sin t - \cos t)\mathbf{i} - (-\sin t + \cos t)\mathbf{k}]$

$\|\mathbf{T}'(t)\| = \frac{\sqrt{6}}{3}$

$\mathbf{N}(t) = \frac{\mathbf{T}'(t)}{\|\mathbf{T}'(t)\|} = \frac{1}{\sqrt{2}}[(-\sin t - \cos t)\mathbf{i}$

$+ (\sin t - \cos t)\mathbf{k}]$

9. $\mathbf{R}'(t) = 2\mathbf{i} + \mathbf{j}; \frac{ds}{dt} = \|\mathbf{R}'(t)\| = \sqrt{5}$

$s = \int_0^4 \sqrt{5}\,dt = 4\sqrt{5}$

10. $\mathbf{R}'(t) = \mathbf{i} + 3\mathbf{j}; \frac{ds}{dt} = \|\mathbf{R}'(t)\| = \sqrt{10}$

$s = \int_0^4 \sqrt{10}\,dt = 4\sqrt{10}$

11. $\mathbf{R}'(t) = 3\mathbf{i} - (3\sin t)\mathbf{j} + (3\cos t)\mathbf{k}$

$\frac{ds}{dt} = \|\mathbf{R}'(t)\| = \sqrt{3^2 + 9\sin^2 t + 9\cos^2 t}$

$= 3\sqrt{2}$

$s = \int_0^{\pi/2} 3\sqrt{2}\,dt = \frac{3\sqrt{2}\pi}{2}$

12. $\mathbf{R}'(t) = t\mathbf{i} + 2\mathbf{j} + 3\mathbf{k}$

$\frac{ds}{dt} = \|\mathbf{R}'(t)\| = \sqrt{1^2 + 2^2 + 3^2} = \sqrt{14}$

$s = \int_0^2 \sqrt{14}\,dt = 2\sqrt{14}$

13. $\mathbf{R}'(t) = -4\sin t\,\mathbf{i} + 4\cos t\mathbf{j} + 5\mathbf{k}$

$\frac{ds}{dt} = \|\mathbf{R}'(t)\| = \sqrt{41}$

$s = \int_0^\pi \sqrt{41}\,dt = \sqrt{41}\pi$

14. $\mathbf{R}'(t) = -3\cos^2 t\sin t\,\mathbf{i} - 2\cos t\sin t\mathbf{k}$

$\frac{ds}{dt} = \|\mathbf{R}'(t)\| = \cos t\sin t\sqrt{4 + 9\cos^2 t}$

$$s = \int_0^{2\pi} \sin t \cos t \sqrt{9 \cos^2 t + 4}\ dt$$

$$= 4 \int_0^{\pi/2} \sin t \cos t \sqrt{9 \cos^2 t + 4}\ dt$$

$$= \frac{4}{18} \int_0^{\sqrt{13}} u^{1/2}\ du \quad \boxed{u = 9 \cos^2 t + 4}$$

$$= \frac{4}{18} \frac{2}{3} u^{3/2} \Big|_0^{\sqrt{13}} = \frac{52}{27} \sqrt{13}$$

15. $\kappa = 0$, as this is a straight line

16. $\kappa = 0$, as this is a straight line

17. $y' = 1 - \frac{2}{9}x;\ y'' = -\frac{2}{9};$

at $x = 3,\ y' = \frac{1}{3}$ and $y'' = -\frac{2}{9}$

$$\kappa = \frac{|y''|}{(1 + y'^2)^{3/2}} = \frac{\frac{2}{9}}{(1 + \frac{1}{9})^{3/2}} = \frac{3\sqrt{10}}{50} \approx 0.19$$

18. $y' = 4x;\ y'' = 4;$ at $x = 1,\ y' = 4$ and $y'' = 4$

$$\kappa = \frac{|y''|}{(1 + y'^2)^{3/2}} = \frac{4}{(1 + 16)^{3/2}} = \frac{4}{17\sqrt{17}}$$

$$\approx 0.06$$

19. $y' = 2ax + b;\ y'' = 2a;$

at $x = c,\ y' = 2ac + b$ and $y'' = 2a$

$$\kappa = \frac{|y''|}{(1 + y'^2)^{3/2}} = \frac{2|a|}{[1 + (2ac + b)^2]^{3/2}}$$

20. $y' = 1 - x^{-2};\ y'' = 2x^{-3};$

at $x = 1,\ y' = 0$ and $y'' = 2$

$$\kappa = \frac{|y''|}{(1 + y'^2)^{3/2}} = \frac{2}{(1 + 0)^{3/2}} = 2$$

21. $y' = \frac{-2x}{2\sqrt{4 - x^2}} = -\frac{x}{\sqrt{4 - x^2}};$

$$y'' = -\frac{\sqrt{4 - x^2} + x^2/\sqrt{4 - x^2}}{4 - x^2};$$

at $x = 1,\ y' = -\frac{1}{\sqrt{3}}$ and

$$y'' = \frac{\sqrt{3} + 1/\sqrt{3}}{3} = \frac{4\sqrt{3}}{9}$$

$$\kappa = \frac{|y''|}{(1 + y'^2)^{3/2}} = \frac{4\sqrt{3}/9}{(1 + \frac{1}{3})^{3/2}} = \frac{1}{2}$$

22. $y' = \frac{-2x}{2\sqrt{r^2 - x^2}} = -\frac{x}{\sqrt{r^2 - x^2}};$

$$y'' = -\frac{\sqrt{r^2 - x^2} + x^2/\sqrt{r^2 - x^2}}{r^2 - x^2};$$

at $x = 0,\ y' = 0$ and $y'' = \frac{1}{r}$

$$\kappa = \frac{|y''|}{(1 + y'^2)^{3/2}} = \frac{1}{r}$$

Note: this solution makes the solution to Problem 21 a special case where $r = 2$.

23. $y' = \cos x;\ y'' = -\sin x;$

at $x = \frac{\pi}{2},\ y' = 0$ and $y'' = -1$

$$\kappa = \frac{|y''|}{(1 + y'^2)^{3/2}} = 1$$

24. $y' = -\sin x;\ y'' = -\cos x;$

at $x = \frac{\pi}{4},\ y' = -\frac{\sqrt{2}}{2}$ and $y'' = -\frac{\sqrt{2}}{2}$

$$\kappa = \frac{|y''|}{(1 + y'^2)^{3/2}} = \frac{\sqrt{2}/2}{(1 + 1/2)^{3/2}} = \frac{2\sqrt{3}}{9}$$

25. $y' = \frac{1}{x};\ y'' = -\frac{1}{x^2};$

at $x = 1,\ y' = 1$ and $y'' = -1$

$$\kappa = \frac{|y''|}{(1 + y'^2)^{3/2}} = \frac{1}{(1 + 1)^{3/2}} = \frac{1}{2\sqrt{2}}$$

26. $y' = e^x;\ y'' = e^x;$

at $x = 0,\ y' = 1$ and $y'' = -1$

$$\kappa = \frac{|y''|}{(1 + y'^2)^{3/2}} = \frac{1}{(1 + 1)^{3/2}} = \frac{1}{2\sqrt{2}}$$

27. If $\mathbf{u}$ and $\mathbf{v}$ are constant vectors, $\mathbf{R}(t) = \mathbf{u} + \mathbf{v}t$, then $\mathbf{V}(t) = \mathbf{v}$, and $\mathbf{A}(t) = \mathbf{0};\ \mathbf{V} \times \mathbf{A}$ will be 0, so κ will be 0.

28. $\frac{d\mathbf{R}}{dt} = \sinh t\ \mathbf{i} + \cosh t\ \mathbf{j};$ at $t = 0,\ \frac{d\mathbf{R}}{dt} = \mathbf{j}$

$\left\| \frac{d\mathbf{R}}{dt} \right\| = 1$, the unit tangent vector is $\mathbf{T} = \mathbf{j}$,

and the unit normal vector is $\mathbf{N} = \pm\mathbf{i}$.

29. $\frac{d\mathbf{R}}{dt} = \frac{\cos t}{\sin t}\mathbf{i} - \frac{\sin t}{\cos t}\mathbf{j} = \cot t\ \mathbf{i} - \tan t\ \mathbf{j}$

$\left\| \frac{d\mathbf{R}}{dt} \right\| = \sqrt{\cos^2 t + \tan^2 t};$ at $t = \frac{\pi}{3}$, the unit tangent vector is

$$\mathbf{T} = \frac{\frac{1}{\sqrt{3}}\mathbf{i} - \sqrt{3}\mathbf{j}}{\sqrt{\frac{10}{3}}} = \frac{1}{\sqrt{10}}(\mathbf{i} - 3\mathbf{j})$$

The unit normal vector is $\mathbf{N} = \pm\dfrac{1}{\sqrt{10}}(3\mathbf{i} + \mathbf{j})$

30. **a.**

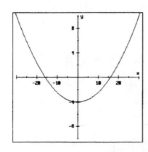

b. $x' = 32$; $y' = 32t$; the unit tangent vector is $\mathbf{T}(3) = \dfrac{1}{\sqrt{10}}(\mathbf{i} + 3\mathbf{j})$

c. At $t = 3$, $x' = 32$, $y' = 96$, $y'' = 32$, $x'' = 0$

$$\kappa = \frac{|x'y'' - y'x''|}{[(x')^2 + (y')^2]^{3/2}} = \frac{|32^2 - 0|}{(32^2 + 96^2)^{3/2}}$$

$$= \frac{1}{320\sqrt{10}};$$

thus, the radius of curvature is

$\rho = 320\sqrt{10}$.

31. **a.** $\mathbf{T} = \dfrac{\mathbf{R}'(t)}{\|\mathbf{R}'(t)\|}$

$$= \frac{(\cos t)\mathbf{i} + (-\sin t)\mathbf{j} + \mathbf{k}}{\sqrt{\cos^2 t + \sin^2 t + 1}}$$

$$= \frac{\sqrt{2}}{2}\,[(\cos t)\mathbf{i} + (-\sin t)\mathbf{j} + \mathbf{k}]$$

$$\mathbf{T}(\pi) = \frac{\sqrt{2}}{2}\,(-\mathbf{i} + \mathbf{k})$$

b. $\mathbf{V}(t) = \mathbf{R}'(t) = (\cos t)\mathbf{i} - (\sin t)\mathbf{j} + \mathbf{k}$

$\mathbf{V}(\pi) = -\mathbf{i} + \mathbf{k}$

$\mathbf{A}(t) = \mathbf{V}'(t) = (-\sin t)\mathbf{i} - (\cos t)\mathbf{j}$

$\mathbf{A}(\pi) = \mathbf{j}$

$$\kappa = \frac{\|\mathbf{V} \times \mathbf{A}\|}{\|\mathbf{V}\|^3} = \frac{\|(-\mathbf{i} + \mathbf{k}) \times \mathbf{j}\|}{(\sqrt{2})^3} = \frac{\sqrt{2}}{2\sqrt{2}} = \frac{1}{2}$$

c. $s = \displaystyle\int_0^\pi \|\mathbf{R}'\|\,dt = \int_0^\pi \sqrt{2}\,dt = \sqrt{2}\pi$

32. **a.** $\mathbf{R}'(t) = (1 - \cos t)\mathbf{i}$

$\qquad\qquad + (\sin t)\mathbf{j} + 2\cos\dfrac{t}{2}\,\mathbf{k}$

$\|\mathbf{R}'(t)\| = \sqrt{(1 - \cos t)^2 + \sin^2 t + 4\cos^2\dfrac{t}{2}}$

$$= \sqrt{2 - 2\cos t + 4\cos^2\frac{t}{2}}$$

$$= \sqrt{4\sin^2\frac{t}{2} + 4\cos^2\frac{t}{2}} = 2$$

$$\mathbf{T}(t) = \frac{\mathbf{R}'(t)}{\|\mathbf{R}'(t)\|}$$

$$= \tfrac{1}{2}[(1 - \cos t)\mathbf{i} + (\sin t)\mathbf{j} + (2\cos\tfrac{t}{2})\mathbf{k}]$$

b. $\mathbf{T}'(t) = \tfrac{1}{2}[(\sin t)\mathbf{i} + (\cos t)\mathbf{j} - (\sin\tfrac{t}{2})\mathbf{k}]$

$\dfrac{d\mathbf{T}}{ds} = \tfrac{1}{4}[(\sin t)\mathbf{i} + (\cos t)\mathbf{j} - (\sin\tfrac{t}{2})\mathbf{k}]$

$\kappa = \left\|\dfrac{d\mathbf{T}}{ds}\right\| = \tfrac{1}{4}\sqrt{\sin^2 t + \cos^2 t + \sin^2\tfrac{t}{2}}$

$$= \tfrac{1}{4}\sqrt{1 + \sin^2\tfrac{t}{2}}$$

33. $9x^2 + 4y^2 = 36$

$\dfrac{x^2}{4} + \dfrac{y^2}{9} = 1$

Parametrize this curve by letting $x = 2\cos t$, $y = 3\sin t$.

$x' = -2\sin t$, $y' = 3\cos t$

$x'' = -2\cos t$, $y'' = -3\sin t$

$$\kappa = \frac{|x'y'' - y'x''|}{[(x')^2 + (y')^2]^{3/2}}$$

$$= \frac{|(-2\sin t)(-3\sin t) - (3\cos t)(-2\cos t)|}{[(-2\sin t)^2 + (3\cos t)^2]^{3/2}}$$

$$= \frac{6}{(4\sin^2 t + 9\cos^2 t)^{3/2}}$$

$\dfrac{d\kappa}{dt} = -\dfrac{6(3)}{2}(4 + 5\cos^2 t)^{-5/2}(-10\cos t \sin t)$

$\dfrac{d\kappa}{dt} = 0$ if $t = 0, \dfrac{\pi}{2}, \pi, \dfrac{3\pi}{2}, 2\pi$

At $t = 0$ and $t = \pi$, $\kappa = \dfrac{6}{(4 + 5)^{3/2}} = \dfrac{2}{9}$

$\qquad\qquad\qquad\qquad\qquad$ (a minimum)

At $t = \dfrac{\pi}{2}$ and $t = \dfrac{3\pi}{2}$, $\kappa = \dfrac{6}{4^{3/2}} = \dfrac{3}{4}$

$\qquad\qquad\qquad\qquad\qquad$ (a maximum)

The points at which a maximum occurs are $P(0, 3)$, and $Q(0, -3)$.

34. $y = e^{2x}$; $y' = 2e^{2x}$; $y'' = 4e^{2x}$

$\kappa = \dfrac{4e^{2x}}{(1 + 4e^{4x})^{3/2}}$

$\dfrac{d\kappa}{dx} = \dfrac{8e^{2x}(1 - 8e^{4x})}{(4e^{4x} + 1)^{5/2}}$; this derivative is 0 when

$$8e^{2x}(1 - 8e^{4x}) = 0$$

$$8e^{4x} = 1$$

$$e^{4x} = \frac{1}{8}$$

$$4x = \ln 2^{-3}$$

$$x = -\tfrac{3}{4}\ln 2$$

The maximum value of κ occurs at this value for x; by substitution we find

$$\kappa = \frac{4e^{2(-3/4)\ln 2}}{(1 + 4e^{4(-3/4)\ln 2})^{3/2}} = \frac{4\sqrt{3}}{9}$$

(At $x = 0$ and at $x = -1$, $\kappa < \kappa_{\max}$)

35. Let $x = t$, then $y = t^6 - 3t^2$
$\mathbf{F} = t\mathbf{i} + (t^6 - 3t^2)\mathbf{j}$; to find the extrema:

$y' = 6x^5 - 6x = 6x(x^2 + 1)(x + 1)(x - 1) = 0$

when $x = 0, 1, -1$ (which are also the t-values). $\mathbf{V} = \langle 1, 6t^5 - 6t \rangle$, $\mathbf{A} = \langle 0, 30t^4 - 6 \rangle$.

$$\mathbf{V} \times \mathbf{A} = \begin{vmatrix} \mathbf{i} & \mathbf{j} & \mathbf{k} \\ 1 & 6t^5 - 6t & 0 \\ 0 & 30t^4 - 6 & 0 \end{vmatrix} = (30t^4 - 6)\mathbf{k}$$

$$\kappa = \frac{\|\mathbf{V} \times \mathbf{A}\|}{\|\mathbf{V}\|^3} = \frac{|30t^4 - 6|}{[1 + (6t^5 - 6t)^2]^{3/2}}$$

$$\rho = \frac{1}{\kappa} = \frac{[1 + (6t^5 - 6t)^2]^{3/2}}{|30t^4 - 6|}$$

$$\tfrac{1}{\kappa}(0) = \tfrac{1}{6}; \quad \tfrac{1}{\kappa}(1) = \tfrac{1}{24}; \quad \tfrac{1}{\kappa}(-1) = \tfrac{1}{24}$$

36. $x' = 1 - \cos t$; $y' = \sin t$; $x'' = \sin t$;
$y'' = \cos t$;

$$\kappa = \frac{1}{4 \sin \frac{t}{4}}$$

37. $\mathbf{R}(t) = 2t\mathbf{i} - t\mathbf{j} + t^2\mathbf{k}$; $P(2, -1, 1)$, so $t = 1$.
$\mathbf{R}'(t) = 2\mathbf{i} - \mathbf{j} + 2t\mathbf{k}$; $\mathbf{R}'(1) = 2\mathbf{i} - \mathbf{j} + 2\mathbf{k}$,
so $x = 2 + 2t$, $y = -1 - t$, $z = 1 + 2t$

38. $\mathbf{R}(t) = e^t\mathbf{i} - 3\mathbf{j} + (1 - t)\mathbf{k}$; $P(1, 3, 1)$, so
$t = 0$. $\mathbf{R}'(t) = e^t\mathbf{i} - \mathbf{k}$; the equations are the
lines $x = 1 + t$, $y = 3$, $z = 1 - t$

39. $\mathbf{R}(t) = t\mathbf{i} + t^2\mathbf{j} + t^3\mathbf{k}$; $\mathbf{v}(t) = \mathbf{i} + 2t\mathbf{j} + 3t^2\mathbf{k}$;
$\mathbf{A}(t) = 2\mathbf{j} + 6t\mathbf{k}$

$$\mathbf{v} \times \mathbf{A} = \begin{vmatrix} \mathbf{i} & \mathbf{j} & \mathbf{k} \\ 1 & 2t & 3t^2 \\ 0 & 2 & 6t \end{vmatrix}$$

$$= (12t^2 - 6t^2)\mathbf{i} - 6t\mathbf{j} + 2\mathbf{k}$$

$\|\mathbf{v} \times \mathbf{A}\| = 2\sqrt{9t^4 + 9t^2 + 1}$;

$\|\mathbf{v}\| = \sqrt{1 + 4t^2 + 9t^4}$; $\kappa = \dfrac{2\sqrt{9t^4 + 9t^2 + 1}}{(1 + 4t^2 + 9t^4)^{3/2}}$

40. $\mathbf{R}(t) = (t - \cos t)\mathbf{i} + \sin t\,\mathbf{j} + 3\mathbf{k}$
$\mathbf{v}(t) = (1 + \sin t)\mathbf{i} + \cos t\,\mathbf{j}$;
$\mathbf{A}(t) = (\cos t)\mathbf{i} - (\sin t)\mathbf{j}$

$$\mathbf{V} \times \mathbf{A} = \begin{vmatrix} \mathbf{i} & \mathbf{j} & \mathbf{k} \\ 1 + \sin t & \cos t & 0 \\ \cos t & -\sin t & 0 \end{vmatrix}$$

$= (-\sin t - \sin^2 t - \cos^2 t)\mathbf{k}$

$= -(1 + \sin t)\mathbf{k}$

$\|\mathbf{v} \times \mathbf{A}\| = 1 + \sin t$; $\|\mathbf{v}\| = \sqrt{2 + 2\sin t}$

$$\kappa = \frac{1 + \sin t}{2^{3/2}(1 + \sin t)^{3/2}} = \frac{1}{2\sqrt{2 + 2\sin t}}$$

41. $y = x^2$; $y' = 2x$; $y'' = 2$; $\kappa = \dfrac{2}{(1 + 4x^2)^{3/2}}$

42. $y = x^3$; $y' = 3x^2$; $y'' = 6x$; $\kappa = \dfrac{6|x|}{(1 + 9x^4)^{3/2}}$

43. $y = x^{-1}$; $y' = -x^{-2}$; $y'' = 2x^{-3}$

$$\kappa = \frac{2}{|x|^3(1 + 1/x^4)^{3/2}} = \frac{2}{|x|^3(x^4 + 1)^{3/2}/x^6}$$

$$= \frac{2|x^3|}{(1 + x^4)^{3/2}}$$

44. $y = \sin x$; $y' = \cos x$; $y'' = -\sin x$;

$$\kappa = \frac{|\sin x|}{(1 + \cos^2 x)^{3/2}}$$

45. $r = r' = r''e^\theta$;

$$\kappa = \frac{e^{2\theta} + 2e^{2\theta} - e^{2\theta}}{(e^{2\theta} + e^{2\theta})^{3/2}} = \frac{1}{\sqrt{2}e^\theta}$$

46. $r = 1 + \cos\theta$; $r' = -\sin\theta$; $r'' = -\cos\theta$

$$\kappa = \frac{(1 + \cos\theta)^2 - 2\sin^2\theta - (1 + \cos\theta)(-\cos\theta)}{(1 + 2\cos\theta + \cos^2\theta + \sin^2\theta)^{3/2}}$$

$$= \frac{1 + 2\cos\theta + \cos^2\theta - 2\sin^2\theta + \cos\theta + \cos^2\theta}{2^{3/2}(1 + \cos\theta)^{3/2}}$$

$$= \frac{1 + 3\cos\theta + 2(\cos^2\theta - \sin^2\theta)}{2^{3/2}(1 + \cos\theta)^{3/2}}$$

47. The function is given (or redefined) as $f(s)$ instead of $f(t)$, where

$$s = \int_0^s \frac{ds}{dt}\,dt \quad \text{and} \quad ds = \sqrt{(x')^2 + (y')^2}\,dt$$

48. Use the arc length parameter form of the formula for curvature if the problem insists on finding a normal vector or if the unit vector **T** is easily differentiated. The vector/derivative or the velocity/acceleration forms are easily used when other forms lead to complications. The parametric form is applicable when the functions are given in terms of a parameter (or x and y can be written in terms of a parameter). The functional form is applied when the dependent variable is given explicitly in terms of the independent variable. The polar form is used when the equation of the curve is given in polar coordinates.

49. a. Let **u** be perpendicular to **v** and $\mathbf{r} = a\mathbf{u} + \mathbf{v}$. Let the x-axis pass through **u** and the y-axis through **v**. The unit vectors **u** and **v** then are **i** and **j**; $\mathbf{r} = a\mathbf{i} + b\mathbf{j}$ is the position of any point in the xy-plane.

b. $\mathbf{w} = \mathbf{p} + \mathbf{n} = (a\mathbf{u} + b\mathbf{v}) + \mathbf{n}$
$\mathbf{w} \cdot \mathbf{u} = a\mathbf{u} \cdot \mathbf{u} + b\mathbf{v} \cdot \mathbf{u} + \mathbf{n} \cdot \mathbf{u} = a$ and
similarly $b = \mathbf{w} \cdot \mathbf{v}$. Hence,
$\mathbf{p} \cdot \mathbf{n} = a\mathbf{n} \cdot \mathbf{u} + b\mathbf{n} \cdot \mathbf{v} = 0$

c. $\mathbf{u} = \frac{1}{\sqrt{3}}(\mathbf{i} + \mathbf{j} + \mathbf{k}); \quad \mathbf{v} = \frac{1}{\sqrt{6}}(2\mathbf{i} - \mathbf{j} - \mathbf{k})$
$\mathbf{w} = \mathbf{j} + 2\mathbf{k};$

$\mathbf{p} = \frac{a}{\sqrt{3}}(\mathbf{i} + \mathbf{j} + \mathbf{k}) + \frac{b}{\sqrt{6}}(2\mathbf{i} - \mathbf{j} - \mathbf{k})$

$a = \mathbf{w} \cdot \mathbf{u} = \frac{1}{\sqrt{3}}(\mathbf{i} + \mathbf{j} + \mathbf{k}) \cdot (\mathbf{j} + 2\mathbf{k})$

$\quad = \frac{1 + 2}{\sqrt{3}} = \sqrt{3}$

$b = \mathbf{w} \cdot \mathbf{v} = \frac{1}{\sqrt{6}}(2\mathbf{i} - \mathbf{j} - \mathbf{k}) \cdot (\mathbf{j} + 2\mathbf{k})$

$\quad = \frac{-1 - 2}{\sqrt{2}\sqrt{6}}(2\mathbf{i} - \mathbf{j} - \mathbf{k})$

$\quad = \frac{1}{2}(2\mathbf{i} + 2\mathbf{j} + 2\mathbf{k} - 2\mathbf{i} - \mathbf{j} - \mathbf{k})$

$\quad = \frac{3}{2}(\mathbf{j} + \mathbf{k})$

d. $\mathbf{w} = a\mathbf{i} + b\mathbf{j} + c\mathbf{k}; \quad \mathbf{p} = a\mathbf{i} + b\mathbf{j};$
$a = \mathbf{w} \cdot \mathbf{i}; \quad b = \mathbf{w} \cdot \mathbf{j}$

e. $\|\mathbf{p}\| = \sqrt{a^2 + b^2}$ is the base of a right triangle (one leg), c is the other leg, and α is the acute angle. The slope is

$$m = \tan \alpha = \frac{c}{\sqrt{a^2 + b^2}}$$

50. a. $\mathbf{R}(t) = e^{-t/3}\cos 3t\, \mathbf{i} + e^{-t/3}\sin 3t\, \mathbf{j}$
$\qquad + \frac{13t}{t^2 + 40}\mathbf{k}$

By computer,

$\mathbf{R}(0) = \mathbf{i}; \quad \mathbf{R}(\frac{\pi}{2}) \approx -0.59\mathbf{j} + 0.48\mathbf{k}$

$\mathbf{R}(\pi) \approx -0.35\mathbf{i} + .82\mathbf{k}$

$\mathbf{R}(\frac{3\pi}{2}) \approx 0.21\mathbf{j} + 0.98\mathbf{k}$

$\mathbf{R}(2\pi) \approx 0.12\mathbf{i} + 1.03\mathbf{k}$

The graph will spiral upward (three cycles) with both x and y getting small, while z rises to about 1.028. The length is about 8 units; that is, a string or wire can be used to verify this value.

b. $\frac{dx}{dt} = e^{-t/3}(-3\sin 3t - \frac{1}{3}\cos 3t)$

$\frac{dy}{dt} = e^{-t/3}(3\cos 3t - \frac{1}{3}\sin 3t)$

$\frac{dz}{dt} = \frac{520 - 13t}{(t^2 + 40)^2}$

$\frac{ds}{dt} = \sqrt{(e^{-t/3})^2(9 + \frac{1}{9}) + \frac{(520 - 13t)^2}{(t^2 + 40)^2}}$

$s = \int_0^{2\pi} \sqrt{(e^{-t/3})^2(9 + \frac{1}{9}) + \frac{(520 - 13t)^2}{(t^2 + 40)^2}}\ dt$

By computer, we approximate this to be 8 (or 8.2246) units of length.

c. $F = \frac{8z'(t)}{\sqrt{(x')^2 + (y')^2}}$

$\quad = \frac{8(520 - 13t)}{(t^2 + 40)^2} \cdot \frac{1}{\sqrt{e^{-2t/3}(\frac{82}{9})}}$

$W = \int_0^{2\pi} \frac{8(520 - 13t)}{(t^2 + 40)^2} \cdot \frac{1}{\sqrt{e^{-2t/3}(\frac{82}{9})}}\ ds$

$\quad \approx 8.2854 \quad$ (integration by computer)

51. $\mathbf{V} = \frac{ds}{dt}\mathbf{T}; \quad \mathbf{A} = \frac{ds}{dt}\frac{d\mathbf{T}}{dt} + \mathbf{T}\frac{d^2s}{dt^2}$

Now $\frac{d\mathbf{T}}{dt} = \frac{d\mathbf{T}/ds}{dt/ds} = \frac{ds}{dt}\frac{d\mathbf{T}}{ds} = \kappa \frac{ds}{dt}\mathbf{N}$

Thus, $\mathbf{A} = \left(\frac{ds}{dt}\right)^2 \kappa\mathbf{N} + \frac{d^2s}{dt^2}\mathbf{T}$

$\mathbf{V} \times \mathbf{A} = \left(\frac{ds}{dt}\right)^3 \kappa\mathbf{T} \times \mathbf{N} + \frac{ds}{dt}\frac{d^2s}{dt^2}\mathbf{T} \times \mathbf{T}$

$\quad = \left(\frac{ds}{dt}\right)^3 \kappa\mathbf{B}$

$\frac{\|\mathbf{V} \times \mathbf{A}\|}{\|\mathbf{V}\|^3} = \frac{(ds/dt)^3}{(ds/dt)^3}\kappa = \kappa$

52. $\mathbf{R}(t) = x(t)\mathbf{i} + y(t)\mathbf{j}; \quad \mathbf{V}(t) = x'(t)\mathbf{i} + y'(t)\mathbf{j}$
$\mathbf{A}(t) = x''(t)\mathbf{i} + y''(t)\mathbf{j}$

$$V \times A = \begin{vmatrix} i & j & k \\ x'(t) & y'(t) & 0 \\ x''(t) & y''(t) & 0 \end{vmatrix}$$

$$= [x'(t)y''(t) - y'(t)x''(t)]k$$

$$\kappa = \frac{\|V \times A\|}{\|V\|^3} = \frac{|x'(t)y''(t) - y'(t)x'(t)|}{[x(t)^2 + y(t)^2]^{3/2}}$$

53. $R = xi + f(x)j$; $V = i + f'(x)j$; $A = y''(x)j$

$$V \times A = \begin{vmatrix} i & j & k \\ 1 & f'(x) & 0 \\ 0 & f''(x) & 0 \end{vmatrix} = f''(x)k$$

$$\kappa = \frac{\|V \times A\|}{\|V\|^3} = \frac{|f''(x)|}{[1 + f'(x)^2]^{3/2}}$$

54. $x = f(\theta)\cos\theta$; $y = f(\theta)\sin\theta$;

$x' = -f(\theta)\sin\theta + f'(\theta)\cos\theta$;

$y' = f(\theta)\cos\theta + f'(\theta)\sin\theta$

$x'' = -f(\theta)\cos\theta - f'(\theta)\sin\theta - f'(\theta)\sin\theta$
$\qquad + f''(\theta)\cos\theta$

$y'' = -f(\theta)\sin\theta + f'(\theta)\cos\theta + f'(\theta)\cos\theta$
$\qquad + f''(\theta)\sin\theta$

$$\kappa = \frac{1}{\|V\|^{3/2}}[f^2(\theta) - f(\theta)f'(\theta) + 2(f'(\theta))^2]$$

$$= \frac{|r^2 + 2(r')^2 - rr''|}{[r^2 + (r')^2]^{3/2}}$$

since $\|V\|^2 = f^2(\theta) + [f'(\theta)]^2$

55. $\|V \times A\| = \|V\|^2\|A\|^2\sin^2\theta$

$\qquad = \|V\|^2\|A\|^2(1 - \cos^2\theta)$

$\qquad = \|V\|^2\|A\|^2 - \|V\|^2\|A\|^2\cos^2\theta$

$\qquad = \|V\|^2\|A\|^2 - (V \times A)^2$

$$\kappa = \frac{\sqrt{\|V\|^2\|A\|^2 - (V \times A)^2}}{\|V\|^3}$$

56. a. $R(t)$ is given; compute $\kappa = \dfrac{\|V \times A\|}{\|V\|^3}$ and $\rho = 1/\kappa$. Obtain **T** and **N**. The center of the osculating circle is along **N** at a distance ρ from the point of contact on the curve.

b. $x(t)i = 32t$; $y(t) = 16t^2 - 4$; $P(32, 12)$, $t = 1$; $R = 32ti + (16t^2 - 4)j$;
$V(t) = 32i + 32tj$; $A(t) = 32j$.
At $t = 1$

$$V \times A = \begin{vmatrix} i & j & k \\ 32 & 32 & 0 \\ 0 & 32 & 0 \end{vmatrix} = 32^2 k$$

$$\|V\| = 32\sqrt{2}$$

$$\kappa = \frac{\|V \times A\|}{\|V\|^3} = \frac{|x'(t)y''(t) - y'(t)x''(t)|}{[(x'(t))^2 + (y'(t))^2]^{3/2}}$$

$$= \frac{32^2}{(32)^3(2)\sqrt{2}} = \frac{1}{64\sqrt{2}}; \ \rho = 32\sqrt{2}$$

$$T = \frac{i + tj}{\sqrt{1 + t^2}};$$

$$\frac{dT}{dt} = \frac{1}{1 + t^2}\left[-\frac{t}{\sqrt{1 + t^2}}i + \left(\sqrt{1 + t^2} - \frac{t^2}{\sqrt{1 + t^2}}\right)j\right]$$

At $t = 1$

$$\frac{dT}{dt} = 0.5[-2^{-1/2}i + (\sqrt{2} - 2^{-1/2})j]$$

$$= \frac{1}{2\sqrt{2}}(-i + j)$$

The unit normal vector is

$$N = \pm\frac{1}{\sqrt{2}}(i - j)$$

Let C be the center of the osculating circle. $PC = 32(-i + j)$ so the coordinates of C are $x = 32 - 32 = 0$ and $y = 12 + 32 = 44$. The equation of the osculating circle is

$$x^2 + (y - 44)^2 = 2{,}048$$

57. a. Since **T** and **B** are orthogonal, $T \cdot B = 0$ and

$$\frac{dT \cdot B}{ds} = \frac{dT}{ds} \cdot B + \frac{dB}{ds} \cdot T$$

Now $\dfrac{dT}{ds} = \kappa N$ and $B \cdot N = 0$ so $\dfrac{dB}{ds} \cdot T = 0$ so $\dfrac{dB}{ds} \cdot T$ which means $\dfrac{dB}{ds}$ is orthogonal to **T**

b. Since $B \cdot B = 1$ and $\dfrac{dB \cdot B}{ds} = 2\dfrac{dB}{ds} \cdot B$, so $\dfrac{dB}{ds} \cdot B = 0$ or $\dfrac{dB}{ds}$ is orthogonal to **B**.

c. Since $\dfrac{dB}{ds}$ is orthogonal to **T** and $\dfrac{dB}{ds}$ is orthogonal to **B**, $\dfrac{dB}{ds}$ is parallel to **N** or

$$\frac{dB}{ds} = -\tau N.$$

11.5 Tangential and Normal Components of Acceleration, Page 726

1. $R(t) = ti + t^2 j$

$V(t) = R'(t) = i + 2tj$

$\mathbf{A}(t) = \mathbf{V}'(t) = 2\mathbf{j}$

$\dfrac{ds}{dt} = \|\mathbf{V}(t)\| = \sqrt{1 + 4t^2}$

$A_T = \dfrac{d^2s}{dt^2} = \dfrac{4t}{\sqrt{1 + 4t^2}};$

$A_N = \sqrt{\|\mathbf{A}\|^2 - A_T{}^2}$

$\quad = \sqrt{4 - \dfrac{16t^2}{1 + 4t^2}} = \dfrac{2}{\sqrt{1 + 4t^2}}$

2. $\mathbf{R}(t) = t\mathbf{i} + e^t\mathbf{j}$

$\mathbf{V}(t) = \mathbf{R}'(t) = \mathbf{i} + e^t\mathbf{j}$

$\mathbf{A}(t) = \mathbf{V}'(t) = e^t\mathbf{j}$

$\dfrac{ds}{dt} = \|\mathbf{V}(t)\| = \sqrt{1 + e^{2t}}$

$A_T = \dfrac{d^2s}{dt^2} = \dfrac{e^{2t}}{\sqrt{1 + e^{2t}}};$

$A_N = \sqrt{\|\mathbf{A}\|^2 - A_T{}^2}$

$\quad = \sqrt{e^{2t} - \dfrac{e^{4t}}{1 + e^{2t}}} = \dfrac{e^t}{\sqrt{1 + e^{2t}}}$

3. $\mathbf{R}(t) = (t\sin t)\mathbf{i} + (t\cos t)\mathbf{j}$

$\mathbf{V}(t) = \mathbf{R}'(t) = (\sin t + t\cos t)\mathbf{i} + (\cos t - t\sin t)\mathbf{j},$

$\mathbf{A}(t) = \mathbf{V}'(t) = (-t\sin t + 2\cos t)\mathbf{i}$

$\qquad\qquad + (-t\cos t - 2\sin t)\mathbf{j}$

$\dfrac{ds}{dt} = \|\mathbf{V}(t)\| = \sqrt{1 + t^2}$

$A_T = \dfrac{d^2s}{dt^2} = \dfrac{t}{\sqrt{1 + t^2}}$

$A_N = \sqrt{\|\mathbf{A}\|^2 - A_T{}^2}\|$

$\quad = \sqrt{t^2 + 4 - \dfrac{t^2}{1 + t^2}}$

$\quad = \dfrac{\sqrt{(t^2 + 4)(1 + t^2) - t^2}}{\sqrt{1 + t^2}}$

$\quad = \dfrac{t^2 + 2}{\sqrt{t^2 + 1}}$

4. $\mathbf{R}(t) = 3\cos t\,\mathbf{i} + 2\sin t\,\mathbf{j}$

$\mathbf{V}(t) = \mathbf{R}'(t) = -3\sin t\,\mathbf{i} + 2\cos t\mathbf{j}$

$\mathbf{A}(t) = \mathbf{V}'(t) = -3\cos t\,\mathbf{i} - 2\sin t\,\mathbf{j}$

$\dfrac{ds}{dt} = \|\mathbf{V}(t)\| = \sqrt{4 + 5\sin^2 t}$

$A_T = \dfrac{d^2s}{dt^2} = \dfrac{5\sin t\cos t}{\sqrt{4 + 5\sin^2 t}}$

$A_N = \sqrt{\|\mathbf{A}\|^2 - A_T{}^2}$

$\quad = \sqrt{\dfrac{(4 + 5\cos^2 t)(4 + 5\sin^2 t) - 25\sin^2 t\cos^2 t}{4 + 5\sin^2 t}}$

$\quad = \dfrac{6}{\sqrt{4 + 5\sin^2 t}}$

5. $\mathbf{R}(t) = t\mathbf{i} + t^2\mathbf{j} + t\mathbf{k}$

$\mathbf{V}(t) = \mathbf{R}'(t) = \mathbf{i} + 2t\mathbf{j} + \mathbf{k}$

$\mathbf{A}(t) = \mathbf{V}'(t) = 2\mathbf{j}$

$\dfrac{ds}{dt} = \|\mathbf{V}(t)\| = \sqrt{2 + 4t^2}$

$A_T = \dfrac{d^2s}{dt^2} = \dfrac{4t}{\sqrt{2 + 4t^2}}$

$A_N = \sqrt{\|\mathbf{A}\|^2 - A_T{}^2}$

$\quad = \sqrt{\dfrac{8 + 16t^2 - 16t^2}{2 + 4t^2}} = \dfrac{2}{\sqrt{1 + 2t^2}}$

6. $\mathbf{R}(t) = \sin t\,\mathbf{i} + \cos t\,\mathbf{j} + \sin t\,\mathbf{k}$

$\mathbf{V}(t) = \mathbf{R}'(t) = \cos t\,\mathbf{i} - \sin t\,\mathbf{j} + \cos t\,\mathbf{k}$

$\mathbf{A}(t) = \mathbf{V}'(t) = -\sin t\,\mathbf{i} - \cos t\,\mathbf{j} + \sin t\,\mathbf{k}$

$\dfrac{ds}{dt} = \|\mathbf{V}(t)\| = \sqrt{1 + \cos^2 t}$

$A_T = \dfrac{d^2s}{dt^2} = -\dfrac{\sin t\cos t}{\sqrt{1 + \cos^2 t}}$

$A_N = \sqrt{\|\mathbf{A}\|^2 - A_T{}^2}$

$\quad = \sqrt{1 + \sin^2 t - \dfrac{\sin^2 t\cos^2 t}{1 + \cos^2 t}}$

$\quad = \sqrt{\dfrac{2}{1 + \cos^2 t}}$

7. $\mathbf{R}(t) = 4\cos t\,\mathbf{i} + \sin t\,\mathbf{k}$

$\mathbf{V}(t) = \mathbf{R}'(t) = -4\sin t\,\mathbf{i} + \cos t\mathbf{k}$

$\mathbf{A}(t) = \mathbf{V}'(t) = -4\cos t\,\mathbf{i} - \sin t\mathbf{k}$

$\dfrac{ds}{dt} = \|\mathbf{V}(t)\| = \sqrt{16\sin^2 t + \cos^2 t}$

$A_T = \dfrac{d^2s}{dt^2} = \dfrac{32\sin t\cos t + 2\cos t\sin t}{2\sqrt{16\sin^2 t + \cos^2 t}}$

$\quad = \dfrac{15\sin t\cos t}{\sqrt{16\sin^2 t + \cos^2 t}} = \dfrac{15\sin t\cos t}{\sqrt{1 + 15\sin^2 t}}$

$A_N = \sqrt{\|\mathbf{A}\|^2 - A_T{}^2}$

$\quad = \sqrt{16\cos^2 t + \sin^2 t - \dfrac{225\sin^2 t\cos^2 t}{16\sin^2 t + \cos^2 t}}$

$\quad = \dfrac{4}{\sqrt{16\sin^2 t + \cos^2 t}} = \dfrac{4}{\sqrt{1 + 15\sin^2 \theta}}$

8. $\mathbf{R}(t) = \frac{5}{13}\cos t\,\mathbf{i} + \frac{12}{13}(1 - \cos t)\mathbf{j} + \sin t\,\mathbf{k}$

$\mathbf{V}(t) = \mathbf{R}'(t) = -\frac{5}{13}\sin t\,\mathbf{i} + \frac{12}{13}\sin t\mathbf{j} + \cos t\mathbf{k}$

$\mathbf{A}(t) = \mathbf{V}'(t) = -\frac{5}{13}\cos t\mathbf{i} + \frac{12}{13}\cos t\mathbf{j} - \sin t\mathbf{k}$

$\frac{ds}{dt} = \|\mathbf{V}(t)\| = \sqrt{\sin^2 t + \cos^2 t} = 1$

$A_T = \frac{d^2 s}{dt^2} = 0$

$A_N = \sqrt{\|\mathbf{A}\|^2 - A_T^{\,2}}$
$\quad\; = \sqrt{1 + \sin^2 t}$

9. $\mathbf{R}(t) = (1 + \cos 2t)\mathbf{i} + 2\cos 2t\,\mathbf{j}$

$\mathbf{V}(t) = \mathbf{R}'(t) = -2\sin 2t\,\mathbf{i} + 2\cos 2t\,\mathbf{j}$

$\mathbf{A}(t) = \mathbf{V}'(t) = -4\cos 2t\,\mathbf{i} - 4\sin 2t\,\mathbf{j}$

$\frac{ds}{dt} = \|\mathbf{V}(t)\| = 2;\ A_T = \frac{d^2 s}{dt^2} = 0$

$A_N = \sqrt{\|\mathbf{A}\|^2 - A_T^{\,2}} = 4$

10. **a.** $\mathbf{R}(t) = r\cos\omega t\mathbf{i} + r\sin\omega t\mathbf{j}$

$\mathbf{V}(t) = \mathbf{R}'(t) = r\omega(-\sin\omega t\mathbf{i} + \cos\omega t\mathbf{j})$

$\mathbf{A}(t) = \mathbf{V}'(t) = -r\omega(\cos\omega t\mathbf{i} + \sin\omega t\mathbf{j})$

$\frac{ds}{dt} = \|\mathbf{V}(t)\| = r\omega$

$A_T = \frac{d^2 s}{dt^2} = 0$

b. $\frac{dx}{dt} = -r\omega\sin\omega t;\ \frac{dy}{dt} = r\omega\cos\omega t$

$\frac{d^2 x}{dt^2} = -r\omega^2\cos\omega t;\ \frac{d^2 y}{dt^2} = -r\omega^2\sin\omega t$

$\kappa = \frac{r^2\omega^2}{(r^2\omega^2)^{3/2}} = \frac{1}{r}$

11. $\mathbf{R}(t) = t\mathbf{i} + 4t^2\mathbf{j}$

$\mathbf{V}(t) = \mathbf{i} + 8t\mathbf{j}$

$\mathbf{A}(t) = 8\mathbf{j};\quad \|\mathbf{A}\| = 8$

$\frac{ds}{dt} = \|\mathbf{V}(t)\| = \sqrt{1 + 64t^2}$

If $\frac{ds}{dt} = 20,\ 1 + 64t^2 = 400,\quad t = \frac{\sqrt{399}}{8}$

$A_T = \frac{d^2 s}{dt^2} = \frac{64\,t}{\sqrt{1 + 64t^2}}$

$A_N = \sqrt{\|\mathbf{A}\|^2 - A_T^{\,2}}$

$\quad = \sqrt{64 - \frac{(64\,t)^2}{1 + 64\,t^2}}$

$\quad = \frac{8}{\sqrt{1 + 64t^2}}$

$A_T\!\left(\frac{\sqrt{399}}{8}\right) = \frac{8\sqrt{399}}{20} \approx 7.98999$

$A_N\!\left(\frac{\sqrt{399}}{8}\right) = \frac{8}{20} = 0.4$

12. Assume that the rope is twirled from a position that only rotates.
$\mathbf{R}(t) = 3\cos\omega t\,\mathbf{i} + 3\sin\omega t\,\mathbf{j}$
$\omega = d\theta/dt = 3/2\pi$ ft/s^2; from Problem 10,
$A_T = 0,\ A_N = r\omega^2 = \frac{27}{4} = 0.6839$.

13. $W = F = 3$ lb ($A_T = 0$), $r = 2$; $F = ma$, so
$a = 3/32$. From Problem 10, $A_N = r\omega^2$,
$2\omega = 32$, so $\omega = 4$ ft/s minimum. The
pressure (on 1 ft^2) is
$P = \frac{F}{A} = mr\omega^2 = \frac{2(3)\omega^2}{32} = \frac{3\omega^2}{16}$ lb/ft^2

14. $9x^2 + 4y^2 = 36$
$\frac{x^2}{4} + \frac{y^2}{9} = 1$

$x = 2\cos\omega t,\ y = 3\sin\omega t;$

$\mathbf{R}(t) = 2\cos\omega t\,\mathbf{i} + 3\sin\omega t\,\mathbf{j}$

$\mathbf{V}(t) = \omega(-2\sin\omega t\,\mathbf{i} + 3\cos\omega t\,\mathbf{j})$

$\mathbf{A}(t) = -\omega^2(2\cos\omega t\,\mathbf{i} + 3\sin\omega t\,\mathbf{j}) = -\omega^2\mathbf{R}$

$\frac{ds}{dt} = \omega\sqrt{4\sin^2\omega t + 9\cos^2\omega t} = 100$ km/h

At $t = 0$, $P(2, 0)$,

$9\omega^2 = 100^2$

$\omega = \frac{100}{3}$

At $t = \pi/2\omega$, $P(0, 3)$

$4\omega^2 = 100^2$

$\omega = 50$

$W = F = mg$ lb

$m = \frac{2{,}700}{(0.0098)(3{,}600)^2} \approx 0.0212584$ kg-h^2 km

At $t = 0$,

$A_T = \frac{\omega^2(4\sin\omega t\cos\omega t - 9\cos\omega t\sin\omega t)}{\sqrt{4\sin^2 t + 9\cos^2\omega t}}$

$\|\mathbf{A}(t)\| = \omega^2\|\mathbf{R}(t)\| = 3\omega^2 = 10{,}000/3$

Thus, $F = ma \approx \frac{(0.0212584)(10{,}000)}{3} \approx 70.9$

15. **a.** We wish to find the force normal to the path of the car.

$F_N = m\kappa\!\left(\frac{ds}{dt}\right)^2$

$m = 3{,}500/32.2;\ \kappa = 1/200,\ \frac{ds}{dt} = 55$ mph
$\approx 242/3$ fps:

$F_N = \frac{3{,}500}{32.2}\!\left(\frac{1}{200}\right)\!\left(\frac{242}{3}\right)^2 \approx 3{,}536.5$

b. $\theta = \tan^{-1}\left(\dfrac{3,500}{3,536.5}\right) \approx 44.7°$

16. $A_T = 0;\ A_N = R\omega^2;\ v = \omega R$

$A_N = 2.4 = \omega^2 R = 66^2/R,$ so

$R = 2.4(66)^2 = 1,815$ ft

17. $r = 15$ ft so the radius of curvature is $\rho = \dfrac{1}{15}$

$$A_N = (2\pi r\omega)^2 \kappa = 4\pi^2 r^2 (\tfrac{1}{r})\omega^2 = 4\pi^2 r\omega^2$$

A person on the wheel "flies off when"

$$mA_N = W$$

$$\frac{W}{g}(4\pi^2 r\omega^2) = W$$

$$\omega^2 = \frac{g}{4\pi^2(15)} = \frac{32}{60\pi^2};$$

$\omega \approx 0.23$ or 14 rpm

18. $\mathbf{R}(t) = t^3\mathbf{i} + t^2\mathbf{j} + t\mathbf{k}$

$\mathbf{V}(t) = \mathbf{R}'(t) = 3t^2\mathbf{i} + 2t\mathbf{j} + \mathbf{k}$

$\mathbf{A}(t) = \mathbf{V}'(t) = \mathbf{R}''(t) = 6t\mathbf{i} + 2\mathbf{j}$

$\dfrac{ds}{dt} = \|\mathbf{R}'(t)\| = \sqrt{9t^4 + 4t^2 + 1}$

$\mathbf{R}' \times \mathbf{R}'' = \begin{vmatrix} \mathbf{i} & \mathbf{j} & \mathbf{k} \\ 3t^2 & 2t & 1 \\ 6t & 2 & 0 \end{vmatrix} = -2\mathbf{i} + 6t\mathbf{j} - 6t^2\mathbf{k}$

$\|\mathbf{R}' \times \mathbf{R}''\| = \sqrt{4 + 36t^2 + 36t^4}$

$A_T = \dfrac{\mathbf{R}' \cdot \mathbf{R}''}{\|\mathbf{R}'\|} = \dfrac{18t^3 + 4t}{\sqrt{9t^4 + 4t^2 + 1}}$

$A_N = \dfrac{\|\mathbf{R}' \times \mathbf{R}''\|}{\|\mathbf{R}'\|} = \dfrac{\sqrt{36t^4 + 36t^2 + 4}}{\sqrt{9t^4 + 4t^2 + 1}}$

19. $\mathbf{R}(t) = t\mathbf{i} + 2t\mathbf{j} + t^2\mathbf{k}$

$\mathbf{V}(t) = \mathbf{R}'(t) = \mathbf{i} + 2\mathbf{j} + 2t\mathbf{k}$

$\mathbf{A}(t) = \mathbf{V}'(t) = \mathbf{R}''(t) = 2\mathbf{k}$

$\dfrac{ds}{dt} = \|\mathbf{R}'(t)\| = \sqrt{1 + 4 + 4t^2} = \sqrt{5 + 4t^2}$

$\mathbf{R}' \cdot \mathbf{R}'' = 4t$

$\mathbf{R}' \times \mathbf{R}'' = \begin{vmatrix} \mathbf{i} & \mathbf{j} & \mathbf{k} \\ 1 & 2 & 2t \\ 0 & 0 & 2 \end{vmatrix} = 4\mathbf{i} - 2\mathbf{j}$

$\|\mathbf{R}' \times \mathbf{R}''\| = \sqrt{16 + 4} = 2\sqrt{5}$

$A_T = \dfrac{\mathbf{R}' \cdot \mathbf{R}''}{\|\mathbf{R}'\|} = \dfrac{4t}{\sqrt{5 + 4t^2}}$

$A_N = \dfrac{\|\mathbf{R}' \times \mathbf{R}''\|}{\|\mathbf{R}'\|} = \dfrac{2\sqrt{5}}{\sqrt{5 + 4t^2}}$

20. $\mathbf{R}(t) = \cos t\mathbf{i} + \sin t\mathbf{j} + \mathbf{k}$

$\mathbf{V}(t) = \mathbf{R}'(t) = -\sin t\mathbf{i} + \cos t\mathbf{j}$

$\mathbf{A}(t) = \mathbf{V}'(t) = \mathbf{R}''(t) = -\cos t\mathbf{i} - \sin t\mathbf{j}$

$\dfrac{ds}{dt} = \|\mathbf{R}'(t)\| = \sqrt{\sin^2 t + \cos^2 t} = 1$

$\mathbf{R}' \cdot \mathbf{R}'' = -\cos t\sin t + \sin t\cos t = 0$

$\mathbf{R}' \times \mathbf{R}'' = \begin{vmatrix} \mathbf{i} & \mathbf{j} & \mathbf{k} \\ -\sin t & \cos t & 0 \\ -\cos t & -\sin t & 0 \end{vmatrix} = \mathbf{k}$

$\|\mathbf{R}' \times \mathbf{R}''\| = 1$

$A_T = \dfrac{\mathbf{R}' \cdot \mathbf{R}''}{\|\mathbf{R}'\|} = 0;\ A_N = \dfrac{\|\mathbf{R}' \times \mathbf{R}''\|}{\|\mathbf{R}'\|} = 1$

21. $\mathbf{R}(t) = e^t\cos t\mathbf{i} + e^t\sin t\mathbf{j} + e^t\mathbf{k}$

$\mathbf{V}(t) = \mathbf{R}'(t)$

$\quad = e^t[(-\sin t + \cos t)\mathbf{i} + (\cos t + \sin t)\mathbf{j} + \mathbf{k}]$

$\mathbf{A}(t) = \mathbf{V}'(t) = \mathbf{R}''(t)$

$\quad = e^t[-2\sin t\mathbf{i} + 2\cos t\mathbf{j} + \mathbf{k}]$

$\dfrac{ds}{dt} = \|\mathbf{R}'(t)\| = \sqrt{3e^{2t}} = \sqrt{3}e^t$

$A_T = \dfrac{\mathbf{R}' \cdot \mathbf{R}''}{\|\mathbf{R}'\|} = \sqrt{3}e^t$

$A_N = \dfrac{\|\mathbf{R}' \times \mathbf{R}''\|}{\|\mathbf{R}'\|} = \sqrt{2}e^t$

22. $\mathbf{B} = \mathbf{T} \times \mathbf{N}$

$\dfrac{d\mathbf{B}}{ds} = \dfrac{d\mathbf{T}}{ds} \times \mathbf{N} + \mathbf{T} \times \dfrac{d\mathbf{N}}{ds}$

$\quad = \mathbf{0} + \mathbf{T} \times \dfrac{d\mathbf{N}}{ds}$ since $\dfrac{d\mathbf{T}}{ds}$ is parallel to $\mathbf{N}$

23. a. $F = \dfrac{W\omega^2 R}{g} = \dfrac{W\omega^2(4\pi^2)(15)}{g} = 60\pi^2\omega^2(\dfrac{W}{g})$

b. $0.12W = \dfrac{60\pi^2\omega^2 W}{32(3,600)}$

$\omega \approx 0.0805727(60) \approx 4.83$ rpm

24. $\dfrac{ds}{dt} = \sqrt{\mathbf{R}' \cdot \mathbf{R}'}$

$A_T = \dfrac{d^2s}{dt^2} = \dfrac{\mathbf{R}' \cdot \mathbf{R}''}{\sqrt{\mathbf{R}' \cdot \mathbf{R}'}} = \dfrac{\mathbf{R}' \cdot \mathbf{R}''}{\|\mathbf{R}'\|}$

$A_N = \kappa\left(\dfrac{ds}{dt}\right)^2 = \dfrac{\|\mathbf{R}' \times \mathbf{R}''\|\|\mathbf{R}' \cdot \mathbf{R}'\|}{\|\mathbf{R}'\|^3} = \dfrac{\|\mathbf{R}' \times \mathbf{R}''\|}{\|\mathbf{R}'\|}$

25. $\mathbf{R}(x) = x\mathbf{i} + f(x)\mathbf{j}$

$\mathbf{V} = \mathbf{i} + f'(x)\mathbf{j};\ \|\mathbf{V}\| = \sqrt{[f'(x)]^2 + 1}$

$\mathbf{A} = f''(x)\mathbf{j}$

$A_T = \dfrac{d}{dx}\{[f'(x)]^2 + 1\} = \dfrac{f'(x)f''(x)}{\sqrt{[f'(x)]^2 + 1}}$

$A_N{}^2 = [f'(x)]^2 - \dfrac{[f'(x)]^2[f''(x)]^2}{[f'(x)]^2 + 1}$

$$= \frac{[f''(x)]^2\{1 + [f'(x)]^2 - [f'(x)]^2\}}{[f'(x)]^2 + 1}$$

$$= \frac{[f''(x)]^2}{[f'(x)]^2 + 1}$$

$$A_N = \frac{|f''(x)|}{\sqrt{1 + [f'(x)]^2}}$$

26. $\mathbf{R}(t) = t\mathbf{i} + f(t)\mathbf{j}$; $\mathbf{V} = \mathbf{i} + f'(t)\mathbf{j}$

$A(t) = f''(t)$

$$\mathbf{V} \times \mathbf{A} = \begin{vmatrix} \mathbf{i} & \mathbf{j} & \mathbf{k} \\ 1 & f'(t) & 0 \\ 0 & f''(t) & 0 \end{vmatrix} = f''(t)\mathbf{k}$$

If $f''(a) = 0$, then $\kappa = \dfrac{\|\mathbf{V} \times \mathbf{A}\|}{\|\mathbf{V}\|^3} = 0$

27. $\mathbf{A} = -g\mathbf{j}$, $\mathbf{V} = (v_0\cos\alpha)\mathbf{i} + (-gt + v_0\sin\alpha)\mathbf{j}$
The **j** component is 0 when $t = (v_0/g)\sin\alpha$ for maximum height. At this maximum height

$$\|\mathbf{V}\| = \sqrt{v_0^2\cos^2\alpha + (-gt + v_0\sin\alpha)^2}$$

$$A_T = \frac{d(v_0\cos\alpha)}{dt} = 0; \quad A_N = \|\mathbf{A}\| = g$$

28. a. With $x^2 + y^2 = r^2$;

$x = r\cos\omega t$ and $y = r\sin\omega t$

$\mathbf{R} = r\cos\omega t\,\mathbf{i} + r\sin\omega t\,\mathbf{j}$

b. $\dfrac{ds}{dt} = \dfrac{d\|\mathbf{R}\|}{dt} = r\omega\| -\sin\omega t\,\mathbf{i} + \cos\omega t\,\mathbf{j}\| = r\omega$

$$A_T = \frac{d^2(r\omega)}{dt^2} = 0,$$

$\mathbf{A} = -r\omega^2(\cos\omega t\,\mathbf{i} + \sin\omega t\,\mathbf{j})$

$A_N = \|\mathbf{A}\| = r\omega^2$

c. Since $A_N = r\omega^2$, doubling ω quadruples A_N. Doubling r results in doubling A_N. If ω is doubled and r is halved, A_N is doubled.

29. a. Let R_s be the distance of the satellite from the center of the earth. From Example 4,

$$v = \sqrt{\frac{GM}{R_s}}\,; \quad T = \frac{2\pi R_s}{v} = \frac{2\pi R_s^{3/2}}{\sqrt{GM}}$$

b. $T = (24)(3,600) = \dfrac{2\pi R_s^{3/2}}{\sqrt{398,600}}$

$$R_s^{3/2} = \frac{(24)(3,600)\sqrt{398,600}}{2\pi} \approx 8,681,655$$

$R_s \approx 42,241$ and $R \approx 42,241 - 6,440$

$\approx 36,000$ km

30. Let x be the horizontal distance from the helicopter to the bunker, h_2 the vertical distance from the helicopter to the bunker, and h_1 the vertical distance from the ground to the bunker.

$h_1 + h_2 = 10,000$; $h_1 = x\tan 15°$; $h_2 = x\tan 20°$

$$x = \frac{10,000}{\tan 15° + \tan 20°} \approx 15,824.8 \text{ ft}$$

$h_2 = (15,824.8)\tan 20° \approx 5,758.8 \text{ ft}$

$y = 16t^2 + 200(\sin 20°)t \approx 5,759.8$

$16t^2 + 68.04t - 5,759.8 = 0$, so

$$t \approx \frac{1}{32}\left[-68.404 + \sqrt{68.404^2 + 64(5,759.8)}\right]$$

≈ 21.23 sec

The horizontal distance after the drop is

$(200)(\cos 20°)(21.23) \approx 3,990.1 \text{ ft}$

The horizontal distance to cover before dropping the canister is

$15,824.8 - 3,990.1 \approx 11,834.7 \text{ ft}$

and the corresponding time is

$\dfrac{11,834.7}{200} \approx 59.1$ or about 1 min

CHAPTER 11 REVIEW

Proficiency Examination, Page 729

1. A vector-valued function (or, simply, a vector function) $\mathbf{F}$ with domain D assigns a unique vector $\mathbf{F}(t)$ to each scalar t in the set D. The set of all vectors $\mathbf{v}$ of the form $\mathbf{v} = \mathbf{F}(t)$ for t in D is the range of $\mathbf{F}$.

2. $\mathbf{F}(t) = f_1(t)\mathbf{i} + f_2(t)\mathbf{j}$ in $\mathbb{R}^2$ (plane) or

$\mathbf{F}(t) = f_1(t)\mathbf{i} + f_2(t)\mathbf{j} + f_3(t)\mathbf{k}$ in $\mathbb{R}^3$ (space)

where f_1, f_2, and f_3 are real-valued (scalar-valued) functions of the real number t defined on the domain set D. In this context f_1, f_2, and f_3 are called the components of $\mathbf{F}$.

3. The graph of $\mathbf{F}(t) = f_1(t)\mathbf{i} + f_2(t)\mathbf{j} + f_3(t)\mathbf{k}$ in $\mathbb{R}^3$ (space) is the graph of the parametric equations $x = f_1(t)$, $y = f_2(x)$, $z = f_3(x)$.

4. The limit of a vector function consists of the vector sum of the limits of its individual components.

5. $\mathbf{F}'(t) = f_1'(t)\mathbf{i} + f_2'(t)\mathbf{j} + f_3'(t)\mathbf{k}$

6. $\int \mathbf{F}(t)\ dt$

$$= \int f_1(t)\ dt\ \mathbf{i} + \int f_2(t)\ dt\ \mathbf{j} + \int f_3(t)\ dt\ \mathbf{k}$$

7. A smooth curve is a curve with no corners. The first derivative is continuous.

8. **a.** Linearity rule
$(a\mathbf{F} + b\mathbf{G})'(t) = a\mathbf{F}'(t) + b\mathbf{G}'(t)$
for constants a, b

b. Scalar multiple
$(h\mathbf{F})'(t) = h'(t)\mathbf{F}(t) + h(t)\mathbf{F}'(t)$

c. Dot product rule
$(\mathbf{F} \cdot \mathbf{G})'(t) = (\mathbf{F}' \cdot \mathbf{G})(t) + (\mathbf{F} \cdot \mathbf{G}')(t)$

d. Cross product rule
$(\mathbf{F} \times \mathbf{G})'(t) = (\mathbf{F}' \times \mathbf{G})(t) + (\mathbf{F} \times \mathbf{G}')(t)$

e Chain rule
$[\mathbf{F}(h(t))]' = h'(t)\mathbf{F}'(h(t))$

9. If the nonzero vector function $\mathbf{F}(t)$ is differentiable and has constant length, then $\mathbf{F}(t)$ is orthogonal to the derivative vector $\mathbf{F}'(t)$.

10. $\mathbf{R}(t) = x(t)\mathbf{i} + y(t)\mathbf{j} + z(t)\mathbf{k}$ is the position;

$$\mathbf{V}(t) = \frac{d\mathbf{R}(t)}{dt} = \mathbf{R}'(t) \text{ is the velocity;}$$

$$\mathbf{A}(t) = \frac{d\mathbf{V}(t)}{dt} = \frac{d^2\mathbf{R}(t)}{dt^2} \text{ is the acceleration}$$

11. The speed is $\|\mathbf{V}(t)\|$.

12. Consider a projectile that travels in a vacuum in a coordinate plane, with the x-axis along level ground. If the projectile is fired from a height of s_0 with initial speed v_0 and angle of elevation α, then at time t $(t \geq 0)$ it will be at the point $\big(x(t),\ y(t)\big)$, where

$$x(t) = (v_0\cos \alpha)t, \ y(t) = -\tfrac{1}{2}gt^2 + (v_0\sin \alpha)t + s_0$$

13. A projectile fired from ground level has time of flight T_f and range R given by the equations

$$T_f = \frac{2}{g}v_0\sin \alpha \text{ and } R = \frac{v_0^2}{g}\sin 2\alpha$$

The maximal range is $R_m = \frac{v_0^2}{g}$, and it occurs when $\alpha = \frac{\pi}{4}$.

14. 1. The planets move about the sun in elliptical orbits, with the sun at one focus.
2. The radius vector joining a planet to the sun sweeps over equal areas in equal intervals of time.
3. The square of the time of one complete revolution of a planet about its orbit is proportional to the cube of the orbit's semimajor axis.

15. $\mathbf{u}_r = (\cos \theta)\mathbf{i} + (\sin \theta)\mathbf{j}$ and
$\mathbf{u}_\theta = (-\sin \theta)\mathbf{i} + (\cos \theta)\mathbf{j}$

16. If $\mathbf{R}(t)$ is a vector function that defines a smooth graph $(\mathbf{T}'(t) \neq 0)$, then at each point a unit tangent is

$$\mathbf{T}(t) = \frac{\mathbf{R}'(t)}{\|\mathbf{R}'(t)\|}$$

and the principal unit normal vector is

$$\mathbf{N}(t) = \frac{\mathbf{T}'(t)}{\|\mathbf{T}'(t)\|}$$

17. Suppose an object moves with displacement $\mathbf{R}(t)$, where $\mathbf{R}'(t)$ is continuous on the interval $[a, b]$. Then the object has speed

$$\|\mathbf{V}(t)\| = \|\mathbf{R}'(t)\| = \frac{ds}{dt} \qquad \text{for } a \leq t \leq b$$

18. If C is a smooth curve defined by $\mathbf{R}(t) = x(t)\mathbf{i} + y(t)\mathbf{j} + z(t)\mathbf{k}$ on an interval $[a, b]$, then the arc length of C is given by

$$s = \int_a^b \|\mathbf{R}'(t)\|\ dt = \int_a^b \sqrt{[x'(t)]^2 + [y'(t)]^2 + [z'(t)]^2}\ dt$$

19. The curvature of a graph is an indication of how quickly the graph changes direction.

$$\kappa = \left\|\frac{d\mathbf{T}}{ds}\right\|$$

where $\mathbf{T}$ is a unit vector.

20. $\kappa = \dfrac{\|\mathbf{V} \times \mathbf{A}\|}{\|\mathbf{V}\|^3}$

21. $\mathbf{T} = \dfrac{d\mathbf{R}}{ds}$ and $\mathbf{N} = \dfrac{1}{\kappa}\left(\dfrac{d\mathbf{T}}{ds}\right)$

22. The radius of curvature is the radius of the osculating circle; $\rho = 1/\kappa$.

23. The acceleration $\mathbf{A}$ of a moving object can be written as $\mathbf{A} = A_T\mathbf{T} + A_N\mathbf{N}$ where

$$A_T = \frac{d^2s}{dt^2} \text{ is the tangential component; and}$$

$$A_N = \kappa\left(\frac{ds}{dt}\right)^2 \text{ is the normal component.}$$

24. This is a helix with radius of 3, climbing in a counter-clockwise direction.

$$s = \int_0^{2\pi} \|\mathbf{R}'\|\ dt$$

$$\mathbf{R} = (3 \cos t)\mathbf{i} + (3 \sin t)\mathbf{j} + t\mathbf{k},$$

$$\mathbf{R}' = (-3 \sin t)\mathbf{i} + (3 \cos t)\mathbf{j} + \mathbf{k},$$

$$\|\mathbf{R}'\| = \sqrt{9 \sin^2 t + 9 \cos^2 t + 1} = \sqrt{10}$$

$$s = \int_0^{2\pi} \sqrt{10}\ dt = 2\pi\sqrt{10}$$

25. $\mathbf{F}' = \dfrac{1}{(1+t)^2}\mathbf{i} + \dfrac{t\cos t - \sin t}{t^2}\mathbf{j} + (-\sin t)\mathbf{k}$

$\mathbf{F}'' = -\dfrac{2}{(1+t)^3}\mathbf{i} + \dfrac{-t^2\sin t - 2t\cos t + 2\sin t}{t^3}\mathbf{j}$

$\qquad - (\cos t)\mathbf{k}$

26. $\begin{vmatrix} \mathbf{i} & \mathbf{j} & \mathbf{k} \\ 3t & 0 & 3 \\ 0 & \ln t & -t^2 \end{vmatrix} = (-3\ln t)\mathbf{i} + (3t^3)\mathbf{j} + (3t\ln t)\mathbf{k}$

$\displaystyle\int_1^2 [(-3\ln t)\mathbf{i} + (3t^3)\mathbf{j} + (3t\ln t)\mathbf{k}]\ dt$

$= -3(t\ln t - t)\mathbf{i} + \dfrac{3t^4}{4}\mathbf{j} + 3\left(\dfrac{t^2}{2}\left[\ln t - \dfrac{1}{2}\right]\right)\mathbf{k}\ \Big|_1^2$

$\boxed{\text{Formulas 499 \& 502}}$

$= (6 - 6\ln 2)\mathbf{i} + 12\mathbf{j} + (6\ln 2 - 3)\mathbf{k}$

$\qquad - (3\mathbf{i} + \tfrac{3}{4}\mathbf{j} - \tfrac{3}{4}\mathbf{k})$

$= (3 - 6\ln 2)\mathbf{i} + \tfrac{45}{4}\mathbf{j} + (6\ln 2 - \tfrac{9}{4})\mathbf{k}$

27. $\mathbf{F}'' = e^t\mathbf{i} - t^2\mathbf{j} + 3\mathbf{k},$

$\mathbf{F}' = e^t\mathbf{i} - \dfrac{t^3}{3}\mathbf{j} + 3t\mathbf{k} + \mathbf{C}$, but $\mathbf{F}'(0) = 3\mathbf{k}$ so

$\mathbf{F}' = (e^t - 1)\mathbf{i} - \dfrac{t^3}{3}\mathbf{j} + (3t + 3)\mathbf{k}$

$\mathbf{F} = (e^t - t)\mathbf{i} - \dfrac{t^4}{12}\mathbf{j} + (\dfrac{3t^2}{2} + 3t)\mathbf{k} + \mathbf{C}$, but

$\qquad \mathbf{F}(0) = \mathbf{i} - 2\mathbf{j}$ so

$\mathbf{F} = (e^t - t)\mathbf{i} - \left(\dfrac{t^4}{12} + 2\right)\mathbf{j} + (\dfrac{3t^2}{2} + 3t)\mathbf{k}$

28. $\mathbf{R} = t\mathbf{i} + 2t\mathbf{j} + te^t\mathbf{k}$

$\mathbf{V} = \mathbf{R}' = \mathbf{i} + 2\mathbf{j} + e^t(t + 1)\mathbf{k}$

$\dfrac{ds}{dt} = \|\mathbf{V}\| = \sqrt{1 + 4 + e^{2t}(t+1)^2}$

$\qquad = \sqrt{5 + e^{2t}(t+1)^2}$

$\mathbf{A} = \mathbf{V}' = e^t(t + 2)\mathbf{k}$

29. $\mathbf{R} = t^2\mathbf{i} + 3t\mathbf{j} - 3t\mathbf{k}$

$\mathbf{T} = \dfrac{\mathbf{R}'}{\|\mathbf{R}'\|} = \dfrac{2t\mathbf{i} + 3\mathbf{j} - 3\mathbf{k}}{\sqrt{4t^2 + 18}}$

$\mathbf{T}' = \dfrac{-4t}{(4t^2 + 18)^{3/2}}(2t\mathbf{i} + 3\mathbf{j} - 3\mathbf{k})$

$\qquad + \dfrac{1}{(4t^2 + 18)^{1/2}}(2\mathbf{i})$

$\qquad = \dfrac{36\mathbf{i} - 12t\mathbf{j} + 12t\mathbf{k}}{(4t^2 + 18)^{3/2}}$

$\mathbf{N} = \dfrac{\mathbf{T}'}{\|\mathbf{T}'\|} = \dfrac{3\mathbf{i} - t\mathbf{j} + t\mathbf{k}}{(2t^2 + 9)^{1/2}}$

$\mathbf{A} = \mathbf{R}'' = 2\mathbf{i}$

$\mathbf{R}' \times \mathbf{R}'' = \begin{bmatrix} \mathbf{i} & \mathbf{j} & \mathbf{k} \\ 2t & 3 & -3 \\ 2 & 0 & 0 \end{bmatrix} = -6\mathbf{j} - 6\mathbf{k}$

$\kappa = \dfrac{\|\mathbf{R}' \times \mathbf{R}''\|}{\|\mathbf{R}'\|^3} = \dfrac{\sqrt{36 + 36}}{(4t^2 + 18)^{3/2}}$

$\qquad = \dfrac{\sqrt{72}}{(4t^2 + 18)^{3/2}} = \dfrac{3}{(2t^2 + 9)^{3/2}}$

$A_T = \dfrac{d^2s}{dt^2} = \dfrac{4t}{(4t^2 + 18)^{1/2}} = \dfrac{2\sqrt{2}t}{\sqrt{2t^2 + 9}}$

$A_N = \dfrac{3}{(2t^2 + 9)^{3/2}}(4t^2 + 18) = \dfrac{6\sqrt{2}}{(4t^2 + 18)^{1/2}}$

$\qquad = \dfrac{6}{\sqrt{2t^2 + 9}}$

30. a. $\mathbf{R} = [(v_0\cos\alpha)t]\mathbf{i}$

$\qquad + [(v_0\sin\alpha)t - \tfrac{1}{2}gt^2 + s_0]\mathbf{j};$

In our case:

$\mathbf{R}(t) = (25\sqrt{3}\ t)\mathbf{i} + (25t - 16t^2)\mathbf{j}$

The maximum height is the value of the coefficient of $\mathbf{j}$ at half the total time of flight, $t = 25/32$ sec.

$\mathbf{R}(\tfrac{25}{32}) = 25(\tfrac{25}{32}) - 16(\tfrac{25}{32})^2 \approx 9.76$ ft

b. $T_f = \dfrac{2}{g}v_0\sin\alpha = \dfrac{1}{16}(50)(\tfrac{1}{2}) = \dfrac{25}{16}$ sec

Range $= \dfrac{v_0^2}{g}\sin 2\alpha = \dfrac{50^2}{32}\sin 60°$

$\qquad = \dfrac{625\sqrt{3}}{16} \approx 67.7$ ft

Supplementary Problems, Page 730

1. $\displaystyle\lim_{t\to 0}\left[t^2\mathbf{i} - 3t\mathbf{j} + \dfrac{\sin 2t}{t}\mathbf{k}\right] = 2\mathbf{k}$

2. $\displaystyle\lim_{t\to 0}\left[\dfrac{\mathbf{i} + t\mathbf{j} - e^{-t}\mathbf{k}}{1 - t}\right] = \mathbf{i} - \mathbf{k}$

3. $\lim\limits_{t\to 0}\ [t\mathbf{i} + 5\mathbf{k}]\cdot[(\sin t)\mathbf{i} + 3t\mathbf{j} - (1 - t)\mathbf{k}]$

    $= (5\mathbf{k})\cdot(-\mathbf{k}) = -5$

4. $\lim\limits_{t\to 0}\ [(1 + t)\mathbf{i} - 3\mathbf{j}] \times [t^2\mathbf{j} + (\cos \pi t)\mathbf{k}]$

    $= \begin{bmatrix} \mathbf{i} & \mathbf{j} & \mathbf{k} \\ 1 & -3 & 0 \\ 0 & 0 & 1 \end{bmatrix} = -3\mathbf{i} - \mathbf{j}$

5. $\lim\limits_{t\to 0}\ [(1 + \frac{1}{t})^t\mathbf{i} - \left(\dfrac{\sin t}{t}\right)\mathbf{j} - t\mathbf{k}]$

    $= \lim\limits_{t\to 0}\Big[\exp[\ln(1 + \frac{1}{t})]\Big]\mathbf{i} - \mathbf{j} - t\mathbf{k}$

    $= \lim\limits_{t\to 0}\ e^{t/(1+t)}\mathbf{i} - \mathbf{j} - t\mathbf{k} = \mathbf{i} - \mathbf{j}$

6. $\lim\limits_{t\to 0}\ \left[\left(\dfrac{1 - \cos t}{t}\right)\mathbf{i} + 4\mathbf{j} + \left(1 + \frac{3}{t}\right)^t\mathbf{k}\right]$

    $= \lim\limits_{t\to 0}[\sin t\mathbf{i} + 4\mathbf{j} + a\mathbf{k}] = 4\mathbf{j} + \mathbf{k}$

    where $a = \lim\limits_{t\to 0}\ \exp\left[\dfrac{\ln(1 + \frac{3}{t})}{\frac{1}{t}}\right]$

    $= \lim\limits_{t\to 0}\ e^{t/(t+3)} = 1$

7. $\mathbf{F}(t) = te^t\mathbf{i} + t^2\mathbf{j}$

    $\mathbf{F}'(t) = (1 + t)e^t\mathbf{i} + 2t\mathbf{j}$

    $\mathbf{F}''(t) = (2 + t)e^t\mathbf{i} + 2\mathbf{j}$

8. $\mathbf{F}(t) = (t \ln 2t)\mathbf{i} + t^{3/2}\mathbf{k}$

    $\mathbf{F}'(t) = (1 + \ln 2t)\mathbf{i} + \frac{3}{2}t^{1/2}\mathbf{k}$

    $\mathbf{F}''(t) = \frac{1}{t}\mathbf{i} + \dfrac{3}{4\sqrt{t}}\mathbf{k}$

9. $\mathbf{F}(t) = 2t^{-1} - 2t\mathbf{j} + te^{-t}\mathbf{k}$

    $\mathbf{F}'(t) = -2t^{-2}\mathbf{i} - 2\mathbf{j} + (1 - t)e^{-t}\mathbf{k}$

    $\mathbf{F}''(t) = 4t^{-3}\mathbf{i} + (t - 2)e^{-t}\mathbf{k}$

10. $\mathbf{F}(t) = [t\mathbf{i} - (1 - t)\mathbf{j}] \times [t^2\mathbf{j} + e^{-t}\mathbf{k}]$

    $= \begin{vmatrix} \mathbf{i} & \mathbf{j} & \mathbf{k} \\ t & -1+t & 0 \\ 0 & t^2 & e^{-t} \end{vmatrix}$

    $= e^{-t}(t - 1)\mathbf{i} - te^{-t}\mathbf{j} + t^3\mathbf{k}$

    $\mathbf{F}'(t) = (2 - t)e^{-t}\mathbf{i} - (1 - t)e^{-t}\mathbf{j} + 3t^2\mathbf{k}$

    $\mathbf{F}''(t) = (t - 3)e^{-t}\mathbf{i} - (t - 2)e^{-t}\mathbf{j} + 6t\mathbf{k}$

11. $\mathbf{F}(t) = (t^2 + e^{at})\mathbf{i} + (te^{-at})\mathbf{j} + (e^{at+1})\mathbf{k}$

    $\mathbf{F}'(t) = (2t + ae^{at})\mathbf{i} + (1 - at)e^{-at}\mathbf{j} + ae^{at+1}\mathbf{k}$

    $\mathbf{F}''(t) = (2 + a^2e^{at})\mathbf{i} + (a^2t - 2a)e^{-at}\mathbf{j} + a^2e^{at+1}\mathbf{k}$

12. $\mathbf{F}(t) = (1 - t)^{-1}\mathbf{i} + (\sin 2t)\mathbf{j} + (\cos^2 t)\mathbf{k}$

$\mathbf{F}'(t) = (1 - t)^{-2}\mathbf{i} + (2\cos 2t)\mathbf{j} + 2\cos t(-\sin t)\mathbf{k}$

$= (1 - t)^{-2}\mathbf{i} + (2\cos 2t\mathbf{j} - \sin 2t\mathbf{k})$

$\mathbf{F}''(t) = 2(1 - t)^{-3}\mathbf{i} - 4\sin 2t\mathbf{j} - 2\cos 2t\mathbf{k}$

13.   14.

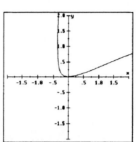   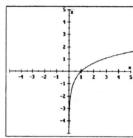

15.   16.

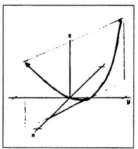   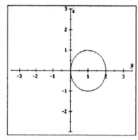

17.   18.

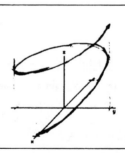   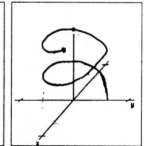

19. $\dfrac{d}{dt}\|\mathbf{F}(t)\| = \dfrac{d}{dt}\sqrt{f_1^2 + f_2^2 + f_3^2}$

    $= \dfrac{f_1f_1' + f_2f_2' + f_3f_3'}{\sqrt{f_1^2 + f_2^2 + f_3^2}}$

    $= \dfrac{\mathbf{F}(t)\cdot\mathbf{F}'(t)}{\|\mathbf{F}(t)\|}$

20. $\dfrac{d}{dt}\Big[\|\mathbf{F}(t)\|\,\mathbf{F}(t)\Big] = \|\mathbf{F}(t)\|\mathbf{F}'(t) + \|\mathbf{F}(t)\|'\mathbf{F}(t)$

21. $\dfrac{d}{dt}\left[\dfrac{\mathbf{F}(t)}{\|\mathbf{F}(t)\|}\right] = \dfrac{d}{dt}\dfrac{\mathbf{F}}{\sqrt{\mathbf{F}\cdot\mathbf{F}}}$

$$= \frac{\sqrt{\mathbf{F} \cdot \mathbf{F}} \ \mathbf{F}' - (\mathbf{F} \cdot \mathbf{F}')\mathbf{F}}{(\mathbf{F} \cdot \mathbf{F})^{3/2}}$$

$$= \frac{\|\mathbf{F}\|^2 \mathbf{F}' - (\mathbf{F} \cdot \mathbf{F}')\mathbf{F}}{\|\mathbf{F}\|^3}$$

22. $\frac{d}{dt}[\mathbf{F}(t) \cdot \mathbf{F}(t)] = 2\mathbf{F} \cdot \mathbf{F}'$

23. $\frac{d}{dt}[\mathbf{F}(t) \times \mathbf{F}(t)] = \mathbf{F} \times \mathbf{F}' + \mathbf{F}' \times \mathbf{F}$

$\quad = \mathbf{F} \times \mathbf{F}' - \mathbf{F} \times \mathbf{F}' = 0$

24. $\frac{d}{dt} \mathbf{F}(e^t) = e^t \mathbf{F}'(e^t)$

25. $\int_{-1}^{1} (e^{-t}\mathbf{i} + t^3\mathbf{j} + 3\mathbf{k}) \, dt$

$\quad = (-e^{-t}\mathbf{i} + \frac{t^4}{4}\mathbf{j} + 3t\mathbf{k}) \Big|_{-1}^{1}$

$\quad = (e - e^{-1})\mathbf{i} + 6\mathbf{k}$

26. $\int_{1}^{2} [(1 - t)\mathbf{i} - t^{-1}\mathbf{j} + e^t\mathbf{k}] \, dt$

$\quad = [(t - \frac{t^2}{2})\mathbf{i} - \ln t \, \mathbf{j} + e^t\mathbf{k}] \Big|_{1}^{2}$

$\quad = -\frac{1}{2}\mathbf{i} - \ln 2 \, \mathbf{j} + (e^2 - e)\mathbf{k}$

27. $\int [te^t\mathbf{i} - (\sin 2t)\mathbf{j} + t^2\mathbf{k}] \, dt$

$\quad \boxed{u = t; \ dv = e^t \, dt}$

$\quad = (te^t - e^t)\mathbf{i} + \frac{\cos 2t}{2}\mathbf{j} + \frac{t^3}{3}\mathbf{k} + \mathbf{C}$

28. $\int e^{2t}[2\mathbf{i} - t\mathbf{j} + (\sin t)\mathbf{k}] \, dt$

$\quad \boxed{u = e^{2t}; \ dv = \sin t \, dt}$

$\quad = e^{2t}\mathbf{i} - \frac{1}{4}(2t-1)e^{2t}\mathbf{j} + \frac{1}{5}(2 \sin t - \cos t)e^{2t}\mathbf{k}$

29. $\int t[e^t\mathbf{i} + (\ln t)\mathbf{j} + 3\mathbf{k}] \, dt$

$\quad \boxed{\text{Formulas 484 and 502}}$

$\quad = (te^t - e^t)\mathbf{i} + \frac{1}{4}t^2(2 \ln t - 1)\mathbf{j} + \frac{3t^2}{2}\mathbf{k} + \mathbf{C}$

30. $\int [e^t\mathbf{i} + 2\mathbf{j} - t\mathbf{k}] \cdot [e^{-t}\mathbf{i} - t\mathbf{j}] \, dt$

$\quad = \int (1 - 2t) \, dt = t - t^2 + C$

31. $\mathbf{R}(t) = t\mathbf{i} + (3 - t)\mathbf{j} + 2\mathbf{k}$

$\quad \mathbf{V}(t) = \mathbf{R}'(t) = \mathbf{i} - \mathbf{j}$

$\quad \frac{ds}{dt} = \|\mathbf{V}(t)\| = \sqrt{2}$

$\quad \mathbf{A}(t) = \mathbf{R}''(t) = \mathbf{0}$

32. $\mathbf{R}(t) = (\sin 2t)\mathbf{i} + 2\mathbf{j} - (\cos 2t)\mathbf{k}$

$\quad \mathbf{V}(t) = \mathbf{R}'(t) = 2 \cos 2t\mathbf{i} + 2 \sin 2t\mathbf{k}$

$\quad \frac{ds}{dt} = \|\mathbf{V}(t)\| = 2$

$\quad \mathbf{A}(t) = \mathbf{R}''(t) = -4 \sin 2t\mathbf{i} + 4 \cos 2t\mathbf{k}$

33. $\mathbf{R}(t) = (t \sin t)\mathbf{i} + (te^{-t})\mathbf{j} - (1 - t)\mathbf{k}$

$\quad \mathbf{V}(t) = \mathbf{R}'(t) = (t\cos t + \sin t)\mathbf{i} + (1 - t)e^{-t}\mathbf{j} + \mathbf{k}$

$\quad \frac{ds}{dt} = \|\mathbf{V}(t)\|$

$\quad = \sqrt{t^2\cos^2 t + t\sin 2t + \sin^2 t + (1 - t)^2 e^{-2t} + 1}$

$\quad \mathbf{A}(t) = \mathbf{R}''(t)$

$\quad = (2 \cos t - t\sin t)\mathbf{i} + e^{-t}(t - 2)\mathbf{j}$

34. $\mathbf{R}(t) = (\ln t)\mathbf{i} + e^t\mathbf{j} - (\tan t)\mathbf{k}$

$\quad \mathbf{V}(t) = \mathbf{R}'(t) = t^{-1}\mathbf{i} + e^t\mathbf{j} - \sec^2 t\mathbf{k}$

$\quad \frac{ds}{dt} = \|\mathbf{V}(t)\| = \sqrt{t^{-2} + e^{2t} + \sec^4 t}$

$\quad \mathbf{A}(t) = \mathbf{R}''(t) = -t^{-2}\mathbf{i} + e^t\mathbf{j} - 2\sec^2 t \tan t\mathbf{k}$

35. $\mathbf{R}(t) = t\mathbf{i} - t^2\mathbf{j}$

$\quad \mathbf{V}(t) = \mathbf{i} - \mathbf{j}$

$\quad \mathbf{T}(t) = \frac{1}{\sqrt{2}}(\mathbf{i} - \mathbf{j})$

$\quad \mathbf{N}(t) = \frac{1}{\sqrt{2}}(\mathbf{i} + \mathbf{j})$

36. $\mathbf{R}(t) = (3 \cos t)\mathbf{i} - (3 \sin t)\mathbf{j}$

$\quad \mathbf{V}(t) = 3(-\sin t \, \mathbf{i} - \cos t \, \mathbf{j})$

$\quad \mathbf{T}(t) = -\sin t \, \mathbf{i} - \cos t \, \mathbf{j}$

$\quad \mathbf{N}(t) = \cos t \, \mathbf{i} - \sin t \, \mathbf{j}$

37. $\mathbf{R}(t) = (4 \cos t)\mathbf{i} - 3t\mathbf{j} + (4 \sin t)\mathbf{k}$

$\quad \mathbf{V}(t) = -4 \sin t \, \mathbf{i} - 3\mathbf{j} + 4 \cos t \, \mathbf{k}$

$\quad \|\mathbf{V}(t)\| = \sqrt{16 + 9} = 5$

$\quad \mathbf{T}(t) = \frac{1}{5}(-4 \sin t \, \mathbf{i} - 3\mathbf{j} + 4 \cos t \, \mathbf{k})$

$\quad \frac{d\mathbf{T}}{dt} = \frac{1}{5}(-4 \cos t \, \mathbf{i} - 4 \sin t \, \mathbf{k})$

$\quad \mathbf{N}(t) = -\cos t \, \mathbf{i} + \sin t \, \mathbf{j}$

38. $\mathbf{R}(t) = (e^t \sin t)\mathbf{i} + (e^t)\mathbf{j} + (e^t \cos t)\mathbf{k}$

$\quad \mathbf{V}(t) = e^t(\cos t + \sin t)\mathbf{i} + e^t\mathbf{j} + e^t(\cos t - \sin t)\mathbf{k}$

$\quad \|\mathbf{V}(t)\| = e^t\sqrt{3}$

$\quad \mathbf{T}(t) = \frac{1}{\sqrt{3}}[(\cos t + \sin t)\mathbf{i} + \mathbf{j} + (\cos t - \sin t)\mathbf{k}]$

$\quad \frac{d\mathbf{T}}{dt} = (\cos t - \sin t)\mathbf{i} - (\cos t + \sin t)\mathbf{k}$

$\quad \frac{d}{dt}\|\mathbf{T}\| = \sqrt{2}$

$\quad \mathbf{N}(t) = \frac{1}{\sqrt{2}}[(\cos t - \sin t)\mathbf{i} - (\sin t + \cos t)\mathbf{k}]$

39. $\mathbf{R}(t) = t^2\mathbf{i} + 2t\mathbf{j} + e^t\mathbf{k}$

$\quad \mathbf{V}(t) = \mathbf{R}'(t) = 2t\mathbf{i} + 2\mathbf{j} + e^t\mathbf{k}$

$\quad \mathbf{A}(t) = \mathbf{R}''(t) = 2\mathbf{i} + e^t\mathbf{k}; \|\mathbf{A}(t)\| = \sqrt{4 + e^{2t}}$

$$\frac{ds}{dt} = \| \mathbf{V}(t) \| = \sqrt{4t^2 + 4 + e^{2t}}$$

$$A_T = \frac{d^2s}{dt^2} = \frac{4t + e^{2t}}{\sqrt{4 + 4t^2 + e^{2t}}}$$

$$A_N = \sqrt{\|\mathbf{A}\|^2 - A_T^2}$$

$$= \sqrt{4 + e^{2t} - \frac{16t^2 + 8te^{2t} + e^{4t}}{4 + 4t^2 + 3^{2t}}}$$

$$= 2\sqrt{\frac{t^2 e^{2t} - 2te^{2t} + 2e^{2t} + 4}{4e^{2t} + 4 + e^{2t}}}$$

$$\mathbf{V} \times \mathbf{A} = \begin{vmatrix} \mathbf{i} & \mathbf{j} & \mathbf{k} \\ 2t & 2 & e^t \\ 2 & 0 & e^t \end{vmatrix}$$

$$= 2e^t\mathbf{i} - (2t-2)e^t\mathbf{j} - 4\mathbf{k}$$

$$\kappa = \frac{\| \mathbf{V} \times \mathbf{A} \|}{\| \mathbf{V} \|^3}$$

$$= \frac{2\sqrt{t^2 e^{2t} - 2te^{2t} + 2e^{2t} + 4}}{(4t^2 + 4 + e^{2t})^{3/2}}$$

40. $\mathbf{R}(t) = t^2\mathbf{i} - 2t\mathbf{j} + (t^2 - t)\mathbf{k}$

$$\mathbf{V}(t) = \mathbf{R}'(t) = 2t\mathbf{i} - 2\mathbf{j} + (2t - 1)\mathbf{k}$$

$$\mathbf{A}(t) = \mathbf{R}''(t) = 2\mathbf{i} + 2\mathbf{k}; \| \mathbf{A}(t) \| = 2\sqrt{2}$$

$$\frac{ds}{dt} = \| \mathbf{V}(t) \| = \sqrt{4t^2 + 4 + 4t^2 - 4t + 1}$$

$$= \sqrt{8t^2 - 4t + 5}$$

$$A_T = \frac{d^2s}{dt^2} = \frac{8t - 2}{\sqrt{8t^2 - 4t + 5}}$$

$$A_N = \sqrt{\|\mathbf{A}\|^2 - A_T^2}$$

$$= \sqrt{8 - \frac{64t^2 - 32t + 4}{8t^2 - 4t + 5}}$$

$$= \frac{\sqrt{64t^2 - 32t + 40 - 64t^2 + 32t}}{8t^2 - 4t + 5}$$

$$= \frac{2\sqrt{10}}{8t^2 - 4t + 5}$$

$$\mathbf{V} \times \mathbf{A} = \begin{vmatrix} \mathbf{i} & \mathbf{j} & \mathbf{k} \\ 2t & -2 & 2t-1 \\ 2 & 0 & 2 \end{vmatrix} = -4\mathbf{i} - 2\mathbf{j} + 4\mathbf{k}$$

$$\kappa = \frac{\| \mathbf{V} \times \mathbf{A} \|}{\| \mathbf{V} \|^3}$$

$$= \frac{\sqrt{(-4)^2 + (-2)^2 + 4^2}}{(4t^2 + 4 + 4t^2 - 4t + 1)^{3/2}}$$

$$= \frac{6}{(8t^2 - 4t + 5)^{3/2}}$$

41. $\mathbf{R}(t) = (4 \sin t)\mathbf{i} + (4 \cos t)\mathbf{j} + 4t\mathbf{k}$

$$\mathbf{V}(t) = \mathbf{R}'(t) = 4 \cos t\,\mathbf{i} - 4 \sin t\,\mathbf{j} + 4\mathbf{k}$$

$$\mathbf{A}(t) = \mathbf{R}''(t) = 4(-\sin t\mathbf{i} - \cos t\mathbf{j}); \| \mathbf{A}(t) \| = 4$$

$$\frac{ds}{dt} = \| \mathbf{V}(t) \| = \sqrt{16 + 16} = 4\sqrt{2}$$

$$A_T = \frac{d^2s}{dt^2} = 0$$

$$A_N = \sqrt{\|\mathbf{A}\|^2 - A_T^2} = 4\sqrt{2}$$

$$\mathbf{V} \times \mathbf{A} = 16 \begin{vmatrix} \mathbf{i} & \mathbf{j} & \mathbf{k} \\ \cos t & -\sin t & 1 \\ -\sin t & -\cos t & 0 \end{vmatrix}$$

$$= 16[\cos t\,\mathbf{i} - \sin t\,\mathbf{j} - \mathbf{k}]$$

$$\kappa = \frac{\| \mathbf{V} \times \mathbf{A} \|}{\| \mathbf{V} \|^3} = \frac{16\sqrt{2}}{64(2\sqrt{2})} = \frac{1}{8}$$

42. $\mathbf{R}(t) = (a\sin 3t)\mathbf{i} + (a + a\cos 3t)\mathbf{j}$
$\qquad + (3a\sin t)\mathbf{k}$

$$\mathbf{V}(t) = \mathbf{R}'(t) = 3a[\cos 3t\mathbf{i} - \sin 3t\mathbf{j} + \cos t\mathbf{k}]$$

$$\mathbf{A}(t) = \mathbf{R}''(t)$$

$$= 3a[-3\sin 3t\mathbf{i} - 3\cos 3t\mathbf{j} - \sin t\mathbf{k}];$$

$$\| \mathbf{A}(t) \| = 9a^2\sqrt{9 + \sin^2 t}$$

$$\frac{ds}{dt} = \| \mathbf{V}(t) \| = 3a\sqrt{1 + \cos^2 t}$$

$$A_T = \frac{d^2s}{dt^2} = \frac{3a(-\sin t \cos t)}{\sqrt{1 + \cos^2 t}} = \frac{3a \sin 2t}{2\sqrt{1 + \cos^2 t}}$$

$$A_N = \sqrt{\|\mathbf{A}\|^2 - A_T^2}$$

$$= \sqrt{9a^2(9 + \sin^2 t) - 9a^2\left(\frac{\sin^2 t \cos^2 t}{1 + \cos^2 t}\right)}$$

$$= \frac{3a\sqrt{9 + 9\cos^2 t + \sin^2 t}}{\sqrt{1 + \cos^2 t}}$$

$$\mathbf{V} \times \mathbf{A} = 9a^2 \begin{vmatrix} \mathbf{i} & \mathbf{j} & \mathbf{k} \\ \cos 3t & -\sin 3t & \cos t \\ -3\sin 3t & -3\cos 3t & -\sin t \end{vmatrix}$$

$$= 9a^2[(\sin 3t \sin t + 3 \cos 3t \cos t)\mathbf{i}$$

$$- (-\sin t \cos 3t + 3 \cos t \sin 3t)\mathbf{j}$$

$$+ (-3 \cos^2 3t - 3 \sin^2 3t)\mathbf{k}]$$

$$\| \mathbf{V} \times \mathbf{A} \| = 9a^2\sqrt{\sin^2 t + 9\cos^2 t + 9}$$

$$\kappa = \frac{\| \mathbf{V} \times \mathbf{A} \|}{\| \mathbf{V} \|^3}$$

$$= \frac{9a^2\sqrt{\sin^2 t + 9\cos^2 t + 9}}{27a^3(1 + \cos^2 t)^{3/2}}$$

43. $r = 4 \cos \theta;\ r' = -4 \cos \theta;\ r'' = -4 \cos \theta;$
at $\theta = \pi/3,\ r = 2,\ r' = -2\sqrt{3},\ r'' = -2$

$$\kappa = \frac{\left|2^2 + 2(-2\sqrt{3})^2 - 2(-2)\right|}{[2^2 + (-2\sqrt{3})^2]^{3/2}} = \frac{1}{2}$$

44. $r = \theta^2$; $r' = 2\theta$; $r'' = 2$;
at $\theta = 2$, $r = 4$, $r' = 4$, $r'' = 2$

$$\kappa = \frac{\left|4^2 + 2(4)^2 - 4(2)\right|}{[4^2 + 4^2]^{3/2}} = \frac{5}{16\sqrt{2}}$$

45. $r = e^{-\theta}$; $r' = -e^{-\theta}$; $r'' = e^{-\theta}$;
at $\theta = 1$, $r = e^{-1}$, $r' = -e^{-1}$, $r'' = e^{-1}$

$$\kappa = \frac{\left|e^{-2\theta} + 2e^{-2\theta} - e^{-2\theta}\right|}{[e^{-2\theta} + e^{-2\theta}]^{3/2}} = \frac{e}{\sqrt{2}} \approx 1.92$$

46. $r = 1 + \cos\theta$; $r' = -\sin\theta$; $r'' = -\cos\theta$;
at $\theta = \pi/2$, $r = 1$, $r' = -1$, $r'' = 0$

$$\kappa = \frac{\left|1^2 + 2(-1)^2 - 1(0)\right|}{[1^2 + (-1)^2]^{3/2}} = \frac{3}{2\sqrt{2}}$$

47. $r = 4\cos 3\theta$; $r' = -12\sin 3\theta$; $r'' = -36\cos 3\theta$
at $\theta = \pi/6$, $r = 0$, $r' = -12$, $r'' = 0$

$$\kappa = \frac{\left|0^2 + 2(-12)^2 - 0(0)\right|}{[0^2 + (-12)^2]^{3/2}} = \frac{1}{6}$$

48. $r = 1 - 2\sin\theta$; $r' = -2\cos\theta$; $r'' = 2\sin\theta$;
at $\theta = \pi/4$, $r = 1 - \sqrt{2}$, $r' = -\sqrt{2}$, $r'' = \sqrt{2}$

$$\kappa = \frac{\left|(1-\sqrt{2})^2 + 2(-\sqrt{2})^2 - (1-\sqrt{2})(\sqrt{2})\right|}{[(1-\sqrt{2})^2 + (-\sqrt{2})^2]^{3/2}}$$

$$= \frac{5 - \sqrt{2}}{(5 - 2\sqrt{2})^{3/2}} \approx 1.12$$

49. $\mathbf{F}(t)$ is continuous for $t \neq 1$.

50. $\mathbf{R}(t) = \left(\frac{t^2 - 2}{2}\right)\mathbf{i} + \frac{(2t+1)^{3/2}}{3}\mathbf{j}$

$$\mathbf{V}(t) = t\mathbf{i} + (2t+1)^{1/2}\mathbf{j}$$

$$\|\mathbf{V}(t)\| = \sqrt{t^2 + 2t + 1} = t + 1$$

$$s = \int_0^6 (t+1)\,dt = \tfrac{1}{2}(t+1)^2\Big|_0^6 = 24$$

51. $\mathbf{F}'''(t) = (\cos t)\mathbf{i} + (\sin t)\mathbf{j} + \frac{t}{\pi}\mathbf{k}$

$$\mathbf{F}''(t) = (\sin t + C_1)\mathbf{i} + (-\cos t + C_2)\mathbf{j} + (\tfrac{t^2}{2\pi} + C_3)\mathbf{k}$$

Since $\mathbf{F}''(0) = 2\mathbf{j} + \mathbf{k}$,

$$\mathbf{F}''(t) = \sin t\,\mathbf{i} + (3 - \cos t)\mathbf{j} + (\tfrac{t^2}{2\pi} + 1)\mathbf{k}$$

$$\mathbf{F}'(t) = (-\cos t + C_4)\mathbf{i} + (3t - \sin t + C_5)\mathbf{j} + (\tfrac{t^3}{6\pi} + t + C_6)\mathbf{k}$$

Since $\mathbf{F}'(0) = \mathbf{i}$,

$$\mathbf{F}'(t) = (2 - \cos t)\mathbf{i} + (3t - \sin t)\mathbf{j} + (\tfrac{t^3}{6\pi} + t)\mathbf{k}$$

$$\mathbf{F}(t) = (2t - \sin t + C_7)\mathbf{i} + (\tfrac{3t^2}{2} + \cos t + C_8)\mathbf{j} + (\tfrac{t^4}{24\pi} + \tfrac{t^2}{2} + C_9)\mathbf{k}$$

Since $\mathbf{F}(0) = \mathbf{i}$,

$$\mathbf{F}(t) = (2t - \sin t + 1)\mathbf{i} + (\tfrac{3t^2}{2} + \cos t - 1)\mathbf{j} + (\tfrac{t^4}{24\pi} + \tfrac{t^2}{2})\mathbf{k}$$

52. **a.** $\mathbf{A}(t) = (-t)\mathbf{i} + 2\mathbf{j} + (2 - t)\mathbf{k}$

$$\mathbf{V}(t) = (-\tfrac{t^2}{2} + C_1)\mathbf{i} + (2t + C_2)\mathbf{j} + (2t - \tfrac{t^2}{2} + C_3)\mathbf{k}$$

Since $\mathbf{V}(0) = 2\mathbf{i} - 4\mathbf{j}$,

$$\mathbf{V}(t) = (-\tfrac{t^2}{2} + 2)\mathbf{i} + (2t - 4)\mathbf{j} + (2t - \tfrac{t^2}{2})\mathbf{k}$$

$$\mathbf{R}(t) = (-\tfrac{t^3}{6} + C_4)\mathbf{i} + (t^2 - 4t + C_5)\mathbf{j} + (t^2 - \tfrac{t^3}{6} + C_6)\mathbf{k}$$

Since $\mathbf{R}(0) = \mathbf{i}$

$$\mathbf{R}(t) = (-\tfrac{t^3}{6} + 2t + 1)\mathbf{i} + (t^2 - 4t)\mathbf{j} + (t^2 - \tfrac{t^3}{6})\mathbf{k}$$

b. $\mathbf{V}(1) = \tfrac{3}{2}\mathbf{i} - 2\mathbf{j} + \tfrac{3}{2}\mathbf{k}$;

$$\|\mathbf{V}(1)\| = \sqrt{(\tfrac{3}{2})^2 + 4 + (\tfrac{3}{2})^2} = \tfrac{1}{2}\sqrt{34}$$

c. The object is stationary when $\mathbf{V}(t) = \mathbf{0}$;
$-\tfrac{t^2}{2} + 2 = 0$; $2t - 4 = 0$, and $2t - \tfrac{t^2}{2} = 0$
There is no value of t simultaneously solving these equations, so the object is not stationary.

53. $\mathbf{R}(s) = a\cos\tfrac{s}{a}\mathbf{i} + a\sin\tfrac{s}{a}\mathbf{j} + 2s\mathbf{k}$

$\mathbf{R}'(s) = -\sin\tfrac{s}{a}\mathbf{i} + \cos\tfrac{s}{a}\mathbf{j} + 2\mathbf{k}$

$\mathbf{T}(s) = \frac{1}{\sqrt{5}}[-\sin\tfrac{s}{a}\mathbf{i} + \cos\tfrac{s}{a}\mathbf{j} + 2\mathbf{k}]$

$\frac{d\mathbf{T}}{ds} = \frac{1}{a}[-\cos\tfrac{s}{a}\mathbf{i} - \sin\tfrac{s}{a}\mathbf{j}]$

$\mathbf{N} = \cos\tfrac{s}{a}\mathbf{i} + \sin\tfrac{s}{a}\mathbf{j}$

54. $y = 1 + \sin x$; $y' = \cos x$; $y'' = -\sin x$

$$\kappa = \frac{|\sin x|}{(1 + \cos^2 x)^{3/2}}$$

a. At $x = \pi/6$, $\kappa = \frac{4}{7^{3/2}}$ and $\rho = \tfrac{7}{4}\sqrt{7}$

b. At $x = \pi/4$, $\kappa = \frac{2}{3\sqrt{3}}$ and $\rho = \tfrac{3}{2}\sqrt{3}$

c. At $x = 3\pi/2$, $\kappa = 1$ and $\rho = 1$

55. $\mathbf{R}(t) = a\cos t\,\mathbf{i} + b\sin t\,\mathbf{j}$

$\mathbf{V}(t) = -a\sin t\,\mathbf{i} + b\cos t\,\mathbf{j}$

$$\|\mathbf{V}(t)\| = \sqrt{a^2\sin^2 t + b^2\cos^2 t}$$

$$\mathbf{V} \times \mathbf{A} = \begin{vmatrix} \mathbf{i} & \mathbf{j} & \mathbf{k} \\ a\sin t & -b\cos t & 0 \\ a\cos t & b\sin t & 0 \end{vmatrix} = ab\mathbf{k}$$

$$\|\mathbf{V} \times \mathbf{A}\| = ab; \quad \rho = \frac{(a^2\sin^2 t + b^2\cos^2 t)^{3/2}}{|ab|}$$

At $t = 0$, $\rho = \dfrac{b^3}{ab} = \dfrac{b^2}{a}$; at $t = \dfrac{\pi}{2}$, $\rho = \dfrac{a^3}{ab} = \dfrac{a^2}{b}$

56. $\mathbf{A} - g\mathbf{j} = -32\mathbf{j}$; $v_0 = \dfrac{180(5,280)}{3,600} = 264$ ft/s

$\mathbf{V} = C_1\mathbf{i} + (-gt + C_2)\mathbf{j}$

where $C_1 = 264$ and $C_2 = 0$

$\mathbf{R} = (264t + C_3)\mathbf{i} + (-16t^2 + C_4)\mathbf{j}$

where $C_3 = 0$ and $C_4 = 4,000$

The ground is reached when $16t^2 = 4,000$ or when $t = 5\sqrt{10}$. The horizontal distance is $x = 264(5\sqrt{10}) \approx 4,174$ ft.

57. a. $\mathbf{R}(t) = (e^{-t}\cos t)\mathbf{i} + (e^{-t}\sin t)\mathbf{j} + e^{-t}\mathbf{k}$

$\mathbf{V}(t) = e^{-t}(-\cos t - \sin t)\mathbf{i}$
$\qquad + e^{-t}(-\sin t + \cos t)\mathbf{j} - e^{-t}\mathbf{k}$

$\|\mathbf{V}(t)\| = e^{-t}\sqrt{2\cos t\sin t + 1 + 1 - 2\sin t\cos t + 1}$

$\mathbf{A}(t) = e^{-t}(2\sin t)\mathbf{i} + e^{-t}(-2\cos t)\mathbf{j} + e^{-t}\mathbf{k}$

b. $\mathbf{V} \times \mathbf{A}$

$$= e^{-2t}\begin{vmatrix} \mathbf{i} & \mathbf{j} & \mathbf{k} \\ -\cos t - \sin t & -\sin t + \cos t & -1 \\ 2\sin t & -2\cos t & 1 \end{vmatrix}$$

$= e^{-2t}[-(\sin t + \cos t)\mathbf{i} - (\sin t - \cos t)\mathbf{j} + 2\mathbf{k}]$

$\|\mathbf{V} \times \mathbf{A}\| = e^{-2t}\sqrt{6}$

$\kappa = \dfrac{\sqrt{6}e^{-2t}}{3\sqrt{3}e^{-3t}} = \dfrac{1}{3}e^t\sqrt{2}$

58. $y = e^{ax}$; $y' = ae^{ax}$; $y'' = a^2 e^{ax}$

$\kappa = \dfrac{a^2 e^{ax}}{(1 + a^2 e^{ax})^{3/2}}$

$\kappa' = \dfrac{a^2}{(1 + a^2 e^{ax})^{3/2}}[(1 + a^2 e^{ax})^{3/2}(ae^{ax})$

$\qquad - e^{ax}(\tfrac{3}{2})(1 + a^2 e^{ax})^{1/2}(a^3 e^{ax})]$

$\kappa' = 0$, if $e^{ax} = 0$ or $2a - a^3 e^{ax} = 0$

which implies that $x = \dfrac{1}{a}\ln\dfrac{2}{a^2}$.

59. $\mathbf{R}(t) = a\cos \omega t\,\mathbf{i} + a\sin \omega t\,\mathbf{j} + \omega^2 t\,\mathbf{k}$

$\mathbf{V}(t) = \mathbf{R}'(t) = \omega(-a\sin \omega t\,\mathbf{i} + a\cos \omega t\,\mathbf{j} + \omega\mathbf{k})$

$\mathbf{A}(t) = \mathbf{R}''(t) = -a\omega^2(\cos \omega\,\mathbf{i} + \sin \omega t\,\mathbf{j})$

$\|\mathbf{V}(t)\| = a^2\omega^2(1 + \omega^2)$; $\dfrac{ds}{dt} = a\omega\sqrt{1 + \omega^2}$

$\|\mathbf{A}(t)\| = a\omega^2$

$A_T = \dfrac{d^2 s}{dt^2} = 0$; $A_N = a\omega^2$

60. $r = 1 + \cos at$; $\theta = e^{-at}$; $\dfrac{dr}{dt} = -a\sin at$

$\dfrac{d^2 r}{dt} = -a^2\cos at$; $\dfrac{d\theta}{dt} = -ae^{-at}$;

$\dfrac{d^2\theta}{dt^2} = a^2 e^{-at}$;

$\mathbf{V}(t) = \dfrac{dr}{dt}\mathbf{u}_r + r\dfrac{d\theta}{dt}\mathbf{u}_\theta$

$\qquad = -a\sin at\,\mathbf{u}_r - ae^{-at}(1 + \cos at)\mathbf{u}_\theta$

$\mathbf{A}(t) = \left[\dfrac{d^2 r}{dt^2} - r\left(\dfrac{d\theta}{dt}\right)^2\right]\mathbf{u}_r + \left[r\dfrac{d^2\theta}{dt^2} + 2\left(\dfrac{dr}{dt}\right)\left(\dfrac{d\theta}{dt}\right)\right]\mathbf{u}_\theta$

$\qquad = [-a^2\cos at - a^2 e^{-2at}(1 + \cos at)\mathbf{u}_r$

$\qquad\quad + [a^2 e^{-at}(1 + \cos at) + 2a^2 e^{-at}\sin at]\mathbf{u}_\theta$

61. $W = 3,000$ lb; $mg = 3,000$, so $m = 3,000/32$

$v = \dfrac{60(5,280)}{3,600} = 88$ ft/s; $r = 40$, $\kappa = 1/40$

$A_N = 88^2/40$;

$\|\mathbf{F}\| = mA_N = \dfrac{88^2}{40}\dfrac{3,000}{32} = 18,150$ lb

62. $x = (v_0\cos\alpha)t$, $y = -16t^2 + v_0\sin\alpha t$

For the range, $y = 0$ or $16t^2 = 50t\sin\alpha$.

$T_f = \dfrac{25\sin \alpha}{16}$ and $t_h = \dfrac{25\sin \alpha}{16}$

$y_{max} = 5$, so

$-16\left(\dfrac{25\sin \alpha}{16}\right)^2 + 50\sin \alpha\left(\dfrac{25\sin \alpha}{16}\right) = 5$

$80 = 625\sin^2\alpha$

$\alpha \approx 21°$

The time of travel is $T_f = \dfrac{25\sin 21°}{8} \approx 1.12$ s

and $x \approx 50(\cos 21°)(1.12) \approx 52.3$ ft

63. $\mathbf{R}(t) = \left(\dfrac{3t}{1 + t^3}\right)\mathbf{i} + \left(\dfrac{3t^2}{1 + t^3}\right)\mathbf{j}$

$\mathbf{T} = \dfrac{d\mathbf{R}}{dt} = \dfrac{3}{(1 + t^3)^2}[(1 + t^3 - 3t^2)\mathbf{i}$

$\qquad - (2t^2 + 2t^5 - 3t^4)\mathbf{j}]$

At $P(0, 0)$, $\mathbf{T} = 3\mathbf{i}$, so the equation of the tangent line are $x = t$, $y = 0$, $z = 0$.

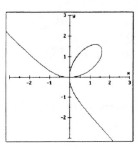

64. $\alpha = 62°$; $t = x/(v_0\cos \alpha)$;

$\quad y = -4.9t^2 + (v_0\sin\alpha)t$

$$= -\frac{4.9x^2}{v_0^2\cos^2\alpha} + \frac{(v_0\sin \alpha)x}{v_0\cos \alpha}$$

$$= -\frac{4.9x^2}{v_0^2\cos^2\alpha} + (\tan \alpha)x$$

$y' = \dfrac{-9.8}{v_0^2\cos^2\alpha}x + \tan \alpha$

$y' = 0$ when $x = \dfrac{v_0^2\sin \alpha \cos \alpha}{9.8}$

The horizontal distance (range) is $2x$. We calculate the necessary values using a spreadsheet and these formulas. The angle measurement at the top is given in both degrees and radians.

Angle:	62	1.08

Vo	x	Horizontal distance
4	0.67677	1.35
5	1.05745	2.11
6	1.52272	3.05
7	2.07259	4.15
8	2.70706	5.41
9	3.42613	6.85
10	4.22978	8.46
11	5.11804	10.24
12	6.09089	12.18
13	7.14834	14.30
14	8.29038	16.58
15	9.51702	19.03
16	10.8282	21.66

We see that the water hits the building for v_0 between 9 and 13 m/s.

65. $v = 10$ m/s; $r = (1/4)$ m; $C = 2\pi r = 0.5\pi$ m

$\omega = 10(2\pi)/(0.5\pi) \approx 40$ rpm or $20/\pi$ rad/s

$\mathbf{R} = r \cos \omega t\, \mathbf{i} + r \sin \omega t\, \mathbf{j}$;

$\mathbf{V} = -r\omega \sin \omega t\, \mathbf{i} + r\omega \cos \omega t\, \mathbf{j}$;

$\|\mathbf{V}\| = \omega r = (\frac{20}{\pi})(\frac{1}{4}) = \frac{5}{\pi}$ m/s

$\mathbf{A} = -r\omega^2 \cos \omega t\, \mathbf{i} - r\omega^2 \sin \omega t\, \mathbf{j} = -r\omega^2\mathbf{R}$

$\mathbf{A}_N = \omega^2 r$ so the point P travels at

$\frac{5}{\pi} + 1 \approx 11.6$ m/s

66. $\mathbf{R} = 10^{-8}\cos t\, \mathbf{i} + 10^{-8}\sin t\, \mathbf{j} + \dfrac{3(10^{-8})t}{2\pi}\, \mathbf{k}$

$\mathbf{V} = 10^{-8}(-\sin t\, \mathbf{i} + \cos t\, \mathbf{j} + \dfrac{3}{2\pi}\, \mathbf{k})$

$\|\mathbf{V}\| = 10^{-8}\sqrt{1 + \dfrac{9}{4\pi^2}}$

Range of t: $0 < t < 6\pi(10^8)$ since there are

3×10^8 turns at 2π radians/turn

$$s = (10^{-8})\int_0^{6\pi \times 10^8} \sqrt{1+\frac{9}{4\pi^2}}\ dt \approx 20.888 \ \mu\text{m}$$

67. $A_T = d^2s/dt^2 = 0$; $ds/dt = \|\mathbf{V}\|$ is constant; if $\|\mathbf{V}\|$ is constant, differentiation produces a zero acceleration, so the converse is true.

68. $r = 10t$; $r' = 10$; $r'' = 0$;

$\theta = 2\pi t$; $\theta' = 2\pi$; $\theta'' = 0$

$\mathbf{V}(t) = \dfrac{dr}{dt}\mathbf{u}_r + r\dfrac{d\theta}{dt}\mathbf{u}_\theta = 10\mathbf{u}_r + 20\pi t\mathbf{u}_\theta$

a. At $t = 0$, $\mathbf{V}(0) = 10\mathbf{u}_r$

b. At $t = 0.25$, $\mathbf{V}(0.25) = 10\mathbf{u}_r + 5\pi\mathbf{u}_\theta$

69. $\mathbf{F}(x) = \displaystyle\int_1^x [(\sin t)\mathbf{i} - (\cos 2t)\mathbf{j} + e^{-t}\mathbf{k}]\ dt$

$\mathbf{F}'(x) = [(\sin t)\mathbf{i} - (\cos 2t)\mathbf{j} + e^{-t}\mathbf{k}]\Big|_{t=x}$

$= \sin x\,\mathbf{i} - \cos 2x\,\mathbf{j} + e^{-x}\mathbf{k}$

70. $\mathbf{F}(x) = \displaystyle\int_1^x [t^2\mathbf{i} + (\sec e^{-t})\mathbf{j} - (\tan e^{2t})\mathbf{k}]\ dt$

$\mathbf{F}'(x) = x^2\mathbf{i} + (\sec e^{-x})\mathbf{j} - \tan e^{2x}\mathbf{k}$

71. $\mathbf{F}(t) = (\frac{t^2}{2} + C_1)\mathbf{i} + (\frac{t^2}{2} + C_2)\mathbf{j} - (\frac{t^2}{2} + C_3)\mathbf{k}$

Since $\mathbf{F}(0) = \mathbf{i} + 2\mathbf{j} - 3\mathbf{k}$,

$\mathbf{F}(t) = (\frac{t^2}{2} + 1)\mathbf{i} + (\frac{t^2}{2} + 2)\mathbf{j} - (\frac{t^2}{2} + 3)\mathbf{k}$

72. $\mathbf{F}(t) = \left[(\frac{3}{2})(\frac{2}{3})(t + 1)^{3/2} + C_1\right]\mathbf{i}$

$\qquad + \left[\ln|t + 1| + C_2\right]\mathbf{j} + \left[e^t + C_3\right]\mathbf{k}$

Since $\mathbf{F}(0) = \mathbf{j} - 3\mathbf{k}$,

$\mathbf{F}(t) = \left[(\frac{3}{2})(\frac{2}{3})(t + 1)^{3/2} - 1\right]\mathbf{i}$

$\qquad + \left[\ln|t + 1| + 1\right]\mathbf{j} + \left[e^t + 1\right]\mathbf{k}$

$= \left[(t+1)^{3/2} - 1\right]\mathbf{i} + \left[\ln|t+1| + 1\right]\mathbf{j} + \left[e^t + 1\right]\mathbf{k}$

73. $\mathbf{F}(t) = \left[(-\frac{1}{2})\cos 2t + C_1\right]\mathbf{i}$

$\qquad + \left[(\frac{e^t}{2})(\sin t + \cos t) + C_2\right]\mathbf{j}$

$\qquad - \left[3\ln|t + 1| + C_3\right]\mathbf{k}$

Since $\mathbf{F}(0) = \mathbf{i} - 3\mathbf{k}$,

$\mathbf{F}(t) = \left[(\frac{1}{2})(3 - \cos 2t)\right]\mathbf{i}$

$\qquad + \left[(\frac{e^t}{2})(\sin t + \cos t) - 1\right]\mathbf{j}$

$\qquad - \left[3\ln|t + 1| + 3\right]\mathbf{k}$

74. $\dfrac{d^2\mathbf{F}}{dt^2} = -32\mathbf{j}$

$$\frac{d\mathbf{F}}{dt} = C_1\mathbf{i} + (-32t + C_2)\mathbf{j} + C_3\mathbf{k}$$

Since $\frac{d\mathbf{F}}{dt} = 5\mathbf{i} + 5\mathbf{k}$ at $t = 0$,

$$\frac{d\mathbf{F}}{dt} = 5\mathbf{i} - 32t\mathbf{j} + 5\mathbf{k}$$

$$\mathbf{F} = (5t + C_4)\mathbf{i} + (-16t^2 + C_5)\mathbf{j} + (5t + C_6)\mathbf{k}$$

Since $\mathbf{F} = 50\mathbf{j}$ at $t = 0$,

$$\mathbf{F} = 5t\mathbf{i} + (-16t^2 + 50)\mathbf{j} + 5t\mathbf{k}$$

75. This is Putnam Problem 6ii of the morning session of 1939. Choose rectangular coordinates with the y-axis vertical, the origin at the position of the gun, and the airplane over a point of the positive x-axis. Then the coordinates of the airplane are (u, h) where $u \geq 0$. If the gun is fired at time $t = 0$ with muzzle velocity V and angle of elevation α, then (neglecting air resistance), the shell's position at time t is given by

$$x = Vt \cos \alpha, \quad y = Vt \sin \alpha - \tfrac{1}{2}gt^2$$

Since it is given that the shell strikes the airplane,

$$u = Vt \cos \alpha \text{ and } h = Vt \sin \alpha - \tfrac{1}{2}gt^2$$

for some t and α. Hence,

$$u^2 + \left(h + \tfrac{1}{2}gt^2\right)^2 = V^2 t^2$$

so that

$$\tfrac{1}{4}g^2 t^4 + (gh - V^2)t^2 + h^2 + u^2 = 0$$

In order for this equation to have a real root, it is necessary that

$$(gh - V^2)^2 \geq g^2(h^2 + u^2)$$
$$g^2 u^2 \leq V^2(V^2 - 2gh)$$
$$V^2 \geq 2gh \text{ and } u \leq \frac{V}{g}\sqrt{V^2 - 2gh}$$

This shows that the gun is within distance

$$\frac{V}{g}\sqrt{V^2 - 2gh}$$

from the point directly below the airplane when it was hit.

76. This is Putnam Problem 4 of the morning session of 1947. We take coordinates with origin at the gun, the y-axis vertical, and the x-axis horizontal in the direction of the fire. For a given angle α and the prescribed initial conditions, the equations motion

$$\frac{d^2 x}{dt^2} = 0, \quad \frac{d^2 y}{dt^2} = -g$$

lead to $x = v_0 t \cos \alpha$, $y = v_0 t \sin \alpha - \tfrac{1}{2}gt^2$.

Elimination of t gives

$$y = x \tan \alpha - \frac{g}{2v_0^2}\sec^2 \alpha$$

For a fixed positive x, this can be written

$$y = \frac{v_0^3}{2g} - \frac{gx^2}{2v_0^2} - \frac{gx^2}{2v_0^2}\left(\tan \alpha - \frac{v_0^2}{gx}\right)^2$$

From this it is clear that we can choose α so as to hit the point (x, y) if and only if

$$y \leq \frac{v_0^2}{2g} - \frac{gx^2}{2v_0^2}$$

To hit a point $(0, y)$, we fire straight up (*i.e.* take $\alpha = 90°$), so that the parametric equation for y can be written

$$y = \frac{v_0^2}{2g} - \frac{v_0^2}{2g}\left(\frac{g}{v_0}t - 1\right)^2$$

and it is clear that we can reach $(0, y)$ if and only if $y \leq v_0^2/(2g)$. Therefore, the desired set H is defined by the inequalities

$$y \leq \frac{v_0^2}{2g} - \frac{gx^2}{2v_0^2}, \quad x \geq 0, \; y > 0$$

Remark: This problem is essentially the same as Problem 6ii of the second competition. A different solution is given there.

77. This is Putnam Problem 6 of the afternoon session of 1946. Suppose the circle has radius r. If Cartesian coordinates are chosen with origin at the center of the circle, then the coordinates of the particle are $(r \cos \theta, r \sin \theta)$ where θ is a function of time t. Differentiating this twice, we see that the acceleration vector is

$$r\frac{d\omega}{dt}(-\sin \theta, \cos \theta) + r\omega^2(-\cos \theta, -\sin \theta)$$

where $\omega = d\theta/dt$ is the angular velocity. Since $(-\sin \theta, \cos \theta)$ and $(-\cos \theta, -\sin \theta)$ are orthogonal unit vectors in the direction of the tangent and the inward normal, we see that the two terms of the acceleration vector are, respectively, the tangential and normal components of the acceleration. Since $\omega \neq 0$ at any time during the motion, the normal component of acceleration, and hence the acceleration vector itself, is never 0. Since $\omega = 0$ at the start and at the finish, by Rolle's theorem there is an intermediate time at which $d\omega/dt = 0$. At that time, the acceleration vector points inward along the radius because $r\omega^2 > 0$.

Remark: We have interpreted "coming to rest" to mean "having velocity 0." If

"coming to rest" means "remaining stationary through some time interval," then the first statement is false, for it is certainly possible that ω and $d\omega/dt$ vanish simultaneously, but not on an interval. The second statement is true, however, because the usual proof of Rolle's theorem shows that $d\omega/dt = 0$ at some point where $\omega \neq 0$, unless $\omega = 0$ identically, which is ruled out.

Cumulative Review for Chapters 7-11, Page 736

1. Step 1. Simplify; Step 2. Use basic formulas; Step 3. Substitute; Step 4. Classify. Step 5: Try again. See Table 7.1, Page 465.

2. Check the following tests for convergence: 1. Divergence test; 2. Limit comparison test; 3. Ratio test; 4. Root test; 5. Integral test; 6. Direct comparison test; 7. Zero-infinity limit. See Table 8.1, Page 527.

3. The graph of
$$Ax^2 + Bxy + Cy^2 + Dx + Ey + F = 0$$
is a quadric surface. Find and sketch the traces of the surface with the coordinate planes to get a clue of the shape of the overall surface. See Table 10.1, Page 645.

4. A vector is a directed line segment. A vector function is a function whose range is a set of vectors for each point in its domain. Vector calculus involves differentiation and integration of vector functions.

5. $y = x\cosh^{-1}x$
$$y' = \frac{x}{\sqrt{x^2 - 1}} + \cosh^{-1}x$$

6. $x\sinh^{-1}y + y\tanh^{-1}x = 0$

$$\frac{xy'}{\sqrt{1+y^2}} + \sinh^{-1}y + \frac{y}{1-x^2} + (\tanh^{-1}x)y' = 0$$

$$y'\frac{x}{\sqrt{1+y^2}} + y'\tanh^{-1}x = -\frac{y}{1-x^2} - \sinh^{-1}y$$

$$\frac{dy}{dx} = \frac{-\dfrac{y}{1-x^2} - \sinh^{-1}y}{\dfrac{x}{\sqrt{1+y^2}} + \tanh^{-1}x}$$

$$= -\frac{y\sqrt{1+y^2} + \sqrt{1+y^2}(1-x^2)\sinh^{-1}y}{x(1-x^2) + \sqrt{1+y^2}(1-x^2)\tanh^{-1}x}$$

7. $\int x \ln \sqrt[3]{x}\, dx = \frac{1}{3}\int x \ln x$ $\boxed{\text{Formula 502}}$

$$= \frac{x^2}{6} \ln x - \frac{x^2}{12} + C$$

8. $\int \sin^2 x \cos^3 x\, dx = \int \sin^2 x(1 - \sin^2 x)\cos x\, dx$
$$= \int \sin^2 x\,(\cos x\, dx) - \int \sin^4 x\,(\cos x\, dx)$$
$$= \frac{1}{3}\sin^3 x - \frac{1}{5}\sin 5x + C$$

9. $\int \tan^2 x \sec x\, dx = \int (\sec^2 x - 1)(\sec x\, dx)$
$$= \int \sec^3 x\, dx - \int \sec x\, dx$$
$$= \left[\frac{1}{2}\sec x \tan x + \frac{1}{2}\int \sec x\, dx\right] - \int \sec x\, dx$$
$$= \frac{1}{2} \sec x \tan x - \frac{1}{2}\ln|\sec x + \tan x| + C$$

10. $\int x\sqrt{16 - x}\, dx$ $\boxed{x = 16 - u^2;\ dx = -2u\, du}$
$$= \int (16 - u^2)u(-2u\, du)$$
$$= -2\int (16u^2 - u^4)\, du$$
$$= -\frac{32u^3}{3} + \frac{2u^5}{5} + C$$
$$= -\frac{32(16 - x^2)^{3/2}}{3} + \frac{2(16 - x^2)^{5/2}}{5} + C$$

11. $\int \frac{\cosh x\, dx}{1 + \sinh^2 x} = \int \frac{du}{1 + u^2} = \tan^{-1}u + C$
$$= \tan^{-1}(\sinh x) + C$$

12. $\int \frac{dx}{1 + \cos x} = \int \frac{1 - \cos x}{1 - \cos^2 x}\, dx$
$$= \int \frac{dx}{\sin^2 x} - \int \frac{\cos x\, dx}{\sin^2 x}$$
$$= \int \csc^2 x\, dx - \int \cot x \csc x\, dx$$
$$= -\cot x + \csc x + C$$

13. $\int \frac{dx}{x^2(x^2 + 5)}$ $\boxed{\text{use partial fractions}}$
$$= \frac{1}{5}\int \left[\frac{1}{x^2} - \frac{1}{x^2 + 5}\right]dx$$
$$= \frac{1}{5}\left[-\frac{1}{x} - \frac{\sqrt{5}}{5}\tan^{-1}\left(\frac{x}{\sqrt{5}}\right)\right] + C$$

14. $\int \frac{dx}{\sqrt{x} - \sqrt[3]{x}}$ $\boxed{\text{let } u^6 = x;\ 6u^5\, du = dx}$
$$= \int \frac{6u^5\, du}{u^3 - u^2} = 6\int \frac{u^3\, du}{u - 1}$$
$$= 6\int [u^2 + u + 1 + (u - 1)^{-1}]\, du$$

$$= 6\left[\frac{u^3}{3} + \frac{u^2}{2} + u + \ln|u - 1|\right] + C$$

$$= 2\sqrt{x} + 3\sqrt[3]{x} + 6\sqrt[6]{x} + 6\ln\left|\sqrt[6]{x} - 1\right| + C$$

15. $\displaystyle\int \frac{dx}{\sqrt{2x - x^2}} = \int \frac{dx}{\sqrt{-(x^2 - 2x + 1) + 1}}$

$$= \int \frac{dx}{\sqrt{1 - (x - 1)^2}} = \int \frac{du}{\sqrt{1 - u^2}}$$

$$= \sin^{-1}u + C = \sin^{-1}(x - 1) + C$$

16. A vector normal to the plane is parallel to the line, so $\mathbf{N} = 2\mathbf{i} - 3\mathbf{j} + \mathbf{k}$,

$$\frac{x + 1}{2} = \frac{y - 2}{-3} = \frac{z - 5}{1}$$

17. Let $A(5, 1, 2)$, $B(3, 1, -2)$, $C(3, 2, 5)$

$\mathbf{AB} = -2\mathbf{i} - 4\mathbf{k}$; $\mathbf{AC} = -2\mathbf{i} + \mathbf{j} + 3\mathbf{k}$

$$\mathbf{N} = \begin{vmatrix} \mathbf{i} & \mathbf{j} & \mathbf{k} \\ -2 & 0 & -4 \\ -2 & 1 & 3 \end{vmatrix} = 4\mathbf{i} + 14\mathbf{j} - 2\mathbf{k}$$

The desired plane is

$$4(x - 5) + 14(y - 1) - 2(z - 2) = 0$$
$$2x + 7y - z - 15 = 0$$

18. $P(-2, -1, 4)$; a vector normal to the plane is parallel to the line, so $\mathbf{N} = 2\mathbf{i} + 5\mathbf{j} - 2\mathbf{k}$.

The desired plane is

$$2(x + 2) + 5(y + 1) - 2(z - 4) = 0$$
$$2x + 5y - 2z + 17 = 0$$

19. $\displaystyle\sum_{k=0}^{\infty} \frac{k^3}{k^4 + 2}$ diverges because $\displaystyle\sum_{k=0}^{\infty} \frac{1}{k}$ diverges

and by the limit comparison test

$$0 < \lim_{k \to 0} \frac{k^4}{k^4 + 2} = 1 < \infty$$

20. $\displaystyle\sum_{k=1}^{\infty} \frac{1}{k \cdot 4^k}$ converges because by the ratio test

$$\sum_{k=1}^{\infty} \frac{k\,4^k}{(k + 1)4^{k+1}} = \frac{1}{4} < 1$$

21. $\displaystyle\sum_{k=2}^{\infty} \frac{1}{k \ln k}$ diverges because by the integral test

$$\lim_{b \to +\infty} \int_{2}^{\infty} (\ln x)^{-1}\frac{dx}{x} = \lim_{b \to +\infty} \ln x \Big|_{2}^{b} = +\infty$$

22. $\displaystyle\sum_{k=0}^{\infty} \frac{3k^2 - 7k + 2}{(2k - 1)(k + 3)}$ diverges because the necessary condition for convergence is not satisfied.

$$\lim_{b \to +\infty} \frac{3k^2 - 7k + 2}{(2k - 1)(k + 3)} = \frac{3}{2} \neq 0$$

23. $\displaystyle\sum_{k=1}^{\infty} \frac{k!}{2^k \cdot k}$ diverges because by the ratio test

$$\sum_{k=1}^{\infty} \frac{(k + 1)!\, 2^k\, k}{(k + 1)2^{k+1}k!} = \sum_{k=1}^{\infty} \frac{(k + 1)k}{2(k + 1)} = \infty > 1$$

24. $\displaystyle\sum_{k=0}^{\infty} \frac{(-1)^{k+1}k}{k^2 + k - 1}$ converges by the Leibniz's alternating series test. The series alternates

$$\lim_{b \to +\infty} \frac{k}{k^2 + k - 1} = 0, \text{ and}$$

$$\frac{k}{k^2 + k - 1} \geq \frac{k + 1}{(k + 1)^2 + (k + 1) - 1}$$

because $f(x) = \dfrac{x}{x^2 + x - 1}$ is decreasing

$(f''(x) < 0)$.

25. $0 < S = 1 + \frac{1}{8} - \frac{1}{27} - \frac{1}{64} + \frac{1}{125} + \frac{1}{216} - \cdots$

$$\leq 1 + \frac{1}{8} + \frac{1}{27} + \frac{1}{64} + \frac{1}{125} + \cdots$$

converges by direct comparison to the convergent p-series with $p = 3$.

26. **a.** $S = \displaystyle\sum_{k=1}^{\infty} \frac{2^{k-1}}{5^{k+3}} = \frac{1}{5^4} \sum_{k=1}^{\infty} \frac{2^{k-1}}{5^{k-1}}$

$$= \frac{1}{625}\left(\frac{1}{1 - \frac{2}{5}}\right) = \frac{1}{375}$$

b. $S = \displaystyle\sum_{k=1}^{\infty} \frac{1}{(3k - 1)(3k + 2)}$

$$S_n = \frac{1}{3} \sum_{k=1}^{n}\left[\frac{1}{3k - 1} - \frac{1}{3k + 2}\right]$$

$$= \frac{1}{3}\left[\left(\frac{1}{2} - \frac{1}{5}\right) + \left(\frac{1}{5} - \frac{1}{8}\right) + \cdots\right]$$

$$= \frac{1}{3}\left[\frac{1}{2} - \frac{1}{3n + 2}\right]$$

$$S = \lim_{k \to +\infty} S_n = \frac{1}{6}$$

27. **a.** $\displaystyle\int_{1}^{\infty} x^2 e^{-x}\, dx$ 　$\boxed{\text{Formula 485}}$

$$= \lim_{t \to +\infty} \int_{1}^{t} x^2 e^{-x}\, dx$$

$$= \lim_{t \to +\infty} \frac{e^{-x}}{-1}(x^2 + 2x + 2)\Big|_{1}^{t}$$

$$= \frac{1}{e}(1 + 2 + 2) = \frac{5}{e}$$

b. $\displaystyle\int_{0}^{2} \frac{dx}{\sqrt{4 - x^2}} = \lim_{t \to 2^-} \int_{0}^{t} \frac{dx}{\sqrt{4 - x^2}}$

$$= \lim_{t \to 2^-} \sin \frac{x}{2}\Big|_{0}^{t} = \frac{\pi}{2}$$

28. $\mathbf{v} \cdot \mathbf{w} = 3 + 6 - 5 = 4$

$$\mathbf{v} \times \mathbf{w} = \begin{vmatrix} \mathbf{i} & \mathbf{j} & \mathbf{k} \\ 3 & -2 & 5 \\ 1 & -3 & -1 \end{vmatrix} = 17\mathbf{i} + 8\mathbf{j} - 7\mathbf{k}$$

29. $\mathbf{F}(t) = 2t\mathbf{i} + e^{-3t}\mathbf{j} + t^4\mathbf{k}$

$\mathbf{F}'(t) = 2\mathbf{i} - 3e^{-3t}\mathbf{j} + 4t^3\mathbf{k}$

$\mathbf{F}''(t) = 9e^{-3t}\mathbf{j} + 12t^2\mathbf{k}$

30. $\int [e^t\mathbf{i} - \mathbf{j} - t\mathbf{k}] \cdot [e^{-t}\mathbf{i} + t\mathbf{j} - \mathbf{k}] \, dt$

$= \int [1 - t + t] \, dt = t + C$

31. $\mathbf{R}(t) = 2(\sin 2t)\mathbf{i} + (2 + 2\cos 2t)\mathbf{j} + 6t\mathbf{k}$

$\mathbf{T}(t) = 4(\cos 2t)\mathbf{i} - (4\sin 2t)\mathbf{j} + 6\mathbf{k}$

$\|\mathbf{T}\| = \sqrt{16 + 36} = 2\sqrt{13}$

$\mathbf{T} = \frac{1}{\sqrt{13}}[2(\cos 2t)\mathbf{i} - 2(\sin 2t)\mathbf{j} + 3\mathbf{k}]$

$\frac{d\mathbf{T}}{dt} = \frac{1}{\sqrt{13}}[-4(\sin 2t)\mathbf{i} - 4(\cos 2t)\mathbf{j}]$

$\mathbf{N} = -\sin 2t\,\mathbf{i} - \cos 2t\,\mathbf{j}$

32. $S = \sum_{k=1}^{\infty} \frac{(-1)^{k+1}}{\sqrt{k}}$ converges by the Leibniz's

alternating series test. The series alternates,

$\lim_{k \to +\infty} \frac{1}{\sqrt{k}} = 0$ and $\frac{1}{\sqrt{k+1}} \le \frac{1}{k}$

$|S_n - S| \le 0.00005$

$\frac{1}{\sqrt{n+1}} < 0.00005$

$\sqrt{n+1} > (0.00005)^{-1}$

$n + 1 > 4(10^8)$

$N > 4(10^8)$

33. a. $f(x) = x^2 e^{-x^2}$

$= x^2 - x^2x^2 + \frac{x^2x^4}{2!} - \frac{x^2x^6}{3!} + \cdots$

$+ \frac{(-1)^k x^2 x^{2k}}{k!}$

$= \sum_{k=0}^{\infty} \frac{(-1)^k x^{2k+2}}{k!}$

b. $\int_0^1 x^2 e^{-x^2} \, dx$

$= \frac{x^3}{3} - \frac{x^5}{5} + \frac{x^7}{7(2!)} - \frac{x^9}{9(3!)} + \cdots \Big|_0^1$

$\approx \frac{1}{3} - \frac{1}{5} + \frac{1}{14} - \frac{1}{54} + \frac{1}{264} - \frac{1}{1,560} + \frac{1}{10,800}$

≈ 0.189

34. $\frac{dy}{dx} + 2y = x^2$; the integrating factor is

$I = e^{\int 2dx} = e^{2x}$

$y = \frac{1}{e^{2x}}\left[\int x^2 e^{2x} \, dx + C\right]$

$= \frac{1}{e^{2x}}\left[\frac{e^{2x}}{4}(2x^2 - 2x + 1) + C\right]$

If $x = 0$, then $y = 2$, so

$y = \frac{x^2}{2} - \frac{x}{2} + \frac{1}{4} + 8e^{-2x}$

35. a. $r = 1 - \frac{1}{2}\sin \theta$ **b.** $r = \frac{9}{2 + 2\cos \theta}$

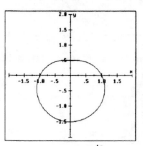
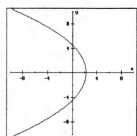

36. $A = 2(\frac{1}{2})\int_0^{\pi/3} (4 - 2\sec \theta) \, d\theta$

$= 4\theta - 2\ln|\sec \theta + \tan \theta|\Big|_0^{\pi/3}$

$= \frac{4\pi}{3} - 2\ln(2 + \sqrt{3}) \approx 1.5549$

37. $x = t^2 - 2t - 1, \; y = t^4 - 4t^2 + 2$

$\frac{dy}{dt} = 4t^3 - 8t; \; \frac{dx}{dt} = 2t - 2$

At $x = -1, \; t = 0, 2$;

Let $t = 0$, then $\frac{dy}{dx} = \frac{4t^3 - 8t}{t-1} = 0$

The equation of the tangent line is

$y - 2 = 0$

Let $t = 2$, then $\frac{dy}{dx} = \frac{2t^3 - 4t}{t-1} = 8$

The equation of the tangent line is

$y - 2 = 8(x + 1)$

$8x - y + 10 = 0$

38. a. $\mathbf{R}(t) = (\sin t)\mathbf{i} - (\cos t)\mathbf{j} + \mathbf{k}$

$\mathbf{V}(t) = (\cos t)\mathbf{i} + (\sin t)\mathbf{j}$

$\mathbf{A}(t) = -(\sin t)\mathbf{i} + (\cos t)\mathbf{j}$

$\|\mathbf{V}(t)\| = 1$

b. $x = \sin t, \; y = -\cos t, \; z = 1$;

$x^2 + y^2 = 1$ and $z = 1$

c. $\frac{ds}{dt} = 1; \; A_T = \frac{d^2s}{dt^2} = 0$

$A_N = \sqrt{1} = 1$

$\mathbf{V} \times \mathbf{A} = \begin{vmatrix} \mathbf{i} & \mathbf{j} & \mathbf{k} \\ \cos t & \sin t & 0 \\ -\sin t & \cos t & 0 \end{vmatrix} = \mathbf{k}$

$\kappa = \frac{1}{1} = 1$

39. $F = \dfrac{GMm}{r^2} = ma$

$A_N = \dfrac{GM}{r^2} = \dfrac{9.56(10^4)}{(5,000)^2} \approx 0.003824$

$v^2 = 19.12; \ v = 4.3726$ mi/s or $v \approx 15,741$ mi/h

40. $\mathbf{V} = v_0\cos\alpha\ \mathbf{i} + (-gt + v_0\sin\alpha)\mathbf{j}$

$\mathbf{R} = (v_0\cos\alpha)t\ \mathbf{i} + [-\dfrac{gt^2}{2} + (v_0\sin\alpha)t + s_0]\mathbf{j}$

$v_0 = 8, \ g = 32, \ s_0 = 120, \ \alpha = \dfrac{\pi}{6}$

$-16t^2 + \dfrac{8t}{2} + 120 = 0$

$4t^2 - t - 30 = 0$

$t = \dfrac{1 \pm \sqrt{1 + 480}}{8}$

$t \approx 2.8665$ sec (disregard negative value);

$x = 4\sqrt{3}(2.8665) \approx 19.86$ ft

CHAPTER 12

Partial Differentiation

12.1 Functions of Several Variables, Page 745

1. $f(x, y) = x^2y + xy^2$
  a. $f(0, 0) = 0$ **b.** $f(-1, 0) = 0$
  c. $f(0, -1) = 0$ **d.** $f(1, 1) = 2$
  e. $f(2, 4) = 48$ **f.** $f(t, t) = 2t^3$
  g. $f(t, t^2) = t^4 + t^5$
  h. $f(1 - t, t) = (1 - t^2)t + (1 - t)t^2$
  $\qquad = t - t^2$

2. $f(x, y) = (1 - \frac{x}{y})^2$
  a. $f(0, 1) = 1$ **b.** $f(5, 5) = 0$
  c. $f(6, 1) = 25$ **d.** $f(1, 2) = \frac{1}{4}$
  e. $f(t, t) = 0$ **f.** $f(5t, t) = 16$
  g. $f(t, 2t) = \frac{1}{4}$
  h. $f(1 + t, t) = \left[1 - \frac{(1 + t)}{t}\right]^2 = \frac{1}{t^2}$

3. $f(x, y, z) = x^2ye^{2x} + (x + y - z)^2$
  a. $f(0, 0, 0) = 0$
  b. $f(1, -1, 1) = 1^2(-1)e^2 + (1-1-1)^2$
  $\qquad = -e^2 + 1$
  c. $f(-1, 1, -1)$
  $\qquad = (-1)^2(1)e^{-2} + (-1 + 1 + 1)^2$
  $\qquad = e^{-2} + 1$
  d. $f(x, x, x) = x^2xe^{2x} + (x + x - x)^2$
  $\qquad = x^3e^{2x} + x^2$
  $\qquad \frac{d}{dx}f(x, x, x) = 2x^3e^{2x} + 3x^2e^{2x} + 2x$
  e. $f(1, y, 1) = 1^2ye^2 + (1 + y - 1)^2$
  $\qquad = e^2y + y^2$
  $\qquad \frac{d}{dy}f(1, y, 1) = e^2 + 2y$
  f. $f(1, 1, z^2) = 1^21e^2 + (1 + 1 - z^2)^2$
  $\qquad = e^2 + (2 - z^2)^2$
  $\qquad \frac{d}{dz}f(1, 1, z^2) = -4z(2 - z^2)$
  $\qquad = 4z(z^2 - 2)$

4. $f(x, y, z) = x \sin y + y \cos z$
  a. $f(0, 0, 0) = 0$
  b. $f(1, \frac{\pi}{2}, \pi) = \sin \frac{\pi}{2} + \frac{\pi}{2} \cos \pi = 1 - \frac{\pi}{2}$
  c. $f(1, \pi, \frac{\pi}{2}) = \sin \pi + \pi \cos \frac{\pi}{2} = 0$
  d. $f(x, x, x) = x \sin x + x \cos x$
  $\qquad \frac{df}{dx} = x(\cos x - \sin x) + \sin x + \cos x$
  e. $f(x, 2x, 3x) = x \sin 2x + 2x \cos 3x$
  $\qquad \frac{df}{dx} = 2x \cos 2x + \sin 2x - 6x \sin 3x + 2 \cos 3x$
  f. $f(y, y, 0) = y \sin y + y$
  $\qquad \frac{df}{dy} = y \cos y + \sin y + 1$

5. A (real) function of two variables associates a real number with each point in the two-dimensional xy-plane

6. $\mathbb{R}^2$ is the domain of a two-dimensional function, while $\mathbb{R}^3$ is the domain of a three-dimensional function.

7. The set $f(x, y, z) = C$ is the set of level curves for different C values. For $z = f(x, y)$, the level curves are the intersections of the surface with a horizontal plane $z = C$. For the volume of a cylinder, $C = \pi rh^2$ are level curves; for the resistance of to two resistors in parallel, $C = 1/r_1 + 1/r^2$ are level curves.

	Domain	Range
8.	$x - y \geq 0$	$f \geq 0$
9.	$x - y > 0$	$f > 0$
10.	$uv > 0$	$f \geq 0$
11.	$\frac{y}{x} \geq 0$	$f \geq 0$
12.	$y - x \geq 0$	$f \neq 0$
13.	$u \sin v \geq 0$	$f \geq 0$
14.	$\mathbb{R}^2$	$f \geq 0$
15.	$y \neq 2$	$f > 0$
16.	$x^2 - y^2 > 0$	$f > 0$
17.	$x^2 + y^2 < 9$	$f > 0$

18.

19.

20.

21.

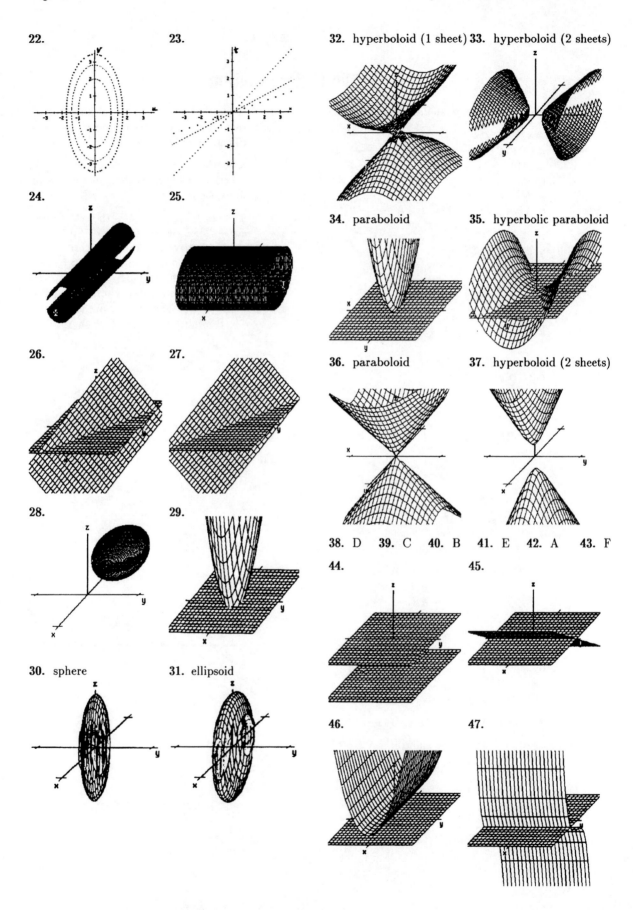

22.

23.

32. hyperboloid (1 sheet) **33.** hyperboloid (2 sheets)

24.

25.

34. paraboloid   **35.** hyperbolic paraboloid

26.

27.

36. paraboloid   **37.** hyperboloid (2 sheets)

28.

29.

38. D **39.** C **40.** B **41.** E **42.** A **43.** F

44.   **45.**

30. sphere   **31.** ellipsoid

46.   **47.**

48.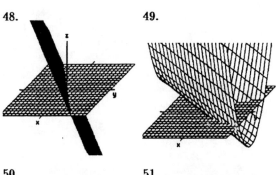

49.

50.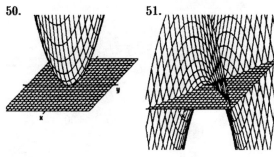

51.

52.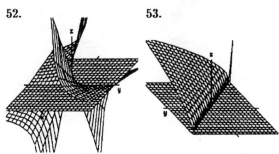

53.

54.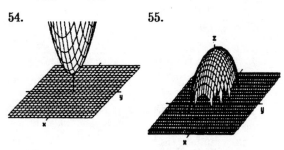

55.

56. If $L = 1$, $\frac{1}{d_0} + \frac{1}{d_i} = 1$

$$d_i = 1 + \frac{1}{d_0 - 1}$$

$$d_i - 1 = \frac{1}{d_0 - 1}$$

If $L = 2$, $d_i - 2 = \frac{4}{d_0 - 2}$

In general, $d_i - L = \frac{L^2}{d_0 - L}$

These level curves are hyperbolas.

57. $f(x, y) = Ax^a y^b$;

$f(2x, 2y) = A(2x)^a(2y)^b = 2^{a+b} Ax^a y^b$

a. Since $a + b > 1$, so the production more than doubles.
b. In this case, $a + b < 1$, so the production less than doubles.
c. In this case, $a + b = 1$, so the production exactly doubles.

58. $10xy = 1,000$ or $y = \frac{100}{x}$

59. $Q(2K, 2L) = C2^r K^r 2^{1-r} L^{1-r} = 2(CK^r L^{1-r})$

This means that the function exactly doubles.

$Q(3K, 3L) = C3^r K^r 3^{1-r} L^{1-r} = 3(CK^r L^{1-r})$

This means that the function exactly triples.

60.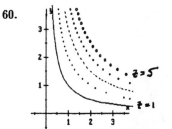

61. If $PV = kT$, then $P = \frac{kT}{V}$;

if $T = 0$, the trace in the PV plane is $P = 0$;

but if $T > 0$, the traces in planes parallel to the PV plane are hyperbolas.

If $V > 0$, the traces in planes parallel to the PT plane are straight lines.

If $P > 0$, the traces in planes parallel to the VT plane are straight lines.

62. The price per unit is p and

$$Q = \frac{r^2}{2,000p} + \frac{s^2}{100} - s$$

The total revenue is

$$R = Qp = \frac{r^2}{2,000} + \frac{s^2 p}{100} - sp$$

63. Let x be the number of units sold at home, and y the number of units sold at the foreign market; then the sales prices at home and in the foreign market, respectively, are

$$p_h = 60 - \frac{x}{5} + \frac{y}{20}; \ p_f = 50 - \frac{x}{10} + \frac{y}{20}$$

The revenue, P, is

$$P = (60 - \frac{x}{5} + \frac{y}{20})x + (50 - \frac{x}{10} + \frac{y}{20})y$$
$$= -\frac{x^2}{5} + \frac{y^2}{20} - \frac{xy}{20} + 60x + 50y$$

12.2 Limits and Continuity, Page 754

1. Pick a positive value ϵ. You are given the surface $z = f(x, y)$. Consider a fixed point (a, b) such that $f(a, b) = L$. Draw a circle around (a, b) with radius $\delta > 0$. If the vertical distance from $z = f(x, y)$ to $z = L$ is less than ϵ for every point in the circle, with the possible exception of the center (a, b), then

$$\lim_{(x,y) \to (a,b)} f(x, y) = L$$

If no such circle exists, then the limit does not exists.

2. Suppose $\lim\limits_{(x,\,y) \to (x_0,\,y_0)} f(x,\,y) = L$ and

$$\lim_{(x,\,y) \to (x_0,\,y_0)} g(x,\,y) = M$$

Then, for a constant a, we have the scalar multiple, sum, product, and quotient rules. See page 750 for details.

3. $\lim\limits_{(x,\,y) \to (-1,\,0)} (xy^2 + x^3 y + 5)$

$$= (-1)0^2 + (-1)^3 0 + 5 = 5$$

4. $\lim\limits_{(x,y) \to (0,\,0)} (5x^2 - 2xy + y^2 + 3)$

$$= 0 - 0 + 0 + 3 = 3$$

5. $\lim\limits_{(x,y) \to (1,\,3)} \dfrac{x + y}{x - y} = \dfrac{1 + 3}{1 - 3} = -2$

6. $\lim\limits_{(x,y) \to (3,\,4)} \dfrac{x - y}{\sqrt{x^2 + y^2}} = \dfrac{3 - 4}{\sqrt{3^2 + 4^2}} = -\dfrac{1}{5}$

7. $\lim\limits_{(x,y) \to (1,\,0)} e^{xy} = 1$

8. $\lim\limits_{(x,y) \to (1,\,0)} (x + y)e^{xy} = (1 + 0)e^0 = 1$

9. $\lim\limits_{(x,y) \to (0,\,1)} [e^{x^2 + x} \ln(ey^2)] = e^0 \ln e = 1$

10. $\lim\limits_{(x,y) \to (e,\,0)} \ln(x^2 + y^2) = \ln e^2 = 2$

11. $\lim\limits_{(x,y) \to (0,\,0)} \dfrac{x^2 - 2xy + y^2}{x - y}$

$$= \lim_{(x,y) \to (0,0)} (x - y) = 0$$

12. $\lim\limits_{(x,y) \to (0,0)} \dfrac{(x^2 - 1)(y^2 - 4)}{(x - 1)(y - 2)}$

$$= \lim_{(x,y) \to (1,2)} (x + 1)(y + 2) = 8$$

13. $\lim\limits_{(x,y) \to (0,0)} \dfrac{e^x \tan^{-1} y}{y} = \lim\limits_{y \to 0} \dfrac{\tan^{-1} y}{y}$

$$= \lim_{y \to 0} \dfrac{1}{1 + y^2} = 1$$

14. $\lim\limits_{(x,y) \to (0,0)} \dfrac{e^x \tan^{-1} y}{x} = 1 \cdot 0 = 0$

15. $\lim\limits_{(x,y) \to (0,0)} \dfrac{\sin(x + y)}{x + y} = \lim\limits_{t \to 0} \dfrac{\sin t}{t} = 1$

16. $\lim\limits_{(x,y) \to (0,0)} (\sin x + \cos y) = 0 + 1 = 1$

17. $\lim\limits_{(x,y) \to (5,5)} \dfrac{x^4 - y^4}{x^2 - y^2}$

$$= \lim_{(x,y) \to (5,5)} \dfrac{(x^2 + y^2)(x^2 - y^2)}{x^2 - y^2}$$

$$= \lim_{(x,y) \to (5,5)} (x^2 + y^2) = 50$$

18. $\lim\limits_{(x,y) \to (a,\,a)} \dfrac{x^4 - y^4}{x^2 - y^2}$

$$= \lim_{(x,y) \to (a,a)} (x^2 + y^2) = 2a^2$$

19. $\lim\limits_{(x,y) \to (2,1)} \dfrac{x^2 - 4y^2}{x - 2y}$

$$= \lim_{(x,y) \to (2,1)} \dfrac{(x + 2y)(x - 2y)}{x - 2y} = 4$$

20. $\lim\limits_{(x,y) \to (0,0)} = \dfrac{x^2 + y^2}{\sqrt{x^2 + y^2 + 4} - 2}$

$$= \lim_{(x,y) \to (0,0)} \dfrac{(x^2 + y^2)\sqrt{x^2 + y^2 + 4} - 2}{x^2 + y^2 + 4 - 4}$$

$$= \lim_{(x,y) \to (0,0)} (\sqrt{4} - 2) = 0$$

21. $\lim\limits_{(x,y) \to (2,\,1)} (xy^2 + x^3 y) = 2(1) + 8(1) = 10$

22. $\lim\limits_{(x,y) \to (1,\,2)} (5x^2 - 2xy + y^2) = 5 - 4 + 4 = 5$

23. $\lim\limits_{(x,y) \to (0,0)} \dfrac{x + y}{x - y}$

$\lim\limits_{(x,y) \to (0,0)} \dfrac{x + y}{x - y}$ along the line $x = 0$ is

$$\lim_{y \to 0} \dfrac{y}{(-y)} = \lim_{y \to 0} (-1) = -1$$

$\lim\limits_{(x,y) \to (0,0)} \dfrac{x + y}{x - y}$ along the line $y = 0$ is

$$\lim_{x \to 0} \dfrac{x}{x} = \lim_{x \to 0} (1) = 1$$

Limit does not exist.

24. $\displaystyle\lim_{(x,y)\to(0,0)} \frac{x-y}{\sqrt{x^2+y^2}}$

Along the line $x = 0$,

$$\lim_{(x,y)\to(0,0)} \frac{x-y}{\sqrt{x^2+y^2}} = \lim_{y\to0} \frac{-y}{\sqrt{y^2}} = -1$$

Along the line $y = 0$

$$\lim_{(x,y)\to(0,0)} \frac{x-y}{\sqrt{x^2+y^2}} = \lim_{x\to0} \frac{x}{\sqrt{x^2}} = 1$$

Limit does not exist.

25. $\displaystyle\lim_{(x,y)\to(0,0)} e^{xy} = e^0 = 1$

26. $\displaystyle\lim_{(x,y)\to(1,\,1)} (x+y)e^{xy} = 2e$

27. $\displaystyle\lim_{(x,y)\to(0,0)} (\sin x - \cos y) = 0 - 1 = -1$

28. $\displaystyle\lim_{(x,y)\to(0,0)} \left[1 - \frac{\sin(x^2+y^2)}{x^2+y^2} \right]$

$$= \lim_{t\to0} \left[1 - \frac{\sin t}{t} \right] = 1 - 1 = 0$$

29. $\displaystyle\lim_{(x,y)\to(0,0)} \frac{\sin(x^2+y^2)}{x^2+y^2} = 1$

30. $f(x,\,y) = \dfrac{x^4 y^4}{(x^2+y^4)^3}$

Along the line $y = mx$, we have

$$\lim_{x\to0} \frac{m^4 x^8}{(x^2+m^4 y^4)^3} = \lim_{x\to0} \frac{m^6 x^8}{x^6(1+m^4 x^2)^3}$$

$$= \lim_{x\to0} \frac{m^4 x^2}{(1+m^4 x^2)^3} = 0$$

Along the line $y = \sqrt{x}$; then

$$\lim_{x\to0} \frac{x^4 x^2}{(x^2+x^2)^3} = \frac{1}{2}$$

Since these limits are not the same, the limit does not exist.

31. $f(x,\,y) = \dfrac{x-y^2}{x^2+y^2}$

Along the line $y = 0$, we have $\displaystyle\lim_{x\to0} \frac{x}{x^2}$, which is not defined, so the limit does not exist.

32. $f(x,\,y) = \dfrac{x^2+y}{x^2+y^2}$

Along the line $x = 0$, we have $\displaystyle\lim_{y\to0} \frac{y}{y^2}$, which is not defined, so the limit does not exist.

33. $f(x,\,y) = \dfrac{x^2 y^2}{x^4+y^4}$

Along the line $y = mx$, we have

$$\lim_{x\to0} \frac{m^2 x^4}{(1+m^4)x^4} = \frac{m^2}{1+m^4} \text{ which is not unique}$$

(the result varies with m). Thus, the limit does not exist.

34. a. Along the line $y = kx$, we have

$$\lim_{x\to0} \frac{k^2 x^3}{x^2+k^4 x^4} = \lim_{x\to0} \frac{k^2 x}{1+k^4 x^2} = 0$$

b. Along the path $x = y^2$, we have

$$\lim_{x\to0} \frac{y^4}{y^4+y^4} = \frac{1}{2}$$

From parts **a** and **b**, we note that the values are not the same so that the limit does not exist.

35. a. Along the line $y = kx$,

$$\lim_{x\to0} \frac{k^3 x^4}{x^2+k^6 x^6} = \lim_{x\to0} \frac{k^3 x^2}{1+k^6 x^4} = 0$$

b. Along the path $x = y^3$,

$$\lim_{x\to0} \frac{y^6}{y^6+y^6} = \frac{1}{2}$$

From parts **a** and **b**, we note that the values are not the same so that the limit does not exist.

36. a. $\displaystyle\lim_{(x,y)\to(2,1)} \frac{x^2-y^2}{x^2+y^2} = \frac{4-1}{4+1} = \frac{3}{5}$

b. Along the line $x = 0$,

$$\lim_{y\to0} \frac{-y^2}{y^2} = -1$$

Along the line $y = 0$

$$\lim_{x\to0} \frac{x^2}{x^2} = 1$$

We note that the values are not the same, so the limit does not exist.

37. a. $\displaystyle\lim_{(x,y)\to(3,1)} \frac{x^2+2y^2}{x^2+y^2} = \frac{9+2}{9+1} = \frac{11}{10}$

b. Along the line $x = 0$,

$$\lim_{y\to0} \frac{2y^2}{y^2} = 2$$

Along the line $y = 0$,

$$\lim_{x\to0} \frac{x^2}{x^2} = 1$$

We note that the values are not the same, so the limit does not exist.

38. Since f is continuous, along the path $y = 0$,

$$\lim_{(x,y)\to(0,0)} \frac{x^3+y^3}{x^2+y^2} = \lim_{x\to0} \frac{x^3}{x^2} = 0$$

Thus, $A = 0$.

39. $\displaystyle\lim_{(x,y)\to(0,0)} \frac{3x^3-3y^3}{x^2-y^2}$

$$= \lim_{(x,y)\to(0,0)} \frac{3(x-y)(x^2+xy+y^2)}{(x+y)(x-y)}$$

$$= \lim_{(x,y)\to(0,0)} \frac{3(x^2+xy+y^2)}{x+y}$$

Along the line $x = y$:

$$\lim_{(x,y)\to(0,0)} \frac{9x^2}{2x} = \lim_{(x,y)\to(0,0)} \frac{9}{2}x = 0$$

Thus, $B = 0$.

40. $\displaystyle\lim_{(x,y)\to(0,0)} \frac{x^2 y^2}{x^2+y^2} = 0$; thus, the function is

continuous at $(0, 0)$.

41. $\displaystyle\lim_{(x,y,z)\to(0,0,0)} \frac{xyz}{x^2+y^2+z^2}$

$$= \lim_{(\rho,\phi,\theta)\to(0,0,0)} \frac{(\rho\sin\phi\cos\theta)(\rho\sin\phi\sin\theta)(\rho\cos\phi)}{\rho^2}$$

$$= \lim_{\rho\to 0} \rho(\sin^2\phi \sin\theta\cos\theta\cos\phi) = 0$$

42. $\left|2x^2 + 3y^2 - 0\right| \le 3\left|x^2+y^2\right| < 3\delta = \epsilon$ where

$\sqrt{x^2+y^2} < \delta$ and $\delta = \frac{\epsilon}{3}$

43. $\left|x + y^2 - 0\right| \le \left|x+y^2\right| < 2\delta = \epsilon$ where

$\sqrt{x^2+y^2} < \delta$ and $\delta = \frac{\epsilon}{2}$

44. $\left|x + y - 0\right| \le |x| + |y| < 2\delta = \epsilon$ where

$\sqrt{x^2+y^2} < \delta$ and $\delta = \frac{\epsilon}{2}$

45. $\left|\frac{x^2-y^2}{x+y} - 2\right| = |x - y - 2|$

$$= \left|x - 1 - (y+1)\right| \le |x-1| + |y+1| < \delta = \epsilon$$

where $\sqrt{(x-1)^2+(y+1)^2} < \delta$ and $\delta = \epsilon$.

46. Let $\epsilon = \dfrac{f(a, b)}{2}$; since f is continuous at (a, b),

there exists a $\delta > 0$ so that

$$\left|f(x, y) - f(a, b)\right| < \epsilon$$

wherever

$$0 < \sqrt{(x-a)^2+(y-b)^2} < \epsilon$$

Thus, if (x, y) is any number inside the punctured disk

$$0 < \sqrt{(x-a)^2+(y-b)^2} < \epsilon$$

we must have

$$-\epsilon < f(x, y) - f(a, b) < \epsilon$$

which implies

$$0 < f(a, b) - \epsilon < f(x, y)$$

since $\epsilon = f(a, b)/2$.

47. Let $\epsilon > 0$ be given. Since $\displaystyle\lim_{(x,y)\to(0,0)} f(x, y) = L$, there exists a $\delta > 0$ so that

$$\left|f(x, y) - L\right| < \frac{\epsilon}{|a|}$$

if $0 < \sqrt{(x-x_0)^2+(y-y_0)^2} < \delta$

which means that

$$\left|af(x, y) - aL\right| < \epsilon$$

if $0 < \sqrt{(x-x_0)^2+(y-y_0)^2} < \delta$

Therefore, $\displaystyle\lim_{(x,y)\to(x_0,y_0)} af(x, y) = aL$.

48. Let $\epsilon > 0$ be given; there exists $\delta_1 > 0$ and $\delta_2 > 0$ so that

$$\left|f(x, y) + g(x, y) - (L + M)\right|$$

$$\le \left|f(x, y) - L\right| + \left|g(x, y) - M\right| < \epsilon_1 + \epsilon_2 = 2\epsilon$$

where $\epsilon = \max(\epsilon_1, \epsilon_2)$. Therefore

$$\lim_{(x,y)\to(x_0,y_0)} [f + g](x, y) = L + M$$

49. Along the path $y = mx$,

$$\lim_{(x,y)\to(0,0)} \frac{x^2 y}{x^2+y^2} = \lim_{x\to 0} \frac{mx^3}{(1+m^2)x^2} = 0$$

Along the path $y = x^2$,

$$\lim_{(x,y)\to(0,0)} \frac{x^2 y}{x^2+y^2} = \lim_{x\to 0} \frac{y^2}{y+y^2} = 0$$

12.3 Partial Derivatives, Page 762

1. Given a function of two (or more) variables, $z = f(x, y)$, hold all but one independent variable(s) constant. The derivative

$$\frac{d}{dx}f(x, y_0) \text{ is a partial derivative}$$

2. Given $w = f(x, y, z)$, then the partial derivatives at (x_0, y_0, z_0) are

$$\frac{\partial w}{\partial x} = \frac{\partial f}{\partial x}(x, y_0, z_0)$$

$$\frac{\partial w}{\partial y} = \frac{\partial f}{\partial x}(x_0, y, z_0)$$

$$\frac{\partial w}{\partial z} = \frac{\partial f}{\partial z}(x_0, y_0, z)$$

3. $f(x, y) = x^3 + x^2 y + xy^2 + y^3$

$f_x = 3x^2 + 2xy + y^2$; $f_y = x^2 + 2xy + 3y^2$;

$f_{xx} = 6x + 2y$; $f_{yx} = 2x + 2y$

4. $f(x, y) = (x + xy + y)^3$

$f_x = 3(x + xy + y)^2(1 + y)$;

$$f_y = 3(x + xy + y)^2(x + 1);$$
$$f_{xx} = 3(1 + y)(2)(x + xy + y)(1 + y)$$
$$= 6(1 + y)^2(x + xy + y)$$
$$f_{yx} = 3[(x + xy + y)^2$$
$$+ (1 + x)(2)(x + xy + y)(1 + y)]$$

5. $f(x, y) = \frac{x}{y}$

$f_x = \frac{1}{y}; f_y = \frac{-x}{y^2}; f_{xx} = 0; f_{yx} = \frac{-1}{y^2}$

6. $f(x, y) = xe^{xy}$

$f_x = e^{xy}(xy + 1); f_y = x^2 e^{xy};$

$f_{xx} = e^{xy}(y + xy^2 + y) = ye^{xy}(xy + 2)$

$f_{yx} = e^{xy}(x^2 y + 2x) = e^{xy}(xy + 2)$

7. $f(x, y) = \ln(2x + 3y)$

$f_x = \dfrac{2}{2x + 3y}; f_y = \dfrac{3}{2x + 3y};$

$f_{xx} = \dfrac{-4}{(2x + 3y)^2}; f_{yx} = \dfrac{-6}{(2x + 3y)^2}$

8. $f(x, y) = \sin x^2 y$

$f_x = (\cos x^2 y)(2xy); f_y = (\cos x^2 y)(x^2)$

$f_{xx} = (2y)[\cos x^2 y - 2x^2 y \sin x^2 y]$

$f_{yx} = 2x \cos x^2 y - 2x^3 y \sin x^2 y$

9. **a.** $f(x, y) = (\sin x^2)\cos y$

$f_x = (2x \cos x^2)(\cos y);$

$f_y = -(\sin x^2)(\sin y)$

b. $f(x, y) = \sin(x^2 \cos y)$

$f_x = \cos(x^2 \cos y)(2x \cos y)$

$f_y = \cos(x^2 \cos y)(-x^2 \sin y)$

10. **a.** $f(x, y) = (\sin\sqrt{x})\ln y^2$

$f_x = \dfrac{\cos\sqrt{x}}{2\sqrt{x}}(\ln y^2); f_y = \dfrac{2\sin\sqrt{x}}{y}$

b. $f(x, y) = \sin(\sqrt{x} \ln y^2)$

$f_x = \cos(\sqrt{x} \ln y^2)\dfrac{\ln y}{\sqrt{x}}$

$f_y = \cos(\sqrt{x} \ln y^2)\dfrac{2\sqrt{x}}{y}$

11. $f(x, y) = \sqrt{3x^2 + y^4}$

$f_x = \dfrac{3x}{(3x^2 + y^4)^{1/2}}; f_y = \dfrac{2y^3}{(3x^2 + y^4)^{1/2}}$

12. $f(x, y) = xy^2 \ln(x + y)$

$f_x = \dfrac{xy^2}{x + y} + y^2 \ln(x + y)$

$f_y = \dfrac{xy^2}{x + y} + 2xy\ln(x + y)$

13. $f(x, y) = x^2 e^{x+y}\cos y$

$f_x = (\cos y)(x^2 e^{x+y} + 2xe^{x+y})$

$\quad = xe^{x+y}(x + 2)\cos y$

$f_y = -x^2 e^{x+y}\sin y + x^2 e^{x+y} \cos y$

$\quad = x^2 e^{x+y}(\cos y - \sin y)$

14. $f(x, y) = xy^3 \tan^{-1} y$

$f_x = y^3 \tan^{-1} y; f_y = \dfrac{xy^3}{1 + y^2} + 3xy^2 \tan^{-1} y$

15. $f(x, y) = \sin^{-1} xy$

$f_x = \dfrac{y}{\sqrt{1 - x^2 y^2}}; f_y = \dfrac{x}{\sqrt{1 - x^2 y^2}}$

16. $f(x, y) = \cos^{-1} xy$

$f_x = \dfrac{-y}{\sqrt{1 - x^2 y^2}}; f_y = \dfrac{-x}{\sqrt{1 - x^2 y^2}}$

17. $f(x, y, z) = xy^2 + yz^3 + xyz$

$f_x = y^2 + yz; f_y = 2xy + z^3 + xz$

$f_z = 3yz^2 + xy$

18. $f(x, y, z) = xye^z$

$f_x = ye^z; f_y = xe^z; f_z = xye^z$

19. $f(x, y, z) = \dfrac{x + y^2}{z}$

$f_x = \dfrac{1}{z}; f_y = \dfrac{2y}{z}; f_z = -\dfrac{x + y^2}{z^2}$

20. $f(x, y, z) = \dfrac{xy + yz}{xz} = \dfrac{y}{z} + \dfrac{y}{x}$

$f_x = -x^{-2}y; f_y = x^{-1} + z^{-1};$

$f_z = -yz^{-2}$

21. $f(x, y, z) = \ln(x + y^2 + z^3)$

$f_x = \dfrac{1}{x + y^2 + z^3}; f_y = \dfrac{2y}{x + y^2 + z^3};$

$f_z = \dfrac{3z^2}{x + y^2 + z^3}$

22. $f(x, y, z) = \sin(xy + z)$

$f_x = y \cos(xy + z); f_y = x \cos(xy + z)$

$f_z = \cos(xy + z)$

23. $\dfrac{2x}{9} + zz_x = 0, z_x = -\dfrac{2x}{9z};$

$-\dfrac{y}{2} + zz_y = 0; z_y = \dfrac{y}{2z}$

24. $6x + 4zz_x = 0, z_x = -\dfrac{3x}{2z};$

$8y + 4zz_y = 0, z_y = -\dfrac{2y}{z}$

25. $6xy + y^3 z_x - 2zz_x = 0$, $z_x = -\dfrac{6xy}{y^3 - 2z}$

$3x^2 + y^3 zy + 3y^2 z - 2zz_y = 0$;

$z_y = -\dfrac{3x^2 + 3y^2 z}{y^3 - 2z}$

26. $3x^2 - y^2 + 2yzz_x = 0$, $z_x = \dfrac{y^2 - 3x^2}{2yz - 3z^2}$

$-2xy + 2yzz_y + z^2 - 3z^2 z_y = 0$; $z_y = \dfrac{2xy - z^2}{2yz - 3z^2}$

27. $\dfrac{1}{2\sqrt{x}} + (\cos xz)(z + xz_x) = 0$;

$z_x = -\dfrac{\dfrac{1}{2\sqrt{x}} + z\cos xz}{x \cos xz}$

$= -\dfrac{1}{2x^{3/2}\cos xz} - \dfrac{z}{x}$

Now, with respect to y:

$2y + (\cos xz)(z + xz_y) = 0$;

$z_y = -\dfrac{2y}{x \cos xz}$

28. $\dfrac{1}{xy + yz + xz}[y + yz_x + xz_x + z] = 0$

$z_x = -\dfrac{y + z}{x + z}$

$\dfrac{1}{xy + yz + xz}[x + yz_y + xz_x + z + xz_y] = 0$

$z_y = -\dfrac{y + z}{x + y}$

29. $f(x, y) = \displaystyle\int_x^y (t^2 + 2t + 1)\, dt$

$= \displaystyle\int_0^y (t^2 + 2t + 1)\, dt - \int_0^x (t^2 + 2t + 1)\, dt$

$f_x = -(x^2 + 2x + 1);\ f_y = y^2 + 2y + 1$

30. $f(x, y) = \displaystyle\int_x^y (e^t + 3t)\, dt$

$= \displaystyle\int_0^y (e^t + 3t)\, dt - \int_0^x (e^t + 3t)\, dt$

$f_x = -(e^x + 3x);\ f_y = e^y + 3y$

31. $2x + x(2y)\dfrac{dy}{dx} + y^2 + 3y^2\dfrac{dy}{dx} = 0$

$(2xy + 3y^2)\dfrac{dy}{dx} = -(2x + y^2)$

$\dfrac{dy}{dx} = -\dfrac{2x + y^2}{2xy + 3y^2}$

At P, $\dfrac{dy}{dx} = -\dfrac{2 + 1}{2 + 3} = -\dfrac{3}{5}$

32. $x\dfrac{dy}{dx} + y - \sin(x^2 + y^2)\left(2x + 2y\dfrac{dy}{dx}\right) = 0$; at P,

$\dfrac{\sqrt{\pi}}{2}\dfrac{dy}{dx} - \sin\dfrac{\pi}{2}\left(\pi + \sqrt{\pi}\dfrac{dy}{dx}\right) = 0$, so $\dfrac{dy}{dx} = -1$

33. $f_x = 6xy;\ f_{xx} = 6y;\ f_y = -3y^2;\ f_{yy} = -6y$;

Thus, $f_{xx} + f_{yy} = 0$.

34. $f_x = \dfrac{2x}{x^2 + y^2}$;

$f_{xx} = \dfrac{2(x^2 + y^2) - 4x^2}{(x^2 + y^2)^2} = \dfrac{2(y^2 - x^2)}{(x^2 + y^2)^2}$

Similarly, $f_{yy} = \dfrac{2(x^2 - y^2)}{(y^2 + x^2)^2}$;

Thus, $f_{xx} + f_{yy} = 0$.

35. $f_x = e^x\sin y;\ f_{xx} = e^x\sin y$

$f_y = e^x\cos y;\ f_{yy} = -e^x\sin y$

Thus, $f_{xx} + f_{yy} = 0$.

36. $f_x = \cos x \cosh y;\ f_{xx} = -\sin x \cosh y$;

$f_y = \sin x \sinh y;\ f_{yy} = \sin x \cosh y$

Thus, $f_{xx} + f_{yy} = 0$.

37. a. y is held constant;

$f_x = y^3 + 3x^2 y$;

$f_x(-1, 1) = -1 - 3 = -4$

b. x is held constant;

$f_y = 3xy^2 + x^3$;

$f_y(1, -1) = 1 + 3 = 4$

38. a. y is held constant;

$f_x = y^{-1} - yx^{-2}$;

$f_x(1, -1) = -1 + 1 = 0$

b. x is held constant;

$f_y = -xy^{-1} + x^{-1}$

$f_y(1, -1) = 1 + 1 = 2$

39. $f_x = -y^2 \sin xy^2$;

$f_{xy} = -y^2(2xy)\cos xy^2 - 2y\sin xy^2$

$= -2xy^3\cos xy^2 - 2y\sin xy^2$

$f_y = -2xy\sin xy^2$;

$f_{yx} = -2y[(\sin xy^2) + (2xy)(\cos xy^2)(y^2)]$

$= -2xy^3\cos xy^2 - 2y\sin xy^2$

They are the same.

40. $f_x = 2\sin x \cos y \sin y = \sin 2x \sin y$

$f_{xy} = \sin 2x \cos y;\ f_y = (\sin^2 x)\cos y$

$f_{yx} = 2 \sin x \cos x \cos y = \sin 2x \cos y$

They are the same.

41. $f_x = 2x - 2y \cos z;\ f_{xz} = 2y \sin z;$

$f_{xzy} = 2 \sin z;$

$f_y = 2y - 2x \cos z;\ f_{yz} = 2x \sin z$

$f_{yxz} = 2x \cos z;$

$f_{xzy} - f_{yzz} = 2(\sin z - x \cos z)$

42. a. $F_x = \dfrac{c\pi x}{2}\sqrt{y - z}$

b. $F_y = \dfrac{c\pi x^2}{4}\dfrac{1}{2\sqrt{y - z}} = \dfrac{c\pi x^2}{8\sqrt{y - z}}$

c. $F_z = \dfrac{c\pi^x}{4}\dfrac{-1}{2\sqrt{y - z}} = \dfrac{-c\pi x}{8\sqrt{y - z}}$

43. a. $C_m = \sigma(T - t)(-0.67 m^{-1.67})$

$= -0.67\sigma(T - t)m^{-1.67}$

b. $C_T = \sigma m^{-0.67}$

c. $C_t = -\sigma m^{-0.67}$

44. a. $S_p = ae^{h/(bT)}$

b. $S_h = ape^{h/(bT)}\left(\dfrac{1}{bT}\right)$

c. $S_T = -ape^{h/(bT)}\left(\dfrac{1}{bT^2}\right)$

45. a. $PV = kT$

$P\dfrac{\partial V}{\partial T} = k$

$\dfrac{\partial V}{\partial T} = \dfrac{k}{P}$

b. $PV = kT$

$P\dfrac{\partial V}{\partial P} + V = 0$

$\dfrac{\partial V}{\partial P} = -\dfrac{V}{P}$

c. $PV = kT$

$V = k\dfrac{\partial T}{\partial T}$

$\dfrac{\partial P}{\partial V}\dfrac{\partial V}{\partial T}\dfrac{\partial T}{\partial P} = -\dfrac{P}{V}\dfrac{k}{P}\dfrac{V}{k} = -1$

46. a. $\dfrac{\partial Q}{\partial K} = 80K^{-1/3}L^{2/5};\ \dfrac{\partial Q}{\partial L} = 48K^{2/3}L^{-3/5}$

b. $\dfrac{\partial^2 Q}{\partial L^2} = -\dfrac{80}{3}K^{-4/3}L^{2/5} < 0$

$\dfrac{\partial Q}{\partial L^2} = -\dfrac{144}{5}K^{2/3}L^{8/5} < 0$

47. a. Moving parallel to the y-axis means y is changing, and x is remaining constant.

$\dfrac{\partial T}{\partial y} = 4xy + 1;\ \dfrac{\partial T}{\partial y}(2, 1) = 9$

b. In this case x is changing and y is constant.

$\dfrac{\partial T}{\partial x} = 3x^2 + 2y^2;$

$\dfrac{\partial T}{\partial x}(2, 1) = 3(2)^2 + 2(1)^2 = 14$

48. a. $z_x = \dfrac{1}{ce^t}(\cos \tfrac{x}{c} - \sin \tfrac{x}{c})$

$z_{xx} = -\dfrac{1}{c^2 e^t}(\sin \tfrac{x}{c} + \cos \tfrac{x}{c})$

$z_t = -e^{-t}(\sin \tfrac{x}{c} + \cos \tfrac{x}{c})$

$z_t = c^2 z_{xx}$, so the given function satisfies the heat equation.

b. $z_x = 3 \sin 3ct \cos 3x;$

$z_{xx} = -9 \sin 3ct \sin 3x$

$z_t = -9c^2 \sin 3ct \sin 3x$

$z_{tt} = c^2 z_{xx}$, so the given function satisfies the wave equation.

c. $z_x = -5 \sin 5t \sin 5x$

$z_{xx} = -25 \sin 5t \cos 5x$

$z_t = 5c \cos 5ct \cos 5x$

$z_{tt} = -25c^2 z_{xx}$, so the given function satisfies the wave equation.

49. $f_x(x, y) = \dfrac{(x^2+y^2)(3x^2 y - y^3) - (x^3 y - xy^3)(2x)}{(x^2 + y^2)^2}$

$f_x(0, y) = \dfrac{y^2(-y^3)}{(y^2)^2} = -y;$

$f_y(x, y) = \dfrac{(x^2+y^2)(x^3 - 3xy^2) - (x^3 y - xy^3)(2y)}{(x^2 + y^2)^2}$

$f_y(x, 0) = -\dfrac{x^5}{x^4} = x$

$f_{yx}(x, 0) = -1$ and $f_{xy}(0, y) = 1;$

continuity fails at $(0, 0)$.

50. Since the sine function has an upper bound of 1, $f(x, y) \le x^2 + y,\ f_x(x, y) \le 2x;\ f_x(0,0) = 0$.

$f_y(x, y) = (x^2 + y) \cos \dfrac{1}{x^2 + y^2}(-1)\dfrac{2y}{(x^2 + y^2)}$

$+ \sin \dfrac{1}{x^2 + y^2}$

This function is not defined because the degree of denominator exceeds that of the numerator, and $(x, y) \to (0, 0)$ shows that the partial is not defined.

51. a. $\dfrac{\partial C}{\partial t} = be^{at+bt};\ \dfrac{\partial C}{\partial x} = ae^{ax+bt};$

$\dfrac{\partial^2 C}{\partial x^2} = a^2 e^{ax+bt}$, so $b = \delta a^2$

b. $\dfrac{\partial C}{\partial t} = e^{-x^2/(4\delta t)}\left[\dfrac{x^2}{4\delta t^{5/2}} - \dfrac{1}{2t^{3/2}}\right]$

$\dfrac{\partial C}{\partial x} = -\dfrac{xe^{-x^2/(4\delta t)}}{2\delta t^{3/2}}$

$\dfrac{\partial^2 C}{\partial x^2} = e^{-x^2/(4\delta t)}\left[\dfrac{x^2}{4\delta^2 t^{5/2}} - \dfrac{1}{2\delta t^{3/2}}\right]$

$\delta\dfrac{\partial^2 C}{\partial x^2} = e^{-x^2/(4\delta t)}\left[\dfrac{x^2}{4\delta t^{5/2}} - \dfrac{1}{2\delta t^{3/2}}\right]$

Thus, the diffusion equation is satisfied.

52. a. $\dfrac{\partial A}{\partial a} = \frac{1}{2}b\sin\gamma;\ \dfrac{\partial A}{\partial b} = \frac{1}{2}a\sin\gamma;$

$\dfrac{\partial A}{\partial \gamma} = \frac{1}{2}ab\cos\gamma$

b. $a = \dfrac{2A}{b\sin\gamma};\ \dfrac{\partial a}{\partial\gamma} = -\dfrac{2A}{b^2\sin\gamma}$

12.4 Tangent Planes, Approximations, and Differentiability, 772

1. $z = (x^2 + y^2)^{1/2};$

$z_x = \frac{1}{2}(x^2 + y^2)^{-1/2}(2x) = \dfrac{x}{\sqrt{x^2 + y^2}}$

$z_y = \dfrac{y}{\sqrt{x^2 + y^2}};$ at $P_0,$

$z_x = \dfrac{3}{\sqrt{10}};\ z_y = \dfrac{1}{\sqrt{10}}$

$z - \sqrt{10} = \dfrac{3}{\sqrt{10}}(x - x) + \dfrac{1}{\sqrt{10}}(y - 1)$

$3x + y - \sqrt{10}z = 0$

2. $z = 10 - x^2 - y^2$

$z_x = -2x;\ z_y = -2y;$ at $P_0,$

$z_x = -4;\ z_y = -4$

$z - 2 = -4(x - 2) - 4(y - 2)$

$4x + 4y + z - 18 = 0$

3. $f(x, y) = x^2 + y^2 + \sin xy$

$f_x = 2x + y\cos xy;$

$f_y = 2y + x\cos xy;$ at $P_0,$

$f_x(x_0, y_0) = 2(0) + 2\cos 0 = 2$

$f_y(x_0, y_0) = 2(2) + 0\cos 0 = 4$

$z - 4 = 2(x - 0) + 4(y - 2)$

$2x + 4y - z - 4 = 0$

4. $f(x, y) = e^{-x}\sin y$

$f_x = -e^{-x}\sin y;\ f_y = e^{-x}\cos y;$ at $P_0,$

$f_x = -1;\ f_y = 0$

$z - 1 = -x$

$x + z - 1 = 0$

5. $z = \tan^{-1}\dfrac{y}{x}$

$z_x = \dfrac{-y/x^2}{1 + (y/x)^2} = \dfrac{-y}{x^2 + y^2};$

$z_y = \dfrac{1/x}{1 + (y/x)^2} = \dfrac{x}{x^2 + y^2};$ at P_0

$z_x = -\frac{2}{8} = -\frac{1}{4};\ z_y = -\frac{2}{8} = \frac{1}{4}$

$z - \dfrac{\pi}{4} = -\frac{1}{4}(x - 2) + \frac{1}{4}(y - 2)$

$x - y + 4z - \pi = 0$

6. $z = \ln|x + y^2|$

$z_x = \dfrac{1}{|x + y^2|};\ z_y = \dfrac{2y}{|x + y^2|};$ at $P_0,$

$z_x = 1,\ z_y = 6$

$z - 0 = 1(x + 3) + 6(y - 2)$

$x + 6y - z - 9 = 0$

7. $f(x, y) = 5x^2y^3$

$df = 10xy^3\,dx + 15x^2y^2\,dy$

8. $f(x, y) = 8x^3y^2 - x^4y^5$

$df = (24x^2y^2 - 4x^3y^5)\,dx$

$\qquad + (16x^3y - 5x^4y^4)\,dy$

9. $f(x, y) = \sin xy$

$df = y(\cos xy)\,dx + x(\cos xy)\,dy$

10. $f(x, y) = \cos x^2y$

$df = -2xy(\sin x^2y)\,dx - x^2(\sin x^2y)\,dy$

11. $f(x, y) = \dfrac{y}{x}$

$df = -\dfrac{y}{x^2}\,dx + \dfrac{1}{x}\,dy$

12. $f(x, y) = \dfrac{x^2}{y}$

$df = \dfrac{2x}{y}\,dx - \dfrac{x^2}{y^2}\,dy$

13. $f(x, y) = ye^x$

$df = ye^x\,dx + e^x\,dy$

14. $f(x, y) = e^{x^2 + y}$

$df = e^{x^2 + y}(2x\,dx + dy)$

15. $f(x, y, z) = 3x^3 - 2y^4 + 5z$

$$df = 9x^2 dx \ - \ 8y^3 \ dy \ + \ 5 \ dz$$

16. $f(x, y, z) = \sin x + \sin y + \cos z$

$$df = \cos x \ dx + \cos y \ dy \ - \ \sin z \ dz$$

17. $f(x, y, z) = z^2 \sin(2x \ - \ 3y)$

$$df = 2z^2 \cos(2x \ - \ 3y) dx$$
$$- \ 3z^2 \cos(2x \ - \ 3y) dy$$
$$+ \ 2z \sin(2x \ - \ 3y) dz$$

18. $f(x, y, z) = 3y^2 z \cos x$

$$df = y^2 z \sin x \ dx + 6yz \cos x \ dy + 2yz \cos x \ dz$$

19. $f(x, y) = xy^3 + 3xy^2$

$$f_x = y^3 + 3y^2; \ f_y = 3xy^2 + 6xy$$

f, f_x, and f_y are all continuous, so the function is continuous.

20. $f(x, y) = x^2 + 4x \ - \ y^2$

$$f_x = 2x + 4; f_y = \ - \ 2y$$

f, f_x, and f_y are all continuous, so the function is continuous.

21. $f(x, y) \ = \ x^3 y^2$

$$f_x = 3x^2 y^2; f_y = 2x^3 y$$

f, f_x, and f_y are all continuous, so the function is continuous.

22. $f(x, y) = x + 4y \ - \ y^2$

$$f_x = 1; f_y = 4 \ - \ 2y$$

f, f_x, and f_y are all continuous, so the function is continuous.

23. Find $f(1, 2) + df$ where $dx = 0.01$ and $dy = 0.03$; $f(1, 2) = 35$

$$df = 12x^3(0.01) + 8y^3(0.03)$$
$$= \ 0.12 + 1.92 \ = \ 2.04$$

$$f(1.01, 2.03) \ \approx \ 35 + 2.04 \ = \ 37.04$$

By calculator,

$$f(1.01, 2.03) \approx 37.08544565$$

24. Find $f(1, 1) + df$ where $dx = \ - 0.02$ and $dy = \ 0.03$; $f(1, 1) = \ - 1$
$$df = 5x^4(\ - 0.02) \ - \ 6y^2(0.03)$$
$$= \ - 0.1 - 1.92 = \ - 0.28$$
$$f(0.98, 1.03) \ \approx \ - 1 \ - \ 0.28 = \ - 1.28$$
By calculator,
$$f(0.98, 1.03) \approx \ - 1.281533203$$

25. Find $f(\frac{\pi}{2}, \frac{\pi}{2}) + df$ where $dx = 0.01$ and $dy = \ - 0.01$; $f(\frac{\pi}{2}, \frac{\pi}{2}) = 0$

$$df = \cos(x + y)(0.01) + \cos(x + y)(\ - 0.01)$$
$$= \ (\ - 1)(0.01) + (\ - 1)(\ - 0.01) = 0$$
$$f(\tfrac{\pi}{2} + 0.01, \tfrac{\pi}{2} \ - \ 0.01) \approx 0 + 0 = 0$$

By calculator, $f(\frac{\pi}{2} + 0.01, \frac{\pi}{2} \ - \ 0.01) = 0$

26. Find $f(\sqrt{\frac{\pi}{2}} + 0.01, \ \sqrt{\frac{\pi}{2}} \ - \ 0.01) + df$ where $dx = 0.01$ and $dy = \ - 0.01$;

$$f(\sqrt{\tfrac{\pi}{2}}, \ \sqrt{\tfrac{\pi}{2}}) = 1$$
$$df = y \cos xy(0.01) + x \cos xy(\ - 0.01)$$
$$= \ \sqrt{\tfrac{\pi}{2}}(0)(0.01) + \sqrt{\tfrac{\pi}{2}}(0)(\ - 0.01) = 0$$
$$f(\sqrt{\tfrac{\pi}{2}} + 0.01, \ \sqrt{\tfrac{\pi}{2}} \ - \ 0.01) \approx 1 + 0 = 1$$

By calculator, $f(\sqrt{\frac{\pi}{2}} + 0.01, \ \sqrt{\frac{\pi}{2}} - 0.01)$
$$\approx 0.9999999995$$

27. Find $f(1 + 0.01, 1 \ - \ 0.02) + df$ where $dx = 0.01$ and $dy = \ - 0.02$; $f(1, 1) = e$

$$df = ye^{xy} \ dx + xe^{xy} \ dy \approx \ e(0.01 \ - \ 0.02)$$
$$\approx \ - 0.03$$

$$f(1.01, 0.98) \ \approx \ 2.718 \ - \ 0.03 \ \approx \ 2.69$$

By calculator, $f(1.01, 0.98) \ \approx 2.6906963$

28. Find $f(1 + 0.01, 1 \ - \ 0.02) + df$ where $dx = 0.01$ and $dy = \ - 0.02$; $f(1, 1) = e$

$$df = e^{x^2 y^2}(2xy^2 \ dx + 2x^2 y \ dy)$$
$$\approx \ e(0.02 \ - \ 0.04) \approx \ - 0.054$$

$$f(1.01, 0.98) \ \approx \ 2.718 \ - \ 0.054 \ \approx \ 2.664$$

By calculator, $f(1.01, 0.98) \ \approx 2.6636426$

29. $z_x = \ - 2x$ is equal to 0 when $x = 0$;

$z_y = \ - 2y + 4$ is equal to 0 when $y = 2$;

Then $z = 5 \ - \ 0^2 \ - \ 2^2 + 4(2) = 9$

The equation of the horizontal plane is $z = 9$.

30. $z_x = 8(x \ - \ 1)$ is equal to 0 when $x = 1$

$z_y = 2(y + 1)$ is equal to 0 when $y = \ - 1$

$z = 4(1 \ - \ 1)^2 + 3(\ - 1 + 1)^2 = 0$

The equation of the plane is $z = 0$.

31. **a.** If f is differentiable there exists ϵ_1 and ϵ_2 that tend to 0 as Δx and Δy tend to 0, where

$$\Delta f = f_x \ dx + f_y \ dy + \epsilon_1 \Delta x + \epsilon_2 \Delta y$$

If x and y are "sufficiently close" to 0,

then

$$f(x, y) \approx f(0, 0) + \Delta f$$

The equation of the horizontal plane at $(0, 0)$ is

$$z - f(0, 0)$$

$$= [f_x(0, 0)](x - 0) + [f(0, 0)](y - 0)$$

Thus, $z = f(x, y)$

$$\approx f(0, 0) + x f_x(0, 0) + y f_y(0, 0)$$

b. $f(x, y) = \dfrac{1}{1 + x - y}$; $f(0, 0) = 1$

$$f_x = \frac{-1}{(1 + x - y)^2}; \, f_x(0, 0) = -1$$

$$f_y = \frac{1}{(1 + x - y)^2}; \, f_y(0, 0) = 1$$

$$f(x, y) \approx f(0, 0) + x f_x(0, 0) + y f_y(0, 0)$$

$$= \tfrac{1}{2}(1 - x - y)$$

$$f(x, y) \approx f(0, 0) + x f_x(0, 0) + y f_y(0, 0)$$

$$\approx 1 + x(-1) + y(1)$$

$$= 1 - x + y$$

c. $f(x, y) = \dfrac{1}{(x + 1)^2 + (y + 1)^2}$

$$f_x = -\frac{2(x + 1)}{[(x + 1)^2 + (y + 1)^2]^2}$$

$$f_y = -\frac{2(y + 1)}{[(x + 1)^2 + (y + 1)^2]^2}$$

$$f(0,0) = \tfrac{1}{2}; \, f_x(0,0) = -\tfrac{1}{2}; \, f_y(0,0) = -\tfrac{1}{2}$$

$$f(x, y) \approx f(0, 0) + x f_x(0, 0) + y f_y(0, 0)$$

$$= \tfrac{1}{2} - \tfrac{1}{2}x - \tfrac{1}{2}y = \tfrac{1}{2}(1 - x - y)$$

32. $$R^{-1} = P^{-1} + Q^{-1}$$

$$-R^{-2} \, dR = -P^{-2} dP - Q^{-2} dQ$$

$$\frac{dR}{R^2} = \frac{dP}{P^2} + \frac{dQ}{Q^2}$$

$$\frac{dR}{R}\left(\frac{Q + P}{PQ}\right) = \frac{dP}{P^2} + \frac{dQ}{Q^2} \text{ since } \frac{1}{R} = \frac{Q + P}{PQ}$$

$$\frac{dR}{R} = \left(\frac{PQ}{Q + P}\right)\left(\frac{dP}{P^2} + \frac{dQ}{Q^2}\right)$$

$$= \frac{1}{Q + P}\left(Q\frac{dP}{P} + P\frac{dQ}{Q}\right)$$

$$\left|\frac{dR}{R}\right| = \left|\frac{1}{Q + P}\right|\left(|Q|\frac{dP}{P} + |P|\frac{dQ}{Q}\right)$$

At $P = 6$, $Q = 10$ with $\left|\dfrac{dP}{P}\right| \leq 0.01$,

$\left|\dfrac{dQ}{Q}\right| \leq 0.01$; The maximum percentage error

is $\left|\dfrac{dR}{R}\right| = \tfrac{1}{16}[(10)(0.01) + (6)(0.01)] = 0.01$

33. Let x, y, and z be the length, and height of the box. Then $|dx| \leq 0.02$, $|dy| \leq 0.02$, and $|dz| \leq 0.02$; the cost per square foot for the top is $C_t = 2$; for side and bottom, it is $C_s = C_b = 1.50$. The cost is

$$C = C_t xy + C_b xy + 2C_s(xz + yz)$$

$$dC = C_t(x \, dy + y \, dx) + C_b(x \, dy + y \, dx)$$

$$+ 2C_s(x \, dz + z \, dx + y \, dz + z \, dy)$$

$$= 2(2 \, dy + 4 \, dx) + 1.5(2 \, dy + 4 \, dx)$$

$$+ 3(6 \, dz + 3 \, dx + 3 \, dy)$$

$$\leq (12 + 9 + 36)(0.02) = 1.14$$

Thus, the maximum possible error will cost $1.14.

34. Let $R = 1$ and $H = 4$ be the radius and height of the cylinder, respectively. Then, $|dR| \leq 0.2$, $|dH| \leq 2(0.2) = 0.04$. The volume is

$$V = \pi R^2 H$$

$$\Delta V \approx dV = \pi(R^2 \, dH + 2RH \, dR)$$

$$= \pi(0.4 + 8(0.2)) \approx 6.3$$

35. Let x and y be the number of units of brands X and Y, respectively. The corresponding prices are $p(x)$ and $q(y)$, respectively.

a. $R(x, y) = x(4,000 - 500x) + y(3,000 - 450y)$

$$= -500x^2 + 4,000x - 450y^2 + 3,000y$$

b. Since the changes are "small" with respect to the current prices, the total differential will give a decent approximation of the change.

$$R = xp(x) + yq(y)$$

$$dR = x \, dp + p(x) \, dx + y \, dq + q(y) \, dy$$

$$|dR| \leq |x| |dp| + |p| |dx| + |y| |dq| + |q| |dy|$$

$$= 7(20) + (500)(-0.04) + 5(18) + (750)(-0.04)$$

$$= 140 - 20 + 90 - 30 = \$180$$

36. a. $K = 500$, $L = 1,500$ and $dK = 0.7$, $dL = 6$;

$$Q(500, 1500) = (150)(500)^{2/3}(1,500)^{1/3}$$

$$\approx 180,169$$

$$Q = 150K^{2/3}L^{1/3}$$

$dQ = 150K^{2/3}(\frac{1}{3})L^{-2/3}dL$

$\qquad + 150(\frac{2}{3})K^{-1/3}L^{2/3}dK$

$\qquad = \frac{1}{3}(150K^{2/3}L^{1/3})\frac{dL}{L}$

$\qquad + \frac{2}{3}(150K^{2/3}L^{1/3})\frac{dK}{K}$

$\qquad = \frac{1}{3}Q\frac{dL}{L} + \frac{2}{3}Q\frac{dK}{K}$

$\qquad = \frac{1}{3}Q\left(\frac{dL}{L} + 2\frac{dK}{K}\right)$

$\qquad = \frac{1}{3}(180,169)(\frac{6}{1,500} + 2\frac{0.7}{500})$

$\qquad \approx 245.183$

b. $dK = 0.5,\ dL = -4$

$dQ = \frac{1}{3}Q\left(\frac{dL}{L} + 2\frac{dK}{K}\right)$

$\qquad = \frac{1}{3}(180,169)(\frac{-4}{1,500} + 2\frac{0.5}{500})$

$\qquad \approx -24.038$

37. $dR = \dfrac{c\,dx}{r^4} - \dfrac{4cx}{r^5}$

$\dfrac{dR}{R} = \dfrac{cr^4 dx}{cxr^4} - \dfrac{4cxr^4 dr}{cxr^5}$

$\qquad = \dfrac{dx}{x} - \dfrac{4dr}{r}$

$\left|\dfrac{dR}{R}\right| \le \left|\dfrac{dx}{x}\right| - \left|\dfrac{4dr}{r}\right| = 0.11$ or 11%

38. $PV = RT;\ T = 400,\ P = 3,000,\ V = 14;$ thus,

$\qquad (3,000)(14) = 400R,$ so $R = 105$

$\qquad PdV + VdP = RdT$

$\qquad (3,000)(0.1) + 14dP = 3(105)$

$\qquad \Delta P \approx dP = \dfrac{15}{14}$ ft/lb

39. $F(x,\ y) = \dfrac{1.786xy}{1.798x + y}$

$f_x = \dfrac{(1.798x + y)(1.786y) - (1.786xy)(1.798)}{(1.798x + y)^2}$

$f_y = \dfrac{(1.798x + y)(1.786x) - 1.786xy(1)}{(1.798x + y)^2}$

$x = 5,\ y = 4,\ dx = 0.1,\ dy = 0.04$

$dF = f_x\,dx + f_y\,dy$

$\qquad = \dfrac{(28.586)(0.1) + (80.2807)(0.04)}{168.7401}$

$\qquad \approx 0.036$

40. $x = 36,\ y = 25,\ dx = 0.5,\ dy = -0.5$

$dQ = Q_x dx + Q_y dy$

$\qquad = 20(\frac{3}{2})x^{1/2}y\,dx + 20x^{3/2}dy$

$\qquad = 2,250 - 2,160 = 90$ copies

41. $x = 50,\ y = 52,\ dx = 1,\ dy = 2$

$dP = (x - 30)(-5\,dx + 4\,dy)$

$\qquad + (70 - 5x + 4y)dx$

$\qquad + (y - 40)(6\,dx - 7\,dy)$

$\qquad + (80 + 6x - 7y)dy$

$\qquad = 24;$ profit is increasing at \$24/week

42. $V(H,\ R) = \pi R^2 H;\ H = 12,\ R = 3,\ dH = -0.2,$
$dR = -0.3$

$dV = \pi(R^2 dH + 2RH\,dR)$

$\qquad = \pi[(9(-0.2) + 2(3)(12)(-0.3)] = -73.5$

43. $T = 2\pi\sqrt{\dfrac{L}{g}};\ L = 4.03,\ g = 32.2,\ dL = -0.03,$
$dg = -0.2$

$f_L = \dfrac{2\pi}{2\sqrt{\frac{L}{g}}}\dfrac{1}{g} = \dfrac{\pi}{\sqrt{Lg}}$

$f_g = \dfrac{2\pi}{2\sqrt{\frac{L}{g}}}\left(-\dfrac{L}{g^2}\right) = -\dfrac{\pi\sqrt{L}}{g^{3/2}}$

$dT = f_L\,dL + f_g\,dg$

$\qquad = \dfrac{\pi}{\sqrt{129.766}}(-0.03) + \left(-\dfrac{\pi\sqrt{4.03}}{32.2\sqrt{32.2}}\right)(-0.2)$

$\qquad \approx -0.0014$

44. $x = 1.2,\ y = 0.5,\ 0 \le dx \le 0.01,\ 0 \le dy \le 0.01$

$dS = \dfrac{x - y - x}{(x - y)^2}\,dx + \dfrac{x}{(x - y)^2}\,dy$

$\qquad = \dfrac{-y\,dx + x\,dy}{(x - y)^2}$

$\qquad \approx \dfrac{0 + (1.2)(0.01)}{(0.7)^2} \approx 0.02449$

45. Approach along the line $y = mx$;

$\displaystyle\lim_{x\to 0}\dfrac{x(mx)}{x^2 + (mx)^2} = \lim_{x\to 0}\dfrac{m}{m^2 + 1} \ne 0$

Not continuous at $(0,\ 0)$.

46. The differential of $x/(x - y)$ is

$\dfrac{x - y - x}{(x - y)^2}\,dx = \dfrac{-y\,dx}{(x - y)^2}$

The differential of $y/(x - y)$ is

$\dfrac{x - y + y}{(x - y)^2}\,dy = \dfrac{x\,dy}{(x - y)^2}$

These results are so much alike because interchanging x and y in the first fraction generates the negative of the second fraction.

47. $A = hb = ab\sin\theta,\ \theta = \dfrac{\pi}{6};$

$da/a = 0.04$, $db/b = 0.03$

$dA = b \sin \theta \, da + a \sin \theta \, db$

$\dfrac{dA}{A} = \dfrac{b \sin \theta \, da}{b \sin \theta \, a} + \dfrac{a \sin \theta \, db}{a \sin \theta \, b}$

$\quad = \dfrac{da}{a} + \dfrac{db}{b}$

$\quad = 0.04 - 0.03 = 0.01$ or 1%

48. $A = hb = ab \sin \theta$, $\theta = \dfrac{\pi}{6}$;

$da/a = 0.04$, $db/b = 0.03$, $d\theta/\theta = 0.02$

$dA = b \sin \theta \, da + a \sin \theta \, db + ab \cos \theta \, d\theta$

$\dfrac{dA}{A} = \dfrac{b \sin \theta \, da}{b \sin \theta \, a} + \dfrac{a \sin \theta \, db}{a \sin \theta \, b} + \dfrac{ab \cos \theta \, d\theta}{ab \sin \theta}$

$\quad = \dfrac{da}{a} + \dfrac{db}{b} + (\cot \theta)\theta \, \dfrac{d\theta}{\theta}$

$\quad = 0.04 - 0.03 + \sqrt{3}(\tfrac{\pi}{6})(2)$

$\quad \approx 0.028$ or 2.8%

12.5 Chain Rules, Page 779

1.   Let $f(x, y)$ be a differentiable function of x and y, and let $x = x(t)$ and $y = y(t)$ be differentiable functions of t. Then $z = f(x, y)$ is a differentiable function of t, and
$$\frac{dz}{dt} = \frac{\partial z}{\partial x}\frac{dx}{dt} + \frac{\partial z}{\partial y}\frac{dy}{dt}$$
Recall the chain rule for a single variable:
$$\frac{dy}{dx} = \frac{dy}{du}\frac{du}{dx}$$
The corresponding rule for two variables is *essentially* the same, the formula involves *both* variables. For two parameters, suppose $z = f(x, y)$ is differentiable at (x, y) and that the partial derivatives of $x = x(u, v)$ and $y = y(u, v)$ exist at (u, v). Then the composite function $z = f[x(u, v), y(u, v)]$ is differentiable at (u, v) with

$$\frac{\partial z}{\partial u} = \frac{\partial z}{\partial x}\frac{\partial x}{\partial u} + \frac{\partial z}{\partial y}\frac{\partial y}{\partial u} \text{ and } \frac{\partial z}{\partial v} = \frac{\partial z}{\partial x}\frac{\partial x}{\partial v} + \frac{\partial z}{\partial y}\frac{\partial y}{\partial v}$$

2.   A schematic representation in the form of a tree lessens the chances of missing a term in the application of the chain rule.

3. $z = f(x, y)$, $x = x(u, v, w)$, $y = y(u, v, w)$,

$\dfrac{\partial z}{\partial u} = \dfrac{\partial f}{\partial x}\dfrac{\partial x}{\partial u} + \dfrac{\partial f}{\partial y}\dfrac{\partial y}{\partial u}$

$\dfrac{\partial z}{\partial v} = \dfrac{\partial f}{\partial x}\dfrac{\partial x}{\partial v} + \dfrac{\partial f}{\partial y}\dfrac{\partial y}{\partial v}$

$\dfrac{\partial z}{\partial w} = \dfrac{\partial f}{\partial x}\dfrac{\partial x}{\partial w} + \dfrac{\partial f}{\partial y}\dfrac{\partial y}{\partial w}$

4.   **a.** $f(t) = 2(-3t^2)(1 + t^3) + (1 + t^3)^2$

$\quad = 2(-3t^2)(1 + t^3) + (1 + t^3)^2$

$\quad = -6t^2 - 6t^5 + 1 + 2t^3 + t^6$

$f'(t) = -12t - 30t^4 + 6t^2 + 6t^5$

b. $f'(t) = \dfrac{\partial}{\partial x}(2xy + y^2)\dfrac{d}{dt}(-3t^2)$

$\qquad\quad + \dfrac{\partial}{\partial y}(2xy + y^2)\dfrac{d}{dt}(1 + t^3)$

$\quad = (2y)(-6t) = (2x + 2y)(3t^2)$

$\quad = -12ty + 6xt^2 + 6yt^2$

$\quad = -12t(1 + t^3) + 6(-3t^2)t^2$

$\qquad + 6(1 + t^3)t^2$

$\quad = -12t - 30t^4 + 6t^2 + 6t^5$

5.   **a.** $z = (4 + e^{6t})e^{2t} = 4e^{2t} + e^{8t}$

$\dfrac{dz}{dt} = 8e^{2t} + 8e^{8t} = 8e^{2t}(1 + e^{6t})$

b. $\dfrac{dz}{dt} = \dfrac{\partial z}{\partial x}\dfrac{dx}{dt} + \dfrac{\partial z}{\partial y}\dfrac{dy}{dt}$

$\qquad = (4 + y^2)2e^{2t} + (2xy)3e^{3t}$

In terms of t:

$(4 + e^{6t})2e^{2t} + (2e^{2t}e^{3t})3e^{3t}$

$\quad = 8e^{2t}(1 + e^{6t})$

6.   **a.** $f(t) = (1 + \cos^2 5t + \sin^2 5t)^{1/2} = \sqrt{2}$

$f'(t) = 0$

b. $f'(t) = \dfrac{\partial}{\partial x}(1 + x^2 + y^2)^{1/2}\dfrac{d}{dt}(\cos 5t)$

$\qquad\quad + \dfrac{\partial}{\partial y}(1 + x^2 + y^2)^{1/2}\dfrac{d}{dt}(\sin 5t)$

$\quad = \dfrac{1}{2\sqrt{1 + x^2 + y^2}}[2x(-5)\sin 5t + 2y(5)\cos 5t]$

$\quad = \dfrac{1}{2\sqrt{1 + x^2 + y^2}}[-10\sin 5t\cos 5t + 10\sin t\cos 5t]$

$\quad = 0$

7.   **a.** $f(t) = (\cos 3t)(\tan^2 3t)$

$f'(t) = \dfrac{3\sin^3 3t}{\cos^2 3t} + 6\sin 3t$

$\quad$ or $3\tan^2 3t\sin 3t + 6\sin 3t$

b. $f'(t) = \dfrac{\partial}{\partial x}(xy^2)\dfrac{d}{dt}(\cos 3t) + \dfrac{\partial}{\partial y}(xy^2)\dfrac{d}{dt}(\tan 3t)$

$\quad = y^2(-3\sin 3t) + 2xy(3\sec^2 t)$

$\quad = -3\sin 3t\tan^2 3t + 6\cos 3t\tan 3t\sec^2 3t$ or

$\qquad \dfrac{-3\sin^3 3t}{\cos^2 3t} + \dfrac{6\sin 3t}{\cos^2 3t}$

8.   **a.** $F(u, v) = u + v + (u + v)^2$

$\dfrac{\partial F}{\partial u} = 1 + 2(u - v)$; $\dfrac{\partial F}{\partial v} = 1 - 2(u - v)$

b. $\dfrac{\partial F}{\partial u} = \dfrac{\partial F}{\partial x}\dfrac{\partial x}{\partial u} + \dfrac{\partial F}{\partial y}\dfrac{\partial y}{\partial u}$

$\qquad = (1)(1) + (2y)(1) = 1 + 2(u - v)$

$\dfrac{\partial F}{\partial v} = \dfrac{\partial F}{\partial x}\dfrac{\partial x}{\partial v} + \dfrac{\partial F}{\partial y}\dfrac{\partial y}{\partial v}$

$\qquad = (1)(1) + (2y)(-1) = 1 - 2(u - v)$

9. a. $F(u, v) = u^2\sin^2 v + u^2 - 4uv + 4v^2$

$\dfrac{\partial F}{\partial u} = 2u\sin^2 v + 2u - 4v$

$\dfrac{\partial F}{\partial v} = 2u^2\sin v \cos v - 4u + 8v$

b. $\dfrac{\partial F}{\partial u} = \dfrac{\partial z}{\partial x}\dfrac{\partial x}{\partial u} + \dfrac{\partial z}{\partial y}\dfrac{\partial y}{\partial u}$

$\qquad = 2x\sin v + 2y(1)$

$\qquad = 2u\sin^2 v + 2(u - 2v)$

$\dfrac{\partial F}{\partial v} = \dfrac{\partial z}{\partial x}\dfrac{\partial x}{\partial v} + \dfrac{\partial z}{\partial y}\dfrac{\partial y}{\partial v}$

$\qquad = 2xu\cos v + 2y(-2)$

$\qquad = 2u^2\sin v \cos v - 4(u - 2v)$

10. a. $F(u, v) = e^{(u+v)(u - v)} = e^{u^2 - v^2}$

$\dfrac{\partial F}{\partial u} = e^{u^2 - v^2}(2u);\ \dfrac{\partial F}{\partial v} = e^{u^2 - v^2}(-2v)$

b. $\dfrac{\partial F}{\partial u} = \dfrac{\partial F}{\partial x}\dfrac{\partial x}{\partial u} + \dfrac{\partial F}{\partial y}\dfrac{\partial y}{\partial u}$

$\qquad = ye^{xy} + xe^{xy}$

$\qquad = e^{u^2 - v^2}(u + v + u - v)$

$\qquad = e^{u^2 - v^2}(2u)$

$\dfrac{\partial F}{\partial v} = \dfrac{\partial F}{\partial x}\dfrac{\partial x}{\partial v} + \dfrac{\partial F}{\partial y}\dfrac{\partial y}{\partial v}$

$\qquad = ye^{xy}(-1) + xe^{xy}$

$\qquad = e^{u^2 - v^2}(-u - v + u - v)$

$\qquad = e^{u^2 - v^2}(-2v)$

11. a. $F(u, v) = \ln e^{uv^2}e^{uv^2 + uv}$

$\qquad = \ln e^{uv(v+1)}$

$\qquad = uv(v + 1)$

$\qquad = uv^2 + uv$

$\dfrac{\partial F}{\partial u} = v^2 + v;\ \dfrac{\partial F}{\partial v} = 2uv + u$

b. $\dfrac{\partial F}{\partial u} = \dfrac{\partial F}{\partial x}\dfrac{\partial x}{\partial u} + \dfrac{\partial F}{\partial y}\dfrac{\partial y}{\partial u}$

$\qquad = \dfrac{1}{x}e^{uv^2}(v^2) + \dfrac{1}{y}e^{uv}(v)$

$\qquad = \dfrac{e^{uv^2}}{e^{uv^2}}v^2 + v$

$\qquad = v^2 + v$

$\dfrac{\partial F}{\partial v} = \dfrac{\partial F}{\partial x}\dfrac{\partial x}{\partial v} + \dfrac{\partial F}{\partial y}\dfrac{\partial y}{\partial v}$

$\qquad = \dfrac{1}{x}e^{uv^2}(2uv) + \dfrac{e^{uv}}{y}u$

$\qquad = 2uv + u$

12. $\dfrac{\partial z}{\partial s} = f_x x_s + f_y y_s;\ \dfrac{\partial z}{\partial t} = f_x x_t + f_y y_t$

13. $\dfrac{\partial w}{\partial s} = \dfrac{\partial w}{\partial x}\dfrac{\partial x}{\partial s} + \dfrac{\partial w}{\partial y}\dfrac{\partial y}{\partial s} + \dfrac{\partial w}{\partial z}\dfrac{\partial z}{\partial s}$

$\dfrac{\partial w}{\partial t} = \dfrac{\partial w}{\partial x}\dfrac{\partial x}{\partial t} + \dfrac{\partial w}{\partial y}\dfrac{\partial y}{\partial t} + \dfrac{\partial w}{\partial z}\dfrac{\partial z}{\partial t}$

14. $\dfrac{\partial t}{\partial x} = \dfrac{\partial t}{\partial u}\dfrac{\partial u}{\partial x} + \dfrac{\partial t}{\partial v}\dfrac{\partial v}{\partial x}$

$\dfrac{\partial t}{\partial y} = \dfrac{\partial t}{\partial u}\dfrac{\partial u}{\partial y} + \dfrac{\partial t}{\partial v}\dfrac{\partial v}{\partial y}$

$\dfrac{\partial t}{\partial z} = \dfrac{\partial t}{\partial u}\dfrac{\partial u}{\partial z} + \dfrac{\partial t}{\partial v}\dfrac{\partial v}{\partial z}$

$\dfrac{\partial t}{\partial w} = \dfrac{\partial t}{\partial u}\dfrac{\partial u}{\partial w} + \dfrac{\partial t}{\partial v}\dfrac{\partial v}{\partial w}$

15. $\dfrac{\partial w}{\partial s} = \dfrac{\partial w}{\partial x}\cdot\dfrac{\partial x}{\partial s} + \dfrac{\partial w}{\partial y}\cdot\dfrac{dy}{\partial s} + \dfrac{\partial w}{\partial z}\cdot\dfrac{\partial z}{\partial s}$

$\dfrac{\partial w}{\partial t} = \dfrac{\partial w}{\partial x}\cdot\dfrac{\partial x}{\partial t} + \dfrac{\partial w}{\partial y}\cdot\dfrac{dy}{\partial t} + \dfrac{\partial w}{\partial z}\cdot\dfrac{\partial z}{\partial t}$

$\dfrac{\partial w}{\partial u} = \dfrac{\partial w}{\partial x}\cdot\dfrac{\partial x}{\partial u} + \dfrac{\partial w}{\partial y}\cdot\dfrac{dy}{\partial u} + \dfrac{\partial w}{\partial z}\cdot\dfrac{\partial z}{\partial u}$

16. $\dfrac{dw}{dt} = \dfrac{\partial w}{\partial x}\dfrac{dx}{dt} + \dfrac{\partial w}{\partial y}\dfrac{dy}{dt} + \dfrac{\partial w}{\partial z}\dfrac{dz}{dt}$

$\qquad = \dfrac{1}{x + 2y - z^2}[(1)(2) + (2)(-\dfrac{1}{t^2})$

$\qquad\quad - (2z)(2\sqrt{t})^{-1}]$

$\qquad = \dfrac{1}{2t - 1 + 2t^{-1}}[2 - 2t^{-2} + 1]$

$\qquad = \dfrac{3 - 2t^{-2}}{t - 1 + 2t^{-1}}$

$\qquad = \dfrac{3t^2 - 2}{t^3 - t^2 + 2t}$

17. $\dfrac{\partial w}{\partial t} = \dfrac{\partial w}{\partial x}\dfrac{\partial x}{\partial t} + \dfrac{\partial w}{\partial y}\dfrac{\partial y}{\partial t} + \dfrac{\partial w}{\partial z}\dfrac{\partial z}{\partial t}$

$\qquad = [yz\cos(xyz)](-3) + [xz\cos(xyz)](-e^{1-t})$

$\qquad\quad + [xy\cos(xyz)](4)$

$\qquad = \cos(xyz)[-3yz - e^{1-t}xz + 4xy]$

18. $\dfrac{\partial w}{\partial t} = \dfrac{\partial w}{\partial x}\dfrac{\partial x}{\partial t} + \dfrac{\partial w}{\partial y}\dfrac{\partial y}{\partial t} + \dfrac{\partial w}{\partial z}\dfrac{\partial z}{\partial t}$

$\qquad = (ze^{xy^2})(y^2)\cos t + (ze^{xy^2})[2xy(-\sin t)$

$\qquad\quad + e^{xy^2}(2\sec^2 2t)]$

$\qquad = e^{xy^2}[y^2z\cos t - 2xyz\sin t + 2\sec^2 2t]$

19. $\dfrac{\partial w}{\partial t} = \dfrac{\partial w}{\partial x}\dfrac{\partial x}{\partial t} + \dfrac{\partial w}{\partial y}\dfrac{\partial y}{\partial t} + \dfrac{\partial w}{\partial z}\dfrac{\partial z}{\partial t}$

$\qquad = e^{x^3 + yz}[3x^2(\dfrac{-2}{t^2}) + \dfrac{2z}{2t - 3} + y(2t)]$

$\qquad = (e^{x^3 + yz})[\dfrac{-6x^2}{t^2} + 2ty + \dfrac{2z}{2t - 3}]$

20. $\dfrac{\partial w}{\partial r} = \dfrac{\partial w}{\partial x}\dfrac{\partial x}{\partial r} + \dfrac{\partial w}{\partial y}\dfrac{\partial y}{\partial r} + \dfrac{\partial w}{\partial z}\dfrac{\partial z}{\partial r}$

$\qquad = e^{2x-y+3z^2}[2(1) + (-1)(2)$

$\qquad\qquad + (6z)(-st \sin rst)]$

$\qquad = e^{2x-y+3z^2} - 6stz \sin rst$

21. $\dfrac{\partial w}{\partial r} = \dfrac{\partial w}{\partial x}\dfrac{\partial x}{\partial r} + \dfrac{\partial w}{\partial y}\dfrac{\partial y}{\partial r} + \dfrac{\partial w}{\partial z}\dfrac{\partial z}{\partial r}$

$\qquad = \dfrac{2s}{2-z} + \dfrac{t\cos(rt)}{2-z} + 0$

$\qquad = \dfrac{2s + t\cos(rt)}{2-z}$

$\dfrac{\partial w}{\partial t} = \dfrac{\partial w}{\partial x}\dfrac{\partial x}{\partial t} + \dfrac{\partial w}{\partial y}\dfrac{\partial y}{\partial t} + \dfrac{\partial w}{\partial z}\dfrac{\partial z}{\partial t}$

$\qquad = 0 + \dfrac{r\cos(rt)}{2-z} + \dfrac{(x+y)(2st)}{(2-z)^2}$

$\qquad = \dfrac{(2-z)[r\cos(rt)] + 2t(x+y)}{(2-z)^2}$

22. **a.** $x^3 + y^2 + z^2 = 5$

$\qquad 3x^2 + 2zz_x = 0$

$\qquad z_x = \dfrac{-3x^2}{2z}$

$\qquad$ Similarly, $z_y = \dfrac{-y}{z}$

$\qquad z_{xy} = -\dfrac{3x^2}{2}\dfrac{z_y}{z^2} = \dfrac{3x^2 y}{2z^3}$

b. $\qquad z_{xx} = -\dfrac{3}{2}\dfrac{z(2x) - x^2 z_x}{z^2}$

$\qquad\qquad = -\dfrac{3}{2z^2}\left(2xz + \dfrac{3x^4}{2z}\right)$

$\qquad\qquad = \dfrac{-12xz^2 - 9x^4}{4z^3}$

c. $z_{yy} = -\dfrac{z - yz}{z^2}$

23. **a.** $z = \dfrac{2}{xy}$; $z_x = -\dfrac{2}{x^2 y}$; $z_{xy} = \dfrac{2}{x^2 y^2} = \dfrac{1}{2}z^2$

b. $z_{xx} = \dfrac{4}{x^3 y}$ $\qquad$ **c.** $z_y = -\dfrac{2}{xy^2}$; $z_{yy} = \dfrac{4}{xy^3}$

24. **a.** $z_x = \dfrac{1}{x+y}$; $z_{xy} = 0$

b. $z_{xx} = -\dfrac{1}{(x+y)^2}$

c. $z_y = \dfrac{1}{x+y} - 2y$; $z_{yy} = -\dfrac{1}{(x+y)^2} - 2$

25. **a.** $-x^{-2} - z^{-2}\dfrac{\partial z}{\partial x} = 0$

$\qquad \dfrac{\partial z}{\partial x} = -\dfrac{z^2}{x^2}$; $\dfrac{\partial z}{\partial y} = -\dfrac{z^2}{y^2}$

$\qquad \dfrac{\partial^2 z}{\partial x \partial y} = -\dfrac{1}{x^2}\left(2z\dfrac{\partial z}{\partial y}\right) = \dfrac{2z^3}{x^2 y^2}$

b. $\dfrac{\partial^2 z}{\partial x^2} = -\dfrac{x^2(2z\frac{\partial z}{\partial x}) - z^2(2x)}{x^4} = \dfrac{2z^2(x+z)}{x^4}$

c. $\dfrac{\partial^2 z}{\partial y^2} = -\dfrac{y^2(2z\frac{\partial z}{\partial y}) - z^2(2y)}{y^4} = \dfrac{2z^2(y+z)}{y^4}$

26. **a.** $\qquad x\cos y = y + z$

$\qquad\qquad \cos y = z_x$

$\qquad\qquad z_{xy} = -\sin y$

b. $z_{xx} = 0$ $\qquad$ **c.** $-x\sin y = 1 + z_y$

$\qquad\qquad\qquad\qquad\qquad -x\cos y = z_{yy}$

27. **a.** $\qquad z^2 + \sin x = \tan y$

$\qquad\qquad 2zz_x + \cos x = 0$

$\qquad\qquad z_x = -\dfrac{\cos x}{2z}$

$\qquad$ Similarly, $z_y = \dfrac{\sec^2 y}{2z}$

$\qquad z_{xy} = \dfrac{\cos x y z_y}{2z^2} = \dfrac{\sec^2 y \cos x}{4z^3}$

b. $z_{xx} = -\dfrac{1}{2}\dfrac{-z\sin x - z_x \cos x}{z^2}$

$\qquad\qquad = \dfrac{2z^2 \sin x - \cos^2 x}{2z^3}$

c. $z_{yy} = \dfrac{1}{2z^2}(2z\sec^2 y \tan y - \sec^2 y\, z_y)$

$\qquad\qquad = \dfrac{4z^2 \sec^2 y \tan y - \sec^4 y}{4z^3}$

28. **a.** $z_r = z_x x_r + z_y y_r = z_x \cos\theta + z_y \sin\theta$

$\qquad z_\theta = z_x x_\theta + z_y y_\theta = -z_x r\sin\theta + z_y r\cos\theta$

b. $z_{rr}^2 + r^{-2}z_{\theta\theta}^2$

$\qquad = z_x^2 \cos^2\theta + 2\sin\theta\cos\theta(z_x z_y) + z_y^2 \sin^2\theta$

$\qquad\quad + z_x^2 \sin^2\theta - 2z_x z_y \sin\theta\cos\theta + z_y^2 \cos\theta$

$\qquad = z_x^2 + z_y^2$

29. $\dfrac{\partial z}{\partial u} = \dfrac{\partial z}{\partial x}\cdot\dfrac{\partial x}{\partial u} + \dfrac{\partial z}{\partial y}\cdot\dfrac{\partial y}{\partial u}$

$\qquad = \dfrac{\partial z}{\partial x}\cdot a + \dfrac{\partial z}{\partial y}\cdot 0 = a\dfrac{\partial z}{\partial x}$

$\dfrac{\partial z}{\partial v} = \dfrac{\partial z}{\partial x}\cdot\dfrac{\partial x}{\partial v} + \dfrac{\partial z}{\partial y}\cdot\dfrac{\partial y}{\partial v}$

$\qquad = \dfrac{\partial z}{\partial x}\cdot 0 + \dfrac{\partial z}{\partial y}\cdot b = b\dfrac{\partial z}{\partial y}$

$\dfrac{\partial^2 z}{\partial u^2} = a\dfrac{\partial}{\partial u}z_x = a[(z_x)_x x_u + (z_x)_y y_u] = a^2 z_{xx}$

$\dfrac{\partial^2 z}{\partial v^2} = b\dfrac{\partial}{\partial v}z_y = b[(z_y)_x x_v + (z_y)_y y_v] = b^2 z_{yy}$

30. $\dfrac{x^2}{a^2} + \dfrac{y^2}{b^2} + \dfrac{z^2}{c^2} = 1$

$$\frac{2x}{a^2} + \frac{2z}{c^2} + \frac{\partial z}{c^2} = 0$$

$$\frac{\partial z}{\partial x} = -\frac{c^2 x}{a^2 z}$$

$$\frac{\partial^2 z}{\partial x^2} = -\frac{c^2}{a^2}\left[\frac{z - x z_x}{z^2}\right]$$

$$= -\frac{c^2}{a^2}\left[\frac{1}{z} - \frac{x}{z^2}\left(-\frac{c^2}{a^2}\right)\frac{x}{z}\right]$$

$$= -\frac{c^2}{a^2}\left[\frac{a^2 z^2 + c^2 x^2}{a^2 z^3}\right]\left(\frac{\partial^2 z}{\partial y \partial x}\right)$$

$$= -\frac{c^2 x z_y}{a^2 z^2}$$

Similarly, $z_y = -\dfrac{c^2 y}{b^2 z}$

$$\frac{\partial^2 z}{\partial x \partial y} = \frac{c^2}{a^2 z^2}\left[-\frac{c^2 y}{b^2 z}\right] = -\frac{c^4 x y}{a^2 b^2 z^3}$$

31. Let $x = \ell(t)$, $y = w(t)$, and $z = z(t)$; then
$x = 10 + 2t$; $y = 8 + 2t$; $z = 20 - 3t$
At $t = 5$ sec, $x = 20$, $y = 28$, and $z = 5$.

$$V = xyz$$

$$\frac{dV}{dt} = xy\frac{dz}{dt} + xz\frac{dy}{dt} + yz\frac{dx}{dt}$$
$$= 20(28)(-3) + (20)(5)(2) + (28)(5)(2)$$

< 0, so the volume is decreasing.

$$S = 2(xy + xz + yz)$$

$$\frac{dS}{dt} = 2\left[x\frac{dy}{dt} + y\frac{dx}{dt} + x\frac{dz}{dt} + z\frac{dx}{dt} + y\frac{dz}{dt} + z\frac{dy}{dt}\right]$$

$$= 2\left[(x + y)\frac{dx}{dt} + (x + z)\frac{dy}{dt} + (x + y)\frac{dz}{dt}\right]$$

$$= 2[(23)(2) + (25)(2) - (48)(3)]$$

< 0, so the surface area is decreasing.

32. $f(x, y) = 10xy^{1/2}$; $x = 30$, $y = 36$, $dx = 1$

$$10xy^{1/2} = C$$

$$x^{1/2}y^{-1/2}\,dy + y^{1/2}\,dx = 0$$

$$dy = \frac{-\sqrt{y}\,dx\,(2)\sqrt{y}}{x}$$

$$= -\frac{2y\,dx}{x}$$

$$= -\frac{2(36)(1)}{30}$$

$$\approx -2.4$$

The manufacturer should decrease the level of unskilled labor by about 2.4 hours.

33. Let $w = uv^2$, so $z = f(w)$

$$\frac{\partial z}{\partial u} = \frac{dz}{dw}\cdot\frac{\partial w}{\partial u} = f'(w)v^2$$

$$\frac{\partial z}{\partial v} = \frac{dz}{dw}\cdot\frac{\partial w}{\partial v} = [f'(w)](2uv)$$

$$2u\frac{\partial z}{\partial u} - v\frac{\partial z}{\partial v} = 2uv^2[f'(w)] - 2uv^2[f'(w)] = 0$$

34. Let $w = u^2v^2$, so $z = u + f(w)$

$$\frac{\partial z}{\partial u} = 1 + \frac{df}{dw}\cdot\frac{\partial w}{\partial u} = 1 + f'(w)(2uv^2)$$

$$\frac{\partial z}{\partial v} = \frac{df}{dw}\cdot\frac{\partial w}{\partial v} = [f'(w)](2u^2v)$$

$$u\frac{\partial z}{\partial u} - v\frac{\partial z}{\partial v} = u + [f'(w)](2u^2v^2 - 2u^2v^2) = u$$

35. $z = f(x, y) = f(u - v, v - u) = f(x, y)$

$x = u - v$, $y = v - u$

$$\frac{\partial z}{\partial u} = \frac{\partial z}{\partial x}\cdot\frac{\partial x}{\partial u} + \frac{\partial z}{\partial y}\cdot\frac{\partial y}{\partial u} = z_x(1) + z_y(-1)$$

$$\frac{\partial z}{\partial v} = \frac{\partial z}{\partial x}\cdot\frac{\partial x}{\partial v} + \frac{\partial z}{\partial y}\cdot\frac{\partial y}{\partial v} = z_x(-1) + z_y(1)$$

$$z_x(1-1) + z_y(1-1) = 0$$

36. Let $w = uv$, so $z = u + f(w)$

$$\frac{\partial z}{\partial u} = 1 + \frac{df}{dw}\cdot\frac{\partial w}{\partial u} = 1 + [f'(w)](v)$$

$$\frac{\partial z}{\partial v} = \frac{df}{dw}\cdot\frac{\partial w}{\partial v} = [f'(w)](u)$$

$$u\frac{\partial z}{\partial u} - v\frac{\partial z}{\partial v} = u + [f'(w)](uv - uv) = u$$

37. Let $u = \dfrac{r - s}{s} = \dfrac{r}{s} - 1$, so $w = f(u)$

$$\frac{\partial w}{\partial r} = \frac{df}{du}\cdot\frac{\partial u}{\partial r} = [f'(u)](s^{-1})$$

$$\frac{\partial w}{\partial s} = \frac{df}{du}\cdot\frac{\partial u}{\partial s} = [f'(u)](-rs^{-2})$$

$$r\frac{\partial w}{\partial r} + s\frac{\partial w}{\partial s} = \frac{r}{s}[f'(u)] - \frac{r}{s}f'(u) = 0$$

38. Let $w = x^2 + y^2$, so $z = f(w)$

$$\frac{\partial w}{\partial x} = \frac{dz}{dw}\cdot\frac{\partial w}{\partial x} = [f'(u)](2x)$$

$$\frac{\partial z}{\partial y} = \frac{dz}{dw}\cdot\frac{\partial w}{\partial y} = [f'(w)](-2y)$$

$$y\frac{\partial z}{\partial x} + x\frac{\partial z}{\partial y} = 2xy[f'(w)] - 2xy[f'(w)] = 0$$

39. Let $u = x^2 + y^2$, so $z = xy + f(w)$

$$\frac{\partial z}{\partial x} = y + \frac{df}{du}\cdot\frac{\partial u}{\partial x} = y + [f'(u)](2x)$$

$$\frac{\partial z}{\partial y} = x + \frac{df}{du}\cdot\frac{\partial u}{\partial y} = x + [f'(u)](2y)$$

$$y\frac{\partial z}{\partial x} - x\frac{\partial z}{\partial y} = y^2 + 2xy[f'(u)] - x^2 - 2xy[f'(u)]$$
$$= y^2 - x^2$$

40. $\dfrac{\partial w}{\partial x} = \dfrac{df}{dt} \cdot \dfrac{\partial t}{\partial x} = [f'(t)](\tfrac{1}{2})(x^2 + y^2 + z^2)^{-1/2}(2x)$

$\dfrac{\partial w}{\partial y} = \dfrac{df}{dt} \cdot \dfrac{\partial t}{\partial y} = [f'(t)](\tfrac{1}{2})(x^2 + y^2 + z^2)^{-1/2}(2y)$

$\dfrac{\partial w}{\partial z} = \dfrac{df}{dt} \cdot \dfrac{\partial t}{\partial z} = [f'(t)](\tfrac{1}{2})(x^2 + y^2 + z^2)^{-1/2}(2z)$

$\left(\dfrac{\partial w}{\partial x}\right)^2 + \left(\dfrac{\partial w}{\partial y}\right)^2 + \left(\dfrac{\partial w}{\partial z}\right)^2$

$\qquad = [f'(t)](x^2 + y^2 + z^2)^{-1}(x^2 + y^2 + z^2)$

$\qquad = [f'(t)]^2 = \left(\dfrac{dw}{dt}\right)^2$

41. Let $u = x^2 + y^2$, then $z = f(u)$

  a. $\dfrac{\partial z}{\partial x} = f'(u)\dfrac{\partial u}{\partial x} = 2xf'(u)$

$\qquad \dfrac{\partial^2 z}{\partial x^2} = 2f'(u) + 2xf''(u)\dfrac{\partial u}{\partial x}$

$\qquad\qquad = 2f'(u) + 4x^2 f''(u)$

$\qquad\qquad = 2f'(x^2 + y^2) + 4x^2 f''(x^2 + y^2)$

  b. $\dfrac{\partial z}{\partial y} = f'(u)\dfrac{\partial u}{\partial y} = 2yf'(u)$

$\qquad \dfrac{\partial^2 z}{\partial y^2} = 2f'(u) + 2yf''(u)\dfrac{\partial u}{\partial y}$

$\qquad\qquad = 2f'(u) + 4y^2 f''(u)$

$\qquad\qquad = 2f'(x^2 + y^2) + 4y^2 f''(x^2 + y^2)$

  c. $\dfrac{\partial z}{\partial x} = 2xf'(u)$

$\qquad \dfrac{\partial^2 z}{\partial x \partial y} = 0[f'(u)] + 2x\left(f''(u)\dfrac{\partial u}{\partial y}\right)$

$\qquad\qquad = 4xyf''(x^2 + y^2)$

42. $\dfrac{dz}{d\theta} = \dfrac{\partial f}{\partial x} \cdot \dfrac{dx}{d\theta} \cdot \dfrac{\partial f}{\partial y} \cdot \dfrac{dy}{d\theta}$

$\qquad = f_x(-\sin \theta) + f_y(\cos \theta)$

$\qquad = -yf_x + xf_y$

$\dfrac{d^2 z}{d\theta^2} = \dfrac{d}{d\theta} \cdot \dfrac{dz}{d\theta}$

$\qquad = \dfrac{d}{d\theta}[-yf_x + xf_y]$

$\qquad = \dfrac{\partial}{\partial x}[-yf_x + xf_y]\dfrac{dx}{d\theta} + \dfrac{\partial}{\partial y}[-yf_x + xf_y]\dfrac{dy}{d\theta}$

$\qquad = -y(f_{xx} + xf_{yx} + f_y)(-\sin \theta)$

$\qquad\qquad + (-yf_{xy} - f_x + xf_{yy})(\cos \theta)$

$\qquad = y^2 f_{xx} - yxf_{yx} - yf_y - xyf_{xy} - xf_x + x^2 f_{yy}$

$\qquad = y^2 f_{xx} + x^2 f_{yy} - 2xyf_{xy} - xf_x - yf_y$

43. $u(x, t) = f(x + ct) + g(x - ct) = f(r, s)$

$\dfrac{\partial u}{\partial x} = \dfrac{\partial u}{\partial r} \dfrac{\partial r}{\partial x} + \dfrac{\partial u}{\partial s} \dfrac{\partial s}{\partial x}$

$\qquad = f'(r)(1) + g(s)(1)$

$\qquad = f'(r) + g'(s)$

$\dfrac{\partial^2 u}{\partial x^2} = \dfrac{\partial u_x}{\partial r} \dfrac{\partial r}{\partial x} + \dfrac{\partial u_x}{\partial s} \dfrac{\partial s}{\partial x}$

$\qquad = f''(r)(1) + g''(s)(1)$

$\qquad = f''(r) + g''(s)$

$\dfrac{\partial u}{\partial t} = \dfrac{\partial u}{\partial r} \dfrac{\partial r}{\partial t} + \dfrac{\partial u}{\partial s} \dfrac{\partial s}{\partial t}$

$\qquad = f'(r)(c) + g'(s)(-c)$

$\qquad = c[f'(r) - g'(s)]$

$\dfrac{\partial^2 u}{\partial t^2} = \dfrac{\partial u_t}{\partial r} \dfrac{\partial r}{\partial t} + \dfrac{\partial u_x}{\partial s} \dfrac{\partial s}{\partial t}$

$\qquad = c\left[f''(r)(c) - g''(s)(-c)\right]$

$\qquad = c^2[f''(r) + g''(s)]$

44. $\dfrac{\partial z}{\partial r} = \dfrac{\partial z}{\partial x} \dfrac{\partial x}{\partial r} + \dfrac{\partial z}{\partial y} \dfrac{\partial y}{\partial r}$

$\qquad = f_x(e^r \cos \theta) + f_y(e^r \sin \theta)$

$\qquad = xf_x + yf_y$

$\dfrac{\partial^2 z}{\partial r^2} = (xf_{xx} + yf_{yx} + f_x)(e^r \cos \theta)$

$\qquad\qquad + (xf_{xy} + f_y + yf_{yy})(e^r \sin \theta)$

$\qquad = x^2 f_{xx} + y^2 f_{yy} + 2xyf_{xy} + xf_x + yf_y$

$\dfrac{\partial z}{\partial \theta} = \dfrac{\partial z}{\partial x} \dfrac{\partial x}{\partial \theta} + \dfrac{\partial z}{\partial y} \dfrac{\partial y}{\partial \theta}$

$\qquad = f_x(-e^r \sin \theta) + f_y(e^r \cos \theta)$

$\qquad = -yf_x + xf_y$

$\dfrac{\partial^2 z}{\partial \theta^2} = (-yf_{xx} + xf_{yx} + f_y)(-e^r \sin \theta)$

$\qquad\qquad + (-yf_{xy} - f_x + xf_{yy})(e^r \cos \theta)$

$\qquad = y^2 f_{xx} + x^2 f_{yy} - 2xyf_{xy} - xf_x - yf_y$

$\dfrac{\partial^2 z}{\partial r^2} + \dfrac{\partial^2 z}{\partial \theta^2} = (x^2 + y^2)f_{xx} + (x^2 + y^2)f_{yy}$

$\qquad + (2xy - 2xy)f_{xy} + (x - x)f_x + (y - y)f_y$

$\qquad\qquad = (x^2 + y^2)f_{xx} + (x^2 + y^2)f_{yy}$

$\qquad\qquad\qquad = e^{-2r}(f_{xx} + f_{yy})$

Thus, $\dfrac{\partial^2 z}{\partial x^2} + \dfrac{\partial^2 z}{\partial y^2} = e^{-2r}\left[\dfrac{\partial^2 z}{\partial r^2} + \dfrac{\partial^2 z}{\partial \theta^2}\right]$

45. $\qquad F(x, y) = C$

$\dfrac{\partial F}{\partial x} dx + \dfrac{\partial F}{\partial y} dy = 0$

$\qquad\qquad \dfrac{dy}{dx} = -\dfrac{F_x}{F_y}$

46.
$$F(x, y, z) = C$$
$$\frac{\partial F}{\partial x} + \frac{\partial F}{\partial y}\frac{\partial y}{\partial x} + \frac{\partial F}{\partial z}\frac{\partial z}{\partial x} = 0$$
$$\frac{\partial z}{\partial x} = -\frac{F_x}{F_z}$$
Note: $\frac{\partial y}{\partial x} = 0$ since x and y are independent.

47.
$$xu + yv - uv = 0$$
$$x\frac{\partial u}{\partial y} + u + y\frac{\partial v}{\partial y} - u\frac{\partial v}{\partial y} - v\frac{\partial u}{\partial y} = 0$$
$$(x - v)\frac{\partial u}{\partial y} + (y - u)\frac{\partial v}{\partial y} = -u$$
Similarly,
$$yu - xv + uv = 0$$
$$y\frac{\partial u}{\partial y} - x\frac{\partial v}{\partial y} - v - u\frac{\partial v}{\partial y} + v\frac{\partial u}{\partial y} = 0$$
$$(y + v)\frac{\partial u}{\partial y} + (-x + u)\frac{\partial v}{\partial y} = v$$

Solve these two equations simultaneously.
$$\frac{\partial u}{\partial x} = \frac{ux - u^2 + uv - yv}{-x^2 + ux + vx - uv - (-uy + y^2 - uv + vy)}$$
$$= \frac{u^2 - uv - ux + vy}{x^2 + y^2 - ux - uy - vx + vy}$$

Similarly,
$$\frac{\partial v}{\partial x} = \frac{v^2 - uv + uy - vx}{x^2 + y^2 - ux - uy - vx + vy}$$

48.
$$xu + yv - uv = 0$$
$$x\frac{\partial u}{\partial y} + y\frac{\partial v}{\partial y} + v - u\frac{\partial v}{\partial y} - v\frac{\partial u}{\partial y} = 0$$
$$(x - v)\frac{\partial u}{\partial y} + (y - u)\frac{\partial v}{\partial y} = -v$$
Similarly,
$$yu - xv + uv = 0$$
$$y\frac{\partial u}{\partial y} + u - x\frac{\partial v}{\partial y} + u\frac{\partial v}{\partial y} + v\frac{\partial u}{\partial y} = 0$$
$$(y + v)\frac{\partial u}{\partial y} + (-x + u)\frac{\partial v}{\partial y} = -u$$

Solve these two equations simultaneously.
$$\frac{\partial u}{\partial y} = \frac{vx + uv - uy + u^2}{x^2 - ux - vx + uv - y^2 - uy - vy + uv}$$
$$= \frac{u^2 + uv - uy - vx}{x^2 + y^2 + 2uv - ux - uy - vx - vy}$$

Similarly,
$$\frac{\partial v}{\partial y} = \frac{v^2 + uv - uy - vx}{x^2 + y^2 + 2uv - ux - uy - vx - vy}$$

49. $\frac{\partial F}{\partial x}\,dx + \frac{\partial F}{\partial y}\,dy + \frac{\partial F}{\partial z}\,dz = 0$

$$\frac{\partial G}{\partial x}\,dx + \frac{\partial G}{\partial y}\,dy + \frac{\partial G}{\partial z}\,dz = 0$$
$$F_x G_z\,dx + F_y G_z\,dy + F_z G_z\,dz = 0$$
$$G_x F_z\,dx + G_y F_z\,dy + G_z F_z\,dz = 0$$
Subtracting,
$$(F_x G_z - G_x F_z)dx + (F_y G_z - G_y F_z)dy = 0$$
From this, we find $\dfrac{dy}{dx} = \dfrac{F_x G_z - G_x F_z}{G_y F_z - F_y G_z}$

Similarly, find $\dfrac{dz}{dx} = \dfrac{F_x G_y - G_x F_y}{G_z F_y - F_z G_y}$

12.6 Directional Derivatives and the Gradient, Page 791

1. $\nabla f = f_x \mathbf{i} + f_y \mathbf{j} = (2x - 2y)\mathbf{i} - 2x\mathbf{j}$

2. $\nabla f = 3\mathbf{i} + 8y\mathbf{j}$

3. $\nabla f = (-\frac{y}{x^2} + \frac{1}{y})\mathbf{i} + (\frac{1}{x} - \frac{x}{y^2})\mathbf{j}$

4. $\nabla f = \dfrac{2x}{x^2 + y^2}\mathbf{i} + \dfrac{2y}{x^2 + y^2}\mathbf{j}$

5. $\nabla f = e^{3 - v}(\mathbf{i} - u\mathbf{j})$

6. $\nabla f = e^{u + v}(\mathbf{i} + \mathbf{j})$

7. $\nabla f = \cos(x + 2y)(\mathbf{i} + 2\mathbf{j})$

8. $\nabla f = yz^2\mathbf{i} + xz^2\mathbf{j} + 2xyz\mathbf{k}$

9. $\nabla f = e^{y + 3z}(\mathbf{i} + x\mathbf{j} + 3x\mathbf{k})$

10. $\nabla f = \dfrac{(xy - 1)(1) - (z + x)(y)}{(z + x)^2}\mathbf{i}$
$$+ \frac{z}{x + z}\mathbf{j} - \frac{xy - 1}{(z + x)^2}\mathbf{k}$$
$$= \frac{1 - xy}{(z + x)^2}\mathbf{i} + \frac{x}{x + z}\mathbf{j} - \frac{xy - 1}{(z + x)^2}\mathbf{k}$$

11. $\mathbf{V} = \dfrac{1}{\sqrt{2}}\mathbf{i} + \dfrac{1}{\sqrt{2}}\mathbf{j};\ \nabla f(1, -2) = \mathbf{j}$
$$D_u f = \frac{1}{\sqrt{2}} = \frac{\sqrt{2}}{2}$$

12. $\mathbf{V} = -\dfrac{1}{\sqrt{2}}\mathbf{i} + \dfrac{1}{\sqrt{2}}\mathbf{j};\ \nabla f(2, -1) = e^{-2}(\mathbf{i} - \mathbf{j})$
$$D_u f = -\frac{1}{\sqrt{2}\,e^2}$$

13. $\mathbf{V} = \dfrac{\sqrt{2}}{2}\mathbf{i} + \dfrac{\sqrt{2}}{2}\mathbf{j};$
$$\nabla f = \frac{2x}{x^2 + 3y}\mathbf{i} + \frac{3}{x^2 + 3y}\mathbf{j}$$
$$\nabla f(1, 1) = \frac{1}{2}\mathbf{i} + \frac{3}{4}\mathbf{j}$$

$$D_u f = \frac{\sqrt{2}}{4} + \frac{3\sqrt{2}}{8} = \frac{5\sqrt{2}}{8}$$

14. $\mathbf{V} = \frac{\sqrt{2}}{2}\mathbf{i} - \frac{\sqrt{2}}{2}\mathbf{j};$

$$\nabla f = \frac{3}{3x + y^2}\mathbf{i} + \frac{2y}{3x + y^2}\mathbf{j}$$

$$\nabla f(0, 1) = 3\mathbf{i} + 2\mathbf{j}$$

$$D_u f = \frac{3\sqrt{2}}{2} - \frac{2\sqrt{2}}{2} = \frac{\sqrt{2}}{2}$$

15. $\mathbf{V} = -\frac{1}{\sqrt{10}}\mathbf{i} - \frac{3}{\sqrt{10}}\mathbf{j};$

$$\nabla f = \sec(xy - y^3)\tan(xy - y^3)[y\mathbf{i}$$
$$+ (x - 3y^2)\mathbf{j}]$$

$$\nabla f(2, 0) = \mathbf{0}$$

$$D_u f = 0$$

16. $\mathbf{V} = \frac{3}{\sqrt{10}}\mathbf{i} - \frac{1}{\sqrt{10}}\mathbf{j};$

$$\nabla f = \cos xy(y\mathbf{i} + x\mathbf{j})$$

$$\nabla f(\sqrt{\pi}, \sqrt{\pi}) = \sqrt{\pi}(\mathbf{i} + \mathbf{j})$$

$$D_u f = \frac{\sqrt{\pi}(-3 + 1)}{\sqrt{10}} = \frac{-2\sqrt{\pi}}{\sqrt{10}}$$

17. $\mathbf{N} = \nabla f = 2x\mathbf{i} + 2y\mathbf{j} + 2z\mathbf{k}$

$$\mathbf{N}(1, -1, 1) = 2\mathbf{i} - 2\mathbf{j} + 2\mathbf{k}$$

$$\mathbf{N}_u = \pm \frac{\mathbf{N}}{\sqrt{12}} = \pm \frac{\sqrt{3}}{3}(\mathbf{i} - \mathbf{j} + \mathbf{k})$$

The tangent plane is: $x - y + z - 3 = 0$

18. $\mathbf{N} = \nabla f = 4x^3\mathbf{i} + 4y^3\mathbf{j} + 4z^3\mathbf{k}$

$$\mathbf{N}(1, -1, -1) = 4\mathbf{i} - 4\mathbf{j} - 4\mathbf{k}$$

$$\mathbf{N}_u = \pm \frac{\mathbf{N}}{4\sqrt{3}} = \pm \frac{\sqrt{3}}{3}(\mathbf{i} - \mathbf{j} + \mathbf{k})$$

The tangent plane is:

$$(x - 1) - (y + 1) - (z + 1) = 0$$
$$x - y - z - 3 = 0$$

19. $\mathbf{N} = \nabla f = \cos(x + y)\mathbf{i} + \cos(x + y)\mathbf{j} + \sin z\mathbf{k}$

$$\mathbf{N}(\tfrac{\pi}{2}, \tfrac{\pi}{2}, \tfrac{\pi}{2}) = -\mathbf{i} - \mathbf{j} + \mathbf{k}$$

$$\mathbf{N}_u = \pm \frac{\mathbf{N}}{\sqrt{3}} = \pm \frac{\sqrt{3}}{3}(-\mathbf{i} - \mathbf{j} + \mathbf{k})$$

The tangent plane is:

$$(x - \tfrac{\pi}{2}) - (y - \tfrac{\pi}{2}) - (z - \tfrac{\pi}{2}) = 0$$
$$x + y - z - \tfrac{\pi}{2} = 0$$

20. $\mathbf{N} = \nabla f = \cos(x + y)\mathbf{i}$
$$+ [\cos(x + y) + \sec^2(y + z)]\mathbf{j}$$
$$+ \sec^2(y + z)\mathbf{k}$$

At $P(\tfrac{\pi}{4}, \tfrac{\pi}{4}, \tfrac{\pi}{4})$ ∇f is not defined

The tangent plane is: $y + z = 0$

21. Write the function as $\ln x - \ln(y - z) = 0$

$$\mathbf{N} = \nabla f = \tfrac{1}{x}\mathbf{i} - \frac{1}{y - z}\mathbf{j} + \frac{1}{y - z}\mathbf{k}$$

$$\mathbf{N}(2, 5, 3) = \tfrac{1}{2}\mathbf{i} - \tfrac{1}{2}\mathbf{j} + \tfrac{1}{2}\mathbf{k} = \tfrac{1}{2}(\mathbf{i} - \mathbf{j} + \mathbf{k})$$

$$\mathbf{N}_u(2, 5, 3) = \pm \frac{\sqrt{3}}{3}(\mathbf{i} - \mathbf{j} + \mathbf{k})$$

The tangent plane is:

$$(x - 2) - (y - 5) + (z - 3) = 0$$
$$x - y + z = 0$$

22. $\ln(x - y) - \ln(y + z) - x + z = 0$

$$\mathbf{N} = \nabla f = \frac{1 - x + y}{x - y}\mathbf{i}$$
$$- [(x - y)^{-1} + (y + z)^{-1}]\mathbf{j}$$
$$+ \frac{-1 + y + z}{y + z}\mathbf{k}$$

$$\mathbf{N}(1, 0, 1) = -2\mathbf{j}$$

$$\mathbf{N}_u = \pm\mathbf{j};$$ The tangent plane is: $y = 0$

23. $\mathbf{N} = \nabla f = e^{x + 2y}(z\mathbf{i} + 2z\mathbf{j} + \mathbf{k})$

$$\mathbf{N}(2, -1, 3) = 3\mathbf{i} + 6\mathbf{j} + \mathbf{k}$$

$$\mathbf{N}_u = \pm \frac{\mathbf{N}}{\sqrt{46}} = \pm \frac{1}{\sqrt{46}}(3\mathbf{i} + 6\mathbf{j} + \mathbf{k})$$

The tangent plane is: $3x + 6y + z - 3 = 0$

24. $\mathbf{N} = \nabla f = 2xze^{x^2}\mathbf{i} - 2y\mathbf{j} + e^{x^2}\mathbf{k}$

$$\mathbf{N}(1, 1, 3) = 2e\mathbf{i} - 2\mathbf{j} + e\mathbf{k}$$

$$\mathbf{N}_u = \pm \frac{\mathbf{N}}{\sqrt{5e^2 + 4}}$$

The tangent plane is:

$$2e(x - 1) - 2(y - 1) + e(z - 3) = 0$$
$$2ex - 2y + ez - 5e + 2 = 0$$

25. $\nabla f = 3z\mathbf{i} + 2y\mathbf{j}$

  a. $\nabla f(1, -1) = 3\mathbf{i} - 2\mathbf{j}; \|\nabla f\| = \sqrt{13}$

  b. $\nabla f(1, 1) = 3\mathbf{i} + 2\mathbf{j}; \|\nabla f\| = \sqrt{13}$

26. $\nabla f = -2x\mathbf{i} - 2y\mathbf{j}$

  a. $\nabla f(1, 2) = -2\mathbf{i} - 4\mathbf{j}; \|\nabla f\| = 2\sqrt{5}$

  b. $\nabla f(0, 0) = \mathbf{0}; \|\nabla f\| = 0$

27. $\nabla f = 3x^2\mathbf{i} + 3y^2\mathbf{j}$

  a. $\nabla f(3, -3) = 27(\mathbf{i} + \mathbf{j}); \|\nabla f\| = 27\sqrt{2}$

  b. $\nabla f(-3, 3) = 27(\mathbf{i} + \mathbf{j}); \|\nabla f\| = 27\sqrt{2}$

28. $\nabla f = a\mathbf{i} + b\mathbf{j}$

$$\nabla f(a, b) = a\mathbf{i} + b\mathbf{j}; \|\nabla f\| = \sqrt{a^2 + b^2}$$

29. $\nabla f = 2axi + 2byj + 2czk$

$\nabla f(a, b, c) = 2(a^2i + b^2j + c^2k)$;

$\|\nabla f\| = 2\sqrt{a^4 + b^4 + c^4}$

30. $\nabla f = 3(ax^2i + by^2j)$

$\nabla f(a, b) = 3(a^3i + b^3j)$; $\|\nabla f\| = 3\sqrt{a^6 + b^6}$

31. $\nabla f = \dfrac{1}{2(x^2 + y^2)}(2xi + 2yj)$

$\nabla f(1, 2) = \frac{1}{5}(i + 2j)$; $\|\nabla f\| = \dfrac{1}{\sqrt{5}}$

32. $\nabla f = (\cos xy)(yi + xj)$

$\nabla f(\dfrac{\sqrt{\pi}}{3}, \dfrac{\sqrt{\pi}}{2}) = \dfrac{\sqrt{3}}{2}\left(\dfrac{\pi}{2}i + \dfrac{\sqrt{3}}{3}j\right)$;

$\|\nabla f\| = \dfrac{\sqrt{39}\pi}{12}$

33. $\nabla f = [2(x + y) + 2(x + z)]i$

$\qquad + [2(x + y) + 2(y + z)]j$

$\qquad + [2(z + y) + 2(x + z)]k$

$\nabla f(2, 1, -2) = 2(5i + 2j + 5k)$;

$\|\nabla f\| = \sqrt{216} = 6\sqrt{6}$

34. $f(x, y, z) = z(\ln y - \ln x)$

$\nabla f = -\dfrac{z}{x}i + \dfrac{z}{y}j + \ln \dfrac{y}{x}k$

$\nabla f(1, e, -1) = i - e^{-1}j + k$

$\|\nabla f\| = \sqrt{2 + e^{-2}}$

35. Let $f(x, y) = ax + by - c = 0$

$\nabla f = ai + bj$; $u = \pm\dfrac{ai + bj}{\sqrt{a^2 + b^2}}$

36. Let $f(x, y) = x^2 + y^2 - a^2 = 0$

$\nabla f = 2(xi + yj)$;

$u = \pm\dfrac{x_0i + y_0j}{\sqrt{x_0^2 + y_0^2}} = \pm\frac{1}{a}(x_0i + y_0j)$

37. Let $f(x, y) = \dfrac{x^2}{a^2} + \dfrac{y^2}{b^2} - 1 = 0$

$\nabla f = \dfrac{2x}{a^2}i + \dfrac{2y}{b^2}j = 2a^{-2}b^{-2}(b^2xi + a^2yj)$

$u = \pm\dfrac{b^2x_0i + ay_0j}{\sqrt{b^4x_0^2 + a^4y_0^2}}$

38. Let $f(x, y) = \dfrac{x^2}{a^2} - \dfrac{y^2}{b^2} - 1 = 0$

$\nabla f = \dfrac{2x}{a^2}i - \dfrac{2y}{b^2}j = 2a^{-2}b^{-2}(b^2xi - a^2yj)$

$u = \pm\dfrac{b^2x_0i - a^2y_0j}{\sqrt{b^4x_0^2 + a^4y_0^2}}$

39. $u = \cos\dfrac{\pi}{6}i + \sin\dfrac{\pi}{6}j = \frac{1}{2}(\sqrt{3}i + j)$

$\nabla f = \nabla(x^2 + y^2) = 2(xi + yj)$

$\nabla f(1, 1) = 2(i + j)$; $D_u f = \sqrt{3} + 1$

40. $u = \dfrac{1}{\sqrt{2}}(-i + j)$

$\nabla f = \nabla(x^2 + xy + y^2) = (2x + y)i + (x + 2y)j$

$\nabla f(1, -1) = i - j$; $D_u f = \dfrac{-2}{\sqrt{2}} = -\sqrt{2}$

41. $\nabla f = e^{x^2y^2}(2xy^2i + 2x^2yj)$

$\nabla(1, -1) = 2e(i - j)$;

$V = \langle 2 - 1, 3 - (-1)\rangle = \langle 1, 4\rangle$;

$u = \langle\dfrac{1}{\sqrt{17}}, \dfrac{4}{\sqrt{17}}\rangle$

$D_u = -2e\left(\dfrac{3}{\sqrt{17}}\right) = -\dfrac{6e}{\sqrt{17}}$

42. **a.** $\nabla f = \nabla(2x^2 - y^2 + 3z^2 - 8x - 4y + 201)$

$\qquad = (4x - 8)i + (-2y - 4)j + (6z)k$

$\nabla f_0(2, -\frac{3}{2}, \frac{1}{2}) = -7j + 3k$

b. $OP_0 = -2i + \frac{3}{2}j - \frac{1}{2}k$

$\cos\theta = \dfrac{OP_0 \cdot \nabla f}{\|OP_0\|\|\nabla f\|}$

$\qquad = \dfrac{-\frac{21}{2} - \frac{3}{2}}{\sqrt{4 + \frac{9}{4} + \frac{1}{4}}\sqrt{49 + 9}} = -\dfrac{12}{\sqrt{377}}$

43. $\nabla f = \nabla xyz = xzi + xzj + xzj$

$\nabla f(1, -1, 2) = -2i + 2j - k$

$v \times w = -i + 7j + 5k$

$u = \dfrac{1}{\sqrt{75}}(-i + 7j + 5k)$

$D_u f = \dfrac{2 + 14 - 5}{5\sqrt{3}} = \dfrac{11\sqrt{3}}{15}$

44. $\nabla f = (ye^{x+z} - ze^{y-x})i + (e^{x+z} + e^{y-x})j$

$\qquad + (ye^{x+z} + e^{y-x})k$

$\nabla f(2, 2, -2) = 4i + 2j + 3k$

$u = \dfrac{1}{\sqrt{29}}(4i + 2j + 3k)$

45. $\nabla T = (y + z)i + (x + z)j + (x + y)k$

$\nabla T(1, 1, 1) = 2(i + j + k)$

$u = \dfrac{\sqrt{3}}{3}(i + j + k)$

$\|\nabla T\| = 2\sqrt{3}$ maximum rate of temperature

change.

46. $E(x, y) = k(x^2 + y^2); \nabla E = 2k(x\mathbf{i} + y\mathbf{j})$

$\nabla E = 2k(3\mathbf{i} + 4\mathbf{j}); \mathbf{u} = \frac{3}{5}\mathbf{i} + \frac{4}{5}\mathbf{j}$

47. $\nabla z = -6x\mathbf{i} - 5y\mathbf{j}$

$\nabla z(\frac{1}{4}, -\frac{1}{2}, \frac{3}{16}) = -\frac{3}{2}\mathbf{i} + \frac{5}{2}\mathbf{j} = -(\frac{3}{2}\mathbf{i} - \frac{5}{2}\mathbf{j})$

The negative sign is for the most rapid decrease; she turns in the direction $\frac{3}{2}\mathbf{i} - \frac{5}{2}\mathbf{j}$.

48. $\nabla f = f_x\mathbf{i} + f_y\mathbf{j}; \mathbf{u} = \frac{3}{5}\mathbf{i} - \frac{4}{5}\mathbf{j}$

$\|\nabla f\| = D_\mathbf{u} f = \frac{3}{5}\mathbf{i} - \frac{4}{5}\mathbf{j} = 100$

$\frac{3}{5}\mathbf{i} - \frac{4}{5}\mathbf{j} = 100$

$3f_x - 4f_y = 500$

$f_y = \frac{3f_x - 500}{4}$

Now,
$$f_x^2 + f_y^2 = 100^2$$

$$f_x^2 + \frac{(3f_x - 500)^2}{16} = 10{,}000$$

$$16f_x^2 + 9f_y^2 - 3{,}00f_x + 250{,}000 = 160{,}000$$

$$f_x^2 - 120f_x + 3{,}600 = 0$$

$$(f_x - 60)^2 = 0$$

$$f_x = 60$$

If $f_x = 60$, then $f_y = \frac{180 - 500}{4} = -80$

Thus, $\nabla f = 60\mathbf{i} - 80\mathbf{j}$

49. To find the directional derivative of f in the direction of u we need f_x and f_y, which we can find with a system of equations.

$$\begin{cases} (f_x\mathbf{i} + f_y\mathbf{j}) \cdot \left(\frac{3\mathbf{i} - 4\mathbf{j}}{5}\right) = 8 \\ (f_x\mathbf{i} + f_y\mathbf{j}) \cdot \left(\frac{12\mathbf{i} + 5\mathbf{j}}{13}\right) = 1 \end{cases}$$

$$\begin{cases} 3f_x - 4f_y = 40 \\ 12f_x + 5f_y = 13 \end{cases}$$

Solving simultaneously: $f_x = 4, f_y = -7$
Now for $\mathbf{V} = 3\mathbf{i} - 5\mathbf{j}$:

$(f_x\mathbf{i} + f_y\mathbf{j}) \cdot \left(\frac{3\mathbf{i} - 5\mathbf{j}}{\sqrt{34}}\right) = (4\mathbf{i} - 7\mathbf{j}) \cdot \left(\frac{3\mathbf{i} - 5\mathbf{j}}{\sqrt{34}}\right)$

$= \frac{12 + 35}{\sqrt{34}} \approx 8.06$

50. $\nabla f = f_x\mathbf{i} + f_y\mathbf{j}$

$v_1 = \frac{1}{\sqrt{13}}(2\mathbf{i} + 3\mathbf{j})$

$\|\nabla f\| = D_{v_1} f = \frac{2f_x + 3f_y}{\sqrt{13}} = 2$

$\frac{2f_x + 3f_y}{\sqrt{13}} = 2$

$2f_x + 3f_y = 2\sqrt{13}$

$v_2 = \frac{1}{\sqrt{34}}(3\mathbf{i} - 5\mathbf{j})$

$\|\nabla f\| = D_{v_2} f = \frac{3f_x - 5f_y}{\sqrt{34}} = -5$

$\frac{3f_x - 5f_y}{\sqrt{34}} = -5$

$3f_x - 5f_y = -5\sqrt{34}$

Solve the system
$$\begin{cases} 2f_x + 3f_y = 2\sqrt{13} \\ 3f_x - 5f_y = -\sqrt{34} \end{cases}$$
to obtain $f_x = -2.7057, f_y = 4.2075$

$\mathbf{v}_3 = \frac{1}{\sqrt{13}}(2\mathbf{i} - 3\mathbf{j})$, so

$$D_{v_3} f = \frac{2(-2.7057) - 4(4.2075)}{\sqrt{13}} \approx -5$$

51. $\mathbf{u} = \frac{1}{\sqrt{10}}(\mathbf{i} - 3\mathbf{j}); D_\mathbf{u} f = \|\nabla f\| = 50$

$\nabla f = \frac{50(\mathbf{i} - 3\mathbf{j})}{\sqrt{10}} = 5\sqrt{10}(\mathbf{i} - 3\mathbf{j})$

52. $T_x = -2x; T_y = -4y$, so $m = \frac{dy}{dx} = \frac{2y}{x}$

Thus, $\frac{dy}{y} = 2\frac{dx}{x}$

$\int y^{-1} dy = \int 2x^{-1} dx$

$\ln|y| = 2\ln|x| + C_1$

$\ln|y| - \ln x^2 = \ln C$

$y = Cx^2$

At $P_0(-1, 1)$, $C = 1$, and the equation of the path is $y = x^2$.

53. $T_x = -2ax; T_y = -2by$, so $m = \frac{dy}{dx} = \frac{by}{ax}$

Thus, $\frac{dy}{y} = \frac{b}{a}\frac{dx}{x}$

$\int y^{-1} dy = \frac{b}{a}\int x^{-1} dx$

$\ln|y| = \frac{b}{a}\ln|x| + C_1$

$\ln|y| - \ln x^{b/a} = \ln C$

$y = Cx^{b/a}$

At $P_0(x_0, y_0)$, $C = \frac{y_0}{x_0^{b/a}}$, and the equation of the path is $y = y_0(x/x_0)^{b/a}$.

54. Answers vary; see the computer manuals that accompany this book; code the function $y = y_0(x/x_0)^{b/a}$ and output a table or generate a plot for the curve.

55. First, note that $T(x, y)$ is minimized when $x = 6$ and $y = 1$; that is, at the point $(6, 1)$. If we let the path he follows be expressed in vector form as

$$\mathbf{R}(t) = x(t)\mathbf{i} + y(t)\mathbf{j},$$

then $\mathbf{R}'(t) = x'(t)\mathbf{i} + y'(t)\mathbf{j}$. Since he always moves in the direction of maximum temperature decrease, we must have

$$\mathbf{R}'(t) = -\nabla T$$

so that

$$\frac{dx}{dt} = -(x - 6) \text{ and } \frac{dy}{dt} = -3(y - 1) \text{ or}$$
$$\frac{dx}{dt} + 6x = 36 \quad \text{and} \quad \frac{dy}{dt} + 3y = 3$$

Solving these first order differential equations, we obtain

$$x(t) = 6 + C_1 e^{-6t}; \quad y(t) = 1 + C_2 e^{-3t}$$

Since $x = 1$ and $y = 5$ when $t = 0$, it follows that $C_1 = -5$ and $C_2 = 4$, so

$$x(t) = 6 - 5e^{-6t}; \quad y(t) = 1 + 4e^{-3t}$$

We can write $e^{-6t} = \frac{6-x}{5}$ and $e^{-3t} = \frac{y-1}{4}$ so that

$$\left(\frac{y-1}{4}\right)^2 = \frac{6-x}{5} \text{ or } x = 6 - \frac{5}{16}(y - 1)^2$$

Thus, the spy walks along the parabola $x = 6 - \frac{5}{16}(y - 1)$ from $(1, 5)$ to $(6, 1)$. The distance (arc length) that he travels is given by the integral

$$L = \int \sqrt{1 + \left(\frac{dx}{dy}\right)^2}\, dy$$

$$= \int_1^5 \sqrt{1 + \left[-\frac{10}{16}(y - 1)\right]^2}\, dy$$

$$= \frac{4}{5}\left[\sqrt{1 + \frac{25}{64}(y - 1)^2}\, \left[\frac{5}{8}(y - 1)\right]\right]$$

$$\ln\left|\sqrt{1 + \frac{25}{64}(y - 1)^2}\right| + \frac{5}{8}(y - 1)\Big|_1^5$$

$$\approx 6.7 \text{ ft}$$

He reaches the hole in about 1 min 40 sec — plenty of time for a cup of tea before Chapter 13.

56. a. Note that $r_1 + r_2 = C$ (a constant) is a level curve of the function $f = r_1 + r_2$. Therefore, by the normal property of the gradient, we know that $\nabla(r_1 + r_2)$ is a normal to $r_1 + r_2 = C$. Thus, $\mathbf{T} \cdot \nabla(r_1 + r_2) = 0$

b. The equation of the ellipse is $r_1 + r_2 = C$

(a constant), which can be regarded a a level curve of the function $f = r_1 + r_2$. But the gradient $\nabla(r_1 + r_2)$ is normal to this level curve. Therefore, we must have

$$\mathbf{T} \cdot \nabla(r_1 + r_2) = 0$$

Let $\mathbf{R}_1 = PF_1$ and $\mathbf{R}_2 = PF_2$ be the vectors from P to the two foci, so that

$$r_1 = \|\mathbf{R}_1\| \text{ and } r_2 = \|\mathbf{R}_2\|$$

By direct computation, it can be shown that

$$\nabla r_1 = \frac{\mathbf{R}_1}{r_1} \text{ and } \nabla r_2 = \frac{\mathbf{R}_2}{r_2}$$

and by substituting into the vector equation $\mathbf{T} \cdot \nabla(r_1 + r_2) = 0$, we have

$$\mathbf{T} \cdot \nabla r_1 = -\mathbf{T} \cdot \nabla r_2$$

$$\mathbf{T} \cdot \left(\frac{\mathbf{R}_1}{r_1}\right) = -\mathbf{T} \cdot \left(\frac{\mathbf{R}_2}{r_2}\right)$$

or

$$\frac{\|\mathbf{T}\|\|\mathbf{R}_1\|}{r_1}\cos(\pi - \theta_1) = -\frac{\|\mathbf{T}\|\|\mathbf{R}_2\|}{r_2}\cos\theta_2$$

so that

$$\cos(\pi - \theta_1) = -\cos\theta_2$$
$$\cos\theta_1 = \cos\theta_2$$
$$\theta_1 = \theta_2$$

57. $r = \sqrt{x^2 + y^2 + z^2}$

a. $\frac{\partial}{\partial x}\left(\frac{1}{r}\right) = -\frac{1}{r^2}\frac{\partial r}{\partial x}$

$$= -\frac{1}{r^2}\left(\frac{2x}{2\sqrt{x^2 + y^2 + z^2}}\right)$$

$$= -\frac{x}{r^3}$$

Likewise for the other partials.

b. $V = -gm_1\frac{1}{r}$

$$-\nabla V = gm_1\left[\frac{\partial}{\partial x}\frac{1}{r}\mathbf{i} + \frac{\partial}{\partial y}\frac{1}{r}\mathbf{j} + \frac{\partial}{\partial z}\frac{1}{r}\mathbf{k}\right]$$

$$= \frac{gm_1}{r^3}(-x\mathbf{i} - y\mathbf{j} - z\mathbf{k}) = \mathbf{F}$$

58. a. $\nabla(cf) = \frac{\partial}{\partial x}(cf)\mathbf{i} + \frac{\partial}{\partial y}(cf)\mathbf{j} + \frac{\partial}{\partial z}(cf)\mathbf{k}$

$$= c(f_x\mathbf{i} + f_y\mathbf{j} + f_z\mathbf{k}) = c\nabla f$$

b. $\nabla(f + g) = \frac{\partial}{\partial x}(f + g)\mathbf{i} + \frac{\partial}{\partial y}(f + g)\mathbf{j}$
$$+ \frac{\partial}{\partial z}(f + g)\mathbf{k}$$

$$= (f_x + g_x)\mathbf{i} + (f_y + g_y)\mathbf{j} + (f_z + g_z)\mathbf{k}$$

$$= f_x\mathbf{i} + f_y\mathbf{j} + f_z\mathbf{k} + g_x\mathbf{i} + g_y\mathbf{j} + g_z\mathbf{k}$$

$$= \nabla f + \nabla g$$

c. $\nabla \dfrac{f}{g} = \dfrac{\partial}{\partial x}\dfrac{f}{g}\mathbf{i} + \dfrac{\partial}{\partial y}\dfrac{f}{g}\mathbf{j} + \dfrac{\partial}{\partial z}\dfrac{f}{g}\mathbf{k}$

$= \dfrac{gf_x - fg_x}{g^2}\mathbf{i} + \dfrac{gf_y - fg_y}{g^2}\mathbf{j}$

$\quad + \dfrac{g_z f - fg_x}{g^2}\mathbf{k}$

$= \dfrac{g(f_x\mathbf{i}+f_y\mathbf{j}+f_z\mathbf{k}) - f(g_x\mathbf{i}+g_y\mathbf{j}+g_z\mathbf{k})}{g^2}$

$= \dfrac{g\nabla f - f\nabla g}{g^2}$

59. $\nabla fg = \dfrac{\partial}{\partial x}(fg)\mathbf{i} + \dfrac{\partial}{\partial y}(fg)\mathbf{j} + \dfrac{\partial}{\partial z}(fg)\mathbf{k}$

$= (fg_x + f_x g)\mathbf{i} + (fg_y + f_y g)\mathbf{j} + (fg_z + f_z g)\mathbf{k}$

$= f(g_x\mathbf{i} + g_y\mathbf{j} + g_z\mathbf{k}) + g(f_x\mathbf{i} + f_y\mathbf{j} + f_z\mathbf{k})$

$= f\nabla g + g\nabla f$

60. $\nabla f = f_x\mathbf{i} + f_y\mathbf{j};\ D_u f = f_x\cos\theta + f_y\sin\theta$

$= e^{x^2 - y^2}[(x + 1)\mathbf{i} + (2y - 2y^2)\mathbf{j}]$

$\nabla f(-1, 3) = 12e^{-7}\mathbf{j};\ D_u f = 6e^{-7}$

61. Let $\mathbf{w} = \mathbf{u} + \mathbf{v}$, then a unit vector in the direction of $\mathbf{w}$ is

$\mathbf{w} = \dfrac{\mathbf{u} + \mathbf{v}}{\|\mathbf{u} + \mathbf{v}\|}$ and

$D_w f = \nabla f \cdot \dfrac{\mathbf{u} + \mathbf{v}}{\|\mathbf{u} + \mathbf{v}\|}$

$= \dfrac{1}{\|\mathbf{u} + \mathbf{v}\|}[\nabla f \cdot \mathbf{u} + \nabla f \cdot \mathbf{v}]$

$= \dfrac{D_u f + D_v f}{\|\mathbf{u} + \mathbf{v}\|}$

62. $\mathbf{R} = x\mathbf{i} + y\mathbf{j} + z\mathbf{k};\ \mathbf{a} = a_1\mathbf{i} + a_2\mathbf{j} + a_3\mathbf{k}$

$\nabla(\mathbf{a}\cdot\mathbf{R}) = \nabla(a_1 x + a_2 y + a_3 z)$

$= a_1\mathbf{i} + a_2\mathbf{j} + a_3\mathbf{k} = \mathbf{a}$

63. a. $\nabla r = \dfrac{2(x\mathbf{i} + y\mathbf{j} + z\mathbf{k})}{\sqrt{x^2 + y^2 + z^2}} = \dfrac{\mathbf{R}}{\|\mathbf{R}\|}$

b. $\nabla r^n = \nabla(x^2 + y^2 + z^2)^{n/2}$

$= \dfrac{n}{2}(r)^{n/2 - 1}(2x\mathbf{i} + 2y\mathbf{j} + 2z\mathbf{k})$

$= nr^{n-2/2}(x\mathbf{i} + y\mathbf{j} + z\mathbf{k})$

$= n(r^{1/2})^{n-2}\mathbf{R}$

12.7 Extrema of Functions of Two Variables, Page 802

1. A critical value of a function f defined on an open set S is a point (x_0, y_0) in S where either one of the following is true:

(1) $f_x(x_0, y_0) = f_y(x_0, y_0) = 0$

(2) $f_x(x_0, y_0)$ or $f_y(x_0, y_0)$ does not exist

(one or both).

2. Solve $z_x = 0$ and $z_y = 0$, and then evaluate $D = z_{xx}z_{yy} - z_{xy}^2$ at the critical points.

3. A regression line is the best fitting line through a set of linearly related data points under the criterion of lease squares.

4. $f(x, y) = 2x^2 - 4xy + y^3 + 2$

$f_x = 4x - 4y = 0$ when $y = x$

$f_y = -4x + 3y^2 = 0$ when $y = 0, \frac{4}{3}$

$f_{xx} = 4, f_{xy} = -4, f_{yy} = 6y$

x	y	f_{xx}	f_{xy}	f_{yy}	D	Classify
0	0	4	-4	0	$-$	saddle point
$\frac{4}{3}$	$\frac{4}{3}$	4	-4	8	$+$	rel minimum

5. $f(x, y) = (x - 2)^2 + (y - 3)^4$

$f_x = 2(x - 2) = 0$ when $x = 2$

$f_y = 4(y - 3)^3 = 0$ when $y = 3$

$f_{xx} = 2, f_{xy} = 0, f_{yy} = 12(y - 3)^2$

x	y	f_{xx}	f_{xy}	f_{yy}	D	Classify
2	3	2	0	0	0	inconclusive

Note that f is the sum of squares, which implies that the minimum is 0, which occurs at $(2, 3)$.

6. $f(x, y) = e^{-x}\sin y$

$f_x = -e^{-x}\sin y = 0$ when $y = n\pi$

$f_y = e^{-x}\cos y = 0$ when $y = (2n + 1)\frac{\pi}{2}$

$f_{xx} = e^{-x}\sin y, f_{xy} = -e^{-x}\cos y,$

$f_{yy} = -e^{-x}\sin y$

x	y	f_{xx}	f_{xy}	f_{yy}	D	Classify
x	$n\pi$	0	∓ 1	0	$-$	saddle point
x	$\frac{(2n+1)\pi}{2}$	e^{-x}	0	$-e^{-x}$	$+$	saddle point

7. $f(x, y) = (1 + x^2 + y^2)e^{1 - x^2 - y^2}$

$f_x = e^{1 - x^2 - y^2}[(-2x)(1 + x^2 + y^2) + 2x]$

$\quad = -2x(x^2 + y^2)e^{1 - x^2 - y^2} = 0$ when $x = 0$

$f_y = e^{1 - x^2 - y^2}[(-2y)(1 + x^2 + y^2) + 2y]$

$\quad = -2y(x^2 + y^2)e^{1 - x^2 - y^2} = 0$ when $y = 0$

$f_{xx} = -2e^{1 - x^2 - y^2}[x(2x) + x(x^2 + y^2)(-2x)$

$\quad + x^2 + y^2]$

$f_{xy} = -2xe^{1-x^2-y^2}[2y + (-2y)(x^2+y^2)]$

$\qquad = -xye^{1-x^2-y^2}(1 - x^2 - y^2)$

$f_{yy} = -2e^{1-x^2-y^2}[y(2y) + y(x^2+y^2)(-2y)$

$\qquad\qquad + x^2 + y^2]$

x	y	f_{xx}	f_{xy}	f_{yy}	D	Classify
0	0	0	0	0	0	inconclusive

$e^{1-x^2-y^2}$ is a maximum at $(0, 0)$; the surface has symmetry with respect to the z-axis. In $\mathbb{R}^2$, $y = (1 + x^2)e^{1-x^2}$ is a bell-shaped curve. Thus, $P(0, 0)$ is a relative maximum.

8. $f(x, y) = \dfrac{9x}{x^2 + y^2 + 1}$

$f_x = \dfrac{9(x^2 + y^2 + 1 - 2x^2)}{(x^2 + y^2 + 1)^2} = \dfrac{9(y^2 + 1 - x^2)}{(x^2 + y^2 + 1)^2}$

$\qquad = 0$ when $y = \pm\sqrt{x^2 + 1}$

$f_y = -\dfrac{9x(2y)}{(x^2 + y^2 + 1)^2} = \dfrac{-18xy}{(x^2 + y^2 + 1)^2}$

$\qquad = 0$ when $x = 0$ or $y = 0$

At $x = 0$, $f_x > 0$, no extremum.

At $y = 0$, $f_x = 0$ if $y = \pm 1$

$f_{xx} = \dfrac{18x^3 - 54xy^2 - 54x}{(x^2 + y^2 + 1)^3}$

$f_{xy} = \dfrac{54x^2y - 18y^3 - 18y}{(x^2 + y^2 + 1)^3}$

$f_{yy} = -\dfrac{18x^3 - 54xy^2 + 18x}{(x^2 + y^2 + 1)^3}$

x	y	f_{xx}	f_{xy}	f_{yy}	D	Classify
1	0	$-\frac{9}{2}$	0	$-\frac{9}{2}$	+	rel max
-1	0	$\frac{9}{2}$	0	$\frac{9}{2}$	+	rel min

9. $f(x, y) = x^2 + xy + y^2$

$f_x = 2x + y = 0$ if $y = -2x$

$f_y = x + 2y = 0$ if $y = -\frac{x}{2}$

$f_{xx} = 2;\ f_{xy} = 1;\ f_{yy} = 2$

x	y	f_{xx}	f_{xy}	f_{yy}	D	Classify
0	0	2	1	2	+	rel min

10. $f(x, y) = xy - x + y$

$f_x = y - 1 = 0$ if $y = 1$

$f_y = x + 1 = 0$ if $x = -1$

$f_{xx} = 0,\ f_{xy} = 1,\ f_{yy} = 0$

x	y	f_{xx}	f_{xy}	f_{yy}	D	Classify
-1	1	0	1	0	1	$-$	saddle point

11. $f(x, y) = -x^3 + 9x - 4y^2$

$f_x = -3x^2 + 9 = 0$ if $x = \pm\sqrt{3}$

$f_y = -8y = 0$ if $y = 0$

$f_{xx} = -6x;\ f_{yy} = -8;\ f_{xy} = 0$

x	y	f_{xx}	f_{xy}	f_{yy}	D	Classify
$\sqrt{3}$	0	$-6\sqrt{3}$	0	-8	+	rel max
$-\sqrt{3}$	0	$6\sqrt{3}$	0	-8	$-$	saddle point

12. $f(x, y) = e^{-(x^2+y^2)}$

$f_x = -2xe^{-(x^2+y^2)} = 0$ if $x = 0$

$f_y = -2ye^{-(x^2+y^2)} = 0$ if $y = 0$

$f_{xx} = -2xe^{-(x^2+y^2)}[x(-2x) + 1]$

$f_{xy} = -2xe^{-(x^2+y^2)}(-2y)$

$f_{yy} = -2e^{-(x^2+y^2)}[y(-2y) + 1]$

x	y	f_{xx}	f_{xy}	f_{yy}	D	Classify
0	0	-2	0	-2	+	rel max

13. $F(x, y) = (x^2 + 2y^2)e^{1-x^2-y^2}$

$F_x = e^{1-x^2-y^2}[(x^2 + 2y^2)(-2x + 2x)]$

$\qquad = 2xe^{1-x^2-y^2}[1 - x^2 - 2y^2] = 0$

$\qquad$ if $x = 0$ or if $x^2 + 2y^2 = 1$

$F_y = e^{1-x^2-y^2}[(x^2 + 2y^2)(-2y) + 4y]$

$\qquad = 2ye^{1-x^2-y^2}[2 - x^2 - 2y^2] = 0$

$\qquad$ if $y = 0$ or $x^2 + 2y^2 = 1$

$F_{xx} = 2e^{1-x^2-y^2}[1 - x^2 - y^2$

$\qquad\qquad + x(-2x)(1 - x^2 - y^2) + x(-2x)]$

$F_{yy} = 2e^{1-x^2-y^2}[2 - x^2 - 2y^2$

$\qquad\qquad + y(-2y)(2 - x^2 - 2y^2) + y(-4y)]$

x	y	f_{xx}	f_{xy}	f_{yy}	D	Classify
0	0	2	0	4	+	rel min
1	0	-2	0	1	$-$	saddle point
-1	0	-2	0	1	$-$	saddle point
0	1	0	0	-4	0	inconclusive
0	-1	0	0	-4	0	inconclusive

Checking points in the vicinity of $(0, \pm 1)$ indicate maximality.

14. $f(x, y) = e^{xy}$

$f_x = ye^{xy} = 0$ if $y = 0$

$f_y = xe^{xy} = 0$ if $x = 0$

$f_{xx} = y^2 e^{xy}$; $f_{xy} = e^{xy}(xy + 1)$; $f_{yy} = x^2 e^{xy}$

x	y	f_{xx}	f_{xy}	f_{yy}	D	Classify
0	0	0	1	0	$-$	saddle point

15. $f(x, y) = x^{-1} + y^{-1} + 2xy$

$f_x = -\dfrac{1}{x^2} + 2y = 0$ when $y = \dfrac{1}{2x^2}$

$f_y = -\dfrac{1}{y^2} + 2x = 0$ when $x = \dfrac{1}{2y^2}$

$f_{xx} = \dfrac{2}{x^3}$; $f_{yy} = \dfrac{2}{y^3}$; $f_{xy} = 2$

$D = \dfrac{4}{x^3 y^3} - 4$

$D\left(\dfrac{1}{\sqrt[3]{2}}, \dfrac{1}{\sqrt[3]{2}}\right) > 0$, and $f_{xx} > 0$, so

$\left(\dfrac{1}{\sqrt[3]{2}}, \dfrac{1}{\sqrt[3]{2}}\right)$ is a relative minimum.

Also consider points where f_x and f_y are undefined: $x = 0$ and/or $y = 0$. However $f(x, y)$ is undefined at these points also.

16. $f(x, y) = (x - 4)\ln xy = (x - 4)(\ln x + \ln y)$

$f_x = (x - 4)\dfrac{1}{x} + \ln x + \ln y$

$f_y = (x - 4)\dfrac{1}{y} = 0$ if $x = 4$; then $y = \dfrac{1}{4}$ makes

$\ln x + \ln y = 0$ if f_x.

$f_{xx} = (x - 4)(-x^{-2}) + 2x^{-1}$

$f_{xy} = y^{-1}$; $f_{yy} = -(x - 4)y^{-2}$

x	y	f_{xx}	f_{xy}	f_{yy}	D	Classify
4	$\frac{1}{4}$	8	4	0	$-$	saddle point

17. $f(x, y) = x^3 + y^3 + 3x^2 - 18y^2 + 81y + 5$

$f_x = 3x^2 + 6x = 3x(x + 2) = 0$ if $x = 0, -2$

$f_y = 3y^2 - 36y + 81 = 3(y - 3)(y - 9) = 0$

$\qquad$ if $y = 3, 9$

$f_{xx} = 6x + 6 = 6(x + 1)$; $f_{xy} = 0$;

$f_{yy} = 6y - 36 = 6(y - 6)$

x	y	f_{xx}	f_{xy}	f_{yy}	D	Classify
0	3	6	0	-18	$-$	saddle point
0	9	6	0	18	$+$	rel min
-2	9	3	-6	-18	$-$	saddle point
-2	3	3	-6	18	$+$	rel max

18. $f(x, y) = 2x^3 + y^3 + 3x^2 - 3y - 12x - 4$

$f_x = 6x^2 + 6x - 12 = 6(x^2 + x - 2)$

$\qquad = 6(x + 2)(x - 1) = 0$ if $x = -2, 1$

$f_y = 3y^2 - 3 = 3(y - 1)(y + 1) = 0$

$\qquad$ if $y = -1$ or $y = 1$

$f_{xx} = 12x + 6 = 6(2x + 1)$; $f_{xy} = 0$; $f_{yy} = 6y$

x	y	f_{xx}	f_{xy}	f_{yy}	D	Classify
-2	-1	$-$	0	$-$	$+$	rel max
-2	1	$-$	0	$+$	$+$	saddle point
1	-1	$+$	0	$-$	$-$	saddle point
1	1	$+$	0	$+$	$+$	rel min

19. $f(x, y) = x^2 + y^2 - 6xy + 9x + 5y + 2$

$f_x = 2x - 6y + 9 = 0$

$f_y = 2y - 6x + 5 = 0$

Solving these equations simultaneously we find $x = \dfrac{3}{2}$, $y = 2$.

x	y	f_{xx}	f_{xy}	f_{yy}	D	Classify
$\frac{3}{2}$	2	2	-6	2	$-$	saddle point

20. $f(x, y) = 2x^2 - y^2$

$f_x = 4x = 0$ if $x = 0$

$f_y = -2y = 0$ if $y = 0$

If $x^2 + y^2 = 1$, then $y^2 = 1 - x^2$ and

$f(x, y) = 2x^2 - (1 - x^2) = 3x^2 - 1 = F(x)$

$F'(x) = 6x = 0$ if $x = 0$ and $y = \pm 1$

Similarly, $x^2 = 1 - y^2$ and

$f(x, y) = 2(1 - y^2) - y^2 = 2 - 3y^2 = G(y)$

$G'(x) = 6y = 0$ if $y = 0$ and $x = \pm 1$.

At $(\pm 1, 0)$, $f(x, y) = 2$;

at $(0, \pm 1)$, $f(x, y) = -1$

Maximum at $(1, 0)$ and $(-1, 0)$; minimum at $(0, 1)$ and $(0, -1)$.

21. $f(x, y) = x^2 + 3y^2 - 4x + 2y - 3$

$f_x = 2x - 4 = 0$ if $x = 2$

$f_y = 6y + 2 = 0$ if $y = -\dfrac{1}{3}$

$f_{xx} = 2$; $f_{xy} = 0$; $f_{yy} = 6$;

$D(2, -\tfrac{1}{3}) = 12$;

$f(0, 0) = -3$; $f(3, 0) = -6$; $f(3, -3) = -9$

$f(0, -3) = 18$; maximum

$f(2, -\frac{1}{3}) = -\frac{22}{3}$; minimum

22. $f(x, y) = 2 \sin x + 5 \cos y$

$f_x = 2 \cos x = 0$ at $x = \frac{\pi}{2}$

$f_y = -5 \sin y = 0$ if $y = 0$, $y = \pi$

$f(\frac{\pi}{2}, 0) = 7$; maximum

$f(\frac{\pi}{2}, \pi) \approx -3.2$; minimum

$f(0, 0) = 5$; $f(2, 0) \approx 6.8$; $f(2, 5) \approx 3.2$;

$f(0, 5) = 5$

23. $f(x, y) = e^{x^2 + 2x + y^2}$

$f_x = (2x + 2)e^{x^2 + 2x + y^2} = 0$ at $x = -1$

$f_y = 2ye^{x^2 + 2x + y^2} = 0$ at $y = 0$

$f(-1, 0) = e^{1-2} \approx 0.36$; minimum

$f(x, y)$ is maximized if the exponent is as large as possible. This will occur when (x, y) is on the circle $x^2 + 2x + y^2$; whenever (x, y) is on the circle, $e^{x^2 + 2x + y^2} = 1$.

24. $f(x, y) = x^2 + xy + y^2$

$f_x = 2x + y = 0$ if $y = -2x$

$f_y = x + 2y = 0$ if $y = -\frac{x}{2}$

$f(0, 0) = 0$; minimum

$f(1, 1) = 3$; maximum

$f(1, -1) = 1$

$f(-1, -1) = 3$; maximum

$f(-1, 1) = 1$

25. $f(x, y) = x^2 - 4xy + y^3 + 4y$

$f_x = 2x - 4y = 0$

$f_y = -4x + 3y^2 + 4 = 0$

Simultaneous solution is found when $x = 4$ or $\frac{4}{3}$ and both of these are outside the boundary of the square.

$f(0, 0) = 0$; minimum

$f(2, 0) = 4$

$f(0, 2) = 16$; maximum

$f(2, 2) = 10$

26. $m = \dfrac{5(25) - (1)(5)}{5(15) - 1^2} = \dfrac{60}{37} \approx 1.62$

$b = \dfrac{15(5) - (1)(25)}{5(15) - 1^2} = \dfrac{25}{37} \approx 0.68$

$y = 1.62x + 0.68$

27. $m = \dfrac{5(40.29) - (10.3)(14.5)}{5(31.45) - (10.3)^2} \approx 1.02$

$b = \dfrac{(31.45)(14.5) - (10.3)(40.29)}{51.16} \approx 0.80$

$y = 1.02x + 0.80$

28. $m = \dfrac{5(155.68) - (28.75)(27.14)}{5(184.47) - (28.75)^2} \approx -0.02$

$b = \dfrac{184.47(27.14) - (28.75)(155.68)}{95.79} \approx 5.54$

$y = -0.02x + 5.54$

29. $m = \dfrac{6(-22) - (-1)(-3)}{6(39) - (-1)^2} \approx -0.58$

$b = \dfrac{39(-3) - (-1)(-22)}{233} \approx -0.60$

$y = -0.58x - 0.60$

30. Let x, y, and z be the dimensions of the box. The volume is $V = xyz$ or $z = V/(xy)$, and the surface area is $S = xy + 2xz + 2yz$. We wish to minimize

$S = xy + 2xz + 2yz$

$= xy + \dfrac{2xV}{xy} + \dfrac{2yV}{xy}$

$= xy + \dfrac{2V}{y} + \dfrac{2V}{x}$

$S_x = y - \dfrac{2V}{x^2} = 0$; $S_y = x - \dfrac{2V}{y^2} = 0$

Solve these equations simultaneously to find

$y = \sqrt[3]{2V}$; $x = \sqrt[3]{2V}$; then $z = \sqrt[3]{0.25V}$

$S_{xx} = \dfrac{4V}{cx^3}$; $S_{xy} = 1$; $S_{yy} = \dfrac{4V}{y^3}$; $D > 0$, so the dimensions for the minimum construction are $x = \sqrt[3]{2V}$, $y = \sqrt[3]{2V}$, $z = \sqrt[3]{0.25V}$

31. To find the area of a triangle when three sides are given use Heron's semi-perimeter formula:

$A = \sqrt{s(s - a)(s - b)(s - c)}$ where $s = \frac{P}{2}$. The values that maximize A^2 will also maximize A.

$A^2 = \dfrac{P}{2}\left(\dfrac{P}{2} - a\right)\left(\dfrac{P}{2} - b\right)\left(\dfrac{P}{2} - c\right)$. To make this a function of two variables substitute $c = P - a - b$.

$16A^2 = P(P - 2a)(P - 2b)(2a + 2b - P)$

$$f_a = P(P - 2b)[(P - 2a)2$$
$$+ (2a + 2b - P)(-2)]$$
$$= P(P - 2b)(4P - 8a - 4b)$$
$$f_b = P(P - 2a)[(P - 2b)2$$
$$+ (2a + 2b - P)(-2)]$$
$$= P(P - 2a)(4P - 8b - 4a)$$

For extrema $f_a = f_b = 0$ at $a = \frac{P}{2}$, $b = \frac{P}{2}$, $c = 0$. $A = 0$, a minimum. Also, solving the system:

$$8a + 4b = 4P, \quad 4a + 8b = 4P,$$
$$a = \frac{P}{3}, \quad b = \frac{P}{3}, \quad c = \frac{P}{3}$$

Verifying that this is a maximum:

$$f_{aa} = P(P - 2b)(-8),$$

$$f_{aa}\left(\frac{P}{3}, \frac{P}{3}\right) = P\left(\frac{P}{3}\right)(-8) = -\frac{8}{3}P^2$$

$$f_{bb} = P(P - 2a)(-8)$$
$$f_{bb}\left(\frac{P}{3}, \frac{P}{3}\right) = P\left(\frac{P}{3}\right)(-8) = -\frac{8}{3}P^2$$
$$f_{ab} = P[(P - 2b)(-4) + (4P - 8a - 4b)(-2)]$$
$$= P(16a + 16b - 12P)$$
$$f_{ab}\left(\frac{P}{3}, \frac{P}{3}\right) = -\frac{4}{3}P^2; \ D = \frac{64}{9}P^4 - \frac{16}{9}P^4 > 0$$

and $f_{aa} < 0$ so $a = b = c = \frac{P}{3}$ is a

maximum. This says that the equilateral triangle has the greatest area.

32. Let x, y, and y be the lengths of the sides of the triangle with a given area A. Let h be the height of the triangle drawn to the base with length x. Then, $\sin \pi/3 = h/y$ so $h = y/2$.

$$A = 2(\tfrac{1}{2}bh) = 2(\tfrac{1}{2})(\tfrac{x}{2})(\tfrac{y}{2}) = \frac{xy}{4}$$

Since A is fixed, write $y = 4A/x$ so that

$$P = x + 2y$$
$$= x + 2(\tfrac{4A}{x})$$
$$= x + 8Ax^{-1}$$

We see $P'(x) = 1 - 8Ax^{-2} = 0$ when

$x = 2\sqrt{2A}$ (we disregard the negative value).

$P''(2\sqrt{2A}) > 0$, so the perimeter is minimized when $x = 2\sqrt{2A}$.

33. $P(x, y) = x(100 - x) + y(100 - y) - x^2 - xy - y^2$
$$= -2x^2 - 2y^2 + 100x + 100y - xy$$

$$P_x = -4x + 100 - y = 0$$
$$P_y = -4y + 100 - x = 0$$

Solving these equations simultaneously, we find $x = 20$, $y = 20$;

$$P_{xx} = -4; \ P_{xy} = -1; \ P_{yy} = -4; \ D > 0$$

$(20, 20)$ is a maximum.

34. $P(x, y) = x(4x - 5) + y(4y - 2) - 2xy - 4$
$$= -5x^2 - 2y^2 + 4x + 4y - 2xy - 4$$
$$P_x = 4 - 10x - 2y - 2 = 0$$
$$P_y = 4 - 4y - 2x = 0$$

Solving these equations simultaneously, we find $x = 0$, $y = 1$;

$$P_{xx} = -10; \ P_{xy} = -2; \ P_{yy} = -4; \ D > 0$$

$(0, 1)$ is a maximum.

35. Use $V_0 = xyz$ to write E as a function of two variables:

$$E(x, y) = \frac{k^2}{8m}\left(\frac{1}{x^2} + \frac{1}{y^2} + \frac{1}{\left(\frac{V_0}{xy}\right)^2}\right)$$

Minimize $\dfrac{8m}{k^2}E = \dfrac{1}{x^2} + \dfrac{1}{y^2} + \dfrac{x^2y^2}{V_0^2}$

$$f_x = -\frac{2}{x^3} + \frac{2xy^2}{V_0^2}; \ f_y = -\frac{2}{y^3} + \frac{2x^2y}{V_0^2}$$

$$-\frac{2}{x^3} + \frac{2xy^2}{V_0^2} = 0 \text{ when } x^4y^2 = V_0^2;$$

$$x^4 = \frac{V_0^2}{y^2}, \ x = \sqrt{\frac{V_0}{y}}$$

Similarly, $y = \sqrt{\dfrac{V_0}{x}}$ and

$$z = \frac{V_0}{xy} = \frac{V_0}{x\sqrt{\frac{V_0}{x}}} = \sqrt{\frac{V_0}{x}} = y$$

$$V_0 = x^3, \ x = y = z = \sqrt[3]{V_0}$$

Verifying that this is a minimum:

$$f_{xx} = \frac{6}{x^4} + \frac{2y^2}{V_0^2}; \ f_{yy} = \frac{6}{y^4} + \frac{2x^2}{V_0^2};$$

$$f_{xy} = \frac{4xy}{V_0^2};$$

$$D = \left(\frac{6}{V_0^{4/3}}\right)^2 - \frac{16V_0^{4/3}}{V_0^4} = \frac{64}{V_0^{8/3}} - \frac{16}{V_0^{8/3}} > 0$$

and $f_{xx} > 0$ so there is a minimum at

$(\sqrt[3]{V_0}, \sqrt[3]{V_0}, \sqrt[3]{V_0})$

36. $R_x = -2x + 2y + 8 = 0$

$R_y = -4y + 2x + 5 = 0$

Solving simultaneously, $x = \frac{21}{2}$, $y = \frac{13}{2}$

$R_{xx} = -2$, $R_{xy} = 2$, $R_{yy} = -4$, $D > 0$, so

the revenue is maximized at $\left(\frac{21}{2}, \frac{13}{2}\right)$.

37. Let x, y, and z be the length, width, and height of the rectangular box, respectively. Then, $xyz = 32$ or $z = 32/(xy)$. The cost of the sides is $C_s = 2xz + 2yz$; the cost of the top is $C_t = 5xy$; the cost of the bottom is $C_b = 3xy$. The total cost of the material is

$C = 8xy + 2xz + 2yz$

$\quad = 8xy + 2x\left(\frac{32}{xy}\right) + 2\left(\frac{32y}{xy}\right)$

$\quad = 8xy + 64y^{-1} + 64x^{-1}$

$C_x = 8y - 64x^{-1} = 0$ when $y = 8x^{-2}$

$C_y = 8x - 64y^{-2} = 0$ when $x = 8y^{-2}$

Solving simultaneously, $x = 2$, $y = 2$, so that $z = 8$.

$C_{xx} = 128x^{-3}$; $C_{xy} = 8$; $C_{yy} = 128y^{-3}$;

$D(2, 2) > 0$, so the cost of the material is minimized for a box that is 2 units long, 2 units wide, and 8 units high.

38. The number of bottles of water from California is $C = 40 - 50x + 40y$, and the number of bottles from New York is $N = 20 + 60x - 70y$. The profit is

$P(x, y) = (x - 2)(40 - 50x + 40y)$

$\qquad\quad + (y - 2)(20 + 60x - 70y)$

$\quad = -50x^2 + 100xy - 70y^2 + 20x + 80y - 120$

$P_x = -100x + 100y + 20$

$P_y = 100x - 140y + 80$

Solving simultaneously, $x = 2.7$, $y = 2.5$

$P_{xx} = -100$; $P_{xy} = 100$; $P_{yy} = -140$;

$D > 0$, so the profit is maximized at $(2.7, 2.5)$; that is, $2.70 for California water and $2.50 for New York water.

39. Profit = Revenue − Cost, or $P = R - C$

$R = (\# \text{ sold})(\text{cost per unit})$

$P = (40 - 8x + 5y)100x + (50 + 9x - 7y)100y$

$\qquad - 1,000(40 - 8x + 5y) - 3,000(50 + 9x - 7y)$

$\frac{P}{100} = -8x^2 - 7y^2 + 14xy - 150x + 210y - 1900$

$P_x = -16x + 14y - 150$;

$P_y = -14y + 14x + 210$

Solving simultaneously, $x = 30$, $y = 45$ Since $P(0, 0) = 0$, $(30, 45)$ must be a maximum. Thus, the first type should be priced at $3,000 and the second type priced at $4,500.

40. Domestic profit $= x(60 - 0.2x + 0.05y - 10)$

$\qquad\qquad\qquad = x(50 - 0.2x + 0.05y)$

Foreign profit $= y(50 - 0.1y + 0.05x - 10)$

$\qquad\qquad\quad = y(40 - 0.1y + 0.05x)$

$P(x, y) = x(50 - 0.2x + 0.05y) + y(40$

$\qquad\qquad - 0.1y + 0.05x)$

$\qquad = -0.2x^2 + 0.1xy - 0.1y^2 + 50x + 40y$

$P_x = -0.4x + 0.1y + 50 = 0$

$P_y = 0.1x - 0.2y + 40 = 0$

Solving simultaneously, $x = 200$, $y = 300$

$P_{xx} = -0.4$; $P_{xy} = 0.1$; $P_{yy} = -0.2$;

$D > 0$, so $(200, 3000)$ is a maximum. That is, 200 machines should be supplied to the domestic market and 300 to the foreign market.

41. $y = 0.78x + 0.19$; GPA is $f(3.75) \approx 3.12$

42. $k \approx m \approx 2.1$

43. **a.** $y = kx^m$; $\ln y = \ln k + m \ln x$ so

$Y = K + mX$; $K = \ln k$; $X = \ln a$;

b. Linear fit is $T \approx -1.6099 + 1.49996A$; so $k = e^K \approx 0.1999 \approx 0.2$ and $m \approx 1.5$ as Kepler predicted.

44. **a.** By assuming that weight-lifters of various sizes are proportioned about the same and that strength is proportional to the cross-section of the muscles involved, we see $m = 2/3$. The linear fit is

$W = 3.42674 + 0.573164X$

so m turns out to be a disappointing 0.573 vs the theoretical value of 2/3.

b. The poor agreement of m with the expected value is largely due to the exceptional performance of the 60 kg lifter.

45. **a.** Only one common zero at $(3, -2)$. In the linear case, regarding common zeros, we have have none (parallel lines), one

(intersecting lines), or infinitely many (the lines are, in fact, one line).

b. In the one-dimensional case, extend the tangent until it intersects the x-axis. In the two dimensional case, extend the tangent plane (for each function) until it intersects the xy-plane.

c. $(x_1, y_1) = (3.37233, 0.130366)$; $(x_2, y_2) = (3.35484, 0.14689)$

d. $f(x_0, y_0) \approx -1.38$; $g(x_0, y_0) \approx 1.04207$
$f(x_1, y_1) \approx 0.140473$;
$g(x_1, y_1) \approx 0.0348811$
For example, $(x_3, y_3) \approx (3.35471, 0.146964)$ so comparing the three computes sets of x and y, it looks like the familiar "second order" convergence.

46. a. By completing the square for the exponential, we have
$$-\left(x - \tfrac{9}{2}\right)^2 - \left(y + \tfrac{1}{4}\right)^2 + \tfrac{325}{16}$$
From this we see the exponential has its maximum at $(4.5, -0.25)$.

b. If $f = H_x$, then $f_x = H_{xx}$ and $f_y = H_{xy}$. If $g = H_y$, then $g_x = H_{yx}$ and $g_y = H_{yy}$. The result follows directly from Problem 45**b**.

c. $(x_1, y_1) \approx (4.48304, -0.292398)$

d. $(x_2, y_2) \approx (4.50007, -0.249822)$; this looks like second-order convergence.

e. It looks like the iterations are diverging.

f. $(x_1, y_1) \approx (1.3819, -0.038193)$ and $(x_2, y_2) \approx (1.03831, -0.0383096)$
The last values are accurate to the number of digits shown.

47. a. $f_x = 4x^3 \neq 0$ and $f_y = 4y^3 = 0$ so that $x = 0$ and $y = 0$.
$$f_{xx} = 12x^2, \ f_{xy} = 0, \ f_{yy} = 12y^2;$$
$D(0, 0) = 0$, so the test is inconclusive. $z = f(0, y) = -y^4$ shows a high point at the origin, while $z = f(x, 0) = x^4$ shows a low point at the origin. Thus, $(0, 0, 0)$ is a saddle point.

b. $g_x = 2xy^2 = 0$ and $g_y = 2x^2y = 0$ so that $x = 0$ and $y = 0$
$$g_{xx} = 2y^2, \ g_{xy} = 4xy, \ g_{yy} = 2x^2;$$
$D(0, 0) = 0$, so the test is inconclusive. $g(0, 0) = 0$ is the minimum because $g(x, y)$ is a product of squares.

c. $h_x = 3x^2 = 0$ and $h_y = 3y^2 = 0$ so that

$x = 0$ and $y = 0$
$$h_{xx} = 6x, \ h_{xy} = 0, \ h_{yy} = 6y;$$
$D(0, 0) = 0$, so the test is inconclusive. $z = h(0, y) = y^3$ shows a point of inflection at the origin, and $z = h(x, 0) = x^3$ also shows a point of inflection at the origin. Thus, $(0, 0, 0)$ is not a high point, not a low point, and not a saddle point.

48. $f_x = (y - x^2)(-4x) + (-2x)(y - 2x^2)$
$$= -2x(3y - 4x^2) = 0 \text{ if } x = 0 \text{ or } y = \tfrac{4}{3}x^2$$
$f_y = y - x^2 + y - 2x^2 = 2y - 3x^2 = 0$
   if $y = \dfrac{3x^2}{2}$

Solving simultaneously, we find $x = 0$, $y = 0$.
$$f_{xx} = -2x(-8x) - 2(3y - 4x^2)$$
$$f_{xy} = -6x; \ f_{yy} = 2;$$
$D = 0$, so the test is inconclusive. $z = f(0, y) = y^2$, which has a low point at the origin. Also, $z = f(x, 0) = 2x4$, which also has a low point at the origin. Thus, $(0, 0)$ is a low point.

49. a. $z = f(x, y)$ has a minimum of 0 at the origin because as we zero in on the origin, the values of f drop toward 0.

b. $z = f(x, y)$ has a saddle point at the origin. In the first and third quadrants, f approaches 0 from above, while in the second and fourth quadrants f approaches 0 from below as $(x, y) \to (0, 0)$. In the plane $y = x$, the cross section is parabola-like with a low point at $(0, 0)$. In the plane $y = -x$, the cross section is parabola-like with a high point at $(0, 0)$.

50. $D_u f = h\left(\dfrac{\partial f}{\partial x + k \dfrac{\partial f}{\partial y}}\right) = F$
$$F_x = hf_{xx} + kf_{xy}; \ F_y = kf_{yy} + hf_{xy}$$
$$D_u F = h(hf_{xx} + kf_{xy}) + k(kf_{yy} + hf_{xy})$$
$$= h^2 f_{xx} + 2hkf_{xy} + k^2 f_{yy}$$
$$= f_{xx}\left(h^2 + \frac{2hkf_{xy}}{f_{xx}} + \frac{k^2 [f_{xy}]^2}{[f_{xx}]^2}\right)$$
$$+ k^2 f_{yy} - k^2 \frac{[f_{xy}]^2}{f_x^2}$$

$$= f_{xx}\left(h + k\frac{f_{xy}}{f_{xx}}\right)^2 + k^2\left(\frac{f_{xx}f_{yy} - f_{xy}^2}{f_{xx}}\right)$$

$$= f_{xx}\left[\left(h + k\frac{f_{xy}}{f_{xx}}\right)^2 + k^2\left(\frac{f_{xx}f_{yy} - f_{xy}^2}{f_{xx}^2}\right)\right]$$

$D_u f > 0$ if $f_{xx} > 0$ and $f_{xx}f_{yy} - f_{xy} > 0$ (all other terms and factors are squares). The curve whose second derivative is positive exhibits a minimum (F is the first derivative).

$D_u f < 0$ if $f_{xx} < 0$ and $f_{xx}f_{yy} - f_{xy} > 0$ (all other terms and factors are squares). The curve whose second derivative is positive exhibits a minimum.

If $f_{xx}f_{yy} - f_{xy} < 0$, $D_u F$ is a difference of squares, which can be positive or negative. The curve whose second derivative varies in sign in the neighborhood of a point exhibits a saddle point.

51. $F_m(m\ h) = 2\sum_{k=1}^{n}[y_k - (mx_k + b)(-x_k)] = 0$

$\sum_{k=1}^{n} x_k y_k = \sum_{k=1}^{n} x_k^2 + b\sum_{k=1}^{n} x_k$ or

$\left(\sum_{k=1}^{n} x_k^2\right)m + \left(\sum_{k=1}^{n} x_k\right)b = \sum_{k=1}^{n} x_k y_k F(m,$

$h) = 2\sum_{k=1}^{n}[y_k - (mx_k + b)(-1)] = 0$

$\sum_{k=1}^{n} y_k = \left(\sum_{k=1}^{n} x_k\right)m + \sum_{k=1}^{n} 1$

$\left(\sum_{k=1}^{n} x_k^2\right)m + \left(\sum_{k=1}^{n} x_k\right)b = \sum_{k=1}^{n} x_k y_k$

Using Cramer's rule, and

$D = n\sum_{k=1}^{n} x^2 - \left(\sum_{k=1}^{n} x\right)^2$

$N_m = n\sum_{k=1}^{n} xy - \sum_{k=1}^{n} x\sum_{k=1}^{n} y$

$N_b = \sum_{k=1}^{n} y\sum_{k=1}^{n} x^2 - \sum_{k=1}^{n} x\sum_{k=1}^{n} xy$

$m = \dfrac{n\sum_{k=1}^{n} x_k y_k - \left(\sum_{k=1}^{n} x_k\right)\left(\sum_{k=1}^{n} y_k\right)}{n\sum_{k=1}^{n} x_k^2 - \left(\sum_{k=1}^{n} x_k\right)^2}$ and

$b = \dfrac{\sum_{k=1}^{n} x_k^2 \sum_{k=1}^{n} y_k - \left(\sum_{k=1}^{n} y_k\right)\left(\sum_{k=1}^{n} x_k y_k\right)}{n\sum_{k=1}^{n} x_k^2 - \left(\sum_{k=1}^{n} x_k\right)^2}$

52. Let the equation of the plane be $y = k_1 x_1 + k_2 x_2$. The (vertical) distance from the actual data point y_k and the corresponding point on the plane is

$d_k = y_k - (k_1 x_1 + k_2 x_2)$

By the method of least squares,

$f(k_1, k_2) = \sum_{k=1}^{n}[y_k - (k_1 x_1 + k_2 x_2)]$

$\dfrac{\partial f}{\partial k_1} = 2\sum_{k=1}^{n}[y_k - (k_1 x_1 + k_2 x_2)](-x_1) = 0$

if $\sum_{k=1}^{n} x_1 y_k = k_1 \sum_{k=1}^{n} x_1^2 + k_2 \sum_{k=1}^{n} x_1 x_2$

$\dfrac{\partial f}{\partial k_2} = 2\sum_{k=1}^{n}[y_k - (k_1 x_1 + k_2 x_2)](-x_2) = 0$

if $\sum_{k=1}^{n} x_1 y_k = k_1 \sum_{k=1}^{n} x_1 x_2 + k_2 \sum_{k=1}^{n} x_2^2$

Use Cramer's rule and

$D = \left(\sum_{k=1}^{n} x_1^2\right)\left(\sum_{k=1}^{n} x_2^2\right) - \left(\sum_{k=1}^{n} x_1 x_2\right)^2$

$N_{k_1} = \left(\sum_{k=1}^{n} x_1 y_k\right)\left(\sum_{k=1}^{n} x_2^2\right)$
$\qquad - \left(\sum_{k=1}^{n} x_1 x_2\right)$

$N_b = \left(\sum_{k=1}^{n} x_1^2\right)\left(\sum_{k=1}^{n} x_2 y_k\right)$
$\qquad - \left(\sum_{k=1}^{n} x_1 x_2\right)\left(\sum_{k=1}^{n} x_1 y_k\right)$

$k_1 = \dfrac{\left(\sum_{k=1}^{n} x_1 y_k\right)\left(\sum_{k=1}^{n} x_2^2\right) - \left(\sum_{k=1}^{n} x_1 x_2\right)\left(\sum_{k=1}^{n} x_2 y_k\right)}{\left(\sum_{k=1}^{n} x_1^2\right)\left(\sum_{k=1}^{n} x_2^2\right) - \left(\sum_{k=1}^{n} x_1 x_2\right)^2}$

$k_2 = \dfrac{\left(\sum_{k=1}^{n} x_1^2\right)\left(\sum_{k=1}^{n} x_2 y_k\right) - \left(\sum_{k=1}^{n} x_1 x_2\right)\left(\sum_{k=1}^{n} x_1 y_k\right)}{\left(\sum_{k=1}^{n} x_1^2\right)\left(\sum_{k=1}^{n} x_2^2\right) - \left(\sum_{k=1}^{n} x_1 x_2\right)^2}$

For the given example, $y = 1.05 x_1 + 1.46 x_2$.

12.8 Lagrange Multipliers, Page 813

1. $f(x, y) = xy;\ g(x, y) = 2x + 2y - 5$

$f_x = y,\ f_y = x,\ g_x = 2,\ g_y = 2$

Solve the system

$$\begin{cases} y = 2\lambda \\ x = 2\lambda \\ 2x + 2y = 5 \end{cases}$$

to find $x = y = \frac{5}{4}$

$f(\frac{5}{4}, \frac{5}{4}) = \frac{25}{16}$ is the constrained maximum.

2. $f(x, y) = xy;\ g(x, y) = x + y - 20$

$f_x = y,\ f_y = x,\ g_x = 1,\ g_y = 1$

Solve the system

$$\begin{cases} y = \lambda \\ x = \lambda \\ x + y = 20 \end{cases}$$

to find $x = y = 10$

$f(10,\ 10) = 100$ is the constrained maximum.

3. $f(x,\ y) = 16 - x^2 - y^2;\ g(x,\ y) = x + 2y - 6$

$f_x = -2x,\ f_y = -2y,\ g_x = 1,\ g_y = 2$

Solve the system

$$\begin{cases} -2x = \lambda \\ -2y = 2\lambda \\ x + 2y = 6 \end{cases}$$

to find $x = \frac{6}{5},\ y = \frac{12}{5}$

$f(\frac{6}{5},\ \frac{12}{5}) = \frac{44}{5}$ is the constrained maximum.

4. $f(x,\ y) = x^2 + y^2;\ x + y - 24$

$f_x = 2x,\ f_y = -2y,\ g_x = 1,\ g_y = 1$

Solve the system

$$\begin{cases} 2x = \lambda \\ 2y = \lambda \\ x + y = 24 \end{cases}$$

to find $x = y = 12$

$f(12,\ 12) = 288$ is the constrained minimum.

5. $f(x,\ y) = x^2 + y^2;\ g(x,\ y) = xy - 1$

$f_x = 2x,\ f_y = 2y,\ g_x = y,\ g_y = x$

Solve the system

$$\begin{cases} 2x = \lambda y \\ 2y = \lambda x \\ xy = 1 \end{cases}$$

to find $x = y = 1$ and $x = y = -1$

$f(\pm 1,\ \pm 1) = 2$ is the constrained minimum.

6. $f(x,\ y) = x^2 - xy + 2y^2;\ g(x,\ y) = 2x + y - 22$

$f_x = 2x - y,\ f_y = -2x + 4y,\ g_x = 2,\ g_y = 1$

Solve the system

$$\begin{cases} 2x - y = 2\lambda \\ -x + 4y = \lambda \\ 2x + y = 22 \end{cases}$$

to find $x = 9,\ y = 4$

$f(9,\ 4) = 77$ is the constrained minimum.

7. $f(x,\ y) = x^2 - y^2;\ g(x,\ y) = x^2 + y^2 - 4$

$f_x = 2x,\ f_y = -2y,\ g_x = 2x,\ g_y = 2y$

Solve the system

$$\begin{cases} 2x = 2\lambda x \\ -2y = 2\lambda y \\ x^2 + y^2 = 4 \end{cases}$$

to find $x = 0,\ y = \pm 2$

$f(0,\ \pm 2) = -4$ is the constrained minimum.

8. $f(x,\ y) = x^2 - 2y - y^2;\ g(x,\ y) = x^2 + y^2 - 1$

$f_x = 2x,\ f_y = -2 - 2y,\ g_x = 2x,\ g_y = 2y$

Solve the system

$$\begin{cases} 2x = 2\lambda x \\ -2 - 2y = 2\lambda y \\ x^2 + y^2 = 1 \end{cases}$$

to find $x = \pm\frac{\sqrt{3}}{2},\ y = -\frac{1}{2}$

$f(\pm\frac{\sqrt{3}}{2},\ -\frac{1}{2}) = \frac{3}{2}$ is the constrained maximum.

9. $f(x,\ y) = \cos x + \cos y;\ g(x,\ y) = y - x - \frac{\pi}{4}$

$f_x = -\sin x,\ f_y = -\sin y,\ g_x = -1,\ g_y = 1$

Solve the system

$$\begin{cases} -\sin x = -\lambda \\ -\sin y = \lambda \\ y + x = \frac{\pi}{4} \end{cases}$$

to find $x = \dfrac{(8n - 1)\pi}{8},\ y = \dfrac{(8n + 1)\pi}{8}$

If $n = 1$, $f(\frac{7\pi}{8},\ \frac{9\pi}{8}) \approx -1.8478$;

If $n = 2$, $f(\frac{15\pi}{8},\ \frac{17\pi}{8}) \approx 1.8478$;

The constrained maximum is approximately 1.8478 for $f(-\frac{\pi}{8} + n\pi,\ \frac{\pi}{8} + n\pi)$ with n even

10. $f(x,\ y,\ z) = xyz;\ g(x,\ y,\ z) = 3x + 2y + z - 6$

$f_x = yz,\ f_y = xz,\ f_z = xy,\ g_x = 3,\ g_y = 2,\ g_z = 1$

Solve the system

$$\begin{cases} yz = 3\lambda \\ xz = 2\lambda \\ xy = \lambda \\ 3x + 2y + z = 6 \end{cases}$$

to find $x = \frac{2}{3}$, $y = 1$, $z = 2$

$f(\frac{2}{3}, 1, 2) = \frac{4}{3}$ is the constrained maximum.

11. $f(x, y, z) = x^2 + y^2 + z^2$;

$g(x, y, z) = x - 2y + 3z - 4$

$f_x = 2x$, $f_y = 2y$, $f_z = 2z$, $g_x = 1$, $g_y = -2$,

$g_z = 3$

Solve the system

$$\begin{cases} 2x = \lambda \\ 2y = -2\lambda \\ 2z = 3\lambda \\ x - 2y + 3z = 4 \end{cases}$$

to find $x = \frac{2}{7}$, $y = -\frac{4}{7}$, $z = \frac{6}{7}$

$f(\frac{2}{7}, -\frac{4}{7}, \frac{6}{7}) = \frac{8}{7}$ is the constrained minimum.

12. $f(x, y, z) = x^2 + y^2 + z^2$;

$g(x, y, z) = 4x^2 + 2y^2 + z^2 - 4$

$f_x = 2x$, $f_y = 2y$, $f_z = 2z$, $g_x = 8x$, $g_y = 4y$,

$g_z = 2z$

Solve the system

$$\begin{cases} 2x = 8\lambda x \\ 2y = 4\lambda y \\ 2z = 2\lambda z \\ x - 2y + 3z = 4 \end{cases}$$

to find solutions $(\pm 1, 0, 0)$, $(0, \pm\sqrt{2}, 0)$, $(0, 0, \pm 2)$; testing each of these leads us to $f(\pm 1, 0, 0) = 1$ is the constrained minimum and $f(0, 0 \pm 2) = 4$.

13. $f(x, y, z) = 2x^2 + 4y^2 + z^2$

$g(x, y, z) = 4x - 8y + 2z - 10$

$f_x = 4x$, $f_y = 8y$, $f_z = 2z$, $g_x = 4$, $g_y = -8$,

$g_z = 2$

Solve the system

$$\begin{cases} 4x = 4\lambda \\ 8y = -8\lambda \\ 2z = 2\lambda \\ 4x - 8y + 2z = 10 \end{cases}$$

to find $x = \frac{5}{7}$, $y = -\frac{5}{7}$, $z = \frac{5}{7}$

$f(\frac{5}{7}, -\frac{5}{7}, \frac{5}{7}) = \frac{25}{7}$ is the constrained minimum.

By using negative values for any two variables in the constraint equation, the third variable can be made arbitrarily large, thus f can be made arbitrarily large and does not have a maximum.

14. $f(x, y, z) = x^2 y^2 z^2$;

$g(x, y, z) = x^2 + y^2 + z^2 - R^2$

$f_x = 2xy^2 z^2$, $f_y = 2x^2 yz^2$, $f_z = 2x^2 y^2 z$,

$g_x = 2x$, $g_y = 2y$, $g_z = 2z$

Solve the system

$$\begin{cases} 2xy^2 z^2 = 2\lambda x \\ 2x^2 yz^2 = 2\lambda y \\ 2x^2 y^2 z = 2\lambda z \\ x^2 + y^2 + z^2 = R^2 \end{cases}$$

to find $x = y = z = \pm\dfrac{R}{\sqrt{3}}$

$f(\dfrac{R}{\sqrt{3}}, \dfrac{R}{\sqrt{3}}, \dfrac{R}{\sqrt{3}}) = \dfrac{R^6}{27}$ is the constrained minimum.

15. $f(x, y, z) = x - y + z$;

$g(x, y, z) = x^2 + y^2 + z^2 - 100$

$f_x = 4$, $f_y = -1$, $f_z = 1$, $g_x = 2x$, $g_y = 2y$,

$g_z = 2z$

Solve the system

$$\begin{cases} 1 = 2\lambda x \\ -1 = 2\lambda y \\ 1 = 2\lambda z \\ x^2 + y^2 + z^2 = 100 \end{cases}$$

to find $x = z = \pm\dfrac{10}{\sqrt{3}}$, $y = \mp\dfrac{10}{\sqrt{3}}$

$f(\dfrac{10}{\sqrt{3}}, -\dfrac{10}{\sqrt{3}}, \dfrac{10}{\sqrt{3}}) = \dfrac{30}{\sqrt{3}} = 10\sqrt{3} \approx 17.3$

is the constrained maximum.

16. $f(x, y, z) = x - y + z;$

$g(x, y, z) = x^2 + y^2 + z^2 - 100$

$f_x = 1, f_y = -1, f_z = 1, g_x = 2x, g_y = 2y,$

$g_z = 2z$

Solve the system

$$\begin{cases} 1 = 2\lambda x \\ -1 = 2\lambda y \\ 1 = 2\lambda z \\ x^2 + y^2 + z^2 = 100 \end{cases}$$

to find $x = z = \pm\dfrac{10}{\sqrt{3}}, y = \mp\dfrac{10}{\sqrt{3}}$

$f\left(-\dfrac{10}{\sqrt{3}}, \dfrac{10}{\sqrt{3}}, -\dfrac{10}{\sqrt{3}}\right) = -\dfrac{30}{\sqrt{3}} = -10\sqrt{3}$

≈ -17.3 is the constrained minimum.

17. The square of the distance is

$f(x, y, z) = x^2 + y^2 + z^2;$

$g(x, y, z) = Ax + By + Cz = 0$

$f_x = 2x, f_y = 2y, f_z = 2z, g_x = A, g_y = B,$

$g_z = C$

Solve the system

$$\begin{cases} 2x = A\lambda \\ 2y = B\lambda \\ 2z = C\lambda \\ Ax + By + Cz = 0 \end{cases}$$

to find $x = y = z = 0$

$f(0, 0, 0) = 0$ is the constrained minimum.

18. The square of the distance is $f(x, y) = x^2 + y^2$

$g(x, y) = 5x^2 - 6xy + 5y^2 - 4$

$f_x = 2x, f_y = 2y, g_x = 10x - 6y,$

$g_y = -6x + 10y$

Solve the system

$$\begin{cases} 2x = \lambda(10x - 6y) \\ 2y = \lambda(-6x + 10y) \\ 5x^2 - 6xy + 5y^2 = 4 \end{cases}$$

to find $x = \pm\frac{1}{2}, \pm 1, y = \pm\frac{1}{2}, \pm 1$

$f(\pm\frac{1}{2}, \pm\frac{1}{2}) = \frac{1}{2}; d = \dfrac{\sqrt{2}}{2}$ is the constrained

minimum distance. $f(\pm 1, \pm 1) = 2;$

$d = \sqrt{2}$ is the constrained maximum distance.

19. The square of the distance is
$f(x, y, z) = x^2 + y^2 + z^2;$
$g(x, y, z) = 2x + y + z - 1$

$f_x = 2x, f_y = 2y, f_z = 2z, g_x = 2, g_y = 1,$

$g_z = 1$

Solve the system

$$\begin{cases} 2x = 2\lambda \\ 2y = \lambda \\ 2z = \lambda \\ 2x + y + z = 1 \end{cases}$$

to find $x = \frac{1}{3}, y = \frac{1}{6}, z = \frac{1}{6}$

$f(\frac{1}{3}, \frac{1}{6}, \frac{1}{6}) = \frac{1}{6}$ is the constrained minimum

and the nearest point is $(\frac{1}{3}, \frac{1}{6}, \frac{1}{6})$

20. $f(x, y, z) = xyz; g(x, y, z) = x + y + z - 24$

$f_x = yz, f_y = xz, f_z = xy, g_x = 1, g_y = 1,$

$g_z = 1$

Solve the system

$$\begin{cases} yz = \lambda \\ xy = \lambda \\ xz = \lambda \\ x + y + z = 24 \end{cases}$$

to find $x = y = z = 8$

The largest product is $f(8, 8, 8) = 512$

21. $f(x, y, z) = xy^2z; g(x, y, z) = x + y + z - 12$

$f_x = y^2z, f_y = 2xyz, f_z = xy^2, g_x = 1, g_y = 1,$

$g_z = 1$

Solve the system

$$\begin{cases} y^2z = \lambda \\ xy = \lambda \\ 2xyz = \lambda \\ x + y + z = 12 \end{cases}$$

to find $x = z = 3, y = 6$

The largest product is $f(3, 6, 3) = 324$

22. Let $x, y,$ and z be the length, width, and
height of the rectangular box, respectively.

$f(x, y, z) = xyz;$

$g(x, y, z) = xy + 2xz + 2yz - 96$

$f_x = yz, f_y = xz, f_z = xy, g_x = y + 2z,$

$g_y = x + 2z, g_z = 2x + 2y$

Solve the system

$$\begin{cases} yz = \lambda(y + 2z) \\ xz = \lambda(x + 2z) \\ xy = \lambda(2x + 2y) \\ xy + 2xz + 2yz = 96 \end{cases}$$

to find $x = y = 4\sqrt{2}, z = 2\sqrt{2}$

The maximum volume is $f(4\sqrt{2}, 4\sqrt{2}, 2\sqrt{2})$

$= 64\sqrt{2}$ cubic units.

23. $T(x, y, z) = 100 - xy - xz - yz;$

$g(x, y, z) = x + y + z - 10;$

$T_x = -y - z, T_y = -x - z, T_z = -y - x,$

$g_x = 1, g_y = 1, g_z = 1$

$$\begin{cases} y + z = \lambda \\ x + z = \lambda \\ x + y = \lambda \\ x + y + z = 10 \end{cases}$$

to find $x = y = z = \frac{10}{3}$

The lowest temperature is $T(\frac{10}{3}, \frac{10}{3}, \frac{10}{3}) = \frac{200}{3}$.

24. Let x (along the river) and y be the sides of the rectangle.

$f(x, y) = x + 2y, g(x, y) = xy - 3,200$

$f_x = 1, f_y = 2, g_x = y, g_y = x$

Solve the system

$$\begin{cases} 1 = \lambda y \\ 2 = \lambda x \\ xy = 3,200 \end{cases}$$

to find $x = 80, y = 40$

The least amount of fencing will be used when there is 80 yd along the river and 40 yd on the sides.

25. $f(x, y) = xy; g(x, y) = 2x + 2y - 320$
$y = \lambda, \quad x = \lambda, \quad x + y = 160, \quad x = y = 80.$
The maximum area is 6,400 sq yd with a square field 80 by 80 yd.

26. Let x and y be the radius and height of the cylinder. $f(x, y) = 2\pi x^2 + 2\pi xy;$
$g(x, y) = \pi x^2 y - 6.89\pi.$

$f_x = 4\pi x + 2\pi y, f_y = 2\pi x, g_x = 2\pi xy, g_y = \pi x^2$

Solve the system

$$\begin{cases} 4\pi x + 2\pi y = 2\lambda\pi xy \\ 2\pi x = \lambda\pi x^2 \\ x^2 y = 6.89 \end{cases}$$

to find $x \approx 1.51, y \approx 3.02$. A real can of Pepsi has a radius of about 1.125 in. (as compared to 1.51) and a height of 4.5 in. (as compared with 3.02).

27. Let x and y be the radius and height of the cylinder. The cost is
$f(x, y) = 2(2\pi x^2) + 2\pi xy;$
$g(x, y) = \pi x^2 y - 4\pi$

$f_x = 8\pi x + 2\pi y, f_y = 2\pi x, g_x = 2\pi xy, g_y = \pi x^2$

Solve the system

$$\begin{cases} 8\pi x + 2\pi y = 2\lambda\pi xy \\ 2\pi x = \lambda\pi x^2 \\ x^2 y = 4 \end{cases}$$

to find the radius $x \approx 1$ in. and the height $y \approx 4$ in.

28. Let x, y, and z be the length, width, and height of one-eight of the rectangular box, respectively.

$f(x, y, z) = xyz$

$g(x, y, z) = x^2 + \frac{y^2}{4} + \frac{z^2}{9} - 1$

$f_x = yz, f_y = xz, f_z = xy, g_x = 2x,$

$g_y = \frac{y}{2}, g_z = \frac{2z}{9}$

Solve the system

$$\begin{cases} yz = 2\lambda x \\ xz = \frac{\lambda y}{2} \\ xy = \frac{2\lambda z}{9} \\ x^2 + \frac{y^2}{4} + \frac{z^2}{9} = 1 \end{cases}$$

to find $x = \frac{1}{\sqrt{3}}, y = \frac{2}{\sqrt{3}}, z = \frac{3}{\sqrt{3}}$

The maximum volume is $f(\frac{1}{\sqrt{3}}, \frac{2}{\sqrt{3}}, \frac{3}{\sqrt{3}})$

$= \frac{2}{\sqrt{3}}$ cubic units.

29. $f(x, y) = 50x^{1/2}y^{3/2}$; $g(x, y) = x + y - 8$

$f_x = 25x^{-1/2}y^{3/2}$, $f_y = 75x^{1/2}y^{1/2}$,

$g_x = 1$, $g_y = 1$

Solve the system

$$\begin{cases} 25x^{-1/2}y^{3/2} = \lambda \\ 75x^{1/2}y^{1/2} = \lambda \\ x + y = 4 \end{cases}$$

to find $x = 2$, $y = 6$. \$2,000 to development and \$6,000 to promotion gives the maximum sales of $f(2, 6) = 50\sqrt{2}(6\sqrt{6}) = 600\sqrt{3}$ units.

30. a. $Q(x, y) = 60x^{1/3}y^{2/3}$;

$g(x, y) = x + y - 120$

$Q_x = 20x^{-2/3}y^{2/3}$; $Q_y = 40x^{1/3}y^{-1/3}$

$g_x = 1$, $g_y = 1$. Hence the three Lagrange

equations are $20x^{-2/3}y^{2/3} = \lambda$,

$40x^{1/3}y^{-1/3} = \lambda$, and $x + y = 120$.

The solution to this system of equations is $x = 40$, $y = 80$. Hence, to generate maximal output, \$40,000 should be spent on labor and \$80,000 should be spent on equipment.

b. It is known that the maximal output will change by λ per unit change of constraint. Substituting in the first Lagrange equation shows $20(40)^{-2/3}(80)^{2/3}$ = 31.75, which implies that the maximal output will increase by approximately \$31,500 if the available money is increased by \$1,000 and allocated optimally.

31. $f(x, y) = 500 - x^2 - 2y^2$

$g(x, y) = 20x + 12y - 236$

$f_x = -2x$, $f_y = -4y$, $g_x = 20$, $g_y = 12$.

Hence the three Lagrange equations are

$-2x = 20\lambda$, $-4y = 12\lambda$, and $20x + 12y = 236$.

Solving, we find $x = 10$, $y = 3$. Thus, the farmer should apply 10 acre-ft of water and 3 acre-ft of fertilizer to maximize the yield.

32. From Problem 31, it is known that the maximal output will change by λ per unit change of constraint. Substituting in the first Lagrange equations shows $\lambda = -x/10 = -1$, which implies that the maximum yield will decrease by 100 units.

33. Maximize $V = xyz$ subject to $2x + 2y + z = 108$.

$f(x, y) = x^2y$; $g(x, y) = 4x + y - 108$.

$f_x = 2xy$, $f_y = x^2$, $g_x = 4$, $g_y = 1$. Hence the three Lagrange equations are $2xy = 4\lambda$, $x^2 = \lambda$, and $4x + y = 108$. Solving, we find $x = y = 18$, $z = 36$. The maximum volume is

$V(18, 18, 36) = 11,664$ cu in or 6.75 cu ft.

34. $P(x, y, z) = \sin x \sin y \sin z$;

$g(x, y, z) = x + y + z - \pi$

$P_x = \cos x \sin y \sin z$; $P_y = \sin x \cos z \sin z$

$P_z = \sin x \sin y \cos z$; $g_x = 1$, $g_y = 1$, $g_z = 1$.

The Lagrange equations are $\cos x \sin y \sin z$

$= \lambda$, $\sin x \cos y \sin z = \lambda$, $\sin x \sin y \cos z$

$= \lambda$, $x + y + z = \pi$. Solving gives $x = y = z$

$= \pi/3$. The maximum product is

$f(\frac{\pi}{3}, \frac{\pi}{3}, \frac{\pi}{3}) = \dfrac{3\sqrt{3}}{8}$.

For $Q(x, y, z) = \cos x \cos y \cos z$;

$g(x, y, z) = x + y + z - \pi = 0$

$Q_x = -\sin x \cos y \cos z$; $Q_y = -\cos x \sin z$

$\cos z$, $Q_z = -\cos x \cos y \sin z$; $g_x = 1$,

$g_y = 1$, $g_z = 1$. The Lagrange equations are

$-\sin x \cos y \cos z = \lambda$, $-\cos x \sin y \cos z$

$= \lambda$, $-\cos x \cos y \sin z = \lambda$, $x + y + z = \pi$.

Solving gives $x = y = z = \pi/3$. The

maximum product is $f(\frac{\pi}{3}, \frac{\pi}{3}, \frac{\pi}{3}) = \dfrac{1}{8}$.

35. $f(x, y, z) = x^2 + y^2 + z^2$;

$g(x, y, z) = x + y - 4$

$h(x, y, z) = y + z - 6$

$f_x = 2x$, $f_y = 2y$, $f_z = 2z$

$g_x = 1$, $g_y = 1$, $g_z = 0$

$h_x = 0$, $h_y = 1$, $h_z = 1$

Solve the system

$$\begin{cases} 2x = \lambda \\ 2y = \lambda + \mu \\ 2y = \mu \\ x + y = 4 \\ y + z = 6 \end{cases}$$

to find $x = 2/3$, $y = 10/3$, $z = 8/3$

The minimum is $f(\frac{2}{3}, \frac{10}{3}, \frac{8}{3}) = \frac{56}{3}$

36. $f(x, y, z) = xyz$

$g(x, y, z) = x^2 + y^2 - 3$

$h(x, y, z) = y - 2z$

$f_x = yz$, $f_y = xz$, $f_z = xy$

$g_x = 2x$, $g_y = 2y$, $g_z = 0$

$h_x = 0$, $h_y = 1$, $h_z = -2$

Solve the system

$$\begin{cases} xy = 2\lambda x \\ xz = 2\lambda y + \mu \\ xy = -2\mu \\ y = 2z \\ x^2 + y^2 = 3 \end{cases}$$

to find $x = \pm 1$, $y = \pm\sqrt{2}$, $z = \pm 1/\sqrt{2}$

The maximum is $\left| f\left(\pm 1, \pm \sqrt{2}, \pm \dfrac{1}{\sqrt{2}} \right) \right| = 1$

37. $f(x, y, z) = xy + xz$

$g(x, y, z) = 2x + 3z - 5$

$h(x, y, z) = xy - 4$

$f_x = y + z, f_y = x, f_z = x$

$g_x = 2x, g_y = 0, g_z = 3$

$h_x = y, h_y = x, h_z = 0$

Solve the system

$$\begin{cases} y + z = 2\lambda + \mu y \\ x = \mu x \\ x = 3\lambda \\ 2x + 3z = 5 \\ xy = 4 \end{cases}$$

to find $x = 5/4$, $y = 16/5$, $z = 5/6$

The maximum is $f(\frac{5}{4}, \frac{16}{5}, \frac{5}{6}) = \frac{121}{24}$

38. $f(x, y, z) = 2x^2 + 3y^2 + 4z^2$;

$g(x, y, z) = x + y + z - 4$

$h(x, y, z) = x - 2y + 5z - 3$

$f_x = 4x, f_y = 6y, f_z = 8z$

$g_x = 1, g_y = 1, g_z = 1$

$h_x = 1, h_y = -2, h_z = 5$

Solve the system

$$\begin{cases} 4x = \lambda + \mu \\ 6y = \lambda - 2\mu \\ 8z = \lambda + 5\mu \\ x + y + z = 4 \\ x - 2y + 5z = 3 \end{cases}$$

to find $x = 24/13$, $y = 125/637$, $z = 71/13$

The minimum is $f(\frac{24}{13}, \frac{125}{637}, \frac{71}{13}) = \frac{3,940,507}{31,213}$

≈ 126.25.

39. a. $P(x, y) = \left(\dfrac{320y}{y + 2} + \dfrac{160x}{x + 4} \right)(150 - 50)$

$\qquad\qquad - 1,000x - 1,000y$

$g(x, y) = x + y - 8$

$P_x = -\dfrac{100(160)(4)}{(x + 4)^2} - 1,000$

$P_y = -\dfrac{100(320)(2)}{(y + 2)^2} - 1,000$

$g_x = 1, g_y = 1$

Since $P_x = P_y = \lambda$,

$-\dfrac{100(160)(4)}{(x + 4)^2} - 1,000$

$= -\dfrac{100(320)(2)}{(y + 2)^2} - 1,000$

or $x + 4 = \pm(y + 2)$

Reject the negative solution as leading to negative spending. Substituting in the constraint equation leads to $x = 3$

thousand dollars for development and $y = 5$ thousand dollars for promotion.

b. $\lambda = P_y = -\dfrac{64,000}{49} - 1,000 \approx 306.122$

(for each \$1,000). Since the change in this promotion/development is \$100, the corresponding change in profit is \$30.61. Remember that the Lagrange multiplier is the change in maximum profit for a 1 (thousand) dollar change in the constraint.

c. To maximize the profit when unlimited funds are available maximize $P(x, y)$ without constraints. To do this, find the critical points by setting $P_x = 0$ and $P_y = 0$, that is

P_x: $\qquad \dfrac{64}{(x + 4)^2} - 1 = 0$

$\qquad\qquad (x + 4)^2 = 64$

$\qquad\qquad\qquad x = 4 \quad$ and

P_y: $\qquad \dfrac{64}{(y + 2)^2} - 1 = 0$

$\qquad\qquad (y + 2)^2 = 64$

$\qquad\qquad\qquad y = 6$

Thus, \$4,000 should be spent on development and \$6,000 should be spent on promotion to maximize profit.

d. If there were a restriction on the amount spent on development and promotion, then constraints would be $g(x, y) = x + y = k$ for some positive constant k. The corresponding Lagrange equations would be

$\dfrac{64}{(x + 4)^2} - 1 = \lambda; \dfrac{64}{(y + 2)^2} - 1 = \lambda;$

and $x + y = k$. To obtain the answer in part a, let $\lambda = 0$. Beginning with the Lagrange equations from part b, set $\lambda = 0$ to obtain $64/(x + 4)^2 - 1 = 0$ or $x = 4$, and $64/(y + 2)^2 - 1 = 0$ or $y = 6$, just as we found in part c.

40. Let h be the height of the cylinder and H the height of the cone. R is the radius.

$S = 2\pi Rh + \pi R\sqrt{h^2 + R^2}$; minimize S

subject to $V = \pi R^2 h + \frac{1}{3}\pi R^2 H - K$. The

Lagrange equations are

$S_R = 2\pi h + \pi\sqrt{R^2 + H^2} + \pi\left(\dfrac{R^2}{\sqrt{H^2 + R^2}} \right)$

$\qquad = \lambda(2\pi Rh + \frac{2}{3}\pi RH) = \lambda V_R$

$S_h = 2\pi = \lambda(\pi R^2) = \lambda V_h$

$\pi R\left(\dfrac{H}{\sqrt{h^2 + R^2}}\right) = \lambda(\tfrac{1}{3})\pi R^2 = \lambda V_H$

Solving these equations simultaneously,

$h = \dfrac{R}{\sqrt{5}},\ H = \dfrac{2R}{\sqrt{5}},\ R = \sqrt[3]{\dfrac{3K}{\pi\sqrt{5}}}$

Thus, the minimum surface area is

$$S = 2\pi R\left(\dfrac{R}{\sqrt{5}}\right) + \pi R\sqrt{\dfrac{4R^2}{5} + R^2}$$

$$= \dfrac{2\pi\sqrt{5}}{5}R^2 + \pi R^2\left(\dfrac{3R}{\sqrt{5}}\right)$$

$$= \sqrt{5}\pi R^2$$

$$= \sqrt{5}\pi\left(\dfrac{3K}{\sqrt{5}\pi}\right)^{2/3}$$

41. Let x, y, z be the length, width, and height of one-eighth of the rectangular box, respectively.

$f(x, y, z) = xyz$

$g(x, y, z) = \dfrac{x^2}{a^2} + \dfrac{y^2}{b^2} + \dfrac{z^2}{c^2} - 1$

$f_x = yz,\ f_y = xz,\ f_z = xy$

$g_x = \dfrac{2x}{a^2},\ g_y = \dfrac{2y}{b^2},\ g_z = \dfrac{2z}{c^2}$

The Lagrange equations are:

$yz = \dfrac{2x}{a^2}\lambda,\ \ xz = \dfrac{2y}{b^2}\lambda,\ \ xy = \dfrac{2z}{c^2}\lambda$

Solving simultaneously,

$x = \dfrac{a}{\sqrt{3}},\ y = \dfrac{b}{\sqrt{3}},\ z = \dfrac{c}{\sqrt{3}}$

The maximum volume is

$f\left(\dfrac{a}{\sqrt{3}}, \dfrac{b}{\sqrt{3}}, \dfrac{c}{\sqrt{3}}\right) = \dfrac{abc}{3\sqrt{3}}$ cubic units

42. Let x be the number of units of labor and y the number of units of capital. With the unit cost of labor and capital p and q respectively, the cost is $C(x, y) = px + qy$. The goal is to minimize cost subject to the fixed production function $f(x, y) = Q(x, y) = c$. Since $C_x = p$ and $C_y = q$, the three Lagrange equations are $p = \lambda f_x$, $q = \lambda f_y$, and $Q(x, y) = c$. Solving this system leads to

$\dfrac{f_x}{p} = \dfrac{f_y}{q}$

43. Let $Q(x, y)$ be the production level curve subject to $px + qy = k$. The three Lagrange equations then are $Q_x = \lambda p$, $Q_y = \lambda q$, and $px + qy - k = 0$. From the first two equations

$\dfrac{Q_x}{p} = \dfrac{Q_y}{q}$

44. With $Q(x, y) = cx^\alpha y^\beta$ and $g(x, y) = px + qy - k$

$Q_x = \alpha x^{\alpha - 1},\ Q_y = \beta x^\alpha y^{\beta - 1}$,

$g_x = p,\ g_y = q$

The three Lagrange equations are:

$\alpha x^{\alpha - 1} y^\beta = \lambda p,\ \beta x^\alpha y^{\beta - 1} = \lambda q,\ px + qy = k$

Solving this system leads to

$x = \dfrac{k}{p(\alpha + \beta p)/\alpha} = \dfrac{k\alpha}{p};\ y = \dfrac{\beta}{\alpha}\dfrac{p}{q}\dfrac{k\alpha}{p} = \dfrac{k\beta}{q}$

If $\alpha + \beta \neq 1$, then

$x = \dfrac{k\alpha}{p(\alpha + \beta)};\ y = \dfrac{k\beta}{q(\alpha + \beta)}$

If $\Delta k = 1$, $dQ/dk = \lambda$,

$\lambda = \dfrac{\alpha x^{\alpha - 1} y^\beta}{p} = \dfrac{\alpha}{p}\dfrac{(k\alpha)^{\alpha - 1}}{p^{\alpha - 1}}\dfrac{(k\beta)^\beta}{p^\beta} = \dfrac{\alpha^\alpha \beta^\beta}{kp}$

45. With $P = Ax^\alpha y^\beta - k;\ \alpha + \beta = 1$, and

$C(x, y) = px + qy$

$P_x = \lambda A\alpha x^{\alpha - 1} y^\beta,\ P_y = \lambda A\beta x^\alpha y^{\beta - 1}$

$C_x = p,\ C_y = q$

The three Lagrange equations are

$p = \alpha Ax^{\alpha-1} y^\beta \lambda,\ \ q = \beta Ax^\alpha y^{\beta-1}\lambda$, and

$Ax^\alpha y^\beta = k$. Solving this system, we find

$x = \dfrac{k}{A}\left(\dfrac{\alpha q}{\beta p}\right)^\beta \qquad y = \dfrac{k}{A}\left(\dfrac{\beta p}{\alpha q}\right)^\alpha$

CHAPTER 12 REVIEW

Proficiency Examination, Page 815

1. A function of two variables is a rule that assigns to each ordered pair (x, y) in a set D a unique number $f(x, y)$.

2. The set D in the answer to Problem 1 is called the domain of the function, and the corresponding values of $f(x, y)$ constitute the range of f.

3. The plane $z = C$ intersects the surface $z = f(x, y)$, the result is the space curve with the equation $f(x, y) = C$. Such an intersection is called the trace of the graph of f in the plane $z = C$. The set of points (x, y) in the xy-plane that satisfy $f(x, y) = C$ is called the level curve of f at C, and an entire family of level curves (or contour curves) is generated as C varies over the range of f.

4. The notation

$$\lim_{(x,\,y)\to(x_0,\,y_0)} f(x,\,y) = L$$

means that the functional values $f(x,\,y)$ can be made arbitrarily close to L by choosing a point $(x,\,y)$ sufficiently close (but not equal) to the point $(x_0,\,y_0)$. In other words, given some $\epsilon > 0$, we wish to find a $\delta > 0$ so that for any point $(x,\,y)$ in the punctured disk of radius δ centered at $(x_0,\,y_0)$, the functional value $f(x,\,y)$ lies between $L + \epsilon$ and $L - \epsilon$.

5. Suppose $\lim\limits_{(x,\,y)\to(x_0,\,y_0)} f(x,\,y) = L$ and

$$\lim_{(x,\,y)\to(x_0,\,y_0)} g(x,\,y) = M$$

Then, for a constant a,

a. $\lim\limits_{(x,\,y)\to(x_0,\,y_0)} [af](x,\,y) = aL$

b. $\lim\limits_{(x,\,y)\to(x_0,\,y_0)} [f + g](x,\,y)$

$$= \left[\lim_{(x,\,y)\to(x_0,\,y_0)} f(x,\,y) \right]$$

$$+ \left[\lim_{(x,\,y)\to(x_0,\,y_0)} g(x,\,y) \right] = L + M$$

c. $\lim\limits_{(x,\,y)\to(x_0,\,y_0)} [fg](x,\,y)$

$$= \left[\lim_{(x,\,y)\to(x_0,\,y_0)} f(x,\,y) \right]$$

$$\times \left[\lim_{(x,\,y)\to(x_0,\,y_0)} g(x,\,y) \right]$$

$$= LM$$

d. $\lim\limits_{(x,\,y)\to(x_0,\,y_0)} \left[\dfrac{f}{g}\right](x,\,y)$

$$= \frac{\lim\limits_{(x,\,y)\to(x_0,\,y_0)} f(x,\,y)}{\lim\limits_{(x,\,y)\to(x_0,\,y_0)} g(x,\,y)} = \frac{L}{M} \;\; (M \neq 0)$$

6. The function $f(x,\,y)$ is continuous at the point $(x_0,\,y_0)$ if and only if
    1. $f(x_0,\,y_0)$ is defined;
    2. $\lim\limits_{(x,y)\to(x_0,y_0)} f(x,\,y)$ exists;
    3. $\lim\limits_{(x,\,y)\to(x_0,\,y_0)} f(x,\,y) = f(x_0,\,y_0)$.

    Also, f is continuous on a set S in its domain if it is continuous at each point in S.

7. If $z = f(x,\,y)$, then the (first) partial derivatives of f with respect to x and y are the functions f_x and f_y, respectively, defined by

$$f_x(x,\,y) = \lim_{\Delta x\to 0} \frac{f(x + \Delta x,\,y) - f(x,\,y)}{\Delta x}$$

$$f_y(x,\,y) = \lim_{\Delta y\to 0} \frac{f(x,\,y + \Delta y) - f(x,\,y)}{\Delta y}$$

provided the limits exist.

8. The line parallel to the xz-plane and tangent to the surface $z = f(x,\,y)$ at the point $P_0(x_0,\,y_0,\,z_0)$ has slope $f_x(x_0,\,y_0)$. Likewise, the tangent line to the surface at P_0 that is parallel to the yz-plane has slope $f_y(x_0,\,y_0)$.

9. $z = f(x,\,y);\; \dfrac{\partial^2 f}{\partial x^2}, \dfrac{\partial^2 f}{\partial y^2}, \dfrac{\partial^2 f}{\partial x \partial y}, \dfrac{\partial^2 f}{\partial y \partial x}$

10. Let $z = f(x,\,y);\; \Delta z = \dfrac{\partial f}{\partial x}\Delta x + \dfrac{\partial f}{\partial y}\Delta y$

    where $\Delta x = dx$ and $\Delta y = dy$.

11. Suppose $f(x,\,y)$ is defined at each point in a circular disk that is centered at $(x_0,\,y_0)$ and contains the point $(x_0 + \Delta x,\,y_0 + \Delta y)$. Then f is said to be differentiable at $(x_0,\,y_0)$ if, the increment of f can be expressed as

$$\Delta f = f_x(x_0,\,y_0)\Delta x + f_y(x_0,\,y_0)\Delta y + \epsilon_1\Delta x + \epsilon_2\Delta y$$

    where $\epsilon_1 \to 0$ and $\epsilon_2 \to 0$ as both $\Delta x \to 0$ and $\Delta y \to 0$ (and $\epsilon_1 = \epsilon_2 = 0$ when $\Delta x = \Delta y = 0$). Also, $f(x,\,y)$ is said to be differentiable on the region R of the plane if f is differentiable at each point in R.

12. If $f(x,\,y)$ and its partial derivatives f_x and f_y are defined in an open region R containing the point $P(x_0,\,y_0)$ and f_x and f_y are continuous at P, then

$$\Delta f = f(x_0+\Delta x,\,y_0 +\Delta y) - f(x_0,\,y_0)$$

$$\approx f_x(x_0,\,y_0)\Delta x + f_y(x_0,\,y_0)\Delta y$$

    so that

$$f(x_0 + \Delta x,\,y_0 + \Delta y)$$

$$\approx f(x_0,\,y_0) + f_x(x_0,\,y_0)\Delta x + f_y(x_0,\,y_0)\Delta y$$

13. If $z = f(x,\,y)$ and Δx and Δy are increments of x and y, respectively, and if we let $dx = \Delta x$ and $dy = \Delta y$ be differentials for x and y, respectively, then the total differential of $f(x,\,y)$ is

$$df = \frac{\partial f}{\partial x}\,dx + \frac{\partial f}{\partial y}\,dy = f_x(x,\,y)\,dx + f_y(x,\,y)\,dy$$

14. Let $f(x,\,y)$ be a differentiable function of x and y, and let $x = x(t)$ and $y = y(t)$ be differentiable functions of t. Then $z = f(x,\,y)$ is a differentiable function of t, and

$$\frac{dz}{dt} = \frac{\partial z}{\partial x}\frac{dx}{dt} + \frac{\partial z}{\partial y}\frac{dy}{dt}$$

15. Suppose $z = f(x,\,y)$ is differentiable at $(x,\,y)$

and that the partial derivatives of $x = x(u, v)$ and $y = y(u, v)$ exist at (u, v). Then the composite function $z = f[x(u, v), y(u, v)]$ is differentiable at (u, v) with

$$\frac{\partial z}{\partial u} = \frac{\partial z}{\partial x}\frac{\partial x}{\partial u} + \frac{\partial z}{\partial y}\frac{\partial y}{\partial u} \text{ and } \frac{\partial z}{\partial v} = \frac{\partial z}{\partial x}\frac{\partial x}{\partial v} + \frac{\partial z}{\partial y}\frac{\partial y}{\partial v}$$

16. Let f be a function of two variables, and let $\mathbf{u} = u_1\mathbf{i} + u_2\mathbf{j}$ be a unit vector. The directional derivative of f at $P_0(x_0, y_0)$ in the direction of $\mathbf{u}$ is given by
$$D_{\mathbf{u}}f(x_0, y_0)$$
$$= \lim_{h \to 0} \frac{f(x_0 + hu_1, y_0 + hu_2) - f(x_0, y_0)}{h}$$
provided the limit exists.

17. Let f be a differentiable function at (x, y) and let $f(x, y)$ have partial derivatives $f_x(x, y)$ and $f_y(x, y)$. Then the gradient of f, denoted by ∇f, is a vector given by

$$\nabla f(x, y) = f_x(x, y)\mathbf{i} + f_y(x, y)\mathbf{j}$$

18. Let f and g be differentiable functions. Then

   a. $\nabla c = \mathbf{0}$ for any constant c

   b. $\nabla(af + bg) = a\nabla f + b\nabla g$

   c. $\nabla(fg) = f\nabla g + g\nabla f$

   d. $\nabla\left(\dfrac{f}{g}\right) = \dfrac{g\nabla f - f\nabla g}{g^2}$ $g \neq 0$

   e. $\nabla(f^n) = nf^{(n-1)}\nabla f$

19. If f is a differentiable function of x and y, then the directional derivative at the point $P_0(x_0, y_0)$ in the direction of the unit vector $\mathbf{u}$ is $D_{\mathbf{u}}f(x, y) = \nabla f \cdot \mathbf{u}$

20. Suppose f is differentiable and let ∇f_0 denote the gradient at P_0. Then if $\nabla f_0 \neq \mathbf{0}$:
   (1) The largest value of the directional derivative of $D_{\mathbf{u}}f$ is $\|\nabla f_0\|$ and occurs when the unit vector $\mathbf{u}$ points in the direction of ∇f_0.
   (2) The smallest value of $D_{\mathbf{u}}f$ is $-\|\nabla f_0\|$ and occurs when $\mathbf{u}$ points in the direction of $-\nabla f_0$.

21. Suppose the function f is differentiable at the point P_0 and that the gradient at P_0 satisfies $\nabla f_0 \neq \mathbf{0}$. Then ∇f_0 is orthogonal to the level surface $f(x, y, z) = K$ at P_0.

22. Suppose the surface S has a nonzero normal vector $\mathbf{N}$ at the point P_0. Then the line through P_0 parallel to $\mathbf{N}$ is called the normal line to S at P_0, and the plane through P_0 with normal vector $\mathbf{N}$ is the tangent plane to

S at P_0.

23. The function $f(x, y)$ is said to have an absolute maximum at (x_0, y_0) if $f(x_0, y_0) \geq f(x, y)$ for all (x, y) in the domain D of f. Similarly, f has an absolute minimum at (x_0, y_0) if $f(x_0, y_0) \leq f(x, y)$ for all (x, y) in D. Collectively, absolute maxima and minima are called absolute extrema.

24. Let f be a function defined at (x_0, y_0). Then $f(x_0, y_0)$ is a relative maximum if $f(x, y) \leq f(x_0, y_0)$ for all (x, y) in an open disk containing (x_0, y_0). $f(x_0, y_0)$ is a relative minimum if $f(x, y) \geq f(x_0, y_0)$ for all (x, y) in an open disk containing (x_0, y_0). Collectively, relative maxima and minima are called relative extrema.

25. A critical point of a function f defined on an open set S is a point (x_0, y_0) in S where either one of the following is true:
   (1) $f_x(x_0, y_0) = f_y(x_0, y_0) = 0$.
   (2) $f_x(x_0, y_0)$ or $f_y(x_0, y_0)$ does not exist (one or both).

26. Let $f(x, y)$ have a critical point at $P_0(x_0, y_0)$ and assume that f has continuous partial derivatives in a disk centered at (x_0, y_0). Let

   $$D = f_{xx}(x_0, y_0)f_{yy}(x_0, y_0) - [f_{xy}(x_0, y_0)]^2$$

   Then, a relative maximum occurs at P_0 if

   $$D > 0 \text{ and } f_{xx}(x_0, y_0) < 0$$

   A relative minimum occurs at P_0 if

   $$D > 0 \text{ and } f_{xx}(x_0, y_0) > 0$$

   A saddle point occurs at P_0 if $D < 0$.

   If $D = 0$, then the test is inconclusive.

27. A function of two variables $f(x, y)$ assumes an absolute extremum on any closed, bounded set S in the plane where it is continuous. Moreover, all absolute extrema must occur either on the boundary of S or at a critical point in the interior of S.

28. Given a set of data points (x_k, y_k), a line $y = mx + b$, called a regression line, is obtained by minimizing the sum of squares of distances $y_k - (mx_k + b)$.

29. Assume that f and g have continuous first partial derivatives and that f has an extremum at $P_0(x_0, y_0)$ on the smooth constraint curve $g(x, y) = c$. If

$\nabla g(x_0, y_0) \neq \mathbf{0}$, there is a number λ such that $\nabla f(x_0, y_0) = \lambda \nabla g(x_0, y_0)$.

30. Suppose f and g satisfy the hypotheses of Lagrange's theorem, and suppose that $f(x)$ has an extremum (minimum and/or a maximum) subject to the constraint $g(x, y) = c$. Then to find the extreme values, proceed as follows:
1. Simultaneously solve the following three equations:
$$f_x(x, y) = \lambda g_x(x, y)$$
$$f_y(x, y) = \lambda g_y(x, y)$$
$$g(x, y) = c$$
2. Evaluate f at all points found in Step 1. The largest of these values is the maximum value of f and the smallest of these values is the minimum value of f.

31. Suppose E is an extreme value (maximum or minimum) of f subject to the constraint $g(x, y) = c$. Then the Lagrange multiplier λ is the rate of change of E with respect to c; that is, $\lambda = dE/dc$.

32. $f(x, y) = \sin^{-1} xy$

Recall $\dfrac{d}{dx} \sin^{-1} u = \dfrac{1}{\sqrt{1 - u^2}} \dfrac{du}{dx}$;

$f_x = \dfrac{y}{\sqrt{1 - x^2 y^2}}; \ f_y = \dfrac{x}{\sqrt{1 - x^2 y^2}}$

$f_{xy} = \dfrac{\sqrt{1 - x^2 y^2}\,(1) - \dfrac{y(-2x^2 y)}{2\sqrt{1 - x^2 y^2}}}{1 - x^2 y^2}$

$= \dfrac{1 - x^2 y^2 + x^2 y^2}{(1 - x^2 y^2)^{3/2}} = \dfrac{1}{(1 - x^2 y^2)^{3/2}}$

$f_{yx} = \dfrac{\sqrt{1 - x^2 y^2}\,(1) - \dfrac{x(-2xy^2)}{2\sqrt{1 - x^2 y^2}}}{1 - x^2 y^2}$

$= \dfrac{1 - x^2 y^2 + x^2 y^2}{(1 - x^2 y^2)^{3/2}} = \dfrac{1}{(1 - x^2 y^2)^{3/2}}$

33. $\dfrac{dw}{dt} = \dfrac{dw}{dx} \cdot \dfrac{dx}{dt} + \dfrac{dw}{dy} \cdot \dfrac{dy}{dt} + \dfrac{dw}{dz} \cdot \dfrac{dz}{dt}$

$= 2xy(t \cos t + \sin t)$
$\quad + (x^2 + 2yz)(-t \sin t + \cos t) + y^2(2)$

$\dfrac{dw}{dt}(\pi) = 2xy(-\pi) + (x^2 + 2yz)(-1) + 2y^2$

$= -2\pi xy + 2y^2 - x^2 - 2yz$

Now if $t = \pi$, $x = 0$, $y = -\pi$, $z = 2\pi$

$\dfrac{dw}{dt}(\pi) = 0 + 2\pi^2 - 0 - 2(-\pi)(2\pi) = 6\pi^2$

34. $f(x, y, z) = xy + yz + xz$ at $(1, 2, -1)$

a. $\nabla f = (y + z)\mathbf{i} + (x + z)\mathbf{j} + (y + x)\mathbf{k}$

$= \mathbf{i} + 3\mathbf{k}$

b. $V_u = \dfrac{-2\mathbf{i} - \mathbf{j}}{\sqrt{5}}$

$df = \nabla f \cdot V_u = \dfrac{-2}{\sqrt{5}} = \dfrac{-2\sqrt{5}}{5}$

c. The directional derivative has its greatest value in the direction of the gradient:

$\dfrac{\mathbf{i} + 3\mathbf{k}}{\sqrt{10}}$. The magnitude is $\|\nabla f\| = \sqrt{10}$

35. $\displaystyle\lim_{(x, y) \to (0, 0)} f(x, y)$ along the line $y = x$ is

$\displaystyle\lim_{x \to 0} \dfrac{x^3}{x^3 + x^3} = \dfrac{1}{2}$.

The limit does not equal $f(x, y)$ so the function is not continuous.

36. $f(x, y) = \ln \dfrac{y}{x} = \ln y - \ln x$

$f_x = -\dfrac{1}{x}, \ f_y = \dfrac{1}{y}, \ f_{yy} = -\dfrac{1}{y^2}, \ f_{xy} = 0$

37. $f(x, y, z) = x^2 y + y^2 z + z^2 x$

$f_x = 2xy + z^2; \ f_y = x^2 + 2yz; \ f_z = y^2 + 2zx$

$f_x + f_y + f_z = 2xy + z^2 + x^2 + 2yz + y^2 + 2zx$

$= (x + y + z)^2$

38. $f(x, y) = x^4 + 2x^2 y^2 + y^4$

$\nabla f = (4x^3 + 4xy^2)\mathbf{i} + (4x^2 y + 4y^3)\mathbf{j}$

$\nabla f(2, -2) = 64(\mathbf{i} - \mathbf{j})$ in the direction of $\dfrac{2\pi}{3}$

with the positive x-axis:

$V = -\dfrac{1}{2}\mathbf{i} + \dfrac{\sqrt{3}}{2}\mathbf{j};$

$\nabla f \cdot V = -32 - 32\sqrt{3}$

$= -32(1 + \sqrt{3}) \approx -87.4$

39. $f_x = e^{-(x^2 + y^2)}[2x + (x^2 + y^2)(-2x)]$

$= 2xe^{-(x^2 + y^2)}(1 - x^2 - y^2)$

$f_y = e^{-(x^2 + y^2)}[2y + (x^2 + y^2)(-2y)]$

$= 2y\, e^{-(x^2 + y^2)}(1 - x^2 - y^2)$

$f_x = f_y = 0$ if $x = y = 0$ or if $x^2 + y^2 = 1$

40. $f(x, y) = x^2 + 2y^2 + 2x + 3$ subject to

$g(x, y) = x^2 + y^2 - 4$

$2x + 2 = 2x\lambda, \ 4y = 2y\lambda$, and $x^2 + y^2 = 4$

From the second equation: $\lambda = 2$ or $y = 0$. If $\lambda = 2$, $2x + 2 = 4x$, $x = 1$, and from the constraint equation $y = \pm\sqrt{3}$. If $y = 0$ the

constraint equation gives $x = \pm 2$. The set of candidates: $(1, \sqrt{3})$, $(1, -\sqrt{3})$, $(2, 0)$, $(-2, 0)$. $f(1, \sqrt{3}) = 12$; $f(1, -\sqrt{3}) = 12$; $f(2, 0) = 11$; $f(-2, 0) = 3$. $f(x, y)$ has a maximum of 12 at $(1, \pm\sqrt{3})$ and a minimum of 3 at $(-2, 0)$.

Supplementary Problems, Page 816

1. $f(x, y) = \sqrt{16 - x^2 - y^2}$

   $16 - x^2 - y^2 \geq 0$; $x^2 + y^2 \leq 16$

   The domain consists of the circle with center at the origin, radius 4, and its interior.

2. $f(x, y) = \dfrac{x^2 - y^2}{x - y} = x + y$, but $x \neq y$ is the domain.

3. $f(x, y) = \sin^{-1} x + \cos^{-1} y$, the sum of two angles. $-1 \leq x \leq 1$ and $-1 \leq y \leq 1$ is the domain.

4. $f(x, y) = e^{x+y}\tan^{-1}\left(\dfrac{y}{x}\right)$

   There is no limitation on the exponent of the exponential or the range of the tangent function (although $x \neq 0$ is appropriate).

5. $f(x, y) = \dfrac{x^2 - y^2}{x + y} = x - y$

   $f_x = 1$; $f_y = -1$

6. $f(x, y) = x^3 e^{3y/2x}$

   $f_x = e^{3y/(2x)}[x^3(\tfrac{3y}{2})(-x^{-2}) + 3x^2]$

   $= \tfrac{3}{2}x(2x - y)e^{3y/(2x)}$

   $f_y = \tfrac{3}{2}x^2 e^{3y/(2x)}$

7. $f(x, y) = x^2 y + \sin\dfrac{y}{x}$

   $f_x = 2xy - \dfrac{y}{x^2}\cos\dfrac{y}{x}$; $f_y = x^2 + \dfrac{1}{x}\cos\dfrac{y}{x}$

8. $f(x, y) = \ln\left(\dfrac{xy}{x + 2y}\right) = \ln x + \ln y - \ln(x + 2y)$

   $f_x = \dfrac{1}{x} - \dfrac{1}{x + 2y}$; $f_y = \dfrac{1}{y} - \dfrac{2}{x + 2y}$

9. $f(x, y) = 2x^3 y + 3xy^2 + \dfrac{y}{x}$

   $f_x = 6x^2 y + 3y^2 - \dfrac{y}{x^2}$; $f_y = 2x^3 + 6xy + \dfrac{1}{x}$

10. $f(x, y) = xye^{xy}$

    $f_x = ye^{xy}(xy + 1)$; $f_y = xe^{xy}(xy + 1)$

11. For $c = 2$, $x^2 - y = 2$ is a parabola opening up, with vertex at $(0, -2)$. For $c = -2$, $x^2 - y = -2$ is a parabola opening up, with vertex at $(0, 2)$.

12. For $c = 0$, $6x + 2y = 0$ is a line through the origin with slope $m = -3$. For $c = 1$,

$6x + 2y = 1$ is a line through the point $(0, \tfrac{1}{2})$ with slope $m = -3$. For $c = 2$, $6x + 2y = 2$ is a line through the point $(0, 1)$ with slope $m = -3$.

13. For $c = 0$, $\sqrt{x^2 + y^2} = 0$ is the origin only. For $c = 1$, $\sqrt{x^2 + y^2} = 1$ is a semicircle (to the right of the y-axis). For $c = -1$, $|y| = 1$ is a pair of half-lines, 1 unit above or below the x-axis, to the left of the y-axis.

14. For $c = 16$, $x^2 + y^2 + z^2 = 16$ is a sphere with center at the origin and radius $r = 4$. For $c = 0$, $x^2 + y^2 + z^2 = 0$ is the origin. For $c = -25$, there are no points.

15. For $c = 1$, $x^2 + \dfrac{y^2}{2} + \dfrac{z^2}{9} = 1$ is an ellipsoid.

    For $c = 2$, $x^2 + \dfrac{y^2}{2} + \dfrac{z^2}{9} = 2$ is an ellipsoid.

16. $\displaystyle\lim_{(x,y)\to(1,1)} \dfrac{xy}{x^2 + y^2} = \dfrac{1}{2}$

17. $\displaystyle\lim_{(x,y)\to(1,1)} \dfrac{x^2 - y^2}{x^4 - y^4}$

    $= \displaystyle\lim_{(x,y)\to(1,1)} \dfrac{x^2 - y^2}{(x^2 - y^2)(x^2 + y^2)} = \dfrac{1}{2}$

18. $\displaystyle\lim_{(x,y)\to(1,1)} \dfrac{x^3 - y^3}{x^3 + y^3} = 0$

19. $\displaystyle\lim_{(x,y)\to(0,0)} \dfrac{x + ye^{-x}}{1 + x^2} = 0$

20. $\displaystyle\lim_{(x,y)\to(0,0)} \dfrac{x^3 - y^3}{x^3 + y^3}$

    Along the line $x = 0$

    $\displaystyle\lim_{y\to 0} = \dfrac{-y^3}{y^3} = -1$

    Along the line $y = 0$

    $\displaystyle\lim_{x\to 0} = \dfrac{x^3}{x^3} = 1$

    Thus, the limit does not exist.

21. $\displaystyle\lim_{(x,y)\to(0,0)} \dfrac{x^4 - y^4}{x^4 + y^4}$

    Along the line $x = 0$

    $\displaystyle\lim_{y\to 0} = \dfrac{-y^4}{y^4} = -1$

    Along the line $y = 0$

    $\displaystyle\lim_{x\to 0} = \dfrac{x^4}{x^4} = 1$

    Thus, the limit does not exist.

22. $\displaystyle\lim_{(x,y)\to(0,0)} \dfrac{x^2 - 2y^2}{2x^2 + y^2}$

    Along the line $x = 0$

$$\lim_{y \to 0} = \frac{-2y^2}{y^2} = -2$$

Along the line $y = 0$

$$\lim_{x \to 0} = \frac{x^2}{2x^2} = \frac{1}{2}$$

Thus, the limit does not exist.

23. $\quad \lim_{(x,y) \to (0,0)} \frac{x^2 - y^3}{x^2 + y^3}$

Along the line $x = 0$

$$\lim_{y \to 0} = \frac{-y^3}{y^3} = -1$$

Along the line $y = 0$

$$\lim_{x \to 0} = \frac{x^2}{x^2} = 1$$

Thus, the limit does not exist.

24. $\quad \dfrac{dz}{dt} = \dfrac{\partial z}{\partial x}\dfrac{dx}{dt} + \dfrac{\partial z}{\partial y}\dfrac{dy}{dt}$

$\qquad = (-y)(-6t) + (-x + 3y^2)(3t^2)$

$\qquad = 6yt - 3xt^2 + 9y^2t^2$

25. $\quad \dfrac{dz}{dt} = \dfrac{\partial z}{\partial x}\dfrac{dx}{dt} + \dfrac{\partial z}{\partial y}\dfrac{dy}{dt}$

$\qquad = ye^t(-t^{-2} + t^{-1}) + (x + 2y)\sec^2 t$

26. $\quad \dfrac{dz}{du} = \dfrac{\partial z}{\partial x}\dfrac{dx}{du} + \dfrac{\partial z}{\partial y}\dfrac{dy}{du}$

$\qquad = 2x(1) + (-2y)(1)$

$\qquad = 2(x - y)$

$\quad \dfrac{dz}{dv} = \dfrac{\partial z}{\partial x}\dfrac{dx}{dv} + \dfrac{\partial z}{\partial y}\dfrac{dy}{dv}$

$\qquad = (2x)(2) + (-2y)(-2)$

$\qquad = 4(x + y)$

27. $\quad \dfrac{dz}{du} = \dfrac{\partial z}{\partial x}\dfrac{dx}{du} + \dfrac{\partial z}{\partial y}\dfrac{dy}{du}$

$\qquad = (\tan \frac{x}{y} + \frac{x}{y}\sec^2 \frac{x}{y})v + (-\frac{x^2}{y^2}\sec^2 \frac{x}{y})v^{-1}$

$\quad \dfrac{dz}{dv} = \dfrac{\partial z}{\partial x}\dfrac{dx}{dv} + \dfrac{\partial z}{\partial y}\dfrac{dy}{dv}$

$\qquad = (\tan \frac{x}{y} + \frac{x}{y}\sec^2 \frac{x}{y})u$

$\qquad \quad + (-\frac{x^2}{y^2}\sec^2 \frac{x}{y})(-uv^{-2})$

28. $\quad x^2 + 6y^2 + 2z^2 = 5$

$\qquad 2x + 4z\dfrac{\partial z}{\partial x} = 0$

$\qquad \dfrac{\partial z}{\partial x} = -\dfrac{x}{2z}$

Also, $\quad 12y + 4z\dfrac{\partial z}{\partial y} = 0$

$\qquad \dfrac{\partial z}{\partial y} = -\dfrac{3y}{z}$

29. $\quad e^x + e^y + e^z = 3$

$\qquad e^x + e^z + 4z\dfrac{\partial z}{\partial x} = 0$

$$\dfrac{\partial z}{\partial x} = -\dfrac{e^x}{e^z} = -e^{x-z}$$

Also, $\qquad e^y + e^z + 4z\dfrac{\partial z}{\partial y} = 0$

$$\dfrac{\partial z}{\partial y} = -\dfrac{e^y}{e^z} = -e^{y-z}$$

30. $\quad x^3 + 2xz - yz^2 - z^3 = 1$

$\quad 3x^2 + 2x\dfrac{\partial z}{\partial x} + 2z - 2yz\dfrac{\partial z}{\partial x} - 3z^2\dfrac{\partial z}{\partial x} = 0$

$\qquad \dfrac{\partial z}{\partial x} = -\dfrac{-3x^2 - 2z}{2x - 2yz - 3z^2}$

Also,

$\quad 2x\dfrac{\partial z}{\partial y} + 2z - 2yz\dfrac{\partial z}{\partial y} - z^2 - 3z^2\dfrac{\partial z}{\partial y} = 0$

$\qquad \dfrac{\partial z}{\partial y} = -\dfrac{z^2}{2x - 2yz - 3z^2}$

31. $\quad \ln(2x - z^2) - (x - y)e^{2z} = 0$

$\dfrac{2 - 2z\frac{\partial z}{\partial x}}{2x - z^2} - (x - y)e^{2z}(2)\dfrac{\partial z}{\partial x} - e^{2z} = 0$

$2 - 2z\dfrac{\partial z}{\partial x} - (x - y)(2x - z^2)e^{2z}(2)\dfrac{\partial z}{\partial x} - (2x - z^2)e^{2z} = 0$

$\qquad \dfrac{\partial z}{\partial x} = \dfrac{(2x - z^2)e^{2z} - 2}{-2z - 2(x - y)(2x - z^2)e^{2z}}$

Similarly,

$$\dfrac{\partial z}{\partial y} = \dfrac{(2x - z^2)e^{2z}}{2z + 2(x - y)(2x - z^2)e^{2z}}$$

32. $\quad x^3 + y^3 + z^3 = 6$

$\qquad 3x^2 + 3z^2\dfrac{\partial z}{\partial x} = 0$

$$\dfrac{\partial z}{\partial x} = -\dfrac{x^2}{z^2}$$

Similarly,

$$\dfrac{\partial z}{\partial y} = -\dfrac{y^2}{z^2}$$

33. $\quad x + 2y - 3z = \ln z$

$\quad 1 - 3\dfrac{\partial z}{\partial x} - \dfrac{1}{z}\dfrac{\partial z}{\partial x} = 0$

$\quad z - 3z\dfrac{\partial z}{\partial x} - \dfrac{\partial z}{\partial x} = 0$

$$\dfrac{\partial z}{\partial x} = \dfrac{z}{3z + 1}$$

Similarly, $\qquad \dfrac{\partial z}{\partial y} = \dfrac{2z}{3z + 1}$

34. $\quad f(x, y) = \tan^{-1} xy$

$\quad f_x = \dfrac{y}{1 + x^2y^2}; f_y = \dfrac{x}{1 + x^2y^2};$

$\quad f_{xx} = -\dfrac{y(2xy^2)}{[1 + (xy)^2]^2} = -\dfrac{2xy^3}{(1 + x^2y^2)^2}$

$\quad f_{yx} = -\dfrac{1 + x^2y^2 - x(2xy^2)}{[1 + (xy)^2]^2} = \dfrac{1 - x^2y^2}{(1 + x^2y^2)^2}$

35. $f(x, y) = \sin^{-1} xy$

$f_x = \dfrac{y}{\sqrt{1 - x^2 y^2}}$; $f_y = \dfrac{x}{\sqrt{1 - x^2 y^2}}$

$f_{xx} = \dfrac{y(2xy^2)}{2[1 - (xy)^2]^{3/2}} = \dfrac{xy^3}{(1 - x^2 y^2)^{3/2}}$

$f_{yx} = \dfrac{\sqrt{1 - x^2 y^2} - \dfrac{x(-2xy^2)}{2(1 - x^2 y^2)^{1/2}}}{1 - (xy)^2}$

$\quad = \dfrac{1 - x^2 y^2 + x^2 y^2}{(1 - x^2 y^2)^{3/2}} = (1 - x^2 y^2)^{-3/2}$

36. $f(x, y) = x^2 + y^3 - 2xy^2$

$f_x = 2x - 2y^2$; $f_y = 3y^2 - 4xy$;

$f_{xx} = 2$; $f_{yx} = -4y$

37. $f(x, y) = e^{x^2 + y^2}$

$f_x = 2x e^{x^2 + y^2}$; $f_y = 2y e^{x^2 + y^2}$;

$f_{xx} = 2e^{x^2 + y^2}(2x^2 + 1)$; $f_{yx} = 4xy e^{x^2 + y^2}$

38. $f(x, y) = x \ln y$

$f_x = \ln y$; $f_y = \dfrac{x}{y}$; $f_{xx} = 0$; $f_{yx} = y^{-1}$

39. $f(x, y, z) = x^3 e^{3y/z^2}$

$f_x = 3x^2 e^{3y/z^2}$; $f_y = \dfrac{3x^2}{z^2} e^{3y/z^2}$

$f_{xx} = 6x e^{3y/z^2}$; $f_{yx} = \dfrac{9x^2}{z^2} e^{3y/z^2}$

40. $f(x, y) = \displaystyle\int_x^y \sin(\cos t)\, dt = -\int_y^x \sin(\cos t)\, dt$

$f_x = -\sin(\cos x)$; $f_y = \sin(\cos y)$

$f_{xx} = [-\cos(\cos x)](-\sin x) = \sin x \cos(\cos x)$

$f_{yx} = 0$

41. $f(x, y, z) = x^3 y^2 \sqrt{z} + \sin(x^2 + y \ln z)$

$f_x = 3x^2 y^2 \sqrt{z} + [\cos(x^2 + y \ln z)](2x)$

$f_y = 2x^3 y \sqrt{z} + [\cos(x^2 + y \ln z)](\ln z)$

$f_{xx} = 6xy^2 \sqrt{z} + [2\cos(x^2 + y \ln z)]$

$\qquad - 4x^2 [\sin(x^2 + y \ln z)]$

$f_{yx} = 6x^2 y \sqrt{z} - 2x(\ln z)[\sin(x^2 + y \ln z)]$

42. $f(x, y, z) = x^2 y^3 z - 8$

$\nabla f = 2xy^3 z \mathbf{i} + 3x^2 y^2 z \mathbf{j} + x^2 y^3 \mathbf{k}$

$\nabla f(2, -1, -2) = 8\mathbf{i} - 48\mathbf{j} - 4\mathbf{k}$

Normal line: $\dfrac{x - 2}{2} = \dfrac{y + 1}{-12} = \dfrac{z + 2}{-1}$

Tangent plane:

$2(x - 2) - 12(y + 1) - (z + 2) = 0$

43. $f(x, y, z) = x^3 + 2xy^2 - 7x^3 + 3y + 1$

$\nabla f = (3x^2 + 2y^2 - 21x^2)\mathbf{i} + (4xy + 3)\mathbf{j}$

$\nabla f(1, 1, 1) = -16\mathbf{i} + 7\mathbf{j}$

Normal line: $\dfrac{x - 1}{16} = \dfrac{y - 1}{-7}$ and $z = 1$

Tangent plane: $-16(x - 1) + 7(y - 1) = 0$

44. $f(x, y, z) = 2z + x^2 z + y^2 z + 5$

$\nabla f = 2xz \mathbf{i} + 2yz \mathbf{j} + (2 + x^2 + y^2)\mathbf{k}$

$\nabla f(1, 1, -1) = -2\mathbf{i} - 2\mathbf{j} + 4\mathbf{k}$

Normal line: $\dfrac{x - 1}{1} = \dfrac{y - 1}{1} = \dfrac{z + 1}{2}$

Tangent plane:

$(x - 1) + (y - 1) - 2(z + 1) = 0$

45. $f(x, y) = x^2 - 6x + 2y^2 + 4y - 2$

$f_x = 2x - 6 = 0$ if $x_1 = 3$

$f_y = 4y + 4 = 0$ if $y = -1$

$f_{xx} = 2$; $f_{xy} = 0$; $f_{yy} = 4$; $D > 0$, so there is a

relative minimum at $(3, -1)$.

46. $f(x, y) = x^3 + y^3 - 6xy$

$f_x = 3x^2 - 6y = 0$; $f_y = 3y^2 - 6x = 0$

Solving this system, we find the solution
$(0, 0)$ and $(2, 2)$.

$f_{xx} = 6x$; $f_{xy} = -6$; $f_{yy} = 6y$;

$D(0, 0) < 0$, so $(0, 0)$ is a saddle point.

$D(2, 2) > 0$, so $(2, 2)$ is a relative minimum.

47. $f(x, y) = (x - 1)(y - 1)(x + y - 1)$

$f_x = (y - 1)[(x - 1)(1) + (x + y - 1)(1)]$

$\quad = (y - 1)(2x + y - 2) = 0$

$f_y = (x - 1)(2y + x - 2) = 0$

Solving this system, we find the solution

$(1, 0), (0, 1), (1, 1), (\tfrac{2}{3}, \tfrac{2}{3})$

$f_{xx} = 2(y - 1)$; $f_{xy} = y - 1 + 2x + y - 2$;

$f_{yy} = 2(x - 1)$

x	y	f_{xx}	f_{xy}	f_{yy}	D	Classify
1	0	-2	-1	0	$-$	saddle point
0	1	0	-1	-2	$-$	saddle point
$\tfrac{2}{3}$	$\tfrac{2}{3}$	$-\tfrac{2}{3}$	$-\tfrac{1}{3}$	$-\tfrac{2}{3}$	$+$	rel max
1	1	0	1	0	0	inconclusive

For $(1, 1)$, note that $f(0.9, 1.1, 1) < 0$ and $f(1.1, 0.9, 1) < 0$, $f(1.1, 1.1, 1) > 0$, so $(1, 1, 1)$ is a saddle point.

48. $f(x, y) = x^2 + y^3 + 6xy - 7x - 6y$

$f_x = 2x + 6y - 7 = 0$

$f_y = 3y^2 + 6x - 6 = 0$

Solving this system, we find the solution $(\frac{1}{2}, 1)$, $(-\frac{23}{2}, 5)$

$f_{xx} = 2$; $f_{xy} = 6$; $f_{yy} = 6y$

x	y	f_{xx}	f_{xy}	f_{yy}	D	Classify
$\frac{1}{2}$	1	2	6	6	$-$	saddle point
$-\frac{23}{2}$	5	2	6	30	$+$	rel min

49. $f(x, y) = x^3 + y^3 - 3x^2 - 18y^2 + 81y + 5$

$f_x = 3x^2 - 6x = 3x(x - 2) = 0$ when $x = 0, 2$

$f_y = 3y^2 - 36y + 81 = 3(y^2 - 12y + 27)$

$\quad = 3(y - 3)(y - 9) = 0$ when $y = 3$, $y = 9$

$f_{xx} = 6(x - 1)$; $f_{xy} = 0$; $f_{yy} = 6(y - 6)$

x	y	f_{xx}	f_{xy}	f_{yy}	D	Classify
0	9	-6	0	18	$-$	saddle point
0	3	-6	0	-18	$+$	rel max
2	9	6	0	18	$+$	rel min
2	3	6	0	-18	$-$	saddle point

50. $f(x, y) = \sin(x + y) + \sin x + \sin y$

$f_x = \cos(x + y) + \cos x$

$f_y = \cos(x + y) + \cos y$

$f_x = f_y = 0$ leads to $\cos y = \cos x$ or $y = x$

This reduces to finding the extrema of

$f(x) = \sin 2x + \sin x$

$f'(x) = 2 \cos 2x + 2 \cos x$

$\quad 2 \cos 2x + 2 \cos x = 0$

$\quad 2(2 \cos^2 x + \cos x - 1) = 0$

$\quad 2(2 \cos x + 1)(\cos x - 1) = 0$

In the given interval, $x = \frac{\pi}{3}$

$f(\frac{\pi}{3}, \frac{\pi}{3})$ is a relative maximum.

51. $\frac{dz}{dt} = (2x)(2) - (3y^2)(2) - (6xy)(2t)$

$\quad = 4x - 6y^2 - 12xyt$

52. $\frac{dz}{dt} = (\ln y)(2) + \frac{x}{y} e^t = 2 \ln y + \frac{xe^t}{y}$

53. $\frac{\partial z}{\partial x} = e^{u^2 - v^2}[(2u^2)(4x) + 4x]$

$\quad + e^{u^2 - v^2}(-2v)(6x)$

$\quad = e^{u^2 - v^2}(8xu^2 + 4x - 12vx)$

$\frac{\partial z}{\partial y} = e^{u^2 - v^2}[(2u^2 + 1)(6y) + 4x]$

$\quad + e^{u^2 - v^2}(-2v)(-4y)$

$\quad = e^{u^2 - v^2}(12yu^2 + 6y + 8vy)$

54. $3x^2 + 2x \frac{\partial z}{\partial x} + 2z - 2yz \frac{\partial z}{\partial x} - 3z^2 \frac{\partial z}{\partial x} = 0$

$\frac{\partial z}{\partial x} = \frac{3x^2 + 2z}{2yz + 3z^2 - 2x}$

$2x \frac{\partial z}{\partial y} - 2yz \frac{\partial z}{\partial y} - z^2 - 3z^2 \frac{\partial z}{\partial y} = 0$

$\frac{\partial z}{\partial y} = \frac{z^2}{2x - 2yz - 3z^2}$

55. $F(x, y) = x^2 - y^2 - 2 = 0$

$F_x = 2x$; $F_y = 2y$; $\frac{dy}{dx} = -\frac{x}{y} = -1$ at $(1, 1)$

56. $F(x, y) = xe^y - 2 = 0$

$F_x = xe^y$; $F_y = e^y$; $\frac{dy}{dx} = -\frac{1}{x} = -\frac{1}{2}$ at $(2, 0)$

57. $f(x, y, z) = \sin x + e^{xy} + 2y - 3 - z = 0$

$\nabla f = (\cos x + ye^{xy})\mathbf{i} + (xe^{xy} + 2)\mathbf{j} - \mathbf{k}$

$\nabla f(0, 1, 0) = 2\mathbf{i} + 2\mathbf{j} - \mathbf{k}$

Normal line: $\frac{x}{2} = \frac{y - 1}{2} = \frac{z - 3}{-1}$

Tangent plane:

$2(x - 0) + 2(y - 1) - (z - 3) = 0$

or $2x + 2y - z + 1 = 0$

58. $f(x, y, z) = x^2 + y^2 < 4$

$V = 2(4 - x^2 - y^2)^{-1/2} = c$

$\frac{2}{4 - x^2 - y^2} = c$

$4 - x^2 - y^2 = 2c^{-1}$

$x^2 + y^2 = 4 - 2c^{-1}$

We recognize these equiponential curves as circles with center $(0, 0)$ and radius $4 - 2c^{-1}$.

If $c = 8$, $r \approx 1.984$; $c = 6$, $r \approx 1.972$

$c = 5$, $r \approx 1.960$ $\qquad c = 2$, $r \approx 1.732$

The four circles are indistinguishable as shown in the following graphs:

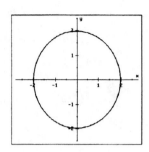

59. $\dfrac{\partial f}{\partial x} = 3x^2 y + z^3; \quad \dfrac{\partial f}{\partial y} = x^3 + 3y^2 z$

$\dfrac{\partial f}{\partial z} = y^3 + 3z^2 x$

$\dfrac{\partial f}{\partial x} + \dfrac{\partial f}{\partial y} + \dfrac{\partial f}{\partial z}$

$= 3x^2 y + z^3 + x^3 + 3y^2 z + y^3 + 3z^2 x$

$= x^3 + y^3 + z^3 + 3(x^2 y + y^2 z + z^2 x)$

so $g(x, y, z) = x^2 y + y^2 z + z^2 x$

60. $\dfrac{\partial u}{\partial x} = \dfrac{1}{y} \cos \dfrac{x}{y} + \dfrac{1}{x}; \quad \dfrac{\partial u}{\partial y} = -\dfrac{x}{y^2} \cos \dfrac{x}{y} + \dfrac{1}{y}$

$y\left(-\dfrac{x}{y^2} \cos \dfrac{x}{y} + \dfrac{1}{y} \right) + x\left(\dfrac{1}{y} \cos \dfrac{x}{y} - \dfrac{1}{x} \right)$

$= -\dfrac{x}{y} \cos \dfrac{x}{y} + 1 + \dfrac{x}{y} \cos \dfrac{x}{y} - 1 = 0$

61. $\dfrac{\partial w}{dt} = \dfrac{(2x)\dfrac{2t}{1 + t^2} + 2ye^r t}{1 + x^2 + y^2} - \dfrac{2e^t}{1 + y^2}$

$= \left[\dfrac{2}{1 + x^2 + y^2} \right]\left(\dfrac{2xt}{1 + t^2} \right)$

$+ \left[\dfrac{2y}{1 + x^2 + y^2} - \dfrac{2}{1 + y^2} \right]e^t$

62. $\nabla f = \dfrac{1}{1 + \dfrac{y^2}{x^2}}\left(-\dfrac{y}{x^2} \right)\mathbf{i} + \dfrac{1}{1 + \dfrac{y^2}{x^2}}(\tfrac{1}{x})\mathbf{j}$

$\nabla f(1, 2) = \tfrac{1}{5}(-4\mathbf{i} + \mathbf{j}); \quad \mathbf{u} = \tfrac{1}{2}(\mathbf{i} + \sqrt{3}\mathbf{j})$

$D_u f = \tfrac{1}{10}(-4 + \sqrt{3})$

63. $\nabla f = y^x \ln y\, \mathbf{i} + xy^{x-1}\mathbf{j}$

$\nabla f(3, 2) = 8 \ln 2\, \mathbf{i} + 12\mathbf{j}; \quad \mathbf{u} = \dfrac{1}{\sqrt{5}}(-2\mathbf{i} - \mathbf{j})$

$D_u f = \dfrac{1}{\sqrt{5}}(-16 \ln 2 - 12) = -\dfrac{4}{\sqrt{5}}(3 + 4 \ln 2)$

64. $\nabla f = yx^{y-1} y^y \mathbf{i} + x^y y^y (1 + \ln y)\mathbf{j}$

$\nabla f(2, 2) = 2^4 \mathbf{i} + 2^4 (1 + \ln 2)\mathbf{j}; \quad \mathbf{u} = \dfrac{1}{\sqrt{2}}(\mathbf{i} - \mathbf{j})$

$D_u f = \dfrac{2^4 \ln 2}{\sqrt{2}}$

65. a. $\nabla f = [5z(x - y)^4 + y^2 z^3]\mathbf{i}$

$\qquad + [-5z(x - y^4 + 2xyz^3)]\mathbf{j}$

$\qquad + [(x - y)^5 + 3xy^2 z^2]\mathbf{k}$

$\nabla f(2, 1, -1) = -6\mathbf{i} + \mathbf{j} + 17\mathbf{k};$

$\mathbf{u} = \dfrac{1}{\sqrt{6}}(2\mathbf{i} + \mathbf{j} + \mathbf{k});$

$D_u f = \dfrac{-12 + 1 - 7}{\sqrt{6}} = -\dfrac{18}{\sqrt{6}} = -3\sqrt{6}$

b. $\mathbf{u} = \dfrac{1}{\sqrt{36 + 1 + 49}}(-6\mathbf{i} + \mathbf{j} + 7\mathbf{k})$

$\qquad = \dfrac{1}{\sqrt{86}}(-6\mathbf{i} + \mathbf{j} + 7\mathbf{k})$

66. $f(x, y, z) = xyz; \; g(x, y, z) = x + y + z = 1$

$f_x = yz; \; f_y = xz; \; f_z = xy; \; g_x = 1; \; g_y = 1; \; g_z = 1$

The Lagrange equations are $yz = 2\lambda x, \; yz = \lambda,$

$xz = \lambda, \; xy = \lambda, \; x + y + z = 1;$ solving this

system of equations gives $x = y = z = 1/3$.

67. $f(x, y, z) = x^2 yz; \; g(x, y, z) = x + y + z = 12$

$f_x = 2xyz; \; f_y = x^2 z; \; f_z = x^2 y; \; g_x = 1; \; g_y = 1;$

$g_z = 1;$ the Lagrange equations are

$2xyz = \lambda x; \; x^2 z = \lambda; \; x^2 y = \lambda; \; x + y + z = 12;$

solving this system of equations gives $x = 6,$

$y = z = 3$.

68. We minimize the square of the distance:

$f(x, y, z) = x^2 + y^2 + z^2;$

$g(x, y, z) = y^2 - z^2 - 10$

$f_x = 2x; \; f_y = 2y; \; f_z = 2z; \; g_x = 0; \; g_y = 2y;$

$g_z = -2z;$ the Lagrange equations are $2x = 0;$

$2y = 2\lambda y; \; 2z = -2\lambda z; \; y^2 - z^2 = 10;$ solving

this system of equations gives $x = 0,$

$y = \pm\sqrt{10}, \; z = 0,$ so $d = \sqrt{10}$.

69. We minimize the square of the distance:

$f(x, y, z) = x^2 + y^2 + z^2;$

$g(x, y, z) = xy - z^2 + 3$

$f_x = 2x; \; f_y = 2y; \; f_z = 2z; \; g_x = y; \; g_y = x;$

$g_z = -2z;$ the Lagrange equations are

$2x = \lambda y; \; 2y = \lambda x; \; 2z = -2\lambda z; \; xy - z^2 = 3;$

solving this system of equations gives $x =$

$\sqrt{3}; \; y = \sqrt{3}; \; z = 0,$ so $d = \sqrt{6}$.

70. $T(x, y) = \dfrac{64}{x^2 + y^2 + 4}$

a. $\nabla T = -64(x^2 + y^2 + 4)^{-2}(2x\mathbf{i} + 2y\mathbf{j})$

$\nabla T(3, 4) = -\dfrac{384}{841}\mathbf{i} - \dfrac{512}{841}\mathbf{j}; \; \mathbf{u} = \dfrac{1}{\sqrt{5}}(2\mathbf{i} + \mathbf{j})$

$D_u T = -\dfrac{128}{841\sqrt{5}}(6 + 4) = -\dfrac{1,280}{29^2 \sqrt{5}}$

b. $\mathbf{u} = \tfrac{1}{5}(3\mathbf{i} + 4\mathbf{j}); \; |D_u T|_{\max} = \dfrac{640}{841}$

71. a.

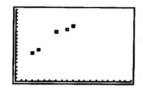

b.

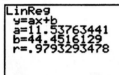

The least squares line is $y = 11.54x + 44.45$

If $x = 5,000$, the sales are

$y = 11.54(5) + 44.45$

≈ 102.15

72. $f_x = 2xy^3 + 3x^2y^2;\ f_{xx} = 2y^3 + 6xy^2$

$f_y = 3x^2y^2 + 2x^3y;\ f_{yy} = 6x^2y + 2x^3$

$f_{xy} = 6xy = 6xy^2 + 6x^2y$

$f_{xx} + f_{xy} + f_{yy} = 2x^3 + 2y^3 + 12xy^2 + 12x^2y$

73. $f_x = -2x\sin(x^2 + y^3 + z^4)$

$f_{xy} = [-2x\cos(x^2 + y^3 + z^4)](2y^2)$

$f_{xyz} = 6xy^2(4z^3)\sin(x^2 + y^2 + z^4)$

$= 24xy^2z^3\sin(x^2 + y^2 + z^4)$

74. a. $\dfrac{dz}{dt} = f_x(1 - \sin t) + f_y e^t$

When $t = 0$, $\dfrac{dz}{dt} = 4(1 - 0) + (-3) = 1$

b. $\dfrac{\partial z}{\partial r} = f_x\cos\theta + f_y\sin\theta$

When $\theta = \dfrac{\pi}{2}$, $\dfrac{\partial z}{\partial r} = (-1)(0) + 3(1) = 3$

$\dfrac{\partial z}{\partial\theta} = f_x(-r\sin\theta) + f_y(r\cos\theta)$

When $\theta = \dfrac{\pi}{2}$ and $r = 2$,

$\dfrac{\partial z}{\partial\theta} = (-1)[-2(1)] + 3[2(0)] = 2$

75. $\dfrac{dy}{dx} = -\dfrac{f_x}{f_y} = -\dfrac{2}{3}$

76. $\nabla f = \lambda(x\mathbf{i} + y\mathbf{j} + z\mathbf{k})$

$f(x,\ y,\ z) = \dfrac{x^2}{2} + \dfrac{y^2}{2} + \dfrac{z^2}{2} + C$

$f(0,\ 0,\ a) = \dfrac{a^2}{2};\ f(0,\ 0,\ -a) = \dfrac{a^2}{2}$

Thus, $f(0,\ 0,\ a) = f(0,\ 0,\ -a)$.

77. $\nabla f = (\cos x + ye^x)\mathbf{i} + (xe^{xy} + 2)\mathbf{j} - \mathbf{k}$

$\mathbf{N} = \pm\dfrac{1}{3}(2\mathbf{i} + 2\mathbf{j} - \mathbf{k})$

78. $\nabla f = 6(x - 2)\mathbf{i} - 10(y + 1)\mathbf{j}$

The equation of the plane is $2x + 2y - 1 = 0$

$\mathbf{N} = 2\mathbf{i} + 2\mathbf{j};\ \nabla f = \lambda\mathbf{N}$

$6(x - 2) = 2 \qquad\qquad -10(y + 1) = 2$

$\qquad x = \dfrac{7}{3} \qquad\qquad\qquad y = \dfrac{4}{5}$

79. $-\sin(x + y) - \sin(x + z)\left(1 + \dfrac{\partial z}{\partial x}\right) = 0$

$\dfrac{\partial z}{\partial x} = -\dfrac{\sin(x + y) + \sin(x + z)}{\sin(x + z)}$

Similarly, $\dfrac{\partial z}{\partial y} = -\dfrac{\sin(x + y)}{\sin(x + z)}$

$\dfrac{\partial^2 z}{\partial x\partial y} = -\dfrac{[\cos(x + y) + \cos(x + z)]}{\sin^2(x + z)}\sin(x + z)\dfrac{\partial z}{\partial y}$

$= -\dfrac{\cos(x+y)\sin^2(x+z) + \sin^2(x+y)\cos(x+z)}{\sin^3(x + z)}$

80. a. Let $t = x^2 + y^2;\ \nabla f = F'(t)(2x\mathbf{i} + 2y\mathbf{j}) = C$

b. $\nabla f(a,\ b) = F'(a,\ b)[2a\mathbf{i} + 2b\mathbf{j}]$

$\mathbf{u} = \dfrac{a\mathbf{i} + b\mathbf{j}}{\sqrt{a^2 + b^2}}$, so $\nabla f(a,\ b)$ is parallel to $\mathbf{u}$

81. $f_x = -12x^{-2} + y = 0$ if $y = 12x^{-2}$;

$f_y = -18y^{-2} + x = 0$ if $x = 18y^{-2}$

Solving simultaneously leads to $(2, 3)$.

$f_{xx} = 24x^{-3};\ f_{xy} = 1;\ f_{yy} = 36y^{-3}$

At $P(2, 3)$, $f_{xx} = 192$, $f_{xy} = 1$, $f_{yy} = \dfrac{4}{3}$, and $D > 0$, so $f(2, 3) = 18$ is a relative minimum.

82. If $y = mx$, $f(x) = 3x^4 - 4mx^3 + m^2x^2$

$f'(x) = 12x^3 - 12mx^2 + 2m^2x = 0$ if $x = 0$

$f''(x) = 36x^2 - 24mx + 2m^2 > 0$ if $m \neq 0$

Thus, $(0, 0)$ is a minimum on $y = mx$.

If $y = mx^{1/2}$, $f(x) = 3x^4 - 4mx^{5/2} + mx$

$f'(x) = 12x^3 - 10mx^{3/2} + m;\ f'(0) = m$ and

$m > 0$ makes f increasing, while $m < 0$ makes it decreasing. Thus, $(0, 0)$ is not a minimum.

83. $f(x,\ y,\ z) = x^2 + y^2 + z^2;$

$g(x,\ y,\ z) = ax + by + cz = 1$

$f_x = 2x;\ f_y = 2y;\ f_z = 2z;\ g_x = a;\ g_y = b;\ g_z = c$

The Lagrange equations are $2x = a\lambda$, $2y = b\lambda$,

$2z = c\lambda$, $ax + by + cz = 1$; solving

simultaneously we find $x = \dfrac{a}{a^2 + b^2 + c^2}$;

$y = \dfrac{b}{a^2 + b^2 + c^2};\ z = \dfrac{c}{a^2 + b^2 + c^2}$

$f_{xx} = 2;\ f_{xy} = 0;\ f_{yy} = 2;\ D > 0$, so

$f\left(\dfrac{a}{a^2 + b^2 + c^2},\ \dfrac{b}{a^2 + b^2 + c^2},\ \dfrac{c}{a^2 + b^2 + c^2}\right)$

is a relative minimum.

84. $f(x,\ y) = x^a y^{1-a}\ ;\ g(x,\ y) = ax + (1 - a)y - 1$

$f_x = ax^a y^{1-a}$; $f_y = (1-a)x^a y^{1-a}$; $g_x = a$;

$g_y = 1 - a$; the Lagrange equations are

$ax^a y^{1-a} = a\lambda$, $(1-a)x^a y^{1-a} = (1-a)\lambda$

Solving this system we find $x = 1$, $y = 1$
$f(1, 1) = 1$ is a relative minimum.

85. $f(x, y, z) = x^{1/3} y^{1/3} z^{1/3}$;

$g(x, y, z) = \frac{1}{3}(x + y + z) - C$

Maximize $P = xyz$ subject to $x + y + z = C$

The Lagrange equations are

$yz = \lambda(1)$; $xz = \lambda(1)$; $xy = \lambda(1)$

Solving this system we find $x = y = c = C/3$

so $xyz = \frac{1}{27}C^3$ which implies

$\frac{1}{27}(x + y + z)^3 \geq xyz$

so that

$$\frac{x + y + z}{3} \geq \sqrt[3]{xyz}$$

Thus, $A(x, y, z) \leq G(x, y, z)$.

86. $\left|\frac{dr}{r}\right| = 0.01$; $\left|\frac{dL}{L}\right| = 0.01$; $\left|\frac{dV}{V}\right| = 0.02$; $\left|\frac{dP}{P}\right| = 0.03$

$\ln a = \ln \pi + \ln P + 4\ln r - \ln 8 - \ln V - \ln L$

$\left|\frac{da}{a}\right| = \left|\frac{dP}{P}\right| + 4\left|\frac{dr}{r}\right| + \left|\frac{dV}{V}\right| + \left|\frac{dL}{L}\right|$

$= 0.03 + 0.04 + 0.02 + 0.01$

$= 0.1$ or 10%

87. **a.** $f_x = g_y$; $f_y = -g_x$; $f = c$, $g = d$

$\nabla f = f_x \mathbf{i} + f_y \mathbf{j}$; $\nabla g = g_x \mathbf{i} + g_y \mathbf{j}$

$\nabla f \cdot \nabla g = f_x g_x + f_y g_y$

$= f_x(-f_y) + f_y(f_x) = 0$

Thus ∇f and ∇g intersect at right angles.

b. $f_{xx} = g_{yx} = g_{xy}$; $f_{yy} = -g_{xy}$ so

$f_{xx} + f_{yy} = g_{xy} - g_{xy} = 0$

$g_{xx} = -f_{yx} = -f_{xy}$, $g_{yy} = f_{xy}$ so

$g_{xx} + g_{yy} = -f_{xy} + f_{xy} = 0$

88. $\theta = \tan^{-1}\frac{y}{x}$; $r_x = \frac{x}{r} = \cos\theta$, $r_y = \sin\theta$

$\theta_x = \frac{-y}{x^2 + y^2} = \frac{-y}{r^2} = \frac{-\sin\theta}{r}$

$\theta_y = \frac{x}{r^2} = \frac{\cos\theta}{r}$

$f_x = f_r r_x + f_\theta \theta_x = f_r \cos\theta + f_\theta(-\frac{\sin\theta}{r})$

$f_{xx} = [f_{rr}\cos\theta - f_{r\theta}\frac{\sin\theta}{r} + f_\theta\frac{\sin\theta}{r^2}]\cos\theta$

$+ [f_{r\theta}\cos\theta - f_r\sin\theta - f_{\theta\theta}\frac{\sin\theta}{r} - f_\theta\frac{\cos\theta}{r}](-\frac{\sin\theta}{r})$

$= f_{rr}\cos^2\theta - f_{r\theta}\frac{\sin\theta\cos\theta}{r} + f_\theta\frac{\sin\theta\cos\theta}{r^2}$

$- f_{r\theta}(\sin\theta\cos\theta) + f_r\frac{\sin^2\theta}{r} + f_{\theta\theta}\frac{\sin^2\theta}{r^2}$

$+ f_\theta\frac{\sin\theta\cos\theta}{r^2}$

$f_y = f_r r_y + f_\theta \theta_y = f_r \sin\theta + f_\theta\frac{\cos\theta}{r}$

$f_{yy} = [f_{rr}\sin\theta + f_{r\theta}\frac{\cos\theta}{r} - f_\theta\frac{\cos\theta}{r^2}]\sin\theta$

$+ [f_{r\theta}\sin\theta + f_{\theta\theta}\frac{\cos\theta}{r} - f_\theta\frac{\sin\theta}{r} + f_r\cos\theta]\frac{\cos\theta}{r}$

$= f_{rr}\sin^2\theta + f_{r\theta}\frac{\sin\theta\cos\theta}{r} - f_\theta\frac{\sin\theta\cos\theta}{r^2}$

$+ f_{r\theta}\frac{\sin\theta\cos\theta}{r} + f_r\frac{\cos^2\theta}{r} + f_{\theta\theta}\frac{\cos^2\theta}{r^2}$

$- f_\theta\frac{\sin\theta\cos\theta}{r^2}$

$f_{xx} + f_{yy} = f_{rr} + \frac{1}{r}f_r + f_{\theta\theta}\frac{1}{r^2}$

89. The perimeter is $f(r, \theta) = 2r + r\theta$; the area is
$g(r, \theta) = \frac{1}{2}r^2\theta - C$

$f_{rr} = 2 + \theta$; $f_\theta = r$; $g_{rr} = r\theta$; $g_\theta = \frac{1}{2}r^2$

The Lagrange equations are $2 + \theta = \lambda r\theta$,

$r = \frac{\lambda r^2}{2}$, $r = 2c$; solving simultaneously we

find $\theta = 2$ and $r = \sqrt{A}$.

90. $Q(x, y) = x^a y^b$; $Q_x = ax^{a-1}y^b$; $Q_y = bx^a y^{b-1}$

$xQ_x + yQ_y = ax^a y^b + bx^a y^b$

$= (a + b)x^a y^b$

$= (a + b)Q$

91. **a.** $dD = dH = 0.5$;
$D_0 = 4$; $H_0 = 8$; $dr = 0.25$

$V = \pi r^2 H$

$dV = \pi(2rH\, dr + \pi r^2\, dH)$

$= \pi(4)(8)(0.25) + \pi(4)(0.5)$

$= 10\pi$

b. $S = 2\pi rH = 2DH$

$dS = \pi[D\, dH + H\, dD]$

$= \pi[(4)(0.5) + (8)(0.5)]$

$= 6\pi$

92. $\left|\frac{dr}{r}\right| = \left|\frac{dh}{h}\right| = 0.02$

$V = \frac{\pi}{3}r^2 h$

$\ln V = \ln\frac{\pi}{3} + 2\ln r + \ln h$

$\left|\frac{dV}{V}\right| = \left|\frac{2\, dr}{r}\right| + \left|\frac{dh}{h}\right|$

$$= 0.04 + 0.02 = 0.06; \; 6\%$$

93. $V = \pi r^2 h; \; dr = -0.02; \; dh = -0.04$

$$dV = \pi[r^2 \, dh + 2rh \, dr]$$

$$= \pi[4(-0.04) + 12(2)(-0.02)]$$

$$\approx -0.64$$

Approximately 0.64 cm^3 has been removed.

94. $\mathbf{u} = a\mathbf{i} + b\mathbf{j}; \; D_u f = af_x + bf_y$

$$\nabla D_u f = (af_{xx} + bf_{xy})\mathbf{i} + (af_{xy} + bf_{yy})\mathbf{j}$$

$$D_u[D_u f] = D_u{}^2 f(x, y)$$

$$= a^2 f_{xx}(x, y) + 2abf_{xy}(x, y) + b^2 f_{yy}(x, y)$$

95. $f(x, y, z) = \sqrt{x} + \sqrt{y} + \sqrt{z} - \sqrt{k}$

The equation of the tangent plane is

$$\frac{1}{2\sqrt{x_0}}(x - x_0) + \frac{1}{2\sqrt{y_0}}(y - y_0) + \frac{1}{2\sqrt{z_0}}(z - z_0) = 0$$

Multiplying by $\sqrt{x_0 y_0 z_0}$ gives

$$\sqrt{y_0 z_0}(x - x_0) + \sqrt{x_0 z_0}(y - y_0) + \sqrt{x_0 y_0}(z - z_0) = 0$$

$$\sqrt{y_0 z_0}\,x + \sqrt{x_0 z_0}\,y + \sqrt{x_0 y_0}\,z$$

$$= x_0\sqrt{y_0 z_0} + y_0\sqrt{x_0 z_0} + z_0\sqrt{x_0 y_0}$$

$$\frac{x}{\sqrt{x_0}} + \frac{y}{\sqrt{y_0}} + \frac{z}{\sqrt{z_0}} = \sqrt{x_0} + \sqrt{y_0} + \sqrt{z_0} = \sqrt{k}$$

The intercepts are $\sqrt{kx_0}$, $\sqrt{ky_0}$, and $\sqrt{kz_0}$, and the sum of the intercepts is

$$(\sqrt{x_0} + \sqrt{y_0} + \sqrt{z_0})\sqrt{k} = k$$

96. $V = \pi r^2 \ell + \frac{4}{3}\pi r^3; \; S = 2\pi r\ell + 4\pi r^2$

$$R = \frac{S}{V}$$

$$= \frac{2\pi r\ell + 4\pi r^2}{\pi r^2 \ell + \frac{4}{3}\pi r^3}$$

$$= \frac{6\ell + 12r}{3r\ell + 4r^2}$$

$$\frac{\partial R}{\partial r} = \frac{(3r\ell + 4r^2)(12) - (6\ell + 12r)(3\ell + 8r)}{(3r\ell + 4r^2)^2}$$

$$= \frac{6(-8r\ell - 10r^2 - 3\ell^2)}{(3r\ell + 4r2)^2} < 0$$

$$\frac{dR}{d\ell} = \frac{(3r\ell + 4r2)(6) - (6\ell + 12r)(3r)}{(3r\ell + 4r^2)^2}$$

$$= \frac{-12r^2}{(3r\ell + 4r^2)^2} < 0$$

97.   If $f(x, y, z) = x^2 + y^2 + z^2$;

$$g(x, y, z) = x^2 + y^2 + z - 4$$

$$f_x = 2x, \; f_y = 2y, \; f_z = 2z;$$

$g_x = 2x, \; g_y = 8y, \; g_z = 1$

The Lagrange equations are $2x = 2x\lambda$,

$2y = 8y\lambda, \; 2z = \lambda, \; x^2 + 4y^2 + z^2 = 4$

Solving this system, we have $(0, 0, 4)$, $(0, \pm\sqrt{31/32}, 1/8), (\pm\sqrt{7/2}, 0, 1/2)$. All points on the ellipse $x^2 + 4y^2 = 4$ are acceptable. One of these is $(0, 1, 0)$.

98.   **a.** $z = \ln A - \ln B + \ln\cos y - \ln\cos x$

$$z_x = \frac{\sin x}{\cos x} = \tan x; \; z_y = \frac{-\sin y}{\cos y} = -\tan y$$

$$z_{xx} = \sec^2 x; \; z_{yy} = -\sec^2 y; \; z_{xy} = 0$$

$$(1 + z_y{}^2)z_{xx} - zz_{xy} + (1 + z_x{}^2)z_{yy}$$

$$= (1 + \tan^2 y)\sec^2 x + (-\ln A + \ln B)$$

$$- [\ln\cos y + \ln\cos x](0)$$

$$+ (1 + \tan^2 x)(-\sec^2 y) = 0$$

As you can see, A and B are immaterial if $AB > 0$ and $AB\cos x\cos y > 0$

b. $z_x = C(\frac{\cos x}{2}) = C\cot x; \; z_y = D\cot y$

$$z_{xx} = -\csc^2 x; \; z_{yy} = -D\csc^2 y; \; z_{xy} = 0$$

$$(1 + z_y{}^2)z_{xx} - zz_{xy} + (1 + z_x{}^2)z_{yy}$$

$$= (1 + D^2\cot^2 y)(C\csc^2 x) - z(0)$$

$$+ (1 + C^2\cot^2 x)(D\csc^2 y)$$

If $D = C = 1$, then

$$(1 + \cot^2 y)(\csc^2 x) + (1 + \cot^2 x)\csc^2 x \neq 0$$

If $C = 1, D = -1$, or $C = -1, D = 1$

$$(1 + \cot^2 y)(\csc^2 x) - (1 + \cot^2 x)\csc^2 x = 0$$

So yes, it is possible to find C and D such that $z = C\ln(\sin x) + D(\sin y)$ is a minimal surface.

99.   This is Putnam Problem 6 in the afternoon session of 1967. Consider the function g whose values are defined by
$g(x, y) = f(x, y) + 2(x^2 + y^2)$
On the circumference of the unit circle, $g(x, y) \geq 1$, and at the origin, $g(0, 0) \leq 1$. Hence, either $g(x, y)$ is a constant and $f(x, y)$ is also a constant $- 2(x^2 + y^2)$, or there is a minimum value for $g(x, y)$ at some interior point. In this case, $g(x, y) = a$ constant and the result is immediate. Otherwise, let (x_0, y_0) be the coordinates of a point where $g(x, y)$ has a minimum. Then, $g_x = g_y = 0$ at (x_0, y_0) and $|f_x(x_0, y_0)| \leq 4|x_0|$, $|f_y(x_0, y_0)| \leq 4|y_0|$. Thus, the conclusion

follows.

100. This is Putnam Problem 5 the morning session of 1946. The tangent plane to the given ellipsoid at the point (x_1, y_1, z_1) has equation

$$\frac{xx_1}{a^2} + \frac{yy_1}{b^2} + \frac{zz_1}{c^2} = 1$$

Its intercepts on the x, y, and z-axes, respectively, are a^2/x_1, b^2/y_1, c^2/z_1. Its volume is cut off by the tangent plane, and the three coordinate planes are:

$$V = \frac{1}{6}\left|\frac{a^2 b^2 c^2}{x_1 y_1 z_1}\right| \quad \text{(If } x_1 y_1 z_1 = 0\text{, then the four}$$
planes do not bound a finite region.)

$$V^2 = \tfrac{1}{36}a^2 b^2 c^2 (x_1{}^2 a^{-2} y_1{}^2 b^{-2} z_1{}^2 c^{-2})^{-1}$$

$$(x_1{}^2 a^{-2} y_1{}^2 b^{-2} z_1{}^2 c^{-2})^{1/3}$$

$$\le \tfrac{1}{3}(x_1{}^2 a^{-2} + y_1{}^2 b^{-2} + z_1{}^2 c^{-2})$$

$$= \tfrac{1}{3}$$

Equality will hold if and only if

$$\frac{x_1{}^2}{a^2} = \frac{y_1{}^2}{b^2} = \frac{z_1{}^2}{c^2} = \frac{1}{3}$$

Hence

$$V^2 = \tfrac{27}{36}a^2 b^2 c^2 \text{ and } V \ge \tfrac{1}{2}\sqrt{3}\,abc$$

with equality if and only if (x_1, y_1, z_1) is one of the eight points for which

$$\frac{x_1{}^2}{a^2} = \frac{y_1{}^2}{b^2} = \frac{z_1{}^2}{c^2} = \frac{1}{3}$$

holds, namely

$$\left(\pm\frac{a}{\sqrt{3}}, \pm\frac{b}{\sqrt{3}}, \frac{c}{\sqrt{3}}\right)$$

Note: It is also possible to minimize V straightforward by maximizing the product $x_1 y_1 z_1$ under the constraint (x_1, y_1, z_1) is a point of the ellipsoid.

101. This is Putnam Problem 13 in the afternoon session of 1938. If the given plane fails to intersect the ellipsoid, then the shortest distance is the distance between the given plane and the nearer of the two tangent planes to the ellipsoid that are parallel to the given plane. The tangent plane to the ellipsoid at the point (x_0, y_0, z_0) is

$$\frac{x_0 x}{a^2} + \frac{y_0 y}{b^2} + \frac{z_0 z}{c^2} = 1$$

If this plane is parallel to $Ax + By + Cz + 1 = 0$, then

$$\frac{x_0}{a^2} = kA; \frac{y_0}{b^2} = kB; \frac{z_0}{c^2} = kC$$

where k is a constant.

$$1 = \frac{x_0}{a^2} + \frac{y_0}{b^2} + \frac{z_0}{c^2}$$

$$= k^2[a^2 A^2 + b^2 B^2 + c^2 C^2]$$

We obtain $|k| = 1/(my)$. The distance from the origin to the given plane is

$$\frac{1}{\sqrt{A^2 + B^2 + C^2}} = h$$

Since the parallel tangent plane can be written in the form $k(Ax + By + Cz) = 1$, the distance from the origin to either parallel plane is

$$\frac{1}{|k|\sqrt{A^2 + B^2 + C^2}} = hm$$

Hence, if $m < 1$, the given plane lies farther from the origin than the tangent planes, and it does not cut the ellipsoid. The distance from the ellipsoid to the given plane in this case is $h(1 - m)$. But if $m \ge 1$, the given plane either lies between the tangent planes or coincides with one of them, so it cuts the ellipsoid and the distance is 0.

CHAPTER 13

Multiple Integration

13.1 Double Integration Over Rectangular Regions, Page 828

1. $\displaystyle\int\int_R 4\ dA = 4(2)(4) = 32$

2. $\displaystyle\int\int_R 5\ dA = (5)(5-2)(3-1) = 30$

3. $\displaystyle\int\int_R (4-y)\ dA = \frac{1}{2}(4)(4)(3) = 24$

4. $\displaystyle\int\int_R (4-2y)\ dA = \frac{1}{2}(4)(2)(4) = 16$

5. $\displaystyle\int\int_R \frac{y}{2}\ dA; \ R = \frac{1}{2}(4)(2)(6) = 24$

6. $\displaystyle\int\int_R \frac{y}{4}\ dA = \frac{1}{2}(8)(2)(2) = 16$

7. $\displaystyle\int\int_R x^2 y\ dA = \int_1^2\int_0^1 x^2 y\ dy\ dx$

$\displaystyle = \int_1^2 \frac{x^2 y^2}{2}\Big|_0^1 dx = \int_1^2 \frac{x^2}{2}\ dx = \frac{x^3}{6}\Big|_1^2 = \frac{7}{6}$

8. $\displaystyle\int\int_R (x+2y)\ dA = \int_1^3\int_{-1}^1 (x+2y)\ dy\ dx$

$\displaystyle = \int_2^3 (xy + y^2)\Big|_{-1}^1 dx = \int_2^3 2x\ dx = x^2\Big|_2^3 = 5$

9. $\displaystyle\int\int_R 2xe^y dA = \int_{-1}^0\int_0^{\ln 2} 2xe^y\ dy\ dx$

$\displaystyle = \int_0^2 2x(e^{\ln 2} - 1)\ dx = -1$

10. $\displaystyle\int\int_R x^2 e^{xy}\ dA = \int_0^1\int_0^1 x^2 e^{xy}\ dy\ dx$

$\displaystyle = \int_0^1 (xe^x - x)\ dx = \frac{1}{2}$

11. $\displaystyle\int\int_R \frac{2xy\ dA}{x^2+1} = \int_0^1\int_1^3 \frac{2xy}{x^2+1}\ dy\ dx$

$\displaystyle = 4\int_0^1 \frac{2x}{x^2+1}\ dx = 4\ln 2$

12. $\displaystyle\int\int_R y\sqrt{1-y^2}\ dA = \int_1^5\int_0^1 y\sqrt{1-y^2}\ dy\ dx$

$\displaystyle = \frac{1}{3}\int_1^5 dx = \frac{4}{3}$

13. $\displaystyle\int\int_R \sin(x+y)\ dA = \int_0^{\pi/4}\int_0^{\pi/2} \sin(x+y)\ dy\ dx$

$\displaystyle = -\int_0^{\pi/4} [\cos(x+\tfrac{\pi}{2}) - \cos x]\ dx = 1$

14. $\displaystyle\int\int_R x\sin xy\ dA = -\int_0^\pi (\cos x - 1)\ dx = \pi$

15. $\displaystyle\int_0^1\int_0^2 (x+3y) = dy\ dx = \int_0^1 (4x+6)\ dx = 8$

16. $\displaystyle\int_0^1\int_0^5 (5x+2y)\ dy\ dx = \int_0^1 (10x^2+4)\ dx = 9$

17. $\displaystyle\int_0^a\int_0^b (ax+by)\ dy\ dx = \int_0^a (abx + \tfrac{1}{2}b^3)\ dx$

$\displaystyle = \frac{a^3 b}{2} + \frac{ab^3}{2}$

18. $\displaystyle\int_0^b\int_0^c axy\ dy\ dx = \frac{ac^2}{2}\int_0^b x\ dx = \frac{ab^2 c^2}{4}$

19. $\displaystyle\int_0^1\int_0^4 x^{1/2} y^{1/2}\ dy\ dx = \frac{2}{3}\int_0^1 x^{1/2}(8)\ dx = \frac{32}{9}$

20. $\displaystyle\int_0^a\int_0^b x^{1/2} y^{1/2}\ dy\ dx = \frac{2}{3}\int_0^a x^{1/2} b^{3/2}\ dx = \frac{4}{9}(ab)^{3/2}$

21. $\displaystyle\int_0^2\int_0^{\ln 2} xe^y\ dy\ dx = \int_0^2 x(2-1)\ dx = 2$

22. $\displaystyle\int_0^1\int_0^{\ln 3} xe^{xy} = \int_0^1 (1 - e^{x(\ln 3)})\ dx = 1 - \frac{2}{\ln 3}$

23. $\displaystyle\int_0^1\int_0^1 (x+y)^5\ dy\ dx = \frac{1}{6}\int_0^1 [(x+1)^6 - x^6]\ dx = 3$

24. $\displaystyle\int_0^1 \int_0^1 \sqrt{x+y} = \frac{3}{2}\int_0^1 [(x+1)^{3/2} - x^{3/2}]\, dx$

$\displaystyle= \frac{15}{4}(2^{5/2} - 1)$

25. $\displaystyle\int_0^1 \int_0^1 (x+y)^n = \frac{1}{n+1}\int_0^1 [(x+1)^{n+1} - x^{n+1}]\, dx$

$\displaystyle= \frac{2^{n+2} - 1}{(n+1)(n+2)}$

26. Divide an area into a grid horizontally and vertically. In the narrow representative rectangle $dy\, dx$ sum up elements vertically (integrate) holding x constant. The limits of integration are from the lower curve to the upper curve. Then sum up (integrate) horizontally (from constant to constant).

27. Either integrate as suggested in Problem 26, or sum up horizontally first, and then vertically.

28. $\displaystyle\int_0^{200} \int_0^{3,000} dy\, dx = 600,000$

29. $\displaystyle\int_0^a \int_0^c f(x,y)\, dy\, dx$

30. $\displaystyle\int_a^b \int_c^d 12 e^{-0.07\sqrt{x^2+y^2}}\, dy\, dx$

31. $\displaystyle 1,400 \int_a^b \int_c^d f(x,y)\, dy\, dx$

32. $\displaystyle\int_0^2 \int_0^1 x(1 - x^2)^{1/2} e^{3y}\, dx\, dy = \frac{1}{3}\int_0^2 e^{3y}\, dy$

$\displaystyle= \frac{e^6}{9} - \frac{1}{9}$

33. $\displaystyle\int_1^4 \int_1^2 \frac{\ln\sqrt{y}}{xy}\, dy\, dx = \frac{1}{4}\int_1^4 x^{-1}\, dx = \frac{1}{2}\ln 2$

34. $\displaystyle\int_1^3 \int_1^2 \frac{xy}{x^2 + y^2}\, dy\, dx$

$\displaystyle= \frac{1}{2}\int_1^3 [x\ln(x^2 + 4) - x\ln(x^2 - 1)]\, dx$

$\displaystyle= \frac{1}{2}[x\ln(x^2 + 4) - 2x + \tan^{-1}\frac{x}{2}]\Big|_1^3$

$\displaystyle= \frac{1}{2}[3\ln\frac{13}{10} + \tan^{-1}\frac{3}{2} + 2\tan^{-1}3$

$\displaystyle= \tan^{-1}\frac{1}{2} - 2\tan^{-1}1] \approx 1.1168$

35. $z = f(x,y)$ is a paraboloid opening downward with vertex at $z = 4$ and intercepts of $(\pm 2, 0, 0)$ and $(0, \pm 2, 0)$. The integral represents the volume above the unit square in the first octant. The minimum value for $z = f(1,1) = 2$. Since $z \geq 2$ over the given R, the value of the integral will be greater than the volume of the prism with unit base and height of 2.

36. Approximations may vary depending on the the evaluation of z_k over each cell of the grid.
$\displaystyle A \approx (0.25)^2 \sum_{k=1}^{16} (0.0625)(56.75) \approx 3.5469$

37. Approximations may vary depending on the the evaluation of z_k over each cell of the grid.
$\displaystyle A \approx (0.25)^2 \sum_{k=1}^{16} (0.0625)(23) \approx 1.44$

38. The integral is the volume of a cylinder of radius 3, height 3 minus the volume of a cone of radius 3 and height 3: $27\pi - 9\pi = 18\pi$

39. The integral is the volume of a cylinder of radius 1, height 1 minus the volume of a cone with radius 1 and height 1:
$$\pi r^2 h - \frac{1}{3}\pi r^2 h = \frac{2}{3}\pi r^2 h = \frac{2\pi}{3}$$

40. The integral is the volume of a cylinder of radius 4, height 4 minus the volume of a cone with radius 4 and height 4:
$$\frac{2\pi}{3}(4^2)(4) = \frac{128\pi}{32}$$

41. $\displaystyle\int\int_R \frac{\partial}{\partial x}\frac{\partial f(x,y)}{\partial y}\, dA = \int_{x_1}^{x_2}\int_{y_1}^{y_2} \frac{\partial}{\partial x}\frac{\partial f(x,y)}{\partial y}\, dy\, dx$

$\displaystyle= \int_{x_1}^{x_2} \frac{\partial}{\partial x} f(x,y)\Big|_{y=y_1}^{y=y_2}$

$\displaystyle= \int_{x_1}^{x_2}\left[\frac{\partial f(x,y_2)}{\partial x} - \frac{\partial f(x,y_1)}{\partial x}\right]\, dx$

$\displaystyle= f(x,y_2) - f(x,y_1)\Big|_{x=x_1}^{x=x_2}$

$\displaystyle= f(x_2,y_2) - f(x_1,y_2) - f(x_2,y_1) + f(x_1,y_1)$

42. $\displaystyle\int\int_R f(x,y)\, dA = \lim_{\|P\|\to 0} \sum_{k=0}^N A(x_k)\Delta x_k$

$\displaystyle A(x) = \int_c^d f(x,y)\, dy,$ so

$\displaystyle\int\int_R f(x,y)\, dA = \int_a^b \underbrace{\int_c^d f(x,y)\, dy}_{A(x)}\, dx$

13.2 Double Integration Over Nonrectangular Regions, Page 836

1. Given $z = f(x, y)$, start with a column $f(x, y)$ $dy\ dx$ over a cell $dy\ dx$ in the xy-plane. Sum up (integrate) over the area in the xy-plane.

2. Sketch the region in the xy-plane. Sum up vertically (integrate) instead of horizontally (or vice-versa) from curve to curve, then horizontally from constant to constant.

3. Use vertical strips:

$$\int_0^4 \int_0^{4-x} xy\ dy\ dx$$

$$= \int_0^4 \frac{x}{2}(4-x)^2\ dx$$

$$= \frac{32}{3}$$

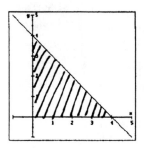

4. Use vertical strips:

$$\int_0^4 \int_{x^2}^{4x} dy\ dx$$

$$= \int_0^4 (4x - x^2)\ dx$$

$$= \frac{32}{3}$$

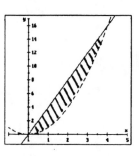

5. Use vertical strips:

$$\int_0^1 \int_{-x^2}^{x^2} dy\ dx$$

$$= \int_0^4 2x^2\ dx = \frac{2}{3}$$

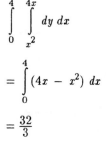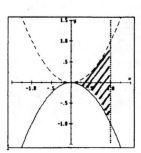

6. Use horizontal strips:

$$\int_1^e \int_0^{\ln x} xy\ dy\ dx$$

$$= \int_0^1 \int_{e^y}^e xy\ dx\ dy$$

$$= \int_0^1 \frac{y}{2}(e^2 - e^{2y})\ dy = \frac{e^2 - 1}{8} \approx 0.7986$$

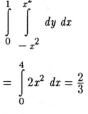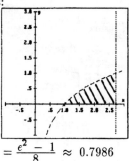

7. Use horizontal strips:

$$\int_0^{2\sqrt{3}} \int_{y^2/4}^{\sqrt{12-y^2}} dx\ dy$$

$$= \int_0^{2\sqrt{3}} \left(\sqrt{12-y^2} - \frac{y^2}{4}\right) dy$$

$$= \left(\frac{y}{2}\sqrt{12-y^2} + 6\sin^{-1}\frac{y}{2\sqrt{3}} - \frac{y^3}{12}\right)\Big|_0^{2\sqrt{3}}$$

(Formula #231)

$$= 3\pi - 2\sqrt{3} \approx 5.9607$$

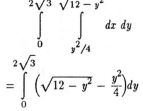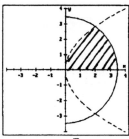

8. Use horizontal strips:

$$\int_{-2}^1 \int_{y^2+4y}^{3y+2} dx\ dy$$

$$= \int_{-2}^1 (-y^2 - y + 2)\ dy$$

$$= \frac{9}{2}$$

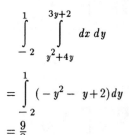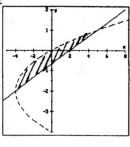

9. Use vertical strips:

$$\int_0^3 \int_1^{4-x} (x+y)\ dy\ dx$$

$$= \int_0^3 -\frac{1}{2}(x^2 + 2x - 15)\ dx$$

$$= \frac{27}{2}$$

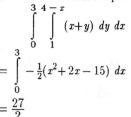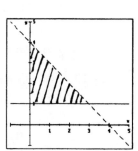

10. Use vertical strips:

$$\int_0^1 \int_{x^3}^1 (x + y^2)\ dy\ dx$$

$$= \int_0^1 \left(x + \frac{1}{3} - x^4 - \frac{x^9}{3}\right) dx$$

$$= \frac{3}{5}$$

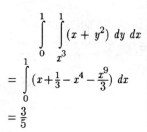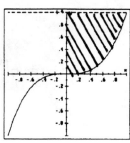

11. Use vertical strips:

$$\int_0^1 \int_0^x (x^2+2y^2)\,dy\,dx$$

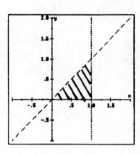

$$= \int_0^1 \left(x^3 + \frac{2x^3}{3}\right)\,dx$$

$$= \frac{5}{12}$$

12. Use vertical strips:

$$\int_{-1}^1 \int_{-1}^x (3x+2y)\,dy\,dx$$

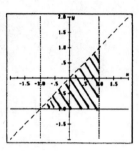

$$= \int_{-1}^1 (4x^2+3x-1)\,dx$$

$$= \frac{2}{3}$$

13. Use vertical strips:

$$\int_0^2 \int_0^{\sin x} y \cos x \, dy \, dx$$

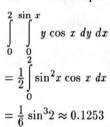

$$= \frac{1}{2}\int_0^2 \sin^2 x \cos x \, dx$$

$$= \frac{1}{6}\sin^3 2 \approx 0.1253$$

14. Use vertical strips:

$$\int_0^{\pi/2} \int_0^{\sin x} e^y \cos x \, dy \, dx$$

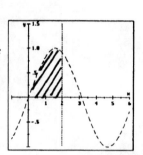

$$= \int_0^{\pi/2} (e^{\sin x}\cos x - \cos x)\,dx$$

$$= (e^{\sin x} - \sin x)\Big|_0^{\pi/2} = e - 2 \approx 0.7183$$

15. Use vertical strips:

$$\int_0^1 \int_x^1 (x + y)\,dy\,dx = \int_0^1 \left(x + \frac{1}{2} - \frac{3x^2}{2}\right)\,dx = \frac{1}{2}$$

16. Use vertical strips:

$$\int_0^1 \int_0^{-2x+2} (x+2y)\,dy\,dx = \int_0^2 (2x^2 - 6x + 4)\,dx = \frac{5}{3}$$

17. Use vertical strips:

$$\int_0^1 \int_{x^2}^{\sqrt{x}} 48 \, xy \, dy \, dx = 24\int_0^1 (x^2 - x^7)\,dx = 5$$

18. Use vertical strips:

$$\int_0^2 \int_{x^2}^{2x} (2y - x)\,dy\,dx = 24\int_0^2 (2x^2 - x^4 + x^3)\,dx = \frac{44}{15}$$

19. Use horizontal strips:

$$\int_0^1 \int_{y^2}^{2-y} y \, dx \, dy = \int_0^1 (2y - y^2 - y^3)\,dy = \frac{5}{12}$$

20. Use vertical strips:

$$\int_0^1 \int_{3x}^{4-x^2} 4x \, dy \, dx = 4\int_0^1 (4x - x^3 - 3x^2)\,dx = 3$$

21. Use horizontal strips:

$$\int_0^3 \int_{y/3}^{\sqrt{4-y}} 4x \, dx \, dy = -\frac{2}{9}\int_0^3 (y^2 + 9y - 36)\,dy = 13$$

22. Use horizontal strips:

$$\int_0^1 \int_{y-1}^{-y+1} (2x + 1)\,dx\,dy = 2\int_0^1 (-y + 1)\,dy = 1$$

23. Use vertical strips:

$$\int_0^1 \int_0^x 2x \, dy \, dx + \int_1^2 \int_0^{1/x^2} 2x \, dy \, dx$$

$$= \int_0^1 2x^2 \, dx + \int_1^2 \frac{2}{x}\,dx = \frac{2}{3} + \ln 4 \approx 2.0530$$

24. Use horizontal strips:

$$\int_0^2 \int_{-y}^{2y} \frac{1}{y^2+1}\,dx\,dy = \int_0^2 \frac{3y}{y^2+1}\,dy = \frac{3}{2}\ln 5 \approx 2.4142$$

25. Use horizontal strips:

$$\int_0^1 \int_y^{y^{1/3}} 12x^2 e^{y^2}\,dx\,dy = 4\int_0^1 (y - y^3)e^{y^2}\,dy$$

$$= 2e - 4 \approx 1.4366$$

26. Use vertical strips:

$$\int_{1}^{e}\int_{0}^{\ln 5} y\,dy\,dx = \frac{1}{2}\int_{1}^{e}\ln^2 x\,dx$$

$$= [x\ln^2 x - 2x\ln x + 2x]\Big|_{1}^{e} = \frac{e-2}{2} \approx 0.3591$$

27.

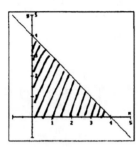

a. vertical strips:

$$\int_{0}^{4}\int_{0}^{4-x} xy\,dy\,dx = \frac{1}{2}\int_{0}^{4}(x^3 - 8x^2 + 16x)\,dx = \frac{32}{3}$$

b. horizontal strips:

$$\int_{0}^{4}\int_{0}^{4-y} xy\,dx\,dy = \frac{1}{2}\int_{0}^{4}(y^3 - 8y^2 + 16y)\,dy = \frac{32}{3}$$

28.

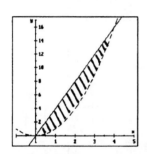

a. vertical strips

$$\int_{0}^{4}\int_{x^2}^{4x} dy\,dx = \int_{0}^{4}(4x - x^2)\,dx = \frac{32}{3}$$

b. horizontal strips

$$\int_{0}^{16}\int_{y/4}^{\sqrt{y}} dx\,dy = \int_{0}^{16}(y^{1/2} - \frac{y}{4})\,dy = \frac{32}{3}$$

29.

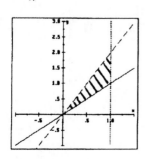

a. vertical strips

$$\int_{0}^{1}\int_{x}^{2x} e^{y-x}\,dy\,dx = \int_{0}^{1}(e^x - 1)\,dx = e - 2 \approx 0.7183$$

b. horizontal strips

$$\int_{0}^{1}\int_{y/2}^{y} e^{y-x}\,dx\,dy + \int_{1}^{2}\int_{y/2}^{1} e^{y-x}\,dx\,dy$$

$$= -\int_{0}^{1}(1 - e^{y/2})\,dy - \int_{1}^{2}(e^{y-1} - e^{y/2})\,dy = e - 2$$

30.

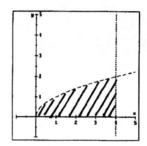

a. vertical strips

$$\int_{0}^{4}\int_{0}^{\sqrt{x}} 3x^5\,dy\,dx = 3\int_{0}^{4} x^{11/2}\,dx = \frac{3(2^{14})}{13} \approx 3{,}781$$

b. horizontal strips

$$\int_{0}^{2}\int_{y^2}^{4} 3x^5\,dx\,dy = \frac{1}{2}\int_{0}^{2}(4^6 - y^{12})\,dy = 2^{12} - \frac{2^{12}}{13} \approx 3{,}781$$

31.

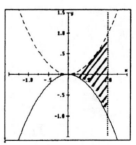

a. vertical strips

$$\int_{0}^{1}\int_{-x^2}^{x^2} dy\,dx = 2\int_{0}^{1} x^2\,dx = \frac{2}{3}$$

b. horizontal strips

$$\int_{0}^{1}\int_{\sqrt{y}}^{1} dx\,dy + \int_{-1}^{0}\int_{\sqrt{-y}}^{1} dx\,dy$$

$$= 2\int_{0}^{1}\int_{\sqrt{y}}^{1} dx\,dy = 2\int_{0}^{1}(1 - \sqrt{y})\,dy = \frac{2}{3}$$

32.

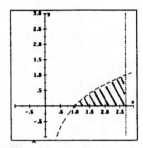

a.   vertical strips

$$\int_1^e \int_0^{\ln x} xy \, dy \, dx = \frac{1}{2}\int_1^e (x \ln^2 x) \, dx$$

$$= \frac{1}{2}(x^2 \ln^2 x)\Big|_1^e - \int_1^e x \ln x \, dx$$

$$= \frac{e^2}{2} - \frac{e^2}{4} + \frac{e^2}{8} - \frac{1}{8} = \frac{e^2 - 1}{8} \approx 0.7986$$

b.   horizontal strips

$$\int_0^1 \int_{e^y}^e xy \, dx \, dy = \frac{1}{2}\int_0^1 (e^2 y - ye^{2y}) \, dy = \frac{e^2 - 1}{8} \approx 0.7986$$

33.

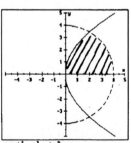

a.   vertical strips

$$\int_0^{2\sqrt{3}} \int_{y^2/6}^{\sqrt{16 - y^2}} dx \, dy = \int_0^{2\sqrt{3}} (\sqrt{16 - y^2} - \frac{y^2}{6}) \, dy$$

$$= \left[8 \sin^{-1}\frac{y}{4} + \frac{y}{2}\sqrt{16 - y^2} - \frac{y^3}{18}\right]\Big|_0^{2\sqrt{3}}$$

$$= \frac{8\pi}{3} + \frac{2\sqrt{3}}{3} \approx 9.5323$$

b.   horizontal strips

$$\int_0^2 \int_0^{\sqrt{6x}} dy \, dx + \int_2^4 \int_0^{\sqrt{6 - x^2}} dy \, dx$$

$$= \int_0^2 x^{1/2} \, dx + \int_2^4 \sqrt{16 - x^2} \, dx$$

$$= \frac{2\sqrt{3}}{3} + \frac{8\pi}{3} \approx 9.5323$$

34.

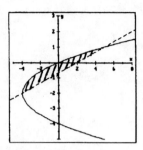

a.   horizontal strips

$$\int_{-2}^1 \int_{y^2+4y}^{3y+2} dx \, dy = -\int_{-2}^1 (y^2 + y - 2) \, dy = \frac{9}{2}$$

b.   vertical strips

$$\int_{-4}^5 \int_{1/3(x - 2)}^{-2+\sqrt{x+4}} dy \, dx = \int_{-4}^5 [-2+\sqrt{x + 4} - \frac{1}{3}x + \frac{2}{3}] \, dx$$

$$= \left[\frac{2}{3}(x + 4)^{3/2} - \frac{1}{6}x^2 - \frac{4}{3}x\right]\Big|_{-4}^5 = \frac{9}{2}$$

35.

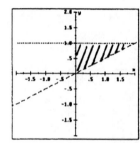

$$\int_0^2 \int_{x/2}^1 f(x, y)\,dy \, dx$$

36.

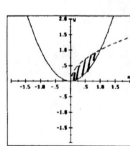

$$\int_0^1 \int_{y^2}^{\sqrt{y}} f(x, y) \, dx \, dy$$

37.

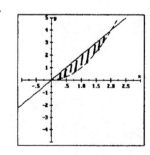

$$\int_0^2 \int_{x^2}^{2x} f(x, y) \, dy \, dx$$

38.

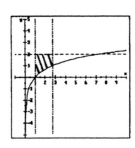

$$\int_0^2 \int_1^{e^y} f(x,\,y)\; dx\; dy$$

42.

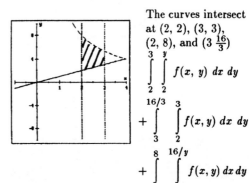

The curves intersect at $(2,\,2)$, $(3,\,3)$, $(2,\,8)$, and $(3\,\frac{16}{3})$

$$\int_2^3 \int_2^y f(x,\,y)\; dx\; dy$$

$$+ \int_3^{16/3} \int_2^3 f(x,y)\; dx\; dy$$

$$+ \int_{16/3}^8 \int_2^{16/y} f(x,y)\; dx\; dy$$

39.

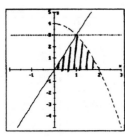

The curves intersect when

$$\frac{y}{3} = \sqrt{4-y}$$

$$y^2 + 9y - 36 = 0$$

$$(y-3)(y+12) = 0$$

Intersection: $(1,\,3)$

$$\int_0^1 \int_0^{3x} f(x,\,y)\; dy\; dx \; + \int_1^2 \int_0^{4-x^2} f(x,\,y)\; dy\; dx$$

43.

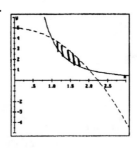

We find the intersection points:

$$\frac{4}{x^2} = 5 - x^2$$

$$x^4 - 5x^2 - 4 = 0$$

$$(x^2-4)(x^2-1) = 0$$

$$x = \pm 1,\; \pm 2$$

Intersections:

$(1,\,4)$, $(2,\,1)$.

Using horizontal strips: $\displaystyle\int_1^4 \int_{2/\sqrt{y}}^{\sqrt{5-y}} dx\; dy$

Using vertical strips:

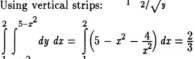

$$\int_1^2 \int_{4/x^2}^{5-x^2} dy\; dx = \int_1^2 \left(5 - x^2 - \frac{4}{x^2}\right) dx = \frac{2}{3}$$

40.

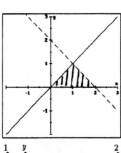

The curves intersect when $x = 2 - x$ or at $(1,\,1)$

$$\int_0^1 \int_0^y f(x,\,y)\; dx\; dy + \int_1^2 \int_1^{2-y} f(x,\,y)\; dx\; dy$$

44.

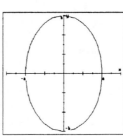

We find the limits of integration:

$$\frac{x^2}{a^2} + \frac{y^2}{b^2} = 1$$

or $y = \frac{b}{a}\sqrt{a^2 - x^2}$

Using vertical strips: $\displaystyle 4\int_0^a \int_0^{(b/a)\sqrt{a^2 - x^2}} dy\; dx$

Using horizontal strips: $\displaystyle 4\int_0^b \int_0^{(b/a)\sqrt{b^2-y^2}} dx\; dy$

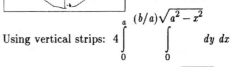

$$= \frac{4a}{b}\int_0^b \sqrt{b^2 - y^2}\; dy = \frac{4a}{b}\left[\frac{b^2}{2}\sin^{-1}\frac{y}{b} + \frac{y}{2}\sqrt{b^2-y^2}\right]\Big|_0^b$$

$$= \frac{4a}{b}\left[\frac{\pi b^2}{4}\right] = \pi ab$$

41.

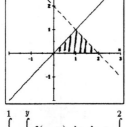

The curves intersect when

$$x^2 = 6 - x$$

$$x^2 + x - 6 = 0$$

$$(x+3)(x-2) = 0$$

Intersections:

$(2,\,4)$ and $(-3,\,9)$

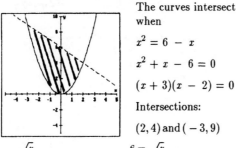

$$\int_0^4 \int_{-\sqrt{y}}^{\sqrt{y}} f(x,\,y)\; dx\; dy + \int_4^9 \int_{-\sqrt{y}}^{6-\sqrt{y}} f(x,\,y)\; dx\; dy$$

45.

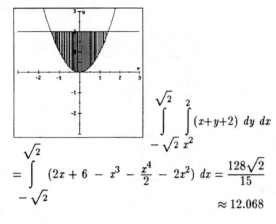

$$\int\limits_{-\sqrt{2}}^{\sqrt{2}} \int\limits_{x^2}^{2} (x+y+2) \; dy \; dx$$

$$= \int\limits_{-\sqrt{2}}^{\sqrt{2}} \left(2x + 6 \; - \; x^3 \; - \; \frac{x^4}{2} \; - \; 2x^2\right) \; dx = \frac{128\sqrt{2}}{15}$$

$$\approx 12.068$$

46.

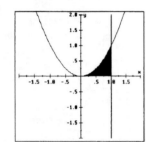

$$\int\limits_{0}^{1} \int\limits_{0}^{x^2} 4x \; dy \; dx$$

$$= 4 \int\limits_{0}^{1} x^3 \; dx = 1$$

47.

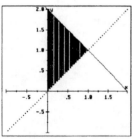

Reversing the order of integration gives a single region:

$$\int\limits_{0}^{1} \int\limits_{x}^{2-x} (x^2 \; + \; y^2) \; dy \; dx = \int\limits_{0}^{1} \left(x^2y \; + \; \frac{y^3}{3}\right)\Bigg|_{x}^{2-x} \; dx$$

$$= \int\limits_{0}^{1} \left(2x^2 - \frac{7x^3}{3} + \frac{(2-x)^3}{3}\right) \; dx = \frac{4}{3}$$

48.

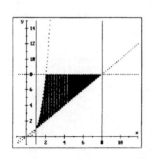

$$\int\limits_{1}^{8} \int\limits_{y^{1/3}}^{y} f(x, \; y) \; dx \; dy$$

49.

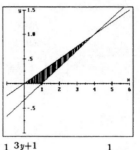

$$\int\limits_{0}^{1} \int\limits_{4y}^{3y+1} xy \; dx \; dy = \frac{1}{2}\int\limits_{0}^{1} (-7y^3 + 6y^2 + y) \; dy = \frac{3}{8}$$

50.

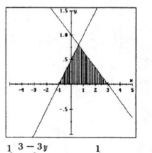

$$\int\limits_{0}^{1} \int\limits_{2y-2}^{3-3y} dx \; dy = \int\limits_{0}^{1} (5 \; - \; 5y^2) \; dy = \frac{5}{2}$$

51.

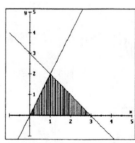

The point of intersection is $(1, 2)$. Horizontal strips will give a single region.

$$\int\limits_{0}^{2} \int\limits_{y/2}^{3-y} (x+2y+4) \; dx \; dy = \int\limits_{0}^{2} \left(-\frac{21}{8}y^2 - 3y + \frac{33}{2}\right) \; dy = 20$$

52.

$$\int\limits_{0}^{3} \int\limits_{0}^{(1/3)\sqrt{36-4x^2}} (3x \; + \; y) \; dy \; dx$$

$$= \int\limits_{0}^{3} \left[2x\sqrt{9 \; - \; x^2} + \frac{9}{2}(x^2 \; - \; 9)\right] \; dx = \frac{40}{3}$$

53. $\displaystyle 4\int\limits_{0}^{1} \int\limits_{0}^{1} (x^2 + y^2) \; dy \; dx = 4\int\limits_{0}^{1} \left(x^2 + \frac{1}{3}\right) \; dx = \frac{8}{3}$

54. $$\int\int_D m \, dA \le \int\int_D f(x, y) \, dA \le \int\int_D M \, dA$$

$$m\int\int_D dA \le \int\int_D f(x, y) \, dA \le M\int\int_D dA$$

$$m \le \int\int_D f(x, y) \, dA \le M$$

55. $f(x, y) = e^{y \sin x}$; $A = \frac{3}{2}$; $m = e^{-0.5(0.5)} \approx$ 0.78, so $mA \approx 1.17$; $M \approx 1.54$ (which in the vicinity of $(0.855, 0.5725)$). A graphing calculator or computer will be required to find these values. Now since the base region is $\frac{3}{2}$ instead of 1 as specified in Problem 54,

$$\frac{3}{2}(0.78) \le \int\int f(x, y) \le \frac{3}{2}(1.54)$$

The value of the integral is between 1.17 and 2.31.

13.3 Double Integrals in Polar Coordinates, Page 844

1. $$\int_0^{\pi/2} \int_0^{2\sin\theta} dr \, d\theta$$

$$= \int_0^{\pi/2} 2 \sin \theta \, d\theta$$

$$= 2$$

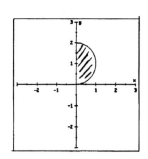

2. $$\int_0^{\pi} \int_0^{1+\sin\theta} dr \, d\theta$$

$$= \int_0^{\pi} (1 + \sin \theta) \, d\theta$$

$$= \pi + 2$$

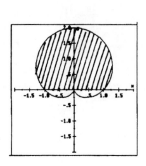

3. $$\int_0^{\pi/2} \int_1^{3} re^{-r^2} \, dr \, d\theta$$

$$= -\frac{1}{2}\int_0^{\pi/2} (e^{-9} - e^{-1}) \, d\theta$$

$$= -\frac{\pi}{4}(e^{-9} - e^{-1})$$

$$= \frac{\pi}{4}\left(\frac{1}{e} - \frac{1}{e^9}\right)$$

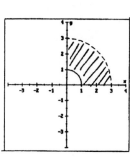

4. $$\int_0^{\pi/2} \int_1^{2} \sqrt{4 - r^2} \, r \, dr \, d\theta$$

$$= \frac{3^{3/2}}{3}\int_0^{\pi/2} d\theta$$

$$= \frac{\pi\sqrt{3}}{2}$$

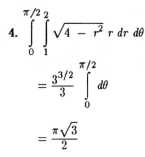

5. $$\int_0^{\pi} \int_0^{4} r^2 \sin^2\theta \, dr \, d\theta$$

$$= \frac{4^3}{3}\int_0^{\pi} \sin^2\theta \, d\theta$$

$$= \frac{32\pi}{3}$$

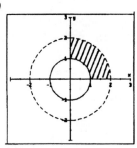

6. $$\int_0^{\pi/2} \int_1^{3} r^2 \cos^2\theta \, dr \, d\theta$$

$$= \frac{26}{3}\int_0^{\pi/2} \cos^2\theta \, d\theta$$

$$= \frac{13\pi}{6}$$

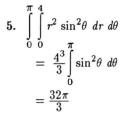

7. $$\int_0^{2\pi} \int_0^{4} 2r^2 \cos \theta \, dr \, d\theta$$

$$= \frac{128}{3}\int_0^{2\pi} \cos \theta \, d\theta$$

$$= 0$$

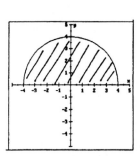

8. $\displaystyle\int_0^{2\pi}\int_0^{1-\sin\theta}\cos\theta\;dr\;d\theta$

$\displaystyle=\int_0^{2\pi}(\cos\theta-\sin\theta\cos\theta)\,d\theta$

$=0$

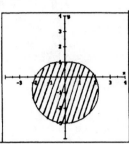

9. $\displaystyle\int_0^{2\pi}\int_0^4 r\;dr\;d\theta=8\int_0^{2\pi}d\theta=16\pi$

10. $\displaystyle 2\int_0^{\pi/2}\int_0^{2\cos\theta}r\;dr\;d\theta=4\int_0^{\pi/2}\cos^2\theta\;d\theta=\pi$

11. $\displaystyle 2\int_0^{\pi}\int_0^{2(1-\cos\theta)}r\;dr\;d\theta$

$\displaystyle=4\int_0^{\pi}\left[1-2\cos\theta+\frac{(1+\cos2\theta)^2}{2}\right]d\theta=6\pi$

12. $\displaystyle 2\int_{-\pi/2}^{\pi/2}\int_0^{1+\sin\theta}r\;dr\;d\theta=\int_{-\pi/2}^{\pi/2}(1+\sin\theta)^2\;d\theta=\frac{3\pi}{2}$

13. $\displaystyle 6\int_0^{\pi/6}\int_0^{4\cos\theta}r\;dr\;d\theta=48\int_0^{\pi/6}\cos^23\theta\;d\theta=4\pi$

14. $\displaystyle 4\int_0^{\pi/2}\int_0^{5\sin2\theta}r\;dr\;d\theta=50\int_0^{\pi/2}\sin^22\theta\;d\theta=\frac{25\pi}{2}$

15. $\displaystyle 8\int_0^{\pi/4}\int_0^{\cos2\theta}r\;dr\;d\theta=4\int_0^{\pi/4}\cos^22\theta\;d\theta=\frac{\pi}{2}$

16. $\displaystyle 2\left[\int_0^{\pi/6}\int_0^{2\sin\theta}r\;dr\;d\theta+\int_{\pi/6}^{\pi/2}\int_0^1 r\;dr\;d\theta\right]$

$\displaystyle=2\left[\int_0^{\pi/6}2\sin^2\theta\;d\theta+\int_{\pi/6}^{\pi/2}r\;dr\;d\theta\right]$

$\displaystyle=\frac{2\pi}{3}-\frac{8\pi-3\sqrt3}{12}$

17. $\displaystyle 2\int_{\pi/6}^{\pi/2}\int_1^{2\sin\theta}r\;dr\;d\theta=2\int_{\pi/6}^{\pi/2}(4\sin^2\theta-1)\;d\theta$

$\displaystyle=\tfrac16(2\pi+3\sqrt3)$

18. $\displaystyle 2\int_0^{\pi/2}\int_1^{1+\cos\theta}r\;dr\;d\theta=2\int_0^{\pi/2}[(1+\cos\theta)^2-1]\;d\theta$

$\displaystyle=2+\frac{\pi}{4}$

19. Use symmetry about the polar axis and the equation of the quarter circle.

$\displaystyle 2[\tfrac14\pi(1)^2]+2\int_{\pi/2}^{\pi}\int_0^{1+\cos\theta}r\;dr\;d\theta$

$\displaystyle=\frac{\pi}{2}+2\int_{\pi/2}^{\pi}\left(\tfrac12\cos^2\theta+\cos\theta+\tfrac12\right)d\theta$

$\displaystyle=\frac{\pi}{2}+\frac{3\pi-8}{4}=\frac{5\pi}{4}-2$

20. $\displaystyle 2\int_{\pi/2}^{\pi}\int_{1+\cos\theta}^{1}r\;dr\;d\theta=\int_{\pi/2}^{\pi}[1-\cos\theta]^2\;d\theta=2-\frac{\pi}{4}$

21. $\displaystyle 2\int_0^{\pi/3}\int_{1+\cos\theta}^{3\cos\theta}r\;dr\;d\theta$

$\displaystyle=\int_0^{\pi/3}[9\cos^2\theta-(1+\cos\theta)^2]\;d\theta$

$\displaystyle=\int_0^{\pi/3}(3+4\cos2\theta-2\cos2\theta)\;d\theta=\pi$

22. $\displaystyle 2\left[\int_0^{\pi/3}\int_0^{2-\cos\theta}r\;dr\;d\theta+\int_{\pi/3}^{\pi/2}\int_0^{3\cos\theta}r\;dr\;d\theta\right]$

$\displaystyle=\int_0^{\pi/3}(2-\cos\theta)^2\;d\theta+\int_{\pi/3}^{\pi/2}9\cos\theta^2\theta\;d\theta$

$\displaystyle=\frac{9\pi}{4}-3\sqrt3$

23. $\displaystyle\int_0^{2\pi}\int_0^a r^2\sin^2\theta\,r\;dr\;d\theta=\frac{a^4}{4}\int_0^{2\pi}\frac{1-\cos2\theta}{2}\;d\theta=\frac{a^4\pi}{4}$

24. $\displaystyle 4\int_0^{\pi/2}\int_0^a(r^2)r\;dr\;d\theta=a^4\int_0^{\pi/2}d\theta=\frac{a^4\pi}{2}$

25. $\displaystyle 4\int_0^{\pi/2}\int_0^a\frac{r}{a^2+r^2}\;dr\;d\theta=2\pi\int_0^{\pi/2}\frac{r\;dr}{r^2+a^2}=\pi\ln2$

26. $4 \displaystyle\int_0^{\pi/2} \int_0^a r e^{-r^2} \, dr \, d\theta = -2 \int_0^{\pi/2} (e^{-a^2} - 1) \, d\theta$

$= \pi(1 - e^{-a^2})$

27. $\displaystyle\int_0^{2\pi} \int_0^a \frac{1}{a+r} \, r \, dr \, d\theta = \int_0^{2\pi} \int_0^a \left(1 - \frac{a}{r+a}\right) dr \, d\theta$

$= \displaystyle\int_0^{2\pi} (a - a \ln 2a + a \ln a) \, d\theta$

$= 2\pi a(1 + \ln \tfrac{1}{2})$ or $2\pi a(1 - \ln 2)$

28. $4 \displaystyle\int_0^{\pi/2} \int_0^a r \ln(a^2 + r^2) \, dr \, d\theta$

$= 4 \displaystyle\int_0^{\pi/2} \left[-a^2 \ln \tfrac{1}{2} + a^2 \ln a - \frac{a^2}{2} \right] d\theta$

$= 2\pi \left(a^2 \ln 2 + a^2 \ln a - \frac{a^2}{2} \right)$

$= \pi a^2 (\ln 4 + 2 \ln a - 1)$

29. $\displaystyle\int_D \int y \, dA = \int_0^{2\pi} \int_0^2 r^2 \sin \theta \, dr \, d\theta$

$= \displaystyle\int_0^{2\pi} \frac{8}{3} \sin \theta \, d\theta = 0$

30. $\displaystyle\int_D \int (x^2 + y^2) \, dA = \int_0^{\pi/4} \int_0^1 r^3 \, dr \, d\theta$

$= \frac{1}{4} \displaystyle\int_0^{\pi/4} d\theta = \frac{\pi}{16}$

31. $\displaystyle\int_D \int e^{x^2+y^2} \, dA = 4 \int_0^{\pi/2} \int_0^3 r e^{r^2} \, dr \, d\theta$

$= 4 \displaystyle\int_0^{\pi/2} \tfrac{1}{2}(e^9 - 1) = \pi(e^9 - 1)$

32. $(x-1)^2 + y^2 = 1$ becomes $-2x + x^2 + y^2$
$= 0$ or $2 \cos \theta$

$\displaystyle\int_D \int \sqrt{x^2 + y^2} \, dA = 4 \int_0^{\pi/2} \int_0^{2\cos\theta} r^2 \, dr \, d\theta$

$= \frac{8}{3} \displaystyle\int_0^{\pi/2} \cos^3\theta \, d\theta = \frac{8}{3} \left[\sin \theta - \frac{\sin^3\theta}{3} \right] \Big|_0^{\pi/2} = \frac{16}{9}$

33. $\displaystyle\int_D \int \ln(x^2 + y^2 + 2) \, dA = \int_0^{\pi/2} \int_0^2 r \ln(r^2 + 2) \, dr \, d\theta$

$= \displaystyle\int_0^{\pi/2} (3 \ln 3 + 2 \ln 2 - 2) \, d\theta$

$= \frac{3}{2}\pi \ln 3 + \pi \ln 2 - \pi \approx 4.2131$

34. $\displaystyle\int_D \int \sin(x^2 + y^2) \, dA = \int_0^{\pi/6} \int_1^2 r \sin r^2 \, dr \, d\theta$

$= \displaystyle\int_0^{\pi/6} \tfrac{1}{2}(\cos 1 - \cos 4) \, d\theta = \frac{\pi}{12}(\cos 1 - \cos 4)$

≈ 0.3126

35. $\displaystyle\int_0^{2\pi} \int_0^2 r^2 \, r \, dr \, d\theta = 4 \int_0^{2\pi} d\theta = 8\pi$

36. $4 \displaystyle\int_0^{\pi/2} \int_0^a r^2 \, dr \, d\theta = 4 \int_0^{\pi/2} \frac{a^3}{3} \, d\theta = 4\pi a^3$

37. The curves intersect at $\theta = \pi/4$.

$\displaystyle\int_0^{\pi/4} \int_0^{4\sin\theta} r^3 \sin \theta \cos \theta \, dr \, d\theta$

$+ \displaystyle\int_{\pi/4}^{\pi/2} \int_0^{4\cos\theta} r^3 \sin \theta \cos \theta \, dr \, d\theta$

$= 4^4 \left[\displaystyle\int_0^{\pi/4} \sin^5\theta \cos \theta \, d\theta + \int_{\pi/2}^{\pi/4} \cos^5\theta \sin \theta \, d\theta \right]$

$= \frac{8}{3}$

38. a. $\displaystyle\int_0^1 \int_0^x dy \, dx$ **b.** $\displaystyle\int_0^{\pi/4} \int_0^{\sec\theta} r \, dr \, d\theta$

39. The curves intersect at $\theta = \pi/4$. Making use of the symmetry we will integrate the $\sin \theta$ curve from 0 to $\pi/4$ to get half of the desired area.

$A = \displaystyle\int_0^{\pi/4} \int_0^{a \sin \theta} r \, dr \, d\theta = 2 \int_0^{\pi/4} \frac{a^2 \sin^2\theta}{2} \, d\theta$

$= a^2 \displaystyle\int_0^{\pi/4} \frac{1 - \cos 2\theta}{2} \, d\theta = \frac{a^2}{8}(\pi - 2)$

40. The curves intersect at $\theta = \pi/3$

$$\int_0^{\pi/3} \int_0^{2+\cos\theta} r \, dr \, d\theta + 2\int_{\pi/3}^{\pi/2} \int_0^{5\cos\theta} r \, dr \, d\theta$$

$$= \int_0^{\pi/3} \left[4 + 4\cos\theta + \frac{1 + \cos 2\theta}{2}\right] d\theta$$

$$+ \frac{25}{2}\int_{\pi/3}^{\pi/2} (1 + \cos 2\theta) \, d\theta$$

$$= \frac{9\pi}{6} + \frac{4\sqrt{3}}{2} + \frac{\sqrt{3}}{8} + \frac{25}{2}\left(\frac{\pi}{6} - \frac{\sqrt{3}}{4}\right) \approx 9.52$$

41. $8\int_0^{\pi/2} \int_3^5 r\sqrt{25 - r^2} \, dr \, d\theta = 8\int_0^{\pi/2} \frac{64}{3} \, d\theta = \frac{256\pi}{3}$

42. $\displaystyle\lim_{n\to\infty} \int_0^{\pi/2} \int_0^n r(r^2 + 1)^{-3/2} \, dr \, d\theta$

$$= \lim_{n\to\infty} \int_0^{\pi/2} \frac{\sqrt{n^2 + 1} - 1}{\sqrt{n^2 + 1}} \, d\theta = \frac{\pi}{2}$$

43. $\displaystyle\lim_{n\to\infty} \int_0^{\pi/2} \int_0^n \frac{1}{(1 + r^2)^2} r \, dr \, d\theta$

$$= -\frac{1}{2} \lim_{n\to\infty} \int_0^{\pi/2} \left(\frac{1}{1 + n^2} - 1\right) d\theta = \frac{\pi}{4}$$

44. $\displaystyle\lim_{n\to\infty} \int_0^{\pi/2} \int_0^n r(r^2 + 1)^{-3} \, dr \, d\theta$

$$= \frac{1}{4} \lim_{n\to\infty} \int_0^{\pi/2} \left[1 - \frac{1}{(n^2 + 1)^2}\right] d\theta = \frac{\pi}{8}$$

45. $\displaystyle\lim_{n\to\infty} \int_0^{\pi/2} \int_0^n re^{-2r^2} \, dr \, d\theta$

$$= \frac{1}{4} \lim_{n\to\infty} \int_0^{\pi/2} (1 - e^{-2n^2}) \, d\theta = \frac{\pi}{8}$$

46. $\displaystyle\lim_{n\to\infty} \int_0^{\pi/2} \int_0^n re^{-4r^2} \, dr \, d\theta$

$$= \frac{1}{8} \lim_{n\to\infty} \int_0^{\pi/2} (1 - e^{-4n^2}) \, d\theta = \frac{\pi}{16}$$

47. The curves intersect at $\theta = \pi/4$.

$$V = 2\int_0^{\pi/4} \int_0^{2\sin\theta} r^2 r \, dr \, d\theta$$

$$= 8\int_0^{\pi/4} \left(\frac{1 - \cos 2\theta}{2}\right)^2 d\theta$$

$$= 2\int_0^{\pi/4} \left(1 - 2\cos 2\theta + \frac{1 + \cos 4\theta}{2}\right) d\theta$$

$$= \frac{3\pi}{4} - 2$$

48. $\displaystyle\int_0^{\pi} \int_1^2 r^3 \, dr \, d\theta = \int_0^{\pi} \frac{15}{4} \, d\theta = \frac{15\pi}{4}$

49. The answer is not 0 because of symmetry the alternate development (with answer 32/9) is correct.

50. a. Let $\displaystyle I = \int_0^\infty e^{-x^2} \, dx = \int_0^\infty e^{-y^2} \, dy$
   with x and y independent. Then
$$I^2 = \left[\int_0^\infty e^{-x^2} \, dx\right]\left[\int_0^\infty e^{-y^2} \, dy\right]$$
$$= \int_0^\infty \int_0^\infty e^{-(x^2 + y^2)} \, dy \, dx$$

   b. Let $\displaystyle I = \int_0^\infty \int_0^\infty e^{-(x^2 + y^2)} \, dy \, dx$
$$= \lim_{n\to\infty} \int_0^{\pi/2} e^{-2x^2} \, dx \int_0^n re^{-r^2} \, dr \, d\theta$$
$$= -\frac{\pi}{4} \lim_{n\to\infty} e^{-r^2}\Big|_0^n = \frac{\pi}{4}$$

   c. $\displaystyle I = \sqrt{I^2} = \frac{\sqrt{\pi}}{2}$

51. Let $u = \sqrt{2}\, x$, $du = \sqrt{2}\, dx$ then our integral becomes:

$$\int_0^\infty e^{-u^2} \frac{du}{\sqrt{2}} = \frac{1}{\sqrt{2}} \int_0^\infty e^{-u^2} \, du = \frac{1}{\sqrt{2}}\left(\frac{\sqrt{\pi}}{2}\right) = \frac{\sqrt{2\pi}}{4}$$

by Problem 50c.

52. In the intersection of the plane $z = R - a$ and the sphere $r^2 + z^2 = R^2$ is

$$r^2 = R^2 - (R - a)^2 = a(2R - a)$$

Thus, $r = R^2 - (R - a)^2 = a(2R - a)$ or
$r = \sqrt{a} \sqrt{2R - ax}$

$$V = \int_0^{\pi/2} \int_0^{\sqrt{a(2R - a)}} r(R^2 - r^2)^{1/2} \, dr \, d\theta$$

$$= -\frac{2\pi}{3}[(R^2 - 2aR + a^2)^{3/2} - R^3]^{1/2}$$

$$= \frac{2\pi}{3}[R^3 - (R - a)^3]$$

13.4 Surface Area, Page 853

1. $z = f(x, y) = 2 - \frac{1}{2}x - \frac{1}{4}y$
$f_x = -\frac{1}{2}; f_y = -\frac{1}{4}$
$\sqrt{f_x^2 + f_y^2 + 1} = \sqrt{\frac{1}{4} + \frac{1}{16} + 1} = \frac{\sqrt{21}}{4}$

$$S = \int_0^4 \int_0^{-2x+8} \frac{\sqrt{21}}{4} \, dy \, dx = \frac{\sqrt{21}}{4} \int_0^4 (-2x + 8) \, dx$$

$$= \frac{16\sqrt{21}}{4} = 4\sqrt{21}$$

2. $z = f(x, y) = 9 - 4x - y$

$f_x = -4; f_y = -1$
$\sqrt{f_x^2 + f_y^2 + 1} = \sqrt{16 + 1 + 1} = \sqrt{18}$

$$S = \int_0^{9/4} \int_0^{9 - 4x} \sqrt{18} \, dy \, dx = \sqrt{18} \int_0^{9/4} (9 - 4x) \, dx$$

$$= \frac{81\sqrt{18}}{8}$$

3. $z = f(x, y) = 4 - x^2 - y^2$
$f_x = -2x; f_y = -2y$
$\sqrt{f_x^2 + f_y^2 + 1} = \sqrt{4x^2 + 4y^2 + 1} = \sqrt{4r^2 + 1}.$

$$S = \int_0^{2\pi} \int_0^2 \sqrt{4r^2 + 1} \, r \, dr \, d\theta$$

$$= \frac{1}{8} \int_0^{2\pi} \int_0^2 \sqrt{4r^2 + 1} \, (8r \, dr) \, d\theta$$

$$= \frac{1}{12} \int_0^{2\pi} (17\sqrt{17} - 1) \, d\theta$$

$$= \frac{\pi}{6} (17\sqrt{17} - 1)$$

4. $z = f(x, y) = x^2 + y^2 - 9$
$f_x = 2x; f_y = 2y$
$\sqrt{f_x^2 + f_y^2 + 1} = \sqrt{4x^2 + 4y^2 + 1} = \sqrt{4r^2 + 1}.$

$$S = \int_0^{2\pi} \int_0^3 r\sqrt{4r^2 + 1} \, dr \, d\theta$$

$$= \frac{1}{12} \int_0^{2\pi} (37\sqrt{37} - 1) \, d\theta$$

$$= \frac{\pi}{6}(37\sqrt{37} - 1)$$

5. $z = f(x, y) = \frac{1}{2}(12 - 3x - 6y)$
$f_x = -\frac{3}{2}; f_y = -3$
$\sqrt{f_x^2 + f_y^2 + 1} = \sqrt{\frac{9}{4} + 9 + 1} = \frac{7}{2}$

$$S = \int_0^1 \int_0^x \frac{7}{2} \, dy \, dx$$

$$= \frac{7}{2} \int_0^1 x \, dx = \frac{7}{4}$$

6. $z = f(x, y) = 2x + 2y - 12$
$f_x = 2; f_y = 2$
$\sqrt{f_x^2 + f_y^2 + 1} = \sqrt{4 + 4 + 1} = 3$

$$S = \int_0^1 \int_0^1 3 \, dy \, dx = 3$$

7. $z = f(x, y) = x^2 - 9; \; f_x = 2x; f_y = 0$
$\sqrt{f_x^2 + f_y^2 + 1} = \sqrt{4x^2 + 1}$

$$S = \int_0^2 \int_0^2 \sqrt{4x^2 + 1} \, dy \, dx = \int_0^2 \sqrt{4x^2 + 1}(2 \, dx)$$

$$= \left(x\sqrt{4x^2 + 1} + \frac{1}{2}\ln|2x + \sqrt{4x^2 + 1}| \right)\Big|_0^2$$
(Formula #168)

$$= 2\sqrt{17} + \frac{1}{2} \ln(4 + \sqrt{17})$$

8. $z = f(x, y) = x^2; \; f_x = 2x; f_y = 0$
$\sqrt{f_x^2 + f_y^2 + 1} = \sqrt{4x^2 + 1}$

$$S = \int_0^1 \int_0^{1 - x} \sqrt{4x^2 + 1} \, dy \, dx$$

$$= \int_0^1 (1 - x)\sqrt{1 + 4x^2}\, dx$$

$$= \sqrt{5} + \tfrac{1}{2}\ln(2 + \sqrt{5}) - \tfrac{1}{12}(5\sqrt{5} + 1)$$

9. $z = f(x, y) = x^2;\ f_x = 2x;\ f_y = 0$

$$\sqrt{f_x^2 + f_y^2 + 1} = \sqrt{4x^2 + 1}$$

$$S = \int_0^4 \int_0^4 \sqrt{4x^2 + 1}\, dy\, dx$$

$$= 4\int_0^4 \sqrt{1 + 4x^2}\, dx$$

$$= 8\sqrt{65} + \ln(8 + \sqrt{65})$$

10. $z = f(x, y) = 2x + y^2;\ f_x = 2;\ f_y = 2y$

$$\sqrt{f_x^2 + f_y^2 + 1} = \sqrt{4 + 4y^2 + 1}$$

$$S = \int_0^3 \int_0^3 \sqrt{5 + 4y^2}\, dx\, dy$$

$$= 3\int_0^3 \sqrt{5 + 4y^2}\, dy$$

$$= \frac{9}{2}\sqrt{41} + \frac{15}{4}\ln\!\left(\frac{6 + \sqrt{41}}{\sqrt{5}}\right) - \frac{15\ln 5}{8}$$

$$\approx 35.2387$$

11. $z = f(x, y) = x^2 + y^2;\ f_x = 2x;\ f_y = 2y$

$$\sqrt{f_x^2 + f_y^2 + 1} = \sqrt{4x^2 + 4y^2 + 1}$$

$$S = 4\int_0^{\pi/2} \int_0^1 r\sqrt{4r^2 + 1}\, dr\, d\theta$$

$$= 4\int_0^{\pi/2} \frac{5\sqrt{15} - 1}{12}\, d\theta$$

$$= \frac{\pi}{6}(5^{3/2} - 1)$$

12. $z = f(x, y) = \sqrt{25 - x^2 - y^2}$

$$f_x = \frac{-x}{\sqrt{25 - x^2 - y^2}};\ f_y = \frac{-y}{\sqrt{25 - x^2 - y^2}}$$

$$\sqrt{f_x^2 + f_y^2 + 1} = \sqrt{\frac{x^2 + y^2 + 25 - x^2 - y^2}{25 - x^2 - y^2}}$$

$$= \sqrt{\frac{1}{25 - r^2}}$$

$$S = \int_0^{2\pi} \int_0^4 \frac{5r\, dr}{\sqrt{25 - r^2}} = -10\int_0^{2\pi} (-2)\, dr = 40\pi$$

13. $z = f(x, y) = \sqrt{4 - x^2};\ f_x = \dfrac{-x}{\sqrt{4 - x^2}};\ f_y = 0$

$$\sqrt{f_x^2 + f_y^2 + 1} = \sqrt{\frac{x^2 + 4 - x^2}{4 - x^2}} = \frac{2}{\sqrt{4 - x^2}}$$

$$S = \int_0^2 \int_0^2 \frac{2}{\sqrt{4 - x^2}}\, dy\, dx = \int_0^2 \frac{4}{\sqrt{4 - x^2}}\, dx$$

$$= 4\sin^{-1}\frac{x}{2}\Big|_0^2 = 2\pi$$

14. $z = f(x, y) = 4 - x - y;\ f_x = -1;\ f_y = -1$

$$\sqrt{f_x^2 + f_y^2 + 1} = \sqrt{1 + 1 + 1} = \sqrt{3}$$

$$S = \int_0^{2\pi} \int_0^4 \sqrt{3}\, r\, dr\, d\theta = \sqrt{3}\int_0^{2\pi} 8\, d\theta = 16\sqrt{3}\pi$$

15. $z = f(x, y) = \sqrt{4 - x^2 - y^2};$

$$f_x = \frac{-x}{\sqrt{4 - x^2 - y^2}};\ f_y = \frac{-y}{\sqrt{4 - x^2 - y^2}}$$

$$\sqrt{f_x^2 + f_y^2 + 1} = \sqrt{\frac{x^2 + y^2 + (4 - x^2 - y^2)}{4 - x^2 - y^2}}$$

$$= \frac{2}{\sqrt{4 - x^2 - y^2}}$$

$$S = 4\int_0^{\pi/2} \int_0^{2\sin\theta} \frac{2r}{\sqrt{4 - r^2}}\, dr\, d\theta$$

$$= -4\int_0^{\pi/2} \int_0^{2\sin\theta} \frac{-2r\, dr}{\sqrt{4 - r^2}}\, d\theta$$

$$= -8\int_0^{\pi/2} \left(\sqrt{4 - 4\sin^2\theta} - 2\right) d\theta$$

$$= -8\int_0^{\pi/2} (2\cos\theta - 2)\, d\theta$$

$$= 8\pi - 16$$

16. The intersection of the two planes (which is also the projection onto the z-plane) is found by eliminating the z: $x^2 + y^2 = 4$

$z = f(x, y) = \sqrt{8 - x^2 - y^2};$

$$f_x = \frac{-x}{\sqrt{8 - x^2 - y^2}};\ f_y = \frac{-y}{\sqrt{8 - x^2 - y^2}}$$

$$\sqrt{f_x^2 + f_y^2 + 1} = \sqrt{\frac{x^2 + y^2 + 8 - x^2 - y^2}{8 - x^2 - y^2}}$$

$$S = 8 \int_0^{\pi/2} \int_0^2 \frac{2\sqrt{2}\,r}{\sqrt{8 - r^2}}\, dr\, d\theta$$

$$= 8 \int_0^{\pi/2} (8 - 4\sqrt{2})\, d\theta = 16\pi(1 - \sqrt{2})$$

17. $z = f(x, y) = x^2 + y;\ f_x = 2x;\ f_y = 1$

$$\sqrt{f_x^2 + f_y^2 + 1} = \sqrt{4x^2 + 2}$$

$$S = \int_0^2 \int_2^5 \sqrt{4x^2 + 2}\, dy\, dx$$

$$= 5\sqrt{2} \int_0^2 \sqrt{1 + 2x^2}\, dx$$

$$= 15\sqrt{2} + \tfrac{5}{2} \ln(2\sqrt{2} + 3)$$

18. $z = f(x, y) = x^2 - y^2;\ f_x = 2x;\ f_y = -2y$

$$\sqrt{f_x^2 + f_y^2 + 1} = \sqrt{4x^2 + 4y^2 + 1}$$

$$S = 4 \int_0^{\pi/2} \int_0^3 r\sqrt{4r^2 + 1}\, dr\, d\theta$$

$$= 4 \int_0^{\pi/2} \left[\frac{37\sqrt{37}}{12} - \frac{1}{12} \right] d\theta$$

$$= \tfrac{\pi}{6}(37\sqrt{37} - 1)$$

19. $z = f(x, y) = 9 - x^2 - y^2;\ f_x = -2x;\ f_y = -2y$

$$\sqrt{f_x^2 + f_y^2 + 1} = \sqrt{4x^2 + 4y^2 + 1}$$

$$S = \int_0^{2\pi} \int_0^3 \sqrt{4r^2 + 1}\, r\, dr\, d\theta$$

$$= \tfrac{1}{12} \int_0^{2\pi} (37\sqrt{37} - 1)\, d\theta$$

$$= \tfrac{\pi}{6}(37\sqrt{37} - 1)$$

20. Obtain a rectangular approximating element of surface area, project onto a coordinate plane (the xy-plane) and then sum up (integrate). In particular find f_x, f_y, and then find

$$dS = \sqrt{f_x^2 + f_y^2 + 1}$$

Then integrate to find S.

21. Let dA be a rectangular element of surface area of **N** normal to this element. Project

onto the xy-plane, which means trace out the shadow in the xy-plane when rays of light parallel to the z-axis hit the surface and then the coordinate plane.

22. $f_x = 0;\ f_y = \tfrac{1}{5};\ dA = \sqrt{\tfrac{1}{25} + 1}\, dy\, dx$

$$S = \frac{120{,}000\sqrt{26}}{5} = 24{,}000\sqrt{26}$$

23. $z = f(x, y) = \tfrac{1}{C}(D - Ax - By)$

$$f_x = -\frac{A}{C};\ f_y = -\frac{B}{C}$$

$$\sqrt{f_x^2 + f_y^2 + 1} = \sqrt{\frac{A^2}{C^2} + \frac{B^2}{C^2} + 1}$$

$$= \frac{\sqrt{A^2 + B^2 + C^2}}{C}$$

$$S = \int_0^{D/A} \int_0^{-\frac{A}{B}x + \frac{D}{B}} \frac{\sqrt{A^2 + B^2 + C^2}}{C}\, dy\, dx$$

$$= \frac{\sqrt{A^2 + B^2 + C^2}}{C} \int_0^{D/A} \left(-\frac{A}{B}x + \frac{D}{B} \right) dx$$

$$= \frac{D^2}{2ABC} \sqrt{A^2 + B^2 + C^2}$$

24. $z = f(x, y) = \tfrac{1}{3}(12 - x - 2y);\ f_x = \tfrac{1}{3};\ f_y = \tfrac{2}{3}$

$$\sqrt{f_x^2 + f_y^2 + 1} = \sqrt{\frac{1 + 4 + 9}{9}} = \frac{\sqrt{14}}{3}$$

$$S = \int_0^a \int_0^x \frac{\sqrt{14}}{3}\, dy\, dx = \int_0^a \frac{x}{3}\sqrt{14}\, d\theta = \frac{a^2}{6}\sqrt{14}$$

25. $z = f(x, y) = a - x - y;\ f_x = -1;\ f_y = -1$

$$\sqrt{f_x^2 + f_y^2 + 1} = \sqrt{3}$$

$$S = \int_0^{2\pi} \int_{a/2}^a \sqrt{3}\, r\, dr\, d\theta = \int_0^{2\pi} \frac{3\sqrt{3}\,a^2}{8}\, d\theta = \frac{3\sqrt{3}\pi a^2}{4}$$

26. $z = f(x, y) = \sqrt{9 - x^2};\ f_x = \frac{-x}{\sqrt{9 - x^2}};\ f_y = 0$

$$\sqrt{f_x^2 + f_y^2 + 1} = \sqrt{\frac{x^2}{9 - x^2} + 0 + 1} = \frac{3}{\sqrt{9 - x^2}}$$

$$S = \int_0^3 \int_0^3 \frac{3}{\sqrt{9 - x^2}}\, dy\, dx = 24 \int_0^3 dx = 72$$

27. $z = f(x, y) = \sqrt{a^2 - x^2 - y^2};$

$$f_x = \frac{-x}{\sqrt{a^2 - x^2 - y^2}};\ f_y = \frac{-y}{\sqrt{a^2 - x^2 - y^2}};$$

$$\sqrt{f_x^{\,2} + f_y^{\,2} + 1} = \sqrt{\frac{x^2 + y^2 + (a^2 - x^2 - y^2)}{a^2 - x^2 - y^2}}$$

$$= \frac{a}{\sqrt{a^2 - x^2 - y^2}}$$

$$S = 8 \int_0^{\pi/2} \int_0^a \frac{a}{\sqrt{a^2 - r^2}} \, r \, dr \, d\theta$$

$$= -4a \int_0^{\pi/2} \int_0^a \frac{(-2r) \, dr}{\sqrt{a^2 - r^2}} \, d\theta$$

$$= -8a \int_0^{\pi/2} (-a) \, d\theta = 8a^2\left(\frac{\pi}{2}\right) = 4\pi a^2$$

28. $z = f(x, y) = \sqrt{r^2 - x^2};$

$$f_x = \frac{-x}{\sqrt{r^2 - x^2}}; \; f_y = 0$$

$$\sqrt{f_x^{\,2} + f_y^{\,2} + 1} = \sqrt{\frac{x^2 + r^2 - x^2}{r^2 - x^2}}$$

$$S = 4 \int_0^r \int_0^h \frac{r \, dy \, dx}{\sqrt{r^2 - x^2}} = 4 \int_0^r \frac{hr}{\sqrt{r^2 - x^2}} \, dx$$

$$= 4rh \sin^{-1} \frac{x}{r} \Big|_0^r = 4\pi rh\frac{\pi}{2} = 2\pi rh$$

29. $z = f(x, y) = \sqrt{x^2 + y^2};$

$$f_x = \frac{x}{\sqrt{x^2 + y^2}}; \; f_y = \frac{y}{\sqrt{x^2 + y^2}}$$

$$\sqrt{f_x^{\,2} + f_y^{\,2} + 1} = \sqrt{\frac{x^2 + y^2}{x^2 + y^2} + 1} = \sqrt{2}$$

$$S = \sqrt{2} \int_0^{2\pi} \int_0^h r \, dr \, d\theta = \frac{\sqrt{2}}{2} \int_0^{2\pi} h^2 \, d\theta = \sqrt{2}\pi h^2$$

30. $z = f(x, y) = 4\sqrt{x^2 + y^2};$

$$f_x = \frac{4x}{\sqrt{x^2 + y^2}}; \; f_y = \frac{4y}{\sqrt{x^2 + y^2}}$$

$$\sqrt{f_x^{\,2} + f_y^{\,2} + 1} = \sqrt{\frac{4x^2 + 4y^2 + x^2 + y^2}{x^2 + y^2}} = \sqrt{5}$$

$$S = \int_0^{2\pi} \int_{h_1/4}^{h_2/4} \sqrt{5} \, r \, dr \, d\theta = \frac{\sqrt{5}}{32} \int_0^{2\pi} (h_2^{\,2} - h_1^{\,2}) \, d\theta$$

$$= \frac{\sqrt{5}\pi}{16}(h_2^{\,2} - h_1^{\,2})$$

31.   Find the intersection of the two surfaces by substituting $x^2 + y^2 = 4z$: $4z + z^2 = 9z$, $z^2 - 5z = 0$, $z = 0, 5$. When $z = 0$ the paraboloid and the sphere are tangent. (The sphere has center at $(0, 0, \frac{9}{2})$ and a radius of

$\frac{9}{2}$.) When $z = 5$ the intersection is the circle $x^2 + y^2 = 20$. The surface we wish to find lies above $x^2 + y^2 = 20$. In polar form $r = 2\sqrt{5}$. $x^2 + y^2 + (z - \frac{9}{2})^2 = \frac{81}{4}$

$$z = \sqrt{\frac{81}{4} - x^2 - y^2} + \frac{9}{2};$$

$$f_x = \frac{-x}{\sqrt{\frac{81}{4} - x^2 - y^2}}$$

$$f_y = \frac{-y}{\sqrt{\frac{81}{4} - x^2 - y^2}}$$

$$\sqrt{f_x^{\,2} + f_y^{\,2} + 1} = \sqrt{\frac{x^2 + y^2 + \frac{81}{4} - x^2 - y^2}{\frac{81}{4} - x^2 - y^2}}$$

$$= \frac{9}{\sqrt{81 - 4x^2 - 4y^2}}.$$

$$S = \int_0^{2\pi} \int_0^{2\sqrt{5}} \frac{9}{\sqrt{81 - 4r^2}} \, r \, dr \, d\theta$$

$$= -\frac{9}{8} \int_0^{2\pi} \int_0^{2\sqrt{5}} \frac{(-8r) \, dr}{\sqrt{81 - 4r^2}} \, d\theta$$

$$= -\frac{9}{4} \int_0^{2\pi} (1 - 9) \, d\theta = 18(2\pi) = 36\pi$$

32. $z = f(x, y) = \sqrt{4 - x^2}; \; f_x = \frac{-x}{\sqrt{4 - x^2}}; \; f_y = 0$

$$\sqrt{f_x^{\,2} + f_y^{\,2} + 1} = \sqrt{\frac{x^2 + 4 - x^2}{4 - x^2}} = \frac{2}{\sqrt{4 - x^2}}$$

$$S = \int_0^1 \int_0^x \frac{2}{\sqrt{4 - x^2}} \, dy \, dx = \int_0^1 \frac{2x}{\sqrt{4 - x^2}} \, dx$$

$$= 2(2 - \sqrt{3})$$

33. $z = f(x, y) = e^{-x}\sin y;$

$$f_x = -e^{-x}\sin y; \; f_y = e^{x}\cos y$$

$$\sqrt{f_x^{\,2} + f_y^{\,2} + 1} = \sqrt{e^{-2x}\sin^2 y + e^{-2x}\sin^2 y + 1}$$

$$S = \int_0^1 \int_0^y \sqrt{e^{-2x} + 1} \, dx \, dy$$

34. $z = f(x, y) = x^3 - xy + y^3;$

$$f_x = 3x^2 - y; \; f_y = 3y^2 - x$$

$$\sqrt{f_x^{\,2} + f_y^{\,2} + 1} = \sqrt{(3x^2 - y)^2 + (3y^2 - x)^2 + 1}$$

$$S = \int_0^2 \int_0^2 \sqrt{(3x^2 - y)^2 + (3y^2 - x)^2 + 1} \, dy \, dx$$

35. $z = \cos(x^2 + y^2); \; f_x = -2x\sin(x^2 + y^2);$

$$f_y = -2y\sin(x^2 + y^2)$$

$$\sqrt{f_x^2 + f_y^2 + 1} = \sqrt{(4x^2 + 4y^2)\sin^2(x^2 + y^2) + 1}$$

$$S = \int_0^{2\pi} \int_0^{\sqrt{\pi/2}} \sqrt{4r^2\sin^2 r^2 + 1} \; r \; dr \; d\theta$$

36. $z = f(x, y) = e^{-x}\cos y$

$$f_x = -e^{-x}\cos y; \; f_y = -e^{-x}\sin y$$

$$\sqrt{f_x^2 + f_y^2 + 1} = \sqrt{e^{-2x}\cos^2 y + e^{-2y}\sin^2 y + 1}$$

$$S = \int_0^{2\pi} \int_0^3 r\sqrt{e^{-2r\cos x} + 1} \; dr \; d\theta$$

37. $z = f(x, y) = x^2 + 5xy + y^2$

$$f_x = 2x + 5y; \; f_y = 2y + 5x$$

$$\sqrt{f_x^2 + f_y^2 + 1} = \sqrt{(2x + 5y)^2 + (2y + 5x)^2 + 1}$$

$$S = \int_0^3 \int_0^x \sqrt{(2x + 5y)^2 + (2y + 5x)^2 + 1} \; dy \; dx$$

38. $z = f(x, y) = x^2 + 3xy + y^2;$

$$f_x = 2x + 3y; \; f_y = 2y + 3x$$

$$\sqrt{f_x^2 + f_y^2 + 1} = \sqrt{(2x + 3y)^2 + (2y + 3x)^2 + 1}$$

$$S = \int_0^4 \int_0^x \sqrt{(2x + 3y)^2 + (2y + 3x)^2 + 1} \; dy \; dx$$

39. $R_u \times R_v = \begin{vmatrix} \mathbf{i} & \mathbf{j} & \mathbf{k} \\ 2\sin v & 2\cos v & 2u \\ 2u\cos v & -2u\sin v & 0 \end{vmatrix}$

$$= 4u^2\sin v\, \mathbf{i} + 4u^2\cos v\, \mathbf{j}$$
$$+ (-4u\sin^2 v - 4u\cos^2 v)\, \mathbf{k}$$
$$= 4u^2\sin v\, \mathbf{i} + 4u^2\cos v\, \mathbf{j} - 4u\, \mathbf{k}$$

$$\|R_u \times R_v\| = \sqrt{16u^4\sin^2 v + 16u^4\cos^2 v + 16u^2}$$

$$= \sqrt{16u^4 + 16u^2}$$

$$= 4|u|\sqrt{u^2 + 1}$$

40. $R_u \times R_v$

$$= \begin{vmatrix} \mathbf{i} & \mathbf{j} & \mathbf{k} \\ 4\cos u \cos v & 4\cos u \sin v & -5\sin u \\ 4\sin u \sin v & 4\sin u \cos v & 0 \end{vmatrix}$$

$$= (20\sin^2 u \cos v)\mathbf{i} + (20\sin^2 u \sin v)\mathbf{j}$$
$$+ (16\sin u \cos u)\, \mathbf{k}$$

$$\|R_u \times R_v\| = 4|\sin u|\sqrt{9\sin^2 u + 16}$$

41. $R_u \times R_v = \begin{vmatrix} \mathbf{i} & \mathbf{j} & \mathbf{k} \\ 1 & 0 & 3u^2 \\ 0 & 1 & 0 \end{vmatrix} = -3u^2\mathbf{i} + \mathbf{k}$

$$\|R_u \times R_v\| = \sqrt{9u^4 + 1}$$

42. $R_u \times R_v = \begin{vmatrix} \mathbf{i} & \mathbf{j} & \mathbf{k} \\ 2\sin v & 2\cos v & 2u\sin 2v \\ 2u\cos v & -2u\sin v & 2u^2\cos 2v \end{vmatrix}$

$$= (4u^2\cos v \cos 2v + 4u^2\sin v \sin 2v)\mathbf{i}$$
$$+ (4u^2\cos v \sin 2v - 4u^2\sin v \cos 2v)\mathbf{j}$$
$$- 4u\, \mathbf{k}$$

$$\|R_u \times R_v\| = 4|u|\sqrt{u^2 + 1}$$

43. $R_u \times R_v = \begin{vmatrix} \mathbf{i} & \mathbf{j} & \mathbf{k} \\ v & 1 & 1 \\ u & -1 & 1 \end{vmatrix}$

$$= 2\mathbf{i} + (u - v)\mathbf{j} - (u + v)\mathbf{k}$$

$$\|R_u \times R_v\| = \sqrt{4 + (u - v)^2 + (-u - v)^2}$$

$$= \sqrt{4 + 2u^2 + 2v^2}$$

$$S = \int_0^{2\pi} \int_0^1 \sqrt{4 + 2r^2} \; r \; dr \; d\theta$$

$$= \frac{1}{4}\int_0^{2\pi} \int_0^1 \sqrt{4 + 2r^2}\,(4r\,dr)\; d\theta$$

$$= \frac{1}{6}\int_0^{2\pi} (6\sqrt{6} - 8) \; d\theta$$

$$= \frac{2\pi}{3}(3\sqrt{6} - 4) \approx 7.01$$

44. $R_u \times R_v = \begin{vmatrix} \mathbf{i} & \mathbf{j} & \mathbf{k} \\ \cos v & \sin v & 1 \\ -u\sin v & u\cos v & 0 \end{vmatrix}$

$$= -u\cos v\, \mathbf{i} - u\sin v\, \mathbf{j} + (u\cos^2 v + u\sin^2 v)\mathbf{k}$$

$$\|R_u \times R_v\| = \sqrt{2}\,u$$

$$S = \int_0^1 \int_0^\pi \sqrt{2}\,u \; dv \; du = \frac{\sqrt{2}\pi}{2}$$

45. $R_u \times R_v$

$$= \begin{vmatrix} \mathbf{i} & \mathbf{j} & \mathbf{k} \\ -(a+b\cos v)\sin u & (a+b\cos v)\cos u & 0 \\ -b\sin v \cos u & -b\sin v \sin u & b\cos v \end{vmatrix}$$

$$= (b^2\cos^2 v + ab\cos v)(\cos u)\mathbf{i}$$
$$+ (b^2\cos^2 v + ab\cos v)(\sin u)\mathbf{j}$$
$$+ (b^2\sin v \cos v + ab\sin v)\mathbf{k}$$

$$\|\boldsymbol{R}_u \times \boldsymbol{R}_v\| = \left| ab + b^2\cos v \right|$$

$$S = \int_0^{2\pi} \int_0^{2\pi} \left| ab + b^2\cos v \right| \, du \, dv$$

$$= \int_0^{2\pi} 2\pi(ab + b^2 \cos v) \, dv$$

$$= 4\pi^2 ab$$

46. $f(x, y, z) = x^2 + y^2 + z^2 - R^2$

$f_x = 2x;\ f_y = 2y;\ f_z = 2z$

$$\frac{\sqrt{f_x^2 + f_y^2 + f_z^2}}{|f_z|} = \frac{2R}{2z} = \frac{R}{\sqrt{R^2 - x^2 - y^2}}$$

$$S = \int_0^R \int_0^{\sqrt{R^2 - x^2}} \frac{R}{\sqrt{R^2 - x^2 - y^2}} \, dy \, dx$$

$$= 8R \int_0^{\pi/2} \int_0^R \frac{r}{\sqrt{R^2 - r^2}} \, dr \, d\theta$$

$$= 8R \int_0^{\pi/2} \frac{R}{2} \, d\theta = 4\pi R^2$$

47. $\nabla f = f_x \mathbf{i} + f_y \mathbf{j} + f_z \mathbf{k};$

$$\|\nabla f\| = \sqrt{f_x^2 + f_y^2 + f_z^2}$$

The unit vector orthogonal to the surface is
$$\mathbf{u} = \frac{\nabla f}{\|\nabla f\|}$$
and the unit vector orthogonal to the projected plane is $\mathbf{k}$, so
$$\cos \gamma = \mathbf{N} \cdot \mathbf{k} = \frac{|\nabla f \cdot \mathbf{u}|}{\|\nabla f\|}$$

Since $dA = \sec \gamma \, dy \, dx$, we have
$$S = \int \int_R \frac{\|\nabla f\|}{|\nabla f \cdot \mathbf{u}|} \, dA$$

13.5 Triple Integrals, Page 863

1. Start with a cube, $dV = dz \, dy \, dx$. Sum up along the z-axis to create a column. Project onto the xy-plane and continue with double integration.

2. $\displaystyle \int_0^4 \int_0^{\sqrt{x}} \int_0^{2-x} dz \, dy \, dx;\qquad \int_0^2 \int_{y^2}^4 \int_0^{2-x} dz \, dx \, dy$

$$\int_0^4 \int_0^{2-x} \int_0^{\sqrt{x}} dy \, dz \, dx \ ;\qquad \int_0^2 \int_{2-z}^4 \int_0^{\sqrt{x}} dy \, dx \, dz$$

$$\int_0^2 \int_0^{\sqrt{2-z}} \int_{y^2}^4 dx \, dy \, dz;\qquad \int_0^2 \int_0^{2-y^2} \int_{y^2}^4 dx \, dz \, dy$$

3. $\displaystyle \int_1^4 \int_{-2}^3 \int_2^5 dx \, dy \, dz = (3)(5)(3) = 45$

4. $\displaystyle \int_{-1}^3 \int_0^2 \int_{-2}^2 dy \, dz \, dx = (4)(2)(4) = 32$

5. $\displaystyle \int_1^2 \int_0^1 \int_{-1}^2 8x^2yz^3 \, dx \, dy \, dz$

$$= 8 \cdot \frac{x^3}{3}\bigg|_{-1}^2 \frac{y^2}{2}\bigg|_0^1 \frac{z^4}{4}\bigg|_1^2 = 8(3)(\tfrac{1}{2})(\tfrac{15}{4}) = 45$$

6. $\displaystyle \int_4^7 \int_{-1}^2 \int_0^3 x^2y^2z^2 \, dx \, dy \, dz$

$$= \frac{x^3}{3}\bigg|_4^7 \frac{y^3}{3}\bigg|_{-1}^2 \frac{z^3}{3}\bigg|_0^3 = 2{,}511$$

7. $\displaystyle \int_0^2 \int_0^x \int_0^{x+y} xyz \, dz \, dy \, dx$

$$= \frac{1}{2} \int_0^2 \int_0^x xy(x + y)^2 \, dy \, dz$$

$$= \frac{1}{2} \int_0^2 \left(\frac{x^5}{2} + 2\frac{x^5}{3} + \frac{x^5}{4} \right) dx$$

$$= \frac{1}{24} \int_0^2 (6 + 8 + 3)x^5 \, dx = \frac{68}{9}$$

8. $\displaystyle \int_0^1 \int_{\sqrt{x}}^{\sqrt{1+x}} \int_0^{xy} y^{-1}z \, dz \, dy \, dx$

$$= \frac{1}{2} \int_0^1 \int_{\sqrt{x}}^{\sqrt{x+1}} x^2y \, dy \, dx$$

$$= \frac{1}{4} \int_0^1 x^2(1 + x - x) \, dx = \frac{1}{12}$$

9. $\displaystyle\int_{-1}^{2}\int_{0}^{\pi}\int_{1}^{4} yz\cos xy \, dz \, dx \, dy$

$\displaystyle= \frac{15}{2}\int_{-1}^{2}\int_{0}^{\pi} y\cos xy \, dx \, dy$

$\displaystyle= \frac{15}{2}\int_{-1}^{2}\sin \pi y \, dy = -\frac{15}{\pi}$

10. $\displaystyle\int_{0}^{\pi}\int_{0}^{1}\int_{0}^{1} x^2 y\cos xyz \, dz \, dy \, dx$

$\displaystyle= \int_{0}^{1}\int_{0}^{1} x\sin xy \, dy \, dx$

$\displaystyle= -\int_{0}^{1}(\cos x - 1) \, dx$

$\displaystyle= 1 - \sin 1$

11. $\displaystyle\int_{0}^{1}\int_{0}^{y}\int_{0}^{\ln y} e^{z+2x} \, dz \, dx \, dy$

$\displaystyle= \int_{0}^{1}\int_{0}^{y} e^{2x}(y - 1) \, dx \, dy$

$\displaystyle= \frac{1}{2}\int_{0}^{1}(e^{2y} - 1)(y - 1) \, dy$

$\displaystyle= \frac{1}{8}(5 - e^2)$

12. $\displaystyle\int_{1}^{3}\int_{0}^{2z}\int_{0}^{\ln y} y\,e^{-x} \, dx \, dy \, dz$

$\displaystyle= -\int_{1}^{3}\int_{0}^{2z} y(e^{\ln y^{-1}} - 1) \, dy \, dz$

$\displaystyle= -\int_{1}^{3}\left(3z - \frac{9z^2}{2}\right) dz = 27$

13. $\displaystyle\int_{1}^{4}\int_{-1}^{2z}\int_{0}^{\sqrt{3}\,x} \frac{x - y}{x^2 + y^2} \, dy \, dx \, dz$

$\displaystyle= \int_{1}^{4}\int_{-1}^{2z}\left[\frac{\pi}{3} - \ln 2\right] dx \, dz$

$\displaystyle= \int_{1}^{4}\left[\frac{2\pi z}{3} - (2\ln 2)z + \frac{\pi}{3} - \ln 2\right] dz$

$\displaystyle= 6\pi - 18\ln 2$

14. $\displaystyle\int_{0}^{1}\int_{x-1}^{x^2}\int_{-x}^{y} (x + y) \, dz \, dy \, dx$

$\displaystyle= \int_{0}^{1}\int_{x-1}^{x^2} (x + y)^2 \, dy \, dx$

$\displaystyle= \frac{1}{3}\int_{0}^{1}\left[(x + x^2)^3 - (2x - 1)^3\right] dx$

$\displaystyle= \frac{1}{3}\int_{0}^{1}(x^6 + 3x^5 + 3x^4 - 7x^3 + 12x^2 - 6x + 1) \, dx$

$\displaystyle= \frac{209}{420} \approx 0.4976$

15. $\displaystyle\int_{0}^{3}\int_{-1}^{1}\int_{2}^{4} (x^2 y + y^2 z) \, dz \, dy \, dx$

$\displaystyle= \int_{1}^{3}\int_{-1}^{1} (2x^2 y + 6y^2) \, dy \, dx$

$\displaystyle= \int_{1}^{3} 4 \, dx = 8$

16. $\displaystyle\int_{2}^{4}\int_{1}^{3}\int_{-2}^{4} (xy + 2yz) \, dz \, dy \, dx$

$\displaystyle= \int_{2}^{4}\int_{1}^{3} (6xy + 12y) \, dy \, dx$

$\displaystyle= 8\int_{2}^{4} (3x + 6) \, dx = 240$

17. $\displaystyle\int_{0}^{1}\int_{0}^{1-x}\int_{0}^{1-x-y} xyz \, dz \, dy \, dx$

$\displaystyle= \frac{1}{2}\int_{0}^{1}\int_{0}^{1-x} xy(1 - x - y)^2 \, dy \, dx$

$\displaystyle= \frac{1}{24}\int_{0}^{1} x(x - 1)^4 \, dx = \frac{1}{720}$

18. $\displaystyle\int_{0}^{3}\int_{0}^{2(1-x/3)}\int_{0}^{1-x/3-y/2} x^2 y \, dz \, dy \, dx$

$\displaystyle= -\frac{1}{6}\int_{0}^{3}\int_{0}^{2(1-x/3)} x^2 y(2x + 3y - 6) \, dy \, dx$

$$= -\frac{1}{81} \int_0^3 2x^2(3 - x)^3 \, dx = \frac{3}{10}$$

19. $$\int_1^3 \int_0^{\sqrt{1-z^2}} \int_{-\sqrt{1-x^2-y^2}}^{\sqrt{1-x^2-y^2}} xyz \, dx \, dy \, dz$$

$$= \int_1^3 \int_0^{\sqrt{1-z^2}} \frac{yzx^2}{2} \bigg|_{-\sqrt{1-x^2-y^2}}^{\sqrt{1-x^2-y^2}} = 0$$

20. $$\int_{-1}^1 \int_{-\sqrt{1-y^2}}^{\sqrt{1-y^2}} \int_0^{x^2+y^2} x \, dz \, dx \, dy$$

$$= \int_{-1}^1 \int_{-\sqrt{1-y^2}}^{\sqrt{1-y^2}} x(x^2 + y^2) \, dx \, dy$$

$$= \int_{-1}^1 \left[\frac{x^4}{2} + \frac{x^2 y^2}{2}\right]\bigg|_{-\sqrt{1-y^2}}^{\sqrt{1-y^2}} dy = 0$$

21. $$\int_0^1 \int_0^x \int_0^{x+y} e^z \, dz \, dy \, dx$$

$$= \int_0^1 \int_0^x (e^{x+y} - 1) \, dy \, dx$$

$$= \int_0^1 (e^{2x} - x - e^x) \, dx$$

$$= \frac{e^2}{2} - e$$

22. $$\int_0^3 \int_0^{\sqrt{9-x^2}} \int_0^{\sqrt{9-x^2-y^2}} yz \, dz \, dy \, dx$$

$$= \frac{1}{2} \int_0^3 \int_0^{\sqrt{9-x^2}} y(9 - x^2 - y^2) \, dy \, dx$$

$$= \frac{1}{8} \int_0^3 (x^2 - 9)^2 \, dx = \frac{81}{5}$$

23. $$\int_0^1 \int_0^{-x+1} \int_0^{1-x-y} dz \, dy \, dx$$

$$= \int_0^1 \int_0^{-x+1} (1 - x - y) \, dy \, dx$$

$$= \int_0^1 \left(\frac{x^2}{2} - x + \frac{1}{2}\right) dx = \frac{1}{6}$$

This result is easily verified with the formula for the volume of a pyramid.

24. $$\int_0^3 \int_0^{9-x^2} \int_0^x dz \, dy \, dx$$

$$= \int_0^3 \int_0^{9-x^2} x \, dy \, dx$$

$$= \int_0^3 x(9 - x^2) \, dx = \frac{729}{4}$$

25. $$8 \int_1^2 \int_2^A \int_3^B dz \, dy \, dx \text{ where}$$

$$A = 2 + \sqrt{1 - (x - 1)^2} \text{ and}$$

$$B = 3 + \sqrt{1 - (x - 1)^2 - (y - 2)^2}$$

$$= 8 \int_1^2 \int_2^A \sqrt{1 - (x-1)^2 - (y-2)^2} \, dy \, dx$$

$$= 8 \int_1^2 \frac{\pi}{4}(2x - x^2) \, dx = \frac{4\pi}{3}$$

This result is easily verified with the formula for a sphere.

26. $$4 \int_0^1 \int_0^{\sqrt{1-x^2}} \int_0^{4[1-x^2-y^2]} dz \, dy \, dx$$

$$= -16 \int_0^1 \int_0^{\sqrt{1-x^2}} (x^2 + y^2 - 1) \, dy \, dx$$

$$= \frac{32}{3} \int_0^1 (1 - x^2)^{3/2} \, dx = 2\pi$$

27. Find the intersection of the parabolic cylinder and the elliptic paraboloid:

$$4 - y^2 = x^2 + 3y^2$$

$$x^2 + 4y^2 = 4$$

$$x = 2\sqrt{1 - y^2}$$

Using the first octant volume and symmetry:

$$V = 4 \int\limits_0^1 \int\limits_0^{2\sqrt{1-y^2}} \int\limits_{x^2+3y^2}^{4-y^2} dz\ dx\ dy$$

$$= 4 \int\limits_0^1 \int\limits_0^{2\sqrt{1-y^2}} (4 - x^2 - 4y^2)\ dx\ dy$$

$$= 4 \int\limits_0^1 \left(8\sqrt{1-y^2} - \frac{8(1-y^2)^{3/2}}{3} - 8y^2\sqrt{1-y^2}\right) dy$$

$$= 4 \int\limits_0^{\pi/2} \left(8\cos\theta - \tfrac{8}{3}\cos^3\theta - 8\sin^2\theta\cos\theta\right)\cos\theta\ d\theta$$

$$= 32 \int\limits_0^{\pi/2} \left(\cos^2\theta - \tfrac{1}{3}\cos^4\theta - \sin^2\theta\,\cos^2\theta\right) d\theta$$

$$= 32 \int\limits_0^{\pi/2} \left((1 - \sin^2\theta)\cos^2\theta - \tfrac{1}{3}\cos^4\theta\right) d\theta$$

$$= 32 \int\limits_0^{\pi/2} \tfrac{2}{3}\cos^4\theta\ d\theta$$

$$= \frac{64}{3} \int\limits_0^{\pi/2} \left(\frac{1 + \cos 2\theta}{2}\right)^2 d\theta$$

$$= \frac{16}{3} \int\limits_0^{\pi/2} \left(1 + 2\cos 2\theta + \frac{1 + \cos 4\theta}{2}\right) d\theta = 4\pi$$

28. $$4 \int\limits_0^{\pi/2} \int\limits_0^3 \int\limits_0^{\sqrt[3]{9-(x^2+y^2)}} dz\ dx\ dy$$

$$= \int\limits_0^{2\pi} \int\limits_0^3 r\sqrt[3]{9 - r^2}\ dr\ d\theta$$

$$= \int\limits_0^{2\pi} \frac{27\sqrt[3]{9}}{8}\ d\theta = \frac{27\sqrt[3]{9}\,\pi}{4}$$

29. $$\int\limits_0^1 \int\limits_y^1 \int\limits_0^y f(x,\ y,\ z)\ dz\ dx\ dy$$

30. $$\int\limits_0^1 \int\limits_1^{x+1} \int\limits_0^x f(x,\ y,\ z)\ dy\ dz\ dx$$

31. $$\int\limits_0^2 \int\limits_0^{\sqrt{4-z^2}} \int\limits_0^{\sqrt{4-y^2-z^2}} f(x,\ y,\ z)\ dx\ dy\ dz$$

32. $\displaystyle\int_0^3 \int_0^{2z/3} \int_0^{\sqrt{4-y^2}} f(x,\,y,\,z)\;dx\;dy\;dz + \int_0^3 \int_{2z/3}^{2} \int_{\sqrt{4z^2-9y^2/3}}^{\sqrt{4-y^2}} f(x,\,y,\,z)\;dx\;dy\;dz$

33. $\displaystyle 8\int_0^2 \int_0^A \int_0^B dz\;dy\;dx = 32\int_0^2 \int_0^A \sqrt{1 - \frac{x^2}{4} - \frac{y^2}{9}}\;dy\;dx = 24\pi\int_0^2 \left(1 - \frac{x^2}{4}\right)\;dx = 32\pi$

where $A = \frac{3}{2}\sqrt{4 - x^2}$ and $B = 4\sqrt{1 - \frac{x^2}{4} - \frac{y^2}{9}}$

34. $\displaystyle 8\int_0^6 \int_0^{\sqrt{4-x^2/9}} \int_0^{4-x^2/9-y^2} dz\;dy\;dx$

$\displaystyle = 8\int_0^6 \int_0^{\sqrt{4-x^2/9}} \left(4 - \frac{x^2}{9} - y^2\right)\;dy\;dx$

$\displaystyle = \frac{16}{81}\int_0^6 (36 - x^2)^{3/2}\;dx$

$\displaystyle = 48\pi$

35. $\displaystyle 4\int\int_D \int_{3x^2+2y^2}^{16-x^2-2y^2} dz\;dy\;dx = 4\int\int_D -4(x^2 + y^2 - 4)\;dy\;dx$

$\displaystyle = 4\int_0^{\pi/2} \int_0^2 4r(4 - r^2)\;dr\;d\theta = 4\int_0^{\pi/2} 16\;d\theta = 32\pi$

36. $\displaystyle 4\int_0^3 \int_0^{(1/3)\sqrt{9-x^2}} \int_0^{x^2/9+y^2} dz\;dy\;dx = 4\int_0^3 \int_0^{(1/3)\sqrt{9-x^2}} \left(\frac{x^2}{9} + y^2\right)\;dy\;dx$

$\displaystyle = 4\int_0^3 \left[\frac{x^2}{27}\sqrt{9 - x^2} + \frac{1}{81}(9 - x^2)^{3/2}\right]\;dx$

$\displaystyle = \frac{4}{81}\int_0^{\pi/2} [18\sin^2\theta(3\cos\theta)(3\cos\theta)]\;d\theta = \frac{7\pi}{4}$

37. $\displaystyle \int_0^{3/2} \int_0^{3-2x} \int_0^{8+2x+y} dz\;dy\;dx = \int_0^{3/2} \int_0^{3-2x} (8 + 2x + y)\;dy\;dx$

$\displaystyle = \int_0^{3/2} \left[(8 + 2x)(3 - 2x) + \frac{(3 - 2x)^2}{2}\right]\;dx$

$\displaystyle = \int_0^{3/2} \left(\frac{57}{2} - 16x - 2x^2\right)\;dx = \frac{45}{2}$

38. Let the origin be the center of the base, with the vertices of the square base on the x and y axes. The x- and y-intercepts are at $755/\sqrt{2}$ from the origin. The equation of the plane through these intercepts and the top of the pyramid is

$$\frac{\sqrt{2}x}{755} + \frac{\sqrt{2}y}{755} + \frac{z}{480} = 1$$

$$V = 4 \int_0^{755/\sqrt{2}} \int_0^{755/\sqrt{2}(1-\sqrt{2}x/755)} \int_0^{A} dz \, dy \, dx$$

where $A = 480(1 - \dfrac{\sqrt{2}x}{755} - \dfrac{\sqrt{2}y}{755})$

$$= 4 \int_0^{755/\sqrt{2}} \int_0^{755/\sqrt{2}(1-\sqrt{2}x/755)} 480(1 - \frac{\sqrt{2}\,x}{755} - \frac{\sqrt{2}\,y}{755}) \, dy \, dx$$

$$= \frac{1,920(755)}{\sqrt{2}} \int_0^{755/\sqrt{2}} \left[(1 - \frac{\sqrt{2}\,x}{755})^2 - \frac{755}{\sqrt{2}(755)\sqrt{2}}(1 - \frac{\sqrt{2}\,x}{755})^2 \right] dx$$

$$= \frac{960(755)^2}{6} \approx 9.12 \times 10^7 \text{ ft}$$

39.
$$V = 8 \int_0^{\pi/2} \int_0^{R} \int_0^{\sqrt{R^2-r^2}} dz \, r \, dr \, d\theta$$

$$= 8 \int_0^{\pi/2} \int_0^{R} r\sqrt{R^2 - r^2} \, dr \, d\theta$$

$$= \frac{8}{3} \int_0^{\pi/2} R^3 \, d\theta = \frac{4}{3}\pi R^3$$

40. Let the origin be the the center of the base, with the vertices of the square base on the x and y axes. The x- and y-intercepts are at $S/\sqrt{2}$ from the origin. The equation of the plane through these intercepts and the top of the pyramid is
$$\frac{\sqrt{2}x}{S} + \frac{\sqrt{2}y}{S} + \frac{z}{H} = 1$$

$$V = 4 \int_0^{S/\sqrt{2}} \int_0^{S/\sqrt{2}(1-\sqrt{2}x/S)} \int_0^{A} dz \, dy \, dx \quad \text{where } A = H(1 - \frac{\sqrt{2}x}{S} - \frac{\sqrt{2}y}{S})$$

$$= 4 \int_0^{S/\sqrt{2}} \int_0^{S/\sqrt{2}(1-\sqrt{2}x/S)} H(1 - \frac{\sqrt{2}x}{S} - \frac{\sqrt{2}y}{S}) dy \, dx$$

$$= 4H \int_0^{S/\sqrt{2}} \left(\left[\frac{S}{\sqrt{2}} - \frac{\sqrt{2}x}{S} \right]2 - \frac{1}{\sqrt{2}S} \frac{S^2}{2} \left[1 - \frac{\sqrt{2}x}{S} \right]2 \right) dx$$

$$= \frac{2SH}{\sqrt{2}} \int_0^{S/\sqrt{2}} \left(1 - \frac{\sqrt{2}x}{S} \right)2 \, dx = \frac{HS^2}{3}$$

41. $z = c\left(1 - \dfrac{x^2}{a^2} - \dfrac{y^2}{b^2} \right)^{1/2}$; $y = b\left(1 - \dfrac{x^2}{a^2} \right)^{1/2}$

$$V = 4\int_0^a \int_0^{b\sqrt{1-x^2/a^2}} \int_0^{c\sqrt{(1-x^2/a^2-y^2/b^2)}} dz\, dy\, dx$$

$$= 8c\int_0^a \int_0^{b\sqrt{1-x^2/a^2}} \left(1 - \frac{x^2}{a^2} - \frac{y^2}{b^2}\right)^{1/2} dy\, dx$$

$$= 4c\int_0^a \left[\left(1 - \frac{x^2}{a^2}\right)\sin^{-1}\frac{y}{b\sqrt{1-\frac{x^2}{a^2}}} + \frac{y}{b}\sqrt{1 - \frac{x^2}{a^2} - \frac{y^2}{b^2}}\right]\Bigg|_0^{b/\sqrt{1-x^2/a^2}} $$

$$= 4bc\int_0^a \left[(1 - \frac{x^2}{a^2})\frac{\pi}{2}\right] dx = \frac{4\pi abc}{3}$$

42. $z = k(x^2 + y^2)$ intersects $x^2 + y^2 + z^2 = \frac{2}{k^2}$ when $\frac{z}{k} + z^2 = \frac{2}{k^2}$

Projected region, D, is the circle with radius z_0 where $k^2 z_0^2 + kz_0 = 2$

$$z_0 = \frac{-k \pm \sqrt{k^2 - 4(k^2)(-2)}}{2k^2} = \frac{1}{k}$$

$$V = \int\int_D \int_{kr^2}^{\sqrt{2/k^2 - r^2}} dA = \int_0^{2\pi}\int_0^{1/k} r\left[\sqrt{\frac{2}{k^2} - r^2} - kr^2\right] dr\, d\theta$$

$$= \int_0^{2\pi}\left[-\frac{1}{3}(\frac{2}{k^2} - r^2)^{3/2} - \frac{kr^4}{4}\right]_0^{1/k} d\theta = \left(\frac{8\sqrt{2} - 7}{12k^3}\right)(2\pi) \approx 2.259\left(\frac{1}{k^3}\right)$$

43. $z = c(1 - \frac{x}{a} - \frac{y}{b})$; $y = b(1 - \frac{x}{a})$

$$V = \int_0^a \int_0^{b(1-x/a)} \int_0^{c(1-x/a-y/b)} dz\, dy\, dx$$

$$= -\frac{c}{ab}\int_0^a \int_0^{b(1-x/a)} (bx + ay - ab)\, dy\, dx$$

$$= \frac{bc}{2a^2}\int_0^a (x - a)^2\, dx = \frac{abc}{6}$$

$$\int_0^a \int_0^{b(1-x/a)} \int_0^{c(1-\frac{x}{a}-\frac{y}{b})} dz\, dy\, dx$$

Similarly, we have

$$V = \int_0^b \int_0^{a(1-y/b)} \int_0^{c(1-x/a-y/b)} dz\, dx\, dy; \qquad V = \int_0^a \int_0^{c(1-x/a)} \int_0^{b(1-x/a-z/c)} dy\, dz\, dx;$$

$$V = \int_0^c \int_0^{a(1-z/c)} \int_0^{b(1-x/a-z/c)} dy\, dx\, dz; \qquad V = \int_0^c \int_0^{b(1-z/c)} \int_0^{a(1-y/b-z/c)} dx\, dy\, dz;$$

$$V = \int_0^b \int_0^{c(1-y/b)} \int_0^{a(1-y/b-z/c)} dx\, dz\, dy$$

44. It is generally true;

$$\int_a^b \int_c^d \int_r^s f(x)g(y)h(z) \ dz \ dy \ dx$$

$$= \int_a^b f(x) \int_c^d g(y) \int_r^s h(z) \ dz \ dy \ dx$$

$$= \left[\int_a^b f(x) \ dx \right]\left[\int_c^d g(y) \ dy \right]\left[\int_r^s h(z) \ dz \right]$$

45. Let u be the horizontal axis, v be the vertical axis, and consider the triangular region determined by $u = 0$, $v = x$, and $v = u$:

$$\int_0^x \int_0^v f(u) \ du \ dv$$

Switching the order of integration this

becomes: $\displaystyle\int_0^x \int_u^x f(u) \ dv \ du = \int_0^x (x - u)f(u) \ du.$

But u is a dummy variable, so this can be written as:

$$\int_0^x (x \ - \ t) \ f(t) \ dt.$$

In like manner, the tetrahedral region determined by he above triangle and the line $u = w$:

$$\int_0^x \int_0^v \int_0^u f(w) \ dw \ du \ dv$$

46. Using Problem 45,

$$\iiint_S \sin(\pi \ - \ z)^3 \ dz \ dy \ dx$$

$$= \int_0^\pi \int_0^y \int_0^x \sin(\pi \ - \ z)^3 \ dz \ dx \ dy$$

$$= \frac{1}{2}\int_0^x (\pi \ - \ t)^2 \sin(\pi \ - \ t)^3 \ dt$$

$$= \frac{\cos(\pi \ - \ t)^3}{6} \Big|_0^\pi = \frac{1 \ - \ \cos \pi^3}{6}$$

47. $\displaystyle\int_1^2 \int_{-1}^1 \int_0^2 \int_0^1 xyz^2 w^2 \ dx \ dy \ dz \ dw$

$$= \int_1^2 \int_{-1}^1 \int_0^2 \frac{w^2 yz^2}{2} \ dy \ dz \ dw = \int_1^2 \int_{-1}^1 w^2 z^2 \ dz \ dw$$

$$= \int_1^2 \frac{2w^2}{3} \ dw = \frac{14}{9}$$

48. $\displaystyle\int_0^4 \int_0^{4 \ - \ x} \int_0^{4 \ - \ x \ - \ y} \int_0^{4 \ - \ x \ - \ y \ - \ z} e^{x \ - \ 2y + z + w} \ dw \ dz \ dy \ dx$

$$= \int_0^4 \int_0^{4 \ - \ x} \int_0^{4 \ - \ x \ - \ y} e^{4 \ - \ 3y} - e^{x \ - \ 2y + z} \ dz \ dy \ dx$$

$$= \int_0^4 \int_0^{4 \ - \ x} (e^{x \ - \ 2y} - e^{4 \ - \ 3y})(x + y \ - \ 3) \ dy \ dx$$

$$= \int_0^4 \left[\frac{e^4(8 \ - \ 3x)}{9} \ - \ \frac{e^{3x - 8}}{18} + \frac{e^x}{2} \right] dx$$

$$= \frac{74e^{12} \ - \ 27e^8 + 1}{54e^8} \approx 74.3197$$

13.6 Mass, Moments, and Probability Density Functions, Page 872

1. Find the total mass. Compute the first moment with respect to the x-axis and the first moment with respect to the y-axis. Divide the first moments by the mass. The resulting coordinates are those of the center of mass.

2. Find the total mass. Computer the first moment with respect to the xy-plane, the first moment with respect to the xz-plane, and the first moment with respect to the yz-plane. Divide the first moments by the mass. The resulting coordinates are those of the center of mass.

3. The center of mass, or centroid, is the point at which the total mass of the region can be considered concentrated so that the effect of the total region is mathematically the same as that of the point. This is an example of mathematical modeling.

4. The moment of inertia is the second moment.

5. $m = \displaystyle\int_0^3 \int_0^4 dy \ dx = 60$

$$M_x = \int_0^3 \int_0^4 y \, dy \, dx = \frac{5}{2}\int_0^3 16 \, dx = 120$$

$$M_y = \int_0^3 \int_0^4 x \, dy \, dx = 20\int_0^3 x \, dx = 90$$

$$(\overline{x}, \overline{y}) = \left(\frac{90}{60}, \frac{120}{60}\right) = \left(\frac{3}{2}, 2\right)$$

6. $$m = 4\int_0^4 \int_0^{\sqrt{x}} dy \, dx = 4\int_0^4 \sqrt{x} \, dx = \frac{64}{3}$$

$$M_x = 4\int_0^4 \int_0^{\sqrt{x}} y \, dy \, dx = 2\int_0^4 x \, dx = 16$$

$$M_y = 4\int_0^4 \int_0^{\sqrt{x}} x \, dy \, dx = 4\int_0^4 x^{3/2} \, dx = \frac{256}{5}$$

$$(\overline{x}, \overline{y}) = \left(\frac{256(3)}{5(64)}, \frac{64}{48}\right) = \left(\frac{12}{5} \ \frac{4}{3}\right)$$

7. $$m = 2\int_0^2 \int_{x^2}^{2x} dy \, dx = 2\int_0^2 (2x - x^2) \, dx = \frac{8}{3}$$

$$M_x = 2\int_0^2 \int_{x^2}^{2x} y \, dy \, dx = \int_0^2 (4x^2 - x^4) \, dx = \frac{8}{5}$$

$$M_y = 2\int_0^2 \int_{x^2}^{2x} x \, dy \, dx = 2\int_0^2 x(2x - x^2) \, dx = \frac{8}{3}$$

$$(\overline{x}, \overline{y}) = \left(\frac{8(3)}{3(8)}, \frac{64(3)}{15(8)}\right) = \left(1, \frac{8}{5}\right)$$

8. $$m = 4\int_0^{1/2} \int_0^{\sin(\pi x/2)} dy \, dx = 4\int_0^{1/2} \sin\frac{\pi x}{2} \, dx$$

$$= \frac{8}{\pi}\left(1 - \frac{\sqrt{2}}{2}\right) \approx 0.7459$$

$$M_x = 4\int_0^{1/2} \int_0^{\sin(\pi x/2)} y \, dy \, dx$$

$$= 2\int_0^{1/2} \sin\frac{\pi x}{2} \, dx = \frac{1}{2} - \frac{1}{\pi} \approx 0.1817$$

$$M_y = 4\int_0^{1/2} \int_0^{\sin(\pi x/2)} x \, dy \, dx$$

$$= 4\int_0^{1/2} x \sin\frac{\pi x}{2} \, dx$$

$$= 4\left[-\frac{\sqrt{2}}{2\pi} + \frac{2\sqrt{2}}{\pi^2}\right] \approx 0.246$$

$$(\overline{x}, \overline{y}) \approx (0.3298, 0.2436)$$

9. $y = 2 - 3x^2$ intersects $3x + 2y = 1$ when

$$6x^2 - 3x - 3 = 0$$

$$3(x - 1)(2x + 1) = 0$$

$$x = 1, \ -\frac{1}{2}$$

$$m = \int_{-1/2}^1 \int_{(-1/2)(1-3x)}^{2-3x^2} dy \, dx$$

$$= \int_{-1/2}^1 \left(-3x^2 + \frac{3}{2}x + \frac{3}{2}\right) dx = \frac{27}{6}$$

$$M_x = \int_{-1/2}^1 \int_{(-1/2)(1-3x)}^{2-3x^2} y \, dy \, dx$$

$$= \frac{1}{2}\int_{-1/2}^1 \left[(2-3x^2)^2 - \frac{1}{4}(1-3x)^2\right] dx = \frac{27}{20}$$

$$M_y = \int_{-1/2}^1 \int_{(-1/2)(1-3x)}^{2-3x^2} x \, dy \, dx$$

$$= \int_{-1/2}^1 x[2 - 3x^2 - \frac{1}{2}(1-3x)] \, dx = \frac{27}{64}$$

$$(\overline{x}, \overline{y}) = \left(\frac{27(16)}{64(27)}, \frac{27(16)}{20(27)}\right) = \left(\frac{1}{4}, \frac{4}{5}\right)$$

10. $$m = 2\int_0^{\pi/2} \int_0^3 \int_0^{\sqrt{9-r^2}} r \, dz \, dr \, d\theta$$

$$= 2\int_0^{\pi/2} \int_0^3 r\sqrt{9 - r^2} \, dr \, d\theta$$

$$= 2\int_0^{\pi/2} 9 \, d\theta = 9\pi$$

$$M_{yz} = \int_0^{\pi/2} \int_0^3 \int_0^{\sqrt{9-r^2}} r^2 \cos\theta \, dz \, dr \, d\theta$$

$$= 2 \int_0^{\pi/2} \int_0^3 r^2 \cos\theta \sqrt{9 - r^2} \, dr \, d\theta$$

$$= 2 \int_0^{\pi/2} \frac{3^4 \pi}{16} \cos\theta \, d\theta = \frac{81\pi}{8}$$

$\bar{x} = \dfrac{81\pi}{8(9\pi)} = \dfrac{9}{8}$; by symmetry $(\bar{x}, \bar{y}, \bar{z}) = \left(\frac{9}{8}, \frac{9}{8}, \frac{9}{8}\right)$

11. $\bar{x} = \dfrac{M_{yz}}{M} = \dfrac{4 \displaystyle\int_0^4 \int_0^{-x+4} \int_0^{4-x-y} x \, dz \, dy \, dx}{4 \displaystyle\int_0^4 \int_0^{-x+4} \int_0^{4-x-y} dz \, dy \, dx}$

$$= \frac{\displaystyle\int_0^4 \int_0^{-x+4} (4x - x^2 - xy) \, dy \, dx}{\text{Vol. of a tetrahedron}}$$

$$= \frac{\displaystyle\int_0^4 \left[(4x - x^2)(-x + 4) - \frac{x(-x+4)^2}{2} \right] dx}{\frac{32}{3}}$$

$$= \frac{3}{32} \int_0^4 \left(\frac{x^3}{2} - 4x^2 + 8x \right) dx = 1$$

By symmetry $\bar{x} = \bar{y} = \bar{z} = 1$ The centroid is at $(1, 1, 1)$.

12. $m = \displaystyle\int_0^{\pi} \int_0^{\sin x} \int_0^{1-z} dy \, dz \, dx$

$$= \int_0^{\pi} \int_0^{\sin x} (1 - z) \, dz \, dx$$

$$= \int_0^{\pi} (\sin x - \tfrac{1}{2} \sin^2 x) \, dx$$

$$= \frac{8 - \pi}{4}$$

$M_{xy} = \displaystyle\int_0^{\pi} \int_0^{\sin x} \int_0^{1-z} z \, dy \, dz \, dx$

$$= \int_0^{\pi} \int_0^{\sin x} z(1 - z) \, dz \, dx$$

$$= \int_0^{\pi} (\tfrac{1}{2} \sin^2 x - \tfrac{1}{3} \sin^3 x) \, dx$$

$$= \frac{9\pi - 16}{36}$$

$M_{yz} = \displaystyle\int_0^{\pi} \int_0^{\sin x} \int_0^{1-z} x \, dy \, dz \, dx$

$$= \int_0^{\pi} \int_0^{\sin x} x(1 - z) \, dz \, dx$$

$$= \frac{1}{2} \int_0^{\pi} (2x \sin x - x \sin x) \, dx$$

$$= \frac{8\pi - \pi^2}{8}$$

$M_{xz} = \displaystyle\int_0^{\pi} \int_0^{\sin x} \int_0^{1-z} y \, dy \, dz \, dx$

$$= \frac{1}{2} \int_0^{\pi} \int_0^{\sin x} (1 - z)^2 \, dz \, dx$$

$$= \frac{1}{6} \int_0^{\pi} (\sin^3 x - 3\sin^2 x + 3 \sin x) \, dx$$

$$= \frac{44 - 9\pi}{36}$$

$(\bar{x}, \bar{y}, \bar{z}) = \left(\dfrac{\pi}{2}, \dfrac{9\pi - 44}{9(\pi - 8)}, \dfrac{9\pi - 16}{9(8 - \pi)} \right)$

$$\approx (1.5708, 0.3596, 0.2807)$$

13. $m = 2 \displaystyle\int_0^3 \int_0^{\sqrt{9 - x^2}} (x^2 + y^2) \, dy \, dx$

$$= 2 \int_0^3 [x^2 \sqrt{9 - x^2} + \tfrac{1}{3}(9 - x^2)^{3/2}] \, dx$$

$$= \frac{4}{3} \int_0^{\pi/2} [(9 \sin^2\theta)(9 \cos^2\theta) + \tfrac{27\pi}{2}] \, d\theta = \frac{81\pi}{4}$$

$\bar{x} = 0$ (by symmetry)

$M_x = 2 \displaystyle\int_0^3 \int_0^{\sqrt{9 - x^2}} y(x^2 + y^2) \, dy \, dx$

$$= 2 \int_0^3 [\tfrac{1}{2} x^2 (9 - x^2) + \tfrac{1}{4}(81 - 18x^2 + x^4)] \, dx$$

$$= \frac{486}{5}$$

$\bar{y} = \dfrac{486(4)}{5(81\pi)} = \dfrac{24}{5\pi}$; $(\bar{x}, \bar{y}) = \left(0, \dfrac{24}{5\pi} \right)$

14. $m = \displaystyle\int_0^{\pi/2} \int_0^a k r^2 \, dr \, d\theta = \frac{k\pi a^4}{8}$

$\overline{x} = 0$ (by symmetry)

$$M_x = \int_0^{\pi/2} \int_0^a kr^2(r \sin \theta)\, r\, dr\, d\theta$$

$$= \frac{1}{5} \int_0^{\pi/2} a^5 k \sin \theta\, d\theta$$

$$= \frac{ka^5}{5}$$

$$\overline{y} = \frac{8ka^5}{5k\pi a^4} = \frac{8a}{5\pi};\ (\overline{x}, \overline{y}) = \left(0, \frac{8a}{5\pi}\right)$$

15.
$$m = \int_0^6 \int_0^{5x/6} 7x\, dy\, dx + \int_6^{12} \int_0^{-5x/6+10} 7x\, dy\, dx$$

$$= \int_0^6 \frac{35}{6} x^2\, dx + \int_6^{12} \left(-\frac{35x^2}{6} + 70x\right) dx$$

$$= 1{,}260$$

$$M_x = \int_0^6 \int_0^{5x/6} 7xy\, dy\, dx + \int_6^{12} \int_0^{-5x/6+10} 7xy\, dy\, dx$$

$$= \int_0^6 \frac{7x}{2}\left(\frac{5x}{6}\right)^2 dx + \int_6^{12} \frac{7x}{2}\left(-\frac{5x}{6} + 10\right)^2 dx$$

$$= \int_0^6 \frac{175}{72} x^3 + \int_6^{12} \left(\frac{175}{72} x^3 - \frac{175}{3} x^2 + 350x\right) dx$$

$$= 2{,}100$$

$$M_y = 7 \int_0^5 \int_{6y/5}^{-(6/5)y+12} x^2\, dx\, dy$$

$$= \frac{7(6^3)}{3} \int_0^5 \left(-\frac{2y^3}{5^3} + \frac{6y^2}{5^2} - \frac{12y}{5} + 8\right) dy$$

$$= 8{,}820$$

$$(\overline{x}, \overline{y}) = \left(\frac{8{,}820}{1{,}260}, \frac{2{,}100}{1{,}260}\right) = \left(7, \tfrac{5}{3}\right)$$

16.
$$m = 3 \int_0^6 \int_0^{x^2} x\, dy\, dx = 3 \int_0^6 x^3\, dx = 972$$

$$M_x = 3 \int_0^1 xy\, dy\, dx = \frac{3}{2} \int_0^6 x^5\, dx = 11{,}664$$

$$M_y = 3 \int_0^6 x^2\, dy\, dx = 3 \int_0^6 x^4\, dx = 4{,}665.6$$

$$(\overline{x}, \overline{y}) = \left(\frac{4{,}665.6}{972}, \frac{11{,}664}{972}\right) \approx (4.8,\ 12)$$

17.
$$m = \int_1^2 \int_0^{\ln x} x^{-1}\, dy\, dx = \int_1^2 \frac{\ln x}{x}\, dx = \frac{(\ln x)^2}{2}$$

$$M_x = \int_1^2 \int_0^{\ln x} \frac{y}{x}\, dy\, dx = \frac{1}{2} \int_1^2 \frac{(\ln x)^2}{x}\, dx = \frac{(\ln 2)^3}{6}$$

$$M_y = \int_1^2 \int_0^{\ln x} \ln x\, dx = 2 \ln 2 - 1$$

$$(\overline{x}, \overline{y}) = \left(\frac{(2 \ln 2 - 1)(2)}{(\ln 2)^2}, \frac{(\ln 2)^3(2)}{6(\ln 2)^2}\right)$$

$$= \left(\frac{4 \ln 2 - 2}{\ln^2 2}, \frac{\ln 2}{3}\right)$$

$$\approx (1.608,\ 0.231)$$

18.
$$m = \int_0^2 \int_0^{e^{-x}} y\, dy\, dx$$

$$= \frac{1}{2} \int_0^2 e^{-2x}\, dx = \frac{1}{4}(1 - e^{-4})$$

$$M_x = \int_0^2 \int_0^{e^{-x}} y^2\, dy\, dx$$

$$= \frac{1}{3} \int_0^2 e^{-3x}\, dx = \frac{1}{9}(1 - e^{-6})$$

$$M_y = \int_0^2 \int_0^{e^{-x}} xy\, dy\, dx$$

$$= \frac{1}{2} \int_0^2 xe^{-2x}\, dx = \frac{1}{8}(1 - 5e^{-4})$$

$$(\overline{x}, \overline{y}) = \left(\frac{2(1 - 5e^{-4})}{1 - e^{-4}}, \frac{4(1 - e^{-6})}{9(1 - e^{-4})}\right)$$

$$\approx (0.4629,\ 0.4399)$$

19. a. By symmetry $\overline{x} = 0$. $\rho = kr$.

$$m = k \int_0^\pi \int_0^a (r)\, r\, dr\, d\theta$$

$$= \frac{ka^3}{3} \int_0^\pi d\theta = \frac{k\pi a^3}{3}$$

$$M_x = k \int_0^\pi \int_0^a (r)(r\sin\theta)\, r\, dr\, d\theta$$

$$= \frac{ka^4}{4} \int_0^\pi \sin\theta\, d\theta$$

$$= \frac{ka^4}{4}(1+1) = \frac{ka^4}{2}$$

$$\bar{y} = \frac{\frac{ka^4}{2}}{\frac{k\pi a^3}{3}} = \frac{3a}{2\pi}$$

The centroid is at $(0, \frac{3a}{2\pi})$

b. $\rho = k\theta$ (use $\rho = \theta$ as the k will cancel in the quotient).

$$m = \int_0^\pi \int_0^a (\theta)\, r\, dr\, d\theta = \frac{a^2}{2} \int_0^\pi \theta\, d\theta = \frac{\pi^2 a^2}{4}$$

$$M_x = \int_0^\pi \int_0^a (\theta)(r\sin\theta)\, r\, dr\, d\theta$$

$$= \frac{a^3}{3} \int_0^\pi \theta\sin\theta\, d\theta = \frac{\pi a^3}{3}$$

$$M_y = \int_0^\pi \int_0^a (\theta)(r\cos\theta)\, r\, dr\, d\theta$$

$$= \frac{a^3}{3} \int_0^\pi \theta\cos\theta\, d\theta = -\frac{2a^3}{3}$$

$$(\bar{x}, \bar{y}) = \left(\frac{-\frac{2a^3}{3}}{\frac{\pi^2 a^2}{4}}, \frac{\frac{\pi a^3}{3}}{\frac{\pi^2 a^2}{4}} \right)$$

$$= \left(-\frac{8a}{3\pi^2}, \frac{4a}{3\pi} \right)$$

20. $\rho = kr^2$

$$m = \int_0^\pi \int_0^a kr^3\, dr\, d\theta = \frac{ka^4\pi}{4}$$

$$M_x = 1 \int_0^\pi \int_0^a kr^4\sin\theta\, dr\, d\theta = \frac{ka^5}{5} \int_0^\pi \sin\theta\, d\theta$$

$$= \frac{2ka^5}{5}$$

$$M_y = \int_0^\pi \int_0^a kr^4\cos\theta\, dr\, d\theta = 0;$$

$$(\bar{x}, \bar{y}) = \left(0, \frac{8a}{5\pi} \right)$$

21. $$m = \int_1^{a^2} \int_0^{\ln x} dy\, dx = \int_1^{a^2} \ln x\, dx = e^2 + 1$$

$$M_x = \int_1^{e^2} \int_0^{\ln x} y\, dy\, dx = \frac{1}{2} \int_1^{e^2} (\ln x)^2\, dx$$

$$= \frac{1}{2} \left[x\ln^2 x - 2 \int_1^{e^2} \ln x\, dx \right] = e^2 - 1$$

$$M_y = \int_1^{e^2} \int_0^{\ln x} x\, dy\, dx = \int_1^{e^2} x\ln x\, dx$$

$$= e^4 - \frac{e^4}{4} + \frac{1}{4} = \frac{1}{4}(3e^4 + 1)$$

$$(\bar{x}, \bar{y}) = \left(\frac{3e^4+1}{4(e^2+1)}, \frac{e^2-1}{e^2+1} \right) \approx (4.911, 0.762)$$

22. $\rho = x^2 + y^2 + z^2 = r^2 + z^2;$
$\bar{x} = \bar{y} = 0$ (by symmetry)

$$m = \int_0^{2\pi} \int_0^3 \int_0^{r^2} r(r^2 + z^2)\, dz\, dr\, d\theta$$

$$= \int_0^{2\pi} \int_0^3 \frac{r^5(r^3 + 3)}{3}\, dr\, d\theta$$

$$= \frac{3,159}{8} \int_0^{2\pi} d\theta = \frac{3,159\pi}{4}$$

$$M_{xy} = \int_0^{2\pi} \int_0^3 \int_0^{r^2} rz(r^2 + z^2)\, dz\, dr\, d\theta$$

$$= \int_0^{2\pi} \int_0^3 \frac{r^7(r^2 + 2)}{4}\, dr\, d\theta$$

$$= \int_0^{2\pi} \frac{150,903}{80}\, d\theta = \frac{150,903\pi}{40}$$

$$(\bar{x}, \bar{y}, \bar{z}) = \left(0, 0, \frac{150,903\pi(4)}{40(3,159\pi)} \right) \approx (0, 0, 4.78)$$

23. $I_x = 2\displaystyle\int_0^1 \int_0^{1-x^2} x^2 y^2 \, dy \, dx$

$= \dfrac{2}{3}\displaystyle\int_0^1 x^2 (1 - x^2)^3 \, dx$

$= \dfrac{2}{3}\displaystyle\int_0^1 (- x^8 + 3x^6 - 3x^4 + x^2) \, dx$

$= \dfrac{32}{945}$

24. $I_z = 4\displaystyle\int_0^1 \int_0^1 (x^2 + y^2) \, x^2 y^2 \, dy \, dx$

$= 4\displaystyle\int_0^1 \int_0^1 (x^4 y^2 + x^2 y^4) \, dy \, dx$

$= \dfrac{4}{15}\displaystyle\int_0^1 (5x^4 + 3x^2) \, dx = \dfrac{8}{15}$

25. $m = 2\displaystyle\int_{-\pi/2}^{\pi/2} \int_0^{1+\sin\theta} r^2 \, dr \, d\theta$

$= \dfrac{2}{3}\displaystyle\int_{-\pi/2}^{\pi/2} (1 + \sin\theta)^3 \, d\theta = \dfrac{5\pi}{3}$

$\bar{x} = 0$ (by symmetry)

$M_y = 2\displaystyle\int_{-\pi/2}^{\pi/2} \int_0^{1+\sin\theta} r^3 \sin\theta \, dr \, d\theta$

$= \dfrac{1}{2}\displaystyle\int_{-\pi/2}^{\pi/2} (1 + \sin\theta)^4 \sin\theta \, d\theta$

$= \dfrac{1}{2}\displaystyle\int_{-\pi/2}^{\pi/2} [\sin\theta + 2 - 2\cos 2\theta + 6\sin^3\theta + (1 - \cos 2\theta)^2 + \sin^5\theta] \, d\theta$

$= \dfrac{7\pi}{4}$

$\bar{y} = \dfrac{7\pi(3)}{4(5\pi)} = \dfrac{21}{20}; \ (\bar{x}, \bar{y}) = (0, 1.05)$

26. $m = \displaystyle\int_0^{\pi/2} \int_0^{\sqrt{2\sin 2\theta}} r \, dr \, d\theta$

$= \displaystyle\int_0^{\pi/2} \sin 2\theta \, d\theta = 1$

$M_x = \displaystyle\int_0^{\pi/2} \int_0^{\sqrt{2\sin 2\theta}} (r\cos\theta) \, r \, dr \, d\theta$

$= \displaystyle\int_0^{\pi/2} \dfrac{2\sqrt{2}}{3} \cos\theta (\sin 2\theta)^{3/2} \, d\theta$

≈ 0.55360367077

(numerical computer approximation)

$\bar{x} \approx 0.55, \ \bar{y} \approx 0.55$ (by symmetry);

$(\bar{x}, \bar{y}) = (0.55, 0.55)$

27. By symmetry $\bar{y} = 0$. $\bar{x}$ for the upper half and the lower half are equal, so consider the upper half; $1 + 2\cos\theta = 0$ at $\cos\theta = -1/2$, $\theta = 2\pi/3$. Also, the subtraction is necessary because $r \le 1 + 2\cos\theta$ is negative for $2\pi/3 \le \theta \le \pi$.

$m = 2\displaystyle\int_0^{2\pi/3} \int_0^{1+2\cos\theta} r \, dr \, d\theta - \int_{2\pi/3}^{\pi} \int_0^{1+2\cos\theta} r \, dr \, d\theta$

$= \displaystyle\int_0^{2\pi/3} (1 + 2\cos\theta)^2 \, d\theta$

$\qquad - \displaystyle\int_{2\pi/3}^{\pi} (1 + 2\cos\theta)^2 \, d\theta$

$= \displaystyle\int_0^{2\pi/3} (1 + 4\cos\theta + 4\cos^2\theta) \, d\theta$

$\qquad - \displaystyle\int_{2\pi/3}^{\pi} (1 + 4\cos\theta + 4\cos^2\theta) \, d\theta$

$= \displaystyle\int_0^{2\pi/3} (1 + 4\cos\theta + 2 + 2\cos 2\theta) \, d\theta$

$\qquad - \displaystyle\int_0^{2\pi/3} (1 + 4\cos\theta + 2 + 2\cos 2\theta) \, d\theta$

$= [3\theta + 4\sin\theta + \sin 2\theta]\Big|_0^{2\pi/3}$

$\qquad - [3\theta + 4\sin\theta + \sin 2\theta]\Big|_{2\pi/3}^{\pi} = \pi + 3\sqrt{3}$

$M_y = 2\displaystyle\int_0^{2\pi/3} \int_0^{1+2\cos\theta} (r\cos\theta) \, r \, dr \, d\theta$

$\qquad - 2\displaystyle\int_{2\pi/3}^{\pi} \int_0^{1+2\cos\theta} (r\cos\theta) \, r \, dr \, d\theta$

$= \dfrac{2}{3}\displaystyle\int_0^{2\pi/3} (\cos\theta)(1 + 2\cos\theta)^3 \, d\theta$

$$-\frac{2}{3}\int_{2\pi/3}^{\pi}(\cos\theta)(1+2\cos\theta)^3\,d\theta$$

$$=\frac{2}{3}\int_{0}^{2\pi/3}(\cos\theta)(1+6\cos\theta+12\cos^2\theta+8\cos^3\theta)\,d\theta$$

$$-\frac{2}{3}\int_{2\pi/3}^{\pi}(\cos\theta)(1+6\cos\theta+12\cos^2\theta+8\cos^3\theta)\,d\theta$$

$$=\frac{2}{3}\Big[\sin\theta+3\theta+\frac{3}{2}\sin 2\theta+12\sin\theta-\frac{12\sin^3\theta}{3}$$

$$+2\Big(\frac{3}{2}\theta+\sin 2\theta+\frac{\sin 4\theta}{8}\Big)\Big]\Big|_{0}^{2\pi/3}$$

$$-\frac{2}{3}\Big[\sin\theta+3\theta+\frac{3}{2}\sin 2\theta+12\sin\theta-\frac{12\sin^3\theta}{3}$$

$$+2\Big(\frac{3}{2}\theta+\sin 2\theta+\frac{\sin 4\theta}{8}\Big)\Big]\Big|_{2\pi/3}^{\pi}$$

$$=\frac{4}{3}\pi+\frac{9}{2}\sqrt{3}$$

$$\bar{x}=\frac{M_y}{M}=\frac{\frac{4}{3}\pi+\frac{9}{2}\sqrt{3}}{\pi+3\sqrt{3}}\approx 1.4372$$

$$(\bar{x},\bar{y})\approx(1.3472,\,0)$$

28.
$$m=\int_{0}^{a}\int_{0}^{bx/a}kx\,dy\,dx=\frac{kb}{a}\int_{0}^{a}x^2\,dx=\frac{a^2bk}{3}$$

$$M_x=\int_{0}^{a}\int_{0}^{bx/a}kxy\,dy\,dx=\frac{kb^2}{2a^2}\int_{0}^{a}x^3\,dx=\frac{ka^2b^2}{8}$$

$$M_y=\int_{0}^{a}\int_{0}^{bx/a}kx^2\,dy\,dx=\frac{kb}{a}\int_{0}^{a}x^3\,dx=\frac{kba^3}{4a}$$

$$(\bar{x},\bar{y})=\Big(\frac{3ka^3b}{4a^2bk},\frac{3ka^2b^2}{8a^2bk}\Big)=\Big(\frac{3a}{4},\frac{3b}{8}\Big)$$

29. a. Since there is symmetry with respect to all axes and $\rho=0$, $(0,0,0)$ is the centroid. For I_z, the average distance squared is $(a^2+b^2)/4$ from the z-axis. If this distance is used as an approximation for the radius of gyration, and the density is 1, I_z should be approximately

$$\frac{(a^2+b^2)(a^3b^3c^3)}{4}$$

b.
$$m=\int_{-a}^{a}\int_{-b}^{b}\int_{-c}^{c}x^2y^2z^2\,dz\,dy\,dx$$

$$=\frac{3c^3}{3}\frac{2b^3}{3}\frac{2a^3}{3}=\frac{8a^3b^3c^3}{27}$$

$$M_{xy}=\int_{-a}^{a}\int_{-b}^{b}\int_{-c}^{c}x^2y^2z^3\,dz\,dy\,dx=0$$

Thus, $\bar{x}=0$ and similarly, $\bar{y}=\bar{z}=0$.

$$I_z=\int_{-a}^{a}\int_{-b}^{b}\int_{-c}^{c}x^2y^2z^2(x^2+y^2)\,dz\,dy\,dx$$

$$=\int_{-a}^{a}\int_{-b}^{b}\frac{2c^3x^2y^2(x^2+y^2)}{3}\,dy\,dx$$

$$=\int_{-a}^{a}\frac{4b^3c^3x^2(5x^2+3b^2)}{45}\,dx$$

$$=\frac{8a^3b^3c^3(a^2+b^2)}{45}$$

Note that our guess is 0.25 times $a^3b^3c^3(a^2+b^2)$ and the calculation gives 0.18 times that same number.

30.
$$m=\int_{0}^{a}\int_{0}^{b}xy\,dy\,dx=\frac{b^2}{2}\int_{0}^{a}x\,dx=\frac{a^2b^2}{4}$$

$$M_x=\int_{0}^{a}\int_{0}^{b}xy^2\,dy\,dx=\frac{b^3}{3}\int_{0}^{a}x\,dx=\frac{a^2b^3}{6}$$

$$M_y=\int_{0}^{a}\int_{0}^{b}x^2y\,dy\,dx=\frac{b^2}{2}\int_{0}^{a}x^2\,dx=\frac{a^3b^2}{6}$$

$$(\bar{x},\bar{y})=\Big(\frac{4a^3b^2}{6a^2b^2},\frac{4a^2b^3}{6a^2b^2}\Big)=\Big(\frac{2a}{3},\frac{2b}{3}\Big)$$

31.
$$I_x=\int_{0}^{\pi}\int_{0}^{a\cos\theta}(r\sin\theta)^2\,r\,dr\,d\theta$$

$$=\frac{a^4}{4}\int_{0}^{\pi}\cos^4\theta\sin^2\theta\,d\theta$$

$$=\frac{a^4}{4}\int_{0}^{\pi}\Big(\frac{1+\cos 2\theta}{2}\Big)^2\Big(\frac{1-\cos 2\theta}{2}\Big)\,d\theta$$

$$=\frac{a^4}{32}\int_{0}^{\pi}(1+\cos 2\theta-\cos^2 2\theta-\cos^3 2\theta)\,d\theta$$

$$=\frac{a^4\pi}{64}$$

32. $m = \int\limits_{0}^{2\pi}\int\limits_{0}^{a} r\, dr\, d\theta = a^2\pi,$

$I_x = \int\limits_{0}^{2\pi}\int\limits_{0}^{a} r^3\sin^2\theta\, dr\, d\theta$

$\quad = \dfrac{a^4}{8}\int\limits_{0}^{2\pi}\dfrac{1 - \cos 2\theta}{2}\, d\theta = \dfrac{a^4\pi}{4}$

Note, $I_x = \dfrac{a^4\pi}{4} = (a^2\pi)\dfrac{a^2}{4} = \dfrac{ma^2}{4}$

33. $m = \int\limits_{0}^{2\pi}\int\limits_{0}^{a} r\, dr\, d\theta = a^2\pi,$

$I_y = \int\limits_{0}^{2\pi}\int\limits_{0}^{a} r^3\cos^2\theta\, dr\, d\theta$

$\quad = \dfrac{a^4}{8}\int\limits_{0}^{2\pi}\dfrac{1 + \cos 2\theta}{2}\, d\theta = \dfrac{a^4\pi}{4}$

Note, $I_y = \dfrac{a^4\pi}{4} = (a^2\pi)\dfrac{a^2}{4} = \dfrac{ma^2}{4}$

34. $m = \int\limits_{0}^{2\pi}\int\limits_{0}^{a} r\, dr\, d\theta = a^2\pi,$

$I_z = \int\limits_{0}^{2\pi}\int\limits_{0}^{a} r^3\, dr\, d\theta = \dfrac{a^4\pi}{2}$

Note, $I_x = \dfrac{a^4\pi}{2} = (a^2\pi)\dfrac{a^2}{2} = \dfrac{ma^2}{2}$

35. If $\rho = 1$ then $m = A = \pi ab.$ Using symmetry:

$I_x = 4\int\limits_{0}^{a}\int\limits_{0}^{\frac{b}{a}\sqrt{a^2 - x^2}} y^2\, dy\, dx$

$\quad = \dfrac{4}{3}\int\limits_{0}^{a}\left(\dfrac{b}{a}\sqrt{a^2 - x^2}\right)^3 dx$

$\quad = \dfrac{4b^3}{3a^3}\int\limits_{0}^{a}(a^2 - x^2)^{3/2}dx \quad \text{(Formula \#245)}$

$\quad = \dfrac{b^3}{3a^3}\left(\dfrac{3a^4\pi}{4}\right) = \dfrac{ab^3\pi}{4}$

Substituting $m = \pi ab$: $I_x = \dfrac{mb^2}{4}$

36. $m = c\int\limits_{0}^{a}\int\limits_{0}^{b(1 - x/a)} x(1 - x/a - y/b)\, dy\, dx$

$\quad = \dfrac{bc}{2}\int\limits_{0}^{a}(x - 2x^2/a + x^3/a^2)\, dx = \dfrac{a^2bc}{24}$

$M_{yz} = bc\int\limits_{0}^{a}\int\limits_{0}^{b(1 - x/a)}\left[x^2(1 - \tfrac{x}{a})^2 - \dfrac{bx^2}{2}(1 - \tfrac{x}{a})^2\right] dx$

$\quad = \dfrac{bc}{2}\int\limits_{0}^{a}\left(x^2 - \dfrac{2x^3}{a} + \dfrac{cx^4}{a}\right) dx = \dfrac{a^2bc^2}{120}$

$M_{xy} = bc\int\limits_{0}^{a}\int\limits_{0}^{b(1 - x/a)}\int\limits_{0}^{c(1 - x/a - y/b)} zx\, dz\, dy\, dx$

$\quad = \dfrac{c^2}{2}\int\limits_{0}^{a}\int\limits_{0}^{b(1 - x/a)}\left(1 - \tfrac{x}{a} - \tfrac{y}{b}\right)^2 x\, dy\, dx$

$\quad = \dfrac{bc^2}{6}\int\limits_{0}^{a} x(1 - \tfrac{x}{a})^3\, dx = \dfrac{a^2b^2c}{120}$

$M_{xz} = c\int\limits_{0}^{a}\int\limits_{0}^{b(1 - x/a)} xy(1 - \tfrac{x}{a} - \tfrac{y}{b})\, dy\, dx$

$\quad = \dfrac{b^2}{6}\int\limits_{0}^{a}\left(x - \dfrac{3x^2}{a} - \dfrac{3x^3}{a} + \dfrac{3x^3}{a^2} - \dfrac{x^4}{a^3}\right) dx$

$\quad = \dfrac{a^2b^2c}{120}$

$(\overline{x},\, \overline{y},\, \overline{z}) = \left(\dfrac{a^3bc(24)}{60a^2bc},\, \dfrac{a^2b^2c(24)}{120a^2bc},\, \dfrac{a^2bc^2(24)}{120a^2bc}\right)$

$\quad = (\tfrac{2a}{5},\, \tfrac{b}{5},\, \tfrac{c}{5})$

37. $m = c(\tfrac{4\pi}{3})(\tfrac{3a^3}{8}) = \dfrac{\pi a^3}{6}$

$M_{yz} = \int\limits_{0}^{a}\int\limits_{0}^{\pi/2} r\cos\theta\sqrt{a^2 - r^2}\; r\, d\theta\, dr$

$\quad = \int\limits_{0}^{a} r^2\sqrt{a^2 - r^2}\, dr$

$\quad = a^4\int\limits_{0}^{\pi/2}\sin^2\alpha\,\cos^2\alpha\, d\alpha = \dfrac{\pi a^4}{16}$

$\overline{x} = \dfrac{6\pi a^4}{16\pi a^3} = \dfrac{3a}{8};$ by symmetry,

$\overline{x} = \overline{y} = \overline{z} = \dfrac{3a}{8}$

38. $P = \int\limits_{0}^{1}\int\limits_{0}^{1 - x} 2e^{-2x}e^{-y}\, dy\, dx$

$$= -2 \int_0^1 e^{-2x}(e^{x-1} - 1) \, dx$$

$$= 2e^{-2} - e^{-1} - e^{-2} + 1 \approx 0.3995$$

39.
$$P = \int_0^1 \int_0^1 xe^{-x}e^{-y} \, dy \, dx$$

$$= -\int_0^1 (xe^{-1} + xe^{-x}) \, dx$$

$$= 1 - 2e^{-1} - \tfrac{1}{2}e^{-1}$$

$$= 1 - \tfrac{5}{2}e^{-1} \approx 0.0803$$

40.
$$P = \tfrac{1}{6} \int_0^3 \int_0^{3-x} e^{-x/2}e^{-y/3} \, dy \, dx$$

$$= \tfrac{1}{2} \int_0^3 (e^{-x/2} - e^{-x/6-1}) \, dx$$

$$= -3e^{-1} + 2e^{-3/2} + 1 \approx .3426$$

41.
$$P = \tfrac{1}{4} \int_0^1 \int_0^{1-x} e^{-x/2}e^{-y/2} \, dy \, dx$$

$$= -\tfrac{1}{2} \int_0^1 e^{-x/2}(e^{-1/2+x/2} - 1) \, dx$$

$$= -\tfrac{1}{2}e^{-1/2} - e^{-1/2} + 1$$

$$= 1 - \tfrac{3}{2}e^{-1/2} \approx 0.0902$$

42.
$$P = \tfrac{1}{8} \int_0^8 \int_0^{8-x} e^{-x/2}e^{-y/4} \, dy \, dx$$

$$= -\tfrac{1}{4} \int_0^8 e^{-x/2}(e^{-2+x/4}) \, dx$$

$$= -\tfrac{1}{4}[-4e^{-4} + 6e^{-2} - 2] \approx 0.3153$$

43. $P(< 30) = \dfrac{1}{300} \displaystyle\int_0^{30}\int_0^{30-x} e^{-x/30}e^{-y/10} \, dy \, dx$

$$= -\tfrac{1}{30} \int_0^{30} e^{-x/30}\left(e^{(x-30)/10} - 1\right) \, dx$$

$$= -\tfrac{1}{30} \int_0^{30} \left(e^{x/15-3} - e^{-x/30}\right) \, dx$$

$$= -\tfrac{1}{2}\left(\tfrac{3}{e} - \tfrac{1}{e^3} - 2\right) \approx 0.4731$$

44. $A = \displaystyle\int_0^1\int_0^{x^2} dy \, dx = \int_0^1 x^2 \, dx = \tfrac{1}{3}$

$$AV = \tfrac{1}{A} \int_0^1\int_0^{x^2} e^{x^3} \, dx = 3\int_0^1 e^{x^3}(x^2) \, dx = e - 1$$

45. $A = \displaystyle\int_0^1\int_{x^2}^1 dy \, dx = \int_0^1 (1 - x^2) \, dx = \tfrac{2}{3}$

$$AV = \tfrac{1}{A} \int_0^1\int_{x^2}^1 e^x y^{-1/2} \, dx = \tfrac{3}{2}\int_0^1 (2e^x - 2xe^x) \, dx$$

$$= \tfrac{3}{2}(2e - 4) = 3(e - 2)$$

46. $V = \displaystyle\int_0^1 \int_0^{1-x} \int_0^{1-x-y} dz \, dy \, dx$

$$= \int_0^1 \int_0^{1-x} (1 - x - y) \, dy \, dx$$

$$= \int_0^1 [(1 - x)^2 - \tfrac{1}{2}(1 - x)^2] \, dx = \tfrac{1}{6}$$

$$AV = \tfrac{1}{V} \int_0^1 \int_0^{1-x} \int_0^{1-x-y} (x + 2y + 3z) \, dz \, dy \, dx$$

$$= 3 \int_0^1 \int_0^{1-x} (x^2 - 4x - y^2 - 2y + 3) \, dy \, dx$$

$$= \int_0^1 (-2x^3 + 9x^2 - 12x + 5) \, dx = \tfrac{3}{2}$$

47. Since the sphere is symmetric in each of the eight octants, and xyz is positive in four and negative in four, the average value is 0.

48. $m = \dfrac{ab}{2}$

$$I_z = \int_0^a \int_0^{-bx/a+b} (x^2 + y^2) \, dy \, dx$$

$$= \int_0^a \left[-\frac{bx^3}{a} - \frac{b^3x^3}{3a^3} + \frac{b^3x^2}{a^2} + bx^2 - \frac{b^3x}{a} + \frac{b^3}{3} \right]$$

$$= \frac{ab(a^2 + b^2)}{12}$$

$$d^2 = \frac{2ab(a^2 + b^2)}{12ab}; \quad d = \sqrt{\frac{a^2 + b^2}{6}}$$

49.

$$m = k \int_0^\pi \int_0^a r^2 \sin\theta \; dr \; d\theta$$

$$= k \int_0^\pi \frac{a^3 \sin\theta}{3} \; d\theta = \frac{2ka^3}{3}$$

$$I_x = \int_0^\pi \int_0^a kr^4 \sin^3\theta \; dr \; d\theta$$

$$= \int_0^\pi \int_0^a kr^4(1 - \cos^2\theta)\sin\theta \; dr \; d\theta$$

$$= \int_0^\pi \frac{ka^5}{5}(1 - \cos^2\theta)\sin\theta \; d\theta = \frac{4ka^5}{15}$$

$$d^2 = \frac{4a^5}{10a^3} = \frac{2a^2}{5}; \; d = \frac{\sqrt{10}\,a}{5}$$

50.

$$m = \int_1^2 \int_1^{x^2} x^2 y \; dy \; dx$$

$$= \frac{1}{2} \int_1^2 x^2(x^4 - 1) \; dx = \frac{332}{42}$$

$$I_x = \int_1^2 \int_1^{x^2} x^2 y^3 \; dy \; dx$$

$$= \frac{1}{4} \int_1^2 x^2(x^8 - 1) \; dx = \frac{1,516}{33}$$

$$d^2 = \frac{1,516(42)}{33(332)}; \; d = \sqrt{\frac{5,306}{913}} \approx 2.4107$$

51. a. $dF = \rho \; dV; \; d = h - z;$

$dW = -\rho(h - z) \; dV,$ so

$$W = \rho \int \int \int (h - z) \; dV$$

b. The volume is considered concentrated at $(0, 0, 0); \; d = h, \; F = \rho V; \; W = \rho h V$

c. $W = \rho(1.5)\pi(36)(10) = 540\pi\rho$

$$W = 4\rho \int_0^6 \int_0^{\sqrt{36 - x^2}} \int_{-.5}^5 (1.5 - z) \; dz \; dy \; dx$$

$$= -2\rho \int_0^6 \int_0^{\sqrt{36 - x^2}} 5(3 - 2x) \; dy \; dx$$

$$= -2\rho \int_0^6 5(3 - 2x)\sqrt{36 - x^2} \; dx$$

$$= -2\rho(135\pi - 720)$$

52. a. Note that $(h - z)/p$ is the cosine of the angle the force vector makes with the z-axis, so it will project this vector onto the z-axis.

b. Just sum up above ΔF.

c. $F = \dfrac{GmM}{p^2} = 5Gm$

d. Numerical integration gives

$F = 4.39741\,Gm$ vs $5.0\,Gm$

e. Changing h to 200 makes $F = 0.008$ and the integral 0.0079982. The center of mass approximation is quite accurate for a large separation (both from a numerical and a geometric point of view).

53.

$$I = 4 \int_0^{\ell/2} \int_0^{h/2} (x^2 + y^2) \; dy \; dx$$

$$= \int_0^{\ell/2} \frac{h(12x^2 + h^2)}{6} \; dx$$

$$= \frac{h\ell(h^2 + \ell^2)}{12}$$

54. Assume that the curve C is smooth and is described by $y = f(x)$ for $a \le x \le b$. Assume also that the axis of revolution is the y-axis. The distance traveled by the centroid $(\overline{x}, \overline{y})$ is $d = 2\pi h$ where $h = \overline{x}$. The area of the surface of revolution is

$$S = 2\pi \int_a^b 2\pi x\sqrt{1 + [f'(x)]^2} \; dx$$

But,

$$h = \overline{x} = \frac{1}{L} \int_a^b x \; ds$$

$$= \frac{1}{L} \int_a^b x\sqrt{1 + [f'(x)]^2} \; dx$$

Thus, $S = 2\pi L h$

55. $S = 2\pi a(2\pi b) = 4\pi^2 ab$

56. Suppose R is the region between $y = f(x)$ and $y = g(x)$ from $x = a$ to $x = b$. Assume the axis of revolution is the y-axis. The distance traveled by the centroid $(\overline{x}, \overline{y})$ is $d = 2\pi\overline{x}$. By the method of cylindrical shells, the volume generated is

$$V = \int_a^b 2\pi x[f(x) - g(x)]\,dx$$

$$= 2\pi \int_a^b x[f(x) - g(x)]\,dx$$

$$= 2\pi(\text{moment of } R \text{ about the } y\text{-axis})$$

$$= 2\pi\overline{x}(\text{area of } R)$$

$$= 2\pi\delta A$$

13.7 Cylindrical and Spherical Coordinates, Page 882

1. Cylindrical coordinates are best used with cylinders, spherical coordinates with spheres, and rectangular coordinates with rectangles.

2. Usually if the solid has symmetry with respect to the z-axis, use cylindrical or spherical coordinates.

3. **a.** Exact: $(\sqrt{2}, \pi/4, 1)$
   Approximate: $(1.41, 0.79, 1.00)$
   b. Exact: $(\sqrt{3}, \pi/4, \cos^{-1}(1/\sqrt{3}))$
   Approximate: $(1.73, 0.79, 0.96)$

4. **a.** Exact: $(4, \pi/2, \sqrt{3})$
   Approximate: $(4,00, 1.57, 1.73)$
   b. Exact: $(\sqrt{19}, \pi/2, \cos^{-1}(\sqrt{3}/\sqrt{19}))$
   Approximate: $(4.36, 1.57, 1.16)$

5. **a.** Exact: $(\sqrt{6}, \tan^{-1}(-2/\sqrt{2}), \sqrt{3})$
   Approximate: $(2.45, -0.96, 1.73)$
   b. Exact: $(3, \tan^{-1}(-\sqrt{3}/2), \cos^{-1}(\sqrt{3}/3))$
   Approximate: $(3.00, -0.96, 0.96)$

6. **a.** Exact: $(\sqrt{5}, \tan^{-1}2, 3)$
   Approximate: $(2.24, 1.11, 3.00)$
   b. Exact: $(\sqrt{14}, \tan^{-1}2, \cos^{-1}(3/\sqrt{14}))$
   Approximate: $(3.74, 1.11, 0.64)$

7. **a.** Exact: $(\sqrt{5}, \tan^{-1}2, 3)$
   Approximate: $(2.24, 1.11, 3.00)$
   b. Exact: $(\sqrt{14}, \tan^{-1}(3/2), \cos^{-1}(3/\sqrt{14}))$
   Approximate: $(3.74, 1.11, 0.64)$

8. **a.** Exact: $(\sqrt{2}, \pi/4, \cos^{-1}(1/\sqrt{3}))$
   Approximate: $(1.41, 0.79, 3.14)$
   b. Exact: $(\sqrt{3}\pi, \pi/4, \cos^{-1}(1/\sqrt{3}))$
   Approximate: $(3.07, 0.79\ 0.96)$

9. **a.** Exact: $(-3/2, 3\sqrt{3}/2, -3)$
   Approximate: $(-1.50, 2.60, -3.00)$
   b. Exact: $(\sqrt{18}\pi, 2\pi/3, \cos^{-1}(-3/\sqrt{18}))$
   Approximate: $(4.24, 2.09, 2.60)$

10. **a.** Exact: $(2\sqrt{3}, 2, -2)$
    Approximate: $(2.83, 2.00, -2.00)$
    b. Exact: $(2\sqrt{5}, \pi/6, \cos^{-1}(1/\sqrt{5}))$
    Approximate: $(4.47, 0.52, 2.03)$

11. **a.** Exact: $(\sqrt{2}, \sqrt{2}, \pi)$
    Approximate: $(1.41, 1.41, 3.14)$
    b. Exact: $(\sqrt{\pi^2+4}, \pi/4, \cos^{-1}\pi/\sqrt{\pi^2+4})$
    Approximate: $(3.72, 0.79, 0.57)$

12. **a.** Exact: $(0, -1, 3)$
    Approximate: $(0.00, -1.00, 3.00)$
    b. Exact: $(\sqrt{10}, \pi/2, \cos^{-1}(3/\sqrt{10}))$
    Approximate: $(3.16, 1.57, 0.32)$

13. **a.** Exact: $(\cos 2, \sin 2, 3)$
    Approximate: $(-0.42, 0.91, 3.00)$
    b. Exact: $(\sqrt{10}, 2, \cos^{-1}(3/\sqrt{10}))$
    Approximate: $(3.16, 2.00, 0.32)$

14. **a.** Exact: $(-\pi, 0, \pi)$
    Approximate: $(-3.14, 0.00, 3.14)$
    b. Exact$(\sqrt{2}\pi, \pi, \cos^{-1}(1/\sqrt{2}))$
    Approximate: $(2.51, 3.14, 0.79)$

15. **a.** Exact: $(3/2, \sqrt{3}/2, -1)$
    Approximate: $(1.50, 0.87, -1.00)$
    b. Exact: $(\sqrt{3}, \frac{\pi}{6}, -1)$
    Approximate: $(1.73, 0.52, -1.00)$

16. **a.** Exact: $(3\sqrt{6}/4, 3\sqrt{6}/4, \sqrt{2}/2)$
    Approximate: $(1.84, 1.84, 0.71)$
    b. Exact: $(1.5, \pi/4, 3\sqrt{3}/2)$
    Approximate: $(1.50, 0.79, 2.60)$

17. **a.** Exact: $(1/2, 0, 1)$
    Approximate: $(0.50, 0.00, 1.00)$
    b. Exact: $(0, \pi/6, 1)$
    Approximate: $(0.00, 0.53, 1.00)$

18. **a.** Exact: $(3\sqrt{2}, \sqrt{6}/2, \sqrt{2})$
    Approximate: $(4.24, 1.22, 1.41)$
    b. Exact: $(\sqrt{2}, \pi/3, \sqrt{2})$
    Approximate: $(1.41, 1.05, 1.41)$

19. **a.** Exact: $(\sin 3 \cos 2, \sin 3 \sin 2, \cos 3)$
    Approximate: $(-0.06, 0.13, -0.99)$
    b. Exact: $(\sin 3, 2, \cos 3)$
    Approximate: $(0.14, 2.00, -0.99)$

20. **a.** Exact: $(0, 0, -\pi)$
    Approximate: $(0.00, 0.00, -3.14)$
    b. Exact: $(0, \pi, -\pi)$
    Approximate: $(0.00, 3.14, -3.14)$

21. $z = r^2\cos 2\theta$

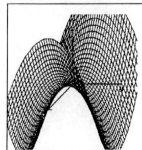

22. $r^2 \cos 2\theta = 1$

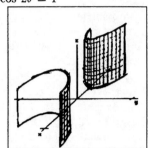

23. $9r^2\cos^2\theta - 4r^2\sin^2\theta + 36z^2 = 0$

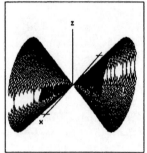

24. $z = r^2$

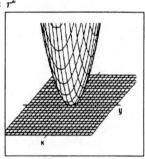

25. $9 - r^2\sin^2\theta = z^2$

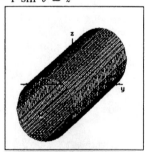

26. $r = 3$

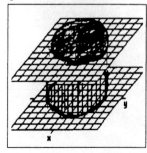

27. $\phi = \dfrac{\pi}{4}$

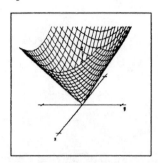

28. $\rho = \csc \phi$

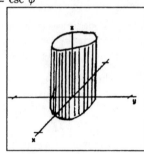

29. $\rho = 1/\sqrt{2}$

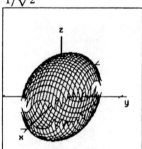

30. $4\rho \cos \phi = \rho^2\sin^2\phi(\cos^2\theta + 3\sin^2\theta)$

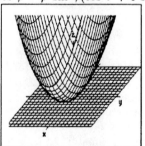

31. $\rho = 2$

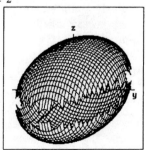

32. $\rho^2 = \dfrac{1}{\sin^2\phi - 4\cos^2\phi}$

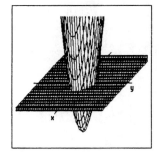

33. $z = 2xy$

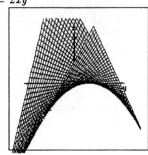

34. $x^2 + y^2 = y$

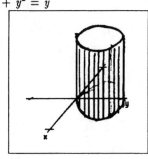

35. $z = x^2 - y^2$

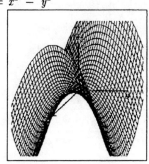

36. $x^2 + y^2 = 1$

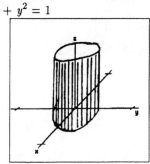

37. $xz = 1$

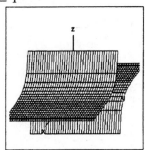

38. $x^2 + y^2 + z^2 = x$

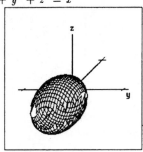

39. $\displaystyle\int_0^\pi \int_0^2 \int_0^{\sqrt{4-r^2}} r\sin\theta \; dz \; dr \; d\theta$

$$= -\frac{1}{2}\int_0^\pi \int_0^2 (\sin\theta)\sqrt{4-r^2}\,(-2r\,dr)\;d\theta$$

$$= -\frac{1}{3}\int_0^\pi (\sin\theta)(-8)\;d\theta = \frac{16}{3}$$

40. $\displaystyle\int_0^{\pi/4} \int_0^1 \int_0^{\sqrt{r}} r^2\sin\theta \; dz \; dr \; d\theta$

$$= \int_0^{\pi/4} \int_0^1 r^{5/2}\sin\theta \; dr \; d\theta$$

$$= \frac{2}{7}\int_0^{\pi/4} \sin\theta \; d\theta = \frac{2-\sqrt{2}}{7}$$

41. $\displaystyle\int_0^{\pi/2} \int_0^{2\pi} \int_0^2 \cos\phi\sin\phi \; d\rho \; d\theta \; d\phi$

$$= 2\int_0^{\pi/2} \int_0^{2\pi} \cos\phi\sin\phi \; d\theta \; d\phi$$

$$= 4\pi\int_0^{\pi/2} \cos\phi\sin\phi \; d\phi = 2\pi$$

42. $\int_0^{\pi/2}\int_0^{\pi/4}\int_0^{\cos\phi} \rho^2 \sin\phi \, d\rho \, d\theta \, d\phi$

$= \frac{1}{3}\int_0^{\pi/2}\int_0^{\pi/4} \cos^3\phi \sin\phi \, d\theta \, d\phi$

$= \frac{\pi}{12}\int_0^{\pi/2} \cos^3\phi \sin\phi \, d\phi = \frac{\pi}{48}$

43. $\int_0^{2\pi}\int_0^4\int_0^1 zr \, dz \, dr \, d\theta$

$= \frac{1}{2}\int_0^{2\pi}\int_0^4 r \, dr \, d\theta$

$= 4\int_0^{2\pi} d\theta = 8\pi$

44. $\int_{-\pi/4}^{\pi/3}\int_0^{\sin\theta}\int_0^{4\cos\theta} r \, dz \, dr \, d\theta$

$= \int_{-\pi/4}^{\pi/3}\int_0^{\sin\theta} 4r\cos\theta \, dr \, d\theta$

$= 2\int_{-\pi/4}^{\pi/3} \sin^2\theta \cos\theta \, d\theta$

$= \frac{1}{12}(3\sqrt{3} + 2\sqrt{2})$

45. $\frac{x^2}{a^2} + \frac{y^2}{b^2} = r^2\sin^2\phi$ and $\frac{z^2}{c^2} = r^2\cos^2\phi$ so

$\frac{x^2}{a^2} + \frac{y^2}{b^2} + \frac{z^2}{c^2} = r^2$

46. $\int_0^{2\pi}\int_0^1\int_0^{15} r^3 \sin\theta \cos\theta \, dz \, dr \, d\theta$

$= 15\int_0^2\int_0^1 r^3 \sin\theta\cos\theta \, dr \, d\theta$

$= \frac{15}{4}\int_0^2 \sin\theta \cos\theta \, d\theta = 0$

47. $\iiint_R (x^2 + y^2)^2 \, dx \, dy \, dz$

This is the four octants where $z > 0$.

$4\int_0^{1/\pi}\int_0^{\pi/2}\int_0^a r^4 \, r \, dr \, d\theta \, dz = \frac{4a^6}{6}\int_0^{1/\pi}\int_0^{\pi/2} d\theta \, dz$

$= \frac{4a^6}{6}\left(\frac{\pi}{2}\right)\left(\frac{1}{\pi}\right) = \frac{a^6}{3}$

48. $\iiint_S z(x^2 + y^2)^{-1/2} \, dx \, dy \, dz$

$= \int_0^{2\pi}\int_0^2\int_0^2 z(r^2)^{-1/2} r \, dz \, dr \, d\theta$

$= \frac{1}{2}\int_0^{2\pi}\int_0^2 \left(4 - \frac{r^4}{4}\right) dr \, d\theta$

$= \frac{16}{5}\int_0^{2\pi} d\theta = \frac{32\pi}{5}$

49. $m = \int_0^{2\pi}\int_0^9\int_r^9 r \, dz \, dr \, d\theta$

$= \int_0^{2\pi}\int_0^9 (9r - r^2) \, dr \, d\theta$

$= \frac{243}{2}\int_0^{2\pi} d\theta = 243\pi$

$\overline{x} = \overline{y} = 0$ (by symmetry)

$M_{xy} = \int_0^{2\pi}\int_0^9\int_r^9 rz \, dz \, dr \, d\theta$

$= \frac{1}{2}\int_0^{2\pi}\int_0^9 r(9^2 - r^2) \, dr \, d\theta$

$= \frac{6{,}561}{8}\int_0^{2\pi} d\theta = \frac{6{,}561\pi}{4}$

$\overline{z} = \frac{6{,}561\pi}{4(243\pi)} = \frac{27}{4}$;

thus, the centroid is $(0, 0, \frac{27}{4})$

50. $\iiint_R \sqrt{x^2 + y^2 + z^2} \, dx \, dy \, dz$

$= 8\int_0^{\pi/2}\int_0^{\pi/2}\int_0^{\sqrt{2}} \rho^3 \sin\phi \, d\rho \, d\theta \, d\phi$

$= 8\int_0^{\pi/2}\int_0^{\pi/2} \sin\phi \, d\theta \, d\phi$

$$= 4\pi \int_0^{\pi/2} \sin\phi \, d\phi = 4\pi$$

51. $\displaystyle\iint\limits_R\int (x^2 + y^2 + z^2) \, dx \, dy \, dz$

$$= 8 \int_0^{\pi/2} \int_0^{\pi/2} \int_0^{\sqrt{2}} \rho^4 \sin\phi \, d\rho \, d\theta \, d\phi$$

$$= \frac{32\sqrt{2}}{5} \int_0^{\pi/2} \int_0^{\pi/2} \sin\phi \, d\theta \, d\phi$$

$$= \frac{16\sqrt{2}\pi}{5} \int_0^{\pi/2} \sin\phi \, d\phi = \frac{16\pi\sqrt{2}}{5}$$

52. $\displaystyle\iint\limits_S\int z^2 \, dx \, dy \, dz$

$$= 4 \int_0^{\pi/2} \int_0^{\pi/2} \int_0^1 \rho^4 \cos^2\phi \sin\phi \, d\rho \, d\theta \, d\phi$$

$$= \frac{4}{5} \int_0^{\pi/2} \int_0^{\pi/2} \sin\phi \cos^2\phi \, d\theta \, d\phi$$

$$= \frac{2\pi}{5} \int_0^{\pi/2} \sin\phi \cos^2\phi \, d\phi = \frac{2\pi}{15}$$

53. $\displaystyle\iint\limits_S\int \frac{dx \, dy \, dz}{\sqrt{x^2 + y^2 + z^2}}$

$$= 8 \int_0^{\pi/2} \int_0^{\pi/2} \int_0^{\sqrt{3}} \rho \sin\phi \, d\rho \, d\theta \, d\phi$$

$$= 12 \int_0^{\pi/2} \int_0^{\pi/2} \sin\phi \, d\theta \, d\phi$$

$$= 6\pi \int_0^{\pi/2} \sin\phi \, d\phi = 6\pi$$

54. $\displaystyle V = 4 \int_0^{\pi/2} \int_0^{1/2} r(1 - 4r^2) \, dr \, d\theta$

$$= \frac{1}{4} \int_0^{\pi/2} d\theta = \frac{\pi}{8}$$

55. $\displaystyle V = \int_0^{2\pi} \int_0^1 \int_0^{4-r^2} dz \, r \, dr \, d\theta$

$$= \int_0^{2\pi} \int_0^1 (4 - r^2) \, r \, dr \, d\theta$$

$$= \int_0^{2\pi} \frac{7}{4} \, d\theta = \frac{7\pi}{2}$$

56. $\displaystyle V = 8 \int_0^{\pi/2} \int_0^1 \int_0^{\sqrt{9-r^2}} r \, dz \, dr \, d\theta$

$$= 8 \int_0^{\pi/2} \int_0^1 r\sqrt{9 - r^2} \, dr \, d\theta$$

$$= \frac{216 - 128\sqrt{2}}{3} \int_0^{\pi/2} d\theta$$

$$= \frac{4\pi}{3}(27 - 16\sqrt{2})$$

57. $\displaystyle V = 2 \int_0^{\pi/2} \int_0^{2\sin\theta} \int_0^{4-r^2} r \, dz \, dr \, d\theta$

$$= \int_0^{\pi/2} \int_0^{2\sin\theta} 2r(4 - r^2) \, dr \, d\theta$$

$$= \int_0^{\pi/2} 8\sin^2\theta \cos^2\theta + 8\sin^2\theta \, d\theta = \frac{5\pi}{2}$$

58. $\displaystyle V = \left[\frac{4}{3}\pi \, 4^3\right] - 4 \int_0^{\pi/2} \int_0^{\pi/6} \int_0^4 \rho^2 \sin\phi \, d\rho \, d\phi \, d\theta$

$$= \frac{4^4}{3}\pi - \frac{256}{3} \int_0^{\pi/2} \int_0^{\pi/6} \sin\phi \, d\phi \, d\theta$$

$$= \frac{4^4}{3}\pi - \frac{128}{3} \int_0^{\pi/2} (2 - \sqrt{3}) \, d\theta$$

$$= \frac{4^4\pi}{3} - \frac{64\pi}{9}(2 - \sqrt{3}) = \frac{64\pi}{3}(2 + \sqrt{3})$$

59. The cave is a cardioid of revolution about the z-axis. The spy will begin to drown when the water level reaches 3 ft above the floor (the xy plane); that is when $z = 3$.

$$\rho \cos\phi = 3$$

$$4(1 + \cos\phi) \cos\phi = 3$$

$$4 \cos^2\phi + 4 \cos\phi - 3 = 0$$

$$(2\cos\phi - 1)(2\cos\phi + 3) = 0$$

$$\cos\phi = \tfrac{1}{2}, -\tfrac{3}{2}$$

so in this application $\phi = \frac{\pi}{3}$. The critical water level is $z = 3 = \rho\cos\theta$ or $\rho = 3\sec\phi$. We need to compute the amount of water in the cave when the depth is 3 ft and use the rate of 25 cu ft/min to determine the time necessary for it to reach that level. In spherical coordinates the volume can be found as the sum of two integrals:

$$V = \int_0^{2\pi}\int_0^{\pi/3}\int_0^{3\sec\phi} \rho^2\sin\phi\; d\rho\; d\phi\; d\theta$$

$$+ \int_0^{2\pi}\int_{\pi/3}^{\pi/2}\int_0^{4(1+\cos\phi)} \rho^2\sin\phi\; d\rho\; d\phi\; d\theta$$

$$= \int_0^{2\pi}\int_0^{\pi/3}\frac{(3\sec\phi)^3}{3}\sin\phi\; d\phi\; d\theta$$

$$+ \int_0^{2\pi}\int_{\pi/3}^{\pi/2}\frac{[4(1+\cos\phi)]^3}{3}\sin\phi\; d\phi\; d\theta$$

$$= \tfrac{9}{2}(3)(2\pi) - \tfrac{64}{3}\left(\tfrac{1}{4} - \tfrac{81}{64}\right)(2\pi)$$

$$= \pi\left(27 + \tfrac{130}{3}\right) \approx 221 \text{ cu ft of water}$$

At the incoming rate of 25 cu ft/min ≈ 8.84 min. So he drowns, you say – nonsense! Any spy worth his salt can hold his breath for a little more than a minute. He frees his hands, stands (to buy more time), hops to the door, pulls up the lever (to stop the water), and opens the door. As the water drains from the room, he unties his feet, and prepares to pursue Purity.

60.   **a.**   This is a routine calculation using the law of cosines.

     b.   Note that
$$\frac{R - z}{p} = \frac{R - \rho\cos\phi}{p}$$
is the cosine of the angle the force vector makes with the z-axis, so it will project this vector onto the z-axis.

     c.   Just sum up the above ΔF.

61.   **a.**   Force $= \dfrac{GmM}{R^2} = \dfrac{Gm\rho(4\pi a^3)}{3R^2}$

     b.   With $R = 4$ and $a = 3$, we obtain
$$\text{Force} = \frac{Gm\rho(9\pi)}{4}$$

     c.   With a rectangular mass m, we got a

poor approximation using the center of mass if the separating distance was small better as the distance increases. With the sphere we always get perfect agreement. Apparently the symmetry of the sphere play the key role. Either a computer with symbolic integration capability or determined work by hand will show that for the sphere the center of mass is perfect.

62. $I_z = 4\displaystyle\int_0^{\pi/2}\int_0^{R} hr^3\; dr\; d\theta = \dfrac{hR^4\pi}{2} = \dfrac{mR^2}{2}$

63. $I = I_x + I_y + I_z$

$$= \iiint_S (y^2 + z^2)\; dV + \iiint_S (x^2 + z^2)\,dV$$

$$+ \iiint_S (x^2 + y^2)\; dV$$

$$= \iiint_S (2x^2 + 2y^2 + 2z^2)\; dV$$

$$= 2\int_0^{\pi}\int_0^{2\pi}\int_0^{1} \rho^2\,\rho^2\sin\phi\; d\rho\; d\theta\; d\phi$$

$$= \int_0^{2\pi}\int_0^{\pi}\int_0^{1} \rho^4\sin\phi\; d\rho\; d\phi\; d\theta$$

$$= \tfrac{2}{5}\int_0^{2\pi}\int_0^{\pi} \sin\phi\; d\phi\; d\theta$$

$$= \tfrac{4}{5}\int_0^{2\pi} d\theta = \tfrac{8\pi}{5}$$

64. $V = 2\displaystyle\int_{\sin^{-1}(b/a)}^{\pi/2}\int_0^{2\pi}\int_{b/\sin\phi}^{a} \rho^2\sin\phi\; d\rho\; d\theta\; d\phi$

$$= \tfrac{2}{3}\int_{\sin^{-1}(b/a)}^{\pi/2}\int_0^{2\pi} \left(a^3\sin\phi - \frac{b^3}{\sin^2\phi}\right) d\theta\; d\phi$$

$$= \tfrac{4\pi}{3}\int_{\sin^{-1}(b/a)}^{\pi/2} (a^3 - b^3\csc^3\phi)\sin\phi\; d\phi$$

$$= \tfrac{4\pi}{3}(a^2 - b^2)^{3/2}$$

13.8 Jacobians: Change of Variables, Page 889

1. $x = u + v, \ y = uv$

$$\frac{\partial(x, y)}{\partial(u, v)} = \begin{vmatrix} \dfrac{\partial x}{\partial u} & \dfrac{\partial x}{\partial v} \\ \dfrac{\partial y}{\partial u} & \dfrac{\partial y}{\partial v} \end{vmatrix} = \begin{vmatrix} 1 & 1 \\ v & u \end{vmatrix} = u - v$$

2. $x = u^2, \ y = u + v$

$$\frac{\partial(x, y)}{\partial(u, v)} = \begin{vmatrix} \dfrac{\partial x}{\partial u} & \dfrac{\partial x}{\partial v} \\ \dfrac{\partial y}{\partial u} & \dfrac{\partial y}{\partial v} \end{vmatrix} = \begin{vmatrix} 2u & 0 \\ 1 & 1 \end{vmatrix} = 2u$$

3. $x = u - v, \ y = u + v$

$$\frac{\partial(x, y)}{\partial(u, v)} = \begin{vmatrix} \dfrac{\partial x}{\partial u} & \dfrac{\partial x}{\partial v} \\ \dfrac{\partial y}{\partial u} & \dfrac{\partial y}{\partial v} \end{vmatrix} = \begin{vmatrix} 1 & -1 \\ 1 & 1 \end{vmatrix} = 2$$

4. $x = u^2 - v^2, \ y = 2uv$

$$\frac{\partial(x, y)}{\partial(u, v)} = \begin{vmatrix} \dfrac{\partial x}{\partial u} & \dfrac{\partial x}{\partial v} \\ \dfrac{\partial y}{\partial u} & \dfrac{\partial y}{\partial v} \end{vmatrix} = \begin{vmatrix} 2u & -2v \\ 2v & 2u \end{vmatrix} = 4(u^2 + v^2)$$

5. $x = u^2 v^2, \ y = v^2 - u^2$

$$\frac{\partial(x, y)}{\partial(u, v)} = \begin{vmatrix} \dfrac{\partial x}{\partial u} & \dfrac{\partial x}{\partial v} \\ \dfrac{\partial y}{\partial u} & \dfrac{\partial y}{\partial v} \end{vmatrix} = \begin{vmatrix} 2uv^2 & 2u^2v \\ -2u & 2v \end{vmatrix}$$

$$= 4uv^3 + 4u^3v = 4uv(v^2 + u^2)$$

6. $x = u \cos v, \ y = u \sin v$

$$\frac{\partial(x, y)}{\partial(u, v)} = \begin{vmatrix} \dfrac{\partial x}{\partial u} & \dfrac{\partial x}{\partial v} \\ \dfrac{\partial y}{\partial u} & \dfrac{\partial y}{\partial v} \end{vmatrix} = \begin{vmatrix} \cos v & -u \sin v \\ \sin v & u \cos v \end{vmatrix} = u$$

7. $x = e^{u+v}, \ y = e^{u-v}$

$$\frac{\partial(x, y)}{\partial(u, v)} = \begin{vmatrix} \dfrac{\partial x}{\partial u} & \dfrac{\partial x}{\partial v} \\ \dfrac{\partial y}{\partial u} & \dfrac{\partial y}{\partial v} \end{vmatrix} = \begin{vmatrix} e^{u+v} & e^{u+v} \\ e^{u-v} & -e^{u-v} \end{vmatrix}$$

$$= -2e^{u+v}e^{u-v} = -2e^{2u}$$

8. $x = e^u \sin v, \ y = e^u \cos v$

$$\frac{\partial(x, y)}{\partial(u, v)} = \begin{vmatrix} \dfrac{\partial x}{\partial u} & \dfrac{\partial x}{\partial v} \\ \dfrac{\partial y}{\partial u} & \dfrac{\partial y}{\partial v} \end{vmatrix} = \begin{vmatrix} e^u \sin v & e^u \cos v \\ e^u \cos v & -e^u \sin v \end{vmatrix}$$

$$= -e^{2u}$$

9. $x = u + v - w, \ y = 2u - v + 3w,$
$z = -u + 2v - w$

$$\frac{\partial(x, y, z)}{\partial(u, v, w)} = \begin{vmatrix} 1 & 1 & -1 \\ 2 & -1 & 3 \\ -1 & 2 & -1 \end{vmatrix} = -9$$

10. $x = 2u - w, \ y = u + 3v, \ z = v + 2w$

$$\frac{\partial(x, y, z)}{\partial(u, v, w)} = \begin{vmatrix} 2 & 0 & -1 \\ 1 & 3 & 0 \\ 0 & 1 & 2 \end{vmatrix} = 11$$

11. $x = u \sin v, \ y = u \sin v, \ z = we^{uv}$

$$\frac{\partial(x, y, z)}{\partial(u, v, w)} = \begin{vmatrix} \cos v & -u \sin v & 0 \\ \sin v & u \cos v & 0 \\ vwe^{uv} & uwe^{uv} & e^{uv} \end{vmatrix} = ue^{uv}$$

12. $x = \dfrac{u}{v}, \ y = \dfrac{v}{w}, \ z = \dfrac{w}{u}$

$$\frac{\partial(x, y, z)}{\partial(u, v, w)} = \begin{vmatrix} \dfrac{1}{v} & -\dfrac{u}{v^2} & 0 \\ 0 & \dfrac{1}{w} & -\dfrac{v}{w^2} \\ -\dfrac{w}{u^2} & 0 & \dfrac{1}{u} \end{vmatrix} = \frac{w-1}{uvw^2}$$

13. $A(0, 5) \rightarrow (5, -5); \ B(6, 5) \rightarrow (11, 1);$
$C(6, 0) \rightarrow (6, 6); \ O(0, 0) \rightarrow (0, 0)$

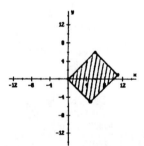

14. $A(0, 5) \rightarrow (10, 5) \ B(4, 6) \rightarrow (8, 10);$
$O(0, 0) \rightarrow (0, 0)$

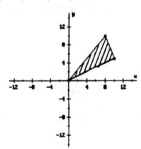

15. $A(5, 0) \to (25, 0);\ B(9, 4) \to (33, 56);$
$C(2, 4) \to (-12, 16);\ O(0, 0) \to (0, 0)$

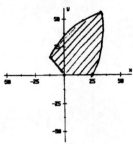

16. $A(4, 1) \to (16, 5);\ B(5, 5) \to (25, 10);$
$C(1, 4) \to (1, 5);\ O(0, 0) \to (0, 0)$

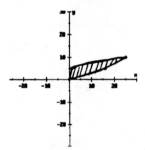

17. $x = u - uv,\quad y = uv$

$$\frac{\partial(x, y)}{\partial(u, v)} = \begin{vmatrix} \dfrac{\partial x}{\partial u} & \dfrac{\partial x}{\partial v} \\ \dfrac{\partial y}{\partial u} & \dfrac{\partial y}{\partial v} \end{vmatrix} = \begin{vmatrix} 1 - v & -u \\ v & u \end{vmatrix}$$

$$= u - uv + uv = u$$

$$dx\ dy = u\ du\ dv$$

18. $x = u^2 - v^2,\quad y = 2uv$

$$\frac{\partial(x, y)}{\partial(u, v)} = \begin{vmatrix} \dfrac{\partial x}{\partial u} & \dfrac{\partial x}{\partial v} \\ \dfrac{\partial y}{\partial u} & \dfrac{\partial y}{\partial v} \end{vmatrix} = \begin{vmatrix} 2u & -2v \\ 2v & 2u \end{vmatrix}$$

$$= 4(u^2 + v^2) = u$$

$$dx\ dy = 4(u^2 + v^2)\ du\ dv$$

For Problems 19-22,
$u = x - y,\ v = x + y,$ so $x = \dfrac{u + v}{2},\ y = \dfrac{v - u}{2}$
and $x + y = 1$ *becomes* $v = 1.$

$$\frac{\partial(x, y)}{\partial(u, v)} = \begin{vmatrix} \dfrac{\partial x}{\partial u} & \dfrac{\partial x}{\partial v} \\ \dfrac{\partial y}{\partial u} & \dfrac{\partial y}{\partial v} \end{vmatrix} = \begin{vmatrix} \dfrac{1}{2} & \dfrac{1}{2} \\ -\dfrac{1}{2} & \dfrac{1}{2} \end{vmatrix} = \dfrac{1}{2}$$

$$dy\ dx = \tfrac{1}{2}\ du\ dv$$

19. $\displaystyle\int_D\int \left(\frac{x - y}{x + y}\right)^5 dy\ dx = \int_0^1\int_{-u}^{u} \frac{u^5}{v^5}\,\frac{1}{2}\ dv\ du$

$$= -\frac{1}{8}\int_0^1 (u - u)\ du = 0$$

20. $\displaystyle\int_D\int \left(\frac{x + y}{x - y}\right)^4 dy\ dx = \int_0^1\int_{-u}^{u} \frac{v^4}{u^4}\,\frac{1}{2}\ dv\ du$

$$= \frac{1}{10}\int_0^1 \left(\frac{u^5}{u^4} + u\right)\ du = \frac{1}{5}\int_0^1 u\ du = \frac{1}{10}$$

21. $\displaystyle\int_D\int (x - y)^5 (x + y)^3\ dy\ dx$

$$= \frac{1}{2}\int_0^1\int_{-u}^{u} u^5 v^3\ dv\ du = -\frac{1}{8}\int_0^1 (u^9 - u^9)\ du = 0$$

22. $\displaystyle\int_D\int (x - y)e^{x^2+y^2}\ dy\ dx$

$$= \frac{1}{2}\int_{-1}^{0} \int_{-v}^{1} ue^{(u^2+v^2)/2}\ du\ dv$$

$$+ \frac{1}{2}\int_{0}^{1} \int_{v}^{1} ue^{(u^2+v^2)/2}\ du\ dv$$

$$= \frac{1}{2}\int_{-1}^{0} \left[e^{(1+v^2)/2} - e^{v^2}\right]\ dv$$

$$+ \frac{1}{2}\int_{0}^{1} \left[e^{(1+v^2)/2} - e^{v^2}\right]\ dv$$

$$\approx 0.5075$$

(numerical or computer approximation)

23. Find the Jacobian, map the region, then integrate.

24. Without the Jacobian, the element of area would not be mapped properly.

For Problems 25-30,
$u = \dfrac{2x + y}{5},\quad v = \dfrac{x - 2y}{5}$ or $y = u - 2v$ and $x = 2u + v$

$$\frac{\partial(x, y)}{\partial(u, v)} = \begin{vmatrix} \dfrac{\partial x}{\partial u} & \dfrac{\partial x}{\partial v} \\ \dfrac{\partial y}{\partial u} & \dfrac{\partial y}{\partial v} \end{vmatrix} = \begin{vmatrix} 2 & 1 \\ 1 & -2 \end{vmatrix} = 5$$

$A(0, 0) \to (0, 0),\quad B(1, -2) \to (0, 1),$
$C(2, 1) \to (1, 0),\quad D(3, -1) \to (1, 1).$
R is the unit square.

25. $\displaystyle\int_S\int \left(\frac{2x + y}{x - 2y + 5}\right)^2 dy\ dx = \int_0^1\int_0^1 \left(\frac{u}{v+1}\right)^2 5\ du\ dv$

$$= \frac{5}{3}\int_0^1 (v + 1)^{-2}\ dv = \frac{5}{6}$$

26. $\displaystyle\int\int_S (2x + y)(x - 2y)^2 \, dy \, dx$

$\displaystyle = \int_0^1\int_0^1 (5u)(5v)^2 \; 5 \, du \, dv$

$\displaystyle = \frac{5^4}{2}\int_0^1 u^2 v^2 \, dv = \frac{625}{6}$

27. $\displaystyle\int\int_S (2x + y)^2(x - 2y) \, dy \, dx$

$\displaystyle = \int_0^1\int_0^1 (5u)^2(5v) \; 5 \, du \, dv$

$\displaystyle = \frac{5^4}{3}\int_0^1 u^3 v \, dv = \frac{625}{6}$

28. $\displaystyle\int\int_S \sqrt{(2x + y)(x - 2y)} \, dy \, dx$

$\displaystyle = \int_0^1\int_0^1 (5u)^{1/2}(5v)^{1/2} \; 5 \, dv \, du$

$\displaystyle = \frac{50}{3}\int_0^1 u^{1/2} \, du = \frac{100}{9}$

29. $\displaystyle\int\int_S (2x + y)\tan^{-1}(x - 2y) \, dy \, dx$

$\displaystyle = \int_0^1\int_0^1 (5u)\tan^{-1}(5v) \; 5 \, dv \, du$

$\displaystyle = \frac{25}{2}\left[v \tan^{-1} 5v \Big|_0^1 - 5\int_0^1 \frac{v \, dv}{1 + 25v^2}\right]$

$\displaystyle = \frac{25}{2}\left[\tan^{-1}5 - \frac{1}{10}\ln(1 + 25v^2)\right]\Big|_0^1$

$\displaystyle = \frac{25}{2}\left[\tan^{-1}5 - \frac{1}{10}\ln 26\right] \approx 13.0949$

Note: computer evaluation gives a form

$\displaystyle -\frac{25}{2}\tan^{-1}\frac{1}{5} + \frac{25\pi}{4} - \frac{5}{4}\ln 26$

30. $\displaystyle\int\int_S \cos(2x + y) \sin(x - 2y) \, dy \, dx$

$\displaystyle = \int_0^1\int_0^1 \cos 5u \sin 5v \; 5 \, dv \, du$

$\displaystyle = \int_0^1 \left[\cos 5u - \cos 5 \cos 5u\right] du$

$\displaystyle = -\frac{1}{5}\sin 5(\cos 5 - 1) \approx -0.1374$

31. $x = u + v$, $y = v$; so $u = x - y$, $v = y$

$\displaystyle \frac{\partial(x, y)}{\partial(u, v)} = \begin{vmatrix} \dfrac{\partial x}{\partial u} & \dfrac{\partial x}{\partial v} \\ \dfrac{\partial y}{\partial u} & \dfrac{\partial y}{\partial v} \end{vmatrix} = \begin{vmatrix} 1 & 1 \\ 0 & 1 \end{vmatrix} = 1$

$A(0, 0) \to (0, 0), \quad B(1, 1) \to (0, 1),$
$C(1, 0) \to (1, 0), \quad D(2, 1) \to (1, 1).$
$dy \, dx = du \, dv$

$\displaystyle \int_R\int (2x - y) \, dy \, dx = \int_0^1\int_0^1 (2u + v) \, dv \, du$

$\displaystyle = \int_0^1 (1 + v) \, dv = \frac{1}{2}$

32. By looking at the function we see a suitable transformation can be obtained when $a = b = s = 1$ and $r = -1$.
$u = x + y$, $v = -x + y$; so $y = \frac{1}{2}(u + v)$,
$x = \frac{1}{2}(u - v)$

$\displaystyle \frac{\partial(x, y)}{\partial(u, v)} = \begin{vmatrix} \dfrac{\partial x}{\partial u} & \dfrac{\partial x}{\partial v} \\ \dfrac{\partial y}{\partial u} & \dfrac{\partial y}{\partial v} \end{vmatrix} = \begin{vmatrix} \frac{1}{2} & -\frac{1}{2} \\ \frac{1}{2} & \frac{1}{2} \end{vmatrix} = \frac{1}{2}$

$A(0, 0) \to (0, 0), \quad B(1, 1) \to (2, 0),$
$C(-1, 1) \to (0, 2), \quad D(2, 0) \to (2, 2).$
$dy \, dx = \frac{1}{2} du \, dv$

$\displaystyle \int_R\int \left(\frac{x + y}{2}\right)^2 e^{(y - x)/2} \, dy \, dx$

$\displaystyle = \frac{1}{8}\int_1^2\int_0^2 (2u + v)u^2 e^{-v/2} \, dv \, du$

$\displaystyle = -\frac{1}{4}(e^{-1} - 1)\int_0^2 u^2 \, du = \frac{2(e - 1)}{3e}$

33. $x = s^2 - t^2$, $y = 2st$; so
$x^2 + y^2 = s^4 - 2s^2 t^2 + t^4 + 4s^2 t^2$
$\qquad = (s^2 + t^2)^2$

$\displaystyle \frac{\partial(x, y)}{\partial(u, v)} = \begin{vmatrix} \dfrac{\partial x}{\partial u} & \dfrac{\partial x}{\partial v} \\ \dfrac{\partial y}{\partial u} & \dfrac{\partial y}{\partial v} \end{vmatrix} = \begin{vmatrix} 2s & -2t \\ 2t & 2s \end{vmatrix} = 4(s^2 + t^2)$

$A(0, 0) \to (0, 0), \quad B(0, 1) \to (-1, 0),$
$C(1, 0) \to (1, 0); \quad dy \, dx = 4(s^2 + t^2) \, ds \, dt$

$$\iint_S \frac{dy\, dx}{\sqrt{x^2+y^2}} = 4\int_0^1 \int_0^{\sqrt{1-t^2}} \frac{s^2+t^2}{s^2+t^2}\, ds\, dt$$

$$= 4\int_0^1 \sqrt{1-t^2}\, dt = \pi$$

34. $x = ar\cos\theta,\ y = br\sin\theta,\ \dfrac{x^2}{a^2}+\dfrac{y^2}{b^2} = r^2$

$$\frac{\partial(x,\,y)}{\partial(u,\,v)} = \begin{vmatrix} \dfrac{\partial x}{\partial u} & \dfrac{\partial x}{\partial v} \\[6pt] \dfrac{\partial y}{\partial u} & \dfrac{\partial y}{\partial v} \end{vmatrix} = \begin{vmatrix} a\cos\theta & -ar\sin\theta \\ b\sin\theta & br\cos\theta \end{vmatrix} = abr$$

$$dy\, dx = abr\, du\, dv$$

$$\iint_S \exp\!\left(-\frac{x^2}{a^2}-\frac{y^2}{b^2}\right) dy\, dx = ab\int_0^{\pi/2}\int_0^1 re^{-r}\, dr\, d\theta$$

$$= \frac{ab}{2}\int_{0}^{\pi/2}(1-e^{-1})\, d\theta = \frac{ab\pi}{4}(1-e^{-1})$$

35. a. $x = u\cos\theta - v\sin\theta,\ y = u\sin\theta + v\cos\theta$

$$\frac{x^2}{a^2}+\frac{y^2}{b^2} = r^2$$

$$\frac{\partial(x,\,y)}{\partial(u,\,v)} = \begin{vmatrix} \dfrac{\partial x}{\partial u} & \dfrac{\partial x}{\partial v} \\[6pt] \dfrac{\partial y}{\partial u} & \dfrac{\partial y}{\partial v} \end{vmatrix} = \begin{vmatrix} \cos\theta & \sin\theta \\ -\sin\theta & \cos\theta \end{vmatrix} = 1$$

$$dy\, dx = du\, dv$$

b. $x^2 + xy + y^2 = 3$; a rotation of $\theta = \pi/2$ will eliminate the xy (or rather the uv) term.

$$x = \frac{\sqrt{2}}{2}(u - v),\quad y = \frac{\sqrt{2}}{2}(u+v)$$

$$\tfrac{1}{2}[u^2 - 2uv + v^2 + u^2 - v^2 + u^2 + 2uv + v^2] = 3;$$

$$\frac{3u^2}{2} + \frac{v^2}{2} = 3;\quad u^2 = 2\!\left(1 - \frac{v^2}{6}\right) = \tfrac{1}{3}(6 - v^2)$$

$$\iint_E y\, dy\, dx = \frac{\sqrt{2}}{2}\int_{-\sqrt{6}}^{\sqrt{6}}\int_{-\sqrt{(6-v^2)/3}}^{\sqrt{(6-v^2)/3}} (u+v)\, du\, dv$$

$$= \frac{\sqrt{2}}{\sqrt{3}}\int_{-\sqrt{6}}^{\sqrt{6}} v\sqrt{6-v^2}\, dv = 0$$

36. Let $u = x + y,\ v = 2y,\ x = u - v/2,\ y = v/2$. The boundary $x + y = 0$ becomes $u = 0$; $x + y = 2$ becomes $u = 2$; $y = 0$ becomes $v = 0$; and $y = 1$ becomes $v = 2$.

$$dy\, dx = \tfrac{1}{2}\, du\, dv$$

$$\iint_R f(x+y)\, dy\, dx = \frac{1}{2}\int_0^1\int_0^2 f(u)\, dv\, du$$

$$= \int_0^2 f(u)\!\left(\frac{v}{2}\right)^2 du = \int_0^2 f(u)\, du = \int_0^2 f(t)\, dt$$

37. $x = r\cos\theta,\ y = r\sin\theta,\ z = z$

$$\frac{\partial(x,\,y,\,z)}{\partial(r,\,\theta,\,z)} = \begin{vmatrix} \cos\theta & \sin\theta & 0 \\ -r\sin\theta & r\cos\theta & 0 \\ 0 & 0 & 1 \end{vmatrix} = r$$

So $dx\, dy\, dz$ becomes $r\, dr\, d\theta\, dz$.

38. $x = au,\ y = bv,\ z = cw$.

$$\frac{\partial(x,\,y,\,z)}{\partial(r,\,\theta,\,z)} = \begin{vmatrix} a & 0 & 0 \\ 0 & b & 0 \\ 0 & 0 & c \end{vmatrix} = abc$$

$$dy\, dx\, dz = abc\, du\, dv\, dw$$

$$\frac{x^2}{a^2} + \frac{y^2}{b^2} + \frac{z^2}{c^2} = u^2 + v^2 + w^2$$

Thus, volume of the ellipsoid $= abc(\text{volume of sphere}) = abc(\frac{4}{3}\pi(1)) = \frac{4}{3}\pi abc$.

39. $u = x - y,\ v = x + y;\ u + v = 8,\ 2u - 1 = 1$, and $v = 1$ are the boundaries.

$$\frac{\partial(x,\,y)}{\partial(u,\,v)} = \begin{vmatrix} \dfrac{\partial x}{\partial u} & \dfrac{\partial x}{\partial v} \\[6pt] \dfrac{\partial y}{\partial u} & \dfrac{\partial y}{\partial v} \end{vmatrix} = \begin{vmatrix} \tfrac{1}{2} & \tfrac{1}{2} \\[4pt] -\tfrac{1}{2} & \tfrac{1}{2} \end{vmatrix} = \tfrac{1}{2}$$

$A(1, 0) \to (1, 1),\quad B(4, -3) \to (7, 1),$
$C(4, 1) \to (3, 5)$
$dy\, dx = 2\, du\, dv$

$$\iint_R \ln\!\left(\frac{x-y}{x+y}\right) dy\, dx$$

$$= \frac{1}{2}\int_1^5 \int_{(v+1)/2}^{8-v} \ln\!\left(\frac{u}{v}\right) du\, dv$$

$$= \frac{1}{4}\int_1^5 \left[-(v+1)\ln\frac{v+1}{2v} + 2(8-v)\ln\frac{8-v}{v} + 3(v-5)\right] dv$$

$$= \frac{1}{4}\!\left(49\ln 7 - \frac{75}{2}\ln 5 - 27\ln 3 + 6\right) \approx 2.8333$$

40. $u = ax + by,\ v = cx + dy$; solve simultaneously to obtain

$$x = \frac{du - bv}{ad - bc};\quad y = \frac{av - cu}{ad - bc}$$

41. $x = u$, $y = v/u$, $z = z$

$$\frac{\partial(x,\, y,\, z)}{\partial(r,\, \theta,\, z)} = \begin{vmatrix} 1 & 0 & 0 \\ -\dfrac{v}{u^2} & \dfrac{1}{u} & 0 \\ 0 & 0 & 1 \end{vmatrix} = \frac{1}{u}$$

$$\int_1^5 \int_1^5 \left(\frac{1}{u}\right)\left(\frac{v}{1+v^2}\right)\, dv\, du = \frac{1}{2}\int_1^5 \frac{1}{u}(\ln 26 - \ln 2)(4)\, du$$

$$= \tfrac{1}{2}\ln 5 \ln 13 \approx 2.064$$

CHAPTER 13 REVIEW

Proficiency Examination, Page 891

1. If f is defined on a closed, bounded region R in the xy-plane, then the double integral of f over R is defined by

$$\int_R \int f(x,\, y)\, dA = \lim_{\|P\| \to 0}\sum_{k=1}^{N} f(\overset{*}{x}_k,\, \overset{*}{y}_k)\Delta A_k$$

provided this limit exists. If the limit exists, we say that f is integrable over R.

2. If $f(x,\, y)$ is continuous over the rectangle R: $a \le x \le b$, $c \le y \le d$, then the double integral

$$\int_R \int f(x,\, y)\, dA$$

may be evaluated by either iterated integral; that is,

$$\int_R \int f(x,\, y)\, dA = \int_c^d \int_a^b f(x,\, y)\, dx\, dy = \int_a^b \int_c^d f(x,\, y)\, dy\, dx$$

3. A type I region contains points $(x,\, y)$ such that for each fixed x between constants a and b, y varies from $g_1(x)$ to $g_2(x)$, where g_1 and g_2 are continuous functions. Think of a vertical strip.

$$\int_D \int f(x,\, y)\, dA = \int_a^b \int_{g_1(x)}^{g_2(x)} f(x,\, y)\, dy\, dx$$

whenever both integrals exist.

4. A type II region contains points $(x,\, y)$ such that for each fixed y between constants c and d, x varies from $h_1(y)$ to $h_2(y)$, where h_1 and h_2 are continuous functions. Think of a horizontal strip.

$$\int_D \int f(x,\, y)\, dA = \int_c^d \int_{h_1(y)}^{h_2(y)} f(x,\, y)\, dx\, dy$$

whenever both integrals exist.

5. The area of the region D in the xy-plane is given by $A = \displaystyle\int_D \int dA$

6. If f is continuous and $f(x,\, y) \ge 0$ on the region D, the volume of the solid under the surface $z = f(x,\, y)$ above the region D is given by $V = \displaystyle\int_D \int f(x,\, y)\, dA$

7. **a.** Linearity rule: for constants a and b,

$$\int_D \int [af(x,\, y) + bg(x,\, y)]\, dA$$
$$= a\int_D \int f(x,\, y)\, dA + b\int_D \int g(x,\, y)\, dA$$

b. Dominance rule: If $f(x,\, y) \ge g(x,\, y)$ throughout a region D, then

$$\int_D \int f(x,\, y)\, dA \ge \int_D \int g(x,\, y)\, dA$$

c. Subdivision rule: If the region of integration D can be subdivided into two subregions D_1 and D_2, then

$$\int_D \int f(x,\, y)\, dA$$
$$= \int_{D_1} \int f(x,\, y)\, dA + \int_{D_2} \int f(x,\, y)\, dA$$

8. If f is continuous in the polar region D described by $a \le r \le b$ ($a,\, b \ge 0$), $\alpha \le \theta \le \beta$ ($0 \le \beta - \alpha < 2\pi$), then

$$\int_D \int f(r,\, \theta)\, dA = \int_\alpha^\beta \int_a^b f(r,\, \theta)\; r\, dr\, d\theta$$

9. Let Q denote the first quadrant of the Cartesian plane, and let C_n denote the quarter circular region described by $r \le n$, $0 \le \theta \le \frac{\pi}{2}$. Then the improper integral

$$\int_Q \int f(x,\, y)\, dA$$

is defined in polar coordinates as

$$\lim_{n \to \infty}\int_{C_n} \int f(r\cos\theta,\, r\sin\theta)r\, dr\, d\theta$$
$$= \lim_{n \to \infty}\int_0^{\pi/2} \int_0^n f(r\cos\theta,\, r\sin\theta)r\, dr\, d\theta$$

If the limit in this definition exists and is equal to L, we say that the improper integral converges to L. Otherwise, we say that the improper integral diverges.

10. Assume that the function $f(x, y)$ has continuous partial derivatives f_x and f_y in a region R of the xy-plane. Then the portion of the surface $z = f(x, y)$ that lies over R has surface area

$$S = \int\int_R \sqrt{[f_x(x, y)]^2 + [f_y(x, y)]^2 + 1}\ dA$$

11. Let D be a region in the xy-plane on which x, y, z and their partial derivatives with respect to u and v are continuous. Also, let S be a surface defined by a vector function

$$\mathbf{R}(u, v) = x(u, v)\mathbf{i} + y(u, v)\mathbf{j} + z(u, v)\mathbf{k}$$

Then the surface area is defined by

$$S = \int\int_D \|\mathbf{R}_u(u, v) \times \mathbf{R}_v(u, v)\|\ du\ dv$$

12. If $f(x, y, z)$ is continuous over a rectangular solid R: $a \le x \le b$, $c \le y \le d$, $r \le z \le s$, then the triple integral may be evaluated by the iterated integral

$$\int\int\int_R f(x, y, z)\ dV = \int_r^s \int_c^d \int_a^b f(x, y, z)\ dx\ dy\ dz$$

The iterated integration can be performed in any order (with appropriate adjustments) to the limits of integration: $dx\ dy\ dz$, $dx\ dz\ dy$, $dz\ dx\ dy$, $dy\ dx\ dz$, $dy\ dz\ dx$, $dz\ dy\ dx$.

13. If V is the volume of the solid region S, then

$$V = \int\int\int_S dV$$

14. If ρ is a continuous density function on the lamina corresponding to a plane region R, then the mass m of the lamina is given by

$$m = \int\int_R \rho(x, y)\ dA$$

15. If ρ is a continuous density function on a lamina corresponding to a plane region R, then the moments of mass with respect to the x-axes is

$$M_x = \int\int_R y\ \rho(x, y)\ dA$$

16. If m is the mass of the lamina, the center of mass is $(\bar{x}, \bar{y})$, where

$$\bar{x} = \frac{M_y}{m} \quad \text{and} \quad \bar{y} = \frac{M_x}{m}$$

If the density ρ is constant, the point $(\bar{x}, \bar{y})$ is called the centroid of the region.

17. The moments of inertia of a lamina of variable density ρ about the x- and y-axes, respectively, are

$$I_x = \int\int_R y^2 \rho(x, y)\ dA$$

and $I_y = \int\int_R x^2 \rho(x, y)\ dA$

18. A joint probability density function for the variables x and y is a function $f(x, y)$ such that $f(x, y) \ge 0$ and

$$\int_{-\infty}^{\infty}\int_{-\infty}^{\infty} f(x, y)\ dx\ dy = 1$$

The probability that the point (x, y) lies in the region R is defined by

$$P[(x, y)\ \text{in}\ R] = \int\int_R f(x, y)\ dA$$

19. Rectangular to cylindrical:

$r = \sqrt{x^2 + y^2}$; $\tan\theta = \frac{y}{x}$; $z = z$

Rectangular to spherical:

$\rho = \sqrt{x^2 + y^2 + z^2}$; $\tan\theta = \frac{y}{x}$

$\phi = \cos^{-1}\left(\frac{z}{\sqrt{x^2 + y^2 + z^2}}\right)$

Cylindrical to rectangular:

$x = r\cos\theta$; $y = r\sin\theta$; $z = z$

Cylindrical to spherical:

$\rho = \sqrt{r^2 + z^2}$; $\theta = \theta$; $\phi = \cos^{-1}\left(\frac{z}{\sqrt{r^2 + z^2}}\right)$

Spherical to rectangular:

$x = \rho\sin\phi\cos\theta$; $y = \rho\sin\phi\sin\theta$; $z = \rho\cos\phi$

Spherical to cylindrical:

$r = \rho\sin\phi$; $\theta = \theta$; $z = \rho\cos\phi$

20. If f is a continuous function of ρ, θ, and ϕ on a bounded, solid region S, the triple integral of f over S is, in spherical coordinates,

$$\int\int\int_S f(\rho, \theta, \phi)\ dV$$

$$= \int \int_S \int f(\rho, \theta, \phi) \, \rho^2 \sin \phi \; d\rho \; d\theta \; d\phi$$

21. $\left| \dfrac{\partial(x, y)}{\partial(u, v)} \right| = \begin{vmatrix} \dfrac{\partial x}{\partial u} & \dfrac{\partial x}{\partial v} \\[6pt] \dfrac{\partial y}{\partial u} & \dfrac{\partial y}{\partial v} \end{vmatrix} = \dfrac{\partial x}{\partial u}\dfrac{\partial y}{\partial v} - \dfrac{\partial y}{\partial u}\dfrac{\partial x}{\partial v}$

22. Let f be a continuous function on a region D, and let T be a one-to-one transformation that maps the region D in the uv-plane onto a region D^* in the xy-plane under the change of variable $x = g(u, v)$, $y = h(u, v)$ where g and h are continuously differentiable on D^*. Then

$$\int_{D^*} \int f(x, y) \; dy \; dx$$

$$= \int_D \int f[g(u, v), h(u, v)] \, |J(u, v)| \; du \; dv$$

23. $\displaystyle\int_0^{\pi/3} \int_0^{\sin y} e^{-x} \cos y \; dx \; dy$

$$= \int_0^{\pi/3} \left((e^{-\sin y})(-\cos y) + \cos y \right) dy$$

$$= e^{-\sqrt{3}/2} + \frac{\sqrt{3}}{2} - 1$$

24. $\displaystyle\int_{-1}^{1} \int_0^{z} \int_y^{y-z} (x + y - z) \; dx \; dy \; dz$

$$= \int_{-1}^{1} \int_0^{z} \left(\frac{(y-z)^2}{2} + (y-z)y - (y-z)z \right.$$
$$\left. - \frac{y^2}{2} - y^2 + yz \right) dy \; dz$$

$$= \int_{-1}^{1} \int_0^{z} \left(-2yz + \frac{3z^2}{2} \right) dy \; dz = 0$$

25. $\displaystyle 2 \int_0^{3} \int_0^{9-x^2} dy \; dx = 2 \int_0^{3} (9 - x^2) \; dx = 36$

26. $A = \displaystyle\int_0^{\pi/2} \int_0^{1} \cos r^2 \, r \; dr \; d\theta$

$$= \frac{1}{2} \int_0^{\pi/2} \int_0^{1} (\cos r^2)(2r \; dr) \; d\theta$$

$$= \frac{1}{2} \int_0^{\pi/2} \sin 1 \; d\theta = \frac{\pi}{4} \sin 1$$

27. This is the four z positive octants of the sphere with radius of 2. Using polar coordinates:

$$S = \int_0^{2\pi} \int_0^{2} \sqrt{f_x^2 + f_y^2 + 1} \; r \; dr \; d\theta$$

$$z = \sqrt{4 - x^2 - y^2}; \; f_x = \frac{-x}{\sqrt{4 - x^2 - y^2}};$$

$$f_y = \frac{-y}{\sqrt{4 - x^2 - y^2}}$$

$$\sqrt{f_x^2 + f_y^2 + 1}$$

$$= \sqrt{\frac{x^2}{4 - x^2 - y^2} + \frac{y^2}{4 - x^2 - y^2} + \frac{4 - x^2 - y^2}{4 - x^2 - y^2}}$$

$$= \frac{2}{\sqrt{4 - x^2 - y^2}}$$

$$S = \int_0^{2\pi} \int_0^{2} \frac{2}{\sqrt{4 - r^2}} \, r \; dr \; d\theta$$

$$= -2(-2) \int_0^{2\pi} d\theta = 8\pi$$

28. Finding the intersection of the plane and the paraboloid:

$$x^2 + 2y^2 = 4x$$

$$(x - 2)^2 + 2y^2 = 4$$

$$\frac{(x - 2)^2}{4} + \frac{y^2}{2} = 1$$

R is a translated ellipse with x intercepts of 0 and 4.

$$V = \int_0^{4} \int_0^{\sqrt{4x - x^2}/\sqrt{2}} \int_{x^2 + 2y^2}^{4x} dz \; dy \; dx$$

$$= \int_0^{4} \int_0^{\sqrt{4x - x^2}/\sqrt{2}} (4x - x^2 - 2y^2) \; dy \; dx$$

$$= \int_0^{4} \left\{ \frac{(4x - x^2)^{3/2}}{\sqrt{2}} - \frac{2}{3} \left(\frac{(4x - x^2)^{3/2}}{2\sqrt{2}} \right) \right\} dx$$

$$= \frac{2}{3\sqrt{2}} \int_0^{4} (4x - x^2)^{3/2} \, dx$$

$$= \frac{\sqrt{2}}{3} \int_0^{4} \left(2^2 - (x - 2)^2 \right)^{3/2} \, dx$$

Now using formula #245:

$$= \frac{\sqrt{2}}{3}\left(\frac{(x-2)(4x-x^2)^{3/2}}{4}\right.$$
$$+ \frac{12(x-2)(4x-x^2)^{1/2}}{8}$$
$$+ \left.\frac{3}{8}(16)\sin^{-1}\frac{x-2}{2}\right)\Big|_0^4$$
$$= \frac{\sqrt{2}}{3}\left(0 + 0 + 6(\tfrac{\pi}{2}) - 0\right)$$
$$= \frac{\sqrt{2}}{3}(6\pi) = 2\pi\sqrt{2}$$

29.
$$m = \int_0^{2\pi}\int_0^2\int_{r^2}^4 (r)\, r\, dz\, dr\, d\theta$$

$$= \int_0^{2\pi}\int_0^2 r^2(4 - r^2)\, dr\, d\theta$$

$$= \frac{64}{15}\int_0^{2\pi} d\theta = \frac{128\pi}{15}$$

30. Let $u = x + y$ and $v = x - 2y$. Then
$u - v = 3y$, $y = (v - u)/3$ and
$2u + v = 3x$, $x = (2u + v)/3$. For the
region transformation $(0, 0)$ becomes $(0, 0)$,
$(2, 0)$ becomes $(2, 2)$ and $(1, 1)$ becomes
$(2, -1)$. The equations of the boundary lines
for the uv region triangle are: $v = u$ and
$v = -\frac{1}{2}u$ or $u = -2v$. The Jacobian:

$$\frac{\partial(x,y)}{\partial(u,v)} = \begin{vmatrix} \frac{2}{3} & \frac{1}{3} \\ \frac{1}{3} & -\frac{1}{3} \end{vmatrix} = \left|-\tfrac{1}{3}\right| = \frac{1}{3}$$

For a single region use vertical slices.

$$\int_0^2\int_{-u/2}^u ue^v(\tfrac{1}{3})\, dv\, du = \frac{1}{3}\int_0^2\left(ue^u - ue^{-u/2}\right) du$$
$$= \frac{1}{3}(e^2 + \tfrac{8}{e} - 3)$$

Supplementary Problems, Page 892

1.

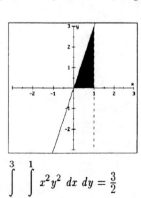

$$\int_0^3\int_{y/3}^1 x^2y^2\, dx\, dy = \frac{3}{2}$$

2.

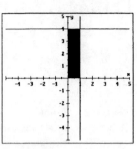

$$\int_0^1\int_0^4 \sqrt{\frac{y}{x}}\, dy\, dx = \frac{32}{3}$$

3.

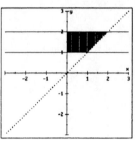

$$\int_0^1\int_1^2 \frac{1}{x^2 + y^2}\, dy\, dx + \int_1^2\int_x^2 \frac{1}{x^2 + y^2}\, dy\, dx$$

Evaluate given integral:
$$\int_1^2\int_0^y \frac{1}{x^2 + y^2}\, dx\, dy = \int_1^2 \frac{\pi}{4y}\, dy = \frac{\pi}{4}\ln 2$$

4.

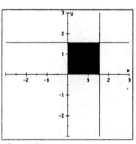

$$\int_0^{\pi/2}\int_0^{\pi/2} \cos(x + y)\, dy\, dx$$
$$= \int_0^{\pi/2} [\sin(\tfrac{\pi}{2} + x) - \sin x]\, dx = 0$$

5.

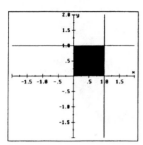

$$\int\limits_0^1 \int\limits_0^1 x\sqrt{x^2 + y} \ dy \ dx$$

Evaluate original integral:

$$\int\limits_0^1 \int\limits_0^1 x\sqrt{x^2 + y} \ dx \ dy$$

$$= \frac{1}{3}\int\limits_0^1 [(1 + y^2)^{3/2} - y^{3/2}] \ dy = \frac{8\sqrt{2} - 4}{15}$$

6.

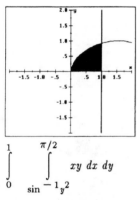

$$\int\limits_0^1 \int\limits_{\sin^{-1}y^2}^{\pi/2} xy \ dx \ dy$$

Evaluate original integral:

$$\int\limits_0^{\pi/2} \int\limits_0^{\sqrt{\sin x}} xy \ dy \ dx = \frac{1}{2}\int\limits_0^{\pi/2} x \sin x \ dx = \frac{1}{2}$$

7.

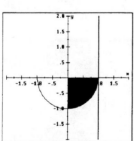

$$\int\limits_{-1}^0 \int\limits_0^{\sqrt{1 - y^2}} y\sqrt{x} \ dx \ dy$$

$$= \frac{2}{3}\int\limits_{-1}^0 y(1 - y^2)^{3/4} \ dy = -\frac{4}{21}$$

8.

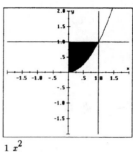

$$\int\limits_0^1 \int\limits_0^{x^2} \sqrt{1 - x^3} \ dy \ dx$$

$$= \int\limits_0^1 x^2\sqrt{1 - x^3} \ dx = \frac{2}{9}$$

9.

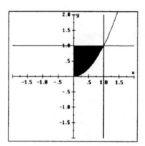

$$\int\limits_0^1 \int\limits_0^{\sqrt{y}} x^3 \sin y^3 \ dx \ dy = \frac{1}{4}\int\limits_0^1 y^2 \sin y^3 \ dy$$

$$= \frac{1 - \cos 1}{12}$$

10.

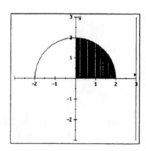

$$\int\limits_0^2 \int\limits_0^{\sqrt{4 - x^2}} \sqrt{4 - x^2 - y^2} \ dx \ dy$$

$$= \int\limits_0^{\pi/2} \int\limits_0^2 r\sqrt{4 - r^2} \ dr \ d\theta = \frac{8}{3}\int\limits_0^{\pi/2} d\theta = \frac{4\pi}{3}$$

11. $\displaystyle\int_0^1 \int_{\sqrt{x}}^1 e^{y^3}\, dy\, dx = \int_0^1 \int_0^{y^2} e^{y^3}\, dx\, dy$

$\displaystyle = \int_0^1 y^2 e^{y^3}\, dy = \tfrac{1}{3}(e - 1)$

12. $\displaystyle\int_0^2 \int_x^2 \frac{y}{(x^2 + y^2)^{3/2}}\, dy\, dx$

$\displaystyle = -\int_0^2 \left(\frac{1}{\sqrt{4 + x^2}} - \frac{1}{\sqrt{2}\,x}\right) dx$

$\displaystyle = \sqrt{2} - \ln\left|\sqrt{2} + 1\right|$

13. $\displaystyle\int_0^1 \int_{-1}^4 \int_x^y z\, dz\, dy\, dx = \tfrac{1}{2}\int_0^1 \int_{-1}^4 (y^2 - x^2)\, dy\, dx$

$\displaystyle \tfrac{1}{2}\int_0^1 (21 - 3x^2)\, dx = 10$

14. $\displaystyle\int_1^2 \int_0^1 \int_0^{\sqrt{1-x^2}} e^{\sqrt{x^2 + y^2}}\, dy\, dx\, dz$

$\displaystyle\int_1^2 \int_0^{2\pi/2} \int_0^1 r e^r\, dr\, d\theta = \int_1^2 \int_0^{2\pi/2} dr\, d\theta = \int_1^2 \frac{\pi}{2}\, d\theta = \frac{\pi}{2}$

15. $\displaystyle\int_0^{\pi/4} \int_0^{2\pi} \int_0^\theta r^2 \sin\phi\, dr\, d\theta\, d\phi = \tfrac{1}{3}\int_0^{\pi/4} \int_0^{2\pi} \theta^3 \sin\phi\, d\theta\, d\phi$

$\displaystyle = \frac{2\pi^4}{3}(2 - \sqrt{2})$

16. $\displaystyle\int_0^1 \int_0^x \int_0^y x^2 y\, dz\, dy\, dx = \int_0^1 \int_0^x x^2 y^2\, dy\, dx$

$\displaystyle = \tfrac{1}{3}\int_0^1 x^5\, dx = \frac{1}{18}$

17. $\displaystyle\int_1^2 \int_x^{x^2} \int_0^{\ln x} x e^z\, dz\, dy\, dx = \int_1^2 \int_x^{x^2} x(x - 1)\, dy\, dx$

$\displaystyle = \int_1^2 x^2(x - 1)^2\, dx = \frac{31}{30}$

18. $\displaystyle\int_0^1 \int_{1-x}^{1+x} \int_0^{xy} xz\, dz\, dy\, dx = \tfrac{1}{2}\int_0^1 \int_{1-x}^{1+x} x^3 y^2\, dy\, dx$

$\displaystyle = \tfrac{1}{3}\int_0^1 x^3(3 + x^2)\, dx = \frac{13}{36}$

19. $\displaystyle\int_{-\sqrt{3}}^{\sqrt{3}} \int_{-\sqrt{3-y^2}}^{\sqrt{3-y^2}} \int_{(x^2+y^2)^2}^9 y^2\, dz\, dx\, dy$

$\displaystyle = \int_0^{2\pi} \int_0^{\sqrt{3}} r^2 \sin^2\theta\, (9 - r^4)\, r\, dr\, d\theta$

$\displaystyle = \int_0^{2\pi} \int_0^{\sqrt{3}} \sin^2\theta(9r^3 - r^7)\, dr\, d\theta$

$\displaystyle = \int_0^{2\pi} (1 - \cos 2\theta)\, d\theta = \frac{81\pi}{8}$

20. $\displaystyle\int\int_D x^2 y\, dA = \int_0^{2\pi} \int_0^2 r^2\cos^2\theta\, r\sin\theta\, r\, dr\, d\theta$

$\displaystyle = \frac{32}{5}\int_0^{2\pi} \sin\theta\,\cos^2\theta\, d\theta = 0$

21. $\displaystyle\int\int_D (x^2 + y^2 + 1)\, dA = \int_0^{2\pi} \int_0^2 (r^2 + 1)\, r\, dr\, d\theta$

$\displaystyle = 2\pi\int_0^2 (r^3 + r)\, dr = 12\pi$

22. $\displaystyle\int\int_D e^{x^2+y^2}\, dA = \int_0^{2\pi} \int_0^2 r e^{r^2}\, dr\, d\theta = \pi(e^4 - 1)$

23. $\displaystyle\int\int_D (x^2 + y^2)^n\, dA = \int_0^{2\pi} \int_0^2 r^{2n+1}\, dr\, d\theta$

$\displaystyle = \int_0^{2\pi} \lim_{s \to 0} \int_s^2 r^{2n+1}\, dr\, d\theta$

$\displaystyle = 2\pi\left[\lim_{x \to 0} \frac{2^{2n+2} - s^{2n+2}}{2(n + 1)}\right] = 2\pi\left[\frac{2^{2n+2}}{2(n + 1)}\right]$

$\displaystyle = \frac{2^{2n+2}\pi}{n + 1}$

24. $\displaystyle\int\int_D \frac{2y}{x}\, dA = \int_0^1 \int_{1-x}^{\sqrt{1-x^2}} 2y x^{-1}\, dy\, dx$

$\displaystyle = \int_0^1 (2 - 2x)\, dx = 2$

25. $\displaystyle\iint_D x^2\sqrt{4-y}\; dA$

$$= \int_{-2}^{2} \int_{-\sqrt{4-x^2}}^{\sqrt{4-x^2}} x^2\sqrt{4-y}\; dy\; dx$$

$$= -\frac{2}{3}\int_{-2}^{2} x^2[x^3 - x^3]\; dx = 0$$

26. $u = y - x,\; v = y + x,$
$y = \frac{1}{2}(u+v),\; x = \frac{1}{2}(v-u);$

$$\frac{\partial(x,y)}{\partial(u,v)} = \begin{vmatrix} \dfrac{\partial x}{\partial u} & \dfrac{\partial x}{\partial v} \\[2mm] \dfrac{\partial y}{\partial u} & \dfrac{\partial y}{\partial v} \end{vmatrix} = \frac{1}{2}\begin{vmatrix} -1 & 1 \\ 1 & 1 \end{vmatrix} = -\frac{1}{2}$$

$$dy\; dx = \frac{1}{2}\; du\; dv$$

$$\iint_D \exp\!\left(\frac{y-x}{y+x}\right)\; dy\; dx = \frac{1}{2}\int_0^2\int_{-v}^{v} e^{u/v}du\; dv$$

$$= \frac{e - e^{-1}}{2}\int_0^2 v\; dv = e - e^{-1}$$

27. $\displaystyle\iint_D x^3\; dA = 0$ (by symmetry)
Alternately,

$$\int_0^1 \int_{-\sqrt{1-x^2}}^{1} x^3\; dy\; dx + \int_0^1 \int_{-1}^{1-\sqrt{1-x^2}} x^3\; dy\; dx$$

$$+ \int_{-1}^{0}\int_{\sqrt{1-x^2}}^{1} x^3\; dy\; dx + \int_{-1}^{0}\int_{-1}^{\sqrt{1-x^2}} x^3\; dy\; dx$$

$$= \int_0^1 x^3(1-\sqrt{1-x^2})\; dx + \int_0^1 x^3(-\sqrt{1-x^2}+1)\; dx$$

$$+ \int_{-1}^{0} x^3(1-\sqrt{1-x^2})\; dx - \int_{-1}^{0} x^3(-\sqrt{1-x^2}+1)\; dx$$

$$= 0$$

28. $\displaystyle\iint_D xy^2\; dA = \int_{-1}^{1}\int_{(y-1)/2}^{1} xy^2\; dx\; dy$

$$= \frac{1}{2}\int_{-1}^{1}\left[\frac{3y^2}{4} - \frac{y^4}{4} + \frac{y}{2}\right]dy = \frac{1}{5}$$

29. $\displaystyle\iint_P \frac{dA}{(a^2+x^2+y^2)^m} = \lim_{N\to\infty}\int_0^{2\pi}\int_0^{N}\frac{r\; dr\; d\theta}{(a^2+r^2)^m}$

$$= 2\pi\lim_{N\to\infty}\left[\frac{a^{2-2m}}{2(m-1)} + \frac{(a^2+N^2)^{1-m}}{2(1-m)}\right]$$

$$= \frac{\pi a^{2-2m}}{m-1}$$

30. $\displaystyle\iint_P e^{-(a|x|+b|y|)}\; dA$

$$= 4\lim_{N\to\infty}\int_0^N\int_0^N e^{-ax}e^{-by}\; dx\; dy$$

$$= -\frac{4}{a}\lim_{N\to\infty}\int_0^N e^{-by}\; dy = \frac{4}{ab}$$

31. $\displaystyle\iint_R \frac{\partial^2 f}{\partial x\partial y}\; dx\; dy = \int_c^d\int_a^b \frac{\partial}{\partial x}\frac{\partial f}{\partial y}\; dx\; dy$

$$= \int_c^d\left[\frac{\partial f(b,y)}{\partial y} - \frac{\partial f(a,y)}{\partial y}\right]dy$$

$$= f(b,d) - f(b,c) - f(a,d) + f(a,c)$$

$$= 5 - 1 + 3 + 4 = 11$$

32. $\displaystyle\iiint_S \sqrt{x^2+y^2+z^2}\; dV$

$$= \int_0^{\pi/2}\int_0^{\pi/2}\int_0^1 \rho^3 \sin\phi\; d\rho\; d\theta\; d\phi$$

$$= \frac{1}{4}\int_0^{\pi/2}\int_0^{\pi/2}\sin\phi\; d\theta\; d\phi$$

$$= \frac{\pi}{8}\int_0^{\pi/2}\sin\phi\; d\phi = \frac{\pi}{8}$$

33. $\displaystyle\iiint_S (x^2+y^2+z^2)\; dV$

$$= \int_0^{\pi/2}\int_0^{\pi/2}\int_0^1 \rho^4 \sin\phi\; d\rho\; d\theta\; d\phi$$

$$= \frac{1}{5}\int_0^{\pi/2}\int_0^{\pi/2}\sin\phi\; d\theta\; d\phi$$

$$= \frac{\pi}{10}\int_0^{\pi/2}\sin\phi\; d\phi = \frac{\pi}{10}$$

34. $\displaystyle\iiint_H z^2 \, dV$

$$= \int_0^{\pi/2} \int_0^{\pi} \int_0^1 \rho^4 \cos^2\phi \sin\phi \, d\rho \, d\theta \, d\phi$$

$$= \frac{1}{5}\int_0^{\pi/2} \int_0^{\pi} \cos^2\phi \sin\phi \, d\theta \, d\phi$$

$$= \frac{\pi}{5}\int_0^{\pi/2} \cos^2\phi \sin\phi \, d\phi = \frac{\pi}{15}$$

35. $\displaystyle\iiint_H \frac{dV}{\sqrt{x^2 + y^2 + z^2}}$

$$= 2\int_0^{\pi/2} \int_0^{\pi} \int_0^1 \rho \sin\phi \, d\rho \, d\theta \, d\phi$$

$$= \int_0^{\pi/2} \int_0^{\pi} \sin\phi \, d\theta \, d\phi$$

$$= \pi\int_0^{\pi/2} \sin\phi \, d\phi = \pi$$

36. $\displaystyle\iiint_S \frac{dV}{x^2 + y^2 + z^2}$

$$= \int_{\pi/4}^{\pi/2} \int_0^{2\pi} \int_{2\cot\phi\csc\phi}^{2\sqrt{2}} \frac{\rho^2}{\rho^2} \sin\phi \, d\rho \, d\theta \, d\phi$$

$$= \int_0^{\pi/2} \int_0^{\pi/2} (2\sqrt{2} - 2\cot\phi\csc\phi) \sin\phi \, d\theta \, d\phi$$

$$= \int_0^{\pi/2} \int_0^{\pi/2} (\sqrt{2}\sin\phi - \cot\phi) \, d\theta \, d\phi$$

$$= 2\pi\int_0^{\pi/2} (\sqrt{2}\sin\phi - \cot\phi) \, d\phi$$

$$= 2\pi(2 + \ln 2)$$

37. $\displaystyle\iiint_S z(x^2 + y^2)^{-1/2} \, dV$

$$= \int_0^{2\pi} \int_0^2 \int_{(1/2)(x^2+y^2)}^2 \frac{zr}{r} \, dz \, dr \, d\theta$$

$$= \int_0^{2\pi} \int_0^2 (x^2 + y^2)^{-1/2}(\tfrac{1}{2})[4 - \tfrac{1}{4}(x^2 + y^2)^2] \, dr \, d\theta$$

$$= \int_0^{2\pi} \int_0^2 (2 - \tfrac{1}{8}r^4) \, dr \, d\theta = \int_0^{2\pi} \frac{16}{5} \, d\theta = \frac{32\pi}{5}$$

38. $\displaystyle\iiint_S (x^4 + 2x^2y^2 + y^4) \, dV$

$$= \int_0^{2\pi} \int_0^a \int_0^{1/\pi} r^5 \, dz \, dr \, d\theta$$

$$= \frac{1}{\pi}\int_0^{2\pi} \int_0^a r^5 \, dr \, d\theta$$

$$= \frac{a^6}{6\pi}\int_0^{2\pi} d\theta = \frac{a^6}{3}$$

39. $\displaystyle\int_0^1 \int_0^x \int_0^{\sqrt{xy}} f(x, y, z) \, dz \, dy \, dx$

$$\int_0^1 \int_z^1 \int_{z^2/x}^x f(x, y, z) \, dy \, dx \, dz$$

40. $\displaystyle\int_0^{\pi/2} \int_0^{2a\cos\theta} r \sin 2\theta \, dr \, d\theta$

$$= \int_0^{\pi/2} \int_0^{2a\cos\theta} 2\sin\theta\cos\theta \, (r \, dr \, d\theta)$$

$$= 2\int_0^{2a} \int_0^{\sqrt{2ax - x^2}} \frac{xy}{x^2 + y^2} \, dy \, dx$$

41. $\displaystyle\int_1^2 \int_{y^2}^{y^5} e^{x/y^2} \, dx \, dy = \int_1^2 y^2(e - e^{y^3}) \, dy$

$$= \frac{1}{3}(e^8 - 8e)$$

42. $\displaystyle\int_0^1 \int_0^y f(x, y) \, dx \, dy + \int_1^4 \int_0^{(4-y)/3} f(x, y) \, dx \, dy$

$$= \int_0^1 \int_x^{4-3x} f(x, y) \, dy \, dx$$

43. $\displaystyle\int_0^1 \int_{-\sqrt{y}}^{\sqrt{y}} f(x, y) \, dx \, dy + \int_1^3 \int_{-1}^{2-y} f(x, y) \, dx \, dy$

$$\int_{-1}^1 \int_{x^2}^{2-x} f(x, y) \, dy \, dx$$

44. When evaluating $\displaystyle\int_{-\infty}^{\infty} e^{-x^2}\, dx$ we would

change to polar coordinates. Using this line of thinking, if $a = c$ and $b = 0$,

$$\int_{-\infty}^{\infty}\int_{-\infty}^{\infty} \exp[-(ax^2 + bxy + cy^2)]\, dA$$

$$= \int_{-\infty}^{\infty}\int_{-\infty}^{\infty} e^{-a(x^2+y^2)}\, dy\, dx$$

$$= \lim_{N\to\infty}\int_{-N}^{N}\int_{-N}^{N} re^{-ar^2}\, dr\, d\theta$$

$$= -\pi \lim_{N\to\infty}\int_{0}^{N} (-2r\, dr)\, e^{-ar^2}\, dr$$

$$= -\frac{\pi}{a} \lim_{N\to\infty} e^{-ar^2}\Big|_{0}^{N} = \frac{\pi}{a}$$

This integral is equal to 1 when $a = b = \pi$ and $b = 0$.

45. $\displaystyle 2\int_{0}^{\pi/3}\int_{1-\cos\theta}^{\cos\theta} r\, dr\, d\theta = \int_{0}^{\pi/3} [\cos^2\theta(1 - \cos\theta)]\, d\theta$

$$= \sqrt{3} - \frac{\pi}{3}$$

46. $\displaystyle \int_{\pi/4}^{5\pi/4}\int_{1+\cos\theta}^{1+\sin\theta} r\, dr\, d\theta$

$$= \frac{1}{2}\int_{\pi/4}^{5\pi/4} [(1+\sin\theta)^2 - (1+\cos\theta)^2]\, d\theta = 2\sqrt{2}$$

47. $\displaystyle \int_{\pi/2}^{7\pi/6}\int_{0}^{1+2\sin\theta} r\, dr\, d\theta - \int_{7\pi/2}^{3\pi/2}\int_{0}^{1+2\sin\theta} r\, dr\, d\theta$

$$= \int_{\pi/2}^{7\pi/6} (1 + 2\sin\theta)^2\, d\theta - \int_{7\pi/2}^{3\pi/2} (1+2\sin\theta)^2\, d\theta$$

$$= \pi + 3\sqrt{3}$$

48. $\displaystyle \int_{0}^{2}\int_{0}^{6-3x}\int_{0}^{(6-3x-y)/2} dz\, dy\, dx$

$$= \frac{1}{2}\int_{0}^{2}\int_{0}^{6-3x} dy\, dx = \frac{1}{4}\int_{0}^{2} (6 - 3x)^2\, dx = 6$$

49. $\displaystyle \int_{0}^{a}\int_{0}^{b(1-b/a)}\int_{0}^{c(1-x/a-y/b)} dz\, dy\, dx$

$$= c\int_{0}^{a}\int_{0}^{b(1-b/a)} \left(1 - \frac{x}{a} - \frac{y}{b}\right) dy\, dx$$

$$= \frac{bc}{2}\int_{0}^{a} \left(1 - \frac{x}{a}\right)^2 dx = \frac{abc}{6}$$

50. $\displaystyle \int_{0}^{1}\int_{0}^{1} (x^2 + y^2)\, dy\, dx = \frac{4}{3}\int_{0}^{1} x^2\, dx = \frac{4}{9}$

51. $\displaystyle \int_{0}^{a}\int_{0}^{\sqrt{a^2 - x^2}} xy\, dy\, dx = \int_{0}^{\pi/2}\int_{0}^{a} r^3 \sin\theta \cos\theta\, dr\, d\theta$

$$= \frac{a^4}{4}\int_{0}^{\pi/2} \sin\theta \cos\theta\, d\theta = \frac{a^4}{8}$$

52. $\displaystyle \int_{-\pi/2}^{\pi/2}\int_{0}^{2\cos\theta} r(4 - r^2 - 4 + 2r\cos\theta)\, dr\, d\theta$

$$= \int_{-\pi/2}^{\pi/2} \left[-4\cos^4\theta + \frac{16\cos^4\theta}{3}\right] d\theta = \frac{\pi}{2}$$

53. $f(x, y, z) = \sqrt{x^2 + y^2} - z;$

$$\nabla f = \frac{x\mathbf{i} + y\mathbf{j}}{\sqrt{x^2 + y^2}} - \mathbf{k}$$

$$\mathbf{N} = \frac{x\mathbf{i} + y\mathbf{j} - \sqrt{x^2 + y^2}\,\mathbf{k}}{\sqrt{x^2 + y^2 + z^2}}$$

$$\|\mathbf{N}\cdot\mathbf{k}\| = \frac{1}{\sqrt{2}}$$

$$S = \sqrt{2}\int_{0}^{2\pi}\int_{0}^{1} r\, dr\, d\theta = \sqrt{2}\pi$$

54. $f(x, y, z) = x^2 + y^2 - z;$

$$\nabla f = 2x\mathbf{i} + 2y\mathbf{j} - \mathbf{k}$$

$$\mathbf{N} = \frac{2x\mathbf{i} + 2y\mathbf{j} - \mathbf{k}}{\sqrt{4(x^2 + y^2) + 1}}$$

$$\|\mathbf{N}\cdot\mathbf{k}\| = \frac{1}{\sqrt{4r^2 + 1}}$$

$$S = \int_0^{2\pi} \int_0^3 r\sqrt{4r^2 + 1} \; dr \; d\theta$$

$$= \frac{37\sqrt{37} - 1}{12} \int_0^{2\pi} d\theta = \frac{\pi}{6}(37^{3/2} - 1)$$

55. $f(x, y, z) = x^2 + z^2 - 4; \; \nabla f = 2x\mathbf{i} + 2z\mathbf{k}$

$$\mathbf{N} = \frac{x\mathbf{i} + z\mathbf{k}}{\sqrt{x^2 + z^2}}; \; \|\mathbf{N} \cdot \mathbf{k}\| = \frac{2}{z}$$

$$S = 2 \int_1^2 \int_0^2 \frac{2}{\sqrt{4 - x^2}} \; dy \; dx = \frac{4\pi}{3}$$

56. $y^2 + z^2 = 2x; \; x + y = 2;$ solving these simultaneously, we find $z^2 + (y + 1)^2 = 3$. If $z = 0, \; y = -1 \pm \sqrt{3}$.

$$2 \int_{1-\sqrt{3}}^{-1+\sqrt{3}} \int_0^{\sqrt{3-(y+1)^2}} \int_{(y^2+z^2)/2}^{1-y} dx \; dz \; dy$$

$$2 \int_{1-\sqrt{3}}^{-1+\sqrt{3}} \int_0^{\sqrt{3-(y+1)^2}} \left(1 - y - \frac{y^2}{2} - \frac{z^2}{2}\right) dz \; dy$$

$$= \frac{2}{3} \int_{1-\sqrt{3}}^{-1+\sqrt{3}} [3 - (1 + y^2)^2]^{3/2} \; dy$$

$$= \frac{2}{3} \int_{-\sqrt{3}}^{\sqrt{3}} (3 - u^2)^{3/2} \; du = \frac{9\pi}{4}$$

57. $z = 4 - x^2 - y^2; \; z = 4 - 2x;$ projection on the xy-plane is $4 - x^2 - y^2 = 4 - 2x$ or $y^2 + (x - 1)^2 = 1$. That is is, $r = 2\cos\theta$.

$$2 \int_0^{\pi/2} \int_0^{2\cos\theta} (4 - r^2 - 4 + 2r\cos\theta) \; r \; dr \; d\theta$$

$$= \frac{32}{12} \int_0^{\pi/2} \cos^4\theta \; d\theta$$

$$= \frac{2}{3} \int_0^{\pi/2} \left[1 + \cos 2\theta + \frac{1 + \cos 4\theta}{3}\right] d\theta = \frac{\pi}{2}$$

58. $8 \int_0^{\pi/2} \int_0^{\cos\theta} r\sqrt{1 - r^2} \; dr \; d\theta = \frac{16}{3} \int_0^{\pi/2} (1 - \sin^3\theta) \; d\theta$

$$= \frac{16}{3} \int_0^{\pi/2} [1 - (1 - \cos^2\theta) \sin\theta] \; d\theta = \frac{8(3\pi - 4)}{9}$$

59. $4 \int_0^{\pi/2} \int_0^1 \int_0^{\sqrt{9-r^2}} r \; dz \; dr \; d\theta = 4 \int_0^{\pi/2} \int_0^1 \sqrt{9 - r^2} \; r \; dr \; d\theta$

$$= \frac{4}{3} \int_0^{\pi/2} (27 - 8^{3/2}) \; d\theta = \frac{2\pi}{3}(27 - 16\sqrt{2})$$

60. $\int_0^{\pi/4} \int_0^{2\pi} \int_0^{2a\cos\phi} \rho^2 \sin\phi \; d\rho \; d\theta \; d\phi$

$$= \frac{8a^3}{3} \int_0^{\pi/4} \int_0^{2\pi} \sin\phi \cos^3\phi \; d\theta \; d\phi$$

$$= \frac{16\pi a^3}{3} \int_0^{\pi/4} \sin\phi \cos^3\phi \; d\phi = \pi a^3$$

61. $\int_0^{\phi_0} \int_0^{2\pi} \int_0^a \rho^2 \sin\phi \; d\rho \; d\theta \; d\phi$

$$= \int_0^{\phi_0} \int_0^{2\pi} \frac{a^3 \sin\phi}{3} \; d\theta \; d\phi = \frac{2\pi a^3}{3} \int_0^{\phi_0} \sin\phi \; d\phi$$

$$= \frac{2\pi a^3}{3}(1 - \cos\phi_0)$$

62. $\int_0^{\phi_0} \int_0^{2\pi} \int_0^a \rho^4 \sin^3\phi \; d\rho \; d\theta \; d\phi$

$$= \int_0^{\phi_0} \int_0^{2\pi} \frac{a^5 \sin^3\phi}{20} \; d\theta \; d\phi = \frac{2\pi a^3}{3} \int_0^{\phi_0} \sin^3\phi \; d\phi$$

$$= \frac{2\pi a^3}{9}(2 - \sin^2\phi_0 \cos\phi_0 - 2\cos\phi_0)$$

63. **a.** $I_z = 5k \int_0^{2\pi} \int_1^2 r^3 \; dr \; d\theta = \frac{75k}{4} \int_0^{2\pi} d\theta = \frac{75\pi k}{2}$

b. $I_z = 5 \int_0^{2\pi} \int_1^2 r^5 \; dr \; d\theta = \frac{105}{2} \int_0^{2\pi} d\theta = 105\pi$

c. $\rho = x^2 y^2 = r^4 \sin^2\theta \cos^2\theta = \frac{r^4}{4} \sin^2 2\theta$

$$= \frac{r^4}{8}(1 - \cos 4\theta)$$

$$I_z = 5 \int_0^{2\pi} \int_1^2 r^2 \frac{r^4}{8}(1 - \cos 4\theta) \, r \, dr \, d\theta$$

$$= \frac{5}{64}(2^8 - 1) \int_0^{2\pi} (1 - \cos 4\theta) \, d\theta$$

$$= \frac{1{,}275\pi}{32} \approx 125.1728$$

64. $m = \int_0^1 \int_0^{1-x} (x^2 + y^2) \, dy \, dx$

$$= \int_0^1 [x^2(1 - x) + \tfrac{1}{3}(1 - x)^3] \, dx = \frac{1}{6}$$

$$M_y = \int_0^1 \int_0^{1-x} x(x^2 + y^2) \, dy \, dx$$

$$= \int_0^1 (-\tfrac{4}{3}x^4 + 2x^3 - x^2 + \tfrac{1}{3}x) \, dx = \frac{1}{15}$$

$$\bar{x} = \frac{6}{15} = \frac{2}{5}$$

65. $\phi_0 = \tan^{-1}\frac{R}{H}$; let k be the constant of proportionality

$$m = k \int_0^{2\pi} \int_0^{\phi_0} \int_0^{H\sec\phi} \rho^3 \sin\phi \, d\rho \, d\phi \, d\theta$$

$$= \frac{H^4 k}{4} \int_0^{2\pi} \int_0^{\phi_0} (\cos\phi)^{-4} \sin\phi \, d\phi \, d\theta$$

$$= \frac{H^4 k}{12} \int_0^{2\pi} [(\cos\phi_0)^{-3} - 1] \, d\theta$$

$$= \frac{H^4 k \pi}{12}(\cos^{-3}\phi_0 - 1)$$

$$= \frac{\pi}{6}kH(R^2 + H^2)^{3/2} - \frac{\pi}{6}kH^4$$

Thus, $m = \frac{\pi k H}{6}[(R^2 + H^2)^{3/2} - H^3]$

$= \frac{\pi k H^4}{6}(\sec^3\phi_0 - 1)$ where $\phi_0 = \tan^{-1}\frac{R}{H}$

is half the vertex angle of the cone.

66. By symmetry, $\bar{x} = \bar{y}$, $m = V = \frac{\pi R^2 H}{3}$

$$M_{xy} = \int_0^{2\pi} \int_0^R \int_{Hr\sin\theta/R}^H rz \, dz \, dr \, d\theta$$

$$= \frac{H}{2} \int_0^{2\pi} \int_0^R \left[r - \frac{r^3 \sin^2\theta}{R^2} \right] dr \, d\theta$$

$$= \frac{H}{2} \int_0^{2\pi} \left[\frac{R^2}{2} - \frac{R^2(1 - \cos^2\theta)}{8} \right] d\theta = \frac{3\pi H^2 R^2}{8}$$

67.

$$\frac{\partial(u, v, w)}{\partial(x, y, z)} = \begin{vmatrix} \dfrac{\partial u}{\partial x} & \dfrac{\partial u}{\partial y} & \dfrac{\partial u}{\partial z} \\[2mm] \dfrac{\partial v}{\partial x} & \dfrac{\partial v}{\partial y} & \dfrac{\partial v}{\partial z} \\[2mm] \dfrac{\partial w}{\partial x} & \dfrac{\partial w}{\partial y} & \dfrac{\partial w}{\partial z} \end{vmatrix}$$

$$= \begin{vmatrix} 2 & -3 & 1 \\ 0 & 2 & -1 \\ 0 & 0 & 2 \end{vmatrix} = 8$$

68.

$$\frac{\partial(u, v, w)}{\partial(x, y, z)} = \begin{vmatrix} \dfrac{\partial u}{\partial x} & \dfrac{\partial u}{\partial y} & \dfrac{\partial u}{\partial z} \\[2mm] \dfrac{\partial v}{\partial x} & \dfrac{\partial v}{\partial y} & \dfrac{\partial v}{\partial z} \\[2mm] \dfrac{\partial w}{\partial x} & \dfrac{\partial w}{\partial y} & \dfrac{\partial w}{\partial z} \end{vmatrix}$$

$$= \begin{vmatrix} 2x & 2y & 2z \\ 0 & 4y & 2z \\ 0 & 0 & 4z \end{vmatrix} = 32xyz$$

69. $u = 2x - y$, $v = x + 2y$; thus,

$x = \tfrac{1}{5}(v + 2u)$, $y = \tfrac{1}{5}(2v - u)$

$y = 0$ becomes $u - 2v = 0$;

$y = 1$ becomes $2v - u = 5$;

$x = 0$ becomes $v + -2u = 0$

$x = 1$ becomes $v + 2u = 5$

70. $u = x - y$, $v = y$;

$y = cx + 1$ becomes $v = c(u + v) + 1$

71. **a.** $\dfrac{\partial(u, v)}{\partial(x, y)} = \begin{vmatrix} \dfrac{\partial u}{\partial x} & \dfrac{\partial u}{\partial y} \\[2mm] \dfrac{\partial v}{\partial x} & \dfrac{\partial v}{\partial y} \end{vmatrix} = \begin{vmatrix} \dfrac{x}{2} & \dfrac{x}{2} \\[2mm] \dfrac{y}{2} & -\dfrac{y}{2} \end{vmatrix} = -2xy$

b. $\dfrac{\partial(x, y)}{\partial(u, v)} = \begin{vmatrix} \dfrac{\partial x}{\partial u} & \dfrac{\partial x}{\partial v} \\[2mm] \dfrac{\partial y}{\partial u} & \dfrac{\partial y}{\partial v} \end{vmatrix} = \begin{vmatrix} \dfrac{1}{4x} & \dfrac{1}{4x} \\[2mm] \dfrac{1}{4y} & -\dfrac{1}{4y} \end{vmatrix} = -\dfrac{1}{2xy}$

c. $\dfrac{\partial(x, y)}{\partial(u, v)} \dfrac{\partial(u, v)}{\partial(x, y)} = (-2xy)(-2xy)^{-1} = 1$

72. **a.** $u = x = y$, $v = x - y$;

$u \geq 0$ becomes $y \geq -x$,

$v \geq 0$ becomes $y \leq x$; thus, $-x \leq y \leq x$

$u + v \geq 1$ becomes $2x \geq 1$ or $x \geq \tfrac{1}{2}$;

$u + v \leq 2$ becomes $2x \leq 2$ or $x \leq 1$; thus $\tfrac{1}{2} \leq x \leq 1$

b. $\dfrac{\partial(u, v)}{\partial(x, y)} = \begin{vmatrix} \dfrac{\partial u}{\partial x} & \dfrac{\partial u}{\partial y} \\ \dfrac{\partial v}{\partial x} & \dfrac{\partial v}{\partial y} \end{vmatrix} = \begin{vmatrix} \frac{1}{2} & \frac{1}{2} \\ \frac{1}{2} & -\frac{1}{2} \end{vmatrix} = -2$

$$\iint_D (u + v)\, du\, dv = 2 \int_{1/2}^{1} \int_{-x}^{x} 2x\, dy\, dx$$

$$= 4 \int_{1/2}^{1} x(x + x)\, dx = \frac{7}{3}$$

73. $\dfrac{\partial(x, y, z)}{\partial(u, v, w)} = \begin{vmatrix} \dfrac{\partial x}{\partial u} & \dfrac{\partial x}{\partial v} & \dfrac{\partial x}{\partial w} \\ \dfrac{\partial y}{\partial u} & \dfrac{\partial y}{\partial v} & \dfrac{\partial y}{\partial w} \\ \dfrac{\partial z}{\partial u} & \dfrac{\partial z}{\partial v} & \dfrac{\partial z}{\partial w} \end{vmatrix}$

$$= \begin{vmatrix} A & 0 & 0 \\ 0 & B & 0 \\ 0 & 0 & C \end{vmatrix} = ABC$$

With $u = x$, $v = 2y$, $w = 3z$, $A = 1$, $B = \frac{1}{2}$, and $C = \frac{1}{3}$, so $\dfrac{\partial(x, y, z)}{\partial(u, v, w)} = \frac{1}{6}$; $\sqrt{u} + \sqrt{v} + \sqrt{w} = 1$

$$V = \frac{1}{6} \int_0^1 \int_0^{(1 - \sqrt{u})^2} \int_0^{(1 - \sqrt{u} - \sqrt{v})^2} dw\, dv\, du$$

$$= \frac{1}{6} \int_0^1 \int_0^{(1 - \sqrt{u})^2} (1 - \sqrt{u} - \sqrt{v})^2\, dv\, du$$

$$= \frac{1}{36} \int_0^1 (1 - 4u^{1/2} + 6u - 4u^{3/2} + u^2)\, du = \frac{1}{540}$$

74. $f(x, y, z) = Ax + By + Cz - 1$

$\nabla f = A\mathbf{i} + B\mathbf{j} + C\mathbf{k}$, and the unit normal vector is $\mathbf{N} = \dfrac{A\mathbf{i} + B\mathbf{j} + C\mathbf{k}}{\sqrt{A^2 + B^2 + C^2}}$

$\|\mathbf{N} \cdot \mathbf{k}\| = \dfrac{C}{A^2 + B^2 + C^2}$

$m = A = \dfrac{\sqrt{A^2 + B^2 + C^2}}{C} \iint_S dS$

$= \dfrac{\sqrt{A^2 + B^2 + C^2}}{2ABC}$

$M_{yz} = \dfrac{\sqrt{A^2+B^2+C^2}}{C} \int_0^{1/A} \int_0^{(-Ax+1)/B} x\, dx\, dy$

$= \dfrac{\sqrt{A^2+B^2+C^2}}{BC} \int_0^{1/A} (-Ax^2 + x)\, dx$

$= \dfrac{\sqrt{A^2 + B^2 + C^2}}{6A^2BC}$

$\overline{x} = \dfrac{2ABC}{6A^2BC} = \dfrac{1}{3A}$; similarly, $\overline{y} = \dfrac{1}{3B}$, $\overline{z} = \dfrac{1}{3C}$

75. $m = A = \frac{\pi}{8}$ since this is a quarter circle.

$M_y = \int_{\pi/4}^{\pi/2} \int_0^1 r^2 \cos\theta\, dr\, d\theta = \frac{1}{3} \int_{\pi/4}^{\pi/2} \cos\theta\, d\theta$

$= \dfrac{2 - \sqrt{2}}{6}$

$\overline{x} = \dfrac{4(2 - \sqrt{2})}{3\pi}$

76. a. $y = 1 - x^2$, $\rho = xy$

$m = \int_0^1 \int_0^{1-x^2} xy\, dy\, dx$

$= \frac{1}{2} \int_0^1 x(1 - x^2)^2\, dx = \frac{1}{12}$

$M_x = \int_0^1 \int_0^{1-x^2} xy^2\, dy\, dx$

$= \frac{1}{3} \int_0^1 x(1 - x^2)^3\, dx = \frac{1}{24}$

$M_y = \int_0^1 \int_0^{1-x^2} x^2 y\, dy\, dx$

$= \frac{1}{2} \int_0^1 x^3(1 - 2x^2 + x^4)^2\, dx = \frac{4}{105}$

$(\overline{x}, \overline{y}) = \left(\dfrac{4(12)}{105}, \dfrac{12}{24} \right) = \left(\dfrac{16}{35}, \dfrac{1}{2} \right)$

b. $I_z = \int_0^1 \int_0^{1-x^2} xy(x^2 + y^2)\, dy\, dx$

$= \frac{1}{4} \int_0^1 (x^9 - 2x^7 + 2x^5 - 2x^3 + x)\, dx$

$= \dfrac{11}{240}$

77. **a.** $m = b \displaystyle\int_0^{2\pi} \int_0^{\sqrt{a}} \int_0^b r^3 \, dr \, d\theta = \dfrac{\pi a^2 b}{2}$

b. $M_{xy} = b \displaystyle\int_0^{2\pi} \int_0^{\sqrt{a}} \int_0^b z r^3 \, dz \, dr \, d\theta = \dfrac{\pi a^2 b^2}{4}$

$\overline{x} = \overline{y} = 0$ by symmetry; $\overline{z} = \dfrac{2\pi a^2 b^2}{4\pi a^2 b} = \dfrac{b}{2}$

The centroid is $\left(0, 0, \frac{b}{2}\right)$.

78. $V = 2 \displaystyle\int_0^{\pi/2} \int_0^{2a\cos\theta} \int_0^r r \, dz \, dr \, d\theta$

$= 2 \displaystyle\int_0^{\pi/2} \int_0^{2a\cos\theta} r^2 \, dr \, d\theta$

$= \dfrac{2a^3}{3} \displaystyle\int_0^{\pi/2} \cos^3\theta \, d\theta = \dfrac{32a^3}{9}$

79. $I_1 = \displaystyle\int_a^b f(x) \, dx = \int_a^b \int_a^x u(x, y) \, dy \, dx$

$I_2 = \displaystyle\int_a^b g(y) \, dy = \int_a^b \int_y^b u(x, y) \, dx \, dy$

The plane region is the triangle bounded by $y = a$, $x = b$ and $y = x$. Thus, $I_1 = I_2$.

80. $x^{2/3} + y^{2/3} + z^{2/3} = a^{2/3}$;

$x^{2/3} = a^{2/3} r^2$; $y^{2/3} = a^{2/3} s^2$; $z^{2/3} = a^{2/3} t^2$ so that $r^2 + s^2 + t^2 = 1$.

$\dfrac{\partial(x, y, z)}{\partial(r, s, t)} = \begin{vmatrix} \dfrac{\partial x}{\partial r} & \dfrac{\partial x}{\partial s} & \dfrac{\partial x}{\partial t} \\ \dfrac{\partial y}{\partial r} & \dfrac{\partial y}{\partial s} & \dfrac{\partial y}{\partial t} \\ \dfrac{\partial z}{\partial r} & \dfrac{\partial z}{\partial s} & \dfrac{\partial z}{\partial t} \end{vmatrix}$

$= \begin{vmatrix} 3ar^2 & 0 & 0 \\ 0 & 3as^2 & 0 \\ 0 & 0 & 3at^2 \end{vmatrix} = 27 r^2 s^2 t^2$

Now the spherical coordinates are
$r = \rho \sin\phi \cos\theta$, $s = \rho \sin\phi \sin\theta$, $t = \rho \cos\phi$

$V = 8(27a^3) \displaystyle\int_0^{\pi/2} \int_0^{\pi/2} \int_0^1 (\rho\sin\phi\cos\theta)^2 (\rho\sin\phi\sin\theta)^2$

$\cdot (\rho \cos\phi)^2 \rho^2 \sin\phi \, d\rho \, d\theta \, d\phi$

$= 216 a^3 \displaystyle\int_0^{\pi/2} \int_0^{\pi/2} \int_0^1 \rho^8 \sin^5\phi \cos^2\phi \cos^2\theta \sin^2\theta \, d\rho \, d\theta \, d\phi$

$= 3a^3 \displaystyle\int_0^{\pi/2} \int_0^{\pi/2} (1 - \cos 4\theta) \sin^5\phi \cos^2\theta \, d\theta \, d\phi$

$= \dfrac{3a^3 \pi}{2} \displaystyle\int_1^0 u^2(1 - u^2)^2 \, du$

$= -\dfrac{3a^3 \pi}{2} \displaystyle\int_1^0 (u^2 - 2u^4 + u^6) \, du = \dfrac{32 a^3 \pi}{35}$

81. **a.** $\dfrac{1}{A(R)} \displaystyle\int\!\!\int_{A(R)} dA = \dfrac{A(R)}{A(R)} = 1$

b. $\displaystyle\int_0^\infty \int_0^\infty \lambda\mu \exp[-\lambda x - \mu y] \, dy \, dx$

$= \displaystyle\lim_{N\to\infty} \int_0^N \int_0^N \lambda\mu \exp[-\lambda x - \mu y] \, dy \, dx$

$= \displaystyle\lim_{N\to\infty} \int_0^N e^{-\lambda x}\left[\dfrac{1}{\mu} - \dfrac{e^{-\mu N}}{\mu}\right] dx$

$= \dfrac{1}{\lambda\mu} \displaystyle\lim_{N\to\infty} [e^{-\lambda N}(e^{-\mu N} - 1)$

$\qquad - e^{-\mu N} + 1]$

$= \dfrac{1}{\lambda\mu}(\lambda\mu) = 1$

82. By Problem 37, Section 13.6, $z_s = \frac{3}{8}$, and by symmetry, $z_c = -1$.

$m = m_s + m_c = \dfrac{4\pi}{6} + 8 = \dfrac{2\pi + 24}{3}$

$M_{xy} = \left(\dfrac{2\pi}{3}\right)\left(\dfrac{3}{8}\right) + 8(-1) = \dfrac{\pi - 32}{4}$

$\overline{z} = \dfrac{3(\pi - 32)}{4(2\pi + 24)} \approx -0.7147$

By symmetry, $\overline{x} = \overline{y} = 0$, so the center is $(0, 0, -0.71)$.

83. The curves intersect when $2 = 4 \sin 2\theta$ or $\theta = \pi/12$; let δ be the density.

$m = \displaystyle\int_{\pi/12}^{\pi/4} \int_{\sqrt{2a}}^{2a\sqrt{\sin 2\theta}} \delta r \, dr \, d\theta$

84. Let δ be the density.

$m = 6 \displaystyle\int_0^{\pi/6} \int_0^{\cos 3\theta} \delta r \, dr \, d\theta$

85.
$$m = 8 \int_0^1 \int_{-1}^1 \int_0^{\sqrt{1-y^2}} dz\, dy\, dx$$

$$= 8 \int_0^1 \int_{-1}^1 \sqrt{1-y^2}\, dy = 2\pi$$

$$M_{yx} = \int_{-2}^1 \int_{-1}^1 \int_{-\sqrt{1-y^2}}^{\sqrt{1-y^2}} z\, dz\, dy\, dx$$

$$= \int_{-2}^1 \int_{-1}^1 [(1-y^2) - (1-y^2)]\, dy\, dx = 0$$

Similarly, $M_{xz} = M_{yz} = 0$. Thus,
$(\overline{x}, \overline{y}, \overline{z}) = (0, 0, 0)$.

86. $x = r \cos\theta;\ F(r, \theta, z) = f(r, \theta) - z$

$y = r \sin\theta;$

$$F_x = F_r r_x + F_\theta \theta_x$$

$$= f_r \frac{2x}{2\sqrt{x^2+y^2}} + f_\theta \frac{-y/x^2}{1 + y^2/x^2}$$

$$= f_r \cos\theta - f_\theta$$

$$F_y = F_r r_y + F_\theta \theta_y$$

$$= f_r \sin\theta + f_\theta \frac{x}{x^2+y^2}$$

$$= f_r \sin\theta + \frac{\cos\theta}{r} f_\theta$$

$$\|\nabla f\|^2 = f_r^2 - 2\left(\frac{\sin\theta\cos\theta}{r}\right) f_r f_\theta + f_\theta^2\left(\frac{1}{r^2}\right)$$

$$+ \frac{2\sin\theta\cos\theta}{r} f_r f_\theta + 1$$

$$= f_r^2 + \frac{1}{r^2} f_\theta^2 + 1$$

$$\cos\gamma = \frac{1}{\|\nabla f\|}$$

$$A = \iint_R \sqrt{1 + f_r^2 + \frac{1}{r^2} f_\theta^2}\ r\, dr\, d\theta$$

87.
$$m = \int_0^{\pi/2} \int_0^{\pi/2} \int_0^1 \frac{\rho \sin\phi}{\rho^2 + 1}\, d\rho\, d\theta\, d\phi$$

$$= \frac{\pi}{2} \int_0^{\pi/2} \int_0^1 \left(1 - \frac{1}{\rho^2+1}\right) \sin\phi\, d\rho\, d\phi$$

$$= \frac{\pi}{2} \int_0^{\pi/2} \left(1 - \frac{\pi}{4}\right) \sin\phi\, d\theta = \frac{\pi}{2}\left(1 - \frac{\pi}{4}\right)$$

$$M_{xy} = \int_0^{\pi/2} \int_0^{\pi/2} \int_0^1 \frac{\rho^3 \cos\phi \sin\phi}{\rho^2 + 1}\, d\rho\, d\theta\, d\phi$$

$$= \frac{\pi}{2} \int_0^{\pi/2} \int_0^{\pi/2} \left(\rho - \frac{\rho}{\rho^2+1}\right) \sin\phi\cos\phi\, d\rho\, d\phi$$

$$= \frac{\pi}{2} \int_0^{\pi/2} \left[\frac{1}{2} - \frac{\ln 2}{2}\right] \sin\phi\cos\phi\, d\phi$$

$$= \frac{\pi}{8}(1 - \ln 2)$$

$$\overline{z} = \frac{\pi(1 - \ln 2)}{\pi(4 - \pi)} \approx 0.3575$$

By symmetry, $\overline{x} = \overline{y} = 0.3575$.

88. First find the volume of the "cap" above $z = c/2$.

$$V_c = 4 \int_0^{\pi/6} \int_0^{\pi/2} \int_{c\sqrt{3}/2}^c \rho^2 \sin\phi\, d\rho\, d\theta\, d\phi$$

$$= 4 \int_0^{\pi/6} \int_0^{\pi/2} c^3\left(\frac{1}{3} - \frac{\sqrt{3}}{8}\right) \sin\phi\, d\theta\, d\phi$$

$$= 2\pi \int_0^{\pi/6} c^3\left(\frac{1}{3} - \frac{\sqrt{3}}{8}\right) \sin\phi\, d\phi$$

$$= \frac{\pi c^3}{24}(8 - 3\sqrt{3})(2 - \sqrt{3})$$

For two such caps, $2V_c = \frac{\pi c^3}{12}(25 - 14\sqrt{3})$

The remainder of the removed region is a prism $V_p = \sqrt{3}c^2$ so that the remaining volume is

$$V = \frac{4}{3}\pi c^3 - 2V_c - V_p$$

$$= \frac{4}{3}\pi c^3 - \frac{\pi c^3}{12}(25 - 14\sqrt{3}) - \sqrt{3}c^2$$

89. This is Putnam Problem 5 in the morning session of 1942. Choose the axes so that the generated circle starts in the xz-plane and is revolved about the z-axis. The generated circle is half of a torid (*i.e.* a solid bounded by a torus). It is clear from symmetry that the centroid lies at the point $(0, \overline{y}, 0)$ on the y-axis, that the requirement of the problem is that $y = b - a$. To find the centroid, we introduce polar coordinates in the xy-plane. Corresponding to the element of area $r\, dr\, d\theta$

in the plane there is the element of volume

$$2\sqrt{a^2 - (r - b)^2}\, r\, dr\, d\theta$$

which contributes

$$2r \sin\theta\sqrt{a^2 - (r - b)^2}\, r\, dr\, d\theta$$

to the moment M of the solid in the y-direction. We have $\overline{y} = M_y/V$ where V is the volume of the semi-torid.

$$V = \int_0^\pi \int_{b-a}^{b+a} 2\sqrt{a^2 - (r - b)^2}\, r\, dr\, d\theta$$

$$= 2\pi \int_{b-a}^{b+a} 2\sqrt{a^2 - (r - b)^2}\, r\, dr$$

$$= 2\pi a^2 \int_{-\pi/2}^{\pi/2} \cos\phi(b + a\sin\phi)\cos\phi\, d\phi$$

$$= \pi^2 a^2 b$$

$$M_y = \int_0^\pi \int_{b-a}^{b+a} 2\sqrt{a^2 - (r - b)^2}\, r^2\, dr\, d\theta$$

$$= 4\pi \int_{b-a}^{b+a} \sqrt{a^2 - (r - b)^2}\, r^2\, dr$$

$$= 4a^2 \int_{-\pi/2}^{\pi/2} \cos\phi(b + a\sin\phi)^2 \cos\phi\, d\phi$$

$$= 2\pi a^2 b^2 + \frac{\pi a^4}{2}$$

In both integrals, we used the substitution $r = b + a\sin\phi$. Hence

$$\overline{y} = \frac{M_y}{V} = \frac{a^2 + 4b^2}{2\pi b}$$

We require $\overline{y} = b - a$, so

$$2\pi b^2 - 2\pi ab = a^2 + 4b^2$$

If $c = b/a$, then

$$(2\pi - 4)c^2 - 2\pi c - 1 = 0$$

$$c = \frac{\pi + \sqrt{\pi^2 + 2\pi - 4}}{2\pi - 4}$$

We chose the positive sign since c must be positive. Remark: the volume of the semi-torid could have been obtained from Pappus' theorem. The critical ratio is 2.91.

90. This is Putnam Problem 3 in the afternoon session of 1957. The desired inequality is equivalent to

$$\int_0^1 [f(x)]^2\, dx \int_0^1 yf(y)\, dy - \int_0^1 y[f(y)]^2\, dy \int_0^1 f(x)\, dx \geq 0$$

which can be rewritten

$$\int_0^1 \int_0^1 f(x)f(y)y[f(x) - f(y)]\, dx\, dy \geq 0$$

This is what we need to prove. Denote the left member of the inequality by I. Then

$$2I = \int_0^1 \int_0^1 f(x)f(y)(y - x)[f(x) - f(y)]\, dx\, dy$$

Because f is decreasing,

$$(y - x)[f(x) - f(y)] \geq 0$$

for all x and y. Then, since f is everywhere positive, it is clear that $2I \geq 0$, and the proof is complete.

91. This is Putnam Problem 5 in the morning session of 1956. Suppose there are two continuous solutions and let g be their difference. Then g is continuous and

$$g(x, y) = \int_0^1 \int_0^y g(u, v)\, du\, dv$$

Since g is continuous, it is bounded on the given square. Let M be a bound. Then

$$\left|g(x, y)\right| \leq \int_0^1 \int_0^y \left|g(u, v)\right|\, du\, dv \leq \int_0^1 \int_0^y M\, du\, dv = Mxy$$

for $0 \leq x \leq 1; 0 \leq y \leq 1$. We now prove

$$\left|g(x, y)\right| \leq M\frac{x^n y^n}{n!\,n!}$$

for any integer n. This has been proved for $n = 1$. Assume that it is true for $n = k$; then

$$\left|g(x, y)\right| \leq \int_0^1 \int_0^y \left|g(u, v)\right|\, du\, dv \leq \int_0^1 \int_0^y M\, du\, dv$$

$$= M\frac{x^{k+1} y^{k+1}}{(k + 1)!(k + 1)!}$$

Thus, the proposition is true by mathematical induction. But for any fixed x and y

$$\lim_{n\to\infty} M\frac{x^n y^n}{n!\,n!} = 0$$

Hence, $\left|g(x, y)\right| \leq 0$; that is, $g(x, y) = 0$. Thus, there cannot be two different continuous solutions. Remark: This problem generalizes immediately to n-dimensions. See Problem 4885, *American Mathematical Monthly*, Vol. 67 (1960), p. 87. Solution is given in Vol. 68 (1961), p. 73.

CHAPTER 14

Vector Analysis

14.1 Properties of a Vector Field: Divergence and Curl, Page 905

1. div $\mathbf{v} = \nabla \cdot \mathbf{v}$, which is applicable to a vector function only, generates a scalar.
Curl $\mathbf{v} = \nabla \times \mathbf{v}$, which is applicable to a vector function only, generates a vector. Note that the words divergence and dot start with the letter "d" and the words curl and cross start with the letter "c."

2. The operator $\nabla = \frac{\partial}{\partial x}\mathbf{i} + \frac{\partial}{\partial y}\mathbf{j} + \frac{\partial}{\partial z}\mathbf{k}$ is applicable to a vector function only.

Answers for Problems 3-8 vary.

9. div $\mathbf{F} = \nabla \cdot \mathbf{F}$
$= \left(\frac{\partial}{\partial x}\mathbf{i} + \frac{\partial}{\partial y}\mathbf{j} + \frac{\partial}{\partial z}\mathbf{k}\right) \cdot (x^2\mathbf{i} + xy\mathbf{j} + z^3\mathbf{k})$
$= 2x + x + 3z^2 = 3x + 3z^2$
$$\text{curl } \mathbf{F} = \nabla \times \mathbf{F} = \begin{vmatrix} \mathbf{i} & \mathbf{j} & \mathbf{k} \\ \frac{\partial}{\partial x} & \frac{\partial}{\partial y} & \frac{\partial}{\partial z} \\ x^2 & xy & z^3 \end{vmatrix}$$
$= 0\mathbf{i} - 0\mathbf{j} + (y - 0)\mathbf{k} = y\mathbf{k}$

10. div $\mathbf{F} = \nabla \cdot \mathbf{F}$
$= \left(\frac{\partial}{\partial x}\mathbf{i} + \frac{\partial}{\partial y}\mathbf{j}\right) \cdot (\mathbf{i} + (x^2 + y^2)\mathbf{j}) = 2y$
$$\text{curl } \mathbf{F} = \nabla \times \mathbf{F} = \begin{vmatrix} \mathbf{i} & \mathbf{j} & \mathbf{k} \\ \frac{\partial}{\partial x} & \frac{\partial}{\partial y} & \frac{\partial}{\partial z} \\ 1 & x^2+y^2 & 0 \end{vmatrix} = \mathbf{0}$$

11. div $\mathbf{F} = \nabla \cdot \mathbf{F}$
$= \left(\frac{\partial}{\partial y}\mathbf{j}\right) \cdot (2y\mathbf{j}) = 2$
$$\text{curl } \mathbf{F} = \nabla \times \mathbf{F} = \begin{vmatrix} \mathbf{i} & \mathbf{j} & \mathbf{k} \\ \frac{\partial}{\partial x} & \frac{\partial}{\partial y} & \frac{\partial}{\partial z} \\ 0 & 2y & 0 \end{vmatrix} = \mathbf{0}$$

12. div $\mathbf{F} = \nabla \cdot \mathbf{F}$
$= \left(\frac{\partial}{\partial x}\mathbf{i} + \frac{\partial}{\partial y}\mathbf{j} + \frac{\partial}{\partial y}\mathbf{k}\right) \cdot (z\mathbf{i} - \mathbf{j} + 2y\mathbf{k}) = 0$
$$\text{curl } \mathbf{F} = \nabla \times \mathbf{F} = \begin{vmatrix} \mathbf{i} & \mathbf{j} & \mathbf{k} \\ \frac{\partial}{\partial x} & \frac{\partial}{\partial y} & \frac{\partial}{\partial z} \\ z & -1 & 2y \end{vmatrix} = \mathbf{i} + \mathbf{j}$$

13. div $\mathbf{F} = \nabla \cdot \mathbf{F}$
$= \left(\frac{\partial}{\partial x}\mathbf{i} + \frac{\partial}{\partial y}\mathbf{j} + \frac{\partial}{\partial z}\mathbf{k}\right) \cdot (\mathbf{i} + \mathbf{j} + \mathbf{k}) = 0$
$$\text{curl } \mathbf{F} = \nabla \times \mathbf{F} = \begin{vmatrix} \mathbf{i} & \mathbf{j} & \mathbf{k} \\ \frac{\partial}{\partial x} & \frac{\partial}{\partial y} & \frac{\partial}{\partial z} \\ 1 & 1 & 1 \end{vmatrix} = \mathbf{0}$$

14. div $\mathbf{F} = \nabla \cdot \mathbf{F}$
$= \left(\frac{\partial}{\partial x}\mathbf{i} + \frac{\partial}{\partial y}\mathbf{j} + \frac{\partial}{\partial z}\mathbf{k}\right) \cdot (xz\mathbf{i} + y^2z\mathbf{j} + xz\mathbf{k})$
$= z + 2yz + x$
$$\text{curl } \mathbf{F} = \nabla \times \mathbf{F} = \begin{vmatrix} \mathbf{i} & \mathbf{j} & \mathbf{k} \\ \frac{\partial}{\partial x} & \frac{\partial}{\partial y} & \frac{\partial}{\partial z} \\ xz & y^2z & xz \end{vmatrix}$$
$= -y^2\mathbf{i} - (z - x)\mathbf{j} + \mathbf{k}$
At $(1, -1, 2)$, div $\mathbf{F} = 2 - 4 + 1 = -1$,
curl $\mathbf{F} = -\mathbf{i} - \mathbf{j}$

15. div $\mathbf{F} = \nabla \cdot \mathbf{F}$
$= \left(\frac{\partial}{\partial x}\mathbf{i} + \frac{\partial}{\partial y}\mathbf{j} + \frac{\partial}{\partial z}\mathbf{k}\right) \cdot (xyz\mathbf{i} + yj + x\mathbf{k})$
$= yz + 1$
$$\text{curl } \mathbf{F} = \nabla \times \mathbf{F} = \begin{vmatrix} \mathbf{i} & \mathbf{j} & \mathbf{k} \\ \frac{\partial}{\partial x} & \frac{\partial}{\partial y} & \frac{\partial}{\partial z} \\ xyz & y & x \end{vmatrix}$$
$= 0\mathbf{i} - (1 - xy)\mathbf{j} - xz\mathbf{k}$
At $(1, 2, 3)$, div $\mathbf{F} = 6 + 1 = 7$,
curl $\mathbf{F} = \mathbf{j} - 3\mathbf{k}$

16. div $\mathbf{F} = \nabla \cdot \mathbf{F}$
$= \left(\frac{\partial}{\partial x}\mathbf{i} + \frac{\partial}{\partial y}\mathbf{j} + \frac{\partial}{\partial z}\mathbf{k}\right) \cdot [(\cos y)\mathbf{i} + (\sin y)\mathbf{j} + \mathbf{k}]$
$= \cos y$
$$\text{curl } \mathbf{F} = \nabla \times \mathbf{F} = \begin{vmatrix} \mathbf{i} & \mathbf{j} & \mathbf{k} \\ \frac{\partial}{\partial x} & \frac{\partial}{\partial y} & \frac{\partial}{\partial z} \\ \cos y & \sin y & 1 \end{vmatrix}$$
$= 0\mathbf{i} - 0\mathbf{j} + \sin y\mathbf{k}$
At $(\frac{\pi}{4}, \pi, 0)$, div $\mathbf{F} = -1$; curl $\mathbf{F} = \mathbf{0}$

17. $\text{div } \mathbf{F} = \nabla \cdot \mathbf{F}$

$= \left(\dfrac{\partial}{\partial x}\mathbf{i} + \dfrac{\partial}{\partial y}\mathbf{j} + \dfrac{\partial}{\partial z}\mathbf{k} \right) \cdot (e^{-xy}\mathbf{i} + e^{xz}\mathbf{j} + e^{yz}\mathbf{k})$

$= -ye^{-xy} + ye^{-xy}$

$\text{curl } \mathbf{F} = \nabla \times \mathbf{F} = \begin{vmatrix} \mathbf{i} & \mathbf{j} & \mathbf{k} \\ \dfrac{\partial}{\partial x} & \dfrac{\partial}{\partial y} & \dfrac{\partial}{\partial z} \\ e^{-xy} & e^{xz} & e^{yz} \end{vmatrix}$

$= (ze^{yz} - xe^{xz})\mathbf{i} - 0\mathbf{j} + (ze^{xz} + xe^{-xy})\mathbf{k}$

At $(3, 2, 0)$, $\text{div } \mathbf{F} = 2 - 2e^{-6}$

$\text{curl } \mathbf{F} = -3\mathbf{i} + 3e^{-6}\mathbf{k}$

18. $\text{div } \mathbf{F} = \nabla \cdot \mathbf{F}$

$= \left(\dfrac{\partial}{\partial x}\mathbf{i} + \dfrac{\partial}{\partial y}\mathbf{j} + \dfrac{\partial}{\partial z}\mathbf{k} \right) \cdot (e^{-x}\sin y)\mathbf{i} + (e^{-x}\cos y)\mathbf{j} + \mathbf{k}$

$= -e^{-x}\sin y - e^{-x}\sin y = -2e^{-x}\sin y$

$\text{curl } \mathbf{F} = \nabla \times \mathbf{F}$

$= \begin{vmatrix} \mathbf{i} & \mathbf{j} & \mathbf{k} \\ \dfrac{\partial}{\partial x} & \dfrac{\partial}{\partial y} & \dfrac{\partial}{\partial z} \\ e^{-x}\sin y & e^{-x}\cos y & 1 \end{vmatrix}$

$= 0\mathbf{i} - 0\mathbf{j} - 2e^{-x}\cos y\, \mathbf{k}$

At $(1, 3, -2)$, $\text{div } \mathbf{F} = -2e^{-1}\sin 3$

$\text{curl } \mathbf{F} = (e^{-1}\cos 3)\mathbf{k}$

19. $\text{div } \mathbf{F} = \nabla \cdot \mathbf{F}$

$= \left(\dfrac{\partial}{\partial x}\mathbf{i} + \dfrac{\partial}{\partial y}\mathbf{j} \right) \cdot [(\sin x)\mathbf{i} + (\cos y)\mathbf{j}]$

$= \cos x - \sin y$

$\text{curl } \mathbf{F} = \nabla \times \mathbf{F} = \begin{vmatrix} \mathbf{i} & \mathbf{j} & \mathbf{k} \\ \dfrac{\partial}{\partial x} & \dfrac{\partial}{\partial y} & \dfrac{\partial}{\partial z} \\ \sin y & \cos y & 0 \end{vmatrix} = \mathbf{0}$

20. $\text{div } \mathbf{F} = \nabla \cdot \mathbf{F}$

$= \left(\dfrac{\partial}{\partial x}\mathbf{i} + \dfrac{\partial}{\partial y}\mathbf{j} \right) \cdot [(-\cos x)\mathbf{i} + (\sin y)\mathbf{j}]$

$= \sin x + \cos y$

$\text{curl } \mathbf{F} = \nabla \times \mathbf{F} = \begin{vmatrix} \mathbf{i} & \mathbf{j} & \mathbf{k} \\ \dfrac{\partial}{\partial x} & \dfrac{\partial}{\partial y} & \dfrac{\partial}{\partial z} \\ -\cos y & \sin y & 0 \end{vmatrix} = \mathbf{0}$

21. $\text{div } \mathbf{F} = \nabla \cdot \mathbf{F}$

$= \left(\dfrac{\partial}{\partial x}\mathbf{i} + \dfrac{\partial}{\partial y}\mathbf{j} \right) \cdot (x\mathbf{i} - y\mathbf{j}) = 0$

$\text{curl } \mathbf{F} = \nabla \times \mathbf{F} = \begin{vmatrix} \mathbf{i} & \mathbf{j} & \mathbf{k} \\ \dfrac{\partial}{\partial x} & \dfrac{\partial}{\partial y} & \dfrac{\partial}{\partial z} \\ x & -y & 0 \end{vmatrix} = \mathbf{0}$

22. $\text{div } \mathbf{F} = \nabla \cdot \mathbf{F}$

$= \left(\dfrac{\partial}{\partial x}\mathbf{i} + \dfrac{\partial}{\partial y}\mathbf{j} \right) \cdot (-x\mathbf{i} + y\mathbf{j}) = 0$

$\text{curl } \mathbf{F} = \nabla \times \mathbf{F} = \begin{vmatrix} \mathbf{i} & \mathbf{j} & \mathbf{k} \\ \dfrac{\partial}{\partial x} & \dfrac{\partial}{\partial y} & \dfrac{\partial}{\partial z} \\ -x & y & 0 \end{vmatrix} = \mathbf{0}$

23. $\text{div } \mathbf{F} = \nabla \cdot \mathbf{F}$

$= \left(\dfrac{\partial}{\partial x}\mathbf{i} + \dfrac{\partial}{\partial y}\mathbf{j} \right) \cdot \left(\dfrac{x}{\sqrt{x^2 + y^2}}\mathbf{i} + \dfrac{y}{\sqrt{x^2 + y^2}}\mathbf{j} \right)$

$= \dfrac{\sqrt{x^2 + y^2} - \dfrac{x^2}{\sqrt{x^2 + y^2}}}{x^2 + y^2}$

$+ \dfrac{\sqrt{x^2 + y^2} + \dfrac{x^2}{\sqrt{x^2 + y^2}}}{x^2 + y^2}$

$= \dfrac{1}{\sqrt{x^2 + y^2}}$

$\text{curl } \mathbf{F} = \nabla \times \mathbf{F} = \begin{vmatrix} \mathbf{i} & \mathbf{j} & \mathbf{k} \\ \dfrac{\partial}{\partial x} & \dfrac{\partial}{\partial y} & \dfrac{\partial}{\partial z} \\ \dfrac{x}{\sqrt{x^2+y^2}} & \dfrac{y}{\sqrt{x^2+y^2}} & 0 \end{vmatrix} = \mathbf{0}$

24. $\text{div } \mathbf{F} = \nabla \cdot \mathbf{F}$

$= \left(\dfrac{\partial}{\partial x}\mathbf{i} + \dfrac{\partial}{\partial y}\mathbf{j} \right) \cdot (x^2\mathbf{i} - y^2\mathbf{j})$

$= 2x - 2y$

$\text{curl } \mathbf{F} = \nabla \times \mathbf{F} = \begin{vmatrix} \mathbf{i} & \mathbf{j} & \mathbf{k} \\ \dfrac{\partial}{\partial x} & \dfrac{\partial}{\partial y} & \dfrac{\partial}{\partial z} \\ x^2 & -y^2 & 0 \end{vmatrix} = \mathbf{0}$

25. $\text{div } \mathbf{F} = \nabla \cdot \mathbf{F}$

$= \left(\dfrac{\partial}{\partial x}\mathbf{i} + \dfrac{\partial}{\partial y}\mathbf{j} + \dfrac{\partial}{\partial z}\mathbf{k} \right) \cdot (ax\mathbf{i} + by\mathbf{j} + c\mathbf{k})$

$= a + b$

$\text{curl } \mathbf{F} = \nabla \times \mathbf{F} = \begin{vmatrix} \mathbf{i} & \mathbf{j} & \mathbf{k} \\ \dfrac{\partial}{\partial x} & \dfrac{\partial}{\partial y} & \dfrac{\partial}{\partial z} \\ ax & by & c \end{vmatrix} = \mathbf{0}$

26. $\text{div } \mathbf{F} = \nabla \cdot \mathbf{F}$

$= \left(\dfrac{\partial}{\partial x}\mathbf{i} + \dfrac{\partial}{\partial y}\mathbf{j} + \dfrac{\partial}{\partial z}\mathbf{k} \right) \cdot [(e^x\sin y)\mathbf{i} + (e^x\cos y)\mathbf{j} + \mathbf{k}]$

$= e^x\sin y - e^x\sin y = 0$

$\text{curl } \mathbf{F} = \nabla \times \mathbf{F} = \begin{vmatrix} \mathbf{i} & \mathbf{j} & \mathbf{k} \\ \dfrac{\partial}{\partial x} & \dfrac{\partial}{\partial y} & \dfrac{\partial}{\partial z} \\ e^x\sin y & e^x\cos y & 1 \end{vmatrix}$

$= (e^x\cos y - e^x\cos y)\mathbf{k} = \mathbf{0}$

27. $\text{div } \mathbf{F} = \nabla \cdot \mathbf{F}$

$$= \left(\frac{\partial}{\partial x}\mathbf{i} + \frac{\partial}{\partial y}\mathbf{j} + \frac{\partial}{\partial z}\mathbf{k}\right) \cdot (x^2\mathbf{i} + y^2\mathbf{j} + z^2\mathbf{k})$$

$$= 2x + 2y + 2z = 2(x + y + z)$$

$$\text{curl } \mathbf{F} = \begin{vmatrix} \mathbf{i} & \mathbf{j} & \mathbf{k} \\ \frac{\partial}{\partial x} & \frac{\partial}{\partial y} & \frac{\partial}{\partial z} \\ x^2 & y^2 & z^2 \end{vmatrix} = \mathbf{0}$$

28. $\text{div } \mathbf{F} = \nabla \cdot \mathbf{F}$

$$= \left(\frac{\partial}{\partial x}\mathbf{i} + \frac{\partial}{\partial y}\mathbf{j} + \frac{\partial}{\partial z}\mathbf{k}\right) \cdot (y\mathbf{i} + z\mathbf{j} + x\mathbf{k}) = 0$$

$$\text{curl } \mathbf{F} = \begin{vmatrix} \mathbf{i} & \mathbf{j} & \mathbf{k} \\ \frac{\partial}{\partial x} & \frac{\partial}{\partial y} & \frac{\partial}{\partial z} \\ y & z & x \end{vmatrix}$$

$$= -\mathbf{i} - \mathbf{j} - \mathbf{k}$$

29. $\text{div } \mathbf{F} = \nabla \cdot \mathbf{F}$

$$= \left(\frac{\partial}{\partial x}\mathbf{i} + \frac{\partial}{\partial y}\mathbf{j} + \frac{\partial}{\partial z}\mathbf{k}\right) \cdot (xy\mathbf{i} + yz\mathbf{j} + xz\mathbf{k})$$

$$= x + y + z$$

$$\text{curl } \mathbf{F} = \begin{vmatrix} \mathbf{i} & \mathbf{j} & \mathbf{k} \\ \frac{\partial}{\partial x} & \frac{\partial}{\partial y} & \frac{\partial}{\partial z} \\ xy & yz & xz \end{vmatrix}$$

$$= -y\mathbf{i} - z\mathbf{j} - x\mathbf{k}$$

30. $\text{div } \mathbf{F} = \nabla \cdot \mathbf{F}$

$$= \left(\frac{\partial}{\partial x}\mathbf{i} + \frac{\partial}{\partial y}\mathbf{j} + \frac{\partial}{\partial z}\mathbf{k}\right) \cdot (2xz\mathbf{i} + 2yz^2\mathbf{j} - \mathbf{k})$$

$$= 2z + 2z^2$$

$$\text{curl } \mathbf{F} = \begin{vmatrix} \mathbf{i} & \mathbf{j} & \mathbf{k} \\ \frac{\partial}{\partial x} & \frac{\partial}{\partial y} & \frac{\partial}{\partial z} \\ 2xz & 2yz^2 & -1 \end{vmatrix}$$

$$= -4yz\mathbf{i} + 2x\mathbf{j}$$

31. $\text{div } \mathbf{F} = \nabla \cdot \mathbf{F}$

$$= \left(\frac{\partial}{\partial x}\mathbf{i} + \frac{\partial}{\partial y}\mathbf{j} + \frac{\partial}{\partial z}\mathbf{k}\right) \cdot (xyz\mathbf{i} + x^2y^2z^2\mathbf{j} + y^2z^3\mathbf{k})$$

$$= yz + 2x^2yz^2 + 3y^2z^2$$

$$\text{curl } \mathbf{F} = \begin{vmatrix} \mathbf{i} & \mathbf{j} & \mathbf{k} \\ \frac{\partial}{\partial x} & \frac{\partial}{\partial y} & \frac{\partial}{\partial z} \\ 2xyz & x^2y^2z^2 & y^2z^3 \end{vmatrix}$$

$$= (2yz^3 - 2x^2y^2z)\mathbf{i} + xy\mathbf{j} + (2xy^2z^2 - xz)\mathbf{k}$$

32. $\text{div } \mathbf{F} = \nabla \cdot \mathbf{F}$

$$= \left(\frac{\partial}{\partial x}\mathbf{i} + \frac{\partial}{\partial y}\mathbf{j} + \frac{\partial}{\partial z}\mathbf{k}\right) \cdot (-z^3\mathbf{i} + 3\mathbf{j} + 2y\mathbf{k}) = 0$$

$$\text{curl } \mathbf{F} = \nabla \times \mathbf{F} = \begin{vmatrix} \mathbf{i} & \mathbf{j} & \mathbf{k} \\ \frac{\partial}{\partial x} & \frac{\partial}{\partial y} & \frac{\partial}{\partial z} \\ -z^3 & 3 & 2y \end{vmatrix}$$

$$= 2\mathbf{i} + 3z^2\mathbf{j}$$

33. $\text{div } \mathbf{F} = \nabla \cdot \mathbf{F}$

$$= \left(\frac{\partial}{\partial x}\mathbf{i} + \frac{\partial}{\partial y}\mathbf{j} + \frac{\partial}{\partial z}\mathbf{k}\right) \cdot (x - y)\mathbf{i} + (y - z)\mathbf{j} + (z - x)\mathbf{k}$$

$$= 3$$

$$\text{curl } \mathbf{F} = \begin{vmatrix} \mathbf{i} & \mathbf{j} & \mathbf{k} \\ \frac{\partial}{\partial x} & \frac{\partial}{\partial y} & \frac{\partial}{\partial z} \\ x - y & y - z & z - x \end{vmatrix}$$

$$= \mathbf{i} + \mathbf{j} + \mathbf{k}$$

34. $\text{div } \mathbf{F} = \nabla \cdot \mathbf{F}$

$$= \left(\frac{\partial}{\partial x}\mathbf{i} + \frac{\partial}{\partial y}\mathbf{j} + \frac{\partial}{\partial z}\mathbf{k}\right) \cdot \left(\frac{x\mathbf{i} + y\mathbf{j} + z\mathbf{k}}{\sqrt{x^2 + y^2 + z^2}}\right)$$

$$= \frac{\sqrt{x^2 + y^2 + z^2} - \dfrac{x^2}{\sqrt{x^2 + y^2 + z^2}}}{x^2 + y^2 + z^2}$$

$$+ \frac{\sqrt{x^2 + y^2 + z^2} - \dfrac{y^2}{\sqrt{x^2 + y^2 + z^2}}}{x^2 + y^2 + z^2}$$

$$+ \frac{\sqrt{x^2 + y^2 + z^2} - \dfrac{z^2}{\sqrt{x^2 + y^2 + z^2}}}{x^2 + y^2 + z^2}$$

$$= \frac{2(x^2 + y^2 + z^2)}{(x^2 + y^2 + z^2)^{3/2}} = \frac{2}{\sqrt{x^2 + y^2 + z^2}}$$

$$\text{curl } \mathbf{F} = \nabla \times \mathbf{F}$$

$$= \begin{vmatrix} \mathbf{i} & \mathbf{j} & \mathbf{k} \\ \frac{\partial}{\partial x} & \frac{\partial}{\partial y} & \frac{\partial}{\partial z} \\ \dfrac{x}{\sqrt{x^2+y^2+z^2}} & \dfrac{y}{\sqrt{x^2+y^2+z^2}} & \dfrac{z}{\sqrt{x^2+y^2+z^2}} \end{vmatrix}$$

$$= \frac{-2yz + 2yz}{2(x^2 + y^2)^{3/2}}\mathbf{i} + \frac{-2xz + 2xz}{2(x^2 + y^2)^{3/2}}\mathbf{j} + \frac{-2xy + 2xy}{2(x^2 + y^2)^{3/2}}\mathbf{k}$$

$$= \mathbf{0}$$

35. $u_x = -e^{-x}(\cos y - \sin y)$

$\quad u_{xx} = e^{-x}(\cos y - \sin y)$

$\quad u_y = e^{-x}(-\sin y - \cos y)$

$\quad u_{yy} = e^{-x}(-\cos y + \sin y)$

$\quad u_{xx} + u_{yy} = 0; \quad u$ is harmonic

36. $v_x = \dfrac{x}{\sqrt{x^2 + y^2 + z^2}}$

$$v_{xx} = \dfrac{\sqrt{x^2 + y^2 + z^2} - \dfrac{x}{\sqrt{x^2 + y^2 + z^2}}}{\sqrt{x^2 + y^2 + z^2}}$$

$$= \dfrac{y^2 + z^2}{(x^2 + y^2 + z^2)^{3/2}}$$

Similarly,

$$v_{yy} = \dfrac{x^2 + z^2}{(x^2 + y^2 + z^2)^{3/2}}; \quad v_{zz} = \dfrac{x^2 + y^2}{(x^2 + y^2 + z^2)^{3/2}}$$

$v_{xx} + v_{yy} + v_{zz} \neq 0$, so the function is not harmonic.

37. $v_x = \dfrac{-x}{(x^2 + y^2 + z^2)^{3/2}}$

$$v_{xx} = \dfrac{(x^2 + y^2 + z^2)^{3/2} - 3x^2(x^2 + y^2 + z^2)^{3/2}}{(x^2 + y^2 + z^2)^3}$$

$$= \dfrac{2x^2 - y^2 - z^2}{(x^2 + y^2 + z^2)^{5/2}}$$

Similarly,

$$v_{yy} = \dfrac{2y^2 - x^2 - y^2}{(x^2 + y^2 + z^2)^{5/2}}; \quad v_{zz} = \dfrac{2z^2 - x^2 - y^2}{(x^2 + y^2 + z^2)^{5/2}}$$

$v_{xx} + v_{yy} + v_{zz} = 0$, so the function is harmonic.

38. $r_x = yz; \ r_{xx} = 0; \ r_{yy} = 0; \ r_{zz} = 0$

$v_{xx} + v_{yy} + v_{zz} = 0$, so the function is harmonic.

39. $\mathbf{F} \times \mathbf{G} = \begin{vmatrix} \mathbf{i} & \mathbf{j} & \mathbf{k} \\ 2 & 2x & 3y \\ x & -y & z \end{vmatrix}$

$$= (2xz + 3y^2)\mathbf{i} + (3xy - 2z)\mathbf{j} + (-2y - 2x^2)\mathbf{k}$$

$$\text{curl}(\mathbf{F} \times \mathbf{G}) = \begin{vmatrix} \mathbf{i} & \mathbf{j} & \mathbf{k} \\ \dfrac{\partial}{\partial x} & \dfrac{\partial}{\partial y} & \dfrac{\partial}{\partial z} \\ 2xz+3y^2 & 3xy-2z & -2y-2x^2 \end{vmatrix}$$

$$= (-2 + 2)\mathbf{i} - (-2x - 4x)\mathbf{j} + (3y - 6y)\mathbf{k}$$

$$= 6x\mathbf{j} - 3y\mathbf{k}$$

40. $\mathbf{F} \times \mathbf{G} = \begin{vmatrix} \mathbf{i} & \mathbf{j} & \mathbf{k} \\ xy & yz & z^2 \\ x & y & -z \end{vmatrix}$

$$= (-2yz^2)\mathbf{i} + (xyz + xz^2)\mathbf{j} + (xy^2 - xyz)\mathbf{k}$$

$$\text{curl}(\mathbf{F} \times \mathbf{G}) = \begin{vmatrix} \mathbf{i} & \mathbf{j} & \mathbf{k} \\ \dfrac{\partial}{\partial x} & \dfrac{\partial}{\partial y} & \dfrac{\partial}{\partial z} \\ -2yz^2 & xyz+xz^2 & xy^2 - xyz \end{vmatrix}$$

$$= (2xy - xz - xy)\mathbf{i} - (y^2 - yz + 4yz)\mathbf{j} + (yz + z^2 + 2z^2)\mathbf{k}$$

$$= (xy - xz)\mathbf{j} - (y^2 + 3yz)\mathbf{j} + (yz + 3z^2)\mathbf{k}$$

41. From Problem 39, $\mathbf{F} \times \mathbf{G}$

$$= (2xz + 3y^2)\mathbf{i} + (3xy - 2z)\mathbf{j} + (-2y - 2x^2)\mathbf{k}$$

$$\text{div}(\mathbf{F} \times \mathbf{G})$$

$$= \dfrac{\partial(2xz+3y^2)}{\partial x} + \dfrac{\partial(3xy-2z)}{\partial y} + \dfrac{\partial(-2y-2x^2)}{\partial z}$$

$$= 2z + 3x$$

42. From Problem 40, $\mathbf{F} \times \mathbf{G}$

$$= (-2yz^2)\mathbf{i} + (xyz + xz^2)\mathbf{j} + (xy^2 - xyz)\mathbf{k}$$

$$\text{div}(\mathbf{F} \times \mathbf{G}) =$$

$$= \dfrac{\partial(-2yz^2)}{\partial x} + \dfrac{\partial(xyz + xz^2)}{\partial y} + \dfrac{\partial(xy^2 - xyz)}{\partial z}$$

$$= xz - xy$$

43. $\mathbf{F} = \nabla f = y^3 z^2 \mathbf{i} + 3xy^2 z^2 \mathbf{j} + 2xy^3 z \mathbf{k}$

$\text{div } \mathbf{F} = 6xyz^2 + 2xy^3$

44. $\mathbf{F} = \nabla f = 2x^2 yz^3 \mathbf{i} ++ x^2 z^3 \mathbf{j} + 3x^2 yz^2 \mathbf{k}$

$\text{div } \mathbf{F} = 2yz^3 + 6x^2 yz$

45. $\text{div } \mathbf{B} = \dfrac{\partial(y^3 z^2)}{\partial x} + \dfrac{\partial y^2 z}{\partial y} + \dfrac{\partial xz^3}{\partial y^2 z^2} = 0$

Thus, $\mathbf{B}$ is solenoidal.

46. $\text{curl } \mathbf{F} = \begin{vmatrix} \mathbf{i} & \mathbf{j} & \mathbf{k} \\ \dfrac{\partial}{\partial x} & \dfrac{\partial}{\partial y} & \dfrac{\partial}{\partial z} \\ f(r)x & f(r)y & f(r)z \end{vmatrix}$

$$= \left[\dfrac{\partial(zf(r))}{\partial x} - \dfrac{\partial(yf(r))}{\partial z} \right]\mathbf{i}$$

$$- \left[\dfrac{\partial(zf(r))}{\partial x} - \dfrac{\partial(xf(r))}{\partial z} \right]\mathbf{j}$$

$$+ \left[\dfrac{\partial(yf(r))}{\partial x} - \dfrac{\partial(xf(r))}{\partial y} \right]\mathbf{i}$$

$$= \left[zf'(r)\dfrac{\partial\sqrt{x^2+y^2+z^2}}{\partial y} - yf'(r)\dfrac{\partial\sqrt{x^2+y^2+z^2}}{\partial z} \right]\mathbf{i}$$

$$+ \left[zf'(r)\dfrac{\partial\sqrt{x^2+y^2+z^2}}{\partial x} - xf'(r)\dfrac{\partial\sqrt{x^2+y^2+z^2}}{\partial z} \right]\mathbf{j}$$

$$+ \left[yf'(r)\frac{\partial\sqrt{x^2+y^2+z^2}}{\partial x} - xf'(r)\frac{\partial\sqrt{x^2+y^2+z^2}}{\partial y} \right]\mathbf{k}$$

$$= [f'(r)]\left[\frac{yz - yz}{r}\mathbf{i} - \frac{xz - xz}{r}\mathbf{j} + \frac{xy - xy}{r}\mathbf{k}\right] = 0$$

47. Let $\mathbf{A} = a\mathbf{i} + b\mathbf{j} + c\mathbf{k}$

$$\mathbf{A}\times\mathbf{B} = \begin{vmatrix} \mathbf{i} & \mathbf{j} & \mathbf{k} \\ a & b & c \\ x & y & z \end{vmatrix}$$

$$= (bz - cy)\mathbf{i} - (az - cx)\mathbf{j} + (ay - bx)\mathbf{k}$$

$$\text{div}(\mathbf{A}\times\mathbf{B}) = \frac{\partial}{\partial x}(bz - cy) - \frac{\partial}{\partial y}(az - cx)$$
$$+ \frac{\partial}{\partial z}(ay - bx) = 0$$

48. For $\mathbf{A}\times\mathbf{R}$ see Problem 47;

$$\text{curl } \mathbf{A}\times\mathbf{R} = \begin{vmatrix} \mathbf{i} & \mathbf{j} & \mathbf{k} \\ \dfrac{\partial}{\partial x} & \dfrac{\partial}{\partial y} & \dfrac{\partial}{\partial z} \\ bz - cy & cx - az & ay - bx \end{vmatrix}$$

$$= 2a\mathbf{i} + 2b\mathbf{j} + 2c\mathbf{k} = 2\mathbf{A}$$

49. $\text{curl } \mathbf{F} = \begin{vmatrix} \mathbf{i} & \mathbf{j} & \mathbf{k} \\ \dfrac{\partial}{\partial x} & \dfrac{\partial}{\partial y} & \dfrac{\partial}{\partial z} \\ u(x,y) & v(x,y) & 0 \end{vmatrix}$

$$= -\frac{\partial}{\partial z}v(x,y)\mathbf{i} + \frac{\partial}{\partial z}u(x,y)\mathbf{j}$$
$$+ \left(\frac{\partial}{\partial x}v(x,y) - \frac{\partial}{\partial y}u(x,y)\right)\mathbf{k}$$

$$= \mathbf{0} \text{ if and only if the coefficient of } \mathbf{k} = 0$$

50. $\mathbf{F} = -\dfrac{x\mathbf{i} + y\mathbf{j} + z\mathbf{k}}{\sqrt{x^2 + y^2 + z^2}}$; then $\text{curl } \mathbf{F} = 0$

(see Problem 34)

51. If the angular velocity is ω, let the speed be ω_0, then

$$\omega = \omega_0\left(\frac{dx}{dt}\mathbf{i} + \frac{dy}{dt}\mathbf{j} + 0\mathbf{k}\right)$$

a. $\mathbf{V} = \omega\times\mathbf{R} = \begin{vmatrix} \mathbf{i} & \mathbf{j} & \mathbf{k} \\ 0 & 0 & \omega_0 \\ x & y & z \end{vmatrix}$

$$= -\omega_0 y\mathbf{i} + \omega_0 x\mathbf{j}$$

b. $\text{div } \mathbf{V} = z - z + 0 = 0$

$$\text{curl } \mathbf{V} = \begin{vmatrix} \mathbf{i} & \mathbf{j} & \mathbf{k} \\ \dfrac{\partial}{\partial x} & \dfrac{\partial}{\partial y} & \dfrac{\partial}{\partial z} \\ -\omega_0 y & \omega_0 x & 0 \end{vmatrix} = 2\omega_0\mathbf{k} = 2\omega$$

Since $\text{curl } \mathbf{V} = 2\omega$, it follows that $\text{curl } \mathbf{V}$ can be used to measure the rotational effect on the fluid flow with velocity $\mathbf{V}$.

52. Let $\mathbf{F} = M\mathbf{i} + N\mathbf{j} + P\mathbf{k}$

$$\text{curl } \mathbf{F} = \begin{vmatrix} \mathbf{i} & \mathbf{j} & \mathbf{k} \\ \dfrac{\partial}{\partial x} & \dfrac{\partial}{\partial y} & \dfrac{\partial}{\partial z} \\ M & N & P \end{vmatrix}$$

$$\text{curl(curl } \mathbf{F}) = \begin{vmatrix} \mathbf{i} & \mathbf{j} & \mathbf{k} \\ \dfrac{\partial}{\partial x} & \dfrac{\partial}{\partial y} & \dfrac{\partial}{\partial z} \\ P_y - N_z & M_z - P_x & N_x - M_y \end{vmatrix}$$

$$= (N_{xy} - M_{yy} - M_{zz} + P_{xz})\mathbf{i}$$
$$+ (P_{yz} - N_{zz} - N_{xx} + m_{yx})\mathbf{j}$$
$$+ (M_{zx} - P_{xx} - P_{yy} + N_{zy})\mathbf{k}$$

53. Let $\mathbf{F} = f_1\mathbf{i} + f_2\mathbf{j} + f_3\mathbf{k}$ and

$$\mathbf{G} = g_1\mathbf{i} + g_2\mathbf{j} + g_3\mathbf{k}$$

$$\mathbf{F}\times\mathbf{G} = \begin{vmatrix} \mathbf{i} & \mathbf{j} & \mathbf{k} \\ f_1 & f_2 & f_3 \\ g_1 & g_2 & g_3 \end{vmatrix}$$

$$= (f_2 g_3 - f_3 g_2)\mathbf{i} + (f_3 g_1 - f_1 g_3)\mathbf{j}$$
$$+ (f_1 g_2 - f_2 g_1)\mathbf{k}$$

$$\text{div}(\mathbf{F}\times\mathbf{G}) = f_2(g_3)_x + (f_2)_x g_3 - f_3(g_2)_x$$
$$- (f_3)_x g_2 - f_1(g_3)_y - (f_1)_y g_3$$
$$+ f_3(g_1)_y + (f_3)_y g_1 + f_1(g_2)_z$$
$$+ (f_1)_z g_2 - f_2(g_1)_z - (f_2)_z g_1$$

Only II is the same.

54. Let $\mathbf{F} = M\mathbf{i} + N\mathbf{j} + P\mathbf{k}$

$$\text{div}(c\mathbf{F}) = (cM)_x + (cN)_y + (cP)_z$$
$$= c(M_x + N_y + P_z)$$
$$= c\,\text{div } \mathbf{F}$$

55. $\text{div } \mathbf{F} + \text{div } \mathbf{G}$

$$= [(f_1)_x + (f_2)_y + (f_3)_z] + [(g_1)_x + (g_2)_y + (g_3)_z]$$
$$= [(f_1)_x + (g_1)_x] + [(f_2)_x + (g_2)_y] + [(f_3)_x + (g_3)_y]$$
$$= \text{div}(\mathbf{F} + \mathbf{G})$$

56. $\text{curl}(\mathbf{F} + \mathbf{G}) = \begin{vmatrix} \mathbf{i} & \mathbf{j} & \mathbf{k} \\ \dfrac{\partial}{\partial x} & \dfrac{\partial}{\partial y} & \dfrac{\partial}{\partial z} \\ f_1 + g_1 & f_2 + g_2 & f_3 + g_3 \end{vmatrix}$

$$= \begin{vmatrix} \mathbf{i} & \mathbf{j} & \mathbf{k} \\ \dfrac{\partial}{\partial x} & \dfrac{\partial}{\partial y} & \dfrac{\partial}{\partial z} \\ f_1 & f_2 & f_3 \end{vmatrix} + \begin{vmatrix} \mathbf{i} & \mathbf{j} & \mathbf{k} \\ \dfrac{\partial}{\partial x} & \dfrac{\partial}{\partial y} & \dfrac{\partial}{\partial z} \\ g_1 & g_2 & g_3 \end{vmatrix}$$

$$= \text{curl } \mathbf{F} + \text{curl } \mathbf{G}$$

57. $\text{curl}(c\mathbf{F}) = \begin{vmatrix} \mathbf{i} & \mathbf{j} & \mathbf{k} \\ \dfrac{\partial}{\partial x} & \dfrac{\partial}{\partial y} & \dfrac{\partial}{\partial z} \\ cf_1 & cf_2 & cf_3 \end{vmatrix} = c\begin{vmatrix} \mathbf{i} & \mathbf{j} & \mathbf{k} \\ \dfrac{\partial}{\partial x} & \dfrac{\partial}{\partial y} & \dfrac{\partial}{\partial z} \\ f_1 & f_2 & f_3 \end{vmatrix}$

58. See Problem 53, part II.

59. Let $\mathbf{F} = M\mathbf{i} + N\mathbf{j} + P\mathbf{k}$

$$\text{curl}(f\mathbf{F}) = \begin{vmatrix} \mathbf{i} & \mathbf{j} & \mathbf{k} \\ \dfrac{\partial}{\partial x} & \dfrac{\partial}{\partial y} & \dfrac{\partial}{\partial z} \\ fM & fN & fP \end{vmatrix}$$

$$= f\begin{vmatrix} \mathbf{i} & \mathbf{j} & \mathbf{k} \\ \dfrac{\partial}{\partial x} & \dfrac{\partial}{\partial y} & \dfrac{\partial}{\partial z} \\ M & N & P \end{vmatrix}$$

$$= (Pf_x - Nf_z)\mathbf{i} - (Pf_x - Mf_z)\mathbf{j}$$
$$\quad + (Nf_x - M_y)\mathbf{k}$$
$$= -(Pf_x - Mf_z)\mathbf{j} + (Nf_x - Mf_y)\mathbf{k}$$
$$= f\,\text{curl}\,\mathbf{F} + \nabla f \times \mathbf{F}$$

60. Let $\mathbf{F} = M\mathbf{i} + N\mathbf{j} + P\mathbf{k}$

$$\text{div}(f\mathbf{F}) = (fM)_x + (fN)_y + (fP)_x$$
$$= f(M_x + N_y + P_z) + Mf_x + Nf_y + Pf_z$$
$$= f\,\text{div}\,\mathbf{F} + [f_x\mathbf{i} + f_y\mathbf{j} + f_z\mathbf{k}] \cdot [M\mathbf{i} + N\mathbf{j} + P\mathbf{k}]$$
$$= f\,\text{div}\,\mathbf{F} + \nabla f \cdot \mathbf{F}$$

61. Let $\mathbf{F} = M\mathbf{i} + N\mathbf{j} + P\mathbf{k}$

$$\text{curl}(\nabla f + \text{curl}\,\mathbf{F})$$
$$= (f_x + P_y - N_z)\mathbf{i} + (f_y + M_z - P_x)\mathbf{j}$$
$$\quad + (f_z + N_x - M_y)\mathbf{k}$$
$$= \text{curl}(f_x\mathbf{i} + f_y\mathbf{j} + f_z\mathbf{k})$$
$$\quad + \text{curl}[(P_y - N_z)\mathbf{i} + (M_z - P_x)\mathbf{j} + (N_x - M_y)\mathbf{k}]$$
$$= \text{curl}(\nabla f) + \text{curl}(\text{curl}\,\mathbf{F})$$

62. Follows immediately from the definition of divergence.

63. $$\nabla \times \nabla f = \begin{vmatrix} \mathbf{i} & \mathbf{j} & \mathbf{k} \\ \dfrac{\partial}{\partial x} & \dfrac{\partial}{\partial y} & \dfrac{\partial}{\partial z} \\ f_x & f_y & f_z \end{vmatrix}$$

$$= (f_y f_z - f_z f_y)\mathbf{i} + (f_z f_x - f_x f_z)\mathbf{j} + (f_x f_y - f_y f_x)\mathbf{k}$$

64. Let $\mathbf{F} = M\mathbf{i} + N\mathbf{j} + P\mathbf{k}$

$$\text{div}(\text{curl}\,\mathbf{F}) = \text{div}\begin{vmatrix} \mathbf{i} & \mathbf{j} & \mathbf{k} \\ \dfrac{\partial}{\partial x} & \dfrac{\partial}{\partial y} & \dfrac{\partial}{\partial z} \\ M & N & P \end{vmatrix}$$

$$= \text{div}[(P_y - N_z)_x\mathbf{i} + (M_y - P_x)_y\mathbf{j} + (N_x - M_y)_z\mathbf{k}]$$
$$= P_{xy} - N_{xz} + M_{yz} - P_{xy} + N_{xz} - M_{yz}$$
$$= 0$$

14.2 Line Integrals, Page 914

1. A line (actually a curve) integral is the summation (integration) of the products of a function evaluated at a point on the curve times an infinitesimal element of arc length.

2. A line (actually a curve) integral is a single (not double or triple) integral involving at least two variables. These variables must be made dependent, so that we can (at least theoretically) integrate by using just one variable.

3. Let $x = t$, $y = 4t^2$ on $0 \le t \le 1$

$$(-y\,dx + x\,dy) = -4t^2\,dt + t(8t)dt = 4t^2\,dt$$
$$\int_C (-y\,dx + x\,dy) = 4\int_0^1 t^2\,dt = \frac{4}{3}$$

4. Let $x = t^2$ and $y = t$ on $1 \le t \le 3$

$$(-y\,dx + 3x\,dy) = -t(2t\,dt) + 3t^2(dt) = t^2\,dt$$
$$\int_C (-y\,dx + 3x\,dy) = \int_1^3 t^2\,dt = \frac{26}{3}$$

5. Let $x = t$ and $y = \frac{1}{4}(2t - 1)$ on $4 \le t \le 8$

$$(x\,dy - y\,dx) = \frac{1}{2}t\,dt - \frac{1}{4}(2t - 1)\,dt = \frac{1}{4}$$
$$\int_C (x\,dy - y\,dx) \int_4^8 \frac{1}{4}\,dt = 1$$

6. Let $x = t^3$ and $y = t^2$ on $-1 \le t \le 1$

$$[(y - x)\,dx + x^2 y\,dy]$$
$$= (t^3 - t^2)(2t\,dt) + (t^4)(t^3)(3t^2\,dt)$$
$$= (2t^4 - 2t^3 + 2t^9)\,dt$$
$$\int_C [(y - x)\,dx + x^2 y\,dy]$$
$$= \int_{-1}^1 (2t^4 - 2t^3 + 2t^9)\,dt = \frac{4}{5}$$

7. C needs to be considered as two regions: let $x = t$ then $y = -2t$ on $-1 \le t \le 0$ and $y = 2t$ on $0 \le t \le 1$

On $[-1, 0]$: $[(x + y)^2\,dx - (x - y)^2\,dy]$
$$= (-t)^2\,dt - (3t)^2(-2\,dt) = 19t^2\,dt$$

On $[0, 1]$: $[(x + y)^2\,dx - (x - y)^2\,dy]$
$$= (3t)^2\,dt - (-t)^2(2\,dt) = 7t^2\,dt$$

$$\int_C [(x + y)^2\,dx - (x - y)^2\,dy]$$

$$= \int_{-1}^{0} 19t^2 \, dt + \int_{0}^{1} 7t^2 \, dt = \frac{19}{3} + \frac{7}{3} = \frac{26}{3}$$

8. Let $x = 2 \sin \theta$ and $y = 2 \cos \theta$ on $0 \leq \theta \leq \frac{\pi}{2}$

$(y^2 - x^2) \, dx - x \, dy$

$$= (4 \cos^2\theta - 4 \sin^2\theta)(2 \cos \theta \, d\theta)$$
$$- (2 \sin \theta)(-3 \sin \theta \, d\theta)$$
$$= (8 \cos^3\theta - 8 \sin^2\theta \cos \theta + 4 \sin^2\theta) \, d\theta$$

$$\int_C [(y^2 - x^2) \, dx - x \, dy]$$
$$= \int_{0}^{\pi/2} [8 \cos^3\theta - 8 \sin^2\theta \cos \theta + 4 \sin^2\theta] \, d\theta$$
$$= \frac{9\pi + 40}{8}$$

9. **a.** Let $x = \cos \theta$ and $y = \sin \theta$ on $0 \leq \theta \leq \frac{\pi}{2}$

$(x^2 + y^2) \, dx + 2xy \, dy$

$$= (1)(-\sin \theta \, d\theta) + 2 \sin \theta \cos^2\theta \, d\theta$$

$$\int_C [(x^2 + y^2) \, dx + 2xy \, dy]$$
$$= \int_{0}^{\pi/2} (-\sin \theta + 2 \sin \theta \cos^2\theta) \, d\theta$$
$$= -\frac{1}{3}$$

b. Let $x = 1 - t$ and $y = t$ on $0 \leq t \leq 1$

$(x^2 + y^2) \, dx + 2xy \, dy$

$$= -(1-t)^2 \, dt - t^2 \, dt + 2t(1-t) \, dt$$
$$= (-4t^2 + 4t - 1) \, dt$$

$$\int_C [(x^2 + y^2) \, dx + 2xy \, dy]$$
$$= \int_{0}^{2} (-4t^2 + 4t - 1) \, dt = -\frac{1}{3}$$

10. **a.** Let $x = t$ and $y = t^2$ on $0 \leq t \leq 2$

$x^2 y \, dx + (x^2 - y^2) \, dy$

$$= t^2(t^2 \, dt) + (t^2 - t^4)(2t \, dt)$$
$$= (t^4 + 2t^3 - 2t^5) \, dt$$

$$\int_C [x^2 y \, dx + (x^2 - y^2) \, dy]$$
$$= \int_{0}^{2} (t^4 + 2t^3 - 2t^5) \, dt = -\frac{104}{15}$$

b. Let $x = t$ and $y = 2t$ on $0 \leq x \leq 2$

$x^2 y \, dx + (x^2 - y^2) \, dy$

$$= t^2(2t \, dt) + (t^2 - 4t^2)(2 \, dt)$$

$$= (2t^3 - 6t^2) \, dt$$

$$\int_C [x^2 y \, dx + (x^2 - y^2) \, dy]$$
$$= \int_{0}^{2} (2t^3 - 6t^2) \, dt = -8$$

11. Since this is not a smooth curve use two regions.

On $[0, 2]$: $x = t$, $y = 0$

$$[x^2 y dx + (x^2 - y^2) dy] = 0$$

On $[2, 4]$ $x = 2$. $y = t$

$$[x^2 y dx + (x^2 - y^2) dy] = (4 - t^2) dt$$

$$\int_C [x^2 y dx + (x^2 - y^2) dy]$$
$$= \int_{0}^{4} (4 - t^2) dt = -\frac{16}{3}$$

12. We must use two regions.

On $[-1, 0]$: $x = t$, $y = 1$

$$(-xy^2 \, dx + x^2 \, dy) = -t \, dt$$

On $[0, \frac{\pi}{2}]$: $x = \sin \theta$ and $y = \cos \theta$

$(-xy^2 \, dx + x^2 \, dy)$

$$= -\sin \theta \cos^2\theta (\cos \theta \, d\theta) + \sin^2\theta(-\sin \theta \, d\theta)$$
$$= (-\sin \theta \cos^3\theta - \sin^3\theta) \, d\theta$$

$$\int_C (-xy^2 \, dx + x^2 \, dy)$$
$$= \int_{-1}^{0} (-t) \, dt + \int_{0}^{\pi/2} (-\sin \theta \cos^3\theta - \sin^3\theta) \, d\theta$$
$$= \frac{1}{2} + \left(-\frac{11}{12}\right) = -\frac{5}{12}$$

13. We must use two regions.

On $[-1, 0]$: $x = t$, $y = 2$

$$-y^2 \, dx + x^2 \, dy = -4 \, dt$$

On $[-1, 0]$: $x = t$ and $y = 2(t - 1)^2$

$$-y^2 \, dx + x^2 \, dy = -4(t-1)^4 \, dt + 4t^2(t-1) dt$$

$$\int_C (-y^2 \, dx + x^2 \, dy)$$
$$= \int_{0}^{-1} -4 \, dt + \int_{1}^{0} [-4(t-1)^4 + (4t^3 - 4t^2)] \, dt$$
$$= 4 + \frac{4}{5} + \frac{4}{12} = \frac{77}{15}$$

14. Let $x = 2\cos\theta$ and $y = 2\sin\theta$ on $[0, 2\pi]$

$$(x^2 - y^2)dx + x\,dy$$

$$= (4\cos^2\theta - 4\sin^2\theta)(-2\sin\theta\,d\theta)$$
$$+ 2\cos\theta(2\cos\theta\,d\theta)$$

$$= [8(2\cos^2\theta - 1)(-\sin\theta) + 2(1 + \cos 2\theta)]\,d\theta$$

$$\int_C [(x^2 - y^2)dx + x\,dy]$$

$$= \int_0^{2\pi} [8(2\cos^2\theta - 1)(-\sin\theta)\,d\theta$$

$$+ \int_0^{2\pi} 2(1 + \cos 2\theta)]\,d\theta = 4\pi$$

15. For the first piece let $0 \le t \le 1$, $x = t$, $y = t^2$

$$x^2 y\,dx - xy\,dy = t^4 dt - 2t^4 dt = -t^4 dt$$

On the second piece let $0 \le t \le 1$, $x = 1 - t$, $y = 1 - t$

$$x^2 y\,dx - xy\,dy = -(1 - t)^3 dt + (1 - t)^2 dt$$
$$= (t^3 - 2t^2 + t)\,dt$$

$$\int_C (x^2 y\,dx - xy\,dy)$$

$$= \int_0^1 (-t^4)\,dt + \int_1^0 (t^3 - 2t^2 + t)\,dt = -\frac{7}{60}$$

16. Let $x = a\cos\theta$ and $y = b\sin\theta$ on $0 \le \theta \le 2\pi$

$$\frac{x\,dx - y\,dy}{\sqrt{x^2 + y^2}} = \frac{-a^2\cos\theta\sin\theta - a^2\sin\theta\cos\theta}{a}\,d\theta$$

$$= -2a\sin\theta\cos\theta\,d\theta$$

$$\int_C \frac{x\,dx - y\,dy}{\sqrt{x^2 + y^2}}$$

$$= -2a\int_0^{2\pi}\cos\theta\sin\theta\,d\theta = -\frac{a}{2}$$

17. Let $x = 2t$ and $y = t$ on $0 \le t \le 1$

$$\mathbf{F} \cdot d\mathbf{R} = [(10t + t)\mathbf{i} + 2t\mathbf{j}] \cdot (2\mathbf{i} + \mathbf{j}) = 24t$$

$$\int_0^1 24t\,dt = 12$$

18. Let $x = 2t$ and $y = t$ on $0 \le t \le 1$

$$\mathbf{F} \cdot d\mathbf{R} = [(10t + t)\mathbf{i} + 2t\mathbf{j}] \cdot (2\mathbf{i} + \mathbf{j}) = 24t$$

$$\int_0^1 24t\,dt = 12$$

19. Since C is not smooth use two regions.

On $(0, 0)$ to $(0, 1)$: $\mathbf{R} = t\mathbf{j}$, for $0 \le t \le 1$

$d\mathbf{R} = \mathbf{j}$, $\mathbf{F} \cdot d\mathbf{R} = x = 0$

On $(0, 1)$ to $(2, 1)$: $\mathbf{R} = t\mathbf{i} + \mathbf{j}$, for $0 \le t \le 2$

$d\mathbf{R} = \mathbf{i}$, $\mathbf{F} \cdot d\mathbf{R} = 5x + y = 5t + 1$

$$\int_C \mathbf{F} \cdot d\mathbf{R} = \int_0^1 0\,dx + \int_0^2 (5t + 1)\,dt = 12$$

20. a. $$\int_{C} (y\,dx - x\,dy + dz)$$

$$= \int_0^{\pi/3} [3\cos t(3\cos t) - 3\sin t(-3\sin t) + 1]\,dt$$

$$= 5\pi$$

b. $$\int_C (y\,dx - x\,dy + dz)$$

$$= [a\cos t(a\cos t) - a\sin t(-a\sin t) + 1]\,dt$$

$$= \frac{\pi}{2}(a^2 + 1)$$

21. a. $$\int_{C} (x\,dx + y\,dy + z\,dz)$$

$$= \int_0^{\pi/2} [-\cos t(\sin t) + \sin t(\cos t) + t]\,dt = \frac{\pi^2}{8}$$

b. $$\int_C (x\,dx + y\,dy + z\,dz)$$

$$= \int_0^1 \left[(1 - t)(-1) + t + \left(\frac{\pi}{2}\right)^2(t)\right]\,dt = \frac{\pi^2}{8}$$

22. a. $$\int_C (-y\,dx + x\,dy + xz\,dz)$$

$$= \int_0^{2\pi} [-\sin t(-\sin t) + \cos^2 t + t\cos t]\,dt$$

$$= 2\pi$$

b. $$\int_C (-y\,dx + x\,dy + xz\,dz)$$

$$= \int_0^{2\pi} [-\sin t(-\sin t) + \cos^2 t + 0]\,dt = 2\pi$$

23. a. For the arc: $x = t^2$, $y = t$, $z = 0$, $0 \le t \le 1$

$$5xy\,dx + 10yz\,dy + z\,dz = 10t^4\,dt$$

For the line: $x = 1$, $y = 1$, $z = t$, $0 \le t \le 1$

$$5xy\,dx + 10yz\,dy + z\,dz = t\,dt$$

$$\int_C (5xy\,dx + 10yz\,dy + z\,dz)$$

$$= \int_0^1 (10t^4 + t)\, dt = \frac{5}{2}$$

b. $x = y = z = t$ for $0 \le t \le 1$

$$\int_C (5xy\, dx + 10yz\, dy + z\, dz)$$

$$= \int_0^1 (5t^2\, dt + 10t^2\, dt + t\, dt)$$

$$= \int_0^1 (15t^2 + t)\, dt = \frac{11}{2}$$

24. C_1: $x + y = 1$ or $y = t$, $x = 1 - t$; $0 \le t \le 1$
C_2: $-x + y = 1$ or $x = -t$, $y = 1 - t$;
$\qquad 0 \le t \le 1$
C_3: $x + y = 1$ or $x = t$, $y = t - 1$; $0 \le t \le 1$
C_4: $x - y = 1$ or $x = t$, $y = t - 1$; $0 \le t \le 1$

$$\int_C \frac{dx + dy}{|x| + |y|} = \int_0^1 \frac{-1 + 1}{1 - t + t}\, dt + \int_0^1 \frac{-1 - 1}{t + 1 - t}\, dt$$

$$+ \int_0^1 \frac{1 - 1}{t - 1 - t}\, dt + \int_0^1 \frac{1 + 1}{t - (t - 1)}\, dt$$

$$= 0 - 2\int_0^1 dt + 0 + 2\int_0^1 dt = 0$$

25. Let $x = \cos t$, $y = \sin t$, $z = 2$; $0 \le t \le 2\pi$

$$\int_C [(y + z)\, dx + (x + z)\, dy + (x + y)\, dz]$$

$$= \int_0^{2\pi} (-\sin^2 t - 2\sin t + \cos^2 t + 2\cos t + 0)\, dt$$

$$= \int_0^{2\pi} (\cos 2t - 2\sin t + 2\cos t)\, dt = 0$$

26. Let $x = \sin t$, $y = -\cos t$; $0 \le t \le \pi$

$\mathbf{F} = (-\cos t - 3)\mathbf{i} + \sin t\mathbf{j}$

$d\mathbf{R} = \cos t\mathbf{i} + \sin t\mathbf{j}$

$$\int_C \mathbf{F} \cdot d\mathbf{R} = \int_0^\pi (-\cos t - 3\cos t + \sin^2 t)\, dt = 0$$

27. Let $x = t$, $y = t^2$, $z = -1$; $1 \le t \le 2$

$\mathbf{F} = (t^2 + 2)\mathbf{i} + t\mathbf{j} - 2t^3\mathbf{k}$; $d\mathbf{R} = \mathbf{i} + 2t\mathbf{j}$

$$\int_C \mathbf{F} \cdot d\mathbf{R} = \int_1^2 [(t^2 + 2) + 2t^2 + 0]\, dt = 9$$

28. Let $x = \cos t$, $y = 1 + \frac{1}{2}\sin t$, $z = 0$; $0 \le t \le 2\pi$

$\mathbf{F} = \cos t\,\mathbf{i} + \cos t(1 + \frac{1}{2}\sin t)\mathbf{j}$;

$\mathbf{R} = \cos t\,\mathbf{i} + (1 + \frac{1}{2}\sin t)\mathbf{j}$;

$d\mathbf{R} = -\sin t\,\mathbf{i} - \frac{1}{2}\cos t\mathbf{j}$

$$\int_C \mathbf{F} \cdot d\mathbf{R} = \int_1^{2\pi} [-\sin t \cos t + \frac{1}{4}\cos^2 t(2 + \sin t)]\, dt$$

$$= \frac{1}{4}\int_0^{2\pi}\int_0^{2\pi} (1 + \cos 2t)\, dt = \frac{\pi}{2}$$

29. C_1: $x = y = t$, $z = 0$; $0 \le t \le 1$; $\mathbf{R} = t\mathbf{i} + t\mathbf{j}$

C_2: $x = 1$, $y = 1 - t$, $z = 0$; $0 \le t \le 1$

$\qquad \mathbf{R} = \mathbf{i} + (1 - t)\mathbf{j}$

C_3: $x = 1 - t$, $y = 0$, $z = 0$; $0 \le t \le 1$

$\qquad \mathbf{R} = (1 - t)\mathbf{i}$

$$\int_C \mathbf{F} \cdot d\mathbf{R} = \int_0^1 2t^2\, dt - \int_0^1 dt + \int_0^1 0\, dt = -\frac{1}{3}$$

30. $x = t$, $y = t$, $z = 1$; $0 \le t \le 1$; $\mathbf{R} = t\mathbf{i} + t\mathbf{j}$

$\mathbf{F} = -3t\mathbf{i} + 3t\mathbf{j} + 3t\mathbf{k}$; $\mathbf{T} = -\frac{1}{\sqrt{2}}(\mathbf{i} + \mathbf{j})$

$ds = \sqrt{1^2 + 1^2 + 1^2} = \sqrt{3}\, dt$

$$\int_C \mathbf{F} \cdot \mathbf{T}\, ds = \int_0^1 \left(\frac{3t}{\sqrt{2}} - \frac{3t}{\sqrt{2}}\right)\sqrt{3}\, dt = 0$$

31. $\mathbf{F} = -x\mathbf{i} + 2\mathbf{j}$, for $0 \le t \le 1$:

$(0, 0)$ to $(0, 1)$: $x = 0$, $y = t$,
$\qquad \mathbf{R} = t\mathbf{j}$, $d\mathbf{R} = \mathbf{j}$, $\mathbf{F} = 2\mathbf{k}$

$(0, 1)$ to $(2, 1)$; $x = 2t$, $y = 1$,
$\qquad \mathbf{R} = 2t\mathbf{i}$, $d\mathbf{R} = 2\mathbf{i}$, $\mathbf{F} = 2t\mathbf{i} + 2\mathbf{k}$

$(2, 1)$ to $(1, 0)$: $x = 2 - t$, $y = 1 - t$,
$\qquad \mathbf{R} = (2 - t)\mathbf{i} + (1 - t)\mathbf{j}$, $d\mathbf{R} = -\mathbf{i} - \mathbf{j}$,
$\qquad \mathbf{F} = (t - 2)\mathbf{i} + 2\mathbf{k}$

$(1, 0)$ to $(0, 0)$, $x = 1 - t$, $y = 0$,
$\qquad \mathbf{R} = (1 - t)\mathbf{i}$, $d\mathbf{R} = -\mathbf{i}$, $\mathbf{F} = (t - 1)\mathbf{i} + 2\mathbf{k}$

$$\int_C \mathbf{F} \cdot \mathbf{T}\, ds = \int_C \mathbf{F} \cdot d\mathbf{R}$$

$$= \int_0^1 [0 - 4t + (2 - t) + (1 - t)]\, dt = 0$$

32. $x = t$, $y = 2t^3$; $0 \le t \le 2$;

$\mathbf{R} = t\mathbf{i} + 2t^3\mathbf{j}$; $ds = \sqrt{1^2 + (6t^2)^2} = \sqrt{1 + 36t^4}\, dt$

$$\int_C y \, ds = \int_0^2 2t^3 \sqrt{1+36t^4} \, dt$$

$$= \frac{577^{3/2} - 1}{108} \approx 128.32$$

33. $\frac{dx}{dt} = 2 \cos t(-\sin t); \frac{dy}{dt} = 2 \cos t \sin t$

$ds = \sqrt{4 \cos^2 t \sin^2 t + 4 \cos^2 t \sin^2 t} \, dt$

$\quad = 2\sqrt{2} \sin t \cos t \, dt$

$$\int_C (x + y) \, ds = 2\sqrt{2} \int_{-\pi/4}^{0} \sin t \cos t(\cos^2 t + \sin^2 t) \, dt$$

$$= -\frac{\sqrt{2}}{2}$$

34. For $x = \cos t, y = \sin t, z = -1; 0 \le t \le 2\pi$

$\frac{dx}{dt} = -\sin t; \frac{dy}{dt} = \cos t; \frac{dz}{dt} = 0$

$\frac{ds}{dt} = 1 \, dt$

$$\int_C \frac{x^2 + xy + y^2}{z^2} \, ds$$

$$= \int_0^{2\pi} (\cos^2 t + \cos t \sin t + \sin^2 t) \, dt$$

$$= \int_0^{2\pi} (1 + \frac{\sin 2t}{2}) \, dt = 2\pi$$

35. For $0 \le t \le 2\pi$, $x = \cos t, y = \sin t$

$\frac{x \, dy - y \, dx}{x^2 + y^2} = \frac{\cos^2 t + \sin^2 t}{\cos^2 t + \sin^2 t} \, dt = 1 \, dt$

$$\int_C \frac{x \, dy - y \, dx}{x^2 + y^2} = \int_0^{2\pi} 1 \, dt = 2\pi$$

36. $x = t, y = at; \mathbf{R}(t) = t\mathbf{i} + at\mathbf{j}$

$dW = \mathbf{F} \cdot d\mathbf{R} = (a\mathbf{i} + \mathbf{j}) \cdot (\mathbf{i} + a\mathbf{j}) = 2a \, dt$

$W = 2a \int_a^0 dt = -2a^2$

37. $C_1: y = 0, \mathbf{R} = x\mathbf{i}; d\mathbf{R} = \mathbf{i} \, dx$

$C_2: x = 2, \mathbf{R} = 2\mathbf{i} + y\mathbf{j}; d\mathbf{R} = \mathbf{j} \, dy$

$C_3: y = 2, \mathbf{R} = x\mathbf{i} + 2\mathbf{j}; d\mathbf{R} = \mathbf{i} \, dx$

$C_4: x = 0, \mathbf{R} = y\mathbf{j}; d\mathbf{R} = \mathbf{j} \, dy$

$$W = \int_C \mathbf{F} \cdot d\mathbf{R}$$

$$= \int_0^2 x^2 \, dx + 4 \int_0^2 y \, dy + \int_2^0 (-4 + x^2) \, dx + 0$$

$$= \frac{8}{3} + 8 - (\frac{8}{3} - 8) = 16$$

38. $C_1: \mathbf{R} = \cos t\mathbf{i} + \sin t\mathbf{j}; 0 \le t \le \pi$

$C_2: \mathbf{R} = t\mathbf{i}; -1 \le t \le 1$

$$\int_0^{\pi} [(\cos^2 t + \sin^2 t)(-\sin t) + (\cos t + \sin t)(\cos t)] \, dt$$

$$+ \int_{-1}^{1} t^2 \, dt = -2 + \frac{\pi}{2}$$

39. For $0 \le t \le 1$, $x = 1 - t$, $y = t$;

$\mathbf{R} = (1 - t)\mathbf{i} + t\mathbf{j}, d\mathbf{R} = -\mathbf{i} + \mathbf{j}$

$\mathbf{F} = y\mathbf{i} + 2x\mathbf{j} = t\mathbf{i} + 2(1 - t)\mathbf{j}$;

$\mathbf{F} \cdot d\mathbf{R} = -t + 2(1 - t) = -3t + 2$

$$W = \int_C \mathbf{F} \cdot d\mathbf{R} = \int_0^1 (-3t + 2) \, dt = \frac{1}{2}$$

40. $$W = \int_C \mathbf{F} \cdot d\mathbf{R}$$

$$= \int_0^1 [(t^2)^2 - t^2 + 2(t^2)(t^3)(2t) - t^2(3t^2)] \, dt$$

$$= \int_0^1 (t^4 - t^2 + 4t^6 - 3t^4) \, dt = -\frac{17}{105}$$

41. $\mathbf{v} = (3 - 1)\mathbf{i} + (4 - 0)\mathbf{j} + (1 - 2)\mathbf{k}$

$\quad = 2\mathbf{i} + 4\mathbf{j} - \mathbf{k}; d\mathbf{R} = (2\mathbf{i} + 4\mathbf{j} - \mathbf{k}) \, dt$

$\mathbf{R} = (1 + 2t)\mathbf{i} + 4t\mathbf{j} + (2 - t)\mathbf{k}$

$\mathbf{F} = 2xy\mathbf{i} + (x^2 + 2)\mathbf{j} + y\mathbf{k}$

$W = \int_C \mathbf{F} \cdot d\mathbf{R}$

$$= \int_0^1 [2(1 + 2t)(4t)(2) + 4(1 + 2t)^2 + 8 + (4t)(-1)] \, dt$$

$$= \int_0^1 (48t^2 + 28t + 12) \, dt = 42$$

42. $\mathbf{R} = 2t\mathbf{i} + t\mathbf{j} + 2t\mathbf{k}; 0 \le t \le 1$

$$W = \int_C \mathbf{F} \cdot d\mathbf{R} = \int_0^1 [(2t)(2) + t + 2(4t^2 - t)] \, dt$$

$$= \int_0^1 (8t^2 - 3t) \, dt = \frac{25}{6}$$

43. $\mathbf{R} = t^2\mathbf{i} + 2t\mathbf{j} + 4t^2\mathbf{k}; d\mathbf{R} = 2t\mathbf{i} + 2\mathbf{j} + 12t^2\mathbf{k}$;

For $0 \le t \le 1$, $x = t^2; y = 2t, z = 4t^3$

$\mathbf{F} = x\mathbf{i} + y\mathbf{j} + (xz - y)\mathbf{k}$

$\quad = t^2\mathbf{i} + 2t\mathbf{j} + (4t^5 - 2t)\mathbf{k}$

$$\mathbf{F} \cdot d\mathbf{R} = 2t^3 + 4t + 48t^7 - 24t^3$$

$$W = \int_C \mathbf{F} \cdot d\mathbf{R} = \int_0^1 (48t^7 - 22t^3 + 4t)dt = \frac{5}{2}$$

44. $\mathbf{F} = (150 + 40)\mathbf{k};$

$\mathbf{R} = 40 \cos t\mathbf{i} + 40 \sin t\mathbf{j} + \frac{6t}{2\pi}\mathbf{k};\ 0 \le t \le 25(2\pi)$

$d\mathbf{R} = -40 \sin t\mathbf{i} + 40 \cos t\mathbf{j} + \frac{6}{2\pi}\mathbf{k}$

$ds = \| d\mathbf{R} \| = \sqrt{1{,}600 + \frac{9}{\pi^2}}\ dt \approx 40.0114\ dt$

$\mathbf{T} \approx -0.9997\sin t\mathbf{i} + 0.9997\cos t\mathbf{j} + 0.02387\mathbf{k}$

$\mathbf{F} \cdot \mathbf{T} = 190\mathbf{k} \cdot (-0.9997\sin t\mathbf{i} + 0.9997\cos t\mathbf{j}$
$\qquad\qquad + 0.02387\mathbf{k}) \approx 4.5353$

The work done in one revolution is

$$W = \int_0^{2\pi} \mathbf{F} \cdot d\mathbf{R} = \int_0^{2\pi} \mathbf{F} \cdot \mathbf{T}\ ds = 4.5353(40.0114\ dt)$$

$$\approx 181.4637(2\pi) \approx 1{,}140$$

For 25 revolutions,

$$W \approx 25(181.4647)(2\pi) \approx 28{,}500\ \text{ft} \cdot \text{lb}$$

45. Work only depends on what happens in the
vertical direction; $\mathbf{R} = 10t\mathbf{k},\ 0 \le t \le 10$
$\mathbf{F} = (190 - t)\mathbf{k}$

$$W = \int_0^{10} \mathbf{F} \cdot d\mathbf{R} = \int_0^{10} [(15)(190 - t)]\ dt$$

$$= 27{,}750\ \text{ft-lb}$$

46. $\mathbf{R} = \cos t\mathbf{i} + \sin t\mathbf{j};\ d\mathbf{R} = -\sin t\mathbf{i} + \cos t\mathbf{j}$

$\mathbf{T} = -\sin t\mathbf{i} + \cos t\mathbf{j};$

$\mathbf{F} = 5{,}000(-\sin t\mathbf{i} + \cos t\mathbf{j}),\ 0 \le t \le 10$

$$W = \int_0^{2\pi} \mathbf{F} \cdot d\mathbf{R} = 5{,}000 \int_0^{2\pi} dt = 10{,}000\pi\ \text{mi} \cdot \text{lb}$$

47. $\int_C \mathbf{E} \cdot d\mathbf{R} = \int_C \mathbf{E} \cdot \frac{d\mathbf{R}}{dt}\ dt$, but $\mathbf{E} \cdot \frac{d\mathbf{R}}{dt} = \frac{m}{Q}\frac{d\mathbf{V}}{dt} \cdot \mathbf{V}$

$$\mathbf{F} = m\mathbf{A} = m\frac{d\mathbf{V}}{dt}$$

$$\mathbf{F} \cdot \mathbf{V} = Q(\mathbf{E} + \mathbf{V} \times \mathbf{B}) \cdot \mathbf{V}$$

$$= Q\mathbf{E} \cdot \mathbf{V} + (\mathbf{V} \times \mathbf{B}) \cdot \mathbf{V}$$

$$= Q\mathbf{E} \cdot \mathbf{V}$$

$$= Q\mathbf{E} \cdot \frac{d\mathbf{R}}{dt}$$

So $\mathbf{F} \cdot \mathbf{V} = m\frac{d\mathbf{V}}{dt} = Q\mathbf{E} \cdot \frac{d\mathbf{R}}{dt}$

48. $dm = \rho(x,\ y,\ z)\ ds$ so

$$m = \int_C \rho(x,\ y,\ z)\ ds;\ M_{xy} = \int_C z\ \rho(x,\ y,\ z)\ ds$$

$$\bar{z} = \frac{\displaystyle\int_C z\ \rho(x,\ y,\ z)\ ds}{\displaystyle\int_C \rho(x,\ y,\ z)\ ds}$$

Similarly, $\bar{x} = \dfrac{\displaystyle\int_C x\ \rho(x,\ y,\ z)\ ds}{\displaystyle\int_C \rho(x,\ y,\ z)\ ds}$ and

$$\bar{y} = \frac{\displaystyle\int_C y\ \rho(x,\ y,\ z)\ ds}{\displaystyle\int_C \rho(x,\ y,\ z)\ ds}$$

14.3 Independence of Path, Page 922

1. $\mathbf{F}$ is conservative if $\mathbf{F} = \nabla f$ where f is
continuously differentiable. Then

$$\int_A^B \mathbf{F}\ d\mathbf{R} = f(B) - f(A)$$

2. With f continuously differentiable

$$\int_A^B \mathbf{F}\ d\mathbf{R} = f(B) - f(A)$$

3. $\mathbf{F}$ is conservative if $\mathbf{F} = \nabla f$ where f is
continuously differentiable.

$$\int_A^B \mathbf{F}\ d\mathbf{R} = f(B) - f(A);\ \int_C^A \mathbf{F}\ d\mathbf{R} = 0$$

where C is a closed path.

4. $\mathbf{E} = y^2\mathbf{i} + 2xy\mathbf{j};\ \dfrac{\partial f}{\partial x} = y^2$

$f(x,\ y) = xy^2 + c(y)$, so
$\dfrac{\partial f}{\partial y} = 2y + c'(y) = 2xy$

$c'(y) = 0$, so $c(y) = K$. If we pick $K = 0$,
then $f(x,\ y) = xy^2$. The field is conservative.

5. $\mathbf{E} = 2xy^3\mathbf{i} + 3x^2y^2\mathbf{j};\ \dfrac{\partial f}{\partial x} = 2xy^3$

$f(x,\ y) = x^2y^3 + c(y)$, so
$\dfrac{\partial f}{\partial y} = 3x^2y^2 + c'(y) = 3x^2y^2$

$c'(y) = 0$, so $c(y) = K$. If we pick $K = 0$,
then $f(x,\ y) = x^2y^3$. The field is conservative.

6. $\mathbf{E} = xe^{xy}\sin y\mathbf{i} + (e^{xy}\cos xy + y)\mathbf{j};$

$u = xe^{xy}\sin y;\ \dfrac{\partial u}{\partial y} = xe^{xy}(\cos y + y)$

$v = ye^{xy}\cos y;\ \dfrac{\partial v}{\partial x} = ye^{xy}\cos y$

$u_y \neq v_x.$ The field is not conservative.

7. $\mathbf{E} = (-y + e^x\sin y)\mathbf{i} + (x + 2)e^2\cos y\mathbf{j}$

$u = -y + e^x\sin y;\ \dfrac{\partial u}{\partial y} = -1 + e^x\cos y$

$v = (x + 2)e^2\cos y;\ \dfrac{\partial v}{\partial x} = e^2\cos y$

$u_y \neq v_x.$ The field is not conservative.

8. $\mathbf{E} = (y - x^2)\mathbf{i} + (2x + y^2)\mathbf{j}$

$u = y - x^2;\ \dfrac{\partial u}{\partial y} = 1;\ v = 2x + y^2;\ \dfrac{\partial v}{\partial x} = 2$

$u_y \neq v_x.$ The field is not conservative.

9. $\mathbf{E} = e^{2x}\sin y\mathbf{i} + e^{2x}\cos y\mathbf{j}$

$u = e^{2x}\sin y;\ \dfrac{\partial u}{\partial y} = e^{2x}\cos y;$

$v = e^{2x}\cos y;\ \dfrac{\partial v}{\partial x} = 2e^{2x}\cos y$

$u_y \neq v_x.$ The field is not conservative.

10. $\displaystyle\int_C (3x + 2y)\,dx + (2x + 3y)\,dy$

$\dfrac{\partial u}{\partial y} = 2;\ \dfrac{\partial v}{\partial x} = 2;\ u_y = v_x,$ so the line integral is path independent.

    a. $x = \cos t,\ y = \sin t;\ 0 \leq t \leq \pi$

$$\int_0^\pi [(3\cos t + 2\sin t)(-\sin t) + (2\cos t + 3\sin t)(\cos t)]\,dt$$
$$= 2\int_0^\pi (\cos^2 t - \sin^2 t)\,dt = 0$$

    b. The integral is 0 because the path could be closed and the effect on the x-axis is 0. Thus, the value is 0.

    c. The integral is 0 because the path is closed.

11. $\displaystyle\int_C (3x + 2y)\,dx - (2x + 3y)\,dy$

$\dfrac{\partial u}{\partial y} = 2;\ \dfrac{\partial v}{\partial x} = -2;\ u_y \neq v_x,$ so the line integral is not path independent.

    a. $x = \cos t,\ y = \sin t;\ 0 \leq t \leq \pi$

$$\int_0^\pi [(3\cos t + 2\sin t)(-\sin t) - (2\cos t + 3\sin t)(\cos t)]\,dt$$
$$= \int_0^\pi (-6\cos t\sin t - 2)\,dt = -2\pi$$

    b. $C_1\!: x = 1 - t,\ y = t;\ 0 \leq t \leq 1$
       $C_2\!: x = -t,\ y = 1 - t;\ 0 \leq t \leq 1$

$$\int_0^1 [3(1 - t) + 2t](-dt) - \int_0^1 [2(1 - t) + 3t]\,dt$$
$$+ \int_0^1 [-3t + 2(1 - t)](-dt)$$
$$- \int_0^1 [2(-t)3(1 - t)](-dt) = -4$$

    c. $\displaystyle -2\pi + \int_{-1}^1 3x\,dx$ (Note: -2π from part **a**.)
$$= -2\pi + 0 = -2\pi$$

12. $\displaystyle\int_C 2x^2y\,dx + x^3\,dy$

$\dfrac{\partial u}{\partial y} = 2x^2;\ \dfrac{\partial v}{\partial x} = 3x^2;\ u_y \neq v_x,$ so the line integral is not path independent.

    a. $\displaystyle\int_{-\pi/2}^{\pi/2} [(2\cos^2 t\sin t)(-\cos t) + \cos^2 t\cos t]\,dt$

$$= 2\int_{-\pi/2}^{\pi/2} \cos^3 t(-\sin t\,dt)$$
$$+ \frac{1}{4}\int_{-\pi/2}^{\pi/2} [1 + 2\cos t + \tfrac{1}{2}(1 + 4t)]\,dt = \frac{3\pi}{8}$$

    b. $\displaystyle\int_0^1 [2t^2(2t - 1) + 2t^3]\,dt + \int_0^1 [2(1 - t)^2](-dt)$

$$= \int_0^1 (6t^3 - 4t^2 + 4t - 2)\,dt = \frac{1}{6}$$

    c. $C_1\!: x = -t,\ y = -1;\ -1 \leq t \leq 1$
       $C_2\!: x = -1,\ y = t;\ -1 \leq t \leq 1$
       $C_3\!: x = t,\ y = 1;\ -1 \leq t \leq 1$
       $C_4\!: x = 1,\ y = -t;\ -1 \leq t \leq 1$

$$\int_{-1}^1 [2(-t)^2(-1)(-1) + t^3 + 2t^2 + (-t)^3(-1)]\,dt$$
$$= \int_{-1}^1 (2t^2 + t^3 + 2t^2 + t^3)\,dt = \frac{8}{3}$$

13. $\displaystyle\int_C 2xy\,dx + x^2\,dy$

$\dfrac{\partial u}{\partial y} = 2x;\ \dfrac{\partial v}{\partial x} = 2x;\ u_y = v_x,$ so the line integral

is path independent.

a. The integral is 0 because the path is closed.

b. $C: x = -t, \; y = t^2; \; 0 \le t \le 2$

$$\int_0^2 8t^3 \, dt = 32$$

c. $C_1: x = -t, \; y = t^2; \; 0 \le t \le 2$
$C_2: x = t, \; y = t + 6; \; -2 \le t \le 2$

$$\int_0^2 [2t(t^2) + t^2(2t)] \, dt + \int_{-2}^2 [2t(t+6) + t^2] \, dt$$
$$= 32$$

14. $\dfrac{\partial u}{\partial y} = 2; \; \dfrac{\partial v}{\partial x} = 2; \; \mathbf{F}$ is conservative.

$$f(x, y) = \frac{x^2}{2} + 2xy + c(y)$$

$$\frac{\partial f}{\partial y} = 2x + c'(y) = 2x + y;$$

$$c'(y) = y, \text{ so } c(y) = \frac{y^2}{2}$$

$$f(x, y) = \frac{x^2}{2} + 2xy + \frac{y^2}{2};$$

$$\int_A^B \mathbf{F} \, d\mathbf{R} = f(1, 1) - f(0, 0) = 3 - 0 = 3$$

15. $\dfrac{\partial u}{\partial y} = 2x; \; \dfrac{\partial v}{\partial x} = 2x; \; \mathbf{F}$ is conservative.

$$f(x, y) = x^2 y + c(y)$$

$$\frac{\partial f}{\partial y} = x^2 + c'(y) = x^2;$$

$$c'(y) = 0, \text{ so } c(y) = K = 0$$

$$f(x, y) = x^2 y;$$

$$\int_A^B \mathbf{F} \, d\mathbf{R} = f(1, 1) - f(0, 0) = 1 - 0 = 1$$

16. $\dfrac{\partial u}{\partial y} = 1; \; \dfrac{\partial v}{\partial x} = 1; \; \mathbf{F}$ is conservative.

$$f(x, y) = xy - \frac{x^3}{3} + c(y)$$

$$\frac{\partial f}{\partial y} = x + c'(y) = x + y^2;$$

$$c'(y) = y^2, \text{ so } c(y) = \frac{y^3}{3}$$

$$f(x, y) = xy - \frac{x^3}{3} + \frac{y^3}{3}$$

$$\int_A^B \mathbf{F} \, d\mathbf{R} = f(1, 1) - f(0, 0) = 1 - 0 = 1$$

17. $\dfrac{\partial u}{\partial y} = -1; \; \dfrac{\partial v}{\partial x} = -1; \; \mathbf{F}$ is conservative

$$f(x, y) = x^2 - yx + c(y)$$

$$\frac{\partial f}{\partial y} = -x + c'(y) = -x + y^2$$

$$c'(y) = y^2, \text{ so } c(y) = \frac{y^3}{3}$$

$$f(x, y) = x^2 - xy + \frac{1}{3}y^3$$

$$\int_C \mathbf{F} \cdot d\mathbf{R} = f(1, 1) - f(0, 0) = 1 - 0 = \frac{1}{3}$$

18. $\dfrac{\partial u}{\partial y} = -3x^2 \sin y + \dfrac{1 + x^2 y^2 - 2x^2 y^2}{(1 + x^2 y^2)^2}$

$\dfrac{\partial v}{\partial x} = -3x^2 \sin y + \dfrac{1 - x^2 y^2}{(1 + x^2 y^2)^2}$

$\mathbf{F}$ is conservative.

$$f(x, y) = x^3 \cos y + \tan^{-1} xy + c(y)$$

$$\frac{\partial f}{\partial y} = -3x^2 \sin y + \frac{x}{1 + x^2 y^2} + c'(y);$$

$$c'(y) = 0, \text{ so } c(y) = K = 0$$

$$f(x, y) = x^3 \cos y + \tan^{-1} xy$$

$$\int_A^B \mathbf{F} \, d\mathbf{R} = f(1, 1) - f(0, 0)$$

$$= \cos 1 + \frac{\pi}{4} - 0 = \cos 1 + \frac{\pi}{4}$$

19. $\dfrac{\partial u}{\partial y} = -e^{-x}(-x \sin xy + \sin xy + xy \cos xy)$

$\dfrac{\partial v}{\partial x} = -e^{-x}(-x \sin xy + \sin xy + xy \cos xy)$

$\mathbf{F}$ is conservative.

$$f(x, y) = e^{-x}[-\cos xy - y \sin xy - x] + c(x)$$

$$\frac{\partial f}{\partial y} = e^{-x}[-\cos xy - y \sin xy - x] + c'(x)$$

$$c'(x) = 2 \cos 2x, \text{ so } c(x) = \sin 2x$$

$$f(x, y) = e^{-x} \cos xy + \sin 2x$$

$$\int_A^B \mathbf{F} \, d\mathbf{R} = f(1, 1) - f(0, 0)$$

$$= e^{-1} \cos 1 + \sin 2 - 1 \approx 0.1081$$

20. $\dfrac{\partial u}{\partial y} = 12xy; \; \dfrac{\partial v}{\partial x} = 12xy; \; \mathbf{F}$ is conservative.

$$f(x, y) = x^3 + 3x^2 y^2 + c(y)$$

$$\frac{\partial f}{\partial y} = 6x^2 y + c'(y)$$

$$c'(y) = 4y^2, \text{ so } c(y) = \frac{4}{3}y^3$$

$f(x, y) = x^3 + 3x^2y^2 + \frac{4}{3}y^3$

$\int_A^B \mathbf{F}\, d\mathbf{R} = f(0, 1) - f(1, 1) = \frac{4}{3} - 1 = \frac{1}{3}$

21. $\frac{\partial u}{\partial y} = 2y; \frac{\partial v}{\partial x} = 2y;$ **F** is conservative

$f(x, y) = x^3 + x^2 + xy^2 + c(y)$

$\frac{\partial f}{\partial y} = 2xy + c'(y)$

$c'(y) = y^3$, so $c(x) = \frac{1}{4}y^4$

$f(x, y) = x^3 + x^2 + y^2x + \frac{y^4}{4}$

$\int_C (3x^2 + 2x + y^2)\,dx + (2xy + y^3)\,dy$

$\qquad = f(1, 1) - f(0, 0) = \frac{13}{4}$

22. $\frac{\partial u}{\partial y} = -x^2y \sin xy + 2x \cos y;$

$\frac{\partial v}{\partial x} = -x^2y \sin xy + 2x \cos xy$

F is conservative.

$f(x, y) = x \sin xy + c(x)$

$\frac{\partial f}{\partial x} = xy \cos xy + \sin xy + c'(x)$

$c'(x) = 0$, so $c(x) = K = 0$

$f(x, y) = x \sin xy$

$\int_C (xy \cos xy + \sin xy)\,dx + (x^2 \cos xy)\,dy$

$\qquad = f(1, \frac{\pi}{6}) - f(0, \frac{\pi}{18}) = \frac{1}{2}$

23. $\frac{\partial u}{\partial y} = 1; \frac{\partial v}{\partial x} = 1;$ **F** is conservative.

$f(x, y) = xy - \frac{x^3}{3} + c(y)$

$\frac{\partial f}{\partial x} = x + c'(y); c'(y) = y^2$, so $c(y) = \frac{y^3}{3}$

$f(x, y) = xy - \frac{x^3}{3} + \frac{y^3}{3}$

$\int_C (y - x^2)\,dx + (x + y^2)\,dy$

$\qquad = f(0, 3) - f(-1, -1) = 8$

24. $\frac{\partial u}{\partial y} = 3x^2 + 2y; \frac{\partial v}{\partial x} = 3x^2 + 2y$

F is conservative.

$f(x, y) = x^3y + xy^2 + c(y)$

$\frac{\partial f}{\partial y} = x^3 + 2xy + c'(y);$

$c'(y) = 0$, so $c(y) = K = 0$

$f(x, y) = x^3 + 2xy$

When $t = 0$, $(x, y) = (0, -2)$

when $t = 2$, $(x, y) = (2, 4)$

$\int_C (3x^2y + y^2)\,dx + (x^3 + 2xy)\,dy$

$\qquad = f(2, 4) - f(0, -2) = 24$

25. $\frac{\partial u}{\partial y} = \cos y; \frac{\partial v}{\partial x} = \cos y$

F is conservative.

$f(x, y) = x \sin y + c(y)$

$\frac{\partial f}{\partial y} = x \cos y + c'(y)$

$c'(y) = 3$, so $c(y) = 3y$

$f(x, y) = x \cos y + 3y$

When $t = 0$, $(x, y) = (0, 0)$

when $t = 1$, $(x, y) = (-2, \pi/2)$

$\int_C (\sin y)\,dx + (3 + x \cos y)\,dy$

$\qquad = f(-2, \pi/2) - f(0, 0) = \frac{1}{2}(3\pi - 4)$

26. $\frac{\partial u}{\partial y} = -e^x \sin y; \frac{\partial v}{\partial x} = -e^x \sin y$

F is conservative.

$f(x, y) = e^x \cos y + c(y)$

$\frac{\partial f}{\partial y} = -xe^x \sin y + c'(y)$

$c'(y) = 0$, so $c(y) = K = 0$

$f(x, y) = e^x \cos y$

When $t = 0$, $(x, y) = (1, 0)$

when $t = \pi/2$, $(x, y) = (0, 1)$

$\int_C (e^x \cos y)\,dx + (-e^x \sin y)\,dy$

$\qquad = f(0, 1) - f(1, 0) = \cos 1 - e$

27. $\frac{\partial u}{\partial y} = 1; \frac{\partial v}{\partial x} = 1;$ **F** is conservative.

$f(x, y) = xy + c(y)$

$\frac{\partial f}{\partial y} = x + c'(y); c'(y) = 0$, so $c(y) = K = 0$

$f(x, y) = xy$

$$\int_C (y\mathbf{i} + x\mathbf{j}) \cdot d\mathbf{R} = f(2, 4) - f(0, 0) = 8$$

28. $\dfrac{\partial u}{\partial y} = 2xy; \dfrac{\partial v}{\partial x} = 2xy;$ **F** is conservative.

$f(x, y) = \dfrac{x^2 y^2}{2} + c(y)$

$\dfrac{\partial f}{\partial y} = x^2 y + c'(y); \ c'(y) = 0,$ so $c(y) = K = 0$

$f(x, y) = \dfrac{x^2 y^2}{2}$

$$\int_C (xy^2\mathbf{i} + x^2 y\mathbf{j}) \cdot d\mathbf{R} = f(0, 0) - f(4, 1) = -8$$

29. $\dfrac{\partial u}{\partial y} = 2; \dfrac{\partial v}{\partial x} = 2;$ **F** is conservative.

$f(x, y) = 2xy + c(y)$

$\dfrac{\partial f}{\partial y} = 2x + c'(y); \ c'(y) = 0,$ so $c(y) = K = 0$

$f(x, y) = 2xy$

$$\int_C (2y\, dx + 2x\, dy) = f(4, 4) - f(0, 0) = 32$$

30. $\dfrac{\partial u}{\partial y} = e^x \cos y; \dfrac{\partial v}{\partial x} = e^x \cos y;$

F is conservative.

$f(x, y) = e^x \sin y + c(y)$

$\dfrac{\partial f}{\partial y} = e^x \cos y + c'(y);$

$c'(y) = 0,$ so $c(y) = K = 0$

$f(x, y) = e^x \sin y$

$$\int_C (e^x \sin y\, dx + e^x \cos y\, dy), \text{ where } C \text{ is}$$

$$= f(0, 2\pi) - f(0, 0) = 0$$

31. $\dfrac{\partial u}{\partial y} = \dfrac{2x(x^2 + y^2) - 2x^3}{(x^2 + y^2)^2} = \dfrac{2xy^2}{(x^2 + y^2)^2};$

$\dfrac{\partial v}{\partial x} = \dfrac{1/x}{1 + y^2/x^2} - \dfrac{x^3 - xy^2}{(x^2 + y^2)^2} = \dfrac{2xy^2}{(x^2 + y^2)^2}$

F is conservative.

$f(x, y) = \tan^{-1}\dfrac{y}{x} + (y - 1)e^{-y} + e^{-y} + c(x)$

$\dfrac{\partial f}{\partial x} = x\tan^{-1}\dfrac{y}{x} + c'(x)$

$c'(y) = 0,$ so $c(y) = K = 0$

$f(x, y) = x\tan^{-1}\dfrac{y}{x} + ye^{-y}$

When $t = 0$, $(x, y) = (1, 0)$;
when $t = 1$, $(x, y) = (1, -\sin 1)$

$$\int_C \left[\tan^{-1}\dfrac{y}{x} - \dfrac{xy}{x^2 + y^2}\right] dx$$

$$+ \left[\dfrac{x^2}{x^2 + y^2} + e^{-y}(1 - y)\right] dy$$

$$= f(\cos 1, -\sin 1) - f(1, 0) = -2.492$$

32. $\text{curl } \mathbf{F} = \begin{vmatrix} \mathbf{i} & \mathbf{j} & \mathbf{k} \\ \dfrac{\partial}{\partial x} & \dfrac{\partial}{\partial y} & \dfrac{\partial}{\partial z} \\ yz^2 & xz^2 & 2xyz \end{vmatrix}$

$= (2xz - 2xz)\mathbf{i} - (2yz - 2yz)\mathbf{j} + (z^2 - z^2)\mathbf{k}$
$= 0$

33. $\text{curl } \mathbf{F} = \begin{vmatrix} \mathbf{i} & \mathbf{j} & \mathbf{k} \\ \dfrac{\partial}{\partial x} & \dfrac{\partial}{\partial y} & \dfrac{\partial}{\partial z} \\ yze^{xy} & xze^{xy} & e^{xy} \end{vmatrix}$

$= (xe^{xy} - xe^{xy})\mathbf{i} - (ye^{xy} - ye^{xy})\mathbf{j}$
$\quad + (ze^{xy} - ze^{xy})\mathbf{k} = 0$

34. $\text{curl } \mathbf{F} = \begin{vmatrix} \mathbf{i} & \mathbf{j} & \mathbf{k} \\ \dfrac{\partial}{\partial x} & \dfrac{\partial}{\partial y} & \dfrac{\partial}{\partial z} \\ \dfrac{y}{z} & \dfrac{x}{z} & -\dfrac{xy}{z^2} \end{vmatrix}$

$= \left(-\dfrac{x}{z^2} + \dfrac{x}{z^2}\right)\mathbf{i} - \left(-\dfrac{y}{z^2} - \dfrac{y}{z^2}\right)\mathbf{j} + \left(\dfrac{1}{z} - \dfrac{1}{z}\right)\mathbf{k}$

$= 0$

35. $\text{curl } \mathbf{F} = \begin{vmatrix} \mathbf{i} & \mathbf{j} & \mathbf{k} \\ \dfrac{\partial}{\partial x} & \dfrac{\partial}{\partial y} & \dfrac{\partial}{\partial z} \\ rx & ry & rz \end{vmatrix}$

$\text{where } r = x^2 + y^2 + z^2$

$= (2yz - 2yz)\mathbf{i} - (2xz - 2xz)\mathbf{j} + (2xy - 2xy)\mathbf{k}$
$= 0$

36. $\text{curl } \mathbf{F} = \begin{vmatrix} \mathbf{i} & \mathbf{j} & \mathbf{k} \\ \dfrac{\partial}{\partial x} & \dfrac{\partial}{\partial y} & \dfrac{\partial}{\partial z} \\ y\sin z & x\sin z + 2y & xy\cos z \end{vmatrix}$

$= (x\cos z - x\cos z)\mathbf{i} - (y\cos z - y\cos z)\mathbf{j}$
$\quad + (\sin z - \sin z)\mathbf{k}$
$= 0$

37. $\operatorname{curl} \mathbf{F} = \begin{vmatrix} \mathbf{i} & \mathbf{j} & \mathbf{k} \\ \dfrac{\partial}{\partial x} & \dfrac{\partial}{\partial y} & \dfrac{\partial}{\partial z} \\ xy^2+yz & x^2y+xz+3y^2z & xy+y^3 \end{vmatrix}$

$= (x + 3y^2 - x - 3y^2)\mathbf{i} - (y - y)\mathbf{j}$
$\quad + (2xy + z - 2xy - z)\mathbf{k}$
$= 0$

38. a. $\operatorname{curl} \mathbf{F} = \begin{vmatrix} \mathbf{i} & \mathbf{j} & \mathbf{k} \\ \dfrac{\partial}{\partial x} & \dfrac{\partial}{\partial y} & \dfrac{\partial}{\partial z} \\ -yx^{-2} & x^{-1} & 0 \end{vmatrix}$

$= 0\mathbf{i} - 0\mathbf{j} + (x^{-2} - x^{-2})\mathbf{k} = 0$

This integral is independent of path if it does not cross the x-axis.

b. $f_x = -\dfrac{y}{x^2} + \dfrac{1}{x}; \ f(x, y) = \dfrac{y}{x} + \ln|x| + c(y)$

$\dfrac{\partial f}{\partial y} = \dfrac{1}{y} + c'(y) = \dfrac{1}{x}; \ c(y) = K = 0$

$f(x, y) = \dfrac{y}{x} + \ln|x|$

With $\mathbf{R} = \cos^3 t\mathbf{i} + 2\sqrt{\sec t}\,\mathbf{j}$;
when $t = 0$, $\mathbf{R} = \mathbf{i} + 2\mathbf{j}$;
when $t = \dfrac{\pi}{6}$, $\mathbf{R} = \dfrac{8}{3\sqrt{3}}\mathbf{i} + 2\sqrt{\dfrac{2}{\sqrt{3}}}\,\mathbf{j}$

$\displaystyle\int_C [(-yx^{-2} + x^{-1})\,dx + x^{-1}dy]$

$= f\left(\dfrac{8}{3\sqrt{3}}, 2\sqrt{\dfrac{2}{\sqrt{3}}}\right) - f(1, 2)$

$= \dfrac{3\sqrt{6}}{\sqrt[4]{3}} + \ln 8 - \dfrac{3}{2}\ln 3 - 2$

39. a. $\operatorname{curl} \mathbf{F}$

$= \left[\dfrac{3kmMyz}{(x^2+y^2+z^2)^{5/2}} - \dfrac{3kmMyz}{(x^2+y^2+z^2)^{5/2}}\right]\mathbf{i}$

$= \left[\dfrac{3kmMxz}{(x^2+y^2+z^2)^{5/2}} - \dfrac{3kmMxz}{(x^2+y^2+z^2)^{5/2}}\right]\mathbf{j}$

$= \left[\dfrac{3kmMxy}{(x^2+y^2+z^2)^{5/2}} - \dfrac{3kmMxy}{(x^2+y^2+z^2)^{5/2}}\right]\mathbf{k}$

$= 0$

$g_x = kmM\dfrac{x}{(x^2 + y^2 + z^2)^{3/2}}$

$g(x, y, z) = \dfrac{kmM}{(x^2 + y^2 + z^2)^{1/2}} + c(y, z)$

$g_y = kmM\dfrac{y}{(x^2 + y^2 + z^2)^{3/2}} + \dfrac{\partial c(y, z)}{\partial y}$

Let $b(z) = c(y, z)$

$g_z = kmM\dfrac{z}{(x^2 + y^2 + z^2)^{3/2}} + b'(z)$

$b(z) = K = 0$, so $c(y, z) = 0$

$g(x, y, z) = \dfrac{kmM}{(x^2 + y^2 + z^2)^{1/2}}$

$\qquad = kmM\left(\dfrac{1}{r}\right)$

Note: Physicists sometimes use $\phi = -\dfrac{kmM}{r}$ as the gravitational component.

b. $W = \displaystyle\int_P^Q \mathbf{G} \cdot d\mathbf{R}$

$= g(a_2, b_2, c_2) - g(a_1, b_1, c_1)$

$= \dfrac{kmM}{(a_2{}^2 + b_2{}^2 + c_2{}^2)^{1/2}}$

$\quad - \dfrac{kmM}{(a_1{}^2 + a_2{}^2 + a_3{}^2)^{1/2}}$

40. a. Let $x = \cos t$ and $y = \sin t$; $0 \le t \le \pi$

$\displaystyle\int_C \dfrac{-y\,dx + x\,dy}{x^2 + y^2} = \int_0^\pi dt = \pi$

on the upper semicircle. For $\pi \le t \le 2\pi$, we have the lower semicircle:

$\displaystyle\int_C \dfrac{-y\,dx + x\,dy}{x^2 + y^2} = \int_\pi^{2\pi} dt = \pi$

b. $M = -\dfrac{y}{x^2 + y^2}$, $N = \dfrac{x}{x^2 + y^2}$

$M_y = \dfrac{-x^2 + y^2}{(x^2 + y^2)^2}$, $N_x = \dfrac{x^2 - y^2}{(x^2 + y^2)^2}$

$M_y = N_x$, but $\mathbf{F}$ is not conservative because it is not continuous at $(0, 0)$.

41. $\mathbf{F}$ is constant, and, therefore, conservative. $\mathbf{R}$ is a simple closed curve, so $W = 0$. If you want to work it out;
$\mathbf{R} = (3\cos t)\mathbf{i} + (3\sin t)\mathbf{j}$,

$F = \dfrac{mv^2}{r} = \dfrac{\frac{30}{32}[3(2\pi)]^2}{3} = \dfrac{45\pi^2}{4}$,

$\mathbf{F}$ is always in a direction that is normal to $\mathbf{R}$, so $\mathbf{F} = \dfrac{45\pi^2}{4}(-\mathbf{R})$. So $\mathbf{F} \cdot \mathbf{R}$, and the line integral, are both equal to zero.

42. a. The path is immaterial since
$\mathbf{F} = -a\mathbf{i} - b\mathbf{j}$ and the partial of

constants are 0. Also, $W = 2a + b$.

b. The path is immaterial since
$\mathbf{F} = -a\mathbf{i} - be^{-y}\mathbf{j}$ and
$$\frac{\partial - a}{\partial y} \neq \frac{\partial - be^{-y+x/9}}{\partial x}$$

43. a. These observations suggest one "bend" the path above the straight line path. That is, to take advantage of the e^{-y} term.

b. Using the parabolic path, $y = x(1 - x/4)$ yields Work $\approx 2.6719a$.
If $y = \sin(\pi x/4)$, we find
Work $\approx 2.67616a$.
Both of these answers are slightly better than the straight line path.

c. For example, using parabolas of the form $y = x(b - ax)$, the condition $y(2) = 1$ leads to $b = (1 + 4a)/2$; and a study of work as a function of a shows that the larger the a, the smaller the amount of work. But large a sends the path way "north" which is possibly unrealistic.

d. The optimal path, if $y \leq 1$ is imposed, seems to be "up to $(1, 0)$ then over to $(2, 1)$." The parabolic paths can improve on this value only if a is allowed to become large making the path swing way north. If $y \leq 1$ is imposed on these paths, the optimal is $y = x(1 - x/4)$. This is found by making a as large as possible subject to $y'(x) \geq 0$ on $[0, 2]$, which insures the path does not exceed $y = 1$. Comparing these results with the crude "up and over" path suggests the family of parabolas was not a good choice; and that basically we want a smooth version of "up, then over." For example, consider $y = 1 - (1 - t)^n$ for large n.

44. If $\mathbf{F}$ is conservative, then $\mathbf{F} = \nabla\phi$ and
$\text{curl } \mathbf{F} = \nabla \times \mathbf{F} = \nabla \times \nabla\phi$
$$= \left[\frac{\partial^2\phi}{\partial y\partial z} - \frac{\partial^2\phi}{\partial z\partial y}\right]\mathbf{i} - \left[\frac{\partial^2\phi}{\partial x\partial z} - \frac{\partial^2\phi}{\partial z\partial x}\right]\mathbf{j}$$
$$+ \left[\frac{\partial^2\phi}{\partial x\partial z} - \frac{\partial^2\phi}{\partial z\partial x}\right]\mathbf{k} = \mathbf{0}$$

Thus, $\text{curl } \mathbf{F} = \left(\frac{\partial P}{\partial y} - \frac{\partial N}{\partial z}\right)\mathbf{i} + \left(\frac{\partial M}{\partial z} - \frac{\partial P}{\partial x}\right)\mathbf{j}$
$$+ \left(\frac{\partial N}{\partial x} - \frac{\partial M}{\partial y}\right)\mathbf{k} = \mathbf{0}$$

Thus, $\frac{\partial P}{\partial y} = \frac{\partial N}{\partial z}$; $\frac{\partial M}{\partial z} = \frac{\partial P}{\partial x}$; $\frac{\partial N}{\partial x} = \frac{\partial M}{\partial y}$

45. a. We use the result from Problem 44,
$$M_y = 2y = N_x; \ M_z = -2x = P_x;$$
$$P_y = 1 = N_z$$

Thus, $\mathbf{F}$ is conservative.
$$f_x = y^2 - 2xz$$
$$f = xy^2 - x^2z + c(y, z)$$

b. $f_y = 2xy + c_y = 2xy + z$
$c(x, y) = yz + c(z)$
$f = xy^2 - x^2z + yz + c(z)$

c. $f_z = -x^2 + y + b'(z) = -x^2 + y$
$b(z) = K = 0$, so
$$f(x, y, z) = xy^2 - x^2z + yz$$

46. We use the result from Problem 44.
$M_y = 1 = N_x$, so $\mathbf{F}$ is conservative.
$h_x = f_x + y$; $h_y = x + c'(y)$
$$c(y) = \int g(y) \, dy \text{ and}$$
$$h(x, y) = \int f(x) \, dx + xy + \int g(y) \, dy$$

47. $\int_C \mathbf{F} \cdot d\mathbf{R} = 0$

$$\int_{C_1} \mathbf{F} \cdot d\mathbf{R} + \int_{C_2} (-\mathbf{F} \cdot d\mathbf{R}) = 0$$

Thus, $\int_{C_1} \mathbf{F} \cdot d\mathbf{R} = \int_{C_2} \mathbf{F} \cdot d\mathbf{R} = 0$

$C_1 \cup C_2 = C$ and $C_1 \cap C_2$ is just P and Q.

14.4 Green's Theorem, Page 931

1. $\int_C (y^2 \, dx + x^2 \, dy) = \int_0^1 \int_0^1 (2x - 2y) \, dy \, dx$

$$= \int_0^1 (2x - 1) \, dx = 0$$

Alternatively, on C:

C_1: $y = 0$; C_2: $x = 1$; C_3: $y = 1$; C_4: $x = 0$
$$\int_0^1 0 \, dx + \int_0^1 dy + \int_1^0 0 \, dx + \int_1^0 0 \, dy = 0$$

2. $\displaystyle\int_C (y^3\,dx - x^3\,dy) = \iint_D (-3x^2 - 3y^2)\,dA$

$\displaystyle = -3\int_0^{2\pi}\int_0^1 r^3\,dr\,d\theta = -\frac{3\pi}{2}$

Alternatively, on C:

$\displaystyle\int_0^{2\pi} (-\sin^3\theta\,\sin\theta - \cos^3\theta\,\cos\theta)\,d\theta$

$\displaystyle = -\frac{1}{2}\int_0^{2\pi}(1 - \cos^2 2\theta)\,d\theta = -\frac{3\pi}{2}$

3. $\displaystyle -\int_C [(2x^2 + 3y)\,dx - 3y^2\,dy]$

$\displaystyle = -\iint_D (0\,dx + 3\,dy) = -3$

Note the negative sign is due to the clockwise path. Alternatively, on C:

C_1: $x = 1 - t,\ y = 1 - t;\ 0 \le t \le 1$
C_2: $x = 2 - t,\ y = t;\ 0 \le t \le 1$
C_3: $y = 0,\ x = 2t;\ 0 \le t \le 1$

$\displaystyle\int_0^1 \big[[2(1-t)^2 + 3(1-t)](-dt) - 3(1-t)^2(-dt)\big]$

$\displaystyle + \int_0^1 \big[[2(2-t)^2 + 3t](-dt) - 3t^2(dt)\big]$

$\displaystyle + \int_0^1 \big[[2(2t)^2 + 3(0)](2\,dt) - 3(0)^2(0)\big]$

$\displaystyle = \int_0^1 (t-1)(t+2)\,dt + \int_0^1 (-5t^2 + 5t - 8)\,dt + \int_0^1 16t^2\,dt$

$\displaystyle = -\frac{7}{6} - \frac{43}{6} + \frac{16}{3} = -3$

4. $\displaystyle\int_C (y^2\,dx + 3xy^2\,dy) = 3\iint_D (3y^2 - 2y)\,dA$

$\displaystyle = \int_0^1\int_y^{2-y} (3y^2 - 2y)\,dy\,dx$

$\displaystyle = \int_0^1 (6y^2 - 6y^3 - 4y + 4y^2)\,dy = -\frac{1}{6}$

Alternatively, on C:

C_1: $x = y = 1 - t;\ 0 \le t \le 1$
C_2: $x = 2 - t,\ y = t;\ 0 \le t \le 1$
C_3: $x = 2t,\ y = 0;\ 0 \le t \le 1$

$\displaystyle\int_0^1 \big[(1-t)^2(-dt) + 3(1-t)(1-t)^2(-dt)\big]$

$\displaystyle + \int_0^1 t^2(-dt) + 3(2-t)t^2\,dt + \int_0^1 0\,dt$

$\displaystyle = -\frac{13}{12} + \frac{11}{12} + 0 = -\frac{1}{6}$

5. $\displaystyle\int_C 4xy\,dx = \iint_D (-4x)\,dA = -4\int_0^{2\pi}\int_0^1 r^2\cos\theta\,d\theta = 0$

Alternatively, on C: $x = \cos\theta,\ y = \sin\theta,$
$dx = -\sin\theta\,d\theta,\ dy = \cos\theta\,d\theta$

$\displaystyle -4\int_0^{2\pi}\cos^2\theta\,\sin\theta\,d\theta = 0$

6. $\displaystyle\int_C (4y\,dx - 3x\,dy) = \iint_D (-4y\,dx - 3x\,dy)$

$\displaystyle = \iint_D -7\,dA = -7\pi(2)(\sqrt{2}) = -14\sqrt{2}\pi$

Recall that the area of an ellipse with semimajor axis a and semiminor axis b is πab. In this example, $a = 2$ and $b = \sqrt{2}$.

Now on C: $x = \sqrt{2}\cos\theta,\ y = 2\sin\theta,$
$dy = 2\cos\theta\,d\theta,\ dx = -\sqrt{2}\sin\theta\,d\theta$

$\displaystyle\int_0^{2\pi} [4(2\sin\theta)(-\sqrt{2}\sin\theta) - 3\sqrt{2}\cos\theta(2\cos\theta)]\,d\theta$

$\displaystyle = \int_0^{2\pi} [-8\sqrt{2}\sin^2\theta - 6\sqrt{2}\cos^2\theta]\,d\theta = -14\sqrt{2}\pi$

7. $\displaystyle\int_C (2y\,dx - x\,dy) = \iint_D (-1 - 2)\,dA$

$\displaystyle = -3(2\pi) = -6\pi$

8. $\displaystyle\int_C (e^x\,dx - \sin x\,dy) = \int_0^1\int_0^1 (-\cos x)\,dy\,dx$

$\displaystyle = -\sin 1$

9. $\displaystyle\int_C (x\sin x\,dx - \tan y\,dy) = \int_{-1}^5\int_{(y-1)/2}^{(y+7)/6} 0\,dx\,dy$

$= 0$

10. $\displaystyle\int_C [(x + y)\,dx - (3x - 2y)\,dy]$

$\displaystyle = \iint_D (-3 - 1)\,dA = -\frac{4(2+3)(4)}{2} = -40$

11. $\int_C [(x - y^2)\,dx + 2xy\,dy]$

$= \int\int_D (2y + 2y)\,dA = \int_0^2 \int_0^2 4y\,dy\,dx$

$= \int_0^2 8\,dx = 16$

12. $\int_C (y^2\,dx + x\,dy) = \int_0^2 \int_0^2 (1 + 2y)\,dy\,dx$

$= \frac{1}{4}\int_0^2 (25 - 1)\,dx = 12$

13. $\int_C [(3y - 4x)\mathbf{i} + (4x - y)\mathbf{j}] \cdot [\mathbf{i}\,dx + \mathbf{j}\,dy]$

$= \int\int_D (4 - 3)\,dA = 2(1)\pi = 2\pi$

since the semimajor and semiminor axes of the ellipse are 2 and 1, respectively.

14. $\int_C (y^2\mathbf{i} + x^2\mathbf{j}) \cdot (\mathbf{i}\,dx + \mathbf{j}\,dy)$

$= 2\int\int_D (x - y)\,dA$

$= 2\int_0^{2\pi} \int_0^1 r^2(\cos\theta - \sin\theta)\,dr\,d\theta = 0$

15. $A = \frac{1}{2}\int_C (-y\,dx + x\,dy)$

$= \frac{1}{2}\int_C [(-2\sin t)(-2\sin t) + (2\cos\theta)(2\cos t)]\,dt$

$= \frac{1}{2}\int_C 4\,dt = \frac{1}{2}\int_0^{2\pi} 4\,dt = 4\pi$

Area of a circle with radius of 2 is $\pi r^2 = 4\pi$.
Check: $A = \pi r^2 = \pi(2)^2 = 4\pi$.

16. C_1: $x = 0,\ y = t,\ dx = 0\ dy = 1;\ 0 \le t \le 1$
C_2: $x = t,\ y = 2 - t;\ 0 \le t \le 1$
C_3: $x = 2 - t,\ y = 0;\ 0 \le t \le 1$

$A = \frac{1}{2}\int_C (-y\,dx + x\,dy)$

$= \frac{1}{2}\int_0^1 (-t\,dt + t\,dt) = 1$

Check: $A = \frac{1}{2}bh = \frac{1}{2}(2)(1) = 1$

17. C_1: $y = 0,\ dy = 0;\ 0 \le x \le 4$
C_2: $x = 4 - 3t,\ y = 3t;\ 0 \le t \le 1$
C_3: $y = 3,\ dy = 0;\ 0 \le x \le 1$
C_4: $x = dx = 0;\ 0 \le t \le 1$

$A = \int_C x\,dy = 0 + 3\int_0^1 (4 - 3t)\,dt + 0 + 0 = \frac{15}{2}$

Check: $A = \frac{1}{2}(b_1 + b_2)h = \frac{1}{2}(1 + 4)(3) = \frac{15}{2}$

18. C_1: $x = 2\cos t,\ y = 2\sin t;\ 0 \le t \le \pi$
C_2: $y = dy = 0;\ -2 \le x \le 2$

$A = \int_C x\,dy = 4\int_0^1 \cos^2 t\,dt + 0$

$= 2\int_0^\pi (1 + \cos 2t)\,dt = 2\pi$

Check: $A = \frac{1}{2}\pi r^2 = \frac{1}{2}\pi(2)^2 = 2\pi$

19. $\int_C (x^2 y\,dx - y^2 x\,dy) = \int\int_D (-y^2 - x^2)\,dA$

$= -\int\int_D r^2\,dA = -\int_0^\pi \int_0^a r^3\,dr\,d\theta = -\frac{\pi a^4}{4}$

20. $\int_C (3y\,dx - 2x\,dy) = \int\int_D (-2 - 3)\,dA$

$= -(5)(2)\int_{-\pi/2}^{\pi/2} \int_0^{1+\sin\theta} r\,dr\,d\theta$

$= -5\int_{-\pi/2}^{\pi/2} [1 + 2\sin\theta + \sin^2\theta]\,dt = -\frac{15\pi}{2}$

21. $I = \int_C (5 - xy - y^2)\,dx - (2xy - x^2)\,dy$

$= \int\int_D (-2y + 2x + x + 2y)\,dA$

$= 3\int\int_D x\,dA = 3M_y$

$m = A = 1$, so

$$\bar{x} = \frac{\displaystyle\iint_D x \, dA}{\displaystyle\iint_D dA} = \frac{1/3}{1} = \frac{1}{3}; \text{ thus } I = 3\bar{x}$$

22. Let $u = x - 2$, $v = y$, $x = u + 2$, $y = v$

$$\frac{\partial(x, y)}{\partial(u, v)} = \begin{vmatrix} \dfrac{\partial x}{\partial u} & \dfrac{\partial x}{\partial v} \\ \dfrac{\partial y}{\partial u} & \dfrac{\partial y}{\partial v} \end{vmatrix} = \begin{vmatrix} 1 & 0 \\ 0 & 1 \end{vmatrix} = 1$$

$$\int_C (x + 2y^2) \, dy = \iint_D (-1) \, dA = -\pi$$

23. $\displaystyle\int_C x^2 \, dy = \iint_D (2x - 0) dA = 2 \iint_D x \, dA = 2\bar{y}A$

Thus, $A\bar{y} = \dfrac{1}{2} \displaystyle\int_C x^2 \, dy$.

24. $C_1: x = t$, $y = (\tan\theta_1)t$; $0 \le t \le a$
$C_2: x = r\cos\theta$, $y = r\sin\theta$, $\theta_1 \le \theta \le \theta_2$
$C_3: x = b - t$, $y = (\tan\theta_2)(b - t)$; $0 \le t \le 1$

$$A = \frac{1}{2}\int_C (-y \, dx + x \, dy)$$

$$= \frac{1}{2}\int_0^a [-(\tan\theta_1)t + t(\tan\theta_1)] \, dt$$

$$+ \frac{1}{2}\int_{\theta_1}^{\theta_2} r^2(\cos^2\theta + \sin^2\theta) \, d\theta$$

$$+ \frac{1}{2}\int_0^1 [(b - t)(\tan\theta_2) + (b - t)(-\tan\theta_2)] \, dt$$

$$= \frac{1}{2}\int_{\theta_1}^{\theta_2} r^2 \, d\theta = \frac{1}{2}\int_{\theta_1}^{\theta_2} [g(\theta)]^2 \, d\theta$$

25. $\displaystyle\int_C \left[\left(\frac{-y}{x^2} + \frac{1}{x}\right)dx + \frac{1}{x} dy\right] = \iint_D \left[-\frac{1}{x^2} + \frac{1}{x^2}\right] dA = 0$

26. $\displaystyle\int_C \frac{x \, dx + y \, dy}{x^2 + y^2} = \iint_D \frac{-y(2x) + x(2y)}{(x^2 + y^2)^2} \, dA = 0$

27. $\displaystyle\int_C \frac{-y \, dx + (x - 1) \, dy}{(x - 1)^2 + y^2}$

$$= \iint_D \frac{(x - 1)(2y) + 2y(x - 1)}{[(x - 1)^2 + y^2]^2} \, dA = 0$$

28. $\displaystyle\int_C \frac{-(y + 2)dx + (x - 1)dy}{(x - 1)^2 + (y + 2)^2}$

$$= \iint_D \frac{2(x - 1)(y + 2) + 2(y + 2)(x - 1)}{[(x - 1)^2 + (y + 2)^2]^2} \, dA = 0$$

29. $\displaystyle\int_C \frac{\partial z}{\partial n} \, ds = \int_C (\nabla z \cdot \mathbf{N}) \, ds = \int_C z_x \, dy - z_y \, dx$

$$= \int_C [4x \, dy - 6y \, dx] = \iint_{\text{circle}} [4 - (-6)] \, dy \, dx$$

$$= 10(\text{area of circle } x^2 + y^2 = 16)$$

$$= 10(\pi 4^2) = 160\pi$$

30. $\nabla f = (2xy - 2y)\mathbf{i} + (x^2 - 2x + 2y)\mathbf{j}$

$$= M\mathbf{i} + N\mathbf{j}$$

$$\int_C \frac{\partial z}{\partial n} \, ds = \int_C (\nabla z \cdot \mathbf{N}) \, ds$$

$$= \int_C (M \, dy + N \, dx)$$

$$= \int_C \nabla f \cdot d\mathbf{R}$$

$$= \iint_D [(2x - 2) - (2x - 2)] \, dA = 0$$

31. The normal derivative, $\dfrac{\partial f}{\partial n} = \nabla f \cdot \mathbf{N}$, so

$$\frac{\partial x}{\partial n} = \nabla x \cdot \mathbf{N} = \mathbf{i} \cdot \left(\frac{dy}{ds}\mathbf{i} - \frac{dx}{ds}\mathbf{j}\right)$$

$$\int_C x \frac{\partial x}{\partial n} \, ds = \int_C x \frac{dy}{ds} \, ds = \int_C x \, dy$$

$$= \iint_D 1 \, dA \text{ (by Green's theorem)}$$

$$= 1 \text{ since } A = 1$$

32. $\displaystyle\int_C [(x - 3y) \, dx + (2x - y^2)] \, dy$

$$= \iint_A (2 + 3) \, dA = 5A$$

33. To the spy's surprise and delight, he finds that the Death force is a conservative field:

$$\frac{\partial}{\partial x}\left[\frac{x}{(x^2 + y^2)} - \frac{1}{y} - \cos y\right] = \frac{y^2 - x^2}{(x^2 + y^2)^2}$$

$$= \frac{\partial}{\partial y}\left[\frac{-y}{(x^2 + y^2)} + \frac{1}{x}\right]$$

It doesn't matter which path he takes, the work will be the same! Since he wants to get out of the room as quickly as possible, he walks in a straight line from $(0, 0)$ to $(5, 4)$ and leaves. As he rests outside the room, recovering his strength, he wonders — was it an accident that the Death Force turned out to be conservative, or did Purity design it that way? Just wait for the second edition!!

34. Part (1): $\mathbf{F}(x, y) = u\mathbf{i} + v\mathbf{j}$, then
$$\frac{\partial u}{\partial y} = \frac{\partial v}{\partial x}$$
This is the meaning of conservative.

Part (2): If $\frac{\partial u}{\partial y} = \frac{\partial v}{\partial x}$, then
$\mathbf{F}(x, y) = f_x\mathbf{i} + f_y\mathbf{j} = \nabla f$. Thus, $\mathbf{F}$ is conservative.

35. $\iint\limits_D (f_x^2 + f_y^2)\, dA$

$$= \iint\limits_D \left[\frac{\partial(ff_x)}{\partial x} - \frac{\partial(-ff_y)}{\partial y}\right] dA$$

$$= \int_C [(-ff_x)\, dx + (ff_x)\, dy]$$

$$= \int_C [ff_x\mathbf{i} + ff_y\mathbf{j}]\cdot(\mathbf{i}\, dy - \mathbf{j}\, dx)\, ds$$

$$= \int_C f\nabla f\cdot\mathbf{N}\, ds = \int_C f\frac{\partial f}{\partial n}\, ds$$

36. $\int_C \left(f\frac{\partial g}{\partial n} - g\frac{\partial f}{\partial n}\right) ds$

$$= \int_C (f\nabla g\cdot\mathbf{N} - g\nabla f\cdot\mathbf{N})\, ds$$

$$= \int_C [f(g_x\mathbf{i} + g + y\mathbf{j})] - g(f_x\mathbf{i} + f_y\mathbf{j})\cdot(dy\mathbf{i} - dx\mathbf{j})$$

$$= \int_C [(fg_x - gf_x)\, dy + (fg_y - gf_y)(-dx)]$$

$$= \iint\limits_D \left[\frac{\partial}{\partial x}(fg_x - gf_x) + \frac{\partial}{\partial y}(fg_y - gf_y)\right] dA$$

$$= \iint\limits_D [fg_{xx} + g_x f_x - gf_{xx} - f_x g_x + fg_{yy}$$

$$+ f_y g_y - gf_{yy} - g_y f_y]\, dA$$

$$= \iint\limits_D [f(g_{xx} + g_{yy}) - g(f_{xx} + f_{yy})]\, dA$$

$$= \iint\limits_D [f\nabla^2 g - g\nabla^2 f]\, dA$$

37. $\int_C f\frac{\partial g}{\partial n}\, ds = \int_C f\frac{\partial g}{\partial n}\, ds = \int_C f\nabla g\cdot\mathbf{N}\, ds$

$$= \int_C f(g_x\mathbf{i} + g_y\mathbf{j})\cdot(dy\,\mathbf{i} - dx\,\mathbf{j})$$

$$= \int_C [fg_x\, dy - fg_y\, dx]$$

$$= \iint\limits_D \left[\frac{\partial}{\partial x}(fg_x) + (\frac{\partial}{\partial y})(fg_y)\right] dA$$

$$= \iint\limits_D [f(g_{xx} + g_{yy})$$

$$+ (f_x\mathbf{i} + f_y\mathbf{j})\cdot(g_x\mathbf{i} + g_y\mathbf{j})]\, dA$$

$$= \iint\limits_D [f\nabla^2 g + \nabla f\cdot\nabla g]\, dA$$

38. $I = \iint\limits_D \left(\frac{\partial N}{\partial x} - \frac{\partial M}{\partial y}\right) dA = \int_C (M\, dx + N\, dy)$

$$= \iint\limits_{D_1} \left(\frac{\partial N}{\partial x} - \frac{\partial M}{\partial y}\right) dA + \iint\limits_{D_2} \left(\frac{\partial N}{\partial x} - \frac{\partial M}{\partial y}\right) dA$$

Let C_3 be the horizontal portion of the path (the line segment separating D_1 and D_2 with $c \leq x \leq d$). Then

$$I = \int_{C_1 - C_3} (M\, dx + N\, dy) + \int_c^d M\, dx$$

$$\int_{C_2 - C_3} (M\, dx + N\, dy) + \int_c^d M\, dx$$

$$= \int_C (M\, dx + N\, dy)$$

14.5 Surface Integrals, 939

For Problems 1-6, $z = \sqrt{4 - x^2 - y^2}$;
$f(x, y, z) = x^2 = y^2 + z^2 - 4$

$\nabla f = 2(x\mathbf{i} + y\mathbf{j} + z\mathbf{k}); \mathbf{N} = \frac{1}{2}(x\mathbf{i} + y\mathbf{j} + z\mathbf{k})$

$\mathbf{N} \cdot \mathbf{k} = \frac{z}{2}; \ dS = \frac{2}{z} dA_{xy} = \frac{2r \ dr \ d\theta}{\sqrt{4 - r^2}}.$

1. $\displaystyle\iint_S z \ dS = 2 \iint_R dA_{xy}$

$\displaystyle = \int_0^{2\pi} \int_0^2 r \ dr \ d\theta = 8\pi$

2. $\displaystyle\iint_S z^2 \ dS = 2 \iint_R z \ dA_{xy}$

$\displaystyle = 2 \int_0^{2\pi} \int_0^2 r\sqrt{4 - r^2} \ dr \ d\theta = \frac{32\pi}{3}$

3. $\displaystyle\iint_S (x - 2y) dS$

$\displaystyle = \int_0^{2\pi} \int_0^2 (2 \cos \theta - 4 \sin \theta) \frac{2}{\sqrt{4 - r^2}} r \ dr \ d\theta$

$\displaystyle = 8 \int_0^{2\pi} (\cos \theta - 2 \sin \theta) \ d\theta = 0$

4. $\displaystyle\iint_S (5 - 2x) dS$

$\displaystyle = 2 \int_0^{2\pi} \int_0^2 (5r - 2r^2 \cos \theta) \frac{dr \ d\theta}{\sqrt{4 - r^2}}$

$\displaystyle = 4 \int_0^{2\pi} (5 - \pi \cos \theta) \ d\theta$

$\displaystyle = 40\pi$

5. $\displaystyle\iint_S (x^2 + y^2)z \ dS = \iint_R (x^2 + y^2)z(\frac{2}{z}) \ dA_{xy}$

$\displaystyle = 2 \int_0^{2\pi} \int_0^r r^3 \ dr \ d\theta = 16\pi$

6. $\displaystyle\iint_S (x^2 + y^2) \ dS = \iint_R (x^2 + y^2)(\frac{2}{z}) \ dA_{xy}$

$\displaystyle = 2 \int_0^{2\pi} \int_0^2 \frac{2r^2 dr d\theta}{\sqrt{4 - r^2}} = 2\pi \int_0^{2\pi} \frac{8 \sin^2\alpha \cos \alpha \ d\alpha}{\cos \alpha}$

$\displaystyle = 2\pi(\pi - 2)$

7. $f(x, y, z) = 2 - y - z; \nabla f = -\mathbf{j} - \mathbf{k}$

$\mathbf{N} = \frac{1}{\sqrt{2}}(-\mathbf{j} - \mathbf{k}); \mathbf{N} \cdot \mathbf{k} = 1/\sqrt{2};$

$dS = \sqrt{2} \ dA_{xy}$

$\displaystyle\iint_S xy \ dS = \int_0^2 \int_0^2 xy \sqrt{2} \ dy \ dx$

$\displaystyle = 2\sqrt{2} \int_0^2 x \ dx = 4\sqrt{2}$

8. $f(x, y, z) = 4 - x - y - z;$

$\nabla f = -(\mathbf{i} + \mathbf{j} + \mathbf{k});$

$\mathbf{N} = -\frac{1}{\sqrt{3}}(\mathbf{i} + \mathbf{j} + \mathbf{k}); |\mathbf{N} \cdot \mathbf{k}| = \frac{1}{\sqrt{3}};$

$dS = \sqrt{3} \ dA_{xy}$

$\displaystyle\iint_S xy \ dS = \sqrt{3} \iint_R xy \ dy \ dx$

$\displaystyle = \frac{\sqrt{3}}{4}(4^2)(4^2) = 64\sqrt{2}$

9. $z = 5$, which is immaterial.

$\displaystyle\int_0^{2\pi} \int_0^1 r^3 \cos \theta \sin \theta \ dr \ d\theta = 0$

10. $z = 10$, which is immaterial.

$\displaystyle\int_0^2 \int_{-\sqrt{4 - x^2}/2}^{\sqrt{4 - x^2}/2} xy \ dy \ dx = 0$

11. $f(x, y, z) = 4 - x - 2y - z;$

$\nabla f = -(\mathbf{i} + 2\mathbf{j} + \mathbf{k}); \mathbf{N} = -\frac{1}{\sqrt{6}}(\mathbf{i} + 2\mathbf{j} + \mathbf{k})$

$|\mathbf{N} \cdot \mathbf{k}| = \frac{1}{\sqrt{6}}; \ dS = \sqrt{6} \ dA_{xy}$

$\displaystyle\iint_S (x^2 + y^2) dS = \sqrt{6} \int_0^4 \int_0^2 (x^2 + y^2) \ dy \ dx$

$\displaystyle = \sqrt{6} \int_0^4 \left(2x^2 + \frac{8}{3}\right) dx = \frac{160\sqrt{6}}{3}$

12. $f(x, y, z) = 4 - x - z; \nabla f = -(\mathbf{i} - \mathbf{k})$

$\mathbf{N} = -\frac{1}{\sqrt{2}}(\mathbf{i} + \mathbf{k}); |\mathbf{N} \cdot \mathbf{k}| = \frac{1}{\sqrt{2}};$

$dS = \sqrt{2} \ dA_{xy}$

$\displaystyle\iint_S (x^2 + y^2) dS = \sqrt{2} \int_0^2 \int_0^2 (x^2 + y^2) \ dy \ dx$

$\displaystyle = \sqrt{2} \int_0^2 \left(x^2 + \frac{4}{3}\right) dx = 0$

13. $f(x, y, z) = 4 - z$, which is immaterial.

$$\int_0^{2\pi} \int_0^1 r^3 \, dr \, d\theta = \frac{\pi}{2}$$

14. $f(x, y, z) = xy - z; \nabla f = y\mathbf{i} + x\mathbf{j} - \mathbf{k};$

$\mathbf{N} = \dfrac{y\mathbf{i} + x\mathbf{j} - \mathbf{k}}{\sqrt{1 + x^2 + y^2}}; \mathbf{N} \cdot \mathbf{k} = \dfrac{1}{\sqrt{1 + r^2}}$

$$\int\int_S (x^2 + y^2) \, dS$$

$$= \int\int_D (x^2 + y^2)\sqrt{1 + x^2 + y^2} \, dA_{xy}$$

$$= \int_0^{2\pi} \int_0^2 r^3\sqrt{1 + r^2} \, dr \, d\theta = \int_0^{2\pi} \left[\frac{10\sqrt{5}}{3} + \frac{2}{15}\right] d\theta$$

$$= 2\pi\left[\frac{10\sqrt{5}}{3} + \frac{2}{15}\right] \approx 47.6699$$

For Problems 15- 18, $f(x, y, z) = x^2 + y^2 - z$

$\nabla f = 2x\mathbf{i} + 2y\mathbf{j} - \mathbf{k}; \mathbf{N} = \dfrac{2x\mathbf{i} + 2y\mathbf{j} - \mathbf{k}}{\sqrt{4x^2 + 4y^2 + 1}}$

$\mathbf{N} \cdot \mathbf{k} = \dfrac{1}{\sqrt{1 + 4r^2}}$

15. $\displaystyle\int\int_S z \, dS = \int_0^{2\pi} \int_0^2 r^2 \sqrt{4r^2 + 1} \, r \, dr \, d\theta$

$$= 4\pi \int_0^2 r^3\sqrt{r^2 + (\tfrac{1}{2})^2} \, dr \quad \text{(Formula \#171)}$$

$$= 4\pi \left[\frac{(r^2 + \frac{1}{4})^{5/2}}{5} - \frac{\frac{1}{4}(r^2 + \frac{1}{4})^{3/2}}{3}\right]\Bigg|_0^2$$

$$= 4\pi\left[\left(\frac{4}{5} - \frac{1}{30}\right)\frac{17}{4}\sqrt{\frac{17}{4}} + \frac{1}{30}\left(\frac{1}{8}\right)\right]$$

$$= 4\pi\left(\frac{391\sqrt{17}}{240} + \frac{1}{240}\right)$$

$$= \frac{\pi}{60}\left(391\sqrt{17} + 1\right) \approx 84.4635$$

16. $\displaystyle\int\int_S (4 - z) \, dS = \int_0^{2\pi} \int_0^2 (4 - r^2)r\sqrt{1 + 4r^2} \, dr \, d\theta$

$$= \int_0^{2\pi}\left[\frac{209\sqrt{17}}{120} - \frac{41}{120}\right]$$

$$= 2\pi\left[\frac{209\sqrt{17}}{120} - \frac{41}{120}\right] \approx 60.2441$$

17. $\displaystyle\int\int_S \sqrt{1 + 4z} \, dS = \int_0^{2\pi} \int_0^2 r^2(1 + 4r^2) \, dr \, d\theta$

$$= 2\pi(2 + 16) = 36\pi$$

18. $\displaystyle\int\int_S \frac{dS}{\sqrt{1 + 4z}} = \int_0^{2\pi} \int_0^2 r \frac{\sqrt{1 + 4r^2}}{\sqrt{1 + 4r^2}} \, dr \, d\theta = 4\pi$

19. $f(x, y, z) = x^2 + y^2 + z^2 - 1;$

$\nabla f = 2(x\mathbf{i} + y\mathbf{j} + z\mathbf{k}); \mathbf{N} = x\mathbf{i} + y\mathbf{j} + \mathbf{k};$

$\mathbf{N} \cdot \mathbf{k} = \dfrac{1}{\sqrt{1 - r^2}}$

$$\int\int_S (x^2 + y^2) \, dS = \int_0^{2\pi} \int_0^1 \frac{r^2}{\sqrt{1 - r^2}} \, r \, dr \, d\theta$$

$$= 2\pi \int_0^1 \frac{r^3 dr}{\sqrt{1 - r^2}} \quad \text{(Formula \#227)}$$

$$= 2\pi\left[\frac{(1 - r^2)^{3/2}}{3} - (1 - r^2)^{1/2}\right]\Bigg|_0^1 = \frac{4\pi}{3}$$

20. $f(x, y, z) = 1 - x - y - z$

$\nabla f = -(\mathbf{i} + \mathbf{j} + \mathbf{k}); \mathbf{N} = -\dfrac{1}{\sqrt{3}}(\mathbf{i} + \mathbf{j} + \mathbf{k})$

$\mathbf{N} \cdot \mathbf{k} = \dfrac{1}{\sqrt{3}}$

$$\int\int_S 2x \, dS = \frac{2}{\sqrt{3}}\int_0^1 \int_0^1 x \, dy \, dx$$

$$= \frac{2}{\sqrt{3}}\int_0^1 x(1 - x) \, dx = \frac{\sqrt{3}}{9}$$

21. $\displaystyle\int\int_S (x^2 + y^2 + z^2) \, dS$

$$= \sqrt{2}\int_0^{2\pi} \int_0^1 [r^2 + (r\cos\theta + 1)^2] \, dr \, d\theta$$

$$= \sqrt{2}\int_0^{2\pi} \int_0^1 [r^3 + r^3\cos^2\theta + 2r^2\cos\theta + r] \, dr \, d\theta$$

$$= \frac{7\sqrt{2}}{8}\int_0^{2\pi} d\theta = \frac{7\pi\sqrt{2}}{4}$$

22. $f(x, y, z) = x + 2y + z - 1$

$\nabla f = \mathbf{i} + 2\mathbf{j} + \mathbf{k}; \mathbf{N} = \dfrac{1}{\sqrt{6}}(\mathbf{i} + 2\mathbf{j} + \mathbf{k})$

$\mathbf{F} = x\mathbf{i} + 2y\mathbf{j} + z\mathbf{k}; \mathbf{N} \cdot \mathbf{k} = \dfrac{1}{\sqrt{6}}(x + 4y + z)$

$$\int\int_S (x + 4y + z) \, dS$$

$$= \int_0^1 \int_0^{(1-x)/2} (x + 4y + 1 - x - 2y) \, dy \, dx$$

$$= \frac{1}{4} \int_0^1 (x^2 - 4x + 3) \, dx = \frac{5}{12}$$

23. $\displaystyle\iint_S \mathbf{F} \cdot \mathbf{N} \, dS$

$dS = \sqrt{(-5)^2 + 4^2 + 1} = \sqrt{42}$

$g(x, y, z) = z + 5x - 4y - 2,$

$\mathbf{N} = \dfrac{5\mathbf{i} - 4\mathbf{j} + \mathbf{k}}{\sqrt{42}}; \ \mathbf{F} \cdot \mathbf{N} = \dfrac{5x - 8y - 3z}{\sqrt{42}}$

$\displaystyle\iint_S \mathbf{F} \cdot \mathbf{N} \, dS = \int_0^1 \int_0^1 \dfrac{5x - 8y - 3z}{\sqrt{42}} \sqrt{42} \, dy \, dx$

$= \displaystyle\int_0^{2/5} (10x - 8) \, dx$

$+ \, 2 \displaystyle\int_{2/5}^1 \left[10x - 8 - \left(\frac{5x}{4} - \frac{1}{2}\right)\left\{10x - 5\left(\frac{5x}{4} - \frac{1}{2}\right) - 3\right\}\right] dx$

$= -\dfrac{24}{5} + 2\displaystyle\int_{2/5}^1 \left(10x - 8 - \frac{75x^2}{16} + \frac{5x}{2} - \frac{1}{4}\right) dx$

$= -\dfrac{24}{5} + 2\left(-\dfrac{25}{16} - 2 + \dfrac{12}{5}\right) = -\dfrac{57}{8}$

24. $\mathbf{F} = x\mathbf{i} + y\mathbf{j} + 2z\mathbf{k}$

Front: $x = 1$; $\mathbf{N} = \mathbf{i}$; $\mathbf{F} \cdot \mathbf{N} = x$; $dS = dy \, dz$

$\displaystyle\iint_S 2 \, dA = 1$

Back: $x = 1$; $\mathbf{N} = -\mathbf{i}$; $\mathbf{F} \cdot \mathbf{N} = -x$; $dS = dy \, dz$

$\displaystyle\iint_S 0 \, dA = 0$

Top: $z = 1$; $\mathbf{N} = \mathbf{k}$; $\mathbf{F} \cdot \mathbf{N} = 2z$; $dS = dx \, dy$

$\displaystyle\iint_S 2 \, dA = 2$

Bottom: $z = 0$; $\mathbf{N} = -\mathbf{k}$; $\mathbf{F} \cdot \mathbf{N} = -2z$; $dS = dx \, dy$

$\displaystyle\iint_S 0 \, dA = 0$

Right: $y = 1$; $\mathbf{N} = \mathbf{j}$; $\mathbf{F} \cdot \mathbf{N} = y$; $dS = dx \, dz$

$\displaystyle\iint_S dA = 1$

Left: $y = 0$; $\mathbf{N} = -\mathbf{j}$; $\mathbf{F} \cdot \mathbf{N} = -y$; $dS = dx \, dz$

$\displaystyle\iint_S 0 \, dA = 1$

$\displaystyle\iint_S \mathbf{F} \cdot \mathbf{N} \, dS = 1 + 2 + 1 = 4$

25. $x^2 + y^2 + z^2 = 1$; $\nabla f = x\mathbf{i} + y\mathbf{j} + z\mathbf{k}$;

$dS = \dfrac{1}{z} \, dA_{xy}$; $\mathbf{N} = x\mathbf{i} + y\mathbf{j} + z\mathbf{k}$

$\mathbf{F} \cdot \mathbf{N} = x^2 + y^2$

$\displaystyle\iint_S \mathbf{F} \cdot \mathbf{N} \, dS = \int_0^{2\pi} \int_0^1 \dfrac{r^3 \, dr \, d\theta}{\sqrt{1 - r^2}}$

$= 2\pi \displaystyle\int_0^1 r^2 \, \dfrac{r \, dr}{\sqrt{1 - r^2}}$

$= \dfrac{4\pi}{3}$

26. $z = 1$ intersects $x^2 + y^2 + z^2 = 5$ at $r = 2$

$\mathbf{F} = 2x\mathbf{i} - 3y\mathbf{j}$; $z = \sqrt{5 - x^2 - y^2}$, $z \ge 1$

$\nabla f = 2(x\mathbf{i} + y\mathbf{j} + z\mathbf{k})$; $dS = \dfrac{\sqrt{5}}{z} \, dA_{xy}$;

$\mathbf{N} = \dfrac{1}{\sqrt{5}}(x\mathbf{i} + y\mathbf{j} + z\mathbf{k})$;

$\mathbf{F} \cdot \mathbf{N} \, dS = \dfrac{1}{\sqrt{5}}(2x^2 - 2y^2)$

$\displaystyle\iint_S \mathbf{F} \cdot \mathbf{N} \, dS = \int_0^{2\pi} \int_0^2 \dfrac{2r^2\cos^2\theta - 3r^2\sin^2\theta}{\sqrt{5 - r^2}} \, r \, dr \, d\theta$

$= 5\displaystyle\int_0^{2\pi} \int_0^2 r^2 \dfrac{r \, dr}{\sqrt{5 - r^2}}\cos^2\theta \, dr \, d\theta - 3\int_0^{2\pi} \int_0^2 \dfrac{r^3 \, d\theta}{\sqrt{5 - r^2}}$

$= \displaystyle\int_0^{2\pi}\left[\dfrac{50\sqrt{5}}{3} - \dfrac{70}{3}\right]\cos^2\theta \, d\theta - \int_0^{2\pi}(10\sqrt{5} - 14) \, d\theta$

$= \dfrac{5\pi}{3}[10\sqrt{5} - 14] - \pi(20\sqrt{5} - 28)$

$= \dfrac{2\pi}{3}[7 - 5\sqrt{5}] \approx -8.75528$

27. $\mathbf{F} = x^2\mathbf{i} + y^2\mathbf{j} + z^2\mathbf{k}$; $z = y + 1$, $dS = \sqrt{2}$

$g(x, y, z) = z - y - 1$; $\mathbf{N} = \dfrac{-\mathbf{j} + \mathbf{k}}{\sqrt{2}}$

$\mathbf{F} \cdot \mathbf{N} = \dfrac{-y^2 + z^2}{\sqrt{2}} = \dfrac{2y + 1}{\sqrt{2}}$

Now $R: x^2 + y^2 = 1$, since the plane intersects the cylinder as a tangent in the xy-plane. Using polar coordinates:

$\displaystyle\iint_S \mathbf{F} \cdot \mathbf{N} \, dS = \int_0^{2\pi} \int_0^1 (2r \sin\theta + 1) \, r \, dr \, d\theta$

$$= \int_0^{2\pi} \left(\frac{2}{3}\sin\theta + \frac{1}{2}\right) d\theta = \pi$$

28. $\mathbf{R} = u\mathbf{i} + v\mathbf{j} - v\mathbf{k}; \; \mathbf{R}_u = \mathbf{i} + \mathbf{j}; \; \mathbf{R}_v = -\mathbf{k}$

$$\mathbf{R}_u \times \mathbf{R}_v = \begin{bmatrix} \mathbf{i} & \mathbf{j} & \mathbf{k} \\ 1 & 1 & 0 \\ 0 & 0 & -1 \end{bmatrix} = -\mathbf{i} + \mathbf{j}$$

$$\|\mathbf{R}_u \times \mathbf{R}_v\| = \sqrt{2}; \; dx\,dy = \sqrt{2}\,du\,dv$$

$$\int\!\!\int_S (3x - y + 2z)\,dS$$

$$= \int_0^1 \int_1^2 (3u - u - 2v)\,dv\,du$$

$$= \int_0^1 (2u - 3)\,du = -2$$

29. $\mathbf{R} = u^2\mathbf{i} + v\mathbf{j} - u\mathbf{k}; \; \mathbf{R}_u = 2u\mathbf{i} + \mathbf{k}; \; \mathbf{R}_v = \mathbf{k}$

$$\mathbf{R}_u \times \mathbf{R}_v = \begin{vmatrix} \mathbf{i} & \mathbf{j} & \mathbf{k} \\ 2u & 0 & 1 \\ 0 & 1 & 0 \end{vmatrix} = -\mathbf{i} + 2u\mathbf{k}$$

$$\|\mathbf{R}_u \times \mathbf{R}_v\| = \sqrt{1 + 4u^2};$$

$$\int\!\!\int_S (x - y^2 + z)\,dS, \text{ where}$$

$$= \int_0^1 \int_0^1 (u^2 - v^2 + u)\sqrt{1 + 4u^2}\,dv\,du$$

$$= \int_0^1 (u^2 - \frac{1}{3} + u)\sqrt{1 + 4u^2}\,du$$

$$= -\frac{19\ln(\sqrt{5} + 2)}{192} + \frac{17\sqrt{5}}{32} - \frac{1}{12}$$

$$\approx 0.9617$$

30. $\mathbf{R} = u\mathbf{i} + v^2\mathbf{j} - v\mathbf{k}; \; \mathbf{R}_u = \mathbf{i}; \; \mathbf{R}_v = 2v\mathbf{j} - \mathbf{k}$

$$\mathbf{R}_u \times \mathbf{R}_v = \begin{vmatrix} \mathbf{i} & \mathbf{j} & \mathbf{k} \\ 1 & 0 & 0 \\ 0 & 2v & -1 \end{vmatrix} = \mathbf{j} + 2v\mathbf{k}$$

$$\|\mathbf{R}_u \times \mathbf{R}_v\| = \sqrt{1 + 4v^2};$$

$$\int\!\!\int_S (\tan^{-1}x + y - z^2)\,dS, \text{ where}$$

$$= \int_0^1 \int_0^1 (\tan^{-1}u + v^2 - v^2)\sqrt{1 + 4v^2}\,dv$$

$$= (\frac{\pi}{4} - \frac{1}{2}\ln 2)\int_0^1 \sqrt{1 + 4v^2}\,dv$$

$$= \left(\frac{\pi}{4} - \frac{1}{2}\ln 2\right)\left[\frac{1}{2}\sqrt{5} + \frac{1}{4}\ln(\sqrt{5} + 2)\right]$$

$$= \frac{1}{16}(\pi - 2\ln 2)(2\sqrt{5} + \ln(\sqrt{5} + 2))$$

31. $\mathbf{R} = u\mathbf{i} - u^2\mathbf{j} + v\mathbf{k}; \; \mathbf{R}_u = \mathbf{i} - 2u\mathbf{j}; \; \mathbf{R}_v = \mathbf{k}$

$$\mathbf{R}_u \times \mathbf{R}_v = \begin{bmatrix} \mathbf{i} & \mathbf{j} & \mathbf{k} \\ 1 & -2u & 0 \\ 0 & 0 & 1 \end{bmatrix} = -2u\mathbf{j} - \mathbf{j}$$

$$\|\mathbf{R}_u \times \mathbf{R}_v\| = \sqrt{1 + 4u^2};$$

$$\int\!\!\int_S (x^2 + y - z)\,dS$$

$$= \int_0^2 \int_0^1 \left[u^2 + (-u^2 - v\sqrt{1 + 4u^2})\right] dv\,du$$

$$= -\frac{1}{2}\int_0^2 \sqrt{1 + 4u^2}\,du$$

$$= -\frac{1}{8}[4\sqrt{17} + \ln(4 + \sqrt{17})] \approx -2.3234$$

32. $m = A; \; f(x, y, z) = 4 - x - 2y - z$

$$\nabla f = -\mathbf{i} - 2\mathbf{j} - \mathbf{k}; \; \mathbf{N}\cdot\mathbf{k} = -\frac{1}{16};$$

$$dS = \sqrt{6}\,A_{xy}$$

$$m = \sqrt{6}\int_0^4 \int_0^{(4-x)/2} dy\,dx$$

$$= \frac{\sqrt{6}}{2}\int_0^4 (4 - x)\,dx = 4\sqrt{6}$$

33. $f(x, y, z) = 10 - 2x - y - z;$

$$\nabla f = -2\mathbf{i} - \mathbf{j} - \mathbf{k}; \; \mathbf{N}\cdot\mathbf{k} = -\frac{1}{\sqrt{6}};$$

$$dS = \sqrt{6}\,dA_{xy}$$

$$m = \sqrt{6}\int_0^5 \int_0^{10-2x} dy\,dx$$

$$= \sqrt{6}\int_0^5 (10 - 2x)\,dx = 25\sqrt{6}$$

34. $f(x, y, z) = x^2 + y^2 - z; \; \nabla f = 2x\mathbf{i} + 2y\mathbf{j} - \mathbf{k}$

$$|\mathbf{N}\cdot\mathbf{k}| = \frac{1}{\sqrt{1 + 4r^2}}; \; dS = \sqrt{1 + 4r^2}\,r\,dr\,d\theta$$

$$m = \int_0^{2\pi} \int_0^1 r(1 + 4r^2)^{1/2}\,dr\,d\theta$$

$$= \frac{1}{12}\int_0^{2\pi} [5\sqrt{5} - 1]\, d\theta$$

$$= \frac{\pi}{6}(5\sqrt{5} - 1)$$

35. $f(x, y, z) = 1 - x^2 - y^2 - z,$

$dS = \sqrt{4x^2 + 4y^2 + 1}$

$$m = \int_0^{2\pi}\int_0^1 1\ \sqrt{4r^2 + 1}\ r\, dr\, d\theta$$

$$= \frac{1}{8}\int_0^{2\pi}\int_0^1 \sqrt{4r^2 + 1}\ (8r\, dr)\, d\theta$$

$$= \frac{1}{8}\int_0^{2\pi}\left(\frac{10\sqrt{5}}{3} - \frac{2}{3}\right) d\theta$$

$$= \frac{\pi}{6}\left(5\sqrt{5} - 1\right)$$

36. $f(x, y, z) = x^2 + y^2 + z^2 - 5$

$\nabla f = 2(x\mathbf{i} + y\mathbf{j} + \mathbf{k}); \mathbf{N}\cdot\mathbf{k} = \frac{z}{\sqrt{5}};$

$dS = \frac{\sqrt{5}}{z}\, dA_{xy}$

$$m = \int_0^{2\pi}\int_2^{\sqrt{5}} \frac{\sqrt{5}\ r}{\sqrt{5 - r^2}}\, dr\, d\theta = 2\sqrt{5}\ \pi$$

37. $f(x, y, z) = 1 - x - y - z$

$\nabla f = -(\mathbf{i} + \mathbf{j} + \mathbf{k}); \mathbf{N}\cdot\mathbf{k} = \frac{1}{\sqrt{3}};$

$dS = \sqrt{3}\, dA_{xy}$

$$m = \sqrt{3}\iint_{R_{xy}} dA_{xy} = \frac{\sqrt{3}}{2}$$

38. $f(x, y, z) = x^2 + y^2 + z^2 - a^2$

$\nabla f = 2(x\mathbf{i} + y\mathbf{j} + z\mathbf{k}); \mathbf{N}\cdot\mathbf{k} = \frac{z}{a};$

$dS = \frac{ar\, dr\, d\theta}{\sqrt{a^2 - r^2}}$

$$m = a\int_0^{2\pi}\int_0^{a\sin\phi} \frac{r\, dr\, d\theta}{\sqrt{a^2 - r^2}}$$

$$= 2\pi a(a - a\sqrt{1 - \sin^2\phi})$$

$$= 2\pi a^2(1 - \cos\phi)$$

39. $f(x, y, z) = z^2 - c^2(x^2 + y^2)$

$\nabla f = 2c^2(x\mathbf{i} + y\mathbf{j} - 2z\mathbf{k})$

$|\mathbf{N}\cdot\mathbf{k}| = \dfrac{z}{\sqrt{c^2(x^2 + y^2) + z^2}};$

$$dS = \frac{\sqrt{2}\,cr^2\, dr\, d\theta}{cr}$$

$$m = \sqrt{2}\int_0^{2\pi}\int_0^a r(1)\, dr\, d\theta = \sqrt{2}\pi a^2$$

$$I_z = \sqrt{2}\int_0^{2\pi}\int_0^a r^3\, dr\, d\theta = \sqrt{2}(2\pi)\frac{a^4}{4} = \frac{ma^2}{2}$$

40. $f(x, y, z) = x^2 + y^2 + z^2 - a^2$

$\nabla f = 2(x\mathbf{i} + y\mathbf{j} + z\mathbf{k}); |\mathbf{N}\cdot\mathbf{k}| = \frac{z}{a}$

$dS = \dfrac{ar\, dr\, d\theta}{\sqrt{a^2 - r^2}}$

$$m = 8a\int_0^{\pi/2}\int_0^a r(a^2 - r^2)^{-1/2}\, dr\, d\theta = 4\pi a^2$$

$$I_z = 8a\int_0^{\pi/2}\int_0^a r^3(a^2 - r^2)^{-1/2}\, dr\, d\theta$$

$$= \frac{16a^4}{3}\int_0^{\pi/2} d\theta = \frac{8\pi a^4}{3} = \frac{2}{3}ma^2$$

41. Consider a long narrow rectangular strip with vertices A, B along one narrow side, and C, D along the opposite narrow side. AC and BD are the long sides. Join A and C and also B and D. The outside unit normal moves continuously around the cylindrical shell and always points outward when moving from point A around to point C (one complete revolution). Now start over with the long narrow strip, twist once so that A is superimposed on D, and B joins C. This is a Möbius strip. The normal points outward at A, but inward at C. It needs to jump across the strip to point outward. This jump implies discontinuity.

14.6 Stokes's Theorem, Page 497

1. Evaluating the line integral $\displaystyle\int_C \mathbf{F}\cdot d\mathbf{R}$

$$\int_C \mathbf{F}\cdot d\mathbf{R} = \int_C (z\, dx + 2x\, dy + 3y\, dz)$$

$$= 2\int_0^{2\pi} 9\cos^2\theta\, d\theta - 3\int_0^{2\pi} 0 = 18\pi$$

Evaluating the integral $\displaystyle\iint_S \text{curl}\ \mathbf{F}\cdot\mathbf{N}\, dS$

$C: x = 3\cos\theta,\ y = 3\sin\theta,\ z = 0;\ dz = 0;$
$0 \le \theta \le 2\pi$

$$\text{curl } \mathbf{F} = \begin{vmatrix} \mathbf{i} & \mathbf{j} & \mathbf{k} \\ \dfrac{\partial}{\partial x} & \dfrac{\partial}{\partial y} & \dfrac{\partial}{\partial z} \\ z & 2x & 3y \end{vmatrix} = 2(x\mathbf{i} + y\mathbf{j} + z\mathbf{k})$$

$$\mathbf{N} = \tfrac{1}{3}(x\mathbf{i} + y\mathbf{j} + z\mathbf{k}); \; dS = \dfrac{3r\,dr\,d\theta}{\sqrt{9 - r^2}}$$

$$\iint_S \text{curl } \mathbf{F} \cdot \mathbf{N} \, dS = \iint_R \tfrac{1}{3}(3x + y + 2z) \dfrac{3r\,dr\,d\theta}{\sqrt{9 - r^2}}$$

$$= \tfrac{1}{3} \int_0^{2\pi} \int_0^3 (3r\cos\theta + r\sin\theta + 2\sqrt{9 - r^2}) \dfrac{3r\,dr}{\sqrt{9 - r^2}}\, d\theta$$

$$= \int_0^{2\pi}\int_0^3 \dfrac{3r^2\cos\theta\,dr\,d\theta}{\sqrt{9 - r^2}} + \int_0^{2\pi}\int_0^3 \dfrac{r^2\sin\theta\,dr\,d\theta}{\sqrt{9 - r^2}}$$

$$+ \int_0^{2\pi}\int_0^3 2r\,dr\,d\theta$$

$$= 0 + 0 + 18\pi = 18\pi$$

2. Evaluating the line integral $\displaystyle\int_C \mathbf{F} \cdot d\mathbf{R}$

C_1: $x = 3 - 3t; \; y = \dfrac{3t}{2}; \; 0 \le t \le 1$

$$I_1 = \int_0^1 \dfrac{3t}{2}(-3\,dt) + \int_0^1 (3 - 3t)(\tfrac{3}{2})\,dt = 0$$

C_2: $z = 3t; \; y = (1 - t)/2; \; 0 \le t \le 1$

$$I_2 = \int_0^1 (3t)(3\,dt) = \dfrac{9}{2}$$

C_3: $z = 3(1 - t); \; x = 3t; \; 0 \le t \le 1$

$$I_3 = \int_0^1 3(1 - t)(3\,dt) + \int_0^1 (3 - 6t)(-3\,dt) = \dfrac{9}{2}$$

$$\int_C \mathbf{F} \cdot d\mathbf{R} = \int_C [(y+z)\,dx + x\,dy + (z - x)\,dz]$$

$$= I_1 + I_2 + I_3 = 9$$

Evaluating the integral $\displaystyle\iint_S \text{curl } \mathbf{F} \cdot \mathbf{N} \, dS$

$$\text{curl } \mathbf{F} = \begin{vmatrix} \mathbf{i} & \mathbf{j} & \mathbf{k} \\ \dfrac{\partial}{\partial x} & \dfrac{\partial}{\partial y} & \dfrac{\partial}{\partial z} \\ y + z & x & z - x \end{vmatrix} = 2\mathbf{j}$$

$$f(x, y, z) = x + 2y + z - 3$$

$$\nabla f = \mathbf{i} + 2\mathbf{j} + \mathbf{k}; \; \mathbf{N} = \dfrac{1}{\sqrt{6}}(\mathbf{i} + 2\mathbf{j} + \mathbf{k})$$

$$dS = \sqrt{6}\, A_{xy}; \; \text{curl } \mathbf{F} \cdot \mathbf{N} = 4\,dA_{xy}$$

$$\iint_S \text{curl } \mathbf{F} \cdot \mathbf{N} \, dS = 4\iint_R dA_{xy} = 4(\tfrac{1}{2})(3)(\tfrac{3}{2}) = 9$$

3. Evaluating the line integral $\displaystyle\int_C \mathbf{F} \cdot d\mathbf{R}$

C_1: $x + 2y = 3$; $\; C_2$: $2y + z = 3$; $\; E_3$: $x + z = 3$.

Parametrizing all three with $0 \le t \le \tfrac{3}{2}$:

C_1: $x = 3 - 2t, \; y = t, \; z = 0,$
  $\mathbf{R} = (3 - 2t)\mathbf{i} + t\mathbf{j},$
  $d\mathbf{R} = -2\mathbf{i} + \mathbf{j},$
  $\mathbf{F} \cdot d\mathbf{R} = -2(x + 2z) + (y - x)$
    $= -3x + y - 4z$
    $= 7t - 9$

$$I_1 = \int_0^{3/2} (7t - 9)\,dt = -\dfrac{45}{8}$$

C_2: $x = 0, \; y = \dfrac{3}{2} - t, \; z = 2t,$
  $\mathbf{R} = (\tfrac{3}{2} - t)\mathbf{j} + 2t\mathbf{k},$
  $d\mathbf{R} = -\mathbf{j} + 2\mathbf{k},$
  $\mathbf{F} \cdot d\mathbf{R} = -(y - x) + 2(z - y)$
    $= x - 3y + 2z$
    $= 7t - \dfrac{9}{2}$

$$I_2 = \int_0^{3/2} \left(7t - \dfrac{9}{2}\right)dt = \dfrac{9}{8}$$

C_3: $x = 2t, \; y = 0, \; z = 3 - 2t,$
  $\mathbf{R} = 2t\mathbf{i} + (3 - 2t)\mathbf{k},$
  $d\mathbf{R} = 2\mathbf{i} - 2\mathbf{k},$
  $\mathbf{F} \cdot d\mathbf{R} = 2(x + 2z) - 2(z - y)$
    $= 2x + 2y + 2z$
    $= 6$

$$I_3 = \int_0^{3/2} 6\,dt = 9$$

$$\int_C \mathbf{F} \cdot d\mathbf{R} = I_1 + I_2 + I_3 = -\dfrac{45}{8} + \dfrac{9}{8} + 9 = \dfrac{9}{2}$$

Evaluating the integral $\displaystyle\iint_S \text{curl } \mathbf{F} \cdot \mathbf{N} \, dS$

$$\text{curl } \mathbf{F} = \begin{vmatrix} \mathbf{i} & \mathbf{j} & \mathbf{k} \\ \dfrac{\partial}{\partial x} & \dfrac{\partial}{\partial y} & \dfrac{\partial}{\partial z} \\ x + 2z & y - x & z - y \end{vmatrix} = -\mathbf{i} + 2\mathbf{j} - \mathbf{k}$$

$$\mathbf{N} = \dfrac{1}{\sqrt{6}}(\mathbf{i} + 2\mathbf{j} + \mathbf{k}); \; dS = \sqrt{6}\, dy\,dx$$

$$\iint_S \text{curl } \mathbf{F} \cdot \mathbf{N} \, dS = \iint_R \dfrac{-1 + 4 - 1}{\sqrt{6}} \sqrt{6}\, dy\,dx$$

$$= 2(\text{the area of the } \triangle)$$

$$= 2\left(\dfrac{9}{4}\right) = \dfrac{9}{2}$$

4. Evaluating the line integral $\int\limits_C \mathbf{F} \cdot d\mathbf{R}$

C: $x = r \cos\theta$, $y = 4$, $z = r \sin\theta$;
$dA = dx\,dz = r\,dr\,d\theta$

$$\int\limits_C \mathbf{F} \cdot d\mathbf{R} = \int\limits_C (2xy\,dx + z^2\,dz)$$

$$= \int\limits_0^{2\pi} [2(2\cos\theta)(4)(-2\sin\theta) + (4\sin^2\theta)(4\cos\theta)]\,d\theta$$

$$= [-4\cos^2\theta + \frac{16}{3}\sin^3\theta]\Big|_0^{2\pi} = 0$$

Evaluating the integral $\iint\limits_S \text{curl } \mathbf{F} \cdot \mathbf{N}\,dS$

$$\text{curl } \mathbf{F} = \begin{vmatrix} \mathbf{i} & \mathbf{j} & \mathbf{k} \\ \frac{\partial}{\partial x} & \frac{\partial}{\partial y} & \frac{\partial}{\partial z} \\ 2xy & 0 & z^2 \end{vmatrix} = -2x\mathbf{k}$$

$f(x, y, z) = x^2 - y + z^2$;

$$\mathbf{N} = \frac{2x\mathbf{i} - \mathbf{j} + 2z\mathbf{k}}{\sqrt{4(x^2 + y^2) + 1}}; \quad dS = \sqrt{1 + 4r^2}\, r\,dr\,d\theta$$

$\text{curl } \mathbf{F} \cdot \mathbf{N}\,dS = -2x\,dS$

$$\iint\limits_S \text{curl } \mathbf{F} \cdot \mathbf{N}\,dS$$

$$= \int\limits_0^{2\pi}\int\limits_0^2 (-2)r^2\cos\theta\sqrt{1 + 4r^2}\,dr\,d\theta$$

$$= 0 \text{ since } \int\limits_0^{2\pi}\cos\theta\,d\theta = 0$$

5. Evaluating the line integral $\int\limits_C \mathbf{F} \cdot d\mathbf{R}$

C: $x = 2\cos\theta$; $y = 2\sin\theta$; $z = dz = 0$

$$\int\limits_C \mathbf{F} \cdot d\mathbf{R} = \int\limits_C (2y\,dx - 6z\,dy + 3x\,dx)$$

$$= \int\limits_0^{2\pi} 2(2\sin\theta)(-2\sin\theta)\,d\theta = -8\pi$$

Evaluating the integral $\iint\limits_S \text{curl } \mathbf{F} \cdot \mathbf{N}\,dS$

$$\text{curl } \mathbf{F} = \begin{vmatrix} \mathbf{i} & \mathbf{j} & \mathbf{k} \\ \frac{\partial}{\partial x} & \frac{\partial}{\partial y} & \frac{\partial}{\partial z} \\ 2y & -6z & 3x \end{vmatrix} = 6\mathbf{i} - 3\mathbf{j} - 2\mathbf{k}$$

$f(x, y, z) = x^2 + y^2 + z - 4$;

$$\nabla f = 2x\mathbf{i} + 2y\mathbf{j} + \mathbf{k}; \quad \mathbf{N} = \frac{2x\mathbf{i} + 2y\mathbf{j} + \mathbf{k}}{\sqrt{4(x^2 + y^2) + 1}}$$

$dS = \sqrt{1 + 4r^2}\, r\,dr\,d\theta$

$$\text{curl } \mathbf{F} \cdot \mathbf{N}\,dS = \frac{12x - 6y - 2}{\sqrt{4r^2 + 1}}\sqrt{1 + 4r^2}\, r\,dr\,d\theta$$

$$\iint\limits_S \text{curl } \mathbf{F} \cdot \mathbf{N}\,dS$$

$$= \int\limits_0^{2\pi}\int\limits_0^2 (12r\cos\theta - 6r\sin\theta - 2)r\,dr\,d\theta$$

$$= \int\limits_0^{2\pi} (32\cos\theta - 16\sin\theta - 4)\,d\theta = -8\pi$$

6. $\mathbf{F} = x^3 y^2\mathbf{i} + \mathbf{j} + z^2\mathbf{k}$

$$\text{curl } \mathbf{F} = \begin{vmatrix} \mathbf{i} & \mathbf{j} & \mathbf{k} \\ \frac{\partial}{\partial x} & \frac{\partial}{\partial y} & \frac{\partial}{\partial z} \\ x^3 y^2 & 1 & z^2 \end{vmatrix} = \frac{-2y}{\sqrt{4r^2 + 1}}\mathbf{k}$$

$f(x, y, z) = x^2 + y^2 - z$;

$$\nabla f = 2x\mathbf{i} + 2y\mathbf{j} - \mathbf{k}; \quad \mathbf{N} = \frac{2x\mathbf{i} + 2y\mathbf{j} - \mathbf{k}}{\sqrt{4(x^2 + y^2) + 1}}$$

$dS = \sqrt{1 + 4r^2}\, r\,dr\,d\theta$

$\text{curl } \mathbf{F} \cdot \mathbf{N}\,dS = -2r^2\sin\theta\,dr\,d\theta$

$$\int\limits_C (x^3 y^2\,dx + dy + z^2\,dz)$$

$$= \iint\limits_S \text{curl } \mathbf{F} \cdot \mathbf{N}\,dS \text{ (Stokes's theorem)}$$

$$= -2\int\limits_0^{2\pi}\int\limits_0^1 r^2\sin\theta\,dr\,d\theta$$

$$= -2\int\limits_0^{2\pi} \sin\theta\,d\theta = 0$$

7. $\mathbf{F} = z\mathbf{i} + x\mathbf{j} + y\mathbf{k}$

$$\text{curl } \mathbf{F} = \begin{vmatrix} \mathbf{i} & \mathbf{j} & \mathbf{k} \\ \frac{\partial}{\partial x} & \frac{\partial}{\partial y} & \frac{\partial}{\partial z} \\ z & x & y \end{vmatrix} = \mathbf{i} + \mathbf{j} + \mathbf{k}$$

$$\mathbf{N} = \frac{2\mathbf{i} + \mathbf{j} + 3\mathbf{k}}{\sqrt{14}}; \quad dS = \frac{\sqrt{14}}{3}\,dy\,dx$$

$$\text{curl } \mathbf{F} \cdot \mathbf{N} = \frac{2 + 1 + 3}{\sqrt{14}} = \frac{6}{\sqrt{14}}$$

$$\int_C (z\ dx + x\ dy + y\ dz)$$

$$= \iint_S \text{curl } \mathbf{F} \cdot \mathbf{N}\ dS \quad \text{(Stokes's theorem)}$$

$$= \iint_R \frac{6}{\sqrt{14}} \frac{\sqrt{14}}{3}\ dy\ dx$$

$$= -2(\text{the area of the } \triangle \text{ region}) = -18$$

8. $\mathbf{F} = y\mathbf{i} - 2x\mathbf{j} + z\mathbf{k}; \ f(x,\ y,\ z) = x^2 + y^2 - z$

$$\text{curl } \mathbf{F} = \begin{vmatrix} \mathbf{i} & \mathbf{j} & \mathbf{k} \\ \dfrac{\partial}{\partial x} & \dfrac{\partial}{\partial y} & \dfrac{\partial}{\partial z} \\ y & -2x & z \end{vmatrix} = -3\mathbf{k}$$

$$dS = \sqrt{(2x)^2 + (2y)^2 + 1}\ dy\ dx$$

$$= \sqrt{4(x^2 + y^2) + 1}\ dy\ dx$$

$$\mathbf{N} = \frac{-2x\mathbf{i} - 2y\mathbf{j} + \mathbf{k}}{\sqrt{4(x^2 + y^2) + 1}}$$

$$\int_C (y\ dx - 2x\ dy + z\ dz)$$

$$= \iint_S \text{curl } \mathbf{F} \cdot \mathbf{N}\ dS \quad \text{(Stokes's theorem)}$$

$$= \iint_R (-3\mathbf{k}) \cdot [-2x\mathbf{i} - 2y\mathbf{j} + \mathbf{k}]\ dy\ dx$$

$$= \iint_R -3\ dy\ dx$$

$$= -3(\text{area of circle})$$

$$= -3\left[\pi(\frac{\sqrt{3}}{2})^2\right] = -\frac{9\pi}{2}$$

9. $\mathbf{F} = 2xy^2\mathbf{i} + 2x^2yz\mathbf{j} + (x^2y^2 - 2z)\mathbf{k};$

$$\text{curl } \mathbf{F} = \begin{vmatrix} \mathbf{i} & \mathbf{j} & \mathbf{k} \\ \dfrac{\partial}{\partial x} & \dfrac{\partial}{\partial y} & \dfrac{\partial}{\partial z} \\ 2xy^2z & 2x^2yz & x^2y^2 - z \end{vmatrix} = 0$$

$$\int_C [2xy^2z\ dx + 2x^2yz\ dy + (x^2y^2 - 2z)dz]$$

$$= \iint_S \text{curl } \mathbf{F} \cdot \mathbf{N}\ dS \quad \text{(Stokes's theorem)}$$

$$= 0$$

10. $\mathbf{F} = y\mathbf{i} + z\mathbf{j} + y\mathbf{k};$

$f(x,\ y,\ z) = x + y + z$

$$\text{curl } \mathbf{F} = \begin{vmatrix} \mathbf{i} & \mathbf{j} & \mathbf{k} \\ \dfrac{\partial}{\partial x} & \dfrac{\partial}{\partial y} & \dfrac{\partial}{\partial z} \\ y & -2x & y \end{vmatrix} = -\mathbf{k}$$

$$\int_C (y\ dx + z\ dy + y\ dz)$$

$$= \iint_S \text{curl } \mathbf{F} \cdot \mathbf{N}\ dS \quad \text{(Stokes's theorem)}$$

$$= -\iint_R dA_{xy} = -4\pi$$

(Because the intersection is a circle of radius 2.)

11. The path is the curve which is the intersection of $x^2 + y^2 = 2$ and $x + y = 2$; it is a circle with radius 2.

$\mathbf{F} = y\mathbf{i} + z\mathbf{j} + x\mathbf{k}$

$$\text{curl } \mathbf{F} = \begin{vmatrix} \mathbf{i} & \mathbf{j} & \mathbf{k} \\ \dfrac{\partial}{\partial x} & \dfrac{\partial}{\partial y} & \dfrac{\partial}{\partial z} \\ y & z & x \end{vmatrix} = -(\mathbf{i} + \mathbf{j} + \mathbf{k})$$

$$\text{curl } \mathbf{F} \cdot \mathbf{N} = -\frac{1}{\sqrt{2}}; \ dS = \sqrt{2}$$

$$\int_C (y\ dx + z\ dy + x\ dz)$$

$$= \iint_S \text{curl } \mathbf{F} \cdot \mathbf{N}\ dS \quad \text{(Stokes's theorem)}$$

$$= \iint_R (-\sqrt{2})\ \sqrt{2}\ dy\ dx$$

$$= -2 \ (\text{area of a circle with } r = 1)$$

$$= -2\pi$$

12. The path is the curve $r = \sqrt{2}$.

$\mathbf{F} = (z + \cos x)\mathbf{i} + (x + y^2)\mathbf{j} + (y + e^z)\mathbf{k}$

$\mathbf{N} = \mathbf{k}; \ dS = r\ dr\ d\theta$

$$\text{curl } \mathbf{F} = \begin{vmatrix} \mathbf{i} & \mathbf{j} & \mathbf{k} \\ \dfrac{\partial}{\partial x} & \dfrac{\partial}{\partial y} & \dfrac{\partial}{\partial z} \\ z + \cos x & x + y^2 & y + e^z \end{vmatrix}$$

$$= \mathbf{i} + \mathbf{j} + \mathbf{k}$$

$$\text{curl } \mathbf{F} \cdot \mathbf{N}\ dS = r\ dr\ d\theta$$

$$\int_C [(z + \cos x)dx + (x + y^2)dy + (y + e^z)dz]$$

$$= \iint_S \text{curl } \mathbf{F} \cdot \mathbf{N} \ dS \quad \text{(Stokes's theorem)}$$

$$= \frac{2\pi r^2}{2}\Big|_0^{\sqrt{2}} = 2\pi$$

13. The path is the curve $r = 1$.

$\mathbf{F} = 3y\mathbf{i} + 2z\mathbf{j} - 5z\mathbf{k}$; $\mathbf{N} = \mathbf{k}$; $dS = r \ dr \ d\theta$

$$\text{curl } \mathbf{F} = \begin{vmatrix} \mathbf{i} & \mathbf{j} & \mathbf{k} \\ \frac{\partial}{\partial x} & \frac{\partial}{\partial y} & \frac{\partial}{\partial z} \\ 3y & 2z & -5x \end{vmatrix} = \mathbf{i} + \mathbf{j} + \mathbf{k}$$

curl $\mathbf{F} \cdot \mathbf{N} \ dS = -3r \ dr \ d\theta$

$$\int_C (3y \ dx + 2z \ dy - 5x \ dz), \text{ where } C \text{ is the}$$

$$= \iint_S \text{curl } \mathbf{F} \cdot \mathbf{N} \ dS \quad \text{(Stokes's theorem)}$$

$$= -3\pi$$

14. The path is the curve $z = 0$, $x = 2 \cos \theta$, $y = 2 \sin \theta$.

$\mathbf{F} = x\mathbf{i} + y^2\mathbf{j} + z\mathbf{k}$; $\mathbf{N} = \mathbf{k}$; $dS = r \ dr \ d\theta$

$$\text{curl } \mathbf{F} = \begin{vmatrix} \mathbf{i} & \mathbf{j} & \mathbf{k} \\ \frac{\partial}{\partial x} & \frac{\partial}{\partial y} & \frac{\partial}{\partial z} \\ x & y^2 & z \end{vmatrix} = 0$$

curl $\mathbf{F} \cdot \mathbf{N} \ dS = 0$

$$\int_C (x \ dx + y^2 \ dy + z \ dz)$$

$$= \int_0^{2\pi} [2 \cos t(-2 \sin t \ dt)$$
$$+ (4 \sin^2 t)(2 \cos t \ dt) + 0]$$

$$= \int_0^{2\pi} [-4 \cos t \sin t + 8 \sin^2 t \cos t] \ dt = 0$$

15. $\mathbf{F} = xy\mathbf{i} - z\mathbf{j}$,

$$\text{curl } \mathbf{F} = \begin{vmatrix} \mathbf{i} & \mathbf{j} & \mathbf{k} \\ \frac{\partial}{\partial x} & \frac{\partial}{\partial y} & \frac{\partial}{\partial z} \\ xy & -z & 0 \end{vmatrix} = \mathbf{i} - x\mathbf{k}$$

The surface is not smooth, so there are 5 different representations for the outward unit normal: $\mathbf{k}, \mathbf{i}, -\mathbf{i}, \mathbf{j}, -\mathbf{j}$. Curl $\mathbf{F} \cdot \mathbf{N}$ for each of these is: $-x, 1, -1, 0, 0$. The 2nd and the 3rd are inverses, and $dS = 1 \ dy \ dx$, so

$$\iint_S \text{curl } \mathbf{F} \cdot \mathbf{N} \ dS = \int_0^1 \int_0^1 (-x) \ dy \ dx = -\frac{1}{2}$$

16. $\mathbf{F} = z\mathbf{i} + z\mathbf{j} + z\mathbf{k}$

C_1: $z = 1 - x$, $y = 0$; $0 \leq x \leq 1$
C_2: $y = 1 - x$, $z = 0$; $0 \leq x \leq 1$
C_3: $z = 1 - y$, $x = 0$; $0 \leq y \leq 1$

$$\iint_S \text{curl } \mathbf{F} \cdot \mathbf{N} \ dS = \int_0^1 [0 + 0 + (1 - z) \ dz]$$

$$+ \int_1^0 (1 - x) \ dx + \int_0^1 (1 - y) \ dy = -\frac{3}{2}$$

17. $\mathbf{F} = xy\mathbf{i} + x^2\mathbf{j} + z^2\mathbf{k}$; the intersection projects onto the xz-plane as $x^2 = z - y^2 = z - z^2$

$$\int_C (xy \ dx + x^2 \ dy + z^2 \ dz)$$

$$= \int_0^1 \sqrt{z - z^2}\left[\frac{-z(1 - 2z)}{2\sqrt{z - z^2}} + (z - z^2) + z^2\right] dz$$

$$+ \int_1^0 \left[-\sqrt{z - z^2}\left(\frac{-z(1 - 2z)}{2\sqrt{z - z^2}}\right) + (z - z^2) + z^2\right] dz$$

$$= 0$$

18. $\mathbf{F} = xy\mathbf{i} + y^2\mathbf{j} + x^2\mathbf{k}$; the intersection projects onto the xz-plane, $y^2 = 9(1 - x^2)$;

$$z = 3 - x - (\pm 3\sqrt{1 - x^2})$$

$$\int_C (xz \ dx + y^2 \ dy + x^2 \ dz)$$

$$= \int_{-1}^1 [x(3 - x - 3\sqrt{1 - x^2}) + 91$$

$$- x^2 \frac{-6x}{2\sqrt{1 - x^2}} + x^2(-1 - \frac{-6x}{2\sqrt{1 - x^2}})] \ dx$$

$$+ \int_{-1}^1 x(3 - x - 3\sqrt{1 - x^2}) \ dx$$

$$+ 9\int_{-1}^1 (1 - x^2)\left(\frac{3x}{\sqrt{1 - x^2}}\right)$$

$$+ x^2\left[-1 + \frac{3x}{\sqrt{1 - x^2}}\right]\right) dx = 0$$

19. $\mathbf{F} = 4y\mathbf{i} + z\mathbf{j} + 2y\mathbf{k}$; the path is the curve $z = 0$, $x = 2 \cos \theta$, $y = 2 \sin \theta$.

$$\text{curl } \mathbf{F} = \begin{vmatrix} \mathbf{i} & \mathbf{j} & \mathbf{k} \\ \frac{\partial}{\partial x} & \frac{\partial}{\partial y} & \frac{\partial}{\partial z} \\ 4y & z & 2y \end{vmatrix} = \mathbf{i} - 4\mathbf{k}$$

$\mathbf{N} = \mathbf{k}$, curl $\mathbf{F} \cdot \mathbf{N} = -4$, R is the circle in the xy-plane with radius of 2. Using polar coordinates:

$$\iint_S \text{curl } \mathbf{F} \cdot \mathbf{N} \, dS = \int_0^{2\pi} \int_0^2 (-4) \, r \, dr \, d\theta$$

$$= -8\pi \int_0^2 r \, dr \, d\theta = -16\pi$$

20. $\mathbf{F} = (1+y)z\mathbf{i} + (1+z)x\mathbf{j} + (1+x)y\mathbf{k}$. The path is the curve $z = 1$; $0 \le \theta \le 2\pi$.
$x = 2\cos\theta$, $y = \sin\theta$, or
$f(x, y, z) = \frac{1}{4}x^2 + y^2 + z^2 - 2$

$$\mathbf{N} = \frac{x\mathbf{i} + 4y\mathbf{j} + 4z\mathbf{k}}{\sqrt{x^2 + 4y^2 + 4z^2}} = \frac{x\mathbf{i} + 4y\mathbf{j} + 4z\mathbf{k}}{\sqrt{8}};$$

$$dS = \frac{\sqrt{2}}{2z} \, dA_{xy}$$

$$\text{curl } \mathbf{F} = \begin{vmatrix} \mathbf{i} & \mathbf{j} & \mathbf{k} \\ \frac{\partial}{\partial x} & \frac{\partial}{\partial y} & \frac{\partial}{\partial z} \\ (1+y)z & (1+z)x & (1+x)y \end{vmatrix}$$

$$= \mathbf{i} + \mathbf{j} + \mathbf{k}$$

Consider the transformation $u = \frac{x}{2}$, $v = y$, $t = z$.

$$\frac{\partial(x, y, z)}{\partial(u, v, t)} = \begin{vmatrix} \frac{1}{2} & 0 & 0 \\ 0 & 1 & 0 \\ 0 & 0 & 1 \end{vmatrix} = \frac{1}{2}$$

The region of integration is now the sphere of radius 2.

$$\iint_S \text{curl } \mathbf{F} \cdot \mathbf{N} \, dS$$

$$= \iint_R \left[\frac{x + 4y + 4z}{4z} \right] dA_{xy}$$

$$= \frac{1}{8} \int_{-\sqrt{21}}^{\sqrt{2}} \int_{-\sqrt{2-u^2}}^{\sqrt{2-u^2}} \frac{u + 4v + 4t}{t} \, dv \, du$$

$$= \frac{1}{8} \int_{-\sqrt{21}}^{\sqrt{2}} \int_{-\sqrt{2-u^2}}^{\sqrt{2-u^2}} \frac{u + 4v + 4t}{t} \, dv \, du$$

Now substitute $t = \sqrt{4 - u^2 - v^2}$ and switch to polar coordinates.

$$= \frac{1}{8} \int_0^{2\pi} \int_0^{\sqrt{2}} \left[\frac{r\cos\theta + 4r\sin\theta}{\sqrt{2 - r^2}} + 4 \right] r \, dr \, d\theta$$

$$= \frac{1}{8} \int_0^{2\pi} \int_0^{\sqrt{2}} \frac{r\cos\theta + 4r\sin\theta}{\sqrt{2 - r^2}} \, r \, dr \, d\theta$$

$$+ \frac{1}{8} \int_0^{2\pi} \int_0^{\sqrt{2}} 4r \, dr \, d\theta$$

$$= 0 + \frac{1}{2} \int_0^{2\pi} d\theta = \pi$$

21. $\mathbf{F} = (1+y)z\mathbf{i} + (1+z)x\mathbf{j} + (1+x)y\mathbf{k}$

$f(x, y, z) = x + y - z - 1$

$\mathbf{N} = \frac{1}{\sqrt{3}}(\mathbf{i} + \mathbf{j} + \mathbf{k})$; $dS = \sqrt{3} \, dA_{xy}$

$$\text{curl } \mathbf{F} = \begin{vmatrix} \mathbf{i} & \mathbf{j} & \mathbf{k} \\ \frac{\partial}{\partial x} & \frac{\partial}{\partial y} & \frac{\partial}{\partial z} \\ (1+y)z & (1+z)x & (1+x)y \end{vmatrix} = \mathbf{i} + \mathbf{j} + \mathbf{k}$$

curl $\mathbf{F} \cdot \mathbf{N} \, dS = 3 \, dA_{xy}$

$$\iint_S \text{curl } \mathbf{F} \cdot \mathbf{N} \, dS$$

$$= \iint_R [(1+y)z \, dx + (1+z)x \, dy + (1+x)y \, dz]$$

$$= \iint_R dA_{xy} = \frac{3}{2}$$

22. $\mathbf{F} = (1+y)z\mathbf{i} + (1+z)x\mathbf{j} + (1+x)y\mathbf{k}$

$f(x, y, z) = 2x - 3y + z - 1$

$\mathbf{N} = \frac{1}{\sqrt{14}}(2\mathbf{i} - 3\mathbf{j} + \mathbf{k})$; $dS = \sqrt{14} \, dA_{xy}$

$$\text{curl } \mathbf{F} = \begin{vmatrix} \mathbf{i} & \mathbf{j} & \mathbf{k} \\ \frac{\partial}{\partial x} & \frac{\partial}{\partial y} & \frac{\partial}{\partial z} \\ (1+y)z & (1+z)x & (1+x)y \end{vmatrix}$$

$$= \mathbf{i} + \mathbf{j} + \mathbf{k}$$

curl $\mathbf{F} \cdot \mathbf{N} \, dS = (2 - 3 + 1) \, dA_{xy} = 0$

$$\iint_S \text{curl } \mathbf{F} \cdot \mathbf{N} \, dS = 0$$

23. $\mathbf{F} = y^2\mathbf{i} + xy\mathbf{j} + xz\mathbf{k}$, $\mathbf{R} = x^2\mathbf{i} + y^2\mathbf{j}$,
$d\mathbf{R} = 2x\mathbf{i} + 2y\mathbf{j}$
$\mathbf{F} \cdot d\mathbf{R} = 2xy^2 + 2xy^2 = 4xy^2$

C: $z = 0$, $x = \cos\theta$, $y = \sin\theta$

$$\int_S \int \text{curl } \mathbf{F} \cdot \mathbf{N} \, dS = \int_C \mathbf{F} \cdot d\mathbf{R}$$

$$\int_C \mathbf{F} \cdot d\mathbf{R} = \int_0^{2\pi} 4(\cos \theta)(\sin^2\theta) \, d\theta = 0$$

24. $\mathbf{F} = z\mathbf{i} + x\mathbf{j} + y\mathbf{k}$;

   $C: x = 2 \cos \theta, \ y = 3 \sin \theta, \ z = \sin \theta$

$$\int_S \int \text{curl } \mathbf{F} \cdot \mathbf{N} \, dS$$

$$\int_0^{2\pi} [\sin \theta(-2 \sin \theta) + (2 \cos \theta)(3 \cos \theta)$$

$$+ (3 \sin \theta)(\cos \theta)] \, d\theta$$

$$= -\int_0^{2\pi} [(1 - \cos 2\theta)\sin \theta + 3(1 + \cos 2\theta)] \, d\theta + 0$$

$$= 4\pi$$

25. Consider the z-positive half of the ellipsoid, S_1, and the z-negative half, S_2, separately. By the symmetry of the surface, $\mathbf{N}(S_1) = -\mathbf{N}(S_2)$.

$$\int_S \int \text{curl } \mathbf{F} \cdot \mathbf{N} \, dS$$

$$= \int_{S_1} \int \text{curl } \mathbf{F} \cdot \mathbf{N} \, dS_1 + \int_{S_2} \int \text{curl } \mathbf{F} \cdot (-\mathbf{N}) \, dS_2$$

$$= 0$$

If $\mathbf{F}$ is a vector field whose component functions have continuous partial derivatives and S can be broken into a number of simple curves which are symmetric about the xy-plane, then

$$\int_S \int \text{curl } \mathbf{F} \cdot \mathbf{N} \, dS = 0$$

26. $\int_C \mathbf{E} \cdot d\mathbf{R} = -\dfrac{\partial \phi}{\partial t} = -\dfrac{\partial}{\partial t} \int_S \int \mathbf{B} \cdot \mathbf{N} \, dS$

$$= \int_S \int \left(-\dfrac{\partial \mathbf{B}}{\partial t}\right) \cdot \mathbf{N} \, dS$$

By Stokes's theorem,

$$\int_C \mathbf{E} \cdot d\mathbf{R} = \int_S \int (\text{curl } \mathbf{E} \cdot \mathbf{N}) \, dS, \text{ so}$$

$$\text{curl } \mathbf{E} = -\dfrac{\partial \mathbf{B}}{\partial t}.$$

27. By Stokes's theorem

$$\int_C \mathbf{B} \cdot d\mathbf{R} = \int_S \int \text{curl } \mathbf{B} \cdot \mathbf{N} \, dS,$$

given curl $\mathbf{B} = k\mathbf{J}$

$$= \int_S \int k\mathbf{J} \cdot \mathbf{N} \, dS = k \int_S \int \mathbf{J} \cdot \mathbf{N} \, dS$$

Also $I = \int_C \mathbf{H} \cdot d\mathbf{R} = \int_S \int \text{curl } \mathbf{H} \cdot \mathbf{N} \, dS$

$$= \int_S \int \mathbf{J} \cdot \mathbf{N} \, dS \text{ from Example 5}$$

Substitute for the desired result.

14.7 Divergence Theorem, Page 955

1. Evaluating the surface integral $\displaystyle\int_S \int \mathbf{F} \cdot \mathbf{N} \, dS$

   For the top portion:

   $\mathbf{F} = xz\mathbf{i} + y^2\mathbf{j} + 2z\mathbf{k}$; $z = \sqrt{4 - x^2 - y^2}$

   $z_x = -\dfrac{x}{z}, \ z_y = -\dfrac{y}{z}$;

   (upward normal): $\mathbf{N} = \dfrac{x}{z}\mathbf{i} + \dfrac{y}{z}\mathbf{j} + \mathbf{k}$

$$\int_{S_T} \int \mathbf{F} \cdot \mathbf{N} \, dS = \int_R \int \left(x^2 + \dfrac{y^3}{z} + 2z\right) dA_{xy}$$

$$= \int_R \int \left(x^2 + \dfrac{y^3}{\sqrt{4 - x^2 - y^2}} + 2\sqrt{4 - x^2 - y^2}\right) dy \, dx$$

$$= \int_0^{2\pi} \int_0^2 \left(r^3 \cos^2\theta + \dfrac{r^4\sin^3\theta}{\sqrt{4 - r^2}} + 2r\sqrt{4 - r^2}\right) dr \, d\theta$$

$$= \int_0^{2\pi} \left(4 \cos^2\theta + 3\pi \sin^3\theta + \dfrac{16}{3}\right) d\theta = \dfrac{44\pi}{3}$$

For the bottom portion:

$\mathbf{F} = xz\mathbf{i} + y^2\mathbf{j} + 2z\mathbf{k}$; $z = -\sqrt{4 - x^2 - y^2}$

$z_x = \dfrac{x}{\sqrt{4 - x^2 - y^2}}; \ z_y = \dfrac{y}{\sqrt{4 - x^2 - y^2}}$

(downward normal):

$$\mathbf{N} = \dfrac{x}{\sqrt{4 - x^2 - y^2}}\mathbf{i} + \dfrac{y}{\sqrt{4 - x^2 - y^2}}\mathbf{j} - \mathbf{k}$$

$$\int_{S_B} \int \mathbf{F} \cdot \mathbf{N} \, dS$$

$$= \int_R \int -x^2 + \dfrac{y^3}{\sqrt{4 - x^2 - y^2}} + 2\sqrt{4 - x^2 - y^2} \, dA_{xy}$$

$$= \int_0^{2\pi} \int_0^2 \left[-r^3\cos^2\theta + \dfrac{r^4\sin^3\theta}{\sqrt{4 - r^2}} + 2r\sqrt{4 - r^2}\right] dr \, d\theta$$

$$= \frac{20\pi}{3}$$

Thus $\iint\limits_{S} \mathbf{F} \cdot \mathbf{N}\, dS = \frac{44\pi}{3} + \frac{20\pi}{3} = \frac{64\pi}{3}$

Evaluating the integral $\iiint\limits_{D} \operatorname{div} \mathbf{F}\, dV$

$\iiint\limits_{D} (z + 2y + 2)\, dV$

$$= \int_0^\pi \int_0^{2\pi} \int_0^2 \left(\rho\cos\phi + 2\rho\sin\phi\sin\theta + 2\right)\rho^2\sin\phi\, d\rho\, d\theta\, d\phi$$

$$= \int_0^\pi \int_0^{2\pi} \left(4\,\sin\phi\cos\phi + 8\sin^2\phi\sin\theta + \frac{16}{3}\sin\phi\right) d\theta\, d\phi$$

$$= \int_0^\pi \left(8\pi\,\sin\phi\,\cos\phi + \frac{32\pi}{3}\,\sin\phi\right) d\phi = \frac{64\pi}{3}$$

2. Evaluating the surface integral $\iint\limits_{S} \mathbf{F} \cdot \mathbf{N}\, dS$

$\mathbf{F} = x\mathbf{i} - 2y\mathbf{j}$; div $\mathbf{F} = -1$

$f(x, y, z) = x^2 + y^2 - z$

$\nabla f = 2(x\mathbf{i} + y\mathbf{j} - \mathbf{k})$; $\mathbf{N} = \dfrac{2x\mathbf{i} + 2y\mathbf{j} - \mathbf{k}}{\sqrt{4x^2 + 4y^2 + 1}}$

$\mathbf{F} \cdot \mathbf{N} = \dfrac{2x^2 - 4y^2}{\sqrt{4x^2 + 4y^2 + 1}} = \dfrac{2r^2\cos^2\theta - 4r^2\sin^2\theta}{\sqrt{4r^2 + 1}}$

$dS = \sqrt{4r^2 + 1}\ r\, dr\, d\theta$

$\iint\limits_{S} \mathbf{F} \cdot \mathbf{N}\, dS$

$$= \int_0^{2\pi} \int_0^3 \left(2r^2\cos^2\theta - 4r^4\sin^2\theta\right) r\, dr\, d\theta$$

$$= \frac{2(3^4)}{4} \int_0^{2\pi} \left(\cos^2\theta - 2\sin^2\theta\right) d\theta = -\frac{81\pi}{2}$$

Evaluating the integral $\iiint\limits_{D} \operatorname{div} \mathbf{F}\, dV$

$$\iiint\limits_{D} (-1)\, dV = -\int_0^{2\pi} \int_0^3 \int_{r^2}^9 r\, dz\, dr\, d\theta$$

$$= -\int_0^\pi \int_0^3 (9r - r^3)\, dr\, d\theta$$

$$= -\frac{81}{4} \int_0^\pi d\theta = -\frac{81\pi}{2}$$

3. Evaluating the surface integral $\iint\limits_{S} \mathbf{F} \cdot \mathbf{N}\, dS$

$\mathbf{F} = 2y^2\mathbf{j}$; div $\mathbf{F} = 4y$;

$f(x, y, z) = x + 4y + z - 8$

$\nabla f = \mathbf{i} + 4\mathbf{j} + \mathbf{k}$

$\mathbf{N} = \dfrac{\mathbf{i} + 4\mathbf{j} + \mathbf{k}}{3\sqrt{2}}$; $\mathbf{F} \cdot \mathbf{N} = \dfrac{8y^2}{3\sqrt{2}}$

$dS = 3\sqrt{2}\ dA_{xy}$

$$\iint\limits_{S} \mathbf{F} \cdot \mathbf{N}\, dS = \int_0^2 \int_0^{8-4y} \frac{8y^2}{3\sqrt{2}}\, 3\sqrt{2}\ dx\, dy$$

$$= 8\int_0^2 y^2(8 - 4y)\, dy = \frac{128}{3}$$

Evaluating the integral $\iiint\limits_{D} \operatorname{div} \mathbf{F}\, dV$

$$\iiint\limits_{D} 4y\, dV = 4\int_0^2 \int_0^{8-4y} \int_0^{8-x-4y} y\, dz\, dx\, dy$$

$$= 4\int_0^2 \int_0^{8-4y} (8y - xy - 4y^2)\, dx\, dy$$

$$= 4\int_0^2 \left(32y - 32y^2 + 8y^3\right) dy$$

$$= 32\int_0^2 (4y - 4y^2 + y^3)\, dy = \frac{128}{3}$$

4. Evaluating the surface integral $\iint\limits_{S} \mathbf{F} \cdot \mathbf{N}\, dS$

$\mathbf{F} = 3x\mathbf{i} + 5y\mathbf{j} + 6z\mathbf{k}$; div $\mathbf{F} = 3 + 5 + 6 = 14$

$f(x, y, z) = 2x + y + z - 4 = 0$

$\nabla f = (2\mathbf{i} + \mathbf{j} + \mathbf{k})$; $\mathbf{N} = \dfrac{1}{\sqrt{6}}(2\mathbf{i} + \mathbf{j} + \mathbf{k})$

$\mathbf{F} \cdot \mathbf{N} = \dfrac{1}{\sqrt{6}}(6x + 5y + 6z)$; $dS = \sqrt{6}\ dA_{xy}$

For the top: $z = 4 - 2x - y$

$I_t = \iint\limits_{S} \mathbf{F} \cdot \mathbf{N}\, dS$

$$= \iint\limits_{R} \left[6x + 5y + 6(4 - 2x - y)\right] dA_{xy}$$

$$= \int_0^2 \int_0^{4-2x} (24 - 6x - y)\, dy\, dx$$

$$= \int_0^2 (10x^2 - 64x + 88)\, dx = \frac{224}{3}$$

For the bottom: $z = 0$; $\mathbf{F} \cdot \mathbf{N} = 0$, so $I_b = 0$
For the back side: $x = 0$; $\mathbf{F} \cdot \mathbf{N} = 0$, so $I_s = 0$
For the left side: $y = 0$; $\mathbf{F} \cdot \mathbf{N} = 0$, so $I_l = 0$

$$\iint_S \mathbf{F} \cdot \mathbf{N}\, dS = \frac{224}{3} + 0 + 0 + 0 = \frac{224}{3}$$

Evaluating the integral $\iiint_D \operatorname{div} \mathbf{F}\, dV$

$$\iiint_D 14\, dV = 14(\tfrac{1}{6})(4)(2)(4) = \frac{224}{3}$$

5. $\operatorname{div} \mathbf{F} = 3$; $\displaystyle\iint_S \mathbf{F} \cdot \mathbf{N}\, dS = \iiint_D \operatorname{div} \mathbf{F}\, dV = 3V = 3$

6. $\mathbf{F} = xyz\mathbf{j}$; $\operatorname{div} \mathbf{F} = xz$;

$$\iint_S \mathbf{F} \cdot \mathbf{N}\, dS = \iiint_D \operatorname{div} \mathbf{F}\, dV$$

$$= \int_0^{2\pi} \int_0^3 \int_0^5 zr^2 \cos\theta\, dz\, dr\, d\theta$$

$$= \frac{25}{2} \int_0^{2\pi} \int_0^3 r^2 \cos\theta\, dr\, d\theta = \frac{9(25)}{2} \int_0^{2\pi} \cos\theta\, d\theta = 0$$

7. $\mathbf{F} = (\cos yz)\mathbf{i} + e^{xz}\mathbf{j} + 3z^2\mathbf{k}$; $\operatorname{div} \mathbf{F} = 6z$;

$$\iint_S \mathbf{F} \cdot \mathbf{N}\, dS = \iiint_D \operatorname{div} \mathbf{F}\, dV$$

$$= 6 \int_0^{2\pi} \int_0^2 \int_0^{\sqrt{4-r^2}} z\, dz\, r\, dr\, d\theta$$

$$= 3 \int_0^{2\pi} \int_0^2 (4 - r^2)\, r\, dr\, d\theta = 3 \int_0^{2\pi} (8 - 4)\, d\theta = 24\pi$$

8. $\mathbf{F} = \begin{vmatrix} \mathbf{i} & \mathbf{j} & \mathbf{k} \\ \dfrac{\partial}{\partial x} & \dfrac{\partial}{\partial y} & \dfrac{\partial}{\partial z} \\ e^{xz} & -4 & \sin xyz \end{vmatrix}$

$= xz \cos xyz\, \mathbf{i} - (yz \cos xyz - xe^{xz})\mathbf{j} - xe^{xz}\mathbf{k}$

$\operatorname{div} \mathbf{F} = x(\cos xyz - xyz \sin xyz)$

$\qquad\qquad - z(\cos xyz - xyz \sin xyz) = 0$

9. Since the surface is not a closed surface, we use the divergence theorem for the closed surface and then subtract the contribution of the bottom face.

$$\iint_S \mathbf{F} \cdot \mathbf{N}\, dS = \iiint_D \operatorname{div} \mathbf{F}\, dV$$

$$= \int_0^1 \int_0^1 \int_0^1 (2x + x^2 + 3)\, dz\, dy\, dx = \frac{13}{3}$$

Since S is not included, the contribution of the bottom face, $z = 0$, is to be subtracted from the above result.

$$\iint_S \mathbf{F} \cdot \mathbf{N}\, dS = \iint_S (-3z)\, dA_{xy} = 0$$

10. Since the surface is not a closed surface, we use the divergence theorem for the closed surface and then subtract the contribution of the bottom face.

Since $\operatorname{div} \mathbf{F} = 0$,

$$\iint_S \mathbf{F} \cdot \mathbf{N}\, dS = \iiint_D \operatorname{div} \mathbf{F}\, dV = 0$$

The contribution of the bottom face, $z = 0$, is to be subtracted from the above result.

$$\iint_S \mathbf{F} \cdot \mathbf{N}\, dS = \iint_s (-3x)\, dA_{xy}$$

$$= -\int_0^1 \int_0^1 \int_0^1 (2x + x^2 + 3)\, dz\, dy\, dx = -\frac{3}{2}$$

Thus, $I = 0 - (-\frac{3}{2}) = \frac{3}{2}$

11. Since this is not a closed surface we will use the divergence theorem for the closed surface of the paraboloid and its disk, then subtract the disk.

$\mathbf{F} = x\mathbf{i} + y\mathbf{j} + z\mathbf{k}$; $\operatorname{div} \mathbf{F} = 1 + 1 + 1 = 3$,

$$\iint_S \mathbf{F} \cdot \mathbf{N}\, dS = \iiint_D \operatorname{div} \mathbf{F}\, dV$$

$$= 3 \int_0^{2\pi} \int_0^3 \int_0^{r^2} dz\, r\, dr\, d\theta = 3 \int_0^{2\pi} \int_0^3 r^2\, r\, dr\, d\theta$$

$$= 6\pi \int_0^3 r^3\, dr = \frac{243\pi}{2}$$

For the disk: $x^2 + y^2 = 9$

$g(x, y, z) = 9 - x^2 - y^2$; $\mathbf{F} = -x^2\mathbf{i} - y^2\mathbf{j}$

and $\mathbf{N} = \mathbf{k}$, $\mathbf{F} \cdot \mathbf{N} = 0$, so

$\iint_S \mathbf{F} \cdot \mathbf{N}\, dS = 0$, so our value remains $\dfrac{243\pi}{2}$

12. $\mathbf{F} = \begin{vmatrix} \mathbf{i} & \mathbf{j} & \mathbf{k} \\ \dfrac{\partial}{\partial x} & \dfrac{\partial}{\partial y} & \dfrac{\partial}{\partial z} \\ y & x & -z \end{vmatrix} = 0$

$\iint_S \mathbf{F} \cdot \mathbf{N}\, dS = \iiint_D \operatorname{div} \mathbf{F}\, dV = 0$

13. $\operatorname{div} \mathbf{F} = 2(x + y + z)$;

$f(x, y, z) = x^2 + y^2 + z^2 - 4$;

$\mathbf{N} = \tfrac{1}{2}(x\mathbf{i} + y\mathbf{j} + z\mathbf{k})$; $dS = \dfrac{2}{z}\, dA_{xy}$

$\iint_S \mathbf{F} \cdot \mathbf{N}\, dS = \iiint_D \operatorname{div} \mathbf{F}\, dV$

$= 2\int_{-2}^{2} \int_{-\sqrt{4-x^2}}^{\sqrt{4-x^2}} \int_{-\sqrt{4-x^2-y^2}}^{\sqrt{4-x^2-y^2}} (x + y + z)\, dz\, dy\, dx$

$= 4\int_{-2}^{2} \int_{-\sqrt{4-x^2}}^{\sqrt{4-x^2}} (x + y)\sqrt{4 - x^2 - y^2}\, dy\, dz$

$= 4\int_{0}^{2\pi} \int_{0}^{2} (r\cos\theta + r\sin\theta)\sqrt{4 - r^2}\, r\, dr\, d\theta$

$= 4\pi \int_{0}^{2\pi} (\cos\theta + \sin\theta)\, d\theta = 0$

14. $\operatorname{div} \mathbf{F} = yz + xz + xy$

$\iint_S \mathbf{F} \cdot \mathbf{N}\, dS = \iiint_D \operatorname{div} \mathbf{F}\, dV$

$= \int_{0}^{1} \int_{0}^{2} \int_{0}^{3} (yz + xz + xy)\, dz\, dy\, dz$

$= \int_{0}^{1} \int_{0}^{2} \left(\dfrac{9y}{2} + \dfrac{9x}{2} + 3xy \right) dy\, dx$

$= \int_{0}^{1} (9 + 9x + 6x)\, dx = \dfrac{33}{2}$

15. $\operatorname{div} \mathbf{F} = 1 + 1 + 2z = 2(z + 1)$

$\iint_S \mathbf{F} \cdot \mathbf{N}\, dS = \iiint_D \operatorname{div} \mathbf{F}\, dV$

$= 2\int_{0}^{2\pi} \int_{0}^{2} \int_{0}^{1} (z + 1)\, dz\, r\, dr\, d\theta$

$= 3\int_{0}^{2\pi} \int_{0}^{2} 3r\, dr\, d\theta = 6\int_{0}^{2\pi} d\theta = 6(2\pi) = 12\pi$

16. $\operatorname{div} \mathbf{F} = 5(x^4 + y^4 + z^4) + 10(y^2 z^2 + x^2 z^2 + x^2 y^2)$

$= 5(x^2 + y^2 + z^2)^2$

S is the hemisphere $z = \sqrt{1 - x^2 - y^2}$

together with the disk $x^2 + y^2 \leq 1$.

$\iint_{\text{closed sphere}} \mathbf{F} \cdot \mathbf{N}\, dS = \iint_{\text{hemisphere}} \mathbf{F} \cdot \mathbf{N}\, dS + \iint_{\text{disk}} \mathbf{F} \cdot \mathbf{N}\, dS$

$= \iiint_{\text{closed sphere}} \operatorname{div} \mathbf{F}\, dV$

$\iint_{\text{disk}} \mathbf{F} \cdot \mathbf{N}\, dS = \iint_{\substack{\mathbf{N} = -\mathbf{k} \text{ on disk}}} \mathbf{F} \cdot \mathbf{N}\, dS$

$= \iint_{x^2+y^2 \leq 1} (z^5 + 10x^2 y^2)\, dS$

$= 0$ since $z = 0$

$\iint_{\text{hemisphere}} \mathbf{F} \cdot \mathbf{N}\, dS = \iiint 5(x^2 + y^2 + z^2)^2\, dV$

$= \int_{0}^{2\pi} \int_{0}^{\pi/2} \int_{0}^{1} 5\rho^4 (\rho^2 \sin\phi)\, d\rho\, d\phi\, d\theta$

$= \int_{0}^{2\pi} \int_{0}^{\pi/2} \dfrac{5}{7} \sin\phi\, d\phi\, d\theta$

$= \int_{0}^{2\pi} -\dfrac{5}{7}\, d\theta = \dfrac{10\pi}{7}$

17. $\operatorname{div} \mathbf{F} = y^2$

$\iint_S \mathbf{F} \cdot \mathbf{N}\, dS = \iiint_D \operatorname{div} \mathbf{F}\, dV$

$= \int_{0}^{2\pi} \int_{0}^{1} \int_{0}^{1} y^2\, dx\, r\, dr\, d\theta$

$$= \int\limits_{0}^{2\pi} \int\limits_{0}^{1} r^3 \cos^2\theta \ dr \ d\theta = \int\limits_{0}^{2\pi} \cos^2\theta \ d\theta = \frac{\pi}{4}$$

18. div $\mathbf{F} = x^2 + y^2 + z^2$

$$\iint\limits_{S} \mathbf{F} \cdot \mathbf{N} \ dS = \iiint\limits_{D} \text{div } \mathbf{F} \ dV$$

$$= \int\limits_{0}^{\pi/4} \int\limits_{0}^{2\pi} \int\limits_{0}^{2} \rho^2(\rho^2 \sin\phi) \ d\rho \ d\theta \ d\phi$$

$$= \frac{32}{5} \int\limits_{0}^{\pi/4} \int\limits_{0}^{2\pi} \sin\phi \ d\theta \ d\phi = \frac{64\pi}{5} \int\limits_{0}^{\pi/4} \sin\phi \ d\phi$$

$$= \frac{32\pi}{5}(2 - \sqrt{2})$$

19. div $\mathbf{F} = 3x^2 + 3y^2 + 3a^2 = 3(r^2 + a^2)$

$$\iint\limits_{S} \mathbf{F} \cdot \mathbf{N} \ dS = \iiint\limits_{D} \text{div } \mathbf{F} \ dV$$

$$= \int\limits_{0}^{2\pi} \int\limits_{0}^{a} \int\limits_{0}^{1} 3(r^2 + a^2) \ dz \ r \ dr \ d\theta$$

$$= 3 \int\limits_{0}^{2\pi} \int\limits_{0}^{a} (r^3 + a^2 r) \ dr \ d\theta$$

$$= 3 \int\limits_{0}^{2\pi} \frac{3a^4}{4} \ d\theta = \frac{9\pi a^4}{2}$$

20. **a.** $\mathbf{F} = x\mathbf{i} + y\mathbf{j} + z\mathbf{k}$; div $\mathbf{F} = 3$

$$V(D) = \iiint\limits_{D} \text{div } \mathbf{F} \ dV = \frac{1}{3} \iiint\limits_{D} dV$$

$$= \frac{1}{3} \iint\limits_{S} \mathbf{F} \cdot \mathbf{N} \ dS$$

b. $f = x^2 + y^2 + z^2 - R^2$;

$\mathbf{N} = \frac{1}{a}(x\mathbf{i} + y\mathbf{j} + z\mathbf{k})$;

$\mathbf{F} \cdot \mathbf{N} = x^2 + y^2 + z^2$; $dS = \frac{a}{z} \ dA_{xy}$

$$\mathbf{F} \cdot \mathbf{N} \ dS = \frac{x^2 + y^2 + z^2}{z} \ dA_{xy}$$

$$= \frac{R}{\sqrt{R^2 - r^2}} \ dA_{xy}$$

From part **a**,

$$V = \frac{1}{3} \iint\limits_{S} \mathbf{F} \cdot \mathbf{N} \ dS = \frac{1}{3} \iint\limits_{S} \frac{R \ r \ dr \ d\theta}{\sqrt{R^2 - r^2}}$$

$$= \frac{1}{3} \int\limits_{0}^{2\pi} \int\limits_{0}^{R} \frac{R^2 \ r \ dr \ d\theta}{\sqrt{R^2 - r^2}}$$

$$= \frac{R^3}{3} \int\limits_{0}^{2\pi} d\theta = \frac{2}{3}\pi R^3$$

21. $\|\mathbf{R}\| = \sqrt{x^2 + y^2 + z^2}$; $\mathbf{N} = \frac{1}{a}(x\mathbf{i} + y\mathbf{j} + z\mathbf{k})$

$\mathbf{R} \cdot \mathbf{N} = \frac{1}{a}(x^2 + y^2 + z^2) = a$;

$\|\mathbf{R}\| \ \mathbf{R} \cdot \mathbf{N} = a(a) = a^2$

Thus, $\iint\limits_{S} \|\mathbf{R}\| \ \mathbf{R} \cdot \mathbf{N} \ dS = a^2 \iint\limits_{S} dS = 4\pi a^4$

22. $\iiint\limits_{T} (x^2 + y^2) \ dV = \frac{1}{3} \iiint\limits_{T} \text{div}(x^3\mathbf{i} + y^3\mathbf{j}) \ dV$

$$= \frac{1}{3} \iiint\limits_{T} (x^3\mathbf{i} + y^3\mathbf{j}) \cdot \mathbf{N} \ dS$$

23. **a.** $\iint\limits_{S} \frac{\partial u}{\partial n} \ dS = \iint\limits_{S} \nabla u \cdot \mathbf{N} \ dS$

$$= \iiint\limits_{T} \text{div}(\nabla u) \ dV$$

$$= \iiint\limits_{T} \nabla \cdot \nabla u \ dV$$

$$= \iiint\limits_{T} \nabla^2 u \ dV$$

b. $\iint\limits_{S} (u\nabla v) \cdot \mathbf{N} \ dS = \iiint\limits_{V} \text{div}(u\nabla v) \ dV$

$$= \iiint\limits_{V} \text{div}(x^2 + y^2 + z^2) \ dV$$

$$= \iiint\limits_{V} (\nabla u \cdot \nabla v + u^2 v) dV$$

$$= \int\limits_{0}^{1} \int\limits_{0}^{1} \int\limits_{0}^{1} [x + y + z + 3(x + y + z)] dz \ dy \ dx$$

$$= 6$$

24. $\iint\limits_{S} \left(f\frac{\partial g}{\partial n} - g\frac{\partial f}{\partial n} \right) dS$

$$= \iint\limits_{S} (f\nabla g - g\nabla f) \cdot \mathbf{N} \ dS$$

$$= \iiint\limits_{D} \text{div}(f\nabla g - g\nabla f) \ dV \quad \text{(divergence thm)}$$

$$= \iiint\limits_{D} (\nabla f \cdot \nabla g + f\nabla^2 g - \nabla g \cdot \nabla f - g\nabla^2 f) \ dV$$

$$= \int\!\!\int\!\!\int_D (f\nabla^2 g - g\nabla^2 f)\ dV$$

25. Recall that the Laplacian of f equal to 0 is defined as harmonic. (p. 904) Use this definition with the result of Problem 23a.

$$\int\!\!\int_S \frac{\partial g}{\partial n}\ dS = \int\!\!\int_S \nabla g \cdot \mathbf{N}\ dS = \int\!\!\int\!\!\int_V \operatorname{div}(\nabla g)\ dV$$

$$= \int\!\!\int\!\!\int_V \nabla^2 g\ dV = \int\!\!\int\!\!\int_V 0\ dV = 0$$

26. Let $\mathbf{U} = u_1\mathbf{i} + u_2\mathbf{j} + u_3\mathbf{k}$; and suppose $\mathbf{F} = \operatorname{curl} \mathbf{U} = 0$; then

$$\operatorname{curl} \mathbf{U} = \nabla \times \mathbf{U}$$

$$= \left(\frac{\partial u_3}{\partial y} - \frac{\partial u_2}{\partial z}\right)\mathbf{i} + \left(\frac{\partial u_1}{\partial z} - \frac{\partial u_3}{\partial x}\right)\mathbf{j}$$
$$+ \left(\frac{\partial u_2}{\partial x} - \frac{\partial u_1}{\partial y}\right)\mathbf{k}$$

$$\operatorname{div}(\operatorname{curl} \mathbf{U}) = \frac{\partial^2 u_3}{\partial x \partial y} - \frac{\partial^2 u_2}{\partial x \partial z} + \frac{\partial^2 u_1}{\partial y \partial z} - \frac{\partial^2 u_3}{\partial y \partial x}$$
$$+ \frac{\partial^2 u_2}{\partial z \partial x} - \frac{\partial^2 u_1}{\partial z \partial y}$$

Thus, $\operatorname{div} \mathbf{F} = 0$; thus,

$$\int\!\!\int_S \mathbf{F} \cdot \mathbf{N}\ dS = \int\!\!\int\!\!\int_V \operatorname{div} \mathbf{F}\ dV = 0$$

27.
$$\int\!\!\int_S \mathbf{F} \cdot \mathbf{N}\ dS = \int\!\!\int\!\!\int_D \operatorname{div}(-k\nabla T)\ dV$$

$$= \int\!\!\int\!\!\int_D (-1)(k\nabla^2 T + \nabla k \cdot \nabla T)\ dV$$

$$= \int\!\!\int\!\!\int_D (-1)(\sigma \rho \frac{\partial T}{\partial t})\ dV$$

For a k a constant,

$$(-1)(k\nabla^2 T + \nabla k \cdot \nabla T)\ dV = (-1)(\sigma \rho \frac{\partial T}{\partial t})$$

Thus, $k\nabla^2 T + \nabla k \cdot \nabla T = \sigma \rho \frac{\partial T}{\partial t}$

28. a. $\mathbf{E} = \frac{q}{4\pi\epsilon}\left[\frac{x\mathbf{i} + y\mathbf{j} + z\mathbf{k}}{(x^2 + y^2 + z^2)^{3/2}}\right]$

$\operatorname{div} \mathbf{E} = \frac{q}{4\pi\epsilon}(x^2 + y^2 + z^2)^3[3(x^2 + y^2 + z^2)^{3/2}$
$\quad - 3y^2(x^2 + y^2 + z^2)^{1/2}$
$\quad - 3z^2(x^2 + y^2 + z^2)^{1/2}] = 0$

Thus, by the divergence theorem,

$$\int\!\!\int_S \mathbf{E} \cdot \mathbf{N}\ dS = \int\!\!\int\!\!\int_V \operatorname{div} \mathbf{E}\ dV = 0$$

b. Enclose the origin in a sphere S' that is

small enough to lie entirely within S. Then the region S'' between S' and S satisfies the conditions of part **a**, so

$$\int\!\!\int_S \mathbf{E} \cdot \mathbf{N}\ dS + \int\!\!\int_{S'} \mathbf{E} \cdot \mathbf{N}\ dS$$

$$= \int\!\!\int_{S''} \mathbf{E} \cdot \mathbf{N}\ dS = 0$$

Thus,

$$\int\!\!\int_S \mathbf{E} \cdot \mathbf{N}\ dS = -\int\!\!\int_{S'} \mathbf{E} \cdot \mathbf{N}\ dS$$

We can evaluate $\int\!\!\int_{S'} \mathbf{E} \cdot \mathbf{N}\ dS$ directly since S' is a sphere. In particular, if S' has radius ρ, then $(0 \leq \theta \leq 2\pi, 0 \leq \phi \leq \pi)$
$$x = \rho \cos \theta \sin \phi$$
$$y = \rho \sin \theta \sin \phi$$
$$z = \rho \cos \phi$$
so that an outward normal on S'' (that is, an inner normal on S') is

$$\mathbf{N} = -\sin\phi[(\rho^2 \cos \theta \sin \phi)\mathbf{i}$$
$$+ (\rho^2 \sin \theta \sin \phi)\mathbf{j} + (\rho^2 \cos \phi)\mathbf{k}]$$

We find
$$\mathbf{E} \cdot \mathbf{N} = \frac{q\mathbf{R}}{4\pi\epsilon\|\mathbf{R}\|^3} \cdot \mathbf{N} = \frac{q}{4\pi\epsilon}\frac{1}{\rho^3}(-\rho^3 \sin \phi)$$
$$= \frac{-q}{4\pi\epsilon} \sin \phi$$

so that

$$\int\!\!\int_S \mathbf{E} \cdot \mathbf{N}\ dS = -\int\!\!\int_{S'} \mathbf{E} \cdot \mathbf{N}\ dS$$

$$= -\int_0^\pi\!\!\int_0^{2\pi} \frac{-q}{4\pi\epsilon} \sin\phi\ d\theta\ d\rho = \frac{q}{4\pi\epsilon}(4\pi) = \frac{q}{\epsilon}$$

29. Suppose the surface S encloses the origin. Then by Gauss' law (part **b** of Problem 28), we have

$$\int\!\!\int_S \mathbf{D} \cdot \mathbf{N}\ dS = \int\!\!\int_S (\epsilon \mathbf{E}) \cdot \mathbf{N}\ dS = \frac{q}{\epsilon}(\epsilon)$$

$$= \int\!\!\int\!\!\int_V Q\ dV$$

Let S' be a sphere centered at the origin that is entirely contained within S, and let S'' be the surface of the region inside S but outside S' (as in the proof of part **b** of Problem 28). Then the divergence theorem applies to S'', and we have

$$\int\int\int_V Q \; dV = \int\int_{S''} \mathbf{D} \cdot \mathbf{N} \; dS = \int\int_{\text{interior of } S}\text{div } \mathbf{D} \; dV$$

If we take S' to be smaller and smaller ($\rho \to 0$, where ρ is the radius of S'), then in the limit, we have

$$\int\int\int_V Q \; dV = \int\int\int_V \text{div } \mathbf{D} \; dV$$

Finally, since this equation holds for *every* region contained within a surface that encloses the origin, it follows that the integrands must be equal; that is, $Q = \text{div } \mathbf{D}$.

30. a. Start with $\nabla \times \mathbf{E} = -\dfrac{\partial \mathbf{B}}{\partial t}$, and take the curl of both sides:

$$\text{curl(curl } \mathbf{E}) = \nabla \times (\nabla \times \mathbf{E}) = \nabla \times \left(-\frac{\partial \mathbf{B}}{\partial t}\right)$$

$$= -\frac{\partial}{\partial t}(\nabla \times \mathbf{B})$$

We can interchange curl and $\partial/\partial t$ because curl involves only the spatial variables x, y, z, and $\partial/\partial t$ involves only time, t.

b. Using the results of part **a**, we obtain

$$\text{curl(curl } \mathbf{E}) = -\frac{\partial}{\partial t}(\nabla \times \mathbf{B})$$

$$= -\frac{\partial}{\partial t}(\nabla \times \mu\mathbf{H}) = -\mu\frac{\partial}{\partial t}(\nabla \times \mathbf{H})$$

$$= -\mu\frac{\partial}{\partial t}(\text{curl } \mathbf{H})$$

It can be shown that

$$\text{curl(curl } \mathbf{E}) = \nabla \times (\nabla \times \mathbf{E})$$

$$= \nabla(\nabla \cdot \mathbf{E}) - (\nabla \cdot \nabla)\mathbf{E}$$

Since $\text{curl } \mathbf{H} = \nabla \times \mathbf{H} = \sigma\mathbf{E} + \epsilon\dfrac{\partial \mathbf{E}}{\partial t}$ we have

$$\nabla(\text{div } \mathbf{E}) - (\nabla \cdot \nabla)\mathbf{E} = \text{curl(curl } \mathbf{E})$$

$$= -\mu\frac{\partial}{\partial t}(\text{curl } \mathbf{H}) = -\mu\frac{\partial}{\partial t}\left(\sigma\mathbf{E} + \epsilon\frac{\partial \mathbf{E}}{\partial t}\right)$$

c. If $Q = 0$, then from Problem 29, we have

$$0 = Q = \nabla \cdot \mathbf{V} = \nabla(\epsilon\mathbf{E}) = \epsilon\nabla \cdot \mathbf{E}$$

so $\nabla \cdot \mathbf{E} = 0$. Substituting into the equation in part **b**, we obtain

$$\nabla(0) - (\nabla \cdot \nabla)\mathbf{E} = -\mu\frac{\partial}{\partial t}\left(\sigma\mathbf{E} + \epsilon\frac{\partial \mathbf{E}}{\partial t}\right)$$

$$-(\nabla \cdot \nabla)\mathbf{E} = -\mu\sigma\frac{\partial E}{\partial t} - \mu\epsilon\frac{\partial^2 \mathbf{E}}{\partial t^2}$$

$$(\nabla \cdot \nabla)\mathbf{E} = \mu\sigma\frac{\partial \mathbf{E}}{\partial t} + \mu\epsilon\frac{\partial^2 \mathbf{E}}{\partial t^2}$$

CHAPTER 14 REVIEW

Proficiency Examination, Page 961

1. A vector field is a collection S of points in space together with a rule that assigns to each point (x, y, z) in S exactly one vector $\mathbf{V}(x, y, z)$.

2. The divergence of a vector field $\mathbf{V}(x, y, z) = u(x, y, z)\mathbf{i} + v(x, y, z)\mathbf{j} + w(x, y, z)\mathbf{k}$ is denoted by div $\mathbf{V}$ and is defined by

$$\text{div } \mathbf{V} = \frac{\partial u}{\partial x}(x, y, z) + \frac{\partial v}{\partial y}(x, y, z) + \frac{\partial w}{\partial z}(x, y, z)$$

3. The curl of a vector field $\mathbf{V}(x, y, z) = u(x, y, z)\mathbf{i} + v(x, y, z)\mathbf{j} + w(x, y, z)\mathbf{k}$ is denoted by curl $\mathbf{V}$ and is defined by

$$\text{curl } \mathbf{V} = \left(\frac{\partial w}{\partial y} - \frac{\partial v}{\partial z}\right)\mathbf{i} + \left(\frac{\partial u}{\partial z} - \frac{\partial w}{\partial x}\right)\mathbf{j} + \left(\frac{\partial v}{\partial x} - \frac{\partial u}{\partial y}\right)\mathbf{k}$$

4. The del operator is defined by

$$\nabla = \frac{\partial}{\partial x}\mathbf{i} + \frac{\partial}{\partial y}\mathbf{j} + \frac{\partial}{\partial z}\mathbf{k}$$

5. The Laplacian of f is

$$\nabla^2 f = \nabla \cdot \nabla f = \frac{\partial^2 f}{\partial x^2} + \frac{\partial^2 f}{\partial y^2} + \frac{\partial^2 f}{\partial z^2}$$

$$= f_{xx} + f_{yy} + f_{zz}$$

The equation $\nabla^2 f = 0$ is called Laplace's equation.

6. A line integral involves two or three variables tied together by one or two additional equations. A Riemann sum is an integral of a single variable. (See p. 960.)

7. Let $\mathbf{F}(x, y, z) = u(x, y, z)\mathbf{i} + v(x, y, z)\mathbf{j} + w(x, y, z)\mathbf{k}$ be a vector field, and let C be the curve with parametric representation $\mathbf{R}(t) = x(t)\mathbf{i} + y(t)\mathbf{j} + z(t)\mathbf{k}$ for $a \le t \le b$ Using $d\mathbf{R} = dx\,\mathbf{i} + dy\,\mathbf{j} + dz\,\mathbf{k}$, we denote the line integral of $\mathbf{F}$ over C by $\int_C \mathbf{F} \cdot d\mathbf{R}$ and is define it by

$$\int_C \mathbf{F} \cdot d\mathbf{R} = \int_C u \; dx + v \; dy + w \; dz$$

$$= \int_a^b u[x(t), y(t), z(t)]\frac{dx}{dt} + v[x(t), y(t), z(t)]\frac{dy}{dt}$$

$$+ w[x(t), y(t), z(t)]\frac{dz}{dt}dt$$

8. Let $\mathbf{F}$ be a continuous force field over a domain D. Then the **work** W done by $\mathbf{F}$ as an object moves along a smooth curve C in D is given by the line integral

$$W = \int_C \mathbf{F} \cdot d\mathbf{R}$$

9. Let f be continuous on a smooth curve C. If

C is defined by

$\mathbf{R}(t) = x(t)\mathbf{i} + y(t)\mathbf{j} + y(t)\mathbf{k}$, where $a \le t \le b$,

then $\displaystyle\int_C f(x,y,z)\,ds$

$$= \int_a^b f[x(t),\, y(t), z(t)] \sqrt{[x'(t)]^2 + [y'(t)]^2 + [z'(t)]^2}\ dt$$

10. Let $\mathbf{F}$ be a conservative vector field on the region D and let f be a scalar potential function for $\mathbf{F}$; that is, $\nabla f = \mathbf{F}$. Then, if C is any piecewise smooth curve lying entirely within D, with initial point P and terminal point Q, we have

$$\int_C \mathbf{F} \cdot d\mathbf{R} = f(Q) - f(P)$$

Thus, the line integral $\int_C \mathbf{F} \cdot d\mathbf{R}$ is independent of path in D.

11. A vector field $\mathbf{F}$ is said to be conservative in a region D if it can be represented in D as the gradient of a continuously differentiable function f, which is then called a scalar potential of $\mathbf{F}$. That is, $\mathbf{F} = \nabla f$ for (x, y) in D.

12. The answer to Problem 11. With $\mathbf{F}$ in D represented as the gradient of a continuously differential function f, f is called a scalar potential of $\mathbf{F}$.

13. A Jordan curve is a closed curve with no self intersections.

14. Let D be a simply connected region with a positively oriented piecewise-smooth boundary C. Then if the vector field
$\mathbf{F}(x, y) = M(x, y)\mathbf{i} + N(x, y)\mathbf{j}$
is continuously differentiable on D, we have

$$\int_C (M\ dx + N\ dy) = \int_D \int \left(\frac{\partial N}{\partial x} - \frac{\partial M}{\partial y}\right) dA$$

15. $A = \displaystyle\int_C (-y\ dx)$ or $A = \displaystyle\int_C (x\ dy)$ or

$A = \dfrac{1}{2}\displaystyle\int_C (-y\ dx + x\ dy)$

16. The normal derivative of f is denoted by $\partial f/\partial n$ and is the directional derivative of f in the direction of the normal pointing to the exterior of the domain of f. The normal derivative of f satisfies the property

$$\frac{\partial f}{\partial n} = \nabla f \cdot \mathbf{N}$$

where $\mathbf{N}$ is an outer unit normal.

17. Let S be a surface defined by $z = f(x, y)$ and R_{xy} its projection on the xy-plane. If f, f_x, and f_y are continuous in R_{xy} and g is continuous on S, then the surface integral of g over S is

$$\int_S \int g(x,y,z)\,dS$$

$$= \int_{R_{xy}} \int g(x,y,f(x,y)) \sqrt{[f_x(x,y)]^2 + [f_y(x,y)]^2 + 1}\ dA_{xy}$$

18. If a surface S is defined parametrically by the vector function
$\mathbf{R}(u, v) = x(u, v)\mathbf{i} + y(u, v)\mathbf{j} + z(u, v)\mathbf{k}$
over a region D in the uv-plane, the surface area of S is given by the integral

$$\int_D \int \|\mathbf{R}_u \times \mathbf{R}_v\|\ du\ dv$$

Then if f is continuous on D, the surface integral of f over D is given by

$$\int_S \int f(x, y, z)\ dS = \int_D \int f(\mathbf{R})\ \|\mathbf{R}_u \times \mathbf{R}_v\|\ du\ dv$$

19. The flux integral of a vector field $\mathbf{F}$ across a surface S is given by

$$\int_S \int \mathbf{F} \cdot \mathbf{N}\ dS$$

20. Let S be an oriented surface with unit normal vector $\mathbf{N}$, and assume that S is bounded by a closed, piecewise smooth curve C whose orientation is compatible with that of S. If $\mathbf{F}$ is a vector field that is continuously differentiable on S, then

$$\int_C \mathbf{F} \cdot d\mathbf{R} = \int_S \int (\text{curl } \mathbf{F} \cdot \mathbf{N})\ dS$$

21. $\mathbf{F}$ is conservative in D if and only if curl $\mathbf{F} = \mathbf{0}$

22. Let D be a region in space bounded by a smooth, orientable closed surface S. If $\mathbf{F}$ is a continuous vector field whose components have continuous partial derivatives in D, then

$$\int_S \int \mathbf{F} \cdot \mathbf{N}\ dS = \int_D \int \int \text{div } \mathbf{F}\ dV$$

where $\mathbf{N}$ is an outward unit normal to the surface S.

23. $\mathbf{F} = yz\mathbf{i} + xz\mathbf{j} + xy\mathbf{k} = M\mathbf{i} + N\mathbf{j} + P\mathbf{k}$

$\dfrac{\partial M}{\partial y} = z = \dfrac{\partial N}{\partial x},\ \dfrac{\partial M}{\partial z} = y = \dfrac{\partial P}{\partial x},\ \dfrac{\partial N}{\partial z} = z = \dfrac{\partial P}{\partial y},$

so **F** is conservative

$\dfrac{\partial f}{\partial x} = yz$, so $f = xyz + a(y, z)$ and

$\dfrac{\partial f}{\partial y} = xz + \dfrac{\partial a}{\partial y}$, so from **F**, $\dfrac{\partial a}{\partial y} = 0$,

$a(y, z) = b(z);\ \ f = xyz + b(z)$, but

$\dfrac{\partial f}{\partial z} = xy + \dfrac{\partial b}{\partial z}$, so from **F**, $\dfrac{\partial b}{\partial z} = 0$,

$b(z) = C$ (a constant) and $f = xyz + C$

24. $\mathbf{F} = x^2 y\mathbf{i} - e^{yz}\mathbf{j} + \dfrac{x}{2}\mathbf{k}$

div $\mathbf{F} = 2xy - ze^{yz}$

$\text{curl } \mathbf{F} = \begin{vmatrix} \mathbf{i} & \mathbf{j} & \mathbf{k} \\ \dfrac{\partial}{\partial x} & \dfrac{\partial}{\partial y} & \dfrac{\partial}{\partial z} \\ x^2 y & -e^{yz} & \dfrac{x}{2} \end{vmatrix}$

$= ye^{yz}\mathbf{i} - \dfrac{1}{2}\mathbf{j} - x^2\mathbf{k}$

25. By Green's theorem

$\displaystyle\int_C \mathbf{F} \cdot d\mathbf{R} = \int_C (M\, dx + N\, dy)$

$\displaystyle = \int\int_D \left(\dfrac{\partial N}{\partial x} - \dfrac{\partial M}{\partial y}\right) dA = \int\int_D (-1)\, dA$

$= -(\text{the area of the triangle}) = -2$

26. By Stokes's theorem

$\displaystyle\int_C \mathbf{F} \cdot d\mathbf{R} = \int\int_S (\text{curl } \mathbf{F}) \cdot \mathbf{N}\, dS$

$\text{curl } \mathbf{F} = \begin{vmatrix} \mathbf{i} & \mathbf{j} & \mathbf{k} \\ \dfrac{\partial}{\partial x} & \dfrac{\partial}{\partial y} & \dfrac{\partial}{\partial z} \\ 2y & z & y \end{vmatrix} = (1-1)\mathbf{i} - 2\mathbf{k} = -2\mathbf{k}$

The intersection of the sphere and the plane:

$$x^2 + y^2 + (x + 2)^2 = 4(x + 2)$$
$$2x^2 + y^2 = 4$$
$$\dfrac{x^2}{2} + \dfrac{y^2}{4} = 1$$

$g(x, y, z) = x - z + 2;\ \ \mathbf{N} = \dfrac{\mathbf{i} - \mathbf{k}}{\sqrt{2}};$

$\text{curl } \mathbf{F} \cdot \mathbf{N} = \dfrac{2}{\sqrt{2}};\ \ dS = \sqrt{2}\, dy\, dx$

$\displaystyle\int\int_S (\text{curl } \mathbf{F}) \cdot \mathbf{N}\, dS = \int\int_S \sqrt{2}\sqrt{2}\, dy\, dx$

$= 2(\text{the area of the ellipse, } \pi ab)$

$= 2(\pi 2\sqrt{2}) = 4\pi\sqrt{2}$

27. By the divergence theorem

$\displaystyle\int_S\int \mathbf{F} \cdot \mathbf{N}\, dS = \int\int\int_D \text{div } \mathbf{F}\, dV$

$\mathbf{F} = x^2\mathbf{i} + (y + z)\mathbf{j} - 2z\mathbf{k};$

div $\mathbf{F} = 2x + 1 - 2 = 2x - 1$

$\displaystyle\int\int\int_D \text{div } \mathbf{F}\, dV = \int\int\int_D (2x - 1)\, dV$

$\displaystyle = \int_0^1\int_0^1\int_0^1 (2x - 1)\, dz\, dy\, dx$

$\displaystyle = \int_0^1 (2x - 1)\, dx = 0$

28. $\mathbf{F} = \dfrac{x\, dx}{(x^2 + y^2)^2} + \dfrac{y\, dy}{(x^2 + y^2)^2} = M\, dx + N\, dy$

$\dfrac{\partial N}{\partial x} = \dfrac{-2y(2x)}{(x^2 + y^2)^3} = \dfrac{-4xy}{(x^2 + y^2)^3};$

$\dfrac{\partial M}{\partial y} = \dfrac{-2x(2y)}{(x^2 + y^2)^3} = \dfrac{-4xy}{(x^2 + y^2)^3}$

$\dfrac{\partial N}{\partial x} = \dfrac{\partial M}{\partial y}$ so **F** is conservative and

independent of path. Since C is a closed path the value of the line integral is 0.

29. Since $m\omega^2$ is a scalar **F** is conservative if **R** is conservative. Testing **R**:

$\dfrac{\partial M}{\partial y} = 0 = \dfrac{\partial N}{\partial x};\ \dfrac{\partial M}{\partial z} = 0 = \dfrac{\partial P}{\partial x};\ \dfrac{\partial N}{\partial z} = 0 = \dfrac{\partial P}{\partial y};$

so **F** is conservative. Ignore for the moment the factor $m\omega^2$.

$\dfrac{\partial f}{\partial x} = x$, so $f = \dfrac{x^2}{2} + g(y, z)$ and $\dfrac{\partial f}{\partial y} = \dfrac{\partial g}{\partial y}$, so

from **F**, $\dfrac{\partial g}{\partial y} = y$, $g = \dfrac{y^2}{2} + h(z)$.

$f = \dfrac{x^2}{2} + \dfrac{y^2}{2} + h(z)$, but $\dfrac{\partial f}{\partial z} = \dfrac{\partial h}{\partial z}$, so from **F**,

$\dfrac{\partial h}{\partial z} = z$, $h = \dfrac{z^2}{2} + C$ (a constant) and

$f = \dfrac{x^2}{2} + \dfrac{y^2}{2} + \dfrac{z^2}{2} + C$

Inserting the scalar factor:

$f = \dfrac{m\omega^2}{2}(x^2 + y^2 + z^2) + C$

30. $W = \displaystyle\int_C \mathbf{F} \cdot d\mathbf{R}$ where

$\mathbf{F} = m\omega^2(x\mathbf{i} + y\mathbf{j} + z\mathbf{k})$ and **R** is the circular path from $(3, 0, 2)$ to $(-3, 0, 2)$.

Parametrizing: $x = 3\cos t$, $y = 3\sin t$,

$\mathbf{R} = (3 \cos t)\mathbf{i} + (3 \sin t)\mathbf{j}$ for $0 \le t \le \pi$.

$d\mathbf{R} = (-3 \sin t)\mathbf{i} + (3 \cos t)\mathbf{j}$,

$\mathbf{F} \cdot d\mathbf{R} = m\omega^2(-9 \sin t \cos t + 9 \sin t \cos t)$

$= 0$

Therefore, $W = 0$.

This result should have been anticipated as z is constant and $\mathbf{R}$ is symmetric about the y-axis.

Supplementary Problems, Page 962

1. curl $\mathbf{F} = 0$, so $\mathbf{F}$ is conservative and $f = 2x - 3y + C$

2. curl $\mathbf{F} = \begin{vmatrix} \mathbf{i} & \mathbf{j} & \mathbf{k} \\ \dfrac{\partial}{\partial x} & \dfrac{\partial}{\partial y} & \dfrac{\partial}{\partial z} \\ xy^{-2} & x^{-2}y & 0 \end{vmatrix}$

$= (-2x^{-3}y + 2xy^{-3})\mathbf{k} \ne 0$

$\mathbf{F}$ is not conservative.

3. curl $\mathbf{F} = \begin{vmatrix} \mathbf{i} & \mathbf{j} & \mathbf{k} \\ \dfrac{\partial}{\partial x} & \dfrac{\partial}{\partial y} & \dfrac{\partial}{\partial z} \\ y^{-3} & -3xy^{-4}+\cos y & 0 \end{vmatrix}$

$= 0$

$\mathbf{F}$ is conservative.

$f_x = y^{-3}; \ f = xy^{-3} + c(y); \ c'(y) = \cos y$, so

$c(y) = \sin y + C$; thus, $f = xy^{-3} + \sin y + C$

4. curl $\mathbf{F} = \begin{vmatrix} \mathbf{i} & \mathbf{j} & \mathbf{k} \\ \dfrac{\partial}{\partial x} & \dfrac{\partial}{\partial y} & \dfrac{\partial}{\partial z} \\ y^2 & 2xy & 0 \end{vmatrix} = 0$

$\mathbf{F}$ is conservative.

$f = xy^2 + c(y, z); \ \dfrac{\partial c(y, z)}{\partial y} = 0; \ c(y, z) = b(z);$

$f_z = 0; \ b(z) = C$; thus, $f(x, y, z) = xy^2 + C$

5. curl $\mathbf{F} = \begin{vmatrix} \mathbf{i} & \mathbf{j} & \mathbf{k} \\ \dfrac{\partial}{\partial x} & \dfrac{\partial}{\partial y} & \dfrac{\partial}{\partial z} \\ \dfrac{1}{y}+\dfrac{y}{x^2} & -\dfrac{x}{y^2}+\dfrac{1}{x} & 0 \end{vmatrix}$

$= \left(-\dfrac{1}{y} - \dfrac{1}{x^2} + \dfrac{1}{y^2} + \dfrac{1}{x^2}\right)\mathbf{k} \ne 0$

$\mathbf{F}$ is not conservative.

6. $\dfrac{\partial}{\partial y}[2x \tan^{-1}\dfrac{y}{x} - y] = -1 + \dfrac{2y^2}{x^2 + y^2}$

$\dfrac{\partial}{\partial x}[2y \tan^{-1}\dfrac{y}{x} + x] = 1 - \dfrac{2y^2}{x^2 + y^2}$

$\mathbf{F}$ is not conservative.

For Problems 7-12, $d\mathbf{R} = (\mathbf{i} + 2t\mathbf{j}) \, dt$

7. $\displaystyle\int_C \mathbf{F} \cdot d\mathbf{R} = \int_1^2 (2 - 6t) = -7$

8. $\mathbf{F} = t^{-3}\mathbf{i} + \mathbf{j}$

$\displaystyle\int_C \mathbf{F} \cdot d\mathbf{R} = \int_1^2 (t^{-3} + 2t) \, dt = \dfrac{27}{8}$

9. $\mathbf{F} = t^{-6}\mathbf{i} + (-3t^{-7} + \cos t^2)\mathbf{j}$

$\displaystyle\int_C \mathbf{F} \cdot d\mathbf{R} = \int_1^2 [t^{-6} + (-3t^{-7} + \cos t^2)(2t)] \, dt$

$= -\dfrac{31}{32} + \sin 4 - \sin 1$

10. $\mathbf{F} = t^4\mathbf{i} + 2t^3\mathbf{j}$

$\displaystyle\int_C \mathbf{F} \cdot d\mathbf{R} = \int_1^2 (t^4 + 4t^4) \, dt = 31$

11. $\mathbf{F} = (1 + \dfrac{1}{t^2})\mathbf{i} - (\dfrac{1}{t^3} - \dfrac{1}{t})\mathbf{j}$

$\displaystyle\int_C \mathbf{F} \cdot d\mathbf{R} = \int_1^2 (\dfrac{1}{t^2} + 1 - \dfrac{2}{t^2} + 2) \, dt = \dfrac{5}{2}$

12. $\mathbf{F} = (2t \tan^{-1}t - t^2)\mathbf{i} + (2t^2 \tan^{-1}t + t)\mathbf{j}$

$\displaystyle\int_C \mathbf{F} \cdot d\mathbf{R}$

$= \displaystyle\int_1^2 (2t\tan^{-1}t - t^2 + 4t^3 \tan^{-1}t + 2t^2) \, dt$

$= \displaystyle\int_1^2 2t\tan^{-1}t \, dt + \int_1^2 4t^3\tan^{-1}t \, dt + \int_1^2 t^2 \, dt$

$= (-5 \tan^{-1}\dfrac{1}{2} + 2\pi - 1) + (-15 \tan^{-1}\dfrac{1}{2}$

$+ \dfrac{15\pi}{2} - \dfrac{4}{3}) + \dfrac{7}{3}$

$= \dfrac{19\pi}{2} - 20 \tan^{-1}\dfrac{1}{2} \approx 20.5722$

13. div $\mathbf{F} = 1 + 1 + 1 = 3$

curl $\mathbf{F} = \begin{vmatrix} \mathbf{i} & \mathbf{j} & \mathbf{k} \\ \dfrac{\partial}{\partial x} & \dfrac{\partial}{\partial y} & \dfrac{\partial}{\partial z} \\ x & y & z \end{vmatrix} = \mathbf{0}$

The line integral is path independent.

$\dfrac{\partial f}{\partial x} = yz;\ f(x,\,y,\,z) = xyz + a(y,\,z)$

$\dfrac{\partial f}{\partial y} = xz + \dfrac{\partial a}{\partial y} = xz;\ a(y,\,z) = b(z)$

$f(x,\,y,\,z) = xyz + b(z)$

$\dfrac{\partial f}{\partial z} = xy + b'(z) = xy + 2;\ b(z) = 2z + C$

$f(x,\,y,\,z) = xyz + 2z + C$

14. div $\mathbf{F} = \dfrac{-\,y/x}{1 + (y/x)^2} + 0 + 2z = \dfrac{-\,y}{x^2 + y^2} + 2z$

$\text{curl } \mathbf{F} = \begin{vmatrix} \mathbf{i} & \mathbf{j} & \mathbf{k} \\ \dfrac{\partial}{\partial x} & \dfrac{\partial}{\partial y} & \dfrac{\partial}{\partial z} \\ \tan^{-1}\dfrac{y}{x} & -3 & z^2 \end{vmatrix}$

$= -\dfrac{1/x}{1 + y^2/x^2}\,\mathbf{k} = -\dfrac{x}{x^2 + y^2}\,\mathbf{k}$

15. div $\mathbf{F} = \dfrac{1}{r}\Big([r^{1/2} - 2x^2(2r^{1/2})]$

$+\ r^{1/2}(-2y^2)/(2r^{1/2}) + r^{1/2}(-2z^2)/(2r^{1/2})\Big)$

$= \dfrac{3r^{1/2}}{x^2 + y^2} + 2z - \dfrac{r}{r^{3/2}2} = \dfrac{2}{r}$

$= \dfrac{2}{\sqrt{x^2 + y^2 + z^2}}$

$\text{curl } \mathbf{F} = \begin{vmatrix} \mathbf{i} & \mathbf{j} & \mathbf{k} \\ \dfrac{\partial}{\partial x} & \dfrac{\partial}{\partial y} & \dfrac{\partial}{\partial z} \\ \dfrac{x}{\sqrt{r}} & \dfrac{y}{\sqrt{r}} & \dfrac{z}{\sqrt{r}} \end{vmatrix}$

$= \Big[-\dfrac{2yz}{2r^{3/2}} + \dfrac{2xz}{2r^{3/2}}\Big]\mathbf{i} - \Big[-\dfrac{2xz}{2r^{3/2}} + \dfrac{2xz}{2r^{3/2}}\Big]\mathbf{j}$

$+ \Big[-\dfrac{2xy}{2r^{3/2}} + \dfrac{2xy}{2r^{3/2}}\Big]\mathbf{k} = \mathbf{0}$

16. div $\mathbf{F} = y \sin z - x^2 z \sin yz + \sin xy$

$\text{curl } \mathbf{F} = \begin{vmatrix} \mathbf{i} & \mathbf{j} & \mathbf{k} \\ \dfrac{\partial}{\partial x} & \dfrac{\partial}{\partial y} & \dfrac{\partial}{\partial z} \\ xy \sin z & x^2\cos yz & z \sin xy \end{vmatrix}$

$= (xz \cos xy + x^2 y \sin yz)\mathbf{i}$

$-\ (yz \cos xy - x \cos z)\mathbf{j}$

$-\ (2x \cos yz - x \sin z)\mathbf{k}$

17. $C_1:\ y = 0,\ x = t;\ 0 \le t \le \pi$

$C_2:\ x = \pi,\ y = t;\ 0 \le t \le \pi$

$\displaystyle\int_C [(\sin \pi y)\ dx + (\cos \pi x)\ dy]$

$\displaystyle = \int_0^\pi \int_0^\pi [0 + \cos \pi^2\ dy] = \pi \cos \pi^2$

18. $C:\ x = 3 \sin t,\ y = 3 \cos t,\ z = t;\ 0 \le t \le \dfrac{\pi}{4}$

$\displaystyle\int_C (z\ dx - x\ dy + dz)$

$\displaystyle = \int_0^{\pi/4} [3t \cos t - (3 \sin t)(-3 \sin t) + 1]\ dt$

$\displaystyle = 3\int_0^{\pi/4} t \cos t\ dt + 9\int_0^{\pi/4} \sin^2 t\ dt + \int_0^{\pi/4} dt$

$= \dfrac{3\pi\sqrt{2}}{8} + \dfrac{11\pi}{8} + \dfrac{3\sqrt{2}}{2} - \dfrac{21}{4} \approx 2.8571$

19. $\mathbf{F} = yz\mathbf{i} + xz\mathbf{j} + (xy + 2)\mathbf{k}$

$C:\ x = \tan t,\ y = t^2,\ z = -3t;\ 0 \le t \le 1$

$\text{curl } \mathbf{F} = \begin{vmatrix} \mathbf{i} & \mathbf{j} & \mathbf{k} \\ \dfrac{\partial}{\partial x} & \dfrac{\partial}{\partial y} & \dfrac{\partial}{\partial z} \\ yz & xz & xy+2 \end{vmatrix} = \mathbf{0}$

The line integral is path independent.

$\displaystyle\int_C [yz\ dx + xz\ dy + (xy + 2)\ dz]$

$= f(\tfrac{\pi}{4},\ 1,\ -3) - f(0,\,0,\,0)$

$= -\dfrac{3\pi}{4} - 6$

20. $\mathbf{F} = (x^2 + y)\mathbf{i} + xz\mathbf{j} - (y + z)\mathbf{k}$

$C:\ x = t,\ y = t^2,\ z = 2;\ 0 \le t \le 1$

$\text{curl } \mathbf{F} = \begin{vmatrix} \mathbf{i} & \mathbf{j} & \mathbf{k} \\ \dfrac{\partial}{\partial x} & \dfrac{\partial}{\partial y} & \dfrac{\partial}{\partial z} \\ x^2+y & xz & -y-z \end{vmatrix} \ne \mathbf{0}$

The line integral is path dependent.

$\displaystyle\int_C [(x^2 + y)\ dx + xz\ dy - (y + z)\ dz]$

$\displaystyle = \int_0^1 [t^2 + t^2 + 2t(2t)]\ dt = 6\int_0^1 t^2\ dt = 2$

21. $\mathbf{R}_u = \mathbf{i} + \mathbf{j} + \mathbf{k}$; $\mathbf{R}_v = \mathbf{j}$

$$\mathbf{R}_u \times \mathbf{R}_v = \begin{vmatrix} \mathbf{i} & \mathbf{j} & \mathbf{k} \\ 1 & 1 & 1 \\ 0 & 1 & 0 \end{vmatrix} = \mathbf{i} + \mathbf{k}$$

$$dS = \|\mathbf{R}_u \times \mathbf{R}_v\| \, du \, dv = \sqrt{2} \, du \, dv$$

$$\iint_S (3x^2 + y - 2z) \, dS$$

$$= \int_0^1 \int_0^1 (3u + u + v - 2u) \sqrt{2} \, dv \, du$$

$$= \sqrt{2} \int_0^1 (3u^2 - u + \tfrac{1}{2}) \, du = \sqrt{2}$$

22. C: $x = 2\cos\theta$, $y = 2\sin\theta$, $z = 0$; $0 \le \theta \le 2\pi$

$\mathbf{N} = -\mathbf{k}$; div $\mathbf{F} = 1$

$$\iint_S \mathbf{F} \cdot \mathbf{N} \, dS = \iiint_D \operatorname{div} \mathbf{F} \, dV - \iint_{\text{circle}} \mathbf{F} \cdot \mathbf{N} \, dS$$

$$= \frac{32\pi}{3} - \int_0^{2\pi} \int_0^2 (-2r\sin\theta) r \, dr \, d\theta$$

$$= \frac{32\pi}{3} + 2 \int_0^{2\pi} \int_0^2 r^2 \sin\theta = \frac{32\pi}{3}$$

23. $\mathbf{F} = x\mathbf{i} + x\mathbf{j} - y\mathbf{k}$

$$\operatorname{curl} \mathbf{F} = \begin{vmatrix} \mathbf{i} & \mathbf{j} & \mathbf{k} \\ \frac{\partial}{\partial x} & \frac{\partial}{\partial y} & \frac{\partial}{\partial z} \\ x & x & y \end{vmatrix} \ne \mathbf{0}$$

The line integral is path dependent.

$$\int_C (x \, dx + x \, dy - y \, dz) = \int_0^1 [t + t(2t) - t^2] \, dt$$

$$= \int_0^1 (t + t^2) \, dt = \frac{5}{6}$$

24. $\mathbf{F} = x^2\mathbf{i} + 0\mathbf{j} - 3y^2\mathbf{k}$

A vector determined by the points is $\mathbf{v} = \mathbf{i} + \mathbf{k}$ and the line passing through those points is L: $x = z - 1$, $y = 1$

$$\int_C (x^2 \, dx - 3y^2 \, dz) = \int_0^1 [x^2 \, dx - 3 \, dx] = -\frac{8}{3}$$

25. $\mathbf{F} = x^2\mathbf{i} + y\mathbf{j}$

$\frac{\partial N}{\partial x} = 0 = \frac{\partial M}{\partial y}$; the integral is path independent

$f_x = x^2$; $f = \frac{x^3}{3} + a(y)$

For a vector parallel to the line determined by

$\mathbf{v} = \mathbf{i} + \mathbf{k}$: L: $z - 1$, $y = 1$

$\frac{\partial f}{\partial y} = a'(y) = y$; $f(x, y) = \frac{x^3}{3} + \frac{y^2}{2} + C$

$$\int_C (x^2 \, dx + y \, dy) = f(0, 1 - 2\pi) - f(0, 1)$$

$$= 2\pi(\pi - 1)$$

26. $\displaystyle \int_C (xy \, dx - x^2 \, dy) = \iint_D \left[\frac{\partial(-x^2)}{\partial x} - \frac{\partial xy}{\partial y} \right] dy \, dx$

$$= \int_{-1}^0 \int_{-x-1}^{x+1} (-2x - x) \, dy \, dx + \int_0^1 \int_{x-1}^{-x+1} (-3x) \, dy \, dx$$

$$= -3 \left[\int_{-1}^0 x(2x + 2) \, dx + \int_0^1 x(-x + 2) \, dx \right] = 0$$

27. $\mathbf{F} = y\mathbf{i} + x\mathbf{j} - 2\mathbf{k}$

$$\operatorname{curl} \mathbf{F} = \begin{vmatrix} \mathbf{i} & \mathbf{j} & \mathbf{k} \\ \frac{\partial}{\partial x} & \frac{\partial}{\partial y} & \frac{\partial}{\partial z} \\ y & x & -2 \end{vmatrix} = \mathbf{0}$$

The line integral is path independent.

$$\int_C (y \, dx + x \, dy - 2 \, dz) = 0$$

28. $\mathbf{F} = (y + z)\mathbf{i} + (x + z)\mathbf{j} + (x + y)\mathbf{k}$

$$\operatorname{curl} \mathbf{F} = \begin{vmatrix} \mathbf{i} & \mathbf{j} & \mathbf{k} \\ \frac{\partial}{\partial x} & \frac{\partial}{\partial y} & \frac{\partial}{\partial z} \\ y+z & x+z & x+y \end{vmatrix} = \mathbf{0}$$

The line integral is path independent.

$$\int_C [(y + z) \, dx + (x + z) \, dy + (x + y) \, dz] = 0$$

29. $\mathbf{F} = -2y\mathbf{i} + 2x\mathbf{j} + \mathbf{k}$; $\mathbf{N} = \mathbf{k}$; $dS = dA_{xy}$

$$\operatorname{curl} \mathbf{F} = \begin{vmatrix} \mathbf{i} & \mathbf{j} & \mathbf{k} \\ \frac{\partial}{\partial x} & \frac{\partial}{\partial y} & \frac{\partial}{\partial z} \\ -2y & 2x & 1 \end{vmatrix} = 4\mathbf{k}$$

$$\int_C (-2y \, dx + 2x \, dy + dz) = 4 \iint_S dA = 4A = 4\pi$$

30. $\mathbf{F} = \operatorname{curl} y\mathbf{i}$; $\mathbf{N} = x\mathbf{i} + y\mathbf{j} + z\mathbf{k}$

$$\operatorname{curl} \mathbf{F} = \begin{vmatrix} \mathbf{i} & \mathbf{j} & \mathbf{k} \\ \frac{\partial}{\partial x} & \frac{\partial}{\partial y} & \frac{\partial}{\partial z} \\ y & 0 & 0 \end{vmatrix} = -\mathbf{k}$$

$$\int\int_S (\text{curl } y\mathbf{i}) \cdot \mathbf{N} \ dS = -\int\int_S z \ dS = -A = -\pi$$

31. $\quad I = \int\int_S (2x^3\mathbf{i} + y^3\mathbf{j} + z^3\mathbf{k}) \cdot \mathbf{N} \ dS$

$$= \int\int\int_D \text{div}(2xz^3\mathbf{i} + y^3\mathbf{j} + z^3\mathbf{k}) \ dV$$

$$= \int\int\int_D (6x^2 + 3y^2 + 3z^2) \ dV$$

Let $\sqrt{2}x = u$, $y = v$, $z = t$, then
$u^2 + v^2 + t^2 = 1$ is a sphere.

$$\frac{\partial(x, y, z)}{\partial(u, v, t)} = \begin{vmatrix} \frac{\partial x}{\partial u} & \frac{\partial x}{\partial v} & \frac{\partial x}{\partial t} \\ \frac{\partial y}{\partial u} & \frac{\partial y}{\partial v} & \frac{\partial y}{\partial t} \\ \frac{\partial z}{\partial u} & \frac{\partial z}{\partial v} & \frac{\partial z}{\partial t} \end{vmatrix} = \begin{vmatrix} \frac{1}{\sqrt{2}} & 0 & 0 \\ 0 & 1 & 0 \\ 0 & 0 & 1 \end{vmatrix}$$

$$= \frac{1}{\sqrt{2}}$$

$$I = \frac{3}{\sqrt{2}} \int\int\int_D (u^2 + v^2 + t^2) \ dV$$

$$= \frac{3}{\sqrt{2}} \int_0^{2\pi} \int_0^{\pi} \int_0^1 \rho^2 (\rho^2 \sin\phi) \ d\rho \ d\phi \ d\theta$$

$$= \frac{3}{\sqrt{2}}(2\pi) \int_0^{\pi} \frac{\sin\phi}{5} \ d\phi = \frac{12\pi}{5\sqrt{2}} = \frac{6\sqrt{2}\pi}{5}$$

32. $\quad \int\int_S (y^2\mathbf{i} + y^2\mathbf{j} + yz\mathbf{k}) \cdot \mathbf{N} \ dS$

$$= \int\int\int_D \text{div}(y^2\mathbf{i} + y^2\mathbf{j} + yz\mathbf{k}) \ dV$$

$$= \int\int\int_D (2y + y) \ dV$$

$$= 3 \int_0^{1/2} \int_0^{1/3} \int_0^{1 - 2x - 3y} y \ dz \ dy \ dx$$

$$= 3 \int_0^{1/2} \int_0^{1/3} [y(1 - 2x) - 3y^2] \ dy \ dx$$

$$= \frac{3}{54} \int_0^{1/2} (1 - 2x)^3 \ dx = \frac{1}{144}$$

33. $\quad \nabla\phi\nabla(2x + 3y) = 2\mathbf{i} + 3\mathbf{j};$

$\qquad f(x, y, z) = ax + by + cz - 1;$

$$\mathbf{N} = \frac{a\mathbf{i} + b\mathbf{j} + c\mathbf{k}}{\sqrt{a^2 + b^2 + c^2}}$$

$$\int\int_S \nabla\phi \cdot \mathbf{N} \ dS = \int\int_S \frac{(2a + 3b)\sqrt{a^2 + b^2 + c^2}}{c(a^2 + b^2 + c^2)}$$

$$= \frac{2a + 3b}{c}(\text{area of triangle})$$

$$= \frac{2a + 3b}{c}(\tfrac{1}{2})(\tfrac{1}{a})(\tfrac{1}{b}) = \frac{2a + 3b}{2abc}$$

34. $\quad \text{div } \mathbf{F} = 2(x + y)$

$$\int\int_S \mathbf{F} \cdot \mathbf{N} \ dS = \int\int\int_D \text{div } \mathbf{F} \ dV$$

$$= \int_0^1 \int_0^1 \int_0^1 2(x + y) \ dz \ dy \ dx$$

$$= 2 \int_0^1 \int_0^1 (x + y) \ dy \ dx$$

$$= 2 \int_0^1 (x + \tfrac{1}{2}) \ dx = 2$$

35. $\quad \text{div } \mathbf{F} = 0$, so

$$\int\int_S \mathbf{F} \cdot \mathbf{N} \ dS = \int\int\int_D \text{div } \mathbf{F} \ dV = 0$$

36. $\quad \text{div } \mathbf{F} = 1$

$$\int\int_S \mathbf{F} \cdot \mathbf{N} \ dS = \int\int\int_D \text{div } \mathbf{F} \ dV - \int\int_S \mathbf{F} \cdot \mathbf{j} \ dS$$

$$= \int_0^{2\pi} \int_0^3 \int_0^{r^2} r \ dy \ dr \ d\theta + 42\int\int_S dA_{xy}$$

$$= 2\pi \int_0^3 r^3 \ dr + 36\pi = \frac{153\pi}{2}$$

37. $\quad \text{div } \mathbf{F} = yz + xz + xy$

$$\int\int_S \mathbf{F} \cdot \mathbf{N} \ dS = \int\int\int_D \text{div } \mathbf{F} \ dV - \int\int_S \mathbf{F} \cdot (-\mathbf{j}) \ dS$$

$$= \int_0^1 \int_0^1 \int_0^1 (yz + xz + xy) \ dz \ dy \ dx - \int\int_S dA_{xy}$$

$$= \int_0^1 \int_0^1 (\tfrac{y}{2} + \tfrac{x}{2} + xy) \ dy \ dx - 0$$

$$= \int_0^1 (\tfrac{1}{4} + \tfrac{x}{2} + \tfrac{x}{2}) \ dx = \frac{3}{4}$$

38. $u = x = z;\ v = y$

$\mathbf{R}(x,\ y) = x\mathbf{i} + y\mathbf{j} + x\mathbf{k};\ \mathbf{T} = \mathbf{i} + \mathbf{j} + 0\mathbf{k};$

$\mathbf{N} = \dfrac{1}{\sqrt{2}}(\mathbf{i} + \mathbf{j});\ \text{div } \mathbf{F} = y$

$$\iint\limits_{S} \mathbf{F} \cdot \mathbf{N}\ dS = \iiint\limits_{D} \text{div } \mathbf{F}\ dV$$

$$= \int_0^1\int_0^1\int_0^1 y\ dz\ dy\ dx = \int_0^1\int_0^1 y\ dy\ dx = \frac{1}{2}$$

39. $\dfrac{\partial}{\partial x}(cx^2 + 4y) = \dfrac{\partial}{\partial y}(\sqrt{x} + 3xy)$

$\qquad 2cx = 3x$

$\qquad c = \dfrac{3}{2}\quad (x \neq 0)$

40. $\dfrac{\partial}{\partial x}(x^{-2} - x^{-1}) = \dfrac{\partial}{\partial y}(cyx^{-3} + x^{-2}y)$

$\qquad \dfrac{-2}{x^3} + \dfrac{1}{x^2} = \dfrac{c}{x^3} + \dfrac{1}{x^2}$

$\qquad c = -2$

41. $\mathbf{F} = e^{yz/x}\left[\left(\dfrac{cyz}{x^2}\right)\mathbf{i} + \left(\dfrac{z}{x}\right)\mathbf{j} + \left(\dfrac{y}{x}\right)\mathbf{k}\right]$

$\text{curl } \mathbf{F} = \left[e^{yz/x}\left(\dfrac{1}{x}\right) + \dfrac{zy}{x^2}e^{yz/x} - \dfrac{1}{x}e^{yz/x} - \dfrac{yz}{x^2}e^{yz/x}\right]\mathbf{i}$

$\qquad + \left(\dfrac{cy}{x^2}e^{yz/x} + \dfrac{cyz}{x^2}\left[\dfrac{y}{x}\right]e^{yz/x}\right.$

$\qquad - \left[\dfrac{-y}{x^2}\right]e^{yz/x} - \left.\dfrac{y}{x}e^{yz/x}\left[\dfrac{-yz}{x^2}\right]\right)\mathbf{j}$

$\qquad + \left(e^{yz/x}\left[\dfrac{-z}{x^2}\right] + \dfrac{z}{x}e^{yz/x}\left[\dfrac{-yz}{x^2}\right]\right.$

$\qquad - \left.e^{yz/x}\left[\dfrac{cz}{x^2}\right] - \dfrac{cyz}{x^2}e^{yz/x}\left[\dfrac{z}{x}\right]\right)\mathbf{k}$

$\qquad = 0 \text{ if } c = -1$

42. $\text{curl } \mathbf{F} = (1 - 1)\mathbf{i} + (0 - 0)\mathbf{j}$

$\qquad + \left(2y + \dfrac{e^{x/y}}{x^2} - \dfrac{1}{xy}e^{x/y} - 2y\right.$

$\qquad - \dfrac{e^{x/y}}{x^2} - \left.\dfrac{y}{x^2}e^{x/y}\left[\dfrac{-x}{y^2}\right]\right)$

$\qquad = 0, \text{ so } \mathbf{F} \text{ is conservative.}$

43. $x = t^2,\ y = t^2 - t,\ z = 3$

$\text{curl } \mathbf{F} = \begin{vmatrix} \mathbf{i} & \mathbf{j} & \mathbf{k} \\ \dfrac{\partial}{\partial x} & \dfrac{\partial}{\partial y} & \dfrac{\partial}{\partial z} \\ 2x & -x-z & y-x \end{vmatrix} \neq \mathbf{0}$

The integral depends on the path.

$W = \displaystyle\int_{C_1} [2x\ dx + (x + z)\ dy + (y - x)\ dz]$

$\qquad = \displaystyle\int_0^1 [(2t^2)(2t) - (t^2 + 3)(2t - 1)]\ dt = \frac{5}{6}$

44. $f_x = yz^2;\ f = xyz^2 + a(y,\ z)$

$f_y = xz^2 + \dfrac{\partial a(y,\ z)}{\partial y} = xz^2 - 1;$

$a(y,\ z) = -y + b(z)$

$f = xyz^2 + y + b(z)$

$f_z = 2xyz + b'(z) = 2xyz - 1;\ b(z) = -z + C$

$f(x,\ y,\ z) = xyz^2 - y - z + C$

$W = \displaystyle\int_C \mathbf{F} \cdot d\mathbf{R} = f(1,\ 0,\ 1) - f(0,\ 0,\ 0) = -1$

45. $(f_x)_y = -(x + y)^{-2} = (f_y)_x$

$f_y = (x + y)^{-1} + a'(y) = (x + y)^{-1};\ a(y) = 0$

$f(x,\ y) = \ln|x + y|$

$W = \displaystyle\int_C \mathbf{F} \cdot d\mathbf{R} = f(c,\ d) - f(a,\ b) = \ln\left(\dfrac{c + d}{a + b}\right)$

46. $\mathbf{F} = (2y \sin x \cos x + y^2\sin x)\mathbf{i}$

$\qquad + (\sin^2 x - 2y \cos x)\mathbf{j}$

$(f_x)_y = 2 \sin x \cos x + 2y \sin x$

$(f_y)_x = 2 \sin x \cos x + 2y \sin x$

Thus, $\mathbf{F}$ is conservative.

$f_x = 2y \sin x \cos x + y^2 \sin x$

$f = y \sin^2 x - y^2 \cos x + a(y)$

$f_y = \sin^2 x - 2y\cos x + a'(y) = \sin^2 x - 2y \cos x$

$a(y) = C$ and $f = y \sin^2 x - y^2\cos x = C$

If $y(0) = 1,\ C = -1$, so

$y \sin^2 x - y^2\cos x + 1 = 0$ is the solution to the given differential equation.

47. $\dfrac{\partial}{\partial y}(6xy + 2y^2 - 5) = 6x + 4y$

$\dfrac{\partial}{\partial x}(2x^2 + 4xy - 6) = 6x + 4y$

$f = 6xy + 2y^2 - 5$

$f = 2x^2 y + 2xy^2 - 5x + a(y)$

$f_y = 3x^2 + 4xy + a'(y) = 3x^2 + 4xy - 6$

$a(y) = -6y + C$, so

$f(x,\ y) = 3x^2 y + 2xy^2 - 5x - 6y + C = 0$

If $y(0) = 0$, then $C = 0$

$3x^2y + 2xy^2 - 5x - 6y = 0$ is the solution to the given differential equation.

48. Let $\mathbf{F} = M\mathbf{i} + N(x + z)\mathbf{j} + P\mathbf{k}$

$$\operatorname{curl} u\mathbf{F} = \begin{vmatrix} \mathbf{i} & \mathbf{j} & \mathbf{k} \\ \dfrac{\partial}{\partial x} & \dfrac{\partial}{\partial y} & \dfrac{\partial}{\partial z} \\ uM & uN & uP \end{vmatrix}$$

$$= [uP_y + u_yP - uN_z - u_zN]\mathbf{i}$$
$$+ [uM_z + u_zM - uP_x - u_xP]\mathbf{j}$$
$$+ [uN_x + u_xN - uM_y - u_yM]\mathbf{k}$$

$$= u[(P_y - N_x)\mathbf{i} + (m_z - P_x)\mathbf{j} + (N_x - M_y)\mathbf{k}]$$
$$+ [(u_yP - u_zN)\mathbf{i} + (u_xP - u_zM)\mathbf{j}$$
$$+ (u_xN - u_yM)\mathbf{k}]$$

$$= u\begin{vmatrix} \mathbf{i} & \mathbf{j} & \mathbf{k} \\ \dfrac{\partial}{\partial x} & \dfrac{\partial}{\partial y} & \dfrac{\partial}{\partial z} \\ M & N & P \end{vmatrix} + \begin{vmatrix} \mathbf{i} & \mathbf{j} & \mathbf{k} \\ \dfrac{\partial}{\partial x} & \dfrac{\partial}{\partial y} & \dfrac{\partial}{\partial z} \\ M & N & P \end{vmatrix}$$

$$= u \operatorname{curl} \mathbf{F} + (\nabla u \times \mathbf{F})$$

49. Let $\mathbf{F} = f_1\mathbf{i} + f_2(x + z)\mathbf{j} + f_3\mathbf{k}$;

$$\mathbf{G} = g_1\mathbf{i} + g_2(x + z)\mathbf{j} + g_3\mathbf{k}$$

$$\mathbf{F} \times \mathbf{G} = \begin{vmatrix} \mathbf{i} & \mathbf{j} & \mathbf{k} \\ f_1 & f_2 & f_3 \\ g_1 & g_2 & g_3 \end{vmatrix}$$

$$= [f_2g_3 - f_3g_2]\mathbf{i} - [f_1g_3 - f_3g_1]\mathbf{j}$$
$$+ [f_1g_2 - f_2g_1]\mathbf{k}$$

$$\operatorname{div}(\mathbf{F} \times \mathbf{G}) = f_2(g_3)_x - f_3(g_2)_x - f_1(g_3)_y$$
$$+ f_3(g_1)_y + f_1(g_2)_z - f_2(g_1)_z$$
$$+ (f_2)_xg_3 - (f_3)_xg_2 - (f_1)_yg$$
$$+ (f_3)_yg_1 + (f_1)_zg_2 - (f_2)_zg_1$$

$$\mathbf{G} \cdot \operatorname{curl} \mathbf{F} - \mathbf{F} \cdot \operatorname{curl} \mathbf{G}$$

$$= \begin{vmatrix} \mathbf{i} & \mathbf{j} & \mathbf{k} \\ \dfrac{\partial}{\partial x} & \dfrac{\partial}{\partial y} & \dfrac{\partial}{\partial z} \\ f_1 & f_2 & f_3 \end{vmatrix} - \mathbf{F} \cdot \begin{vmatrix} \mathbf{i} & \mathbf{j} & \mathbf{k} \\ u_x & u_y & u_z \\ g_1 & g_2 & g_3 \end{vmatrix}$$

$$= \mathbf{G} \cdot \{[(f_3)_y - (f_2)z]\mathbf{i} - [(f_3)_x - (f_1)_z]\mathbf{j}$$
$$+ [(f_2)_x - (f_1)_y]\mathbf{k}\}$$
$$- \mathbf{F} \cdot \{[(g_3)_y - (g_2)_z]\mathbf{i}$$
$$- [(g_3)_x - (g_1)_z]\mathbf{j}$$
$$+ [(g_2)_x - (g_1)_y]\mathbf{k}\}$$

$$= (f_3)_yg_1 - (f_2)_zg_1 - (f_3)_xg_2$$
$$+ (f_1)_z(g_2) + (f_2)_xg_3 - (f_1)_yg_3$$
$$- f_1(g_3)_y + f_1(g_3)_x - f_2(g_1)_z$$
$$- f_3(g_2)_x + f_3(g_1)_y$$

Thus, $\operatorname{div}(\mathbf{F} \times \mathbf{G}) = \mathbf{G} \cdot \operatorname{curl} \mathbf{F} - \mathbf{F} \cdot \operatorname{curl} \mathbf{G}$.

50. $\mathbf{F}$ and $\mathbf{G}$ are conservative.

$$\nabla f = \begin{vmatrix} \mathbf{i} & \mathbf{j} & \mathbf{k} \\ \dfrac{\partial}{\partial x} & \dfrac{\partial}{\partial y} & \dfrac{\partial}{\partial z} \\ f_x & f_y & f_z \end{vmatrix} = 0$$

$$\operatorname{curl} \mathbf{G} = \begin{vmatrix} \mathbf{i} & \mathbf{j} & \mathbf{k} \\ \dfrac{\partial}{\partial x} & \dfrac{\partial}{\partial y} & \dfrac{\partial}{\partial z} \\ g_1 & g_2 & g_3 \end{vmatrix} = 0$$

$\mathbf{F} = \nabla f$ and $\mathbf{G} = \nabla g$; By Problem 49,

$$\operatorname{div}(\mathbf{F} \times \mathbf{G}) = \mathbf{G} \cdot \operatorname{curl} \mathbf{F} - \mathbf{F} \cdot \operatorname{curl} \mathbf{G} = 0$$

so that $\mathbf{F} \times \mathbf{G}$ is solenoidal.

51. $\operatorname{div} \mathbf{F} = \operatorname{div}(\operatorname{curl} \mathbf{G}) = 0$ so $\mathbf{F}$ is solenoidal (See Problem 50.)

52. $\mathbf{F} = f(x, y, z)\mathbf{A}$ with $\mathbf{A} = a_1\mathbf{i} + a_2\mathbf{j} + a_3\mathbf{k}$

$$\operatorname{curl} \mathbf{F} = \begin{vmatrix} \mathbf{i} & \mathbf{j} & \mathbf{k} \\ \dfrac{\partial}{\partial x} & \dfrac{\partial}{\partial y} & \dfrac{\partial}{\partial z} \\ fa_1 & fa_2 & fa_3 \end{vmatrix}$$

$$= (a_3f_y - f_za_2)\mathbf{i} - (f_xa_3 - f_za_1)\mathbf{j}$$
$$+ (f_xa_2 - f_ya_1)\mathbf{k}$$

$$= a_1(f_y\mathbf{j} - f_y\mathbf{k}) + a_2(-f_z\mathbf{i} - f_x\mathbf{k})$$
$$+ a_3(f_y - f_x\mathbf{j})$$

$$\mathbf{A} \cdot \operatorname{curl} \mathbf{F} = a_1a_2f_z - a_1a_3f_y - a_2a_1f_z$$
$$+ a_2a_3f_x + a_3a_1f_y - a_3a_2f_x = 0$$

Thus, $\operatorname{curl} \mathbf{F}$ is orthogonal to $\mathbf{A}$.

$$\nabla f \cdot \operatorname{curl} \mathbf{F} = a_1f_zf_x - a_1ff - a_2f_zf_x + a_2f_zf_x$$
$$+ a_3f_xf_y - a_3f_xf_y = 0$$

Thus, $\operatorname{curl} \mathbf{F}$ is orthogonal to ∇f.

53. $\operatorname{div} \mathbf{A} \times \mathbf{F} = \operatorname{div} \begin{vmatrix} \mathbf{i} & \mathbf{j} & \mathbf{k} \\ \dfrac{\partial}{\partial x} & \dfrac{\partial}{\partial y} & \dfrac{\partial}{\partial z} \\ f_1 & f_2 & f_3 \end{vmatrix}$

$$= \operatorname{div}[(a_2f_3 - a_3f_2)\mathbf{i} - (a_1f_3 - a_3f_1)\mathbf{j}$$
$$+ (a_1f_2 - a_2f_1)\mathbf{k}]$$

$$= a_2(f_3)_x - a_3(f_2)_x - a_1(f_3)_y + a_3(f_1)_y$$
$$+ a_1(f_2)_z - a_2(f_1)_z$$

$$= a_1[-(f_3)_y + (f_2)_z] - a_2[(f_1)_z - (f_3)_x]$$
$$+ a_3[-(f_2)_x + (f_1)_y]$$

$$= -\mathbf{A} \cdot \operatorname{curl} \mathbf{F}$$

54. If $f = \ln\left| x^2 - y^2 \right|$; $\nabla f = \dfrac{2x\mathbf{i} - 2y\mathbf{j}}{x^2 - y^2}$

$\displaystyle\int_C \frac{x\,dx - y\,dy}{x^2 - y^2} = \frac{1}{2}\int \nabla f \cdot d\mathbf{R}$

$= f(5, 4) - f(2, 0) = \frac{1}{2}[\ln 9 - \ln 4] = \ln \frac{3}{2}$

55. $\displaystyle\int_C \left(\frac{-y}{x^2}\,dx + \frac{1}{x}\,dy\right)$

$= \displaystyle\iint_D \left[\frac{\partial x^{-1}}{\partial x} - \frac{\partial(-y/x^2)}{\partial y}\right]\,dy\,dx$

$= \displaystyle\iint_D \left[-\frac{1}{x^2} + \frac{1}{x^2}\right]\,dy\,dx = 0$

56. $\dfrac{\partial u}{\partial y} = \dfrac{\partial v}{\partial x}$; let uxy^2 and $v = x^2 y$, then $u_y = 2xy$

and $v_x = 2xy$; $u_x = y^2$, $u_{xx} = 0$, $u_y = 2xy$,

$u_{yy} = 2x$; thus $u_{xx} + u_{yy} \neq 0$, so u and v are

not necessarily harmonic.

57. a. C is any region that does not contain the y-axis.

$f_x = (1 + y^2)x^{-3}$; $f = -\dfrac{1 + y^2}{2x^2} + a(y)$

$f_y = -\dfrac{2y}{2x^2} + a'(y) = -\dfrac{y}{x^2} - y$

$a(y) = -\dfrac{y^2}{2} + C$;

$f = -\dfrac{1 + y^2}{2x^2} - \dfrac{y^2}{2} + C$

b. $\displaystyle\int_C \mathbf{F} \cdot d\mathbf{R} = \int_C \nabla f \cdot d\mathbf{R} = f(3, 4) - f(1, 1)$

$= \left[-\dfrac{1 + y^2}{2x^2} - \dfrac{y^2}{2} + C\right]\Big|_{(1,1)}^{(3,4)} = -\dfrac{134}{18}$

58. $u_y = (2ye^x + y^2)_x = 2ye^x + 3y^2 e^{2x}$

$u = y^2 e^x + y^3 e^3 x + c(x)$

59. Evaluating the line integral directly:

C_1: $x = t$, $y = (5 - t)/2$; $1 \le t \le 2$

$I_1 = \displaystyle\int_1^2 \left[\frac{2}{5 - t} - \frac{1}{2t}\right]\,dt = \ln\frac{16}{9} - \frac{1}{2}\ln 2$

C_2: $x = 2$, $y = t$; $\frac{3}{2} \le t \le 4$

$I_2 = \displaystyle\int_{3/2}^4 \frac{dt}{2} = \frac{5}{4}$

C_3: $x = 2 - t$, $y = 2(2 - t)$; $0 \le t \le 1$

$I_3 = \displaystyle\int_0^1 \left[\frac{-1}{4 - 2t} - \frac{2}{2 - t}\right]\,dt = -\frac{5}{2}\ln 2$

$\displaystyle\int_C \left(\frac{dx}{y} + \frac{dy}{x}\right) = I_1 + I_2 + I_3 = \frac{5}{4} + \ln\frac{2}{9}$

Using Green's theorem:

$\displaystyle\int_C \left(\frac{dx}{y} + \frac{dy}{x}\right) = \iint_D \left(-\frac{1}{x^2} + \frac{1}{y^2}\right)\,dA_{xy}$

$= \displaystyle\int_1^2 \int_{(5-x)/2}^{2x} \left(-\frac{1}{x^2} + \frac{1}{y^2}\right)\,dy\,dx$

$= \displaystyle\int_1^2 \frac{5(2x^2 - 7x + 5)}{2x^2(5 - x)}\,dx$

$= \displaystyle\int_1^2 \left[-\frac{2}{x - 5} + \frac{5}{2x^2} - \frac{3}{x}\right]\,dx$

$= \dfrac{5}{4} + \ln\dfrac{2}{9}$

60. $\operatorname{div}(\operatorname{curl}\mathbf{F}) = \operatorname{div}\begin{vmatrix} \mathbf{i} & \mathbf{j} & \mathbf{k} \\ \frac{\partial}{\partial x} & \frac{\partial}{\partial y} & \frac{\partial}{\partial z} \\ f_1 & f_2 & f_3 \end{vmatrix} = 0$

$\displaystyle\iint_S (\operatorname{curl}\mathbf{F}\cdot\mathbf{N})\,dS = \iiint_D \operatorname{div}\operatorname{curl}\mathbf{F}\,dV = 0$

61. $c\mathbf{k}\times\mathbf{R} = \mathbf{F} = \begin{vmatrix} \mathbf{i} & \mathbf{j} & \mathbf{k} \\ 0 & 0 & c \\ x & y & z \end{vmatrix} = -cy\mathbf{i} + cx\mathbf{j}$

$\operatorname{curl}\mathbf{F} = \begin{vmatrix} \mathbf{i} & \mathbf{j} & \mathbf{k} \\ \frac{\partial}{\partial x} & \frac{\partial}{\partial y} & \frac{\partial}{\partial z} \\ -cy & cx & 0 \end{vmatrix} = -2c\mathbf{k}$

$\nabla f = 2\mathbf{i} + 2\mathbf{j} + \mathbf{k}$; $\mathbf{N} = \frac{1}{3}(2\mathbf{i} + 2\mathbf{j} + \mathbf{k})$

$\operatorname{curl}\mathbf{F}\cdot\mathbf{N}\,dS = \frac{1}{3}\,dS$

$\displaystyle\int_C (c\mathbf{k}\times\mathbf{R})\cdot d\mathbf{R} = \iint_D \operatorname{curl}(c\mathbf{k}\times\mathbf{R})\cdot\mathbf{N}\,dS$

$= \frac{1}{3}(3)(2\pi) = 2\pi c$

62. a. $\nabla^2 w = 0$;

$\nabla\cdot(w\nabla w) = \left(\frac{\partial}{\partial x}\mathbf{i} + \frac{\partial}{\partial y}\mathbf{j} + \frac{\partial}{\partial z}\mathbf{k}\right)\cdot[ww_x\mathbf{i} + ww_y\mathbf{j} + ww_z\mathbf{k}]$

$= (ww_x)_x + (ww_y)_y + (ww_z)_z$

$= [ww_{xx} + ww_{yy} + ww_{zz}]$

$$+ [(w_x)^2 + (w_y)^2 + (w_z)^2]$$
$$= 0 + \|\nabla w\|^2$$
$$= \|\nabla w\|^2$$

b. $w = x - y + 2z$; $f = x^2 + y^2 + z^2 - 9 = 0$

$w \dfrac{\partial w}{\partial n} = w \nabla w \cdot \mathbf{N}$; $\nabla w = \mathbf{i} - \mathbf{j} + 2\mathbf{k}$; $\nabla^2 w = 0$

so w is harmonic. $\|\nabla w\| = \sqrt{6}$

$$\iint\limits_S w \frac{\partial w}{\partial n}\, dS = \iint\limits_S w \nabla w \cdot \mathbf{N}\, dS$$

$$= \iiint\limits_D \operatorname{div}(w \nabla w)\, dV$$

$$= \iiint\limits_D \|\nabla w\|^2\, dV$$

$$= 6 V = 6(\tfrac{4\pi}{3})(3^3) = 324\pi$$

63. $\mathbf{R}(t) = x(t)\mathbf{i} + y(t)\mathbf{j} + z(t)\mathbf{k}$;

$\dfrac{d\mathbf{R}}{dt} = \dfrac{dx}{dt}\mathbf{i} + \dfrac{dy}{dt}\mathbf{j} + \dfrac{dz}{dt}\mathbf{k}$

Since $\mathbf{F}$ is always perpendicular to path C, we must have

$$\mathbf{F} \cdot \frac{d\mathbf{R}}{dt} = 0$$

Thus, $W = \displaystyle\int_C \mathbf{F} \cdot d\mathbf{R} = 0$

64. $\mathbf{D} = 2x^2\mathbf{i} + 3y^2\mathbf{j} - 2x^2\mathbf{k}$; $\operatorname{div} \mathbf{D} = 4x + 6y$

The equation of the plane is $x + 3y = 0$

65. $\mathbf{F} = m\omega^2\mathbf{R}$; $\omega = 9.8m$, $m = 10,000/9.8$;

$\omega = d\theta/dt$; $0 \le t \le \pi$

$$dW = \mathbf{F} \cdot d\mathbf{R} = -\frac{10,000}{9.8}\omega^2 \int_C \mathbf{R} \cdot d\mathbf{R}$$

$$= -\frac{10,000}{9.8}\omega^2 \int_C (x\, dx + y^2\, dy)$$

$$= -\frac{10,000}{9.8}\omega^2(7,000)(10^3)\int_0^\pi (-\cos\theta \sin\theta + \sin\theta \cos\theta)\, d\theta$$

$$= 0$$

66. The difference is potential energy is

$$\Delta p = p_2 - p_1 = mv_2{}^2 - mv_1{}^2$$
$$= 10[(2.5)^2 - 3^2] = -27.5 \text{ g cm}^2/\text{s}^2$$

67. $\mathbf{F}$ is conservative with scalar potential

$$f(x, y, z) = x^2yz - \frac{y}{z}\tan^{-1}\frac{y}{z} + \frac{1}{2}\ln(1 + \frac{y^2}{z^2})$$

The endpoints are $t = 0$: $(0, 1, 1)$ and

$t = \tfrac{1}{2}$: $(1, 0, 2)$, so the work is

$$W = \int_C \mathbf{F} \cdot d\mathbf{R} = f(1, 0, 2) - f(0, 1, 1)$$

$$= [(0 - 0 + \tfrac{1}{2}\ln(1 + 0)]$$

$$\quad - [0 - \tfrac{1}{1}\tan^{-1}(\tfrac{1}{1}) + \tfrac{1}{2}\ln(1 + 1^2/1^2)]$$

$$= \tan^{-1}1 - \tfrac{1}{2}\ln 2 = \frac{\pi}{4} - \frac{1}{2}\ln 2$$

68. C: $x^2 + \dfrac{y^2}{k^2} = 1$; $x = \cos\theta$, $y = k\sin\theta$; $0 \le \theta \le \pi$

$$W = \int_C [4xy\, dx + (3x + y)\, dy]$$

$$= \int_0^\pi [4k\cos\theta \sin\theta(-\sin\theta) + 3k\cos\theta(\cos\theta) + k^2\sin\theta \cos\theta]\, d\theta$$

$$= \frac{3}{2}k\int_0^\pi (1 + \cos 2\theta)\, d\theta = \frac{3k\pi}{2}$$

This is minimized when $k = 0$ (the ellipse degenerates to a line segment).

69.
$$\iint\limits_S \mathbf{E} \cdot \mathbf{N}\, dS = \iiint\limits_D \operatorname{div} \mathbf{F}\, dV$$

$$= \frac{1}{\epsilon}\iiint\limits_D Q\, dV = \frac{q}{\epsilon}$$

70.
$$\frac{\partial}{\partial x}\left(\frac{x}{x^2 + y^2}\right) = \frac{-x^2 + y^2}{(x^2 + y^2)^2}$$

$$\frac{\partial}{\partial y}\left(\frac{-y}{x^2 + y^2}\right) = \frac{-x^2 + y^2}{(x^2 + y^2)^2}$$

Thus, the integral path is independent, so

$$\int_C \frac{-y\, dx + x\, dy}{x^2 + y^2} = 0$$

71. a. The normal derivative $\partial g/\partial n$ is defined as the directional derivative in the normal direction, that is

$$\frac{\partial g}{\partial n} = \nabla g \cdot \mathbf{N}$$

Applying the divergence theorem to $\mathbf{F} = f\nabla g$, we obtain

$$\iint\limits_S f\frac{\partial g}{\partial n}\, dS = \iint\limits_S f(\nabla g \cdot \mathbf{N})\, dS$$

$$= \iiint\limits_D \operatorname{div}(f\nabla g)\, dV$$

$$= \iiint_D \nabla \cdot (f \nabla g) \, dV$$

$$= \iiint_D \underbrace{[f\nabla^2 g + (\nabla f) \cdot (\nabla f)]}_{\text{expansion of } \nabla \cdot (f\nabla g) \text{ by properties of gradient}} dV$$

Similarly, we have

$$\iint_S g \frac{\partial f}{\partial n} \, dS = \iiint_D [gf\nabla^2 f + (\nabla g) \cdot (\nabla g)] \, dV$$

But f *and* g are given to be harmonic, so $\nabla^2 f = \nabla^2 g = 0$ and we have

$$\iint_S f \frac{\partial g}{\partial n} \, dS = \iiint_D (\nabla f) \cdot (\nabla g) \, dV$$

$$= \iint_S g \frac{\partial f}{\partial n} \, dS$$

b. Using the results of part **a**,

$$\iint_S f \frac{\partial f}{\partial n} \, dS = \iiint_D (\nabla f) \cdot (\nabla g) \, dV$$

$$= \iiint_D \| \nabla f \|^2 \, dV$$

72. $\text{curl}(\text{curl } \mathbf{F}) = \text{curl}\{[(f_3)_y - (f_2)_z]\mathbf{i}$

$- [(f_3)_x - (f_1)_z]\mathbf{j} + [(f_2)_x - (f_1)_y]\mathbf{k}\}$

$= [(f_2)_{xy} - (f_1)_{yy} + (f_3)_{xz} - (f_1)_{zz}]\mathbf{i}$

$\quad - [(f_2)_{xx} - (f_1)_{xy} - (f_3)_{yz} + (f_2)_{zz}]\mathbf{j}$

$\quad + [-(f_3)_{xx} + (f_1)_{xz} - (f_3)_{yy} + (f_2)_{yz}]\mathbf{k}$

$\nabla(\text{div } \mathbf{F}) - \nabla^2 \mathbf{F}$

$\quad = \nabla[(f_1)_x + (f_2)_y + (f_3)_z] - \nabla \cdot \nabla \mathbf{F}$

$\quad = [(f_1)_{xx} + (f_2)_{xy} + (f_3)_{xz}]\mathbf{i}$

$\qquad + [(f_1)_{xy} + (f_2)_{yy} + (f_3)_{yz}]\mathbf{j}$

$\qquad + [(f_1)_{xz} + (f_2)_{yz} - (f_3)_{zz}]\mathbf{k}$

$\qquad - \left(\frac{\partial}{\partial x}\mathbf{i} + \frac{\partial}{\partial y}\mathbf{j} + \frac{\partial}{\partial x}\mathbf{k}\right) \cdot \left(\frac{\partial}{\partial x}\mathbf{i} + \frac{\partial}{\partial y}\mathbf{j}\right.$

$\qquad + \left.\frac{\partial}{\partial x}\mathbf{k}\right)(f_1\mathbf{i} + f_2\mathbf{j} + f_3\mathbf{k})$

$\quad = [(f_2)_{xy} - (f_1)_{yy} + (f_3)_{xz}]\mathbf{i}$

$\qquad - [(f_2)_{xx} - (f_1)_{xy} - (f_3)_{yz} + (f_2)_{zz}]\mathbf{j}$

$\qquad + [-(f_3)_{xx} + (f_1)_{xz} - (f_3)_{yy} + (f_2)_{yz}]\mathbf{k}$

Thus, $\text{curl}(\text{curl } \mathbf{F}) = \nabla(\text{div } \mathbf{F}) - \nabla^2 \mathbf{F}$.

73. $\text{div}(\nabla f) = \nabla \cdot (f_x \mathbf{i} + f_y \mathbf{j} + f_z \mathbf{k})$

$$= f_{xx} + f_{yy} + f_{zz}$$

$\text{div}(f \nabla f)$

$= \left(\frac{\partial}{\partial x}\mathbf{i} + \frac{\partial}{\partial y}\mathbf{j} + \frac{\partial}{\partial z}\mathbf{k}\right) \cdot (ff_x \mathbf{i} + ff_y \mathbf{j} + ff_z \mathbf{k})$

$= f(f_{xx} + f_{yy} + f_{zz}) + f_x^2 + f_y^2 + f_z^2$

$= f(\text{div } \nabla f) + \| \nabla f \|^2$

We have $7f = f(\text{div}\nabla f) + 3f$, $\text{div}\nabla f = 4$

$$\iint_S \frac{\partial f}{\partial n} \, dS = \iint_S \nabla f \cdot \mathbf{N} \, dS$$

$$= \iiint_D (\nabla f) \, dV$$

$$= \iiint_D 4 \, dV = 4\left(\frac{4\pi}{3}\right)(1) = \frac{16\pi}{3}$$

74. $\mathbf{R}_u = [(a + b \cos v)(-\sin u)\mathbf{i}$

$\quad + (a + b \cos v)\cos u \, \mathbf{j}]$

$\mathbf{R}_v = [-b \sin v \cos u \mathbf{i} - b \sin v \sin u \mathbf{j}$

$\quad + b \cos v \mathbf{k}]$

$\mathbf{R}_u \times \mathbf{R}_v = [b(a + b \cos v) \cos u \cos v \, \mathbf{i}$

$\quad - v(a + b \cos v) \sin u \cos v \, \mathbf{j}$

$\quad + b(a + b \cos v)\sin v \, \mathbf{k}]$

$\| \mathbf{R}_u \times \mathbf{R}_v \| = b(a + b \cos v)$

Thus,

$$\iint dS = \iint \| \mathbf{R}_u \times \mathbf{R}_v \| \, dv \, du$$

$$= \int_0^{2\pi} \int_0^{2\pi} b(a + b \cos v) \, dv \, du$$

$$= \int_0^{2\pi} 2ab\pi \, du = 4\pi^2 ab$$

75. $\iint_S \mathbf{F} \cdot \mathbf{N} \, dS = \iiint_D \text{div}(x\mathbf{i} + y\mathbf{j} + z\mathbf{k}) \, dV$

$= 3V = 3(8 - 1) = 21$

76. div $\mathbf{F}$ is independent of coordinates, so it does not matter which corner is removed. From Problem 75, the value is 21.

77.
$$I = \frac{1}{3}\int_C (-y^3\ dx + x^3\ dy)$$
$$= \frac{1}{3}\int\int_S (3x^2 + 3y^2)\ dA$$
$$= \int\int_A r^2\ dm = I_z$$

78. By symmetry, $\bar{x} = \bar{y} = 0$. To find $\bar{z}$, we first compute the surface area of the cone.

$$S = \pi(3)(3\sqrt{2}) = 9\sqrt{2}\pi$$
$$\bar{z} = \frac{1}{9\sqrt{2}\pi}\int\int_S z\ dS$$
$$= \frac{1}{9\sqrt{2}\pi}\int\int_{\text{disk } x^2+y^2 \le 9} \sqrt{x^2 + y^2}\ A\ dy\ dx$$

where $A = \sqrt{\dfrac{x^2}{x^2 + y^2} + \dfrac{y^2}{x^2 + y^2} + 1}\ dy\ dx$

$$= \frac{1}{9\sqrt{2}\pi}\int_0^{2\pi}\int_0^3 r\sqrt{2}(r\ dr\ d\theta)$$
$$= \frac{1}{9\sqrt{2}\pi}\int_0^{2\pi} 9\sqrt{2}\ d\theta = 2$$

The centroid of the conical shell is $(0, 0, 2)$.

79. Once the Green Purity makes her choice of N (say $N = \bar{N}$), the "head start function" $S(x, N)$ becomes a function x alone; specifically,

$$\bar{S}(x) = S(x, \bar{N}) = \frac{3e^{4\bar{N}+22}}{(e^{40} + e^{8\bar{N}})}\left[\frac{e^{(x+\bar{N})} - 1}{(x + \bar{N})^2}\right]$$

and $\dfrac{d\bar{S}(x)}{dx} = 0$ when $(x + \bar{N})^2 = 1$

Thus, the minimum value of $\bar{S}(x)$ — the most damage the Red Purity can do — occurs when $(x + \bar{N})^2 = 1$, and this minimum is given by

$$M(\bar{N}) = S(-\bar{N} \pm 1, \bar{N})$$
$$= \frac{3e^{4\bar{N}+22}}{(e^{40} + e^{8\bar{N}})}\left[\frac{e^{1^2}}{(1)^2}\right]$$
$$= \frac{3e^{4\bar{N}+23}}{e^{40} + e^{8\bar{N}}}$$
$$= 3e^{22}\left[\frac{e^{4\bar{N}}}{e^{40} + e^{8N}}\right]$$

This is the *worst* the Green Purity can expect once she chooses $N = \bar{N}$, so she wants to

maximize $M(\bar{N})$. Differentiating $M(\bar{N})$ with respect to $\bar{N}$, we obtain

$$\frac{dM}{d\bar{N}} = 3(e^{23})\left[\frac{(e^{40} + e^{8\bar{N}})(4e^{4N}) - (e^{4\bar{N}})(8e^{8\bar{N}})}{(e^{40} + e^{8\bar{N}})^2}\right]$$
$$= 3e^{23}(4e^{4\bar{N}})\left[\frac{e^{40} + e^{8\bar{N}} - 2e^{8\bar{N}}}{(e^{40} + e^{8\bar{N}})}\right]$$

This is 0 when $e^{40} = e^{8\bar{N}}$; this is when $\bar{N} = 5$

Thus, the *minimum* value $M(\bar{N})$ is *maximized* when N is chosen to be 5. If both Puritys "play" properly, the Green Purity chooses $\bar{N} = 5$, the Red Purity chooses $x = -\bar{N} \pm 1 = -4, -6$ and the resulting head start time is

$$S(-4, 6) = \frac{3e^{[(1)^2 + 4(5) + 22]}}{(1)^2(e^{40} + e^{40})} = \frac{3e^{43}}{2e^{40}}$$
$$= 1.5e^3 \approx 30 \text{ min}$$

However, the spy reneges on his deal and subdues a rather surprised Red Purity (in a politically correct way, of course). The Green Purity cries, "How did you know?" The spy smiles smugly and says, "The woman who set those traps to kill me knows some math. You don't and your sister does." The Red Purity snarls, "So she wins because I'm smarter! I'll remember that next time, and believe me, there *will* be a next time."

80. This is Putnam Problem 6i in the morning session of 1948. Let $\rho = \rho(s)$ be the parametric vector equation of the given curve, where s is the arc length. We assume that ρ is a periodic function of class C^2 (that is, f^2 is continuous) and period L, the length of the curve, so that ρ describes the curve once as s varies from 0 to L. Then $d\rho/ds = \mathbf{T} = \mathbf{T}(s)$ is the tangent vector to the curve at $\rho(s)$, and

$$\frac{d\mathbf{T}}{ds} = \kappa\mathbf{N}$$

where their concave side of the curve and $\kappa = \kappa(s)$ is the curvature. The radius of curvature is $r = 1/\kappa$. It is given that the force on an element ds of the curve has magnitude ds/r and direction $-\mathbf{N}$; that is,

$$d\mathbf{F} = -\frac{\mathbf{N}\ ds}{r}$$

so that $d\mathbf{F} = -dt$. The condition for equilibrium is that the total force should be zero and that the total moment of the force with respect to some point should be zero. Thus, we must show

$$\int_C d\mathbf{F} = 0 \text{ and } \int_C \rho \times d\mathbf{F} = \mathbf{0}$$

Now

$$\int_C d\mathbf{F} = -\int_C d\mathbf{T} = -\mathbf{T}(s)\Big|_0^L = 0$$

For the second requirement, note that

$$\frac{d}{ds}(\rho + \mathbf{T}) = \mathbf{T} \times \mathbf{T} + \rho \times \frac{d\mathbf{T}}{ds}$$

Hence

$$\int_C \rho \times d\mathbf{F} = -\int_C \rho \times d\mathbf{T} = -\rho \times \mathbf{T}\Big|_0^L = 0$$

Cumulative Review for Chapters 12-14, Page 969

1. Calculus is the study of dynamic processes (rather than the static). It is the study of infinitesimals, the behavior of functions at or near a point. The are three fundamental ideas of calculus; the notion of a limit, derivatives (the limit of difference quotients), and integrals (the sum of infinitesimal quantities). Calculus is also the study of transformations of reference frames (coordinate systems), and motion with respect to reference frames (vector analysis).

2. The notation $\lim_{x \to c} f(x) = L$ is read "the limit of $f(x)$ as x approaches c is L" and means that the functional values $f(x)$ can be made arbitrarily close to L by choosing x sufficiently close to c. See Chapter 6 of the *Mathematics Handbook* for a summary of all the limit ideas in this book. The derivative of f at x is given by

$$f'(x) = \lim_{\Delta x \to 0} \frac{f(x + \Delta x) - f(x)}{\Delta x}$$

provided this limit exists. See Chapter 7 of the *Mathematics Handbook* for a summary of the derivative ideas in this book. If f is defined on the closed interval $[a, b]$ we say f is integrable on $[a, b]$ if

$$I = \lim_{\|P\| \to 0} \sum_{k=1}^{n} f(\overset{*}{x}_k)\Delta x_k$$

exists. This limit, if it exists, is called the definite integral of f from a to b. The definite integral is denoted by

$$I = \int_a^b f(x)\, dx$$

See Chapter 8 of the *Mathematics Handbook* for a summary of the integral ideas in this

book.

3. Multivariable calculus involves functions whose domain consists of more than one independent variable.

4. $f(x, y) = 2x^2 + xy - 5y^3$
$f_x = 4x + y;\ f_y = x - 15y^2;\ f_{xy} = 1$

5. $f(x, y) = x^2 e^{y/x}$
$f_x = e^{y/x}(2x - y);\ f_y = xe^{y/x};$
$f_{xy} = e^{x/y}(1 - y/x)$

6. $f(x, y) = \dfrac{x^2 - y^2}{x - y} = x + y$ (if $x \neq y$)
$f_x = 1;\ f_y = 1;\ f_{xy} = 0$

7. $f(x, y) = y\sin^2 x + \cos xy$
$f_x = 2y\sin x\cos x - y\sin xy$
$f_y = \sin^2 x - x\sin xy$
$f_{xy} = \sin 2x - xy\cos xy - \sin xy$

8. $f(x, y) = e^{x+y}$
$f_x = e^x e^y;\ f_y = e^x e^y;\ f_{xy} = e^x e^y$

9. $f(x, y) = \dfrac{x^2 + y^2}{x - y}$

$$f_x = \frac{(x-y)(2x) - (x^2 + y^2)}{(x-y)^2} = \frac{x^2 - 2xy - y^2}{(x-y)^2}$$

$$f_y = \frac{(x-y)(2y) + (x^2 + y^2)}{(x-y)^2} = \frac{x^2 + 2xy - y^2}{(x-y)^2}$$

$$f_{xy}$$
$$= \frac{(x-y)^2(-2x - 2y) + (x^2 - 2xy - y^2)(2)(x-y)}{2}$$
$$= \frac{-4xy}{(x-y)^3}$$

10. $\displaystyle\int_0^1 \int_x^{2x} e^{y-x}\, dy\, dx = \int_0^1 (e^x - 1)\, dx = e - 2$

11. $\displaystyle\int_0^4 \int_0^{\sqrt{x}} 3x^5\, dy\, dx = \int_0^4 3x^{11/2}\, dx = \dfrac{49{,}152}{13}$

12. $\displaystyle\int_0^1 \int_0^z \int_y^{y-z} (x + y + z)\, dx\, dy\, dz$

$$= \int_0^1 \int_0^z -\tfrac{1}{2}z(4y + z)\, dy\, dz$$

$$= \int_0^1 -\tfrac{3}{2}z^3\, dz = -\tfrac{3}{8}$$

13. $$\int_0^{15\pi}\int_0^{\pi}\int_0^{\sin\phi} \rho^3 \sin\phi \, d\rho \, d\theta \, d\phi$$

$$= \frac{1}{4}\int_0^{15\pi}\int_0^{\pi} \sin^5\phi \, d\theta \, d\phi$$

$$= \frac{\pi}{4}\int_0^{15\pi} \sin^5\phi \, d\pi \, d\phi = \frac{4\pi}{15}$$

14. $$\iint_R e^{x+y} \, dA = \int_0^1\int_0^1 e^y e^x \, dx \, dy$$

$$= (e - 1)\int_0^1 e^y \, dy = (e - 1)^2$$

15. $$\iint_R ye^{xy} \, dA = \int_0^1\int_0^2 ye^{xy} \, dy \, dx$$

$$\int_0^2\int_0^1 ye^{xy} \, dx \, dy = \int_0^2 (e^y - 1) \, dy = e^2 - 3$$

16. $$\iint_R \sin(x + y) \, dA = \int_0^{\pi/2}\int_0^{\pi/4} \sin(x + y) \, dy \, dx$$

$$= -\int_0^{\pi/2} [\cos(x + \tfrac{\pi}{4}) - \cos x] \, dx = 1$$

17. $$\iint_R x \sin xy \, dA = \int_0^{\pi}\int_0^{1} x \sin xy \, dy \, dx$$

$$= -\int_0^{\pi} (\cos x - 1) \, dx = \pi$$

18. $$\int_C (y^2 z \, dx + 2xyz \, dy + xy^2 \, dz) = \int_0^1 t^3 \, dt = 1$$

19. $$\int_C (5 \, xy \, dx + 10 \, yz \, dy + z \, dz)$$

$$= \int_0^1 [5(t^2)(t)(2t) + 10(t)(2t^3) + (2t^3)(6t^2)] \, dt$$

$$= \int_0^1 (30t^4 + 12t^5) \, dt = 8$$

20. $\int_C \mathbf{F} \cdot d\mathbf{R}$, where $\mathbf{F} = yz\mathbf{i} - x\mathbf{k}$; curl $\mathbf{F} \neq 0$ so the the line integral depends on the path.

C_1: $z = 1$, $x = y = t$; $0 \le t \le 1$

C_2: $z = 1$, $y = 1 - t$, $x = 1$; $0 \le t \le 1$

C_3: $z = 1$, $y = 0$, $x = 1 - t$; $0 \le t \le 1$

$$\int_C \mathbf{F} \cdot d\mathbf{R} = \int_0^1 [t(1) \, dt - t(0)] \, dt$$

$$+ \int_0^1 [(1 - t)(1)(0) - (1)(0)] \, dt$$

$$+ \int_0^1 [(0)(1)(-dt) - (1 - t)(0)] \, dt$$

$$= \int_0^1 t \, dt = \frac{1}{2}$$

You can check this result by using Stokes's theorem.

21. $$\frac{ds}{dt} = \sqrt{\left(\frac{dx}{dt}\right)^2 + \left(\frac{dy}{dt}\right)^2}$$

$$= \sqrt{[3(\cos^2 t)(-\sin t)]^2 + [3(\sin^2 t)(\cos t)]^2}$$

$$= 3 \sin t \cos t\sqrt{\cos^2 t + \sin^2 t}$$

$$\int_C (x^3 + y^3) \, ds$$

$$= \int_0^{2\pi} [\cos^9 t(3 \sin t \cos t) + \sin^9 t(3 \sin t \cos t)] \, dt$$

$$= 0$$

22. $\nabla f = \mathbf{N} = (2x + y\cos xy)\mathbf{i} + (2y + x\cos xy)\mathbf{j} - \mathbf{k}$

$\nabla f(0, 2, 4) = 2\mathbf{i} + 4\mathbf{j} - \mathbf{k}$

The equation of the plane is

$$2x + 4(y - 2) + (z - 4) = 0$$

$$2x + 4y + z - 12 = 0$$

or $\frac{x}{2} = \frac{y - 2}{4} = \frac{z - 4}{-1}$

23. $V_0 = xyz$; $S = 2(xy + xz + yz)$

$yz = 2\lambda(y + z)$; $xz = 2\lambda(x + z)$; $xy = 2\lambda(x + y)$

Solving this system simultaneously, we find $x = y = z = \sqrt[3]{V_0}$

24. **a.** $T(x, y, z) = \frac{1}{10}(x^2 + y + z^3)$;

$\nabla T = \frac{1}{10}[2x\mathbf{i} + \mathbf{j} + 3z^2\mathbf{k}]$

$\nabla T(-2, 9, 1) = \frac{1}{10}(-4\mathbf{i} + \mathbf{j} + 2\mathbf{k})$

$\mathbf{u} = \mathbf{P_0Q} = 3\mathbf{i} - 12\mathbf{j} + 4\mathbf{k} = \dfrac{3\mathbf{i} - 12\mathbf{j} + 9\mathbf{k}}{\sqrt{9+144+81}}$

$= \dfrac{1}{\sqrt{234}}(3\mathbf{i} - 12\mathbf{j} + 9\mathbf{k})$

$\dfrac{dT}{ds} = \nabla T \cdot \mathbf{u} = \dfrac{1}{10\sqrt{234}}(-12 - 12 + 27)$

$$= \frac{9}{10\sqrt{26}}$$

b $\nabla T = \mathbf{N}; \mathbf{N} = \frac{1}{\sqrt{26}}(-4\mathbf{i} + \mathbf{j} + 3\mathbf{k})$

c. $\frac{dT}{ds} = \|\nabla T\| = \sqrt{26}$

25. $A = 2\int_{\pi/2}^{\pi} \int_{1}^{1-\cos\theta} r \, dr \, d\theta = \int_{\pi/2}^{\pi} (1 - \cos\theta)^2 \, d\theta$

$$= \int_{\pi/2}^{\pi} (\cos^2\theta - 2\cos\theta) \, d\theta = \frac{\pi}{4} + 2$$

26. $V = \int_{0}^{2\pi} \int_{0}^{1} \int_{0}^{1+r^2} r \, dz \, dr \, d\theta = \int_{0}^{2\pi} \int_{0}^{1} r(1 + r^2) \, dr \, d\theta$

$$= \int_{0}^{2\pi} \frac{3}{4} \, d\theta = \frac{3\pi}{2}$$

27. $S: f(x, y, z) = x^2 + y^2 - z = 0;$

$\nabla f = 2x\mathbf{i} + 2y\mathbf{j} - \mathbf{k}; \mathbf{N} = \frac{2x\mathbf{i} + 2y\mathbf{j} - \mathbf{k}}{\sqrt{4r^2 + 1}};$

$dS = r\sqrt{4r^2 + 1} \, dr \, d\theta$

$$S = \int_{0}^{2\pi} \int_{0}^{4} r(4r^2 + 1)^{1/2} \, dr \, d\theta = 2\pi\left(\frac{65\sqrt{65} - 1}{12}\right)$$

$$= \frac{\pi}{6}(65\sqrt{65} - 1)$$

28. $\delta = x + y;$

$dm = (x + y) \, dA = r(r\cos\theta + r\sin\theta) \, dr \, d\theta$

$$m = \int_{0}^{\pi/2} \int_{0}^{1} r^2(\cos\theta + \sin\theta) \, dr \, d\theta$$

$$= \frac{1}{3}\int_{0}^{\pi/2} (\cos\theta + \sin\theta) \, d\theta = \frac{2}{3}$$

$$M_x = \int_{0}^{\pi/2} r^3(\cos\theta\sin\theta + \sin^2\theta) \, dr \, d\theta$$

$$= \frac{1}{4}\int_{0}^{\pi/2} \int_{0}^{1} [\cos\theta\sin\theta + \frac{1 - \cos 2\theta}{2}] \, d\theta$$

$$= \frac{\pi - 1}{8}$$

$(\bar{x}, \bar{y}) = \left(\frac{3\pi - 3}{16}, \frac{3\pi - 3}{16}\right) \approx (0.4155, 0.4155)$

29. $u(x, y) = x - 2y; \frac{\partial u}{\partial y} = -2; v(x, y) = y - 2x;$

$\frac{\partial v}{\partial x} = -2$, so $\mathbf{F}$ is conservative.

$f_x = x - 2y; f = \frac{x^2}{2} - 2xy + a(y)$

$f_y = -2x + a'(y) = -2x + y; a(y) = \frac{y^2}{2} + C$

$f(x, y) = \frac{x^2}{2} + \frac{y^2}{2} - 2xy + C$

$W = \int_{C} \mathbf{F} \cdot d\mathbf{R} = f(1, 0) - f(0, 1) = 0$

30. $x = \cos t, y = \sin t, z = t$

$$W = \int_{C} \mathbf{F} \cdot d\mathbf{R} = \int_{0}^{2\pi} (-w) \, dt = -2\pi w$$